Handbook of Pharmaceutical Wet Granulation

Handbook of Pharmaceutical Wet Granulation
Theory and Practice in a Quality by Design Paradigm

Second Edition

Edited by

Ajit S. Narang
Pharmaceutical Sciences, ORIC Pharmaceuticals, Inc.,
South San Francisco, CA, United States

Sherif I.F. Badawy
Drug Product Science & Technology, Bristol-Myers
Squibb Co., New Brunswick, NJ, United States

 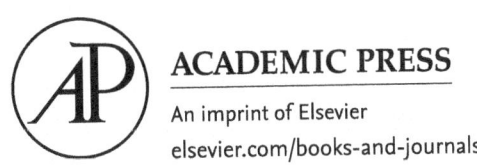

Academic Press is an imprint of Elsevier

125 London Wall, London EC2Y 5AS, United Kingdom
50 Hampshire Street, 5th Floor, Cambridge, MA 02139, United States

Copyright © 2026 Elsevier Inc. All rights are reserved, including those for text and data mining, AI training, and similar technologies.

Publisher's note: Elsevier takes a neutral position with respect to territorial disputes or jurisdictional claims in its published content, including in maps and institutional affiliations.

Books and Journals published by Elsevier comply with applicable product safety requirements. For any product safety concerns or queries, please contact our authorised representative, Elsevier B.V., at productsafety@elsevier.com.

For accessibility purposes, images in electronic versions of this book are accompanied by alt text descriptions provided by Elsevier. For more information, see https://www.elsevier.com/about/accessibility.

No part of this publication may be reproduced or transmitted in any form or by any means, electronic or mechanical, including photocopying, recording, or any information storage and retrieval system, without permission in writing from the publisher. Details on how to seek permission, further information about the Publisher's permissions policies and our arrangements with organizations such as the Copyright Clearance Center and the Copyright Licensing Agency, can be found at our website: www.elsevier.com/permissions.

This book and the individual contributions contained in it are protected under copyright by the Publisher (other than as may be noted herein).

Notices

Knowledge and best practice in this field are constantly changing. As new research and experience broaden our understanding, changes in research methods, professional practices, or medical treatment may become necessary.

Practitioners and researchers must always rely on their own experience and knowledge in evaluating and using any information, methods, compounds, or experiments described herein. In using such information or methods they should be mindful of their own safety and the safety of others, including parties for whom they have a professional responsibility.

To the fullest extent of the law, neither the Publisher nor the authors, contributors, or editors, assume any liability for any injury and/or damage to persons or property as a matter of products liability, negligence or otherwise, or from any use or operation of any methods, products, instructions, or ideas contained in the material herein.

ISBN: 978-0-443-29817-2

For Information on all Academic Press publications visit our website at https://www.elsevier.com/books-and-journals

Publisher: Mica H. Haley
Acquisitions Editor: Andre G. Wolff
Editorial Project Manager: Ashi Jain
Production Project Manager: Kiruthigadevi
Cover Designer: Christian J. Bilbow

Typeset by Aptara, New Delhi, India

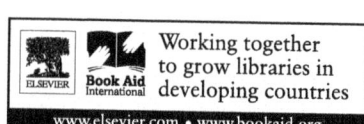

Dedication

To the path of innovation amidst our busy daily lives …
    I dedicate this work to the sacrifices of my parents, Mr. Tirath Singh and Mrs. Gurdip Kaur, and my wife, Swayamjot; to the love of my brother, Supreet, and sons, Manvir and Arjun; and to the confidence and critique bestowed upon me by my teachers and mentors.

Ajit S. Narang

To my beloved family, my wife Irene and sons, Ramy and Daniel, in gratitude for their continued support. To the memory of my parents: my father, who instilled in me the love for science, and my mother, in gratitude for her love and sacrifice.

Sherif I.F. Badawy

Contents

Contributors xxiii
About the Editors xxix
Foreword xxxi
Preface xxxiii
Acknowledgments xxxvii

SECTION I Process design and product quality attributes

CHAPTER 1 Physicochemical principles governing agglomeration and growth kinetics 3
Jonathan B. Wade

1.1 General introduction 3
    1.1.1 Overview of common agglomeration techniques 3
    1.1.2 Introduction to wet granulation 3
    1.1.3 Wetting and nucleation 4
    1.1.4 Consolidation and coalescence 11
    1.1.5 Alternative wet granulation methods 26
    References 28

CHAPTER 2 Microstructure and mechanical properties of granules formed in high-shear wet granulation 35
Leon Farber

2.1 Introduction 35
2.2 Characterization techniques 36
    2.2.1 Microstructure visualization and quantification 36
    2.2.2 Mechanical properties 39
2.3 Microstructure of granules 42
    2.3.1 Granule formation 42
    2.3.2 Pores 43
    2.3.3 Distribution of components 64
    2.3.4 Material transformation during granulation: Microcrystalline cellulose 69
2.4 Connection between granule microstructure and granule strength 73
    2.4.1 Theoretical models 73
    2.4.2 Evolution of microstructure and strength development of interparticle material bridges (Model systems) 74
    2.4.3 Simulation of granule deformation and breakage behavior: From understanding to design 75

2.5 Connection between granule microstructure and granule dissolution 81
2.6 Granule microstructure and granulation process 83
2.7 Concluding remarks 84
   Acknowledgments 86
   References 86

CHAPTER 3 Mechanistic basis for the effects of process parameters on quality attributes in high shear wet granulation 93
Sherif I.F. Badawy, Ajit S. Narang, Keirnan R. LaMarche, Ganeshkumar A. Subramanian and Sailesh A. Varia

3.1 Introduction 93
3.2 Lactose study 94
   3.2.1 Granule size distribution 96
   3.2.2 Compaction properties 97
   3.2.3 Bulk and tapped density of the granulation 102
3.3 Active pharmaceutical ingredient (API) studies 103
   3.3.1 Characterization of granule and tablet attributes 104
   3.3.2 Process parameters 110
   3.3.3 Final product critical quality attributes and granulation bulk powder properties 114
3.4 Summary and conclusions 118
   References 119

CHAPTER 4 Structures and properties of granules prepared by high-shear wet granulation 123
Tianxiang Gao, Jiangnan Dun and Changquan Calvin Sun

4.1 Introduction 123
4.2 Material sciences tetrahedron 124
4.3 Topics of practical importance in high-shear wet granulation 125
   4.3.1 High-shear wet granulation in batch mode 125
   4.3.2 High-shear wet granulation in continuous mode 126
   4.3.3 End point determination of high-shear wet granulation process 127
4.4 Evolution of granule structure during high-shear wet granulation 128
   4.4.1 Wetting and nucleation 128
   4.4.2 Growth and consolidation 130
   4.4.3 Granule breakage during high-shear wet granulation 133
   4.4.4 Granule breakage during fluidized bed drying 136
4.5 Applications of materials science tetrahedron in high-shear wet granulation 136
   4.5.1 Effects of initial water content on granule structure and property 136
   4.5.2 Effects of massing time on structure and properties of granule 138

  4.5.3 Effects of surface smoothing and granule size enlargement on flowability and tabletability 141
  4.5.4 Effects of screw profiles on granule structures and properties in twin-screw wet granulation 141
  4.5.5 Effects of material properties on granule structures and properties in twin-screw wet granulation 142
 4.6 A formulation strategy for solving overgranulation problem during high-shear wet granulation 144
 4.7 Conclusion 147
  References 147

CHAPTER 5 Wet granulation and chemical stability of drug products 155
Sherif I.F. Badawy

 5.1 Introduction 155
 5.2 Reduced stability by wet granulation 156
 5.3 Enhanced stability by wet granulation 161
  5.3.1 Triazine derivative (CPD-1) 162
  5.3.2 DMP 754 165
  5.3.3 Salt form and stability in wet granulation 173
 5.4 Summary and conclusions 175
  References 175

CHAPTER 6 Material physical modifications induced by wet granulation 179
Sherif I.F. Badawy

 6.1 Introduction 179
 6.2 Solid state transformation in wet granulation 179
  6.2.1 Transformation to a more stable polymorphic form 179
  6.2.2 Formation of a high energy metastable form 180
  6.2.3 Transformation of a salt to the free form 180
  6.2.4 Formation of a hydrate/solvate 181
  6.2.5 Control of solid state transformations in wet granulation .. 191
 6.3 Micromeritic changes in wet granulation 192
  6.3.1 Microcrystalline cellulose 192
 6.4 Summary and conclusions 203
  References 203

CHAPTER 7 Current practices in wet granulation-based generic product development 207
Rajan Verma, Maneesha Patil and Carlos O. Paz

 7.1 Introduction 207
 7.2 Mechanism of granule formation 208

  7.2.1 Wetting and nucleation 208
  7.2.2 Consolidation and growth 209
  7.2.3 Attrition and breakage 210
 7.3 Granulation methodologies 210
  7.3.1 High-shear wet granulation process 210
  7.3.2 Fluid-bed wet granulation process 216
  7.3.3 Low-shear wet granulation process 221
  7.3.4 Single-pot granulation 222
  7.3.5 Moisture-activated dry granulation (MADG) 224
  7.3.6 Melt granulation 225
 7.4 Application of QbD in wet granulation process development 225
  7.4.1 Case study IV 229
  7.4.2 Case study V 240
 7.5 Granulation end point determination 250
 7.6 Scaleup of wet granulation process 254
  7.6.1 High-shear granulator 254
  7.6.2 Fluid bed granulation 258
 7.7 Conclusions and recommendations 259
  Acknowledgments 260
  References 261

SECTION II Excipients and input material attributes

CHAPTER 8 Material attributes and their impact on wet granulation process performance 267

Praveen Hiremath, Kalyan Nuguru and Vivek Agrahari

 8.1 Granulation technology 267
  8.1.1 Dry granulation 267
  8.1.2 Wet granulation 268
 8.2 Excipients in wet granulation process 269
  8.2.1 Excipient variability 273
  8.2.2 Excipient functionality 275
  8.2.3 Role of active substances 277
  8.2.4 Diluents 278
  8.2.5 Binders 285
  8.2.6 Lubricants 296
  8.2.7 Disintegrants 299
 8.3 Conclusion and future directions 305
  References 306
  Further reading 312

CHAPTER 9 Binders in wet granulation **313**
Thomas Dürig and Kapish Karan

9.1 Introduction 313
9.2 Physical–chemical properties of common wet granulation binders 314
    9.2.1 Hydroxypropyl cellulose 315
    9.2.2 Methyl cellulose 315
    9.2.3 Hypromellose 318
    9.2.4 Ethyl cellulose 318
    9.2.5 Sodium carboxymethyl cellulose 319
    9.2.6 Povidone 319
    9.2.7 Copovidone 319
    9.2.8 Starch and modified starches 320
    9.2.9 Gum acacia 321
9.3 Important considerations for binder selection and use 321
    9.3.1 Use levels and binder efficiency 321
    9.3.2 Stability and compatibility considerations for binders 322
    9.3.3 Binder water content 322
    9.3.4 Stability considerations 323
    9.3.5 Properties impacting binder mechanisms in wet granulation 324
9.4 Regulatory acceptance and supplier reliability 344
    References 344

CHAPTER 10 Effect of binder attributes on granule growth and densification **347**
Ajit S. Narang, Li Tao, Junshu Zhao, Rekha Keluskar, Shruti Gour, Tim Stevens, Kevin Macias, Brenda Remy, Preetanshu Pandey, Richard D. LaRoche, Anna Sosnowska, Stephen Cole, Atul Dubey, Rohit Ramachandran, Jinjiang Li and Dilbir Bindra

10.1 Functional excipients and critical material attributes 347
10.2 Impact of binder attributes on granulation outcomes 348
    10.2.1 Quality attributes of interest 348
    10.2.2 Mechanistic basis of binder effects 349
    10.2.3 Binder selection 351
10.3 Impact of binders on granulation outcomes 352
10.4 Simulation of particle velocities and impaction forces 357
10.5 Modeling binder mode of addition effects 359
10.6 Granule consolidation kinetics 371
    10.6.1 Binder hydration 371
    10.6.2 Consolidation rate studies 372
    10.6.3 Granule consolidation 375

10.7 Conclusion	376
Acknowledgments	376
Abbreviations	376
References	376

CHAPTER 11 Role of drug substance material properties in the processibility and performance of wet granulated products 379
Chandra Vemavarapu and Sherif I.F. Badawy

11.1 Introduction	379
11.2 Multivariate study of input material properties	379
11.2.1 Input material properties	380
11.2.2 Statistical model for wet-granulated product attributes	383
11.3 Effect of particle size	393
11.3.1 Particle size and wet granule mechanical properties	394
11.3.2 Particle size and granule porosity	400
11.3.3 Particle size and granule growth	401
11.4 Surface area	404
11.5 Contact angle	405
11.6 Solubility	405
11.7 Compaction properties	406
11.8 Flow properties	408
11.9 Summary and conclusions	409
References	409

CHAPTER 12 Critical material attributes in wet granulation 413
Arvind K. Bansal, Garima Balwani and Sneha Sheokand

12.1 Introduction	413
12.2 Basics of wet granulation: Understanding the process, the equipment, and the materials	414
12.2.1 Rate processes	415
12.2.2 Types of wet granulation	415
12.2.3 Continuous wet granulation	417
12.2.4 Endpoint	417
12.2.5 Excipients in wet granulation	417
12.2.6 Functionality of excipients	418
12.3 Quality by design paradigm in wet granulation	419
12.3.1 Critical quality attributes of granules	420
12.3.2 Critical process parameters of wet granulation	420
12.3.3 Critical material attributes for granules	422
12.4 Execution of quality by design methodology in assessing criticality of material attributes	434

	12.4.1 Quality function deployment	436
12.5	Conclusions	437
	References	437

CHAPTER 13 Critical material attributes during continuous twin-screw wet granulation 443

Valerie Vanhoorne, Phaedra Denduyver and Chris Vervaet

13.1	Continuous twin-screw wet granulation	443
13.2	Formulation adjustments: Batch-wise versus continuous manufacturing	445
13.3	Solubility	448
13.4	Active pharmaceutical ingredient	450
13.5	Fillers	451
	13.5.1 Lactose	451
	13.5.2 Microcrystalline cellulose	452
	13.5.3 Mannitol	453
13.6	Binders	453
	13.6.1 Binder addition method	454
	13.6.2 Binder attributes	457
13.7	Surfactants	459
13.8	Disintegrants	459
13.9	Controlled release formulations	460
13.10	Conclusion	462
	References	462

SECTION III PAT, scale-up, control strategy

CHAPTER 14 Inline focused beam reflectance measurement during wet granulation 471

Ajit S. Narang, Tim Stevens, Srinivasa Paruchuri, Kevin Macias, Zhihui Gao, Sherif I.F. Badawy, Dilbir Bindra and Mario Hubert

14.1	Introduction	471
14.2	Focused beam reflectance measurement probe setup and operation	472
	14.2.1 Position	472
	14.2.2 Orientation	472
	14.2.3 Operation	473
14.3	Understanding chord length distribution	474
	14.3.1 Effect of impeller speed on the inline chord length distribution of dry microcrystalline cellulose	480
	14.3.2 Effect of water on chord length distribution of dry microcrystalline cellulose	482

14.4 Granulations that predominantly increase the particle size 483
    14.4.1 Effect of binder concentration on the chord length distribution profile 484
    14.4.2 Effect of water concentration on the chord length distribution profile 489
    14.4.3 Reproducibility of chord length distribution measurement 494
14.5 Granulations that predominantly reduce fines 495
    14.5.1 Effect of water concentration 496
    14.5.2 End-point indicator algorithm 502
    14.5.3 Scale-up 504
14.6 Conclusion 506
Acknowledgments 506
References 506

CHAPTER 15 Principles and applications of drag force flow sensor **509**
Valery Sheverev, Vadim Stepaniuk and Ajit S. Narang

15.1 Introduction 509
15.2 In-line rheometry as process analytical technology 510
    15.2.1 Impeller torque and power consumption 511
    15.2.2 Flow force measured locally inside granulator 512
15.3 Measurement principle of the drag force flow sensor 514
    15.3.1 Sensing model 517
    15.3.2 Temperature compensation 519
    15.3.3 Drag force flow probe in a uniform fluid flow 519
    15.3.4 Drag force flow probe in a powder flow 519
15.4 Drag force flow sensor data interpretation 522
    15.4.1 Raw data 522
    15.4.2 Force pulse magnitude 525
    15.4.3 Amplitude at process frequency 530
15.5 Granule densification and scale-up in brivanib alaninate granulation 531
    15.5.1 Design space of process parameters 533
    15.5.2 Wet mass consistency by drag force flow sensor 537
15.6 Mean force pulse magnitude and CVFPM compared with basic flowability energy measured by FT4 powder rheometer 544
    15.6.1 Comparison of drag force flow sensor and FT4 measurements during high-shear wet granulation 545
    15.6.2 Off-line dry powder characterization by drag force flow sensor and FT4 548
15.7 Comparison of mean force pulse magnitude and CVFPM measured by drag force flow sensor with shaft amperage 549
15.8 Conclusions 553
References 554

CHAPTER 16 An introduction to powder characterization 557
Jamie Clayton

- 16.1 The relevance of powder flow for wet granulation 557
- 16.2 Considerations and challenges in characterizing powder flow 558
  - 16.2.1 Why do we test powders? 558
  - 16.2.2 A complex problem 558
  - 16.2.3 Powder flowability 559
  - 16.2.4 Understanding the impact of the process environment 566
- 16.3 Defining a powder—Fundamental principles of particles 568
  - 16.3.1 Friction 569
  - 16.3.2 Mechanical interlocking 569
  - 16.3.3 Liquid bridging 570
  - 16.3.4 Inter-particulate forces of cohesion 571
  - 16.3.5 Gravity 571
- 16.4 Methods available to the modern scientist 573
  - 16.4.1 Traditional powder characterization techniques 573
  - 16.4.2 Uniaxial testing 575
  - 16.4.3 Biaxial shear cell testing 576
  - 16.4.4 Dynamic powder testing 578
- 16.5 Practical process relevance—Three case studies 579
  - 16.5.1 Predicting feeder performance from powder flow measurements 579
  - 16.5.2 Developing a design space for a die filling operation 585
  - 16.5.3 A QbD approach to continuous tablet manufacture 589
- 16.6 Conclusion 594
- References 597

CHAPTER 17 A quality by design approach to scale-up of high-shear wet granulation process 601
Preetanshu Pandey and Sherif I.F. Badawy

- 17.1 Introduction 601
- 17.2 High-shear wet granulation process 602
- 17.3 Quality by design in high-shear wet granulation 605
  - 17.3.1 Critical material attributes 605
  - 17.3.2 Critical process parameters 607
- 17.4 Scale-up principles 609
  - 17.4.1 Attribute-based scale-up strategy 612
  - 17.4.2 Parametric-based scale-up strategy 616
- 17.5 Modeling techniques 621
- 17.6 Summary 623
- References 624

CHAPTER 18 Integrated application of quality-by-design principles to drug product and its control strategy development 633

Sherif I.F. Badawy, Ajit S. Narang, Keirnan R LaMarche, Ganeshkumar A. Subramanian, Sailesh A. Varia, Judy Lin, Tim Stevens and Pankaj A. Shah

- 18.1 Introduction 633
- 18.2 QbD methodology 634
- 18.3 Physical characterization 635
- 18.4 Process analytical technology 636
    - 18.4.1 NIR moisture methods 636
    - 18.4.2 NIR tablet potency method 636
- 18.5 Formulation selection and optimization 637
    - 18.5.1 Formulation selection 637
    - 18.5.2 Formulation optimization 639
- 18.6 Critical quality attributes, process parameters, and material attributes 642
    - 18.6.1 In vitro drug release (tablet dissolution) 643
    - 18.6.2 Water content and impurities 647
    - 18.6.3 Potency and content uniformity 653
    - 18.6.4 Tablet appearance 654
- 18.7 Control strategy 655
    - 18.7.1 Process parameters design space 657
    - 18.7.2 Input (raw) material controls 657
    - 18.7.3 In-process attributes control 659
    - 18.7.4 Product release testing 662
- 18.8 Key considerations 662
    - 18.8.1 Building process robustness by formulation design 664
    - 18.8.2 Design space strategy 664
    - 18.8.3 Model-based specification limits 665
- 18.9 Concluding remarks 666
- References 666

CHAPTER 19 Implementation of pharmaceutical quality by design in wet granulation 669

Xiang Yu, Lawrence X. Yu, Yue Teng, Dhaval K. Gaglani, Bhagwant D. Rege and Susan Rosencrance

- 19.1 Introduction 669
- 19.2 Function and popularity of wet granulation 670

19.3 Quality by design and unit operation profiles for wet granulation 672
    19.3.1 From product to unit operation 672
    19.3.2 Process operation procedure in high-shear/low-shear granulation profile 674
    19.3.3 Process operation procedure in fluidized bed granulation profile 675
19.4 Quality characterizations of wet granulation product 676
19.5 Product design and understanding of wet granulation 678
19.6 Process design and understanding of wet granulation 680
    19.6.1 Process parameters for high-shear/low-shear granulation 681
    19.6.2 Process parameters for fluidized bed granulation 686
    19.6.3 Scale-up consideration 689
19.7 Quality by design tools and wet granulation development 690
    19.7.1 Risk assessment 691
    19.7.2 Design of experiment 691
    19.7.3 Process analytical technology 693
19.8 Summary 694
    References 695
    Further reading 697

SECTION IV Process modeling and emerging trends

CHAPTER 20 Numerical modeling for wet granulation processes 701
Satoru Watano and Hideya Nakamura

20.1 Introduction 701
20.2 Random models 702
20.3 Population balance models 703
20.4 Fundamentals of the discrete element method 706
20.5 Analysis and design of granulation processes using the discrete element method 709
    20.5.1 Discrete element method simulation of high-shear mixer granulators 709
    20.5.2 Discrete element method simulation of fluidized bed spray granulators/coaters 712
20.6 Advanced modeling of particle transformation in wet granulation by discrete element method-related methods 716
    20.6.1 Coupling with particle adhesion/coalescence model 716
    20.6.2 Coupling with population balance modeling 718
    20.6.3 Direct numerical simulation of fundamental processes of the particle transformation 719
    References 720

CHAPTER 21 Application of the discrete element method to scale-up of high-shear granulation 725
Hideya Nakamura

- 21.1 Introduction 725
  - 21.1.1 A brief introduction of discrete element method 725
  - 21.1.2 What should we investigate for scale-up of high-shear granulation? 726
  - 21.1.3 Advantages of discrete element modeling in the scale-up issue 726
- 21.2 What can we extract from the discrete element modeling? 727
  - 21.2.1 Geometric properties 727
  - 21.2.2 Kinematic properties 728
  - 21.2.3 Dynamic properties 731
- 21.3 Application examples 732
  - 21.3.1 Similarities between different sizes of vertical mixers at bumping flow regime 733
  - 21.3.2 Kinematic similarity based on relative swept volume and impeller speed-mixer diameter scaling relationship 733
  - 21.3.3 Dynamic similarity based on the blade-particle bed stress 735
  - 21.3.4 Combined kinematic-dynamic scale-up method 737
- 21.4 Summary and outlook 744
- References 744

CHAPTER 22 Advances in computational modeling and simulation of wet granulation processes 747
Nejat Rahmanian and Tony Bediako Arthur

- 22.1 Introduction 747
  - 22.1.1 Wet granulation 748
  - 22.1.2 Wet granulation rate processes 749
- 22.2 Approaches to modeling of wet granulation processes 751
  - 22.2.1 The data-driven models (Empirical) 751
  - 22.2.2 The physical (Mechanistic)-based models 751
  - 22.2.3 Modeling through population balance model 752
  - 22.2.4 Population balance modeling governing equations: 3D population balance models 752
  - 22.2.5 Discrete element method simulation 754
  - 22.2.6 Base models 755
  - 22.2.7 CFD simulation 756
  - 22.2.8 Equations of conservation 756
  - 22.2.9 Approaches for computational fluid dynamics multiphase simulations 758
- 22.3 Simulation of high granulators 759
  - 22.3.1 Discrete element method simulation of high shear granulation 760

22.3.2 CFD simulation of high shear granulation 768
22.3.3 Coupled simulation of high shear granulation 770
22.4 Fluidized bed granulation simulation 774
22.4.1 Discrete element method simulation of fluidized bed granulation 775
22.4.2 Coupled granulation simulations and modeling of fluidized bed granulation 775
22.5 Simulation of twin-screw granulation 782
22.5.1 DEM simulation twin-screw granulation 783
22.5.2 Couple simulation of the twin-screw granulation 786
22.6 DEM simulation of other granulation equipment 788
22.7 Conclusion 790
Acknowledgment 792
References 792

CHAPTER 23 Twin-screw continuous wet granulation 801
Niyati Niranjan Kodange, Tongzhou Liu, Adwait Pradhan, Ankita V. Shah, Abu T. Serajuddin and Feng Zhang

23.1 Introduction 801
23.2 Continuous wet granulation 802
23.3 Twin-screw wet granulation 803
23.4 Effects of formulation parameters on twin-screw wet granulation 805
23.4.1 Active pharmaceutical ingredient properties 805
23.4.2 Binder type, concentration, and its properties 806
23.4.3 Binder form (powder versus solution) 807
23.4.4 Fillers and filler combinations 808
23.4.5 Surfactants 809
23.5 Effects of process parameters on twin-screw wet granulation 810
23.5.1 Powder feed rate 810
23.5.2 Screw speed 811
23.5.3 Screw elements 811
23.5.4 Residence time 814
23.5.5 Liquid-to-solid ratio 815
23.5.6 Barrel temperature 817
23.6 Process monitoring of twin-screw wet granulation 817
23.7 Granulation mechanism in twin-screw wet granulation and its effect on granule properties 819
23.8 Comparison of twin-screw granulation and conventional batch high-shear granulation 822
23.9 Challenges in twin-screw wet granulation and emerging solutions 822
23.9.1 Abrasion 822
23.9.2 Post-granulation milling 823

		23.9.3 Post-granulation drying	823
		23.9.4 Recent advancements to eliminate multiple unit operations	824
	23.10	Scale-up in twin-screw wet granulation	825
	23.11	Summary	827
		References	828

CHAPTER 24 Melt granulation: Granulation mechanisms, formulation and process design for batch and twin-screw systems ... 837
Adwait Pradhan, Niyati Niranjan Kodange, Fengyuan Yang, Kapish Karan, Thomas Durig and Feng Zhang

	24.1	Introduction	837
	24.2	Similarities in the granulation mechanism between wet and melt granulation processes	838
		24.2.1 Nucleation mechanism	841
		24.2.2 Granule growth	841
		24.2.3 Granule attrition	841
	24.3	Critical quality attributes for melt granulation	842
	24.4	Formulation design for melt granulation	842
		24.4.1 Binder as the most important formulation variable	843
		24.4.2 Physicochemical properties of the drug	848
	24.5	Process design for melt granulation	849
		24.5.1 Batch melt granulation (*In-situ* melt granulation)	849
		24.5.2 Continuous melt granulation (twin-screw melt granulation)	851
	24.6	Advantages of twin-screw melt granulation	855
	24.7	Challenges of twin-screw melt granulation	856
	24.8	Expert's opinion on the future of twin-screw melt granulation	857
	24.9	Summary	857
		References	858

CHAPTER 25 The application of the state-of-the-art material library/material database approach to the process understanding and process modeling of wet granulation ... 867
Zichen Liang, Gan Luo and Bing Xu

	25.1	The material library/material database concept and its application in pharmaceutical formulation development	867
	25.2	Wet granulation data fusion modeling based on material library	872
		25.2.1 Data sources	874
		25.2.2 Formulation properties estimation	874
		25.2.3 Process parameters processing	875
		25.2.4 Multivariate process modeling	876
		25.2.5 Process model derived from integrated data (literature data and laboratory data)	877

  25.2.6 Conclusion 885
 25.3 Loss of tabletability after wet granulation based on material library 885
  25.3.1 Wet granulation and tableting 886
  25.3.2 Methodology of the material library 886
  25.3.3 Experimental method for high shear wet granulation 887
  25.3.4 Evaluation of change in tabletability 891
  25.3.5 Comparison of compression behavior of powders and granules 892
  25.3.6 Change of tabletability from powders to granules 898
 25.4 Conclusion 902
  References 903

CHAPTER 26 Emerging paradigms in pharmaceutical wet granulation **911**
 Ajit S. Narang and Sherif I.F. Badawy
 26.1 Introduction 911
  26.1.1 End point control 912
  26.1.2 Design of experiment approaches 912
  26.1.3 Modeling and simulation 912
  26.1.4 Residual challenges 913
 26.2 Moisture-activated granulation 914
  26.2.1 Granulation with different physical states of binder 914
  26.2.2 Moisture-activated dry granulation 915
 26.3 Technologies with concurrent heat and mass flow dynamics 916
  26.3.1 Freeze granulation 916
  26.3.2 Melt granulation 917
  26.3.3 Steam granulation 918
 26.4 Foam granulation 919
  26.4.1 Comparison with wet granulation 919
  26.4.2 Physics of liquid penetration 919
  26.4.3 Liquid marble formation 920
  26.4.4 Advantages and opportunities 921
 26.5 Conclusion and future trends 921
  References 922

Index **927**

Contributors

Vivek Agrahari
Department of Formulation Technology, Bayer Animal Health GmbH, Leverkusen, Germany; Department of Technical Development, Bayer U.S. LLC, Shawnee, KS, United States

Tony Bediako Arthur
Chemical Engineering Program, School of Engineering, Faculty of Engineering and Technologies, University of Bradford, Bradford, United Kingdom

Sherif I.F. Badawy
Drug Product Science & Technology, Bristol-Myers Squibb Co., New Brunswick, NJ, United States

Garima Balwani
Department of Pharmaceutics, National Institute of Pharmaceutical Education and Research (NIPER), Mohali, Punjab, India

Arvind K. Bansal
Department of Pharmaceutics, National Institute of Pharmaceutical Education and Research (NIPER), Mohali, Punjab, India

Dilbir Bindra
Drug Product Science & Technology, Bristol-Myers Squibb Co., New Brunswick, NJ, United States

Jamie Clayton
Freeman Technology Ltd., Gloucestershire, United Kingdom

Stephen Cole
DEM Solutions, Denver, CO, United States

Phaedra Denduyver
Laboratory of Pharmaceutical Technology, Ghent University, Ghent, Belgium

Atul Dubey
Tridiagonal Solutions Pvt. Ltd., Pune, Maharashtra, India; Department of Process Engineering, Aditya Birla Science and Technology Company Pvt. Ltd., Navi Mumbai, Maharashtra, India

Jiangnan Dun
Department of Pharmacy, National University of Singapore, Singapore

Thomas Durig
Barentz North America HQ, Avon, OH, United States

Thomas Dürig
Life Sciences R&D Innovation, Ashland Inc, Wilmington, DE, United States

Dhaval K. Gaglani
Center for Drug Evaluation and Research, Office of Pharmaceutical Quality, Food and Drug Administration, Silver Spring, MD, United States

Tianxiang Gao
Department of Industrial and Molecular Pharmaceutics, Purdue University, West Lafayette, IN, United States

Zhihui Gao
Drug Product Science & Technology, Bristol-Myers Squibb Co., New Brunswick, NJ, United States

Shruti Gour
Drug Product Science & Technology, Bristol-Myers Squibb Co., New Brunswick, NJ, United States

Praveen Hiremath
Department of Formulation Technology, Bayer Animal Health GmbH, Leverkusen, Germany

Mario Hubert
Analytical and Bioanalytical Development, Bristol-Myers Squibb Co., New Brunswick, NJ, United States

Kapish Karan
Life Sciences R&D Innovation, Ashland Inc, Wilmington, DE, United States

Kapish Karan
Pharmaceutical Technology, Ashland Specialty Ingredients, Wilmington, DE, United States

Rekha Keluskar
Drug Product Science & Technology, Bristol-Myers Squibb Co., New Brunswick, NJ, United States

Niyati Niranjan Kodange
Division of Molecular Pharmaceutics and Drug Delivery, College of Pharmacy, University of Texas at Austin, Austin, TX, United States

Keirnan R LaMarche
Patheon Part of Thermo Fisher Scientific, Greenville, NC, United States

Richard D. LaRoche
DEM Solutions, Denver, CO, United States

Jinjiang Li
Wolfe Labs, Woburn, MA, United States

Zichen Liang
Department of Chinese Medicine Informatics, Beijing University of Chinese Medicine, Beijing, China; Beijing Key Laboratory of Chinese Medicine Manufacturing Process Control and Quality Evaluation, Beijing, China

Judy Lin
Global Drug Development, Regulatory Affair – CMC, Novartis Pharmaceutical Corp., East Hanover, NJ, United States

Tongzhou Liu
Department of Product Development, Science and Technology Operations, AbbVie Inc., North Chicago, IL, United States

Gan Luo
Department of Chinese Medicine Informatics, Beijing University of Chinese Medicine, Beijing, China; Beijing Key Laboratory of Chinese Medicine Manufacturing Process Control and Quality Evaluation, Beijing, China

Kevin Macias
Analytical and Bioanalytical Development, Bristol-Myers Squibb Co., New Brunswick, NJ, United States

Hideya Nakamura
Department of Chemical Engineering, Osaka Prefecture University, Osaka, Japan

Ajit S. Narang
Pharmaceutical Sciences, ORIC Pharmaceuticals, Inc., South San Francisco, CA, United States

Kalyan Nuguru
Department of Formulation Technology, Bayer Animal Health GmbH, Leverkusen, Germany; Department of Technical Development, Bayer U.S. LLC, Shawnee, KS, United States

Preetanshu Pandey
Drug Product Development, Kura Oncology, San Diego, CA, United States

Srinivasa Paruchuri
Appco Pharma, Somerset, NJ, United States

Maneesha Patil
Formulation Research and Development, Perrigo Laboratories India Private Limited, Ambernath, India; Innovation & Development (I&D) Center, Abbott India Private Limited, Mumbai, India

Carlos O. Paz
Consumer Health Care (CHC) Research and Development, Perrigo Company Plc, Allegan, MI, United States

Adwait Pradhan
Division of Molecular Pharmaceutics and Drug Delivery, College of Pharmacy, University of Texas at Austin, Austin, TX, United States

Nejat Rahmanian
Chemical Engineering Program, School of Engineering, Faculty of Engineering and Technologies, University of Bradford, Bradford, United Kingdom

Rohit Ramachandran
Department of Chemical & Biochemical Engineering, Rutgers University, Piscataway, NJ, United States

Bhagwant D. Rege
Center for Drug Evaluation and Research, Office of Pharmaceutical Quality, Food and Drug Administration, Silver Spring, MD, United States

Brenda Remy
Drug Product Science & Technology, Bristol-Myers Squibb Co., New Brunswick, NJ, United States

Susan Rosencrance
Center for Drug Evaluation and Research, Office of Pharmaceutical Quality, Food and Drug Administration, Silver Spring, MD, United States

Abu T. Serajuddin
Department of Pharmaceutical Sciences, College of Pharmacy, St. John's University, Queens, NU, United States

Ankita V. Shah
Department of Deerfield Discovery and Development, Deerfield Management, New York, NY, United States

Pankaj A. Shah
Drug Product Science & Technology, Bristol-Myers Squibb Co., New Brunswick, NJ, United States

Sneha Sheokand
Department of Pharmaceutics, National Institute of Pharmaceutical Education and Research (NIPER), Mohali, Punjab, India

Valery Sheverev
Lenterra, Inc., Newark, NJ, United States

Anna Sosnowska
DEM Solutions, Denver, CO, United States

Vadim Stepaniuk
Lenterra, Inc., Newark, NJ, United States

Tim Stevens
Drug Product Science & Technology, Bristol-Myers Squibb Co., New Brunswick, NJ, United States

Ganeshkumar A. Subramanian
Manufacturing Science and Technology, Bristol-Myers Squibb Co., New Brunswick, NJ, United States

Changquan Calvin Sun
Department of Industrial and Molecular Pharmaceutics, Purdue University, West Lafayette, IN, United States

Li Tao
Drug Product Science & Technology, Bristol-Myers Squibb Co., New Brunswick, NJ, United States

Yue Teng
Center for Drug Evaluation and Research, Office of Pharmaceutical Quality, Food and Drug Administration, Silver Spring, MD, United States

Valerie Vanhoorne
Laboratory of Pharmaceutical Technology, Ghent University, Ghent, Belgium

Sailesh A. Varia
Drug Product Science & Technology, Bristol-Myers Squibb Co., New Brunswick, NJ, United States

Chandra Vemavarapu
Global Regulatory Sciences, Bristol-Myers Squibb, Hopewell, NJ, United States

Rajan Verma
Formulation Research and Development, Perrigo Laboratories India Private Limited, Ambernath, India; Innovation & Development (I&D) Center, Abbott India Private Limited, Mumbai, India

Chris Vervaet
Laboratory of Pharmaceutical Technology, Ghent University, Ghent, Belgium

Jonathan B. Wade
Technical Services/Manufacturing Science Division, Eli Lilly & Company, Indianapolis, IN, United States

Satoru Watano
Department of Chemical Engineering, Osaka Prefecture University, Osaka, Japan

Bing Xu
Department of Chinese Medicine Informatics, Beijing University of Chinese Medicine, Beijing, China; Beijing Key Laboratory of Chinese Medicine Manufacturing Process Control and Quality Evaluation, Beijing, China

Fengyuan Yang
Pharmaceutical Technology, Ashland Specialty Ingredients, Wilmington, DE, United States

Xiang Yu
Center for Drug Evaluation and Research, Office of Pharmaceutical Quality, Food and Drug Administration, Silver Spring, MD, United States

Lawrence X. Yu
Center for Drug Evaluation and Research, Office of Pharmaceutical Quality, Food and Drug Administration, Silver Spring, MD, United States

Feng Zhang
Division of Molecular Pharmaceutics and Drug Delivery, The University of Texas at Austin, Austin, TX, United States

Junshu Zhao
Drug Product Science & Technology, Bristol-Myers Squibb Co., New Brunswick, NJ, United States

About the Editors

Ajit S. Narang works for the Department of Pharmaceutical Sciences of ORIC Pharmaceuticals, Inc., in South San Francisco, CA, responsible for the pharmaceutical development of new chemical entities through preclinical and clinical stages. He has served as adjunct faculty at the University of Tennessee, Memphis, TN; University of Phoenix, Phoenix, AZ; University of Nebraska Medical Center, Omaha, NE; University of the Pacific, Stockton, CA; Campbell University, Buies Creek, NC; and Western Michigan University, Kalamazoo, MI, United States. He has served as a panel member of the Biopharmaceutics Technical Committee (BTC) of the Pharmaceutical Quality Research Institute (PQRI) in Arlington, VA; a panel member of the  International Pharmaceutics Excipient Council (IPEC) committees; chair of the Formulation Design and Delivery (FDD) section of the American Association of Pharmaceutical Scientists (AAPS); a member of the Systems-based Pharmaceutics (SBP) alliance of the Process Systems Enterprise, Inc. (PSE) in London, United Kingdom; and a scientific advisor to the editors of *JPharmSci*.

He holds over 20 years of pharmaceutical industry experience in drug development and commercialization of across preclinical, clinical, and commercialization stages for both small and large molecule drugs. Prior to ORIC, he worked for Genentech (a member of the Roche group) in South San Francisco, CA, United States; Bristol-Myers Squibb, Co. in New Brunswick, NJ, United States; Ranbaxy Research Labs (currently a subsidiary of Daiichi Sankyo, Japan) in Gurgaon, India; and Morton Grove Pharmaceuticals (currently Wockhardt USA) in Gurnee, IL, United States. He holds an undergraduate pharmacy degree from the University of Delhi, India, and graduate degrees in pharmaceutics from the Banaras Hindu University, India, and the University of Tennessee Health Science Center (UTHSC) in Memphis, TN, United States.

He has contributed to several preclinical, clinical, and commercialized drug products, including NDAs, ANDAs, and 505B2s. He is credited with 54 peer-reviewed articles, 22 editorial contributions, 5 books, 10 patent applications, 47 invited talks, and 85 presentations at various scientific meetings. His current research interests are translation from preclinical to clinical and commercial drug product design, incorporation of QbD elements in drug product development, and mechanistic understanding of the role of material properties on product performance.

Sherif I.F. Badawy is an Executive Scientific Director in the Drug Product Development department of the Bristol-Myers Squibb Company. He received his B.S. in pharmacy and M.S. in pharmaceutics from Cairo University and his Ph.D. in pharmaceutics from Duquesne University. He has more than 29 years of industrial experience in drug product development. His current responsibilities at Bristol-Myers Squibb include formulation and process development and scale-up of commercial oral solid and liquid dosage forms. His areas of research interest include high-shear wet granulation, tablet compaction, stability of solid dosage forms, and bioavailability enhancement of poorly water-soluble compounds. He has authored more than 46 manuscripts and numerous abstracts and presentations in those areas.

Foreword

Foreword by Courtney V. Fletcher

Education and practice of the pharmaceutical sciences are rapidly evolving in these modern times. The pace of such advancements creates an increasing need for updated and contemporary resources for students and faculty and for practitioners in the fields of pharmaceutics. I am pleased to see this book, which covers that gap by highlighting modern practices and recent advancements in the well-established field of pharmaceutical processing by wet granulation.

This text features outstanding contributions from leading scientists across the globe from all practices of the profession—industry, regulatory, and academic. The topics cover the depth and breadth of the various facets of technology and practice.

I am confident this book will enable a deeper understanding of the fundamental principles and provide a wider perspective of their practice to our next generations of students and scientists—while also fueling innovation in the evolving areas.

Happy reading!
Courtney V. Fletcher, Pharm.D.
Dean and Professor, College of Pharmacy
University of Nebraska Medical Center, Omaha, NE, United States

Foreword by Xiaoling Li

It is my pleasure to write this foreword for this book co-edited by Dr. Ajit Narang and Dr. Sherif Badawy. Topics in this book encompass fundamental principles of the industrial practice of wet granulation. As a crucial step in pharmaceutical product development, the granulation process has evolved over the years of my teaching, consulting, and pharmaceutical product development career. While the fundamental scientific principles remain unchanged, application of technology and cross-disciplinary knowledge have definitely accelerated progress in this specific field. As emerging technologies fuse into pharmaceutical product development, the process design for the development of new drug products has shifted to a new paradigm. Quality by design (QbD) is just one of the important shifts that impact and enhance pharmaceutical product development. This book provides ample insights into these aspects through both the depth and breadth of coverage of content in the field of high-shear wet granulation. The sections on process analytical technologies, process modeling, and emerging trends in pharmaceutical sciences highlight the technological advancements and the multidisciplinary nature of these advancements. The sections on control strategy, material attributes, and process design highlight the evolution of the field itself in the context of QbD.

With diminishing contents of this book in the current undergraduate and graduate curriculum in the United States, this book provides much-needed content on fundamentals, practice, and evolution of an important step in pharmaceutical product development with the contemporary QbD embedded in the contents. This book is a very valuable reference and tool for those entering the field from different disciplines, as well as those seeking systemic and advanced knowledge in wet granulation. The advance of science and practice for wet granulation will be driven by needs in pharmaceutical product development, regulatory requirements, and emerging technologies. Therefore this book will also serve as a launching pad to inspire scientists in this field to continue to innovate in the future.

Xiaoling Li, Ph.D.
Professor and Associate Dean
Thomas J Long School of Pharmacy and Health Sciences, University of the Pacific
Stockton, CA, United States

Foreword by Clive Wilson

The pharmaceutical industry is in a constant flux, with high market growth, new medical entities, and the challenge of different regulations across the world pressuring the manufacturing sector. In addition, and importantly, as the industry focuses on taking risks in the discovery and development of the new molecular entities, the risks involved in every other part of the development pathway and commercialization of new drugs must be minimized and addressed in the contemporary quality by design (QbD) paradigm. The fundamental scientific principles and technologies have stood the test of time, but additionally, in-line instrumentation of process operations allows critical attributes to be determined. The QbD method of drug development empowers developers thorough understanding and practical utilization of tools and technologies that translate the product and process understanding to build robust drug products that will be consistent and of high quality over time.

While QbD has been constantly making its mark in the field of pharmaceutical development, a book that outlines and delves in-depth into the QbD application to one of the core processes of pharmaceutical manufacturing—wet granulation—has long been needed. The authors have tackled the task admirably, aided by excellent editorship from two scientists at the top of this field. The content on very practical applications in wet granulation written by industry leaders is complemented by academia-industry applications in Process Analytical Technology (PAT) and modeling in the final sections. This book fulfills this important role of highlighting how QbD is understood and applied by the pharmaceutics professionals in new drug product development.

I am confident that all members of the pharmaceutical community will find the contents of this book relevant and meaningful to their study and work.

Clive Wilson, Ph.D.
JP Todd Professor of Pharmaceutics
Strathclyde Institute of Pharmacy and Biomedical Sciences
University of Strathclyde, Glasgow, United Kingdom

Preface

The second edition of this handbook is updated and expanded to include latest developments in the field of pharmaceutical high shear wet granulation. Several chapters have been revised and updated, including

- Physicochemical principles governing agglomeration and growth kinetics
- Microstructure and mechanical properties of granules formed in high shear wet granulation
- Structures and properties of granules prepared By high shear wet granulation
- Principles of drag force flow sensor
- Twin screw continuous wet granulation
- Emerging paradigms in pharmaceutical wet granulation

New chapters in this edition include

- Material properties in twin screw granulation
- Advances in computational modeling and simulation of granulation processes
- Melt granulation: a new type of wet granulation
- Material library approach for accelerating the process development of wet granulation

This handbook is designed for pharmaceutical students, researchers, regulators, and practitioners alike. It brings forth a mix of basic principles, current practices, and advancements in the evolving science and technology of wet granulation—making for *an integrated approach* to benefit busy professionals.

Granulation process transforms the shape, size, surface, and density of powders or powder mixtures to improve their physicochemical properties and handling to enable rapid and robust manufacture on high-speed commercial equipment. The success of a granulation operation depends on optimal achievement of several quality attributes, including size enlargement with uniformity of drug and excipient distribution, improved flow, and densification without compromising compactibility, compressibility, tabletability, and drug release. Focusing primarily on pharmaceutical wet granulation, this book provides extensive and in-depth understanding of *current* paradigms and practices.

The industrial practice of pharmaceutical wet granulation was dominated in the last few decades by high-shear wet granulation (HSWG). Fluid bed granulation is not uncommon in the pharmaceutical industry, but its application has been limited by a number of factors—not the least of which is the ability to uniformly fluidize the cohesive initial blend characteristic of many pharmaceutical formulations. The focus of this book, therefore, is the more widely practiced HSWG.

Despite its wide utilization, the practice and evolution of HSWG have been evolving rather slowly, which is inherent to the fundamental concepts and established traditional processes of HSWG. In its fundamental form, HSWG entails the use of a high-shear blender to mix liquid and powder components, creating time-dependent trajectory of the wet mass attributes until the desired attributes are achieved at the process end point. Early practice of HSWG emphasized end point determination as the main goal of process development and paid little attention to fundamental process understanding. Wet granulation processes have been traditionally developed using a trial and error approach, with the main objective of identifying a "magical" tool to detect granulation end point. A major limitation of the end point concept was the use of methods with only empirical and indirect correlations to the wet mass quality attributes critical to downstream process and product performance. In fact, the favorite approach for end point determination in the early days was the "fist test," in which the operator squeezes a sample of the wet mass and subjectively decides, based on the wet mass consistency, if the end point has been reached. This book addresses evolving scientific development and understanding of end point determination methods in several chapters.

While a reliable end point detection technology could be a valuable part of a successful process control strategy, the scope of modern process development and control in the QbD era goes far beyond end point. Several chapters in this book highlight how process dynamics and material properties impact granulation trajectory and end product.

Physical models of wet granulation characterize the rate processes in HSWG, such as nucleation, granule coalescence, and breakage. These concepts are reviewed and embedded in the study of granule structure, critical quality attributes, and the effects of process parameters and material properties throughout this book.

While the physical and computational models have enhanced the fundamental understanding of wet granulation, they are not directly amenable to the design of industrial processes. These models have been generally developed for a particular rate process using simple systems, and hence, they are not directly applicable to the complex pharmaceutical formulations and wet granulation processes, which involve simultaneous rate processes.

This handbook places particular emphasis on the design of experiment (DoE) approach, which is now commonly applied to study the effect of various input variables on HSWG outcomes. Early DoE studies were limited to empirical interpretations. Later studies utilized physical models to interpret the effect of process parameters on primary granule properties (such as granule size and density), granulation bulk powder properties, and final product critical quality attributes (such as flow, compaction properties, and tablet dissolution rate). A similar approach was used for input material properties, where the observed effects of material properties on process performance were linked to granulation mechanisms and physical models of granulation.

Predictive numerical simulation of HSWG has evolved over the last two decades with increasing complexity of the models enabled by the leap in available computational power. The various modeling approaches utilized for HSWG include population balance modeling (PBM), discrete element modeling (DEM), PBM coupled with DEM, PBM coupled with computational fluid dynamics (CFD), PBM with compartmental model and DEM, and PBM coupled with volume of fluid (VoF) models. Early numerical modeling approaches were based on empirical population balance models. The advances in physical models of granulation enabled the evolution of the PBM efforts and the implementation of physically based model parameters. However, the complexity of the HSWG precluded the development of a widely predictive model based on one computational approach. This complexity arises from the number of simultaneous rate processes in the HSWG and the heterogeneity of forces and particle velocities in the granulator. Hybrid modeling approaches, such as the combined PBM and DEM modeling, are emerging as promising advances in HSWG process modeling. This handbook includes in-depth review of the computational modeling of HSWG.

This handbook highlights the substantial progress in the understanding, development, and control of HSWG in the last few decades. Nevertheless, significant challenges remain, which are arguably stemming from the inherent nature of the process itself. It is becoming more apparent that a paradigm shift may be necessary to address residual challenges, rather than incremental improvements in current paradigms. The current state-of-the-art in HSWG, including advances described above, is covered in depth in this book. The final section of this book provides broad perspectives and overviews of *evolving* paradigms and practices in the field of pharmaceutical wet granulation. These topics include continuous wet granulation, moisture-activated dry granulation, and processes that concurrently utilize heat and mass flow dynamics, such as steam granulation, melt granulation, and freeze granulation.

We hope that these perspectives provide the value to our readers.

Ajit S. Narang
Sherif I.F. Badawy

Acknowledgments

We gratefully thank our contributors, who put in tremendous work and dedication to share their knowledge and findings from years of learning and practice in this field.

Many people have contributed to our thinking and have been invisible guides to this book. We are particularly thankful to our academic and industrial mentors and teachers who walked with us through the learning of scientific principles; our colleagues and collaborators who worked with us on the applications of technologies and projects; and the scientific community across time and space through their contributions to the literature.

We especially thank the Elsevier staff who patiently and expertly worked with us through the planning, preparation, and production of this book.

Ajit S. Narang
Sherif I.F. Badawy

SECTION I

Process design and product quality attributes

1. Physicochemical principles governing agglomeration and growth kinetics 3
2. Microstructure and mechanical properties of granules formed in high-shear wet granulation 35
3. Mechanistic basis for the effects of process parameters on quality attributes in high shear wet granulation 93
4. Structures and properties of granules prepared by high-shear wet granulation 123
5. Wet granulation and chemical stability of drug products 155
6. Material physical modifications induced by wet granulation 179
7. Current practices in wet granulation-based generic product development 207

CHAPTER 1

Physicochemical principles governing agglomeration and growth kinetics

Jonathan B. Wade

Technical Services/Manufacturing Science Division, Eli Lilly & Company, Indianapolis, IN, United States

1.1 General Introduction

1.1.1 Overview of Common Agglomeration Techniques

Wet granulation typically involves an initial dry blending of the powders, to give a homogeneous distribution, followed by the addition of a granulation liquid. Continued mixing ensures wetting of the powder surfaces and promotes agglomeration of the particles to form granules. Mixing is continued until the desired end-point is reached. The end-point could be defined by a number of parameters, including mixing time, quantity of binder liquid added, and power or torque reading on the mixing impeller. The wet granules are then passed through a coarse screen to break large lumps, dried to remove the binder liquid, and milled to produce granules of the desired particle size distribution. The milled granules often are blended with a lubricant, and potentially a portion of extra-granular binder and disintegrant, to form the final granule blend that is homogeneous, free flowing, and possesses sufficient compaction properties to form a suitable tablet under compression.

Wet granulation is used to reduce segregation potential and to enhance powder flowability, density, and compaction properties. Additionally wet granulation offers the ability to increase the wettability of poorly soluble drugs. A surface active agent can be readily incorporated into the binder liquid, where it is brought into intimate contact with the surfaces of the drug particles during the granulation process, to enhance dissolution rate (Chowdary & Manjula, 1999; Grace et al., 2011). Approximately 40% of the world's top oral drugs are classified as Biopharmaceutics Classification System (BCS) class II (low solubility, high permeability) and IV (low solubility, low permeability) compounds (Amidon et al., 2009), and that the problem is considered even worse in drug discovery pipelines (Lipinski, 2002). Consequently the use of wet granulation processes to simultaneously form granules of the desired physical properties and enhanced bioavailability remains a popular option.

1.1.2 Introduction to Wet Granulation

Wet granulation processes have been a topic of research for several decades with (Newitt & Conway-Jones, 1958) pioneering some of the earliest work in 1958, where they investigated the agglomeration behavior of sand in a drum granulator. Since then, a substantial volume of work has been published studying a wide range of materials, such as detergents, minerals, and pharmaceuticals, which have

4 Chapter 1 Physicochemical principles governing agglomeration and growth

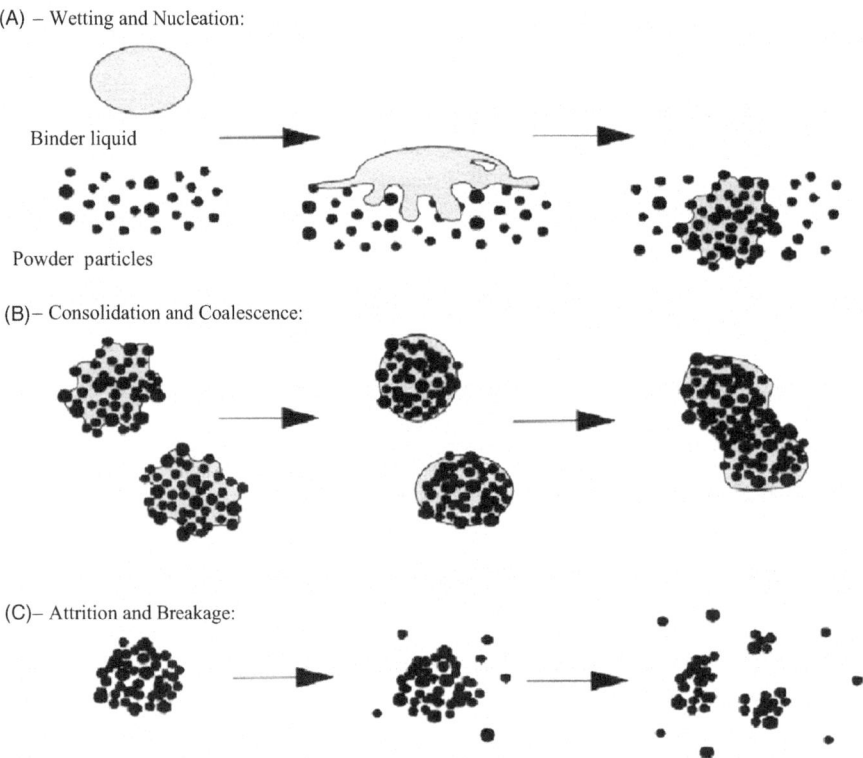

FIGURE 1.1

Representation of wet granulation rate processes: (A) wetting and nucleation, (B) consolidation and coalescence, and (C) attrition and breakage.

From Kristensen (1988).

been granulated in a variety of equipment, including coffee grinders, food processors, rotating drums, fluidized beds, and high shear mixers. Several comprehensive review papers have been written to summarize the state of knowledge in the discipline; notably in the 1970s, when Kapur (1978) reviewed the "balling" process, the 1980s (Kristensen, 1988; Kristensen & Schaefer, 1987; Schubert, 1981) and 1990s (Ennis, 1996), when authors reviewed "agglomeration" and "size enlargement," and the 2000s, when Iveson et al., 2001a reviewed the "nucleation, growth, and breakage phenomena in agitated wet granulation processes." These reviews provide the basis for the current understanding that wet agglomeration processes involve three simultaneous rate processes (Ennis & Litster, 1997) as shown in Fig. 1.1: wetting and nucleation; consolidation and coalescence; and attrition and breakage. Each has received considerable attention in the literature and is discussed separately.

1.1.3 Wetting and Nucleation

Wetting is the process of displacing air from the powder surface with binder liquid. Nucleation is the process of bringing two or more surface wet particles into contact to form nuclei (Fig.1.1A). The area

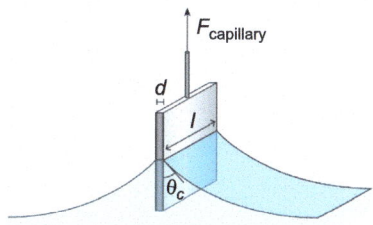

FIGURE 1.2

Illustration depicting surface tension measurement using the Wilhelmy plate method.

From Schaefer and Mathiesen (1996).

where the binder liquid and powder surface come into contact to form initial nuclei is termed the nucleation zone. Nuclei formation and binder liquid dispersion are both important in the nucleation zone. The size of the binder liquid droplet relative to the powder particles onto which it is deposited will influence the nucleation mechanism. Two different nucleation mechanisms have been proposed (Schaefer & Mathiesen, 1996; Scott et al., 2000). If the droplet is large compared to the particles, nucleation will occur by immersion of the smaller particles into the larger droplet to form nuclei with saturated pores. Alternatively nucleation with relatively small droplets will occur by the distribution of the drops on the surface of the particles, which then will start to coalesce.

1.1.3.1 Binder Liquid Characterization

Binder liquid properties need to be characterized and understood before attempting to design a wet granulation process. Properties of interest include liquid density, surface tension, viscosity, and contact angle between the liquid and solid.

1.1.3.1.1 Surface Tension

Surface tension is the attractive force exerted by molecules beneath the surface that draw the surface molecules into the bulk liquid. Several methods can be used to measure surface tension of a liquid, including the capillary rise method, stalagmometer method, pendant drop method, and, commonly, the Wilhelmy plate method. An illustration of the Wilhelmy plate method is shown in Fig.1.2.

In the Wilhelmy plate method, the liquid sample is added to the sample vessel, and the temperature recorded. A platinum plate is thoroughly cleaned with water and heat treated with an oxyacetylene flame prior to use. The plate is assumed to be perfectly wetted such that the contact angle, θ, between the plate and the liquid sample is zero and therefore the term $\cos\theta$ is taken to be 1. The plate then is attached to the instrument microbalance and lowered to a fixed immersion depth where force measurements are recorded at intervals over a fixed period. Surface tension is calculated according to Eq. (1.1).

$$\gamma = \frac{F}{L|\cos|\theta} \tag{1.1}$$

where γ is the surface tension of the liquid (mN m^{-1}), F is the force acting on the microbalance (mN), L is the wetted length (m), and θ is the contact angle between the liquid and the plate (degrees).

6 Chapter 1 Physicochemical principles governing agglomeration and growth

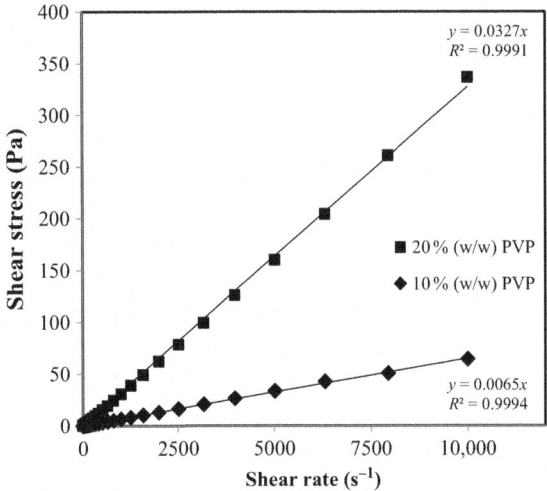

FIGURE 1.3

Example shear rate versus shear stress profile for 10% and 20% (w/w) PVP aqueous solutions generated using a continuous shear rotational rheometer.

From Schaefer and Mathiesen (1996).

Viscosity Viscosity of a liquid is a measure of its resistance to flow and commonly is measured using a continuous shear rotational rheometer. The cone and sample stage typically are cleaned with deionized water prior to each sample measurement. The sample stage temperature should be maintained by means such as a thermostatically controlled water jacket. Viscosity is determined over a wide shear rate range (e.g., $1-10,000$ s^{-1}) with measurements recorded at intervals across the shear rate range. The results are presented as a plot of shear stress as a function of shear rate. An example of the relationship between shear rate and shear stress is presented in Fig.1.3 for 10% and 20% (w/w) poly(vinyl pyrrolidone) (PVP) aqueous binder liquids.

These example data demonstrate that both 10% and 20% (w/w) PVP binder liquids are Newtonian, that is, viscosity is shear rate independent. Newtonian behavior of binder liquids is important because it is likely that a wide range of shear rates exists within the granulator. Newtonian behavior allows the assumption that the contribution of viscous forces will be consistent regardless of flow patterns, impeller speed, etc. The use of a non-Newtonian binder would complicate evaluation of the granulation process because the viscous contributions from the binder liquid can be expected to vary with the experimental conditions used.

Contact Angle Where an interface exists between a droplet of liquid and a solid, the angle between the surface of a liquid and the plane of the solid is referred to as the contact angle. The contact angle is a result of interfacial energies and is a measure of the wettability of a solid by a liquid. A contact angle between the binder liquid and the powder <90 degrees implies that the liquid wets the surface, which will favor binder liquid distribution (Hapgood et al., 2002) and should be the aim when selecting a wet granulation product formulation.

Contact angle between a binder liquid and a powder formulation is commonly determined by sorption measurements using the Washburn equation (Nowak et al., 2013). A stainless steel fritted base sample holder fitted with a filter paper is packed with a known mass powder. Care is taken to present a consistently packed powder column before each measurement. The holder containing the sample is then attached to the instrument microbalance and lowered into a sample of the binder liquid. The mass increase as a function of time is recorded at intervals over a fixed period. The gradient of a linear region of the slope is determined in order to calculate the contact angle according to Eq. (1.2):

$$m^2 = c \frac{\rho^2 |\gamma| \cos\theta}{\mu} T| \tag{1.2}$$

where m is the mass of liquid in the capillary (g), c is a constant (m^{-5}), ρ is the liquid density (kg m^{-3}), γ is the liquid surface tension (N m^{-1}), θ is the contact angle (degrees) between the liquid sample and solid sample, μ is the liquid viscosity (Pa s), and T is the time (s). The constant, c, is determined experimentally from capillary rise measurements using hexane as a completely wetting liquid (Siebold et al., 1997). This technique has been widely reported for a range of liquids and powders, including water, paraffin oil, 3%, 10%, 13%, and 20% (w/w) aqueous PVP and 0.5%, 1%, and 1.5% (w/w) aqueous HPMC solutions and ethylene glycol and powders, including glass beads, aluminum oxide, zirconium dioxide, calcium carbonate, hydroxyapatite, and microcrystalline cellulose (Chitu et al., 2011; Marston et al., 2013; Nowak et al., 2013; Siebold et al., 1997; Wade et al., 2014a). These studies all reported the standard deviation for contact angle measurements in the range of 3% to 10%.

1.1.3.2 Drop Penetration Time

When a binder liquid contacts the powder bed surface, it penetrates into the capillary pores, both within and between particles, to form nuclei. The time taken for the drop to penetrate completely into a powder bed with no liquid remaining on the surface is defined as the drop penetration time. Denesuk et al. (1993) developed a simple drop penetration time model taking into account the liquid surface tension, liquid viscosity, solid–liquid contact angle, and the pore size of the powder bed. They applied the Washburn equation, in which wetting is driven by capillary pressure and resisted by viscous dissipation of the flow, and assumed the powder bed behaved as a bundle of parallel cylindrical capillaries. They also assumed the drop penetration behavior follows the constant drawing area (CDA) model, in which the three-phase contact line remains stationary (i.e., constant drop radius) and the apparent contact angle slowly decreases as the liquid drains from the droplet into the porous surface. The CDA model is used more commonly than the decreasing drawing area model, in which the apparent contact angle at the substrate surface remains constant throughout the penetration process and the liquid contact line retards toward the center of the drop (Hapgood et al., 2002). Hapgood et al. (2002) modified the approach by introducing terms for the effective porosity and effective pore size of the powder bed to account for the fact that the assumption of parallel capillary pores is unlikely to represent a moving powder in which particles are packed loosely, resulting in the following equation:

$$t_p = 1.35 \frac{V_d^{2/3}}{\varepsilon_{\text{eff}}^2 R_{\text{eff}}} \frac{\mu}{|\gamma| \cos\theta} \tag{1.3}$$

where t_p is the predicted drop penetration time, V_d is the drop volume, ε_{eff} is the effective porosity of the powder bed, R_{eff} is the effective pore radius, μ is the liquid viscosity, γ is the liquid surface tension, and θ is the liquid–solid contact angle.

Hapgood et al. (2009) later revised the model to incorporate the effects of hydrophobic powder components and defined drop penetration time as:

$$t_p = 1.37 \frac{V_d^{2/3}}{\left((1-\zeta)\varepsilon_{eff}^*\right)^2 R_{eff}} \frac{\mu}{|\gamma|\cos\theta} \tag{1.4}$$

Where the modified effective porosity, ε^*_{eff}, is defined as:

$$\varepsilon_{eff}^* = \varepsilon_{tap}\left(1 - \varepsilon + \varepsilon_{tap} - \varepsilon \frac{\varepsilon SA_{phobic}}{SA_{blend}}\right)^y \tag{1.5}$$

and the effective pore radius, R_{eff}, is defined as:

$$R_{eff} = \frac{\varphi d_{3,2}}{3} \frac{\varepsilon_{eff}}{(1-\varepsilon_{eff})} \tag{1.6}$$

where ζ is the proportion of the powder surface comprised of hydrophobic particles, ε_{tap} is the tapped porosity of the powder bed, ε is the porosity of the powder bed, fSA_{phobic} is the ratio of hydrophobic particle surface area to the total surface area of the powder blend, SA_{blend}, y is the percolation factor, φ is the shape factor, and $d_{3,2}$ is the surface mean particle size. Short penetration times are facilitated by small droplet size, low binder viscosity, porous nonhydrophobic powders, large powder pore size, high surface tension, and low contact angle.

In order to apply drop penetration time measurements between different formulations and granulation equipment, a dimensionless drop penetration time, τ_p, was proposed (Hapgood et al., 2003):

$$\tau_p = \frac{t_p}{t_c} \tag{1.7}$$

where t_p is the drop penetration time and t_c is the circulation time, which is the time interval between a quantity of powder leaving and re-entering the spray zone. The circulation time is a function of powder flow patterns, the quantity of powder in the granulator, and impeller speed (Hapgood et al., 2003).

1.1.3.3 Binder Liquid Delivery

It has been hypothesized that if all particles contain an equal amount of binder, their physical properties should be the same and produce a narrow size distribution (Mort & Tardos, 1999), provided the primary particles are the same size at the beginning. If the binder liquid is unevenly distributed, some nuclei will be more saturated than others and will grow preferentially, resulting in heterogeneous granule properties. There are three operating variables in binder liquid delivery: droplet size distribution, flow rate, and spray area. When the binder liquid is poured into the granulator, the initial liquid distribution is poor with localized areas of high moisture content, resulting in an initial bimodal size distribution and an increased fraction of coarse granules (Holm et al., 1983; Scott et al., 2000). During the granulation process, a unimodal distribution can result depending upon the mechanical dispersion conditions (Knight et al., 1998).

It is possible to alter the nucleation zone by changing the type, position, and settings of the spray nozzle. Large spray angles and high nozzle to bed distance increase the spray area and decrease the spray density, reducing the likelihood of binder droplets coalescing and reducing the size and spread of the nuclei produced (Rankell et al., 1964).

1.1.3.4 Powder Mixing

Effective powder mixing is critical to binder liquid dispersion in all granulator types. A high rate of fresh powder passing through the nucleation zone facilitates uniform distribution of the binder liquid throughout the powder. Increasing the impeller speed aids binder liquid dispersion by increasing both the shear forces in the granulator, which induces breakage of large wet granules, and promotes powder flux through the nucleation zone (Holm et al., 1983, 1984; Kokubo & Sunada, 1996).

Benali et al. (2009) showed that the degree of mixing in a high shear mixer depends on the competition between the gravitational force, F_g, and the centrifugal force, F_c, as shown in Eqs. (1.8) and (1.9):

$$F_g = mg \qquad (1.8)$$

$$F_c = m\frac{(\pi N D_b)^2}{\left(\frac{D_b}{2}\right)} \qquad (1.9)$$

where m is the wet powder mass, g is the acceleration because of gravity, N is the number of impeller rotations per unit time, and D_b is the diameter of the granulator bowl. The equilibrium between these forces occurs when a critical impeller tip speed, N_c, is reached:

$$N_c = \left(\frac{g}{2\pi^2 D_b}\right)^{1/2} \qquad (1.10)$$

At speeds below the critical value, gravitational forces dominate, and mixing will be less efficient.

1.1.3.5 Dimensionless Spray Flux

Tardos et al. (1997) attempted to standardize the description of nucleation zone conditions across equipment scales. They suggested measuring binder flow rate compared to the size of the spray zone and the powder flux through the spray zone. A decreased flow rate, increased spray zone, or increased flux would be expected to reduce the granule size because there is a lower probability of drop coalescence and a lower liquid volume available for agglomeration per unit of powder (Rankell et al., 1964).

More recently Litster et al. (2001) quantified spray conditions in a high shear mixer through the development of a dimensionless spray flux, Ψ_a, defined as:

$$\psi_a = \frac{3V}{2|d_d|vw_s} \qquad (1.11)$$

where V is the liquid volumetric flow rate (m³ s⁻¹), d_d is the average droplet size (m), v is the powder surface velocity beneath the nozzle (m s⁻¹), and w_s is the width of the spray (m) 90 degrees to powder flow direction. A high Ψ_a, > 1.0, indicates that the ratio of binder liquid addition rate to the powder flux rate is high and the probability of droplets overlapping on the powder surface and coalescing is increased. The surface of the powder bed will become saturated and form large wet areas, and nucleation will occur by immersion, resulting in a wide nuclei size distribution. A low Ψ_a, < 0.1, indicates that the ratio of binder liquid addition rate to the powder flux is sufficiently low that individual droplets are unlikely to coalesce and therefore will form single nuclei that leave the nucleation zone before being re-wet by another droplet.

This hypothesis has been tested for a conventional granulation process scaling from a 10 to a 75 L granulator (Kayrak-Talay et al., 2013). Although the authors predicted no change in the growth regime as a result of maintaining a constant liquid saturation and Stokes deformation number, they observed marked differences in particle size distribution and lump formation. This was attributed to an approximate doubling of the dimensionless spray flux upon scale up, which resulted from the attempt to maintain a constant liquid addition time but inadvertently increased the liquid addition rate. This resulted in a nonhomogenous liquid distribution and demonstrated the importance of dimensionless spray flux.

Some authors have argued that fast dispersion of the binder liquid is taken for granted in high shear granulators and that the binder liquid addition conditions are not critical (Faure et al., 2001). It is proposed that the coalescence into granules is affected mainly by the mixing conditions and the binder liquid amount, and that the system usually "recovers" even when binder liquid is added as a single event because the shear forces in the granulator break the initial nuclei and distribute the binder liquid by mechanical dispersion (Holm et al., 1983; Knight et al., 2000; Terashita et al., 1990). Plank et al. (2003) also highlight several practical limitations in achieving a spray flux <0.1 upon scale up of the process to commercial scales, such as the need to either decrease volumetric binder liquid flow rate or increase impeller tip speed, both of which have significant effects on granule consolidation and growth. Another option considered was to increase the number of spray nozzles. In the scale-up example used, however, this resulted in >14 nozzles in a 300 L scale granulator, which is impractical because of space limitations. As a result, Plank et al. (2003) predict that wet granulation processes executed beyond the laboratory scale routinely operate outside the drop-controlled regime. They do acknowledge the practical benefits of quantifying liquid coverage relative to powder flux, however, they challenge the relevance of using drop size measurements to quantify liquid coverage when routinely operating outside the drop-controlled regime. They propose that the area over which the binder liquid is delivered is more influential, resulting in the development of an empirical equation:

$$\Psi_{alt} = \left. \frac{V}{vA} \right| \tag{1.12}$$

where A is the area of the spray zone. It is proposed that the most reliable method to determine V/A is via direct measurement, such as collecting liquid from a spray nozzle into a collection device such as a grid of adjacent square cuvettes and weighing the amount of liquid collected over a given period of time. Alternatively one could calculate the ratio of the spray volume and spray area directly provided that there is confidence in the spray area estimation.

1.1.3.6 Nucleation Regime Map

Nucleation can be considered a combination of single drop behavior (dimensionless drop penetration time, τ_p) and multiple drop interactions (dimensionless spray flux, Ψ_a) with the specific formulation properties and equipment operating conditions determining the nucleation regime (Iveson et al., 2001a). Three nucleation regimes have been proposed (Litster et al., 2001): drop controlled, intermediate, and mechanical dispersion. Based on these regimes, Hapgood et al. (2003) proposed the nucleation regime

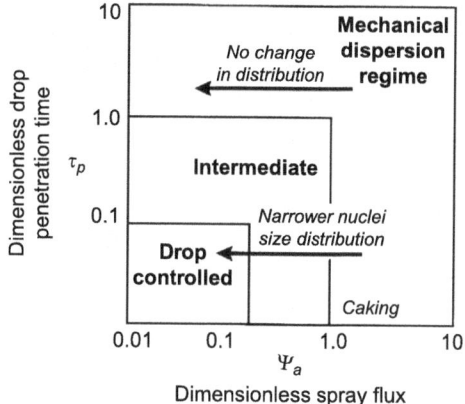

FIGURE 1.4

Nucleation regime map where Ψ_a is the dimensionless spray flux and τ_p is the dimensionless drop penetration time.

From Hapgood et al. (2003).

map (Fig. 1.4) which considers both the dimensionless drop penetration time and the dimensionless spray flux. Although exact values for the regime map boundaries are unknown, the regime map provides a logical basis for investigation and development of granulation processes. Quantification of the regime map transition boundaries is a clear area for future investigations, which would further advance the field of controlled nucleation.

In the drop-controlled regime, binder liquid droplets penetrate the powder bed immediately, and the individual droplet size determines the nuclei size distribution. As the drop penetration time slows and/or the dimensionless spray flux increases, nucleation moves toward the mechanical dispersion regime. In the mechanical dispersion regime, mixing intensity relative to capillary and viscous forces determine nucleation. The retarding effect of viscosity on the nucleation process has been reported frequently (Ennis et al., 1991; Mills et al., 2000; Schaefer & Mathiesen, 1996; Schaefer et al., 2004). In these cases, nucleation and binder liquid distribution occur by mechanical mixing, and the liquid addition method has minimal effect on the particle size distribution. In the intermediate regime, both drop penetration time and mechanical dispersion are influencing factors, making the granulation process difficult to control. If the binder liquid addition rate exceeds the binder liquid dispersion rate, local overwetting will occur. In practice, control of the nucleation process requires measurement and understanding of a number of important variables, such as the wetting behavior of the binder liquid on the powder bed (i.e., formulation), the binder droplet size (i.e., method of addition), binder spray width (i.e., spray nozzle design), and powder surface velocity (i.e., impeller speed and design).

1.1.4 Consolidation and Coalescence

As granules are agitated in a granulator, they experience many collisions with other granules, the walls of the granulator, and the impeller and chopper blades. These collisions can cause either dilation of the granule structure, which creates a more porous assembly, or consolidation (also called densification

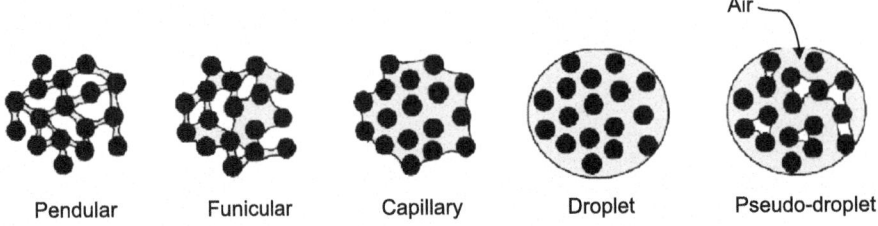

FIGURE 1.5

Representation of the different pore saturation states of liquid-bound granules.

From Iveson and Litster (1998a).

or compaction), which reduces their porosity and size, squeezes out entrapped air, and possibly even squeezes binder liquid to the granule surface (Iveson & Litster, 1998a). Granule porosity controls granule deformability and liquid saturation, both of which have a strong influence on granule growth mechanisms (Kristensen et al., 1985).

1.1.4.1 Liquid Saturation

Newitt and Conway-Jones (1958) first described the effect of increased binder liquid content on the wet granule structure (Fig. 1.5). As the binder liquid content increases, the number of liquid bridges between primary particles increases and the granule moves from a pendular state to a funicular state. At increased binder liquid content, liquid bridges merge into a continuous network leading to the capillary state. Further increase in binder liquid content leads to oversaturation and the static strength of the granules drops to zero (Iveson et al., 2002). It has been suggested that if air is entrapped in the granules it might be feasible for surface liquid to become available at liquid saturation <100% (Iveson et al., 2001b) as represented by the pseudo-droplet state.

The increase in pore liquid saturation as a function of binder liquid content can vary during the granulation process, depending upon formulation and equipment variables that both affect the extent of granule consolidation. Therefore, the maximum granule liquid saturation (S_{max}) is used as a dimensionless measure of binder liquid content (Iveson & Litster, 1998b; Kristensen et al., 1985), which can be used as a means of comparison between different granulation systems:

$$S_{max} = \frac{m\rho_s |1 - \varepsilon_{min}|}{\rho_l |\varepsilon_{min}|} \quad (1.13)$$

where m is the mass ratio of liquid to solid, ρ_s, is the density of solid particles, ρ_l, is the liquid density, and ε_{min} is the minimum porosity the formulation reaches under given operating conditions. The term m should take into account liquid volume changes because of solids dissolution (Iveson et al., 2001b). ε_{min} is a complex function of formulation properties, liquid content, and operating conditions (Iveson & Litster, 1998a; Iveson et al., 1996) and cannot be predicted without performing actual measurements of granule porosity (Iveson & Litster, 1998b). For low-intensity drum granulation, it has been suggested that ε_{min} could be estimated by measuring the in-process wet-tapped porosity of the formulation (Iveson & Litster, 1998b). In high-intensity granulators, such as the high shear granulator, however, formulations are consolidated to a greater extent and ε_{min} is recommended to be measured more directly (Iveson & Litster, 1998b). The determination of granule porosity for use in the calculation of S_{max} has been measured by low-pressure mercury porosimetry (Giesche & Porosimetry, 2006; Iveson et al., 2001b), however, mercury is a toxic substance with associated handling problems, the method is relatively time consuming, and the apparatus is relatively costly. More recently a powder pycnometry has been shown to be robust for determining the porosity of larger bodies, such as rectangular roller compacted ribbons of 10 mm width × 22 mm length × 2 mm thickness (Zinchuk et al., 2004), spherical green iron ore pellets of 10 to 12.5 mm diameter (Forsmo & Vuori, 2005), and pharmaceutical granules (Carvajal & Macias, 2012; Ghadiri et al., 2009; Litster et al., 2008). A thorough evaluation of the technique for determining granule porosity has been reported (Wade et al., 2014c). Powder pycnometry methods measure the granule envelope density, which is used to calculate ε_{min} (Litster et al., 2008):

$$\varepsilon_{min} = 1 - \left(\frac{\rho_e}{\rho_g}\right) \tag{1.14}$$

where ρ_e is the envelope density of the granule sample and ρ_g is the true density of the granule sample. The envelope density represents the density of a granule when its volume is considered on the basis of a perimeter drawn around the envelope of the granule. In this fashion, the envelope density represents the porosity contributions from pores between granules because of surface roughness and imperfect packing with the DryFlo media, and pores that are contained within the granule structure. True density can be determined from granule samples or expressed as a function of the true densities of the constituent single-component powders, ρ_1, ρ_2, ρ_3, etc., as follows:

$$\frac{1}{\rho_g} = \frac{n_1}{\rho_1} + \frac{n_2}{\rho_2} + \frac{n_3}{\rho_3} \tag{1.15}$$

where, n_1, n_2, and n_3 are the weight fractions of the constituent powders, respectively. Such an approach assumes homogeneous mixing of the constituent powder components and representative sampling of the mixture.

Tardos et al. (1997) noted that nucleation is the predominant granulation mechanism at low liquid saturation, with the rate of nucleation being proportional to the strength of the nuclei formed. Kristensen et al. (1985) investigated the granulation of lactose monohydrate and found that growth proceeded slowly when liquid saturation was <0.6. Above a liquid saturation of 0.6, the growth rate increased rapidly because of rapid coalescence of moist agglomerates. Similarly the power consumption of the granulation increased significantly at the same time, thereby providing a basis for an instrumented endpoint where the process could be ended immediately prior to the uncontrolled growth stage. Kristensen (1996) studied the granulation of calcium hydrogen phosphate and showed that as liquid saturation increased so did the geometric mean diameter of the agglomerates, with a rapid increase in growth occurring above a liquid saturation of approximately 0.8.

Granule consolidation, and the resultant liquid-bound granule strength, is controlled by at least three forces: capillary forces, viscous forces, and interparticulate friction (Iveson et al., 2002). Interparticulate friction and viscous forces are dissipative, in that they resist particle motion, whereas capillary forces are conservative, in that they always act to pull particles together and therefore aid consolidation and resist dilation (Iveson et al., 2002). Other forces such as van der Waals and electrostatic forces also can be important in some cases. In liquid-bound systems, however, electrostatic forces are usually minimal because the conductive binder liquid dissipates charge density. Van der Waals forces usually are negligible compared to liquid bridge forces for systems with particle size >10 μm (Rumpf, 1962).

Granule deformation upon collision promotes growth through rearrangement of individual particles, subsequent increases in contact area between colliding granules, and dissipation of collision kinetic energy. The effect of binder liquid content on deformation is complex because it has different effects on interparticulate friction, viscous forces, and capillary forces. When granule deformation occurs, movement of binder liquid can take place from within the granule to the surface, through intragranular capillaries (Kapur, 1978). The contribution of capillary forces, which act to conserve energy and pull particles together via a liquid bridge, increases up to an S_{max} of 1 (Newitt & Conway-Jones, 1958; Rumpf, 1962). Viscous force contribution is increased with an increased binder liquid content because a greater quantity of binder liquid is present within and on the surfaces of granules. When interparticulate friction dominates, increasing the quantity of binder liquid generally increases the extent of consolidation because of a lubrication effect (Iveson et al., 2002). In contrast, when viscous forces dominate, an increase in the quantity of binder liquid can decrease the extent of consolidation. Discrete element modeling (DEM) simulations (Lian et al., 1998) showed that at low collision velocity, viscous effects dissipated the majority of the collision energy because little interparticulate movement occurs. At higher collision velocities, however, energy is dissipated significantly by both viscous and frictional effects. It has been reported that consideration of the effects of binder liquid viscosity further complicates the relationship because increasing binder viscosity resulted in a decrease in the rate of consolidation, and that the effects of liquid content and viscosity were intertwined. Increasing the amount of a binder with a low viscosity of 0.0011 Pa s was found to increase the rate of consolidation because of increased lubrication, whereas increasing the amount of a binder with a high viscosity of 1.1 Pa s decreased the rate of consolidation because of increased viscous forces (Iveson et al., 1996; Iveson & Litster, 1998a). Unless the relative magnitudes of the frictional, viscous, and capillary forces are known, Iveson and Litster (1998a) warn that it is impossible to even qualitatively predict the effect of changing binder liquid content and viscosity.

The effect of increased binder (water) quantity on both the granule size and pore structure, and the subsequent relationship of these properties with drug dissolution rate of mefenamic acid in a formulation containing lactose monohydrate, low-substituted hydroxypropyl cellulose, hydroxypropyl cellulose, and microcrystalline cellulose was reported by Ohno et al. (2007). An increase in binder liquid quantity from 30% to 50% (w/w) resulted in a significant increase in the d_{50} particle size, decrease in the d_{50} pore diameter, and corresponding decrease in the amount of mefenamic acid dissolved after 15 min. These results illustrate the importance of binder liquid amount on both the size and density of granules.

1.1.4.2 Modified Capillary Number

Granule strength and consolidation behavior are controlled by at least three forces: capillary forces, viscous forces, and interparticulate friction (Iveson et al., 2002). In the context of wet granulation, interparticulate friction and viscous forces resist particle motion (i.e., dissipate collision energy), whereas capillary forces always act to pull particles together and aid consolidation and resist dilation (i.e., conserve collision energy) (Iveson et al., 2002). Predicting the effect of variables such as binder liquid content, binder liquid viscosity, and impeller speed on the relative magnitude of these forces is complex (Iveson & Litster, 1998a), and for this purpose a capillary number approach has been developed.

The ratio between viscous forces and static forces has been defined as a viscous capillary number (Ca_{vis}) (Ennis et al., 1991):

$$Ca_{vis} = \frac{\mu v_o}{\gamma} \tag{1.16}$$

where, μ is the binder liquid viscosity (Pa s), γ is the binder liquid surface tension (mN m^{-1}), and v_o is the speed of the particles (m s^{-1}). In a high shear granulator, v_o can be approximated as equal to $\pi N D_b$, where N is the impeller speed and D_b is the granulator bowl diameter. If $Ca_{vis} < 10^{-3}$ viscous forces are negligible when compared to capillary forces and adhesion is the product of interfacial forces (Ennis et al., 1991), whereas if Ca_{vis} is > 1 viscous forces are dominant over capillary forces. The approach was later refined by Benali et al. (2009) through development of a modified capillary number ($Ca*$) that incorporates the work of adhesion of the binder liquid:

$$Ca^* = \frac{\mu v_o}{\omega} \tag{1.17}$$

where ω is the work of adhesion (N m^{-1}) equal to $\gamma(1 + \cos\theta)$ and θ is the contact angle (degrees) of the binder liquid on the powder surface. In general, if $Ca* < 1$ interfacial forces are dominant and growth is not affected by binder viscosity, and if $Ca* > 1.62$, viscous forces dominate and control growth. Experimental data (Benali et al., 2009) suggest a $Ca*$ value of 0.80 represents the threshold between interfacial and viscous forces dominating.

Wade et al. (2016) extended the modified capillary number to the comparison of the conventional and reverse-phase granulation processes using hydroxyapatite and two concentrations of PVP aqueous binder liquid. Fig. 1.6 shows the relationship reported between $Ca*$ and the granule mass mean diameter. An increasing $Ca*$ represents a decrease in the relative contribution of capillary forces and an increase in the viscous contributions of the binder liquid.

The mass mean diameter of granules formed by the conventional granulation process was insensitive to $Ca* < 0.8$. At $Ca* > 0.8$, there was an increase in the mass mean diameter that coincided with a decrease in intragranular porosity and increase in S_{max}. These data suggest that an increase in the contribution of viscous forces facilitates granule growth for the conventional granulation process. In contrast, the mass mean diameter of granules formed by the reverse-phase process decreased as $Ca*$ increased to ~ 0.8, and a relative plateau is reached at $Ca* > 1.1$. These data suggest that granule growth in the reverse-phase process might be driven by capillary forces. This is consistent with the greater degree of capillary wetting of particles that occurs when compared to the conventional process, and the observation that granule growth begins after S_{max} has decreased sufficiently such that capillary forces begin to contribute to wet granule strength.

1.1.4.3 Wet Granule Strength

Rumpf (1962) first calculated wet granule strength, σ, with the assumption that capillary forces in wet granules were dominant:

$$\sigma = 6S_{max} \frac{1-\varepsilon}{\varepsilon} \frac{\gamma \cos\theta}{d_p} \tag{1.18}$$

where S_{max} is the liquid saturation, ε is the granule porosity, γ is the liquid surface tension, θ is the liquid–solid contact angle, and d_p is the particle size of the primary particles.

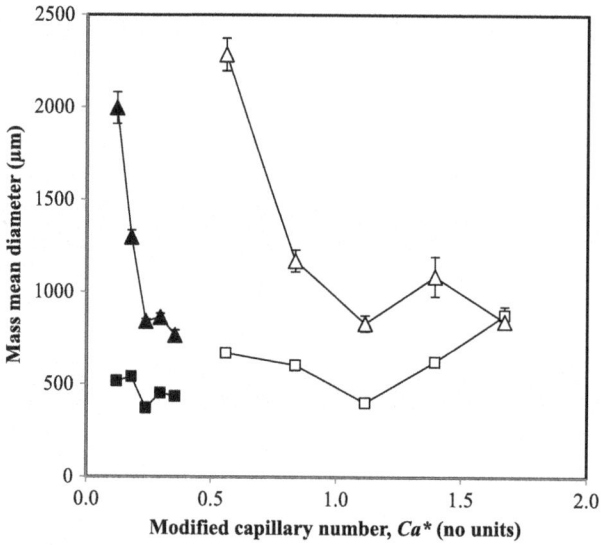

FIGURE 1.6

Hydroxyapatite granule mass mean diameter as a function of modified capillary number, Ca∗. ■, 10% (w/w) PVP by conventional granulation; □, 20% (w/w) PVP by conventional granulation; ▲, 10% (w/w) PVP by reverse-phase granulation; △, 20% (w/w) PVP by reverse-phase granulation. Error bars represent 1 SD, $n = 4$.

From Liu et al. (2000).

Ennis et al. (1991) postulated that in a dynamic situation viscous forces are likely to be dominant and therefore proposed a modified version of the Rumpf equation to calculate granule strength in the wet state:

$$\sigma = \frac{9}{8} \frac{(1-\varepsilon)^2}{\varepsilon^2} \frac{9\pi \mu v_p}{16 d_{3,2}} \quad (1.19)$$

where μ is the liquid viscosity, v_p is the relative velocity of particles within the granulator, and $d_{3,2}$ is the surface mean particle size of the primary particles.

It has been shown, however, that both static and dynamic forces contribute to granule growth (Benali et al., 2009); therefore, an additive model derived by Liu et al. (2009) has been proposed.

$$\sigma = S_{max} \left[6 \frac{1-\varepsilon}{\varepsilon} \frac{\gamma \cos\theta}{d_p} + \frac{9}{8} \frac{(1-\varepsilon)^2}{\varepsilon^2} \frac{9\pi \mu v_p}{16 d_{3,2}} \right] \quad (1.20)$$

This model considers both the capillary strength and dynamic strength of the liquid bridges when calculating granule strength in the wet state.

1.1.4.4 Stokes Deformation Number

The dimensionless Stokes deformation number, St_{def}, first proposed by Tardos et al. (1997) and later adapted by Iveson et al., 2001b, is the ratio between the impact kinetic energy of the wet granules just before collision to the energy dissipated by the liquid bonds between the particles (Iveson et al., 2001b;

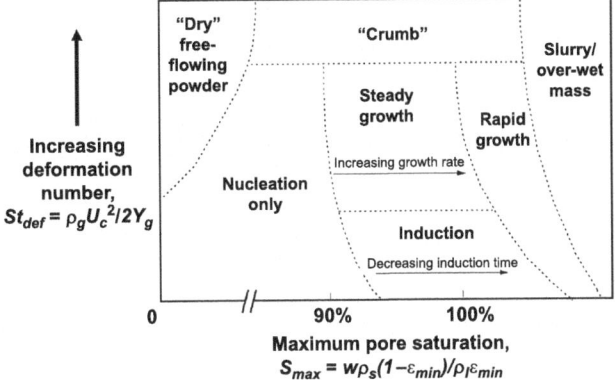

FIGURE 1.7

Growth regime map for liquid bound granules.

From Kristensen et al. (1985).

Tardos et al. 1997), or alternatively the characteristic work done in a collision to plastically deform the granules (Iveson et al., 2001b; Tardos et al. 1997). St_{def} gives a manner in which the relationships among formulation properties, binder liquid properties, and operating conditions can be described.

$$St_{def} = \frac{\rho_g v_c^2}{2\sigma} \quad (1.21)$$

where ρ_g is the granule's density, v_c is the representative collision velocity, and σ is the wet granule strength. A smaller St_{def} implies that significant energy is dissipated in deforming the granule during a collision and granule coalescence is favored.

The use of Stokes deformation number has been reported in the literature for both drum granulators and high shear mixers (Bouwman et al., 2006; Cavinato et al., 2010; Iveson & Page, 2005; Iveson et al., 2001b; Kayrak-Talay et al., 2013).

1.1.4.5 Growth Regime Map

The growth regime map (Fig. 1.7), originally proposed by Iveson and Litster (1998b) and later modified by Iveson et al., 2001b, considers the granule liquid saturation, S_{max}, and the Stokes deformation number, St_{def}, to be the two variables that determine the growth regime within which a given granulation process will operate. The model is considered useful in that it ties easily measurable material properties (e.g., binder viscosity, wet granule strength, granule density, and size) to operating conditions such as particle collision velocity or shear rate (Mort, 2009).

The growth regime map for liquid-bound granules depicts the regime that a given granulation process will operate within to be a function of granule liquid saturation, S_{max}, and Stokes deformation number, St_{def} (Iveson & Litster, 1998b; Iveson et al., 2001b). Several general themes are evident from the growth regime map. An increase in binder liquid quantity will increase S_{max} and move the system along the x-axis from dry to nucleation to induction or steady growth, to rapid growth, and then to an overwet mass regime. An increase in impeller speed will increase the St_{def}, and potentially S_{max}, and therefore move the system from nucleation or induction to steady growth and then to the crumb regime.

The regime map does have limitations. It does not consider wetting and nucleation effects and, by consequence, does not consider control of drop penetration time and dimensionless spray flux (Iveson & Litster, 1998b). These are of particular importance in the nucleation and induction regimes in which they can influence the size distribution and porosity of granules. The regime map also is not able to predict the rate or extent of granule growth, but it does give an indication of the dominant growth mechanism and an informed directional approach to process scale-up, transfer, and process control strategies.

The regime map does not replace experimentation with a given formulation or piece of equipment because variables such as temperature, which will affect binder liquid viscosity, surface tension, and solids dissolution, and the extent of consolidation that a given formulation will reach, cannot currently be predicted (Iveson et al., 2001b). Therefore, some degree of experimentation with a given formulation and piece of granulation equipment are necessary.

The regime map boundaries are not "validated," however, some experimental data are available. The boundary between nucleation and steady or induction growth depends upon the binder quantity required to reach a critical saturation of granule pores and the wet granule consolidation behavior. At $S_{max} < 0.7$, insufficient liquid is present to achieve coalescence and only nuclei are formed (Kayrak-Talay et al., 2013). An S_{max} between 0.8 and 0.9 generally results in a steady growth regime (Butensky & Hyman, 1971; Iveson et al., 2001b; Sherington, 1968), whereas induction growth occurs where granules are strong and do not deform sufficiently to coalesce without the presence of free liquid at the granule surfaces. The boundary between steady growth and induction growth has been reported for St_{def} of 0.001 to 0.003 (Iveson et al., 2001b) and ~0.08 (Wade et al., 2016). Since an increase in S_{max} reduces the induction time, logically at high S_{max} the difference between steady growth and induction growth disappears and both systems exhibit fast steady growth. The boundary between the steady growth and crumb regimes has been reported to be St_{def} of 0.01 (vanden Dries et al., 2003), 0.04 (Iveson & Litster, 1998b), 0.1 (Iveson et al., 2001b), and 0.2 (Tardos et al., 1997; Wade et al., 2016).

Several potential sources of error must be considered when calculating St_{def} including changes in viscosity (Cavinato et al., 2010) and S_{max} (Bouwman et al., 2006) during the course of the process because of dissolution of formulation components, variability in measuring granule porosity (Adepu et al., 2016), lack of consideration for interparticulate frictional and cohesive forces when determining wet granule strength (Bouwman et al., 2006; Kayrak-Talay et al., 2013), and variability in estimation of the representative collision velocity. Notably errors in the estimation of the representative collisions velocity can magnify discrepancies greatly because St_{def} is proportional to the representative collision velocity squared (Iveson et al., 2001b).

The representative collision velocity, v_c, in Eq. (1.23) has been reported to be 15% to 20% of the impeller tip speed (Liu et al., 2009; Kayrak-Talay et al., 2013; Knight et al., 2001), however, this fixed percentage has been criticized (Cavinato et al., 2010). (Wade et al., 2016) employed powder surface velocity measurements to study the effect of liquid saturation, binder liquid viscosity, and impeller tip speed on the granule surface velocity. Figs. 1.8 and 1.9 show the measured surface velocity of granules as a function of S_{max} and impeller speed, respectively.

In the case of S_{max} the surface velocity varied between 5.98% and 20.81% of the impeller tip speed, and, when considering the effect of impeller tip speed, the surface velocity varied between 9.71% and 37.80% of the impeller tip speed. These data indicate that the collision velocity might be expected to vary based on the binder liquid quantity, viscosity, and consolidation properties of the granules and that the fixed percentage representative collision velocity likely does not accommodate the effects of granule

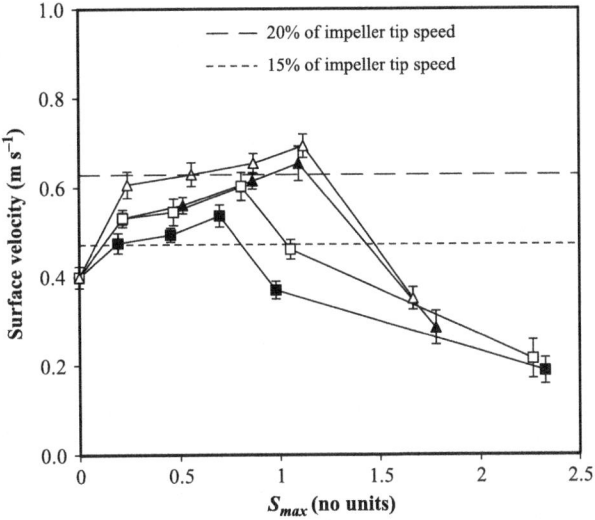

FIGURE 1.8

Hydroxyapatite wet granule surface velocity as a function of liquid saturation, S_{max}, at 3.14 m s^{-1} impeller tip speed. ■, 10% (w/w) PVP by conventional granulation; □, 20% (w/w) PVP by conventional granulation; ▲, 10% (w/w) PVP by reverse-phase granulation; △, 20% (w/w) PVP by reverse-phase granulation. Error bars represent 1 SD, $n = 6$.

From Wade et al. (2016).

and process variables. Data exist only for hydroxyapatite granulation with PVP binder liquid in a small-scale 1 L granulator so broad conclusions should be considered conservatively. Velocity gradients are likely to exist from the center of the granulator bowl to the outside and from the base of the granulator bowl to the powder surface. Therefore, surface velocity measurements are likely to represent a minimum collision velocity experienced.

Rather than calculate the wet granule strength from measured properties, direct measurements of granule yield strength have been pursued. Iveson et al. (2001b) studied direct measurement of wet granule yield strength at strain rates of 0.015 mm s^{-1}. These strain rates, however, were concluded to be too low to be representative of those experienced in a granulator and that at such low strain rates dynamic effects and viscous dissipation might not be participating, resulting in an artificially high St_{def}.

Subsequently dynamic yield strength measurements were performed using a specialized hydraulic load frame capable of high-speed impact velocities up to 15 cm s^{-1}. At these speeds, the yield strength of granules was found to be strain-rate dependent and, for granules prepared with different viscosity binders, the ranking varied with strain rate (Iveson et al, 2003). This strain rate is still significantly less than high-shear granulator impact velocities, which have been suggested up to 10 m s^{-1} (Iveson et al., 2002; Knight et al., 2001). In practicality, granules experience a wide variety of shear and impact stresses dependent upon the flow patterns of the granulator design, formulation, and process variables in question (Mort, 2005). Any *ex granulator* measurements will be only an estimate of the true shear and impact stresses experienced. Recently positron emission particle tracking has been applied to the measurement

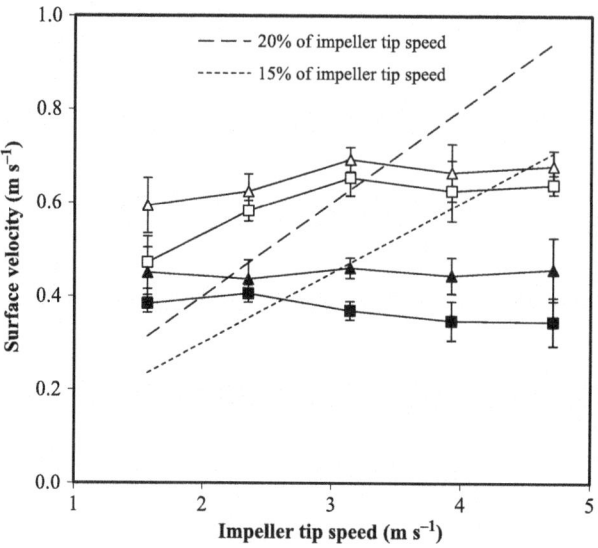

FIGURE 1.9

Hydroxyapatite wet granule surface velocity as a function of impeller tip speed using 200 mL binder liquid. ■, 10% (w/w) PVP by conventional granulation; □, 20% (w/w) PVP by conventional granulation; ▲, 10% (w/w) PVP by reverse-phase granulation; △, 20% (w/w) PVP by reverse-phase granulation. Error bars represent 1 SD, $n = 6$.

From Wade et al. (2016).

of particle motion real time inside granulators, however, this equipment is highly specialized and not widely available (Hassanpour et al., 2009).

Even considering the limitations discussed, it should be emphasized that the application of dimensionless approaches such as spray flux, nucleation regime map, liquid saturation, Stokes deformation number, and the growth regime map have proven effective in the design and characterization of granulation processes, and their further adoption is recommended.

1.1.4.6 Coalescence Models

Granule growth behavior is determined fundamentally by whether colliding granules coalesce or rebound. This simple concept is difficult to model because of the complex nature of granules. Ennis et al. (1991) developed a model for granule coalescence of surface wet nondeformable granules based on the dynamic liquid bridge. They defined a dimensionless Stokes viscous number, St_v, to predict whether a collision between two spherical granules will result in coalescence or rebound, depending upon their kinetic energy and the energy dissipated during the collision:

$$St_v = \frac{8\rho_g v_p r}{9\mu} \tag{1.22}$$

where ρ_g is the granule density, v_p the relative velocity between two spheres estimated as the impeller tip speed, r is the radius of the sphere, and μ is the binder viscosity. The model predicts that the collision will result in coalescence if the St_v is less than the critical Stokes viscous number, St^*_v:

$$St^*_v = \left(1 + \frac{1}{e}\right) ln\left(\frac{h}{h_a}\right) \tag{1.23}$$

where e is the coefficient of restitution, h is the thickness of the surface liquid layer, and h_a is the characteristic height of surface asperities. Three regimes were defined: (1) $St_v \ll St^*_v$—the noninertial regime where all collisions result in coalescence regardless of the size of the colliding granules, (2) $St_v = St^*_v$—the inertial regime where some collisions result in coalescence; collisions between two large granules are less likely to result in coalescence, and (3) $St_v \gg St^*_v$—the coating regime where no collisions result in coalescence.

As shown by the model, the probability of coalescence increases with decreasing particle density, impeller speed, and granule size, and increasing surface liquid layer thickness and binder viscosity. Practical difficulty in applying the model is in the estimation of the coefficient of restitution, thickness of the surface liquid layer, and the height of the surface asperities, as these are experimentally difficult to quantify and are a function of time and binder liquid content and will be dynamic during the course of the granulation process (Liu et al., 2000).

The model of Liu et al. (2000) builds on the Ennis model and is written in terms of bulk parameters of the formulation-binder mixture and of process intensity, including dimensionless groups such as viscous and Stokes deformation numbers, and the ratio of the plastic yield stress to elastic modulus. The model gives the conditions for two types of coalescence: type I and type II (Fig. 1.10).

For type I, coalescence granules coalesce by viscous dissipation in the surface liquid layer before the granule surfaces contact. In type II, coalescence granule surfaces contact and deform. Relative granule velocity is reduced to zero by viscous forces during rebound. The model considers coalescence between two deformable surface wet granules to occur in three stages. The approach stage describes the squeezing of surface liquid from between granules in the collision contact area. The deformation stage describes the collision impact velocity between two granules and their separation distance. In this stage, granules begin to deform and kinetic energy is stored elastically. When the collision velocity is reduced to zero, a contact area has formed between the two granules. The separation stage describes the rebound of the two granules with an initial velocity equal to the stored elastic energy of the impact. Granules will coalesce if the collision kinetic energy is completely dissipated by either viscous loss in the surface liquid layer or plastic deformation in the granule matrix. The model predicts that granules will rebound if the energy attributable to the granule velocity at the end of the separation stage exceeds the strength of the liquid bridge.

1.1.4.7 Growth Behavior

Granule growth behavior can be divided into two broad classes: steady growth and induction growth (Fig. 1.11) (Iveson & Litster, 1998b). The type of growth depends on the deformability and consolidation rate of the granules.

Steady growth occurs when weak, deformable particles form a large contact area during collision and liquid can be squeezed into the contact zone. If this bond is strong enough to resist the separating forces within the granulator, then the pair of granules will coalesce to form a new larger granule (Fig. 1.11A). This behavior leads to a steady increase in granule size and is common in systems with coarse, narrowly sized particles, and low surface tension and/or low viscosity binders (Capes & Danckwerts, 1965; Linkson et al., 1973; Newitt & Conway-Jones, 1958). Induction growth occurs when strong, slowly

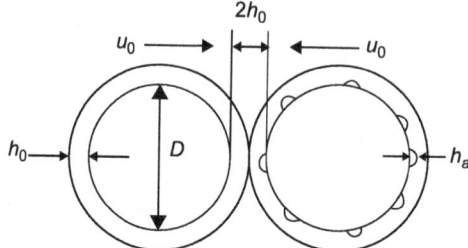
1. Initial approach stage: type 1 coalescence may occur.

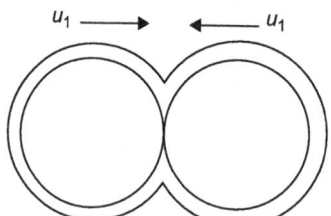
2. Deformation stage: solid layers touch.

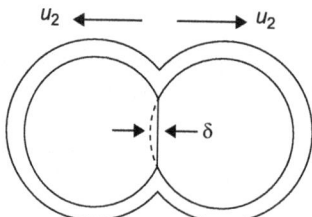
3. Initial separation stage: rebound begins.

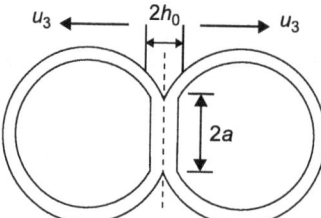
4. Final separation stage: type II coalescence or rebound occurs.

FIGURE 1.10

Representation of type 1 and type 2 wet granule coalescence models.

From Lian et al. (1998).

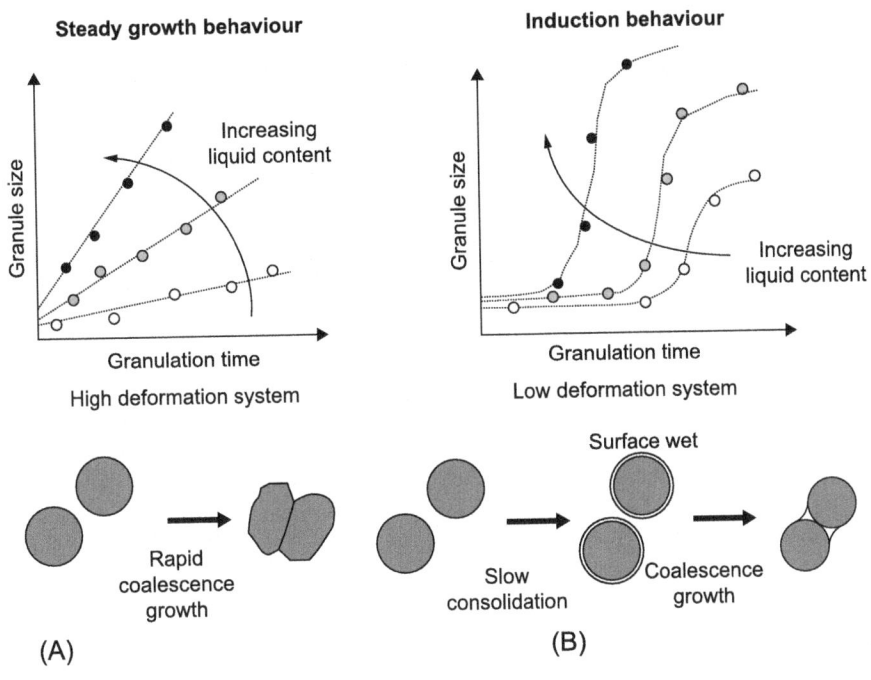

FIGURE 1.11

Schematic of steady and induction growth mechanisms.

From Iveson et al. (2002).

consolidating granules do not deform sufficiently during collision to form a large contact area or strong contact bonds. Pairs of collided granules quickly break apart, and there is a period of little or no growth, so named the induction period (Fig. 1.11B). The length of this induction period decreases with increasing liquid content and will become zero above a critical liquid content. If granules consolidate sufficiently the binder liquid eventually can be squeezed to the surface of the granules. This surface liquid enables bonds to form between granules without the need for large amounts of deformation. This triggers rapid granule growth until granules become so large that further coalescence is prevented by the impact force of the granulator impeller. This type of growth behavior often is seen in systems with fine particles and/or viscous binders (Kapur, 1978; Linkson et al., 1973; Sastry et al., 1977). The boundary between the nucleation and induction growth regime can be characterized by determining the extent of granule consolidation during the wet massing stage, that is, when liquid addition is complete but the impeller is still rotating. Induction growth is characteristic of formulations that undergo significant consolidation (densification) during the wet mass stage.

Other types of behavior include nucleation only when granule nuclei form, but there is insufficient binder liquid to promote further growth; crumb behavior when the formulation is too weak to form permanent granules but instead forms a loose crumb material that cushions a few large granules that are constantly breaking and reforming; and overwetting, in which excess binder liquid is present and the system exhibits uncontrolled growth (Iveson & Litster, 1998b).

1.1.4.8 Effect of Primary Particle Size

Because free surface liquid is necessary in order to obtain agglomeration, the optimum amount of liquid required for granulation will depend upon the accessible surface area of the powders. A smaller particle size results in a larger surface area of the powder. Therefore, for a given constant binder liquid volume, a decreasing particle size will result in a smaller granule size. Alternatively a larger volume of binder liquid will be required to keep granule size constant when powder particle size decreases.

Kristensen et al. (1985), and later Iveson and Litster (1998a), found that dicalcium phosphate granule strength increased with decreasing powder particle size. Because the deformability of the granule determines the probability of coalescence, a higher strength granule deforms less and slows the growth rate. Supporting data were reported by Hsiau et al. (2008), who found that as the primary particle size (d_{50}) of calcium carbonate increased from 1.5 to 82.5 µm, the rate of granule growth increased. They also proposed that granules comprising smaller primary particle size were stronger and suffered a lesser degree of breakage, though the resultant granule size was smaller. Hunter & Ganderton (1972) found a similar trend for lactose granulations. They explained the observations as resulting from the fact that the attractive force for finer agglomerates is greater per unit volume than the larger particles and therefore a stronger granule is obtained. Such nuclei will be brittle and will tend to rebound off each other when they collide, resulting in a lower growth rate when compared to granules composed of larger particles, which will be more plastic and deform readily upon collision and coalesce to form larger granules. These findings most likely are dependent upon material properties and process conditions and shown by the fact that when the experiments were performed with a constant volume of binder liquid but a changing dry powder particle size, the differences were overcome by bringing all granulations to the same liquid saturation (Hunter & Ganderton, 1972).

It has been suggested that a bimodal particle size distribution will favor coalescence and growth through a layering mechanism of smaller particles onto the surface of larger ones (Hounslow et al., 2001; Litster & Waters, 1988; Tardos et al., 1997). Smaller particles will have lower collision energy than larger particles. This lower energy is more likely to be dissipated upon collision for smaller agglomerates resulting in a greater probability of coalescence between a small and a large particle, than between a collision of two large particles.

When the excipients are semisoluble in the binder liquid, the amount of liquid has to be decreased to account for the reduced powder surface area and increased liquid volume caused by dissolution of excipients. Because the smallest particles will dissolve faster, the reduction in surface area can be large and typically is reflected in a reduction in binder liquid amount. When preparing lactose granules that were partially soluble in the binder liquid, the liquid saturation required for significant growth was within the range of 0.3 to 0.6 (Kristensen et al., 1984). However, when experiments were carried out using a polyethylene glycol (PEG) binder liquid, in which the lactose was not soluble, rapid growth by coalescence required liquid saturations in the markedly higher range of 0.8 to 0.9 (Schaefer et al., 1990).

1.1.4.9 Binder Liquid Properties

The powder bed can be considered as a porous surface consisting of a series of capillary pores into which binder liquid will penetrate only when the wetting thermodynamics are favorable, that is, the contact angle between the powder surface and the binder liquid droplet is <90 degrees (Iveson et al., 2001a). A high contact angle between the binder liquid and the dry powder can result in granulation processes that are difficult to control and highly variable (Hapgood et al., 2002, 2009). Provided that wetting thermodynamics are favorable, then the binder liquid properties that will influence the granulation process are viscosity and surface tension.

An increase in binder liquid viscosity has been found to promote granule growth up to a certain critical value above which granule growth decreases (Johansen & Schaefer, 2001). For example, the minimum silicone oil binder liquid viscosity necessary to form granules from various particle sizes of calcium carbonate has been reported (Keningly et al., 1997). A silicone oil viscosity of 10 mPa s was necessary to form granules when considering calcium carbonate primary particles with a mean size of 8 μm, a viscosity of 100 mPa s was necessary for particles of 50 to 80 μm, and a viscosity of 1 mPa s was necessary for particles of 230 μm.

Previous studies have reported on the effects of binder liquid properties on granules prepared using the conventional granulation process. Iveson et al. (1996) showed an interaction between binder liquid content and viscosity. These researchers reported that an increase in content of a low viscosity binder (1 mPa s) decreased the granule porosity, while an increase in content of a high viscosity binder (650 mPa s) increased the granule porosity, yet an increase in content of intermediate viscosity binders (5.4 and 70 mPa s) had minimal effect on porosity. These differences were attributed to the different relative contributions of interparticulate friction, viscous dissipation, and capillary forces in resisting deformation. Johansen and Schaefer (2001) reported that, when using a lower binder viscosity of 66 mPa s, the granulation process proceeded by nucleation and coalescence mechanisms gave steady growth. However, when viscosity was increased above 760 mPa s, immersion of particles in the binder droplets dominated and growth proceeded by layering of particles on the surface of the agglomerates and granule growth was hindered.

A decrease in binder liquid surface tension can decrease granule growth by decreasing the strength of liquid bridges between particles, meaning dilation or breakage is more likely (Iveson et al., 2002). A decrease in surface tension, therefore, typically produces granules of a smaller size and higher porosity. This effect was demonstrated for glass ballotini with surface mean diameters of 19 μm that were granulated with glycerol binder liquids of varying surface tension that was manipulated using sodium dodecylbenzene sulfonate surfactant. As surface tension was increased from 48 to 72 mN m^{-1}, the granule porosity decreased from ~0.39 to ~0.35 with the rate of consolidation also decreasing as the surface tension increased (Iveson & Litster, 1998a).

1.1.4.10 *Effect of Impeller Speed*

Impeller speed will affect granule consolidation (Saleh et al., 2005), growth (Knight et al., 2000), and breakage (Liu et al., 2009).

Knight et al. (2000) reported that as impeller speed increased, there was an initial increase in granule size, followed by a subsequent reduction. At low impeller speeds, the granules had highly spherical shapes, whereas at high impeller speeds they assumed irregular shapes, indicative of breakage phenomena. Saleh et al. (2005) studied the interactive effects of liquid quantity and impeller speed on alumina powder granulation with aqueous PEG binders and reported that increasing liquid content and/or increasing impeller speed resulted in a decrease in granule porosity and increase in pore saturation. The effect of impeller speed on gabapentin granule size distribution was found to be complex (Litster & Kayrak-Talay, 2011) because of the increase in growth and the increase in breakage. These findings make a priori prediction difficult without quantitative knowledge of the growth and breakage regimes.

An increase in impeller speed will increase the frequency and energy of granule collisions and impacts with the impeller and equipment surfaces, resulting in increased consolidation and increased liquid saturation (Ennis et al., 1991; Iveson & Litster, 1998b; Ouchiyama & Tanaka, 1980). For glass ballotini and lactose particles, the rate of breakage was influenced by liquid saturation, binder liquid viscosity, binder liquid surface tension, and primary powder particle size (Liu et al., 2009). Stokes deformation number, St_{def}, gave a good prediction of the breakage probability and a boundary condition of 0.2 was proposed. The increase in impeller speed was proposed to promote a transition through nucleation to steady growth to crumb regimes, which could explain the contrasting reports that an increase in impeller speed can both promote granule growth and granule breakage.

The breakage mechanism is important in wet granulation processes because it influences, and potentially controls, the final granule size distribution (Liu et al., 2009; Wade et al., 2014b). An increase in impeller speed will increase the collision velocity, or energy, between granules (Iveson et al., 2001b). If the collision energy of the granules is greater than their dissipation capacity, the granules will rebound, reducing the probability of coalescence and growth (Eliasen et al., 1998; Hoornaert et al., 1998). Therefore, the effect of impeller speed on granule consolidation and the resultant granule particle size need to be considered in order to gain further insight. For granulation of hydroxyapatite (Wade et al., 2014b), lower impeller speeds resulted in poor binder distribution, presence of large wet granules, and a bimodal particle size distribution. As the impeller speed increased, the degree of breakage and binder liquid distribution increased and the particle size transitioned to a unimodal distribution of significantly lower mass mean diameter. The findings of vanden Dries et al. (2003) are similar, in which an increase in impeller speed increased the extent of granule breakage and improved granule homogeneity.

1.1.5 Alternative Wet Granulation Methods

Several alternative and streamlined wet granulation techniques have been developed, techniques that eliminate variables that currently present barriers to the control of the granulation process or reduce the costs of production. Recent examples include development of melt granulation techniques, in which a solid phase binder is melted during the granulation process to promote granule consolidation and growth, to allow the processing of hygroscopic sodium valproate drug substances (Thies & Kleinebudde, 1999) and to improve moisture stability of dipeptidylpeptidase IV (Kowalski et al., 2009). Fluidized bed granulation, in which the binder liquid is sprayed onto the solid formulation components while they are fluidized, offers more uniform particle characteristics and higher granule porosity compared to conventional high-shear mixer processes (Turton et al., 1999). These two techniques also have been combined in a comelt fluidized bed granulation process developed to enhance ibuprofen bioavailability (Walker et al., 2007). Foam granulation, in which the traditional binder liquid is added to the powder formulation as a dense foam rather than a liquid, has been introduced with the advantages of improved binder distribution, shorter processing times, and elimination of spray nozzles (Keary & Sheskey, 2004).

More recently the feasibility of reverse-phase granulation has been demonstrated (Wade et al., 2014a; Wade et al., 2015a; Wade et al., 2015b; 2016). In this process, dry powder is immersed into the binder liquid, thus eliminating the traditional granule nucleation variables and providing simplified process control. Such a reverse-phase process involves the controlled addition of the powder formulation into the agitated binder liquid to create completely wetted powder particles, which favor granule formation. Addition of further powder is proposed to reduce the S_{max} of the granules, with the desired particle size

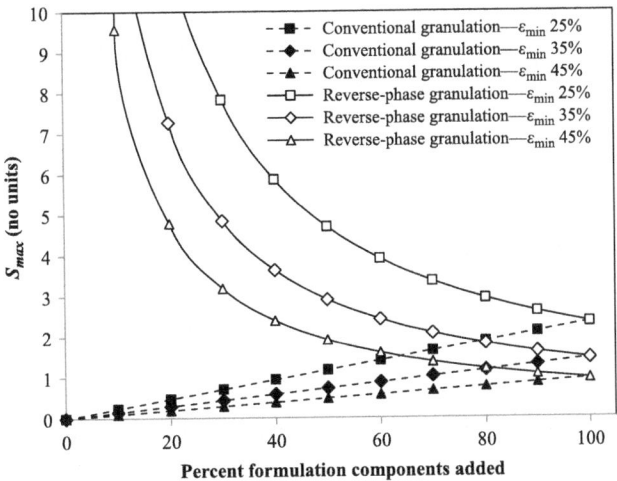

FIGURE 1.12

Plot of liquid saturation as a function of percent formulation components added showing the effect of different minimum intragranular porosity reached. For the conventional process the component added is 250 cm³ 10% (w/w) PVP binder liquid. For the reverse-phase process the formulation component is 600 g hydroxyapatite powder.

From Wade et al. (2014a).

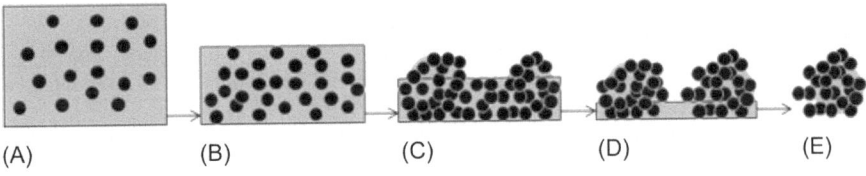

FIGURE 1.13

Proposed mechanism for the reverse-phase granulation process.

From Wade et al. (2014a).

being obtained through controlled breakage. Fig. 1.12 shows the hypothetical progression in S_{max} for both the conventional and reverse-phase granulation processes (Wade et al., 2014a).

The effect of three minimum porosity, ε_{min}, values is simulated in Fig. 12, and it is assumed that the ε_{min} reached is constant regardless of the amount of binder liquid added. The conventional process proceeds through a gradual linear increase in S_{max} as binder liquid is added. In contrast, the reverse-phase process undergoes a steep nonlinear decrease in S_{max} as powder is added. After all formulation components are added, that is, 100% liquid added for the conventional process or 100% powder added for the reverse-phase process, both granulation processes reach identical S_{max} values provided that the same ε_{min} is achieved. As ε_{min} is decreased, that is, granule consolidation occurs, the final S_{max} attained is increased. A proposed mechanism for the reverse-phase granulation process is presented in Fig. 1.13.

The mechanism for growth in the reverse-phase process depicts the start of the process as powder particles suspended in binder liquid (Fig. 1.13A). In this state, only viscous and frictional forces act, and they are insufficient to dissipate the collision energy of the impeller. The concentration of particles in the binder liquid is low, such that the probability of a successful collision and coalescence between two particles is low. As additional dry powder is added, the suspension becomes more concentrated and binder liquid begins to cover and penetrate powder particles (Fig. 1.13B). At a certain point, binder liquid becomes limiting and powder begins to be added onto the surface of liquid-saturated powder rather than directly into binder liquid (Fig. 1.13C). Distribution of the binder liquid is controlled by wet granule breakage and mechanical distribution by the impeller blade. In this region, S_{max} is decreased to a critical point in which capillary forces begin to contribute and granules begin to coalesce (Fig. 1.13D). At this stage, equilibrium is reached between growth and breakage, the balance of which is determined by the strength of the wet granules versus the collision energy (Fig. 1.13E). This is the desired endpoint for the reverse-phase granulation process.

References

Adepu, M., Hate, S., Betard, A., Oka, S., Schongut, M., Sen, M., Sood, Y., Wolf, D., Wieland, S., Stepanek, F., Muzzio, F., Glasser, B., & Ramachandran, R. (2016). Quantitative validation and analysis of the regime map approach for the wet granulation of industrially relevant zirconium hydroxide powders. *Powder Technology, 294*, 177–184. https://doi.org/10.1016/j.powtec.2016.02.026.

Amidon, G. L., Dahan, A., & Miller, J. M. (2009). Prediction of solubility and permeability class membership: Provisional BCS classification of the world's top oral drugs. *AAPS Journal, 11*, 740–746.

Benali, M., Gerbaud, V., & Hemati, M. (2009). Effect of operating conditions and physico-chemical properties on the wet granulation kinetics in high shear mixer. *Powder Technology, 190*(1-2), 160–169. https://doi.org/10.1016/j.powtec.2008.04.082.

Bouwman, A. M., Visser, M. R., Meesters, G. M. H., & Frijlink, H. W. (2006). The use of Stokes deformation number as a predictive tool for material exchange behaviour of granules in the 'equilibrium phase' in high shear granulation. *International Journal of Pharmaceutics, 318*(1-2), 78–85. https://doi.org/10.1016/j.ijpharm.2006.03.038.

Butensky, M., & Hyman, D. (1971). Rotary drum granulation. An experimental study of the factors affecting granule size. *Industrial and Engineering Chemistry Fundamentals, 10*(2), 212–219. https://doi.org/10.1021/i160038a005.

Capes, C. E., & Danckwerts, P. V. (1965). Granule formation by the agglomeration of damp powders: 1. The mechanism of granule growth. *Transactions of the Institution of Chemical Engineers, 43*, 116–122.

Carvajal, M. T., & Macias, K. A. (2012). The influence of granule density on granule strength and resulting compact strength. *Chemical Engineering Science, 72*, 205–213.

Cavinato, M., Franceschinis, E., Cavallari, S., Realdon, N., & Santomaso, A. (2010). Relationship between particle shape and some process variables in high shear wet granulation using binders of different viscosity. *Chemical Engineering Journal, 164*(2-3), 292–298. https://doi.org/10.1016/j.cej.2010.08.029.

Chitu, T. M., Oulahna, D., & Hemati, M. (2011). Wet granulation in laboratory scale high shear mixers: Effect of binder properties. *Powder Technology, 206*, 25–33.

Chowdary, K. P. R., & Manjula, T. (1999). Effect of surfactants on the solubility and dissolution rate of nimesulide from tablets. *Indian Journal of Pharmaceutical Sciences, 62*, 97–101.

References

Denesuk, M., Smith, G. L., Zelinski, B. J. J., Kreidl, N. J., & Uhlmann, D. R. (1993). Capillary penetration of liquid droplets into porous materials. *Journal of Colloid Interface Science, 158*, 114–122.

Eliasen, H., Schaefer, T., & Kristensen, H. (1998). Effects of binder rheology on melt agglomeration in a high shear mixer. *International Journal of Pharmaceutics, 176*, 73–83.

Ennis, B. J. (1996). Agglomeration and size enlargement session summary paper. *Powder Technology, 88*(3), 203–225. https://doi.org/10.1016/S0032-5910(96)03124-5.

Ennis, B.J., & Litster, J.D. (1997). Particle size enlargement. In Perry, R. H., Green, D. W., & Maloney, J. O. (Eds.), *Perry's chemical engineers' handbook.* (7th ed.), pp. 73–117). McGraw-Hill

Ennis, B. J., Tardos, G., & Pfeffer, R. (1991). A microlevel-based characterization of granulation phenomena. *Powder Technology, 65*(1-3), 257–272. https://doi.org/10.1016/0032-5910(91)80189-P.

Faure, A., York, P., & Rowe, R. C. (2001). Process control and scale-up of pharmaceutical wet granulation processes: A review. *European Journal of Pharmaceutics and Biopharmaceutics, 52*, 269–277.

Forsmo, S. P. E., & Vuori, J. P. (2005). The determination of porosity in iron ore green pellets by packing in silica sand. *Powder Technology, 159*(2), 71–77. https://doi.org/10.1016/j.powtec.2005.05.032.

Ghadiri, M., Rahmanian, N., Jia, X., & Stepanek, F. (2009). Characterisation of granule structure and strength made in a high shear granulator. *Powder Technology, 192*, 184–194.

Giesche, H., & Porosimetry, M. (2006). Mercury porosimetry: A general (practical) overview. *Particle and Particle Systems Characterization, 23*, 9–19. doi:10.1002/ppsc.200601009.

Grace, X. F., Latha, S., Shanthi, S., & Reddy, C. U. (2011). Comparative study of different surfactants for solubility enhancement of two class II drugs for type II diabetes mellitus. *International Journal of Pharmacy and Pharmaceutical Sciences, 4*, 377–379.

Hapgood, K. P., Litster, J. D., Biggs, S. R., & Howes, T. (2002). Drop penetration time into porous powder beds. *Journal of Colloid and Interface Science, 253*, 353–366. https://doi.org/10.1006/jcis.2002.8527.

Hapgood, K. P., Litster, J. D., & Smith, R. (2003). Nucleation regime map for liquid bound granules. *AIChE Journal, 49*(2), 350–361. https://doi.org/10.1002/aic.690490207.

Hapgood, K. P., Nguyen, T., & Shen, W. (2009). Drop penetration time in heterogeneous powder beds. *Chemical Engineering Science, 64*, 5210–5221.

Hassanpour, A., Kwan, C. C., Ng, B. H., Rahmanian, N., Ding, Y. L., Antony, S. J., Jia, X. D., & Ghadiri, M. (2009). Effect of granulation scale-up on the strength of granules. *Powder Technology, 189*(2), 304–312. https://doi.org/10.1016/j.powtec.2008.04.023.

Holm, P., Jungerson, O., Schaefer, T., & Kristensen, H. G. (1983). Granulation in high speed mixers part I: Effects of process variables during kneading. *La Pharmacie Industrielle, 45*, 806–811.

Hoornaert, F., Wauters, P., Meesters, G., Pratsinis, S., & Scarlett, B. (1998). Agglomeration behaviour of powders in a lodige mixer granulator. *Powder Technology, 96*, 116–128.

Hounslow, M., Pearson, J., & Instone, T. (2001). Tracer studies of high-shear granulation: II. Population balance modelling. *AICHE Journal, 47*, 1984–1998.

Hsiau, S. S., Tu, W. D., Ingram, A., & Seville, J. (2008). The effect of powder size on induction behaviour and binder distribution during high shear melt agglomeration of calcium carbonate. *Powder Technology, 184*, 298–312.

Hunter, B. M., & Ganderton, D. (1972). The effect of particle size on the granulation of lactose by massing and screening. *The Journal of Pharmacy and Pharmacology, 24*(Suppl), 17–24.

Iveson, S., & Page, N. (2005). Dynamic strength of liquid-bound granular materials: The effect of particle size and shape. *Powder Technology, 152*, 79–89.

Iveson, S. M., Beathe, J. A., & Page, N. W. (2002). The dynamic strength of partially saturated powder compacts: The effect of liquid properties. *Powder Technology, 127*, 149–161.

Iveson, S. M., & Litster, J. D. (1998a). Fundamental studies of granule consolidation part 2: Quantifying the effects of particle and binder properties. *Powder Technology, 99*, 243–250.

Iveson, S. M., & Litster, J. D. (1998b). Growth regime map for liquid-bound granules. *AICHE Journal, 44*, 1510–1518.

Iveson, S. M., Litster, J. D., & Ennis, B. J. (1996). Fundamental studies of granule consolidation part 1: Effects of binder content and binder viscosity. *Powder Technology, 88*, 15–20.

Iveson, S. M., Litster, J. D., Hapgood, H., & Ennis, B. J. (2001a). Nucleation, growth and breakage phenomena in agitated wet granulation processes: a review. *Powder Technology, 117*, 3–39.

Iveson, S. M., Page, N. W., & Litster, J. D. (2003). The importance of wet-powder dynamic mechanical properties in understanding granulation. *Powder Technology, 130*, 97–101.

Iveson, S. M., Wauters, P. A. L., Forrest, S., Litster, J. D., Meesters, G. M. H., & Scarlett, B. (2001b). Growth regime map for liquid-bound granules: Further development and experimental validation. *Powder Technology, 117*, 83–97.

Johansen, A., & Schaefer, T. (2001). Effects of interactions between powder particle size and binder viscosity on agglomerate growth mechanisms in a high shear mixer. *European Journal of Pharmaceutical Sciences, 12*, 297–309.

Kapur, P. C. (1978). Balling and granulation. *Advances in Chemical Engineering, 10*(C), 55–123. https://doi.org/10.1016/S0065-2377(08)60132-5.

Kayrak-Talay, D., Dale, S., Wassgren, C., & Litster, J. (2013). Quality by design for wet granulation in pharmaceutical processing: Assessing models for *a priori* design and scaling. *Powder Technology, 240*, 7–18.

Keary, C. M., & Sheskey, P. J. (2004). Preliminary report of the discovery of a new pharmaceutical granulation process using foamed aqueous binders. *Drug Development and Industrial Pharmacy, 30*(8), 831–845. https://doi.org/10.1081/DDC-200030504.

Keningly, S., Knight, P., & Marson, A. (1997). An investigation into the effects of binder viscosity on agglomeration behaviour. *Powder Technology, 91*, 95–103.

Knight, P. C., Instone, T., Pearson, J. M. K., & Hounslow, M. J. (1998). An investigation into the kinetics of liquid distribution and growth in high shear mixer agglomeration. *Powder Technology, 97*, 246–257.

Knight, P. C., Johansen, A., Kristensen, H. G., Schæfer, T., & Seville, J. P. K. (2000). An investigation of the effects on agglomeration of changing the speed of a mechanical mixer. *Powder Technology, 110*(3), 204–209. https://doi.org/10.1016/s0032-5910(99)00259-4.

Knight, P. C., Seville, J. P. K., Wellm, A. B., & Instone, T. (2001). Prediction of impeller torque in high shear powder mixers. *Chemical Engineering Science, 56*(15), 4457–4471. https://doi.org/10.1016/s0009-2509(01)00114-2.

Kokubo, H., & Sunada, H. (1996). Effect of process variables on the properties and binder distribution of granules prepared by high-speed mixer. *Chemical & Pharmaceutical Bulletin, 44*, 1546–1549.

Kowalski, J., Kalb, O., Joshi, Y. M., & Serajuddin, A. T. M. (2009). Application of melt granulation to enhance stability of a moisture sensitive immediate-release drug product. *International Journal of Pharmaceutics, 381*, 56–61.

Kristensen, H. G. (1988). Agglomeration of powders. *Acta Pharmaceutica Suecica, 25*, 187–204.

Kristensen, H. G. (1996). Particle agglomeration in high shear mixers. *Powder Technology, 88*(3), 197–202. https://doi.org/10.1016/S0032-5910(96)03123-3.

Kristensen, H. G., Holm, P., Jaegerskou, A., & Schaefer, T. (1984). Granulation in high shear mixers, part 4: Effect of liquid saturation on the agglomeration. *La Pharmacie Industrielle, 46*, 763–766.

Kristensen, H. G., Holm, P., & Schaefer, T. (1985). Mechanical properties of moist agglomerates in relation to granulation mechanisms part I. Deformability of moist, densified agglomerates. *Powder Technology, 44*(3), 227–237. https://doi.org/10.1016/0032-5910(85)85004-x.

Kristensen, H. G., & Schaefer, T. (1987). Granulation: A review on pharmaceutical wet-granulation. *Drug Development and Industrial Pharmacy, 13*(4-5), 803–872. https://doi.org/10.3109/03639048709105217.

Lian, G., Thornton, C., & Adams, M. J. (1998). Discrete particle simulation of agglomerate impact coalescence. *Chemical Engineering Science, 53*, 3381–3391.

Linkson, P. B., Glastonbury, J. R., & Duffy, G. J. (1973). The mechanism of granule growth in wet palletisation. *Transactions of the Institution of Chemical Engineers, 51*, 251–257.

Lipinski, C. (2002). Poor aqueous solubility: An industry wide problem in drug discovery. *American Pharmaceutical Review, 5*(3), 82–85.

Litster, J. D., Hapgood, K. P., Michaels, J. N., Sims, A., Roberts, M., Kameneni, S. K., Hsu, T., Doyle, F. J., Stepanek, F., Wang, F.-Y, & Cameron, I. T. (2001). Liquid distribution in wet granulation: Dimensionless spray flux. *Powder Technology, 114*(1-3), 32–39. https://doi.org/10.1016/S0032-5910(00)00259-X.

Litster, J. D., & Kayrak-Talay, D. (2011). *A priori* performance prediction in pharmaceutical wet granulation: Testing the applicability of the nucleation regime map to a formulation with a broad size distribution and dry binder addition. *International Journal of Pharmaceutics, 418*, 254–264.

Litster, J. D., Ramachandran, R., Poon, J., Sanders, C., Glaser, T., Immauel, C., … Cameron, I. T. (2008). Experimental studies on distributions of granule size, binder content and porosity in batch drum granulation: Inferences on process modelling requirements and process sensitivities. *Powder Technology, 188*, 89–101.

Litster, J. D., & Waters, A. (1988). Influence of the material properties of iron ore sinter feed on granulation effectiveness. *Powder Technology, 55*(2), 141–151. https://doi.org/10.1016/0032-5910(88)80097-4.

Liu, L. X., Litster, J. D., Iveson, S. M., & Ennis, B. J. (2000). Coalescence of deformable granules in wet granulation processes. *AIChE Journal, 46*(3), 529–539. https://doi.org/10.1002/aic.690460312.

Liu, L. X., Smith, R., & Litster, J. D. (2009). Wet granule breakage in a breakage only high-hear mixer: Effect of formulation properties on breakage behaviour. *Powder Technology, 189*(2), 158–164. https://doi.org/10.1016/j.powtec.2008.04.029.

Marston, J. O., Sprittles, J. E., Zhu, Y., Li, E. Q., Vakarelski, I. U., & Thoroddsen, S. T. (2013). Drop spreading and penetration into pre-wetted powders. *Powder Technology, 239*, 128–136. https://doi.org/10.1016/j.powtec.2013.01.062.

Mills, P. J. T., Seville, J. P. K., Knight, P. C., & Adams, M. J. (2000). The effect of binder viscosity on particle agglomeration in a low shear mixer/agglomerator. *Powder Technology, 113*, 140–147.

Mort, P. R. (2005). Scale-up of binder agglomeration processes. *Powder Technology, 150*(2), 86–103. https://doi.org/10.1016/j.powtec.2004.11.025.

Mort, P. R. (2009). Scale-up and control of binder agglomeration processes: Flow and stress fields. *Powder Technology, 189*(2), 313–317. https://doi.org/10.1016/j.powtec.2008.04.022.

Mort, P. R., & Tardos, G. I. (1999). Scale-up of agglomeration processes using transformations. *KONA Powder and Particle Journal, 17*(May), 64–75. https://doi.org/10.14356/kona.1999013.

Newitt, D. M., & Conway-Jones, J. M. (1958). A contribution to the theory and practice of granulation. *Transactions of the Institution of Chemical Engineers, 36*, 422–430.

Nowak, E., Combes, G., Stitt, E. H., & Pacek, A. W. (2013). A comparison of contact angle measurement techniques applied to highly porous catalyst supports. *Powder Technology, 233*, 52–64. https://doi.org/10.1016/j.powtec.2012.08.032.

Ohno, I., Hasegawa, S., Yada, S., Kusai, A., Moribe, K., & Yamamoto, K. (2007). Importance of evaluating the consolidation of granules manufactured by high shear mixer. *International Journal of Pharmaceutics, 338*(1-2), 79–86. https://doi.org/10.1016/j.ijpharm.2007.01.030.

Ouchiyama, N., & Tanaka, T. (1980). Stochastic model for compaction of pellets in granulation. *Industrial & Engineering Chemistry Process Design and Development, 19*(4), 555–560. https://doi.org/10.1021/i260076a009.

Plank, R., Diehl, B., Grinstead, H., & Zega, J. (2003). Quantifying liquid coverage and powder flux in high-shear granulators. *Powder Technology, 134*(3), 223–234. https://doi.org/10.1016/s0032-5910(03)00171-2.

Rankell, A. S., Scott, M. W., Lieberman, H. A., Chow, F. S., & Battista, J. V. (1964). Continuous production of tablet granulations in a fluidized bed II. Operation and performance of equipment. *Journal of Pharmaceutical Sciences, 53*, 320–324. https://doi.org/10.1002/jps.2600530316.

Rumpf, H. (1962). The strength of granules and agglomerates. In Knepper, W. A. (Ed.), *Agglomeration* (pp. 379–418). Wiley Interscience.

Saleh, K., Vialatte, L., & Guigon, P. (2005). Wet granulation in a batch high shear mixer. *Chemical Engineering Science, 60*(14), 3763–3775. https://doi.org/10.1016/j.ces.2005.02.006.

Sastry, K. V., Panigraphy, S. C., & Fuerstenau, D. W. (1977). Effect of wet grinding and dry grinding on the batch balling behaviour of particulate materials. *Transactions of the Society of Mining Engineers, 262*, 325–331.

Schaefer, T., Holm, P., & Kristensen, H. G. (1990). Melt granulation in a laboratory scale high shear mixer. *Drug Development and Industrial Pharmacy, 16*, 1249–1277.

Schaefer, T., Johnsen, D., & Johansen, A. (2004). Effects of powder particle size and binder viscosity on intergranular and intra-granular particle size heterogeneity during high shear granulation. *European Journal of Pharmaceutical Sciences, 21*, 525–531.

Schaefer, T., & Mathiesen, C. (1996). Melt pelletization in a high shear mixer VIII: Effects of binder viscosity. *International Journal of Pharmaceutics, 139*, 125–138.

Schubert, H. (1981). Principles of agglomeration. *International Chemical Engineering, 21*(3), 363–377.

Scott, A. C., Hounslow, M. J., & Instone, T. (2000). Direct evidence of heterogeneity during high-shear granulation. *Powder Technology, 113*(1-2), 205–213. https://doi.org/10.1016/s0032-5910(00)00354-5.

Sherington, P. J. (1968). The granulation of sand as an aid to understanding fertilizer granulation. *Chemical Engineer, 220*, 201–215.

Siebold, A., Walliser, A., Nardin, M., Oppliger, M., & Schultz, J. (1997). Capillary rise for thermodynamic characterization of solid particle surface. *Journal of Colloid and Interface Science, 186*(1), 60–70. https://doi.org/10.1006/jcis.1996.4640.

Tardos, G. I., Khan, M. I., & Mort, P. R. (1997). Critical parameters and limiting conditions in binder granulation of fine powers. *Powder Technology, 94*, 245–258. https://doi.org/10.1016/S0032-5910(97)03321-4.

Terashita, K., Watano, S., & Miyanami, K. (1990). Determination of end-point by frequency analysis of power consumption in agitation granulation. *Chemical & Pharmaceutical Bulletin, 38*, 3120–3123.

Thies, R., & Kleinebudde, P. (1999). Melt pelletisation of a hygroscopic drug in a high shear mixer: Part 1. Influence of process variables. *International Journal of Pharmaceutics, 188*(2), 131–143. https://doi.org/10.1016/S0378-5173(99)00214-8.

Turton, R., Tardos, G. I., & Ennis, B. J. (1999). Fluidized bed coating and granulation. In W. C. Yang (Ed.), *Selected topics on fluidization, solids handling and processing* (pp. 331–429). Noyes Publications.

vanden Dries, K., de Vegt, O. M., Girard, V., & Vromans, H. (2003). Granule breakage phenomena in a high shear mixer: Influence of process and formulation variables and consequences on granule homogeneity. *Powder Technology, 133*, 228–236.

Wade, J. B., Martin, G. P., & Long, D. F. (2014a). Feasibility assessment for a novel reverse-phase wet granulation process: The effect of liquid saturation and binder liquid viscosity. *International Journal of Pharmaceutics, 475*(1-2), 450–461. https://doi.org/10.1016/j.ijpharm.2014.09.012.

Wade, J. B., Martin, G. P., & Long, D. F. (2014b). Controlling granule size through breakage in a novel reverse-phase wet granulation process: The effect of impeller speed and binder liquid viscosity. *International Journal of Pharmaceutics, 478*, 439–446.

Wade, J. B., Martin, G. P., & Long, D. F. (2014c). An assessment of powder pycnometry as a means of determining granule envelope density. *Pharmaceutical Development and Technology, 20*, 257–265.

Wade, J. B., Martin, G. P., & Long, D. F. (2015a). The evolution of granule fracture strength as a function of impeller tip speed and granule size for a novel reverse-phase wet granulation process. *International Journal of Pharmaceutics, 488*(1–2), 95–101. https://doi.org/10.1016/j.ijpharm.2015.04.033.

Wade, J. B., Martin, G. P., & Long, D. F. (2015b). Controlling granule size through breakage in a novel reverse-phase wet granulation process; the effect of impeller speed and binder liquid viscosity. *International Journal of Pharmaceutics, 478*(2), 439–446. https://doi.org/10.1016/j.ijpharm.2014.11.067.

Wade, J. B., Martin, G. P., & Long, D. F. (2016). The development of a growth regime map for a novel reverse-phase wet granulation process. *International Journal of Pharmaceutics, 512*(1), 224–233. https://doi.org/10.1016/j.ijpharm.2016.08.050.

Walker, M., Bell, S. E. J., Andrews, G., & Jones, D. (2007). Co-melt fluidised bed granulation of pharmaceutical powders: Improvements in drug bioavailability. *Chemical Engineering Science, 62*, 451–462.

Zinchuk, A., Mullarney, M., & Hancock, B. (2004). Simulation of roller compaction using a laboratory scale compaction simulator. *International Journal of Pharmaceutics, 269*, 403–415.

CHAPTER 2

Microstructure and mechanical properties of granules formed in high-shear wet granulation

Leon Farber
Merck & Co., Inc., Rahway, NJ, United States

2.1 Introduction

The structure of a material or an object usually relates to the arrangement of its internal components. For a granule, the structure would be expected to describe the arrangement of its initial particles–solid regions between particles formed by solidified liquid binder, and voids within the granule. For high-shear wet granulation (HSWG) of pharmaceutical materials, the typical mean particle size of initial powders ranges from microns to tens of microns, with the desired final mean granule size generally being several hundred microns. At that scale, the physical features of the granule are observed via microscopy, and the term "microstructure" is usually used. The length scale of interest for granule microstructure can span from submicron distances (such as the morphology of interparticle pores and solidified bridges) to a scale much larger than the original particles, at which the size and spatial distribution of larger pores or surface morphology are of primary interest. Because HSWG is usually done on a multicomponent system, in which the individual components have some inherent particle size distribution (PSD), agglomeration can lead to a nonuniform distribution of the particles within the granule by both composition and size. The granule microstructure could also describe the nonuniform distribution of components within the granule by composition and size.

The characterization of the internal granule microstructure has not been widely performed in the framework of research on wet granulation. Most of the reported research has focused on granulation mechanisms and the resultant granule size distribution as the attribute of interest. Size distribution is relatively easy to measure and usually allows the author to discuss the kinetics of various rate processes that occur during granulation. In production, a granule size distribution measurement is used as a fast and cost-effective tool to monitor process performance. In contrast, characterization of an internal granule structure requires relatively sophisticated and expensive tools and is relatively slow. Similarly mechanical properties of individual granules have been characterized only in a limited number of published reports, and usually as an attribute that differentiates between outcomes of the individual experiments performed within a study. A review published in 2007 by Barrera-Medrano, Salman, Reynolds, and Hounslow was the first attempt to summarize the data available on the microstructure of granules produced by various techniques. It mostly focused on initial results of the application of X-ray microtomography to granules produced in model systems at a small scale. Since then, significant developments in the field of granulation have been made, and a lot of new data on granule microstructure have become available.

This chapter reviews the specific microstructures reported for granules formed by HSWG and the current understanding of the mechanisms underlying their formation. The chapter discusses the relationship between the properties of the initial powder materials and the liquid binder and the process parameters that result in the formation of the specific microstructure. It also reviews the connection between the microstructure and the resultant mechanical properties of individual granules. The chapter illustrates the differences between the microstructure of granules formed by HSWG versus granules formed by other methods. Lastly the chapter will examine and discuss the importance of granule microstructure within the framework of critical quality attributes and critical process parameters for the HSWG manufacturing process train.

The following section will discuss the various techniques used to characterize granule microstructure and their mechanical properties.

2.2 Characterization techniques

2.2.1 Microstructure visualization and quantification

Various microscopy techniques that are used for the characterization of granule microstructure are described below. Nonmicroscopic techniques exist for quantifying granule total porosity, an important and simplest descriptor of pore structure, but these techniques, such as He/Hg pycnometry or oil impregnation, are not discussed. Mercury intrusion porosimetry is discussed in later sections.

2.2.1.1 *X-ray computed tomography*

X-ray computed tomography (XRCT) became the technique of choice to visualize the microstructure of pharmaceutical granules. It is a nondestructive technique that assesses the three-dimensional distribution of density within the object. The principles and recent advances of the technique can be found in a recent review paper (Maire & Withers, 2014). In most instances, a sample is rotated 180 or 360 degrees inside the instrument, and a series of X-ray transmission images are acquired as shown in Fig. 2.1. Using a sophisticated Fourier transform algorithm, the two-dimensional, side-view projection images create a set of horizontal cross-sectional images which combined, results in a complete three-dimensional map of the sample. The contrast mechanism of the techniques is based on the difference in X-ray linear attenuation coefficient by different materials. The use of the technique for characterization of granule microstructure became feasible in the early 2000s with the development and availability of standalone and table-top systems with micron-sized point sources, which were able to image micron-sized details.

For characterization of pharmaceutical granules, XRCT was applied initially for imaging of pores in granules produced by wet granulation (Farber et al., 2003). The application was expanded for imaging of the arrangement of major excipients in granules produced by fluid-bed granulation (Rajniak et al., 2007) and by melt granulation (Barrera-Medrano et al., 2007), solidified binder bridges in fluidized bed granulation (FBG) granules (Ansari & Stepanek, 2008), and spatial distribution of primary particles within HSWG by size (Rahmanian et al., 2009). XRCT analysis of solidified interparticle bridges in granules produced by wet granulation has been reported only for nonpharmaceutical model systems (Dale et al., 2014).

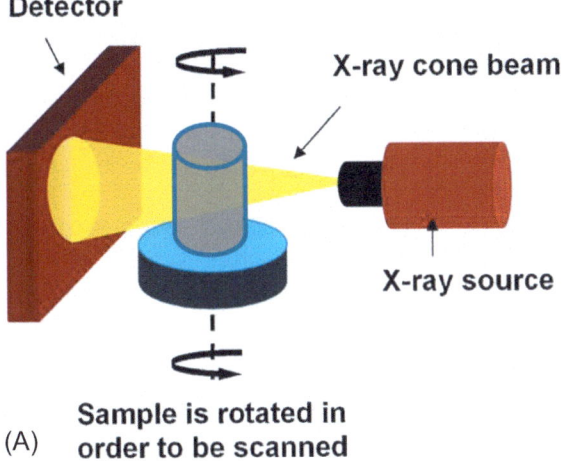
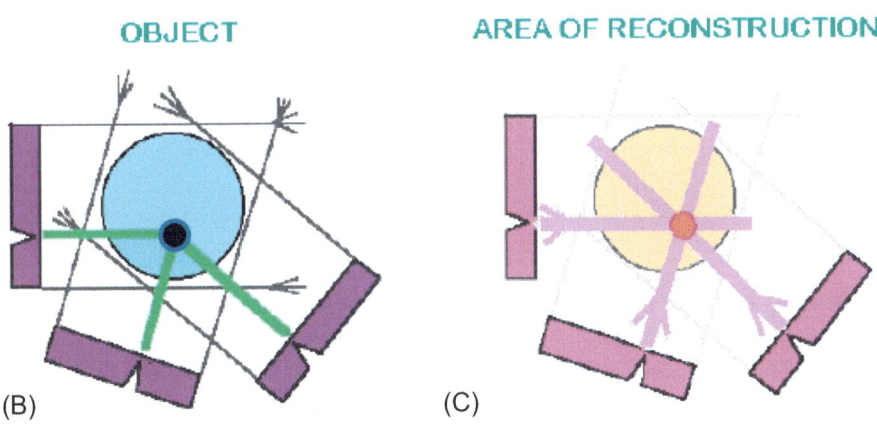

FIGURE 2.1

XRCT schematics: (A) diagram of a XRCT scanning device, (B) illumination of an object containing a dense disk with X-rays and generation of transmission shadow images at different angles of rotation; (C) back-projection calculations using shadow images generated in (B) to recreate the position and the shape of the high-density object inside the body.

Adapted from Moreno-Atanasio et al. (2010). (B and C) Skyscan 1072 Desktop X-Ray microtomograph Instruction manual, Skyscan, 2001.

In addition to standalone micro-XRCT systems, a micro-XRCT attachment that could be fitted inside a standard scanning electron microscopy (SEM) has been developed by Bruker. The system use the primary electron beam of the host SEM to generate the soft X-ray beams needed for CT measurements. The typical resolution of the system is on the order of several microns. This equipment to analyze granules from different stages of the dissolution process allowed, in combination with classical release kinetics measurement, to quantitatively relate porosity with the active pharmaceutical ingredient (API) dissolution kinetics (Kaspar et al., 2013; Oka et al., 2015).

The use of synchrotron radiation is an alternative instrumental approach to achieve relatively high resolution with XRCT. It was applied to study pharmaceutical granules prepared by FBG and by melt granulation (Noguchi et al., 2013), and by spray drying (Guo et al., 2016). Due to its high temporal resolution, synchrotron X-ray micro-computed tomography is preferentially used for studying the dynamics of interaction between binder droplets and powder beds on model systems for better understanding of granule nucleation early stages (Danalou et. al, 2022; Li et. al, 2019).

A novel nano-XRCT unit that operates in Zernike Phase Contrast mode with nominal 2D resolution of 44 and 172 nm has been introduced by Bruker. Various pharmaceutical particles were investigated with this unit (Wong, D'Sa et al., 2014). It was found that only internal voids in micron-sized particles could be imaged with high fidelity. It has been concluded that similarity in linear attenuation coefficient of many pharmaceutical materials–both API and excipients–in real multicomponent systems still remains one of the major challenges in distinguishing between the components of granules and other complex particles. Imaging the topology of solidified interparticle bridges in HSWG granules of pharmaceutical formulations by XRCT has yet to be reported.

Because of the nature of the method, which usually requires physical rotation of a sample and no structural changes during image acquisition, direct observation of dynamic events is challenging. Typical acquisition time for a full set of images spans from several minutes to several hours (for high-resolution datasets), so the dynamic events usually are studied by performing time-lapse imaging (Kaspar et al., 2013).

2.2.1.2 *Scanning electron microscopy*

SEM allows the study of surface morphology at submicron and even nanometer scale resolution. With a typically installed energy dispersive X-ray spectroscopy (EDS) attachment, elemental composition can also be studied at micron and submicron level resolution. SEM is widely used to evaluate the shape and surface topography of pharmaceutical particles and granules. To study the internal structure of granules, however, a cross-section needs to be prepared and imaged.

For relatively hard materials, an object is usually embedded in a resin/epoxy, and then polished using aqueous suspensions of polishing material (diamond or alumina). This approach was used to prepare and study granules of ceramic powders (Eckhard et al., 2014; Eckhard & Nebelung, 2011). "Soft" materials, such as polymers and biological samples, are usually embedded and then cut with a diamond knife using a microtome, with water used as a medium to remove the cut sections. Thus the standard techniques are not directly applicable for the majority of pharmaceutical materials and products that are typically water soluble or otherwise strongly affected by water or moisture. Nevertheless, the approach was used for granules of model pharmaceutical formulation produced by HSWG (Rahmanian et al., 2011). In order to analyze the internal structure in their work, granules were embedded in a resin, sectioned, and polished using a precision lapping and polishing machine, PM2A (Logitech, United Kingdom). The main abrasive was diamond, used in three stages of polishing: 9, 3, and lastly with 1 μm, with ethane diol as lubricant. The samples were washed between each stage using isopropanol. The primary advantages of this approach are a higher resolution than XRCT and a larger number of granules that can be imaged with the higher resolution.

2.2.1.3 Scanning electron microscope-focused ion beam

Dual-beam systems that combine an SEM and a focused ion beam column can acquire a 3D microstructure of microscopic objects by removing material in a controlled manner with an ion beam, producing sequential cross-sections, and imaging the resultant sections with SEM. The technique allows imaging with nm resolution and is used widely in microelectronics and materials science to elucidate the microstructure of materials and devices at nm and micron-size scales. It was applied to study the internal morphology of pharmaceutical particles used for inhalation aerosols (Heng et al., 2007; Zhu et al., 2014). The typical material removal rate, however, is parts of microns per minute, and the technique is destructive. The technique can also introduce artifacts that include heat damage, gallium implantation, and changing crystal structure (Miller et al., 2007). Although the technique could be applicable for studying the microstructure of pharmaceutical granules, it appears to be too slow to be cost-effective for granules of hundreds of microns in size because dual-beam systems are relatively expensive.

2.2.1.4 Optical microscopy

Optical microscopy is traditionally used in materials science to understand the internal microstructure of materials with a resolution down to 0.3 μm. For intensive applications, cross-sections are usually prepared and studied, but, as described above, cross-sectioning of granules of pharmaceutical materials is challenging, so it is not typically used. Optical microscopy was used to qualitatively assess binder distribution within granules (Le et al., 2011; Mangwandi et al., 2010; Nguyen et al., 2010; Schöngut et al., 2015). For this study, a food-grade colorant or a fluorescent dye was added to the binder. Optical microscopy was also used to extract the shape and surface roughness of individual particles (Štěpánek et al., 2009), and this information was used as an input for further simulations of granule growth.

2.2.2 Mechanical properties

The mechanical properties of individual granules are important to their performance in practically all postgranulation unit operations such as drying, milling, pneumatic conveying, transport, blending/lubrication, storage, and compaction. Granules exhibit a rich mechanical behavior that varies from brittle to fully plastic, depending on preparation method, environment (relative humidity, temperature), structure, and loading conditions (i.e., loading rate) (Bika et al., 2001). It is widely accepted that most available practical measures of agglomerate mechanical behavior are not intrinsic, that is, not independent of test specimen geometry and the manner in which stress is applied. Therefore selection and execution of measurements are usually guided by loading conditions and agglomerate size and structure from the process of concern. Practical questions that are often asked in industrial settings pertain to formulation and processing: Measurements need to be able to answer questions such as "Does a vendor source change or PSD shift affect the granule strength or attrition behavior?" or "Can we pneumatically convey this material?" Several methods are used to evaluate the mechanical properties of individual granules. Crushing strength can be used to evaluate the behavior during quasistatic diametrical compression. Single and multiple impact tests, as well as attrition measurements, can be successful in understanding agglomerate behavior during fluid bed drying, pneumatic handling, milling, and coating applications. The methods most often used in applied research and characterization of HSWG granules are described below.

2.2.2.1 *Quasistatic diametrical compression*
Quasistatic diametrical compression is a common test to study the crushing strength of individual granules. The exact procedure for this method is described by ASTM Standard D4179 (1982), Ryu and Saito (1991), and Couroyer et al. (2000). Granules usually are sieved to a narrow cut, and randomly selected granules from this sieve cut are compressed individually between two rigid plates using a testing machine, Fig. 2.2A. A sufficiently large number of granules should be tested per sample, in which the mean and standard deviations do not vary with further testing.

When the granule is brittle, and assuming the granule cross-sectional area to be circular in the plane of the punch, with diameter d_g, the crush strength can be calculated from the load F at breakage as

$$\sigma_{cr} = \frac{4F}{\pi d_g^2}$$

In some studies, an additional coefficient of 0.7 is used to account for irregular shape (Hiramatsu & Oka, 1966).

Single granule diametrical compression measurements also have been used to evaluate the effective elastic modulus of granules and tensile strength (Mangwandi et al., 2010).

The energy that is required to initiate fracture, W_B, can be estimated from the area under the force–displacement data to the point of failure:

$$W_B = \frac{1}{\pi R^2} \int_0^{\Delta max} F d\Delta \tag{2.1}$$

where Δ max is the displacement at fracture.

2.2.2.2 *Single impact tester*
A single-particle normal impact tester was used to investigate attrition and breakage of granular solids by impact (as shown in Fig. 2.2B) (Yuregir et al., 1986). A wide range of particulate materials was investigated using this test, such as solid semibrittle salts (Zhang & Ghadiri, 2002), polymer particles (Papadopoulus & Ghadiri, 1996), weak (binderless, cohesive) agglomerates (Boerefijn et al., 1998), and strong agglomerates (Subero & Ghadiri, 2001). The fragmentation of particles showed qualitatively distinct regimes of breakage that appear clearly connected to particle structure and impact velocity. In the apparatus, individual granules are entrained by air and impact a metal target, Fig. 2.2B. The impact velocity is controlled by varying the vacuum level in the apparatus and is measured optically. Typically 8 to 10 g of granules from a specific mesh cut (850–1000 μm sieve size) are fed into the testing tube,

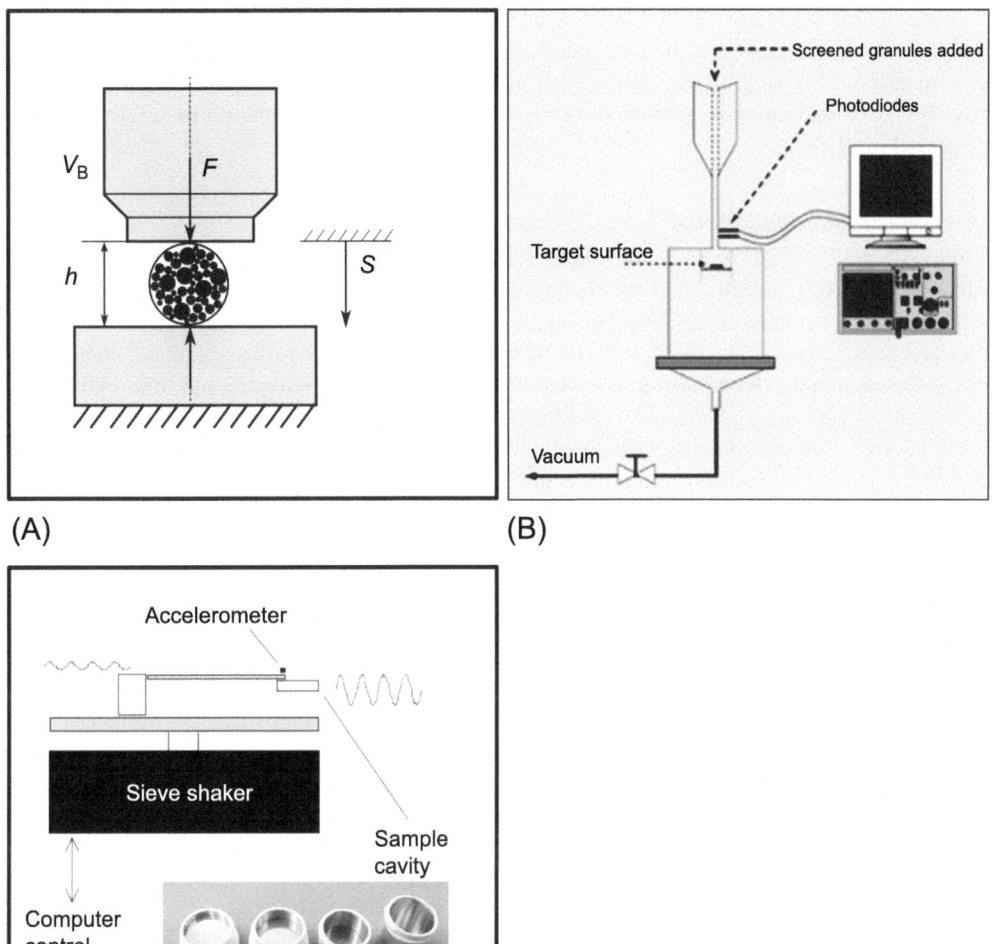

FIGURE 2.2

Schematic representation of different mechanical tests on granules: (A) compression test of a single granule; (B) single impact tester; (C) vibrational impact tester.

Adapted from Dosta et al. (2016); Cai et al. (2013); Gentzler and Michaels (2004).

and their fragments are then collected and sieved. The retention on a specific sieve is measured. The weight loss of the granules could be calculated and plotted. When this test was used to compare the impact strength of HSWG granules produced from API powder and different binders, it showed that breakage behavior on impact is controlled mostly by a binder rather than by particle size or nature of API (Cai et al., 2013).

2.2.2.3 Vibrational impact test

A multiple-impact tester based on the resonant cantilever impactor was developed by Beekman et al. (2002) and modified by Gentzler and Michaels (2004), who also developed a novel technique to experimentally obtain single-particle attrition rates from the data, Fig. 2.2C. For the test, granules are sieved to a narrow cut of 1180 to 1400 µm. Six to eight granules are placed in a resonator chamber of a specific depth and tested at a specific combination of amplitude and time. After testing, all significantly sized material was collected from the impact chamber and sieved through a 600 µm sieve, intentionally chosen to be approximately one half of the particle diameter. Particle damage or loss is defined as the fraction of original mass not recovered on the 600 µm sieve. The particle weight loss rate is measured and then recalculated for weight loss as a function of velocity. The study on pharmaceutical granules showed that a transition from attrition to gross fragmentation is connected to the granule microstructure.

2.2.2.4 Granule friability test

In industrial settings, relatively simple granule friability tests can be used to compare propensity for damage under specific conditions. A description of a small-scale ball milling test is an example (Li et al., 2011). "Approximately 2 g of accurately weighed and recorded sample of granules of the desired size range (425–710 µm) are placed into a 20 mL scintillation vial, followed by the addition of 5 g of 4 mm diameter glass balls. The vial was then placed into a 60-mL glass bottle, supported by article towels, to avoid any movement of the vial within the glass bottle. After supporting the vial within the bottle, the bottle was placed in a Turbula Mixer and rotated for 20 min at 56 rpm. For measuring the loss due to friability, the contents of the vial (granules and glass beads) were poured onto a sieve stack consisting of 20 and 40 mesh screens. The glass balls were retained on the 20 mesh screen and removed. The sieve stack with the sample was placed into the sonic sifter, and run at shift five, pulse five for 5 min. Finally the granules retained on the 40 mesh screen were accurately weighed and recorded. The friability index was taken as $(1 - Y/X)/100$, where X is the original sample retained on the screen (425–710 µm) and Y is the sample retained by a 40 mesh (425 µm) sieve after testing."

2.3 Microstructure of granules

2.3.1 Granule formation

Agglomeration in HSWG processes is usually described as a combination of rate processes that include wetting and nucleation, coalescence and growth, and breakage (Iveson et al., 2001). Regime maps based

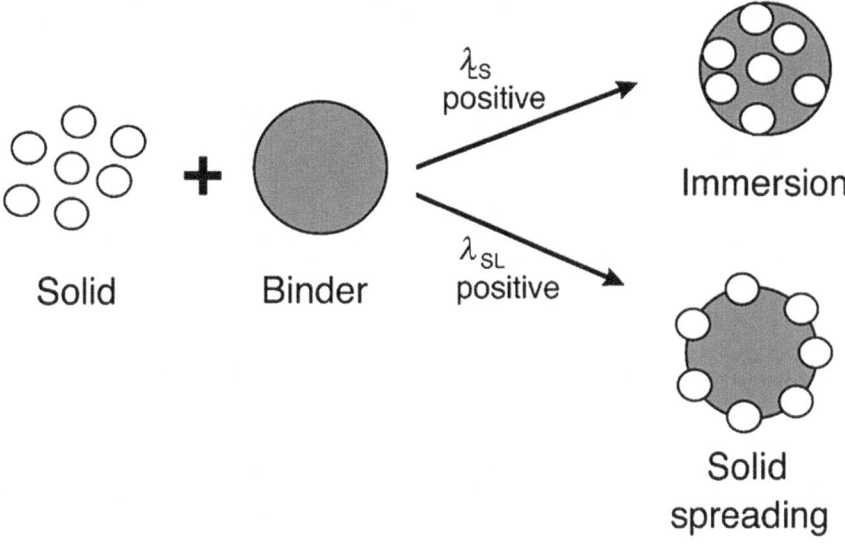

FIGURE 2.3

Nucleation formation mechanisms as a function of spreading coefficient values.

Adapted from Hapgood et al. (2009).

on the dimensionless groups consisting of relevant physical parameters have been proposed to describe those processes.

Wetting and spreading of the granulation liquid in nucleation play a critical role in determining the overall granule microstructure. This behavior can be driven by spreading coefficients (Krycer & Pope, 1983; Planinsek et al., 2000; Rowe, 1989a, 1989b; Wu, 1973) and is schematically shown in Fig. 2.3. In a typical wetting system, granule nuclei are formed by the penetration of liquid droplets into dry powder. When the powder is agitated, these wet nuclei grow into granules by coalescence or layering. The microstructure of granules in such systems is described in Sections 2.3.2.1 and 2.3.2 to 2.3.4. In poorly wetting systems, the granulating liquid cannot penetrate the powder to form nuclei. Instead, powder spreads around the drop during agitation in HSWG. Such behavior can lead to the formation of hollow granules, which are described in Sections 2.3.2.2 and 2.3.2.3. Hollow granules may also form in wetting systems when the liquid can not penetrate the powder due to high viscosity. This is described in Section 2.3.2.3.

2.3.2 Pores

2.3.2.1 Typical wetting systems

A representative microstructure of granules produced by HSWG from a single-component powder in a system with good wetting and a low viscosity binder is shown in Fig. 2.4 (Farber et al., 2003;

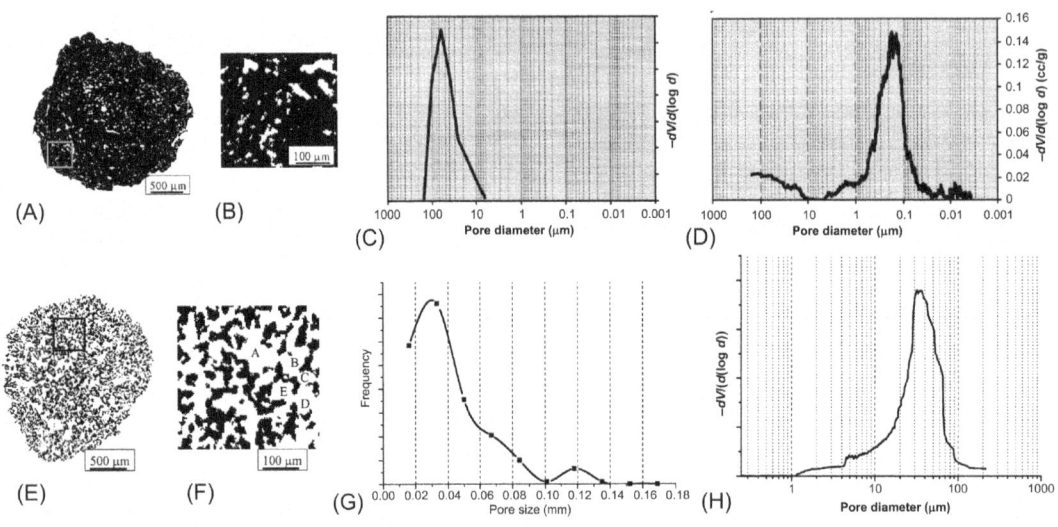

FIGURE 2.4

Tomographic images and pore size distributions of mannitol granules produced under (A–D) high shear and (E–F) low shear in the Fukae mixer: (A), (E) typical reconstructed cross-sectional image (binarized) (B), (F) high-magnification image of area in the box in (A) and (E), respectively (C), (G) pore size distribution calculated from tomographic images; (D), (H) pore size distribution as measured by mercury porosimetry.

Adapted from Farber et al. (2003).

Farber et al., 2003b). Mannitol–a brittle crystalline hydrophilic material–was granulated using a small-scale 2 L granulator at different impeller/chopper speeds with different liquid binders (pure water or HPC aqueous solution of low viscosity). The produced granules can be described as porous bodies. Depending on processing conditions and binder, total porosity can vary significantly, ranging from about 12% to 58%. For a denser granule produced at high shear conditions, most of the pores in the images appear occluded (Fig. 2.4). The overall porosity determined from XRCT data is about 12%. The volume-based pore size distribution from XRCT, calculated from the equivalent projected area in Fig. 2.4, was unimodal with the mode close to $dv = 62$ μm. Mercury porosimetry showed total accessible porosity of about 15% and a unimodal pore size distribution with a mode of 0.1 to 0.2 μm (Fig. 2.4D). This indicates either that the granules indeed have closed pores and the mercury crushed the granules at pressure higher than ca. 10 MPa, or that pores visualized by XRCT in Fig. 2.4 are interconnected (and connected to the surface) by channels with effective diameter of 0.1 to 0.2 μm, which is below the resolution of XRCT, and are accessible, in spite of appearing occluded.

A granule produced under conditions of much lower shear had much larger overall porosity of about 58% as determined from both mercury porosimetry and XRCT, Fig. 2.4E and Fig. 2.4F. The pore morphology is different: relatively large chambers (pores) of 80 to 120 μm in diameter (identified by capital letters from A to D in Fig. 2.4F) are seen to be connected to each other via orifices that appear about 20 to 40 μm in diameter. A diameter-equivalent pore size distribution calculated from the linear distribution showed two peaks that correspond in size to channels and cavities, respectively. Mercury porosimetry showed an unimodal pore size distribution with the mode of about 33.5 µm, failing again to fully characterize the internal pore distribution. XRCT demonstrates clearly that the size of the internal pores does not change dramatically, even though this could easily (but erroneously) be inferred from the mercury intrusion pore size distributions. The major difference in granules produced under different conditions is not the size of the pores, but the connectivity. The work also showed that, although XRCT is inferior in precision and dynamic range to mercury porosimetry for measuring total porosity, it provides morphological information such as pore shape, spatial distribution, and connectivity that cannot be measured by mercury intrusion, which measures the diameter of pore necks only.

For comparison, mannitol granules produced by FBG are shown in Fig. 2.5 (Rajniak et al., 2007). Granules produced with 5% HPC binder appear somewhat similar to the granules produced by HSWG at lower shear. With increases in binder solution concentration/viscosity (to 10% and 15%), however, progressively more open structures are formed in FBG. The observed fluid bed dryer (FBD) granule microstructures were in good agreement with the computed microstructures. An approach for computing microstructures of FBD granules was developed in Štěpánek (2004) and Štěpánek and Ansari (2005). Virtual granules have been generated by random packing of primary particles using the viscosity and wetting properties of a binder, as well as the shape and surface topography of solid particles, and assuming a small binder droplet size versus particle size. The approach allows "growth" of virtual agglomerates and investigation of the effect of process parameters, binder properties, and morphology of primary particles on kinetics of granule growth and granule attributes such as shape and porosity.

The microstructure of granules of a typical multicomponent pharmaceutical formulation produced by HSWG at an industrial scale is qualitatively similar to the microstructure of HSWG mannitol granules produced at high-shear conditions, Fig. 2.6. Large pores that appear isolated are uniformly distributed throughout the volume. Total porosity by XRCT, however, was significantly lower than the porosity measured by mercury porosimetry. XRCT resolution of 3 µm/pixel was sufficient only to image relatively large pores of a typical size of 20 µm, which accounted for less than half of the total porosity. Total porosity calculated from XRCT was 14%, while 31% porosity was measured by mercury porosimetry. Mercury porosimetry also showed bimodal PSD, with a main peak at approximately 2 µm, indicating the presence of smaller channels that interconnect larger pores and form a continuous network that is accessible from the surface.

For other industrial granules, XRCT was helpful in understanding the difference in dissolution properties of granules produced at different Froude numbers (Smrcka et al., 2016), Fig. 2.7. The XRCT analysis of granules produced at lower Froude numbers revealed the existence of a percolating pore network spanning the entire granule and connected to the granule surface. The pores present in granules produced at higher Froude numbers appear to be more isolated in the XRCT cross-sections. Granule strength has not been measured in any of these studies.

Unlike the above studies where the granules seemed to be formed by consolidation and layering, two different types of granules were observed in each sieve fraction of granules produced by HSWG of a starch/lactose blend (Le et al., 2011). In that study, HSWG mechanisms and their effect on granules' microstructure and properties were investigated by using blue- and red-colored HPC binder solutions that were added at two locations simultaneously (opposite sides of the granulator). The approach allowed measurement of binder content in samples by dissolving them and measuring UV absorption.

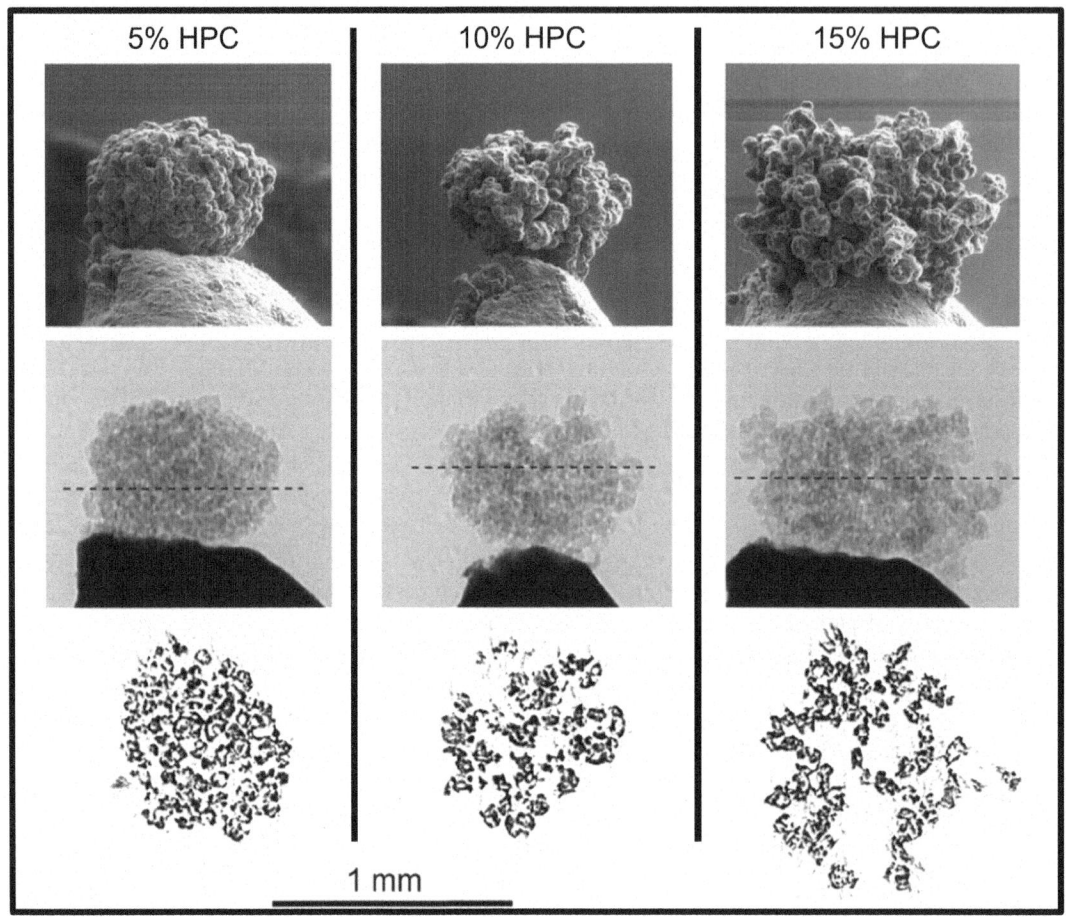

FIGURE 2.5

Morphological analysis of the mannitol granules manufactured via FBG with different binders. HPC concentration in the liquid binder is shown (wt%). Top images, SEM; middle images, X-ray transmission; bottom images, reconstructed cross-sections (binarized) at a height approximately shown as a dashed line on the transmission images

Adapted from Rajniak et al. (2007).

Representative cross-sectional tomography images of granules from the same sample are shown in Fig. 2.8, along with optical images. Granules were taken from the same sieve cut. Some granules (as in Fig. 2.8A and Fig. 2.8B) appeared to be formed by the coalescence of smaller granules with a resulting relatively large aggregate structure. Other granules (Fig. 2.8C and Fig. 2.8D) appeared to be formed directly from primary particles by consolidation and had significantly lower porosity.

2.3 Microstructure of granules

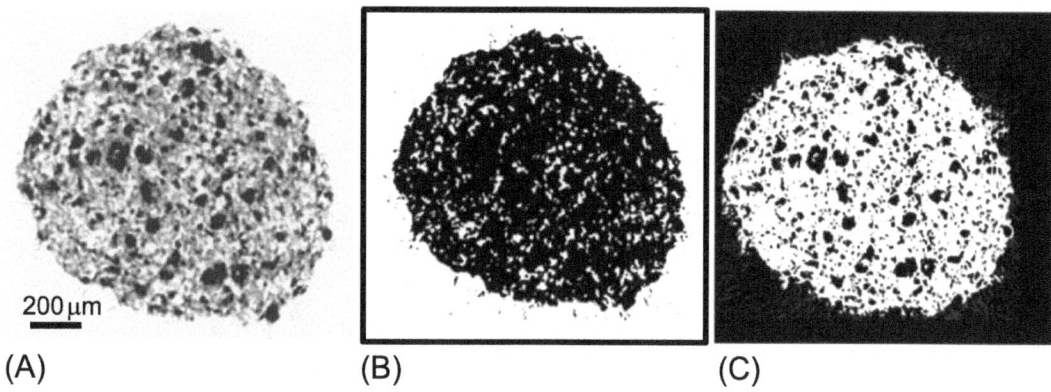

FIGURE 2.6

XRCT results for a typical granule of a commercial HSWG product: (A) reconstructed cross-section, showing presence of pores (*white contrast*) and particles of di-basic calcium phosphate anhydrous (*dark contrast*) within the granule; (B) pores only (*white contrast*); (C) dibasic calcium phosphate only (*dark contrast* in the body of the granule) (granule total porosity calculated from XRCT images is 14%).

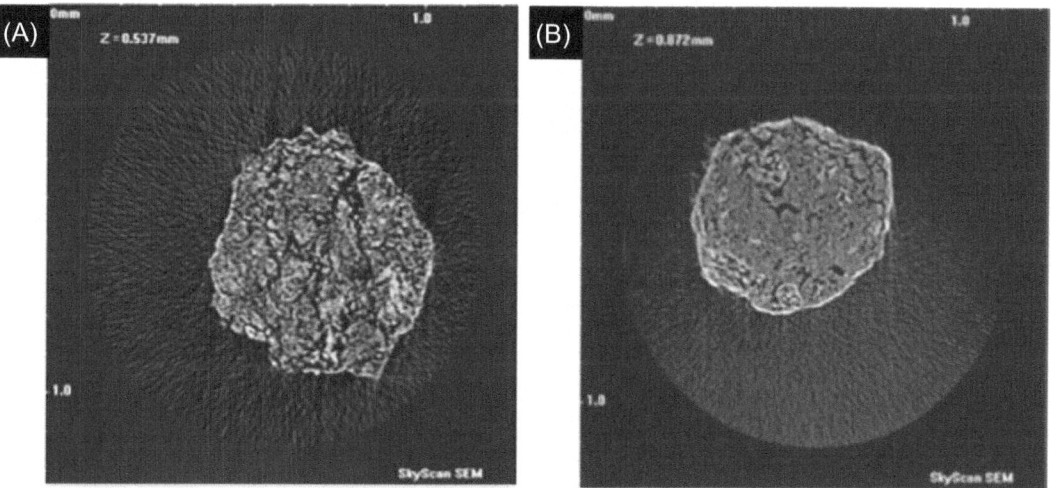

FIGURE 2.7

Representative XRCT images (reconstructed cross-sections) of granules produced at different Froude numbers at 4 L scale: (A) $Fr = 0.26$; (B) $Fr = 1.23$.

Adapted from Smrcka et al. (2016).

Overall granules with consolidation microstructure had higher binder content, lower porosity, and higher strength (see Fig. 2.9A–C). As would be expected, they dissolved more slowly than the granules formed by coalescence (Fig. 2.9F).

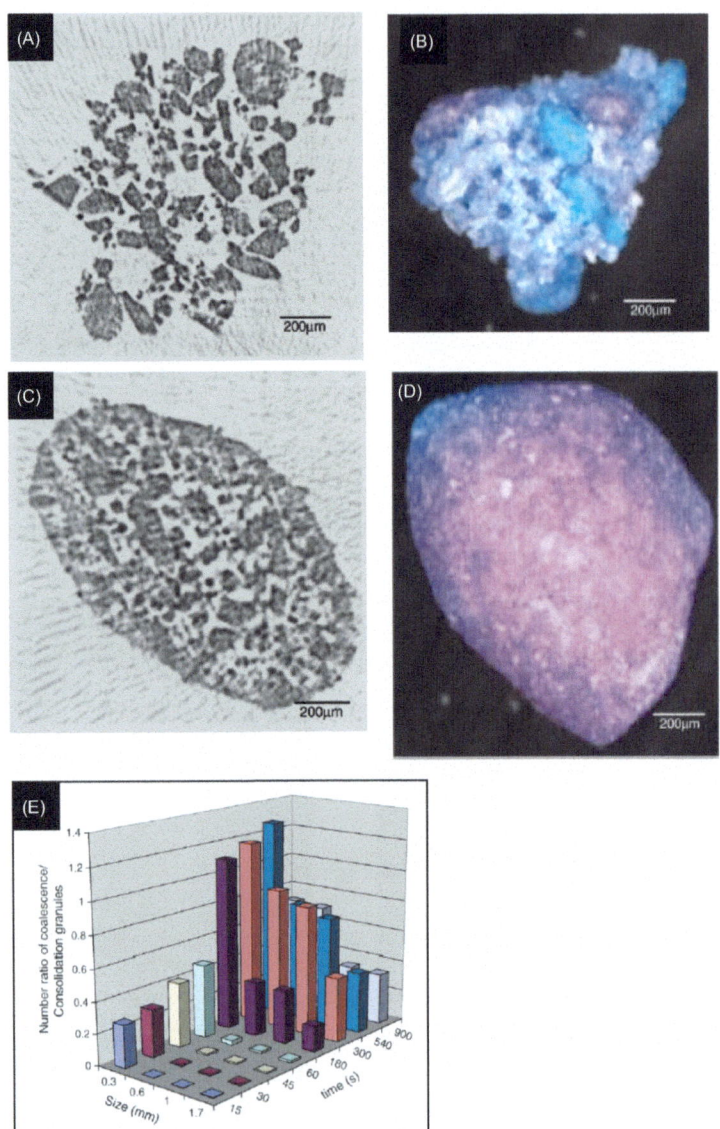

FIGURE 2.8

Representative (A and C) X-ray tomography and (B and D) optical images of two structures of granules–coalescence granules and consolidation granules–in the size range of 1.0–1.4 mm observed in a granulation sample, and (E) number ratio of coalescence granules to consolidation granules at different granulation times and size classes: (A and B) coalescence granule structure; (C and D) consolidation granule structure.

Adapted from Le et al. (2011).

2.3 Microstructure of granules

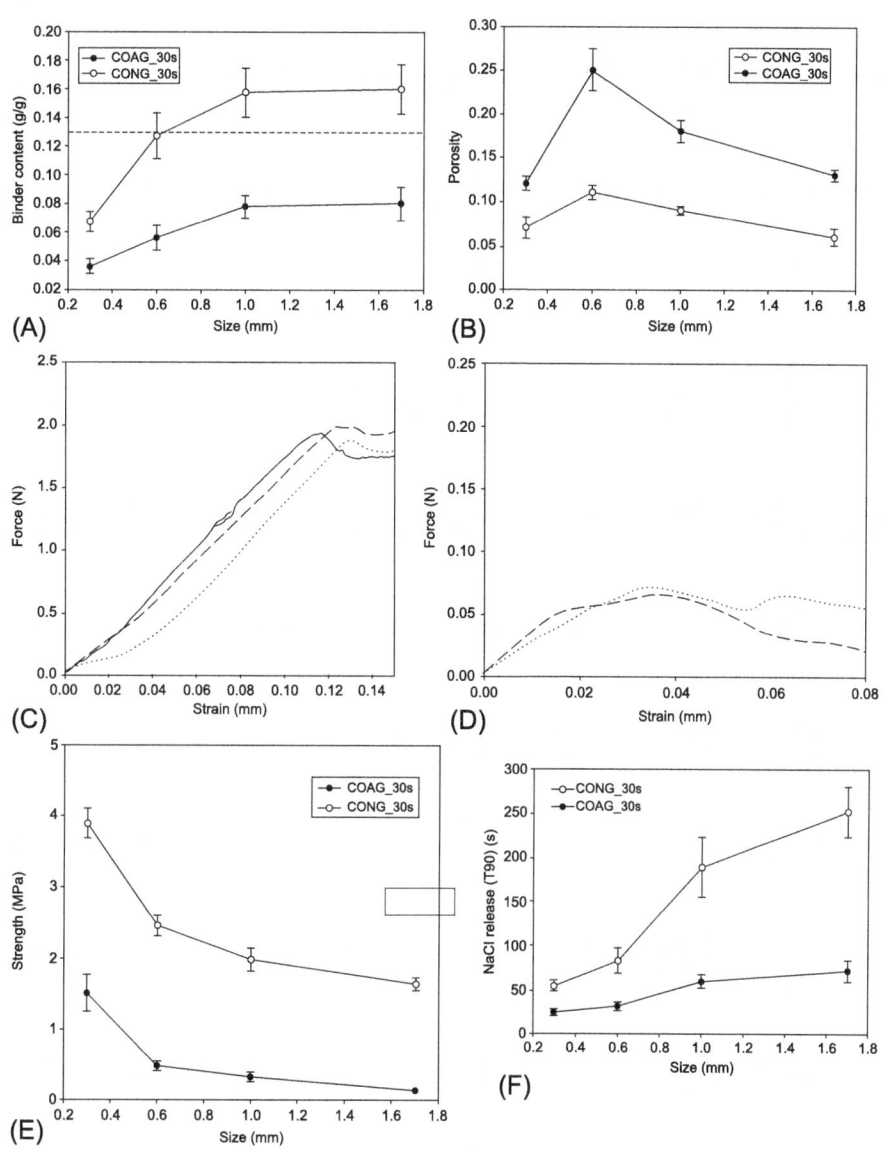

FIGURE 2.9

Properties of granules from two structural classes—coalescence (COAG) and consolidation (CONG): (A) binder content as a function of a size; (B) granule porosity as a function of a size; (C) representative compression behavior of CONG granules; (D) representative compression behavior of COAG granules; (E) granule crush strength as a function of size; (F) characteristic drug release time (T90) as a function of size.

Adapted from Le et al. (2011).

Table 2.1 Formulation, particle diameter, and estimated droplet diameter corresponding to the maximum interparticle distance for a specific excipient.

Component	Wt%	Particle diameter (d, µm)	Droplet diameter (D, µm)
Hydrophobic powder	70	1	n/a
MCC	20	100	296
SLS	1	30	293
Ac-Di-Sol	5	50	264
HPC-LF	4	200	1155

Adapted from Hapgood et al. (2009).
Ac-Di-Sol, Cross-carmelose sodium; HPC-LF, hydroxypropyl cellulose HPC-LF; MCC, microcrystalline cellulose; SLS, sodium lauril sulfate.

Although it appears that granules formed by coalescence (i.e., agglomeration) generally are weaker than the granules formed by layering, the formulation, not surprisingly, still can play a dominant role in determining granule mechanical behavior. For example, when the attrition behavior of HSWG granules of two different formulations–layering and coalescence–was compared, it was reported that granules formed by coalescence had higher strength than the granules formed by layering, although both had approximately similar porosity (Fig 2.10) (Gentzler & Michaels, 2004).

Knowledge about specific microstructure and its evolution during the granulation process allows a better understanding of the interplay between the different granulation rate processes, such as breakage, consolidation, and coalescence. For example, Fig. 2.8E shows that the number of coalescence granules increases initially during the granulation process, but then, at longer times (after 300 s in this study), decreases, apparently because of granule breakage. This information could be used as a calibration tool for the development of population balance models.

2.3.2.2 Poorly wetting systems of pharmaceutical formulations. hollow granules.

Qualitatively different microstructures have been observed for HSWG granules formulated with highly hydrophobic powder (Hapgood et al., 2009). Unlike granules of hydrophilic formulation, in which porosity is distributed relatively uniformly through the volume of granules, robust hollow granules have been produced by HSWG. The formulation is shown in Table 2.1. The binder (HPC) was added as a powder.

Fig. 2.11 shows representative X-ray tomographic reconstructions of the internal structure of granules retained in different sieve fractions. These images show that the majority of finer granules consisted of a consolidated powder shell and an empty core. The wall thickness of the granule ranged from 25 to 50 µm. It is believed that the internal hollow space was originally filled with granulating fluid that evaporated during drying. These results are consistent with granule nuclei formed by solid spreading around water droplets to form a liquid sphere with subsequent evaporation of the fluid to form a hollow granule (Fig. 2.3). The majority of larger granules had hybrid internal structures, containing several hollow nuclei surrounded by less dense material. Fig. 2.11 shows a side view of such a granule and several reconstructed cross-sections. They show that the hollow areas typically are

2.3 Microstructure of granules

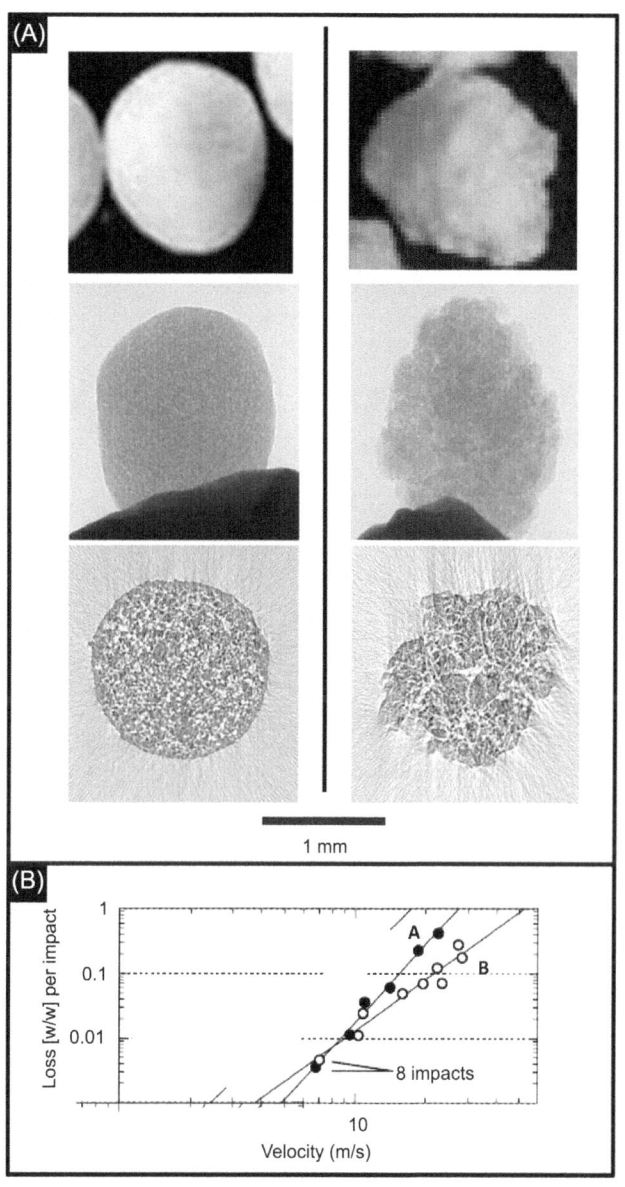

FIGURE 2.10

Representative images of HSWG granules of two commercial products with different dominant microstructure (a—consolidation; b—coalescence) and (B) their attrition behavior; in (A): top images, light microscopy; middle images, transmission X-ray images; bottom images, reconstructed cross-sectional images.

Adapted from Gentzler and Michaels (2004).

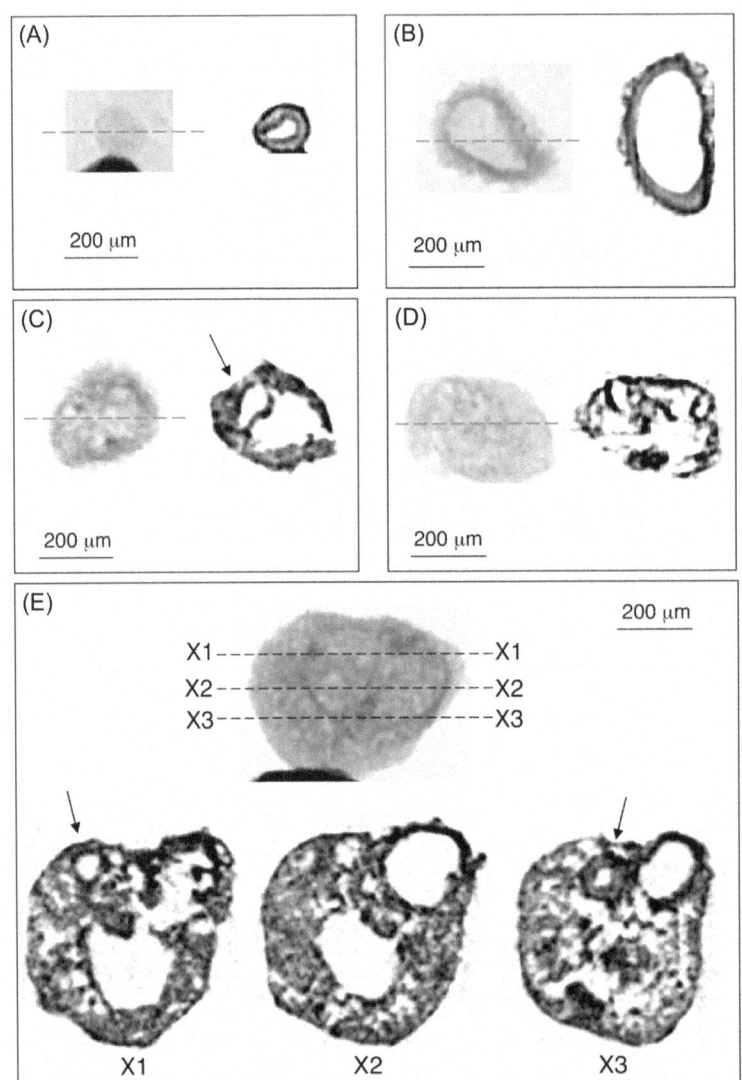

FIGURE 2.11

Hollow granules formed in a formulation of a hydrophobic powder after HSWG. X-ray tomographs of granules in different mesh cuts: (A) − 105 + 150 mm; (B), (C), and (D) − 180 + 300 mm; (E) − 500 + 1000 mm. (A–D) left image and (E) top image: transmission shadow X-ray image; right image on (A)–(D) and bottom images on (E) are reconstructed cross-sections. *Dotted lines* on side view images indicate the position of the reconstructed cross-sections. *Arrow* points to embedded small nuclei that are comparable in size with the nuclei in (A).

Adapted from Hapgood et al. (2009).

Table 2.2 Mass fraction of the components in the formulation.

Component	Mass fraction	Particle diameter (d_{MEAN}, μm)
API	0.58	3.3
Prosolv SMCC90	0.24	63.8
Neosorb PT100	0.07	96.5
Povidone K25	0.05	72.5
SLS	0.01	42.5
Crosspovidone (Polyplasdone XL)	0.05	73.5

Adapted from Smrcka et al. (2016).
SLS, Sodium lauril sulfate.

surrounded by a denser, homogenous layer that is believed to be the wall of a hollow nucleus. It appears that this microstructure was formed by the coalescence of hollow nuclei and subsequent layering with ungranulated fines to form a thick powder shell around the coalesced nuclei. For granules of intermediate sizes, both single- and multinuclei microstructures were observed, as shown in Fig. 2.11B–D.

Although granule strength has not been measured explicitly, the hollow granules had sufficient mechanical strength to withstand dry milling. Milling eliminated granules larger than the mill screen opening of 1.2 mm; however, the distribution was unchanged qualitatively. In particular, the fines fraction did not increase significantly, indicating that the smaller granules largely survived milling, and the large granules broke into fragments rather than disintegrating into primary particles.

Various experiments have demonstrated the critical role of both the formulation and the processing conditions for the formation and survival of stable hollow granules.

The effect of fluid level and wet massing (WM) time on the external morphology of 300 to 500 μm granules is shown in Fig. 2.12. At 1 min of WM, the morphology did not change appreciably between 70% and 78% fluid level. The external surfaces of the granules in all of these cases are composed of small particles, suggesting that they were produced by the layering of fines. WM for 7 min at 70% water did not change PSD or surface morphology; however, the formerly hollow granules appear collapsed and deformed. At 78% water and 7 min of WM, the external microstructure changed significantly; these granules are conglomerates of smaller granules rather than primary particles. The domination of collapsed spheres after prolonged WM indicates an additional rate process during the granulation, which is most likely shear-mediated.

In a single drop experiment with the API from Table 2.2, liquid marble nuclei were easily formed. However, the structure collapsed and formed a hemisphere when dried, consistent with the lack of a binding agent in the dry state. Formation of robust hollow granules from the formulation in Table 2.2 suggests that the shell contains HPC (the only binder in the dry mix) and that water permeated into the shell and dissolved the HPC during processing. On drying, this created solid bridges between particles in the shell, which cemented the shell (Farber et al., 2005).

Calculations also showed the importance of drop size relative to the interparticle distance of hydrophilic excipients dispersed in the hydrophobic matrix. Droplet sizes that were smaller than the

54 Chapter 2 Microstructure and mechanical properties

FIGURE 2.12

Effect of GFL level and wet massing time on granule morphology (SEM) granules produced at 70% and 78% fluid level and 1 and 7 min of wet massing (granules from − 300 + 500 μm mesh cut).

Adapted from Hapgood et al. (2009).

calculated average interparticle distances (see Table 2.1) between the hydrophilic particles should mostly result in the formation of stable liquid marbles. Larger drops may initially form a liquid marble, but they will interact with hydrophilic excipients to produce a drainage path through the powder shell. The results in Table 2.1 predict the maximum drop diameters between 300 and 1000 μm, bracketing the 300–500 μm hollow granule size observed in the granulation experiments (Table 2.1).

Formation of hollow granules in HSWG of another pharmaceutical formulation of a hydrophobic API was reported by Smrcka et al. (2016). The formulation shown in Table 2.2 was granulated

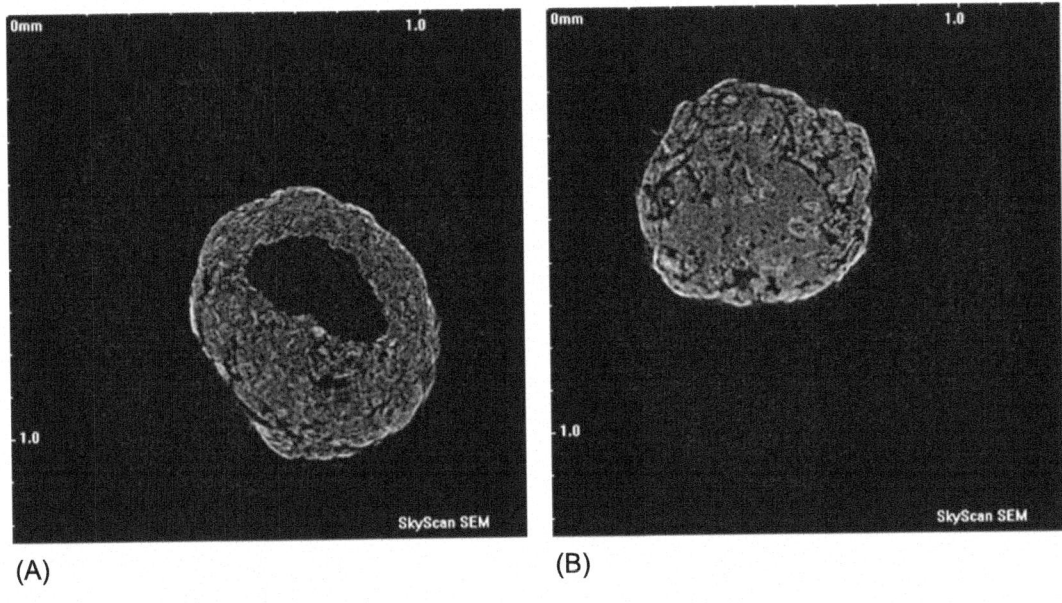

(A) (B)

FIGURE 2.13

Example of a (A) hollow-core and (B) regular granules from the same HSWG batch.

Adapted from Smrcka et al. (2016).

using different agitation intensities, expressed by the Froude number, as well as at two scales—4 and 400 L.

From five randomly chosen granules above 400 μm produced at $Fr = 1.23$ in the 4-L granulator that were analyzed by micro-CT, two contained a hollow core, Fig. 2.13. In other batches, hollow granules were not observed. Apparently the formulation is similar enough to the one in Table 2.2 to allow formation of hollow granules at some processing conditions (which might include a combination of a flow pattern and a specific spray droplet size distribution), but still not sufficiently adequate to allow robust formation of hollow granules for a wide range of conditions, unlike the formulation in Table 2.1.

The microstructure of granules of a hydrophobic powder blend appeared to be somewhat different when a polymeric binder was delivered in solution during HSWG (Fig. 2.14). In this work, aqueous solutions of either hydroxypropylcellulose (HPC) grade E or polyvinylpyrrolidone (PVP) grade K29 were delivered to a small-scale granulator droplet-wise using a peristaltic pump, and relatively large granules were studied using XRCT. The data on binder viscosity has not been provided. Images demonstrate that both granules have a core/shell microstructure. For HPC-containing granulation, the core consists of several hollow granules with relatively dense shells. For PVP-based granules, the core appears to consist of fully solid nuclei that could be formed from collapsed liquid marbles. It was shown in Nguyen et al. (2010) that aqueous solution of PVP might transitionally wet hydrophobic powder,

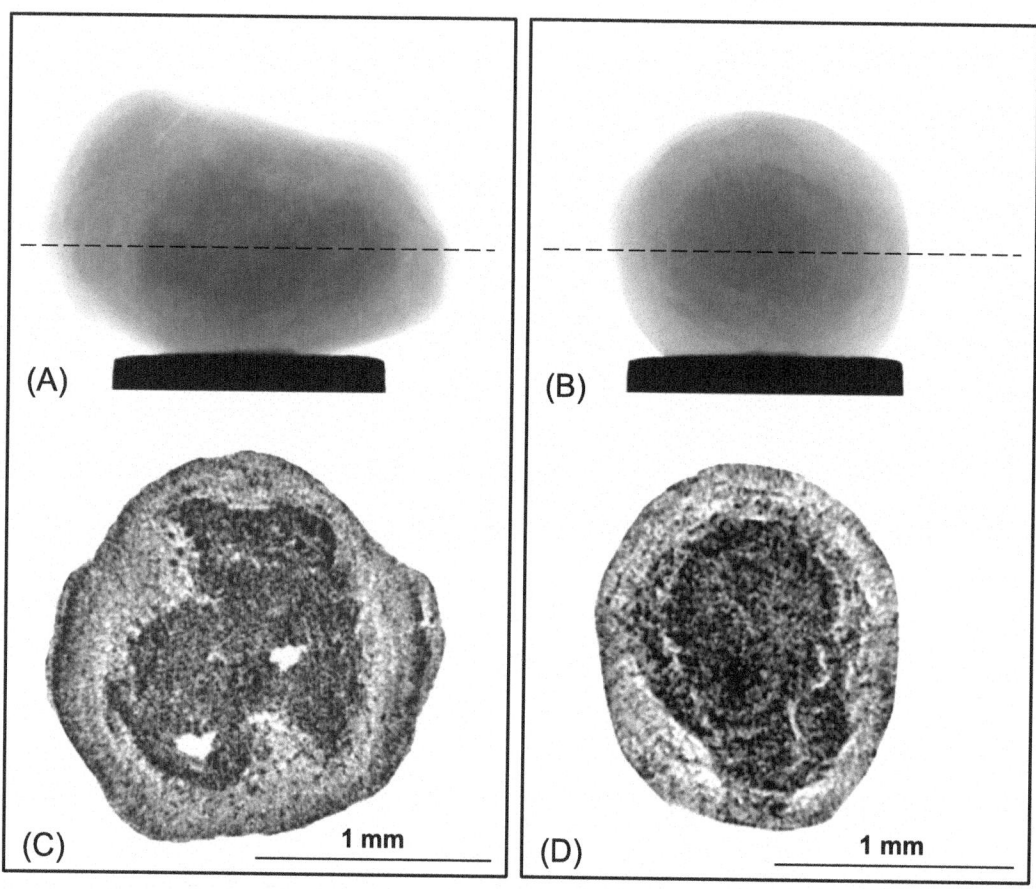

FIGURE 2.14

X-ray tomographs of granules of hydrophobic powder granulated with different polymeric binder solutions: (A), (C): HPC solution; (B), (D) PVP solution; (A), (B): transmission shadow X-ray images; (C), (D): reconstructed cross-sections. *Dotted line* on side view images indicate the position of the reconstructed cross-section.

providing relatively short penetration time into a powder bed of a hydrophobic powder. Outside the core, granules are significantly less dense for both binders, with porosity distributed nonuniformly. Many relatively coarse pores (tens of microns in size) are concentrated in the vicinity of the core/shell interface. The microstructure is consistent with the nucleation and granulation mechanism of mixtures of hydrophilic and hydrophobic components proposed in (Charles-Williams et al., 2013), Fig. 2.15. Mechanical properties of the granules were characterized in this study.

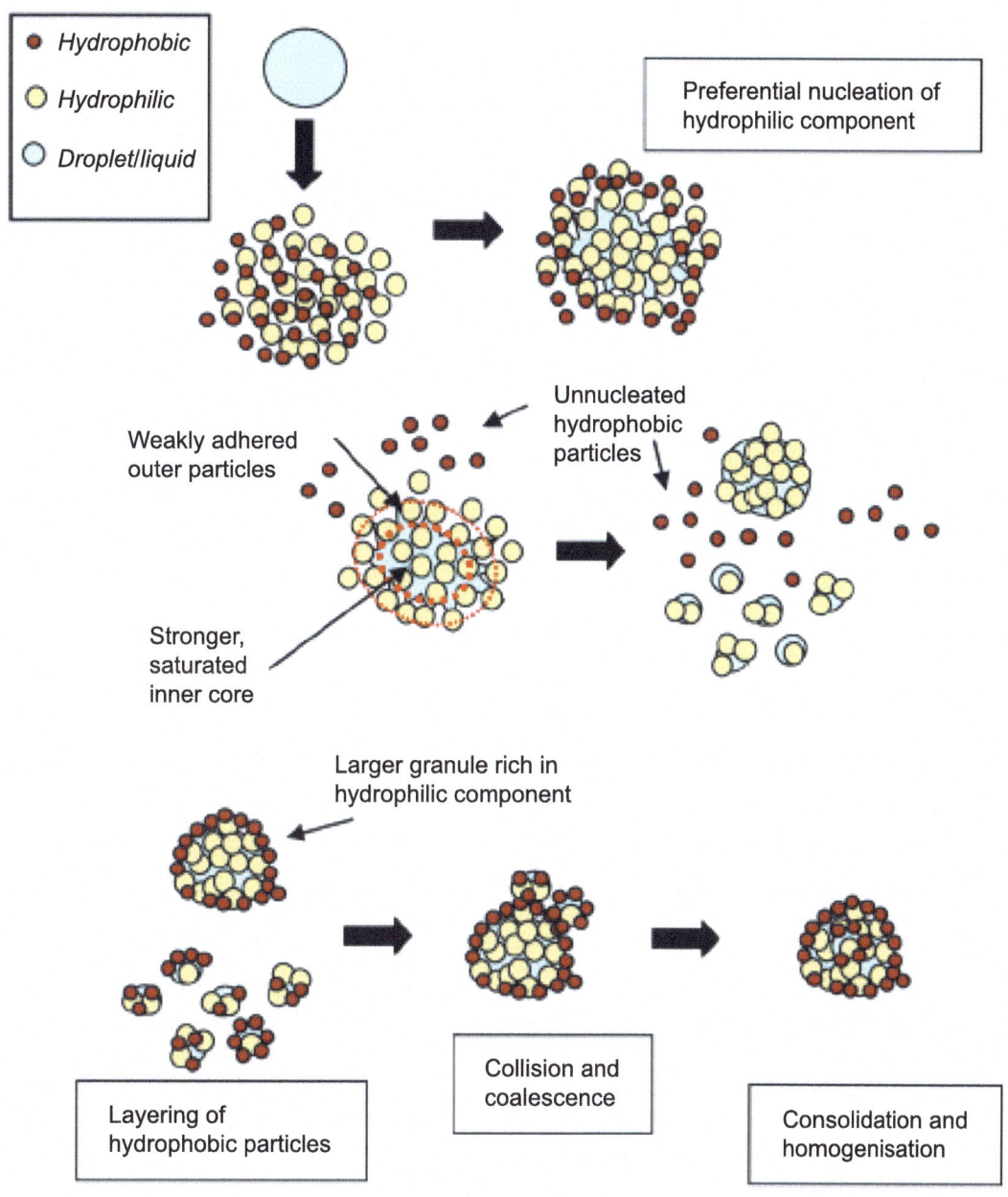

FIGURE 2.15

Proposed nucleation and granulation mechanism for granulation of mixtures of hydrophobic and hydrophilic components.

Adapted from Charles-Williams et al. (2013).

2.3.2.3 *Hollow granules: Model studies*

2.3.2.3.1 Single-component hydrophobic powder and a binder

The formation of hollow granules was investigated for several types of hydrophobic powder using droplets of various binders. The droplets were delivered on a powder bed from a syringe and then lightly agitated on the surface of the powder bed, which resulted in the formation of "liquid marbles" (Eshtiaghi et al., 2010). It has been recognized in this work that the drying of such liquid marbles would be expected to show some similarities with the drying of liquid drops containing solutes that can result in the formation of hollow spherical shells at some drying conditions (Walton & Mumford, 1999). Indeed elevated drying temperature, smaller or nano-size particles, and higher binder concentration tended to promote the formation of spherical hollow granules, which appeared to consist of a solidified binder shell coated with hydrophobic powder particles. Different combinations of drying temperature, binder type, binder concentration, and powder type/grade resulted in large changes in the survival rate, indicating the importance of careful formulation during hollow granule process development. However, attempts to produce hollow granules in the most promising simple binary model system from that study (hydrophobic Aerosil R202 and 5% HPC binder) in HSWG were not very successful (Eshtiaghi et al., 2009). Although some stable hollow granules were formed under some granulation conditions, the granule structure was far more irregular and complex than the spherical single cavity granules produced either in model experiments with light agitation or from the pharmaceutical formulation shown in Table 2.1.

2.3.2.3.2 Wetting and nonwetting binary mixtures of hydrophobic active pharmaceutical ingredient and hydrophilic iron oxide granulated with binders of different viscosities

Nonwetting micron-size pharmaceutical active powder was blended with a submicron wetting excipient (iron oxide) in three different ratios to produce wetting, nonwetting, and marginally wetting powder blends. Dextran solutions with viscosities of *ca* 10, 100, and 1000 mPa•s served as binders (Farber et al., 2020). Model nuclei were prepared from all combinations of these blends and binders. A binder droplet was delivered on a powder bed using a syringe, and the formed nuclei were further granulated in a tumbling drum and a high shear granulator. The effects of blend wettability, binder viscosity, granulation time, and shear forces on granule size and microstructure were investigated. In the tumbling drum, hollow granules were produced at higher binder viscosities of 100 and 1000 mPa•s, regardless of blend wettability, Fig. 2.16. Based on the original drop size, the tumbled marbles have evolved and accreted powder but not undergone significant aggregation or breakage. Unlike that, solid granules with a complex internal microstructure were produced at a low binder viscosity of 10 mPa•s regardless of blend wettability. Overall granulation of the marginally wetting blend with medium viscosity binder in the tumble blender for 5 min resulted in the formation of "perfect" hollow granules: the shell was compositionally uniform, the size was significantly larger than the size of the original droplet, the shape was close to spherical, and there was no rupture from internal gas escape.

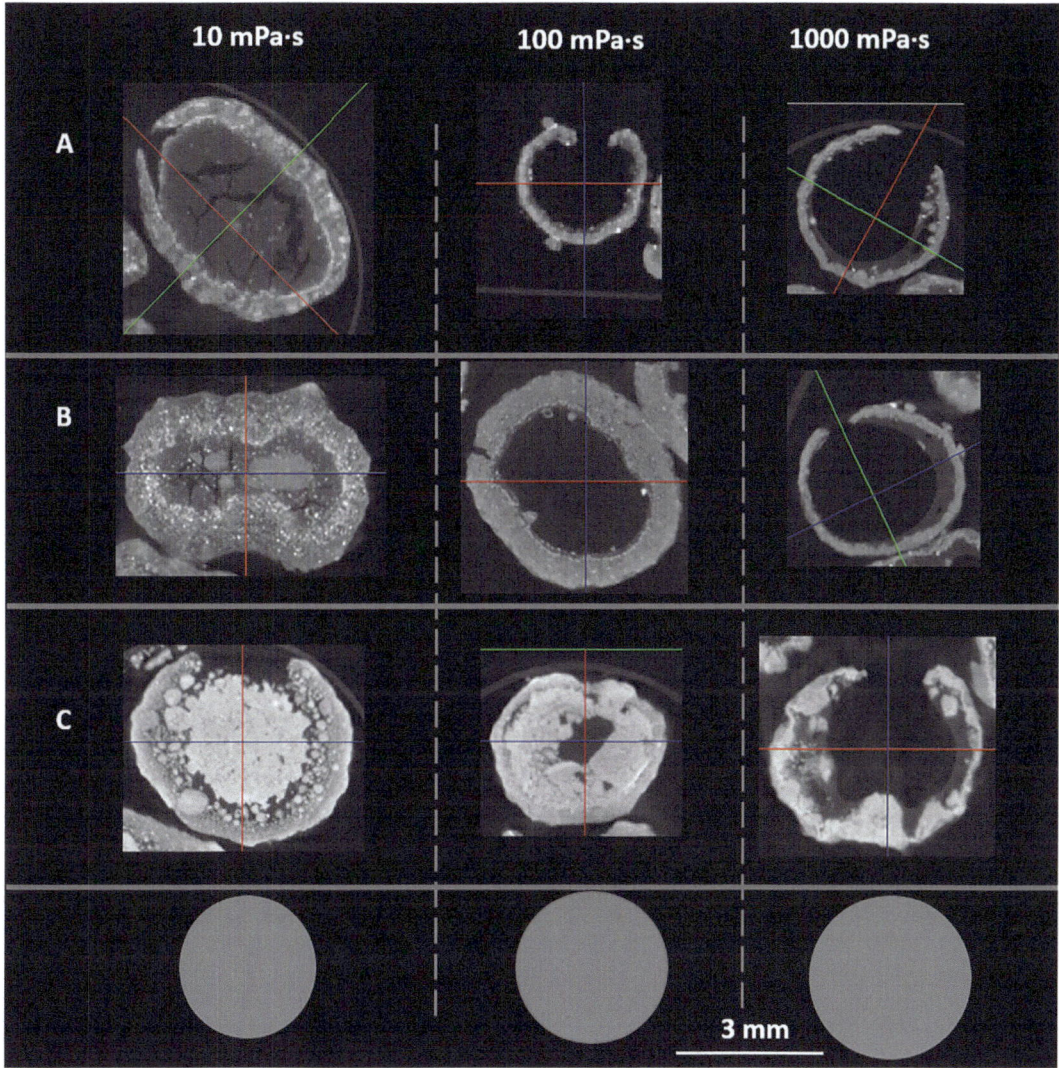

FIGURE 2.16

Microstructure [XRCT] of granules after 5 min. of granulation in the tumble drum; disks represent the initial droplet size; row A: non-wetting blend; row B: marginally wetting blend, row C: wetting blend

Adapted from Farber et al. (2020).

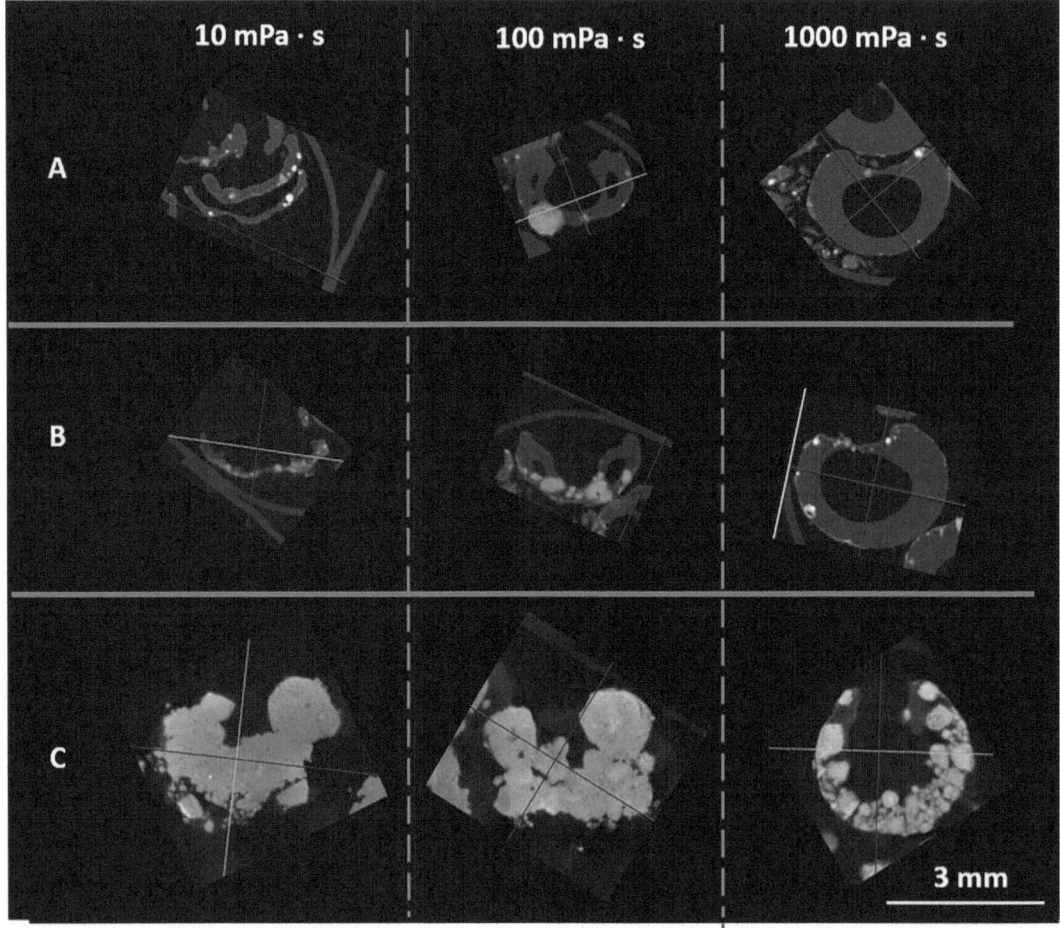

FIGURE 2.17

Microstructure [XRCT] of dried nuclei. (Adapted from Farber et al., [2020] Reproduced with permission from Elsevier); row A: non-wetting blend; row B: marginally wetting blend, row C: wetting blend

Adapted from Farber et al. (2020).

Comparing the microstructure of a nucleus after drying (Fig. 2.17). with the microstructure of the granules (i.e., tumbled nuclei) after drying, the most obvious difference is that the tumbling allowed for more powder incorporation, structural symmetry, and reinforcement. A similar prevalence of shell or hollow particles in Fig. 2.17 indicates that hollow granules primarily arise from the classic drying mechanism in which evaporation increases outer surface viscosity to the point that droplet shrinkage is too slow to prevent nucleation of gas bubbles. This was demonstrated by XRCT images of

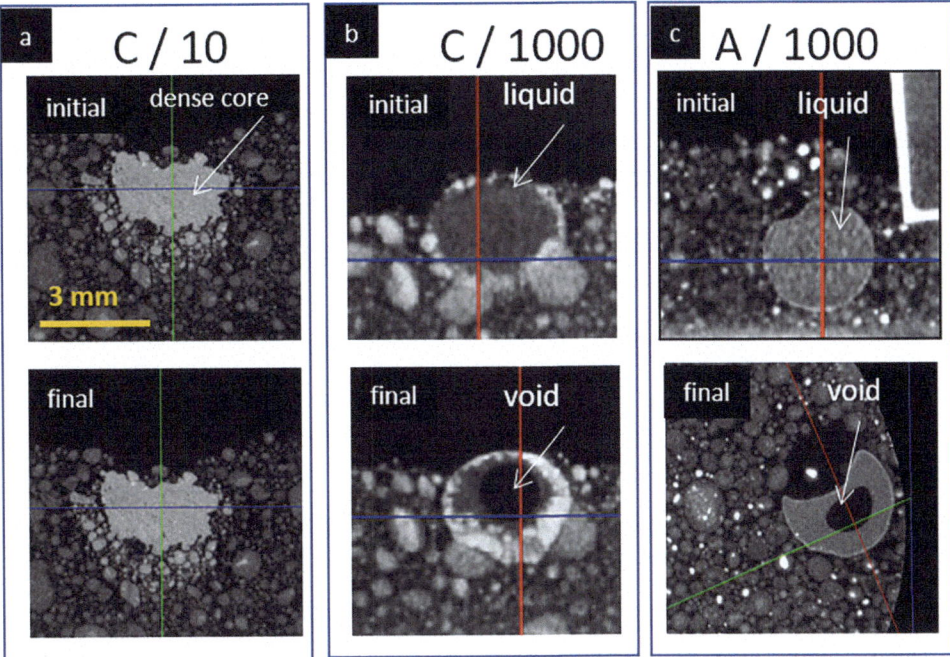

FIGURE 2.18

Formation and evolution of nuclei microstructure [XRCT] after deposition of a binder droplet on a blend (*top images*) and its subsequent drying (bottom images) for (A) low viscosity binder / wetting blend, (B) high viscosity binder / wetting blend, and (C) high viscosity binder / non-wetting blend. (Adapted from Farber et al., [2020] Reproduced with permission from Elsevier)

Adapted from Farber et al. (2020).

nuclei of different blend/binder combinations immediately after deposition and after subsequent drying (Fig. 2.18).

The dried nuclei from the nonwetting blend in Fig. 2.17 agree with the results of Hapgood and Khanmohammadi (2009), who reported that stable liquid marbles formed more with lower binder viscosity than with higher viscosity but that they collapsed into hemispheres after drying. Similarly Eshtiaghi et al. (2010) found that increasing binder concentration/viscosity improved the survivability of hollow granules and concluded that the internal fluid flow needed for collapse during drying is limited by viscosity.

The fact that granulation of the nominally nonwetting blend with high and medium-viscosity binders resulted in the formation and growth of a mixed powder/binder layer on top of the binder droplet surface indicates the constructive and critical roles of both the external forces and the viscous forces. External force results in the immersion of the nonwetting powder into the surface of the liquid, while viscous interactions result in the retention of that material in the binder. The negative contribution from capillary forces is apparently overcome. Reducing external forces (such as in experiments of Eshtiaghi et al. (2010) with rolling the single high viscosity binder droplet on a Petri dish with hydrophobic powder, and in this study) practically eliminated granule growth. Reducing binder viscosity changed the growth mechanism from hollow granules to core/shell. For the highly wetting blend, using high and medium-viscosity binders slowed the propagation of wetting in addition to preventing significant granule deformation. Apparently a similar mechanism led to the formation of hollow granules in melt granulation as reported by Ansari and Stepanek (2006). In their study, mixtures of small API powder particles and relatively large solid binder particles were fluidized in FBG and heated to a temperature slightly above the melting point of the polymeric binder for a limited amount of time. That resulted in the formation of hollow granules from particles sticking to the binder and forming a rigid shell that further grew by layering as the binder front moved outwards.

Five blend/binder combinations were also used for HSWG experiments. A 2L Diosna granulator was run for 25 sec at 300 rpm (no chopper) to get the number of revolutions approximately similar to the number of revolutions of the tumble drum. Destructive high-strain deformation within the mixing time did not occur with the high viscosity binder for both wetting and nonwetting powders or with the low viscosity binder for the wetting powder, Fig. 2.19.

The granules for those three extremes were similar to those from the tumbled marbles in Fig. 2.16, although most observed granules had an oval shape rather than the more spherical shape in the case of low shear tumbling, indicating some deformation in addition to binary aggregation. This demonstrates that deformation resistance by viscosity (binder) is critical for producing hollow granules in high-shear granulation.

Generally, it can be concluded that the formation of hollow granules consists of three stages: (1) initial dispersion of liquid binder if break up under shear is favorable, (2) formation of the liquid marble and growth of the shell, (3) final microstructure evolution during drying. It is clear that the internal microstructure and growth kinetics of hollow granules are controlled by viscous forces, capillary forces, impact, and shear forces. The modified Reynolds and Weber numbers were proposed to rationalize and quantitatively assess the formation and stability of shell microstructure (Farber et al., 2020).

For those seeking to produce hollow or layered granules for product design, the model study suggests that solid particle size should be much smaller than the liquid binder drops, the binder should have large viscosity to reduce shear dispersion, and, in wetting systems, to retard and reduce the drop wetting timescale to under the process time. Finally the binder contact angle should ideally be close to (but smaller than) 90 degrees. In apparent contradiction, the hollow granules in Section 3.2.2. (Hapgood et al. 2009) were formed with water (very low viscosity) and the nonwetting fine powder blend. However, the dry powder blend also contained HPC powder (L-grade) that would have medium viscosity at 4% dissolved in water and would wet the hydrophobic powder used in the paper. Those results also show an alternative, more practical approach to produce hollow granules: start with a hydrophobic powder blend containing solid binder and use water for ease of atomization. When in contact with nonwetting powder, liquid marbles form. As the binder starts to dissolve in water droplets, it makes the liquid viscous and wettable to start the growth of a shell,

2.3 Microstructure of granules

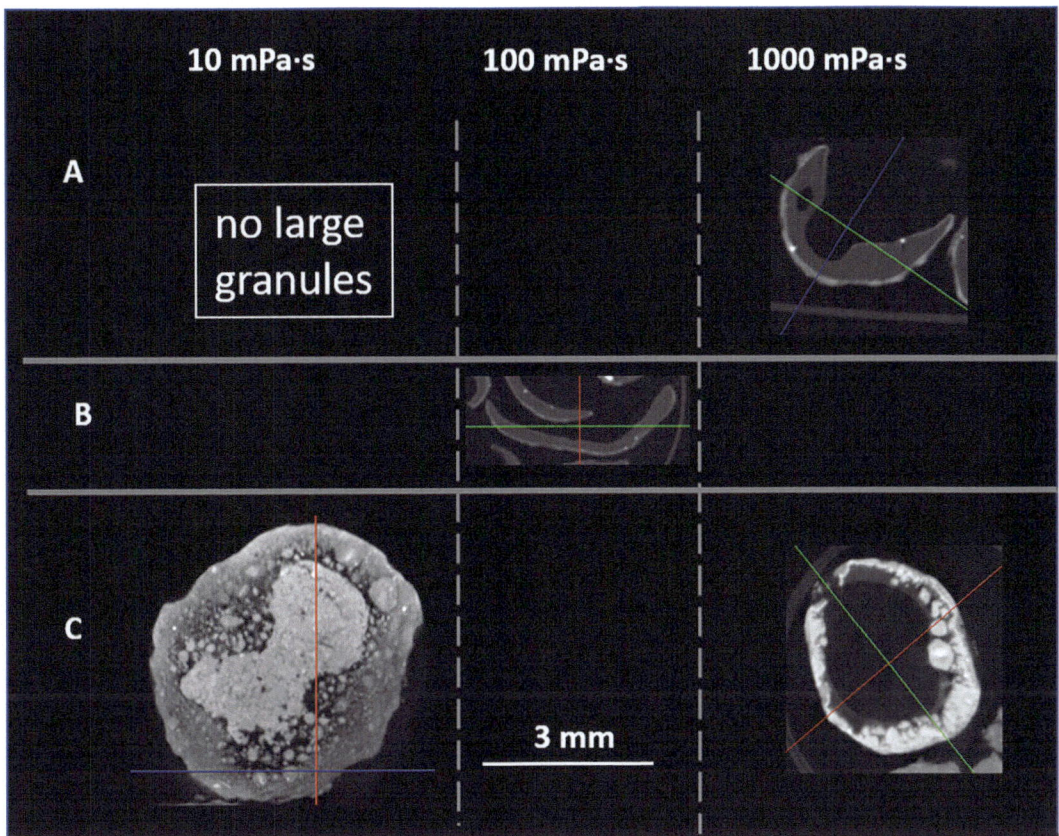

FIGURE 2.19

Microstructure [XRCT] of granules after 25 sec granulation in the high shear mixer; row A: non-wetting blend; row B: marginally wetting blend, row C : wetting blend

Adapted from Farber et al. (2020).

as described in this section for a marginally wetting powder blend and a medium viscosity binder system.

2.3.2.4 Wetting mixture of microcrystalline cellulose with a small amount of marginally wetting active pharmaceutical ingredient

HSWG of binary mixture that contained hydrophilic microcrystalline cellulose (MCC) and 3% to 7% of fine marginally wetting API powder (micronized acetaminophen) conducted with 80% of pure water (by weight of dry powder) resulted in formation of granules that contained relatively large irregular central pores or a group of pores (Oka et al., 2015), Fig. 2.20. Out of the typical granule porosity

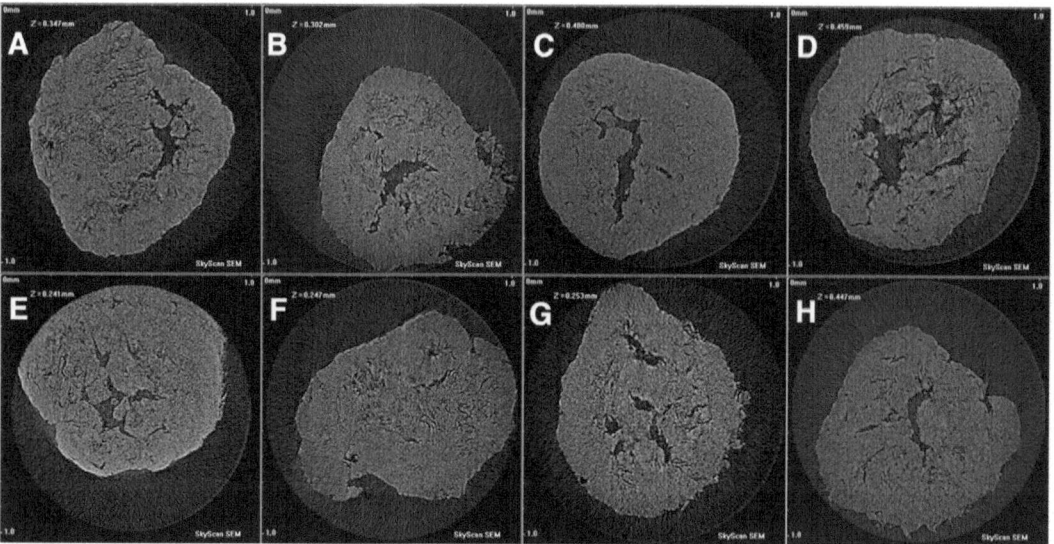

FIGURE 2.20

Representative images of granules with the central pore, obtained by HSWG of MCC- acetaminophen blend using pure water as a binder. Each image represents a batch that has a unique combination of percentage of API (3 or 7), impeller speed (225 or 300 rpm), and wet massing time (5 or 7 min).

Adapted from Oka et al. (2015).

of about 15% to 21% measured by XRCT, 60% to 90% of that volume was located in the largest cavity. (XRCT characterization was only reported for granules larger than 800 microns). The exact formation mechanism of such microstructure and the role of the formulation components have not been investigated in this study. The authors propose that the existence of the central pore suggests the presence of a trapped macroscopic fluid droplet(s) in the granule core at the end of granulation. However, the matching shape of the cavities' surfaces and the overall lower density of the material in the core suggest cavity formation during drying by material rupture due to directional shrinkage. This could happen if the material in the outer layer of the granule was able to form a relatively rigid shell, unlike the soft material in the granule core. The possibility of MCC doing this in HSWG is demonstrated and discussed in Section 2.3.4.

2.3.3 Distribution of components
2.3.3.1 Seeded granules

It is usually expected that granules produced by HSWG have a uniform spatial distribution of primary particles of all sizes, as in Fig. 2.6C. However, granules with very distinct characteristic microstructure were observed for calcium carbonate powder HSWG at different conditions and scales using an aqueous solution of polyethylene glycol (PEG) as a binder. At their core, the granules contain a large primary particle (Fig. 2.21) (Rahmanian et al., 2011). In those studies, it was found that under some conditions, a large particle serves as a seed for the formation of a granule, and the formation of the seeded structure is strongly dependent on the dynamic conditions in the granulator and the primary PSD, regardless of the scale of the granulator. The results were rationalized using the Stokes deformation number approach introduced by Tardos et al. (1997). Stokes deformation number allows for comparison of externally applied kinetic energy to the granules with the energy required for their deformation and reflects both the process conditions and materials properties. If the PSD is wide, the coarse particles serve as a nucleus (seed) at Stokes deformation numbers >0.1 in the granulator, and the granule grows by layering of finer particles on that seed. Practically all granules under those conditions had this seeded microstructure. Occasionally granules with two larger particles at the core were observed. When the conditions were not fulfilled, the probability of observing the seeded granules decreased dramatically. The microstructure of granules in that case varied significantly: There were granules containing $CaCO_3$ particles randomly distributed in the granule volume as well as granules enriched with binder at the core, see Fig. 2.21.

Differences in granule microstructure were directly correlated to differences in mechanical properties (Rahmanian et al., 2009; Rahmanian et al., 2008). It was concluded that granules with a large particle in the core have the highest strength, while granules with randomly distributed particles have the lowest strength. The study also demonstrated that for $CaCO_3$ of wide PSD, the microstructure of granules could be tailored via parameters of the HSWG process.

2.3.3.2 Binder distribution

Visualizing binder spatial distribution in HSWG granules has proven to be quite challenging. Earlier studies on granule microstructure as related to the manufacturing process and final product properties (Seager et al., 1979) used a solvent extraction method to visualize binder distribution in granules. In that work, paracetamol powder was granulated using a gelatin binder solution. It was found that in dried HSWG granules after WM, the binder exists as a single solid sponge-like body that holds together API crystals. The individual plates of the sponge were thin–about a micron or less. It was found that the structure had the same appearance at different binder levels, except for differences in thickness. The authors proposed a formation sequence of such microstructure (see Fig. 2.18 in (Seager et al., 1979)). The results suggest that individual particles in a HSWG granule after WM become coated with a very thin binder-rich layer that forms a continuous matrix phase. Such microstructure is in good agreement with the reconstructed cross-sectional images of HSWG granules consisting of $CaCO_3$ powder and PEG as a binder (Fig. 2.21), where the binder can be observed forming a matrix.

This solvent extraction study also clearly demonstrated that different manufacturing techniques create qualitatively different binder structures. For roller compaction, the binder was present as discrete particles, forming the bonds between the paracetamol crystals. The binder particles showed different morphologies depending on the roll pressure. For spray drying, the granules consisted of spherical particles composed of an outer shell of binder with an inner core of paracetamol particles. Using the same methodology of binder extraction, earlier studies on FBG (Gamlen et al., 1982) found that individual solid particles or aggregates formed by electrostatic attraction become coated or partially covered with the granulation solution, leaving a layer of sticky binder. Other particles collided with

66 Chapter 2 Microstructure and mechanical properties

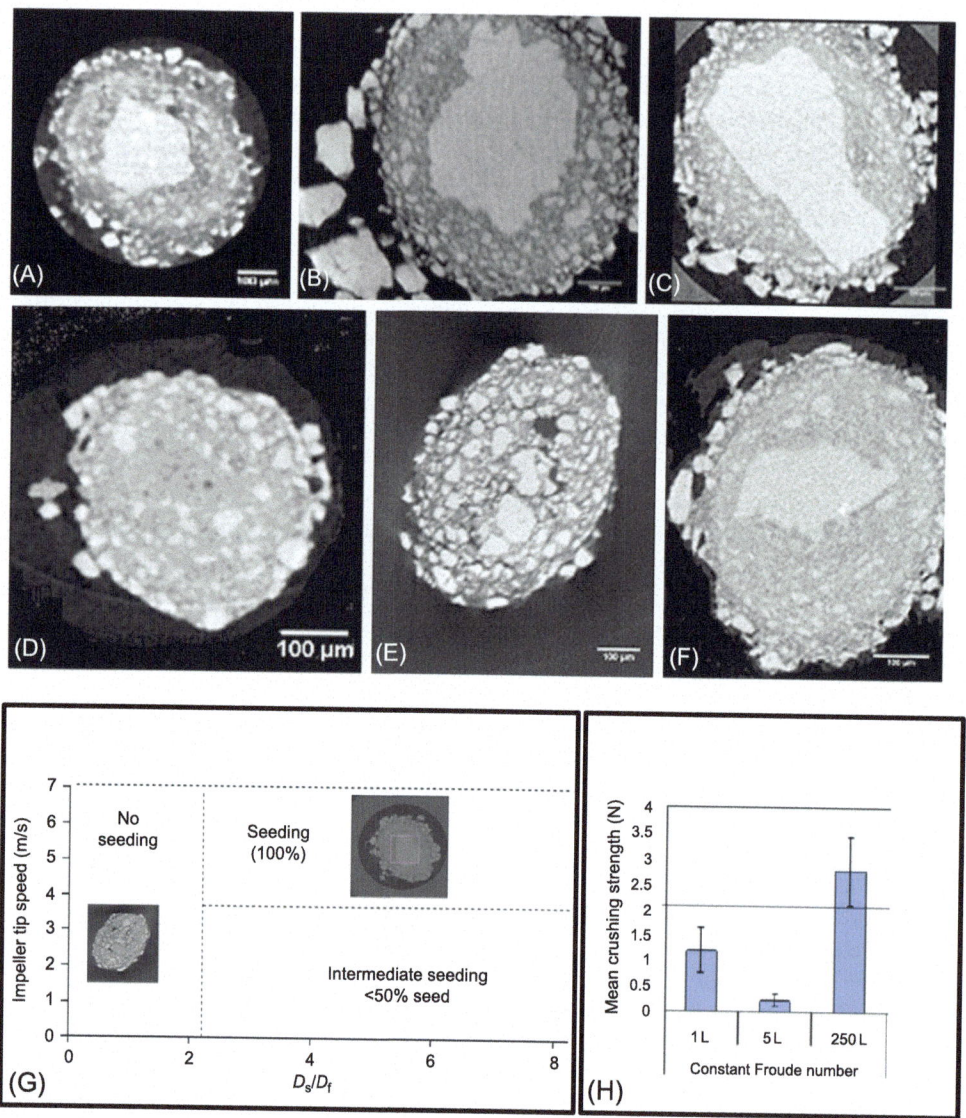

FIGURE 2.21

Seeded granulation. (A–F) Central cross-sections of granules produced under (A–C) the constant tip speed and (D–F) constant Froude number at different scales (A), (D) 1 L; (B), (E) 5 L; (C), (F) 250 L; (G) The regime map for production of seeded granule structure (D_s–seed size; D_f–mean feed particle size); (H) Mean crushing strength of the granules with different microstructures shown in (D–F).

Adapted from Rahmanian et al. (2008), Rahmanian et al. (2009), and Rahmanian et al. (2011).

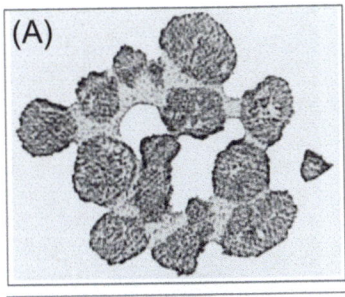
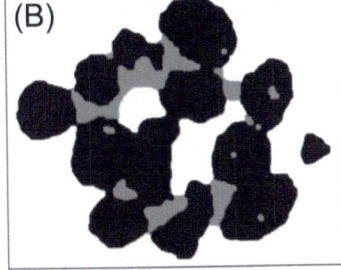
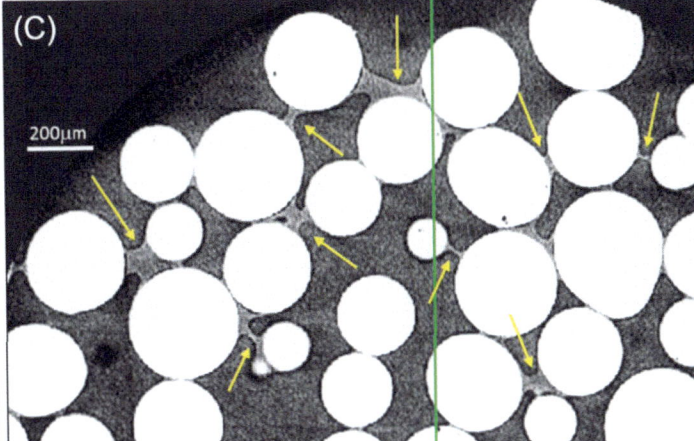

FIGURE 2.22

Binder visualization (by XRCT) in model granules produced via various methods: (A), (B) FBG (Suglet granules (S1-PEG8000, binder content 11%) granule cross-section before (A) and after (B) image filtering and segmentation). (C) nucleation in a static or quasistatic powder bed (droplet of a PVP binder deposited in a glass ballotini powder bed), glass in the image is white, interparticle bridges after drying are pointed at by *arrows*.

(A and B) Adapted from Ansari and Stepanek (2008). (C) Adapted from Dale et al. (2014).

these sticky crystals/agglomerates, forming aggregates that then became coated with further deposits of binder solution. This process continued as the aggregates grew into granules, giving a characteristic open structure. Reviews of other publications suggest that both isolated individual bridges between primary particles and short branch-like bridges are expected (Štěpánek & Ansari, 2005) (Fig. 2.22). Finally XRCT images of model granules formed by nucleation in a static or quasistatic powder bed showed that binder in such granules can be described as individual cylindrical bridges between particles (Dosta et al., 2016) (Fig. 2.22C).

The binder distribution within HSWG granules is not necessarily uniform. A qualitative difference in binder radial macrodistribution within HSWG granules is obvious in Fig. 2.21D, with clear enrichment of the granule core with the binder that was observed at some conditions. These granules showed intermediate yield strength in that study.

Macrodistribution of a polymeric binder in HSWG granules was also visualized in work by Mangwandi et al. (2010) (Fig. 2.23). The granules were produced from a lactose/starch blend using a colored aqueous HPC solution at different impeller speeds. Although a volumetric binder distribution was not studied quantitatively in this work, it is clear from the images that the binder is distributed much more uniformly in granules produced with high impeller speed. Measurements in this study also

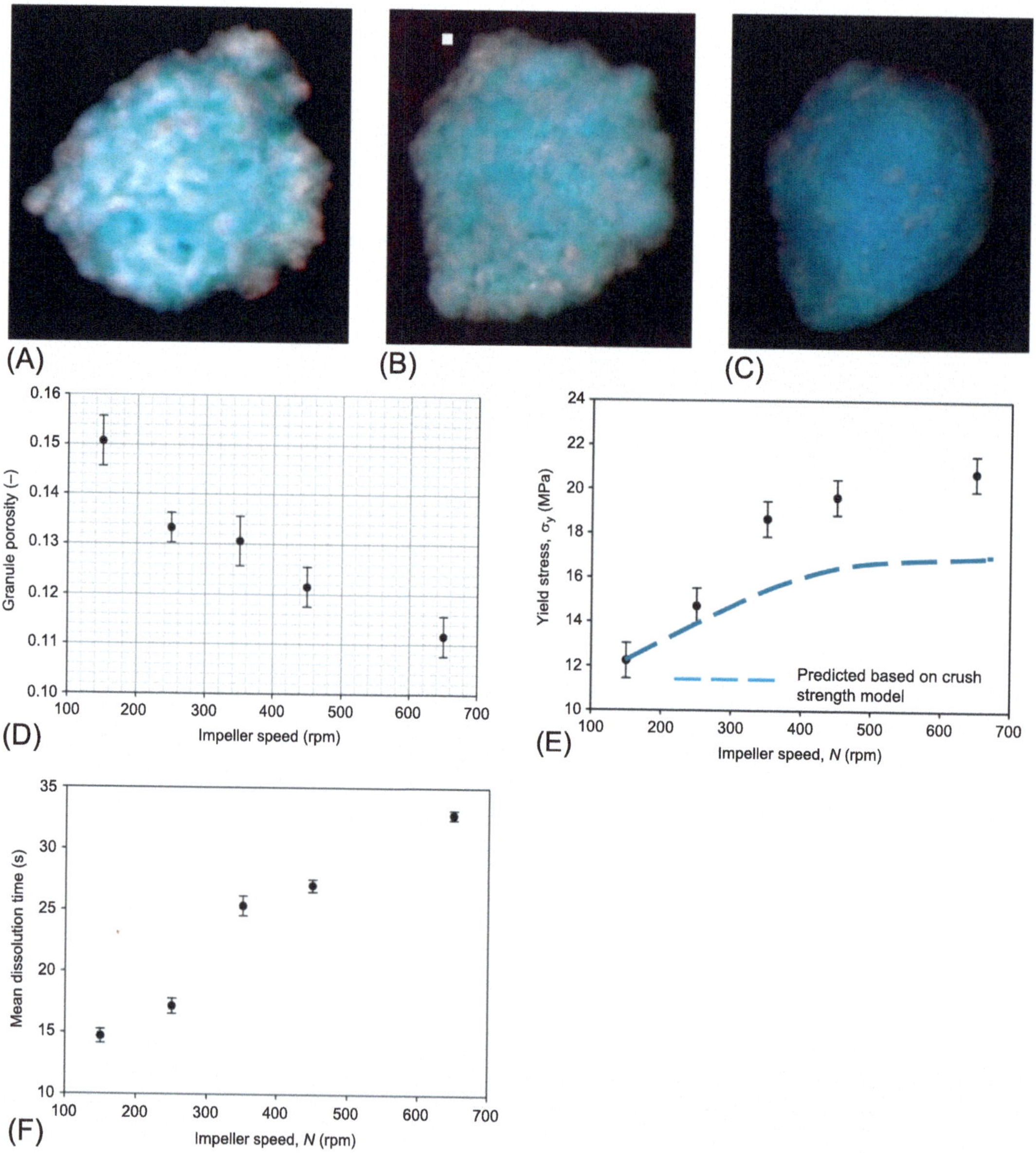

FIGURE 2.23

Effect of impeller speed on surface appearance and properties of lactose–starch HPC granules in the size range 1.0–1.18 mm. (A–C): light microscopy of batches produced at (A) 150 rpm, (B) 250 rpm, (C) 650 rpm; (D) granule porosity; (E) granule yield stress, (F) granule dissolution (predicted values in (E) are recalculated from porosity in (D) using Eq. (2.4) from Bika et al. (2005).

Adapted from Mangwandi et al. (2010).

Table 2.3 Composition of dry powder mix Component.

Component	Wt%
Lactose monohydrate	73
Microcrystalline cellulose (MCC)	20
Hydroxypropyl cellulose (HPC)	3
Cross-carmelose sodium (CCS)	3
Active pharmaceutical ingredient (API)	1
Citric acid (added w/granulating fluid)	1

Adapted from Michaels et al. (2009).
API, *Active pharmaceutical ingredient;* CCS, *cross-carmelose sodium;* HPC, *hydroxypropyl cellulose;* MCC, *microcrystalline cellulose*

showed that porosity and mechanical properties of the granules depend on impeller speed. The porosity of granules produced at low impeller speed is higher than the porosity of granules produced at high impeller speed (ca. 15% vs. 11%), while the yield stress is lower. Higher granule strength is expected when the porosity is lower because of purely geometrical reasons (see Section 2.4). However, the value of the strength is significantly higher than could be expected from the crush strength model, which accounts for porosity (Bika et al., 2005). This excess in strength is most likely associated with the difference in binder distribution throughout the granule, that is, because of the difference in granule microstructure.

2.3.4 Material transformation during granulation: Microcrystalline cellulose
2.3.4.1 "Swiss-cheese" granules

A solid material can experience a transformation during HSWG that profoundly affects the granule microstructure and properties. In the course of the work on steady states in HSWG granulation (Michaels et al., 2009), it was found that extended WM of typical lactose/MCC-based formulations results in a significant and unusual evolution of the granule microstructure. During extended WM, these granules evolve from typical agglomerates composed of primary MCC and lactose particles to a bicontinuous structure with a skeleton of porous cellulose with pores filled with the remaining formulation components. The soluble components can be removed from the pores by dissolution, leaving the porous MCC skeleton (Farber & Michaels, 2018). The effect of WM time, granulating fluid level (GFL), and nature of the liquid, as well as the amount of MCC, on the formation of the microstructure is described below.

The formulation is shown in Table 2.3. The granulation was performed in a 25 L Fielder high shear granulator using 28% and 32% of USP-purified water as GFL, added at a 2%/min rate. Water addition was followed by WM for 40 min. As was demonstrated in a study by (Michaels et al., 2009), the PSD stabilized by approximately 10 min of WM time and practically did not change with longer WM. An external chiller was used to keep the temperature of the granulation vessel at 27°C. One more experimental run was performed using 35.5 w/w% ethanol as a granulating liquid (which has the same volume as 28% of water).

Time evolution of typical granule morphology is shown in Fig. 2.24. Granules produced during water addition without WM are composed of identifiable particles of the excipients attached to each other in a conventional granular structure. These granules disintegrated into primary particles when immersed in water. With WM, however, the microstructure evolves with time. There is a significant change in the appearance of granules after 10 min of WM. With even longer WM, the granule surface continues to change and becomes smoother, and granules become dense spheroids with some individual lactose crystals either attached to or embedded into their surface. It was found that both the initial microstructure (individual particles stuck together) and the final microstructure (dense spheroids) did not depend on the size of the granules. During WM, the transformation of the microstructure occurs on the scale of tens of minutes after the start of WM. The actual time depends on the granule size. For granules of about 100 μm, the transformation occurred within 10 min. For granules of 250 μm, it took about 30 min.

The granules, which remained intact during this experiment, were immersed in water and then dried. No visual changes were observed by the naked eye; however, weight measurements of the granules showed that they had lost about 76% of their initial weight. This agrees well with the weight fraction of lactose and HPC in the mixture (77%). Both lactose and HPC are soluble in water. It could be concluded that lactose and HPC were dissolved by water and removed. The remaining material should be MCC and CCS, which are not soluble.

Representative SEM images of granules at higher magnification before and after immersion into water are shown in Fig. 2.25. Prior to immersion, the granules appear to be a continuous, dense spherical body with some individual lactose crystals either attached to or embedded into its surface. It is impossible to distinguish individual MCC or CCS filaments in or on the granule. After the water immersion experiment, the overall spheroidal shape of the granule is preserved, but now the surface contains holes in the shape of lactose particles. Changes in the internal microstructure and porosity were revealed by examining granules prior to and after immersion by XRCT. Fig. 2.25 shows typical cross-sections of the same granule before and after water immersion. The overall shape and dimensions of the granule did not change; however, there was a significant change in both total porosity and pore size distribution. Internal porosity prior to immersion was about 15%, with the majority of pores of about 15 to 25 μm that appear to be isolated. After immersion, the total porosity was close to 70%, and it was composed of larger interconnected cavities. Further experiments showed that dissolution occurs via a shrinking core mechanism. Granulation with ethanol instead of water resulted in granules of the same microstructure. When the amount of MCC was reduced in the formulation to 10% (via substitution with lactose), the vast majority of granules appear to consist of the initial particles stuck together. This shows that the transformation of cellulose into a continuous porous structure requires a threshold amount of MCC in that formulation: 10% of MCC was insufficient, and 20% was sufficient. Although the mechanism of the MCC transformation was not studied in that work, it was speculated that amorphous regions in semicrystalline MCC were plasticized by hydration, and mobile cellulose chains on the surface of adjacent particles entangled, forming a strong adhesive physical bond. As a result, the individual cellulose particles became a single continuous solid.

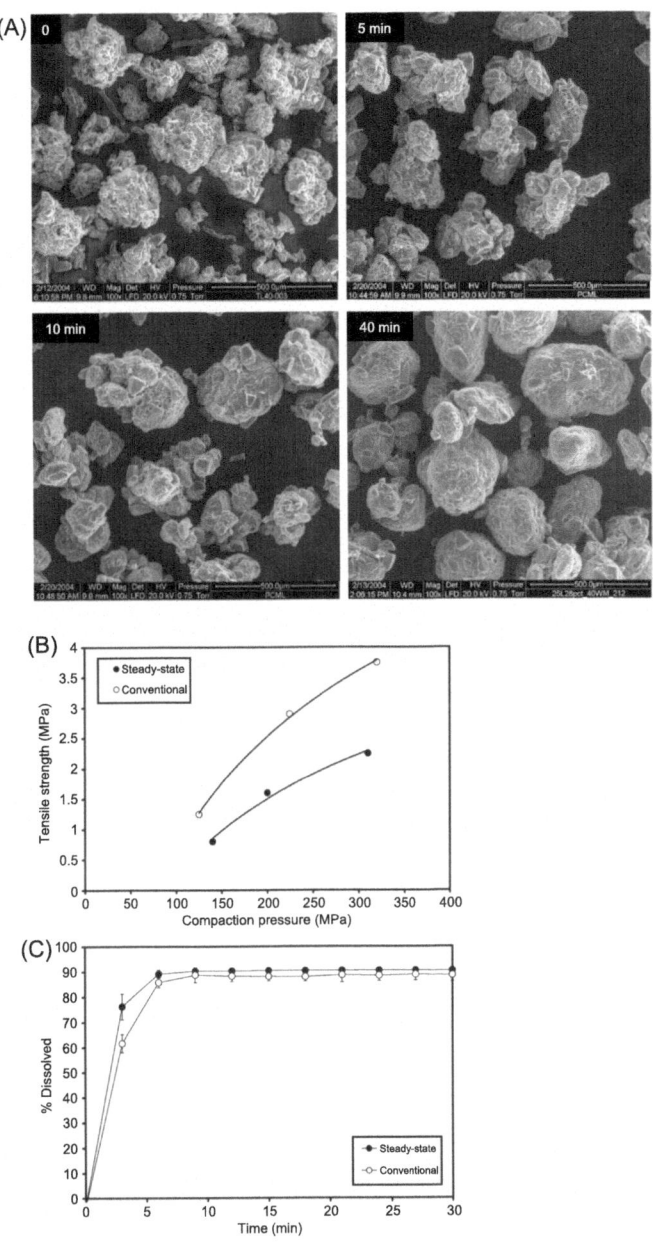

FIGURE 2.24

(A) Time evolution of granule morphology during long wet massing; (B) compaction profiles of granules obtained from steady-state and conventional granulation; (C) dissolution profiles of tablets compressed from steady-state and conventional granulation.

(A) Adapted from Farber and Michaels (2018). (C) Adapted from Michaels et al. (Seager et al., 1979).

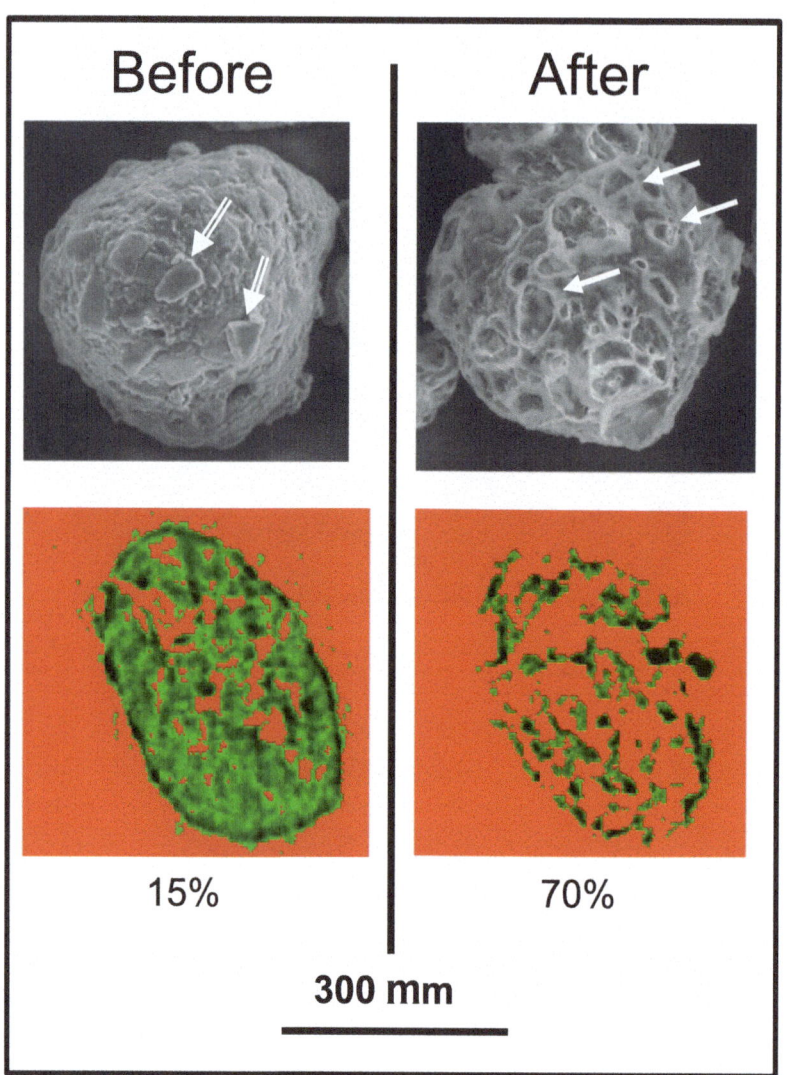

FIGURE 2.25

Granules from the batch produced by 40 min wet massing before and after dissolution experiment. Top images, SEM; lower images, XRCT, reconstructed cross-sections; granule porosity was calculated from XRCT data.

Adapted from Farber and Michaels (2018).

2.4 Connection between granule microstructure and granule strength

Manufacturing of granules via HSWG includes two steps. The first step is granule formation via agglomeration of primary particles, which occurs in a granulator and is facilitated by a liquid. In a wet (sometimes called green) granule, particles are mostly held together by liquid bridges via capillary forces. The second step is drying. During that step, liquid evaporates from the bridges to leave behind solid bridges or necks that impart mechanical strength to the dry granule.

It was recognized early that the strength of solid interparticle bridges might control the tensile strength of agglomerates. Various micromechanical models have been proposed to predict dry granule strength and describe criteria for brittle failure. Three of them are described briefly below: the crush strength model, the recrystallized bridge model, and the autoadhesion model (JKR theory).

2.4.1 Theoretical models

2.4.1.1 Crush strength model

To calculate the crush strength of the agglomerate, σ_{cr}, from the interparticle binding force F, a correlation has been proposed by Rumpf (1975). This applies to an agglomerate of randomly packed monodispersed spheres with the diameter d_p that reads

$$\sigma_{cr} = z(\varepsilon)\frac{F}{d_p^2} \tag{2.2}$$

where z is a porosity (ε)-dependent function given as $z(\varepsilon) = (1 - \varepsilon)/\varepsilon$ by Rumpf and alternatively $z(\varepsilon) = 13.3(1 - \varepsilon)^4$ by (Kendall, 1987). Assuming that particles are connected by bridges, and bridge fracture occurs across the narrowest cross-section of the bridge, the normal force F required to separate two particles after the bridge has formed between them can be calculated as

$$F = \pi r_{sb}^2 \sigma_{sb} \tag{2.3}$$

where r_{sb} is the radius of the narrowest portion of the bridge or neck, σ_{sb} is its strength. The assumption is that the bridge breaks at its narrowest point (and not at the interface with either particle). The force and the corresponding stress can be either tensile (compression or tension) or shear. The difficulty in this computation is the prediction of the neck size r_{sb} and assigning a value to its strength σ_{sb}.

For granules produced by wet granulation, Eqs. (2.2) and (2.3) were extended to express the microscopic crushing strength of granules in terms of the bridge strength while incorporating the concentration of the binder solution and liquid-to-solid ratio (Bika et al., 2005). The resulting equation is given by:

$$\sigma_{cr} = \pi b^2 \frac{1-\varepsilon}{\varepsilon}\left[\frac{4\pi}{3}C_L r_{liq}\right]^{2c}\sigma_{sb} \tag{2.4}$$

$C_L = C_S/\rho_b$ is the solid concentration in the binder solution expressed in g/g, and r_{liq} is the liquid-to-solid ratio, g/g. This approach allows calculation of the strength of an individual bridge if the granule crush strength is measured, or predicting the granule crush strength if some assumption about the materials properties of the bridge can be made.

2.4.1.2 Recrystallized bridge model

Alternatively, if one assumes that the bridge is a brittle solid (for example, formed by recrystallized excipient such as mannitol or lactose), an upper limit to the bridge strength σ_G could be described by the Griffith model extended to the porous body (Bika et al., 2001)

$$\sigma_G = \sqrt{\frac{2E \cdot \gamma}{\pi c}} \tag{2.5}$$

Here, $E = E_0(1 - \varepsilon)^k$, where E_0 is the Young's modulus, ε is the compact's porosity, $k = 2$, γ is the surface energy, and c is a characteristic defect size. The difficulty in this computation is assigning the value to the critical defect size—ideally, the largest void or binder-free region.

2.4.1.3 Autoadhesion model (JKR Theory)

In the absence of dissolved solids in the liquid bridge (no polymeric binder or evaporative recrystallization), dry granules are held together by autoadhesive forces, and the granule strength can be calculated according to Johnson et al. (1971):

$$\sigma_C = 24.7 z(\varepsilon) \frac{\gamma}{d_p} \tag{2.6}$$

assuming that the interfacial fracture energy is equal to the surface energy, γ, and the characteristic defect size is equal to the primary particle size, $c = d_p$. The porosity function in Eq. 2.6 can be taken as either $z(\varepsilon) = 13.3(1 - \varepsilon)^4$ (Kendall, 1987) or $z(\varepsilon) = (1 - \varepsilon)/\varepsilon$ (Pietsch et al., 1967).

In a study by Bika et al. (2005), the strength of solid bridges between primary particles that are composed of granules of lactose and mannitol was measured experimentally and then related to the above models. For lactose and mannitol powders granulated with polymeric binders dissolved in ethanol (in which lactose is practically insoluble and mannitol is only sparingly soluble), the strength of the bridges was found to correspond to the strength of pure polymeric films. In the absence of polymers, dry bridge strength was comparable to the strength of dry crystals (Griffith strength, Eq. 2.5). However, in granulations with aqueous solutions of polymeric binders, bridge strength differs significantly from either the strength of polymeric binder or Griffith strength of the dry crystals, which also depends on the concentration of the binder and presence of other excipients (such as surfactants), cannot be explained or predicted easily, for reasons which are explained in the next section.

2.4.2 Evolution of microstructure and strength development of interparticle material bridges (Model systems)

In the models for granule strength, a solid bridge is usually assumed to be nonporous and of similar chemical composition and physical properties to either the primary powder particles or the polymeric binder material used. This, however, is an overly simplistic view of a very complex problem, especially if the original fine powder is itself soluble in the binder solution. In this case, the liquid should partially dissolve some solid powder and form solid bridges of a complex composition and microstructure.

The evolution and phase composition of drying material bridges of pure lactose and mannitol were studied first in (Farber et al., 2003a, 2003b). In the case of lactose, it was found that bridge solidification at ambient conditions occurs via simultaneous crystallization and vitrification within minutes. As a result, a solid bridge usually contains both crystalline and noncrystalline phases, with the crystalline phase being predominantly α-lactose monohydrate. Most of the noncrystalline phase eventually converts to crystalline β-lactose, but the process can take many hours or even days (depending on humidity). In the case of mannitol, different polymorphic forms crystallize as the drying/crystallization process progresses.

Forces exerted by the drying interparticle bridges were measured in the work that followed (Farber et al., 2005). The bridges were formed from aqueous solutions of common pharmaceutical excipients that are both nonpolymeric (lactose, mannitol) and polymeric (hydroxypropyl cellulose or HPC, hydroxypropyl methylcellulose or HPMC, polyvinylpyrrolidone/povidone or PVP). The work also included a study of the morphology, microstructure, and crystalline structure of solidifying bridges. It was found that the solidifying behavior and final properties of bridges differ dramatically, depending on the composition of the solution. Bridges containing only lactose or mannitol tend to expand upon solidification, pushing the ends of the bridge apart; in contrast, pure HPC, HPMC, or PVP bridges tend to contract. Bridges crystallized from a solution of the pure nonpolymeric excipients are polycrystalline, brittle, and have low strength; bridges from the polymeric excipients are amorphous, strong, and tough. When the polymeric and nonpolymeric excipients are used together, behavior closer to one of the extremes takes place. This depends on the relative amount of polymer in the bridge, see Figs. 2.26 and 2.27. It was also found that the different polymers impart different behaviors on the bridge. The observed differences in solidification behavior have important implications for granule mechanical properties. For example, when comparing HPC- and HPMC-containing bridges, HPMC provided a slightly higher tensile strength, both as a pure binder and in a mixture with excipients. HPMC-containing materials, however, appear to be much more brittle than those containing HPC, thus more prone to mechanical failure under stress. It was also found that the behavior of HPC- and HPMC-containing bridges can be rationalized in terms of a simple linear combination of the behavior of pure polymer and excipient. For PVP, the behavior indicated a strong interaction (such as solution plasticization) between PVP and a nonpolymeric excipient.

The obtained results demonstrate that the properties of the interparticle bridge can be difficult to predict, especially if the original fine powder is itself soluble in the binder solution. The properties potentially could depend on processing conditions, including granulation and WM time, bed temperature during granulation, and drying conditions.

2.4.3 Simulation of granule deformation and breakage behavior: From understanding to design

Numerically, granule breakage has mainly been studied using the discrete element method (DEM). In order to reproduce the agglomerate shape and internal structure, the bonded particle model (BPM) can be used effectively (Potyondy & Cundall, 2004; Shen et al., 2016). By BPM, the granule is represented as a set of primary particles connected with ideally elastic or viscoelastic solid bonds (Kozhar et al., 2015; Spettl et al., 2015). DEM simulations have been used to investigate both the impact breakage (Antonyuk et al., 2006; Metzger & Glasser, 2012; Moreno et al., 2003; Subero et al., 1999;

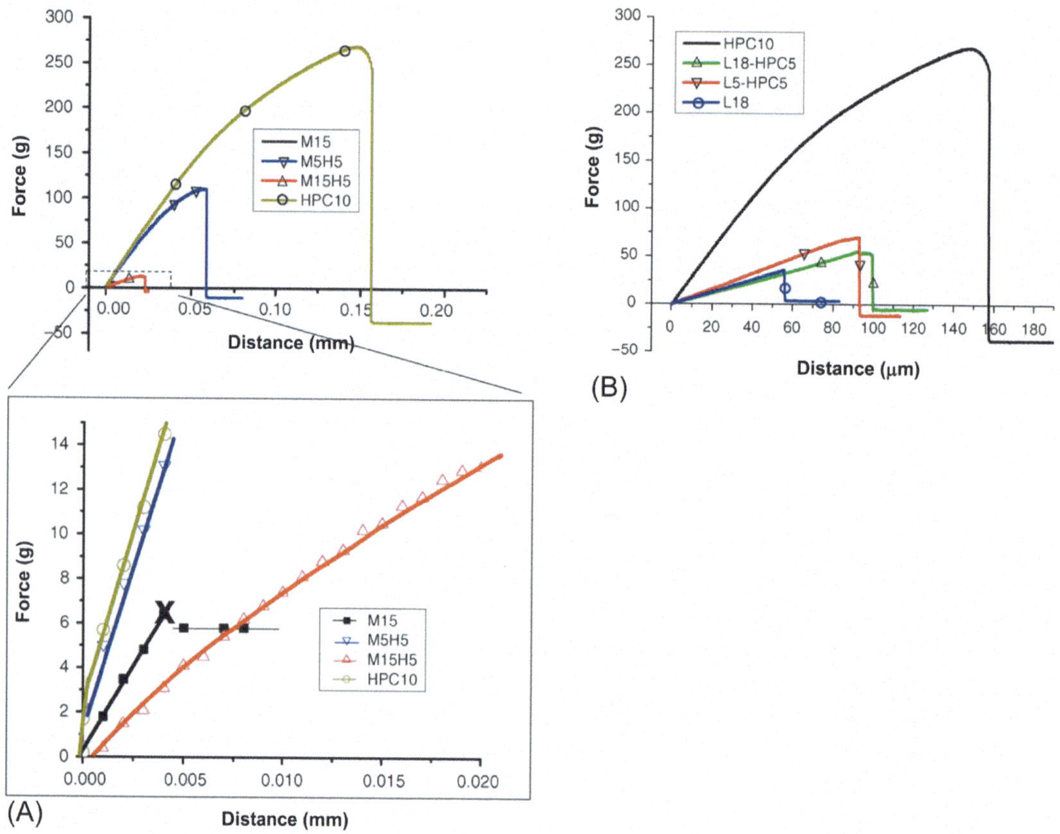

FIGURE 2.26

Response to uniaxial tension of the solidified dried model bridges in: (A) mannitol–HPC systems and (B) lactose–HPC systems. (M5H5 means a bridge formed from an aqueous solution of 5% mannitol and 5% HPC, etc.).

Adapted from Farber et al. (2005).

Tong et al., 2009) and the compression breakage of agglomerates (Golchert et al., 2004; Thornton et al., 2004). Study by Thornton et al. (2004) and Mishra and Thornton (2001) demonstrated that packing density has a large impact on the compression and impact breakage. The failure mechanisms of intermediate packing densities were shown to be different based on the packing structure near the site of impact. This demonstrates that the local structure determines the breakage of agglomerates, thus motivating the measurement of internal structure and the structure variation within granules.

To consider the actual microstructure of granules, the computer simulations can be extended effectively with XRCT (Moreno-Atanasio et al., 2010). Dale et al. (2014) have developed a method

2.4 Connection between granule microstructure and granule strength

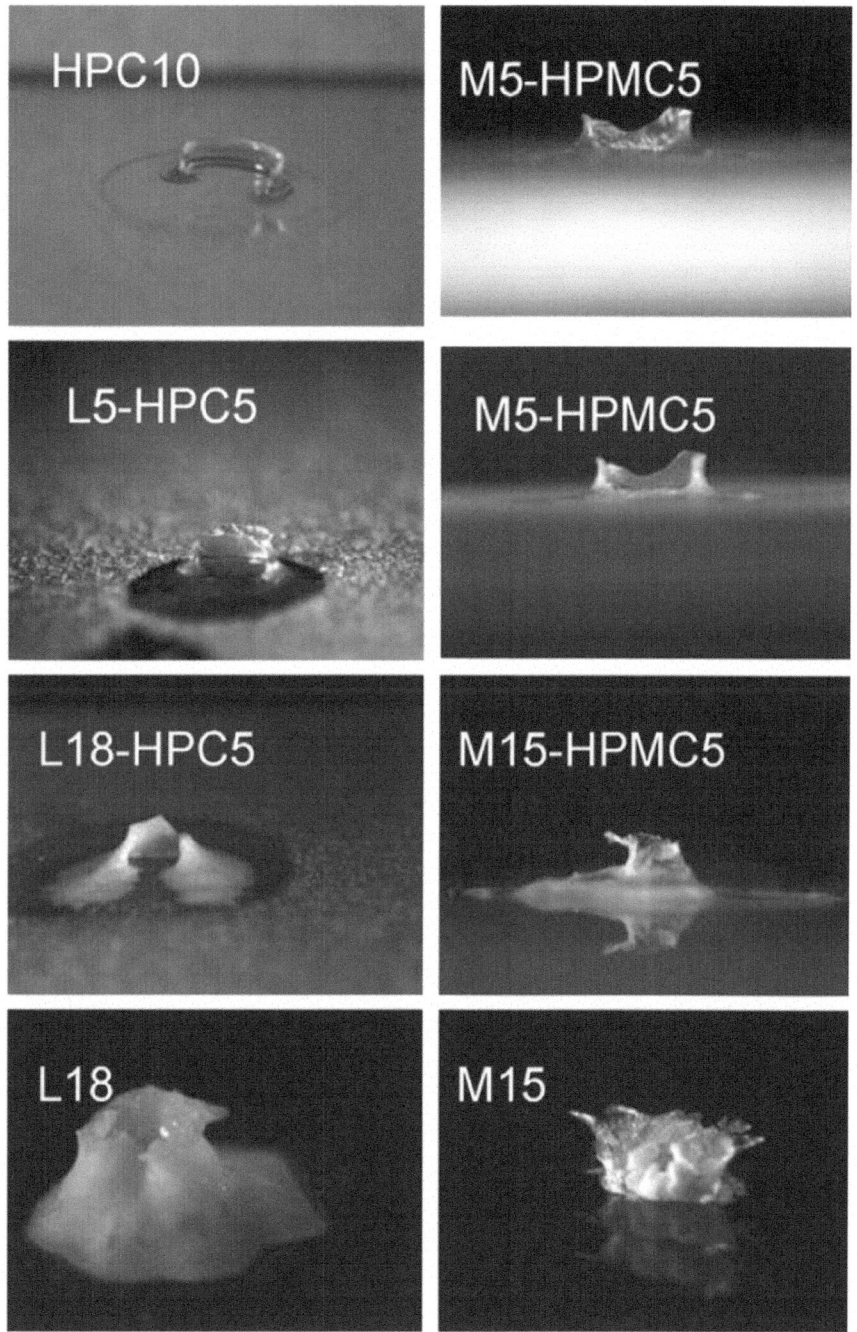

FIGURE 2.27

Optical micrographs of model bridges after failure in the tension experiment.

Adapted from Farber et al. (2003).

to visualize and describe phase distributions within granules and have shown that, for certain granule formation methods, phases are not homogenously distributed. XRCT data were used to generate virtual granules with the same microstructure, including the same binder distribution, as in actual granules using the BPM approach (Dosta et al., 2016), Fig. 2.28. The breakage behavior of simulated granules during diametrical compression has been investigated numerically and compared to the experimental results for the actual granules. The range of granule strength values predicted from simulations overlapped with the range of experimental measurements in most instances (Fig. 2.29A). The comparison between experimental and simulation results showed that simulation results follow the expected trends, and indeed are quantitatively accurate for the spherical fluid bed granules. The BPM approach was also used to simulate different binder distributions within granules. The simulations demonstrated that granules with increased binder concentration toward the perimeter of the granule should have greater attrition resistance, while increasing binder concentration at the core should lead to higher resistance to fragmentation Fig. 2.29B.

It was also found that simulations with the chosen DEM models and material parameters cannot describe experiments exactly. Much smaller gradients and smooth behavior of the force–displacement curve were observed in the experiments than in simulations. This was attributed to the fact that a pure elastic solid bond model was used in the DEM, so incorporation of plastic deformation in future DEM simulations of the agglomerates will be required.

The above studies have been conducted on granules formed from relatively large (255–355 μm) glass ballotini and PVP binder in a static or a fluidized bed. Although the shape and connectivity of the binder phase in granules produced by HSWG could differ significantly from the mostly pillar-like shape of interparticle bridges in model granules that were investigated in the above study, the approach should be applicable for simulation and numerical analysis of HSWG granules as well as granules produced via other methods that use binders. It appears that the main impediments to the application of the described approach for simulating pharmaceutical HSWG granules are the insufficient resolution of XRCT and potentially low contrast between the typical binder and the primary particles. Nevertheless, it could be expected that further improvements in XRCT would be able to overcome those barriers. Coupled with the development of the simulation tools to incorporate elastoplastic behavior of bonds, simulations will allow investigators not only to estimate mechanical properties of produced granules but also to design and predict optimal granule structure, which in turn, could be targeted by the granulation process.

It should be noted that the HSWG process might not be the optimal approach to create various desired microstructures. For example, the creation, tight control, and reproducibility of a microstructure with a radial gradient in porosity or a component distribution (such as a binder) can be hard to realize via HSWG, given the fact that the process is transient, and breakage and coalescence of granules could become dominant mechanisms at some point during the process. At the same time, it appears that such a gradient could be realized more easily via FBG.

As discussed in the next section, however, there is still a significant gap in the clear understanding of the connection between the granule microstructure and the properties of the final product, especially tablets, in which the connection between the microstructure and mechanical properties or functional properties of granules (such as dissolution or drug release rate) is studied, established, and could be simulated.

2.4 Connection between granule microstructure and granule strength

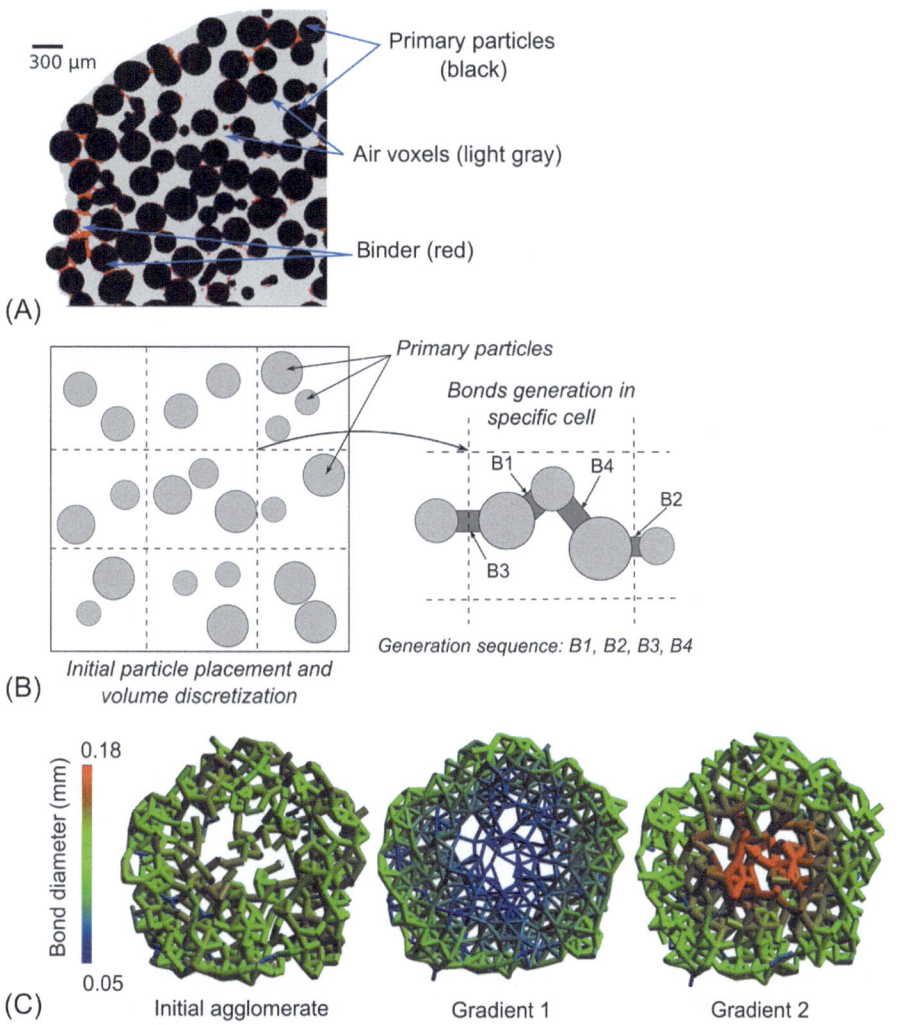

FIGURE 2.28

Generation of virtual granules for numerical analysis: (A) phase segmentation of XRCT dataset from an actual granule; (B) general representation of agglomerate generation algorithm (for two-dimensional case); (C) representation of internal structure of an initial granule (*left*) and of simulated agglomerates with different gradients of binder distributions (Gradient 1–binder volume gradient increases from the center to shel, Gradient 2–binder volume gradient decreased from the center to shel).

Adapted from Dosta et al. (2016).

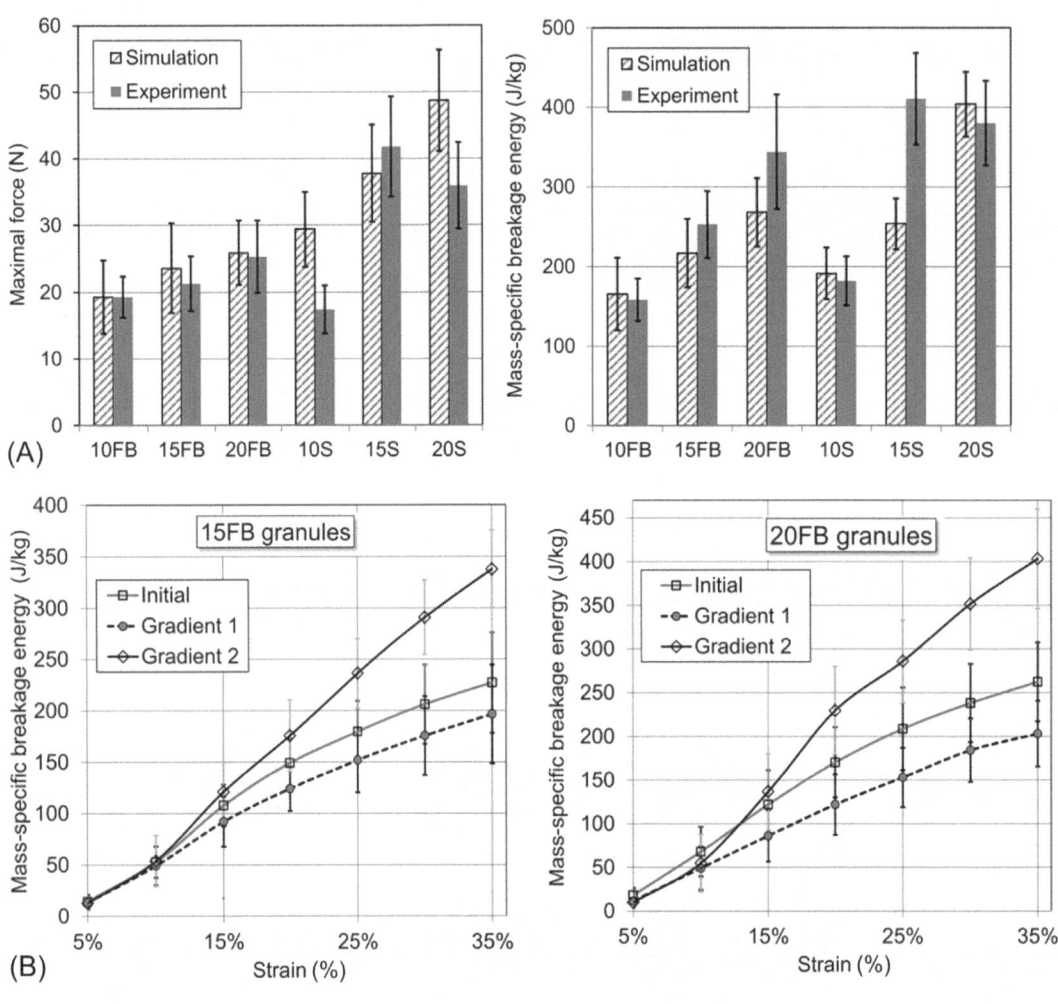

FIGURE 2.29

Mechanical properties of granules. (A) Comparison between simulation and experimental results; (B) Breakage energy calculated for different gradients of binder distribution within granules; FB–granules from fluidized bed; S–granules from static bed; 10, 15, 20–solution concentration of PVP binder (wt%) used to produce granules (amount of liquid binder was kept approximately the same in all experiments).

Adapted from Dosta et al. (2016).

2.5 Connection between granule microstructure and granule dissolution

The rate of drug release by a granule of a certain size in a dissolution medium could depend on the granule microstructure (total porosity, pore size distribution and spatial distribution, distribution of components, most importantly of a binder) and experimental conditions during dissolution. This section presents several examples where the dissolution and/or disintegration behavior of granules produced by HSWG could be connected directly to their microstructure.

Multiple studies have reported that a decrease in total measured granule porosity generally results in slower granule dissolution as observed via dissolution curves. While classic dissolution tests provide information about the kinetics of API release into the bulk solution, they do not enable direct observation of the subparticle-level phenomena that govern the dissolution/release process. The dissolution mechanism and the rate-limiting step had to be inferred indirectly from the shape of the release curve, which could lead to ill-posed inverse problems (Kimber et al., 2013). The progress of advanced imaging methods, however, has made it possible to combine findings from classical dissolution tests with information about the internal distribution of formulation components within the dosage form (pellet, granule, tablet). Wetting front propagation and gel layer formation after contact with the dissolution medium, development of API concentration profiles both inside the tablet and in the external boundary layer, or recrystallisation of the API could be observed both before and during the dissolution process. When combined with appropriate image processing algorithms and physics-based mathematical models of dissolution, a more complete picture of the underlying mechanisms and their parametric sensitivity can be obtained. In Kaspar et al. (2013), a combination of time-resolved micro-CT imaging of individual granules with UV/vis detection of the quantity of dissolved API from those granules made it possible to determine the effective diffusion coefficient of the API through the granule structure, and thus establish a quantitative structure–property relationship for dissolution. Granules were produced by HSWG from an API-based to MCC-based blend. Depending on processing conditions, the porosity of granules varied from approximately 7% to 18%. Relatively fine pores (tens of microns and larger) were distributed relatively uniformly throughout the granule volume. A power-law dependence of the effective diffusivity on porosity (Archie's law) was found to hold (Fig. 2.30). Dissolution of the drug in that case occurred via leaching of the drug from the granule, leaving the granule skeleton intact.

An example of a strong effect of binder distribution within the HSWG granule on granule dissolution time is shown in Fig. 2.23F (Mangwandi et al., 2010). It was found that the mean dissolution time of granules with uniform binder distribution (Fig. 2.23C) is about two times longer than for the granules with less uniform binder distribution (Fig. 2.23A and Fig. 2.23B) or for the granules produced without polymeric binder. Although the porosity of the granules with less uniform binder distribution (produced at a lower impeller speed) is somewhat higher (ca. 15% and 13% at 150 and 250 rpm vs. 11% at 650 rpm, respectively), change in the porosity from 15% to 13% in the granules with less uniformly distributed binder does not affect their dissolution rate, yet practically similar decrease in porosity from 13% to 11% coupled with significant change in binder distribution resulted in significant decrease in the dissolution rate. It suggests that a change in binder distribution rather than a change in total porosity is the major

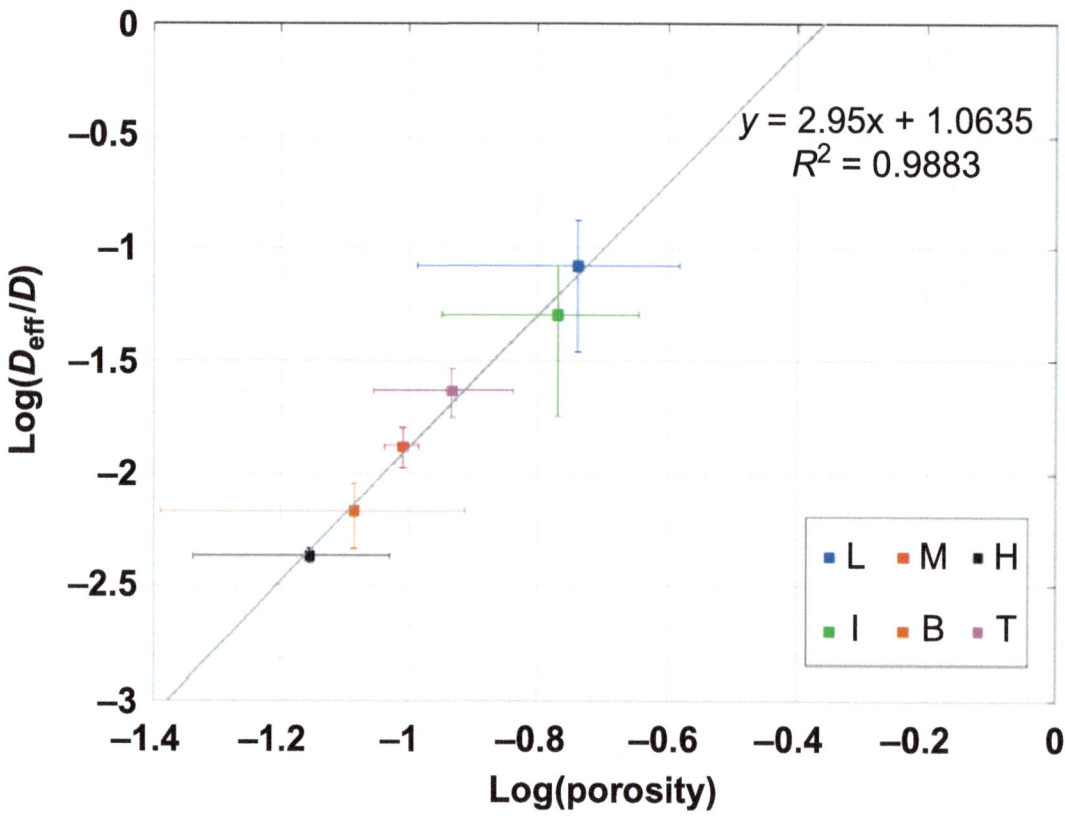

FIGURE 2.30

Double logarithmic plot of the effective diffusion coefficient as function of porosity for all samples, confirming power-law scaling (Archie's law) with a microstructure-dependent exponent.

Adapted from Kaspar et al. (2013).

factor in the dramatically changed dissolution behavior. Most likely, the binder forms a 3D continuous network within the granules with uniform binder distribution (similarly to what is shown in Figs. 2.22D and 2.23). The data indicate that the presence of a well-distributed, slow-dissolving binder results in a decrease in the dissolution rate of granules that consist of fast-dissolving components. At the same time, an opposite effect of a well-distributed binder on granule dissolution was reported in Cai et al. (2013). Well-distributed binders (either HPC or PVP) that had higher solubility than API resulted in a significant increase in the dissolution rate of HSWG granules (which mostly consisted of poorly soluble API).

Differences in dissolution properties also were observed for granules that had different microstructures because of alternative formation mechanisms–layering versus coalescence–described in Section 3.2.1, Fig. 2.9F. Granules with a consolidation microstructure dissolved more slowly than granules formed by coalescence. This is not surprising because granules with consolidation microstructure had both lower porosity and a higher amount of binder per granule weight (see Fig. 2.9A), so both of those factors would be expected to result in slower granule dissolution, although the relative contribution of each has not been studied in that work.

In the examples presented above, the microstructure of produced granules is characterized, dissolution properties are measured in a physical experiment, and the structure–property connection is established. An alternative approach to understand the effect of granule microstructure on dissolution properties is to carry out computational (in silico) experiments, that is, create a population of virtual granules with different microstructures by computer simulation, followed by the solution of convection–diffusion equations governing the dissolution process, which was used in (Ansari & Stepanek, 2008) to study effects of formulation and process variables on dissolution of granules produced by FBG process. Initially the microstructure of actual granules from several key conditions was characterized by XRCT. Then, three-dimensional virtual granules were generated by simulating the key steps of primary particle packing, binder spreading, and binder solidification during the FBG process following the approach described in (Štěpánek, 2004). This approach could be used for understanding and predicting of dissolution properties of HSWG granules. It has not been applied yet, however, most probably because of both the limitation in XRCT resolution to visualize binder distribution in HSWG granules and a lack of adequate understanding of key steps of primary particle packing, binder spreading, and binder solidification during the HSWG process.

In addition to intrinsic factors affecting granule dissolution (i.e., granule microstructure), one needs to be aware of the importance of the conditions in a dissolution apparatus in which granule dissolution might be measured in the pharmaceutical industry. The pharmacopeia paddle method employs a round-bottom unbaffled vessel and a single half-round paddle. This configuration, however, often causes the accumulation of particles near the bottom of the vessel because of insufficient agitation underneath the paddle (Bai & Armenante, 2008; Bai et al., 2007). This phenomenon is often referred to as coning or heap formation. After the coning phenomenon occurred, the dissolution rate of a drug could become slower and more variable compared to well-suspended cases (Gray et al., 2009). The occurrence of coning depends on the particle size, the particle density, the fluid viscosity, the fluid density, the apparatus configurations, and the agitation strength (Higuchi et al., 2014). One can imagine that if the borderline paddle speed was selected for the dissolution experiment, the difference in the granule microstructure that could affect boyency and flow of granules in the liquid (porosity/density, surface roughness, mean size) might result in establishment of a completely different flow regime/coning in the apparatus that would significantly affect the resulting dissolution profile.

2.6 Granule microstructure and granulation process

The microstructure of granules could provide a link between the granulation processing conditions and their resulting properties (such as strength, flowability, compressibility, and dissolution or drug release kinetics). Lactose/MCC-based granules described in Section 3.4.1 are an extreme example. Extending WM time completely transforms the granule microstructure, dramatically affecting its properties: From an agglomerate of individual particles that disintegrates in water within seconds and minutes (depending on size), the granules become a composite consisting of an insoluble MCC-based matrix containing lactose particles. Upon immersion in water, such granules do not change their shape and size for days (although soluble components are leached from the matrix). The nucleated granulation described in Section 3.3.2 provides another example: A change in impeller speed can result in qualitative changes in the microstructure because of differences in the agglomeration mechanism. Alternatively, it could be hypothesized that a change in PSD of an API (for example, reduction in D90) could result in a completely different dominant microstructure in the granulation and affect the mechanical properties.

Although the granule microstructure from the HSWG process is not well characterized in the majority of published studies in the pharmaceutical field, it is well documented that an increase in impeller speed, adding more GFL, or extending WM time will result in lower granule porosity and higher strength. The inverse relationship between granule porosity and strength is expected based on the theoretical models described in Section 2.4.1. Analysis of the actual data (see Section 2.3.3.2), however, demonstrates that, in some cases, the decrease in porosity alone cannot fully account for the increase in strength, suggesting that significant changes in the microstructure were caused by the change in processing conditions.

The results on solidification of liquid bridges presented in Section 2.4.2 demonstrate the potential importance of postgranulation processing steps such as hold time of wet granulate prior to drying and the drying step itself. It is expected that phase composition, morphology, and mechanical properties of solidified interparticle bridges can be affected by differences in hold time as well as drying temperature and drying kinetics. This will affect granule strength significantly, although the apparent microstructure (porosity distribution within the granule and distribution of the main excipients) might remain the same.

As will be discussed in the concluding remarks, however, granule microstructure might be only one of several factors that affect the processability of the granulation, process robustness, and the properties of the final dosage form.

2.7 Concluding remarks

This chapter considers the microstructure and properties of individual granules. It was demonstrated that granule microstructure might affect both the mechanical properties (crush strength, attrition resistance) and transient properties of individual granules (dissolution, disintegration, drug release kinetics). For granules that are filled into capsules, the properties of the granules would determine the properties of the final product. As such, if there is a correlation between the granule microstructure and drug release from granules, this should also be observed for the capsule. If tablet dissolution is controlled by the dissolution of compressed granules, then granule microstructure could be a controlling factor. For example, it was found for hollow granules in Section 2.3.2.2 that in spite of similar tabletability (i.e., ability to form a strong compact when multiple granules are compressed in a tablet) of blends based on both the hollow granules and collapsed hollow granules (Fig. 2.12, 78/1 and 78/7), dissolution of tablets from collapsed granules was significantly slower and directly correlated with slow dissolution of individual granules with collapsed hollow structure. Similarly it was found that the dissolution of tablets produced from resultant insoluble MCC-based granules of steady-state granulation directly correlates with drug release kinetics of individual granules. Such a relationship, however, might not necessarily exist in a general case.

As important as microstructure or strength of individual granules might be for the intermediate processing steps (drying, pneumatic conveying, storage), their input into collective properties of granules–flowability and especially tabletability–needs to be considered. For example, tabletability might be independent of the granule microstructure, as was shown in Cai et al. (2013), who studied HSWG of pure API with and without polymeric binder. Size distribution, granule porosity, and dissolution kinetics for granules of each of the two drugs–simvastatin and etoricoxib–were practically unaffected by the presence of a polymeric binder in the formulation, and for etoricoxib, tabletability was good and independent of the presence of a polymeric binder. For simvastatin, granulation without a polymeric binder was unable to form strong tablets, while binder-containing granulations had excellent tableting properties, resulting in tablets with fast dissolution profiles. The latter case also clearly demonstrates the role of the polymeric binder in the microstructure of simvastatin granules. Not only does it form bridges between the original particles and hold them together in a granule, but it also provides bonding between particles within the granules and between granules during tablet compression, as well as serving as a conduit for dissolution media during the dissolution test.

Significant differences in granule microstructure and tabletability, however, might not necessarily result in different dissolution properties of the resultant tablets. The compaction profile of lubricated steady-state granules discussed in Section 2.3.4 is compared to the compaction profile of milled, lubricated conventional granules in Fig. 2.24B. Although the tensile strength of tablets obtained from granules produced by the steady-state procedure is not as high as that of tablets produced from granules of a conventional granulation, the rates of dissolution of the two types of tablets were found to be comparable (Fig. 2.24C). The dissolution profile of fast disintegrated granules (no WM) mostly likely controlled by the dissolution kinetics of individual API particles, while the dissolution profile of MCC-based matrix granules is probably controlled by other rate-limited processes. Nevertheless, this difference in microstructure and dissolution mechanisms did not result in a difference in tablet dissolution.

Additionally, postgranulation processing steps—milling, blending with other excipients, lubrication, compression, and, potentially, film coating—might strongly affect and even dominate the tabletability of the granulation and properties of a tablet, including its strength, disintegration, and dissolution behavior, regardless of granule microstructure. For example, overlubrication followed by overcompression of fast-dissolving granules could result in very slow disintegration and slow dissolution of a tablet.

Although a number of studies investigate tabletability of the granulation after wet granulation or resulting tablet properties (Sheskey & Williams, 1996) (see, e.g., Bacher et al., 2008; Emori et al., 1997; Murakami et al., 2001; Nguyen et al., 2013; Shi et al., 2010, 2011a, 2011b; Wu et al., 2001), few published studies to date connect tablet properties to granule microstructure. Understanding the potential contribution of the microstructure of individual granules to the collective properties of granulation and to the final drug product could provide a valuable insight into the properties of the drug product and into the factors controlling variability and robustness of the process.

Acknowledgments

The author would like to thank his current and former colleagues at Merck & Co., Inc., for encouragement and fruitful discussions during the writing of this section, as well as for successful collaboration throughout the years of research work. Special thanks go to B. Ricart, M. Gentzler, G. Wong, J.N. Michaels, G. Tardos, and K. Hapgood.

References

Ansari, M. A., & Stepanek, F. (2006). Formation of hollow core granules by fluid bed in-situ melt granulation: Modeling and experiments. *International Journal of Pharmaceutics, 321*, 108–118.

Ansari, M. A., & Stepanek, F. (2008). The effect of granule microstructure on dissolution rate. *Powder Technology, 181*(2), 104–114. doi:10.1016/j.powtec.2006.12.012.

Antonyuk, S., Khanal, M., Tomas, J., Heinrich, S., & Mörl, L. (2006). Impact breakage of spherical granules: Experimental study and DEM simulation. *Chemical Engineering and Processing: Process Intensification, 45*, 838–856.

Bacher, C., Olsen, P. M., Bertelsen, P., & Sonnergaard, J. M. (2008). Compressibility and compactibility of granules produced by wet and dry granulation. *International Journal of Pharmaceutics, 358*, 69–74.

Bai, G., & Armenante, P. M. (2008). Velocity distribution and shear rate variability resulting from changes in the impeller location in the USP dissolution testing apparatus II. *Pharmaceutical Research, 25*(2), 320–336. doi:10.1007/s11095-007-9477-z.

Bai, G., Armenante, P. M., Plank, R. V., Gentzler, M., Ford, K., & Harmon, P. (2007). Hydrodynamic investigation of USP dissolution test apparatus II. *Journal of Pharmaceutical Sciences, 96*(9), 2327–2349. doi:10.1002/jps.20818.

Barrera-Medrano, D., Salman, A. D., Reynolds, G. K., & Hounslow, M. J. (2007). Granule Structure, Chapter 25. In A. D. Salman, M. J. Hounslow, & J. P. K. Seville (Eds.), *Granulation*. Elsevier B.V..

Beekman, W. J., Meesters, G. M. H., Scarlett, B., & Becker, T. (2002). Measurement of granule attrition and fatigue in a vibrating box. *Particle and Particle Systems Characterization, 19*, 5–11.

Bika, D., Tardos, G., Panmai, S., Farber, L., & Michaels, J. N. (2005). Strength and morphology of solid bridges in dry granules of pharmaceutical powders. *Powder Technology, 150*, 104–116.

Bika, D. G., Gentzler, M., & Michaels, J. N. (2001). Mechanical properties of agglomerates. *Powder Technology, 117*, 98–112.

Boerefijn, R., Ning, Z., & Ghadiri, M. (1998). Disintegration of weak lactose agglomerates for inhalation applications. *International Journal of Pharmaceutics, 172*, 199–209.

Cai, L., Farber, L., Zhang, D., Li, F., & Farabaugh, J. (2013). A new methodology for high drug loading wet granulation formulation development. *International Journal of Pharmaceutics, 441*(1-2), 790–800. doi:10.1016/j.ijpharm.2012.09.052.

Charles-Williams, H., Wengeler, R., Flore, K., Feise, H., Hounslow, M. J., & Salman, A. D. (2013). Granulation behaviour of increasingly hydrophobic mixtures. *Powder Technology, 238*, 64–76. doi:10.1016/j.powtec.2012.06.009.

Couroyer, C., Ghadiri, M., Laval, P., Brunard, N., & Kolenda, F. (2000). Methodology for investigating the mechanical strength of reforming catalyst beads. *Oil and Gas Science and Technology: Revue de l'Institute Fracais du Petrole, 55*, 67–85.

References

Dale, S., Wassgren, C., & Litster, J. (2014). Measuring granule phase volume distributions using X-ray microtomography. *Powder Technology, 264*, 550–560. doi:10.1016/j.powtec.2014.06.009.

Danalou, S. Z., Ding, X. F., Zhu, N., Emady, H. N., & Zhang, L. (2022). 4D study of liquid binder penetration dynamics in pharmaceutical powders using synchrotron X-ray micro computed tomography. *International Journal of Pharmaceutics, 627*, 122192.

Dosta, M., Dale, S., Antonyuk, S., Wassgren, C., Heinrich, S., & Litster, J. D. (2016). Numerical and experimental analysis of influence of granule microstructure on its compression breakage. *Powder Technology, 299*, 87–97. doi:10.1016/j.powtec.2016.05.005.

Eckhard, S., Fries, M., Teipel, U., Antonyuk, S., & Heinrich, S. (2014). Modification of the mechanical granule properties via internal structure. *Powder Technology, 258*, 252–264.

Eckhard, S., & Nebelung, M. (2011). Investigations of the correlation between granule structure and deformation behavior. *Powder Technology, 206*, 79–87.

Emori, H., Sakuraba, Y., Takahashi, K., Nishihata, T., & Mayumi, T. (1997). Prospective validation of high-shear wet granulation process by wet granule sieving method. II. Utility of wet granule sieving method. *Drug Development and Industrial Pharmacy, 23*(2), 203–215.

Eshtiaghi, N., Arhatari, B., & Hapgood, K. P. (2009). Producing hollow granules from hydrophobic powders in high-shear mixer granulators. *Advanced Powder Technology, 20*, 558–566.

Eshtiaghi, N., Liu, J. J. S., & Hapgood, K. P. (2010). Formation of hollow granules from liquid marbles: Small scale experiments. *Powder Technology, 197*, 184–195.

Farber, L., Al-aaraj, D. K., Smith, R., & Gentzler, M. (2020). Formation and internal microstructure of granules from wetting and non-wetting efavirenz/iron oxide blends. *Chemical Engineering Science, 227*, 115909.

Farber, L., & Michaels, J. N. (2018). Microstructure of micro-crystalline cellulose based granules produced by high-shear wet granulation with long wet-massing time. *Institution of Chemical Engineers, United States Chemical Engineering Research and Design, 132*, 1054–1059. doi:10.1016/j.cherd.2017.12.025.

Farber, L., Tardos, G., & Michaels, J. N. (2003a). Use of X-ray tomography to study the porosity and morphology of granules. *Powder Technology, 132*(1), 57–63. doi:10.1016/S0032-5910(03)00043-3.

Farber, L., Tardos, G. I., & Michaels, J. N. (2003b). Evolution and structure of drying material bridges of pharmaceutical excipients: Studies on a microscope slide. *Chemical Engineering Science, 58*, 4515–4525.

Farber, L., Tardos, G. I., & Michaels, J. N. (2005). Micro-mechanical properties of drying material bridges of pharmaceutical excipients. *International Journal of Pharmaceutics, 306*(1-2), 41–55. doi:10.1016/j.ijpharm.2005.08.028.

Gamlen, M. J., Seager, H., & Warrack, J. K. (1982). The structure and tablet properties of paracetamol granules prepared in a fluid bed and by wet massing. *International Journal of Pharmaceutical Technology and Product Manufacture, 3*, 108–114.

Gentzler, M., & Michaels, J. N. (2004). Impact attrition of brittle structured particles at low velocities: Rigorous use of a laboratory vibrational impact tester. *Chemical Engineering Science, 59*(24), 5949–5958. doi:10.1016/j.ces.2004.07.038.

Golchert, D., Moreno, R., Ghadiri, M., & Litster, J. (2004). Effect of granule morphology on breakage behaviour during compression. *Powder Technology, 143–144*, 84–96.

Gray, V., Kelly, G., Xia, M., Butler, C., Thomas, S., & Mayock, S. (2009). The science of USP 1 and 2 Dissolution: Present challenges and future relevance. *Pharmaceutical Research, 26*(6), 1289–1302. doi:10.1007/s11095-008-9822-x.

Guo, Z., Yin, X., Liu, C., Wu, L., Zhu, W., Shao, Q., York, P., Patterson, L., & Zhang, J. (2016). Microstructural investigation using synchrotron radiation X-ray microtomography reveals taste-masking mechanism of acetaminophen microspheres. *International Journal of Pharmaceutics, 499*(1-2), 47–57. doi:10.1016/j.ijpharm.2015.12.045.

Hapgood, K. P., Farber, L., & Michaels, J. N. (2009). Agglomeration of hydrophobic powders via solid spreading nucleation. *Powder Technology, 188*, 248–254.

Heng, D., Tang, P., Cairney, J. M., Chan, H. K., Cutler, D. J., Salama, R., & Yun, J. (2007). Focused-ion-beam milling: A novel approach to probing the interior of particles used for inhalation aerosols. *Pharmaceutical Research, 24*(9), 1608–1617. doi:10.1007/s11095-007-9276-6.

Higuchi, M., Yoshihashi, Y., Tarada, K., & Sugano, K. (2014). Minimum rotation speed to prevent coning phenomena in compendium paddle dissolution apparatus. *European Journal of Pharmaceutical Sciences, 65*, 74–78.

Hiramatsu, Y., & Oka, Y. (1966). Determination of the tensile strength of rock by a compression test of an irregular test piece. *International Journal of Rock Mechanics and Mining Sciences, 3*(2), 89–90. doi:10.1016/0148-9062(66)90002-7.

Iveson, S. M., Litster, J. D., Hapgood, K., & Ennis, B. J. (2001). Nucleation, growth and breakage phenomena in agitated wet granulation processes: A review. *Powder Technology, 117*, 3–39.

Johnson, K. L., Kendall, K., & Roberts, A. D. (1971). Surface energy and the contact of elastic solids. *Proceedings of the Royal Society of London A, 324*, 301–313.

Kaspar, O., Tokarovaa, V., Okab, S., Sowrirajanb, K., Ramachandranb, R., & Stepanek, F. (2013). Combined UV/Vis and micro-tomography investigation of acetaminophen dissolution from granules. *International Journal of Pharmaceutics, 458*, 272–281.

Kendall, K. (1987). Tribology in particular technology. In Adams, M. & Briscoe, B. (Eds.), Bristol: IOP Publishing.

Kimber, J. A., Kazarian, S. G., & Stepanek, F. (2013). Formulation design space analysis for drug release from swelling polymer tablets. *Powder Technology, 236*, 179–187.

Kozhar, S., Dosta, M., Antonyuk, S., Heinrich, S., & Bröckel, U. (2015). DEM simulations of amorphous irregular shaped micrometer-sized titania agglomerates at compression. *Advanced Powder Technology, 26*, 767–777.

Krycer, I., & Pope, D. G. (1983). An evaluation of tablet binding agents. Part I. Solution binders. *Powder Technology, 34*, 39–51.

Le, P. K., Avontuur, P., Hounslow, M. J., & Salman, A. D. (2011). A microscopic study of granulation mechanisms and their effect on granule properties. *Powder Technology, 206*(1-2), 18–24. doi:10.1016/j.powtec.2010.06.014.

Li, C., Zhu, N., Emady, H. N., & Zhang, L. (2019). Synchrotron-based X-ray in-situ imaging techniques for advancing the understanding of pharmaceutical granulation. *International Journal of Pharmaceutics, 572*, 118797.

Li, J., Tao, L., Dali, M., Buckley, D., Gao, J., & Hubert, M. (2011). The effect of the physical states of binders on high-shear wet granulation and granule properties: A mechanistic approach toward understanding high-shear wet granulation process. Part II. Granulation and granule properties. *Journal of Pharmaceutical Sciences, 100*(1), 294–310. doi:10.1002/jps.22261.

Maire, E., & Withers, P. J. (2014). Quantitative X-ray tomography. *International Materials Reviews, 59*(1), 1–43. doi:10.1179/1743280413Y.0000000023France.

Mangwandi, C., Adams, M. J., Hounslow, M. J., & Salman, A. D. (2010). Effect of impeller speed on mechanical and dissolution properties of high-shear granules. *Chemical Engineering Journal, 164*(2-3), 305–315. doi:10.1016/j.cej.2010.05.039.

Metzger, M. J., & Glasser, B. J. (2012). Numerical investigation of the breakage of bonded agglomerates during impact. *Powder Technology, 217*, 304–314. doi:10.1016/j.powtec.2011.10.042.

Michaels, J. N., Farber, L., Wong, G. S., Hapgood, K., Heidel, S. J., Farabaugh, J., Chou, J. H., & Tardos, G. I. (2009). Steady states in granulation of pharmaceutical powders with application to scale-up. *Powder Technology, 189*(2), 295–303. doi:10.1016/j.powtec.2008.04.028.

Miller, M. K., Russell, K. F., Thompson, K., Alvis, R., & Larson, D. J. (2007). Review of atom probe FIB-based specimen preparation methods. *Microscopy and Microanalysis, 13*, 428–436.

Mishra, B. K., & Thornton, C. (2001). Impact breakage of particle agglomerates. *International Journal of Mineral Processing, 61*(4), 225–239. doi:10.1016/s0301-7516(00)00065-x.

Moreno, R., Ghadiri, M., & Antony, S. J. (2003). Effect of the impact angle on the breakage of agglomerates: A numerical study using DEM. *Powder Technology, 130*, 132–137.

Moreno-Atanasio, R., Williams, R., & Jia, X. (2010). Combining X-ray tomography with computer simulation for analysis of granular and porous materials. *Particuology, 8*, 81–99.

Murakami, H., Yoneyama, T., Nakajima, K., & Kobayashi, M. (2001). Correlation between loose density and compactibility of granules prepared by various granulation methods. *International Journal of Pharmaceutics, 216*, 159–164.

Nguyen, T. H., Morton, D. A. V., & Hapgood, K. P. (2013). Application of the unified compaction curve to link wet granulation and tablet compaction behavior. *Powder Technology, 240*, 103–115.

Nguyen, T. H., Shen, W., & Hapgood, K. (2010). Effect of formulation hydrophobicity on drug distribution in wet granulation. *Chemical Engineering Journal, 164*(2-3), 330–339. doi:10.1016/j.cej.2010.05.008.

Noguchi, S., Kajihara, R., Iwao, Y., Fujinami, Y., Suzuki, Y., Terada, Y., Uesugi, K., Miura, K., & Itai, S. (2013). Investigation of internal structure of fine granules by microtomography using synchrotron X-ray radiation. *International Journal of Pharmaceutics, 445*(1-2), 93–98. doi:10.1016/j.ijpharm.2013.01.048.

Oka, S., Emady, H., Kašpar, O., Tokárová, V., Muzzio, F., Štěpánek, F., & Ramachandran, R. (2015). The effects of improper mixing and preferential wetting of active and excipient ingredients on content uniformity in high shear wet granulation. *Powder Technology, 278*, 266–277. doi:10.1016/j.powtec.2015.03.018.

Papadopoulus, D. G., & Ghadiri, M. (1996). Impact breakage of poly-methylmethacrylate (PMMA) extrudates: I. Chipping mechanism.. *Advanced Powder Technology, 7*(3), 183–197.

Pietsch, W., Rumpf, H., & Haftkraft, K. (1967). Adhesive force, capillary pressure, liquid volume and critical angle of a liquid bridge between two spheres. *Chemical Engineering Technology, 39*(15), 885–893. doi:10.1002/cite.330391502.

Planinsek, O., Pisek, R., Trojak, A., & Srcic, S. (2000). The utilization of surface free-energy parameters for the selection of a suitable binder in fluidized bed granulation. *International Journal of Pharmaceutics, 207*, 77–88.

Potyondy, D. O., & Cundall, P. A. (2004). A bonded-particle model for rock. *International Journal of Rock Mechanics and Mining Sciences, 41*(8), 1329–1364. doi:10.1016/j.ijrmms.2004.09.011.

Rahmanian, N., Ghadiri, M., & Ding, Y. (2008). Effect of scale of operation on granule strength in high shear granulators. *Chemical Engineering Science, 63*, 915–923.

Rahmanian, N., Ghadiri, M., & Jia, X. (2011). Seeded granulation. *Powder Technology, 206*(1-2), 53–62. doi:10.1016/j.powtec.2010.07.011.

Rahmanian, N., Ghadiri, M., Jia, X., & Stepanek, F. (2009). Characterisation of granule structure and strength made in a high shear granulator. *Powder Technology, 192*(2), 184–194. doi:10.1016/j.powtec.2008.12.016.

Rajniak, P., Mancinelli, C., Chern, R., Stepanek, F., Farber, L., & Hill, B. (2007). Experimental study of wet granulation in fluidized bed: Impact of the binder properties on the granule morphology. *International Journal of Pharmaceutics, 334*(1-2), 92–102. doi:10.1016/j.ijpharm.2006.10.040.

Rowe, R. C. (1989a). Surface free energy and polarity effects in the granulation of a model system. *International Journal of Pharmaceutics, 53*(1), 75–78. doi:10.1016/0378-5173(89)90364-5.

Rowe, R. C. (1989b). Binder–substrate interactions in granulation, a theoretical approach based on surface free energy and polarity. *International Journal of Pharmaceutics, 52*, 149–154.

Rumpf, H. (1975). Particle technology. Chapman and Hall.

Ryu, H. J., & Saito, F. (1991). Single particle crushing of nonmetallic inorganic brittle materials. *Solid State Ionics, 47*(1-2), 35–50. doi:10.1016/0167-2738(91)90177-D.

Schöngut, M., Smrcĭka, D., Gregor, T., & Štelpánek, F. (2015). Investigation of rate-limiting steps during granulation with a chemically reactive binder. *Powder Technology, 270*, 510–519.

Seager, H., Burt, I., Ryder, J., Rue, P., Murray, S., Beal, N., & Warrack, J. K. (1979). The relationship between granule structure, process of manufacture and the tableting properties of the granulated product. Part I. *International Journal of Pharmaceutical Technology and Product Manufacture, 1*, 36–44.

Shen, Z., Jiang, M., & Thornton, C. (2016). DEM simulation of bonded granular material. Part I: Contact model and application to cement sand. *Computers and Geotechnics, 75*, 192–209.

Sheskey, P. J., & Williams, D. M. (1996). Comparison of low-shear and high-shear wet granulation techniques and the influence of percent water addition in the preparation of a controlled-release matrix tablet containing HPMC and a high-dose, highly water-soluble drug. *Pharmaceutical Technology, 20*(3), 80–92.

Shi, L., Feng, Y., & Sun, C. C. (2010). Roles of granule size in over-granulation during high shear wet granulation. *Journal of Pharmaceutical Sciences, 99*(8), 3322–3325.

Shi, L., Feng, Y., & Sun, C. C. (2011a). Massing in high shear wet granulation can simultaneously improve powder flow and deteriorate powder compaction: A double-edged sword. *European Journal of Pharmaceutical Sciences, 43*(1–2), 50–56.

Shi, L., Feng, Y., & Sun, C. C. (2011b). Origin of profound changes in powder properties during wetting and nucleation stages of high-shear wet granulation of microcrystalline cellulose. *Powder Technology, 208*(3), 663–668.

Smrcka, D., Dohnal, J., & Štepánek, F. (2016). Dissolution and disintegration kinetics of high-active pharmaceutical granules produced at laboratory and manufacturing scale. *European Journal of Pharmaceutics and Biopharmaceutics, 106*, 107–116.

Spettl, A., Dosta, M., Antonyuk, S., Heinrich, S., & Schmidt, V. (2015). Statistical investigation of agglomerate breakage based on combined stochastic mocrostructure modeling and DEM simulations. *Advanced Powder Technology, 26*, 1021–1030.

ŠteÏpánek, F., Rajniak, P., Mancinelli, C., Chern, R. T., & Ramachandran, R. (2009). Distribution and accessibility of binder in wet granules. *Powder Technology, 189*, 376–384.

Štěpánek, F. (2004). Computer-aided product design: Granule dissolution. *Chemical Engineering Research and Design, 82*(11), 1458–1466. doi:10.1205/cerd.82.11.1458.52035.

Štěpánek, F., & Ansari, M. A. (2005). Computer simulation of granule microstructure formation. *Chemical Engineering Science, 60*(14), 4019–4029. https://doi.org/10.1016/j.ces.2005.02.030.

Subero, J., & Ghadiri, M. (2001). Breakage patterns of agglomerates. *Powder Technology, 120*, 232–243.

Subero, J., Ning, Z., Ghadiri, M., & Thornton, C. (1999). Effect of interface energy on the impact strength of agglomerates. *Powder Technology, 105*, 66–73.

Tardos, G. I., Khan, I. M., & Mort, P. M. (1997). Critical parameters and limiting conditions in binder granulation of fine powders. *Powder Technology, 94*, 245–258.

Thornton, C., Ciomocos, M. T., & Adams, M. J. (2004). Numerical simulations of diametrical compression tests on agglomerates. *Powder Technology, 140*(3), 258–267. https://doi.org/10.1016/j.powtec.2004.01.022.

Tong, Z. B., Yang, R. Y., Yu, A. B., Adi, S., & Chan, H. K. (2009). Numerical modeling of the breakage of loose agglomerates of fine particles. *Powder Technology, 196*, 213–221.

Walton, D. E., & Mumford, C. J. (1999). The morphology of spray-dried particles: The effect of process variables upon the morphology of spray-dried particles. *Chemical Engineering Research & Design, 77*, 260–442.

Wong, J., D'Sa, D., Foley, M., Chan, J. G. Y., & Chan, H.-K. (2014). NanoXCT: A novel technique to probe the internal architecture of pharmaceutical particles. *Pharmaceutical Research, 31*, 3085–3094.

Wu, J.-S., Ho, H.-O., & Sheu, M.-T. (2001). Influence of wet granulation and lubrication on the powder and tableting properties of codried product of microcrystalline cellulose with β-cyclodextrin. *European Journal of Pharmaceutics and Biopharmaceutics, 51*, 63–69.

Wu, S. (1973). Polar and nonpolar interactions in adhesion. *Journal of Adhesion, 5*, 39–55.

Yuregir, K. R., Ghadiri, M., & Clift, R. (1986). Observations on impact attrition of granular solids. *Powder Technology, 49*, 53–57.

Zhang, Z., & Ghadiri, M. (2002). Impact attrition of particulate solids. Part 2: Experimental work. *Chemical Engineering Science, 57*(17), 3671–3686. https://doi.org/10.1016/s0009-2509(02)00241-5.

Zhu, B., Trainia, D., Lewisc, D. A., & Young, P. (2014). The solid-state and morphological characteristics of particles generated from solution-based metered dose inhalers: Influence of ethanol concentration and intrinsic drug properties. *Colloids and Surfaces A: Physicochemical Engineering Aspects, 443*, 345–355.

CHAPTER 3

Mechanistic basis for the effects of process parameters on quality attributes in high shear wet granulation

Sherif I.F. Badawy[a], Ajit S. Narang[b], Keirnan R. LaMarche[c], Ganeshkumar A. Subramanian[d] and Sailesh A. Varia[a]

[a]*Drug Product Science & Technology, Bristol-Myers Squibb Co., New Brunswick, NJ, United States,*
[b]*Pharmaceutical Sciences, ORIC Pharmaceuticals, Inc., South San Francisco, CA, United States,*
[c]*Patheon part of Thermo Fisher Scientific, Greenville, NC, United States,* [d]*Manufacturing Science and Technology, Bristol-Myers Squibb Co., New Brunswick, NJ, United States*

3.1 Introduction

High shear wet granulation is a particle agglomeration process extensively used in the pharmaceutical industry. During wet granulation, particle size enlargement is achieved by the addition of a liquid to the powder formulation while mixing in a high shear mixer (Kristensen, 1996; Kristensen and Schaefer, 1987; Schaefer et al., 1986). Granulation parameters must be controlled in order to ensure the manufacture of a granulation with the desired particle size. The two main components of the wet granulation process that potentially dictate the granulation outcome include the granulator equipment and granulation liquid factors. Granulator equipment factors include bowl impeller blade and chopper design (Holm, 1987), bowl fill fraction, impeller placement (top or bottom driven), impeller and chopper speed, and process time (Badawy and Pandey, 2017; Oulahna, Cordier, Galet, & Dodds, 2003). The factors associated with granulation liquid include amount, liquid introduction method (spray versus drip), spray rate, spray characteristics (droplet size, velocity, spray area, etc.), liquid addition time, addition location, liquid properties (viscosity, surface tension, etc.), and binder addition method (wet versus dry). Despite its extensive use in the pharmaceutical industry, wet granulation processes traditionally have been developed using a trial-and-error approach. Several studies investigated the effect of granulation parameters in the high shear mixer on product attributes and demonstrated that control of process parameters is necessary to obtain product with the desired attributes (Johansson and Alderborn, 2001; Ritala, Holm, Schaefer, & Kristensen, 1988; Shiraishi, Kondo, Yuasa, & Kanaya, 1994). Design of experiment (DOE) approach has been occasionally applied to study the effect of wet granulation parameters on product attributes, but the results are usually not interpreted in terms of physical mechanisms of granulation (Iskandarani, Shiromani, & Clair, 2001; Ring, Oliveira, & Crean, 2011). In the last two decades, however, efforts have increased to develop physical models that characterize the rate processes in wet granulation such as nucleation, granule coalescence, and breakage (Knight, 1993; Iveson, Litster, & Ennis, 1996;

Iveson, Litster, Hapgood, & Ennis, 2001; Liu, Litster, Iveson, & Ennis, 2000; Tardos, Khan, & Mort, 1997). Although these models have enhanced the fundamental understanding of wet granulation, they are not directly amenable to the design of industrial processes. Those models generally have been developed for a particular rate process using simple systems and they are not directly applicable to the complex pharmaceutical formulations and wet granulation processes that involve simultaneous rate processes. In the quality by design (QbD) era, there is a need to understand the manufacturing process and to define critical process parameters that affect product performance (Badawy et al., 2016; Yu, 2008; Yu et al., 2014). Moreover, it is critical to understand how those parameters affect the critical quality attributes (CQAs) of the drug product. This requires efforts to bridge the gap between physical models of wet granulation and industrial process development practice. While the models are not currently in a position to be used directly in process development or control, they provide the fundamental knowledge necessary to understand the mechanisms involved in the industrial process. Such understanding is evolving, which would allow for a more systematic approach to process development instead of trial and error.

In this chapter, knowledge obtained from wet granulation studies on development compounds and placebo formulations (Badawy, Menning, Gorko, & Gilbert, 2000; Badawy, Narang, LaMarche, Subramanian, & Varia, 2012) are used to discuss the effect of process parameters on primary granule properties (granule size and density), on key granulation bulk powder properties, and on final product critical quality attributes (flow, compaction properties, and tablet dissolution rate). The mechanisms of the effect of process parameters on primary granule properties are reviewed. In addition, the correlation of those primary properties with bulk powder properties and tablet dissolution are identified in order to connect mechanisms involving process parameters to product performance. The findings are compared between the different formulations in order to recognize common mechanistic themes and also explain differences in behavior when they exist. As the performance of those compounds is assessed collectively, some themes are identified that can be viewed as general rules for the pharmaceutical wet granulation process.

3.2 Lactose study

A screening experimental design was used to identify critical parameters that influence granulation characteristics of a lactose-based placebo formulation, including granule size and compaction properties (Badawy et al., 2000). Anhydrous lactose, NF, was the major component of the formulation (approximately 93%, w/w). Among other excipients in the formulation, povidone was used as a binder (2%, w/w), and crospovidone as a disintegrant (2%, w/w); magnesium stearate (1%, w/w) was used as a lubricant.

Evaluated for their effect on granulation characteristics were six variables: impeller speed of the granulator; granulating solution addition rate; total amount of water added in the granulation step; mixing time after complete addition of granulating solution (wet massing time); moisture content of the granulation after drying; and screen size used for milling of the dried granulation. These variables were screened with an eight run Plackett-Burman design shown in Table 3.1. Regression analysis of the data was carried out in the Statistical Analysis System (SAS) software by a linear model with no interactions. Analysis was done on the centered values of the factors (high level set to 1, low level set

Table 3.1 Plackett-Burman design for lactose-based formulation study (Badawy et al., 2000)

Run number	Impeller speed (rpm)	Liquid addition rate[a] (g/min)	Total amount of water (g)	Wet massing time (min)	Moisture level after drying[b] (%)	Screen size for dry milling (in.)
1	600	200	225	1	2.0	0.024
2	300	200	225	3	0.8	0.032
3	300	100	225	3	2.0	0.024
4	600	100	185	3	2.0	0.032
5	300	200	185	1	2.0	0.032
6	600	100	225	1	0.8	0.032
7	600	200	185	3	0.8	0.024
8	300	100	185	1	0.8	0.024

[a] Target values. Actual values ranged from 90.8 to 143 g/min for the low level, and 186.8–216.5 g/min for the high level. Actual values were used for data analysis.
[b] Target values. Actual values ranged from 0.64% to 0.73% for the low level and 1.92%–2.54% for the high level. Actual values were used for data analysis.

to −1). In the case of moisture content of the granulation and liquid addition rate, actual experimental values (not the target values) were used for the regression analysis.

Granulation was carried out in a TK-Fielder Spectrum Processor (SP-1) granulator at a batch size of 2 kg using an M8 impeller and a Christmas tree chopper design. Anhydrous lactose was blended with half of the quantity of crospovidone for 2 min in the granulator operating at impeller speed of 300 or 600 rpm and chopper speed of 3000 rpm. Povidone was dissolved in the granulating solution, which was then added to the blend in the granulator at a rate of 100 or 200 g/min using a pressurized solution pot and a spray nozzle. The amount of water in the granulating solution represented 9.5% or 11.5% w/w of the total solids in the granulator. Mixing was continued for 1 or 3 min after complete addition of the granulating solution (wet massing time). The granulation was dried in a Glatt WSG-3 V fluid bed at 40°C inlet air temperature to a moisture content of approximately 0.8% or 2%. The dried granulation was milled using a Quadro 197S Comil through a screen of 0.032″ or 0.024″ opening diameter at 1000 rpm motor speed. The milled granulation was blended with the remaining quantity of crospovidone for 15 min in a 4 quart V-blender. Magnesium stearate then was added to the V-blender and blended for 5 min. The lubricated granulation was compressed into 60 mg tablets using 7/32″ round standard concave tooling.

In addition to the 2 kg experimental-design batches shown in Table 3.1, three batches were manufactured using a TK Fielder PMA-65 granulator at 20 kg scale using a similar process. Chopper speed was 3000 rpm and impeller speed was 200 rpm, which represents the same impeller tip speed as the 300 rpm for the SP-1 granulator. The granulating solution was added to the granulator at a rate of 1000 g/min using a pressurized solution pot and a spray nozzle. The amount of water in the granulating solution represented 9.5% w/w of the total solids in the granulator. Wet massing time was 0, 1, or 3 min. The granulation was dried in Aeromatic S2 fluid bed at 60°C inlet air temperature to a moisture content of approximately 1%. The dried granulation was milled using a Quadro 197S Comil through a screen of 0.032″ opening diameter at 1000 rpm motor speed. The milled granulation was blended with the remaining quantity of crospovidone and magnesium stearate in a V-blender. The lubricated granulation then was compressed into 240 mg tablets using 11/32″ round standard concave tooling.

Granulation was characterized for particle size distribution, bulk and tapped density, compaction properties, and pore volume distribution (mercury intrusion porosimetry) using methods similar to those described below for the API study. Granulation moisture content was determined by LOD at 105°C. Tablet pore volume distribution also was characterized by mercury intrusion porosimetry.

3.2.1 Granule size distribution

Percent of the dried (unmilled) granulation retained on 25-mesh screen (oversize granulation) and the percent of the final granulation retained on 60-mesh screen were used to compare the extent of granule growth between the various runs. Percent oversize granulation was comparable for all runs except for runs 1 and 6. The percent oversize granulation was 31.4%, and 23.2% for these two runs, respectively, compared to 6.7%–13.9% for the other batches. Mesh analysis results of the lubricated granulation showed a larger fraction of the granulation retained on the 60-mesh screen for runs 1 and 6 compared to other runs. Runs 1 and 6 have high impeller speed (600 rpm), a high level of the total amount of water (225 g) and low wet massing time (1 min). Therefore, this combination of parameters appears to have a pronounced effect on particle agglomeration, resulting in the observed increase in percent oversize granulation and the larger fraction retained on the 60-mesh screen regardless of the screen size used for milling. This conclusion is supported by the results of the regression analysis for the 40–60 mesh granulation fraction and percent oversize data (Table 3.2). Regression analysis showed that impeller speed, total amount of water, and wet massing time have the largest effect on granule size, as shown by the large magnitude of their parameter estimates for the percent oversize granulation and the 40–60 mesh fraction models. The positive sign of the parameter estimates for the impeller speed and total amount of water indicates that the high level of these parameters leads to larger granule size. To the contrary, the negative sign for the wet massing coefficient suggests that shorter wet massing time results in larger granule size.

The effect of impeller speed and the amount of water on granule growth can be explained by their effect on the degree of liquid saturation of the granules during the granulation process. Granule growth by coalescence is dependent on the degree of liquid saturation of the colliding particles. Depending on the degree of liquid saturation, the bonding strength between the colliding granules can be strong enough to resist the separating forces exerted by the impeller of the high shear granulator. Granule liquid saturation takes into account both the volume of available intragranular pores as well as the liquid volume as follows:

$$S = \frac{w\rho_s(1-\varepsilon)}{\rho_l \varepsilon} \quad (3.1)$$

where S is granule liquid saturation, w is the mass ratio of liquid-to-solid, ρ_s is solid density of primary particles, ρ_l is liquid density, and [is granule porosity. As liquid saturation increases, granules become more plastic and the thickness of liquid layer on granule surface increases. Because coalescence is enhanced by both of these factors, granule growth increases at higher liquid saturation (Kristensen and Schaefer, 1987; Ritala et al., 1988; Ritala, Jungersen, Holm, Schæfer, & Kristensen, 1986). The effect of high impeller speed and high level of water on percent oversize granulation can be described, at least partly, as the result of increased liquid saturation of the granulation. High amounts of water increases the magnitude of w in the above equation, while higher impeller speed is expected to increase granule densification and hence decrease the porosity term, [, thus increasing liquid saturation at the same granule liquid content.

During the granulation process in a high-speed mixer, particles or small granules agglomerate because of coalescence, and larger granules are broken because of the mechanical forces acting on them (Iveson et al., 2001). These two effects counteract each other and the net effect depends on several factors such as the nature of the formulation, impeller speed, liquid saturation, and granule size. Under the experimental conditions of this study, the increased wet massing time appears to result in breaking of the large granules, as shown by the decrease in the oversize granulation and the granulation fraction retained on the 60-mesh screen. It is therefore inferred that granule breakage is the more dominant mechanism affecting particle size during wet massing.

3.2.2 Compaction properties

Maximum hardness values were obtained from the compression profiles of the different runs and were used to compare the compaction properties of the different granulations (Table 3.3). All granulations with higher moisture content showed increased compactibility and the maximum hardness values ranged between 8 and 10.4 SCU. This suggests that moisture content of the granulation has a pronounced effect

Table 3.2 Regression analysis data for the Plackett-Burman experimental design of the lactose-based formulation study (Badawy et al., 2000)

Variable	Parameter estimate (P value)					
	Impeller speed	Liquid addition rate	Total amount of water	Wet massing time	Moisture level after drying	Screen size for dry milling
Maximum hardness	−0.729 (0.25)	−0.065 (0.91)	−0.436 (0.39)	−1.035 (0.19)	2.249 (0.11)	−0.213 (0.62)
Weight variation	−0.045 (0.10)	0.060 (0.12)	−0.008 (0.45)	0.001 (0.90)	−0.003 (0.79)	0.001 (0.88)
Friability	−0.039 (0.26)	−0.010 (0.76)	−0.0001 (0.99)	0.040 (0.27)	0.049 (0.27)	0.023 (0.41)
% Oversize	3.430 (0.22)	0.320 (0.89)	4.260 (0.18)	−5.938 (0.14)	0.922 (0.67)	−1.856 (0.38)
Bulk density	0.026 (0.10)	−0.012 (0.31)	0.003 (0.61)	0.015 (0.18)	−0.002 (0.74)	0.001 (0.80)
Tapped density	0.021 (0.02)	−0.009 (0.07)	0.002 (0.19)	0.014 (0.03)	−0.019 (0.03)	0.002 (0.15)
40–60 mesh fraction	9.409 (0.08)	−1.220 (0.63)	7.433 (0.10)	−5.827 (0.13)	4.210 (0.23)	−1.043 (0.55)
80–200 mesh fraction	−7.630 (0.03)	−0.512 (0.56)	−5.053 (0.05)	7.623 (0.04)	−4.589 (0.07)	0.691 (0.34)
> 325 mesh fraction	−1.773 (0.27)	1.657 (0.40)	−2.427 (0.20)	−1.811 (0.27)	0.369 (0.79)	0.345 (0.74)

3.2 Lactose study

Table 3.3 Results of compression experiments for the lactose-based study (Badawy et al., 2000)

Run number	1	2	3	4	5	6	7	8
Maximum hardness for 7/32″ tablets (SCU)	8.8	4.7	9.1	8.0	10.4	4.9	4.1	8.2
Weight variation for 7/32″ tablets of 5 SCU hardness (SD)	0.39	0.46	0.36	0.29	0.47	0.31	0.39	0.42
Friability for 7/32″ tablets of 5 SCU hardness (%)	0.02	0.17	0.27	0.19	0.19	0.06	0.08	0.06

on compactibility. Granulations dried to the low moisture content showed lower compactibility except for run 8, which had maximum hardness of 8.2 SCU, comparable to the tablets with higher moisture content. Therefore, granulations from the different runs appeared to separate into two groups according to the maximum achievable tablet hardness. One group showed maximum hardness between 8 and 10.4 SCU and includes all runs dried to approximately 2% moisture together with run 8. The other group showed maximum hardness values between 4.1 and 4.7 SCU and includes the remaining runs dried to approximately 0.8% moisture. Results of the regression analysis (Table 3.2) showed that moisture content has the largest effect on the maximum hardness, followed by the wet massing time and impeller speed. The negative sign of the coefficients for the wet massing time and impeller speed indicates that a low level of these parameters leads to more compactible granulation. Granulation from run 8, which showed the highest compactibility among granulations dried to low moisture content, was manufactured using low impeller speed and short wet massing time (second and third most important parameters). Run 5, which showed the highest maximum hardness value among granulations dried to a high moisture content also was manufactured with low impeller speed and short wet massing time. Increasing the impeller speed and/or the wet massing time resulted in the decrease in granulation compactibility.

High impeller speed and long wet massing time could decrease the granulation porosity by subjecting the granulation to high-shear forces for longer periods of time (Badawy, Gray and Hussain, 2006). Porosity of the granulation from runs 7 and 8 was determined by mercury intrusion porosimetry on a sample of the granulation with particle size > 150 µm. Pore volume for pores in the 1–8 µm diameter range was diminished for granulation from run 7, which was manufactured using high impeller speed (600 rpm) and long wet massing time (3 min), compared to run 8 (Figure 3.1). These pores appeared to be critical for granulation compactibility as shown by the remarkably higher compactibility of granulation from run 8. Tablets compressed using the more porous granulation showed reduced pore volume in the 1–4 µm pore diameter range compared to tablets compressed using the less porous granulation under the same compression force (Figure 3.2). This finding suggests a higher tendency of the more porous granulation to densify upon application of the compression force resulting in closer packing of the particles and, consequently, diminishing the pores in the 1–4 µm range. In contrast, granulation from run 7 showed a peak at 2 µm, suggesting a lower tendency for densification upon compression. This is consistent with the finding by Wikberg and Alderborn (Wikberg and Alderborn, 1993) that demonstrated wider and bimodal pore-size distribution for the tablets compressed from granulation with low porosity compared to the narrower and smaller pore-size distribution for tablets compressed from the more porous granulation. The reduced porosity of the granulation from run 7 resulted in a decreased fragmentation propensity and volume reduction behavior of the granulation, which led to

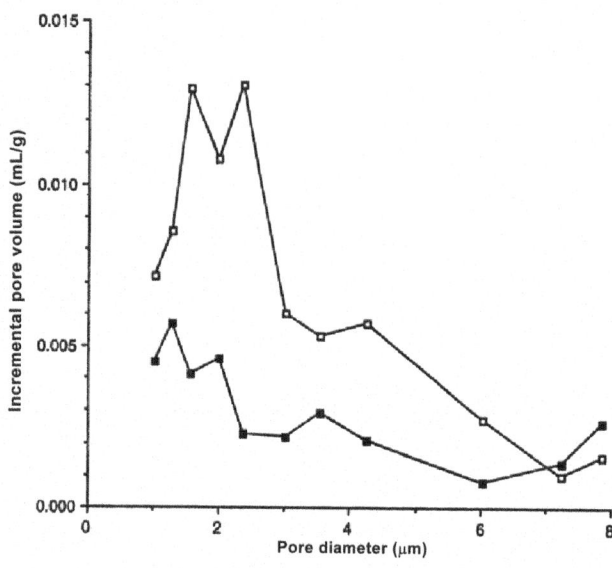

FIGURE 3.1

Effect of wet massing time on compactibility of lactose-base granulation manufactured in the PMA-65 high-shear granulator (20 kg scale) (Badawy et al., 2000).

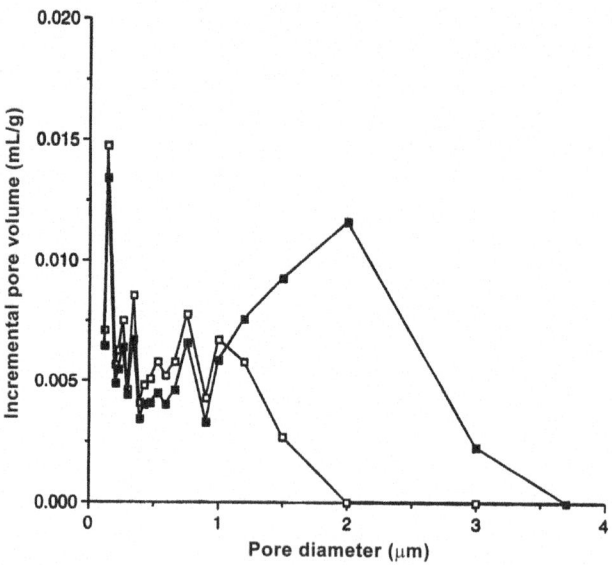

FIGURE 3.2

Pore volume distribution determined by mercury intrusion porosimetry for lactose-based tablets manufactured using different parameters. ■, high impeller speed, long wet massing time; □, low impeller speed, short wet massing time (Badawy et al., 2000).

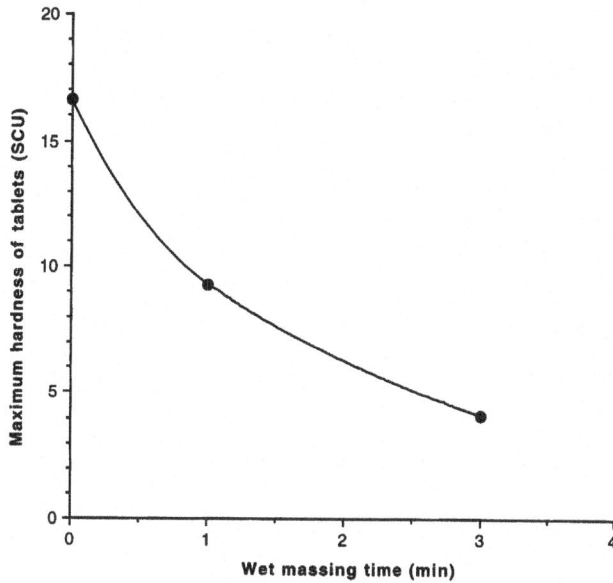

FIGURE 3.3

Effect of wet massing time on compactibility of lactose-base granulation manufactured in the PMA-65 high-shear granulator (20 kg scale) (Badawy et al., 2000).

decreased granulation compactibility. Consequently, tablets compressed from the granulation with lower pore volume showed lower tablet hardness than tablets manufactured using more porous granulation.

The effect of wet massing time on granulation compactibility also was evaluated at 20 kg scale in the 65-Liter (PMA-65) granulator. The experiments were conducted at the low impeller speed based on the results from the SP-1 experiments, which showed that high impeller speed decreased granulation compactibility. The granulation was dried to a target moisture content of approximately 1%, which was found to be the equilibrium moisture content for the granulation at ambient conditions. Other parameters (amount of water, liquid addition rate, and screen size) were selected arbitrarily because the SP-1 experiments showed they have minimal effect on granulation characteristics.

Compactibility of granulation manufactured in the PMA-65 granulator also was dependent on wet massing time. Granulation compactibility decreased as the wet massing time increased from 0 to 3 min (Figure 3.3), indicating that the effect of wet massing time at this scale is similar to the SP-1 batches. Similar to the SP-1 batches, granulation manufactured using long wet massing time in the PMA-65 granulator showed a relatively low fragmentation tendency. Scanning electron microscopy of tablets compressed using granulation manufactured in the PMA-65 granulator with 3 min wet massing time showed a more preserved outline of granules than tablets compressed using the more compactible granulation from run 8 in the above SP-1 experiments (Figure 3.4).

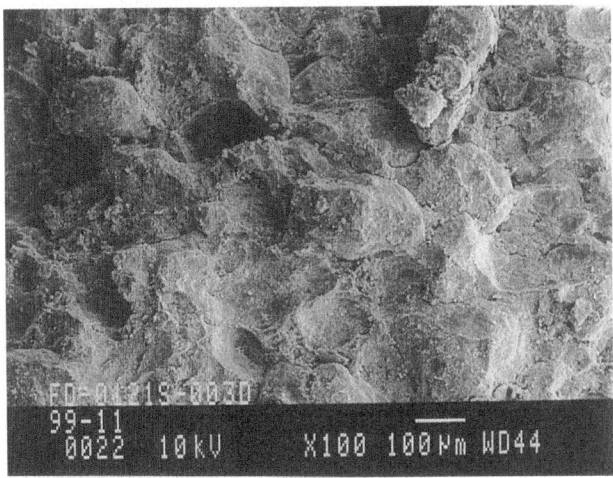

FIGURE 3.4

Scanning electron micrographs of broken surface of lactose-based tablets compressed using more compactible (top) and less compactible granulation (bottom) (Badawy et al., 2000).

3.2.3 Bulk and tapped density of the granulation

Results of the regression analysis showed that impeller speed has the largest effect on bulk density followed by the wet massing time. These parameters showed a positive sign for the estimate of their coefficients. This is consistent with the above discussion that high impeller speed and wet massing time result in lower granulation porosity and, therefore, higher bulk density. In addition, the tapped density of the granulation showed a similar trend for the effect of wet massing time and impeller speed. It is noteworthy that both bulk and tapped density showed a correlation with the maximum hardness values.

Table 3.4 Experimental design[a] for the Brivanib formulation (Badawy et al., 2012)[b]

Water level (w/w%)	Impeller speed (m/s)	Wet massing time (s)	Water addition time (min)
49	6	50	3
44	3.6	10	5
49	6	50	5
44	3.6	10	3
49	6	10	3
46.5	4.8	30	3
44	3.6	50	3
46.5	4.8	30	3
49	3.6	50	3
44	6	50	3
46.5	4.8	30	3
49	3.6	10	3
44	6	10	3

[a] 2^3 Factorial design with three center point replicates for the water level, impeller speed, and wet massing time; effect of liquid addition time was tested at the extreme combinations of the other three parameters.
[b] Intragranular components: microcrystalline cellulose, Brivanib (50%, w/w), hydroxypropyl cellulose, croscarmellose sodium; extragranular: crospovidone, colloidal silicon dioxide, microcrystalline cellulose and magnesium stearate–2.0 kg batch size.

Runs 8 and 5, which showed the highest maximum hardness among granulations with low and high moisture content, respectively, demonstrated the lowest bulk and tapped density among the eight runs.

3.3 Active pharmaceutical ingredient (API) studies

This section reviews studies for three different APIs, which represent typical development compounds in the pharmaceutical industry. The three APIs are weak bases (representing a large fraction of development compounds) differing in intrinsic solubility, pK_a, solid state form (salt versus free form), and melting point. Brivanib alaninate (Huynh et al., 2008) (BMS-582664) (Br) is a weakly basic l-alanine ester prodrug with pK_a of 6.9 and intrinsic solubility of < 10 µg/mL. Br API is a neat crystalline free form with a melting point of 143°C. Razaxaban (BMS-561389) (Ra) is a weak base with very low intrinsic solubility (< 1.0 µg/mL). Ra API is a neat hydrochloride salt form with aqueous solubility of 3.3 mg/mL (resulting solution pH of 3.3) and melting point of 252°C (Badawy et al., 2006). Pexacerfont (BMS-562086) (Broxer et al., 2011) (Pexa) is a weak base with pK_a of 3.9 and intrinsic solubility of ∼ 6 µg/mL. Pexa API is a neat crystalline free form with melting point of 118°C.

The effect of wet granulation process parameters on the product attributes for the three compounds was studied using a design of experiment (DOE) approach. The qualitative composition and experimental designs for the three formulations are shown in Table 3.4, Table 3.5, and Table 3.6. Batches were manufactured according to the experimental design and the resulting granulated materials were characterized as described below. Responses obtained from the characterization experiments were analyzed by regression analysis using SAS JMP (SAS Institute Inc., Cary, NC, USA).

Table 3.5 Experimental design[a] for the Razaxabn formulation (Badawy et al., 2012)[b]

Run#	Water added (g)	Water addition rate (g/min)	Impeller speed (rpm)	Wet massing time (s)
1	675	300	334	30
2	540	300	334	180
3	540	300	668	30
4	540	100	668	180
5	675	300	668	180
6	675	100	668	30
7	540	100	334	30
8	675	100	334	180

[a] 2^{4-1} Fractional factorial design.
[b] Intragranular components: *microcrystalline cellulose, Razaxaban (33%, w/w), hydroxypropyl cellulose, croscarmellose sodium;* extragranular: *magnesium stearate–1.5 kg batch size.*

Granulations were manufactured by mixing the intragranular components of the formulation in the bowl of a 10 L high shear granulator (Fielder PMA-10) using the impeller speeds specified in Table 3.4–3.6. The preblend then was granulated with the specified amount of water added using the target addition rate (Ra) or specified water addition time (Br and Pexa). After water addition is completed, the granulation was wet massed for the specified time per Table 3.4, Table 3.5 and Table 3.6. The wet granules were screened through 8-mesh screen and subsequently dried in a fluid bed drier (Glatt GPCG-1) (Br and Pexa) at inlet air temperature of 70°C or tray dried (Ra) at 50°C to a specified moisture content (NMT 3.0%, 2.0%, or 1.7% for Ra, Pexa and Br, respectively). The dried granulation was milled using a conical mill through 18-mesh (1 mm) screen. The milled granulation was blended with the extragranular components and lubricated with magnesium stearate in a V-blender (Ra) or bin blender (Pexa and Br) for 3, 30, or 10 min for Ra, Pexa and Br, respectively.

3.3.1 Characterization of granule and tablet attributes

Granulated materials manufactured by the wet granulation process were characterized according to the following methods.

The particle size distribution of the final granulation was determined by mesh analysis using an Allen Bradley Sonic Sifter (Allan Bradley, Milwaukee, WI, USA) equipped with a series of six screens and a pan. The geometric mean diameter (GMD) of the granulation was calculated from the granulation weight fraction retained on each screen. The percent fines (< 75 μm) was calculated by the summation of the percent material retained over the 325-mesh screen and dust collected in the pan of the sonic sifter.

Bulk density of the final blend was measured using a graduated cylinder. Granulation was poured slowly into the cylinder up to a specified volume mark. The weight of the granulation then was determined and used to calculate the bulk density. Tap density of the final blend was obtained by subjecting the powder in the graduated cylinder to 400 taps in tap density equipment.

Table 3.6 Experimental design[a] for the Pexacerfont formulation (Badawy et al., 2012)[b]

Water addition time (min)	Water conc. (%w/w)	Wet massing time (s)	Impeller tip speed (m/s)
2	46	10	3.3
2	46	10	6.7
2	46	50	3.3
2	46	50	6.7
2	54	10	3.3
2	54	10	6.7
2	54	50	3.3
2	54	50	6.7
6	46	10	3.3
6	46	10	6.7
6	46	50	3.3
6	46	50	6.7
6	54	10	3.3
6	54	10	6.7
6	54	50	3.3
6	54	50	6.7
2	50	30	5
6	50	30	5
4	46	30	5
4	54	30	5
4	50	10	5
4	50	50	5
4	50	30	3.3
4	50	30	6.7
4	50	30	5
4	50	30	5
4	50	30	5

[a] *Central composite design with three center point replicates.* [b] *Intragranular components: microcrystalline cellulose, Pexacerfont (25%, w/w), hydroxypropyl cellulose, croscarmellose sodium; extragranular: croscarmellose sodium, magnesium stearate–1.8 kg batch size.*

Pore distribution of the Br granulation and tablets was determined by mercury intrusion porosimetry (MIP, AutoPore III, Micromeritics, Norcross, GA, USA). A fraction of the tablet or a granulation sample of the particle size fraction passing through 60-mesh screen and retained on 100-mesh screen was used for the porosity determination. Incremental pore volume was determined at different pressures ranging from 1 to 61,000 psi. The pore diameter corresponding to a given pressure was calculated by the software of the instrument according to the Washburn equation using mercury surface tension value of 485 dyn/cm and mercury contact angle of 130 degrees. Pores in the 0.1–10 μm range were used for the assessment and comparison of the different batches (Badawy, Gray and Hussain, 2006).

Compaction properties of the three formulations were assessed by compressing the granulation at different compression forces and determining tablet hardness obtained at each compression force using a tablet hardness tester (Key International Inc., Cranbury, NJ, USA). The compaction profile was constructed using tablet hardness value in SCU (on the y-axis) as a function of compression force in kN (on the x-axis). A rotary tablet press (Korsch PH106 six station press, Korsch AG, Berlin, Germany) and round standard concave B-tooling were used to compress Ra and Pexa lubricated blends. The tooling diameter was 9/32″ and 3/8″ and tablet weight was 150 and 300 mg for Ra and Pexa, respectively. The maximum tablet hardness obtained from the compaction profile was determined and used for the statistical analysis. For Br, a Presster compaction simulator (Metropolitan Computing Corporation, East Hanover, NJ, USA) was used to generate the compaction profile data at 14 ms dwell time using oval shaped B-tooling and 800 mg tablet weight. A compactibility parameter was calculated as the slope of the initial linear part of the compaction profile and used for the statistical analysis.

Flow properties of the lubricated blend was tested using an Erweka GTB powder flow tester (Erweka GmbH, Heusenstamm, Germany). The time required for 100 g of the blend to flow through 10 mm aperture (Br) or 15 mm aperture (Pexa) was determined and converted to a flow rate in g/s. For Ra, the angle of repose was determined using the Erweka GTB tester and used for the statistical analysis.

Dissolution rate was conducted on tablets compressed to a target hardness of 28, 9, and 13 SCU for Br, Ra and Pexa respectively. Dissolution test was performed using USP Type II apparatus (paddle) at 75 (Br and Pexa) or 60 rpm (Ra). Dissolution media were selected to provide appropriate sink condition for the different compounds. Dissolution medium was pH 6.8 phosphate buffer with 1.0% w/v Triton-X for Br, pH 4.0 citrate buffer with 0.5% w/v sodium lauryl sulfate (SLS) for Pexa and 0.01 N HCl for Ra (dissolution medium temperature set at 37°C).

Regression analysis was used with four factors (water amount, impeller speed, wet massing time, water addition time or rate) as the independent variables. The various in-process critical materials attributes (granule GMD, percent fines, porosity, blend density, flow and compaction properties) and the final product critical quality attribute (dissolution) were the responses used in the regression analysis. Results of the statistical analyses are shown in Table 3.7, Table 3.8 and Table 3.9. Parameter estimates are listed with the corresponding probability (P) values. Only parameter estimates that are statistically significant ($P < 0.1$) are shown in the tables. A parameter estimate represents the coefficient of the regression model associated with a particular model term. Hence, a larger magnitude indicates a more pronounced effect of the model parameter on the response and the parameter estimate sign indicates the direction in which the response changes as the level of the parameter is increased (i.e., a positive sign for the parameter estimate indicates increase in the magnitude of the response as the level of the parameter is increased, while a negative sign indicates an opposite parameter effect).

In the following sections, discussion of the effect of each process parameter on the granule growth and densification as reflected on granule size and density/porosity is presented. This is followed by discussion on the granulation bulk powder properties and tablet dissolution and their link to process parameters and primary attributes (granule size and density). Granule size and density are referred to as primary granule properties because they represent the particle-level properties modulated by the granulation process. Bulk powder properties of the granulation (e.g., flow and compaction) and some final product critical quality attributes (e.g., dissolution) are functions of primary granule properties and therefore can be viewed as derived properties that are dictated by the aforementioned particle-level properties.

Table 3.7 Regression analysis results[a] for Brivanib (Badawy et al., 2012)

Parameter	Parameter estimate (probability, P value)						
	GMD (µm)	Fines (%)	Mean pore diameter (µm)	Tap density (g/cm³)	Flow rate (g/s)	Compactibility (SCU/kN)	% Dissolved in 10 min
Water amount	7.17 (0.047)	−0.772 (0.023)	−0.101 (0.048)	0.007 (0.044)	0.182 (0.037)		−2.67 (0.006)
Impeller speed		−0.436 (0.094)	−0.121 (0.030)	0.012 (0.013)	0.221 (0.022)		−3.83 (0.002)
Wet massing time			−0.090 (0.062)	0.011 (0.014)	0.166 (0.047)		−3.08 (0.004)
Impeller speed * water amount	9.44 (0.023)	−0.920 (0.014)				0.158 (0.048)	
Wet massing time * water amount						0.221 (0.020)	
Impeller speed * wet massing time				0.006 (0.075)	0.136 (0.074)	0.193 (0.029)	

[a] One step full factorial regression analysis.

Table 3.8 Regression analysis results[a] for Pexacerfont (Badawy et al., 2012)

Parameter	Parameter estimate (probability, P value)					
	GMD (μm)	Fines (%)	Tap density (g/cm^3)	Flow rate (g/s)	Maximum hardness (SCU)	% Dissolved in 30 min
Water amount	23.2 (<0.0001)	−4.8 (<0.0001)	0.014 (0.004)	4.64 (<0.0001)	−1.40 (0.0002)	−3.57 (<0.0001)
Impeller speed	−11.2 (<0.0001)	2.3 (0.005)	0.013 (0.007)		−1.14 (0.001)	1.61 (0.009)
Wet massing time	−3.9 (0.027)				−0.89 (0.007)	
Water addition time	−4.9 (0.008)				−0.58 (0.055)	1.94 (0.0003)
Impeller speed * water amount	−7.3 (0.0008)		−0.012 (0.014)	−0.58 (0.095)	0.82 (0.015)	1.13 (0.063)
Wet massing time * water amount						
Impeller speed * wet massing time					−0.56 (0.074)	
Impeller speed * water addition time						
Water amount * water addition time	−3.8 (0.038)	−1.8 (0.021)				
Water massing time * water addition time				2.75 (0.005)		
water amount * water amount						

[a] One step regression analysis of central composite model.

Table 3.9 Regression analysis results[a] for Razaxaban (Badawy et al., 2012)

Parameter	Parameter estimate (probability, P value)					
	GMD (μm)	Fines (%)	Bulk density (g/cm^3)	Angle of repose (θ)	Maximum hardness (SCU)	% Dissolved in 10 min
Water amount	21.5 (0.005)	−10.9 (0.006)	0.048 (0.062)	−2.04 (0.024)	−2.96 (0.035)	−7.80 (0.006)
Impeller speed	8.75 (0.086)		0.052 (0.047)		−2.76 (0.043)	−5.92 (0.017)
Wet massing time						
Impeller speed * water amount	12 (0.036)		0.040 (0.096)		−2.31 (0.070)	−6.08 (0.015)

[a] *Backward stepwise regression of fractional factorial model; then check for two way interactions for selected parameters.*

3.3.2 Process parameters

3.3.2.1 Water amount

Water amount is the most important factor (main effect) for granule growth as shown by the higher parameter estimate of water on GMD for all three compounds. As described for the lactose formulation, increasing water concentration increases liquid saturation of the granules (higher fraction of the void space within the granules is filled with liquid) which was reported to enhance granule coalescence (Ritala et al., 1988). The higher liquid saturation makes the granule more easily deformable and results in more liquid available at the granule surface, both of which increase probability of successful coalescence upon granule collision (Liu et al., 2000).

Water also showed statistically significant increase in blend density for all three compounds and a decrease in granule porosity for Br. Low viscosity granulating liquids such as water act as a lubricant, which reduces interparticulate friction within the granule and facilitates particle movement in response to densifying forces in the high shear granulator. It is noteworthy, however, that an opposite effect was reported when a high viscosity liquid was used (Iveson, Litster and Ennis, 1996). For high viscosity liquids, the disruption of interparticle friction by the liquid was outweighed by the increased strength of the liquid bridges of the viscous liquid, which resists densification.

3.3.2.2 Impeller speed

The literature offers conflicting reports about the effect of impeller speed on granule size. Some studies reported the increase in granule size as the impeller speed was increased (Schaefer et al., 1986), while other studies suggested an opposite effect (Iskandarani, Shiromani and Clair, 2001). Higher impeller speed increases velocity of particle collision in the granulator. The effect of collision velocity on granule growth depends on the mechanical properties of the wet granules (e.g., plastically deformable versus brittle). For plastic granules resulting from high granule liquid saturation, high collision velocity results in more pronounced granule deformation (higher area of contact between colliding granules) and therefore higher probability of successful coalescence and rapid granule growth (Knight, 1993). This appears to be the case for Br and Ra, but not for Pexa, in which increasing impeller speed increased GMD. This effect of impeller speed for Br and Ra, however, was dependent the on water level, as shown by the significant interaction term between water and impeller speed. For both Br and Ra, the increase in GMD at higher impeller speed is predominantly at the high water level where granules have sufficient liquid saturation and plasticity. At low water level and liquid saturation, granules are more brittle and high collision velocity is more likely to result in granule breakage rather than coalescence and growth. In other words, granule behavior when yield stress is exceeded determines the effect of impeller speed (Iveson et al., 2001). For brittle granules that break at their yield stress, a maximum stable granule size is achieved under given conditions according to the model proposed by (Tardos, Khan and Mort, 1997):

$$a_{cr} = \frac{\left(2\tau_y St_{def}/\rho_p\right)^{1/2}}{\gamma} \tag{3.2}$$

where a_{cr} is the is the maximum stable granule size, τ_y is the yield stress, ρ_p is the granule density, γ is the average shear rate in the granulator, and St_{def} is the Stokes deformation number that relates the externally applied kinetic energy to the energy required for granule breakage. Because γ is directly related to the impeller speed, this model predicts the maximum stable granule size to be inversely related to impeller speed.

In contrast to brittle granules, plastic granules, which deform rather than break at their yield stress, demonstrate increase in probability of coalescence with the increase in Stokes deformation number (St_{def}) as indicated by the model of (Liu et al., 2000). St_{def} is expressed as follows:

$$St_{def} = mu_0/2DY_d \tag{3.3}$$

where m and D are the mean granule mass and diameter, respectively, u_0 is the relative velocity of colliding granules and Y_d is the granule yield stress. Because u_0 increases with the increase in impeller speed, Eq. (3.3) predicts St_{def} and the probability of coalescence to be enhanced by higher impeller speed for the plastically deformable granules.

Pexa showed different behavior than Br and Ra with respect to impeller speed as the increase in impeller speed significantly reduced granule size. This could be attributed to the brittle behavior of Pexa granules as described above. A two-way interaction term, however, showed that the effect of impeller speed in reducing particle size is more predominant at the higher water level. This suggests that the mechanism of impeller speed effect might not be limited to the brittle behavior of the Pexa granules, because higher water level is expected to make granules less brittle and therefore less susceptible to breakage. This interaction term might be explained by the effect of impeller on liquid dispersion. As pharmaceutical high shear wet granulation usually operates in the mechanical dispersion regime for nucleation in which mechanical forces are responsible for liquid distribution and nucleation (Hapgood, Litster, & Smith, 2003), a low impeller speed that does not efficiently distribute the liquid can result is localized overwetting and formation of lumps. Earlier reports showed that lumps are formed during liquid addition, suggesting that those oversize granules are formed during nucleation and not as the result of coalescence. Therefore, an initial bimodal granule size distribution was observed and slowly changed to unimodal distribution by mixing in the granulator (Scott, Hounslow, & Instone, 2000). Higher impeller speed improves water distribution and breaks lumps resulting from localized overwetting, which is a likely contributing factor to the reduced GMD at the higher impeller speed for Pexa.

The effect of impeller speed on GMD, therefore, is dependent on two factors: granule mechanical properties, in which more plastic granules are likely to have more pronounced growth as impeller speed is increased, while brittle granules can undergo more attrition at the higher impeller speed and reduction in granule size; and liquid distribution in the granulator because pharmaceutical high shear wet granulation almost always operates in the mechanical dispersion regime. When the low impeller speed does not efficiently distribute the liquid, the result is localized overwetting and formation of lumps. Those lumps or oversize granules are distinct, at least from a mechanistic point of view, from large granules resulting from granule coalescence. Those oversize lumps represent larger particle size than would be achieved if the granulating liquid is uniformly dispersed and granulation process is allowed to proceed to steady state. If this is case, higher impeller speed increases the efficiency of water distribution and reduces oversized lumps and consequently GMD is reduced.

The higher impeller speed also increased blend density of all three formulations and reduced the porosity of Br granules. The higher impeller speed produces higher shear forces in the granulator which increases granule density and reduces porosity (Badawy, Gray and Hussain, 2006; Rahmanian, Ghadiri, Jia, & Stepanek, 2009; Rahmanian, Naji, & Ghadiri, 2011). This also was demonstrated for the lactose-based formulation study (Badawy et al., 2000).

3.3.2.3 Wet massing time

Similar to impeller speed, the increase in wet massing time was reported to increase granule size in some instances (Ohno et al., 2007), while decreasing granule size in other reports (Iskandarani, Shiromani and Clair, 2001). During wet massing, granule coalescence and growth takes place, but large granules also can undergo breakage until a steady state particle size distribution is achieved. In addition, granule densification usually takes place during wet massing (Narang et al., 2015; Pandey et al., 2013; Zhao, Tao, Li, Bindra, & Narang, 2013), which can affect granule liquid saturation and mechanical properties and therefore particle size distribution. According to the regime growth map of (Iveson and Litster, 1998), this increase in liquid saturation shifts the system's behavior from the nucleation to the growth regions, resulting in faster granule growth as the increased consolidation squeezes the liquid to the granule surface, enhancing coalescence.

The range of wet massing times used in the three studies is relatively narrow (within 1 min for Br and Pexa, and up to 3 min for Ra). Despite the short wet massing time used in these studies, its effect on blend density/porosity for Br and Ra was significant, in which increasing wet massing time increased density and decreased porosity. Typically, an exponential decay of porosity is observed during wet massing (Am Ende et al., 2010; Hapgood and Litster, 2010) as shown in Eq. (3.4).

$$\frac{\varepsilon - \varepsilon_{min}}{\varepsilon_0 - \varepsilon_{min}} = exp(-k_c t) \tag{3.4}$$

where ε is the average granule porosity, ε_{min} is the minimum achievable porosity at given process conditions, ε_0 is the initial porosity, k_c is the granulation consolidation rate constant, and t is the wet massing time. Consolidation rate during wet massing usually is dependent on other process conditions, including impeller speed and the amount of water used during granulation.

This effect of wet massing on blend density was not observed for Pexa. This could be because of the decrease of granule size by the longer wet massing time, which could offset the effect of decreased granule porosity on the observed blend density. In other words, smaller granule size tends to reduce blend density (because of entrapped air and reduced packing of the granules in the powder bed) and masks the increase in blend density that would be observed if only granule porosity is decreased. This could be the reason that the effect of wet massing on blend density of Pexa was not statistically significant. Compaction property change for Pexa also suggests that granule density is increased by wet massing.

Unlike its effect on density/porosity, wet massing showed no effect on the GMD for Br and Ra. This suggests that the rate of granule size change is lower than that for granule porosity, resulting in no observable change in granule size during the short wet massing times used in the study. The rapid change in porosity during wet massing with limited or no change in granule size is similar to the lactose-based formulation study (Badawy et al., 2000). An alternative hypothesis is that particle size steady state is achieved almost immediately as water addition is completed and no further particle size change takes place during wet massing. This does not seem to be the case at least for Br, in which an experiment conducted with long wet massing time resulted in further, albeit limited, particle size growth in 25 min.

Wet massing showed significant decrease in GMD for Pexa, which suggests that granule breakage outweighs growth during wet massing. As discussed for impeller speed, water distribution and oversize granules resulting from localized overwetting seem to be important factors for Pexa. Therefore, longer wet massing results in breakage of the oversize granules with the decrease in GMD as granule size distribution continues to approach steady state.

Overall, granule densification is the key effect to be taken into consideration during wet massing. There are negative effects on product quality attributes resulting from granule densification in the high shear mixer. The narrow range of wet massing times used in the studies reported here was selected to avoid excessive densification. The fact that statistically significant reduction in density/porosity still was observed within this narrow range illustrates the need to carefully design high shear wet granulation processes that avoid overdensification. It also leads to the design of wet granulation processes that use sufficient granulating liquid to achieve desired granule growth during liquid addition instead of relying on wet massing to increase granule size. High shear wet granulation processes should therefore be designed with sufficient granulating liquid added to achieve adequate granule liquid saturations to be in the steady growth regime (Iveson and Litster, 1998) at the end of liquid addition phase, which results in coalescence and desired granule growth. Design of induction systems requiring long wet massing times and consolidation to increase liquid saturation and induce growth should be avoided. Increasing the granulating liquid amount to achieve desired growth is a preferred strategy than using lower liquid amount and long wet massing time (Badawy and Pandey, 2017).

3.3.2.4 *Water addition rate (time)*

Water addition time is inversely related to addition rate and therefore both are used for the purpose of this discussion to describe the duration of the water addition phase of wet granulation. Water addition rate showed minimal effect on granule size for Br and Ra. In these studies, water was added to the granulator in the form of a stream using a peristaltic pump and a tube without nozzle. It is likely that nucleation is controlled by mechanical dispersion in these studies. Mixing conditions in the granulator appear to be effective in the dispersion of water for the Br and Ra formulations, and water addition rate did not have a significant effect on granule size. For Pexa, mechanical dispersion of the water is not as effective, as indicated by the effects of impeller speed and wet massing time on granule size. In the case of Pexa, slow water addition aids in the water dispersion resulting in reduced overwetting and lump formation. Consequently, water addition time has a statistically significant effect on GMD for Pexa and the increase in water addition time (slower water addition) results in smaller granule size. The statistically significant term for the two-way interaction between water amount and water addition time, which showed more pronounced effect of water addition time on GMD at the higher water level, also is consistent with this assessment. Batches manufactured at the higher water level are more likely to have a higher content of the overwetted lumps, and the slower water addition is more beneficial in improving water dispersion for those batches.

Water addition rate did not have a significant effect on density/porosity of any of the three formulations despite the reduced porosity/density for Br and Ra resulting from longer wet massing time. In other words, while the increase in mixing time after complete water addition (wet massing time) resulted in a significant effect on granule densification, the increase in mixing time during water addition (water addition time) had minimal effect on granule porosity. This is similar to the lactose-based granulation (Badawy et al., 2000) and indicates that the effect of mixing on granule porosity is dependent on the water level (granule liquid saturation). Mixing after complete water addition is more effective in decreasing granule porosity (because of the lubricant effect of water). On the other hand, mixing during water addition, when granule liquid saturation is still low, is not effective in reducing granule porosity and consequently prolonging water addition time has little or no effect on granule porosity. It is obvious that wet massing time needs to be controlled as a separate parameter distinct of water addition time and not lumped together as one parameter representing total processing time.

Based on these studies, liquid addition rate seems to be the least consequential factor among the four parameters evaluated as it showed minimal impact on granule density and growth. Liquid addition rate affects liquid dispersion only under conditions where mechanical dispersion of the granulating liquid is not very effective. In this case, reducing addition rate aids in the dispersion of the liquid, decreases overwetted lumps, and produces narrower granule size distribution. This is more likely to be the case for high viscosity granulating liquid when the polymeric binder is dissolved in the granulating liquid or if the formulation contains a high concentration of polymeric material that hydrates rapidly when it comes in contact with water (e.g., cellulosic polymer-based controlled release matrix formulations).

3.3.3 Final product critical quality attributes and granulation bulk powder properties

Primary granule attributes (granule size and density) were discussed in conjunction with the effect of granulation parameters. In this section, tablet dissolution and granulation bulk powder properties are examined. Those attributes are functions of primary granule properties and are directly related to final product performance or to the granulation performance in downstream unit operations.

3.3.3.1 Tablet dissolution

Dissolution rate represented by the percentage dissolved in 10 min for Br tablets was reduced significantly by the increase in water amount, impeller speed, or wet massing time ($P = 0.006, 0.002, 0.004$, respectively). Dissolution rate of Br tables thus appears to be affected by the granulation parameters in the same direction as granule porosity (mean pore diameter). A plot of percentage dissolved in 10 min versus granule pore diameter shows a strong correlation between the two attributes ($r^2 = 0.931$) (Figure 3.5A). In contrast, poor correlation exists between percentage dissolved in 10 min and granule GMD ($r^2 = -0.308$) (Figure 3.5B). This is consistent with the hypothesis that granule porosity controls tablets dissolution. Tablet porosity also was tested by MIP, which indicated that granule porosity is reflected in tablet porosity (i.e., granules with smaller pores resulted in lower porosity tablets) (Figure 3.6), suggesting that granule pore structure still is maintained after tablet compression. Tablets with different dissolution rates resulting from differences in granulation parameters showed similar disintegration time, including a batch made with long wet massing time (25 min) that showed very low porosity and slow dissolution. This leads to the conclusion that granule disintegration, rather than tablet disintegration, is the rate controlling step for drug release from the Br tablets. As granule pore diameter is decreased, dissolution medium ingress into the granule and subsequent granule disintegration become slower, resulting in the slower tablet dissolution.

Unlike Br, dissolution rate for Pexa was inversely dependent on granule GMD (Figure 3.7A). Tablet dissolution showed no correlation to blend density (used a surrogate for granule density in this case) (Figure 3.7B). The effect of GMD on dissolution rate also was consistent with the observation that tablet dissolution rate was affected by granulation parameters in a direction opposite to their effect on granule GMD. Therefore, increase in water amount used for granulation increased GMD ($P < 0.0001$) and reduced the percentage dissolved in 30 min ($P < 0.0001$), while the increase in impeller speed reduced GMD ($P = 0.005$) and increased dissolution rate ($P = 0.009$). Because granule particle size rather than porosity (as inferred from blend density data) seems to be the factor controlling dissolution, it is concluded that granule erosion (rather than granule disintegration) might be the more predominant

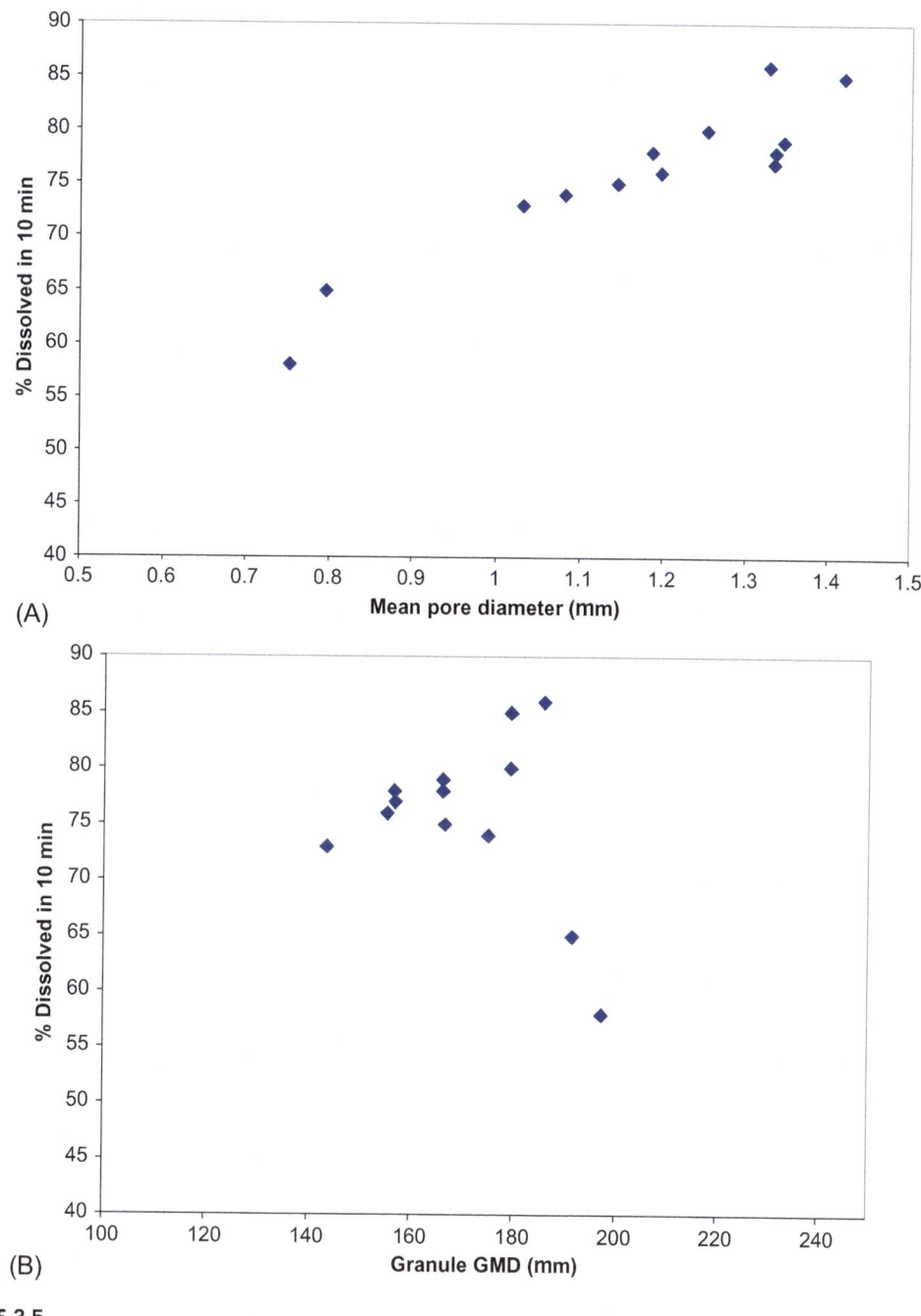

FIGURE 3.5

(A) Effect of granule pore diameter on dissolution rate of Brivanib tablets. (B) Effect of granule size on dissolution rate of Brivanib tablets (Badawy et al., 2012).

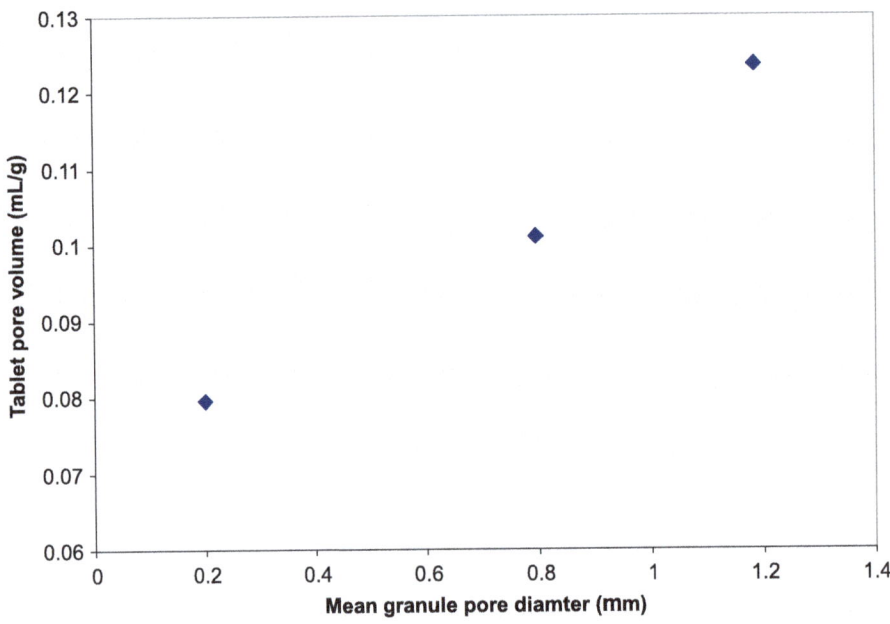

FIGURE 3.6

Effect of Brivanib granule pore size on tablet porosity (Badawy et al., 2012). (Data for DOE center point batch, DOE batch with smallest pore diameter and extreme batch with 25 min wet massing time outside of the DOE.)

mechanism for dissolution in the case of Pexa. A granule disintegration-controlled mechanism is likely to be affected by granule porosity as it was the case for Br.

Percentage dissolved in 30 min for Ra tablets was reduced by the increase in water amount, impeller speed, or wet massing time ($P = 0.006$, 0.017 and 0.015, respectively).

The percentage dissolved in 10 min for Ra tablets showed good correlation with both GMD ($r^2 = -0.946$) and blend density ($r^2 = -0.914$), where the increase in GMD or blend density was associated with decreased percent dissolved in 10 min. The decrease in dissolution rate by the increase in both granule particle size and density suggests a dissolution mechanism controlled by both granule erosion and disintegration.

Data from the three formulations illustrate that tablet dissolution can be affected by granule size or density depending on the rate controlling mechanism for dissolution. The understanding of dissolution mechanism provides the basis for modulating granulation parameters in order to optimize the primary granule quality attribute controlling dissolution and hence maximize dissolution rate (Badawy et al., 2012).

3.3.3.2 Granule compactibity

None of the main effects for the granulation parameters was statistically significant for Br compactibility. While the interaction terms were significant, the magnitude of their effect (parameter estimate) was small. This suggests little effect of granulation parameters on compaction properties of Br within the

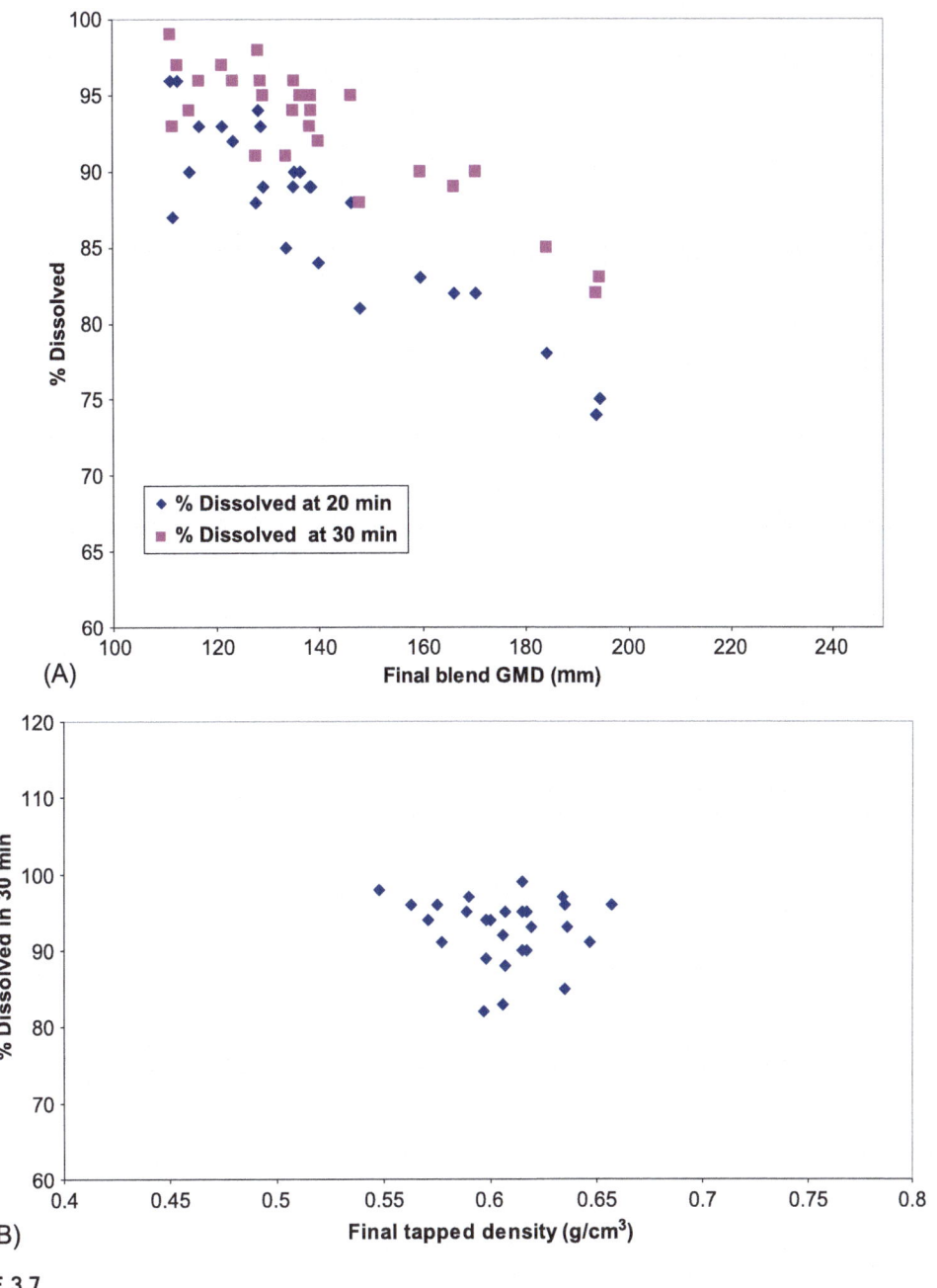

FIGURE 3.7

(A) Effect of final blend granule size on dissolution rate of Pexcerfont tablets. (B) Effect of final blend density on dissolution rate of Pexcerfont tablets (Badawy et al., 2012).

range tested. Thus, observed variation in granule size and porosity induced by change in granulation parameters were not reflected on the compaction behavior of Br.

In contrast, maximum tablet hardness achieved for Ra was reduced significantly by water amount, impeller speed, and wet massing time. As all three parameters reduced blend density, the reduced maximum hardness resulting from the increase in the three granulation parameters can be attributed to their effect on blend (granule) density. Thus, an excellent correlation exists between blend destiny and maximum tablet hardness ($r^2 = -0.997$). The reduced compactibility of denser, less porous granules is consistent with the observation for the lactose-based formulation and other studies (Badawy et al., 2000; Johansson and Alderborn, 2001) and is attributed to reduced fragmentation and/or deformation propensity of those granules during compression.

Maximum tablet hardness for Pexa was significantly reduced by water amount and impeller speed ($P = 0.0002$ and 0.001, respectively) and to a lesser extent by wet massing time and water addition time ($P = 0.007$ and 0.055, respectively). Because blend density also is increased significantly by water amount and impeller speed, it is possible that negative effect of those two parameters on maximum hardness is the result of their effect on density (as it is the case for Ra). Because higher impeller speed reduces granule size while reducing tablet hardness, it is unlikely that reduced compactibility by wet granulation is the result of the increased particle size during granulation but rather a consequence of granule densification.

Based on the analysis of the compaction data for the three formulations, it seems that compaction is more likely to be affected by granule density than granule size, at least within range of granule size achieved in this study.

3.3.3.3 Blend flow

Flow is enhanced by the increased granule size and/or density during granulation. The contribution of increased size and density varied among the three formulations. The enhanced flow rate for Br is attributed mainly to increased granule density with little contribution from particle size enlargement. A good correlation was observed between flow rate and mean pore diameter ($r^2 = -0.841$) while no correlation existed between GMD and flow rate ($r^2 = -0.040$). In addition, flow rate was enhanced significantly by the increase in water amount, impeller speed and wet massing time, all of which reduced granule pore size. Water amount, however, was the only significant main factor for GMD, a further indication that enhanced blend flow properties for Br is not directly related to particle size enlargement.

The increase in both particle size and density appeared to contribute to reduced angle of repose and enhanced flow properties of Ra. Angle of repose showed good correlation with both bulk density ($r^2 = -0.909$) and GMD ($r^2 = -0.877$).

Water amount was the only statistically significant main effect for the Pexa flow rate ($P < 0.0001$). Increase in water amount resulted in increases in both granule GMD and blend density, both of which could enhance flow properties. Higher impeller speed, however, increased blend density but decreased GMD, two effects that are expected to affect blend flow in opposite directions. Therefore, impeller speed did not have a net measurable effect on blend flow rate.

3.4 Summary and conclusions

Wet granulation studies with a lactose-based placebo and model API formulations were used to identify the role of the four main process parameters on primary granule properties (size and density). Trends were interpreted in light of physical models of wet granulation. Mechanisms for effect of process parameters on granulation bulk powder properties and a final product critical quality attribute (tablet dissolution) were reviewed and linked to the primary granule properties. Comparisons of the trends between these formulations were used to establish themes common to pharmaceutical wet granulation.

References

Am Ende, M. T., Bharadwaj, R., García-Muñoz, S., Ketterhagen, W., Prpich, A., & Doshi, P. (2010). Process modeling techniques and applications for solid oral drug products. *Chemical engineering in the pharmaceutical industry* (pp. 633–662). Hoboken, NJ: John Wiley and Sons, Inc.

Badawy, S., & Pandey, P. (2017). Design, development and scale-up of the high shear wet granulation process. In Y. Qiu, Y. Chen, G. Zhang, L. Yu, & R. V. Mantri (Eds.), *Developing solid oral dosage forms* (2nd ed.). Academic Press.

Badawy, S. I., Narang, A. S., LaMarche, K., Subramanian, G., & Varia, S. A. (2012). Mechanistic basis for the effects of process parameters on quality attributes in high shear wet granulation. *International Journal of Pharmaceutics, 439*(1–2), 324–333. https://doi.org/10.1016/j.ijpharm.2012.09.011.

Badawy, S. I. F., Gray, D. B., & Hussain, M. A. (2006). A study on the effect of wet granulation on microcrystalline cellulose particle structure and performance. *Pharmaceutical Research, 23*(3), 634–640. https://doi.org/10.1007/s11095-005-9555-z.

Badawy, S. I. F., Gray, D. B., Zhao, F., Sun, D., Schuster, A. E., & Hussain, M. A. (2006). Formulation of solid dosage forms to overcome gastric pH interaction of the factor Xa inhibitor, BMS-561389. *Pharmaceutical Research, 23*(5), 989–996. https://doi.org/10.1007/s11095-006-9899-z.

Badawy, S. I. F., Menning, M. M., Gorko, M. A., & Gilbert, D. L. (2000). Effect of process parameters on compressibility of granulation manufactured in a high-shear mixer. *International Journal of Pharmaceutics, 198*(1), 51–61. doi:10.1016/S0378-5173(99)00445-7.

Badawy, S. I. F., Narang, A. S., Lamarche, K. R., Subramanian, G. A., Varia, S. A., Lin, J., et al. (2016). Integrated application of quality-by-design principles to drug product development: A case study of Brivanib alaninate film- coated tablets. *Journal of Pharmaceutical Sciences, 105*(1), 168–181. https://doi.org/10.1016/j.xphs.2015.11.023.

Broxer, S., Fitzgerald, M. A., Sfouggatakis, C., Defreese, J. L., Barlow, E., Powers, G. L., et al. (2011). The development of a robust process for a CRF1 receptor antagonist. *Organic Process Research and Development, 15*(2), 343–352. https://doi.org/10.1021/op100270u.

Hapgood, K. P., & Litster, J. D. (2010). Wet granulation processes. *Chemical engineering in the pharmaceutical industry* (pp. 757–780). Hoboken, NJ: John Wiley and Sons, Inc. doi:10.1002/9780470882221.ch39.

Hapgood, K. P., Litster, J. D., & Smith, R. (2003). Nucleation regime map for liquid bound granules. *AICHE Journal, 49*(2), 350–361. https://doi.org/10.1002/aic.690490207.

Holm, P. (1987). Effect of impeller and chopper design on granulation in a high speed mixer. *Drug Development and Industrial Pharmacy, 13*(9–11), 1675–1701. https://doi.org/10.3109/03639048709068687.

Huynh, H., Ngo, V. C., Fargnoli, J., Ayers, M., Khee, C. S., Heng, N. K., et al. (2008). Brivanib alaninate, a dual inhibitor of vascular endothelial growth factor receptor and fibroblast growth factor receptor tyrosine kinases, induces growth inhibition in mouse models of human hepatocellular carcinoma. *Clinical Cancer Research, 14*(19), 6146–6153. https://doi.org/10.1158/1078-0432.CCR-08-0509.

Iskandarani, B., Shiromani, P. K., & Clair, J. H. (2001). Scale-up feasibility in high-shear mixers: Determination through statistical procedures. *Drug Development and Industrial Pharmacy, 27*(7), 651–657. doi:10.1081/DDC-100107321.

Iveson, S. M., & Litster, J. D. (1998). Growth regime map for liquid-bound granules. *AICHE Journal, 44*(7), 1510–1518. https://doi.org/10.1002/aic.690440705.

Iveson, S. M., Litster, J. D., & Ennis, B. J. (1996). Fundamental studies of granule consolidation part 1: Effects of binder content and binder viscosity. *Powder Technology, 88*(1), 15–20. https://doi.org/10.1016/0032-5910(96)03096–03093.

Iveson, S. M., Litster, J. D., Hapgood, K., & Ennis, B. J. (2001). Nucleation, growth and breakage phenomena in agitated wet granulation processes: A review. *Powder Technology, 117*(1–2), 3–39. 2001 https://doi.org/10.1016/S0032-5910(01)00313-8 .

Johansson, B., & Alderborn, G. (2001). The effect of shape and porosity on the compression behaviour and tablet forming ability of granular materials formed from microcrystalline cellulose. *European Journal of Pharmaceutics and Biopharmaceutics, 52*(3), 347–357. https://doi.org/10.1016/S0939-6411(01)00186-2.

Knight, P. C. (1993). An investigation of the kinetics of granulation using a high shear mixer. *Powder Technology, 77*(2), 159–169. https://doi.org/10.1016/0032-5910(93)80053-D.

Kristensen, H. G. (1996). Particle agglomeration in high shear mixers. *Powder Technology, 88*(3), 197–202. 1996 https://doi.org/10.1016/S0032-5910(96)03123-3 .

Kristensen, H. G., & Schaefer, T. (1987). Granulation: A review on pharmaceutical wet-granulation. *Drug Development and Industrial Pharmacy, 13*(4–5), 803–872. doi:10.3109/03639048709105217.

Liu, L. X., Litster, J. D., Iveson, S. M., & Ennis, B. J. (2000). Coalescence of deformable granules in wet granulation processes. *AIChE Journal, 46*(3), 529–539. doi:10.1002/aic.690460312.

Narang, V. A. S, Stepaniuk, V., Badawy, S., Stevens, T., Macias, K., Wolf, A., Pandey, P., Bindra, D., & Varia, S. (2015). Real-time assessment of granule densification in high shear wet granulation and application to scale-up of a placebo and a brivanib alaninate formulation. *Journal of Pharmaceutical Sciences, 104*(3), 1019–1034. Www.interscience.wiley.com/jpages/0022-3549. doi:10.1002/jps.24233.

Ohno, I., Hasegawa, S., Yada, S., Kusai, A., Moribe, K., & Yamamoto, K. (2007). Importance of evaluating the consolidation of granules manufactured by high shear mixer. *International Journal of Pharmaceutics, 338*(1–2), 79–86. https://doi.org/10.1016/j.ijpharm.2007.01.030.

Oulahna, D., Cordier, F., Galet, L., & Dodds, J. A. (2003). Wet granulation: The effect of shear on granule properties. *Powder Technology, 130*(1–3), 238–246. 2003 https://doi.org/10.1016/S0032-5910(02)00272-3 .

Pandey, P., Tao, J., Chaudhury, A., Ramachandran, R., Gao, J. Z., & Bindra, D. S. (2013). A combined experimental and modeling approach to study the effects of high-shear wet granulation process parameters on granule characteristics. *Pharmaceutical Development and Technology, 18*(1), 210–224. https://doi.org/10.3109/10837450.2012.700933.

Rahmanian, N., Ghadiri, M., Jia, X., & Stepanek, F. (2009). Characterisation of granule structure and strength made in a high shear granulator. *Powder Technology, 192*(2), 184–194. 2009 https://doi.org/10.1016/j.powtec.2008.12.016 .

Rahmanian, N., Naji, A., & Ghadiri, M. (2011). Effects of process parameters on granules properties produced in a high shear granulator. *Chemical Engineering Research and Design, 89*(5), 512–518. https://doi.org/10.1016/j.cherd.2010.10.021.

Ring, D. T., Oliveira, J. C. O., & Crean, A. (2011). Evaluation of the influence of granulation processing parameters on the granule properties and dissolution characteristics of a modified release drug. *Advanced Powder Technology, 22*(2), 245–252. https://doi.org/10.1016/j.apt.2011.01.006.

Ritala, M., Holm, P., Schaefer, T., & Kristensen, H. G. (1988). Influence of liquid bonding strength on power consumption during granulation in a high shear mixer. *Drug Development and Industrial Pharmacy, 14*(8), 1041–1060. https://doi.org/10.3109/03639048809151919.

Ritala, M., Jungersen, O., Holm, P., Schæfer, T., & Kristensen, H. G. (1986). A comparison between binders in the wet phase of granulation in a high shear mixer. *Drug Development and Industrial Pharmacy, 12*(1113), 1685–1700. https://doi.org/10.3109/03639048609042603.

Schaefer, T., Bak, H., Jaegerskou, A., Kristensen, A., Svensson, J., Holm, P., et al. (1986). Granulation in different types of high speed mixers. I: Effects of process variables and up-scaling. *Pharmazeutische Industrie, 48*(9), 1083–1089.

Scott, A. C., Hounslow, M. J., & Instone, T. (2000). Direct evidence of heterogeneity during high-shear granulation. *Powder Technology, 113*(1–2), 205–213. https://doi.org/10.1016/S0032-5910(00)00354-5.

Shiraishi, T., Kondo, S., Yuasa, H., & Kanaya, Y. (1994). Studies on the granulation process of granules for tableting with a high speed mixer. Physical properties of granules for tableting. *Chemical and Pharmaceutical Bulletin, 42*(4), 932–936. https://doi.org/10.1248/cpb.42.932.

Tardos, G. I., Khan, M. I., & Mort, P. R. (1997). Critical parameters and limiting conditions in binder granulation of fine powders. *Powder Technology, 94*(3), 245–258. https://doi.org/10.1016/S0032-5910(97)03321-4.

Wikberg, M., & Alderborn, G. (1993). Compression characteristics of granulated materials. VII. The effect of Intragranular binder distribution on the compactibility of some lactose granulations. *Pharmaceutical Research: An Official Journal of the American Association of Pharmaceutical Scientists, 10*(1), 88–94. https://doi.org/10.1023/A:1018929214629.

Yu, L. X. (2008). Pharmaceutical quality by design: Product and process development, understanding, and control. *Pharmaceutical Research, 25*(4), 781–791. https://doi.org/10.1007/s11095-007-9511-1.

Yu, L. X., Amidon, G., Khan, M. A., Hoag, S. W., Polli, J., Raju, G. K., & Woodcock, J. (2014). Understanding pharmaceutical quality by design. *AAPS Journal, 16*(4), 771–783. http://www.springerlink.com/content/1550-7416/. doi:10.1208/s12248-014-9598-3.

Zhao, J., Tao, L., Li, J., Bindra, D., & Narang, A. S. (2013). *Granule consolidation kinetics and the indication on granulation mechanism during high shear wet granulation: Effect of binder type, molecular weight, and method of additionPaper presented at the AAPS annual meeting, San Antonio, TX.*

CHAPTER 4

Structures and properties of granules prepared by high-shear wet granulation

Tianxiang Gao[a], Jiangnan Dun[b] and Changquan Calvin Sun[a]

[a]*Department of Industrial and Molecular Pharmaceutics, Purdue University, West Lafayette, IN, United States,* [b]*Department of Pharmacy, National University of Singapore, Singapore*

4.1 Introduction

The tablet is one of the most widely used pharmaceutical oral solid dosage forms for drug delivery. Advantages of the tablet include high physical and chemical stability, easy administration, low manufacturing cost, and outstanding patient compliance.

Critical quality attribute (CQA) refers to a physical, chemical, biological, or microbiological property or characteristic that should be within an appropriate limit, range, or distribution to ensure the desired product quality (FDA & Tripartite, 2009). Some examples are tablet tensile strength, friability, disintegration time, content uniformity, and dissolution. To manufacture tablets that meet all required CQAs, excipients are selected according to the unique properties of each active pharmaceutical ingredient (API). In addition to formulation development, an appropriate process must be developed and optimized for manufacturing a drug product that meets required CQAs. The three most common manufacturing processes in pharmaceutical tablet manufacturing are direct compression (DC), dry granulation (DG), and wet granulation (WG).

The DC process is the simplest for making pharmaceutical tablets. It involves blending of excipients and APIs, followed by compression. The DC process is attractive mainly because of the lower manufacturing cost associated with the fewer unit operations (Yuan et al., 2013). However, successful DC manufacturing requires good tabletability, satisfying flowability, and acceptable content uniformity of the blends. For APIs with poor tabletability or flowability, DC may not be possible when the drug loading is high. Furthermore, segregation is a main challenge to DC when API loading is very low. The DG process involves milling precompacted slugs or ribbons, prepared by either slugging or roller compaction, to form granules (Kleinebudde, 2004). Roller compaction is currently the leading approach because of its efficiency and flexibility. Since no liquid is involved, water or heat-sensitive drugs can be processed by DG (Teng et al., 2009). Compared to DC and DG can handle a wider range of drug loadings because the formation of larger granules enhances powder flow and improves content uniformity by minimizing segregation between API and excipients. The WG process involves forming agglomerates consisting of small primary particles by agitation in the presence of a liquid, which is removed in subsequent drying. Thus powders granulated by WG exhibit improved flow, reduced dust, and lower tendency to segregation during subsequent unit operations, which improve the manufacturability and content uniformity of the final product (Sherrington & Oliver, 1981). When properly formulated and processed, WG can be used for manufacturing tablets over a wider range of API loadings than both DC and DG, while still maintaining good flowability, tabletability, and content uniformity. Modern WG is usually executed under high-shear. During the HSWG, a liquid is sprayed onto a powder bed as it is vigorously agitated under high-shear in an appropriate equipment, such as a high-shear mixer or twin-screw mixer, to produce agglomerates (Osei-Yeboah et al., 2014). In modern pharmaceutical tablet manufacturing, HSWG remains an excellent choice for tablet manufacturing when an API is not water or heat-sensitive.

Content uniformity is an important CQA, which is often a problem for low-dose tablet formulations. USP has set standards for evaluating drug content uniformity in tablets (Huang et al., 2010). For powders consisting of components with very different particle size distributions, simple blending may not be sufficient for achieving uniform distribution of components due to segregation. For a given formulation, API particle size greatly influences content uniformity, and it is possible to estimate the probability of passing the USP content uniformity test from the particle size distribution of API (Rohrs et al., 2006). Generally, smaller API particles favor better content uniformity. However, size reduction may cause other manufacturing problems, such as inadequate flowability. Segregation is prevented in granules prepared by HSWG because API particles are secured in granules. HSWG is, therefore, very effective for achieving desired content uniformity, even for very low-dose APIs.

Powder flow is another property of great importance for the successful commercial manufacturing of tablets. Poor powder flow usually results in manufacturing problems, such as low efficiency (long manufacturing cycle time), loss of materials, and high tablet weight variability. Powder flowability is influenced by several factors, including but not limited to particle size and size distribution, particle morphology, surface roughness, surface coating, hydration state, and moisture content (Hou & Sun, 2008; Shi et al., 2011c; Sun, 2009a, 2016; Yuan et al., 2013; Zhou et al., 2012). Larger particles usually correspond to better powder flow. However, no cut-off particle size can be suggested for acceptable powder flowability because of the significant effects on powder flow by other particle properties, processing equipment design, for example, hopper angle, and process parameters, for example, emptying rate. Also, smaller particles would correspond to higher cohesion and, therefore, significantly decreased flowability when processed on the same equipment and identical process parameters (Aulton, 2002). A practical system to broadly classify powders is based on particle size and particle density (Geldart, 1973), based on which pharmaceutical powders with mean particle size below 25 μm are usually highly cohesive. It was shown that flow problems tend to occur as particle size falls below 100 μm (Aulton, 2002; Liu et al., 2008). HSWG is effective in improving powder flowability through not only size enlargement but also several other mechanisms, such as surface smoothing, shape rounding, and densification (Shi et al., 2011c; Zeng et al., 2001).

4.2 Material sciences tetrahedron

The development of high-quality drug products, including those by HSWG, requires a fundamental understanding of underlying materials science and relevant engineering principles. It was recognized that drug formulation development should be a science rather than an art, and quality should be designed into a product not by trial-and-error. In this context, the concept of material sciences tetrahedron (MST) plays a prominent role in drug product development (Sun, 2009b).

The MST describes the relationships among structure, properties, process, and performance (Fig. 4.1). Performance of a drug product, the goal of drug development, depends on properties of

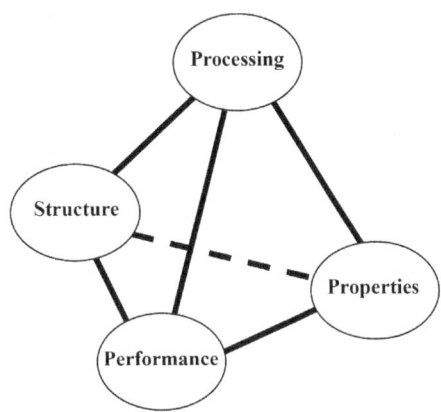

FIGURE 4.1

A diagram illustrating the interconnections between four components of the Material sciences tetrahedron.
From Sun (2009b).

the product, which are dictated by the structure of the drug product. Once the relationships among structure, property, and performance are elucidated, appropriate processes can be designed to realize the desired performance by precisely engineering material structure through various approaches, such as crystal and particle engineering. MST can be applied to solve problems at multiple length scales. Well-known examples include, (1) molecular level—avoiding moisture exposure by moisture-resistant packaging can stabilize drugs that degrade through hydrolysis reaction (Sun, 2009b); (2) particle level—adding polymers to inhibit or slow down nucleation and/or growth of crystals to maintain physical stability of amorphous drugs (Taylor, 2015); (3) crystal level—improving solubility and plasticity by modifying crystal structures of drugs (Chattoraj et al., 2010; Chattoraj et al., 2014; Chow et al., 2012; Sun, 2013; Ullah et al., 2016; Zhou et al., 2016); (4) bulk powder level—nanocoating to improve powder flowability and choice of appropriate excipients for optimum tableting performance (Chattoraj et al., 2011; Zhou et al., 2013).

Research on HSWG can be broadly divided into three topics: (1) effects of process parameters on properties of granulated powders; (2) effects of formulation components on properties of granulated powders; (3) mechanistic investigation into factors that influence granule formation. These have laid a solid foundation for systematic investigations of HSWG under the guidance of MST. Future research in HSWG will benefit tremendously from considering the evolution of granule structures by blends of different compositions under different processing conditions. A clear understanding of structure–property relationship is essential for developing high-quality tablet products by HSWG.

4.3 Topics of practical importance in high-shear wet granulation

4.3.1 High-shear wet granulation in batch mode

A traditional high-shear granulator (Fig. 4.2) mainly consists of a bowl, an impeller (usually mounted in the base of the bowl), and a chopper (mounted at the side of the bowl). The speeds of the impeller and chopper are adjustable in modern high-shear granulators. The impeller shears the powder bed to

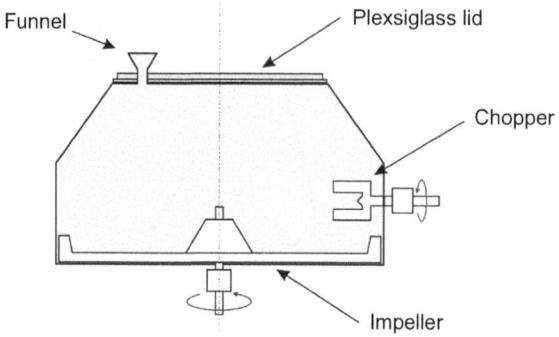

FIGURE 4.2

A diagram showing a standard high shear mixer.

From Žižek et al. (2013).

facilitate agglomeration, while the chopper is intended to break larger agglomerates and to facilitate more uniform mixing. Granulation liquid, such as water and binder solution, could be introduced into the granulator from the top of the instrument by pouring or spraying. Spraying is preferred due to the more uniform distribution of liquid, which allows more robust control of the process and quality of the granules.

4.3.2 High-shear wet granulation in continuous mode

In an effort to reduce time-to-market and to increase the cost-effectiveness of production in pharmaceutical industry, continuous manufacturing has been a topic of great recent interest. A main technique that enables continuous HSWG is the twin-screw wet granulation (TSWG) (Cartwright et al., 2013).

Two main components in a typical twin-screw granulator are feeders and screws (Fig. 4.3). A variety of feeders, including screw feeders, gravity feeders, and vibratory feeders, can be used to feed powder into the barrel inlet. The choice of powder feeder mainly depends on flowability of the powder to be processed. For very cohesive powders with poor flowability, loss in weight feeders should be used to ensure consistent delivery of materials into the TSWG (Cartwright et al., 2013). The screws are typically constructed from matching pairs of individual screw elements on the screw root, including conveying elements and kneading blocks. Conveying elements are designed to generate low mechanical energy and low shear stress with the intention to transport material between mixing zones (Seem et al., 2015). Compared with conveying elements, kneading blocks, which are located after the liquid feeder, are designed to generate high mechanical energy and high-shear force with the goal to compact and distribute the wet mass. Densification and consolidation occur inside the intermeshing region created by a pair of kneading discs. The chopping and shearing motion of kneading block breaks apart large agglomerates and disperses liquid (Seem et al., 2015). Increasing the number of kneading elements leads to decreased proportion of fines, increased proportion of oversized agglomerates, and increased density of granules (Djuric et al., 2009).

Thus the design of TSWG can influence the properties of granules. It was demonstrated, using two different extruders from APV Baker (MP19 TC25, Newcastle-under-Lyme, United Kingdom) and

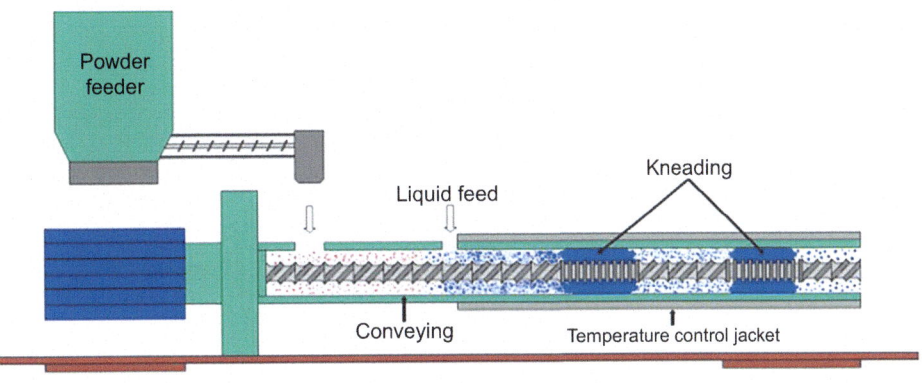

FIGURE 4.3

A schematic of components of a typical twin-screw granulator.

From Seem et al. (2015).

Leistritz Extrusionstechnik GmbH (Micro 27GL/28D, Nuremberg, Germany), that granule friability was inversely related to the total powder feeding rate. In addition, the number of fines and oversized agglomerates were significantly influenced by extruder type. For example, the APV extruder produced almost no oversized agglomerates. In contrast, the Leistritz extruder produced a larger fraction of oversized agglomerates due to its relatively higher densification inside the barrel. Also, increasing the degree of filling in the extruder barrel or reducing the free chamber volume tends to increase the number of oversized agglomerates and decrease the number of fines due to the higher degree of densification (Djuric et al., 2009).

4.3.3 End point determination of high-shear wet granulation process

The end point in HSWG is usually defined as a target mean particle size or particle size distribution. By the "Principle of Equifinality," the granule properties at the real end point and the subsequent tableting performance are essentially unchanged regardless of the granulation process parameters (Emori et al., 1997). For example, granules with the same properties can be obtained by different combinations of granulation parameters, that is, combinations of lower granulating water level and prolonged massing versus higher granulating water level and shorter massing. However, the successful determination of end point of HSWG is a challenge. Limited by the relatively poor reproducibility of the HSWG process, different operators may make quite different determinations of the end point, which usually leads to batch-to-batch inconsistency. Several approaches have been explored to investigate the end point of HSWG. These include monitoring (1) power consumption of mixer (Holm et al., 1985; Levin, 2006), (2) impeller torque and torque rheometer (Corvari et al., 1992; Dan et al., 2023; Levin, 2006; Rowe & Sadeghnejad, 1987), (3) acoustic emission (Whitaker et al., 2000), (4) near-infrared moisture sensor (Miwa et al., 2000; Rantanen et al., 2005), and (5) focused beam reflectance (Huang et al., 2010).

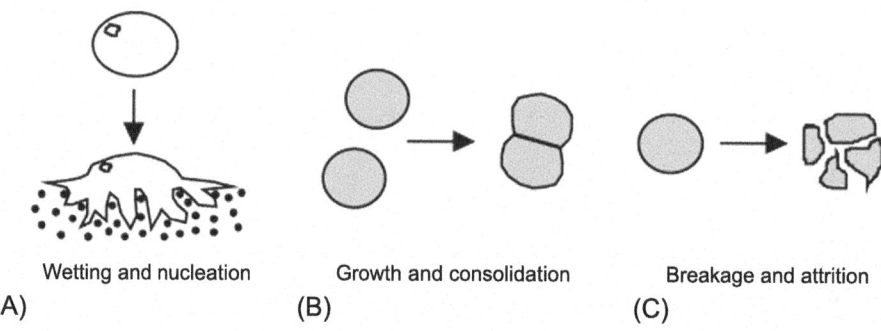

FIGURE 4.4

Diagrams illustrating the three granulation processes: (A) wetting and nucleation, (B) growth and consolidation, and (C) breakage and attrition.

From Litster (2003).

4.4 Evolution of granule structure during high-shear wet granulation

A key step in implementing the MST for the efficient development of high-quality tablet products by HSWG is the understanding of structure evolution during this process. To achieve this, three important stages in HSWG–wetting and nucleation (Fig. 4.4A), growth and consolidation (Fig. 4.4B), and breakage and attrition–must be considered (Ennis & Litster, 1997).

4.4.1 Wetting and nucleation

The wetting and nucleation process involves bringing liquid binder into contact with a dry powder and distributing it evenly throughout the powder (Iveson et al., 2001). The distribution of granulation liquid on the surface of powder bed is the first step of all HSWG processes. "Nuclei" are formed after the initial contact between the liquid binder and powder surface. Depending on both the formulation and process variables, two distinct mechanisms could be in effect: drop-controlled nucleation and mechanical dispersion. When the drop penetrates fast enough and the drops are well separated from each other, nucleation is drop-controlled. This mechanism is often achieved by lowering the spray flux and drop penetration time. If either the drop penetration is slow or the spray flux is too high, the mechanical dispersion mechanism dominates (Iveson et al., 2001).

A nucleation regime map (Fig. 4.5) was proposed to summarize the effects of drop penetration time τ_p, which is affected by formulation composition and binder properties, and dimensionless spray flux, Ψ_a (Hapgood et al., 2003). τ_p is the time required for a single drop to penetrate into the powder bed. Ψ_a, a measure of the nozzle spray zone compared to the renewal flux of powder surface passing through the nozzle spray zone, was a key process parameter used to quantify the effects of the liquid flow rate, binder drop size, and powder flux (Litster et al., 2001).

The nucleation regime map shows three regions of nucleation-based on parameters related to τ_p and Ψ_a. In drop-controlled regime, the controlling variable is the size of binder droplet. To maintain the drop-controlled regime, two consecutive droplets cannot overlap with each other. Therefore the flux

4.4 Evolution of granule structure during high-shear wet granulation

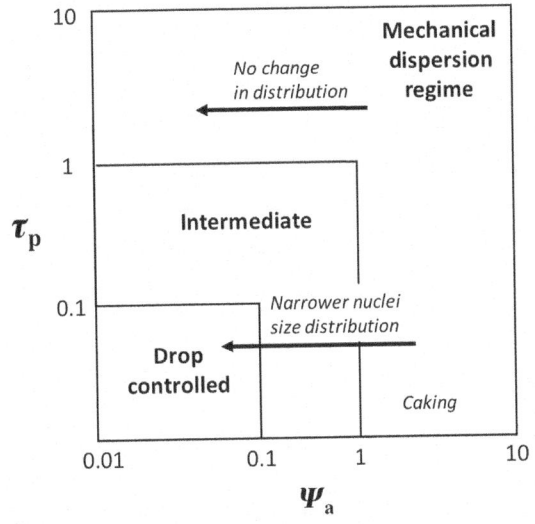

FIGURE 4.5

A graph showing a nucleation regime map.

From Hapgood et al. (2003).

needs to be slow so that a drop has sufficient amount of time to penetrate into the powder bed before meeting with another drop. At such a low spray flux, the droplet size distribution determines the nuclei size distribution.

When binder solution penetration is slow, drop coalescence occurs on the powder surface, which results in a broader nuclei size distribution. In this case, nucleation will enter the mechanical dispersion regime, where capillary pressure and binder viscosity dictate granulation (Iveson et al., 2001). In addition, surface caking may occur when the number of overlapped binder droplets is very high (Hapgood et al., 2003). One factor that influences drop penetration is wettability by the binder solution, which can be characterized by spreading coefficient. A binder liquid with a larger spreading coefficient can penetrate more quickly and tends to form good interparticular liquid bridges, which become solid bridge upon drying. As a result, denser and stronger granules can be obtained. In contrast, binder liquid with a poor spreading coefficient lead to more open-structured granules, which are more friable (Zhang et al., 2002). In the intermediate regime, both drop penetration dynamics and mechanical force dispersion play a significant role, making it more difficult to control the wet granulation process.

In extreme cases where powder mixing in the spray zone is poor, surface caking because of many overlapping drops or liquid pooling due to slow penetration can occur. Viscosity of the binder liquid affects the nucleation mechanism by influencing the drop penetration time. A more viscous binder liquid tends to prolong the drop penetration time and moves the system away from drop-controlled regime. Moreover, the overlap of binder droplets tends to cause different degrees of nucleation, which results in a relatively broad particle size distribution. In a study of granulating lactose monohydrate with water and polyethylene 200 (PEG200), it was found that decreasing the flux narrowed the particle size distribution when water was used, but no significant change in particle size distribution was seen when PEG200 was

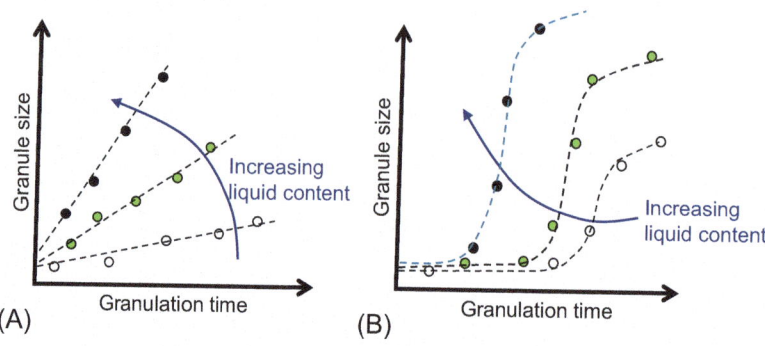

FIGURE 4.6

Graphs showing granule growth mechanisms (granule size vs. granulation time), (A) steady growth and (B) induction growth.

From Iveson and Litster (1998).

used (Hapgood et al., 2003). Thus flux has more pronounced effects on nucleation when less viscous binders are used. Therefore drop-controlled nucleation requires either the binder solution to have low viscosity or the flux of binder liquid to be very slow.

4.4.2 Growth and consolidation

Granule consolidation occurs when a granule is subject to external stress during collisions with another granule, granulator wall, or impeller (Iveson et al., 2001). Consolidation decreases granule porosity and increases its pore saturation by liquid binder. Depending on the deformability of granules, two granule growth mechanisms may result: steady growth and induction growth (Fig. 4.6).

During steady growth, granule size continuously increases with increasing granulation time and the amount of binder solution. This can occur when materials are mechanically deformable, where collisions lead to large contact area. Deformability of granules may be estimated by Stokes deformation number (St_{def}) (Iveson & Litster, 1998). If the collision is sufficiently strong or fast, binder liquid is squeezed out from the points of contact to become available for promoting further growth of granules during collisions. This process is characterized by a linear growth profile (Fig. 4.6A), where granule size enlargement commences immediately after the addition of binder liquid.

Induction growth, however, exhibits a delay in granule size growth after adding the binder liquid (Fig. 4.6). This is more likely to occur in materials that are nondeformable. Negligible degree of particle deformation does not effectively squeeze out liquid binder during collisions. Thus the amount of available binder liquid on the surface is insufficient to sustain steady granule growth during the initial stage. Consequently, the growth of granules is arrested until a sufficient amount of the binder liquid is available on the surface of granules through repeated breakage and consolidation. After a critical point, granules start to grow rapidly. For this growth behavior, an effective approach to decrease the length of induction period is using a larger amount of liquid binder (Fig. 4.6B). In fact, for both steady growth and induction growth, a higher level of binder liquid level favors granule growth because more liquid is available on the surface to promote continued growth of granules through collisions.

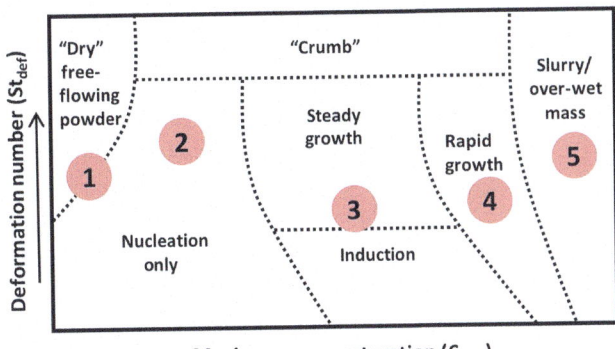

FIGURE 4.7

A granule growth regime map, showing five regimes depending on granule deformation number and granule maximum pore saturation.

From Iveson and Litster (1998).

Following an approach similar to that used for developing the nucleation regime map, a *granule growth regime map* was proposed to classify granule growth behavior by considering binder liquid content in the system and granule deformability (Iveson & Litster, 1998). Here, maximum granule pore saturation (S_{max}) was used to measure the binder liquid content, and St_{def} was used to characterize the granule deformability. The granule growth map may be divided into five regions according to pore saturation, as shown in Fig. 4.7.

In Region 1, the liquid content is so low that it can barely cover the surface of particles. Consequently, no granule growth is possible if the particle deformability is high. Therefore only nucleation occurs when particle deformability is low. Whether the nucleation is controlled by drop introduction or mechanical dispersion depends on the liquid addition rate (Fig. 4.5), which must be considered separately from this map. Region 2 is reached when the binder liquid content is sufficient in the system. In this region, the amount of binder liquid is still less than that required for sustained normal granule growth. Therefore only nucleation occurs if the particle deformation is low. However, when deformability is high, weak granules could form. Such powders appear as "crumbs" since they break easily even under a low stress. When more binder liquid is added, Regions 3 is reached. In this region, normal granule growth takes place. Whether steady growth or induction growth occurs depends on the deformability of materials. Induction growth is observed for poorly deformable systems, because more collisions are required in order to squeeze out sufficient amount of liquid onto surface to sustain granule growth. In contrast, steady granule growth takes place for highly deformable systems, because each collision lead to sufficient deformation and bonding to sustain granule growth. Particle-binder mixtures below the minimum strength only form a "crumb" because they cannot survive the shear stress during the process (Poskart, 1988). In Region 4, the binder liquid content is so high that granules generally grow rapidly except when the deformability is extremely low. In that case, an induction period may still be required before growth is observed. In Region 5, an excessive amount of binder liquid is added. This region should be avoided in HSWG because the system becomes a slurry or wet mass.

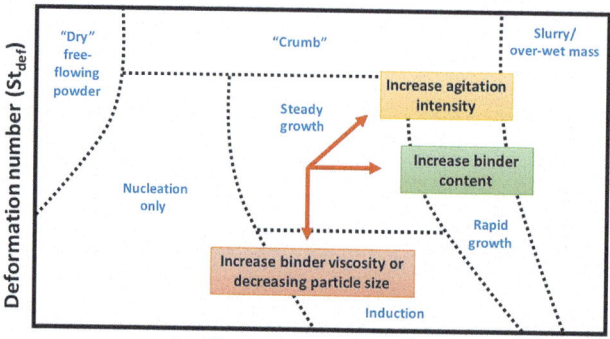

FIGURE 4.8

A graph showing granule growth regime map with effects of common factors indicated.

From Iveson and Litster (1998).

Many other factors also influence the evolution of granule structure during HSWG, such as binder content, impeller speed, and particle size of the solids. Their effects on granule growth could also be explained by the granule growth regime map (Fig. 4.8).

Increasing impeller speed or agitation intensity increases the effective collision velocity during HSWG. Therefore more deformation of the particles is expected during each collision, which is indicated by a larger deformation number. Meanwhile, increasing agitation intensity increases the degree of granule consolidation, which promotes the S_{max} (Iveson & Litster, 1998). Consequently, higher impeller speed will move the system to the upper right corner of the granule growth map (Fig. 4.8).

Higher binder content increases the maximum pore saturation, which also moves the system to the right-hand side on the map (Fig. 4.8). Therefore simple incorporation of more liquid binder can move the system from nucleation only to granule growth or from rapid growth to over-wet mass. This explains the observed effects of binder delivery on HSWG, where the pour-in method resulted in significantly larger variations in binder distribution compared with pump-in and pressure pot methods (Tan et al., 2014). In this case, the portion of powder mass in contact with more water followed different nucleation and growth paths to arrive at granules with a structure and properties that cannot be reversed after redistribution of binder solution by the impeller.

The binder viscosity affects both binder distribution and granule coalescence. Decreasing binder viscosity enhances the size reduction of binder droplets and promotes the binder distribution. Increasing binder viscosity decreases the granule growth by increasing dynamic yield stress and reducing the deformability of the granule (Schæfer & Mathiesen, 1996). Moreover, both smaller particle size of raw materials and higher binder viscosity tend to increase the dynamic yield stress of the granule, which decreases the deformability and lowers rate of granule growth (Iveson & Litster, 1998; Iveson et al., 1996).

Material properties of the formulation also affect the granule growth behavior, through influencing the S_{max} and granule deformability (evaluated by St_{def}) (Muthancheri & Ramachandran, 2020). For binary mixtures with MCC, both S_{max} and St_{def} increase as the loading of a less hydrophobic API, acetaminophen, increases from 10% to 80%. When the loading of the more hydrophobic ibuprofen increases from 10% to 80%, S_{max} increases in the entire range, but St_{def} only increases up to 50% of ibuprofen before decreasing. The different granulation behaviors of the two systems could be attributed to the different nucleation and growth mechanisms. Solid spread nucleation mechanism and aggregation-dominated growth of ibuprofen formulation lead to more uniform and stronger granules that are more resistant to granule breakage. In contrast, the immersion-driven nucleation and faster granule growth by particle layering result in acetaminophen granules that are mechanically weaker and less uniform (Muthancheri & Ramachandran, 2020).

Both the nucleation regime map and the granule growth regime map qualitatively elaborate the HSWG process. Quantitative understanding of HSWG requires detailed characterization of structures and properties of resulting granules as per the MST. This is illustrated by recent work, which systematically investigated the evolution of granule structure and properties of microcrystalline cellulose (MCC, Avicel PH101) and polyvinylpyrrolidone (PVP and K30) blends throughout a HSWG process (Osei-Yeboah et al., 2014).

MCC particles were irregularly shaped porous agglomerates with rough surfaces (Fig. 4.9A). The median volume-based diameter, d_{50}, analyzed by laser diffraction, slightly decreased when increasing granulation water content from 0% to 35%. This observation suggests possible surface smoothing and densification of MCC particles within this period (Osei-Yeboah et al., 2014). This is supported by SEM images (Fig. 4.9A–D), which show diminishing number of observable pores and smoother surfaces. Thus surface smoothing and densification occurred during the initial addition of water up to 35%. After this point, rapid granule growth took place and the size of granules increased dramatically to about 800 μm and fine particles disappeared completely as water content reached 105% (Fig. 4.9D–K).

Surface smoothing and densification were also supported by the continually decreased specific surface area (SSA) and porosity of the granules measured by BET nitrogen adsorption and mercury porosimetry, respectively (Osei-Yeboah et al., 2014). Size enlargement of MCC granules was also accompanied by shape rounding, where granule became more spherical from the originally irregular and elongated shape. All of these changes led to profoundly improved powder flowability (Osei-Yeboah et al., 2014). However, they also led to severely reduced tabletability (Fig. 4.10), due to deteriorated compressibility, as a result of lower plasticity, lower compactibility, and smaller surface area (Xiao et al., 2022). Plasticity of granulated MCC can be reliably quantified by in-die mean yield pressure (P_y) (Vreeman & Sun, 2021). The deteriorated compactibility may be attributed to the smaller surface area, caused by the increased larger particle size and higher density of granules. Thus changes in critical properties of granulated powders could be clearly explained from the evolution of granule structure as the HSWG process proceeds. Similar approach is expected to better understand properties of resulting granules in other systems, where different properties of API and binder type were found to affect granule properties (Dan et al., 2023).

4.4.3 Granule breakage during high-shear wet granulation

In addition to granule nucleation and growth, granule breakage is another important process that can significantly impact final granule properties, which makes it more challenging to accurately predict granule properties (Kayrak-Talay & Litster, 2011). Using colored tracers, it was shown that granule breakage influenced the binder distribution and final granule size distribution (Van Den Dries et al., 2003). During the high-shear granulation process, smaller granules tend to coalescence into bigger

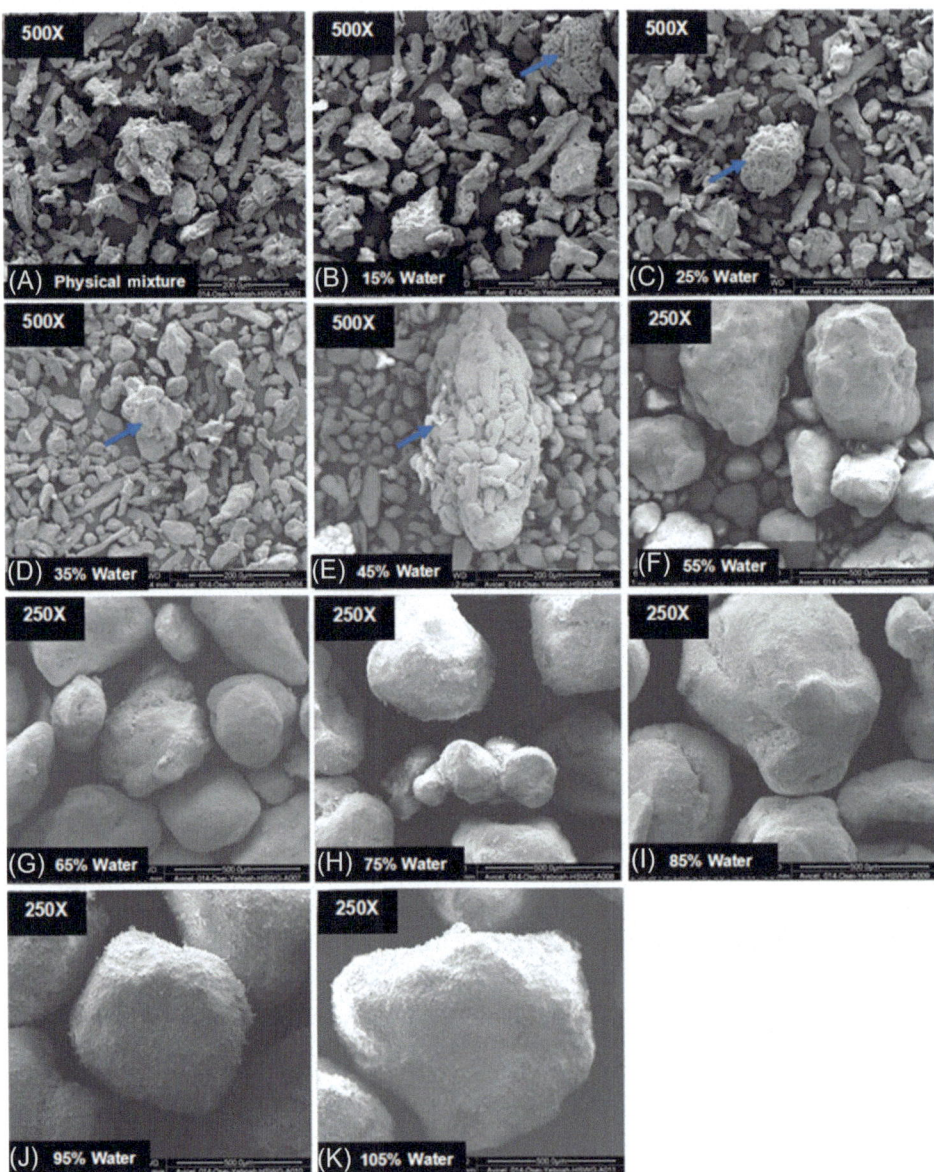

FIGURE 4.9

A series of scanning electron microscope images of physical mixture and granules of Avicel PH101 and polyvinylpyrrolidone K30 at various water levels, magnified at 500 times and 250 times.

From Osei-Yeboah et al. (2014).

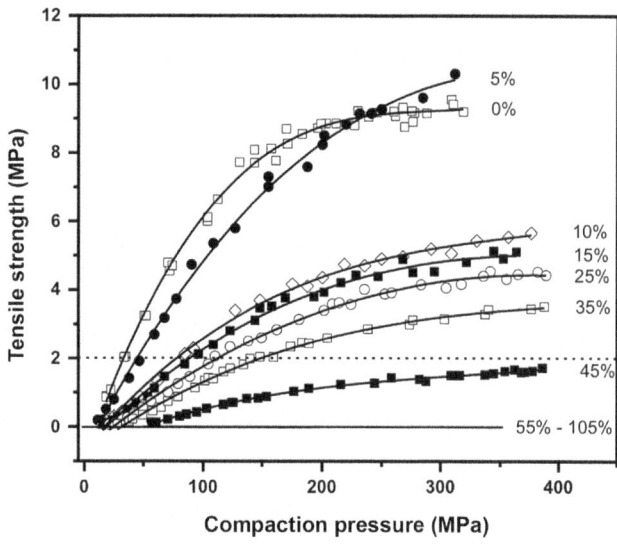

FIGURE 4.10

A graph showing tabletability profiles of a physical mixture of microcrystalline cellulose and polyvinylpyrrolidone and granules prepared with different amounts of water.

From Osei-Yeboah et al. (2014).

agglomerates while larger granules tend to break into smaller particles. The destructive nucleation growth mechanism accounts for the effects of both coalescence and breakage on granule growth (Vonk et al., 1997). On one hand, primary nuclei form and grow by layering of primary particles in the presence of binder liquid. On the other hand, attrition and fragmentation generate secondary nuclei, which can further grow into bigger granules by coalescence.

The extent of granule breakage depends on the process parameters, especially impeller speed. Increasing impeller speed initially increases granule size due to more coalescence events. However, further increasing in the impeller speed may reduce granule size due to more extensive granule breakage (Benali et al., 2009; Knight et al., 2000). Thus higher impeller speed may increase or decrease final granule size, depending on the formulation and other process parameters (Eliasen et al., 1998; Hoornaert et al., 1998). Varying impeller speed can also change granule shape (Logan & Briens, 2012). During reverse-phase wet granulation, mechanical stress introduced by impeller was found to play an essential role in granule breakage. Impact pressure, along with granule tensile strength estimated from the static tensile strength (Wade et al., 2015), may be used to predict the tendency of granule breakage. Impact pressure can eventually exceed the granule strength, when a sufficiently high impeller speed is used, to cause granule breakage (Vonk et al., 1997).

It was suggested that granules elongation due to shear preceded breakage at a material-specific St_{def} (Tardos et al., 1997). For example, lowering St_{def} was shown to eventually lead to breakage of lactose monohydrate granules (Van Den Dries et al., 2003). A capillary number, C_a, which is related to St_{def}, was also found to linearly relate to the reciprocal of the probability of breakage (de Koster et al., 2021). Thus the use of St_{def} or C_a may be an attractive alternative to monitoring impeller speed for predicting granule breakage. This is because the impeller speed alone may not lead to an accurate prediction of granule breakage without considering other factors, such as granulation water content (Badawy et al., 2012; de Koster et al., 2021). Higher water content usually corresponds to higher plasticity of the material, which favors coalescence during granulation. When water content is high, higher impeller speed promotes the coalescence due to the availability of binding liquid. In contrast, when water content is low, higher impeller speed likely causes more granule breakage as the granules are more brittle.

Material properties of a formulation can also significantly influence granule breakage. Impeller speed and processing time affect granule breakage only if the material properties lie in a certain range (Van Den Dries et al., 2003). For example, mechanically strong materials may experience low breakage even at high impeller speed. Binder surface saturation level and binder viscosity also influence granule breakage. Higher binder saturation level favors coalescence and results in stronger granules, thus, less breakage. Meanwhile, higher binder viscosity corresponds to less granule breakage by decreasing granule porosity and increasing granule strength, while large granule size promotes breakage (Van Den Dries et al., 2003).

4.4.4 Granule breakage during fluidized bed drying

Granule breakage can also occur during fluidized bed drying post HSWG. If not controlled, excessive fines generated during drying can cause problems in downstream processing such as inconsistent die filling due to poor powder flowability (Bemrose & Bridgwater, 1987).

Water content is one of the most critical factors that affect granule breakage behavior during drying. At the same air flow rate, granule breakage increases as the water content decreases with time (Nieuwmeyer et al., 2007; Nieuwmeyer et al., 2007; van den Dries & Vromans, 2002). Fines are formed only when the water content falls below a critical value (Nieuwmeyer et al., 2008). Thus fines are generated primarily during the late stage of the fluid bed drying process. Higher air flow rate generally lead to more fines because it causes more rapid water loss as well as higher extent of collision. Granules remain intact until impact stress during collisions exceeds granule strength, at which point granule breakage occurs (Vonk et al., 1997). Weak granules formed using suboptimal amount of water during granulation can break easily even at the beginning of the drying process. In contrast, granules obtained using a sufficient amount of granulation water are usually large and strong, thus, more resistant to fine generation during fluidized bed drying. Therefore granule breakage during fluidized bed drying depends on flow rate of drying air, drying time, granule size, mechanical properties, and amount of water used during granulation. These factors must be carefully considered when designing a fluid bed drying process to avoid an excessive number of fines.

4.5 Applications of materials science tetrahedron in high-shear wet granulation

4.5.1 Effects of initial water content on granule structure and property

For hygroscopic materials, water content can vary significantly when exposed to different relative humidity levels (RH) during storage or handling. RH is expected to swing wildly with seasons.

4.5 Applications of materials science tetrahedron in high-shear wet granulation

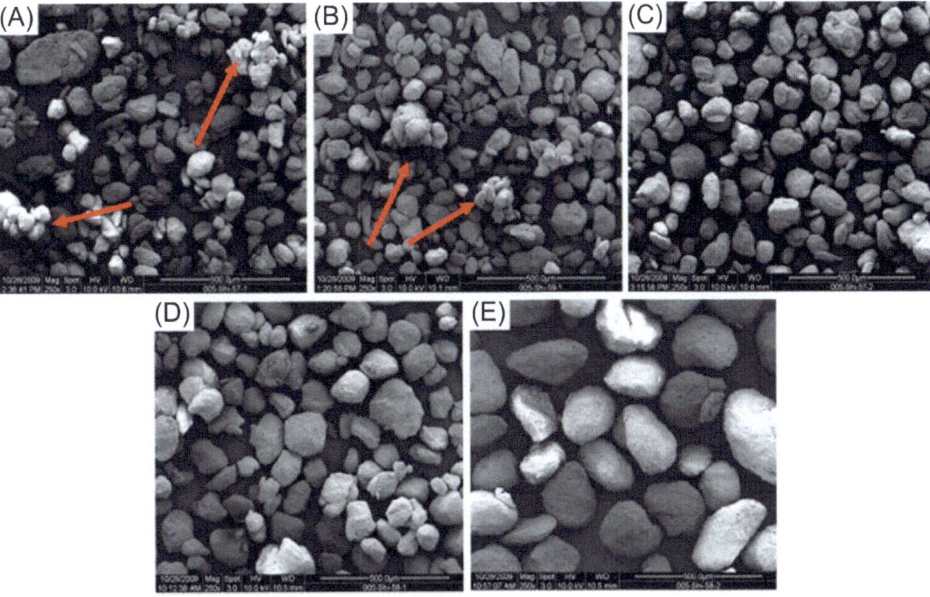

FIGURE 4.11

A series of five SEM images of granules prepared from microcrystalline cellulose with various of initial water contents, (A) 0.9%, (B) 2.6%, (C) 4.6%, (D) 7.2%, (E) 10.5%).

From Shi et al. (2011a).

Depending on locations, RH during summer can be as high as 60% but as low as 2% during winter in uncontrolled laboratory environments. This can have profound impact on the outcome of HSWG.

MCC is one of the most commonly used diluents in pharmaceutical tablets. It is hygroscopic and tends to absorb different amounts of water under various levels of humidity (Sun, 2008). MCC Avicel PH102 was found to exhibit minimum flow properties that should be met for high-speed tableting (Sun, 2010). Thus it can be used as a convenient reference for assessing adequacy of powder flowability of new tablet formulations. It was shown that structures and properties of MCC granules by HSWG were drastically affected by increasing initial moisture content in MCC (Shi et al., 2011a). This led to profound changes in performance of granulated bulk powders, such as increased flowability along with severely deteriorated tabletability. Following the MST, clear understanding of such impact requires a systematic examination of the structures and properties of granules.

When water content in MCC was 0.9% and 2.6% (w/w), only loose agglomerates (red arrow) existed in the powder (Fig. 4.11A and B). This indicates that the HSWG process remained at nucleation stage instead of growth regime. A slight increase in the initial water content to 4.6% brought the system to the granule growth regime, which led to generation of loose agglomerates (Fig. 4.11C). Further increasing the initial water content to 7.2% and 10.5% in MCC resulted in significant size enlargement (Fig. 4.11D and E), indicating the commencement of the rapid growth stage of HSWG. This effect, indicated by SEM data, was confirmed by size distribution data from laser diffraction experiments, which showed a clear trend of granule size changing with the initial water content in MCC (Fig. 4.12A and B).

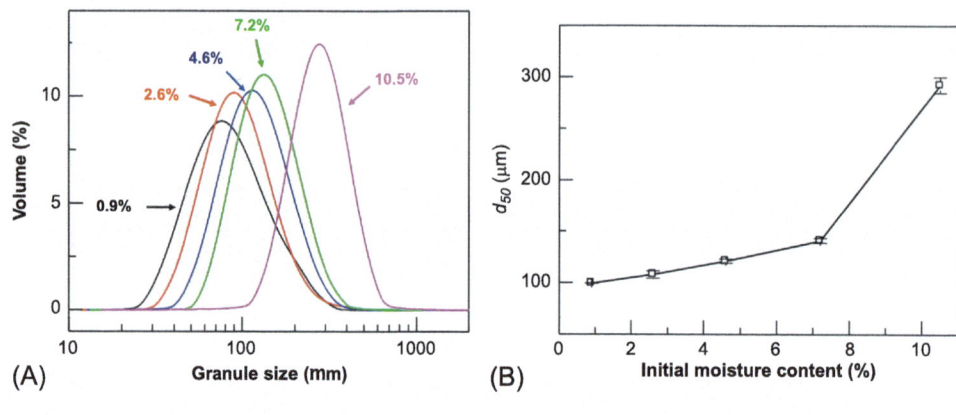

FIGURE 4.12

Graphs showing microcrystalline cellulose granule size as a function of initial water content, (A) volume-based granule size distribution, and (B) median granule size.

From Shi et al. (2011a).

By following the granule porosity in the final granules, a critical initial water content of 2.6% was identified. With increasing initial water content, pore shrinkage and elimination occurred more extensively during HSWG before the critical initial water content was reached. Beyond the critical initial water content, granule porosity did not change with water content (Shi et al., 2011a). Although granule porosity remained unchanged above the critical initial water content, SSA continued to decrease because of granule surface smoothing and size enlargement. These are in agreement with the effects of binder solution on granule growth process illustrated in the growth regime map (Fig. 4.8).

4.5.2 Effects of massing time on structure and properties of granule

One practice during HSWG is to knead the wet mass after all binding liquid has been added to the granulator, which is commonly known as massing. Massing is performed mainly to improve homogeneity of binder distribution and to increase the degree of granule densification. However, massing can also negatively impact structures and properties of granules if not carefully controlled. Again, such changes in powder properties could be explained by structural changes of the granules. It was shown that prolonged massing during HSWG of MCC successfully improved granule densification (Shi et al., 2011a, 2011b). Granules prepared without massing had an irregular shape with highly porous and rough surfaces. Increasing massing time smoothened the surfaces of granules and made them more rounded. This was accompanied by elimination of pores and reduction of SSA, although granule size did not change dramatically (Fig. 4.13).

These changes in granule structure due to prolonged massing led to significantly decreased tabletability over the entire pressure range (Fig. 4.14A), including the pressure range of 150 to 300 MPa commonly applied in pharmaceutical tablet manufacturing. For example, more than 70% of tensile strength drop occurred at 300 MPa pressure within 10 min of massing (Fig. 4.14B). Clearly, caution needs to be exercised when introducing a massing step into an HSWG for processing formulations

4.5 Applications of materials science tetrahedron in high-shear wet granulation

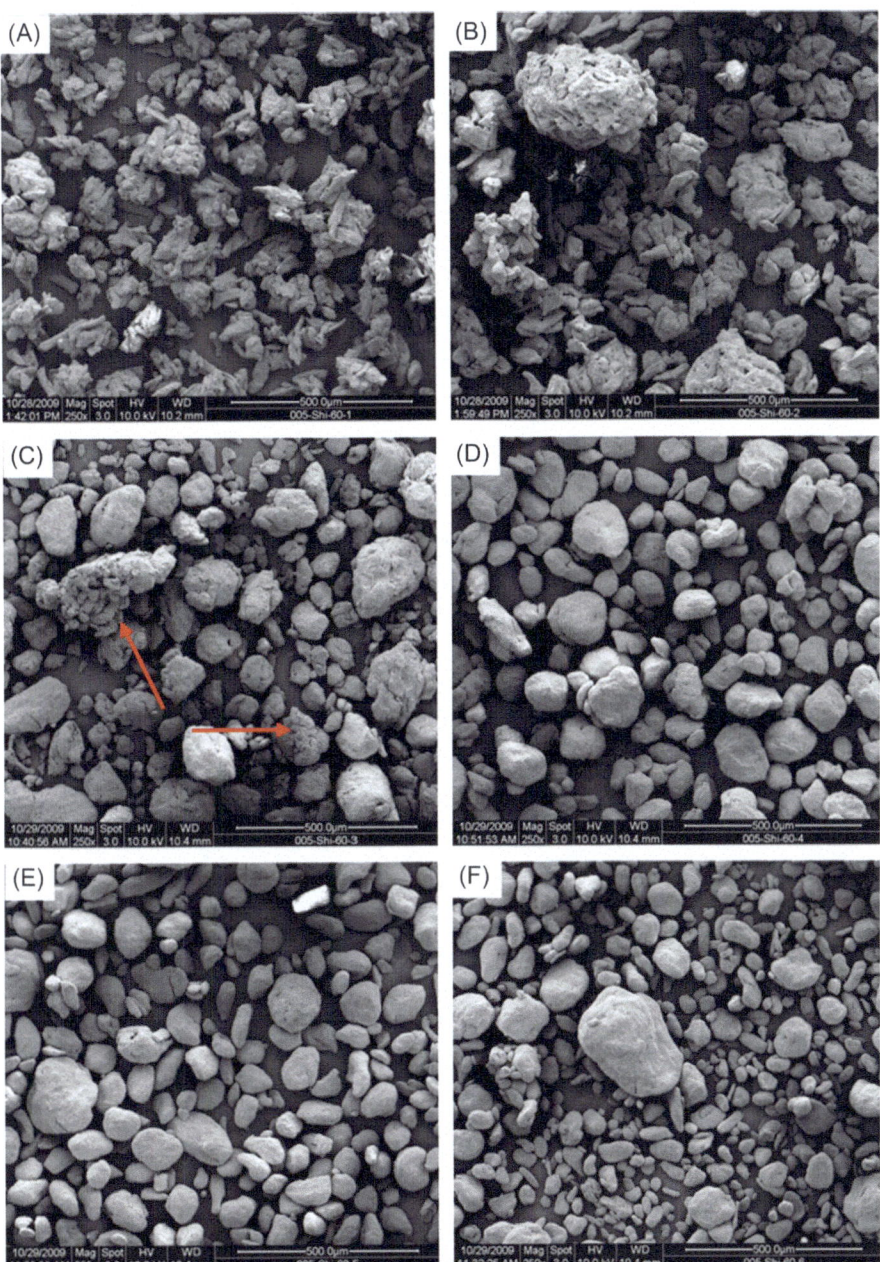

FIGURE 4.13

A series of SEM images of microcrystalline cellulose granules prepared by high shear wet granulation with various massing durations: (A) 0 min, (B) 1 min, (C) 5 min, (D) 10 min, (E) 20 min, and (F) 40 min.

From Shi et al. (2011b).

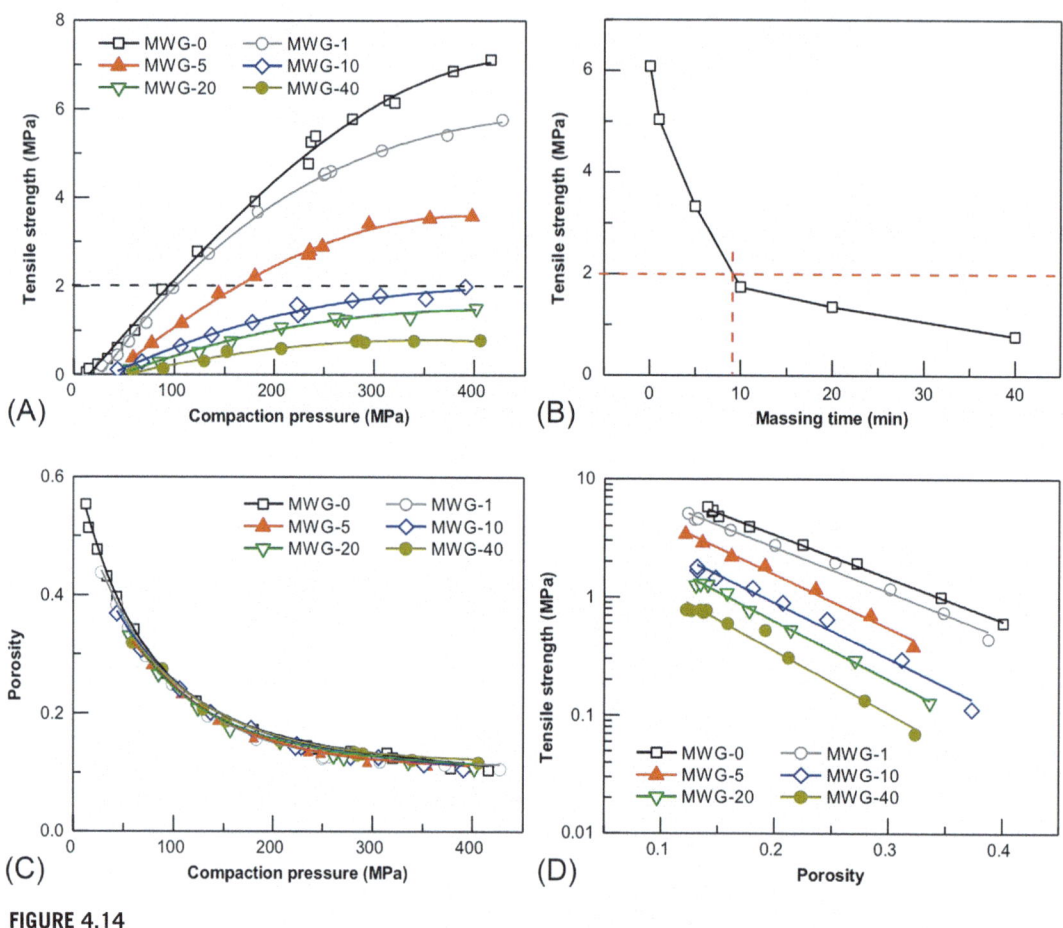

FIGURE 4.14

Graphs showing mechanical properties of microcrystalline cellulose granule prepared by HSWG with various massing times: (A) tabletability, (B) tensile strength at 300 MPa compaction pressure, (C) compressibility, (D) compactibility. The numbers after MWG in the figures indicate massing time in minutes.

From Shi et al. (2011b).

containing MCC. The deterioration in tabletability can be explained based on a consideration of bonding area among granules. It was found that massing time did not impact compressibility (Fig. 4.14C). Thus total tablet porosity was independent of massing time. Since intragranular porosity decreased with longer massing time, intergranular porosity must be higher for granules prepared with longer massing time (Shi et al., 2011b). The larger intergranular porosity corresponds to smaller intergranular bonding area and, consequently, weaker tablets according to the bonding area-bonding strength interplay theory (Osei-Yeboah et al., 2016). Significant deterioration of compactibility was also observed with increasing massing time (Fig. 4.14D). This is because denser granules are more resistant to permanent plastic deformation during compaction. Therefore actual bonding area is lower when compressed to comparable tablet porosity. Similar effects of massing time on granules are also observed for API/excipients mixtures (Macho et al., 2023) and nonpharmaceutical mineral systems (You et al., 2022). For example, when hydralazine mixed with MCC and anhydrous lactose were granulated with different impeller speeds and wet massing times, the tabletability and compactibility of granules produced from longest wet massing time (120 s) are reduced more significantly compared to those of original mixtures. This effect can be explained by the lower plasticity, indicated by the higher P_y, of the granules.

4.5.3 Effects of surface smoothing and granule size enlargement on flowability and tabletability

As discussed above, surface smoothing and densification occur along with the size enlargement during the HSWG for both MCC and its mixture with PVP. Consequently, tablet tensile strength decreased while flowability improved sharply as the granulation water content increased (Osei-Yeboah et al., 2014; Shi et al., 2010, 2011c). Optimum control of these properties for successful tablet manufacturing requires solid understanding of structural basis for both. As explained earlier in the studies of effects of massing time and initial water content during HSWG, surface smoothing, densification, and size enlargement all contributed to smaller bonding area, hence lower tensile strength (Osei-Yeboah & Sun, 2015a; Osei-Yeboah et al., 2014; Osei-Yeboah et al., 2016; Shi et al., 2011c). This structure-property-performance relationship is likely broadly applicable in pharmaceutical manufacturing because MCC is widely used in pharmaceutical tablet formulations of different drugs.

As mentioned earlier in this chapter, enhancing powder flow is the main motivation for HSWG and size enlargement is the main contributor to the desired improvement in powder flowability. However, other mechanisms may also be in effect. For particles with rough surfaces, mechanical interlocking between asperities on the particle surfaces can effectively hinder the relative movement between adjacent particles and, thus, decrease powder flowability (Zeng et al., 2001). Surface smoothing reduces mechanical interlocking. In fact, the enhancement in powder flow of MCC was profound when merely 5% of water was used during HSWG, where laser diffraction data excluded particle size enlargement as a possible mechanism (Shi et al., 2011c). SEM data confirmed that the improved flowability was accompanied by surface smoothing instead. When more water was used during HSWG, densification, size enlargement, and shape rounding all contributed to the much enhanced flowability of granules (Shi et al., 2011a, 2011c). Such effects could also be attained by increasing massing time (Shi et al., 2011b) or using MCC containing higher amount of moisture as discussed earlier (Shi et al., 2011a).

4.5.4 Effects of screw profiles on granule structures and properties in twin-screw wet granulation

As an emerging technique amenable for continuous wet granulation, the TSWG has drawn great interest, as mentioned in Section 4.3.2. The mechanisms of blend mixing and granulation during TSWG is complex (Zhang et al., 2021). Thus understanding factors that affect granules produced is critical to achieve desirable granulation by the TSWG process.

Screw profile is a key factor that plays an important role in granule properties due to its flexibility in geometries and design (Thompson & Sun, 2010). It is shown that different screw profiles affect the granule structures, content uniformity, and tabletability of a hydrophobic drug formulation

FIGURE 4.15

An image showing different screw element types arranged in a row: (1) 20 mm GLC, (2) 20 mm K90, (3) 90 mm GFA30 (30 mm pitch), and (4) 60 mm GFA 20 (20 mm pitch).

From Kashani Rahimi et al. (2020).

(Kashani Rahimi et al., 2020). Unlike hydrophilic formulation, mixtures containing highly hydrophobic ingredients will suffer poor or inhomogeneous wetting with the binder. Extra efforts by mechanical forces may be needed to achieve more homogenous mixture-binder interaction since it falls into the mechanical dispersion regime (Fig. 4.5). The use of a distributive comb mixing element (GLC), compared with kneading blocks (K) and general forwarding action screw element (GFA), results in granules with smaller and narrower particle size distribution, more porosity, lower strength, and higher content uniformity. This can be attributed to the characteristics of the GLC: more cutting actions, open channels, and free volume within the elements (Fig. 4.15). Granules are fragmented and re-granulated into porous structure by the axial breakage and attrition. A "back mixing," where the forward-moving powder meets the backward traveling wet fragments, occurs by the unique "fragmentation-re-agglomeration dispersive cycle" provided by the GLCs. Such repeated mixing favors better mixing between the powder and binder regardless of hydrophobicity. As a result, both the compressibility and compactibility of the granules from GLC are improved due to the higher porosity and rougher surface of granules (Fig. 4.16). It is clear that adopting appropriate screw elements in TSWG is essential to achieve granules with desirable properties (Seem et al., 2015; Thompson, 2015; Zhang et al., 2021).

4.5.5 Effects of material properties on granule structures and properties in twin-screw wet granulation

According to the MST concept, the performance of granules depends on the properties of formulation components as well. Hence, it is expected that the raw materials, including fillers, binders, surfactants, and APIs, may influence granule properties (Franke et al., 2023; Lute et al., 2018; Portier et al., 2020a, 2020b; Portier et al., 2021; Stauffer et al., 2019; Vandevivere et al., 2020). It was shown that excipient fillers with different primary particle morphologies (Lute et al., 2018) and mechanical properties (Khorsheed et al., 2019) impacted granule structures and tabletability. Materials with relatively high plasticity, such as MCC, tend to suffer a tabletability reduction upon granulation with kneading elements. In contrast, tabletability of brittle materials, such as mannitol 160C, is improved after kneading. This is mainly due to the fact that more deformable MCC can form denser and stronger granules than

4.5 Applications of materials science tetrahedron in high-shear wet granulation

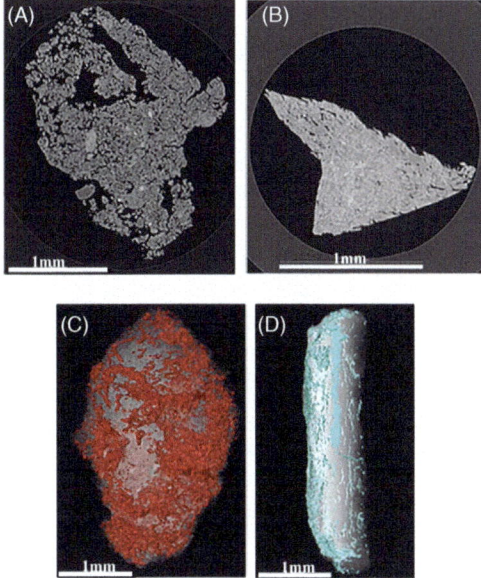

FIGURE 4.16

Micrographs showing cross-section X-ray CT images of granules processed with (A) GLC and (B) K90 elements and 3D void distribution sideways images of granules processed with (C) GLC and (D) K90 elements.

From Kashani Rahimi et al. (2020).

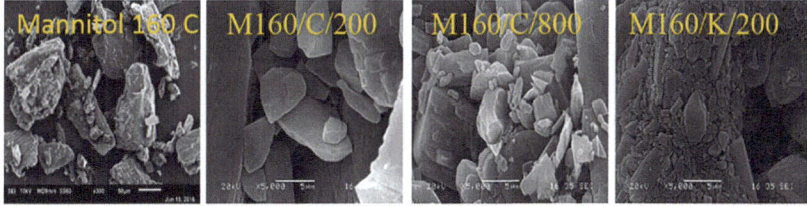

FIGURE 4.17

Micrographs showing scanning electron microscopy images of granules from Mannitol 160C and processing conditions at 5000 × magnification, (C indicates conveying element only, K indicates conveying + kneading elements. Screw speed is 200/800 rpm).

From Khorsheed et al. (2019).

others. On the other hand, smaller crystals were formed after granulation when the brittle mannitol 160C is granulated, which lead to improve the tabletability (Fig. 4.17). The formation of smaller mannitol particles were hypothesized to be a result of the dissolution of mannitol crystals into binder liquid followed by crystallization into smaller crystals upon drying (Khorsheed et al., 2019). Another possible mechanism is the fragmentation of the brittle mannitol by the high-shear force of kneading elements during the TSWG process (Wu & Sun, 2007).

4.6 A formulation strategy for solving overgranulation problem during high-shear wet granulation

An important aspect of MST is that structural understanding is useful for solving problems only when appropriate engineering strategies can be designed to modify relevant material structures and properties (Sun, 2009b). In the context of HSWG, a main problem is the overgranulation, which is characterized by severely deteriorated tabletability or dissolution rate (Osei-Yeboah et al., 2014). It has been shown that the reduced bonding area between granules, due to larger size, higher density, and smoother surface, is responsible for reduced tabletability (Sun, 2011). Based on this understanding, a possible strategy is size reduction. This was confirmed by showing that size reduction of an overgranulated batch of MCC granule significantly recovered tabletability (Shi et al., 2010). Therefore closely monitoring granule size during HSWG is important for avoiding overgranulation. However, the window for obtaining granules with suitable sizes to achieve acceptable flowability without significant loss of tabletability is narrow. Therefore effectively addressing this risk requires another strategy that renders tabletability insensitive of granule size enlargement.

One such strategy is to introduce a brittle material to the formulation. The incorporation of a sufficient amount of brittle material makes granules brittle. Such granules maintain excellent flowability during handling, while they can also undergo extensive fracture during compaction. The in-situ size reduction generates much larger area for bonding and, thus, preserves good tabletability. This formulation strategy was demonstrated by incorporating lactose and dibasic calcium phosphate (Dical) with MCC (Osei-Yeboah et al., 2014).

The tabletability of pure MCC granules decreased sharply as water content increases and eventually reached the overgranulation zone (Fig. 4.18). However, for blends containing lactose or Dical, tabletability decreased initially with increasing granulation water level until reaching a minimum tensile strength (Fig. 4.18). At higher water level, tablet tensile strength rose again. Such effect is more prominent when more brittle excipient is used in the formulation. For formulations containing 80% of either lactose or Dical, tablet tensile strength varied with water level but overall maintained high values (Fig. 4.19B and C). Of course, the higher pressure leads to higher tensile strength. This is in stark contrast to MCC without any brittle excipient, where tensile strength quickly reached zero with increasing granulation water level irrespectively of the compaction pressure applied (Fig. 4.19A).

A comprehensive study to understand the change of tabletability of granules by HSWG for 30 materials, including pharmaceutical excipients and natural product powders, was conducted (Wang et al., 2022). These materials were categorized according to their mechanical properties (plastic, brittle, and elastic types), and tableting performance was classified according to change in tabletability (Type I—loss, Type II—unchanged, and Type III—increased). Among the brittle fillers, mannitol belongs to Type III with increased tabletability after granulation, due to the size reduction and larger bonding area as aforementioned. This observation suggests the use of mannitol as a filler with plastic excipients to maintain tabletability of formulated granules.

Thus incorporating brittle excipients makes granule more resistant against potential overgranulation problem during HSWG. This also highlights the importance of characterizing both brittleness and plasticity of drugs and excipients when developing a formulation for HSWG (Wang et al., 2022, 2023). For example, if the drug is very plastic, the use of more brittle excipient is desired to mitigate any possible overgranulation problem. Brittleness of a material can be quantified using a tablet brittleness index extrapolated to zero porosity (Gong & Sun, 2015; Gong et al., 2015; Paul & Sun, 2017). On

4.6 A formulation strategy for solving overgranulation problem during high-shear

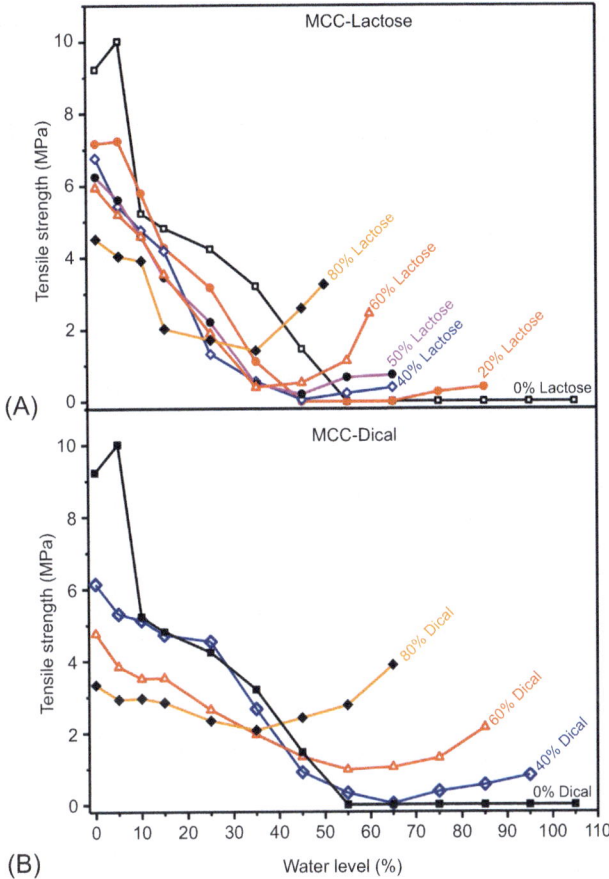

FIGURE 4.18

Line graphs showing the effects of brittle excipient concentration on the tensile strength of microcrystalline cellulose granules at 300 MPa compaction pressure with different water content (A) lactose and (B) Dical.

From Osei-Yeboah et al. (2014).

the other hand, material plasticity may be quantified from analyzing powder compressibility data using an appropriate equation (Sun, 2017), or through determining tablet indentation hardness (Patel & Sun, 2016). Successful implementation of this formulation strategy to address the overgranulation problems requires maintaining a balance between plasticity and brittleness. While a formulation lacking brittle fracture tends to risk overgranulation, completely brittle materials tend to exhibit very high ejection force and high friability even when tabletability may be good (Osei-Yeboah & Sun, 2015a, 2015b). Aided by tablet brittleness index, balance between plasticity and brittleness of formulations can be attained through appropriate excipient selection.

Water-holding capacity, defined as the maximum amount of water added into the system before paste formation, is another property to consider when choosing excipients for a HSWG formulation. It was determined that the water-holding capacity followed the order of: MCC > Dical > lactose (Fig. 4.20)

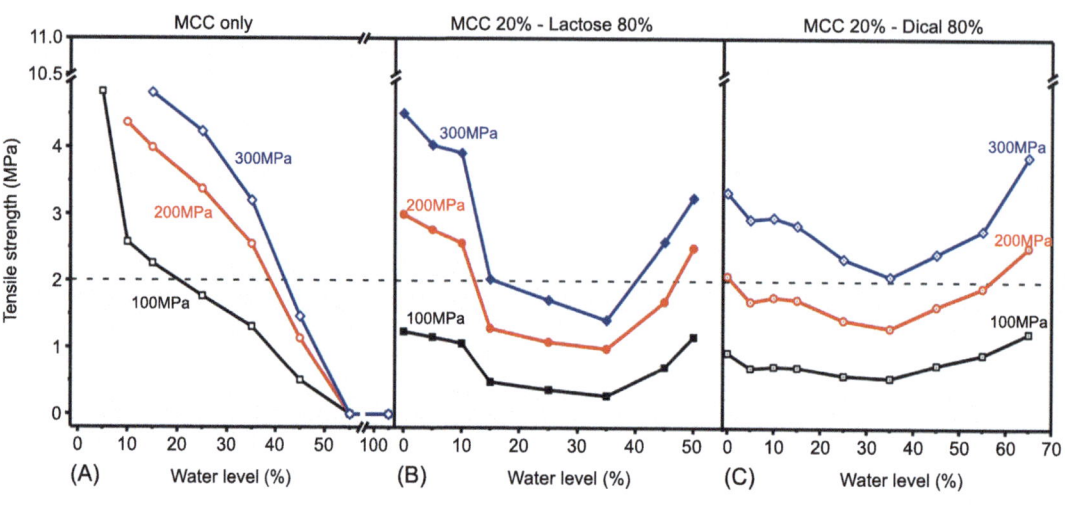

FIGURE 4.19

Line graphs showing the dependence of tablet tensile strength with granulating water content at different compaction pressures for: (A) microcrystalline cellulose (MCC), (B) 20% MCC + 80% lactose; and (C) 20% MCC + 80% Dical.

From Osei-Yeboah et al. (2014).

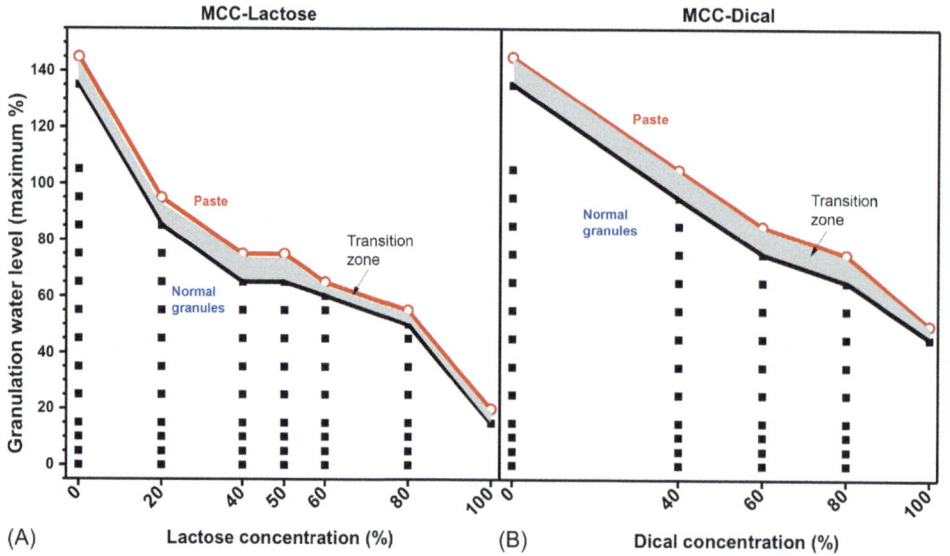

FIGURE 4.20

Graphs showing water-holding capacity of mixtures containing microcrystalline cellulose (MCC) and either (A) lactose or (B) Dical.

From Osei-Yeboah et al. (2014).

(Osei-Yeboah et al., 2014a, 2014b). Water-holding capacity provides formulation scientists important information because they determine the maximum amount of granulation water that can be used for HSWG. The granulation water content must be kept below the water-holding capacity of the formulation mixture in order to avoid entering five regions of the granule growth regime map (Fig. 4.7), where HSWG is no longer feasible.

4.7 Conclusion

HSWG involves spraying binder liquid onto the powder bed while being vigorously sheared by impeller blades or screws. Though HSWG is not suitable for materials that are sensitive to moisture and heat, several advantages of HSWG, including improved flow, improved homogeneity and content uniformity, fine reduction, and improved handling, make it one of the most commonly used powder processing methods in the pharmaceutical industry. Continuous HSWG holds promise because of its ability to decrease the time to market and good reproducibility. Twin-screw granulator is a leading choice for implementing continuous HSWG.

Numerous attempts have been made historically to understand and manipulate the HSWG process. However, their impact is limited by a lack of structural insight of resultant granules. Material science tetrahedron is a useful tool to guide formulation scientists for developing high-quality HSWG products based on a clear understanding of structure–property relationship. Understanding of the evolution of granule structure starts from an examination of nucleation and growth stages of HSWG. Nucleation regime map and granule growth map are useful for qualitatively understanding formulation and process variables on the properties of HSWG granules. It has been recognized that understanding structures of HSWG granules is critical for assessing granule properties and their sensitivity to raw material properties, such as mechanical properties, size, and initial water content, and process variables, such as, massing time, and screw profiles in TSWG. Powder flowability benefits from granule size enlargement, densification, surface smoothing, and shape rounding. Unfortunately, these changes also deteriorate tabletability of plastic materials, which is a symptom of overgranulation. Finally, maintaining an appropriate balance between plasticity and brittleness of the formulation is important for solving the overgranulation problem without impairing the robustness of HSWG formulations.

References

Aulton, M. E. (2002). *Pharmaceutics: The science of dosage form design* (p. 2002). Churchill Livingstone.

Badawy, S. I. F., Narang, A. S., Lamarche, K., Subramanian, G., & Varia, S. A. (2012). Mechanistic basis for the effects of process parameters on quality attributes in high shear wet granulation. *International Journal of Pharmaceutics, 439*(1–2), 324–333. https://doi.org/10.1016/j.ijpharm.2012.09.011.

Bemrose, C. R., & Bridgwater, J. (1987). A review of attrition and attrition test methods. *Powder Technology, 49*(2), 97–126. https://doi.org/10.1016/0032-5910(87)80054-2.

Benali, M., Gerbaud, V., & Hemati, M. (2009). Effect of operating conditions and physico–chemical properties on the wet granulation kinetics in high shear mixer. *Powder Technology, 190*(1–2), 160–169. https://doi.org/10.1016/j.powtec.2008.04.082.

Cartwright, J. J., Robertson, J., D'Haene, D., Burke, M. D., & Hennenkamp, J. R. (2013). Twin screw wet granulation: Loss in weight feeding of a poorly flowing active pharmaceutical ingredient. *Powder Technology, 238*, 116–121. https://doi.org/10.1016/j.powtec.2012.04.034.

Chattoraj, S., Shi, L., Chen, M., Alhalaweh, A., Velaga, S., & Sun, C. C. (2014). Origin of deteriorated crystal plasticity and compaction properties of a 1:1 cocrystal between piroxicam and saccharin. *Crystal Growth & Design, 14*(8), 3864–3874. https://doi.org/10.1021/cg500388s.

Chattoraj, S., Shi, L., & Sun, C. C. (2010). Understanding the relationship between crystal structure, plasticity and compaction behaviour of theophylline, methyl gallate, and their 1:1 co-crystal. *CrystEngComm, 12*(8), 2466–2472. https://doi.org/10.1039/c000614a.

Chattoraj, S., Shi, L., & Sun, C. C. (2011). Profoundly improving flow properties of a cohesive cellulose powder by surface coating with nano-silica through comilling. *Journal of Pharmaceutical Sciences, 100*(11), 4943–4952. https://doi.org/10.1002/jps.22677.

Chow, S. F., Chen, M., Shi, L., Chow, A. H. L., & Sun, C. C. (2012). Simultaneously improving the mechanical properties, dissolution performance, and hygroscopicity of ibuprofen and flurbiprofen by cocrystallization with nicotinamide. *Pharmaceutical Research, 29*(7), 1854–1865. https://doi.org/10.1007/s11095-012-0709-5.

Corvari, V., Fry, W. C., Seibert, W. L., & Augsburger, L. (1992). Instrumentation of a high-shear mixer: Evaluation and comparison of a new capacitive sensor, a watt meter, and a strain-gage torque sensor for wet granulation monitoring. *Pharmaceutical Research: An Official Journal of the American Association of Pharmaceutical Scientists, 9*(12), 1525–1533. https://doi.org/10.1023/A:1015843820526.

Dan, A., Vaswani, H., Šimonová, A., Grząbka-Zasadzińska, A., Li, J., Sen, K., Paul, S., Tseng, Y. C., & Ramachandran, R. (2023). End-point determination of heterogeneous formulations using inline torque measurements for a high-shear wet granulation process. *International Journal of Pharmaceutics: X, 6*. https://doi.org/10.1016/j.ijpx.2023.100188.

de Koster, S. A. L., Liu, L. X., Litster, J. D., & Smith, R. M. (2021). High-shear granulation: An investigation into granule breakage rates. *Advanced Powder Technology, 32*(5), 1390–1398. https://doi.org/10.1016/j.apt.2021.03.006.

Djuric, D., Vanmelkebeke, B., Kleinebudde, P., Remon, J., & Vervaet, C. (2009). Comparison of two twin-screw extruders for continuous granulation. *European Journal of Pharmaceutics and Biopharmaceutics, 71*(1), 155–160. https://doi.org/10.1016/j.ejpb.2008.06.033.

Eliasen, H., Schæfer, T., & Gjelstrup Kristensen, H. (1998). Effects of binder rheology on melt agglomeration in a high shear mixer. *International Journal of Pharmaceutics, 176*(1), 73–83. https://doi.org/10.1016/s0378-5173(98)00306-8.

Emori, H., Sakuraba, Y., Takahashi, K., Nishihata, T., & Mayumi, T. (1997). Prospective validation of high-shear wet granulation process by wet granule sieving method. II. utility of wet granule sieving method. *Drug Development and Industrial Pharmacy, 23*(2), 203–215. https://doi.org/10.3109/03639049709149794.

Ennis, B. J., & Litster, J. D. (1997). *Perry's chemical engineers' handbook* (pp. 20–56). McGraw-Hill.

Franke, M., Riedel, T., Meier, R., Schmidt, C., & Kleinebudde, P. (2023). Scale-up in twin-screw wet granulation: Impact of formulation properties. *Pharmaceutical Development and Technology, 28*(10), 948–961. https://doi.org/10.1080/10837450.2023.2276791.

Geldart, D. (1973). Types of gas fluidization. *Powder Technology, 7*(5), 285–292. https://doi.org/10.1016/0032-5910(73)80037-3.

Gong, X., Chang, S. Y., Osei-Yeboah, F., Paul, S., Perumalla, S. R., Shi, L., Sun, W. J., Zhou, Q., & Sun, C. C. (2015). Dependence of tablet brittleness on tensile strength and porosity. *International Journal of Pharmaceutics, 493*(1–2), 208–213. https://doi.org/10.1016/j.ijpharm.2015.07.050.

Gong, X., & Sun, C. C. (2015). A new tablet brittleness index. *European Journal of Pharmaceutics and Biopharmaceutics, 93*, 260–266. https://doi.org/10.1016/j.ejpb.2015.04.007.

References

Hapgood, K. P., Litster, J. D., & Smith, R. (2003). Nucleation regime map for liquid bound granules. *AIChE Journal, 49*(2), 350–361. https://doi.org/10.1002/aic.690490207.

Holm, P., Schaefer, T., & Kristensen, H. G. (1985). Granulation in high-speed mixers Part VI. effects of process conditions on power consumption and granule growth. *Powder Technology, 43*(3), 225–233. https://doi.org/10.1016/0032-5910(85)80003-6.

Hoornaert, F., Wauters, P. A. L., Meesters, G. M. H., Pratsinis, S. E., & Scarlett, B. (1998). Agglomeration behaviour of powders in a lodige mixer granulator. *Powder Technology, 96*(2), 116–128. https://doi.org/10.1016/S0032-5910(97)03364-0.

Hou, H., & Sun, C. C. (2008). Quantifying effects of particulate properties on powder flow properties using a ring shear tester. *Journal of Pharmaceutical Sciences, 97*(9), 4030–4039. https://doi.org/10.1002/jps.21288.

Huang, C. Y., & Sherry Ku, M. (2010). Prediction of drug particle size and content uniformity in low-dose solid dosage forms. *International Journal of Pharmaceutics, 383*(1–2), 70–80. https://doi.org/10.1016/j.ijpharm.2009.09.009.

Huang, J., Kaul, G., Utz, J., Hernandez, P., Wong, V., Bradley, D., Nagi, A., & O'Grady, D. (2010). A PAT approach to improve process understanding of high shear wet granulation through in-line particle measurement using FBRM C35. *Journal of Pharmaceutical Sciences, 99*(7), 3205–3212. https://doi.org/10.1002/jps.22089.

International Council for Harmonisation of Technical Requirements for Pharmaceuticals for Human Use (ICH). (2009). ICH harmonised tripartite guideline Q8(R2): Pharmaceutical development. https://database.ich.org/sites/default/files/Q8_R2_Guideline.pdf.

Iveson, S. M., & Litster, J. D. (1998). Growth regime map for liquid-bound granules. *AIChE Journal, 44*(7), 1510–1518. https://doi.org/10.1002/aic.690440705.

Iveson, S. M., Litster, J. D., & Ennis, B. J. (1996). Fundamental studies of granule consolidation Part 1: Effects of binder content and binder viscosity. *Powder Technology, 88*(1), 15–20. https://doi.org/10.1016/0032-5910(96)03096-3.

Iveson, S. M., Litster, J. D., Hapgood, K., & Ennis, B. J. (2001). Nucleation, growth and breakage phenomena in agitated wet granulation processes: A review. *Powder Technology, 117*(1–2), 3–39. https://doi.org/10.1016/S0032-5910(01)00313-8.

Kashani Rahimi, S., Paul, S., Calvin Sun, C., & Zhang, F. (2020). The role of the screw profile on granular structure and mixing efficiency of a high-dose hydrophobic drug formulation during twin screw wet granulation. *International Journal of Pharmaceutics, 575*. https://doi.org/10.1016/j.ijpharm.2019.118958.

Kayrak-Talay, D., & Litster, J. D. (2011). A priori performance prediction in pharmaceutical wet granulation: Testing the applicability of the nucleation regime map to a formulation with a broad size distribution and dry binder addition. *International Journal of Pharmaceutics, 418*(2), 254–264. https://doi.org/10.1016/j.ijpharm.2011.04.019.

Khorsheed, B., Gabbott, I., Reynolds, G. K., Taylor, S. C., Roberts, R. J., & Salman, A. D. (2019). Twin-screw granulation: Understanding the mechanical properties from powder to tablets. *Powder Technology, 341*, 104–115. https://doi.org/10.1016/j.powtec.2018.05.013.

Kleinebudde, P. (2004). Roll compaction/dry granulation: Pharmaceutical applications. *European Journal of Pharmaceutics and Biopharmaceutics, 58*(2), 317–326. https://doi.org/10.1016/j.ejpb.2004.04.014.

Knight, P. C., Johansen, A., Kristensen, H. G., Schæfer, T., & Seville, J. P. K. (2000). An investigation of the effects on agglomeration of changing the speed of a mechanical mixer. *Powder Technology, 110*(3), 204–209. https://doi.org/10.1016/s0032-5910(99)00259-4.

Levin, M. (2006). Wet granulation: End-point determination and scale-up. In J. Swarbrick (Ed.), *Encyclopedia of pharmaceutical technology* (3rd ed.) (pp. 4078–4098). Marcel Dekker.

Litster, J. D. (2003). Scaleup of wet granulation processes: Science not art. *Powder Technology, 130*(1–3), 35–40. https://doi.org/10.1016/s0032-5910(02)00222-x.

Litster, J. D., Hapgood, K. P., Michaels, J. N., Sims, A., Roberts, M., Kameneni, S. K., & Hsu, T. (2001). Liquid distribution in wet granulation: Dimensionless spray flux. *Powder Technology, 114*(1–3), 32–39. https://doi.org/10.1016/S0032-5910(00)00259-X.

Liu, L. X., Marziano, I., Bentham, A. C., Litster, J. D., & Howes, T. (2008). Effect of particle properties on the flowability of ibuprofen powders. *International Journal of Pharmaceutics, 362*(1–2), 109–117. https://doi.org/10.1016/j.ijpharm.2008.06.023.

Logan, R., & Briens, L. (2012). Investigation of the effect of impeller speed on granules formed using a PMA-1 high shear granulator. *Drug Development and Industrial Pharmacy, 38*(11), 1394–1404. https://doi.org/10.3109/03639045.2011.653361.

Lute, S. V., Dhenge, R. M., & Salman, A. D. (2018). Twin screw granulation: An investigation of the effect of barrel fill level. *Pharmaceutics, 10*(2). https://doi.org/10.3390/pharmaceutics10020067.

Macho, O., Gabrišová, Ľ., Guštafík, A., Jezso, K., Juriga, M., Kabát, J., & Blaško, J. (2023). The influence of wet granulation parameters on the compaction behavior and tablet strength of a hydralazine powder mixture. *Pharmaceutics, 15*(8), 2148. https://doi.org/10.3390/pharmaceutics15082148.

Miwa, A., Yajima, T., & Itai, S. (2000). Prediction of suitable amount of water addition for wet granulation. *International Journal of Pharmaceutics, 195*(1–2), 81–92. https://doi.org/10.1016/s0378-5173(99)00376-2.

Muthancheri, I., & Ramachandran, R. (2020). Mechanistic understanding of granule growth behavior in bi-component wet granulation processes with wettability differentials. *Powder Technology, 367*, 841–859. https://doi.org/10.1016/j.powtec.2020.04.016.

Nieuwmeyer, F., van der Voort Maarschalk, K., & Vromans, H. (2008). The consequences of granulate heterogeneity towards breakage and attrition upon fluid-bed drying. *European Journal of Pharmaceutics and Biopharmaceutics, 70*(1), 402–408. https://doi.org/10.1016/j.ejpb.2008.03.003.

Nieuwmeyer, F. J. S., Damen, M., Gerich, A., Rusmini, F., Van Der Voort Maarschalk, K., & Vromans, H. (2007a). Granule characterization during fluid bed drying by development of a near infrared method to determine water content and median granule size. *Pharmaceutical Research, 24*(10), 1854–1861. https://doi.org/10.1007/s11095-007-9305-5.

Nieuwmeyer, F. J. S., van der Voort Maarschalk, K., & Vromans, H. (2007b). Granule breakage during drying processes. *International Journal of Pharmaceutics, 329*(1–2), 81–87. https://doi.org/10.1016/j.ijpharm.2006.08.017.

Osei-Yeboah, F., Chang, S. Y., & Sun, C. C. (2016). A critical examination of the phenomenon of bonding area - bonding strength interplay in powder tableting. *Pharmaceutical Research, 33*(5), 1126–1132. https://doi.org/10.1007/s11095-016-1858-8.

Osei-Yeboah, F., Feng, Y., & Sun, C. C. (2014a). Evolution of structure and properties of granules containing microcrystalline cellulose and polyvinylpyrrolidone during high-shear wet granulation. *Journal of Pharmaceutical Sciences, 103*(1), 207–215. https://doi.org/10.1002/jps.23776.

Osei-Yeboah, F., & Sun, C. C. (2015a). Tabletability modulation through surface engineering. *Journal of Pharmaceutical Sciences, 104*(8), 2645–2648. https://doi.org/10.1002/jps.24532.

Osei-Yeboah, F., & Sun, C. C. (2015b). Validation and applications of an expedited tablet friability method. *International Journal of Pharmaceutics, 484*(1–2), 146–155. https://doi.org/10.1016/j.ijpharm.2015.02.061.

Osei-Yeboah, F., Zhang, M., Feng, Y., & Sun, C. C. (2014b). A formulation strategy for solving the overgranulation problem in high shear wet granulation. *Journal of Pharmaceutical Sciences, 103*(8), 2434–2440. https://doi.org/10.1002/jps.24066.

Patel, S., & Sun, C. C. (2016). Macroindentation hardness measurement - Modernization and applications. *International Journal of Pharmaceutics, 506*(1–2), 262–267. https://doi.org/10.1016/j.ijpharm.2016.04.068.

Paul, S., & Sun, C. C. (2017). Lubrication with magnesium stearate increases tablet brittleness. *Powder Technology, 309*, 126–132. https://doi.org/10.1016/j.powtec.2016.12.012.

Portier, C., Pandelaere, K., Delaet, U., Vigh, T., Kumar, A., Di Pretoro, G., De Beer, T., Vervaet, C., & Vanhoorne, V. (2020). Continuous twin screw granulation: Influence of process and formulation variables on granule quality attributes of model formulations. *International Journal of Pharmaceutics, 576*, 118981. https://doi.org/10.1016/j.ijpharm.2019.118981.

Portier, C., Vervaet, C., & Vanhoorne, V. (2021). Continuous twin screw granulation: A review of recent progress and opportunities in formulation and equipment design. *Pharmaceutics, 13*(5), 668. https://doi.org/10.3390/pharmaceutics13050668.

Poskart, M. B. (1988). Evaluation of a New Pelletiser. BS Thesis. Australia: Department of Chemical Engineering, University of Queensland.

Rantanen, J., Wikström, H., Turner, R., & Taylor, L. S. (2005). Use of in-line near-infrared spectroscopy in combination with chemometrics for improved understanding of pharmaceutical processes. *Analytical Chemistry, 77*(2), 556–563. https://doi.org/10.1021/ac048842u.

Rohrs, B. R., Amidon, G. E., Meury, R. H., Secreast, P. J., King, H. M., & Skoug, C. J. (2006). Particle size limits to meet USP content uniformity criteria for tablets and capsules. *Journal of Pharmaceutical Sciences, 95*(5), 1049–1059. https://doi.org/10.1002/jps.20587.

Rowe, R. C., & Sadeghnejad, G. R. (1987). The rheology of microcrystalline cellulose powder/water mixes - measurement using a mixer torque rheometer. *International Journal of Pharmaceutics, 38*(1–3), 227–229. https://doi.org/10.1016/0378-5173(87)90118-9.

Schæfer, T., & Mathiesen, C. (1996). Melt pelletization in a high shear mixer. IX. effects of binder particle size. *International Journal of Pharmaceutics, 139*(1–2), 139–148. https://doi.org/10.1016/0378-5173(96)04548-6.

Seem, T. C., Rowson, N. A., Ingram, A., Huang, Z., Yu, S., de Matas, M., Gabbott, I., & Reynolds, G. K. (2015). Twin screw granulation—A literature review. *Powder Technology, 276*, 89–102. https://doi.org/10.1016/j.powtec.2015.01.075.

Sherrington, P. J., & Oliver, O. (1981). *Granulation*. Heyden.

Shi, L., Feng, Y., & Sun, C. C. (2010). Roles of granule size in over-granulation during high shear wet granulation. *Journal of Pharmaceutical Sciences, 99*(8), 3322–3325. https://doi.org/10.1002/jps.22118.

Shi, L., Feng, Y., & Sun, C. C. (2011a). Initial moisture content in raw material can profoundly influence high shear wet granulation process. *International Journal of Pharmaceutics, 416*(1), 43–48. https://doi.org/10.1016/j.ijpharm.2011.05.080.

Shi, L., Feng, Y., & Sun, C. C. (2011b). Massing in high shear wet granulation can simultaneously improve powder flow and deteriorate powder compaction: A double-edged sword. *European Journal of Pharmaceutical Sciences, 43*(1–2), 50–56. https://doi.org/10.1016/j.ejps.2011.03.009.

Shi, L., Feng, Y., & Sun, C. C. (2011c). Origin of profound changes in powder properties during wetting and nucleation stages of high-shear wet granulation of microcrystalline cellulose. *Powder Technology, 208*(3), 663–668. https://doi.org/10.1016/j.powtec.2011.01.006.

Stauffer, F., Vanhoorne, V., Pilcer, G., Chavez, P.-F., Vervaet, C., & De Beer, T. (2019). Managing API raw material variability during continuous twin-screw wet granulation. *International Journal of Pharmaceutics, 561*, 265–273. https://doi.org/10.1016/j.ijpharm.2019.03.012.

Sun, C. C. (2008). Mechanism of moisture induced variations in true density and compaction properties of microcrystalline cellulose. *International Journal of Pharmaceutics, 346*(1–2), 93–101. https://doi.org/10.1016/j.ijpharm.2007.06.017.

Sun, C. C. (2009a). Improving powder flow properties of citric acid by crystal hydration. *Journal of Pharmaceutical Sciences, 98*(5), 1744–1749. https://doi.org/10.1002/jps.21554.

Sun, C. C. (2009b). Materials science tetrahedron—A useful tool for pharmaceutical research and development. *Journal of Pharmaceutical Sciences, 98*(5), 1671–1687. https://doi.org/10.1002/jps.21552.

Sun, C. C. (2010). Setting the bar for powder flow properties in successful high speed tableting. *Powder Technology, 201*(1), 106–108. https://doi.org/10.1016/j.powtec.2010.03.011.

Sun, C. C. (2011). Decoding powder tabletability: Roles of particle adhesion and plasticity. *Journal of Adhesion Science and Technology, 25*(4–5), 483–499. https://doi.org/10.1163/016942410×525678.

Sun, C. C. (2013). Cocrystallization for successful drug delivery. *Expert Opinion on Drug Delivery, 10*(2), 201–213. https://doi.org/10.1517/17425247.2013.747508.

Sun, C. C. (2016). Quantifying effects of moisture content on flow properties of microcrystalline cellulose using a ring shear tester. *Powder Technology, 289*, 104–108. https://doi.org/10.1016/j.powtec.2015.11.044.

Sun, C. C. (2017). Microstructure of tablet—pharmaceutical significance, assessment, and engineering. *Pharmaceutical Research, 34*(5), 918–928. https://doi.org/10.1007/s11095-016-1989-y.

Tan, B. M. J., Loh, Z. H., Soh, J. L. P., Liew, C. V., & Heng, P. W. S. (2014). Distribution of a viscous binder during high shear granulation—Sensitivity to the method of delivery and its impact on product properties. *International Journal of Pharmaceutics, 460*(1–2), 255–263. https://doi.org/10.1016/j.ijpharm.2013.11.020.

Tardos, G. I., Khan, M. I., & Mort, P. R. (1997). Critical parameters and limiting conditions in binder granulation of fine powders. *Powder Technology, 94*(3), 245–258. https://doi.org/10.1016/S0032-5910(97)03321-4.3.245.258.

Taylor, L. S. (2015). Physical stability and crystallization inhibition. In J. Swarbrick (Ed.), *Pharmaceutical sciences encyclopedia* (pp. 179–217). John Wiley & Sons, Inc. doi:10.1002/9780470571224.pse526.

Teng, Y., Qiu, Z., & Wen, H. (2009). Systematical approach of formulation and process development using roller compaction. *European Journal of Pharmaceutics and Biopharmaceutics, 73*(2), 219–229. https://doi.org/10.1016/j.ejpb.2009.04.008.

Thompson, M. R. (2015). Twin screw granulation-review of current progress. *Drug Development and Industrial Pharmacy, 41*(8), 1223–1231. https://doi.org/10.3109/03639045.2014.983931.

Thompson, M. R., & Sun, J. (2010). Wet granulation in a twin-screw extruder: Implications of screw design. *Journal of Pharmaceutical Sciences, 99*(4), 2090–2103. https://doi.org/10.1002/jps.21973.

Ullah, M., Hussain, I., & Sun, C. C. (2016). The development of carbamazepine-succinic acid cocrystal tablet formulations with improved in vitro and in vivo performance. *Drug Development and Industrial Pharmacy, 42*(6), 969–976. https://doi.org/10.3109/03639045.2015.1096281.

Van Den Dries, K., De Vegt, O. M., Girard, V., & Vromans, H. (2003). Granule breakage phenomena in a high shear mixer; influence of process and formulation variables and consequences on granule homogeneity. *Powder Technology, 133*(1–3), 228–236. https://doi.org/10.1016/S0032-5910(03)00106-2.

van den Dries, K., & Vromans, H. (2002). Relationship between inhomogeneity phenomena and granule growth mechanisms in a high-shear mixer. *International Journal of Pharmaceutics, 247*(1–2), 167–177. https://doi.org/10.1016/s0378-5173(02)00419-2.

Vandevivere, L., Denduyver, P., Portier, C., Häusler, O., De Beer, T., Vervaet, C., & Vanhoorne, V. (2020). Influence of binder attributes on binder effectiveness in a continuous twin screw wet granulation process via wet and dry binder addition. *International Journal of Pharmaceutics, 585*, 119466. https://doi.org/10.1016/j.ijpharm.2020.119466.

Vonk, P., Guillaume, C. P. F., Ramaker, J. S., Vromans, H., & Kossen, N. W. F. (1997). Growth mechanisms of high-shear pelletisation. *International Journal of Pharmaceutics, 157*(1), 93–102. https://doi.org/10.1016/s0378-5173(97)00232-9.

Vreeman, G., & Sun, C. C. (2021). Mean yield pressure from the in-die Heckel analysis is a reliable plasticity parameter. *International Journal of Pharmaceutics: X, 3*, 100094. https://doi.org/10.1016/j.ijpx.2021.100094.

Wade, J. B., Martin, G. P., & Long, D. F. (2015). Controlling granule size through breakage in a novel reverse-phase wet granulation process; the effect of impeller speed and binder liquid viscosity. *International Journal of Pharmaceutics, 478*(2), 439–446. https://doi.org/10.1016/j.ijpharm.2014.11.067.

Wang, L. F., Zhao, L., Hong, Y. L., Shen, L., & Lin, X. (2023). Attribute transmission and effects of diluents and granulation liquids on granule properties and tablet quality for high shear wet granulation and tableting process. *International Journal of Pharmaceutics, 642*. https://doi.org/10.1016/j.ijpharm.2023.123177.

Wang, Y., Cao, J., Zhao, X., Liang, Z., Qiao, Y., Luo, G., & Xu, B. (2022). Using a material library to understand the change of tabletability by high shear wet granulation. *Pharmaceutics, 14*(12). https://doi.org/10.3390/pharmaceutics14122631.

Whitaker, M., Baker, G. R., Westrup, J., Goulding, P. A., Rudd, D. R., Belchamber, R. M., & Collins, M. P. (2000). Application of acoustic emission to the monitoring and end point determination of a high shear granulation process. *International Journal of Pharmaceutics, 205*(1–2), 79–91. https://doi.org/10.1016/S0378-5173(00)00479-8.

Wu, S. J., & Sun, C. (2007). Insensitivity of compaction properties of brittle granules to size enlargement by roller compaction. *Journal of Pharmaceutical Sciences, 96*(5), 1445–1450. https://doi.org/10.1002/jps.20929.

Xiao, B., Zhang, J., Geng, L., Tang, X., Wang, Y., Yin, T., Zhang, Y., Gou, J., & He, H. (2022). Studies on the influence of high-shear granulation process on the compressibility of microcrystalline cellulose. *International Journal of Pharmaceutics, 625*, 122075. https://doi.org/10.1016/j.ijpharm.2022.122075.

You, Y., Guo, J., Li, G., Zheng, Z., Li, Y., & Lü, X. (2022). Effects of process parameters on the growth behavior and granule size distribution of iron ore mixtures in a novel high-shear granulator. *International Journal of Minerals, Metallurgy and Materials, 29*(12), 2152–2161. https://doi.org/10.1007/s12613-021-2407-y.

Yuan, J., Shi, L., Sun, W. J., Chen, J., Zhou, Q., & Sun, C. C. (2013). Enabling direct compression of formulated Danshen powder by surface engineering. *Powder Technology, 241*, 211–218. https://doi.org/10.1016/j.powtec.2013.03.010.

Zeng, X. M., Martin, G. P., & Marriott, C. (2001). *Particulate interactions in dry powder formulations for inhalation*. Taylor & Francis.

Zhang, D., Flory, J. H., Panmai, S., Batra, U., & Kaufman, M. J. (2002). Wettability of pharmaceutical solids: Its measurement and influence on wet granulation. *Colloids and Surfaces A: Physicochemical and Engineering Aspects, 206*(1–3), 547–554. https://doi.org/10.1016/S0927-7757(02)00091-2.

Zhang, Y., Liu, T., Kashani-Rahimi, S., & Zhang, F. (2021). A review of twin screw wet granulation mechanisms in relation to granule attributes. *Drug Development and Industrial Pharmacy, 47*(3), 349–360. https://doi.org/10.1080/03639045.2021.1879844.

Zhou, Q., Shi, L., Chattoraj, S., & Sun, C. C. (2012). Preparation and characterization of surface-engineered coarse microcrystalline cellulose through dry coating with silica nanoparticles. *Journal of Pharmaceutical Sciences, 1*(11), 4258–4266. https://doi.org/10.1002/jps.23301.

Zhou, Q., Shi, L., Marinaro, W., Lu, Q., & Sun, C. C. (2013). Improving manufacturability of an ibuprofen powder blend by surface coating with silica nanoparticles. *Powder Technology, 249*, 290–296. https://doi.org/10.1016/j.powtec.2013.08.031.

Zhou, Z., Li, W., Sun, W. J., Lu, T., Tong, H. H. Y., Sun, C. C., & Zheng, Y. (2016). Resveratrol cocrystals with enhanced solubility and tabletability. *International Journal of Pharmaceutics, 509*(1–2), 391–399. https://doi.org/10.1016/j.ijpharm.2016.06.006.

Žižek, K., Hraste, M., & Gomzi, Z. (2013). High shear granulation of dolomite – I: Effect of shear regime on process kinetics. *Chemical Engineering Research and Design, 91*(1), 70–86. https://doi.org/10.1016/j.cherd.2012.06.014.

CHAPTER 5

Wet granulation and chemical stability of drug products

Sherif I.F. Badawy[a]

[a]*Drug Product Science & Technology, Bristol-Myers Squibb Co., New Brunswick, NJ, United States*

5.1 Introduction

The wet granulation process has been reported to affect chemical stability of drug molecules. In some cases, the effect of the wet granulation process on chemical stability manifests itself by enhancing degradation reactions during processing, whereby an increase in degradant content is observed at the end of manufacturing. Exposure of a moisture-sensitive compound to the high water activity experienced in aqueous wet granulation might be the underlying mechanism. Alternatively, degradation in the solution phase, as fraction of the active dissolves in the granulating liquid, can be the causal mechanism. Actives with high solubility in the granulating liquid are expected to be more susceptible to the latter mechanism. An example of increased degradation during manufacturing was provided by lovastatin, which showed higher degradant levels in tablets manufactured by a wet granulation process in initial testing compared to the dry granulation and direct compression processes (Saquib, Sheikh, Ahmed, Usmanghani, & Khattak, 2015).

Degradation during manufacturing, however, is not observed frequently in wet granulation because of the short processing time relative to reaction rates even for moisture sensitive compounds (Farag Badawy, 2001; Badawy et al., 2016; Badawy & Pandey, 2017; Badawy, Vickery, Shah, & Hussain, 2004; Farag Badawy, Williams, & Gilbert, 1999). Enhanced degradation caused by wet granulation is observed more commonly on storage (accelerated and long-term stability) and not at the end of manufacturing. While this might be attributed in some cases to high level of residual water in the drug product, it is more frequently the result of creation of high-energy disordered states during wet granulation. Rate of the degradation reaction in the high-energy phase is significantly higher than the crystalline phase, and therefore drug degradation in the solid state usually takes place in disordered regions or in crystal defects (Waterman et al., 2002). The higher molecular mobility in those disordered regions results in faster degradation rate than in the crystal lattice. In addition, disordered regions have higher water content because of absorbed water, which further enhances degradation rate in those less ordered regions for reactions in which water acts as a reactant (Narang, Desai, & Badawy, 2012). Amorphous regions of the active can potentially form in wet granulation, and although the fraction of the drug in amorphous state might be small, this still leads to a measurable impact on stability because the acceptable specification limit of a drug product degradant is usually very low (Narang, Desai and Badawy, 2012). In some cases, wet granulation can result in the formation of a high-energy state of one or more excipients, which increases formulation hygroscopicity and has a negative impact on the stability of moisture-sensitive compounds.

The effect of wet granulation on product stability also can be caused by enhanced contact between drug and excipients. Partial dissolution of an excipient and/or the drug substance during wet granulation and subsequent crystallization during drying (albeit to the same initial form) results in higher surface area of drug/excipient contact. This intimate contact of the drug and the excipient enhances rate of drug degradation caused by drug-excipient interaction and results in a less stable drug product. The intimate contact of drug and excipients resulting from wet granulation was leveraged to enhance stability in some cases. For formulations containing a pH modifier, dissolution of the buffer components in the granulating liquid results in better distribution of the pH modifier in the formulation and in a more effective pH control, therefore maximizing drug product stability (Farag Badawy and Hussain, 2007).

5.2 Reduced stability by wet granulation

Reduced stability of wet granulated products was observed in many studies. The rate of impurity formation in crystalline cenicriviroc mesylate tablets stored at 40°C/75% RH was higher in tablets manufactured by wet granulation compared to tablets of similar composition prepared by dry granulation (Menning and Dalziel, 2013). Enhanced degradation of drug products manufactured by wet granulation on stability is frequently the result of phase change during granulation. The reduced stability of Abbott-232 tablets was attributed to solution-mediated phase transformation from the starting anhydrate form to amorphous Abbott-232 during wet granulation. Because of the low drug loading, a wet granulation process was used, using an aqueous solution of Abbott-232 as the granulating liquid. Studies at a higher drug loading showed that the wet granulated product lacked the powder X-ray diffraction (PXRD) peaks characteristic of the crystalline anhydrate drug substance, indicating that the wet processing conditions resulted in predominantly amorphous form in the drug product. The drug did not crystallize during drying of the wet granulated mass and instead formed an amorphous phase, which slowly crystallized to the anhydrate form during the course of the stability study. Product stability was improved when a direct compression process was used, preventing the solution-mediated process induced phase transformation (Wardrop et al., 2006).

In another study, stability of two acidic drugs, piroxicam and lornoxicam, showed enhanced degradation in formulations manufactured by wet granulation. The wet granulation induced the interaction between the acidic drug and basic excipients in the formulation (sodium bicarbonate and dicalcium phosphate). Consequently, solid-state analysis by PXRD and vibrational spectroscopy (infrared and Raman spectroscopy) showed the formation of the sodium salt of the two drugs in the wet granulated product. The enhanced degradation on storage was attributed to the wet granulation process-induced salt formation (Christensen, Nielsen, Rantanen, Cornett, & Bertelsen, 2014).

Stability of the water soluble triazine derivative (compound I; CPD-I) was affected negatively by wet granulation (Badawy et al., 2004). CPD-I is the benzenesulfonate salt of a weakly basic triazine derivative with pK_a of 4.0. Because of its weakly basic nature, solutions of CPD-I (salt of a weak base) have low pH and a solution at a concentration of $\sim$ 3.5 mg/mL has a pH of 2.2. CPD-I has a relatively low melting point of $\sim$ 91°C. The main degradation pathway for CPD-I in solution is through acid catalyzed hydrolysis with pH of maximum stability between 7 and 8. Acid catalyzed hydrolysis of CPD-I also was observed in the solid state. Because of the acidic nature of this salt, the microenvironment pH of the

Table 5.1 Experimental design for the study of formulation and processing variables on tablet stability of CPD-I

Run number	Amount of water used for granulation (g)	Ball milling time (hours)	Concentration of sodium carbonate (%)
1	2	0	5
2	0	0	0
3	0	2	5
4	2	2	0
5	0	0	5
6	2	2	5
7	2	0	0
8	0	2	0
9	0	0	0
10	0	0	5

From Badawy, S., Vickery, R., Shah, K., & Hussain, M. (2004). Effect of processing and formulation variables on the stability of a salt of a weakly basic drug candidate. Pharmaceutical Development and Technology, 9(3), 239–245.

drug particles is lower than the pH of maximum stability for CPD-I. Stability of CPD-I in drug product also was found to dependent on the manufacturing process and similar formulations showed different stability profiles depending on the process selected for dosage form manufacture.

A two-level full factorial design was used to study the effect of three variables on CPD-I tablet stability: aqueous wet granulation, ball milling of the drug substance with microcrystalline cellulose, and inclusion of sodium carbonate in the formulation. The factorial design is shown in Table 5.1. In addition to experimental design batches, additional batches were prepared for stability testing. One batch without pH modifier was manufactured by dissolving the drug substance in water and then using the aqueous drug solution to granulate microcrystalline cellulose. Another batch was similar to run 1 in Table 5.1 but containing 5% sodium bicarbonate instead of sodium carbonate. Wet granulated batches without pH modifier also were manufactured using different amounts of granulating water and another batch was manufactured by preblending the drug substance with colloidal silicon dioxide before wet granulation. Stability of tablets from the different batches was evaluated at 40°C and 40°C/75% RH. Batches were manufactured by mixing CPD-I with microcrystalline cellulose and sodium carbonate, when applicable, in the bowl of the Bohle minigranulator (L.B. Bohle, Bristol, PA) for 2 min at impeller speed of 500 rpm and chopper speed of 2500 rpm. For batches without wet granulation, the blend was discharged from the granulator and then mixed with 0.3 g magnesium stearate for 5 min using a Turbula T2C mixer (Willy A. Bachofen AG, Basel, Switzerland) at 42 rpm. Tablets containing 50 mg equivalent of the free base were compressed using a Carver press (Fred S. Carver Inc., Menomonee Falls, WI) to a hardness of 8–12 SCU. For wet granulated batches, the blend in the high shear granulator was granulated with water using an impeller speed of 500 rpm and chopper speed of 2500 rpm. The granulation then was dried in an oven at 50°C for 2 h. The dried granulation was hand-screened through a 20-mesh screen and then lubricated with magnesium stearate. Results of stability testing for tablets from the experimental design batches are shown in Table 5.2. Preliminary data analysis using a statistical model with all three main effects and

Table 5.2 Total degradant concentration in tablets from the experimental design batches of CPD-I

Run number	% Total degradant at 40°C/75% RH				% Total degradant at 40°C			
	Initial	4 weeks	8 weeks	13 weeks	Initial	4 weeks	8 weeks	13 weeks
1	0	0.01	0	0.02	0	0.04	0.03	0.05
2	0.01	0.78	1.48	2.73	0.01	0.09	0.11	0.11
3	0	0.49	0.92	1.57	0	0.03	0.05	0.07
4	0.01	1.43	3.01	5.44	0.01	0.45	0.72	1.00
5	0	0.13	0.21	0.31	0	ND	0	0
6	0.01	0.01	0.02	0.01	0.01	ND	0.05	0.08
7	0.19	1.47	2.18	3.27	0.19	ND	0.62	0.84
8	0.02	0.55	0.83	1.23	0.02	ND	0.03	0.05
9	0.05	0.66	1.07	1.83	0.05	ND	ND	ND
10	0.02	0.32	0.54	0.85	0.02	ND	ND	ND

From Badawy, S., Vickery, R., Shah, K., & Hussain, M. (2004). Effect of processing and formulation variables on the stability of a salt of a weakly basic drug candidate. Pharmaceutical Development and Technology, 9(3), 239–245.

two-way interactions indicated that the magnitude of the ball milling-wet granulation and ball milling-sodium carbonate interaction terms is insignificant. As a result, data subsequently was analyzed with a model composed of an intercept, main effects, and the wet-granulation-sodium carbonate two-way interaction term.

As expected for compounds undergoing degradation via hydrolysis, more degradation was observed in tablets stored at 40°C/75% RH compared to the 40°C condition. Moreover, wet granulation decreased tablet stability and enhanced degradation of CPD-I in the tablets devoid of sodium carbonate at 40°C ($P < 0.0001$) and 40°C/75% RH ($P = 0.01$) (Figure 5.1). CPD-I does not crystallize from aqueous solution but rather forms an amorphous material when the water is evaporated. Thus, during wet granulation, the fraction of CPD-I dissolving in the granulating liquid is not expected to re-crystallize during drying resulting in partial loss of crystallinity. This may explain the negative effect of wet granulation on tablet stability as amorphous material is often less stable. Molecules in the amorphous regions have higher mobility, which facilitates collision between reacting molecules in a bimolecular reaction such as hydrolysis (Ahlneck and Zografi, 1990; Shalaev and Zografi, 1996; Waterman et al., 2002). The higher energy state of the molecules in the amorphous domains also reduces the energy required by those molecules to reach the transition state compared to molecules in the crystal lattice. The significant degradation in tablets manufactured by completely dissolving the drug substance in the granulating liquid is a further illustration of the negative effect of wet granulation on tablet stability. As expected, X-ray powder diffraction indicated that CDP-I exists in a predominantly amorphous form in the latter granulation (Figure 5.2).

The rate of degradation of CPD-I on accelerated stability also was dependent on the amount of water used for granulation (Figure 5.3) despite the similar granulation moisture content at the end of drying. This was attributed to the higher amorphous content resulting from the increased fraction of the active dissolving in the larger volume of granulating liquid as the fraction dissolving during granulation forms

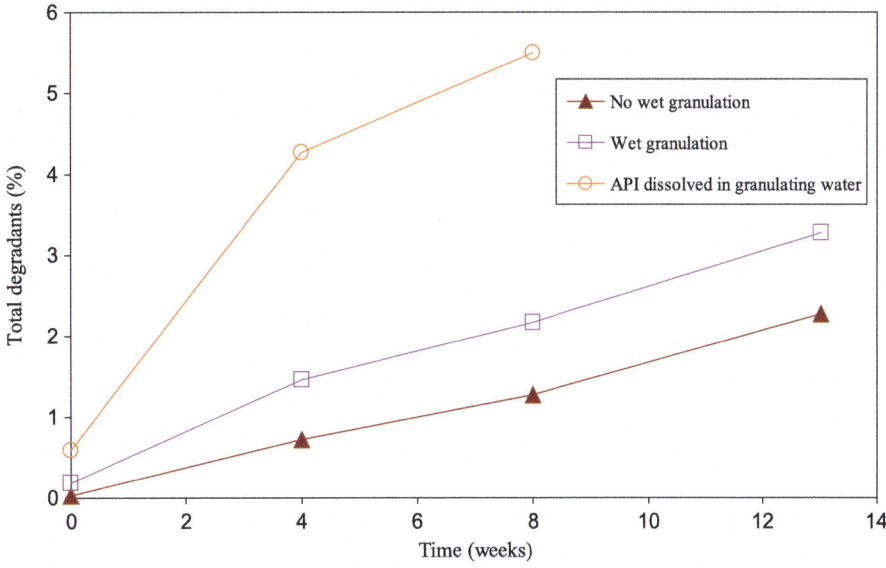

FIGURE 5.1

Effect of wet granulation process on CPD-I stability at 40°C/75% RH.

Adapted from Badawy, S., Vickery, R., Shah, K., & Hussain, M. (2004). Effect of processing and formulation variables on the stability of a salt of a weakly basic drug candidate. Pharmaceutical Development and Technology, 9(3), 239–245.

amorphous material after drying as shown above. Degradation rate also was enhanced by preblending the drug substance with colloidal silicon dioxide prior to wet granulation (Figure 5.4). This is not because of chemical catalysis by colloidal silicon dioxide, as the rate of CPD-I degradation is solution was not affected by the addition of colloidal silicon dioxide (Figure 5.4). The effect of colloidal silicon dioxide was attributed to a physical mechanism. Drug substance particles were covered with the nanosized colloidal silicon dioxide particles during preblending. The hygroscopic colloidal silicon dioxide on the surface of the drug substance particles enhances their interaction with the granulating water, which eventually results in greater formation of amorphous material.

Unlike wet granulation, ball milling of the drug substance with microcrystalline cellulose did not have significant effect on tablet stability at 40°C ($P = 0.59$) or at 40°C/75% RH ($P = 0.23$). This suggests that no significant disruption of crystallinity was caused by ball milling despite the low crystal lattice energy of this drug substance, as disruption of crystallinity is expected to enhance reactivity in the solid dosage form similar to the observation with wet granulation. Mechanical stress in ball milling was not as effective as wet granulation at crystallinity disruption. In the case of CPD-I, the formation of amorphous material during wet granulation is caused by dissolution of the drug substance in the granulating water rather than mechanical stress. Dissolution of the lower energy polymorph in the granulating liquid is a common mechanism for the formation of a high-energy form that is kinetically trapped during drying (Morris, Griesser, Eckhardt, & Stowell, 2001). High aqueous solubility is therefore a risk factor for the reduced stability of a drug product manufactured by wet granulation. In contrast to CPD-I, CPD-II showed similar degradation rate for drug products manufactured by dry or wet granulation

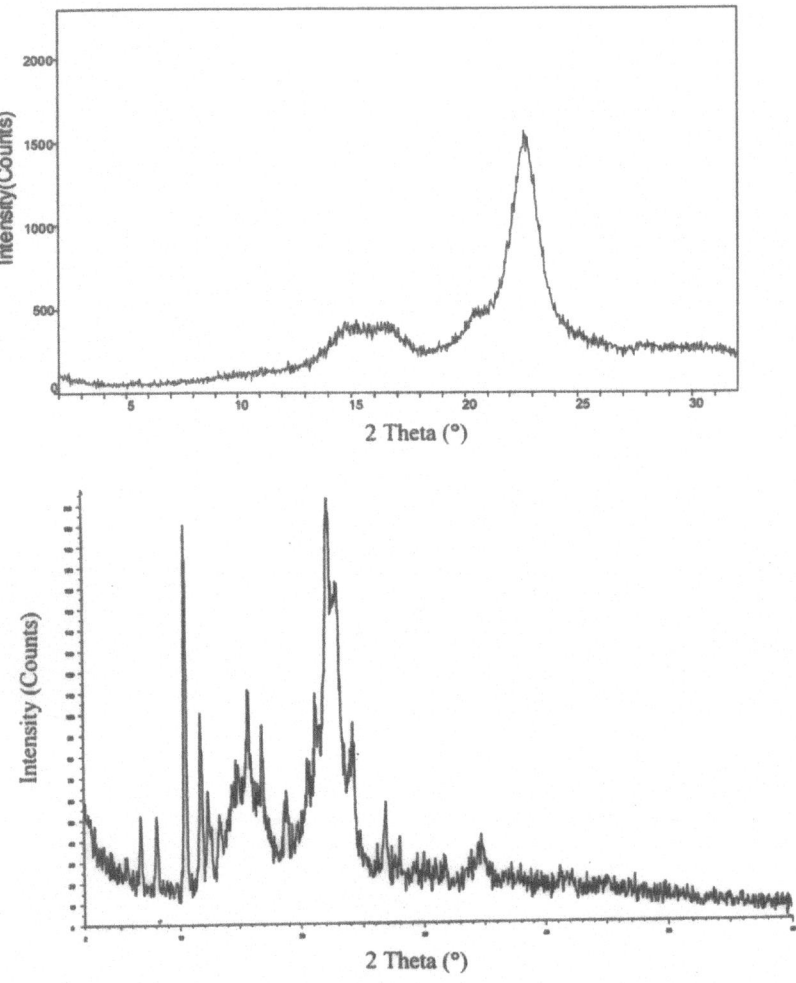

FIGURE 5.2

Powder X-ray diffraction pattern of granulation manufactured by dissolving CPD-I drug substance in granulating water (top) and blend manufactured by dry blending (no wet granulation, bottom).

From Badawy, S., Vickery, R., Shah, K., & Hussain, M. (2004). Effect of processing and formulation variables on the stability of a salt of a weakly basic drug candidate. Pharmaceutical Development and Technology, 9(3), 239–245.

(Figure 5.5). CPD-II is a moisture sensitive ester compound with a higher first order rate constant of hydrolysis in solution than CPD-I around their expected solid state microenvironmental pH (Figure 5.5). CPD-II, however, is a poorly water soluble compound and no solid state transformation was observed during wet granulation. Stable products for moisture sensitive compounds can be manufactured successfully by wet granulation if solid state transformation and formation of high energy solids do not take place during granulation (Badawy et al., 2014, 2016).

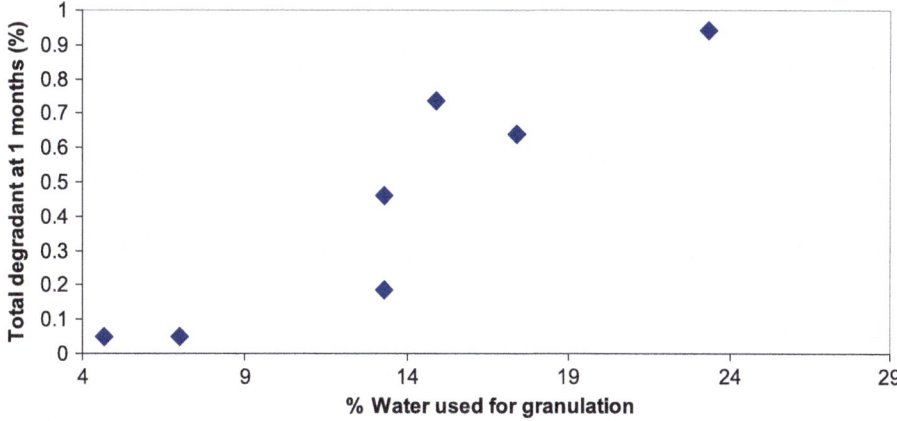

FIGURE 5.3

Effect of water amount used for granulation on stability of CPD-I tablets at 40°C/75% RH.

From Badawy, S., & Pandey, P. (2017). Design, development, and scale-up of the high shear wet granulation process. In Qiu, C., Zhang, Y., & Mantri (Eds.), Developing solid oral dosage forms (2nd ed.): Academic Press.

5.3 Enhanced stability by wet granulation

Wet granulation also was reported to enhance the stability of drug formulations in some cases. Typically, the enhanced stability by wet granulation is stemming from its ability to create more intimate contact between the drug substance and a stabilizing excipient. pH modifiers are commonly used stabilizers for compounds with pH-dependent degradation kinetics. Degradation rate and profile of those compounds are modulated by the microenvironmental pH of the solid dosage form. This provides the opportunity to improve stability by incorporating a pH modifier in the formulation in order to provide the optimal microenvironmental pH for the drug and therefore maximize its stability. The drug microenvironmental pH is determined by the type and the concentration of the pH modifier used in the formulation (Farag Badawy and Hussain, 2007).

Stability of formulations containing pH modifiers also was found to be dependent on the process used for their manufacture. Different degradation rates were observed for moexipril hydrochloride formulations containing alkalinizing (basic) agents and prepared by different manufacturing processes. Alkalinizing agents were found to destabilize moexipril hydrochloride in dry powder mixtures, however, the same alkalinizing agents improved its stability in wet granulated mixtures. The authors attributed the stabilization to the neutralization of the acidic drug by the alkalinizing agent or to the formation of more stable cationic salts during wet granulation (Gu, Strickley, Chi, & Chowhan, 1990). In another example, the rate of peroxide-induced N-oxide formation was found to be reduced by lowering the pH in solution and excipient slurries. Formulations manufactured by wet granulation using citric acid solution as the granulating liquid similarly were shown to be substantially stabilized with respect to peroxide-induced oxidation (Freed et al., 2008).

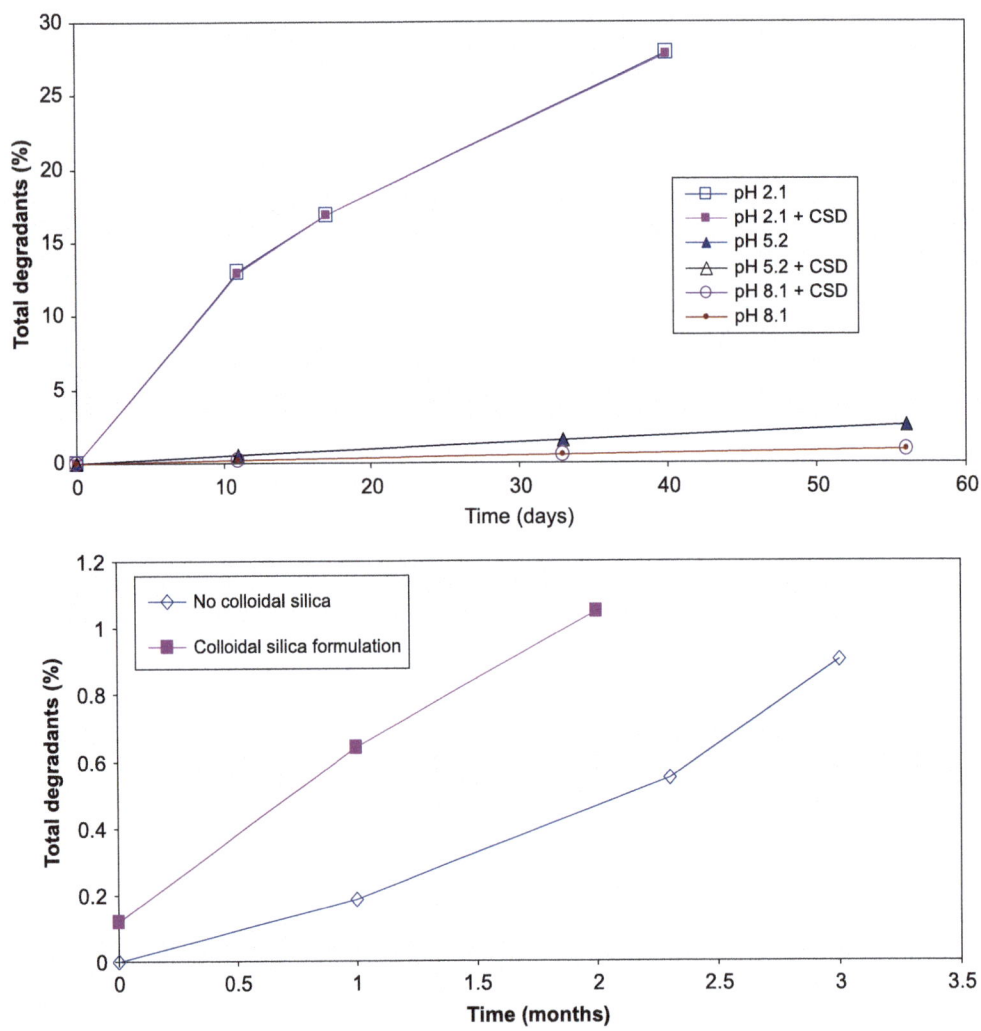

FIGURE 5.4

Effect of colloidal silicon dioxide (CSD) on CPD-I stability at 40°C/75% RH. Top, no effect of CSD on solution stability; bottom, increased degradation rate in tablets with CSD.

5.3.1 Triazine Derivative (CPD-1)

While the rate of acid catalyzed hydrolysis of CPD-I was enhanced by wet granulation compared to a dry blend in formulations without pH modifier, the opposite effect of wet granulation was observed in formulations containing sodium carbonate as a pH modifier. Addition of sodium carbonate in the formulation enhanced tablet stability and decreased the rate of degradants formation at 40°C ($P < 0.0001$) and 40°C/75% RH ($P < 0.0001$) (Figure 5.6) (Badawy et al., 2004). Sodium carbonate increases

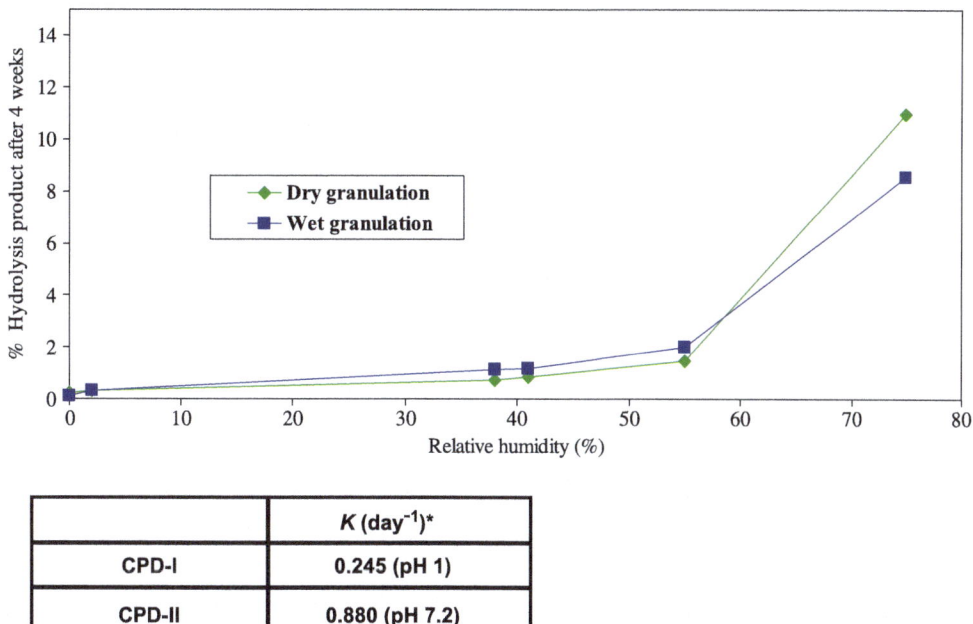

FIGURE 5.5

Stability of tablet formulation of a moisture sensitive ester compound (CPD-II) showing comparable stability of tablets manufactured by wet and dry granulation at 40°C/75% RH.

the microenvironment pH thus diminishing the rate of acid catalyzed hydrolytic degradation. The stabilizing effect of the carbonate was more pronounced when the wet granulation process was used for tablet manufacture as shown by the significant interaction term between sodium carbonate and wet granulation at 40°C ($P < 0.0001$) and 40°C/75% RH ($P < 0.0001$) (Figure 5.4). Tablets containing sodium carbonate and manufactured by a dry process were less stable than those manufactured by wet granulation. The use of wet granulation probably results in a better distribution of the sodium carbonate in the formulation resulting in more effective pH control. Thus, the use of wet granulation enhances the pH modifying effect of the carbonate, which appears to outweigh the undesirable effect of crystallinity disruption caused by wet granulation. In other words, the loss of crystallinity appears to have no negative effect on stability when the microenvironment pH is not sufficiently acidic to promote the acid catalyzed reaction. In addition, the hydrolysis reaction is slowed down considerably by increasing the microenvironment pH to the point that the higher water activity at the 40°C/75% RH condition did not result in any significant degradation in those tablets containing sodium carbonate compared to the 40°C dry condition. Sodium bicarbonate was less effective than sodium carbonate in stabilizing the tablets, suggesting that the higher microenvironment pH expected for the sodium carbonate formulation was needed for enhanced tablet stability (Figure 5.7) (Badawy et al., 2004).

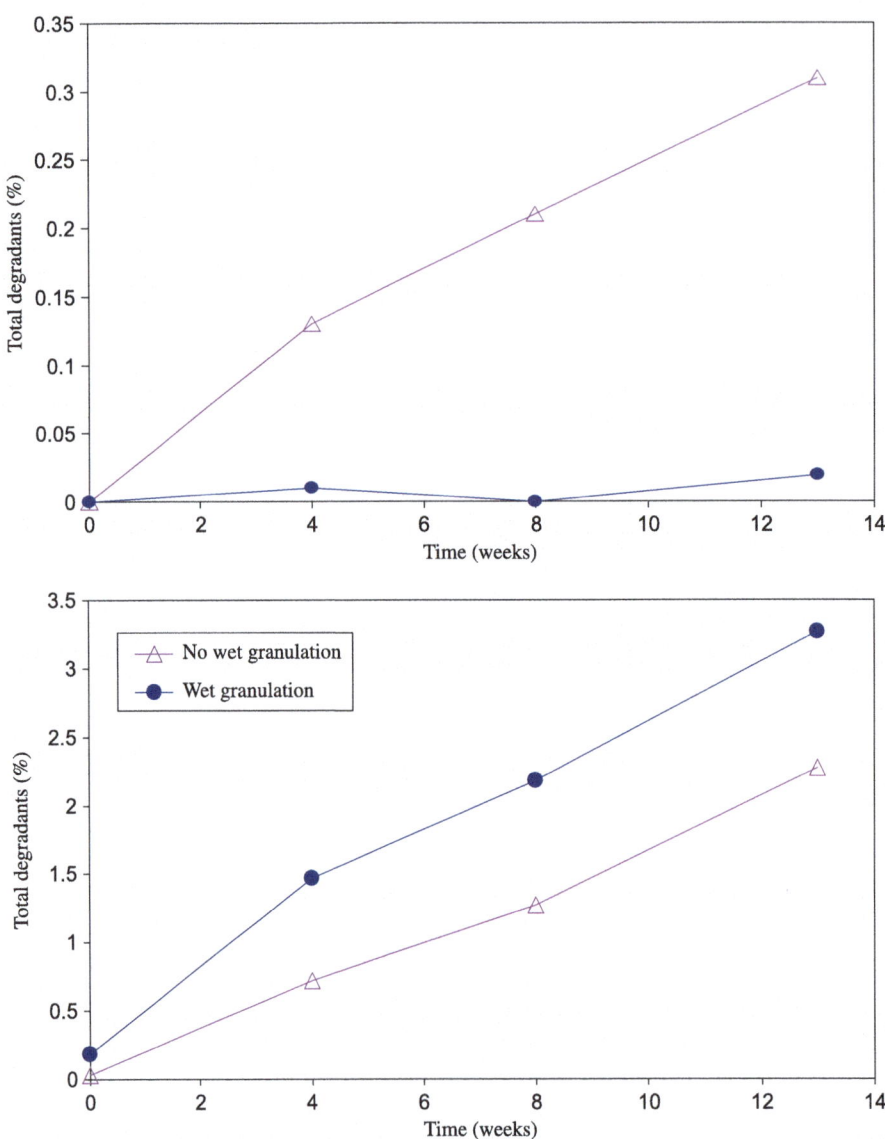

FIGURE 5.6

Formulation-dependent effect of wet granulation on stability of CPD-I tablets at 40°C/75% RH. Formulation containing sodium carbonate (top) and formulation without sodium carbonated (bottom). Wet granulation enhances distribution of the pH modifier, which outweighs its negative effect on crystallinity.

Adapted from Badawy, S., Vickery, R., Shah, K., & Hussain, M. (2004). Effect of processing and formulation variables on the stability of a salt of a weakly basic drug candidate. Pharmaceutical Development and Technology, 9(3), 239–245.

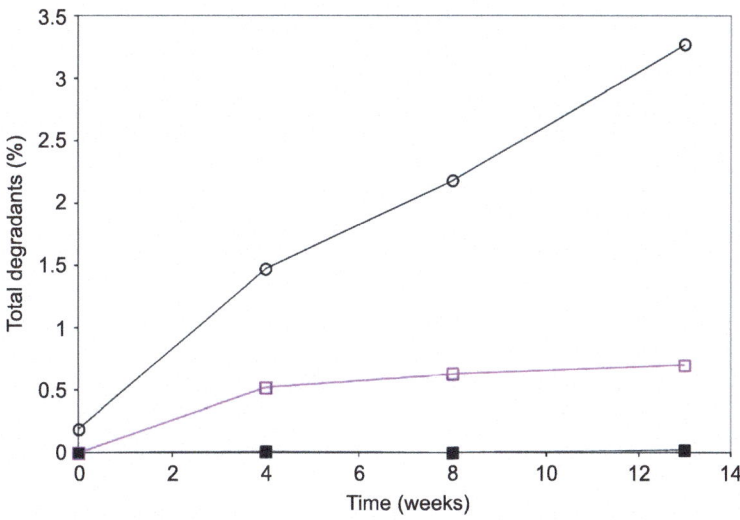

FIGURE 5.7

Effect of pH modifiers on stability of CPD-I tablets manufactured by wet granulation at 40°C/75% RH. 5% sodium carbonate, (■); 5% sodium bicarbonate, (□); no pH modifier, (○).

Adapted from Badawy, S., Vickery, R., Shah, K., & Hussain, M. (2004). Effect of processing and formulation variables on the stability of a salt of a weakly basic drug candidate. Pharmaceutical Development and Technology, 9(3), 239–245.

5.3.2 DMP 754

The improved stability of a formulation containing pH modifier was similarly reported for DMP 754 tablets and capsules when a wet granulation process was used for product manufacture. DMP 754 is an ester prodrug that demonstrated significant instability in the solid state (Badawy, Williams, & Gilbert, 1999).

5.3.2.1 *Stability behavior of DMP-754*

The crystalline DMP 754 drug substance exhibits good stability in the solid state. DMP 754 degradation in the solid state was significantly enhanced in the presence of different excipients, and the rate of degradation was proportional to the excipient-to-drug ratio. The two main degradation products isolated in the solid state (Figure 5.8) were the ester hydrolysis product (XV459) and the amidine hydrolysis product (SJ459) (Farag Badawy, Williams and Gilbert, 1999). Among all the fillers tested, anhydrous lactose showed the lowest rate of DMP 754 degradation. However, DMP 754 still showed significant degradation in the presence of anhydrous lactose at high excipient-to-drug ratios. Enhanced hydrolysis of DMP 754 in the presence of lactose was attributed, at least partly, to lactose catalysis, because lactose was shown to provide concentration-dependent catalysis of ester and amidine degradation in solution. Hydrolysis rate of DMP 754 in the presence of lactose was found to be dependent on the microenvironment pH. Lactose and DMP 754 have a saturated solution pH of approximately 6 and 6.8, respectively. Consequently, the pH of the microenvironment for the drug particles is expected to be in this range, which is 2–3 pH units higher than the pH of maximum stability for DMP 754 in solution (~

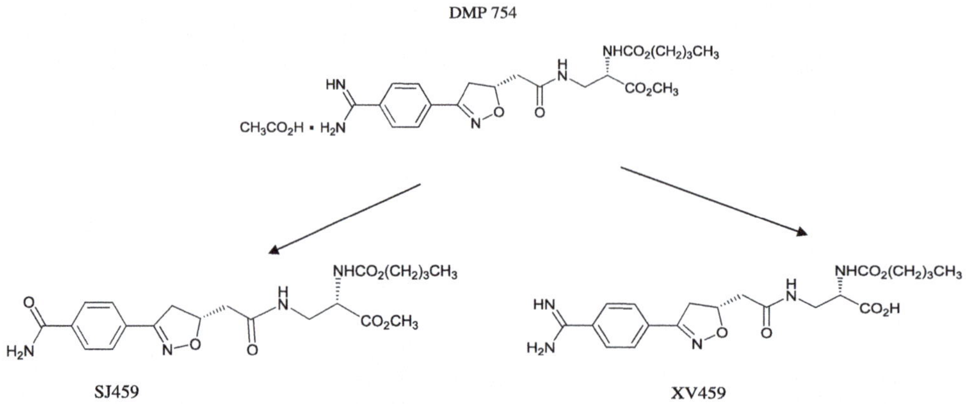

FIGURE 5.8

Structure of DMP 754 degradants.

From Badawy, S. I. F., Williams, R. C., & Gilbert, D. L. (1999). Chemical stability of an ester prodrug of a glycoprotein IIb/IIIa receptor antagonist in solid dosage forms. Journal of Pharmaceutical Sciences, 88(4), 428–433.

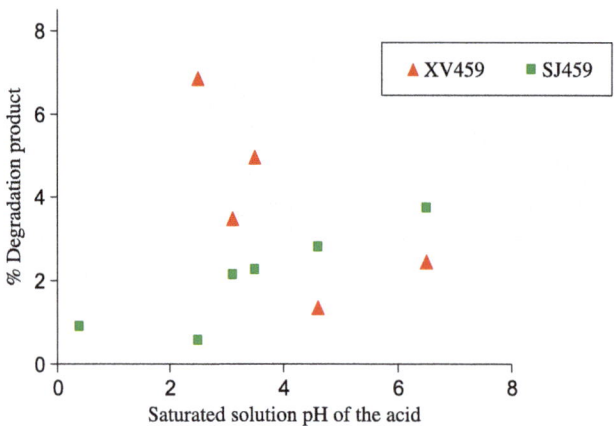

FIGURE 5.9

DMP 754 degradation in lactose blends (0.33% w/w drug loading) after 2 months at 40°C/75% RH.

Adapted from Farag Badawy, S. I., Williams, R. C., & Gilbert, D. L. (1999). Effect of different acids on solid-state stability of an ester prodrug of a IIb/IIIa glycoprotein receptor antagonist. Pharmaceutical Development and Technology, 4(3), 325–331.

pH 4.5). The hydrolysis rates of the ester and amidine groups of DMP 754 in lactose blends were altered by incorporation of acidic components in the blend. The effect of an acid on the microenvironment pH of DMP 754 was predicted by the saturated solution pH of the acid (Farag Badawy, Williams and Gilbert, 1999). The ester group attained maximum stability with acids having saturated solution pH of approximately 4. The amidine group, however, showed increased stability with the more acidic modifiers having saturated solution pH values as low as 0.4 (Figure 5.9). Disodium citrate (saturated solution pH of

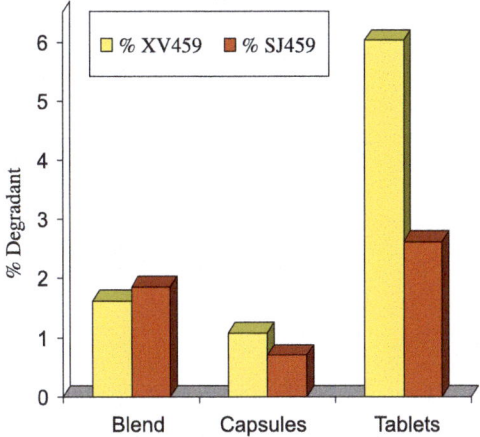

FIGURE 5.10

Effect of encapsulation and compression on the degradation of DMP 754 after 1 month at 40°C/75% RH.

Adapted from Badawy, S. I. F., Williams, R. C., & Gilbert, D. L. (1999). Chemical stability of an ester prodrug of a glycoprotein IIb/IIIa receptor antagonist in solid dosage forms. Journal of Pharmaceutical Sciences, 88(4), 428–433.

4.6) was the only acid tested that improved the stability of both groups. Stability of DMP 754 in the solid state, therefore, can be improved by the use of an appropriate acid that adjusts the microenvironmental pH to approximately 4.

Processing was shown to influence the stability of DMP 754. The effect of compression and encapsulation on DMP 754 stability was examined for binary blends with anhydrous lactose at a low drug loading (0.33% w/w). Stability was evaluated at 40°C/75% RH in HDPE bottles. Encapsulation and compression affected the stability of DMP 754 in the binary blend with anhydrous lactose (Figure 5.10) (Badawy et al., 1999). The encapsulation of the lactose/DMP 754 blend into hard gelatin capsules decreased drug degradation. The stabilizing effect of encapsulation was comparable for the ester and the amidine groups. The encapsulation of the blend reduces the surface area of the blend exposed to the environment and decreases the rate of penetration of water vapor into the powder bed. The capsule shell also can act as a barrier that water vapor has to penetrate before it reaches the blend. This is particularly true if the gelatin shell is more hygroscopic than the blend and consequently can act as a desiccant. Thus, encapsulation can prolong the time that it takes the moisture content of powder bed to equilibrate with water vapor pressure at 75% RH, which might be the reason for the improved stability of blends encapsulated into hard gelatin shells. To the contrary, tableting of the DMP 754/lactose blend enhanced drug degradation (Figure 5.10). Tableting increases the number of contact points between lactose and DMP 754. This would enhance lactose catalysis and increases the rate of moisture transfer between lactose and the drug, thus resulting in an increased rate of drug degradation in the tablets as compared to the blends. Despite the low concentration of moisture associated with anhydrous lactose (approximately 0.5% w/w at 75% RH), this moisture corresponds to high water:drug molar ratio because of the small amount of DMP 754 in the blend (0.33% w/w) and the low molecular weight of water. The destabilizing effect of tableting was more pronounced for ester hydrolysis than amidine hydrolysis.

Table 5.3 DMP 754 formulation composition

Ingredients	Concentration (% w/w)						
	Physical blends		Dry granulation		Wet granulation		
DMP 754	0.33	0.33	0.33	0.33	0.8	1.7	0.33
Disodium citrate	0	2.5	2.5	2.5	2.5	2.5	0
Povidone	0	0	0	2.0	2.0	2.0	20
Lactic acid	0	0	0	0	0	0	0.0083
Magnesium stearate	0	0	1.0	1.0	1.0	1.0	1.0
Anhydrous lactose	99.67	97.17	96.17	94.17	93.7	92.8	96.66
Tablet or capsule strength (mg)	0.1	0.1	0.2	0.2	0.5	1.0	0.1
Weight of tablet or capsule content (mg)	30	30	60	60	60	60	30

From Badawy, S. I. F., Williams, R. C., & Gilbert, D. L. (1999). Chemical stability of an ester prodrug of a glycoprotein IIb/IIIa receptor antagonist in solid dosage forms. Journal of Pharmaceutical Sciences, 88(4), 428–433.

5.3.2.2 Stability of DMP 754 drug product manufactured by wet granulation

The effect of processing on the stability of drug product containing disodium citrate was also evaluated. Composition of DMP 754 formulations are shown in Table 5.3 (Badawy et al., 1999). Stability of DMP 754 tablets and capsules containing disodium citrate and manufactured by a dry granulation (slugging) process or a wet granulation process was assessed. DMP 754 capsules manufactured by a wet granulation process were more stable when stored at 40°C/75% RH compared to capsules manufactured by the dry granulation process in the same packaging configuration (Figure 5.11). The rates of degradation of the ester and amidine groups were lower for the wet granulation capsules than for the dry granulation capsules. As for CPD-I, the higher stability of capsules manufactured by wet granulation was explained by the more uniform distribution of the citrate in this formulation. Adding the citrate to the granulating solution leads to intimate contact of this acidic component with the drug and other formulation components, resulting in a better control of the microenvironment pH. The granulating solution wets the particles, and when the water evaporates, the citrate is in close contact with the formulation constituents. This is shown by the sodium distribution in the blend that was found to be more diffuse in the wet granulation sample compared to localized distribution in the dry granulation sample, thus suggesting a more uniform distribution of the citrate in the former formulation (Figure 5.12) (Badawy et al., 1999).

While capsules were more stable than tablets for the dry granulation process, the opposite was observed for wet granulation. DMP 754 tablets compressed from the granulation manufactured by the wet process were more stable than the capsules filled with the same granulation at 40°C/75% RH (Figure 5.11) and both ester and amidine hydrolysis rates were lower in the tablet dosage form. In the case of the wet granulation formulation with disodium citrate, compression did not demonstrate the destabilizing effect observed for the dry blends. Increasing the number of contact points in the case of granulation manufactured by the wet process did not result in enhanced degradation, probably because of the control

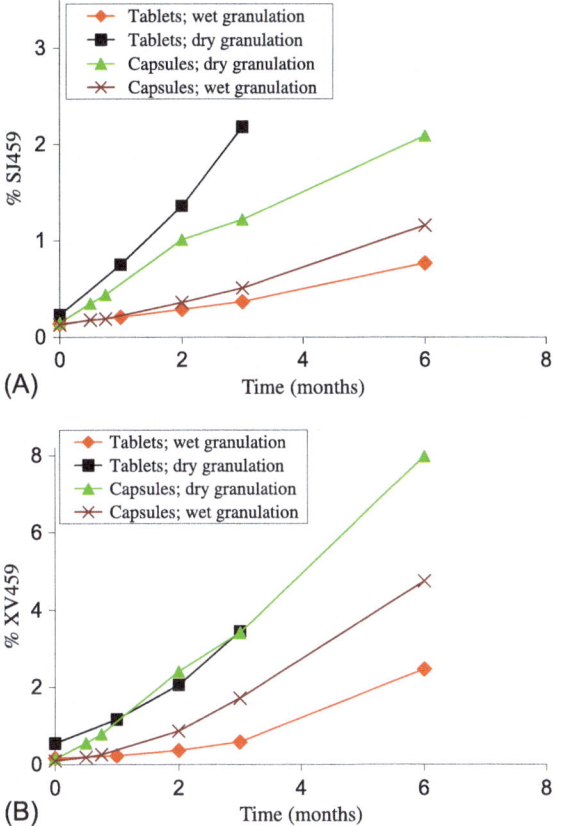

FIGURE 5.11

Hydrolysis in DMP 754 dosage forms stored at 40°C/75% RH in HDPE bottles with desiccant. (A) Ester hydrolysis, (B) Amidine hydrolysis.

Adapted from Badawy, S. I. F., Williams, R. C., & Gilbert, D. L. (1999). Chemical stability of an ester prodrug of a glycoprotein IIb/IIIa receptor antagonist in solid dosage forms. Journal of Pharmaceutical Sciences, 88(4), 428–433.

of microenvironment pH. Stability of tablets manufactured by the wet granulation process increased with the decrease in the excipient-to-drug ratio (higher formulation strength) (Table 5.4).

The stabilizing effect of disodium citrate for product manufactured by wet granulation was demonstrated further by the instability of tablets without disodium citrate. Although the formulation without disodium citrate was manufactured using a granulating solution buffered to pH 4 with lactate, these tablets showed a higher degradation rate than for the tablets with disodium citrate (Figure 5.13). The very low concentration of lactate (0.0083% w/w of the total weight of the formulation) probably was insufficient to control the microenvironment pH in the tablets. This was shown by slurry pH measurement where the pH values of the slurries prepared from the granulations with disodium citrate and lactate were found to be 4.4 and 7.7, respectively (Badawy et al., 1999).

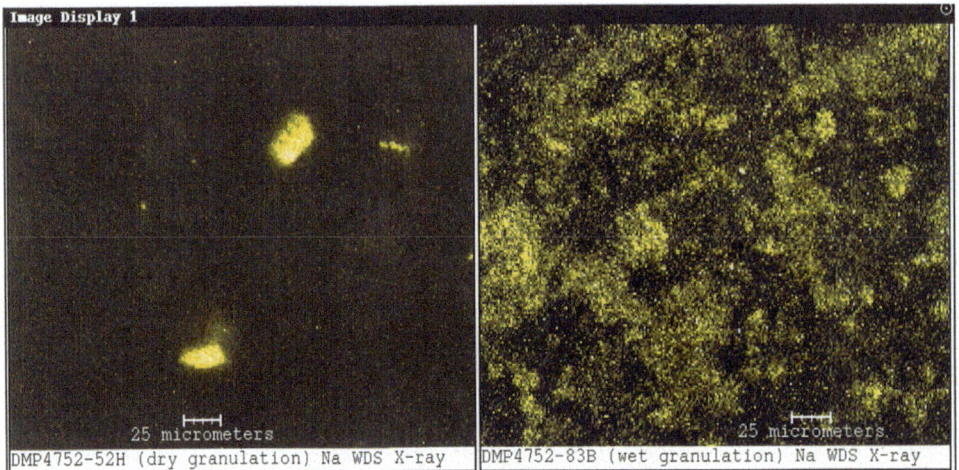

FIGURE 5.12

Distribution of sodium in DMP 754 tablets. Left, dry granulation; right, wet granulation.

From Badawy, S. I. F., Williams, R. C., & Gilbert, D. L. (1999). Chemical stability of an ester prodrug of a glycoprotein IIb/IIIa receptor antagonist in solid dosage forms. Journal of Pharmaceutical Sciences, 88(4), 428–433.

Table 5.4 Degradation of DMP 754 in tablets and capsules after 3 months at 40°C/75% RH					
	Disodium citrate concentration (%)	Strength (mg)	Moisture content (%)[a]	Degradation product	
				XV459	SJ459
Capsules/dry granulation	2.5	0.2	0.7	3.41	1.22
Tablets/dry granulation	2.5	0.2	0.7	3.44	2.18
Capsules/wet granulation	2.5	0.2	2.5	1.72	0.51
Tablets/wet granulation	2.5	0.2	1.5	0.59	0.37
Tablets/wet granulation	2.5	0.5	1.2	0.46	0.32
Tablets/wet granulation	2.5	1.0	1.4	0.28	0.25
Tablets/wet granulation	0[b]	0.1	1.5[c]	1.36[c]	2.00[c]

[a] Moisture of tablet or capsule content determined by Karl Fischer assay after 3 months at 40°C/75% RH.
[b] Contains 0.0083% lactic acid.
[c] Two months timepoint. From Badawy, S. I. F., Williams, R. C., & Gilbert, D. L. (1999). Chemical stability of an ester prodrug of a glycoprotein IIb/IIIa receptor antagonist in solid dosage forms. Journal of Pharmaceutical Sciences, 88(4), 428–433.

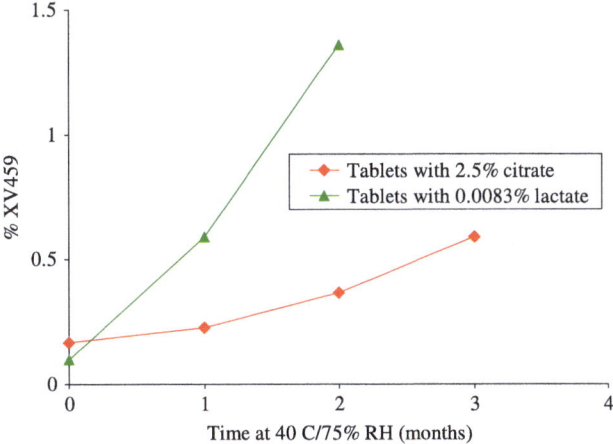

FIGURE 5.13

Effect of concentration of pH modifiers on stability of DMP 754 tablets (pH modifier added in granulating solution; pH adjusted to 4)

From Badawy, S. I. F., Williams, R. C., & Gilbert, D. L. (1999). Chemical stability of an ester prodrug of a glycoprotein IIb/IIIa receptor antagonist in solid dosage forms. Journal of Pharmaceutical Sciences, 88(4), 428–433.

5.3.2.3 *Moisture uptake behavior of DMP 754 formulations*

Moisture content of capsules manufactured by the wet granulation process was higher than those manufactured by the dry granulation process. Capsule moisture content at time zero, determined by a Karl-Fisher titration, was found to be 0.6% and 2.0% for the dry and wet formulations, respectively. The moisture content of capsules manufactured by wet granulation also was higher than those manufactured by dry granulation after three months of storage at 40°C/75% RH (Table 5.4) (Badawy et al., 1999). The higher moisture content of the wet granulation formulation is attributed to two reasons. First, partial conversion to lactose monohydrate during the wet granulation process was observed by X-ray diffraction of the granulation manufactured by the wet process. Second, the formulation manufactured by the wet granulation process was found to be more hygroscopic than the dry granulation formulation as determined by moisture sorption-desorption isotherms for the two formulations. Percent weight gain of the granulation upon the increase of relative humidity from 40% to 90% was 1.7% and 6.5% for the dry and wet formulations, respectively (Figure 5.14A and B). The higher hygroscopicity of the wet granulated blend is attributed to a low level of amorphous lactose formation during wet granulation, although lactose remained predominantly crystalline after wet granulation.

The hygroscopic nature of the formulation manufactured by wet granulation might be a contributing factor to the higher degradation rate in the case of capsules relative to tablets. Because of the hygroscopicity of granulation manufactured by wet granulation, moisture is transferred from the capsule shell to the granulation (which is in direct contact with the gelatin shell) resulting in a higher degradation rate compared to the tablets (Figure 5.11).

Although the increased hygroscopicity of a formulation generally is expected to increase the degradation rate of a moisture-sensitive drug, the effective microenvironment pH control in the case of the wet

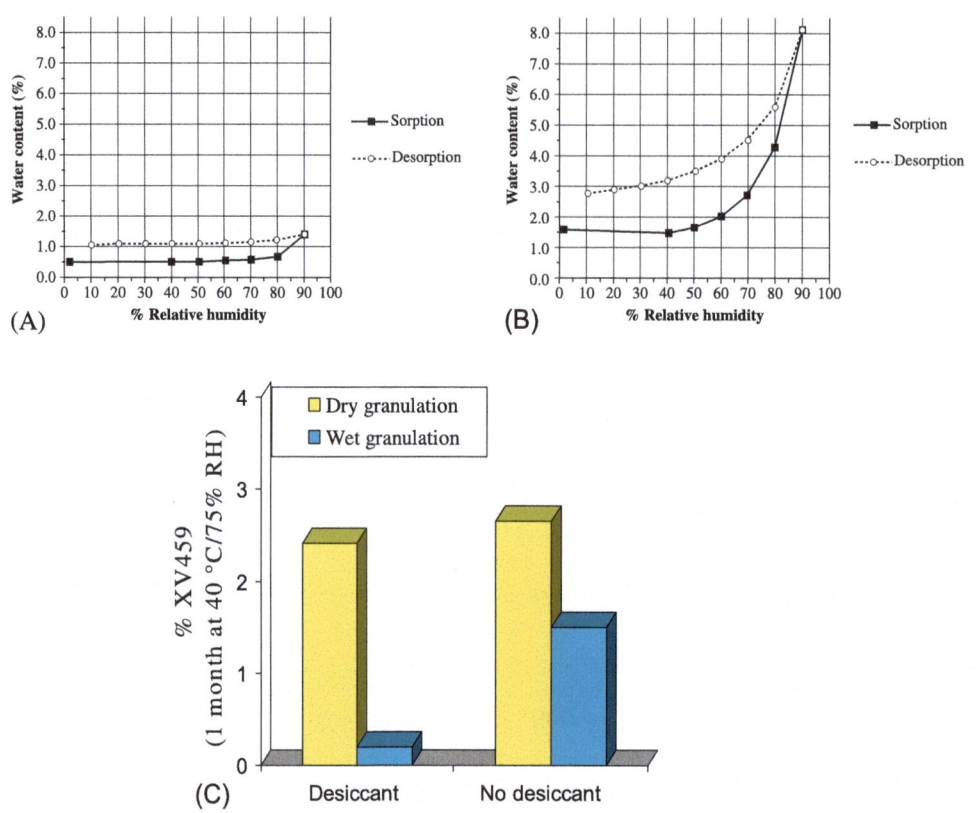

FIGURE 5.14

Effect of manufacturing process on the moisture sensitivity of DMP-754 tablets. Moisture uptake of blend manufactured by dry granulation (A) and wet granulation (B). Stability in HDPE bottles showing higher moisture sensitivity of product manufactured by wet granulation (C).

Adapted from Badawy, S. I. F., Williams, R. C., & Gilbert, D. L. (1999). Chemical stability of an ester prodrug of a glycoprotein IIb/IIIa receptor antagonist in solid dosage forms. Journal of Pharmaceutical Sciences, 88(4), 428–433.

granulation formulation was a key factor for the stability of this formulation. The microenvironment pH control in the case of the wet granulated formulation probably was able to compensate for the increased hygroscopicity, resulting in a more stable dosage form than the less hygroscopic dry granulation formulation, which lacked effective pH control. Nevertheless, the stability of the wet granulated product was more dependent on the presence of desiccant. While the dry granulated formulation was less stable than wet granulated product in both cases, the less hygroscopic dry granulated product showed little dependence on the desiccant (relative humidity inside the bottle) compared to the more hygroscopic wet granulated formulation (Figure 5.14C).

Stability of the wet granulated formulation was enhanced further by wet granulating formulation components without the drug substance (including lactose and disodium citrate) and then adding the drug substance to the extragranular phase by dry blending (Figure 5.15). This modified process exploits

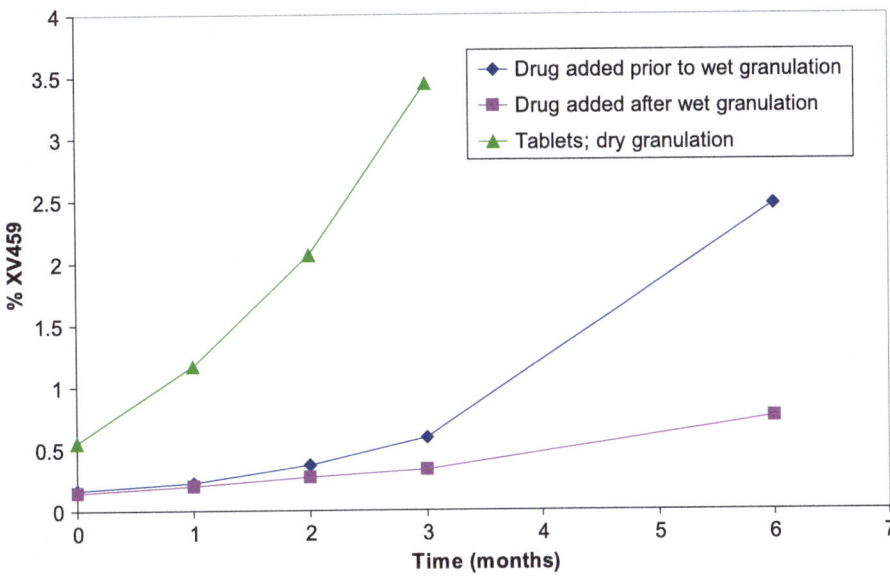

FIGURE 5.15

Ester hydrolysis in DMP 754 tablets manufactured by wet granulation at 40°C/75% RH showing improved stability by wet granulation of excipients with citrate buffer prior to addition of the active.

the wet granulation process to achieve effective pH control of the lactose (thus diminishing its catalytic effect on DMP 754 hydrolysis), while sparing the drug substance from the stress of wet granulation. The modified process also reduces the contact between the lactose and the drug (as the drug is in the extragranular phase), which is beneficial because of the increased hygroscopicity of wet granulated lactose.

5.3.3 Salt form and stability in wet granulation

Solid state hydrolysis rates of the ester and amidine groups on DMP 754 were found to decrease by the use of a salt of a stronger acid. DMP 755 is a crystalline salt of methanesulfonic acid that has a lower pKa than acetic acid (− 1.2 vs. 4.76). As a result, the pH of a concentrated solution of DMP 755 is approximately 2, compared with 6.8 for a saturated solution of the acetate salt (DMP 754) (Farag Badawy, 2001). The use of the methanesulfonic acid (mesylate) salt, therefore, provides a lower microenvironment pH than the acetate salt. This is expected to enhance the stability of the amidine group because it showed continuous increase in stability with the decrease in microenvironment pH. The lower pH also improves the stability of the ester group provided the microenvironment pH does not drop below 4, which is the pH of maximum stability for the ester group. Physical properties of DMP 755 and DMP 754 are summarized in Table 5.5. Similar to DMP 754, DMP 755 drug substance showed minimal or no increase in degradant concentration when stored at 40°C/75% RH in an open dish, which shows that the drug substance for both salt forms is stable at this condition (Table 5.6). Both salt forms are nonhygroscopic and neither one picks up any appreciable quantity of moisture. The two salts showed

Table 5.5 Physical properties of DMP 755 (mesylate salt) and DMP 754 (acetate salt)

	DMP 754	DMP 755
Solubility (mg/mL)	10	> 100
Solution pH	6.8	2
Melting point (°C)	213	158

From Farag Badawy, S. I. (2001). Effect of salt form on chemical stability of an ester prodrug of a glycoprotein IIb/IIIa receptor antagonist in solid dosage forms. International Journal of Pharmaceutics, 223(1–2), 81–87.

Table 5.6 Degradant concentrations in drug substance and binary blend-filled hard gelatin capsules after storage for 4 weeks

	Moisture content (%)[a]	Degradation product (<%)	
		XV459	SJ459
DMP 754 drug substance			
Time zero	0.0	< 0.1	0.11
40°C/75% RH	0.2	0.11	0.13
DMP 754 binary blend			
Time zero	0.3	0.21	0.29
40°C/75% RH	1.3	1.08	0.70
50°C	0.6	1.25	0.99
DMP 755 drug substance			
Time zero	0.1	< 0.1	< 0.1
40°C/75% RH	0.0	< 0.1	< 0.1
DMP 755 binary blend			
Time zero	0.3	< 0.10	< 0.10
40°C/75% RH	0.4	0.42	0.29
50°C	0.2	0.41	0.34

[a] Moisture content determined by a Karl Fischer assay. From Farag Badawy, S. I. (2001). Effect of salt form on chemical stability of an ester prodrug of a glycoprotein IIb/IIIa receptor antagonist in solid dosage forms. International Journal of Pharmaceutics, 223(1–2), 81–87.

higher degradant concentrations in their binary mixtures with lactose compared with the drug substance, despite the fact that these mixtures were stored in sealed HDPE bottles rather than an open dish (Table 5.6). As expected based on the microenvironmental pH, mesylate salt was more stable than the acetate salt in the binary blend after four weeks at 40°C/75% RH and 50°C (Farag Badawy, 2001).

Stability of drug product for DMP 755 was compared with that of DMP 754 manufactured by the same process. Comparison was made between the two salt forms using the more stable dosage form for the given manufacturing process (tablets for wet granulation and capsules for the dry process) (Fig. 16) (Farag Badawy, 2001). DMP 755 capsules manufactured by the dry granulation process and containing disodium citrate showed remarkably lower rate of degradation than DMP 754 capsules manufactured by the same process. Moisture content, determined by a Karl Fischer assay, was comparable for the two capsules (0.7% for the DMP 755 capsules and 0.8% for the DMP 754 capsules). These results were interpreted knowing the effect of microenvironment pH on the stability of this molecule in the solid state. The high degradation rate for DMP 754 capsules manufactured by dry granulation shows the inability of disodium citrate to lower the pH of the formulation in this case. The mesylate salt, however, is expected to have a lower microenvironment pH in the solid state, regardless of the effectiveness of the pH modifier, and therefore demonstrated improved stability of the drug product manufactured by the dry process. DMP 755 capsules manufactured by the dry granulation process without disodium citrate also showed comparable degradation rate to those containing disodium citrate, confirming the minimal contribution of the pH modifying agent to the stability of DMP 755 capsules. It is noteworthy that DMP 755 capsules manufactured by dry granulation showed similar degradation rates to the DMP 754 tablets manufactured by wet granulation. Knowing the sensitivity of this molecule to the microenvironment pH, these results suggest that both formulations have comparable microenvironment pH values. The uniform distribution of the buffer in the case of wet granulation makes the pH of the salt less critical for stability. The effect of the counter ion on the microenvironment pH is the most plausible explanation for the observed differences in the stability of the two salt forms. This is particularly true because DMP 755 is more soluble in water than DMP 754 (> 100 vs. 10 mg/mL) and has a lower melting point (158 vs. 213°C), factors that can negatively affect the stability of the mesylate salt (Gould, 1986). The impact of wet granulation process on stability can be dependent therefore not only on the formulation but also on the salt form of the drug substance.

5.4 Summary and conclusions

The wet granulation process may have a negative impact on long-term stability of some drug products. The negative impact of wet granulation usually is associated with some loss of crystallinity or with increased contact between the drug substance and incompatible excipients. The effect of wet granulation on stability depends on the composition of the formulation and, in some cases, on the salt form of the drug substance. The increased contact of the drug substance and formulation components achieved in wet granulation can be leveraged to enhance the effectiveness of stabilizing excipients, such as pH modifiers. Stability of compounds undergoing hydrolytic degradation is dependent on the microenvironment pH and the manufacturing process. Selection of the proper pH modifier that provides the desired pH for optimum stability, together with the wet granulation process that maximizes its effect, is critical to dosage form stability. In the absence of the pH modifier, degradation rate is significantly higher because of some loss of crystallinity and increased moisture uptake of the formulation caused by wet granulation.

References

Ahlneck, C., & Zografi, G. (1990). The molecular basis of moisture effects on the physical and chemical stability of drugs in the solid state. *International Journal of Pharmaceutics, 62*(2–3), 87–95. https://doi.org/10.1016/0378-5173(90)90221-O.

Badawy, S., & Pandey, P. (2017). Design, development, and scale-up of the high shear wet granulation process. In C. Qiu, Y. Zhang, & Mantri (Eds.), *Developing solid oral dosage forms* (2nd ed.). Academic Press.

Badawy, S., Vickery, R., Shah, K., & Hussain, M. (2004). Effect of processing and formulation variables on the stability of a salt of a weakly basic drug candidate. *Pharmaceutical Development and Technology, 9*(3), 239–245. https://doi.org/10.1081/PDT-200031417.

Badawy, S. I. F., Lin, J., Gokhale, M., Desai, S., Nesarikar, V. V., Lamarche, K. R., et al. (2014). Quality by design development of brivanib alaninate tablets: Degradant and moisture control strategy. *International Journal of Pharmaceutics, 469*(1), 111–120. https://doi.org/10.1016/j.ijpharm.2014.04.059.

Badawy, S. I. F., Narang, A. S., Lamarche, K. R., Subramanian, G. A., Varia, S. A., Lin, J., et al. (2016). Integrated application of quality-by-design principles to drug product development: A case study of brivanib alaninate film-coated tablets. *Journal of Pharmaceutical Sciences, 105*(1), 168–181. https://doi.org/10.1016/j.xphs.2015.11.023.

Badawy, S. I. F., Williams, R. C., & Gilbert, D. L. (1999). Chemical stability of an ester prodrug of a glycoprotein IIb/IIIa receptor antagonist in solid dosage forms. *Journal of Pharmaceutical Sciences, 88*(4), 428–433. https://doi.org/10.1021/js9803297.

Christensen, N. P. A., Nielsen, S., Rantanen, J., Cornett, C., & Bertelsen, P. (2014). Processing-induced salt formation of two oxicams in solid dosage forms affects dissolution behavior and chemical degradation. *Powder Technology, 266*, 175–182. https://doi.org/10.1016/j.powtec.2014.05.054.

Farag Badawy, S. I. (2001). Effect of salt form on chemical stability of an ester prodrug of a glycoprotein IIb/IIIa receptor antagonist in solid dosage forms. *International Journal of Pharmaceutics, 223*(1–2), 81–87. https://doi.org/10.1016/S0378-5173(01)00726-8.

Farag Badawy, S. I., & Hussain, M. A. (2007). Microenvironmental pH modulation in solid dosage forms. *Journal of Pharmaceutical Sciences, 96*(5), 948–959. https://doi.org/10.1002/jps.20932.

Farag Badawy, S. I., Williams, R. C., & Gilbert, D. L. (1999). Effect of different acids on solid-state stability of an ester prodrug of a IIb/IIIa glycoprotein receptor antagonist. *Pharmaceutical Development and Technology, 4*(3), 325–331. https://doi.org/10.1081/PDT-100101368.

Freed, A. L., Strohmeyer, H. E., Mahjour, M., Sadineni, V., Reid, D. L., & Kingsmill, C. A. (2008). pH control of nucleophilic/electrophilic oxidation. *International Journal of Pharmaceutics, 357*(1–2), 180–188. https://doi.org/10.1016/j.ijpharm.2008.01.061.

Gould, P. L. (1986). Salt selection for basic drugs. *International Journal of Pharmaceutics, 33*(1–3), 201–217. https://doi.org/10.1016/0378-5173(86)90055-4.

Gu, L., Strickley, R. G., Chi, L. H., & Chowhan, Z. T. (1990). Drug-excipient incompatibility studies of the dipeptide angiotensin-converting enzyme inhibitor, Moexipril hydrochloride: Dry powder vs wet granulation. *Pharmaceutical Research: An Official Journal of the American Association of Pharmaceutical Scientists, 7*(4), 379–383. https://doi.org/10.1023/A:1015871406549.

Menning, M. M., & Dalziel, S. M. (2013). Fumaric acid microenvironment tablet formulation and process development for crystalline cenicriviroc mesylate, a BCS IV compound. *Molecular Pharmaceutics, 10*(11), 4005–4015. https://doi.org/10.1021/mp400286s.

Morris, K. R., Griesser, U. J., Eckhardt, C. J., & Stowell, J. G. (2001). Theoretical approaches to physical transformations of active pharmaceutical ingredients during manufacturing processes. *Advanced Drug Delivery Reviews, 48*(1), 91–114. https://doi.org/10.1016/S0169-409X(01)00100-4.

Narang, A. S., Desai, D., & Badawy, S. (2012). Impact of excipient interactions on solid dosage form stability. *Pharmaceutical Research, 29*(10), 2660–2683. https://doi.org/10.1007/s11095-012-0782-9.

Saquib, N. U., Sheikh, D., Ahmed, I., Usmanghani, K., & Khattak, S. U. R. (2015). Effect of formulation and process variables on degradation products of lovastatin in tablet dosage form. *Asian Journal of Pharmaceutical and Clinical Research, 8*(1), 131–133.

Shalaev, E. Y., & Zografi, G. (1996). How does residual water affect the solid-state degradation of drugs in the amorphous state? *Journal of Pharmaceutical Sciences, 85*(11), 1137–1141. https://doi.org/10.1021/js960257o.

Wardrop, J., Law, D., Qiu, Y., Engh, K., Faitsch, L., & Ling, C. (2006). Influence of solid phase and formulation processing on stability of Abbott-232 tablet formulations. *Journal of Pharmaceutical Sciences, 95*(11), 2380–2392. https://doi.org/10.1002/jps.20679.

Waterman, K. C., Adami, R. C., Alsante, K. M., Antipas, A. S., Arenson, D. R., Carrier, R., Hong, J., Landis, M. S., Lombardo, F., Shah, J. C., Shalaev, E., Smith, S. W., & Wang, H. (2002). Hydrolysis in pharmaceutical formulations. *Pharmaceutical Development and Technology, 7*(2), 113–146. https://doi.org/10.1081/PDT-120003494.

Material physical modifications induced by wet granulation

Sherif I.F. Badawy[a]

[a]*Drug Product Science & Technology, Bristol-Myers Squibb Co., New Brunswick, NJ, United States*

6.1 Introduction

In the wet granulation process, material experiences mechanical stresses resulting from agitation in the granulator. Mechanical stresses frequently induce physical changes in the material being granulated, particularly in the presence of the granulating solvent. These physical changes include solid state transformations, which involve modification of the crystal structure of the drug substance or excipient, and micromeritic (particle structure/property) changes. Mechanical and solvent stresses also can induce chemical degradation, either directly or as a consequence of the physical changes. Physical changes in wet granulation are addressed in this chapter. The impact of wet granulation on chemical stability is the subject of Chapter 5.

6.2 Solid state transformation in wet granulation

Solid state phase transformation is a common occurrence in wet granulation. Phase transformation in wet granulation can be classified into four categories: transformation to a more stable polymorphic form; formation of a high energy metastable form (e.g., amorphous); transformation of a salt form to the less soluble free form; and formation of a hydrate/solvate. Solid state transformations present quality and regulatory challenges to drug product development and commercial manufacturing. Therefore, careful risk assessment, mechanistic understanding, and appropriate control strategy should be in place if wet granulation is to be used for a drug substance that is prone to such transformations (Badawy & Pandey, 2017).

6.2.1 Transformation to a more stable polymorphic form

Transformation of a metastable polymorph to the more thermodynamically stable polymorph can take place via a solvent-mediated mechanism during wet granulation. For example, transformation of form A to the more stable tautomeric polymorph B was reported to take place during wet granulation of an irbesartan formulation (Pan, Crull, Yin, & Grosso, 2014). Polymorphic conversion of the metastable form I of flufenamic acid to the stable polymorph III also was observed during wet granulation with ethanol (Davis et al., 2003). As the more soluble metastable polymorph dissolves in the granulating liquid, it forms a supersaturated solution with respect to the stable polymorph, which subsequently crystallizes from the supersaturated solution. The less soluble polymorph can have lower dissolution rate and bioavailability compared to the metastable form (Aguiar, Krc, Kinkel, & Samyn, 1967; Kobayashi, Ito, Itai, & Yamamoto, 2000). The extent of conversion to the more stable polymorphic form depends on the kinetics of phase transformation relative to processing time. Rate of phase transformation is a function of dissolution rate of the metastable form, volume of granulating liquid used in the process, and crystallization kinetics of the stable polymorph (Morris, Griesser, Eckhardt, & Stowell, 2001). Extent of transformation, therefore, can depend on granulation parameters that affect these factors. Processing time depends on liquid addition rate and wet massing time, while dissolution rate can be a function of impeller speed. Kinoshita et al. studied the effect of wet granulation process parameters on the crystallization and dissolution behavior of amorphous nilvadipine from a solid dispersion with hypromellose (Kinoshita, Ohta, Shiraki, Higashi, & Moribe, 2017). They showed that crystallization was induced by a higher ethanol ratio in the hydroalcoholic granulating liquid. This was attributed to the higher solubility of nilvadipine in ethanol, which resulted in partial dissolution of nilvadipine in the granulating liquid during the granulation process and subsequent recrystallization during drying. In addition, a lower amount of the granulating liquid reduced the process-induced crystallization of nilvadipine.

6.2.2 Formation of a high energy metastable form

Formation of a high-energy metastable form can take place when the starting lower energy polymorph dissolves in the granulating liquid and is trapped kinetically into a high-energy form during drying (Morris et al., 2001; Badawy & Pandey, 2016). A high-energy disordered phase also can form by a nonsolvent mediated mechanism in which shear forces applied during wet granulation in the presence of water causes disruption of crystallinity and creates crystal defects or amorphous phase. Formation of an amorphous phase also can increase formulation hygroscopicity after wet granulation (Badawy, Williams, & Gilbert, 1999). Wet granulation of BMS-561388 resulted in partial transformation to the less stable amorphous form (Badawy et al., 2009). Similarly, aqueous wet granulation of either polymorph A or B of dexketoprofen trometamol with microcrystalline cellulose caused transformation to the amorphous form regardless of the initial polymorph used in granulation (Blanco, Alcalá, González, & Torras, 2006).

6.2.3 Transformation of a salt to the free form

Salts of weak bases or weak acids can covert to the free form (free base or free acid) during aqueous wet granulation (Badawy & Pandey, 2016). As the salt dissolves in the granulating water, it will form a saturated solution with respect to the free form if the solution pH is higher (for a weak base) or lower (for a weak acid) than the pH of maximum solubility (pH_{max}). The free form subsequently precipitates from this saturated solution, resulting in the salt-to-free form transformation. If this solution pH requirement is not met, the salt would be the thermodynamically stable form in equilibrium with solution and consequently there would be no conversion to the free form. Basic excipients such as croscarmellose sodium were shown to increase the disproportionation risk for the salts of weak bases as they raise the microenvironmental pH. For example, disproportionation of pioglitazone hydrochloride was enhanced because of the presence of croscarmellose sodium in the formulation (Nie, Xu, Taylor, Marsac, & Byrn, 2017). The acidic excipients, citric acid and polyacrylic acid, however, were found to induce the disproportionation of the calcium salt of atorvastatin (weak acid salt). Moreover, it was observed that milling of the formulation blend promoted atorvastatin calcium disproportionation on stability, which was attributed to the formation of close contacts between the drug substance particles and the acidic components (Christensen, Rantanen, Cornett, & Taylor, 2012).

As for polymorphic transformations, wet granulation parameters (amount of granulating liquid, impeller speed, and process duration) can affect the extent of solvent mediated transformation of the salt to the free form.

6.2.4 Formation of a hydrate/solvate

Compounds that exist in a hydrate form can convert to the hydrate if the anhydrous form is used in aqueous wet granulation, because water activity during wet granulation is usually higher than the critical humidity for the formation of the hydrate (Badawy & Pandey, 2016). Hydrate formation also can take place through a solvent-mediated mechanism as described above for polymorphic transformations. Many literature articles exist about the transformation of the anhydrous form to hydrate during wet granulation (Gift, Luner, Luedeman, & Taylor, 2009; Wikström, Carroll, & Taylor, 2008; Wikström, Marsac, & Taylor, 2005). The lower solubility of hydrates can lead to lower dissolution rate and bioavailability (Debnath and Suryanarayanan, 2004; Kobayashi et al., 2000). The hydrate form can partially or completely convert back to the anhydrous form during drying of the granulation so a mixture of forms can exist in the final product. Wikström et al. (2008, 2005) used Raman spectroscopy to follow hydrate formation of theophylline during wet granulation. They concluded that hydrate formation takes place via a solvent mediated mechanism and showed that the rate of transformation increased by the increase in agitation speed. Formation of theophylline monohydrate in the wet mass was inhibited by granulation at elevated temperature (50°C) (Otsuka, Kanai, & Hattori, 2014). Theophylline monohydrate formed by granulation at lower temperatures (27–40°C) was found to convert back to the anhydrous form by drying at 70°C (Otsuka, Kanai and Hattori, 2014).

6.2.4.1 Hydrate formation of lactose in wet granulation

Hydrate formation of an excipient also can take place during wet granulation. Formation of lactose monohydrate upon wet granulation of anhydrous lactose and its impact on excipient performance was studied by Shah, Hussain, Hubert, and Farag Badawy (2008). Lactose is one of the most frequently used fillers in tablet formulations. Lactose is available in different forms depending upon the crystallization conditions and in various grades with different particle sizes and different compaction properties. α-lactose monohydrate and β-anhydrous lactose are among the most commonly used forms of lactose. Even for the same crystal form, different lactose grades from different suppliers exhibit different powder properties and therefore could not be treated as interchangeable in direct compression formulations (Whiteman and Yarwood, 1988). In particular, compaction behavior is different for the various forms of lactose. Roller dried β-anhydrous lactose has significantly superior compaction properties to α-lactose monohydrate because of its particle morphology. Anhydrous lactose particles have high surface roughness and porosity, in contrast to the smooth low-porosity lactose monohydrate particles, resulting in higher tendency of the particles to fracture during compaction

Table 6.1 Full factorial design for the anhydrous lactose wet granulation study (Badawy, Menning, Gorko, & Gilbert, 2000).

RUN#	% w/w of MCC in the formulation	%w/w of granulating water[a]	Drying conditions
1	0	18	Tray
2	20	10	Tray
3	0	10	FBD
4	10	14	Tray
5	20	18	Tray
6	0	10	Tray
7	10	14	FBD
8	20	18	FBD
9	0	18	FBD
10	20	10	FBD

[a] %w/w of granulating water = 100 × weight of granulating water/weight of solids in the granulator.

(Lerk, 1993; Vromans et al., 1985; Vromans, Bolhuis, Lerk, & Kussendrager, 1987). Therefore, anhydrous lactose would be the lactose form of choice in tablet formulation where an improvement in compactibility is required, including wet granulated tablet formulations. Anhydrous lactose, however, can undergo partial form conversion to α-lactose monohydrate in presence of water used during granulation. In addition to its possible impact on compactibility, this form conversion should be monitored closely because it could cause dilution of the active ingredient in the formulation, potentially lowering the potency of the final tablets because of the water associated with monohydrate form. The study by Shah et al. showed that wet granulated blends containing anhydrous lactose maintained better compactibility than lactose monohydrate granulations, despite partial conversion of β-anhydrous lactose to α-lactose monohydrate during granulation. Interestingly, lactose monohydrate particles created from anhydrous lactose particles in wet granulation still retained the morphological features of the initial anhydrous lactose particles and consequently maintained higher compactibility compared to the as-is lactose monohydrate.

In the 2008 study by Shah et al., the effect of three factors on form conversion of anhydrous lactose was evaluated: inclusion of MCC along with anhydrous lactose in the formulation; amount of water used for granulation; and drying conditions of the granulation (tray drying vs. fluid bed drying), because fluidized bed drying has been shown to be significantly faster in both drying and handling time than tray drying (Gao, Gray, Motheram, & Hussain, 2000). MCC is hygroscopic and has the ability to absorb a large quantity of water. Angberg, Nystr, and Castensson (1991) demonstrated that MCC protects anhydrous lactose when water vapor is abundant by slowing down the rate of hydration, because water is preferentially absorbed by MCC. The addition of MCC in the wet granulated tablet formulation along with anhydrous lactose was hypothesized to influence the incorporation of water into the anhydrous lactose crystal and reduce form conversion of anhydrous lactose to monohydrate.

A two-level full factorial design with two center points (Table 6.1) was used to study the effect of three variables: percentage of MCC (low 0% and high 20%), water to intragranular solids ratio (low 0.10 and high 0.18), and drying conditions (tray drying vs. fluid bed drying, FBD). All formulations

Table 6.2 Response data for factorial design evaluating factors affecting percent lactose form conversion during wet granulation (Badawy et al., 2000).

RUN#	Difference between LOD at 105°C and 150°C (%)	Percent lactose conversion
1	1.47	33.1
2	0.92	26.3
3	1.03	23.2
4	1.42	35.8
5	1.32	37.7
6	1.23	27.7
7	1.04	26.2
8	0.95	27.1
9	1.21	27.2
10	0.8	22.8

contained, 3% hydroxypropyl cellulose and 2% croscarmellose sodium (1% intra- and 1% extragranular). Percent lactose conversion (see below) was the response variable evaluated. Statistical analysis of the data was carried out using a regression model. Granulation of the different experimental design batches was carried out in a 10 L high-shear granulator at a batch size of 1.5 kg using water as the granulating liquid. The wet granulation was dried either in a hot air oven at 50°C or in a Glatt fluid bed dryer at an inlet temperature of 65°C–70°C to a moisture content of < 1.5% w/w.

The percent lactose conversion in the milled granulation was measured by determining the loss on drying of the granulation at two different temperatures, 105°C and 150°C using a moisture analyzer. At 105°C, only the unbound water would be lost, as compared to the loss of bound water (water of hydration) at 150°C. As reported in the literature, lactose monohydrate contains $\sim$ 5.2% total moisture, while lactose anhydrous contains < 0.5% moisture (David and Augsburger, 1977; Reier and Shangraw, 1966; Shukla and Price, 1991). The difference between the two LOD values at 105°C and 150°C were calculated, and the percent form conversion from lactose anhydrous to monohydrate was determined by the following formula:

$$\% \text{Lactose form conversion} = \frac{\text{LODat150°C} - \text{LODat105°C} \times 100}{\text{Wt.fraction of lactose in formulation} \times 4.7}$$

where 4.7 is the theoretical % water (w/w) in the crystal of α-lactose monohydrate.

Table 6.2 shows the data for the wet granulation DOE study. The percent form conversion for all the runs ranged from 22.8% to 37.7%. Table 6.3 depicts the regression analysis of the percent form conversion during wet granulation. Statistical analysis (Table 6.3) demonstrated that the formulation granulated with a higher percent of water showed a higher tendency for form conversion during wet granulation as indicated by the positive sign of the parameter estimate ($p = 0.0122$). Higher percent form conversion eventually would lead to dilution of the active ingredient in the formulation and could lower the potency of the tablets. Therefore, it is desirable to use the minimum amount of water needed to achieve acceptable granulation, in order to avoid dilution in formulations containing anhydrous lactose.

Table 6.3 Regression analysis of percent lactose form conversion during wet granulation (Badawy et al., 2000).

Percent lactose form conversion during wet granulation		
Parameter	Parameter estimate	P value
% w/w of MCC	0.3375	0.7165
% w/w of water	3.1375	0.0122
Drying conditions (FBD)	− 3.41	0.0051

Granulation subjected to fluidized bed drying demonstrated a significantly lower percent form conversion than those subjected to convective tray drying ($p = 0.0051$), as indicated by the negative sign of the parameter estimate. A fluid bed dryer significantly reduces drying time compared to tray dryer or vacuum dryer (Gao et al., 2000). The fluid bed dryer took approximately 45 min to dry a granulation to the target moisture content of < 1.5% as compared to 6–8 h for tray drying. Therefore, the residence time during which moisture is available to interact with the anhydrous material is much higher in tray drying as compared to the fluidized bed dryer. Effect of MCC on percent form conversion was not statistically significant ($p = 0.7165$) as per the regression analysis of percent form conversion (Table 6.3). MCC is known to have high affinity for water and therefore its presence in the formulation could make granulating water less available for form conversion. Levels of MCC up to 20% used in this study, however, did not reduce the percent form conversion.

In the second part of the study by Shah et al., the effect of form conversion from anhydrous to monohydrate on compactibility of lactose was investigated, because lactose monohydrate is less compactible than anhydrous lactose. Complete form conversion of lactose anhydrous to monohydrate form was induced by exposure to a high humidity condition. Anhydrous β-lactose at higher humidity incorporates water and undergoes form conversion to crystalline α-lactose monohydrate (Shukla and Price, 1991; Angberg, Nyström, & Castensson, 1991; Berlin, Kliman, Anderson, & Pallansch, 1971). Anhydrous lactose was placed in desiccators filled with saturated solution of potassium sulfate, resulting in an equilibrium humidity of 97% for 4 weeks to induce the form conversion of anhydrous lactose to the monohydrate form. The difference between the LOD values at 105°C and 150°C, as described earlier, was used to calculate the percent conversion in the sample. The as-received samples (anhydrous β-lactose and control α-lactose monohydrate) and the samples stored at high humidity were subjected to several tests to confirm the conversion of anhydrous lactose to the monohydrate form, to characterize powder properties of initial and form converted material, and to assess the impact of form conversion on performance. Thermogravimetric analysis (TGA) and powder X-ray diffraction (PXRD) were used to confirm the conversion of anhydrous lactose to the monohydrate form. The particle size distribution was determined by mesh analysis using a sonic sifter. The surface area for the entire sample and the size fraction between 88 and 149 μm for all the lactose samples was determined using a nitrogen adsorption technique. The amount of nitrogen adsorbed was determined at partial nitrogen pressure (P/P_0) ranging from 0.05 to 0.30. Surface area was determined using the Brunauer-Emmett-Teller's (BET) calculation for the nitrogen data in the P/P_0 range from 0.05 to 0.3. Pore volume distributions of the samples also were determined for a size fraction passed through 100 mesh (149 μm) and retained on 170 mesh (88 μm), by mercury intrusion porosimetry. Incremental pore volume and total pore volume were determined in the pressure range from 1 to 60,000 psi corresponding to pore diameters between 148 to 0.003 µm. Different lactose samples also were characterized by scanning electron microscopy (SEM).

The compaction profiles of the different lactose samples of same size fraction (88–149 µm) were obtained using an ESH tablet compaction simulator. Only the material passing through 100 mesh (149 µm) and retained on 170 mesh (88 µm) screens was used for analysis. The similar size fraction removes a particle size bias and facilitates a comparison of compactibility. A 260 mg sample of each batch was compressed on the simulator using 3/8″ flat face tooling to a pre-determined in-die thickness. Each batch was compressed to five different thickness targets with two-three tablets obtained at each thickness. Actual thickness achieved for each compressed tablet was determined by the linear variable displacement transducer (LVDT) of the compaction simulator. Compression and ejection forces were determined for each tablet. The hardness of each tablet was determined by a diametrical compression test on a hardness tester and converted to a tensile strength. Compaction profiles then were constructed using the compression pressure (obtained by dividing the compression force by punch area) and corresponding compact tensile strength. The compactibility parameter was calculated as the slope of the line obtained by linear regression of the data points in the compaction profile. The other compaction parameters calculated from the compaction simulator data were the yield pressure and bonding. The yield pressure was obtained from the slope of the linear portion of the natural logarithm of the reciprocal tablet porosity versus compression pressure. The bonding parameter was calculated as the Y-intercept of the plot of tensile strength versus tablet porosity and represents theoretical tablet tensile strength at zero porosity.

Anhydrous lactose converted completely to the monohydrate form when stored under high humidity (97% RH/25°C) for 4 weeks. The difference between the LOD values at 105°C and 150°C for the anhydrous lactose material stored under those conditions was approximately 4.8%, which is in close agreement with theoretical water of hydration for lactose monohydrate (Berlin et al., 1971; Brittain et al., 1991). In addition, the unexposed anhydrous lactose (as received) material showed only minimal difference between the two LOD values. Thermogravimetric analysis performed on the anhydrous lactose stored under the same conditions demonstrated a weight loss of 4.721% between 105°C and 150°C (Figure 6.1), which further confirmed the complete conversion of anhydrous lactose to monohydrate under high RH. The X-ray pattern of anhydrous lactose subjected to high RH was similar to the pattern of the lactose monohydrate control and different from the control anhydrous lactose (Figure 6.2). This established the conversion of anhydrous lactose to monohydrate under high RH.

Sieve analysis revealed that anhydrous lactose, when stored under high RH, agglomerates and the mean geometric diameter increased from 81.9 µm for control anhydrous lactose to 129.0 µm. The % fines decreased from 36.8% to 17.8% (Table 6.4). Because the three different lactose materials exhibited different particle size distribution, further physical characterization of these materials, which included BET surface area, mercury porosimetry, and compactibility measurements, were performed on the same size fraction materials (88–149 µm), which would eliminate any bias because of particle size differences.

BET surface area and total intrusion volume measurements of the different lactose materials for similar size fraction (88–149 µm) also are listed in Table 6.4. The BET surface area for the form converted material was 0.906 m^2/g, compared to 0.284 and 0.256 m^2/g for control anhydrous lactose and lactose monohydrate, respectively. Similar surface area measurements also were performed on the entire sample and showed surface area in the following order: Lactose monohydrate < Anhydrous lactose < Form converted material (Table 6.4), which is in agreement with other literature reports stating that β-anhydrous lactose has more pronounced particle surface irregularities than the α-lactose monohydrate, leading to higher surface area.

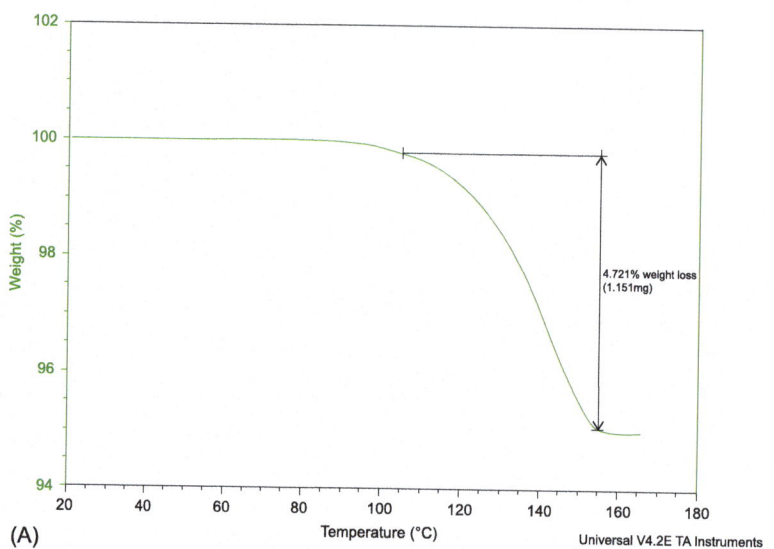

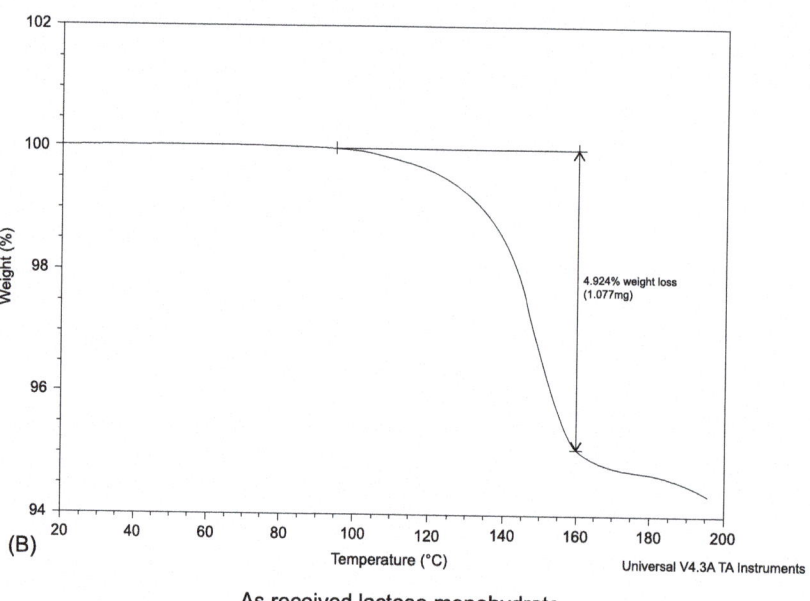

FIGURE 6.1

Thermogravimetric analysis of lactose (Badawy et al., 2000).

6.2 Solid state transformation in wet granulation

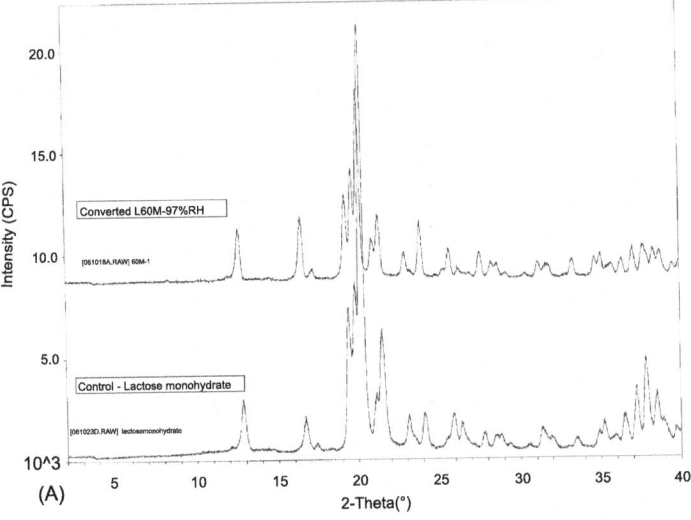

(A) Comparison of PXRD patterns of converted material (top) versus "As received" lactose monohydrate (bottom)

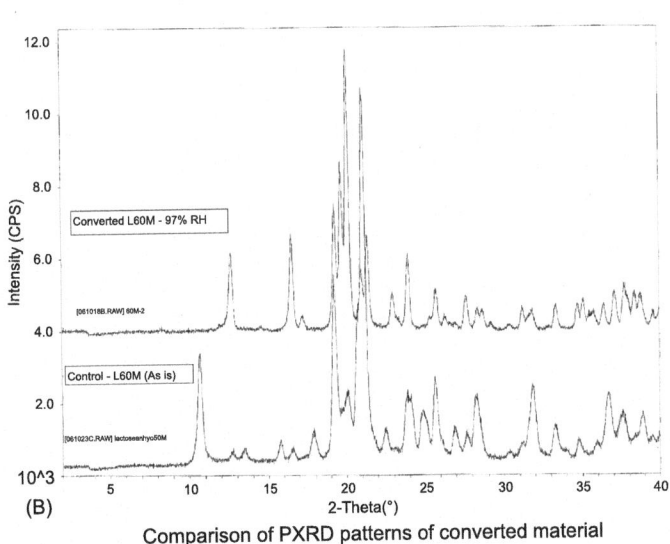

(B) Comparison of PXRD patterns of converted material (top) versus "As received" anhydrous lactose (bottom)

FIGURE 6.2

Powder X-ray diffraction (PXRD) patterns for different grades of lactose (Badawy et al., 2000).

Table 6.4 Physical characterization and compaction data for the different samples of lactose (Badawy et al., 2000).

Sample	Particle size by mesh analysis		Compaction parameters				BET surface area (m²/g)	Mercury porosimetry (total intrusion volume (mL/g) from 0.1 to 10.0 μm pore diameter)
	Geometric mean diameter (μm)	% Fines (< 75 μm)	Compactibility (kPa/MPa)	Bonding (MPa)		Yield pressure (MPa)		
Lactose monohydrate (as received)	57.4[a]	58.8[a]	5.2[b]	2.2[b]		281.3[b]	0.256[b]/0.350[a]	0.0598[b]
Anhydrous lactose (as received)	81.9[a]	36.8[a]	11.0[b]	4.3[b]		182.5[b]	0.284[b]/0.453[a]	0.1383[b]
Form converted material	135.9[a]	17.6[a]	11.8[b]	5.4[b]		258.3[b]	0.906[b]/0.884[a]	0.3318[b]

[a] *Entire sample.*
[b] *Size Fraction between 88 and 149 μm.*

Figure 6.3A and B illustrate the pore distribution of different lactose materials for similar size fractions between 88 and 149 μm. The total intrusion volume (total pore volume) for anhydrous lactose, determined by mercury porosimetry, increased from 0.1383 to 0.3318 mL/g, upon conversion to the monohydrate form. The total intrusion volume for the different samples of lactose is in the following order: Lactose monohydrate < Anhydrous lactose < Form converted material. This is consistent with the results from the BET surface area measurements.

The scanning electron micrograph imaging (Figure 6.4) illustrates the difference in the surface morphology between the different samples of lactose. The pictures confirmed the rougher and irregular surfaces of control anhydrous lactose particles as compared to lactose monohydrate. Anhydrous lactose upon exposure to high RH undergoes form conversion to monohydrate. Upon conversion, the anhydrous lactose particles change their morphology. Each particle of the resulting monohydrate form is an aggregate of small crystals with subsequently higher surface area and higher porosity. Thus, while both have the same crystal form, the commercial and converted lactose monohydrate materials have different particle morphology.

Anhydrous lactose also demonstrated increase in surface area upon wet granulation. For the DOE granulation containing 94.5% anhydrous lactose, granulated with 18% w/w water and dried using a tray dryer (Run 1 of Table 6.1), the percent of lactose conversion was 33.1% (Table 6.2). The surface area for this granulation was 0.756 m^2/g as compared to 0.453 m^2/g for the starting material anhydrous lactose. It was inferred that form conversion of anhydrous lactose during wet granulation also is accompanied by similar morphological changes as the form converted material under high humidity.

Table 6.4 shows the compaction parameters of the three different samples of lactose, for the same size fraction (88–149 μm). Compaction profiles are shown in Figure 6.5. As expected, the compactibility of as-received anhydrous lactose material (11.0 kPa/MPa) was significantly higher than the compactibility of as-received lactose monohydrate (5.2 kPa/MPa), which is in agreement with other literature reports (Lerk, 1993; Vromans et al., 1985; 1987). The compactibility of the form converted material (11.8 kPa/MPa) was comparable to the starting material, despite its conversion from anhydrous to monohydrate form during storage at high relative humidity. Compactibility of a given material depends upon two factors: bonding strength and compressibility, which is the ability to undergo volume reduction with pressure (densification tendency). Even though the compactibility of control anhydrous lactose and the form converted material is similar, considerable differences were observed in bonding and yield pressure. The form converted material demonstrated a higher bonding of 5.4 MPa compared to 4.3 and 2.2 MPa for the control anhydrous lactose and control lactose monohydrate, respectively. The yield pressure for the form converted material, which is inversely related to densification tendency, was 258.3 MPa as compared to 182.5 MPa for control anhydrous lactose. This points out that the form converted material requires more pressure to densify to a given solid fraction during compression than the control anhydrous lactose, indicating a lower compressibility for the converted material. The form converted material being an aggregate of very fine particles, is expected to have higher fragmentation tendency. The fine particles resulting from fragmentation of primary particles of the form converted material appear to be more resistant to densification compared to the anhydrous starting material and can be responsible for higher yield pressure and lower the compressibility. The higher fragmentation tendency, however, leads to a higher bonding surface area per unit volume, and therefore higher bonding strength is observed for the form converted material as compared to the control anhydrous lactose. Although the bonding strength increases, there is no increase in the compactibility because the lower

190 Chapter 6 Material physical modifications induced by wet granulation

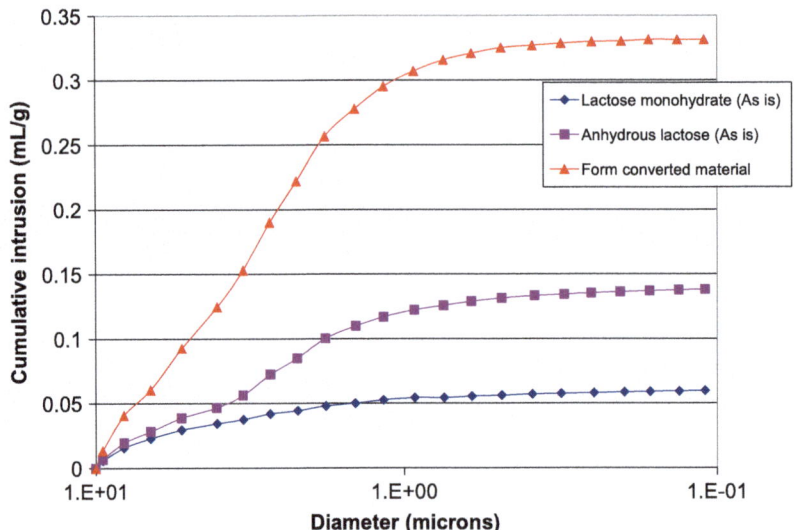

(A) Cumulative intrusion volume vs diameter

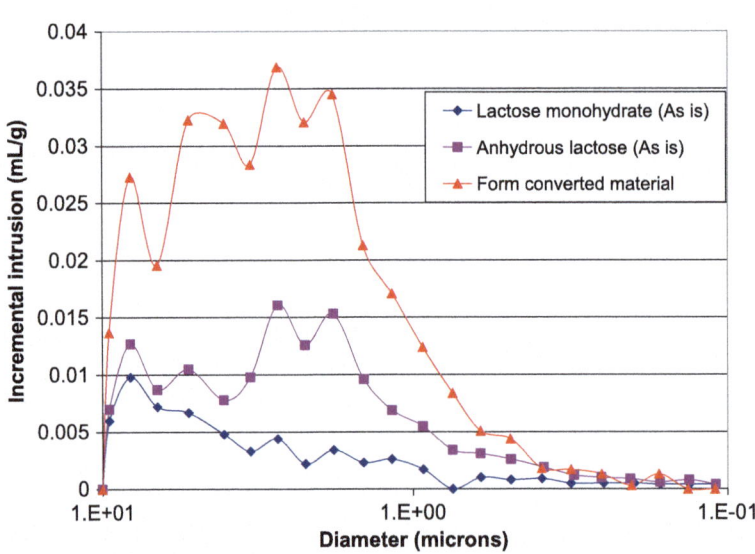

(B) Incremental intrusion volume vs diameter

FIGURE 6.3

Mercury intrusion porosimetry plots for different grades of lactose for size fraction between 88 and 149 μm (Badawy et al., 2000).

FIGURE 6.4

SEM Imaging of different grades of lactose (Badawy et al., 2000).

compressibility tends to act in an opposite direction, leading to similar compactibility to that of starting anhydrous lactose material.

6.2.5 Control of solid state transformations in wet granulation

Excipients can be used to stabilize a drug substance against form conversion during wet granulation. Excipients can inhibit solvent-mediated polymorphic and hydrate transformations by absorption of the solvent (water), thus making it less available to dissolve the active ingredient. In addition, specific interactions between a drug molecule and an excipient can slow down nucleation and/or crystal growth. In a study of the effect of several polymers on hydrate formation of three model compounds in wet granulation (Gift et al., 2009), found varying degrees of inhibitory effect on hydrate formation depending on the compound and polymer used. Cross-linked poly(acrylic) acid was found to completely inhibit caffeine hydrate transformation and both hydroxypropyl methylcellulose and crosslinked poly(acrylic) acid completely inhibited the carabamazepine hydrate formation. Sulfaguanidine transformation was rapid even in the presence of polymers. (Airaksinen et al., 2005) found that hygroscopic partially crystalline excipients hindered hydrate formation of nitrofurantoin in wet granulation, probably because of their significant uptake of granulating water. Form transition from a metastable to a stable form

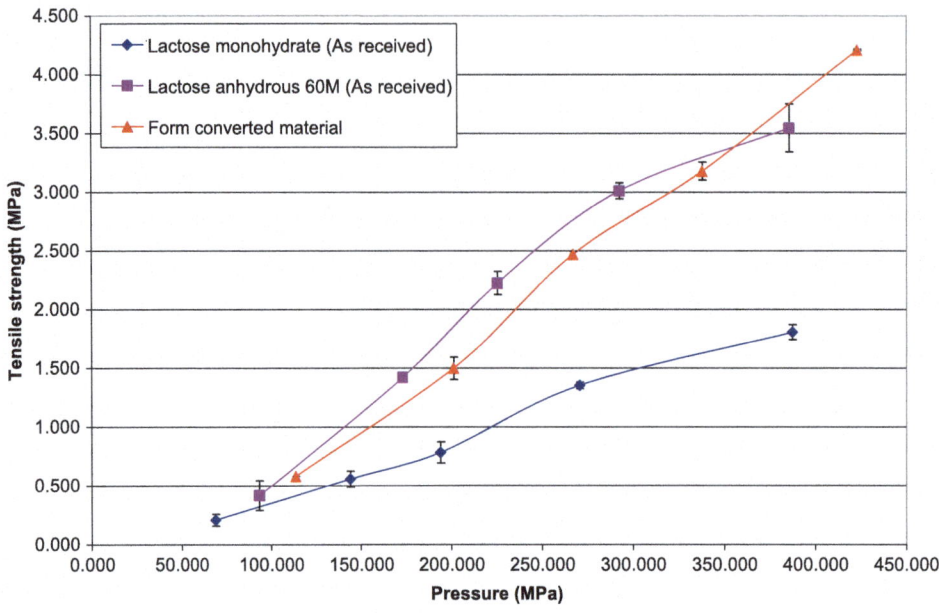

FIGURE 6.5

Compactibility profiles of different lactose materials with similar size fraction (between 88 and 149 μm) (Badawy et al., 2000).

of theophylline during wet granulation also was inhibited by povidone. Higher molecular weight of povidone was more effective in inhibiting the transition than the lower molecular weight polymer (Tantry, Tank, & Suryanarayanan, 2007).

The dispersion of the pioglitazone hydrochloride salt into polymeric matrices was used to prevent direct contact with croscarmellose sodium and reduce salt disproportion in wet granulation (Nie et al., 2017). In the absence of a mitigation strategy, significant disproportionation of the salt was observed in the formulation containing croscarmellose sodium during wet granulation. Dispersion of the salt in the polymeric matrix was found to be more effective than the physical mixture of the salt and the polymer. In addition, hydroxypropyl acetate succinate (HPMC-AS) was found to be more effective as the dispersion polymer in reducing disproportionation than polyvinylpyrrolidone-vinyl acetate (PVPVA) because of the presence of succinic acid groups in the HPMC-AS, which reduces the micro environmental pH during wet granulation.

6.3 Micromeritic changes in wet granulation
6.3.1 Microcrystalline cellulose

Microcrystalline cellulose (MCC) is another widely used pharmaceutical excipient in the manufacture of solid dosage forms because of its desirable attributes. In particular, MCC has excellent compactibility, which makes it a valuable excipient in the formulation of a tablet dosage form. Partially crystalline cellulose fibrils represent the basic structure component of MCC. Those fibrils agglomerate into fibril aggregates, which then are combined to form the particles of commercially available MCC grades (Staniforth, Baichwal, Hart, & Heng, 1988). It has been recognized that MCC compactibility can be reduced significantly upon wet granulation (Gustafsson, Lennholm, Iversen, & Nyström, 2003). The change in the nature of MCC has been suggested as an underlying mechanism for the decreased compactibility. Buckton, Yonemochi, Yoon, and Moffat (1999) suggested a change in the internal bonding within the cellulose upon wet granulation and drying, which produced an altered physical structure and was manifested by the change in enthalpy of water sorption. The change in the extent of hydrogen bonding between the cellulose hydroxy groups also has been suggested to occur upon wet processing (Nakai, Nakajima, Fukuoka, & Hasegawa, 1977). Later reports, however, failed to detect significant differences in the degree of hydrogen bonding because of wet granulation and extrusion of MCC (Millili, Wigent, & Schwartz, 1996). Suzuki, Kikuchi, Yamamura, Terada, and Yamamoto (2001) suggested a change in the crystallinity of MCC upon high-shear wet granulation, where the crystallite size of cellulose determined by an X-ray diffraction technique was found to decrease with increasing granulation time and the amount of granulating water. Other authors have suggested a crystallite-gel model (Kleinebudde, 1997) or a sponge model (Ek, Newton, & Kleinebudde, 1998) to describe the performance of MCC during wet processing.

The agglomeration of primary particles of MCC into larger dense granules also can be a contributing mechanism to its diminished compactibility, even without changes in primary particle structure. The over-granulation of lactose using excessive water and shear forces was found to decrease its compactibility upon wet granulation in a high-shear mixer. In this case, the failure of the large dense granules with reduced surface area to significantly fracture or deform upon application of the compression force was shown to result in the decreased compactibility (Wikberg and Alderborn, 1991; Badawy, et al., 2000). A similar observation was suggested for MCC pellets manufactured by an extrusion/spheronization technique. Tensile strength of compacts formed at a given compression pressure was inversely related to pellet porosity (Johansson, Wikberg, Ek, & Alderborn, 1995).

A study conducted by Badawy, Gray, and Hussain (2006) provided further understanding of the behavior of MCC in wet granulation. The relevance of particle and/or granule porosity to compactibility of MCC based formulations was investigated. The porosity changes in MCC upon wet granulation were characterized in order to distinguish between pores within primary MCC particles and those between particles in a granule. The effect of binder addition and subsequent milling of dried granules on MCC compactibility loss also was evaluated.

In the study by Badawy et al., three granulation batches of microcrystalline cellulose (MCC) were manufactured by high-shear wet granulation. Two of the batches contained 3% hydroxypropyl cellulose (HPC), while the third batch consisted of only MCC. The two batches containing HPC were granulated using two different granulation parameters (overgranulated and undergranulated batches). The overgranulated batch was obtained by using higher impeller speed, higher amount of granulating water, and a longer wet massing time than the undergranulated batch. Formulation without HPC was granulated using identical parameters as the overgranulated batch with HPC.

For the two overgranulated batches, a sample of the dried and screened granulation was ball milled using a micro mill. Lubrication of the granulation was performed by blending the dried and screened granulation with a quantity of magnesium stearate representing 0.5% of the final batch weight in a V-blender.

6.3.1.1 Particle size distribution

Particle size distribution of the dried granulation and MCC starting material was determined by mesh analysis using a sonic sifter. Particle size distribution of the dried granulation and MCC starting material also was determined by a dry laser diffraction technique.

Particle size measurements by mesh analysis and laser diffraction indicated minimal particle size growth for the undergranulated batch with HPC and the overgranulated batch without HPC (Table 6.5). Geometric mean diameter determined by mesh analysis has increased only from 81 to 87 μm and 101 μm for the two batches, respectively. Similarly, laser diffraction data do not suggest an increase in particle size for the two batches compared to the starting material. The overgranulated batch with HPC, however, showed significant increase in particle size compared to the starting material as indicated by the two particle size measurement techniques. Geometric mean diameter by mesh analysis was 214 μm and volume mean diameter by laser diffraction was 324 μm compared to 81 and 115 μm for the starting material determined by the two methods, respectively.

Ball milling of the two overgranulated batches reduced the particle size, resulting in a smaller particle size than the starting material. Volume mean diameter by laser diffraction was 41 and 34 μm for the two ball milled batches with and without HPC, respectively. The geometric mean particle size by mesh analysis was 54 and 89 μm for the two batches, respectively. The higher number for the ball milled batch without HPC in this case is attributed to agglomeration of the fine particles that would not break up during mesh analysis. The presence of agglomerates in this case was corroborated by visual observations.

6.3.1.2 Porosity and surface area

Porosity of the granulation and MCC starting material was tested by mercury intrusion porosimetry at different pressures ranging from 1 to 61,000 psi using a similar method as the lactose study. A particle size fraction passing through 70-mesh screen and retained on 100-mesh screen was used for the porosity determination, except for the ball milled material. Because of the small particle size of the ball milled material, it was not possible to collect a 70–100 mesh fraction and therefore the whole particle size distribution was used. Pore size distribution for tablets compressed using the same compression force for all three granulations also was determined by mercury intrusion porosimetry. In addition to incremental pore volume, incremental pore wall surface area was calculated for each increment. Total pore volume and pore area were calculated for pore diameters ranging from 0.1 to 10 μm. The normalized pore wall surface area per unit pore volume then was calculated by dividing total pore area by pore volume for each tablet.

In a granulation process, primary particles of a starting material agglomerate to form larger aggregates described as granules. Three types of pores can exist in a granulated system: void space between granules (intergranular pores); pores between primary particles within the granules (intragranular pores); and pores within the primary particles (intraparticulate pores).

Pore size distribution by mercury intrusion porosimetry of a granulated system shows intergranular (void) space at lower pressures and intragranular/intraparticulate (pore) space at higher pressures. Void space is a function of particle size/packing and provides little useful information regarding pore structure. Intraparticulate/intragranular space, however, is directly related to pore structure and has been shown to correlate well with important particle performance characteristics such as compactibility (Wikberg and Alderborn, 1991; Badawy et al., 2000).

Table 6.5 Physical characterization of granules and MCC starting material (Badawy et al., 2006).

	BET surface area (m²/g)	BJH pore volume, 0.002–0.07 μm pore diameter (mL/g)	Geometric mean diameter by sieve analysis (μm)	Malvern particle size Volume mean diameter (D[4,3], μm)	Mercury intrusion porosity (mL/g)	
					1–10 μm pore volume	0.1–1 μm pore volume
Avicel PH102	1.09	0.0032	81.3	114.9	0.4081[a]	0.0791[a]
Undergranulated, HPC	0.91	0.0023	87	125.1	0.3639[a]	0.0724[a]
Overgranulated, HPC	0.53	0.0013	214.3	324.1	0.0588[a]	0.0461[a]
Overgranulated, HPC, ball milled	1.33	0.0060	53.9	41.4	0.4358	0.0481
Overgranulated, no HPC	0.34	0.0008	101.1	97.6	0.1324[a]	0.0398[a]
Overgranulated, no HPC, ball milled	1.24	0.0053	88.7	34.3	0.5431	0.0348

[a] 70–100 mesh fraction.

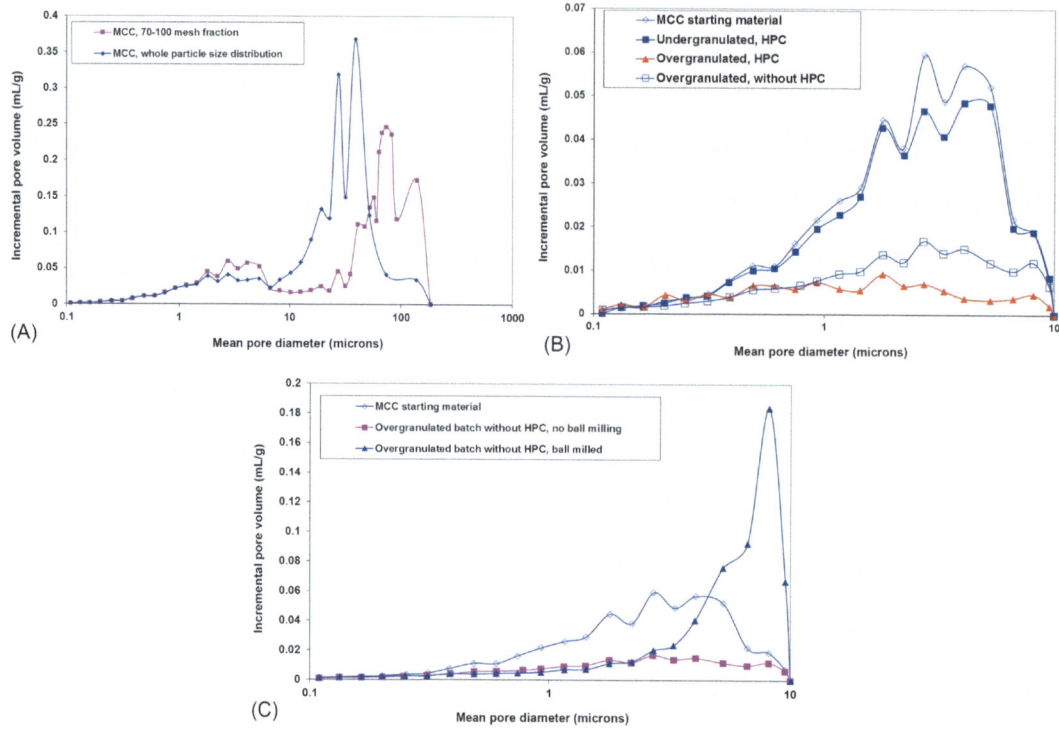

FIGURE 6.6

Mercury intrusion pore size distribution of granulations of (A) MCC starting material, (B) granulation batches, and (C) ball milled batch without HPC. Incremental intrusion volume represents pore volume associated with corresponding mean pore diameter (Badawy et al., 2006).

MIP pore distribution profiles typically are bimodal or multimodal with the peaks at the larger pore sizes representing void space (Figure 6.6A). In this study by Badawy et al., 10 μm was taken as the cut-off point between pores and void space with values below 10 μm representing pores and values above 10 μm representing void space. To enhance resolution, samples were fractionated by sieve and the 70–100 mesh fraction was used for the MIP studies. Results for total pore volume in both the 0.1–1.0 μm and the 1.0–10.0 μm ranges are listed in Table 6.5.

MCC starting material showed higher pore volume at all pore sizes than the granulated samples (Figure 6.6B). Undergranulated material showed a modest decrease in porosity, while overgranulated material (with or without binder) showed a substantial decrease in porosity. Particle size in the overgranulated batch without the binder varies little from MCC indicating minimal granule growth. As a result, particles in the 70–100 mesh fraction are mainly primary MCC particles. The diminished porosity in this case is likely because of pore structure changes between the fibrils and/or fibril aggregates (Ek, Alderborn, & Nyström, 1994) within the MCC primary particles.

In the case of the overgranulated batch containing HPC, the significant particle growth suggests that particles in the 70–100 mesh range are a mixture of large primary particles and granulated small particles. Regardless of the origin of the particles, they appeared to have the lowest pore volume in the 0.1–10 μm range among all samples tested. In addition to the decreased primary particle porosity, it also was inferred that granules formed during processing of this batch are likely to possess low pore volume between the primary particles in the granule.

Ball milling of the overgranulated batches reduced particle size to such an extent that distinguishing between void space and pore structure became difficult in the 1.0–10.0 μm range. In the 0.1–1.0 μm range, however, the contribution from pore structure is enhanced. For this reason, evaluation of the porosity for the ball milled batches was performed by examining pore volumes in the 0.1–1 μm range. For both milled batches, pore volume in this range was similar to pore volume in the unmilled samples suggesting that ball milling does not affect the porosity in this pore size range (Figure 6.6C).

Surface area and porosity of the lubricated granulation and MCC starting material also was determined using a nitrogen adsorption technique. The amount of nitrogen adsorbed was determined at partial nitrogen vapor pressure (P/P_0) ranging between 0.05 and 0.98. Surface area was determined by the (BET) calculation for the nitrogen adsorption data in the P/P_0 range from 0.05 to 0.30. Pore size distribution and total pore volume was calculated by using Barrett, Joyner and Halenda (BJH) analysis for the adsorption data in the pore diameter range of 0.002–0.07 μm.

Granulation of MCC decreased its surface area in all cases (Table 6.5). For the under-granulated HPC-containing batch, there was approximately 17% decrease in surface area compared to ungranulated MCC. The decrease in surface area was even more pronounced for the overgranulated batches. Surface area reduction was approximately 51% and 69% for the overgranulated batches with and without HPC, respectively. The decrease in surface area appears to be more than expected from the increase in particle size alone, particularly in the case of the overgranulated batch without HPC. Total surface area is the result of external surface area, which is inversely related the particle size and pore surface area within the particles or granules. While minimal increase in particle size was observed for the overgranulated batch without HPC, its surface area was significantly lower than MCC and other granulated batches. Thus, it appears that the decrease in surface area in this case is primarily because of decreased porosity. Ball milling increased the surface area of overgranulated batches and the surface area of milled materials was even slightly higher than ungranulated MCC.

Pore volume distribution also was obtained for pores in the 0.002–0.07 μm range from the nitrogen adsorption studies using BJH analysis (Table 6.5). Because of the small pore size in this case, pore structure examined by this technique is almost exclusively limited to pores within the primary particle rather than pores between the primary particles in the granule structure. Similar to 0.1–10 μm pore diameter range examined by MIP, wet granulation also diminished pore volume in the 0.002–0.07 μm range. Significant reduction of the pore volume in the later range was observed for all granulated batches with more pronounced reduction observed for the two overgranulated batches. The decreased porosity resulting from wet granulation was reflected in the surface area reduction of the granulated batches, which was more pronounced than what would be expected from particle size enlargement alone.

Pore volume in the 0.002–0.07 μm range for the ball milled material was substantially higher than the unmilled material from the same batch and higher than ungranulated MCC. This can be attributed to the formation of cracks or fissures during the ball milling process. It is noteworthy that ball milling did not significantly change pore volume in the 0.1–1 μm range as determined by MIP, suggesting that any created cracks are mostly below 0.1 μm. The higher porosity of the ball milled material in the 0.002–0.07 μm range is likely to be a contributing factor to its higher surface area as described above.

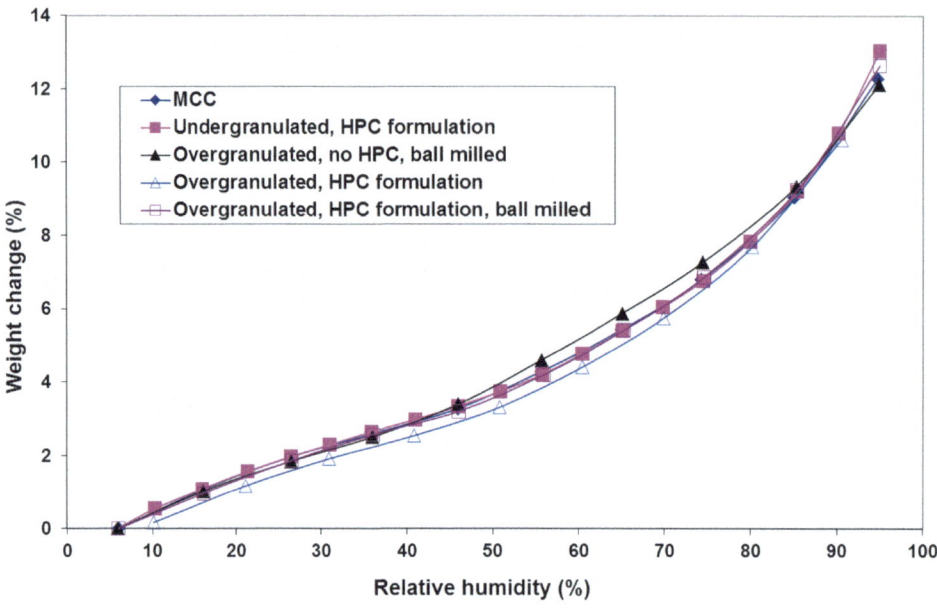

FIGURE 6.7

Moisture uptake isotherms of granulations and of MCC starting material at 25°C (Badawy et al., 2006).

6.3.1.3 Moisture uptake studies

Moisture uptake by the granulations and MCC starting material was determined at 25°C. Moisture uptake by the sample was determined as the relative humidity was increased in increments of 5% or 10%. Equilibrium moisture uptake was determined by measuring weight gain at each relative humidity. Equilibrium was deemed to be achieved when the rate of weight change is < 0.005% per minute.

Adsorption isotherms of the different granulations, starting material, and ball milled granulations were essentially identical (Figure 6.7). Most of the moisture uptake by MCC and its processed formulations results from sorption by the amorphous regions of MCC (Zografi and Kontny, 1986). Water uptake by surface adsorption represents only a small fraction of the total moisture uptake by MCC, which explains the similar hygroscopicity of the different batches evaluated here despite their surface area differences. The similar moisture uptake of those formulations and the starting material, however, suggests very little change in amorphous to crystalline ratio during processing.

6.3.1.4 Compaction properties

Compaction profiles of the lubricated and unlubricated granulations and MCC starting material was obtained using a compaction simulator following the same approach as described for the lactose study. The two compaction parameters calculated from the compaction simulator data were the yield pressure (calculated from the slope of the linear portion of the natural logarithm of the reciprocal tablet porosity versus compression pressure) and bonding parameter (calculated as the Y-intercept of the plot of tensile strength versus tablet porosity and represents theoretical tablet tensile strength at zero porosity).

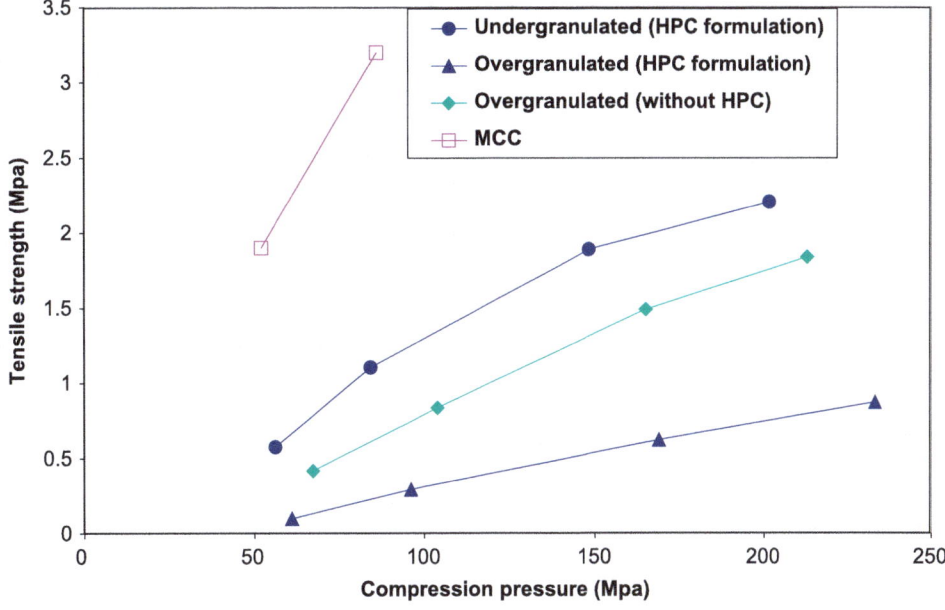

FIGURE 6.8

Compaction profiles of un-lubricated MCC granulation and starting material (Badawy et al., 2006).

As expected, ungranulated MCC showed excellent compactibility. In all cases, granulated formulations were less compactible than MCC. Undergranulated HPC formulation showed higher compactibility than the same formulation processed using higher amount of water and more shear (overgranulated) (Figure 6.8).

Overgranulated formulation without HPC showed higher compactibility than the overgranulated HPC formulation that was manufactured using identical granulation parameters. The incorporation of HPC appeared to make the formulation more susceptible to reduction in compactibility when overgranulated, probably as it enhanced the growth of dense granules under those processing conditions. The overgranulated HPC-free formulation was still less compactible than MCC and the undergranulated HPC formulation. Ball milling of the two overgranulated formulations resulted in significant enhancement of compactibility approaching that of the ungranulated MCC (Figure 6.9).

Wet granulation of MCC resulted in an increase in yield pressure and a decrease in bonding parameter calculated from the compaction simulator studies (Table 6.6). The yield pressure is inversely related to the volume reduction propensity of the material and therefore an increase in yield pressure indicates lower tendency for plastic deformation or fragmentation. The bonding parameter is decreased if bond strength is decreased or if bonding surface area per unit volume is decreased. It is noteworthy that increased fragmentation or plastic deformation also can contribute to enhanced bonding parameter by increasing bonding surface area within a given volume.

The increase in yield pressure upon wet granulation can be the result of the decreased porosity observed for the wet granulated material. In addition to reducing MCC primary particle porosity, the wet

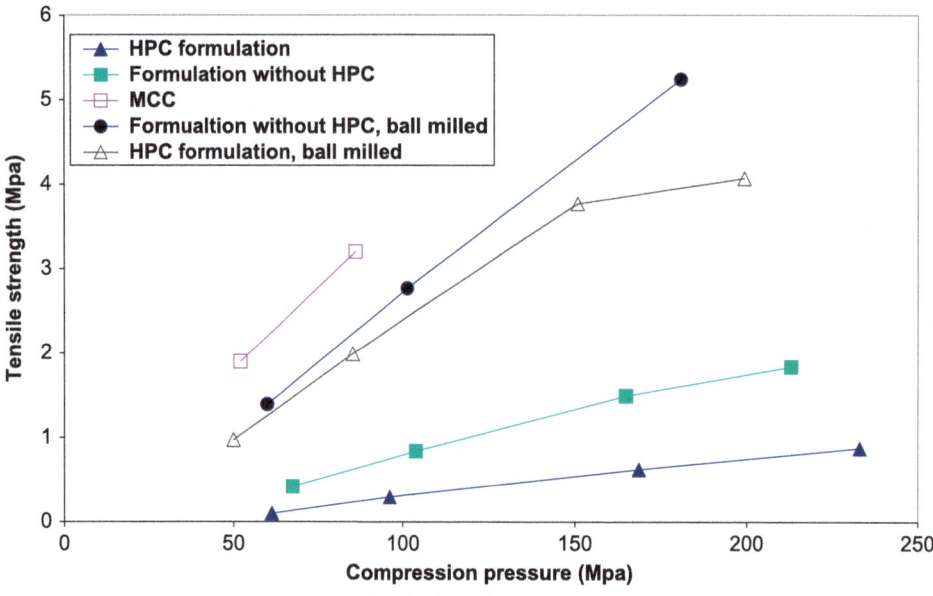

FIGURE 6.9

Effect of milling on compactibility of over-granulated MCC formulations (Badawy et al., 2006).

Table 6.6 Compaction parameters for the different MCC granulations and starting material (Badawy et al., 2006).

	Compactibility (kPa/MPa)	Yield pressure (MPa)	Bonding (MPa)
Avicel PH102	41.03	72.84	5.42
Undergranulated, HPC	11.12	101.03	2.67
Undergranulated, HPC (lubricated)	9.31	106.51	2.16
Overgranulated, HPC	4.47	113.96	1.01
Overgranulated, HPC (lubricated)	0.60	133.45	0.13
Overgranulated, HPC, ball milled	21.54	92.11	4.94
Overgranulated, no HPC	9.85	107.09	2.27
Overgranulated, no HPC, ball milled	31.69	97.63	6.54

granulation process also produced granules with lower porosity, particularly for the overgranulated batch with HPC. Dense primary particles and/or granules with lower porosity are less prone to fragmentation or plastic deformation during compression, resulting in the observed increase in yield pressure.

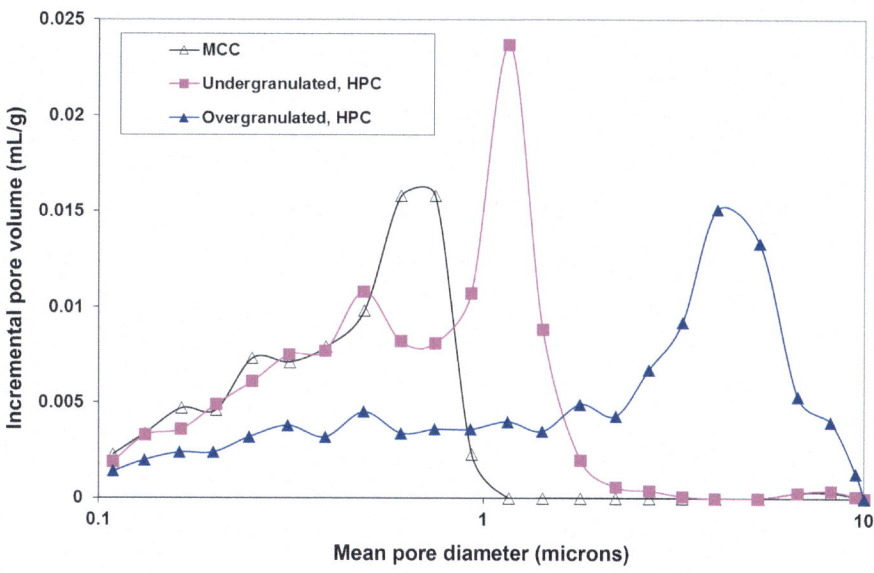

FIGURE 6.10

Mercury intrusion pore size distribution of MCC tablets obtained at 10 kN compression force (Badawy et al., 2006).

The decreased bonding parameter was attributed to the decreased bonding surface area per unit volume (bond volume density). This is supported by tablet pore size distribution, which showed the largest median pore diameter for the tablets compressed from overgranulated HPC batch, followed by the undergranulated HPC batch, and finally the ungranulated MCC (Figure 6.10). A smaller pore diameter of a tablet is reflected in a higher pore surface area (therefore higher bonding surface area) at a given porosity. Normalized pore area was 11.7, 8.9 and 5.2 m^2/mL for the MCC, undergranulated, and overgranulated batches, respectively. Decreased fragmentation or plastic deformation of the densified MCC particles/granules is a contributor to the observed decrease in pore surface area. It cannot be excluded, however, that decreased bonding parameter also might be because of reduced bond strength resulting from particle surface changes induced by wet granulation.

Ball milling of the overgranulated batches decreased their yield pressure compared to the unmilled material. The decrease in yield pressure was associated with the observed increase in pore volume in the 0.002–0.07 μm pore diameter range. Cracks induced by the milling process could be the reason for the increased porosity and the decreased yield pressure. The presence of cracks would decrease particle strength, therefore reducing yield pressure. Ball milling also increased bonding parameter, which could be attributed to increased bonding surface area per unit volume resulting from decreased particle size.

Lubrication with magnesium stearate diminished compactibility of both undergranulated and overgranulated batches containing HPC (Figure 6.11). Compactibility reduction was more pronounced for the overgranulated compared to the undergranulated batch. Magnesium stearate is known to coat granules and particles and interferes with bond formation during compression, diminishing compactibility. Granule and primary particle fragmentation during compression exposes fresh surfaces that are not

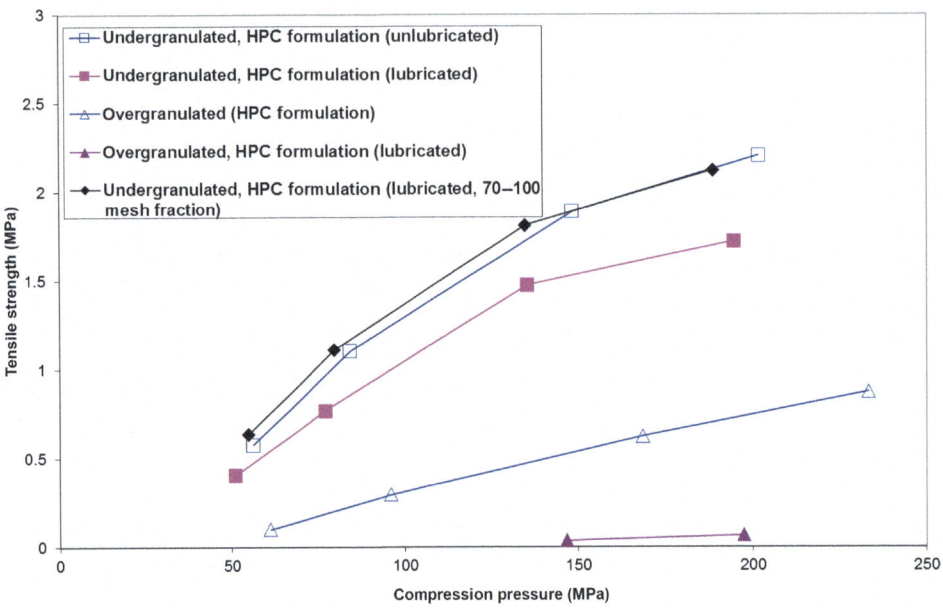

FIGURE 6.11

Compaction profiles of the unlubricated and lubricated MCC granulation batches (Badawy et al., 2006).

coated with magnesium stearate, which will enhance bonding and compactibility. The more porous primary particles and granules from the ungranulated batch might have higher fragmentation tendency than the less porous material from the overgranulated batch and therefore be less susceptible to the negative effect of magnesium stearate on compactibility.

Sieve fractions, 70–100 mesh, of the two lubricated granulations also were evaluated for compaction behavior. Compared to the whole granulation, compaction behavior of this large particle size fraction was different for the two batches. For the overgranulated batch, the 70–100 mesh fraction was less compactible than the whole granulation and intact tablets could not be obtained. On the contrary, the 70–100 mesh fraction of the undergranulated batch was more compactible than the whole granulation demonstrating a compaction profile similar to the unlubricated granulation. The difference in behavior between the two granulations was explained by the different structure and fragmentation propensity of the 70–100 mesh particles for the two granulations. Because of the minimal particle growth during granulation, the larger particle size fraction of the undergranulated batch consists mainly of large primary MCC particles. For this batch, the large primary particles appeared to have higher fragmentation tendency than the smaller primary particles present in the whole granulation resulting in a higher lubricant sensitivity for the whole granulation. The large particle size fraction of the over-granulated batch, however, contains a higher proportion of densified large granules. Those granules have lower fragmentation tendency (because of diminished porosity) compared to the smaller primary particles and/or granules present in the whole granulation resulting in a higher lubricant sensitivity for the 70–100 mesh fraction.

6.4 Summary and conclusions

Mechanical and solvent stresses in wet granulation can induce physical changes in the drug substance or excipients undergoing granulation. These physical changes can be in the form of solid state transformation such as polymorphic change, salt disproportionation, or hydrate formation. Micromeritic (particle structure/property) changes also can take place during wet granulation.

Partial transformation of anhydrous lactose to lactose monohydrate in high shear wet granulation is an example of solid state transformations. Despite form conversion of anhydrous lactose to lactose monohydrate during wet granulation, it retains its inherent particle morphology, surface area, and porosity. The higher compactibility of anhydrous lactose was maintained after wet granulation, indicating that particle structure and not crystal structure is the factor governing performance of lactose in compaction.

The decrease in MCC particle porosity by wet granulation is an example of micromeritic changes in wet granulation. The decrease in MCC primary particle porosity resulted in the decrease in MCC compactibility even without significant particle size growth. Aggregation of MCC particles into large dense granules when the binder is included in the formulation resulted in further decrease in compactibility. Milling was able to counteract the effect of wet granulation on MCC compactibility as particle surface area and porosity are increased by milling.

References

Aguiar, A. J., Krc, J., Kinkel, A. W., & Samyn, J. C. (1967). Effect of polymorphism on the absorption of chloramphenicol from chloramphenicol palmitate. *Journal of Pharmaceutical Sciences, 56*(7), 847–853. https://doi.org/10.1002/jps.2600560712.

Airaksinen, S., Karjalainen, M., Kivikero, N., Westermarck, S., Shevchenko, A., Rantanen, J., et al. (2005). Excipient selection can significantly affect solid-state phase transformation in formulation during wet granulation. *AAPS PharmSciTech, 6*(2), E311–E322. [electronic resource] 10.1208/pt060241.

Angberg, M., Nystr, C., & Castensson, S. (1991). Evaluation of heat-conduction microcalorimetry in pharmaceutical stability studies. IV. The influence of microcrystalline cellulose on the hydration rate of anhydrous lactose. *International Journal of Pharmaceutics, 77*(2–3), 269–277. https://doi.org/10.1016/0378-5173(91)90327-K.

Angberg, M., Nyström, C., & Castensson, S. (1991). Evaluation of heat-conduction microcalorimetry in pharmaceutical stability studies. III. Crystallographic changes because of water vapor uptake in anhydrous lactose powder. *International Journal of Pharmaceutics, 73*(3), 209–220. https://doi.org/10.1016/0378-5173(91)90413-I.

Badawy, S., Hussain, M., Zhao, F., Ye, Q., Huang, Y., & Palaniswamy, V. (2009). Degradation pathways of a corticotropin-releasing factor antagonist in solution and solid states. *Journal of Pharmaceutical Sciences, 98*(8), 2636–2647. https://doi.org/10.1002/jps.21637.

Badawy, S., & Pandey, P. (2017). Design, development and scale-up of the high shear wet granulation process. In Y. Qiu, Y. Chen, G. Zhang, L. Yu, & R. V. Mantri (Eds.), *Developing solid oral dosage forms* (2nd ed.) (pp. 749–776). Academic Press.

Badawy, S. I., Menning, M. M., Gorko, M. A., & Gilbert, D. L. (2000). Effect of process parameters on compressibility of granulation manufactured in a high-shear mixer. *International Journal of Pharmaceutics, 198*(1), 51–61.

Badawy, S. I. F., Gray, D. B., & Hussain, M. A. (2006). A study on the effect of wet granulation on microcrystalline cellulose particle structure and performance. *Pharmaceutical Research, 23*(3), 634–640. https://doi.org/10.1007/s11095-005-9555-z.

Badawy, S. I. F., Williams, R. C., & Gilbert, D. L. (1999). Chemical stability of an ester prodrug of a glycoprotein IIb/IIIa receptor antagonist in solid dosage forms. *Journal of Pharmaceutical Sciences, 88*(4), 428–433. https://doi.org/10.1021/js9803297.

Berlin, E., Kliman, P. G., Anderson, B. A., & Pallansch, M. J. (1971). Calorimetric measurement of the heat of desorption of water vapor from amorphous and crystalline lactose. *Thermochimica Acta, 2*(2), 143–152. https://doi.org/10.1016/0040-6031(71)85043-8.

Blanco, M., Alcalá, M., González, J. M., & Torras, E. (2006). Near infrared spectroscopy in the study of polymorphic transformations. *Analytica Chimica Acta, 567*(2), 262–268. https://doi.org/10.1016/j.aca.2006.03.036.

Brittain, H. G., Bogdanowich, S. J., Bugay, D. E., DeVincentis, J., Lewen, G., & Newman, A. W. (1991). Lactose, Anhydrous. *Analytical Profiles of Drug Substances and Excipients, 20*(C), 369–398. http://www.elsevier.com. 10.1016/S0099-5428(08)60536-5.

Brittain, H. G., Bogdanowich, S. J., Bugay, D. E., DeVincentis, J., Lewen, G., & Newman, A. W. (1991). Lactose, anhydrous. In K. Florey (Ed.), *Analytical profiles of drug substances and excipients* (20, pp. 369–398). Academic Press.

Buckton, G., Yonemochi, E., Yoon, W. L., & Moffat, A. C. (1999). Water sorption and near IR spectroscopy to study the differences between microcrystalline cellulose and silicified microcrystalline cellulose before and after wet granulation. *International Journal of Pharmaceutics, 181*(1), 41–47. https://doi.org/10.1016/S0378-5173(98)00413-X.

Christensen, N. P. A., Rantanen, J., Cornett, C., & Taylor, L. S. (2012). Disproportionation of the calcium salt of atorvastatin in the presence of acidic excipients. *European Journal of Pharmaceutics and Biopharmaceutics, 82*(2), 410–416. https://doi.org/10.1016/j.ejpb.2012.07.003.

David, S. T., & Augsburger, L. L. (1977). Plastic flow during compression of directly compressible fillers and its effect on tablet strength. *Journal of Pharmaceutical Sciences, 66*(2), 155–159. https://doi.org/10.1002/jps.2600660205.

Davis, T. D., Morris, K. R., Huang, H., Peck, G. E., Stowell, J. G., Eisenhauer, B. J., Hilden, J. L., Gibson, D., & Byrn, S. R. (2003). In situ Monitoring of Wet Granulation Using Online X-Ray Powder Diffraction. *Pharmaceutical Research, 20*(11), 1851–1857. 10.1023/B:PHAM.0000003385.20030.9a.

Debnath, S., & Suryanarayanan, R. (2004). Influence of processing-induced phase transformations on the dissolution of theophylline tablets. *AAPS PharmSciTech, 5*(1), 39–49. doi:10.1007/bf02830576.

Ek, R., Alderborn, G., & Nyström, C. (1994). Particle analysis of microcrystalline cellulose: Differentiation between individual particles and their agglomerates. *International Journal of Pharmaceutics, 111*(1), 43–50. https://doi.org/10.1016/0378-5173(94)90400-6.

Ek, R., Newton, J. M., & Kleinebudde, P. (1998). Microcrystalline cellulose as a sponge as an alternative concept to the crystallite-gel model for extrusion and spheronization (multiple letters) [1]. *Pharmaceutical Research, 15*(4), 509–512. https://doi.org/10.1023/A:1011905222168.

Gao, J. Z. H., Gray, D. B., Motheram, R., & Hussain, M. A. (2000). Importance of inlet air velocity in fluid bed drying of a granulation prepared in a high shear granulator. *AAPS PharmSciTech, 1*(4), 3.

Gift, A. D., Luner, P. E., Luedeman, L., & Taylor, L. S. (2009). Manipulating hydrate formation during high shear wet granulation using polymeric excipients. *Journal of Pharmaceutical Sciences, 98*(12), 4670–4683. https://doi.org/10.1002/jps.21763.

Gustafsson, C., Lennholm, H., Iversen, T., & Nyström, C. (2003). Evaluation of surface and bulk characteristics of cellulose I powders in relation to compaction behavior and tablet properties. *Drug Development and Industrial Pharmacy, 29*(10), 1095–1107. https://doi.org/10.1081/DDC-120025867.

Johansson, B., Wikberg, M., Ek, R., & Alderborn, G. (1995). Compression behaviour and compactability of microcrystalline cellulose pellets in relationship to their pore structure and mechanical properties. *International Journal of Pharmaceutics, 117*(1), 57–73. https://doi.org/10.1016/0378-5173(94)00295-G.

Kinoshita, R., Ohta, T., Shiraki, K., Higashi, K., & Moribe, K. (2017). Effects of wet-granulation process parameters on the dissolution and physical stability of a solid dispersion. *International Journal of Pharmaceutics, 524*(1–2), 304–311. https://doi.org/10.1016/j.ijpharm.2017.04.007.

Kleinebudde, P. (1997). The crystallite-gel-model for microcrystalline cellulose in wet-granulation, extrusion, and spheronization. *Pharmaceutical Research, 14*(6), 804–809. https://doi.org/10.1023/A:1012166809583.

Kobayashi, Y., Ito, S., Itai, S., & Yamamoto, K. (2000). Physicochemical properties and bioavailability of carbamazepine polymorphs and dihydrate. *International Journal of Pharmaceutics, 193*(2), 137–146. https://doi.org/10.1016/S0378-5173(99)00315-4.

Lerk, C. F. (1993). Consolidation and compaction of lactose. *Drug Development and Industrial Pharmacy, 19*(17–18), 2359–2398. https://doi.org/10.3109/03639049309047195.

Millili, G. P., Wigent, R. J., & Schwartz, J. B. (1996). Differences in the mechanical strength of dried microcrystalline cellulose pellets are not because of significant changes in the degree of hydrogen bonding. *Pharmaceutical Development and Technology, 1*(3), 239–249.

Morris, K. R., Griesser, U. J., Eckhardt, C. J., & Stowell, J. G. (2001). Theoretical approaches to physical transformations of active pharmaceutical ingredients during manufacturing processes. *Advanced Drug Delivery Reviews, 48*(1), 91–114. https://doi.org/10.1016/S0169-409X(01)00100-4.

Nakai, Y., Nakajima, S., Fukuoka, E., & Hasegawa, J. (1977). Crystallinity and physical characteristics of microcrystalline cellulose. *Chemical and Pharmaceutical Bulletin, 25*(1), 96–101. https://doi.org/10.1248/cpb.25.96.

Nie, H., Xu, W., Taylor, L. S., Marsac, P. J., & Byrn, S. R. (2017). Crystalline solid dispersion: A strategy to slowdown salt disproportionation in solid state formulations during storage and wet granulation. *International Journal of Pharmaceutics, 517*(1–2), 203–215. https://doi.org/10.1016/j.ijpharm.2016.12.014.

Otsuka, M., Kanai, Y., & Hattori, Y. (2014). Real-time monitoring of changes of adsorbed and crystalline water contents in tablet formulation powder containing theophylline anhydrate at various temperatures during agitated granulation by near-infrared spectroscopy. *Journal of Pharmaceutical Sciences, 103*(9), 2924–2936. https://doi.org/10.1002/jps.24006.

Pan, D., Crull, G., Yin, S., & Grosso, J. (2014). Low level drug product API form analysis: Avalide tablet NIR quantitative method development and robustness challenges. *Journal of Pharmaceutical and Biomedical Analysis, 89*, 268–275. https://doi.org/10.1016/j.jpba.2013.11.011.

Reier, G. E., & Shangraw, R. F. (1966). Microcrystalline cellulose in tableting. *Journal of Pharmaceutical Sciences, 55*(5), 510–514. https://doi.org/10.1002/jps.2600550513.

Shah, K. R., Hussain, M. A., Hubert, M., & Farag Badawy, S. I. (2008). Form conversion of anhydrous lactose during wet granulation and its effect on compactibility. *International Journal of Pharmaceutics, 357*(1–2), 228–234. https://doi.org/10.1016/j.ijpharm.2008.02.008.

Shukla, A. J., & Price, J. C. (1991). Effect of moisture content on compression properties of directly compressible high beta-content anhydrous lactose. *Drug Development and Industrial Pharmacy, 17*(15), 2067–2081. https://doi.org/10.3109/03639049109048533.

Staniforth, J. N., Baichwal, A. R., Hart, J. P., & Heng, P. W. S. (1988). Effect of addition of water on the rheological and mechanical properties of microcrystalline celluloses. *International Journal of Pharmaceutics, 41*(3), 231–236. https://doi.org/10.1016/0378-5173(88)90198-6.

Suzuki, T., Kikuchi, H., Yamamura, S., Terada, K., & Yamamoto, K. (2001). The change in characteristics of microcrystalline cellulose during wet granulation using a high-shear mixer. *Journal of Pharmacy and Pharmacology, 53*(5), 609–616. https://doi.org/10.1211/0022357011775938.

Tantry, J. S., Tank, J., & Suryanarayanan, R. (2007). Processing-induced phase transitions of theophylline: Implications on the dissolution of theophylline tablets. *Journal of Pharmaceutical Sciences, 96*(5), 1434–1444. https://doi.org/10.1002/jps.20746.

Vromans, H., Bolhuis, G. K., Lerk, C. F., & Kussendrager, K. D. (1987). Studies of tableting properties of lactose. IX. The relationship between particle structure and compactibility of crystalline lactose. *International Journal of Pharmaceutics, 39*(3), 207–212. https://doi.org/10.1016/0378-5173(87)90218-3.

Vromans, H., De Boer, A. H., Bolhuis, G. K., Lerk, C. F., Kussendrager, K. D., & Bosch, H. (1985). Studies on tableting properties of lactose—Part 2. Consolidation and compaction of different types of crystalline lactose. *Pharmaceutisch Weekblad Scientific Edition, 7*(5), 186–193. https://doi.org/10.1007/BF02307575.

Whiteman, M., & Yarwood, R. J. (1988). The evaluation of six lactose-based materials as direct compression tablet excipients. *Drug Development and Industrial Pharmacy, 14*(8), 1023–1040. https://doi.org/10.3109/03639048809151918.

Wikberg, M., & Alderborn, G. (1991). Compression characteristics of granulated materials. IV. The effect of granule porosity on the fragmentation propensity and the compatibility of some granulations. *International Journal of Pharmaceutics, 69*(3), 239–253. https://doi.org/10.1016/0378-5173(91)90366-V.

Wikström, H., Carroll, W. J., & Taylor, L. S. (2008). Manipulating theophylline monohydrate formation during high-shear wet granulation through improved understanding of the role of pharmaceutical excipients. *Pharmaceutical Research, 25*(4), 923–935. https://doi.org/10.1007/s11095-007-9450-x.

Wikström, H., Marsac, P. J., & Taylor, L. S. (2005). In-line monitoring of hydrate formation during wet granulation using Raman spectroscopy. *Journal of Pharmaceutical Sciences, 94*(1), 209–219. https://doi.org/10.1002/jps.20241.

Zografi, G., & Kontny, M. J. (1986). The interactions of water with cellulose- and starch-derived pharmaceutical excipients. *Pharmaceutical Research: An Official Journal of the American Association of Pharmaceutical Scientists, 3*(4), 187–194. https://doi.org/10.1023/A:1016330528260.

CHAPTER 7

Current practices in wet granulation-based generic product development

Rajan Verma[a,b], Maneesha Patil[a,b] and Carlos O. Paz[c]

[a]*Formulation Research and Development, Perrigo Laboratories India Private Limited, Ambernath, India*, [b]*Innovation & Development (I&D) Center, Abbott India Private Limited, Mumbai, India*, [c]*Consumer Health Care (CHC) Research and Development, Perrigo Company Plc, Allegan, MI, United States*

7.1 Introduction

Among the various routes to deliver drugs to the human body, the oral route is the most preferred because of patient compliance. Drug delivery via the oral route can be achieved using different dosage forms, such as tablets, liquids (solutions, suspensions, etc.), and capsules. Solid dosage forms constitute about two-thirds of all dosage forms available (Augsburger & Hoag, 2008).

Granulation is one of the most important unit operations in the production of pharmaceutical dosage forms. In context to the pharmaceutical industry, granulation refers to a process in which a powder mixture (often consisting of drug and excipients) is mixed and made to agglomerate to form a larger entity, called granules, by the process of dry compaction or by using a binder solution.

Dry granulation or dry compaction is a technique to form granules without using a liquid solution (Kleinebudde, 2004). This process is used mostly for products that are heat and/or moisture sensitive and readily compatible. In dry granulation, high pressure is applied to form agglomerates, e.g., by compression of powders within a cavity to form loose and large tablets (also known as slugs) in a heavy-duty tablet press or by passing the powder mixture into the gap between two counter rotating rollers (i.e., using roller compactor) to produce a sheet of materials (called ribbons or compacts). The resulting agglomerates can be milled and sifted to a target particle size distribution.

Among all the powder agglomeration processes, wet granulation is one of the most commonly used techniques. It consists of the agglomeration of different powders through the addition of a granulating fluid and mixing. The liquid binds the particles together by a combination of capillary and viscous forces until more permanent bonds are formed by subsequent drying (Iveson, Litster, Hapgood, & Ennis, 2001). Granulation is performed to create free-flowing particles having good compressibility. Another reason is to prevent the segregation of critical components in a powder mixture by reducing the difference in size and density between different powders. The granules are processed further, such as tableting, capsule filling, and making dry suspension, to manufacture the finished dosage form or drug product.

(i) Wetting & nucleation

(ii) Consolidation & coalescense

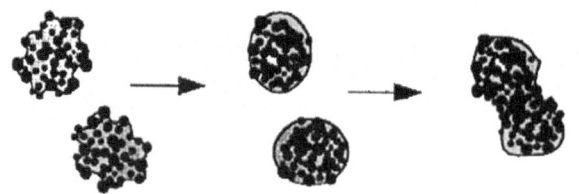

(iii) Attrition & breakage

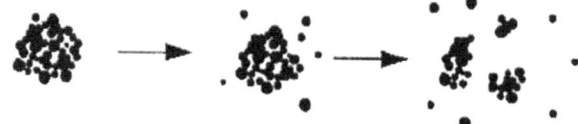

FIGURE 7.1

Schematic view of granulation process.

From Iveson, S. M., Litster, J. D., Hapgood, K., & Ennis, B. J. (2001). Nucleation, growth and breakage phenomena in agitated wet granulation processes: a review. Powder Technology, 3–39.

7.2 Mechanism of granule formation

Three sets of rate processes are important in determining wet granulation behavior: wetting and nucleation; consolidation and growth; and breakage and attrition (Ennis and Litster, 1997). Each of these are mentioned below (Iveson et al., 2001) and shown schematically in Figure 7.1.

7.2.1 Wetting and nucleation

The liquid binder is brought into contact with a dry powder bed and is distributed through the bed to generate a distribution of nuclei granules. This mechanism is regarded as an important stage in granulation processes, however, it is rarely identified and separated from other effects such as coalescence and attrition.

7.2 Mechanism of granule formation

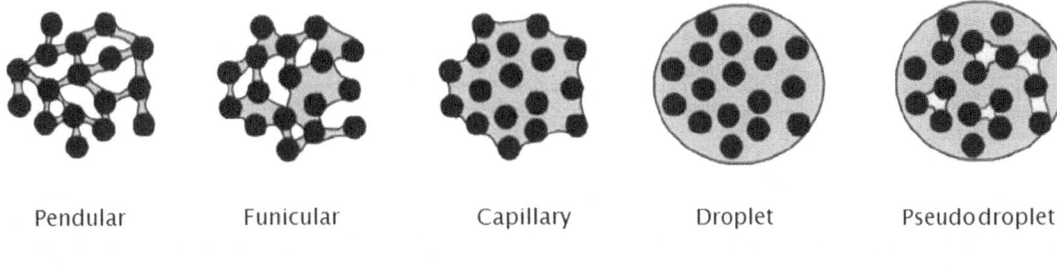

| Pendular | Funicular | Capillary | Droplet | Pseudodroplet |

FIGURE 7.2

The different states of saturation of liquid-bound granules.

From Iveson, S. M., Litster, J. D., Hapgood, K., & Ennis, B. J. (2001). Nucleation, growth and breakage phenomena in agitated wet granulation processes: a review. Powder Technology, 3–39.

Two processes are important in the nucleation zone. The first stage is nuclei formation, which is a function of wetting thermodynamics and kinetics. The second stage is dispersion of binder within the powder mix to ensure effective mixing of the powder and binder, which is a function of process variables. Choosing a poor combination of powder and binder (e.g., a high contact angle) or using an inefficient binder dispersion method (e.g., high liquid flow rate or poor spray characteristics) produces a product that is difficult to control and reproduce.

7.2.2 Consolidation and growth

In this stage, collisions between two granules, granules and feed powder, or a granule and the equipment lead to granule compaction and growth. Granule growth occurs whenever material in the granulator collides and sticks together. For two large granules, this process traditionally is referred to as coalescence, whereas the sticking of fine material onto the surface of large preexisting granules often is termed layering.

These growth processes can begin as soon as liquid is added to an agitated powder mass (i.e., simultaneous with the wetting and nucleation stage) and can continue well after liquid addition has been completed. Whether a collision between two granules results in permanent coalescence depends on a wide range of factors, including the mechanical properties of the granules and the availability of liquid binder at or near the granule surface.

Granules can exist in a number of different states of liquid saturation (Newitt and Conway-Jones, 1958), as shown in Figure 7.2.

- In the *pendular* state, particles are held together by liquid bridges at their contact points (pendular bonds).
- The *capillary* state occurs when a granule is saturated. All the voids are filled with liquid and the surface liquid is drawn back into the pores under capillary action.
- The *funicular* state is a transition between the pendular and capillary state in which the voids are not fully saturated with liquid.
- The *droplet* state occurs when the particles are held within or at the surface of a liquid drop.
- A *pseudodroplet* state occurs when unfilled voids remain trapped inside the droplet.

During granulation, it is possible for the saturation state of the granules to shift from the pendular state through to the droplet state, either because of the continuous addition of liquid binder or because of consolidation reducing the granule porosity.

7.2.3 Attrition and breakage

In this stage, wet or dried granules break because of impact, shear, or compaction in the granulator or during subsequent product handling. Breakage of wet granules influences the final granule size distribution, especially in high-shear granulators. In some circumstances, breakage can limit the maximum granule size or help distribute a viscous binder. On the other hand, attrition of dry granules leads to the generation of dusty fines.

7.3 Granulation methodologies

Quite a number of granulation technologies are available to pharmaceutical manufacturers. The choice of a specific process often depends on the properties of the input materials (primarily the drug for high drug-load formulations), desired final product characteristics (e.g., granule particle size distribution and density, tablet hardness, dissolution), economic reasons (e.g., available equipment at the manufacturing site).

Wet granulation can be done using a high-shear or low-shear granulator, fluid bed granulator, or other techniques (e.g., single pot granulation, melt granulation, extrusion).

7.3.1 High-shear wet granulation process

High-shear wet granulation is one of the most commonly used techniques for the production of granules. A typical high shear granulator consists of a mixing bowl, an impeller, and a secondary mixer/chopper. In a high-shear granulator, mixing, densification, and agglomeration of wetted materials are achieved as a result of shearing and compaction forces exerted by the main impeller (Kristensen and Schaefer, 1987). Most impellers typically rotate at high speed on either a vertical or horizontal axis to create the agitation required for granulation. The vertical lifting of the powder bed against the gravitational pull and the centrifugal movement of the particles creates a well-mixed cascading ribbon movement of the particles within the granulator in which particle momentum is imparted by radial and tangential particle movement, and the weight of the powder bed.

The chopper is much smaller, a distance away from the impeller in the powder bed, and rotates at a much higher speed (1000–3000 rpm). The role of the chopper in granulation is usually to fracture larger agglomerates or cause growth of smaller agglomerates, depending on the feed properties, operating conditions, and the geometry of the mixer, impeller, and chopper.

Binder addition to high-shear granulators can be in the form of a liquid spray through nozzles or by pouring. The granulating liquid typically is added from a port in the lid. Care should be taken to add the granulating liquid away from the location of the chopper.

Typically, a high-shear wet granulation process includes six steps.

1 Loading the drug and excipient powder mixture into the mixer bowl.
2 Dry mixing of powder mixture for a short period of time (typically 5–15 min). Impeller speed and the need of chopper (on, off, or intermittent on) is decided based on the properties of the powder mixture.
3 Addition of granulating liquid (by spraying/pouring) on the powder mixture surface, while the impeller is running. Chopper can be kept on or off based on product requirements. While chopper can help prevent dry lump formation in the dry mixing stage (step 2), during binder addition, the function of the chopper is to prevent formation of large agglomerates by breaking down the wet lumps.
4 After the binder addition is complete, impeller rotation is continued onto a stage of kneading or wet massing. During this stage, both the impeller and the chopper usually are run together.
5 Unloading of wet granules from the mixer followed by wet sieving (optional) and drying (in a tray dryer or a fluid-bed dryer).
6 Milling and sifting of dry granules, followed by further processing (e.g., blending, tableting, capsule filling).

In a high-shear wet granulation process, depending upon the choice of excipients, including binder, granule growth is usually very fast. In a well-controlled process, this is typically an advantage because the processing time is very short. To ensure that a wet granulation process is well-controlled (not undergranulated or overgranulated), it is of paramount importance to select the composition appropriately (right mix of water-soluble and water-insoluble diluent, binder, and other excipients) and to control the process.

7.3.1.1 Types of high-shear granulators

The two main types of batch high-shear granulators are vertical shaft and horizontal shaft. In vertical shaft granulators, an impeller can be on the bottom (Figure 7.3) or top-mounted (Figure 7.4), with a chopper located at or near the side of the bowl (chopper can be side-mounted or top-mounted). The impeller rotates the powder bed under the spray zone of the liquid binder.

In horizontal shaft granulators (Figure 7.5), the impeller lifts and distributes the powder under the liquid binder spray, and the chopper is located near the bottom of the bowl. Depending upon the input material characteristics and processing parameters, granule properties and/or finished product critical quality attributes (CQAs) might be affected by the design of the granulator. Kinetics of granule growth and mechanisms in the horizontal mixers differ from the vertical type, where the granules roll down the inner wall of the cylindrical mixing chamber. Normally, higher densification of the granules can be expected in the vertical mixer, where granules are assumed to be thrown extensively against the wall by centrifugal forces (Kristensen and Schaefer, 1987).

A few studies have compared granules prepared by top-drive and bottom-drive high-shear granulation processes.

In one of the studies (Prescott et al., 2004), it was found that granule properties were similar whether produced by top-drive or bottom-drive high shear granulators. For this study, blends of starch, microcrystalline cellulose, and lactose were granulated with increasing amounts of water using a 25-L top-drive (Vector GMX-25) and the same size bottom-drive granulator (Powrex FM-VG-25). Granulation parameters were optimized and water uptake ranged from 70% to 100% w/w of intragranular mass (110% w/w in the case of top-drive granulator). Post-granulation, the wet granules were fluid-bed dried

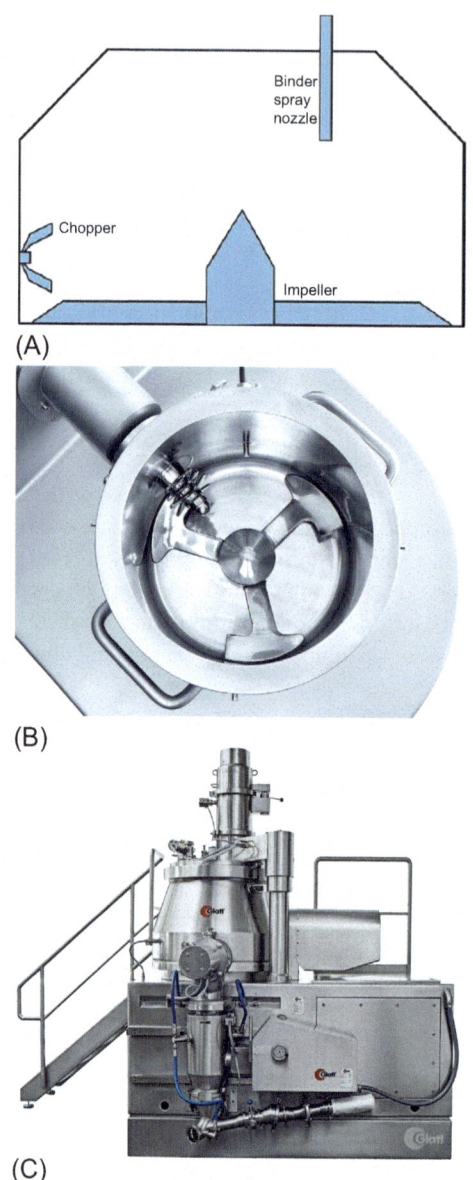

FIGURE 7.3

(A) Schematic diagram of a vertical-bottom mounted high shear granulator. (B) Laboratory model MTGL granulator with impeller and chopper. (C) Glatt VG Pro 600 high-shear granulator.

(A) From Briens, L., & Logan, R. (2011). The effect of the chopper on granules from wet high-shear granulation using a PMA-1 granulator. AAPS PharmSciTech, 1358–1365. (B) Courtesy of Loedige, Germany. (C) Courtesy of Glatt, Germany.

7.3 Granulation methodologies

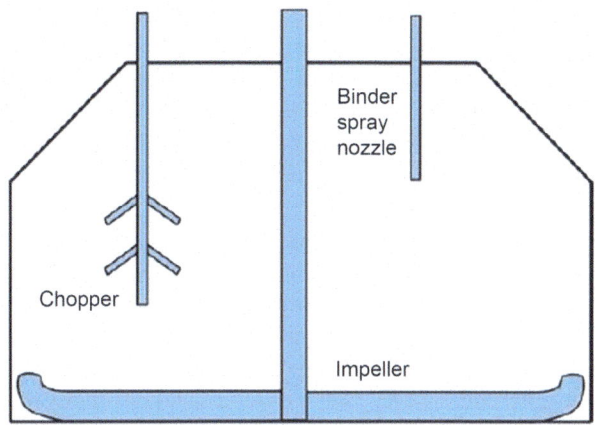

FIGURE 7.4

Schematic diagram of a vertical top-mounted high-shear granulator.

Source: Briens, L., & Logan, R. (2011). The effect of the chopper on granules from wet high-shear granulation using a PMA-1 granulator. AAPS PharmSciTech, 1358–1365.

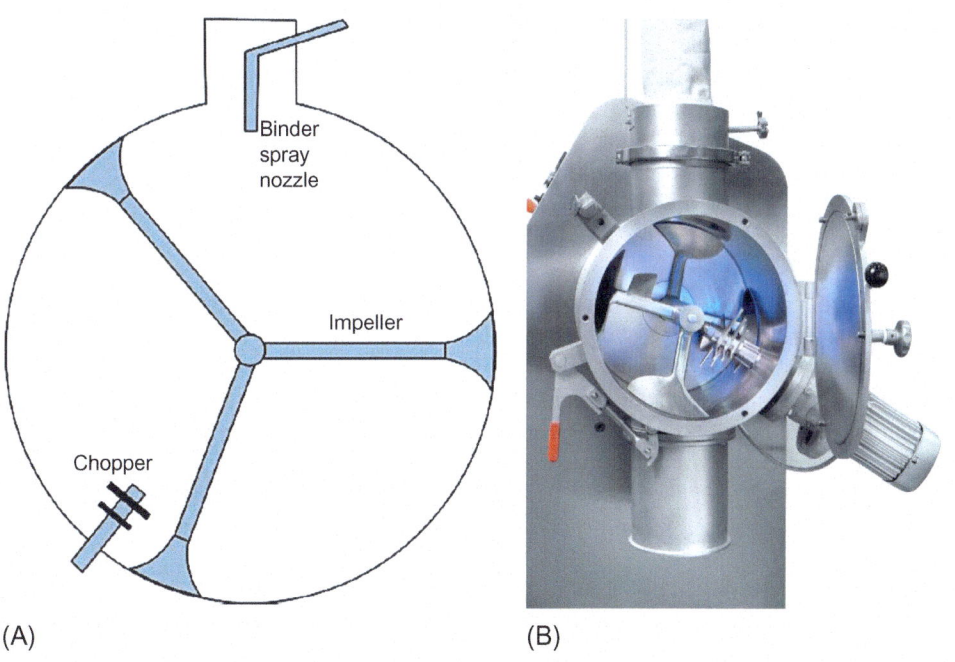

FIGURE 7.5

(A) Schematic diagram of a horizontal high-shear granulator. (B) Ploughshare Mixer L 50.

(A) Source: Briens, L., & Logan, R. (2011). The effect of the chopper on granules from wet high-shear granulation using a PMA-1 granulator. AAPS PharmSciTech, 1358–1365. (B) Source: Courtesy of Loedige, Germany.

and milled using a Fitz mill. The authors concluded that both top-drive and bottom-drive granulators produced granulations that were substantially similar. The enlargement of particle size (measured using Malvern Mastersizer 2000) was maximized at 90% w/w water addition for the bottom-drive and at 100% w/w for the top-drive. The granulations from each granulator had minimal cohesive strength at 80% w/w water addition. The granules were least prone to attrition when produced using 90% w/w water. Bulk density increased with increasing percentage of water, but leveled off at higher water level.

In another study (Smith *et al.*, 2003, 2004), granule properties and dissolution profile from tablets made using top-driven and bottom-driven high shear granulators were compared. Blends of pregelatinized starch, microcrystalline cellulose, and lactose were granulated with increasing amount of water using 25-L top drive (Vector GMX-25) and bottom-drive (Powrex FM-VG-25) high-shear granulators. The wet granules were dried in a fluid-bed dryer and blended with dye (as a marker for drug release during dissolution) and magnesium stearate. For the same amount of water added, mean diameter (measured using sieve analysis) and the percentage of coarse granules generated by the bottom-drive granulator was slightly higher than with the top-drive granulator. This difference was attributed to the shape of the product chamber and the location of the chopper. There was no significant difference in bulk density or the rise in temperature during high-shear granulation. Both granulators produced granulations with no significant differences in Carr's index; however, the flow of dried granules measured through narrow aperture were reduced at higher water levels. Compaction profiles were not dependent on the equipment design for the same level of water addition. The compressed tablets were subjected to dissolution testing. The dissolution profiles were similar (f_2 value ranged from 50 to 96) after the first 5 min for a given water level and compression force. Thus, the dissolution profiles were not dependent on the equipment design. For a given level of water addition, increasing compression force resulted in decreased tablet dissolution rates for both top-drive and bottom-drive high shear granulators.

7.3.1.2 *Case study I*

In one of the studies at Perrigo Labs, batches prepared by high-shear granulation process by top-driven and bottom-driven granulators were compared during the course of product development for an abbreviated new drug application (ANDA) of compound A. Compound A is a low-dose drug (20 mg), a calcium salt having molecular weight of 1209.4 g/mol, Log *P* value of 6.36 (o/w), and pKa of 4.46. During initial development trials, suitable particle size of API (micronized grade), a stable polymorph, and excipients were identified. The drug load for the study was approximately 7% w/w; other excipients were binder (hydroxypropyl cellulose), disintegrant (croscaremllose sodium), diluents (microcrystalline cellulose, lactose, and calcium carbonate), wetting agent (Tween 80), and lubricant (magnesium stearate).

For this study, composition of the batches was kept the same. Two granulators were evaluated: top-driven (GMX; Freund Vector, USA) and bottom-driven (Sams Techno Mech., India). Brief manufacturing steps are listed below:

1. Sifting of all raw materials and drug.
2. Dry mixing followed by wet granulation.
3. Wet milling followed by drying in a fluid-bed dryer.
4. Milling and sizing of granules.

Table 7.1 Processing parameters for bottom-drive and top-drive high-shear granulator.

Parameters	Bottom-driven granulation	Top-driven granulation
Premix time/Impeller speed	4 min/150 rpm	5 min/500 rpm
Binder addition time/Impeller speed	2 min/150 rpm	2 min/500 rpm
Wet mass time/Impeller speed	1 min/150 rpm	1 min/500 rpm
Chopper speed during wet massing	1500 rpm	1500 rpm

Table 7.2 Physical evaluation of the final lubricated blend for compound A made using bottom-drive and top-drive high-shear granulators.

Evaluation parameter	Bottom-driven granulator	Top-driven granulator
BD (bulk density in g/mL)	0.50	0.63
TD (tapped density in g/mL)	0.66	0.73
Compressibility index	24.2	13.9
Hausner ratio	1.32	1.16

5 Blending with extra-granular excipients and lubrication in V shell blender.
6 Compression followed by coating.

Although both the batches followed these steps, there were some differences in impeller speed (standard settings were used) because of different principles of the granulators. These are captured in Table 7.1.

Physical properties of the final lubricated blend are listed in Table 7.2. Hausner ratio and compressibility index is an indication of flow property of powders and is calculated by measuring bulk and tapped density of powders (USP ⟨1174⟩). Based on values, one can assess whether the powder is having good or a poor flow (the lower the value of these two parameters, the better the flow). Dissolution was evaluated in 900 mL of 0.1 N HCl using USP-II apparatus at 75 rpm and sampling was performed at 5, 10, 15, and 30 min. Analysis was carried out using a validated High-pressure liquid chromatography (HPLC) method. The dissolution results are listed in Table 7.3.

There were differences in the characteristics of the final blend of the batches made using two different types of granulators. The granules were denser with top-driven granulator with bulk density of 0.63 g/mL, compared to the granules made by bottom-driven granulation with bulk density of 0.50 g/mL tested as per USP ⟨616⟩. These differences in densities could be explained partly by the differences in equipment parameters. In addition, slower dissolution was observed for the batch made using top-driven granulator, as compared to the bottom-driven granulator. Granule densification correlated well with the dissolution results.

Table 7.3 Dissolution result comparison (of tablets of compound A) in 0.1 N HCl of batches made using bottom-drive and top-drive high-shear granulator.

Time (min)	% Dissolved (%RSD); $n = 6$		
	Innovator product	Batch manufactured with bottom-driven granulator	Batch manufactured with top-driven granulator
5	55 (2.3)	45 (3.0)	17 (5.3)
10	59 (2.6)	48 (2.4)	24 (4.1)
20	59 (3.0)	50 (1.5)	29 (2.2)
30	58 (3.4)	52 (1.1)	37 (3.2)

7.3.1.3 Continuous wet granulation

Traditionally, pharmaceuticals have been produced using a method known as batch manufacturing, which involves multiple discrete steps. After each step in the process, production typically stops so samples can be tested offline for quality. This increases the production times and, in case of increased demands, production might require larger equipment (requiring more physical space, a bigger footprint, and more time and money). Recent advances in manufacturing technology, however, have prompted the pharmaceutical industry to consider moving away from batch manufacturing to a faster, more efficient process known as continuous manufacturing. Pharmaceuticals made using continuous manufacturing are moved nonstop within the same facility, eliminating hold times between steps. Material is fed through an assembly line of fully integrated components. This method saves time, reduces the likelihood for human error, and can respond more quickly to market demands. The FDA is taking proactive steps to facilitate the drug industry's implementation of emerging technologies, including continuous manufacturing, to improve product quality and address many of the underlying causes of drug shortages and recalls (Lee, 2017). A typical continuous manufacturing setup involves a high-shear wet granulation module, a segmented dryer module, and a granule conditioning module (Stahl, 2004). In the granulation module, dry ingredients can be dosed individually and mixed in the continuous high-shear granulator. After a small dry mix section, the granulation liquid is added, so each particle receives the same amount of liquid. The particles follow a granulation track that mimics the granulation in a batch process. The whole wet granulation process takes place in a few seconds with only a few grams of product in process at a given time, resulting in faster startup and no losses. The particle size can be adjusted by changing the working level (the amount of powder) in the granulator. This facilitates continuous flow of wet granules with a constant quality and density. The wet granules then are transferred to the dryer. There are no oversized agglomerates and, thus, no wet milling. Elaborate discussion about continuous manufacturing is beyond the scope of this chapter and reader is referred to other sources (Fonteyne et al., 2013, 2015).

7.3.2 Fluid-bed wet granulation process

Fluid-bed granulation (also known as top-spray granulation) is becoming a popular process because of its all-operations-in-one-equipment concept. Unlike the multiple operations in series of equipment in other wet granulation techniques, this technique has advantages of:

- All the operations (dry mixing, wet granulation, and drying) can be performed in one equipment.

7.3 Granulation methodologies 217

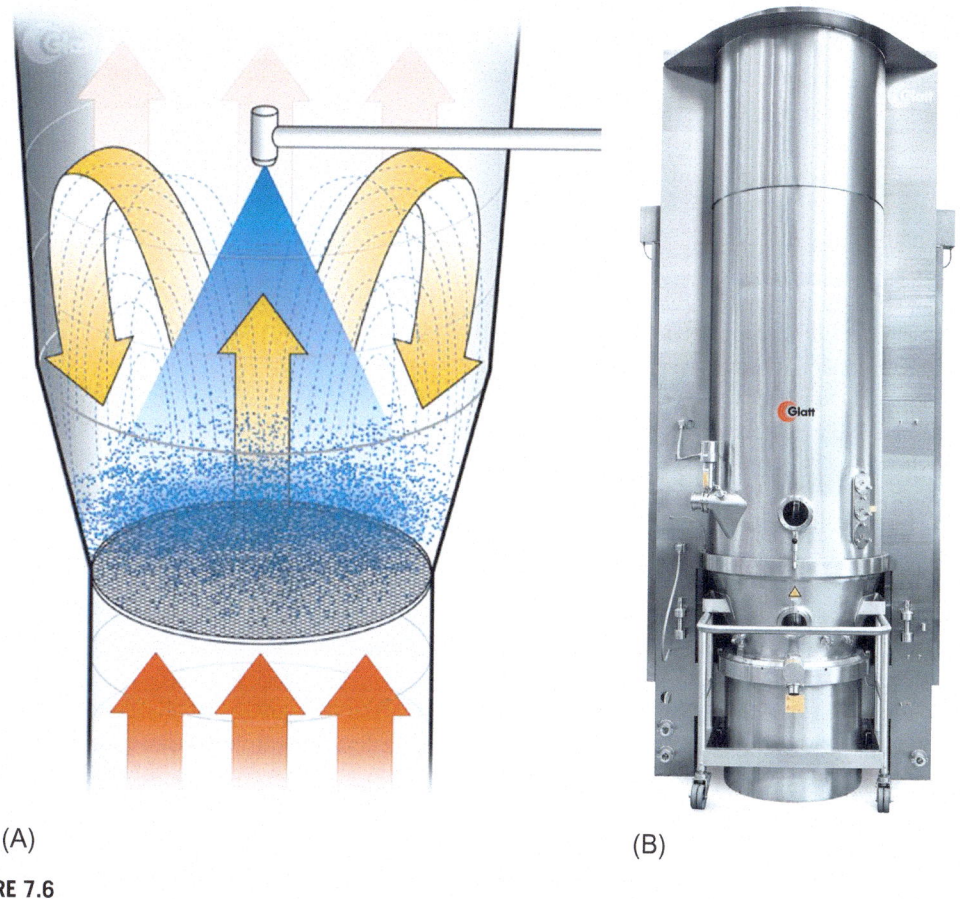

FIGURE 7.6

(A) Schematic diagram of a fluid-bed granulator. (B) Glatt WSG Pro 200.

(A) Source: Courtesy of Glatt, Germany. (B) Source: Courtesy of Glatt, Germany.

- Relatively, it yields more porous granules with narrow particle size distribution, as compared to other granulation techniques. Therefore, this can be choice of process where these key attributes are desired.
- Because this method yields more uniform granule size distribution, milling might not be needed and is likely to be lower intensity, if needed. The consequent reduction in the generation of fines and dust provides advantages in terms of a consistent product and operational safety (especially for high potent compounds).

A typical fluid-bed granulator design is shown in Figure 7.6. The air is drawn by a ventilator through a material container (also known as product container) of a conical shape placed at the bottom of the long vertical cylinder.

- The role of the product container, which contains a distribution plate at the bottom to direct the flow of air, is to hold the blend to be granulated in sufficiently fluidized state so that simultaneous granulation and drying is possible in granulation stage, followed by uniform drying in the drying stage.
- The role of spray gun is mainly to make available the granulating solution or binder in the form of fine mist to the fluidized blend.
- The role of filter bags is mainly to avoid any loss of fluidized blend and to ensure continuous recirculation of the product.

The air flow through the ventilator moves the particles up in the central part of the bed and down by the wall. The fluidized air is heated to a temperature typically ranging from 40 to 80°C. The binder solution is introduced by spraying. When the liquid addition is completed, the granules are dried in the same equipment (Kristensen and Schaefer, 1987). At a more micro level, few studies have been reported using discrete-element-method (DEM) or combination of DEM and computational fluid dynamics (CFD), wherein each particle is tracked individually (Borner, Buck, & Tsotsas, 2017; Fries, Antonyuk, Heinrich, & Palzer, 2011). This technology allows a complete representation of the particle-particle and particle-wall interaction and their influence on the process dynamics. Moreover, these methods can give comprehensive details about particle and fluid dynamics such as residence time distribution in the spray zone and evaluation of process parameters. Going in the details of DEM and CFD is beyond the scope of this chapter and reader can refer to other references on this subject (Gantt and Gatzke, 2005; Goldschmidt & Kuipers, 2003; Heinrich, Ihlow, & Morl, 2003).

In one of the studies, properties of granules prepared by fluid-bed granulation and high-shear granulation process were compared (Morin and Briens, 2014). In this study, the authors prepared placebo granules containing 50% lactose monohydrate, 45% microcrystalline cellulose (Avicel PH 101; FMC Bioploymer, USA), 4% hydroxypropyl methyl cellulose (Pharmacoat 603; Shin Etsu, Japan) as a binder, and 1% w/w croscarmellose sodium (Alfa Aesar) as a disintegrant. High-shear granulation was performed using a Fielder PMA-1 (GEA) granulator. Fluidized-bed granulation was performed in a custom-made top-spray conical fluidized-bed equipment. Authors reported differences in the formation, and subsequent growth, of the granule nuclei by the two methods. The differences in the granule nuclei formation resulted from the different particle flow patterns within the high-shear granulator compared with the fluidized bed, which provided different exposure to the liquid binder spray. Based on the scanning electron microscope (SEM) images of the samples taken from the granulations, authors concluded that the formation of granule nuclei was different for the high-shear and fluidized-bed granulations. For the high-shear granulation, nuclei contained only microcrystalline cellulose. (There was powder segregation during the dry mixing stage and larger lactose particles were forced toward the sides and the bottom of the bowl, while the microcrystalline cellulose fibers segregated to the middle of the bowl and near the top of the powder bed.) As a result, during binder addition, primarily microcrystalline cellulose fibers were exposed initially to the binder spray and these hygroscopic fibers easily absorbed the binder and formed nuclei. For the fluidized-bed granulations, the nuclei consisted primarily of lactose particles with a few attached microcrystalline cellulose fibers. Because there was no segregation during initial fluidization of powder, liquid binder was absorbed by the microcrystalline cellulose and also coated the surface of the lactose monohydrate particles. Collisions between a wetted lactose monohydrate and microcrystalline cellulose resulted in nuclei that included both particles. The differences in the granule nuclei formation pattern, combined with different flow and shear levels, led to induction growth in the high-shear granulation process, but steady state growth was observed in the fluidized-bed apparatus. The SEM images combined with the particle size analysis indicated that the granule growth mechanism was different for the two types of granulation processes. For the high-shear granulation process, granule formation followed the induction growth mechanism. Growth remained minimal until a moisture content of about 20% w/w and then increased rapidly. For the fluidized-bed granulations, granule formation followed a slow but steady growth mechanism. Granule growth occurred through coalescence of nuclei and smaller granules as the binder continued to be sprayed onto the bed.

There were differences in granule properties in the granules prepared using two processes despite composition being the same. Particle size distribution of the granules was narrow for both processes at low granule moisture content; however, at moisture levels just below 30% w/w, the size distribution for the high-shear granulation remained narrow, while it was wider for granules made using fluidized-bed granulations. At even higher moisture levels, the size distribution of the high-shear granulation widened as the formation of oversized agglomerates occurred, while the distribution for the fluidized-bed granulations did not change significantly. The dynamic densities, measured using the revolution analyzer, were much lower for the fluidized-bed granules than for the high-shear granules: 0.32–0.42 g/mL from fluidized-bed granulation vs. 0.59 g/mL from high-shear granulation. The high-shear granules exhibit better flow properties (compressibility index value of 8) compared to the granules produced by the fluidized-bed (compressibility index value ranging from 10 to 20).

7.3.2.1 *Case study II*

In another internal study at Perrigo, feasibility trials of compound A (as presented in case study I; with same qualitative composition but with increased quantity of disintegrant) were conducted in fluid-bed granulator (5 L capacity; FBE 5; Pam Glatt, India) and high-shear granulator (bottom-driven granulator; 10 L capacity; Sams Techno. Mech., India). For this study, compositions of batches were kept same.

The high-level manufacturing process for high-shear granulation process was similar to that presented in case study I; details are captured below.

- Pre-mix time/Impeller speed: 10 min/150 rpm
- Binder addition time/Impeller speed: 3 min/150 rpm (Total fluid uptake was 50% w/w of blend; added by pouring method with an addition rate of 380 g/min)
- Wet mass time/Impeller speed: 1 min/150 rpm with chopper on at 1500 rpm
- Drying was done in fluid bed dryer at 65°C till loss on drying (LOD) of < 3% w/w was achieved.

The brief manufacturing steps for fluid-bed granulation and details are captured below:

1. Mixing and sifting of all raw materials and drug.
2. Loading in a fluid-bed granulator and spraying the binder solution using a top spray gun assembly.
3. Drying in fluid-bed dryer.
4. Milling and sizing of granules.
5. Blending with extragranular excipients and lubrication in V shell blender.
6. Compression followed by coating.

Total fluid uptake was 50% w/w of blend; sprayed using 0.8 mm nozzle with spray rate of 15 g/min. Total spraying time was 80 min and drying process was continued (at 70°C until LOD of < 3% w/w was achieved).

Table 7.4 Physical evaluation of the final lubricated blend for compound A made using high-shear granulation and fluid-bed granulation.

Evaluation parameters	Batch manufactured with high-shear granulation	Batch manufactured with fluid-bed granulation
BD (bulk density in g/mL)	0.55	0.35
TD (tapped density in g/mL)	0.68	0.42
Compressibility index	19.11	16.66
Hausner ratio	1.24	1.20

Table 7.5 Dissolution results comparison (of tablets of compound A) in 0.1 N HCl of batches made using high-shear granulation and fluid-bed granulation.

	% Dissolved (% RSD); $n = 6$		
Time (min)	Innovator product	Batch manufactured with high-shear granulation	Batch manufactured with fluid-bed granulation
5	55 (2.3)	48 (3.5)	45 (1.7)
10	59 (2.6)	51 (2.4)	48 (3.4)
20	59 (3.0)	53 (2.1)	47 (3.7)
30	58 (3.4)	55 (2.0)	47 (3.0)

Physical properties of the final lubricated blend are listed in Table 7.4. Tablets were compressed at an average weight of 300 mg using 12.1 × 6.3 mm capsule shaped tool and target hardness of 12–14 SCU (Strong Cobb Unit). The dissolution profile of tablets was evaluated in 900 mL of 0.1 N HCl using USP-II apparatus at 75 rpm; sampling was performed at 5, 10, 15, and 30 min; and analysis was carried out using a validated HPLC method. The results of the dissolution study are listed in Table 7.5.

The density of the granules made using fluid bed granulation was lower (BD: 0.35 g/mL) than that from the high-shear granulation process (BD: 0.55 g/mL). However, the values of compressibility index and Hausner ratio in both cases were quite similar (Table 7.4) and suggest the flow values to be similar for both the processes. (Flow characteristics to be fair considering USP ⟨1174⟩.) Friability values of tablets (determined as per USP ⟨1216⟩) made using the fluid-bed granulation approach was higher (0.78% w/w) as compared to high-shear granulation process (Friability: Nil), and it was anticipated that the tablets made using fluid-bed granulation process will have increased defects after coating and packaging. The dissolution profile of tablets made from the fluid-bed granulation process was similar to high-shear granulation process (f_2 value of 63) especially during initial phase; however, slightly slower (5%–7% slower) during the later phase than that from high-shear granulation process. These differences could be attributed to the specific product characteristics, choice of excipients, and/or processing parameters, and in no way should it be construed that one equipment is better than the other. In the majority of cases, it is always possible to achieve the desired CQAs by proper selection of excipients and carefully optimizing the processing parameters. In generic product development, though not necessary, it is always advantageous to be as close as possible (in terms of composition and process) to the innovator

Table 7.6 Physical evaluation of the final blend for compound B made using high-shear granulation and fluid-bed granulation.

Parameters	Final blend manufactured with fluid-bed granulation	Final blend manufactured with high-shear granulation
BD (bulk density in g/mL)	0.34	0.41
TD (tapped density in g/mL)	0.49	0.51
Compressibility index	29.41	20.00
Hausner ratio	1.42	1.25

product. Considering the innovator product was made using a high-shear wet granulation process, a similar approach was taken forward for generic product development.

7.3.2.2 Case study III

In another internal Perrigo product development study (for compound B), similar comparison was made for change in manufacturing process using high-shear granulation (bottom-driven granulator) and fluid-bed granulation by keeping the composition the same. Compound B is a high-dose drug (180 mg), a hydrochloride salt having molecular weight of 538.13 g/mol, Log P value of 5.6, and pKa 1 of 8.76 and pKa 2 of 4.28 and shows pH dependent solubility (highly soluble in low pH and solubility decreases as the pH increases). During initial development trials, suitable particle size of API (D_{90} < 20 μm) and excipients were identified. The drug load for the study was approximately 30% w/w and other excipients included were disintegrants (croscarmellose sodium and pregelatinized starch), diluents (microcrystalline cellulose and lactose), and lubricant (magnesium stearate). The manufacturing process for high-shear granulation process was similar to that presented in case study I. The manufacturing process for fluid-bed granulation was similar to that presented in case study II.

Physical properties of the granules are listed in Table 7.6 and were tested as per the methodology presented in case study I. The granules manufactured using fluid-bed granulation had lower bulk density and higher compressibility index value (indicating poor flow) than the granules manufactured using high-shear granulation process (Table 7.6). Granules manufactured with fluid-bed granulation had low bulk density indicating more porous characteristics with smaller particle size (Figure 7.7) with narrower particle size distribution. This is in line with case study II, wherein fluid-bed granulation process also produced granules with lower bulk density.

7.3.3 Low-shear wet granulation process

Low-shear wet granulation is a powder densification and agglomeration technique that works on lower agitator speeds, has lower sweep volumes, and lower pressures on the powder bed than high-shear mixers or fluid-bed granulators. Low-shear granulators primarily are distinguished from high-shear granulators by the geometry and the design of the shear-inducing components. Shear can be induced by rotating impeller, reciprocal kneading action, or convection screw action (Fda, 2014). Some of these low-shear subclasses are listed below.

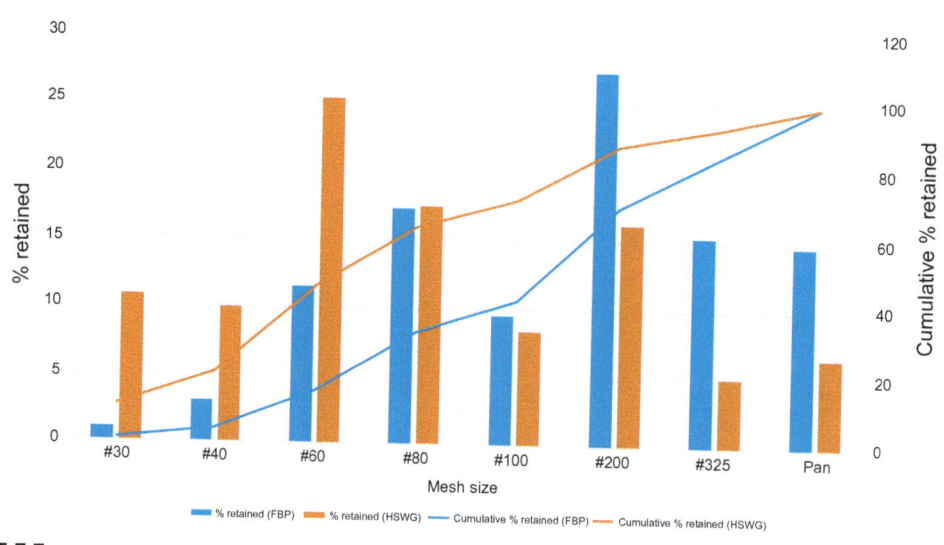

FIGURE 7.7

Particle size distribution of the final blend for compound B made using high-shear granulation and fluid-bed granulation.

7.3.3.1 *Planetary mixer*

The planetary mixer consists of a mixing bowl and a mixer blade attached to a mixing shaft (Gokhale & Trivedi, 2010). In this mixer, both the mixing blade shaft and the mixing blade rotate simultaneously. The planetary mixer can be operated at a variable speed. The slower speed is used to mix powders, and faster speeds are used for the kneading action required during the wet granulation process. One of the advantages of the planetary mixers is that the mixing bowl has no dead spaces. One of the disadvantages, however, is the limited batch size that can be prepared at one time. Several sublots are prepared using this equipment, and the final blend of the sublots is done in a large tumbling mixer.

7.3.3.2 *Screw type mixer*

The basic design consists of a cylindrical screw attached to an arm that rotates around the perimeter of the vessel causing convective mixing and shear (Cantor, Augsburger, & Hoag, 2008). This conveys product from the bottom of the conical vessel to the product surface. The particles at the surface opposite the screw accelerate downward in a mass flow by gravity as the diameter of the vessel decreases. The three actions result in a unique three-dimensional mixing.

7.3.4 Single-pot granulation

In this technique, granulation (high-shear) and drying happens in the same equipment (Stahl, 2004). Single-pot processing is an established pharmaceutical wet granulation and drying technique often used when high containment is required for the production of potent oncology substances or hormones. The basic drying principle relies on the application of a vacuum in the bowl, thereby, drastically lowering the evaporation temperature of the granulation liquid. The traditional heat source comes from the heated dryer (which, in this case, is the same as the granulator) walls. The heat transfer is related to the surface area of the dryer walls and the volume of product. Therefore, this direct heating method is most effective for small scale, where organic solvents or low quantities of binder fluids are used. To enhance the drying process and reduce drying times, particularly for larger-scale operations, additional drying techniques can be implemented, such as use of stripping gas or microwave energy.

Introducing stripping gas (typically, a dry gas, air, or nitrogen) into the pot enhances the vacuum drying and allows lower final moisture content to be achieved. In this technique, a small quantity of dry gas is introduced from the bottom of the equipment and passes through the product bed, improving the efficiency of vapor removal. Because the heated wall is the only source of drying energy, linear scale-up is not possible. Microwave drying relies on additional energy being supplied that is preferentially absorbed by the solvents in the process to enhance evaporation. In a comparative study (Bossche and Vaerenbergh, 2014), authors studied the efficiency of two drying methods (stripping gas and use of microwave energy) and found that drying process is much more efficient when using microwave energy. Microwaves were capable of removing the water, approximately 1 kg in 40 min, whereas gas-stripping required > 3 h to remove the same amount of water.

One of the technologies that can be used to prepare granules in a single-pot fluid-bed rotor processor is using Granurex (Freund Vector, 2017). It consists of a processing chamber with a rotating disc inside a stationary container (also called as stator) that fits within a VFC Flo-coater fluid-bed system. A spray gun is located in the stator, which is immersed in the product bed. The rotor imparts energy to the product for granule formation (similar to high-shear granulation), however, less emphasis is placed on the fluidization for product movement. The geometry of the rotor and stator combined with the angular movement of the rotor produces characteristic helical product movement specific to rotor processing. This helical product movement past the spray gun creates an environment for high efficiency and excellent uniformity (Figure 7.8).

Spherical granulation using Granurex is a hybrid of top-spray and high-shear granulation processes. The process starts with a powder blend inside the rotor disc. A binder solution is sprayed onto the powder to granulate it. The rotor imparts force onto the powder, forming a more spherical granule than a top spray would (Freund Vector, 2017). Process parameters such as spray rate, atomization air pressure, temperature, and speed of the rotor, determine the desired particle size of the granules. Using this technology, it is possible to get narrow particle size of the granules that are also more spherical than those produced using other methodologies (Patil, Khadse, & Ige, 2016). Drug loading also can be very high using this technique, similar to that achieved using high-shear granulation process.

A few studies (Kimura et al., 2010, 2014) investigated the relationships among the operational conditions of Granurex, material attributes such as water content during granulation, and a series of associated micromeritics, including the flowability (measured using compressibility index), granule mean size (measured using sieve analysis), and granule density values. A variety of different operational conditions were tested, including the binder flow rate, atomization pressure, slit air flow rate, rotating speed, and temperature of the inlet air. For the purpose of this study, authors used lactose monohydrate (Pharmatose 200M, DFE Pharma, Netherlands) and corn starch (Nihon, Japan) as fillers (70:30 ratio) and hydroxypropyl cellulose (HPC-L, Nippon, Japan) was used as a binder (26% w/w total dry mix). The results of this study revealed three things: the granule mean size was negatively affected by the atomization pressure, and positively affected by the binder flow rate; the granule flow property was positively affected by the atomization pressure, and negatively affected by the slit air flow rate; and the

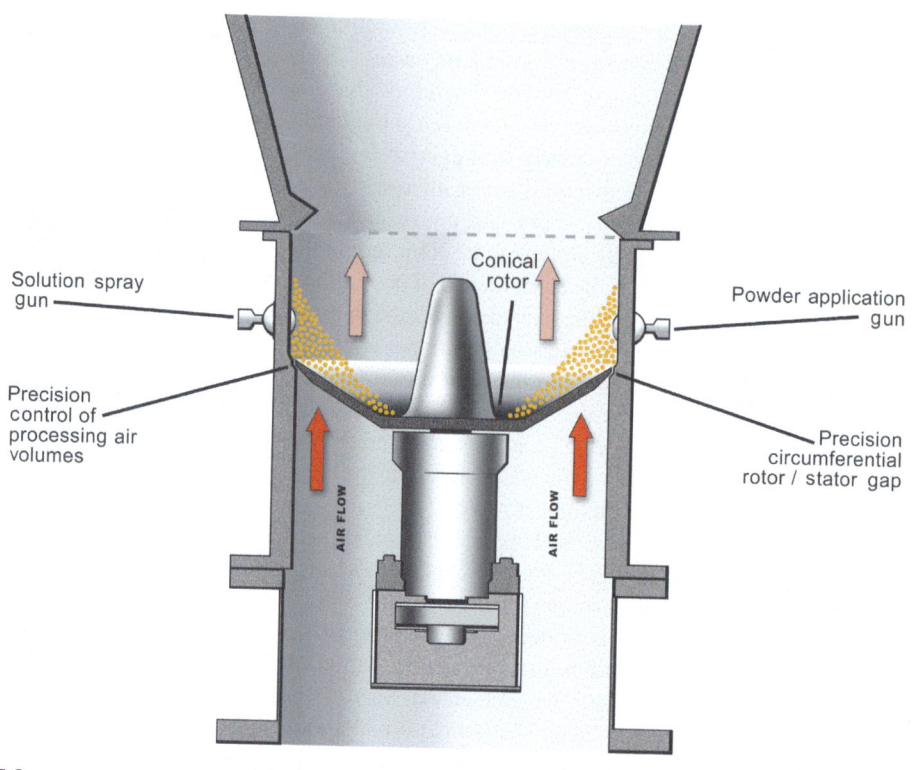

FIGURE 7.8

Courtesy of Freund Inc., USA.

granule density was positively affected by the rotating speed. Also, the roundness of granules (measured by image analysis using WinROOF image analysis software) was positively affected by the binder flow rate.

7.3.5 Moisture-activated dry granulation (MADG)

In MADG, granules are created with water and a granulating binder, as in wet granulation, but are not heat dried or milled (Ullah, 2011). MADG has two stages: agglomeration and moisture distribution. During agglomeration, a major portion of the formulation containing the drug is agglomerated. The drug is blended with filler and binder in the powder form, and this blend constitutes approximately 50%–80% w/w of the formula weight. In the second stage, a small amount (1%–4% w/w) of water is sprayed as small droplets onto the powder mixture (while blending). Water moistens the blend and causes the binder to become tacky, leading to particles, particularly fines, to form moist agglomerates. The process does not create large granules, which would need milling. In addition, the end point is not sensitive to blending because very little water is used in the process. The remaining excipients then are added during blending/extragranular mixing stage that results in dry and free-flowing granulation.

The excipients used in MADG are similar to that of a typical tablet composition made using wet granulation, however, in MADG, these excipients are added in a specific order. For example, following agglomeration, microcrystalline cellulose is added while mixing, which absorbs most of the excess water. Colloidal silica then is added, absorbing any remaining moisture. Other excipients, such as disintegrants and lubricants, are added and blended last. The process is relatively quick (in comparison with high-shear and fluid-bed granulation process) and the final blend looks like a direct blend formulation with fine particle size distribution. MADG could be challenging in case of high drug load moisture absorbing APIs. One of the requirements for this technology is the need to accurately deliver a very small amount of water in spray form. This process also requires an airless spray system that accurately delivers the desired amount of water in small (50–200 μm) droplets (Ullah, 2011). The system should not have drips (peristaltic pumps, in particular, are not suitable). The gear pump or pressure vessel also must provide the right type of spray. Despite sounding simple, less familiarity with the process and some apprehension toward adoption has resulted in limited use of this technology in pharmaceutical industry.

7.3.6 Melt granulation

Another variation of the high-shear wet granulation process is melt granulation, wherein a meltable binder (binder that melts or softens at relatively low temperatures, e.g., 50–100°C) is added (usually in a dry form). Melting is achieved by the heat energy provided by friction in the mixer and the heated jacket of the mixer bowl. This process efficiently agglomerates pharmaceutical powders for use in both immediate and sustained-release solid dosage forms (Royce, Suryawanshi, Shah, & Vishnupad, 1996). The process uses binders that are effective as granulating fluids when they are in the molten state. Some of the binders that can be used for this purpose are Gelucire (melts at 45–50°C) and polyethylene glycols (melts between 42 and 65°C, depending upon molecular weight). Cooling of the agglomerated powders and the resultant solidification of the molten materials completes the granulation process.

Table 7.7 captures different types of granulators and few of their vendors. This list is representative only. Because of wide usage and complexities involved, scope of this chapter is limited to wet granulation process using high-shear granulation and fluid-bed granulation.

7.4 Application of QbD in wet granulation process development

Quality by design (QbD) is defined as "a systematic approach to development that begins with predefined objectives and emphasizes product and process understanding and process control, based on sound science and quality risk management" (Fda, 2012). This systematic approach is illustrated in Figure 7.9 and includes development studies using univariate trials (also known as one factor at a time or OFAT) and multivariate statistical experimental studies (i.e., design of experiments or DoE). This framework represents a move away from the traditional approach in the industry of quality by testing (QbT). In the context of the pharmaceutical manufacturing process, QbT is characterized by extensive testing, limited flexibility in process conditions, and often no clear reasoning for batch failures. QbD, on the other hand, offers operational flexibility as long as product attributes are maintained within an approved design space (Yu et al., 2014).

Table 7.7 List of different wet granulation equipment with few vendors and models.

Type	Vendor	Comments
Vertical-Bottom-driven high-shear granulators	GEA	Capacity 1 to 1800 L PMA Classic, PMA Advanced, and PharmaConnect high-shear granulators
	Glatt	TMG; with swap vessels with capacity of 0.5, 1, 2, 4, and 6 L VG 65/10; with exchangeable product containers for maximum product volume flexibility from 2.5 to 50 L. Even batch sizes from 0.1 L are possible when a TMG adapter is used
	Freund-Vector	GMXB Granumeist bottom-drive high-shear granulator systems, available in different models ranging from 1 to 1800 L
	Diosna	Available in lab, pilot, and production capacities
Vertical-Top-driven high-shear granulators	Glatt	TDG Lab with working vessels with volumes of 1, 2, 4, and 6 L
	Freund-Vector	GMX Granumeist; top-drive high-shear wet granulation mixing systems, available in micro model (1–4 L), lab model (4–25 L), pilot model (75–150 L), and production capacities (300 – 1200 L)
	GEA	10–1200 L capacity UltimaGral
Horizontal-High-shear granulators	Loedige	Horizontal Ploughshare Mixer can be used both for batch processes (Type FKM) and for continuous applications (Type KM).
Low-shear granulators	Ross	Planetary disperser (PowerMix) and dual planetary disperser available in lab and production models.
Fluid-bed granulator	GEA	Capacity 25–1800 L FlexStream fluid-bed processors
	Glatt	GPCG fluid-bed processors available in lab, pilot, and production models
	Freund-Vector	VFC Flo-Coater system available for lab, pilot, and production capacities
	Diosna	Available in mini and midi lab models and also pilot/production capacities
Rotary granulators	Freund-Vector	Granurex Rotor; one-pot fluid-bed rotor processor

(continued on next page)

7.4 Application of QbD in wet granulation process development

Table 7.7 List of different wet granulation equipment with few vendors and models—cont'd

Type	Vendor	Comments
Single-pot processing	GEA	Capacity 10–1200 L UltimaPro single pot Mixing, granulating, and drying options integrated into a single processing vessel Relies on the application of a vacuum within the bowl to dry the wet mass
	Diosna	Mixing, granulation, and vacuum drying in a single system; Single Pot Processor P/VAC-10; P/VAC 10–60, and VAC 150–2000

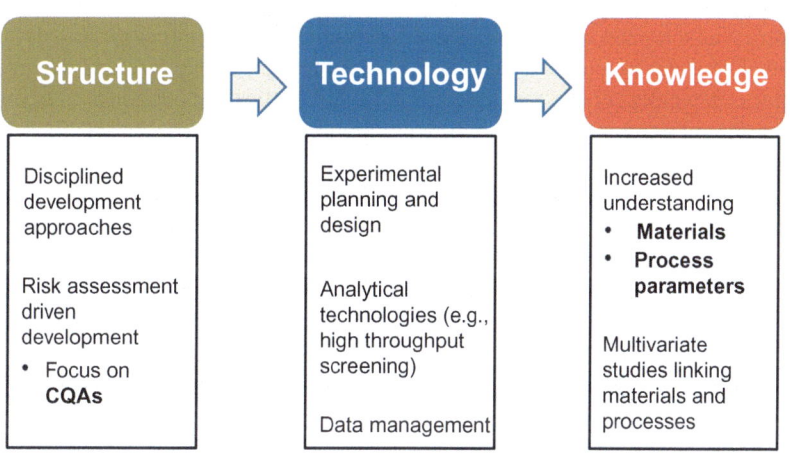

FIGURE 7.9

Application of QbD in product development.

In one study (Wang, Ye, Heng, & Ma, 2008), granulation processes were compared for different formulations using a laboratory-scale high-shear mixer. The effects of critical process parameters or CPPs (impeller speed, chopper speed, and kneading time) on granule characteristics were evaluated. The characteristics of the granules studied included the size distribution, friability, and morphological properties. The flow profiles of the wet mass and material deposition during the process also were studied. The results showed that the effect of the impeller speed was influenced by the starting material system. Chopper speeds from 1200 to 3600 rpm and kneading times from 120 to 240 s had a consistent influence on all formulations. It also was observed that the toroidal flow pattern of the wet mass could be maintained for a longer period and granules with a good spherical shape were obtained by removing the chopper during the last 120 s of the granulation process. It was concluded that the effectiveness of high-shear wet granulation could be improved by choosing a proper combination of starting material and process parameters and by monitoring the mass motion during the process.

During the development of Ramipril tablets (Aksu et al., 2013), CQAs and CPPs were identified based on the quality target product profile (QTPP) using the historical data and risk assessment using failure mode and effect analysis (FMEA). CPPs that affected the product and process were used to establish an experimental design, and the results were used to facilitate definition of the design space using tools of DoE, the response surface method (RSM), and artificial neural networks (ANNs). Authors used this project to discover knowledge associated with the manufacture of Ramipril tablets using a range of ANN-based software, with the intention of establishing a multidimensional design space that ensures consistent product quality. The study was concluded with a design space based on the study data and specifications, and a new formulation was optimized. On the basis of this formulation, a new laboratory batch formulation was prepared and tested. It was confirmed that the explored formulation was within the design space.

In another study, three model compounds were used to study the effect of process parameters on inprocess critical material attributes (CMAs) and final product CQA (Badawy, Narang, LaMarche, Subramanian, & Varia, 2012). The effect of four process parameters was evaluated using DoE approach and batches were characterized for particle size distribution, density (porosity), flow, compaction, and dissolution rate. The water amount showed significant effect on granule size and density. The role of impeller speed was dependent on the granule mechanical properties and efficiency of liquid distribution in the granulator. Blend density was found to increase rapidly during wet massing. The liquid addition rate was the least consequential factor and showed minimal impact on granule density and growth. Correlations of primary properties with granulation bulk powder properties (compaction and flow) and tablet dissolution also were identified. The effects of the process parameters on the bulk powder properties and tablet dissolution were consistent with their proposed link to primary granule properties.

In yet another study, a pharmaceutical compound was used to study the effect of batch wet granulation process parameters in combination with the residual moisture content remaining after drying on granule and tablet quality attributes (Gabbott, Husban, & Reynolds, 2016). The effect of three batch wet granulation process parameters was evaluated using a multivariate experimental design, with a novel constrained design space. Batches were characterized for moisture content, granule density, crushing strength, porosity, disintegration time, and dissolution. Water quantity added during granulation showed a significant effect on granule density and tablet dissolution rate. Mixing time showed a significant effect on tablet crushing strength, and mixing speed showed a significant effect on the distribution of tablet crushing strengths obtained. The residual moisture content remaining after granule drying showed a significant effect on tablet crushing strength. Correlations between the energy input during granulation, the density of granules produced, and the quality attributes of the final tablets also were identified.

In another study (Rahmanian, Naji, & Ghadiri, 2011), influence of process parameters such as impeller speed, granulation time, and binder viscosity on granule strength and properties were reported. A high-shear granulator was used to produce granules using calcium carbonate (Durcal) as a model compound and aqueous polyethylene glycol (PEG) as the binder. It was observed that higher impeller speeds gave rise to more consolidation and compaction of granules, and produced granules with high strength and low porosity. Increasing the granulation time had an effect on granules' strength, until an optimum time was reached. Binder addition methods have showed no considerable effect on granule properties, such as granule strength or particle size distribution.

In one study, effects of high-shear wet granulation process parameters were studied on granule characteristics (Pandey et al., 2013). Authors did full factorial design of experiments on three process parameters, such as water amount, impeller speed, and wet massing time. It was observed that the water amount had the largest impact on granule characteristics, and that the effect of other process variables was more pronounced at a higher water amount. At high-water amounts, an increase in impeller speed and/or wet massing time showed a decrease in granule porosity and compactability, and a strong correlation between granule porosity and compactability. A three-dimensional population balance model was developed that considered agglomeration and consolidation in the granulation process. Using this model, particle size distribution of three other batches were predicted, each of which was manufactured under different process parameters (water amount, impeller speed, and wet massing time) and the model was able to capture and predict successfully the shifts in granule particle size distribution with changes in these process parameters.

Experimental design and multivariate data analysis was used to define design space for granulation and tableting processes in one of the studies (Djuris, Medarevic, Krstic, Djuric, & Ibric, 2013). Acetaminophen was used as the model drug, and authors also investigated the possibility of the development of immediate or extended-release acetaminophen tablets. Granulation experiments were performed in the fluid bed processor using polyethylene oxide polymer as a binder in the direct granulation method. The first set of experiments was organized according to the Plackett-Burman design, followed by the full factorial experimental design. Principal component analysis and partial least squares regression were applied as the multivariate analysis techniques. By using these different methods, CQAs and PPs were identified and quantified. Also, an inline method was developed to monitor the temperature during the fluidized-bed granulation process and control strategies were proposed.

7.4.1 Case study IV

This section presents a Perrigo internal case study of a generic product developed using high-shear wet granulation process. Principles of QbD were used to ensure robust product and process development for compound B (as presented in case study III).

In QbD-based development, the first step is to determine QTPP that is defined as "a prospective summary of the quality characteristics of a drug product that ideally will be achieved to ensure the desired quality, taking into account safety and efficacy of the drug product" (Fda, 2011). In development of drug products, especially in case of generic products, the QTPP should be defined early in development based on the properties of the drug substance (DS), characterization of the reference listed drug (RLD) product (also called the innovator product), and considerations of the RLD label and intended patient population. For generic drug products, the QTPP includes all product attributes that are needed to ensure equivalent safety and efficacy to that of RLD. Some of the elements of QTPPs include dosage form (i.e., tablet) and drug release rate design (i.e., immediate release or IR, modified release or MR, chewable tablet, orally disintegrating tablet or ODT), route of administration, strength, pharmacokinetics and stability requirements, drug product quality attributes (e.g., assay, dissolution, impurity profile), container closure system, product labeling (e.g., administration recommendation, storage conditions, food effect). Based on the clinical and pharmacokinetic characteristics of the RLD and prior knowledge, a QTPP (Table 7.8) was defined and justified to guide the product development of compound B.

After QTPP is identified, the next step is to define CQA that is "a physical, chemical, biological, or microbiological property or characteristic that should be within an appropriate limit, range, or

Table 7.8 QTPP for product containing compound B.

QTPP element	Target	Justification
Dosage form/design	Immediate release (IR) tablet	Pharmaceutical equivalence requirement: Same dosage form
Route of administration	Oral	Pharmaceutical equivalence requirement: Same route administration
Dosage strength	180 mg	Pharmaceutical equivalence requirement: Same strength
Pharmacokinetics/bioequivalence	Bioequivalent to RLD	Bioequivalence requirement
Stability	24-month shelf-life at room temperature	Storage conditions and use pattern are appropriate for commercialization; shelf life meets ICH Q1A (R2) criteria
Drug product quality attributes	Physical attributes (appearance, size, shape, color, friability) Identification Assay Content uniformity (CU) Degradation products Residual solvents Drug release Water content Microbial limits	Pharmaceutical equivalence requirement; meets the compendial or other applicable standards
Container closure system	Suitable container closure system to achieve the target shelf life	Protects product from degradation and contamination. Permits dosing that are acceptable for commercialization
Administration/concurrence with labeling	180 mg once daily with water	Same administration/concurrence with RLD labeling
Storage conditions	Store at controlled room temperature 20–25°C (68–77°F) (see USP controlled room temperature)	Storage conditions and use pattern are appropriate for commercialization

distribution to ensure the desired product quality" (Fda, 2011). The identification of a CQA from the QTPP is based on the severity of harm to a patient should the product fall outside the acceptable range for that attribute. Typical CQAs in case of an IR product could be assay, content uniformity, dissolution, and degradation products. Other CQAs such as identity, residual solvents, and microbial limits are unlikely to be affected by formulation and/or process variables. These, therefore, usually are monitored but not necessarily investigated in depth. For compound B, assay, content uniformity, degradation products, and drug release were identified as the CQAs. Identified CQAs are listed in Table 7.9.

Table 7.9 Drug product CQAs for product containing compound B.

Quality attributes of the drug product		Target	Is this critical?	Justification of criticality
Physical attributes and properties	Appearance (color and shape)	Brown-colored capsule shape film coated tablet	No	Color, shape and appearance are not directly linked to safety and efficacy. Therefore, they are not critical. The target is set to ensure patient acceptability.
	Flavor (odor)	No unpleasant odor	No	Critical for qualitative sensory attributes, but not critical for product performance. The flavor or odor is not directly linked to safety and efficacy.
	Size	Similar to RLD	Yes[a]	Tablet size correlates to acceptability since tablet is to be administered orally; therefore, it is critical.
	Score configuration	Not scored	No	The RLD and generic are both unscored tablets, and thus score configuration is not critical.
	Friability	Not > 1% w/w	No	Friability is a routine test as per compendial requirements for tablets. A target of NMT 1.0% of mean weight loss assures a low impact on patient safety and efficacy and minimizes customer complaints.
Identification		Positive for compound B	Yes[a]	Though identification is critical for safety and efficacy, this CQA can be effectively controlled by the quality management system and will be monitored at drug product release. Formulation and process do not affect identity.
Assay		95.0–105.0% of label claim	Yes	Assay variability will affect safety and efficacy. Process can affect the assay value of the drug product, and therefore assay will be evaluated through product and process development.
Content uniformity		Complies with USP (905) uniformity of dosage units	Yes	Variability in content uniformity will affect safety and efficacy. Both formulation and process affect the uniformity and CU will be evaluated through product and process development.

(*continued on next page*)

Table 7.9 Drug product CQAs for product containing compound B—cont'd

Quality attributes of the drug product	Target	Is this critical?	Justification of criticality
Degradation of products	Meets USP monograph and ICH Q3 requirements	Yes	Degradation of products can affect safety and must be controlled based on compendia/ICH requirements or RLD characterization to limit patient exposure. Formulation and process can affect degradation of products. Therefore, degradation will be assessed during product and process development.
Residual solvents	Complies with USP ⟨467⟩	Yes[a]	Residual solvents can affect safety. However, no solvent is used in the drug product manufacturing process and the drug product complies with USP ⟨467⟩. Therefore, formulation and process variables are unlikely to affect this CQA.
Drug release/dissolution	Meets USP monograph	Yes	Failure to meet the drug release specification can affect bioavailability. Both formulation and process affect the drug release profile. This CQA will be investigated during formulation and process development.
Water content	2%–3%	Yes[a]	Generally, water content can affect degradation and microbial growth in the drug product and can be a potential CQA. Formulation and process can affect water content. However, in this case, API is not sensitive to hydrolysis and moisture will not affect stability. Therefore, this CQA will not be discussed during formulation and process development.
Microbial limits	Meets relevant pharmacopeia criteria	Yes[a]	Noncompliance of microbial limits can affect patient safety. However, in this case, the risk of microbial growth is very low. Microbial testing will be done throughout the process and is effectively controlled by the quality management system. Therefore, this CQA will not be discussed in detail during formulation and process development

[a] *Formulation and process variables are unlikely to affect the CQA. Therefore, the CQA will not be investigated and discussed in detail in subsequent risk assessment and pharmaceutical development. However, the CQA remains a target element of the drug product profile and should be addressed accordingly.*

After QTPP has been set and CQAs identified, the next step is to arrive at a composition. In case of product development of compound B, excipients were selected for further evaluation based on the excipients used in the RLD, intellectual property (IP) landscape, drug-excipient compatibility results, and the known degradation pathways for the API. The manufacturing process was identified during the early stages of product development. Direct compression strategy was evaluated; however, this approach was not pursued further as the blend exhibited poor flow (compressibility index of 36). In the case of fluid-bed granulation, dissolution was found to be fast in one of the key dissolution conditions (900 mL of 0.001 N HCl at 50 rpm using USP apparatus type II). High-shear granulation process was found to be suitable based on physical characterization of blend and dissolution data.

The qualitative compositions in all three processes studied (direct compression, fluid-bed granulation, and high-shear granulation process) was the same and contained API (around 30% w/w), diluent (around 64% w/w), disintegrant (5% w/w), and lubricant (1% w/w). In fluid-bed and high-shear granulation process, granulation was done with water and high-level manufacturing process is described in case study III. In case of direct compression strategy, API and excipients (except lubricant) were blended followed by lubrication and compression. The flow of the lubricated blend was almost similar in case of high-shear granulation process (compressibility index of 20) and fluid bed granulation process (compressibility index of 21), whereas, the flow of lubricated blend was poor in case of direct compression strategy (compressibility index of 36). These observations were further supported with the actual compression trials. For both fluid-bed granulation process and high-shear granulation process, though the flow of blend was satisfactory, dissolution comparison with RLD was better in case of the batch made using high-shear granulation process (f_2 value $>$ 60) than the batch made using fluid bed granulation process (f_2 value on borderline of 50). Therefore, high-shear granulation process was selected for further development.

After a feasible composition was identified, risk assessment was conducted to understand the effect of potential changes in drug substance and formulation variables on drug product CQAs. In an ideal scenario, risk assessment meetings are conducted in a cross-functional setup. Based on the risk assessment, a risk can be identified as low (broadly acceptable risk; no further investigation needed), medium (risk is acceptable; further investigation might be needed in order to reduce the risk), or high (risk is unacceptable; further investigation is needed to reduce the risk). Drug substance and formulation variable risks were identified using Fish-bone diagram (Fda, 2006) and mitigated via appropriate experimentations. They are not discussed in this chapter.

After a composition was identified meeting the desired CQAs, material attributes (MAs) of excipients were evaluated in terms of impact on CQAs. The MAs that affect the CQAs are called CMAs. For the CMAs, more knowledge is gained and mitigation strategy is developed after appropriate experimentations. Next, the impact of different process parameters (PPs) on CQAs is identified. Typically, a complete proposed manufacturing process is evaluated at each step, each of which serves as an input for next step.

The drug product (compound B) was manufactured using a wet granulation process using a bottom-driven, high-shear granulator. The manufacturing steps included sifting of all raw materials, dry mixing and wet granulation in high-shear granulator, wet milling, drying in a fluid-bed dryer, milling and sizing, blending and lubrication, and tablet compression followed by coating. A risk assessment of the granulation process was performed to identify the high and medium risk steps that might affect the CQAs of the final drug product. This risk assessment is presented in Table 7.10. Based on prior knowledge and initial development trials, it was anticipated that granulation process could have a medium impact

Table 7.10 Initial process parameter risk assessment table for drug product (compound B).

Drug product CQAs	Manufacturing steps		
	Granulation	Final mix	Compression
Assay	No impact	Low	Medium
Content uniformity	Low	Low	Medium
Degradation products	No impact	No impact	No impact
Dissolution	Medium	Medium	Medium

Note: For the purpose of this chapter, risk assessment of only the granulation parameter is presented.

on one of the CQAs (i.e., dissolution) and low impact on other CQAs (assay, degradation products, and content uniformity). For this case study, authors have used qualitative risk assessment tools based on prior knowledge, initial development trials, and properties of API; however, other risk assessment tools, including quantitative risk assessment also can be used (Fda, 2006). Risk assessment was performed to evaluate its impact on drug product CQAs. The relative risk that each attribute presents was ranked as high, medium, or low. The high-risk attributes warrant further investigation, whereas low-risk attributes required no further investigation. The medium risk is considered acceptable based on current knowledge.

Initial development trials indicated percentage of fluid uptake (% w/w of dry mix), fluid addition rate, and wet massing time (or kneading time) affect the granule properties and dissolution profile. Granule size was increasing and dissolution was being negatively affected by increasing the percentage of fluid uptake, fluid addition rate, or the wet massing time. These variables were investigated using DoE in order to better understand the granulation process and to develop a design space to reduce the risk. For all these experiments, several parameters were kept constant:

- Granulator: High speed; bottom driven; 150 L scale; (Tapasya Eng., Saizoner mixer granulator, Model SAI 15)
- Impeller speed: 100 rpm
- Chopper speed: 1500 rpm
- Dry mix time: 10 min
- Load: 70% occupancy (based on working volume) corresponding to 33.36 kg or 60,000 tablets.

The responses studied in this DoE included dissolution, the Hausner ratio (for flow), and the compressibility index (also for flow). Compressibility index and Hausner ratio were determined as per USP ⟨1174⟩. A set of eight initial experimental runs were planned, with four augmented runs (Table 7.11) added later to improve the detection of subtle relationships among the three inputs and the outputs (dissolution at Q point, i.e., at 45 min, Hausner ratio, and compressibility index). During initial development, discriminating media (0.1 N HCl; 900 mL; USP-2/50 rpm) was selected for evaluation based on its sensitivity to manufacturing process changes. An independent factor screening design, based on the D-optimality criterion (inbuilt feature of JMP Software; D-Optimal designs are most appropriate for screening experiments and are most efficient for designing experiments where the primary goal is inference), was used to find the inputs that have relationships to the various outputs.

7.4 Application of QbD in wet granulation process development

Table 7.11 Experimental plan for the DoE trials for drug product (compound B).

DoE run no.	1	2	3	4	5	6	7	8	9	10	11	12
	Initial experimentation run								Augmented runs			
Batch no.	231/001A	231/002A	231/003A	231/004A	231/005A	231/006A	231/007A	231/008A	236/006A	233/004A	236/007A	236/008A
Fluid uptake (% w/w of dry mix)	50	60	40	40	60	60	40	40	55	45	45	55
Binder addition rate (kg/min)	4	2	2	6	6	2	2	6	5	3	5	3
Kneading time (min)	2	1	3	3	1	3	1	1	2	2	3	3

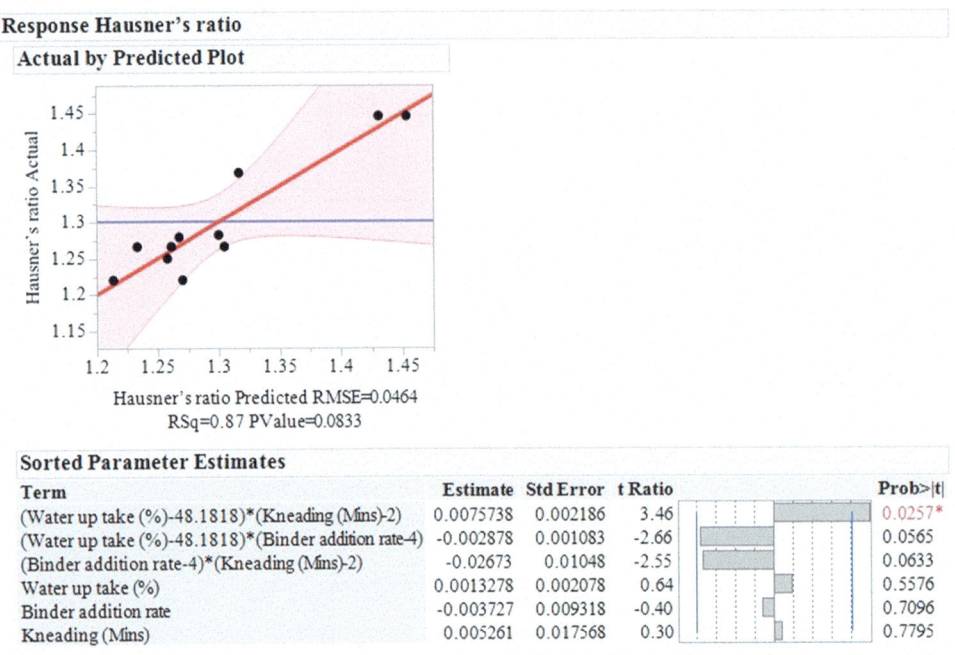

FIGURE 7.10

Model results for Hausner ratio (compound B).

The outputs were measured from samples of material obtained from each of the experimental batches. The JMP-PRO v12.1 software was used to design the experiments and to analyze the results. Results of these batches are shown in Table 7.12.

Batch number 231-006A-180 was found to exert undue influence on all results as identified by Cook's D Influence (> 1), which has been excluded from the analysis to mitigate error. Cook's D Influence is a measure of the influence of an outlier on a statistical model. Based on the data analysis, percentage of fluid uptake and kneading time are likely to have influence on the Hausner ratio and the compressibility index. The kneading time might impact the dissolution at 45 min. An example of the model fit for Hausner ratio is shown in Figure 7.10. For this mathematical model, the changes in input parameters explain 87% of the variability in the Hausner ratio (r-square $= 0.87$). In addition, the model was found to be significant with a P value of 0.08. The interaction of water uptake and kneading time significantly influenced Hausner ratio (Prob $> |t| = 0.0257$).

The relationships among the three inputs and the three outputs of interest can be seen comprehensively in the prediction profiler (Figure 7.11). The steep angle seen for model line of kneading time and the dissolution at 45 min illustrates a strong relationship. The relatively flat angle for the model line of kneading time and the compressibility index (as per USP ⟨1174⟩) illustrates a weak relationship. The profiler is dynamic and allows for the exploring of an infinite number of input levels to make estimates for the three outputs and eventual optimization of the process.

Table 7.12 Blend properties and dissolution results of DoE trials (compound B).

Batch no.	231/001A	231/002A	231/003A	231/004A	231/005A	231/006A	231/007A	231/008A	236/006A	233/004A	236/007A	236/008A
Hausner ratio	1.37	1.25	1.27	1.27	1.22	1.30	1.28	1.45	1.28	1.27	1.22	1.27
Compressibility index %	26.9	20.0	21.0	21.0	18.0	22.9	21.8	30.8	22.0	21.0	18.0	21.0
Dissolution: 0.1 N HCl; USP-2/50 rpm												
Time (min)	% drug dissolved; n = 6											
5	25	11	39	36	15	9	35	31	29	48	32	23
10	41	30	54	47	35	23	46	42	44	62	47	50
15	51	43	61	55	46	40	54	49	54	69	55	62
20	58	51	66	59	53	49	59	54	60	74	61	69
30	68	63	73	67	63	59	66	62	71	81	72	78
45	77	72	81	75	72	69	73	70	80	87	80	86

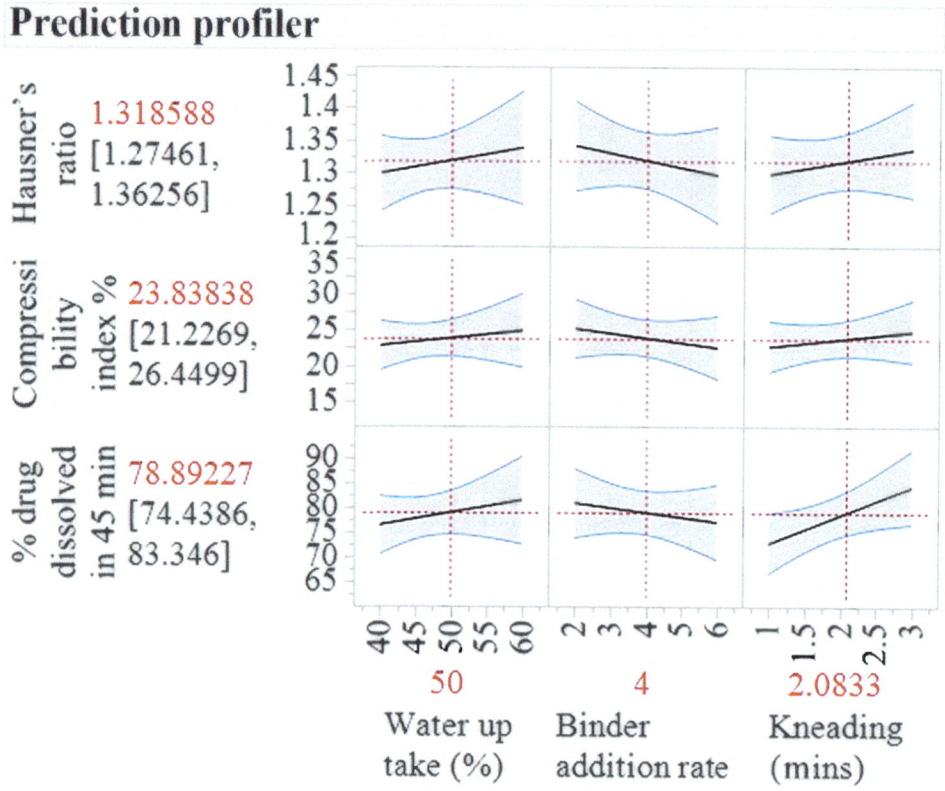

FIGURE 7.11

Dynamic prediction profiler showing impact of various inputs on outputs (compound B).

The experimental modeling was used to identify the settings of the inputs that are likely to produce optimal results for all of the measured outputs. Based on the DoE trials, results, and data analysis, it was decided to manufacture a confirmatory batch with the following parameters (and keeping all other parameters constant as in DoE trials):

- Water uptake: 50% w/w of dry mix
- Kneading (wet massing time): 2 min
- Binder addition rate: 4.17 kg/min (corresponding to binder addition time of 4 min).

The confirmatory batch results were compared to the model estimate to determine how well the actual results can be predicted. Considering composition was finalized and confirmatory batch was prepared, dissolution evaluation was done in release media 0.001 N HCl; 900 mL; USP-2/50 rpm. It was assumed that the design space already is established in the discriminatory media and going forward, the batches will be tested in release media only. Table 7.13 includes 95% confidence interval used to

Table 7.13 Comparison of confirmatory batch results with theoretical estimates (compound B).

Inputs			Outputs		
Water uptake %	Binder addition rate (kg/min)	Kneading time (min)	Hausner ratio	Compressibility index %	45-min dissolution (0.001 N HCl; USP-2/50 rpm)
50	4.17	2	1.2–1.3	16.4–25.4	97.2–99.5
Actual results			1.3	25	98
Hit/Miss			Hit	Hit	Hit

Table 7.14 Updated process parameter risk assessment table for drug product (compound B).

	Manufacturing steps		
Drug product CQAs	Granulation	Final mix	Compression
Assay	No impact	Low*	Low
Content uniformity	Low*	Low*	Low
Degradation products	No impact	No impact	No impact
Dissolution	Low	Low	Low

*No further mitigation was required as the risk was identified as low during initial risk assessment.

estimate average results for each of the outputs. The model correctly estimated the results achieved with the confirmation run.

The above case study exemplifies that how risks were identified and based on the experiments, the medium risk identified for granulation parameters on CQA (i.e., dissolution) was mitigated (Table 7.14).

Based on this data, it was decided to identify the design space for parameters under evaluation such as percentage of water uptake, binder addition rate, and wet massing time (or kneading time) so that CQA-like dissolution will always hit the desired target within specified range.

The contour plots were used to illustrate the contour grid of the percentage of water uptake and binder addition rate inputs (Figure 7.12), binder addition rate and kneading time inputs (Figure 7.13), and the percentage of water uptake and kneading time inputs (Figure 7.14).

From above data, it was concluded that there is no limit to the design space within all the ranges studied for all process parameters because of acceptable results for all the outputs. After the design space was identified for process parameters for this product, the next step was to identify and propose the parameters for scaleup batches for submission/commercial scale (Table 7.15). It is important to understand that these identified/proposed parameters will be the starting point for manufacturing the higher scale batch and, if some modifications need to be done, knowledge gained during development can be used. The scaleup section in this chapter provides more details about the calculation of these parameters.

7.4.2 Case study V

In another Perrigo development case study, excipient levels and CPPs were optimized during development of an extended release product of compound C using a high-shear granulation process (bottom-driven). Compound C is a high-dose drug (600 mg), a base having molecular weight of 198.2 g/mol, Log P vale of 1.39, and shows high water solubility across the gastro-intestinal pH range. The study with multiple excipients levels and CPPs was designed using structured experiments to gain information on the design space, where dissolution targets can be met in a biorelevant media. The multivariate experiments were planned using the definitive screening design (DSD) approach available in JMP-PRO v12.0 software. The design included minimum and maximum treatment levels for inputs as well as a center point for each. Use of DSD assumes that only a small number of inputs, interactions, and squared

Table 7.15 Recommended parameters for scaleup batches for compound B.

Parameters	Identified ranges on 150 L Scale	Proposed ranges for higher scale (e.g., fielder 1800)	Scaleup principle used
Fluid uptake (% w/w of dry mix)	40%–60%	40%–60%	Same as that of development scale
Binder addition rate (kg/min)	2–6	3–9	Constant Froude number
Kneading (min)	1–3	1–4	
Impeller speed	100 rpm	64 rpm	

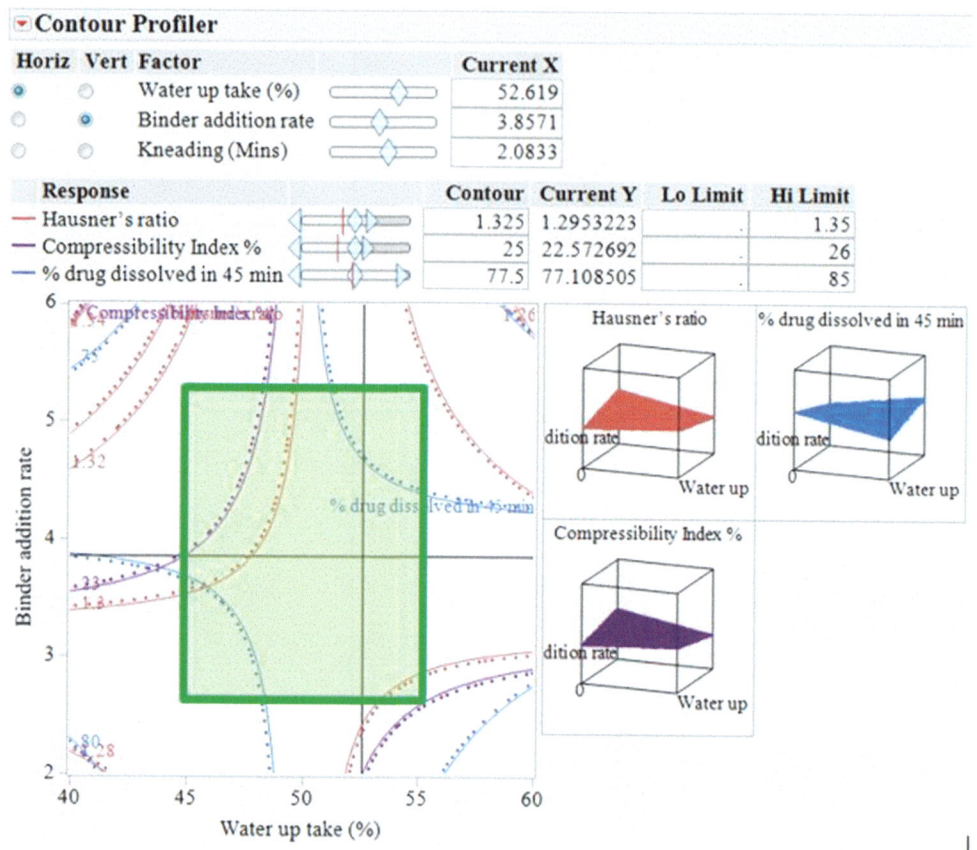

FIGURE 7.12

Contour grid of percentage of water uptake and binder addition rate (compound B).

terms have significant influence on the dissolution output. Based on early development trials, CPPs ranges and also the excipient levels of interest were defined (Table 7.16). This particular design allowed evaluation of six factors in 13 runs as compared to conventional multivariate DOE. A full-factorial design for six inputs would have involved 64 experimental runs. The definitive screening design is exceptionally efficient at analyzing six or more inputs.

The manufacturing process involved granulation in a high-shear granulator (PMA 300) followed by drying in fluid-bed dryer (Glatt WST 60) and milling in Fitz mill. After blending in a 10 cubic foot V-shell mixer, tablets were compressed using a Killian tablet press. All the process inputs other than the wet mass time of the high-shear granulation were controlled at fixed levels. The six inputs mentioned in Table 7.16 were studied with 13 batches in the random order shown in Table 7.17.

Individual inputs of the tablet hardness and wet mass time were found to be very significant in the influence exerted on dissolution at various time points. A complex (squared term) effect and the interaction between wet mass time and diluent levels were significant at influencing dissolution values.

7.4 Application of QbD in wet granulation process development

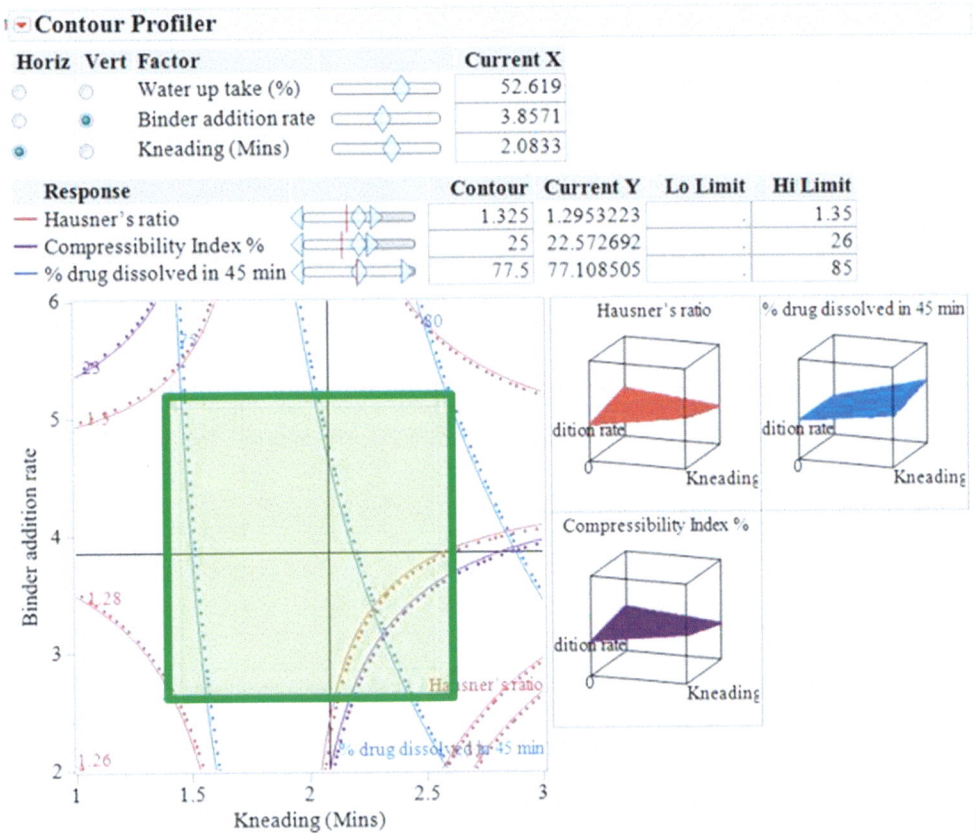

FIGURE 7.13

Contour grid of binder addition rate and kneading time (compound B).

Table 7.16 Experiment inputs for the study design (compound C).			
Factor	Low level (− 1)	Target	High level (+ 1)
Wet massing time (min)	5	8	15
Hardness (strong cobb unit; SCU)	10	15	20
Extended release polymer 1 (mg/tab)	1.5	4.75	8
Extended release polymer 2 (mg/tab)	3	7.5	12
Diluent (mg/tab)	10	18	26
Binder (mg/tab)	15	20	25

Table 7.17 Design of experiments to study excipient levels and CPPs (compound C).

Run No	Wet massing time (min)	Hardness (strong cobb unit; SCU)	Extended release polymer 1 (mg/tab)	Extended release polymer 2 (mg/tab)	Diluent (mg/tab)	Binder (mg/tab)
1	11	10	8	12	10	20
2	5	20	4.75	12	10	15
3	5	10	1.5	12	18	25
4	11	20	8	3	18	15
5	5	10	8	7.5	26	15
6	8	15	4.75	7.5	18	20
7	8	20	8	12	26	25
8	11	10	4.75	3	26	25
9	11	15	1.5	12	26	15
10	5	20	1.5	3	26	20
11	5	15	8	3	10	25
12	8	10	1.5	3	10	10
13	15	20	1.5	7.5	10	25

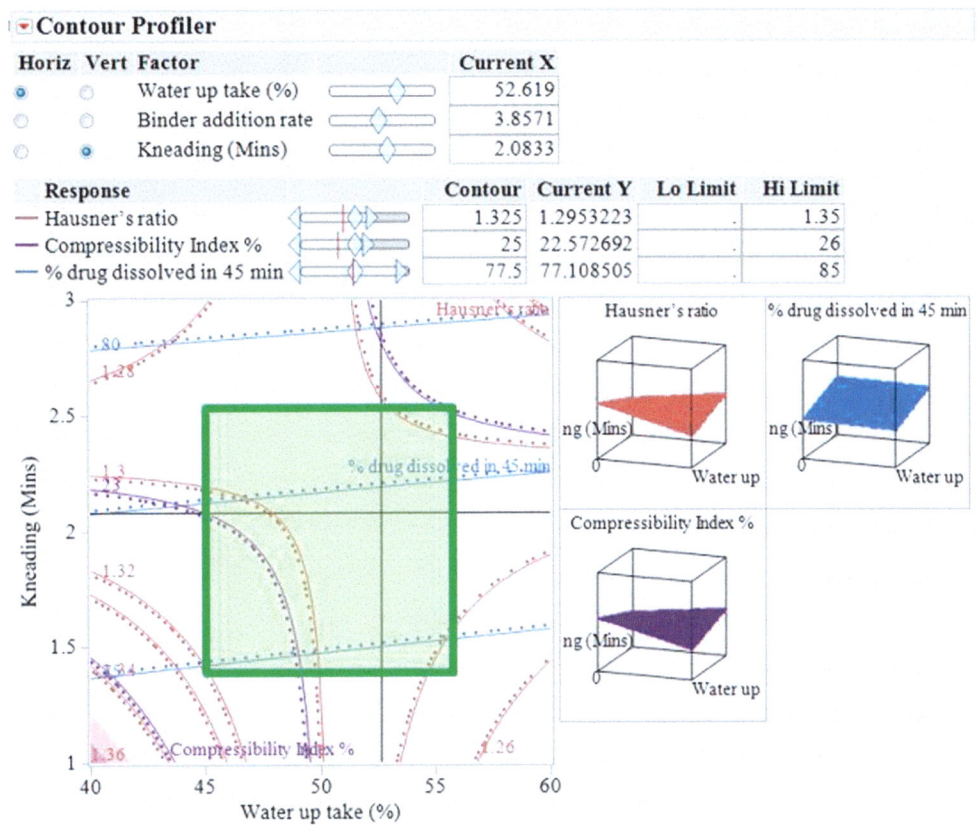

FIGURE 7.14

Contour grid of percentage of water uptake and kneading time (compound B).

Figure 7.15 shows the trends between the inputs and the dissolution outputs (CQA). The interactions among inputs can be identified by the non-parallel effects in dissolution in the comparative space of wet mass time and diluent. A slight interaction also is seen in the space of wet mass time and extended release polymer 1. Based on the data analysis, complex effect also is seen for the binder input indicating a nonlinear influence of the material on dissolution. Increase in binder levels between 10 and 18 results in dramatic drop in dissolution values; however, drop stabilizes and reverses for binder levels between 18 and 24.

Based on the results, a prediction pattern was estimated considering dissolution as a CQA. The theoretical best results and input levels to achieve them are noted in Table 7.18.

A continuous function can be used to better define the dissolution profile above and beyond the specific time points specified as CQAs. Growth rate is the average slope of increase of the release of a drug over time and is a good indicator of how process inputs can affect dissolution. The three-parameter

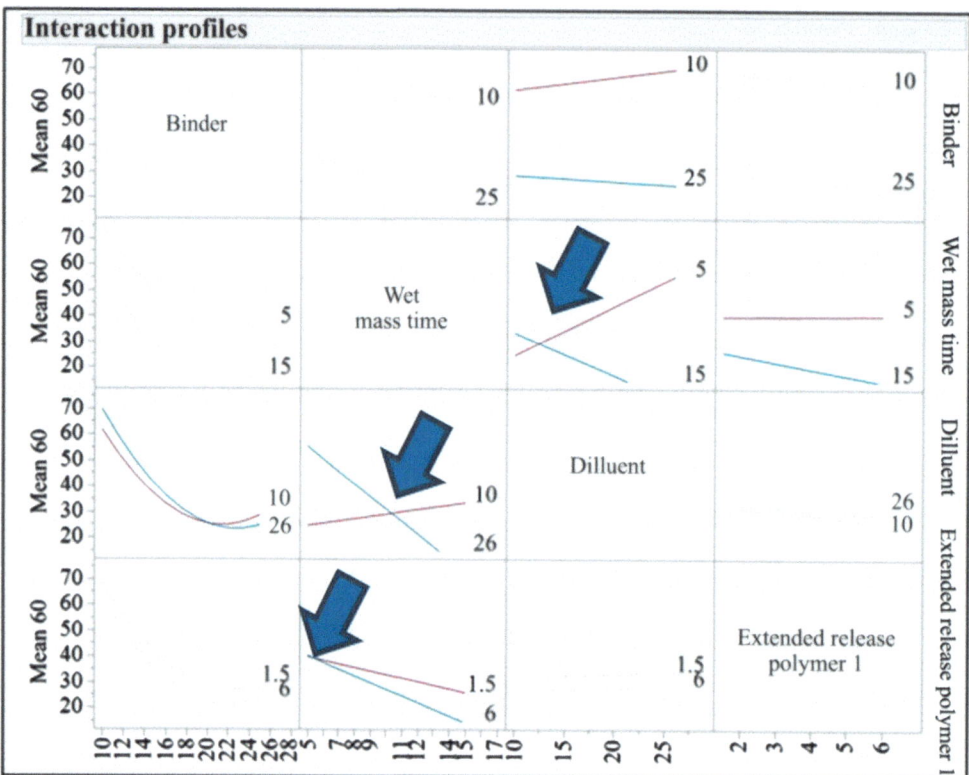

FIGURE 7.15

Interactions plot for the inputs on dissolution (compound C).

Table 7.18 Optimal results with inputs and expected outcomes (compound C).

Inputs		Outputs	
Binder (mg/tab)	18	% dissolution (30 min)	29.2–31.6
Wet mass time (min)	5	% dissolution (60 min)	38.5–46.3
Diluent (mg/tab)	20	% dissolution (120 min)	50.8–64.5
Extended release polymer 1	1.5	Growth rate (3Pe)	− 0.02 to − 0.013
Hardness (SCU)	15		

exponential function explains > 98% of the variation in percentage of release over the scope of time studied (Figure 7.16).

Based on modeling of the dissolution data, a confirmation plan was established to study the control space, and six confirmation batches were run per the levels noted in Table 7.19.

7.4 Application of QbD in wet granulation process development

Table 7.19 Confirmation batch plan (compound C).

Factor	DOE (level) predicted	Batch 1 (DOE predicted batch)	Batch 2	Repeat DOE batches (to augment DOE understanding)			Batch 6 (Independent batch)
				Batch 3	Batch 4	Batch 5	
Binder (mg/tab)	18	18	18	25	20	15	22.5
Wet massing time (min)	5	5	5	5	5	8	5
Diluent (mg/tab)	20	20	10	18	26	10	10
Extended release polymer 1 (mg/tab)	1.5	1.5	1.5	1.5	1.5	1.5	1.5
Extended release polymer 2 (mg/tab)	3	3	3	12	3	3	3
Hardness (SCU)	15	15	15	10	20	10	15

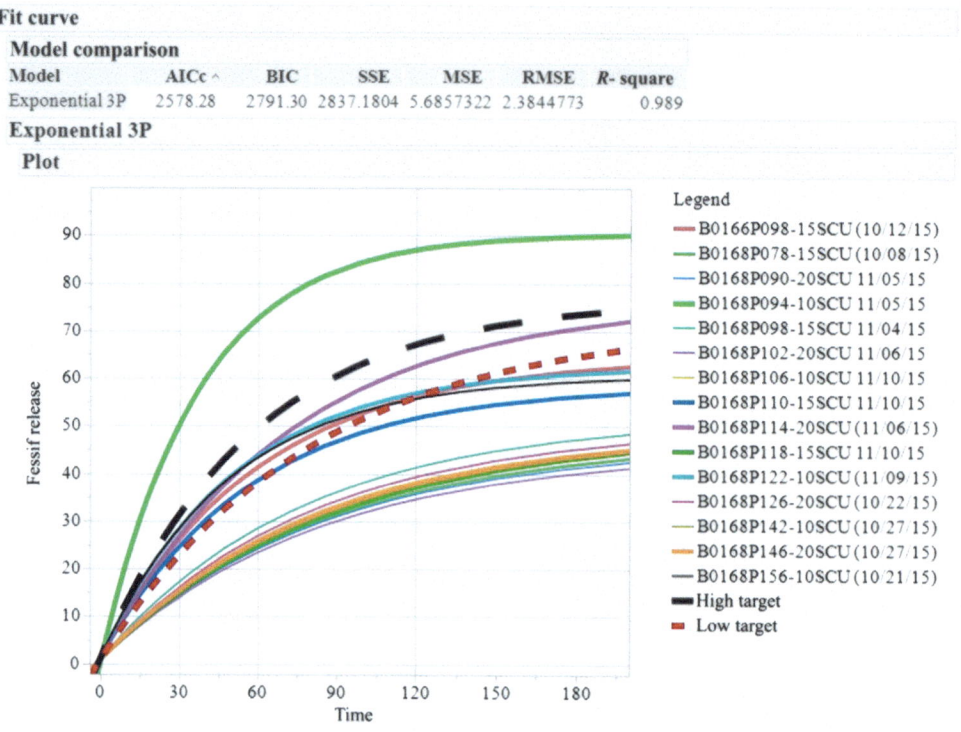

FIGURE 7.16

Nonlinear dissolution profiles (three parameter exponential functions).

Table 7.20 Example of confirmation batch performance to predictions (compound C).				
	30-min dissolution	60-min dissolution	120-min dissolution	3Pe growth rate[a]
Average	29.5	40.8	54.6	65.04
Standard deviation	1.98	2.26	1.68	
RSD	6.7	5.5	3.1	
Low 95% predicted	28.8	39.4	52.4	54.8
High 95% predicted	32.0	45.4	62.9	83.8
Hit/miss	Hit	Hit	Hit	Hit

[a] The growth rate of a 3-parameter exponential function that fits the data. This is how fast the percentage of drug dissolved increases over time.

The dissolution values for four of the six batches were within the prediction intervals for the CQA outputs. Table 7.20 illustrates that the actual results of a confirmation batch that are within the 95% prediction interval from the structured model. Two of the batches with more extreme levels of binder had 30-min and 60-min results that are just outside of the confidence interval prediction; however, the

7.4 Application of QbD in wet granulation process development

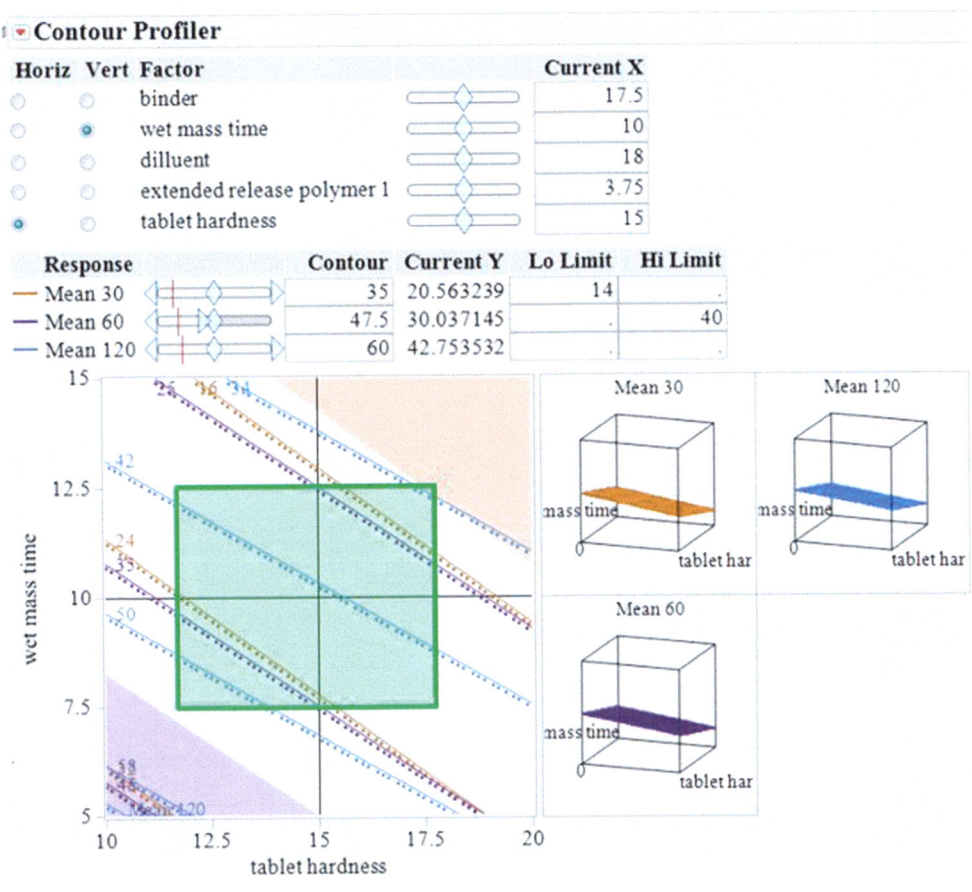

FIGURE 7.17

Contour grid of wet mass time and tablet hardness (compound C).

prediction zone was near the inflection point of the response curve and difficult for precise and accurate estimations. This space is outside of the levels that will be suggested for commercial production.

From the design evaluations, valuable information about the control space for the excipient levels and CPPs was obtained to ensure that the product will meet CQA targets robustly. The model estimations were confirmed by additional batches, which provided the guidelines for operational controls that are likely to constantly produce product that meets the dissolution requirements.

The contour plots (Figure 7.17 and Figure 7.18) were used to illustrate the two most significant inputs of the process to explore the space of the three dissolution pulls. The white space in the middle of the plot illustrates the design space that will provide acceptable dissolution results. A control space is developed from within the design space, which is indicated by blue shaded square over the plot to illustrate that wet mass time of between 7.5 and 12.5, and tablet hardness of between 12.5 and 17.5 and wet mass time of between 7.5 and 12.5 and diluent between 10 and 17 to show a potential set of manufacturing limits that will work in achieving the desired output of dissolution successfully.

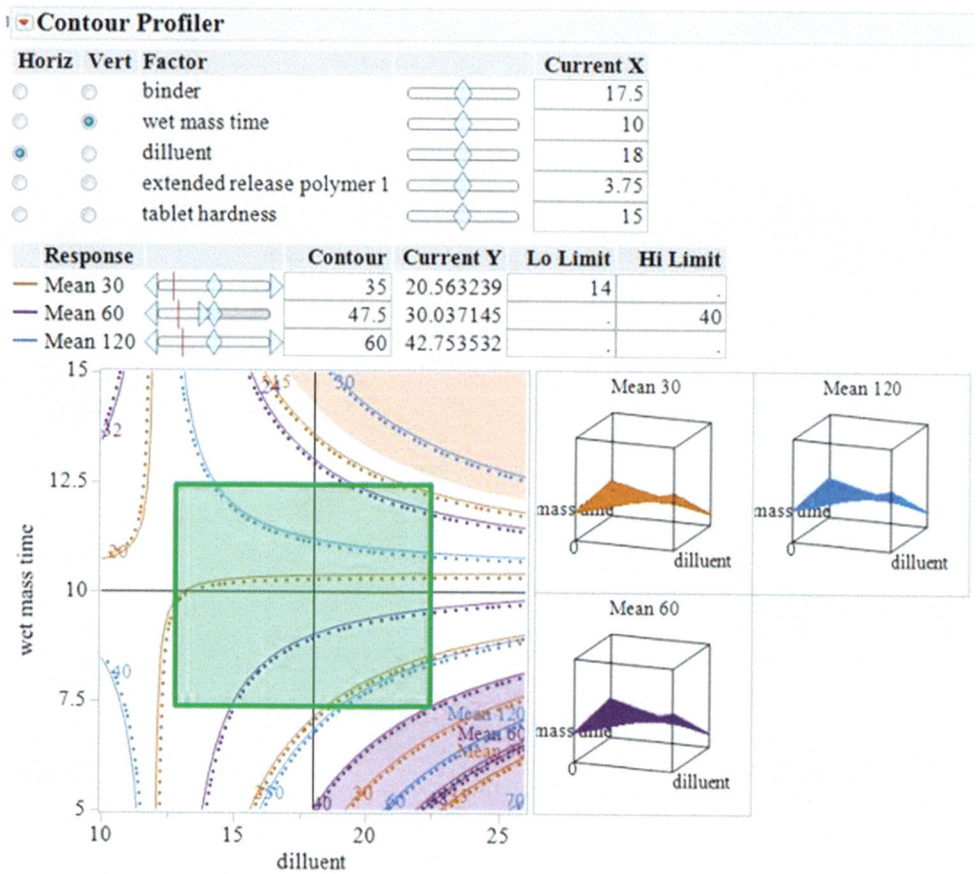

FIGURE 7.18

Contour grid of wet mass time and diluent level (compound C).

These two case studies clearly demonstrate that how use of risk assessment tools and multivariate experiments (DoE) have resulted in identification and mitigation of high-risk items (excipient levels/CPPs) and identification of preferred ranges (i.e., design space). Use of QbD approach helps to define clear targets early on and constant use of risk assessment tools and statistical applications helps mitigate risks and arrive at informed decisions (Figure 7.19).

7.5 Granulation end point determination

In context of pharmaceutical wet granulation, the end point can be defined as reaching a stage during granulation process, wherein granule properties (e.g., target mean particle size or distribution) and subsequent tableting properties (e.g., tablet hardness, dissolution) are expected to be very similar regardless of wet granulation processing parameters (e.g., impeller or chopper speed in high-shear

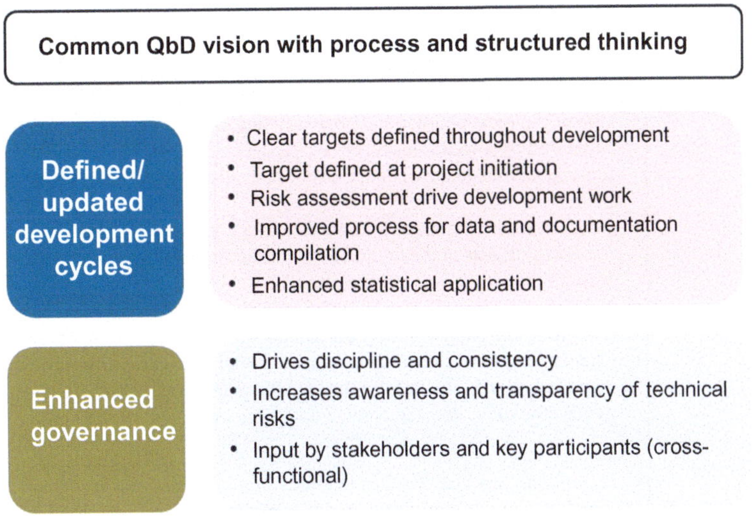

7.5 Granulation end point determination

Common QbD vision with process and structured thinking

Defined/updated development cycles
- Clear targets defined throughout development
- Target defined at project initiation
- Risk assessment drive development work
- Improved process for data and documentation compilation
- Enhanced statistical application

Enhanced governance
- Drives discipline and consistency
- Increases awareness and transparency of technical risks
- Input by stakeholders and key participants (cross-functional)

FIGURE 7.19

QbD approach in development.

granulator, air volume in fluid-bed granulation, and binder addition rate). Granule growth often is difficult to predict because the process is sensitive to raw material properties and operating conditions. Regulatory agencies are encouraging development of process analytical technologies (PAT) to improve process understanding and online monitoring (FDA, 2004).

Broadly, technologies investigated for high-shear wet granulation monitoring include traditional methods such as power consumption, impeller torque, torque rheometer, reaction torque, and capacitance measurements. Also used are emerging technologies such as acoustic, near infrared, focused beam reflectance measurement, Raman spectroscopy, imaging, spatial filter velocimetry, and stress and vibration measurements. These techniques are briefly covered in Table 7.21. Covering all these technologies in detail is outside the scope of this chapter, and interested readers can refer to one of the comprehensive reviews (Hansuld and Briens, 2014).

Apart from the advanced techniques for end-point determination, there are often practical aspects such as wet massing time, which an operator relies upon to stop the granulation. This conventional technology is a reliable technique and should be studied and established early on.

In one study (Narang, Sheverev, et al., 2016), a drag flow force (DFF) sensor that measures the force exerted by wet mass in a granulator on a thin cylindrical probe was shown as a promising process analytical technology (PAT) for real-time inline high-resolution monitoring of wet mass consistency during high-shear wet granulation. In this study, the measurements of flow force by a DFF sensor, taken during wet granulation of three placebo formulations with different binder content, were compared with concurrent at line FT4 Powder Rheometer characterization of wet granules collected at different time points of the processing. The wet mass consistency measured by the DFF sensor correlated well with the granulation's resistance to flow and interparticulate interactions as measured by FT4 Powder Rheometer. Authors concluded that the force pulse magnitude measured by the DFF sensor was indicative of

Table 7.21 Few techniques for determination of granulation end point

S. no.	Technique	Remarks
1	Power consumption	- One of the most popular and relatively inexpensive measurements. - Measured by a watt transducer or a power cell using Hall Effect (a measurable transverse voltage between the two radial sides of a current conductor in a magnetic field). - Correlates with the mean granule size of a granulation, although the correlation is not always linear in the entire range. Intragranular porosity also shows some correlation with power consumption. - Normalized work of granulation (power profile integrated over time) can accurately determine end-points and is correlated well with properties of granulates. - Challenge with this technique is that this variable reflects load on the motor rather than load on the impeller. It relates to the overall mixer performance, depends on the motor efficiency, and can change with time regardless of the load.
2	Impeller torque	- In a mixing process, changes in torque on the blades and power consumption of the impeller occur as a result of change in the cohesive force or the tensile strength of the agglomerates in the moistened powder bed. - Direct torque measurement requires installation of strain gages on the impeller shaft or on the coupling between the motor and impeller shaft. As the shaft is rotating, a device called slip ring is used to transmit the signal to the stationary data acquisition system. - One of the best inline PAT measurement of the load on the main impeller
3	Torque rheometer	- A torque rheometer is a device that provides an offline measurement of torque required to rotate the blades of the device, and this torque can be used to assess rheological properties of the granulation. - Has been used extensively for end-point determination - The torque values thus obtained were termed as measure of wet mass consistency.
4	Reaction torque	- By the third law of Newton, for every force there is a counterforce, collinear, equal, and opposite in direction. - As the impeller shaft rotates, the motor tries to rotate in the opposite direction, but it does not because it is bolted in place. The tensions in the stationary motor base can be measured by a reaction torque transducer.
5	Acoustic	- Use of piezoelectric acoustic emission sensors - Noninvasive, sensitive, and relatively inexpensive.
6	Near infrared	- Use refractive near-infrared (NIR) moisture sensor for end-point determination of wet granulation process - There are technological challenges associated with this approach, as the sensor can measure only the amount of water at the powder surface.

Modified from Levin, M. (2006). Wet granulation: End-point determination and scaleup. In J. Swarbrick (Ed.), Encyclopedia of pharmaceutical technology (3rd ed., pp. 4078–4098). London: Informa Healthcare.

fundamental material properties (e.g., shear viscosity and granule size/density), as they were changing during the granulation process.

In another study (Narang, Stevens, et al., 2016), authors evaluated the resolution and sensitivity of chord length distribution (CLD) for inline measurement inside a high-shear granulator using focused beam reflectance measurement (FBRM) C35 probe using different particle size grades of microcrystalline cellulose (MCC). Also, the impact of water and impeller tip speed on the measurement accuracy as well as correlation with offline particle sizing techniques (FBRM, laser diffraction [Malvern Mastersizer], microscopy [Sympatec QicPic], and nested sieve analysis) was studied. Inline FBRM resolved size differences among different MCC grades, and the data correlated well with offline analyses. Impeller tip speed changed the number density of inline CLD measurements while addition of water reduced the CLD of dry MCC, likely because of deagglomeration of primary particles. Authors concluded that inline FBRM CLD measurement in high-shear granulator provides adequate resolution and reproducible measurements in the pharmaceutically relevant size range both in the presence and in the absence of water. In another study (Narang et al., 2017), authors used FBRM C35 probe to investigate the effect of formulation and process parameters on the granule growth profile over time during the high-shear wet granulation of a high drug load formulation of brivanib alaninate. The probe quantitatively captured changes in the granule CLD with the progress of granulation and delineated the impact of water concentration used during granulation. The results correlated well with offline particle size distribution measured by nested sieve analyses. An end point indication algorithm was developed that was able to successfully track the process time needed to reach the target CLD. Testing of the brivanib alaninate granulation through 25-fold scaleup of the batch process indicated that the FBRM CLD profile can provide a scale-independent granule attribute-based process fingerprint.

An inline particle size analyzer was used in conjunction with multivariate methods in a fluid-bed granulation process in one of the studies (Huang et al., 2010). The evaluation of inline monitoring of quality attributes, particle size, and particle size distribution, was conducted using the Parsum probe that is based on spatial filtering technique. Several granulation batches were manufactured and monitored using a commercial-scale fluid bed granulator. Reference measurements by offline Malvern MasterSizer showed good agreement with those by Parsum at end-of-spray phase. Multivariate/batch statistical process control methods were used to evaluate batch process performance, batch-to-batch variation and develop potential control strategy. The authors concluded that the Parsum analyzer is a viable tool for inline particle size characterization and improved process understanding in combination with multivariate tools.

In another study, qualitative and quantitative chemometric models were evaluated to monitor moisture content of a wet granulation in a fluid-bed dryer using near infrared (NIR) technology (Barla, Kumar, Nalluri, Gandhi, & Venkatesh, 2014). A principal component analysis (PCA) model was evaluated to obtain qualitative information. Multiple linear regression (MLR), principal component regression (PCR), partial least squares regression (PLS-R) and support vector machine regression (SVM-R) were evaluated using The Unscrambler X. The PLS-R method was selected to demonstrate real-time monitoring of the moisture content. An ABB FT-NIR spectrometer with a Galileo direct-contact fiber-optic diffuse reflectance probe was used for NIR measurements. The Unscrambler X Process Pulse was employed to upload the PLS-R model in measuring real-time moisture content. The PCA model successfully projected test batch data for the process signature, and the PLS-R model successfully predicted the moisture content of the granulation during fluid-bed drying.

Table 7.22 Commonly used dimensionless numbers to describe wet granulation process.

Dimensionless number	Equation	Comments
Newton (power) number (N_p)	$Np = P/(\rho \times n^3 \times d^5)$ —(1)	Newton (power) number, which relates the drag force acting on a unit area of the impeller and the inertial stress, represents a measure of power requirement to overcome friction in fluid flow in a stirred reactor. In mixer-granulation applications, this number can be calculated from the power consumption of the impeller or estimated from the power consumption of the motor.
Froude number (F_r)	$F_r = n^2 \times d/g$ —(2)	Froude Number has been described for powder blending and was suggested as a criterion for dynamic similarity and a scale-up parameter in wet granulation. It relates the inertial stress to the gravitational force per unit area acting on the material. It is a ratio of the centrifugal force to the gravitational force. The mechanics of the phenomenon was described as interplay of the centrifugal force (pushing the particles against the mixer wall) and the centripetal force produced by the wall, creating a compaction zone.
Reynolds number (Re)	$Re = d^2 \times n \times p/\eta$ —(3)	Reynolds numbers relate the inertial force to the viscous force. They are frequently used to describe mixing processes and viscous flow, especially in chemical engineering.

Where P = power required by the impeller or motor (W = J/s), dimensional units $[M\ L^2\ T^{-5}]$; ρ = specific density of particles (kg/m³), dimensional units $[M\ L^{-3}]$; n = impeller speed (rev/s), dimensional units $[T^{-1}]$; d = impeller (blade) diameter or radius (m), dimensional units $[L]$; g = gravitational constant (m/s²), dimensional units $[L\ T^{-2}]$; p = specific density of particles (kg/m³), dimensional units $[M\ L^{-3}]$; η = dynamic viscosity (Pa s), dimensional units $[M\ L^{-1}\ T^{-1}]$.
Modified from Levin, M. (2006). Wet granulation: End-point determination and scaleup. In J. Swarbrick (Ed.), Encyclopedia of pharmaceutical technology (3rd ed., pp. 4078–4098). London: Informa Healthcare.

7.6 Scaleup of wet granulation process
7.6.1 High-shear granulator

Traditionally, dimensional analysis has been used for scaleup in chemical engineering, and this approach, which uses process similarities between different scales, had been applied to pharmaceutical granulation since the early days (Leuenberger, Bier, & Sucker, 1979).

The essence of dimensionless analysis is as follows: For any two dynamically similar systems, all the dimensionless numbers necessary to describe the process have the same numerical value. The process is characterized solely by numerical values of the dimensionless variables (numbers). In other words, dimensionless representation of the process is scale invariant. Dimensionless numbers most commonly

Table 7.23 Scaleup considerations for fluid bed granulation process..

Parameters	Comments
Batch size (kg)	A batch size can be calculated by following formula $S_{max} = V \times 0.8 \times BD$—(7) $S_{min} = V \times 0.5 \times BD$—(8) where S = batch size V = volume of the container BD = bulk density of the substrate
Liquid amount (kg)	Calculate total amount of liquid (kg) so as to maintain similar ratio of liquid (kg)/product (kg) ratio in lab scale and scaled-up machine Liquid amount (scale-up machine) = Liquid amount (development machine) × Batch size (scale-up machine)/Batch size (development machine)—(9)
Air temperature (°C)	Use same inlet air temperature in the scaled-up machine and lab scale machine (preferably, maintain similar dew point control or use slightly higher temperature to minimize impact of inlet air humidity).
Air flow (ft^3/min)	There should be a right balance between the air volumes used for fluid-bed granulation as a very high air volume can cause not only spray drying but also cause deceleration of coated particles in filter bags causing reduction on process efficiency. A very low air volume will not allow the particles to come across the granulation cycle affecting the granulation efficiency. Air flow (ft^3/min) = Air flow air velocity (ft/min) × Area of distributor plate (in ft^2)—(10)
Air velocity (ft/min)	Use same velocity in the scaled-up machine as that of lab scale machine. Adjust air flow to keep air velocity constant
Bed depth (ft)	Maintain same bed depth in the scaled-up machine as that of lab scale machine.
Nozzle height above bed (ft)	Maintain same nozzle height above bed in the scaled-up machine as that of lab scale machine.
Liquid droplet size	Droplet size has a great role to play in physical properties of granules. It is always preferred to have smallest droplet size for any fluid bed granulation and this can be achieved by using higher atomization pressure. However, care should be taken to avoid blinding of filter bags because of smaller size of expansion chamber as compared to Wurster process. It is desirable to maintain the droplet size similar across the higher scale equipment. This can be further controlled by using additional or multiheaded nozzles to reduce the spray rate per nozzle port.

(*continued on next page*)

used to describe the wet granulation process are Newton, Froude, and Reynolds and are listed in Table 7.22.

In a high-shear granulation process, several variables should be kept in mind while scaling up.

7.6.1.1 Height of the material and batch size

Ideally, the percentage of occupancy in the small-scale and scaleup equipment should be kept similar to provide equivalent process conditions across the size ranges. The height of the material in the bowl will have an effect on the overall energy being used in the granulation process. As the process is scaled up, the depth of material in the granulator also will change.

Table 7.23 Scaleup considerations for fluid bed granulation process.—cont'd	
Parameters	Comments
Spray rate (kg/min)	This is one of the critical process parameters for fluid bed granulation process. There can be two ways for estimating spray rate. In general, granule size is directly proportional to the humidity in bed during granulation. Typically, spray rate must be proportional to the volume of the drying air which can achieve adequate fluidization pattern. If small-scale and large-scale equipment has air volume indicator, this equation can be used to give adequate fluidization in each machine. $S_1 = \frac{V_1}{V_2} \times S_2$—(11) where V_1 = air volume in the scaled-up machine V_2 = air volume in the lab scale machine S_1 = spray rate in the scaled-up machine S_2 = spray rate in the lab scale machine However, if the equipment does not have air volume indicator, a ratio of cross sectional areas of product bowl screen can be considered. Assumption in this case would be similar fluidization pattern will require similar air velocity. $S_1 = \frac{A_1}{A_2} \times S_2$—(12) where A_1 = cross-sectional area of the scaled-up machine A_2 = cross-sectional area of the lab scale machine Other factors that can impact spray rate includes, tackiness of spraying solution, size of the substrate on which it is sprayed, properties of the liquid being sprayed, temperature, and the coating zone, where the application is taking place.

Modified from Jones, D. M. (1985). Factors to consider in fluid bed processing. Pharmaceutical Technology, 50–55; Mehta, A. M. (1988). Scaleup considerations in the fluid bed process for controlled release products. Pharmaceutical Technology, 46–52.

7.6.1.2 Binder addition rate

Ideally, the granulating fluid should be added at a slow rate to avoid local overwetting. To ensure consistency in the spray rate, pneumatic or binary nozzle or atomization by pressure nozzle can be used. During scaleup, longer liquid addition periods are more beneficial in order to avoid localized overwetting of the product. Some companies also prefer keeping liquid addition time the same as that of smaller scale machine; however, in another approach, large-scale binder addition time can be calculated as

Binder addition rate (scaleup machine) = Small scale binder addition time × small scale rpm/large scale rpm. (4)

7.6.1.3 Chopper speed

The purpose of the chopper is to break loose granules so as to allow further granule growth. Normally, chopper speed has little effect on the granule size; however, one of the literature reports (Briens and Logan, 2011) shows the impact of chopper speed based on impeller speed. In this study, chopper presence and thereafter, chopper speed was varied during wet high-shear granulation of a placebo formulation using a PMA-1 granulator while also varying the impeller speed. The effect of the chopper on the granules varied with impeller speed from no effect at a low impeller speed of 300 rpm to flow interruptions at an impeller speed of 700 rpm to minimal impact at very high impeller speeds as caking at the bowl perimeter obscured the effect of the chopper on the flow pattern. Differences in the granule

7.6 Scaleup of wet granulation process

Table 7.24 Typical process parameters in a wet granulation process

High-shear granulation (top-driven/bottom-driven/low-shear granulation process	Fluid-bed granulation
Scale dependent[a]	Scale dependent[a]
Impeller speed	Atomization air pressure/air flow
Spray rate	Nozzle type
Kneading/mixing time	Location of nozzles
% fluid uptake	Flow/spray rate
Scale independent[b]	Scale independent[b]
Chopper speed (only in case of high shear granulator)	Atomization air pressure/air flow
Nozzle type	Nozzle type
Nozzle size	Location of nozzles
Nozzle location	No. of nozzles
Granulator loading level (% of working volume)	Nozzle dimension
Liquid/binder addition method (gravity feed/pump)	Air flow
	Dew point
	Bottom screen type/size
	Filter type
	Filter bag pores (filter bag efficiency)
	Filter shake time/Interval
	Filter system type (blow back/bag filter)
	Granulation load size
	Loading and unloading methods
	Product temperature
	Inlet air temperature

[a] Scale dependent parameters (factors) require level/quantity adjusted with changes to the scale (either from lab-scale to pilot-scale to commercial-scale; or commercial-scale to lab-scale) to obtain similar CQAs/QTPPs between the different scales of manufacturing.
[b] Scale independent parameters (factors) are not required to change level/quantity with changes to the scale (either from lab-scale to pilot-scale to commercial-scale; or commercial-scale to lab-scale) to obtain similar CQAs/QTPPs between the different scales of manufacturing.

flowability were minimal. It was concluded, however, that the largest fraction of optimal granules would be obtained at an impeller speed of 700 rpm with the chopper at 1000 rpm allowing balances between flow establishment, segregation, and centrifugal forces.

7.6.1.4 Impeller speed

The most critical scaleup variable for high-shear granulators is the impeller speed. For a given equipment design, the force experienced by the particles and the particle velocities mostly are dictated by the impeller speed (Tao, Pandey, Bindra, Gao, & Narang, 2015). Thus, significant effort has been put into establishing the best way to scale up impeller speed. In the past, many production machines were capable of only one or two speeds. With the advent of high-quality variable frequency drives, however, many newer production models also are provided with variable speed impellers allowing much easier scaleup.

Impeller tip speed can be calculated using Eq. (5)

$$\text{Tip speed} = \frac{\pi n d}{t} \tag{5}$$

where n is the impeller speed in revolutions per minute, d is the diameter of the impeller, and t is the time.

The most commonly used scaleup rules related to impeller speed are the power law correlation governed by Eq. (6) (Tao et al., 2015).

$$\frac{n_2}{n_1} = \left(\frac{d_1}{d_2}\right)^n \tag{6}$$

where n is the impeller speed in revolutions per minute, d is the diameter of the impeller and subscripts 1 and 2 represent different scales of granulator. A value of $n=1$ in Eq. (6) corresponds to maintaining a constant impeller tip speed across scales, whereas a value of $n=0.5$ corresponds to maintaining a constant Froude number across scales. A value of $n=0.8$ is an empirically derived exponential parameter that was shown to deliver similar shear stress across different scales, with successful scaleup results in some previous studies and is also known as constant empirical stress method.

Scaling to constant impeller tip speed assumes the entire granulation process is based upon the impaction of the impeller blade against the granules. This excludes any potential forces being created because of the weight of material and from centrifugal/centripetal forces. Scaling to constant Froude number takes into account the greater forces generated in smaller machines because of the effect of the bowl wall. As the Froude number is related to the square of the impeller speed (in rpm), the range of Froude numbers is much greater on the faster running small scale machines. This means it is essential to know the maximum size granulator that the product will be processed on, before choosing the speed of the small machine.

7.6.2 Fluid bed granulation

Granule growth in fluid-bed granulation involves three stages: nucleation, transition, and ball growth. A typical transition to ball growth depends on the humidity in the blend. Some of the parameters that can affect scaleup of fluid-bed granulation are listed in Table 7.23.

7.7 Conclusions and recommendations

This chapter provided an overview of the granulation theory and of various wet granulation methodologies, with major focus on high-shear wet granulation and fluid-bed granulation process. High-shear wet granulation (equipped with either a vertical main impeller shaft or a horizontal main impeller shaft) is one of the most commonly used techniques for production of granules. The various technical references cited found that in majority of cases, granule properties were similar, when produced by top-drive and bottom-drive vertical high shear granulators; however, exceptions could be there based on granule properties. Fluid-bed granulation is a popular alternative process (to high-shear granulation) because of obvious advantages of executing all processing steps (dry mixing, wet granulation, and drying operations) in single equipment.

The application of QbD in wet granulation process development was presented along with two case studies, wherein DoE also was used to identify the design space for the processing parameters. The objectives of the DoEs are to

1. Obtain maximum amount of information using the reasonable amount of resources.
2. Determine which inputs (factors), namely CMAs or CPPs impact the average response (i.e., CQA) and that will have no effect (in other words, separates the critical few from the trivial many).
3. Build mathematical models relating "y" (response; CQA) to "x" (inputs; CMAs, CPPs)
4. The model can be used to simulate the process, if the behavior of each input is known
5. Future changes to the inputs can be evaluated through simulation using the model.
6. Identify input factor settings (CMA, CPP) that optimize the response (CQA) and minimize the cost
7. Provide a historical reference available for the next similar drug formula being developed.

Identification of CPPs and CMAs are always on the critical path of product development, thereby forcing one to be very efficient with QbD development processes by:

1. Leveraging documented prior knowledge
2. Internal data mining
3. Research articles, review papers, patents, or reference books
4. RLD labeling
5. Conducting risk assessment of the drug substance, material attributes, formulation variables, and processing parameters
6. Risk identification through risk assessment process
7. Risk mitigation through experimental studies (univariate, multivariate trials, etc.)
8. Documenting residual risk, if any, through appropriate justification.

A development case study was presented, using multivariate structured experiments (using DSD approach) to gain information on the design space, where dissolution targets can be met in a biorelevant media. From the design evaluations, valuable information on the control space for the excipient levels and CPPs was obtained to ensure that the product will robustly meet CQA targets. The model estimations were confirmed by additional batches, which provided the guidelines for operational controls that are likely to constantly produce product that meets the dissolution requirements. Using QbD approach helps to define clear targets early on and constant use of risk assessment tools along with statistical applications helps in mitigating risks and arriving at informed decisions.

Techniques of granulation end-point determination and scaleup principles were presented for both high shear and fluid-bed based granulation along with the practical aspects. Finally, the following recommendations and practical tips should be useful to the reader for successful development of wet-granulation-based products.

1. Initiate feasibility trials with tentative composition in mind (based on prior knowledge, public domain, information about RLD). Try screening different manufacturing processes and shortlist one based on desired granule and tablet characteristics (e.g., better flow, good hardness, desired dissolution).
2. Finalize the composition and the process that meets the desired CQAs.
3. Identify the proposed commercial scale equipment and assess (based on development trials, prior knowledge, literature) which factor could be critical for the product in question. Few of these parameters that by no means are exhaustive are listed in Table 7.24.
4. Identify the scaleup/scaledown factors (given in previous sections for high shear and fluid bed granulation process) and take batches as per target parameters. Always remember that the trials taken using scaleup/scaledown factors are only the starting point and, based on results from experimentation, the processing parameters can be modified.
5. Based on the trial batch, perform more experimental trials (e.g., DOE trials) to identify the design space that gives desired CQAs.
6. It is always beneficial to identify the failure points during early stage development so as to define the working boundaries.
7. Ensure all development work (using either R&D or commercial scale equipment) is conducted using process parameters within the qualified ranges of the equipment. Sometimes, there could be a risk of going out of qualified ranges when process is scaled up from development scale to production scale. To avoid this, one of the approaches could be checking/using scaledown approach from production scale to development scale equipment at the start of project to select process parameters. This approach can ensure successful scaleup within the qualified range of production scale equipment.
8. Although the entire design space does not have to be reestablished (e.g., DoE) at commercial scale, design spaces should be verified initially, including monitoring or testing of CQAs that are influenced by scale-dependent parameters as suitable prior to commercial manufacturing.

Acknowledgments

The authors would like to acknowledge Bruce D. Johnson for providing the opportunity to contribute in this book chapter and also Sumedha Nadkar for supporting this initiative. Contributions of Rob Lievense in this manuscript, especially, providing statistical analysis of case studies, DoE designs and data analysis and extensive discussions is greatly appreciated. Authors would also like to acknowledge James Jin, Gaurav Gupta, Basavaraj Shidagonnavar, Paritosh Singh, and Pankaj Chhipa for the work carried out and providing all the inputs.

References

Aksu, B., Paradkar, A., De Matas, M., Özer, O., Güneri, T., & York, P. (2013). A quality by design approach using artificial intelligence techniques to control the critical quality attributes of ramipril tablets manufactured by wet granulation. *Pharmaceutical Development and Technology, 18*(1), 236–245. Available from http://www.tandfonline.com/loi/iphd20 . doi:10.3109/10837450.2012.705294.

Augsburger, L. L., & Hoag, S. W. (2008). Forward. In L. L. Augsburger, & S. W. Hoag (Eds.), *Pharmaceutical dosage forms: Tablets* (Vol. 1, pp. V–VII). New York: Informa Healthcare USA, Inc Vol. 1(pp. V–VII).

Badawy, S. I., Narang, A. S., LaMarche, K., Subramanian, G., & Varia, S. A. (2012). Mechanistic basis for the effects of process parameters on quality attributes in high shear wet granulation. *International Journal of Pharmaceutics*, 324–333.

Barla, V. S., Kumar, R., Nalluri, V. R., Gandhi, R. R., & Venkatesh, K. (2014). A practical evaluation of qualitative and quantitative chemometric models for real-time monitoring of moisture content in a fluidized bed dryer using near infrared technology. *Journal of Near Infrared Spectroscopy*, 221–228.

Borner, M., Buck, A., & Tsotsas, E. (2017). DEM-CFD investigation of particle residence time distribution in top-spray fluidized bed granulation. *Chemical Engineering Science*, 187–197.

Bossche, M. V., & Vaerenbergh, G. V. (2014). *Microwave drying: a more efficient technology than gas-stripping*. Retrieved from http://www.gea.com/en/stories/microwave-drying.jsp .

Briens, L., & Logan, R. (2011). The Effect of the Chopper on Granules from Wet High-Shear Granulation Using a PMA-1 Granulator. *AAPS PharmSciTech, 12*(4), 1358–1365. doi:10.1208/s12249-011-9703-1.

Cantor, S. L., Augsburger, L. L., & Hoag, S. W (2008). Pharmaceutical granulation processes, mechanism, and the use of binders. In L. L. Augsburger, & S. W. Hoag (Eds.), *Pharmaceutical dosage forms: Tablets* (Vol. 1, pp. 261–301). New York: Informa Healthcare USA, Inc Vol. 1.

Djuris, J., Medarevic, D., Krstic, M., Djuric, Z., & Ibric, S. (2013). Application of quality by design concepts in the development of fluidized bed granulation and tableting processes. *Journal of Pharmaceutical Sciences*, 1869–1882.

Ennis, B. J., & Litster, J. D. (1997). Particle size enlargement. In R. Perry, & D. Green (Eds.), *Perry's chemical engineers' handbook* (7th ed) (pp. 20–89) Eds.. New York: McGraw-Hill.

FDA. (2004). *Guidance for industry: PAT—A framework for innovative pharmaceutical development, manufacturing, and quality assurance*. Retrieved from (2004). http://www.fda.gov/downloads/drugs/guidances/ucm070305.pdf.

FDA. (2006). *Guidance for industry: Q9 quality risk management*. Retrieved from (2006). https://www.fda.gov/downloads/Drugs/Guidances/ucm073511.pdf.

FDA. (2011). *Guidance for industry: quality by design for ANDAs: an example for modified release dosage forms*. Retrieved from (2011). http://www.fda.gov/downloads/drugs/developmentapprovalprocess/howdrugsaredevelopedandapproved/approvalapplications/abbreviatednewdrugapplicationandagenerics/ucm286595.pdf.

FDA. (2012). *Guidance for industry: quality by design for ANDAs: an example for immediate release dosage forms*. Retrieved from (2012). http://www.fda.gov/downloads/drugs/developmentapprovalprocess/howdrugsaredevelopedandapproved/approvalapplications/abbreviatednewdrugapplicationandagenerics/ucm304305.pdf.

FDA. (2014). *Guidance for industry: SUPAC: manufacturing equipment addendum*. Retrieved from (2014). http://www.fda.gov/downloads/drugs/guidancecomplianceregulatoryinformation/guidances/ucm346049.pdf.

Fonteyne, M., Vercruysse, J., Díaz, D. C., Gildemyn, D., Vervaet, C., Remon, J. P., & Beer, T. D. (2013). Real-time assessment of critical quality attributes of a continuous granulation process. *Pharmaceutical Development and*

Technology, 18(1), 85–97. Available from http://www.tandfonline.com/loi/iphd20 . doi:10.3109/10837450. 2011.627869.

Fonteyne, M., Vercruysse, J., Leersnyder, F. D., Snickb, B. V., Vervaet, C., Remon, J. P., et al. (2015). Process analytical technology for continuous manufacturing of solid-dosage forms. *Trends in Analytical Chemistry*, 159–166.

Fries, L., Antonyuk, S., Heinrich, S., & Palzer, S. (2011). DFM-CFD modeling of a fluidized bed spray granulator. *Chemical Engineering Science*, 2340–2355.

Gabbott, I. P., Husban, F. A., & Reynolds, G. K. (2016). The combined effect of wet granulation process parameters and dried granule moisture content on tablet quality attributes. *European Journal of Pharmaceutics and Biopharmaceutics*, 70–78.

Gantt, J. A., & Gatzke, E. P. (2005). High shear granulation modelling using a discrete element simulation approach. *Powder Technology*, 195–212.

Gokhale, R., & Trivedi, N. R. (2010). Wet granulation in low- and high-shear mixers. In D. M. Parikh (Ed.), *Handbook of pharmaceutical granulation technology* (pp. 183–203). New York: Informa Healthcare USA, Inc.

Goldschmidt, M. J., & Kuipers, J. A. (2003). Discrete element modeling of fluidized bed spray granulation. *Powder Technology*, 39–45.

Hansuld, E. M., & Briens, L. (2014). A review of monitoring methods for pharmaceutical wet granulation. *International Journal of Pharmaceutics, 472*(1-2), 192–201. doi:10.1016/j.ijpharm.2014.06.027.

Heinrich, S., Ihlow, M., & Morl, L. (2003). Particle population modeling in fluidized bed-spray granulation: analysis of the steady state and unsteady behavior. *Powder Technology*, 154–161.

Huang, J., Goolcharran, C., Utz, J., Hernandez-Abad, P., Ghosh, K., & Nagi, A. (2010). A PAT Approach to Enhance Process Understanding of Fluid Bed Granulation Using In-line Particle Size Characterization and Multivariate Analysis. *Journal of Pharmaceutical Innovation, 5*(1-2), 58–68. doi:10.1007/s12247-010-9079-x.

Iveson, S. M., Litster, J. D., Hapgood, K., & Ennis, B. J. (2001). Nucleation, growth and breakage phenomena in agitated wet granulation processes: A review. *Powder Technology, 117*(1-2), 3–39. doi:10.1016/S0032-5910(01)00313-8.

Jones, D. M. (1985). Factors to consider in fluid bed processing. *Pharmaceutical Technology*, 50–55.

Kimura, S. I., Iwao, Y., Ishida, M., Noguchi, S., Itai, S., Uchida, S., et al. (2014). Evaluation of the physicochemical properties of fine globular granules prepared by a multifunctional rotor processor. *Chemical and Pharmaceutical Bulletin*, 309–315.

Kimura, S. I., Iwao, Y., Ishida, M., Uchimoto, T., Miyagishima, A., Sonobe, T., & Itai, S. (2010). Optimal conditions to prepare fine globular granules with a multi-functional rotor processor. *International Journal of Pharmaceutics, 391*(1-2), 244–247. doi:10.1016/j.ijpharm.2010.03.005.

Kleinebudde, P. (2004). Roll compaction/dry granulation: pharmaceutical applications. *European Journal of Pharmaceutics and Biopharmaceutics, 58*(2), 317–326. doi:10.1016/j.ejpb.2004.04.014.

Kristensen, H. G., & Schaefer, T. (1987). Granulation: A review on pharmaceutical wet-granulation. *Drug Development and Industrial Pharmacy, 13*(4-5), 803–872. doi:10.3109/03639048709105217.

Lee, S. L. (2017). *Modernizing the way drugs are made: a transition to continuous manufacturing.* Retrieved from FDA: https://www.fda.gov/Drugs/NewsEvents/ucm557448.htm.

Leuenberger, H., Bier, H., & Sucker, H. (1979). Theory of the granulating-liquid requirement in the conventional granulation process. *Pharmaceutical Technology*, 61–68.

Levin, M. (2006). Wet granulation: end-point determination and scaleup. In J. Swarbrick (Ed.), *Encyclopedia of pharmaceutical technology* (pp. 4078–4098). London: Informa Healthcare 3rd ed.

Mehta, A. M. (1988). Scaleup considerations in the fluid bed process for controlled release products. *Pharmaceutical Technology*, 46–52.

Morin, G., & Briens, L. (2014). A Comparison of Granules Produced by High-Shear and Fluidized-Bed Granulation Methods. *AAPS PharmSciTech, 15*(4), 1039–1048. doi:10.1208/s12249-014-0134-7.

Narang, A. S., Sheverev, V., Freeman, T., Both, D., Stepaniuk, V., Delancy, M., Millington-Smith, D., Macias, K., & Subramanian, G. (2016). Process Analytical Technology for High Shear Wet Granulation: Wet Mass Consistency Reported by In-Line Drag Flow Force Sensor Is Consistent with Powder Rheology Measured by At-Line FT4 Powder Rheometer®. *Journal of Pharmaceutical Sciences, 105*(1), 182–187. Available from www.interscience.wiley.com/jpages/0022-3549 . doi:10.1016/j.xphs.2015.11.030.

Narang, A. S., Stevens, T., Hubert, M., Paruchuri, S., Macias, K., Bindra, D., Gao, Z., & Badawy, S. (2016). Resolution and Sensitivity of Inline Focused Beam Reflectance Measurement During Wet Granulation in Pharmaceutically Relevant Particle Size Ranges. *Journal of Pharmaceutical Sciences, 105*(12), 3594–3602. Available from www.interscience.wiley.com/jpages/0022-3549 . doi:10.1016/j.xphs.2016.09.001.

Narang, A. S., Stevens, T., Macias, K., Paruchuri, S., Gao, Z., & Badawy, S. (2017). Application of In-line Focused Beam Reflectance Measurement to Brivanib Alaninate Wet Granulation Process to Enable Scale-up and Attribute-based Monitoring and Control Strategies. *Journal of Pharmaceutical Sciences, 106*(1), 224–233. Available from www.interscience.wiley.com/jpages/0022-3549 . doi:10.1016/j.xphs.2016.08.025.

Newitt, D. M., & Conway-Jones, J. M. (1958). A contribution to the theory and practice of granulation. *Transactions of the Institution of Chemical Engineers*, 422–441 1958.

Pandey, P., Tao, J., Chaudhury, A., Ramachandran, R., Gao, J. Z., & Bindra, D. S. (2013). A combined experimental and modeling approach to study the effects of high-shear wet granulation process parameters on granule characteristics. *Pharmaceutical Development and Technology, 18*(1), 210–224. Available from http://www.tandfonline.com/loi/iphd20 . doi:10.3109/10837450.2012.700933.

Patil, N., Khadse, S. C., & Ige, P. P. (2016). Review on novel granulation techniques. *World Journal of Pharmaceutical Research*, 1961–1975.

Prescott, J. K., Smith, T. J., Li, J.-X., Carlin, B., Ray, S., Sackett, G. L., et al. (2004). *Comparison of flow properties of granules prepared by top-drive and bottom-drive high-shear granulators 2004 aaps annual meeting and exposition*. Baltimore: Courtesy Freund-Vector Corporation.

Rahmanian, N., Naji, A., & Ghadiri, M. (2011). Effects of process parameters on granules properties produced in a high shear granulator. *Chemical Engineering Research and Design*, 512–518.

Royce, A., Suryawanshi, J., Shah, U., & Vishnupad, K. (1996). Alternative granulation technique: melt granulation. *Drug Development and Industrial Pharmacy*, 917–924.

Smith, T. J., Li, J.-X., Carlin, B., Ray, S., Sackett, G. L., Sheskey, P., et al. (2003). *Top-drive vs. bottom-drive high shear granulation: effect on granule properties of an immediate release formulation 2003 AAPS annual meeting and exposition*. Salt Lake City: Courtesy Freund-Vector Corporation.

Smith, T. J., Peng, Y., Xu, W., Li, J.-X., Ray, S., Sackett, G. L., et al. (2004). Top-drive vs. bottom-drive high-shear granulation: dissolution profile of immediate release tablets. 2004 AAPS annual meeting and exposition. Baltimore Courtesy Freund-Vector Corporation.

Stahl, H. (2004). Comparing different granulation techniques. *Pharmaceutical Technology Europe, 16*(11), 23–33. http://www.pharmtech.com/harald-stahl.

Tao, J., Pandey, P., Bindra, D. S., Gao, J. Z., & Narang, A. S. (2015). Evaluating Scale-Up Rules of a High-Shear Wet Granulation Process. *Journal of Pharmaceutical Sciences, 104*(7), 2323–2333. Available from www.interscience.wiley.com/jpages/0022-3549 . doi:10.1002/jps.24504.

Ullah, I. (2011). *Moisture-activated dry granulation*. Retrieved from Pharmaceutical Technology Europe (2011). http://www.pharmtech.com/moisture-activated-dry-granulation.

Vector Freund Corporation Limited (2017). *Granurex rotor: one-pot fluid bed rotor processor GXR insert*. Retrieved from (2017). https://www.freund-vector.com/products/granurexrotor/.

Wang, S., Ye, G., Heng, P. W., & Ma, M. (2008). Investigation of high shear wet granulation processes using different parameters and formulations. *Chemical and Pharmaceutical Bulletin*, 22–27.

Yu, L. X., Amidon, G., Khan, M. A., Hoag, S. W., Polli, J., Raju, G. K., & Woodcock, J. (2014). Understanding pharmaceutical quality by design. *AAPS Journal, 16*(4), 771–783. Available from http://www.springerlink.com/content/1550-7416/ . doi:10.1208/s12248-014-9598-3.

SECTION II

Excipients and input material attributes

8. Material attributes and their impact on wet granulation process performance 267
9. Binders in wet granulation 313
10. Effect of binder attributes on granule growth and densification 347
11. Role of drug substance material properties in the processibility and performance of wet granulated products 379
12. Critical material attributes in wet granulation 413
13. Critical material attributes during continuous twin-screw wet granulation 443

Material attributes and their impact on wet granulation process performance

Praveen Hiremath[a], Kalyan Nuguru[a,b] and Vivek Agrahari[a,b]

[a]*Department of Formulation Technology, Bayer Animal Health GmbH, Leverkusen, Germany,*
[b]*Department of Technical Development, Bayer U.S. LLC, Shawnee, KS, United States*

8.1 Granulation technology

Granulation technology is commonly used in the pharmaceutical industry during solid oral dosage form development, especially in tablet and capsule manufacturing (Parikh, 2010; Shanmugam, 2015). The process is commonly used for size enlargement, where small particles are produced by the addition of liquid to the powder mixture (excipient/API: active pharmaceutical ingredient) and the massing of the mix to produce granules.

The wet granulation process includes a series of steps, including wetting, nucleation, consolidation, growth, attrition, and breakage. The process starts as a mixture of dry powder followed by wetting, which is the addition and even distribution of liquid throughout the blend. Nucleation takes place when the liquid joins together nearby primary particles in weak structures known as nuclei. Nucleation is followed by granule growth, which typically proceeds according to either a steady growth or an induction mechanism. As granules grow they consolidate because of the agitation forces present with mixing. Consolidation increases granule strength and forces excess liquid to the surface. Agitation can cause breakage in granules that are weak or poorly formed. Dried granules also can undergo attrition and/or breakage during processing or handling. In general granulation is performed mainly to prevent segregation of the constituents of the powder, to improve flow properties, compaction characteristics, appearance, to minimize dust, and to densify the material (Parikh, 2010).

The common granulation processes for solid dosage forms are wet and dry granulation.

8.1.1 Dry granulation

The dry granulation process is used to form granules without adding any liquid to the powder blend or mixture. Because no moisture is involved in the process, it is an ideal way to process moisture-sensitive compounds. Furthermore, it is not necessary to dry the granules, making the process energy efficient. Forming granules without moisture involves compacting under high pressure using suitable tooling, mostly rollers followed by size reduction to produce a granular, free-flowing blend of acceptable size distribution (Parikh, 2010; Shanmugam, 2015).

8.1.2 Wet granulation

In the wet granulation process, granules are produced by the addition of a liquid/dry binder with liquid added to the powder blend or mixture (Parikh, 2010; Shanmugam, 2015). The drug is combined with other excipients and processed with the use of a solvent (aqueous or organic) with subsequent drying and milling to produce granules. Solvent mixed into the powders can form bonds between powder particles that are strong enough to lock them together. After the liquid dries, however, some powders can fall apart. In such cases a liquid solution containing binder is required. Organic solvents, such as ethanol, are used when water-sensitive drugs are involved or when a rapid drying of granules is required. Additionally it is possible to use stabilizing agents such as pH modifiers in close contact with the drug, which could maximize product stability. After the solvent has been removed or dried, the powder blend forms a more densely held mass. In the traditional wet granulation method the wet mass is forced through a sieve to produce presized wet granules that are subsequently dried. A subsequent, postdrying, screening/milling step breaks agglomerates into granules of desired particle size and distribution. Wet granulation can be divided into low-shear, high-shear, and fluid-bed granulation processes. Each process has its advantages and limitations, which can be useful for different formulations.

8.1.2.1 Low shear

Low shear is the traditional granulation process performed using low-speed planetary or trough mixers in which the drug and excipients are granulated with a binder solution. The resulting wet mass is screened to form discrete granules, which typically are dried in a tray dryer. The dried granules are rescreened or milled to the required size, blended with extragranular excipients, lubricated, and compressed. The major limitations of this process are that the equipment might not be a closed vessel, long drying times, potential for migration of soluble components during tray drying, and lack of in-process control. Low-shear (e.g., fluid-bed) granulation is preferred over high-shear granulation when the resultant granules need to be porous and show fast disintegration and dissolution properties. The fluid-bed granulation process results in a narrow particle size distribution compared to a high-shear granulation process. Additionally fluid bed granulation can be used for particle coating (e.g., Wurster process), and the fluid-bed process is a single-pot process.

8.1.2.2 High shear

High-shear mixers/granulators are characterized by the presence of two mixing blades: an impeller that rotates in the base of the mixer and a high-speed chopper that continually breaks up the wet mass (typically large agglomerates) during the granulation process. This combination provides an effective mixing of materials and use of small amounts of water compared to the low-shear granulation. The entire process can be completed in a few minutes, and the systems can be fitted with a variety of devices to monitor and determine the end point of granulation. The high-shear granulator is a closed-vessel system, and the granules produced generally are able to be transferred to a fluid-bed processor/dryer under high containment, thus minimizing the extent of handling.

A key advantage of high-shear granulation is its wide applicability to various formulations. The intensity of the mixing process, however, can lead rapidly to overgranulation that affects granule tabletability. In general any process factor that increases the extent of granulation (water, massing time, or impeller speed) tends to increase granule density and reduce tabletability. It is possible to counteract the effect of overgranulation to some extent by milling the dried granules. An optimum high-shear granulation is achieved by carefully controlling the relative proportions of diluents, the amount of granulating water, and the duration of wet massing. Thus, with high-shear granulation, it is extremely important to determine and control the end-point of the granulation process to achieve critical granule properties such as size and density. A disadvantage of both low- and high-shear granulation is that water can encounter the components for a significant time, leading to changes in the drug and/or excipients.

8.1.2.3 Fluidized bed granulation

Fluidized-bed granulation has three steps.

1. *Blending*: The drug and excipients are blended with a low volume of fluidizing air to achieve homogeneity and to preheat (warm) the dry powder blend. Care needs to be taken in this stage that fine drug particles are not removed from the bed by entrainment in high-velocity air. Effective fluidization depends in part on the particle size and density of the ingredients to be fluidized.
2. *Granulation*: An aqueous/organic solvent or binder solution is sprayed onto the fluidized bed. Granule growth during this phase is dependent on many factors, including granulating fluid viscosity, droplet size, and spray rate. In general increasing any of these factors tends to increase the granule growth rate.
3. *Drying*: After the spraying is stopped, the powder bed continues to be gently fluidized until the granulation is dry. The end point usually is determined by the bed temperature and moisture content (via loss on drying). During this phase, there can be some granule attrition that could lead to a breakdown of granules into powders.

The advantages of fluidized-bed granulation are that it is a one-pot (contained) process, meaning a single piece of equipment can be used for both granulation and drying, and granules are typically of low density and possess better compressibility. Some combinations of drugs and excipients, however, are incapable of being mixed by fluidization before granulation, and, in these cases, a premix must be prepared. Also, milled/micronized drugs can be entrained in the fluidizing airstream and lost through the filter assembly.

Wet granulation has had numerous technological innovations recently, and the significance and limitations of these recent techniques are summarized in Table 8.1. The recent progress of different granulation techniques is represented in Fig. 8.1. Although, several advantages are inherent to wet granulation, the process also has some shortcomings. The liquids used in the granulation process can have negative influence or effects on the drugs and/or excipients. Moreover, the complexity of the process compared to direct compression increases the number of critical process parameters that must be controlled precisely for the desired product performance.

The powders/granules intended for compression into tablets must possess good flowability and compressibility. Flowability is required to produce tablets of consistent weight and uniform strength, whereas compressibility is required to form a compact and intact mass. These two essential properties are achieved by adding excipients to tablet formulation and/or through a granulation process.

8.2 Excipients in wet granulation process

Excipients play a crucial role and are important constituents in a drug formulation (Rowe et al., 2009; Zarmpi et al., 2017). To design a product that is effective, convenient to handle, and facilitates patient

Table 8.1 Summary of recent advances in granulation techniques.

Techniques/technologies	Description	Granule characteristics	Merits	Limitations	Equipment
Pneumatic dry granulation	• Dry granulation • Mild compaction and pneumatic classification	æ Porous, highly compressible æ Taste masking æ Fast disintegration æ Release time modification	↑ Drug loading æ Thermolabile and moisture-sensitive drugs ↑ Product stability ↓ Cost and waste	X Recycled granule quality X Segregation potential X Friability	æ Roller compaction with air stream or vacuum
Reverse wet granulation	• Wet granulation • Water or solvent is granulating liquid	æ Uniform wetting æ Uniform erosion	↓ Particle size æ Spherical shape æ Poorly water-soluble drugs	X Larger particle size[a] X Lower porosity X Many problems similar to conventional wet granulation	æ High-speed mixer
Steam granulation	• Wet granulation Steam is granulating liquid	↑ Diffusion rate ↑ Uniform distribution ↑ Surface area æ Spherical shape	æ Eco-friendly æ Sterility Process time æ No solvent use æ No health hazards	X Local overheating/wetting X High energy inputs X Thermolabile drugs X Limited binders	æ High-speed mixer with steam generator/regulator
Moisture-activated dry granulation	• Wet granulation • 1%–4% water is granulating liquid and moisture-absorbing material	æ Uniform size ↑ Flowability ↑ Compressibility	æ Less energy input æ No drying process æ Wide applicability æ Continuous processing ↓ Shorter process time Process variables	X Moisture-sensitive drugs X Impossible high drug loading X Limited absorbents	æ High-shear mixer coupled with a sprayer
Thermal adhesion granulation	• Wet granulation • Low water/solvent is granulating liquid and heating at 30°C–130°C	æ Flowability æ Friability Tensile strength	↑ Drug loading æ No drying process Dust	X High energy inputs X Thermolabile and moisture sensitive drugs X Limited binders	æ Tumble blender or similar equipment coupled with heating system

(continued on next page)

			æ	X	æ
Melt granulation	• Wet granulation • Meltable binder as granulating liquid, heating at 50°C–90°C	æ Possible modified release ↑ Dissolution	æ No water or solvent æ No drying process ↓ Energy input ↓ Cost and process time æ Water-sensitive drugs	X Thermolabile drugs X Limited binders	æ High-shear mixer æ Fluidized bed
Freeze granulation	• Wet granulation • Spray freezing and subsequent freeze drying for slurry or suspensions	æ Uniform size æ Flowability æ Spherical shape	↑ Granule homogeneity æ Thermolabile drugs æ Granule density control ↓ Material waste	X Limited solvent medium X Only suitable for conversion of liquid slurry or suspension to granules	æ Spray freezer coupled with freeze dryer
Foam granulation	• Wet granulation • Foam as granulating liquid	æ Uniform binder distribution æ No overwetting ↑ Surface area	↓ Water requirement No spray nozzle use æ Low water required ↓ Cost and process time æ Water-sensitive drugs	X Moisture sensitive drugs X Limited binders	æ High-shear mixer or fluidized bed granulator coupled with foam generator/regulator

a At a lower binder concentration. ↓, Reduced or decreased; ↑, increased or high; æ, possibility or suitability or availability; X, unsuitable or not applicable. Reproduced with permission from Shanmugam (2015).

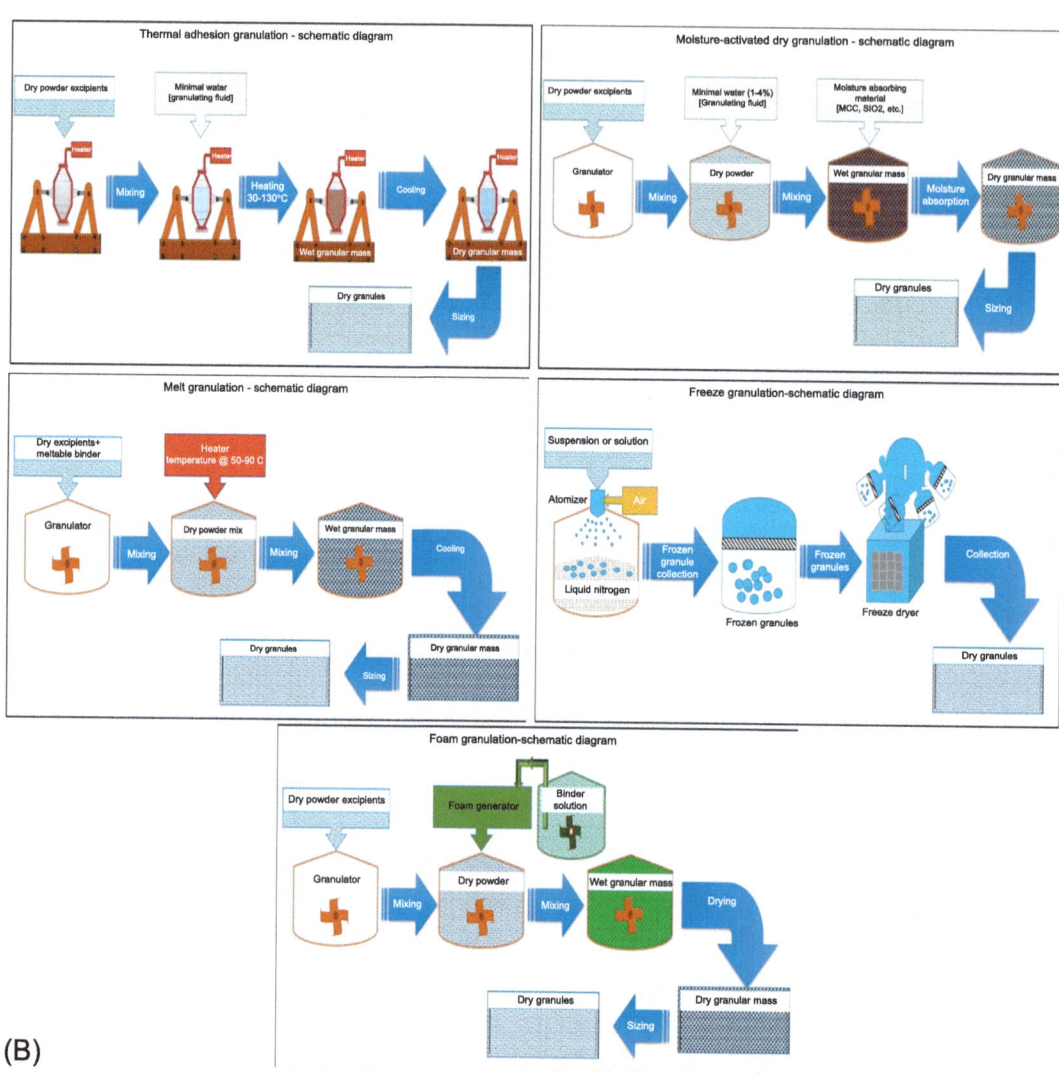

(B)

FIGURE 8.1

(A and B) Schematic representation of different granulation techniques.

Reproduced with permission from Shanmugam (2015).

compliance, the drug substance must be processed with excipients. While excipients are generally pharmacologically inert components, technical advancements have brought an evolution in their roles in enhancing formulation and processing efficiencies, modulating drug release, and optimizing therapeutic outcomes. Critical quality attributes (CQAs) refer to the characteristics of the product, while critical material attributes (CMAs) are related to the input materials (excipients). According to the ICH Q8(R2) guidelines, a CQA is defined as "a physical, chemical, biological, or microbiological property or characteristic that should be within an appropriate limit, range, or distribution to ensure the desired product quality" (https://www.fda.gov/downloads/drugs/guidances/ucm073507.pdf). CMA is not defined in the ICH guidelines, but it can be defined as a physical, chemical, biological or microbiological property or characteristic of an input material that should be within an appropriate limit, range, or distribution to ensure the desired quality of output material (drug substance, excipient, or inprocess material) (Zhang & Mao, 2017).

Factors that can trigger variability in the final product quality, include APIs, excipients, and process parameters. The limited knowledge of the role of physicochemical properties of pharmaceutical excipients hinders the possibility to further optimize the dosage forms from the aspects of manufacturing and drug product performance. Therefore, a thorough understanding of molecular, structural, and overall physicochemical properties of an excipient is highly essential (Rowe et al., 2009; Zarmpi et al., 2017).

8.2.1 Excipient variability

The drug-excipient and excipient–excipient interactions can affect the performance, stability, and the bioavailability of the drug substance in a dosage form significantly. These effects could be because of the direct interaction of the drug with excipients, reactive impurities in the excipients, or indirect effect of excipients, for example, because of their variability in manufacturing, quality, and at the molecular level. Understanding and integrating excipient variability in the product design is of utmost importance to achieve robust manufacturing processes and quality products. Excipients comprise a substantial mass of the formulation and can have a significant contribution to the variability in the CQAs of the final drug product (Dave et al., 2015; Rowe et al., 2009; Zarmpi et al., 2017) as summarized in Fig. 8.2A. It is important, therefore, to understand the impact of excipient variability on the drug product manufacturability and performance.

Excipient variability at the molecular level could be explained by the differences in the polymorphic form, structure, crystalline or amorphous content, and water content variations. Excipient variability at the macroscopic level could be because of differences in the particle morphology and density. The main reasons for excipient variability are environmental variations that can lead to different raw material properties and a manufacturing process that refers to variability caused by a change during excipient production (Dave et al., 2015; Rowe et al., 2009; Zarmpi et al., 2017). The variability (lot-to-lot and supplier-to-supplier) in excipient properties should be considered and appropriate controls in place to ensure consistent performance of a product.

The different approaches used to assess excipient variability at the molecular and macroscopic level are summarized in Fig. 8.2B. Most excipients are naturally derived or produced after chemical modification of a natural product. Variability of excipient attributes could be because of source-to-source or lot-to-lot variability of excipients. The source of raw materials is a major factor for their variability, for example, microcrystalline cellulose (MCC) can be produced from wood (soft or hard) or cotton. Significant variations in chemical structure, crystal structure, and particle size of four different brands of MCC were observed (two brands derived from softwood and two from hardwood) (Landín et al., 1993a). Differences in lignin content could be the cause for the increased dissolution rate of prednisone tablets manufactured with MCC from different sources (Landín et al., 1993b). Supplier-to-supplier variability also could arise either from different natural sourcing or manufacturing processes.

274 Chapter 8 Material attributes and their impact on wet granulation

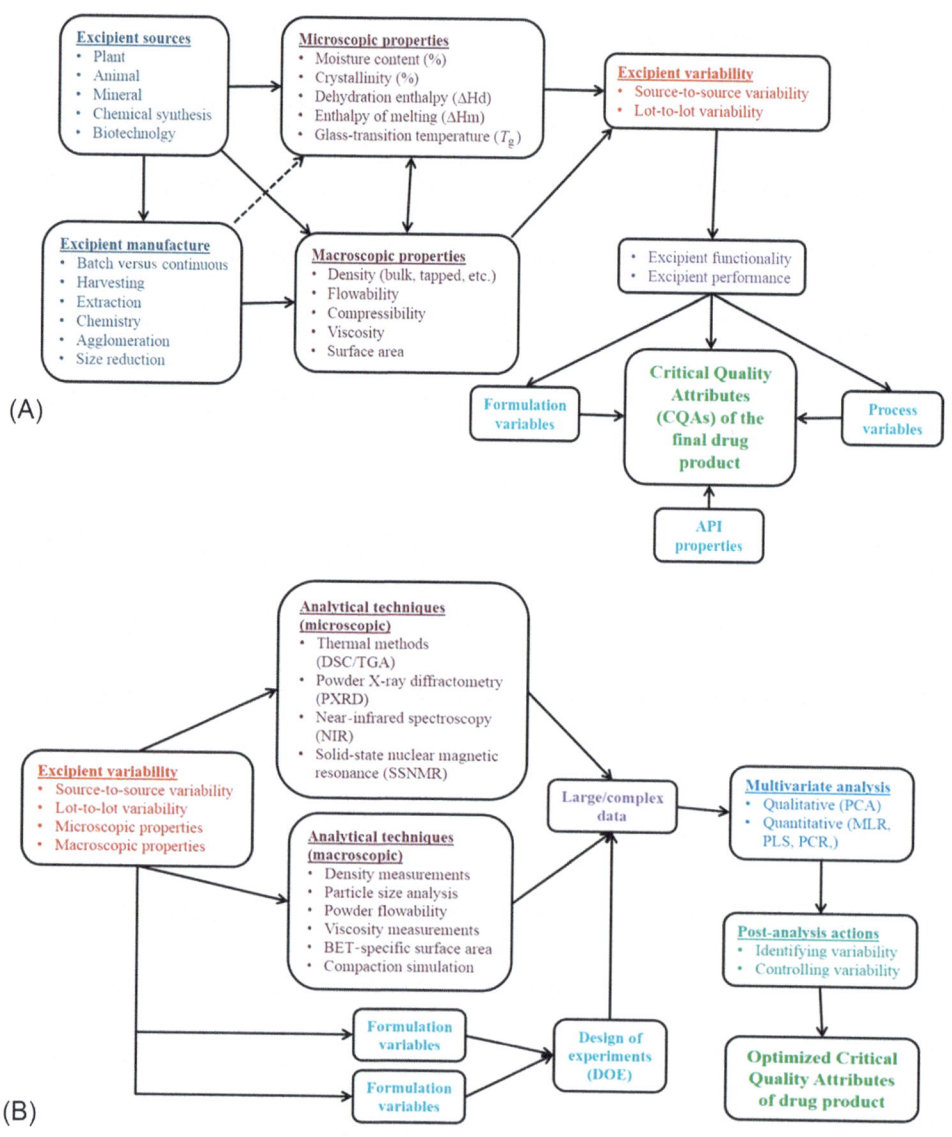

FIGURE 8.2

(A) The impact of excipient variability on the CQAs of the drug product. (B) Analysis of excipient variability.

Reproduced with permission from Dave et al. (2015).

For instance, significant intersupplier variability were observed in anhydrous lactose properties that could affect the final product according to the amount of excipient used (Gamble et al., 2010).

8.2.2 Excipient functionality

The influence of material attributes depends on the amount and function of a material (excipient) within a dosage form. Excipients provide a range of functional roles to a drug product such as to enhance volume or increase the weight of the blend, disintegration, binding, lubrication, and adjust the drug release profile (Narang et al., 2017). Generally the impact is reduced greatly when the component is at low concentration although sometimes excipients even at low concentration (e.g., lubricants) can have a significant impact on the manufacturing process and/or product quality. In the wet granulation process, excipients such as binders, diluents, lubricants, disintegrants, colorants, flavors, and stabilizing agents are used commonly and serve different functions in the formulation. The effects of excipient properties on a drug product, however, are product-specific and are formulation- and process-dependent. The selection of appropriate excipients is critical and is based on the functionality as well as compatibility with the active molecule. The use of excipients in dosage form requires robust manufacturing, quality, and performance. A profound knowledge of excipient functionality is a part of product development strategy. The United States Pharmacopeia (USP) 34-NF 29 chapter, 1059, provides an overview of key functional categories and physical/chemical properties of excipients that could affect the functionality, along with a general description of the mechanisms by which excipients can accomplish their functions, and tests that can relate to excipient performance. A draft of the nonmandatory functionality-related characteristics (FRCs) chapter has been published by the European Pharmacopeia (Ph.Eur.). The FRCs are defined as controllable physical or chemical characteristics of an excipient that affect its functionality. FRCs are critical because excipient functionality can be sensitive to variations (different grades, batch-to-batch, and manufacture-to-manufacture) and can have an impact on product performance. The Ph.Eur. chapter includes each of the excipient monographs based on the function in identifying the right excipient and their CQAs for a specific purpose.

It is possible that one excipient has multiple functionalities in the dosage forms depending on its physical and/or chemical attributes. Polyvinylpyrrolidone (PVP) is used as a binder in the wet granulation process, as well as for stabilization of the amorphous drug in a solid-dispersion system (Narang et al., 2017). Similarly the use of povidone as a solubilizer and stabilizer in liquid formulations include viscosity and molecular mass, although molecular mass has been considered sufficient as an FRC when it is used as a binder in tablet granules. There are cases where the same excipient can affect the performance of the dosage form by having multiple functions. For example, MCCs are used as diluents, flowability aid, binder, and disintegrants in tablets. Another example is the higher specific surface area of magnesium stearate, which is useful in its lubricant property, however, the same attribute, however, negatively affects the disintegration, dissolution and hardness of solid oral dosage forms.

In general it is complicated to identify all the functional attributes of an excipient relevant for the specified dosage form. Thus, for a given product, the critical attributes of excipients must be assessed and controlled to ensure a consistent product performance. This requires a thorough understanding of the physical and chemical attributes of drug and excipients. Some representative examples of wet granulation excipients and their functional categories are given in Table 8.2, and their brief descriptions are provided in the following subsections.

Table 8.2 Summary of wet granulation excipients.

Functional category	Excipients	Typical concentration	Critical material attributes (CMAs)
Diluents	Lactose	Up to 90%	• Particle size • Morphology • Moisture content • bulk density • Surface area • Crystallinity • Wettability • Manufacturing process • Degree of polymerization (DP)
	Microcrystalline cellulose	20%–90%	–
	Dicalcium phosphate	Up to 90%	–
	Mannitol	Up to 90%	–
Binders	Starch paste	5%–25%	• Concentration, • Solvent used • Surface tension • Viscosity • Particle size • Solubility • Molecular weight • Mixing time • Storage conditions
	Microcrystalline cellulose	20%–90%	–
	Pregelatinized starch	5%–75% w/w	–
	Hydroxylpropyl methylcellulose	2%–5%	–
	Hydroxylpropyl cellulose	2%–6%	–
	Polyvinyl Pyrrolidone	0.5%–5% w/w	–
Lubricants	Stearic Acid	1%–3%	• Particle size • Surface area • Chemical composition • Crystallinity • Concentration • Amount • Mixing time
	Magnesium stearate	0.5%–1%	–
	Sodium stearyl fumarate	1%–3%	–

(continued on next page)

Table 8.2 Summary of wet granulation excipients—cont'd

Functional category	Excipients	Typical concentration	Critical material attributes (CMAs)
Disintegrants	Pregelatinized starch	5%–20%	• Particle size • Degree of substitution • Concentration • Methods of incorporation • Moisture content • Swelling property • Porosity • Ionization • Degree of cross-linking
	Sodium starch glycolate	2%–6%	–
	Croscarmellose sodium	2%–6%	–
	Crospovidone	2%–6%	–
Glidants	Colloidal silicon dioxide	0.1%–0.5%	• Specific surface area • Particle size distribution • Hygroscopicity • Chemical composition

8.2.3 Role of active substances

The physicochemical attributes of active molecules (APIs) such as solubility, water content, particle size distribution, particle shape, morphology, density, surface area, and polymorphism can influence the performance of a drug product. The quantity of API within the bulk of the formulation excipients is also critical to product performance. Taking the specific example of the granulation process, when the drug load is low, a uniform distribution is critical to ensure content uniformity, whereas, in case of higher drug amount, coating of the drug particles with the appropriate excipient is critical.

In one study the influence of aqueous solubility, wettability (contact angle with water), water holding capacity, particle size distribution, and surface area of APIs on the processability and performance of the granulations was examined by observing the granule growth, compactability, and flow changes in the high-shear wet granulation process (Fonteyne et al., 2014). The rationale behind selecting these specific responses was based on several factors: that controlled/steady growth of granules is expected to provide robustness to the final drug product, whereas uncontrolled/rapid granule growth is undesirable as it renders the drug product to be highly sensitive to moderate changes in formulation/process parameters; that the width of granule size distribution provides useful information about the granulation events and has significant influence on segregation, flow, and compressibility of the granulations; that changes in compactability upon wet granulation is a collective effect of loss in porosity and size enlargement of the constituent particles; that flow improvement upon granulation is possibly a combined effect of size growth, densification, reduction in the width of powder size distribution, and surface modification. Based on the statistical data the following observations were generated: controlled growth was highest in readily wettable APIs with low surface area; uncontrolled growth was higher in APIs of high solubility and low water holding capacity; polydisperse granulations were produced from the APIs of high contact angle and surface area; increase in compactability was high in the APIs with large surface area and broader particle size distributions; compounds with high water uptake showed reduced compactability upon wet granulation; flow enhancement as a result of wet granulation was highest in APIs of large-size distributions; and large API particles showed a negative effect on granule growth.

In summary, the physicochemical properties of APIs can have significant influence the processability and performance of wet granulation systems and dosage forms. Therefore studies identifying and controlling the CQAs of APIs are equally important as of excipients and should be investigated.

8.2.4 Diluents

Diluents are inorganic or organic inert materials and incorporated into solid dosage forms to act as fillers or bulking agents to increase the weight of the blend for enabling tableting or encapsulation process. Diluents can comprise up to 90% by weight of the dosage form, especially for a low-dose drug, to ensure adequate product processability (Desai et al., 1993; Koo, 2017; Zarmpi et al., 2017). Properties of diluents that can affect the performance of a dosage form includes crystallinity, polymorphism, particle size, particle shape, moisture content, density and specific surface area, degree of polymerization (DP), flow properties, solubility, and compactability.

Particle size is considered one of the most important CMAs. Kushner observed that excipients and their suppliers play a critical role in drug product quality. The author confirmed that variability in particle size can affect tablet hardness, friability, disintegration, and content uniformity (Kushner, 2013). Hlinak et al. (2006) observed that particle size also can affect wetting properties, dissolution, and stability of drug products.

An increase of moisture level up to an optimum percentage increases the tablet strength because bound water vapor layers reduce interparticular surface distances and increase intermolecular attraction forces (Patel et al., 2006). Moisture in a material also can exert Van der Waals' forces. The presence of free water reduces intermolecular attractive forces and allows separation of the particles. The two fundamental forces that can affect powder flow are cohesion and friction. As moisture content increases, frictional forces and electrostatic charges between particles can be reduced. Moisture also can increase cohesion because of the creation of liquid or a solid bridges between particles (Nokhodchi, 2005). A summary of several critical factors that can affect the performance of diluents is given in Table 8.3.

Physicochemical compatibility between diluent and API is also an important factor to consider. Typically used diluents in the wet granulation process include lactose, MCC, starch, mannitol, and dicalcium phosphate (DCP). Some of these diluents and their physicochemical properties are discussed below.

8.2.4.1 Lactose

Lactose is a disaccharide consisting of d-galactose and d-glucose units linked through β (1–4) glycosidic bonds (Zarmpi et al., 2017). Lactose varies in secondary processing methods (sieved, milled, spray-dried, etc.) and therefore has differences in its particle size distribution and surface morphology (Rowe et al., 2009; Zarmpi et al., 2017). Two anomeric forms of lactose (α and β) are present and at equilibrium (0°C–100°C), and the ratio between the two forms is about 60:40 for β- and α-lactose, respectively (Zarmpi et al., 2017). The α- and β-forms differ in their physicochemical attributes, such as melting

Table 8.3 Factors affecting the performance of diluents.

Moisture content	Moisture content influences tensile strength, compaction, flowability, and viscoelastic properties of a diluent.
Particle size and morphology	Variability in excipient particle size may impact not only tablet hardness, friability, and disintegration, but also the content uniformity, wetting properties, dissolution, and stability of drug products. Diluents morphology can significantly influence the tabletability, stability, and product performance of a product.
Bulk density	A porous structure and a relatively low bulk density facilitate compressibility. It can be generalized that a decrease in bulk density improves tabletability; however, it often obstructs flowability.
Specific surface area	High specific surface area of diluent may improve cohesion, but it may negatively impact the powder flow properties.
Crystallinity and polymorphism	Crystallinity influences the water adsorption which may affect flowability, dissolution, solubility, compactibility, disintegration, tabletability, and stability of products.
Manufacturing process	Differences in manufacturing process of diluents play an important role on their functionality and may impact the dissolution efficiency of a tablet.
Degree of polymerization (DP)	DP may have significant effects on diluents properties such as on water absorption and compressibility. However, the effect of DP is applicable to few excipients such as MCC and starch and may not be valid for all other diluents.

point, density, and solubility. These molecular differences affect solid state properties of lactose and subsequently of products with varying attributes (Gamble et al., 2010). Compared to the anhydrous α-form of lactose, the anhydrous β-form has higher aqueous solubility and dissolution rate (van Kamp et al., 1986). A higher dissolution rate and initial solubility was observed for the products consisting mainly of β-lactose when compared to the products with a high α-content (Fig. 8.3). The initial solubility of anhydrous α-lactose was higher than that of α-lactose monohydrate but decreased after reaching a maximum. Of the β-lactoses, the highest initial solubility was found for roller-dried β-lactoses. Monohydrate α- and β-lactose disintegrates faster than anhydrous lactose. Moreover, an increase in compression force resulted in a faster disintegration because of the increased porosity of tablets (because of increased brittle fracture of the lactose molecules at higher compression forces), which allowed quick water penetration (van Kamp et al., 1986). In the case of α-lactose monohydrate (Fig. 8.4A) or crystallized β-lactose, it can be seen that an increase in compression force causes an initial retardation of the penetration rate into tablets (Fig. 8.4C). This effect also can be seen for roller-dried β-lactose at the lowest compression forces. For tablets from anhydrous α-lactose or roller-dried β-lactose, water uptake was poor and incomplete when high compression forces were used (Fig. 8.4B and D).

Spray-dried lactose consists of α-lactose monohydrate and amorphous lactose in a spherical shape with excellent flow properties (Gamble et al., 2010; Zarmpi et al., 2017). Studies showed that increasing

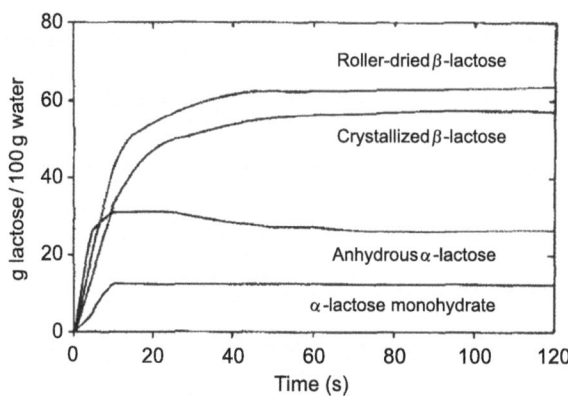

FIGURE 8.3

Initial solubility profiles of different types of crystalline lactose powder in water of 37°C ± 0.5°C.

Reproduced with permission from van Kamp et al. (1986).

the amorphous content of lactose prepared by spray drying increased the crushing strength and disintegration time of tablets prepared without disintegrants (Vromans et al., 1987a, 1987b). A change in crushing strength was attributed to the differences of amorphous (plastic) and crystalline content (brittle) under compression pressure. The spray-dried lactose is characterized by a porous structure and a relatively low bulk density that facilitate compressibility. The porosity and particle properties of lactose can have a significant impact on dissolution. Particle size and size distribution has been shown to affect moisture content, flow, blending, compactability, hardness, friability, and dissolution. Granule porosity was increased as the starting material particle size decreased because of the higher resistance of small particles to densification tendency (Badawy & Hussain, 2004). Narrow particle size ranges (40–75 and 212–250 μm) produced the higher porosity granules because of the lack of fine particles that likely occupy the void volumes in between the larger particles (Badawy & Hussain, 2004).

The concentration of lactose plays an important role. Increasing the percentage of fine lactose in indomethacin formulations (interactive mixture of API and coarse lactose) resulted in improvement of the drug's dissolution rate (Allahham & Stewart, 2007). This is explained by the fact that the rapid dissolution of lactose opened the matrix structure, which enables deagglomeration of indomethacin particles and dissolution. The batch-to-batch variability and intravendor variability also have been observed to have an impact on the processability and/or functionality of anhydrous lactose (Gamble et al., 2010). The presence of impurities such as formic acid, acetic acid, and aldehydes in lactose can catalyze the drug hydrolysis (Narang et al., 2012; Wu et al., 2011). For example, spray-dried lactose is reported to contain furfuraldehyde which was correlated with the discoloration of lactose and the reaction between 5-hydroxymethyl-2-furfuraldehyde and primary amine functional groups in APIs could lead to the formation of Schiff bases (Narang et al., 2012; Wu et al., 2011).

8.2.4.2 Microcrystalline cellulose

Cellulose is a polysaccharide composed of β (1 → 4) linked d-glucose units. MCC frequently is used in wet granulation processes and is obtained through hydrolytic depolymerization of cellulose

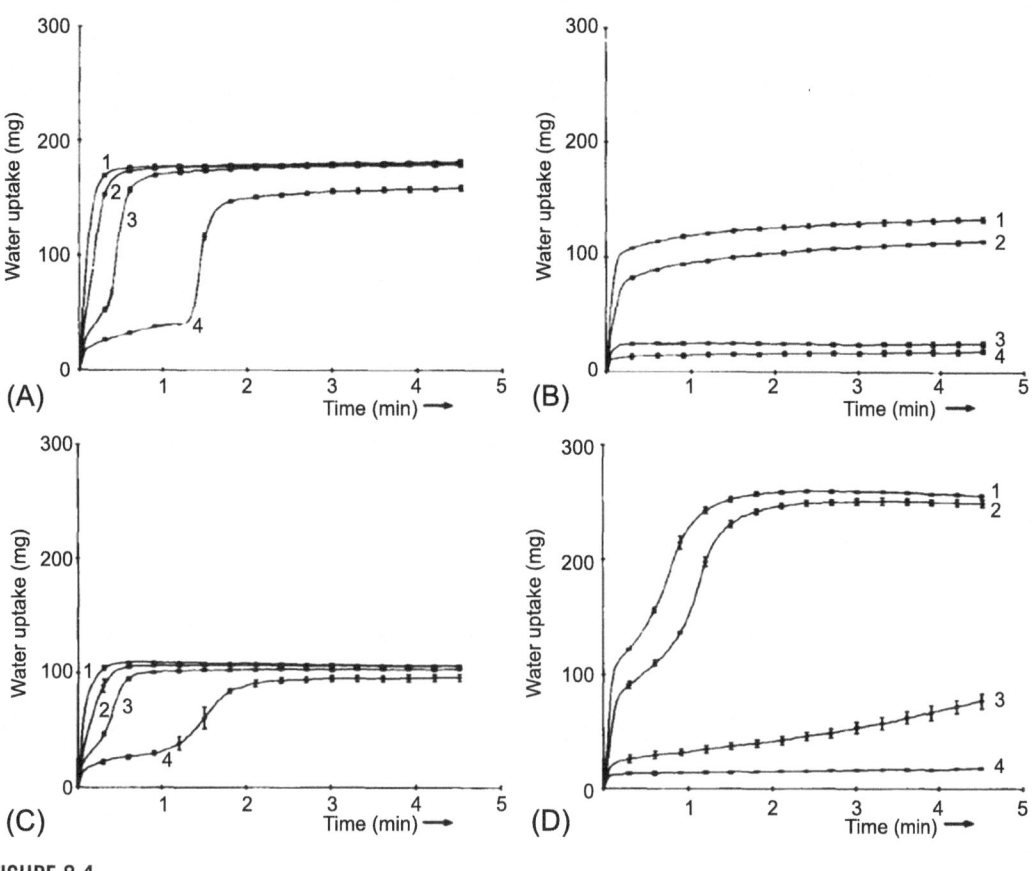

FIGURE 8.4

Water penetration into tablets compressed from different types of crystalline lactose. (A) α-Lactose monohydrate, (B) anhydrous α-lactose, (C) crystallized β-lactose, and (D) roller-dried β-lactose. Compression force: (1) 5 kN, (2) 10 kN, (3) 20 kN, and (4) 40 kN.

Reproduced with permission from van Kamp et al. (1986).

(Rowe et al., 2009). MCC is highly compactable. The wet granulation process, however, reduces the compactability of MCC because of change in its structure and loss of bonding surfaces. The factors that tend to reduce the compactability of MCC are increasing amount of water, longer massing time, and higher mixer speed. Reducing the particle size of MCC affects its flowability, and, because of its increased cohesiveness, results in the production of stronger tablets. The use of coarser particle size MCC improves flowability and reduces tablet weight variation, but, because of the increased risk of segregation, content uniformity can be affected. Obae et al. observed that MCC morphology, described by the length of particles (L) and their width (D) are one of the most important factors influencing tabletability (Obae et al., 1999). Presence of other excipients also can affect the functionality of MCC. For example, because of the ability of MCC to undergo plastic deformation, it is susceptible to a lubricant, such as magnesium stearate (van der Watt, 1987). To decrease the sensitivity of MCC,

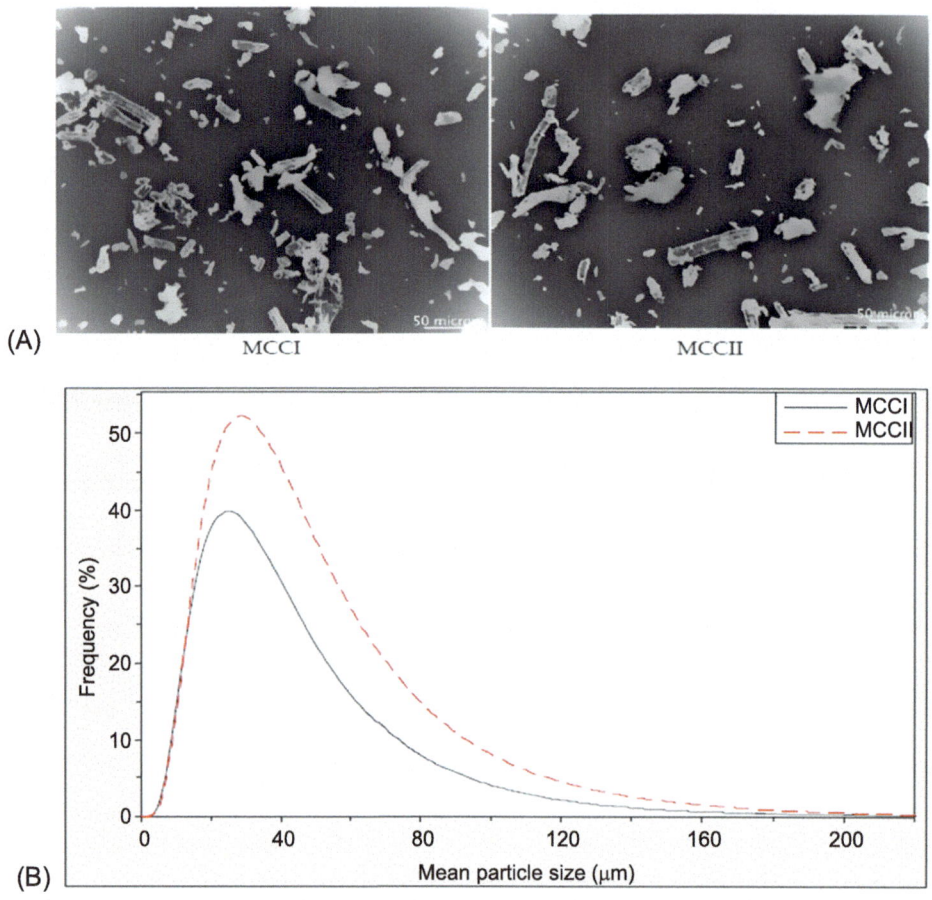

FIGURE 8.5
(A) Optical micropictures of microcrystalline allomorphs. (B) Particle size distribution of microcrystalline celluloses.

Reproduced with permission from Rojas (2013).

blending it with colloidal silica prior to lubrication is recommended (Pingali et al., 2011) because of the preferential binding of colloidal silica to the magnesium stearate.

MCC crystallinity has an impact on dissolution, and an increase in water penetration was observed with decreasing MCC crystallinity (Awa et al., 2015). The moisture within the pores of MCC can reduce frictional forces and facilitate plastic flow within the individual microcrystals. Two MCC polymorphs, MCC-I and MCC-II have been reported (Zarmpi et al., 2017). Both, MCC-I and MCC-II consisted of aggregated and irregularly shaped particles with rough surfaces and sharp edges (Fig. 8.5A). Elongated particles, however, were more predominant for MCC-I (Rojas, 2013). The particle size distribution is depicted in Fig. 8.5B. Both MCC-I and MCC-II showed a positively skewed distribution, but MCC-II had a slightly larger tendency to have high frequencies in the low particle size region (Rojas, 2013).

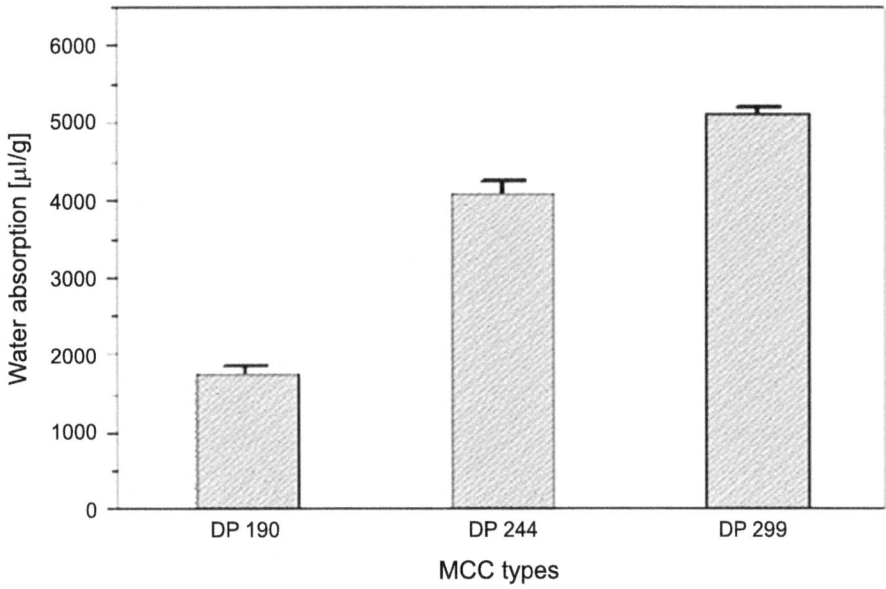

FIGURE 8.6

Water absorption for each degree of polymerization (DP).

Reproduced with permission from Rojas (2013).

The effect of MCC crystallinity on compactibility and dissolution of acetaminophen was studied (Suzuki & Nakagami, 1999). Results showed that the compression energy was lowered with a decrease in crystallinity, however, a lower dissolution was observed as crystallinity decreased up to a level. Studies showed that the MCC with higher DP had greater water absorption and compressibility (Fig. 8.6). This behavior probably resulted from the higher molecular weight of MCC, which can bind more water molecules. The molecular weight distribution of MCC with different DPs also was investigated and a unimodal molecular weight distribution (Fig. 8.7) was correlated. It can be assumed that the DP plays an important role in the mechanical properties of the MCC. Friability data of the MCCs with different DPs also was analyzed (Fig. 8.8). Results from another study showed that the DP did not affect the MCC characteristics, and the differences were related to the origin of raw material and the manufacturing process involved (Dybowski, 1997).

8.2.4.3 Mannitol

Mannitol is a polyol isomer of sorbitol. It is a nonhygroscopic diluent and widely used in pharmaceutical preparations at concentrations of 10% to 90% w/w. Mannitol occurs as a white, odorless, crystalline powder, or free-flowing granules (Rowe et al., 2009). It has a sweet taste and imparts a cooling sensation in the mouth because of its negative heat of solution and therefore is a preferred diluent in chewable tablets. Because of its nonhygroscopic nature, it is a good choice for moisture-sensitive drugs. Mannitol is present in different polymorphic forms with different compression characteristics (Debord et al., 1987). Crystalline grade mannitol generally is used for wet granulation, whereas the spray-dried form

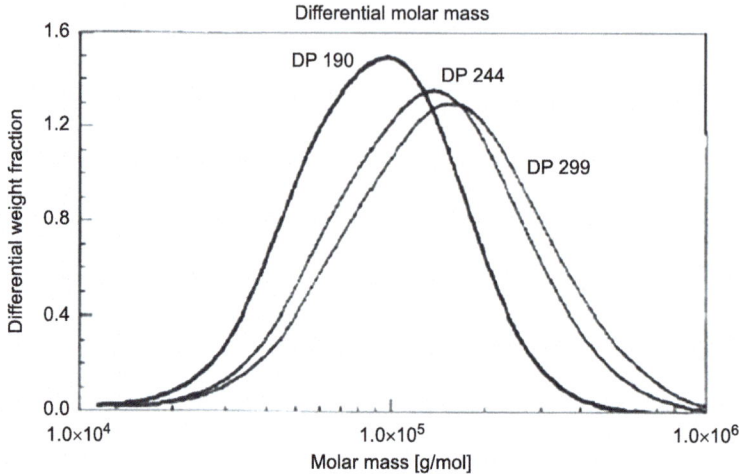

FIGURE 8.7

Molecular weight distributions of MCC with different degree of polymerizations (DPs).

Reproduced with permission from Shlieout et al. (2002).

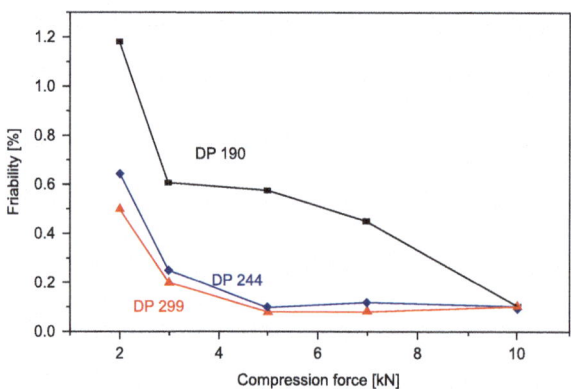

FIGURE 8.8

Friability of MCC tablets with different degree of polymerizations (DPs).

Reproduced with permission from Shlieout et al. (2002).

is used for direct compression. Without spray drying, mannitol has poor flow and binding properties, making it poor as a direct compression excipient. Granulations containing mannitol have the advantage of being dried easily. Higher lubricant quantities were needed for mannitol-containing granulations compared to those made with other diluents (Rowe et al., 2009). The size distribution of granules prepared from lactose, glucose, and mannitol in a high-shear granulator, was studied (Juppo et al., 1992). It was noted that the shapes and the size distribution of the granules produced from the three excipients were different. The size of the granules prepared with lactose, glucose, and mannitol were increased when the amount of granulation liquid increased. Similar results were observed for granulating high-shear granulation mixtures of mannitol and MCC (Westerhuis et al., 1996).

8.2.4.4 Dicalcium phosphate

DCP is an insoluble inorganic diluent. The anhydrous (triclinic crystal) and dihydrate (monoclinic structure) forms of DCP are used in pharmaceutical development (Rowe et al., 2009). The dihydrate form of DCP presents good flow properties and low hygroscopicity. Based on the temperature (40°C–50°C) and humidity (32%–75% relative humidity: % RH), however, the dihydrate form tends to lose water of hydration, which can cause chemical instability of APIs in dosage forms (Landín et al., 1994; Miyazaki et al., 2009). Because of the absence of water in its crystal structure, the anhydrous form of DCP exhibits higher porosity, leading to better compressibility and faster disintegration (Doldán et al., 1995). The dihydrate form transforms (partially or wholly) to the anhydrous form because of the high temperature in the wet granulation process.

Comparison between the two forms of DCP showed that the anhydrous form has better intraparticular porosity and better compressibility than the dihydrate form (Doldán et al., 1995). In addition, because of the higher intraparticular porosity, disintegration of anhydrous DCP is better than the dihydrate form. Both forms of DCP, however, needed a swelling-type disintegrant in the formulation when used (Khan & Rooke, 1976). Different grades of DCP are available, with coarse grade used for direct compression and milled grade for wet granulation or roller compaction. The milled grade has an alkaline pH and cannot be used with API incompatible with basic pH.

8.2.5 Binders

Binders promote cohesiveness and convert powder into granules (Zarmpi et al., 2017). In the wet granulation process, binders can be added to the powder blend as a dry ingredient or as a solution in the solvent (usually ethanol, water, or a mixture of these two). The uniformity of the particle size, hardness, disintegration, and compressibility of the granulation depends on the type and quantity of binder added. Several factors influence the effectiveness of a binder in a formulation (Parikh, 2010) (Table 8.4). Some of these factors are related to the drug and excipients in the formulation and others are related to the binder and solvent system used. The method of addition of the binder is also important.

The type of construction material of a granulator determines the granule size distribution and the volume of binder required (Bouwman et al., 2004). Vessel walls that are wetted easily require a high volume of binders. The use of plastics (mainly in lab-scale granulators) such as polymethylmethacrylate (PMMA) and polytetrafluoroethylene (PTFE) narrow the particle size distribution of granules because of the high contact angle whereby all liquids are forced directly/immediately to the powder bed (Bouwman et al., 2004). Results from this study showed that the size and size distribution of granules after drying was different when different vessels were used (Fig. 8.9). The granulation mechanisms for the different vessels and materials are depicted in Fig. 8.10. With stainless steel material (used in industrial-scale granulators), less contact angle liquid layer is formed in the wall surface, causing nonhomogeneous distribution of the liquid over the powder bed, resulting in broader-size granules.

Storage conditions as well as granulation process parameters can also significantly affect the binder efficiency and physicochemical characteristics. The hardness and disintegration time of Ranitidine tablets prepared using the PVP in wet granulation method decreased when the tablets were stored under elevated moisture levels (Uzunarslan & Akbuğa, 1991). PVP absorbs a significant quantity of water at

Table 8.4 Factors affecting the performance of binders.

Characteristics of drugs and other excipients	Compressibility, particle size, surface area, porosity, hydrophobicity, solubility of drugs in binder are important. Drugs with poor compressibility need stronger binders to be added. Fine/porous drug particles require a higher amount of binder compared to the larger particles. The granule size and friability are also affected by the solubility of drug in the granulation solution.
Binder concentration	Increasing the concentration of binder generally improves the mechanical strength of the tablets. Hard granulations result due to stronger/highly concentrated binder solution and require excessive compression force during tableting. However, weak granulations result due to insufficient quantity of binder which segregates easily. Also there is an increase in the plastic deformation and available bonding area with increasing the binder concentration.
Binder solvent quantity	An optimum quantity of a liquid is needed to achieve a desired granule size to keep the batch-to-batch variations to a minimum. Larger quantities of granulating liquid produce a coarser, narrow particle size distribution, and harder granules.
Particle size	Smaller particles with higher surface area have more contact points to allow the formation of stronger granules compared to the larger particles. However, due to the higher surface area, the solvent and binder requirements increase as the primary particle size decreases. At the same binder and solvent level, excipients with larger particle sizes consolidate easier and produce less porous granules as compared to finer particles.
Solubility	Increasing the excipients solubility in the granulating solvent decreases the solvent requirement and lead to the formation of granules of narrow particle size distribution and reduced friability. Drug solubility in the granulating solvent can affect its distribution in different granule size fractions. Drugs with high solubility in the granulating solvent have a higher tendency to migrate during drying.
Molecular weight (MW)	Increasing the MW of a binder improves tablets strength and influences the drug release rate and dissolution.
Binder solution viscosity	A less viscous binder that spreads easily on particles is superior compared to that shows poor wetting quality. However, when the solution viscosity is very high, problems with binder spreading and distribution take place. A binder of higher viscosity produces larger, denser granules and decreased the amount of binder required to initiate granule growth.

(continued on next page)

Table 8.4 Factors affecting the performance of binders—cont'd

Characteristics of drugs and other excipients	**Compressibility, particle size, surface area, porosity, hydrophobicity, solubility of drugs in binder are important. Drugs with poor compressibility need stronger binders to be added. Fine/porous drug particles require a higher amount of binder compared to the larger particles. The granule size and friability are also affected by the solubility of drug in the granulation solution.**
Binder solution surface tension	Decreasing binder solution surface tension decreases the capillary suction pressure and friction resistance to consolidation. This increases the granule consolidation rate and minimal attainable granule porosity. Decreasing the binder solution surface tension decreases the liquid requirement to attain overwetting. In general when using binders with lower surface tension, a large amount is required to produce the same size granules compared to a binder of higher surface tension. Higher surface tension of binder liquid also results in lower granule porosity.
Mixing time	If wet massing time is higher it produces harder granules.
Mechanical properties of binder	The mechanical properties of a binder determine the strength and deformation behavior of a binder matrix. When the spreading coefficient of binder over substrate is positive, dense granules are expected while negative spreading coefficient leads to the formation of porous granules.
Solvent composition and properties	Granule properties can be significantly affected by the granulating solvent. The porosity, friability, and tablet strength can be significantly affected by the types of the solvent system used. Changing the solvent system can also affect the excipients wettability and binder distribution.
Storage conditions	Storage conditions can severely alter the tablet properties since it changes the binder's physicochemical characteristics. Such changes can affect tablet hardness, friability, disintegration, and drug release.
Specific surface area	Increase in surface area causes an increase in massing time and in the water or binder quantity.
Impurity	Impurities that increase wettability may cause a decrease in massing time or decrease in the quantity of the binder required. Impurities also affect drug stability by inducing hydrolysis.
Polarity	The surface polarity calculated from contact angle measurement indicated that binders and powders with similar polarities produce stronger granules.

(continued on next page)

Table 8.4 Factors affecting the performance of binders—cont'd	
Characteristics of drugs and other excipients	Compressibility, particle size, surface area, porosity, hydrophobicity, solubility of drugs in binder are important. Drugs with poor compressibility need stronger binders to be added. Fine/porous drug particles require a higher amount of binder compared to the larger particles. The granule size and friability are also affected by the solubility of drug in the granulation solution.
Granulator: material of construction	Vessel wall made up of stainless steel require higher volume of binder as compared to vessel made up of plastic materials such as polymethylmethacrylate (PMMA) and polytetrafluoroethylene (PTFE). However, plastic materials are primarily used in lab-scale granulators. The use of plastics narrows the particle size distribution of granules due to the high contact angle whereby all liquids are forced directly/immediately to the powder bed.

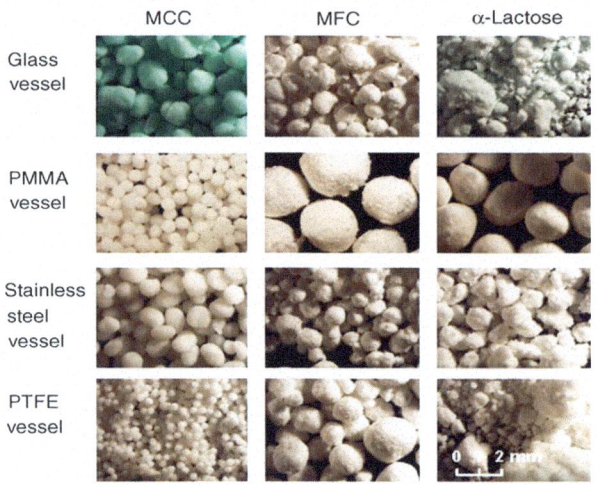

FIGURE 8.9

Granules produced in the different vessels. Bar represents 2 mm.

Reproduced with permission from Bouwman et al. (2004).

elevated humidity, which can act as a plasticizer, and decreases the polymer glass transition temperature (Fitzpatrick et al., 2002). The binder solution viscosity and surface tension are also important factors in their functionality. Increasing the binder solution viscosity could lead to increasing granule size and decreased amount of binder required to initiate granule growth in high-shear and fluid-bed granulation processes (Hoornaert et al.,1998; Sakr et al., 2012). This can be better understood by the mechanism of granule formation. When two particles collide during granulation, they either coalesce or spring back

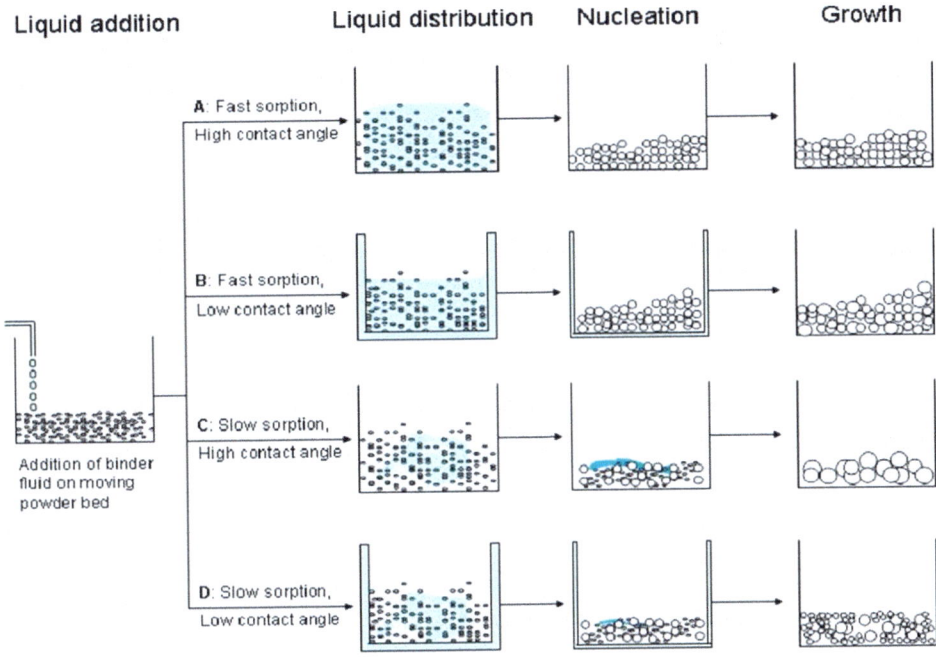

FIGURE 8.10

The different granulation mechanisms. Liquid binder is added to the moving powder bed, the liquid is distributed over the bed and nucleation starts. Addition to the powder bed is similar for all situations. A difference in nucleation originates in differences in the contact angle of the vessel wall material and the liquid sorption rate of the powder.

Reproduced with permission from Bouwman et al. (2004).

from each other. Collision energy that keeps the colliding particles together includes surface tension and viscous and interparticle friction forces. Surface tension plays a critical role for low-viscosity binders such as water in a high-shear wet granulation process (Cantor et al., 2008). Spreading efficiency of the binder solution between solid particles is the main controlling factor in wetting and granulation (Rowe, 1990). Dense nonfriable granules were produced when the spreading coefficient of binder over substrate was positive, while a negative spreading coefficient led to the formation of porous granules (Rowe, 1990). Also the higher spreading coefficient of binders resulted in stronger tablets.

Binder solvent is one important variable that could significantly influence granule properties as well as the mechanism of granule consolidation. In pharmaceutical manufacturing, aqueous, hydroethanolic, and ethanolic solvent systems are used widely. The selection of a solvent system is important because it could affect the formulation wettability and binder distribution. For example, PVP distribution in low-shear mixers was improved when water was replaced with a hydroalcoholic solution (Shah et al., 1996). Lactose granulations prepared using ethanolic PVP solution had higher porosity and friability compared to those prepared using water as a granulating solvent (Wikberg & Alderborn, 1993). Also increasing the ethanol content in hydroalcoholic solvent system increases the tablet strength because of the increased fragmentation tendency of granules increased with ethanol content, which could have further enhanced the consolidation (Alderborn, 1988). Binder-substrate interactions are significant determinants of the granule and tablet strength. The significant determinants of the granule are wettability of the substrate by the binder solution, binder cohesion, and binder-substrate adhesion (Reading & Spring, 1984). The impact of binder physicochemical properties on granule friability and strength of different hydrophilic and hydrophobic powders prepared by wet granulation method has been studied (Horisawa et al., 1993). The surface polarities calculated from contact angle measurement indicated that binders and powders with similar polarities produced stronger granules.

Binder concentration impacts the effectiveness of a binder (Becker et al., 1997). Increasing the binder concentration generally improves the mechanical strength of tablets. Solubility of the binder in the granulating solvent was found to decrease the solvent requirement and produced granules of uniform particle size distribution with reduced friability (Dias & Pinto, 2002). Drug or excipient particle size affects granule strength, porosity, and consolidation rate in granulation. Formation of stronger granules was achieved when smaller particles with higher surface areas were used because of availability of more contact points between colliding particles (van den Dries & Vromans, 2002). It should be noted that the solvent and binder requirements increase as the primary particle size decreases because of increased surface area. It has been shown that at the same binder and solvent level, an excipient with larger particle sizes consolidates more easily and produced less porous granules compared to finer particles (Sakr et al., 2012). The degree of drug solubility in the binder solution could affect its distribution in different granule size fractions. Drugs with higher solubility in the binder solution exhibited migration during the drying process, which led to higher drug concentration at the outer granular surfaces (van den Dries & Vromans, 2002).

The common excipients used as binders are starch, MCC, hydroxypropyl methylcellulose (HPMC), hydroxypropyl cellulose (HPC), PVP, hydroxyethyl cellulose, and sodium carboxymethyl cellulose.

8.2.5.1 Starch

Starch is a carbohydrate consisting of amylose (soluble) and amylopectin (insoluble). Amylose is a linear α1–4 linked polymer chain of glucose subunits, whereas amylopectin is composed of larger, branched polymer chains of α-glucose units with α1–4 and α1–6 linkages (Zarmpi et al., 2017). Amylose and amylopectin consist of semicrystalline and intercrystalline amorphous phases in alternating layers. Native starches do not possess ideal binder properties, such as good compressibility, and can lead to softer tablets with higher friability. Moreover, a high viscosity of starch paste (5%–25% w/w) makes the distribution of binder difficult and leads to its uneven distribution in granules. Therefore starch has been replaced by pregelatinized starch that can be used as a free-flowing powder. Pregelatinized starches are chemically and/or mechanically processed starches and typically contain 5% of free amylose, 15% of free amylopectin, and 80% unmodified starch. Pregelatinized starch is used in the concentration range of 5% to 10% w/w for the wet granulation processes. The concentration of starch and presence of other excipients in formulation play a significant role in its effectiveness and granules properties. Increasing starch content (in lactose: starch-based formulation) led to small granules with wider particle size distribution (Holm et al., 2001). Results were explained in terms of water absorption by starch particles and the formation of weaker granules that could not grow to the same extent as when starch was absent.

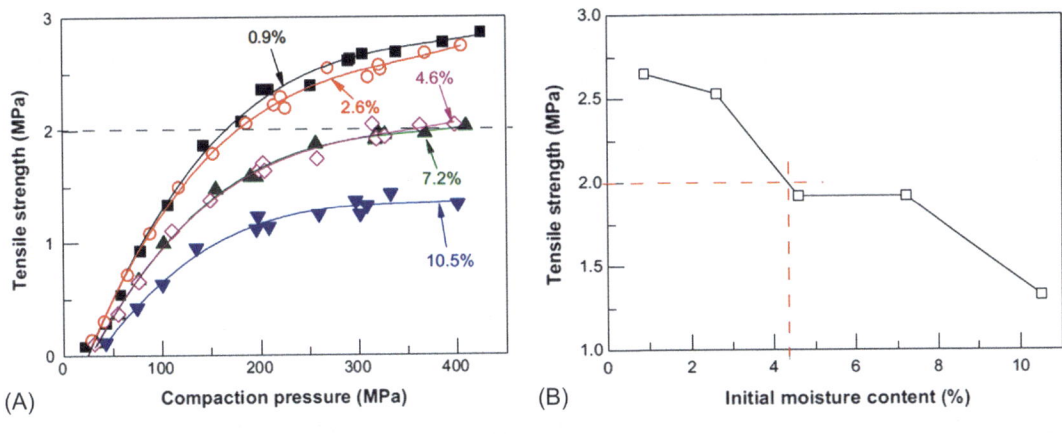

FIGURE 8.11

(A) Tabletability of granules prepared with MCC containing different amounts of water. (B) Effect of initial moisture content of starting MCC on tablet tensile strength at 300 MPa compaction pressure.

Reproduced with permission from Shi et al. (2011).

8.2.5.2 Microcrystalline cellulose

MCC functionality as a binder is related to its ability to deform plastically when compression force is applied (Thoorens, et al., 2014). MCC particles come in closer contact and form hydrogen bonds leading to strong compacts. The critical properties of MCC with respect to its functionality as a binder include moisture content, particle size, bulk density, specific surface area, DP, and crystallinity (Thoorens et al., 2014). DP affects tabletability, with highly polymerized MCC leading to powders with small particle size and smooth surface. Tablets containing MCC with a DP of 244 and 299 were twice as strong as those produced with a DP of 199 (Shlieout et al., 2002). In general DP favors the fibrous structure of MCC; improving tabletability but compromising powder flowability (Thoorens et al., 2014). Typically MCC is incorporated into solid dosage forms at the concentration of 20% to 90% w/w of the tablet when used as a binder or diluent (Rowe et al., 2009). High MCC concentrations can increase tablet hardness, which affects the disintegration and dissolution profiles. Inclusion of a disintegrant, therefore, is recommended when MCC is used as a binder. The increased viscous layer caused by swelling of MCC in aqueous solutions can affect the drug release when MCC is used at high concentrations.

It is critical to control the moisture level in MCC because of its high hygroscopicity, which can affect moisture-sensitive drugs. An optimum moisture level is critical, however, because low and high moisture can compromise compactability. Changes in moisture content can affect different stages during tablet manufacturing. Shi et al. (2011) studied the high-shear wet granulation of MCC with different initial moisture levels from 0.9% to 10.5% w/w. The obtained granules were designated as MWG-X, with X corresponding to the initial moisture content in the starting MCC powders. It was observed that the granule tabletability deteriorated with increasing initial moisture content in the starting materials (Fig. 8.11A). The tabletability of MWG-0.9 was highest, with that of MWG-2.6 slightly lower but MWG-2.6 was higher than 2 MPa when compaction pressure is ≥200 MPa. When initial moisture content was increased from 2.6% to 4.6%, tabletability dropped sharply. When the initial moisture content increased from 7.2% to 10.5%, tabletability also dropped sharply, however, tablets formed from these granules

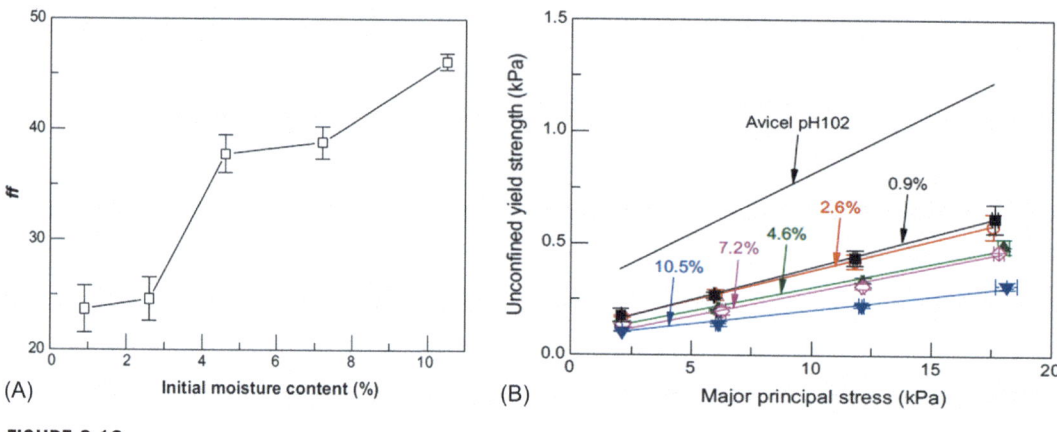

FIGURE 8.12

(A) Granule flow factor, at 10 kPa major principal stress, as a function of initial moisture content of starting MCC. (B) Effect of moisture content of starting MCC on granule flow performance.

Reproduced with permission from Shi et al. (2011).

had significantly lower tensile strength. The trend of change in tabletability was shown in Fig. 8.11B. Variations in initial moisture content in starting materials significantly affected the flow properties of the resulting granules (Figs. 8.12A and B). Increasing moisture content from 0.9% to 2.6% did not cause significant change in the flow factor of the granules. The flow factor increased significantly, however, when the initial moisture content increased from 2.6% to 4.6% and then stayed constant up to 7.2% (Fig. 8.12A). The granule size was increased with increasing the initial moisture content, which led to an improved flowability. For MWG-0.9, most of the granules appeared to be round in shape and smooth in surface texture without pores (Fig. 8.13A), indicating that these granules were well-consolidated and dense. The granules also contained a small fraction of loose agglomerates (indicated by arrows shown in Fig. 8.13B). Granules in MWG-4.6 were more uniform in size and loose agglomerates were essentially absent (Fig. 8.13C). With a further increase in initial moisture content from 4.6%, the only noticeable effect was larger granule size (Fig. 8.13D and E). The qualitative SEM observation of granule size induced by variations in initial moisture content in starting MCC was supported by the quantitative size measurements using laser diffraction (Fig. 8.14A). All granules showed unimodal size distributions. Granule size (d_{50}) increased from 99.4 ± 0.2 to 140.9 ± 2.3 μm when the initial moisture content increased from 0.9% to 7.2%. However, d_{50} increased sharply to 292.2 ± 7.7 μm when initial moisture content was 10.5%. This corresponded to approximately 200% and 100% increase in granule size when compared to MWG-0.9 and MWG-7.2, respectively (Fig. 8.14B). Another important material property, SSA, decreased with increasing initial moisture content in MCC (Fig. 8.15). SSA of MWG-0.9 (0.189 ± 0.003 m²/g) was more than twice that of MWG-10.5 (0.087 ± 0.002 m²/g). The reduction in SSA was consistent with both the increase in granule size and reduction in granule porosity. Other factors such as impurities in MCC (e.g., presence of reducing sugars) can affect drug stability by inducing hydrolysis in addition to the drug adsorption onto the polymer (Narang et al., 2012; Wu et al., 2011).

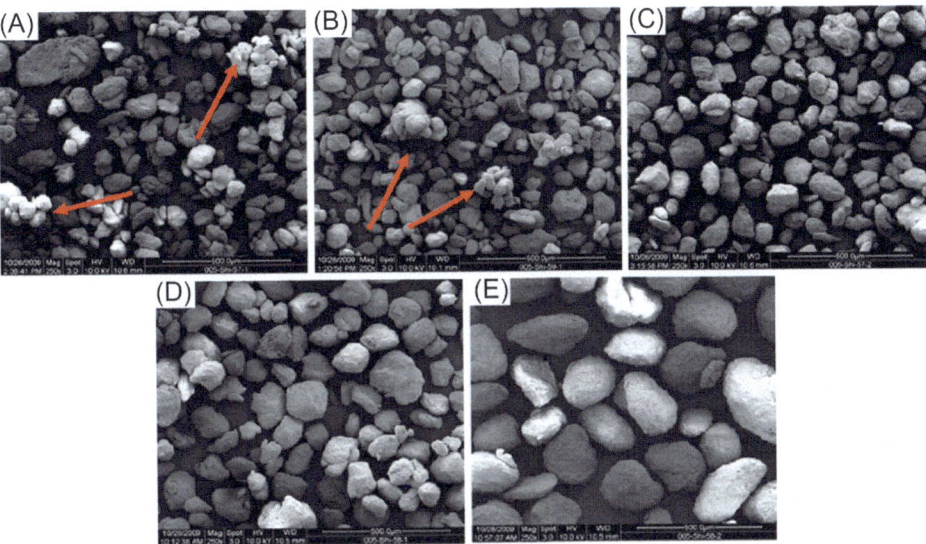

FIGURE 8.13

Evolution of granule morphology with increasing initial moisture content of starting MCC. (A) 0.9%, (B) 2.6%, (C) 4.6%, (D) 7.2%, and (E) 10.5%.

Reproduced with permission from Shi et al. (2011).

8.2.5.3 *Polyvinylpyrrolidone*

PVP or povidone is a hygroscopic, amorphous, synthetic polymer consisting of linear 1-vinyl-2-pyrrolidinone groups. As a binder, PVP is used in the concentration range of 0.5% to 5% w/w. Different degrees of polymerization of PVP resulted in polymers of various molecular weights. It is generally characterized by its viscosity in aqueous solution relative to that of water and expressed as a K value in the range of 10 to 120. Povidones with K-values ≤ 30 are manufactured by spray drying as spheres, whereas povidones with higher K-values are manufactured by drum drying as plates (Chakraborty et al., 2015). Wet granulation with povidone K25/30/90 generally gives harder granules with better flow properties than with other binders with lower friability and higher binding strength. Moreover, povidone also promotes the dissolution of APIs. For example, the drug release was faster in paracetamol tablets with 4% povidone K90 compared to tablets with gelatin or HPMC as binder (Jun et al., 1989). It has been shown that PVP was more efficient than HPMC owing to the lower work of cohesion and adhesion of HPMC. It could be further attributed to the better adhesion of PVP, especially to hydrophilic surfaces. Using PVP solution as granulating agent, it was observed that the addition of MCC as an insoluble excipient to a lactose-based formulation led to increase in solvent requirement and produced larger granules (Rohera & Zahir, 1993).

8.2.5.4 *Hydroxylpropyl methylcellulose*

HPMC is a water-soluble nonionic cellulosic polymer in which some of the hydroxyl groups are substituted with methoxy and hydroxypropyl groups (Zarmpi et al., 2017). As a binder, HPMC is

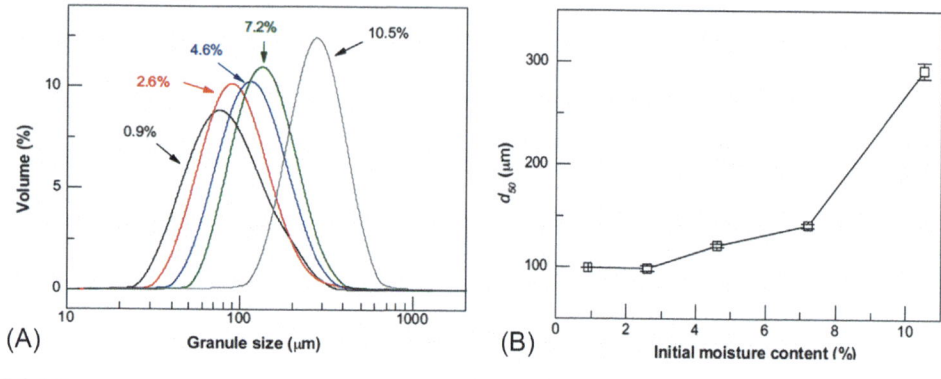

FIGURE 8.14

(A) Typical size distribution profiles of granules prepared with MCC containing different amounts of water. (B) Granule size as a function of initial moisture content of starting MCC.

Reproduced with permission from Shi et al. (2011).

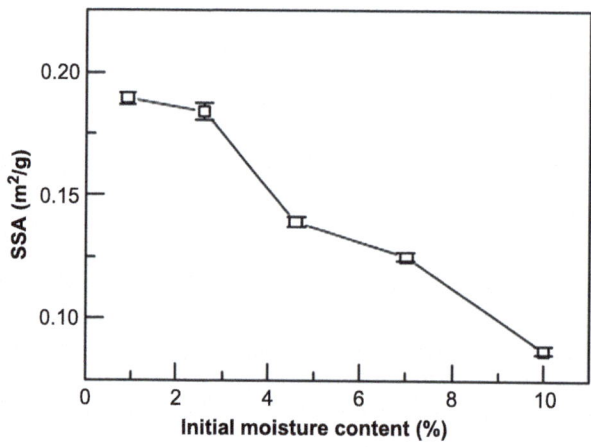

FIGURE 8.15

Effect of the initial moisture content of starting MCC on granule specific surface area (SSA).

Reproduced with permission from Shi et al. (2011).

used at the concentration of 2% to 5% w/w, however, it has been commonly used as tablet film coating polymer. The polymer chain length, size and degree of branching determine the viscosity of the polymer in solution. In general a tablet film coating requires low-viscosity polymers. Using low-viscosity polymers, the solid content in the coating formulation can be increased with a lesser amount of water, which can increase coating speed and efficiency. HPMC has many of the desired coating polymer properties. It provides aqueous soluble films; easy processing because of its nontacky nature; a transparent, tough, and flexible film that protects fragile tablets; improved appearance; and resistance to abrasion. The lower viscosity HPMC, however, produced film with lower tensile strength. The higher viscosity grades of HPMC provide film with good tensile strength, but their films have poor adhesion to the core surface and can easily peel off the tablet surface. When used alone HPMC has the tendency to bridge or fill the debossed tablet surfaces. Therefore a mixture of HPMC with other polymers or plasticizers is used to improve its binding to the tablet surface and eliminate bridging or filling problems.

Different grades of HPMC are available according to their particle size distribution, viscosity, molecular weights, and substitution of methoxy and hydroxypropyl groups. Because water penetration affects drug release and dissolution, not only the hydroxypropyl group and the degree of substitution but also the substitution pattern affects the release and dissolution (Zarmpi et al., 2017). Heterogeneity in the substitution pattern alters the release of the polymer because of hydrophobic interactions between the substituent, and a subsequent drug release alteration (Viriden et al., 2010; Zhou et al., 2014). Variation in substitution pattern causes batch-to-batch variability (Dahl et al., 1990). On aqueous solutions interaction, HPMC hydrates and forms a viscous gel layer that thickened when more water penetrated. These characteristics of HPMC affect its functionality. The MW and chain length have a significant effect on the viscosity of HPMC aqueous solutions and affects drug release and dissolution (Zarmpi et al., 2017). The hydrogen-bonding between oxygen atoms in ether groups of HPMC and water molecules leads to extension of the polymer and formation of a coil-shaped structure (Zarmpi et al., 2017). Coiled structures tend to form more hydrogen-bonds, entrap water, and form entanglements with other coiled molecules resulting in increased resistance to flow. Therefore HPMC with high MW tends to swell faster and forms viscous layers.

Particle size of HPMC affects drug release and dissolution through its impact on tablet hardness and water penetration. HPMC of smaller particle size formed stronger tablets because of increased surface area and interparticle cohesiveness, whereas HPMC of larger particles enhanced the dissolution because they do not fully occupy the space around each particle, leaving voids for water penetration (Mohamed et al., 2015). Drug release was caused by disintegration, diffusion, and a combination of diffusion and erosion for large, medium, and small particle sizes of HPMC, respectively. These effects were attributed to the proximity of polymer particles and the differences in the porosity of the formed hydrogel. Faster dissolution observed with higher HPMC particle sizes because of their porous arrangement, was counterbalanced via high concentrations because more polymer chains were present leaving no spaces for water penetration. The drug release rate from HPMC matrices was influenced by the drug/HPMC ratio, drug solubility, compression force, and viscosity grade of the HPMC. Lower viscosity grade HPMCs are more sensitive to the effect of compression force and tend to provide erosion-based release, compared to higher viscosity grade HPMCs for predominantly diffusion-based drug release (Hiremath & Saha, 2008a, 2008b). Mechanical properties of HPMC also affect its functionality. According to Rowe, dense granules are expected when the spreading coefficient of binder over substrate was positive, while negative spreading coefficient led to the formation of porous granules (Rowe, 1990). HPC is also one of the widely used binders (Parikh, 2010). HPC is nonionic, water-soluble, and pH-insensitive cellulose ether polymer. The high level of substitutions makes HPC more thermoplastic and less hygroscopic than other water-soluble cellulose ethers. Although, HPC has a good film formation property and excellent plasticity, it is not commonly used as an aqueous film coating polymer compared to HPMC because of its strong binding force. Various MW grades of HPC are available; however, low MW grades are typically used as binders. To overcome the above bridging or filling issues of HPMC, the combination of HPMC and HPC is used. HPC provides better film adhesion; however, the cost of HPC is much higher than HPMC.

Table 8.5 Factors affecting the performance of lubricants

Surface area and particle size	Increasing the surface area of a lubricant (e.g., magnesium stearate) due to particle size reduction leads to a thin evenly distributed layer on granules surfaces, whereas decreasing the surface area results in enrichment on the surface of granule particles. The high coverage of particle surfaces with the lubricant during blending reduces the interparticle bonding leading to weak tablets.
Chemical composition	Commercially available lubricants exhibit significant batch-to-batch variation and their lubricant properties. The factors mainly responsible for this variation are differences in chemical composition, specific surface area, and crystal structure.
Crystal structure	Different crystalline structures have different strengths of attraction between adjacent lamellae thereby affecting its relative ability to delaminate and subsequently coat the adjacent particles.
Concentration	The higher the concentration of lubricants used or the longer the blending times, the more complete this coating of the adjacent particles will become. The hydrophobic coating interferes with wetting phenomena leading to an increase in the time required for the tablet to disintegrate and/or drug to dissolve. Additionally a complete coating of lubricant may affect tablet hardness by interfering with the interparticle bonding required by formulations.
Amount and the mixing time	The use of proper quantities and homogeneous distribution of lubricants is important. An excess amount of lubricants or overmixing leads to increase in disintegration time and the decrease in hardness of compressed tablets. Over mixing can prevent the bonding of powder blend during tablet compression, resulting in softer tablets.
Process parameters	The process parameters such as blending time, speed, can exaggerate the deleterious effects of lubricants.

8.2.6 Lubricants

Friction or cohesiveness between particles enforces a barrier in powder or granule flowability. This affects content uniformity, compaction, tablet hardness, weight uniformity, and, ultimately, product performance. Lubricants address flowability issues by adherence to die wall or particle surfaces (Li & Wu, 2014; Wang et al., 2010b). Lubrication creates a film between surfaces or interfaces to reduce cohesion between particles or adhesion of particles onto surfaces (Zarmpi et al., 2017). The type, concentration, and method of incorporating lubricants into the process affect tablet compression in terms of compactibility and tablet weight variation. Factors affecting the performance of lubricants are summarized in Table 8.5.

The two major types of lubricants are as follows:

1. *Hydrophilic*: Generally poor lubricants, no glidant and antiadherent properties.
2. *Hydrophobic*: Most widely used lubricants, usually effective at low concentration, have antiadherent and glidant properties.

Three major roles are identified with lubricants:

1. *Lubricant role*: To decrease friction at the interface between a tablet's surface and the die wall during ejection and reduce wear on punches and dies.
2. *Antiadherent role*: Prevent sticking to punch faces or, in the case of encapsulation, lubricants prevent sticking to machine dosators, tamping pins, etc.
3. *Glidant role*: Enhance product flow by reducing interparticulate friction.

There are two ways in which lubricants reduce friction. A liquid lubricant can form a thin, continuous fluid layer between the tablet and the metal die surface or lubricant particles can form a boundary layer on the particles or metal die surfaces (Morin & Briens, 2013).

Lubricants levels and the way they are added into a formulation are critical. If the concentration is very low or distribution and mixing times are inadequate, problems such as punch filming, picking, sticking, capping, and binding in the die cavity can arise. If concentrations, distribution, and mixing times are too high, potential problems are a decrease in tablet hardness, the inability to compress into tablets, an increase in disintegration times, and a decrease in the rate of dissolution of tablets. Compared to other excipients in solid oral dosage forms, lubricants are required in low concentrations. Overall, lubricants affect hardness, friability, disintegration, and dissolution of the tablet.

Commonly used lubricants include stearates (magnesium, calcium, and sodium), stearic acid, sodium stearyl fumarate, talc, and sodium lauryl sulfate. Lubricants such as magnesium stearate and stearic acid are almost insoluble in water and that property of their powder blend coating can decrease dissolution of the formulation (Desai et al., 1993; Koo, 2017). The water-proofing effect is dependent on the solubility of the APIs, and the maximum effect is seen with the low-solubility APIs (Desai et al., 1993; Koo, 2017).

8.2.6.1 Stearic acid

Stearic acid is a saturated monobasic acid with 18 carbon-chain lengths. It is synthesized by the hydrolysis of animal fat or from hydrogenation of cottonseed or vegetable oil. Commercial stearic acid is a mixture of stearic acid with palmitic and myristic acid. Depending on the ratio of the stearic to palmitic acid, it can vary from macrocrystalline (45:55 w/w) to microcrystalline (between 50:50 and 90:10 w/w) structure (Li & Wu, 2014). Stearic acid polymorphic forms A, B, and C (most stable) are made using different organic solvents and crystallization conditions (Garti et al., 1980). Thermal studies indicated that stearic acid from different suppliers showed little batch-to-batch or manufacturer-to-manufacturer variability (Garti et al., 1980; Inaoka et al., 1988).

Because of its lower surface area, stearic acid is used at 1% to 3% w/w concentration. Because magnesium stearate at a concentration of 0.25% w/w is reported to soften the tablets made with pregelatinized starch and potentially affects tablet strength and dissolution, stearic acid is the preferred lubricant for pregelatinized starch. The starch undergoes plastic deformation during tableting and therefore has higher sensitivity to the concentration of magnesium stearate. Also, as reported by Fouda et al. (1998), although magnesium stearate accelerated the degradation of aspirin, stearic acid can protect drugs (aspirin) against degradation. In addition, stearic acid also can play a role in the polymorphic phase transformation of drugs, which subsequently resulted in a slowing down of the dissolution of tablets (Wang et al., 2010). Tablet dissolution was slow because of the transformation of polymorphic forms (Form II to Form I) of the drug, facilitated by stearic acid (Wang et al., 2010).

8.2.6.2 Magnesium stearate

Magnesium stearate is the most commonly used metallic salt boundary lubricant containing two equivalents of a fatty acid (usually stearic and palmitic acid) and a charged magnesium (Zarmpi et al., 2017). It is relatively inexpensive, chemically stable, has a high melting point and lubrication property. A concentration of 0.25% to 5.0% w/w magnesium stearate was used in formulation development (Rowe et al., 2009). Its lubricant effect relates to the adherence of the polar moiety on granules/powders, while the lipophilic moiety is oriented outward from the particle's surface (Jójárt et al., 2012). Its capacity to form a hydrophobic (waxy) layer around particles leads to reduced water penetration, which compromises the dissolution profile.

Commercially available magnesium stearates are generally a mixture of crystalline forms (anhydrate, monohydrate, dihydrate, and trihydrate) (Li & Wu, 2014). The crystal structures identified for the magnesium stearate hydrates include the plate-shaped dihydrate and needle-shaped monohydrate and trihydrate forms (Ertel & Carstensen, 1988). The different crystalline forms have different strengths of attraction between adjacent lamellae, which affect its relative ability to delaminate and subsequently coat adjacent particles. The monohydrate form produces tablets with lower variability (Li & Wu, 2014). The dihydrate form acts as a better lubricant because of its lamellar shape as it shears readily under applied tangential forces and because it has a lower tendency to cause overlubrication (Rajala & Laine, 1995; Zarmpi et al., 2017). The irregularity of the shapes of commercially available magnesium stearates compared to pure magnesium stearate relates to their reduced lubricant effects (Miller & York, 1985).

The amount and the mixing time of magnesium stearate in the formulation are critical variables. A higher level and longer mixing time reduce the drug dissolution. Depending upon the source, magnesium stearate differs in morphology, crystallinity, batch-to-batch variability in particle size, surface area, bulk strength, and fatty acid composition (Li & Wu, 2014). These differences can result in different compression profiles and lubrication efficiency that lead to differences in hardness and tablet friability. Three factors: differences in chemical composition, specific surface area, and crystal structure, have considered being mainly responsible for these variations.

A composition consisting of magnesium stearate to palmitate in a ratio of 25% to 75%, respectively, is optimum for its lubrication and shear properties, but this composition is generally not found in commercial samples. The surface area of magnesium stearate is also an important variable and the greater the surface area, the higher the ability to coat other particles in the formulation, ultimately leading to an effective lubrication.

Magnesium stearate with impurities of magnesium oxide creates an alkaline environment, causing drug degradation, especially for base-sensitive molecules (Li & Wu, 2014; Nokhodchi, 2005). Kararli et al. (1989) reported that MgO reacts with ibuprofen at certain temperatures and humidity to form the magnesium salt of ibuprofen. Magnesium stearate also induced an oxidation reaction, and the decomposition of drotaverine HCl was accelerated when magnesium stearate and talc were present in a formulation (Pawelczyk & Opielewicz, 1978). Specifically drotaverine HCl was degraded to drotaveraldine by an oxidative degradation pathway, which can be inhibited using an antioxidant or an acidic auxiliary material. Degradation of drugs also was mediated by the presence of magnesium ions. Upon an accelerated stress treatment, fosinopril sodium was degraded into a β-ketoamide (III) and a phosphoric acid (IV) in a tablet formulation with magnesium stearate, mediated by magnesium metal ions (Thakur et al., 1993).

8.2.6.3 Sodium stearyl fumarate

SSF can be used in solid dosage forms, generally at the concentration of 1% to 3% (Li & Wu, 2014; Wang et al., 2010). It is less hydrophobic and has a less retardant effect on tablet dissolution than magnesium stearate. As reported, compared with magnesium stearate, SSF has the same lubrication efficiency and influence on tablet strength and disintegration (Hölzer & Sjögren, 1979). Prolonged mixing, however, improved the lubrication efficiency of SSF. It was indicated that the SSF should be used as the tablet lubricant where magnesium stearate cannot be used because of compaction, lubrication, stability issues (Li & Wu, 2014).

8.2.7 Disintegrants

Disintegration is needed in dosage forms to achieve a quick onset of action and immediate drug release. The three major mechanisms of tablet disintegration process are (Desai et al., 2016) as follows:

1. *Swelling*: Swelling of the disintegrant can overcome the adhesion forces of other formulation excipients/components, causing the tablet fragmentation.
2. *Porosity and capillary action (wicking)*: Disintegrants that do not swell impart their disintegrating action through this mechanism. Tablet porosity provides pathways for the penetration of fluid into tablets. The disintegrant particles enhance porosity and provide these pathways into the tablet. Liquid is drawn up or wicked into these pathways through capillary action, and rupture the interparticulate bonds causing the tablet to break apart.
3. *Deformation*: Fragmentation of tablets can be caused by elastic deformation of disintegrants under pressure and release of high energy upon contact to water because of the ability of particles to recover their initial structure.

The first and often the rate-determining step in disintegration is the liquid penetration in the porous powder compact (Markl & Zeitler, 2017). Liquid penetration does not directly build up the pressure necessary to rupture the particle–particle bonds, but it is a prerequisite to initiate other mechanisms such as swelling. An overview of mechanisms involved in the disintegration of powder compacts is depicted in Fig. 8.16.

Factors influencing the performance of a disintegrant include its particle size, incorporation methods, applied compression force, moisture content, and degree of substitution (Desai et al., 2016). A disintegrant used in granulation can be more effective if used both intragranularly and extragranularly thereby acting to break the tablet up into granules that further disintegrate the granules to release the drug into solution. Disintegrants added intragranularly (in wet granulation) are usually not as effective as those added extragranularly because disintegrants are exposed to wetting and drying as part of the granulation process, which reduces their activity. In a study by Khattab et al. (1993), the disintegration and dissolution were enhanced for tablets prepared with extragranular and intragranular disintegrants in comparison with formulations with disintegrants added either in extragranular or intragranular phases. Gordon et al. (1990), however, showed that the disintegrant promoted the dissolution of poorly

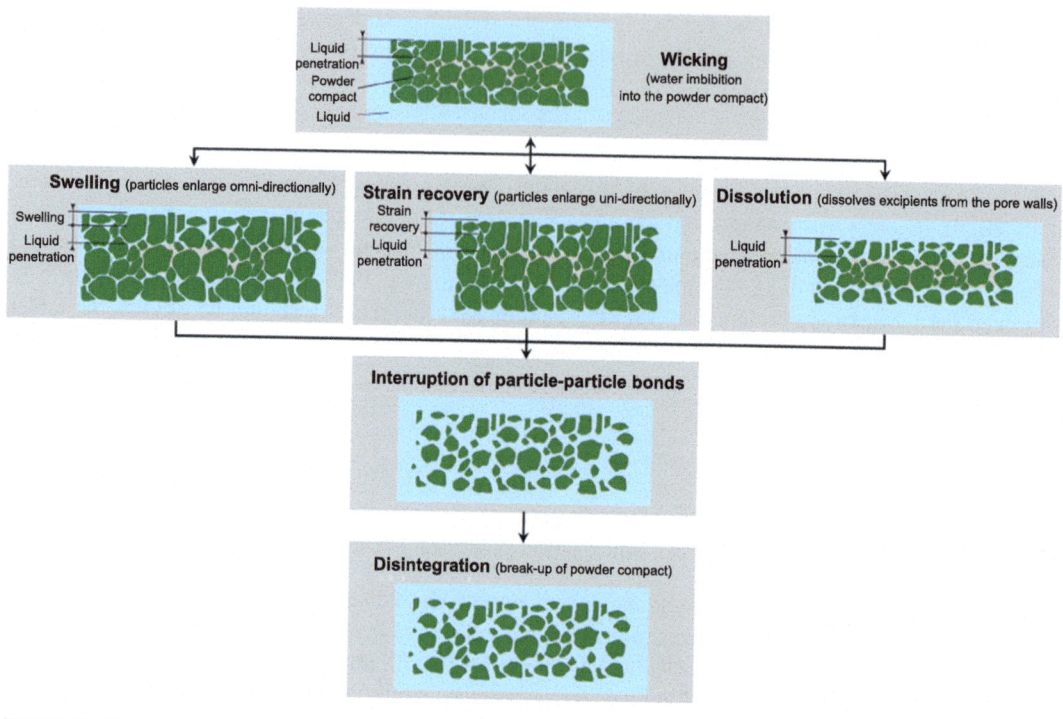

FIGURE 8.16

Overview of mechanisms involved in disintegration of pharmaceutical powder compacts.

Reproduced with permission from Markl and Zeitler (2017).

soluble drug when added intragranularly. He et al. (2008) reported that disintegrants incorporated intragranularly reduced the dispersibility of a poorly wettable drug because of the formation of dense granules; however, extragranular disintegrants provided the best dispersibility.

Chemical modifications of starch, cellulose, and povidone produced more efficient disintegrants, classified as superdisintegrants (Desai et al., 2016). Superdisintegrants can provide the disintegration within a few minutes and enhance the dissolution rate of solid dosage forms. Superdisintegrants provide improved compressibility and compatibility and have no negative effect on the mechanical strength of dosage forms containing high-dose drugs. The most common superdisintegrants are sodium starch glycolate (SSG), croscarmellose sodium (CCS), and crospovidone. It is also reported that the mode of incorporation of superdisintegrants had no significant effect on tablet dissolution, but related to the granulation methods, levels, and types of disintegrants used, and the physicochemical properties of tablet components (van Kamp et al., 1983).

Quodbach and Kleinebudde (2015) performed a systematic study to understand the impact of storage conditions (different RH) and relative tablet density on the functionality of tablet disintegrants. They observed that the storage conditions significantly affected the water uptake for SSG, but not as much for crospovidone and CCS. The swelling capacity of some disintegrants is pH dependent. The disintegration rate of tablets containing SSG or CCS decreased in acidic media but that of tablets containing crospovidone were unaffected (Chen et al., 1998). The reduced swelling of superdisintegrants caused a decrease in the dissolution of hydrochlorothiazide tablets in acidic medium in the presence of lactose as a diluent (Zhao & Augsburger, 2005). Substituting lactose to DCP dihydrate led to higher drug release in acidic media compared to water, and the effect of superdisintegrant swelling to reduce dissolution was not observed.

The common disintegrants used in wet granulation process are starch and cellulose-based excipients and are summarized below.

8.2.7.1 *Starch*

Starch and its derivatives are multifunctional excipients and disintegrants in tablet formulation. The versatility of starch as an excipient is mainly based on its derivatizations. Swelling is the primary mechanism of starch-based disintegrants. Native starches do not have ideal properties to act as an efficient tablet disintegrant, and a large amount (10%–15%) is generally required. At high concentration, however, native starch can have a negative impact on the tablet hardness in addition to flow and compactability issues. Moreover, intra-granular starch is not effective as disintegrant, so modified starches (produced through physical and chemical modifications) to produce pregelatinized starches are processed (Rowe et al., 2009). Pregelatinization of the starch has significant impact on its physical attributes. Partial pregelatinization improves the flowability and compressibility of starch and offers some disintegrant property. Full pregelatinization makes the starch soluble in cold water and not effective as a disintegrant, but it can be used as a binder (Rowe et al., 2009).

The functionality of starch also has been affected by the presence of other excipients. For example, reduced disintegration time of formulation was observed with an increase in the particle size of starch in the formulation without a lubricant (Smallenbroek et al., 1981). Disintegration time, however, was decreased with a decrease in particle size of the disintegrant when used with a lubricant. This is because as lubricant coverage decreased with an increase in the surface area of disintegrant, a faster disintegration was resulted.

8.2.7.2 *Sodium starch glycolate*

SSG is the sodium salt of cross-linked carboxymethyl starch. SSG is derived from starch with two chemical modifications: substitution (to increase hydrophilicity) and cross-linking (to reduce solubility and gel formation upon contact with water) (Shah & Augsburger, 2002). SSG acts as a superdisintegrant through rapid swelling because of the adsorption of large amounts of water leading to faster disintegration (Desai et al., 2014). Also, because of its spherical shape, SSG can have good flow properties.

Different grades of SSG are available according to particle size distribution, sodium chloride percentage, and pH. The amount of sodium in SSG is established between 2.8% and 4.2% in USP, whereas the degree of substitution (carboxymethyl groups) is not specified (*Sodium Starch Glycolate. USP 39-NF 34*, Rockville, MD, USA, 2016). Values for the degree of substitution between 0.23 and 0.32 have been reported (Zarmpi et al., 2017). Hydration and swelling of SSG are related to the degree of substitution. An increase in swelling and water uptake was observed as substitution increased from 0.20 to 0.29 and opposite effects were seen at higher substitution (Rudnic et al., 1985). A higher degree of substitution also can lead to increasing drug-excipient interactions of weakly basic drugs because the drug can be adsorbed onto the polymer (Narang et al., 2012).

The differences in the disintegration and dissolution properties of SSG were attributed to changes in the degree of cross-linking and extent of carboxymethylation (induces hydrophilicity by disrupting hydrogen bonding and allowing water access to the molecules). Disintegration efficiency has been inversely related to its level of cross-linking. Because carboxymethylation and cross-linking have opposing effects on water solubility, water access, and viscosity, a balance between the two is important for optimal performance. The phosphate group-based cross-linking in SSG leads to a high spacing between SSG chains, facilitating water penetration and swelling, and reduces the gel formation (Rojas et al., 2012). The extended swelling of SSG is attributed to phosphate cross-linking compared to the esterification in CCS, which does not allow this high spacing between the polymer chains. An increase of 25% to 35% of crosslinking leads to powders with increased swelling and water uptake, but further increase in crosslinking leads to lower swelling and water uptake (Zarmpi et al., 2017).

Particle size also affects the functionality of SSG with larger particles being more efficient. A threefold increase in particle size of SSG resulted in a proportional decrease in the disintegration time (Smallenbroek et al., 1981). As the particle size decreased, a more viscous layer is formed because of increased interaction with water, creating a barrier for drug diffusion and leading to delayed dissolution. A typical amount of SSG used in drug formulation ranges between 2% and 8% w/w (Rowe et al., 2009). SSG at higher levels (>8% of tablet weight) causes an increase in disintegration time because of the formation of a viscous layer that hinders water penetration in the formulation (Desai et al., 2014). The purity of SSG is also important and affects its disintegration efficiency. SSG has the impurities of sodium chloride, sodium glycolate, sodium citrate, and sodium acetate as the byproducts of the synthesis of SSG. Monochloroacetate, nitriles, and nitrates are potentially reactive impurities in SSG. Weakly basic drugs can compete with the sodium counter ion in SSG, adsorbing onto the surface of the disintegrant particles (Narang et al., 2012; Wu et al., 2011).

8.2.7.3 *Croscarmellose sodium*

CCS is cross-linked carboxymethylcellulose sodium and used as a superdisintegrant (Zarmpi et al., 2017). CCS is hydrophilic, but, made insoluble and fibrous by cross-linking of sodium salt of carboxymethylcellulose. The source of cellulose can result in differences in the physical properties of CCS. CCS derived from wood pulp has lower MW, higher water solubility, a slightly lower pH, decreased water capacity and swelling rate compared to CCS made from cotton linters (Koo, 2017). The functionality of CCS as a superdisintegrant is related to its fluid uptake and swellability characteristics. Swelling, wicking, and strain recovery are proposed mechanisms for CCS (Desai et al., 2016). Because swelling of CCS is attributed to the hydration of the carboxymethyl group, the degree of substitution determines CCS functionality.

The degree of substitution of CCS, as per the United States National Formulary monograph is 0.60 to 0.85 (Narang & Boddu, 2015). Zhao and Augsburger (2006) investigated the function-related properties of CCS and identified that particle size, degree of substitution, and the ratio of acidic to basic substituents are major factors influencing the swelling of CCS . The degree of cross-linking and the hydrophilicity of CCS are determined by the oxygen in the backbone as ether linkages, hydroxyls, and carboxylic acid functional groups (Narang & Boddu, 2015). In addition, ionized carboxylic acid groups interact with amine groups of basic substances, which can result in binding of the drug leading to slow and/or incomplete drug release (Narang & Boddu, 2015). The carboxymethyl substitution increases the swelling ability of CCS, with basic substituents having a better tendency to swell than acidic. A larger particle size of CCS leads to enhanced swelling and faster disintegration because CCS forms a viscous layer upon contact with water, and a smaller particle size forms a more viscous layer because of enhanced interactions with water (Zarmpi et al., 2017). Similarly morphologies that favor moisture absorption or have more sites for moisture uptake enhances the wicking ability of CCS.

In tablet formulations, CCS is used at the concentration of 0.5% to 5.0% w/w (Rowe et al., 2009). Any further increase in its level may result in the formation of a viscous gel layer acting as a barrier for disintegration. As in SSG, monochloroacetate, nitriles, and nitrates are potentially reactive impurities in CCS. Adsorption of weakly basic drugs and their salts to CCS has been reported to cause incomplete in vitro dissolution (Chien et al., 1981; Narang et al., 2012; Wu et al., 2011). The differences in the disintegration rate because of the presence of soluble or insoluble components in tablet were studied (Caramella et al., 1987, 1988; Desai et al., 2016). In general disintegrants acted better when formulated with insoluble excipients (e.g., DCP) than with soluble (e.g., lactose). Bindra et al. studied the impact of alkalinity of excipients (sodium carbonate) on tablets formulated with CCS (Bindra, et al., 2014). There was a significant slowdown of dissolution upon storage, with the decrease being proportional to the increase in alkalinity. This change was attributed to partial or complete hydrolysis of the ester cross-links in CCS, led to byproducts of increasing hydrogel characteristics. This led to a viscous barrier on tablets that delayed dissolution. Disintegrants with higher water affinity tend to show a greater slowdown in disintegration, and the impact was greater when tablets contain soluble excipients.

8.2.7.4 Crospovidone

Crospovidone is a water-insoluble synthetic cross-linked PVP. Crospovidone acts via different mechanisms including the swelling, and wicking followed by secondary swelling as proposed by Kornblum and Stoopak (1973). Strain recovery was also proposed as the disintegration mechanism of crospovidone (Desai et al., 2016; Kornblum & Stoopak, 1973). Crospovidone swells without gelling, a property that is advantageous for developing orally disintegration tablets and where gelling can delay the dissolution process. When compaction force is applied, the polymer deforms. Upon contact with water, it absorbs water via capillary action and regains its normal structure releasing an amount of energy capable to break the tablet. The particle size of crospovidone strongly affects the disintegration process, and larger particles provide a faster disintegration. As size increases, the intraparticular porosity increases, leading to larger water uptake and faster disintegration. Rudnic et al. concluded that the increase in crospovidone particle size improved the disintegration and dissolution of tablets (Rudnic et al., 1980). Several grades of crospovidone are available, differing in particle size distribution (standard, fine, superfine, and micronized), bulk density, hydration capacity, and peroxide levels.

Generally crospovidone is used at the concentration of 2% to 5% w/w (Rowe et al., 2009). Crospovidone levels higher than 8% of tablet weight produce weaker tablets with a faster disintegration (Desai et al., 2014). Being nonionic in nature, the disintegration efficiency of crospovidone is independent of the pH of media and thus, a potentially suitable disintegrant for cationic drugs (Zhao & Augsburger, 2005). Molecular properties of crospovidone, therefore, are unlikely to affect the excipient's functionality. The residual peroxides in crospovidone, however, can affect the stability of oxygen-sensitive molecules (Narang, Desai & Badawy, 2012). Hartauer et al. (2000) observed an oxidative degradation product of raloxifene hydrochloride in a tablet formulation that was identified as an N-oxide derivative. Results from drug-excipient studies and a formulation spiking study showed that residual peroxides in crospovidone promoted the formation of the N-oxide. Effect of sorbed moisture on the functionality of disintegrant has been studied. Differential scanning calorimetry and dynamic vapor sorption analysis were performed for crospovidone with different particle sizes to evaluate their differences in water uptake, water distribution characteristics, and water–polymer interactions

Table 8.6 Factors affecting the performance of disintegrants.

Particle size	In general an increase in particle size improves the disintegration and dissolution of tablets. This attributes to the efficient formation of hydrophilic networks by larger size particles.
Degree of substitution of superdisintegrants	Degree of substitution relates to the total acidic (acid form) and basic (sodium salt) components of superdisintegrants. Brands with higher basic substituents exhibit larger settling volumes, higher water uptake and size increase. Higher degree of substitution may also induce drug-excipient interactions.
Methods of incorporation	Disintegrants can be more effective if used both intragranularly and extragranularly. In general disintegrants added intragranularly are not as effective as that added extragranularly. However, the method of addition of disintegrant alone cannot determine its efficiency.
Applied compression force	When superdisintegrants, SSG, CCS, and crospovidone, were used extragranularly, recompression improved dissolution. However, when used intragranularly, recompression decreased dissolution of tablets containing CCS and crospovidone but increased for SSG.
Moisture content (during the storage)	Storage conditions affect the water uptake and force development for disintegrants such as SSG; however, it does not significantly affect the crospovidone, and CCS.
Swelling property and porosity	The swelling of disintegrants depends on several factors such as chemical structure, degree of crosslinking, and porosity. A porous tablet matrix with large void spaces could reduce the swelling of disintegrants and impede their efficiency. Conversely low porosity compacts prepared using very high compression forces could hinder liquid entry and prolong the disintegration time. Thus optimal porosity is needed to provide an adequate mechanical integrity without compromising the disintegratability.
Presence of other excipients	Presence of other excipients such as lubricants can impact the efficiency of disintegrants. Coating of particles by lubricants can interfere with the wetting of tablets. The solubility of excipients can also impact the efficiency of disintegrants and they are more effective in the presence of insoluble excipients. In case of soluble excipients, tablet dissolves rather than disintegrate in the presence of water. Tablet excipient dissolving can lead to generation of the higher viscosity of the liquid and generation of more porous matrix thus reducing the efficiency of disintegrants that act through swelling mechanism. Disintegrants with higher water affinity show a greater slowdown in disintegration and the impact is greater in the presence of soluble excipients.

(continued on next page)

Table 8.6 Factors affecting the performance of disintegrants—cont'd

Particle size	In general an increase in particle size improves the disintegration and dissolution of tablets. This attributes to the efficient formation of hydrophilic networks by larger size particles.
Concentration	At high concentration, some disintegrants can negatively impact the tablet hardness, flow, and compactability issues in addition to producing weaker tablets with a faster or slower disintegration.
Other factors	Factors such as pH can affect the efficiency of disintegrants by affecting their swelling capacity. This is true especially for CCS and SSG and has been attributed to lower liquid holding capacity of unionized forms of cellulosic and modified starch polymers. The impact of alkalinity of excipients was studied and a partial or complete hydrolysis of the ester cross-links in CCS was observed. These generated by-products of increasing hydrogel characteristics and led to a viscous barrier on tablets to delay the dissolution.

(Saripella et al., 2014). Despite the differences in particle sizes of crospovidone samples, they exhibited similar water interactions. Glass transition temperature of PVP was reduced with an increase in moisture content indicating that water plasticizes the disintegrant (Fitzpatrick et al., 2002). Although the particle sizes differed, their abilities for water uptake and water distribution were not significantly different. Dynamic vapor sorption measurements confirmed that the water interaction profiles were similar for all crospovidone grades. Factors affecting the performance of disintegrants are summarized in Table 8.6.

8.3 Conclusion and future directions

Granulation is used commonly in the pharmaceutical industry during solid oral dosage form development and manufacturing. The wet granulation process is frequently employed to obtain homogeneous granules to improve the flow, content uniformity, compression characteristics, drug release profile, and to reduce segregation and the potential for dusting (contamination/safety). The properties of the granules such as porosity, hygroscopicity, hardness, size, shape, texture, surface area, density, elasticity, and plasticity can be controlled through manipulation of several material attributes. Excipients play a crucial role and are important constituents in a drug formulation. Excipients constitute a major component of solid dosage forms, and their use covers a variety of functions, from dosage manufacturing to dosage performance. Although excipients are considered as an inert material, they can significantly alter the therapeutic response of a drug, making their amount and uniformity in a dosage form as important. The limited knowledge of the role of physicochemical properties of excipients hinders the possibility to further optimizing the dosage forms from the aspects of manufacturing and drug product performance. Therefore understanding and integrating excipient variability in the product design is important to achieve quality products. It is critical to understand the impact of excipient variability on the drug product manufacturability and performance. This chapter has attempted to evaluate and summarize the

CMAs of some of the excipients on the CQAs of granules. In the wet granulation process, excipients such as binders, diluents, lubricants, and disintegrants, are commonly used. Despite the rising awareness on excipient roles in dosage forms, not many reports are available. Hence, a great deal of effort still needs to be made to delineate the complex effects of excipient physicochemical factors on product performance.

References

Alderborn, G. (1988). Granule properties of importance to tableting. *Acta Pharmaceutica Suecica, 25*, 1988.

Allahham, A., & Stewart, P. J. (2007). Enhancement of the dissolution of indomethacin in interactive mixtures using added fine lactose. *European Journal of Pharmaceutics and Biopharmaceutics, 67*(3), 732–742. https://doi.org/10.1016/j.ejpb.2007.04.013.

Awa, K., Shinzawa, H., & Ozaki, Y. (2015). The effect of microcrystalline cellulose crystallinity on the hydrophilic property of tablets and the hydrolysis of acetylsalicylic acid as active pharmaceutical ingredient inside tablets. *AAPS PharmSciTech, 16*(4), 865–870. https://doi.org/10.1208/s12249-014-0276-7.

Badawy, S. I., & Hussain, M. A. (2004). Effect of starting material particle size on its agglomeration behavior in high shear wet granulation. *AAPS PharmSciTech, 5*(3), e38. https://doi.org/10.1208/pt050338.

Becker, D., Rigassi, T., & Bauer-Brandl, A. (1997). Effectiveness of binders in wet granulation: A comparison using model formulations of different tabletability. *Drug Development and Industrial Pharmacy, 23*(8), 791–808. https://doi.org/10.3109/03639049709150550.

Bindra, D. S., Stein, D., Pandey, P., & Barbour, N. (2014). Incompatibility of croscarmellose sodium with alkaline excipients in a tablet formulation. *Pharmaceutical Development and Technology, 19*(3), 285–289. https://doi.org/10.3109/10837450.2013.778869.

Bouwman, A. M., Visser, M. R., Eissens, A. C., Wesselingh, J. A., & Frijlink, H. W. (2004). The effect of vessel material on granules produced in a high-shear mixer. *European Journal of Pharmaceutical Sciences, 23*(2), 169–179. https://doi.org/10.1016/j.ejps.2004.07.008.

Cantor, S., Augsburger, L., & Gerhardt, A. (2008). Pharmaceutical granulation processes, mechanism and the use of binders. *Pharmaceutical dosage forms: Tablets* (3rd ed., pp. 261–301). CRC Press.

Caramella, C., Colombo, P., Conte, U., Ferrari, F., Gazzaniga, A., La Manna, A., et al. (1987). The mechanisms of disintegration of compressed particulate systems. *Polymer Bulletin, 18*(6), 541–544. https://doi.org/10.1007/bf00255339.

Caramella, C., Colombo, P., Conte, U., Ferrari, F., Gazzaniga, A., LaManna, A., et al. (1988). A physical analysis of the phenomenon of tablet disintegration. *International Journal of Pharmaceutics, 44*(1), 177–186. https://doi.org/10.1016/0378-5173(88)90114-7.

Chakraborty, P., Ghosh, A., & Chakraborty, D. D. (2015). Polymeric systems in quick dissolving novel films. *Handbook of polymers for pharmaceutical technologies* (pp. 143–165). John Wiley and Sons, Inc./Scrivener Publishing LLC.

Chen, C. R., Cho, S. L., Lin, C. K., Lin, Y. H., Chiang, S. T., & Wu, H. L. S. (1998). Dissolution difference between acidic and neutral media of acetaminophen tablets containing a super disintegrant and a soluble excipient. II. *Chemical and Pharmaceutical Bulletin, 46*(3), 478–481. https://doi.org/10.1248/cpb.46.478.

Chien, Y. W., Van Nostrand, P., Hurwitz, A. R., & Shami, E. G. (1981). Drug-disintegrant interactions: Binding of oxymorphone derivatives. *Journal of Pharmaceutical Sciences, 70*(6), 709–710.

Dahl, T. C., Calderwood, T., Bormeth, A., Trimble, K., & Piepmeier, E. (1990). Influence of physicochemical properties of hydroxypropyl methylcellulose on naproxen release from sustained release matrix tablets. *Journal of Controlled Release, 14*(1), 1–10. https://doi.org/10.1016/0168-3659(90)90055-X.

Dave, V. S., Saoji, S. D., Raut, N. A., & Haware, R. V. (2015). Excipient variability and its impact on dosage form functionality. *Journal of Pharmaceutical Sciences, 104*(3), 906–915. https://doi.org/10.1002/jps.24299.

Debord, B., Lefebvre, C., Guyot-Hermann, A. M., Hubert, J., Bouche, R., & Cuyot, J. C. (1987). Study of different crystalline forms of mannitol: Comparative behavior under compression. *Drug Development and Industrial Pharmacy, 13*(9–11), 1533–1546. https://doi.org/10.3109/03639048709068679.

Desai, D. S., Rubitski, B. A., Varia, S. A., & Newman, A. W. (1993). Physical interactions of magnesium stearate with starch-derived disintegrants and their effects on capsule and tablet dissolution. *International Journal of Pharmaceutics, 91*(2), 217–226. https://doi.org/10.1016/0378-5173(93)90341-C.

Desai, P. M., Er, P. X., Liew, C. V., & Heng, P. W. (2014). Functionality of disintegrants and their mixtures in enabling fast disintegration of tablets by a quality by design approach. *AAPS PharmSciTech, 15*(5), 1093–1104. https://doi.org/10.1208/s12249-014-0137-4.

Desai, P. M., Liew, C. V., & Heng, P. W. (2016). Review of disintegrants and the disintegration phenomena. *Journal of Pharmaceutical Sciences, 105*(9), 2545–2555. https://doi.org/10.1016/j.xphs.2015.12.019.

Dias, V. H., & Pinto, J. F. (2002). Identification of the most relevant factors that affect and reflect the quality of granules by application of canonical and cluster analysis. *Journal of Pharmaceutical Sciences, 91*(1), 273–281. https://doi.org/10.1002/jps.10015.

Doldán, C., Souto, C., Concheiro, A., Martínez-Pacheco, R., & Gómez-Amoza, J. L. (1995). Dicalcium phosphate dihydrate and anhydrous dicalcium phosphate for direct compression: a comparative study. *International Journal of Pharmaceutics, 124*(1), 69–74. https://doi.org/10.1016/0378-5173(95)00077-V.

Dybowski, U. (1997). Does polymerisation degree matter? *Manufacturing Chemist*, 19–21.

Ertel, K. D., & Carstensen, J. T. (1988). An examination of the physical properties of pure magnesium stearate. *International Journal of Pharmaceutics, 42*(1), 171–180. https://doi.org/10.1016/0378-5173(88)90173-1.

Fitzpatrick, S., McCabe, J. F., Petts, C. R., & Booth, S. W. (2002). Effect of moisture on polyvinylpyrrolidone in accelerated stability testing. *International Journal of Pharmaceutics, 246*(1–2), 143–151. https://doi.org/10.1016/S0378-5173(02)00375-7.

Fonteyne, M., Wickstr om, H., Peeters, E., Vercruysse, J., Ehlers, H., Peters, B. H., et al. (2014). Influence of raw material properties upon critical quality attributes of continuously produced granules and tablets. *European Journal of Pharmaceutics and Biopharmaceutics, 87*(2), 252–263. https://doi.org/10.1016/j.ejpb.2014.02.011.

Fouda, M. A., Mady, O. Y., & El-Azab, G. A. (1998). Stabilization and control of aspirin release via solid dispersion systems. *Mansoura Journal of Pharmaceutical Sciences, 14*, 36–70.

Gamble, J. F., Chiu, W. S., Gray, V., Toale, H., Tobyn, M., & Wu, Y. (2010). Investigation into the degree of variability in the solid-state properties of common pharmaceutical excipients-anhydrous lactose. *AAPS PharmSciTech, 11*(4), 1552–1557. https://doi.org/10.1208/s12249-010-9527-4.

Garti, N., Wellner, E., & Sarig, S. (1980). Stearic acid polymorphs in correlation with crystallization conditions and solvents. *Kristall und Technik, 15*(11), 1303–1310. https://doi.org/10.1002/crat.19800151112.

Gordon, M. S., Chatterjee, B., & Chowhan, Z. T. (1990). Effect of the mode of croscarmellose sodium incorporation on tablet dissolution and friability. *Journal of Pharmaceutical Sciences, 79*(1), 43–47.

Hartauer, K. J., Arbuthnot, G. N., Baertschi, S. W., Johnson, R. A., Luke, W. D., Pearson, N. G., et al. (2000). Influence of peroxide impurities in povidone and crospovidone on the stability of raloxifene hydrochloride in tablets: Identification and control of an oxidative degradation product. *Pharmaceutical Development and Technology, 5*(3), 303–310. https://doi.org/10.1081/PDT-100100545.

He, X., Barone, M. R., Marsac, P. J., & Sperry, D. C. (2008). Development of a rapidly dispersing tablet of a poorly wettable compound: Formulation DOE and mechanistic study of effect of formulation excipients on wetting of celecoxib. *International Journal of Pharmaceutics, 353*(1–2), 176–186. https://doi.org/10.1016/j.ijpharm.2007.11.045.

Hiremath, P. S., & Saha, R. N. (2008a). Oral controlled release formulations of rifampicin. Part II: Effect of formulation variables and process parameters on in vitro release. *Drug Delivery, 15*(3), 159–168. https://doi.org/10.1080/10717540801952498.

Hiremath, P. S., & Saha, R. N. (2008b). Oral matrix tablet formulations for concomitant controlled release of antitubercular drugs: Design and in vitro evaluations. *International Journal of Pharmaceutics, 362*(1–2), 118–125. https://doi.org/10.1016/j.ijpharm.2008.06.019.

Hlinak, A. J., Kuriyan, K., Morris, K. R., Reklaitis, G. V., & Basu, P. K. (2006). Understanding critical material properties for solid dosage form design. *Journal of Pharmaceutical Innovation, 1*(1), 12–17. https://doi.org/10.1007/bf02784876.

Holm, P., Schaefer, T., & Larsen, C. (2001). End-point detection in a wet granulation process. *Pharmaceutical Development and Technology, 6*(2), 181–192. https://doi.org/10.1081/PDT-100000739.

Holzer, A. W., & Sj ogren, J. (1979). Evaluation of sodium stearyl fumarate as a tablet lubricant. *International Journal of Pharmaceutics, 2*(3), 145–153. https://doi.org/10.1016/0378-5173(79)90015-2.

Hoornaert, F., Wauters, P. A. L., Meesters, G. M. H., Pratsinis, S. E., & Scarlett, B. (1998). Agglomeration behavior of powders in a L odige mixer granulator. *Powder Technology, 96*(2), 116–128. https://doi.org/10.1016/S0032-5910(97)03364-0.

Horisawa, E., Komura, A., Danjo, K., & Otsuka, A. (1993). Effect of binder characteristics on the strength of agglomerates prepared by the wet method. *Chemical and Pharmaceutical Bulletin, 41*(8), 1428–1433. https://doi.org/10.1248/cpb.41.1428.

Inaoka, K., Kobayashi, M., Okada, M., & Sato, K. (1988). Stability, occurrence and step morphology of polymorphs and polytypes of stearic acid. *Journal of Crystal Growth, 87*(2). https://doi.org/10.1016/0022-0248(88)90171-6.

Jójárt, I., Sovány, T., Pintye-Hódi, K., & Kása, P. (2012). Study of the behavior of magnesium stearate with different specific surface areas on the surface of particles during mixing. *Journal of Adhesion Science and Technology, 26*(24), 2737–2744. https://doi.org/10.1080/01694243.2012.701481.

Jun, Y. B. M., Kim, S. I., & Kim, Y. I. J. (1989). Preparation and evaluation of acetaminophen tablets. *Korean Pharmaceutical Sciences, 19*, 123–128.

Juppo, A. M., Yliruusi, J., Kervinen, L., & Str om, P. (1992). Determination of size distribution of lactose, glucose and mannitol granules by sieve analysis and laser diffractometry. *International Journal of Pharmaceutics, 88*(1), 141–149. https://doi.org/10.1016/0378-5173(92)90310-X.

Kararli, T. T., Needham, T. E., Seul, C. J., & Finnegan, P. M. (1989). Solid-state interaction of magnesium oxide and ibuprofen to form a salt. *Pharmaceutical Research, 6*(9), 804–808.

Khan, K. A., & Rooke, D. J. (1976). Effect of disintegrant type upon the relationship between compressional pressure and dissolution efficiency. *Journal of Pharmacy and Pharmacology, 28*(8), 633–636. https://doi.org/10.1111/j.2042-7158.1976.tb02816.x.

Khattab, I., Menon, A., & Sakr, A. (1993). Effect of mode of incorporation of disintegrants on the characteristics of fluid-bed wet-granulated tablets. *The Journal of Pharmacy and Pharmacology, 45*(8), 687–691.

Koo, O. M. Y. (2017). *Pharmaceutical excipients: Properties, functionality, and applications in research and industry*. John Wiley and Sons, Inc.

Kornblum, S. S., & Stoopak, S. B. (1973). A new tablet disintegrating agent: Cross-linked polyvinylpyrrolidone. *Journal of Pharmaceutical Sciences, 62*(1), 43–49. https://doi.org/10.1002/jps.2600620107.

Kushner, J. T. (2013). Utilizing quantitative certificate of analysis data to assess the amount of excipient lot-to-lot variability sampled during drug product development. *Pharmaceutical Development and Technology, 18*(2), 333–342. https://doi.org/10.3109/10837450.2011.604784.

Landín, M., Martínez-Pacheco, R., Gómez-Amoza, J. L., Souto, C., Concheiro, A., & Rowe, R. C. (1993a). Effect of country of origin on the properties of microcrystalline cellulose. *International Journal of Pharmaceutics, 91*(2), 123–131. https://doi.org/10.1016/0378-5173(93)90331-9.

Landín, M., Martínez-Pacheco, R., Gómez-Amoza, J. L., Souto, C., Concheiro, A., & Rowe, R. C. (1993b). Influence of microcrystalline cellulose source and batch variation on the tabletting behavior and stability of prednisone formulations. *International Journal of Pharmaceutics, 91*(2), 143–149. https://doi.org/10.1016/0378-5173(93)-90333-B.

Landín, M., Martínez-Pacheco, R., Gómez-Amoza, J. L., Souto, C., Concheiro, A., & Rowe, R. C. (1994). Dicalcium phosphate dihydrate for direct compression: Characterization and intermanufacturer variability. *International Journal of Pharmaceutics, 109*(1), 1–8. https://doi.org/10.1016/0378-5173(94)90115-5.

Li, J., & Wu, Y. (2014). Lubricants in pharmaceutical solid dosage forms. *Lubricants, 2*(1), 21–43. https://doi.org/10.3390/lubricants2010021.

Markl, D., & Zeitler, J. A. (2017). A review of disintegration mechanisms and measurement techniques. *Pharmaceutical Research, 34*(5), 890–917. https://doi.org/10.1007/s11095-017-2129-z.

Miller, T. A., & York, P. (1985). Frictional assessment of magnesium stearate and palmitate lubricant powders. *Powder Technology, 44*(3), 219–226. https://doi.org/10.1016/0032-5910(85)85003-8.

Miyazaki, T., Sivaprakasam, K., Tantry, J., & Suryanarayanan, R. (2009). Physical characterization of dibasic calcium phosphate dihydrate and anhydrate. *Journal of Pharmaceutical Sciences, 98*(3), 905–916. https://doi.org/10.1002/jps.21443.

Mohamed, F. A., Roberts, M., Seton, L., Ford, J. L., Levina, M., & Rajabi-Siahboomi, A. R. (2015). The effect of HPMC particle size on the drug release rate and the percolation threshold in extended-release mini-tablets. *Drug Development and Industrial Pharmacy, 41*(1), 70–78. https://doi.org/10.3109/03639045.2013.845843.

Morin, G., & Briens, L. (2013). The effect of lubricants on powder flowability for pharmaceutical application. *AAPS PharmSciTech, 14*(3), 1158–1168. https://doi.org/10.1208/s12249-013-0007-5.

Narang, A. S., & Boddu, S. H. (2015). *Excipient applications in formulation design and drug delivery* (pp. 1–681). Springer International Publishing. https://doi.org/10.1007/978-3-319-20206-8.

Narang, A. S., Desai, D., & Badawy, S. (2012). Impact of excipient interactions on solid dosage form stability. *Pharmaceutical Research, 29*(10), 2660–2683. https://doi.org/10.1007/s11095-012-0782-9.

Narang, A. S., Mantri, R. V., & Raghavan, K. S. (2017). Excipient compatibility and functionality. *Developing solid oral dosage forms* (2nd ed., pp. 151–179). Academic Press.

Nokhodchi, A. (2005). An overview of the effect of moisture on compaction and compression. *Pharmaceutical Technology*, 46–66. https://doi.org/10.1002/jps.20805.

Obae, K., Iijima, H., & Imada, K. (1999). Morphological effect of microcrystalline cellulose particles on tablet tensile strength. *International Journal of Pharmaceutics, 182*(2), 155–164.

Parikh, D. M. (2010). *Handbook of pharmaceutical granulation technology* (3rd ed.). Informa Healthcare.

Patel, S., Kaushal, A. M., & Bansal, A. K. (2006). Compression physics in the formulation development of tablets. *Critical Reviews in Therapeutic Drug Carrier Systems, 23*(1), 1–65.

Pawelczyk, E., & Opielewicz, M. (1978). Drug decomposition kinetics. XLIX. Kinetics of autooxidation of drotaverine hydrochloride in the solid phase. *Acta Poloniae Pharmaceutica—Drug Research, 35*(3), 311–319.

Pingali, K., Mendez, R., Lewis, D., Michniak-Kohn, B., Cuitino, A., & Muzzio, F. (2011). Mixing order of glidant and lubricant influence on powder and tablet properties. *International Journal of Pharmaceutics, 409*(1–2), 269–277. https://doi.org/10.1016/j.ijpharm.2011.02.032.

Quodbach, J., & Kleinebudde, P. (2015). Performance of tablet disintegrants: Impact of storage conditions and relative tablet density. *Pharmaceutical Development and Technology, 20*(6), 762–768. https://doi.org/10.3109/10837450.2014.920357.

Rajala, R., & Laine, E. (1995). The effect of moisture on the structure of magnesium stearate. *Thermochimica Acta, 248*(C), 177–188. https://doi.org/10.1016/0040-6031(94)01950-l.

Reading, S. J., & Spring, M. S. (1984). The effects of binder film characteristics on granule and tablet properties. *Journal of Pharmacy and Pharmacology, 36*(7), 421–426. https://doi.org/10.1111/j.2042-7158.1984.tb04417.x.

Rohera, B. D., & Zahir, A. (1993). Granulations in a fluidized-bed: Effect of binders and their concentrations on granule growth and modeling the relationship between granule size and binder concentration. *Drug Development and Industrial Pharmacy, 19*(7), 773–792. https://doi.org/10.3109/03639049309062982.

Rojas, J. (2013). Effect of polymorphism on the particle and compaction properties of microcrystalline cellulose. Eds. T. V. D. Ven, & L. Godbout (Eds.), *Cellulose: medical, pharmaceutical and electronic applications*. InTech.

Rojas, J., Guisao, S., & Ruge, V. (2012). Functional assessment of four types of disintegrants and their effect on the spironolactone release properties. *AAPS PharmSciTech, 13*(4), 1054–1062. https://doi.org/10.1208/s12249-012-9835-y.

Rowe, R. C. (1990). Correlation between predicted binder spreading coefficients and measured granule and tablet properties in the granulation of paracetamol. *International Journal of Pharmaceutics, 58*(3), 209–213. https://doi.org/10.1016/0378-5173(90)90197-C.

Rowe, R. C. S., Paul, J., & Quinn, M. E. (2009). *Handbook of pharmaceutical excipients*: 6. Pharmaceutical Press; American Pharmacists Association.

Rudnic, E. M., Kanig, J. L., & Rhodes, C. T. (1985). Effect of molecular structure variation on the disintegrant action of sodium starch glycolate. *Journal of Pharmaceutical Sciences, 74*(6), 647–650.

Rudnic, E. M., Lausier, J. M., Chilamkurti, R. N., & Rhodes, C. T. (1980). Studies of the utility of cross linked polyvinlpolypyrrolidine as a tablet disintegrant. *Drug Development and Industrial Pharmacy, 6*(3), 291–309. https://doi.org/10.3109/03639048009051943.

Sakr, W. F., Ibrahim, M. A., Alanazi, F. K., & Sakr, A. A. (2012). Upgrading wet granulation monitoring from hand squeeze test to mixing torque rheometry. *Saudi Pharmaceutical Journal, 20*(1), 9–19. https://doi.org/10.1016/j.jsps.2011.04.007.

Saripella, K. K., Mallipeddi, R., & Neau, S. H. (2014). Crospovidone interactions with water. II. Dynamic vapor sorption analysis of the effect of polyplasdone particle size on its uptake and distribution of water. *International Journal of Pharmaceutics, 475*(1–2), 174–180. https://doi.org/10.1016/j.ijpharm.2014.08.040.

Shah, N., et al. (1996). *Effect of processing techniques in controlling the release rate and mechanical strength of hydroxypropyl methylcellulose based hydrogel matrices*: 42. Elsevier.

Shah, U., & Augsburger, L. (2002). Multiple sources of sodium starch glycolate, NF: Evaluation of functional equivalence and development of standard performance tests. *Pharmaceutical Development and Technology, 7*(3), 345–359. https://doi.org/10.1081/PDT-120005731.

Shanmugam, S. (2015). Granulation techniques and technologies: recent progresses. *BioImpacts, 5*(1), 55–63. https://doi.org/10.15171/bi.2015.04.

Shi, L., Feng, Y., & Sun, C. C. (2011). Initial moisture content in raw material can profoundly influence high shear wet granulation process. *International Journal of Pharmaceutics, 416*(1), 43–48. https://doi.org/10.1016/j.ijpharm.2011.05.080.

Shlieout, G., Arnold, K., & Muller, G. (2002). Powder and mechanical properties of microcrystalline cellulose with different degrees of polymerization. *AAPS PharmSciTech, 3*(2), E11. https://doi.org/10.1208/pt030211.

Smallenbroek, A. J., Bolhuis, G. K., & Lerk, C. F. (1981). The effect of particle size of disintegrants on the disintegration of tablets. *Pharmaceutisch Weekblad, 3*(1), 1048–1051. https://doi.org/10.1007/bf02193321.

Suzuki, T., & Nakagami, H. (1999). Effect of crystallinity of microcrystalline cellulose on the compactability and dissolution of tablets. *European Journal of Pharmaceutics and Biopharmaceutics, 47*(3), 225–230.

Thakur, A. B., Morris, K., Grosso, J. A., Himes, K., Thottathil, J. K., Jerzewski, R. L., Wadke, D. A., & Carstensen, J. T. (1993). Mechanism and kinetics of metal ion-mediated degradation of fosinopril sodium. *Pharmaceutical Research, 10*(6), 800–809. https://doi.org/10.1023/A:1018940623174.

Thoorens, G., Krier, F., Leclercq, B., Carlin, B., & Evrard, B. (2014). Microcrystalline cellulose, a direct compression binder in a quality by design environment: A review. *International Journal of Pharmaceutics, 473*(1–2), 64–72. https://doi.org/10.1016/j.ijpharm.2014.06.055.

Uzunarslan, K., & AkbugÆa, J. (1991). The effect of moisture on the physical characteristics of ranitidine hydrochloride tablets prepared by different binders and techniques. *Drug Development and Industrial Pharmacy, 17*(8), 1067–1081. https://doi.org/10.3109/03639049109043845.

van den Dries, K., & Vromans, H. (2002). Relationship between inhomogeneity phenomena and granule growth mechanisms in a high-shear mixer. *International Journal of Pharmaceutics, 247*(1-2), 167–177. https://doi.org/10.1016/s0378-5173(02)00419-2.

van der Watt, J. G. (1987). The effect of the particle size of microcrystalline cellulose on tablet properties in mixtures with magnesium stearate. *International Journal of Pharmaceutics, 36*(1), 51–54. https://doi.org/10.1016/0378-5173(87)90235-3.

van Kamp, H. V., Bolhuis, G. K., Kussendrager, K. D., & Lerk, C. F. (1986). Studies on tableting properties of lactose. IV. Dissolution and disintegration properties of different types of crystalline lactose. *International Journal of Pharmaceutics, 28*(2-3), 229–238. https://doi.org/10.1016/0378-5173(86)90249-8.

van Kamp, H. V., Bolhuis, G. K., & Lerk, C. F. (1983). Improvement by super disintegrants of the properties of tablets containing lactose, prepared by wet granulation. *Pharmaceutisch Weekblad, Scientific Edition, 5*(4), 165–171.

Viriden, A., Larsson, A., & Wittgren, B. (2010). The effect of substitution pattern of HPMC on polymer release from matrix tablets. *International Journal of Pharmaceutics, 389*(1–2), 147–156. https://doi.org/10.1016/j.ijpharm.2010.01.029.

Vromans, H., Bolhuis, G. K., Lerk, C. F., & Kussendrager, K. D. (1987a). Studies on tableting properties of lactose. VIII. The effect of variations in primary particle size, percentage of amorphous lactose and addition of a disintegrant on the disintegration of spray-dried lactose tablets. *International Journal of Pharmaceutics, 39*(3), 201–206. https://doi.org/10.1016/0378-5173(87)90217-1.

Vromans, H., Bolhuis, G. K., Lerk, C. F., van de Biggelaar, H., & Bosch, H. (1987b). Studies on tableting properties of lactose. VII. The effect of variations in primary particle size and percentage of amorphous lactose in spray dried lactose products. *International Journal of Pharmaceutics, 35*(1), 29–37. https://doi.org/10.1016/0378-5173(87)90071-8.

Wang, J., Davidovich, M., Desai, D., Bu, D., Hussain, M., & Morris, K. (2010a). Solid-state interactions of a drug substance and excipients and their impact on tablet dissolution: a thermal-mechanical facilitated process-induced transformation or PIT. *Journal of Pharmaceutical Sciences, 99*(9), 3849–3862. https://doi.org/10.1002/jps.22222.

Wang, J., Wen, H., & Desai, D. (2010b). Lubrication in tablet formulations. *European Journal of Pharmaceutics and Biopharmaceutics, 75*(1), 1–15. https://doi.org/10.1016/j.ejpb.2010.01.007.

Watt (1987). The effect of the particle size of microcrystalline cellulose on tablet properties in mixtures with magnesium stearate. *International Journal of Pharmaceutics, 36*(1), 90235. https://doi.org/10.1016/0378-51731987.

Westerhuis, J. A., de Haan, P., Zwinkels, J., Jansen, W. T., Coenegracht, P. J. M., & Lerk, C. F. (1996). Optimization of the composition and production of mannitol/microcrystalline cellulose tablets. *International Journal of Pharmaceutics, 143*(2), 151–162. https://doi.org/10.1016/S0378-5173(96)04699-6.

Wikberg, M., & Alderborn, G. (1993). Compression Characteristics of Granulated Materials. VII. The effect of intragranular binder distribution on the compactibility of some lactose granulations. *Pharmaceutical Research, 10*(1), 88–94. https://doi.org/10.1023/A:1018929214629.

Wu, Y., Levons, J., Narang, A. S., Raghavan, K., & Rao, V. M. (2011). Reactive impurities in excipients: Profiling, identification and mitigation of drug-excipient incompatibility. *AAPS PharmSciTech, 12*(4), 1248–1263. https://doi.org/10.1208/s12249-011-9677-z.

Zarmpi, P., Flanagan, T., Meehan, E., Mann, J., & Fotaki, N. (2017). Biopharmaceutical aspects and implications of excipient variability in drug product performance. *European Journal of Pharmaceutics and Biopharmaceutics, 111*, 1–15. https://doi.org/10.1016/j.ejpb.2016.11.004.

Zhang, L., & Mao, S. (2017). Application of quality by design in the current drug development. *Asian Journal of Pharmaceutical Sciences, 12*(1), 1–8. https://doi.org/10.1016/j.ajps.2016.07.006.

Zhao, N., & Augsburger, L. L. (2005). The influence of swelling capacity of superdisintegrants in different pH media on the dissolution of hydrochlorothiazide from directly compressed tablets. *AAPS PharmSciTech, 6*(1), E120. https://doi.org/10.1208/pt060119.

Zhao, N., & Augsburger, L. L. (2006). The influence of product brand-to-brand variability on superdisintegrant performance: A case study with croscarmellose sodium. *Pharmaceutical Development and Technology, 11*(2), 179–185. https://doi.org/10.1080/10837450600561281.

Zhou, D., Law, D., Reynolds, J., Davis, L., Smith, C., Torres, J. L., et al. (2014). Understanding and managing the impact of HPMC variability on drug release from controlled release formulations. *Journal of Pharmaceutical Sciences, 103*(6), 1664–1672. https://doi.org/10.1002/jps.23953.

Further reading

FMC BioPolymer Technical Bulletin (2015). *Impact of cellulose source on properties and disintegration performance of croscarmellose sodium.*

CHAPTER 9

Binders in wet granulation

Thomas Dürig and Kapish Karan

Life Sciences R&D Innovation, Ashland Inc, Wilmington, DE, United States

9.1 Introduction

Wet granulation is the most commonly practiced unit process for the agglomeration of pharmaceutical powders and has long been a central part of solid dosage form manufacturing. It is an essential enabling technology allowing poorly flowable, poorly compressible, or poorly dissolving active ingredients to be incorporated in elegant, safe, and effective drug products such as tablets and capsules.

Mostly this is achieved by intimately combining a plastic binder in solution form with the active pharmaceutical ingredient (API) and other excipients, thus improving powder flowability and tablet compactibility. Wet granulation is also useful to ensure uniform drug distribution and to prevent powder segregation, and therefore is frequently employed in the processing of low-dose, high-potency drugs. Lastly a traditional use of wet granulation, frequently practiced in the food and nutraceutical industry, is instantizing or hydrophilizing. Here wet granulation with a suitable hydrophilic binder improves the solubility or dispersibility of powders or tablets in water.

As a technological concept, wet granulation has remained relevant for more than a century and has undergone significant innovation and adaptation to evolving industry needs. Traditionally wet granulation was practiced as a low-shear technique using planetary, oscillator, or ribbon blender granulators. Beginning in the 1980s, however, more efficient, high-shear granulation processes were introduced, while simultaneously fluid bed–granulation became more prominent. Most recently wet granulation process improvements have involved continuous processing and twin-screw extrusion.

Although it is possible to perform wet granulations with only a solvent and the active ingredients with or without a filler (for example, glucosamine and chondroitin can be granulated successfully with only water), most active ingredients and fillers lack both wet binding ability to form a coherent mass and the high degree of plasticity needed to ensure homogeneous, dense, dry granules with low friability, and excellent compactibility. In most cases it is necessary to add a wet binder, or granulating agent, with appropriate surface wetting ability to ensure excellent adhesion and cohesion between interparticulate surfaces in the wet state, as well as excellent plasticity, compactibility, and binding ability after the granules are dried and reduced in size.

The ideal binder therefore will have a high degree of surface wetting and spreadability and a high degree of wet adhesion (strong liquid bridges in the wet granules) to allow the formation of agglomerates, while also possessing plasticity in the dry state to overcome unfavorable powder flow and mechanical properties. Such a binder should yield dense, uniform granules with low friability and

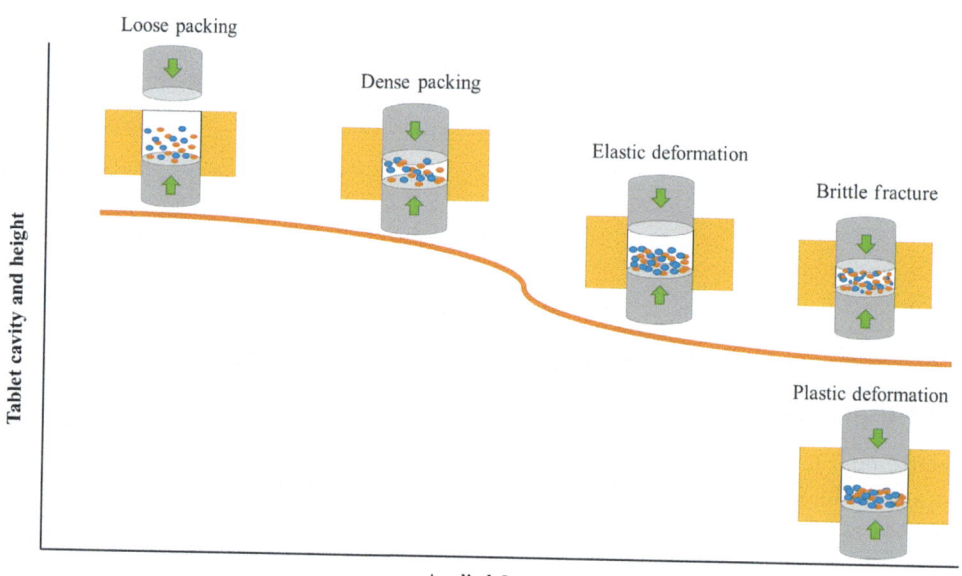

FIGURE 9.1

Mechanisms of consolidation for tableting materials.

From Rudnic and Schwartz (1995).

a high degree of compactibility, while minimizing the amount of applied force required to form strong, dense tablets.

An ideal binder would enable the packing and deformation of granulated powder through the various stages of powder consolidation during tablet compaction, are shown in Fig. 9.1.

Traditionally binder selection has been empirical, often based on the formulators' experience and subjective preferences. Pharmaceutical powder technology, however, has made significant advances in mechanistic and quantitative particle engineering and materials science, therefore facilitating a more logical approach to binder and excipient selection.

The purpose of this chapter is to review the physical–chemical properties of the major binders in current pharmaceutical and commercial use, and to provide a practical and logical guide to wet binder selection and use in the context of physical–chemical properties of the substrate and the binder. Recent advances in the mechanistic understanding of granulation technology will be discussed in detail, specifically the importance of selecting binders with a focus on surface energetics, wetting, spreadability, and thermomechanical properties of the system.

9.2 Physical–chemical properties of common wet granulation binders

Traditional materials that have been used as wet binders include natural polymers such as gelatin, gum acacia, gum tragacanth, starch, guar gum, and sugars (such as sucrose and glucose). Of these, only gum acacia and sucrose are still used occasionally, with gum acacia being found frequently in nutraceutical granulations.

For the most part current pharmaceutical practice favors the use of less variable synthetic and semisynthetic materials, such as various cellulose derivatives, vinyl pyrrolidone derivatives, and modified starches. Among the most frequently used binders are povidone (also known as polyvinyl pyrrolidone [PVP]), copovidone (PVP-polyvinyl acetate [PVA] copolymer), partially gelatinized starch (PGS), and various cellulose ethers such as hydroxypropyl cellulose (HPC), methyl cellulose (MC), hypromellose (also known as hydroxypropylmethyl cellulose [HPMC]), and occasionally ethyl cellulose (EC) and sodium carboxymethyl cellulose (NaCMC). Table 9.1 lists some of the most frequently used wet binders, typical use levels, and suitable solvents.

9.2.1 Hydroxypropyl cellulose

HPC is a cellulose ether in which hydroxyl groups on the cellulose backbone have been hydroxypropylated. It is manufactured by reacting alkali cellulose with propylene oxide at elevated pressure and temperature to yield a highly substituted cellulose ether, with 3.4 to 4.1 mol of hydroxypropyl substituent per mole of anhydro glucose backbone units (Ashland, 2001). Because of the high levels of hydroxypropylation (~70%), HPC is more plastic and relatively hydrophobic as compared to other water-soluble cellulose ethers. It is fully soluble in water and polar organic solvents, such as methanol, ethanol, isopropyl alcohol (IPA), and acetone. Solubility of HPC in water is temperature dependent; it is readily soluble at temperatures below the cloud point (the temperature below which the polymer starts to phase-separate, and two phases appear, thus becoming cloudy), which is around 45°C.

HPC is available commercially in different viscosity grades, with an average molecular weight (MW) grades ranging from 20 to 1500 kDa. The low MW grades most typically are used as binders (Table 9.2). HPC is a superior binder and has shown equivalent binder efficiency and good compactibility when added as a solution or in dry powdered form (Skinner & Harcum, 1998). Additionally for dry addition, fine particle size grades are preferred because of faster hydration and uniformity of mixing and distribution. The coarse grades are preferred for solution addition because they disperse more easily without lumping.

To prepare a lump-free dispersion, HPC can be dispersed in 50% of the required final volume of hot water (>60°C), and after 10 min of hydration the remaining water can be added cold while continuing to stir. Because of its high binder efficiency, HPC tends to be particularly well-suited for high-dose, difficult-to-compress tablets, in which only small amounts of binder can be added. In general use levels above 8% are inadvisable as they tend to cause excessive slowing of disintegration and dissolution times.

HPC is used widely in pharmaceutical applications (as a binder), in film coating, controlled-release matrix (higher MW grades), and melt extrusion. HPC has harmonized compendial status in the National Formulary of the United States Pharmacopeia (USP/NF), the European Pharmacopeia (pH.Eur.), Japanese Pharmacopeia (JP), and the Food Chemicals Codex (FCC).

9.2.2 Methyl cellulose

MC is the methyl ether of cellulose, produced by reacting methyl chloride and alkali cellulose. It contains 27.58% to 31.5% of methoxy groups. When compared to HPC, MC is less substituted by the weight of the polymer.

Table 9.1 Commonly used wet binders.

Binder	Typical use level	Comments	Inactive ingredient database IID Limits[a] (mg/dose)
Hydroxypropylcellulose (HPC)	2%–6%	Used with water, hydroalcoholic and neat polar organic solvents	95.0 mg
Methylcellulose (MC)	2%–10%	Used with water or hydroalcoholic solvents	68.0 mg
Sodium carboxymethylcellulose (NaCMC)	2%–5%	Used with water	50.0 mg
Hypromellose (HPMC)	2%–10%	Used with water or hydroalcoholic solvents	480.0 mg
Ethylcellulose (EC)	2%–10%	Used with polar and nonpolar organic solvents, not soluble if water exceeds 20% of total solvent. Hydrophobic coating can slow down drug release for low-soluble drugs, best used for high-dose, highly soluble drugs and moisture-sensitive drugs	188.8 mg
Povidone (PVP)	2%–10%	Used with water, hydroalcoholic and neat polar organic solvents. Ultralow viscosity grades, allow high solution concentrations (20%)	300.0 mg
Copovidone (PVA–PVP)	2%–8%	Used with water and hydroalcoholic solvents. More plastic than PVP	853.8 mg
Pregelatinized starch (PGS)	5%–15%	Can be used only with water, also acts as a disintegrant, effective use levels are mostly higher than other binders	345.95 mg

[a] Data accessed in July 2017. For the latest IID or IIG limits: https://www.accessdata.fda.gov/scripts/cder/iig/.

9.2 Physical–chemical properties of common wet granulation binders

Table 9.2 Selected commercial binder grades.

Binder	Trade name/grade/supplier	Nominal viscosity
Hydroxypropylcellulose	Klucel Hydroxypropylcellulose ELF, EF and LF pharm, also available as fine particle grades EXF and LXF pharm	2% viscosities 5, 8, and 12 cps, respectively
	Nisso HPC SL and L also available as (F) fine grades	2% viscosities 5 and 8 cps, respectively
Hypromellose	Methocel E3, E5, E6 and E15 Premium LV Hypromellose	2% viscosities 3, 5, 6, and 15 cps, respectively
	Pharmacoat 603, 605, 606, and 615 hypromellose	2% viscosities 3, 5, 6, and 15 cps, respectively
Methylcellulose	Methocel A15 premium LV methylcellulose	2% viscosity 15 cps
	Benecel A15 LV pharm methylcellulose	2% viscosity 15 cps
Ethylcellulose	Aqualon ethylcellulose N7, N10, N14 and N22 pharm	5% viscosities 4, 7, 10, 14, and 22 cps, respectively
	Ethocel standard premium ethylcellulose NF	5% viscosities 4, 7, 10, and 20 cps, respectively
Sodium carboxymethylcellulose	Aqualon sodium carboxymethylcellulose 7L2P and 7LF pharm	2% viscosities 20 and 50 cps, respectively
	Blanose sodium carboxymethylcellulose 7L2P and 7LF pharm	2% viscosities 20 and 50 cps, respectively
Povidone	Kollidon 25, 30, and 90F pharm	5% viscosities 2, 2.5, and 55 cps, respectively
	Plasdone K12, K17, K25, K29/32 and K90 povidone	5% viscosities 1, 1.8, 2.0, 2.5, and 55, respectively
Copovidone	Kollidon VA 64 copovidone	5% viscosity 2.5 cps
	Plasdone S630 copovidone	5% viscosity 2.5 cps
Pregelatinized starch	Starch 1500 partially pregelatinized starch	N/A
	Lycatab PGS partially pregelatinized starch	N/A

Klucel, Benecel, Aqualon, and Blanose are registered trademarks of Ashland LLC. Nisso is a registered trademark of Nippon Soda Company. Pharmacoat is a registered trademark of Shin-Etsu Corporation. Methocel and Ethocel are trademarks of the Dow Chemical Company. Kollidon is a registered trademark of BASF Corporation. Plasdone is a registered trademark of ISP. Starch 1500 is a registered trademark of BPSI. Lycatab is a registered trademark of Roquette Frères.

MC is listed in the USP/NF, pH.Eur., JP, and FCC. MC is used widely in oral solid pharmaceutical formulations as a binder, coating agent, and as a controlled release matrix.

MC is soluble in water (up to 55°C). This indicates slightly higher water solubility than HPC. Its aqueous solution exhibits thermal gelation properties at elevated temperatures. A hydroalcoholic solution of MC can be prepared with polar organic solvents such as ethanol, methanol, and IPA, as long as a small amount of water (10% v/v) is added as a cosolvent.

MC is available in a wide range of MW grades, but almost exclusively the low MW grade with a nominal viscosity of 15 cps at 2% w/v concentration is used as a tablet binder (Table 9.2). Although MC can be added as a dry powder or as a solution, it generally is more effective when predissolved and added as a solution (Skinner & Harcum, 1998). Low MW MC is a versatile binder with good plastic flow and wetting ability. It produces granulations that compress easily and tablets with moderate hardness. Aqueous solutions of MC can be prepared in analogous fashion as described above for HPC.

9.2.3 Hypromellose

HPMC is a partly *O*-methylated and O-2-hydroxypropylated cellulose ether. Manufactured by reacting alkali cellulose with methyl chloride and propylene oxide, this polymer is available in various substitution ratios and MW grades. HPMC is widely used in solid dosage forms as a binder, a film coating agent, and a controlled release matrix.

The binder properties of HPMC are comparable to MC, with the exception that it is somewhat more hydrophilic. As a binder, primarily low viscosity grades with substitution type 2910 (28%–30% methoxy groups by weight and 4%–12% hydroxypropyl groups by weight, often described as the *E*-grade) are used (Table 9.1). Concentrations of 2%–5% (w/w) can be used as a binder in either wet or dry addition processes, but is less efficient in the latter form (Skinner & Harcum, 1998).

HPMC type 2910 has a cloud-point of about 65°C, which necessitates higher water temperatures for solution preparation. To prepare an aqueous solution, HPMC is first hydrated in the required amount of hot water (>65°C) with vigorous stirring, followed by addition of cold water to make up the volume. Hydroalcoholic solutions, with a minimum of 10% v/v water, or mixtures of other water-miscible solvents such as glycerin also can be used to solvate HPMC. HPMC is listed in the USP/NF, pH.Eur., JP, and the FCC.

9.2.4 Ethyl cellulose

Ethyl cellulose (EC) is a partly *O*-ethylated cellulose ether derivative. It is available in a variety of grades, which differ in viscosity. EC is prepared by the reaction of ethyl chloride with alkali cellulose. During this reaction the substitution of ethoxyl groups is controlled to a range of 44.5% and higher, which is slightly lower than (2.0 and higher) the theoretical maximum of 3.0. Within this range, four ethyoxyl types are defined for the EC-G-type (44.5%–45.5%), K-type (45.5%–46.8%), N-type (47.5%–49.0%), and T-type (49.0% and higher).

It is insoluble in water but soluble in a variety of solvents. EC is widely used in pharmaceutical formulations as a binder, coating agent for modified release tablets and multiparticulates, taste masking agent, and in matrix tablets to achieve modified release.

Low viscosity grades of N (48.0%–49.5% wt. ethoxyl) and T (49.6%–51.0% wt. ethoxyl) types are used as binders in concentrations of 2% to 10%. Granulations using ethyl cellulose result in softer granules, which compress into tablets that disintegrate easily. The dissolution of the active ingredient, however, can be retarded because EC is insoluble in water.

EC can be used in the dry form or in a hydroalcoholic solution for wet granulation. Ethyl cellulose is a good nonaqueous binder for water-sensitive formulations but has been largely replaced by other environmentally friendly binder systems.

9.2.5 Sodium carboxymethyl cellulose

NaCMC is the sodium salt of carboxymethyl cellulose, an anionic derivative. It is widely used in oral, ophthalmic, injectable, and topical pharmaceutical formulations. For solid dosage forms it is used primarily as a binder or matrix former.

Pharmaceutical grades of NaCMC are available commercially at degree of substitution (DS) values of 0.7, 0.9, and 1.2, with a corresponding sodium content of 6.5% to 12% wt. It is also available in several different viscosity grades. NaCMC is highly soluble in water at all temperatures, forming clear solutions. Its solubility depends on its DS.

NaCMC, when used as a binder, yields softer granules that have good compressibility, forming tough tablets of moderate strength. NaCMC, being highly hygroscopic, can absorb a large quantity of water (>50%) at elevated relative humidity conditions. Hence, the tablets tend to harden with age.

9.2.6 Povidone

PVP is a linear polymer consisting of N-vinyl-2 pyrrolidone monomers. It is made with different degrees of polymerization, which results in a wide range of MWs ranging from 2.5 to 3000 kDa.

PVP is recognized as a versatile excipient and is used as a tablet binder, complexing agent, film former, suspending agent, and solubilizer. PVP is water-soluble, with the maximum concentration being limited only by the solution viscosity. Being nonionic in nature, the viscosity of povidone in an aqueous solution is unaffected by pH. The low MW grades are often used as tablet binders. Nonetheless, high MW grades have also been reported to have very high binder efficiency.

In tablet formulations, povidone is used as a wet granulation binder, mostly added in solution, at recommended use levels of 2% to 5% wt. (Skinner & Harcum, 1998), although higher use levels of up to 10% wt. have been reported for poorly compactible and challenging formulations. Povidone is highly soluble in water and freely soluble in many polar organic solvents, such as ethanol, methanol, IPA, and butanol. It is insoluble in nonpolar organic solvents. PVP generally is used in the form of a solution, where its low viscosity allows solid concentrations as high as 15% to 20%.

PVP is highly hygroscopic. At 50% RH, the typical equilibrium moisture content exceeds 15% by weight (Fig. 9.2). It is advisable, therefore, to take precautions against uncontrolled and unnecessary exposure to atmospheric moisture. Tablets made with PVP as a binder tend to show hardening and dissolution slowdown on storage. PVP is listed in the USP/NF, pH.Eur., and JP (BASF, 2008).

9.2.7 Copovidone

Copovidone (PVP–PVA) is the random, linear copolymer of N-vinyl-2-pyrrolidone and vinyl acetate in a 60:40 ratio, making it a derivative of PVP. The addition of vinyl acetate to the polymer chain reduces

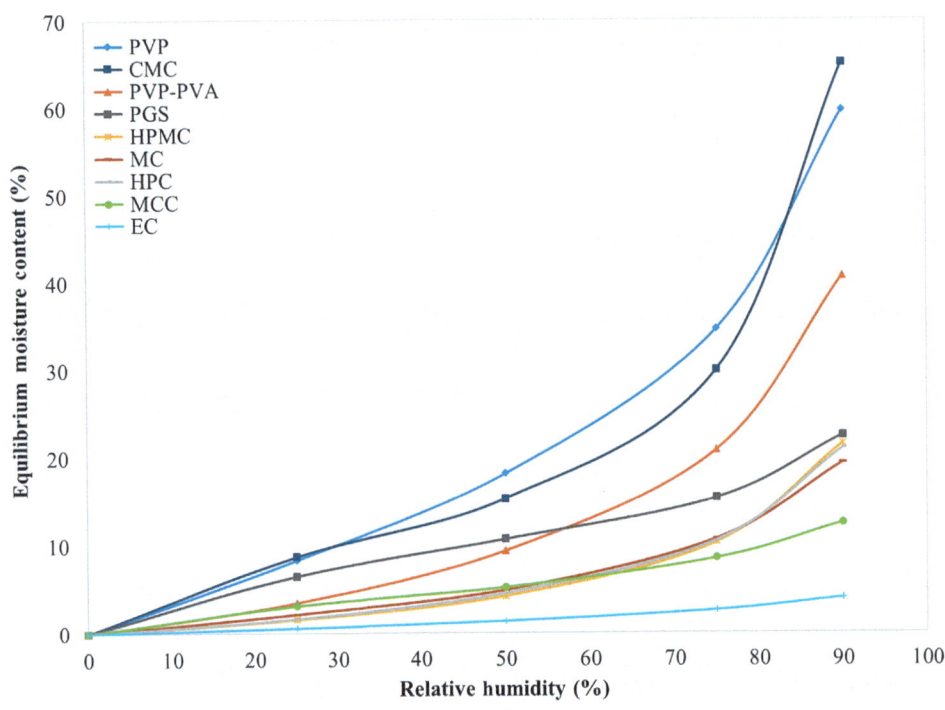

FIGURE 9.2

Equilibrium moisture isotherms for selected common wet granulation binders and excipients.

the hydrophilicity and hygroscopicity of the PVP homopolymer. At 50% RH, the typical equilibrium moisture content is approximately 10% (Fig. 9.2).

PVP–PVA can be used in wet granulation either in dissolved form or added dry to the powder blend, followed by wet massing. Binder effectiveness is approximately equivalent for these two methods of incorporation.

Copovidone is soluble in water and polar organic solvents and is listed in the USP/NF, pH.Eur., and JP. The addition of vinyl acetate also has a positive effect on the plasticity of the polymer, thus lowering the glass transition temperature and improving compactibility and adhesiveness. In addition to being used as a wet and dry binder, PVA–PVP can also be incorporated into film coating formulations together with HPMC (ISP, 2009).

9.2.8 Starch and modified starches

Starch is a polymeric carbohydrate consisting of glucose monomers linked by glycosidic bonds. It is a generally recognized as safe (GRAS) listed material with monographs in the USP/NF, pH.Eur., and JP. Traditionally, it was one of the most widely used tablet binders, but in today's industrial practice, pregelatinized starch or other binders are used in its place.

Starch is obtained from plant sources, such as maize, potato, wheat, rice, and tapioca. Maize and potato are major sources of excipient-grade starch. It is not soluble in either cold water or alcohol.

Traditionally it is used by gelatinizing in hot water to form a paste. Starch paste can be prepared by wetting starch with cold water, followed by the addition of boiling water with continuous stirring until a translucent paste is formed. This paste can be diluted further with cold water to achieve an appropriate concentration.

Binder use levels for starch are relatively higher (5%–25%) as compared to synthetic binders. The high viscosity of starch paste can make efficient binder distribution and substrate wetting problematic, generating soft and friable granules. However it produces tablets that tend to disintegrate easily.

9.2.8.1 Pregelatinized starch

PGS is a modified starch with multifunctional benefits. It can be used as a binder, disintegrant, and diluent in tablet formulations. PGS is manufactured by rupturing all or part of the native starch granules using chemical and mechanical processing. This process enhances cold-water solubility and improves compactibility and flowability. It is available in fully or partially pregelatinized forms (Colorcon, 1999). The degree of pregelatinization determines its solubility in cold water.

As a binder in wet granulation, PGS is typically used in solution. It can also be used dry, but this reduces its efficiency significantly. Furthermore, its use levels (15%–20% w/w) are usually higher relative to synthetic binders.

It is incompatible with organic solvents and, thus is used only in aqueous binder systems. PGS can be used as a stabilizer or moisture sequestrant because it holds water in two states: bound (e.g., water of hydration) and free water, with only a portion of the sorbed water available as free water.

Partially pregelatinized starch is the most frequently used form of PGS, but fully pregelatinized starch is also available. Commercial, partially pregelatinized starch typically has around 10%–20% pregelatinized or water-soluble content, which makes it useful for wet granulation. The cold water-soluble part acts as a binder, while the remainder aids tablet disintegration. PGS is found in the USP/NF, pH.Eur., and JP.

9.2.9 Gum acacia

Gum acacia, also called gum arabic, is a naturally occurring gum extracted from the hardened exudates of plants *Acacia senegal* and *Acacia seyal*. Commercially available gum acacia is largely sourced from trees in the Sahel region of Africa. It is available in powdered, granular, and spray-dried forms. It is a complex mixture of sugars and hemicellulose. It is commonly used as an emulsion stabilizer in the food industry.

In granulation, gum acacia is used in an aqueous solution or added dry. It is an exceptional binder as it forms very strong tablets and granules. Disintegration and dissolution, however, are often impeded. In today's market, acacia is used in nutritional supplements. Because of its incompatibilities with a number of substances, large variability, and sporadic supply, however it is seldom used in the pharmaceutical industry.

9.3 Important considerations for binder selection and use

9.3.1 Use levels and binder efficiency

Typical binder use levels are given in Table 9.1; however binder use levels tend to be drug and formulation-specific and can differ significantly from the typical stated levels. In general for all binders, increased binder levels and higher binder solution concentrations lead to more viscous solutions, lower surface tension, lower wetting angles, and larger droplet size. This, in turn, leads to high wet adhesion forces and faster granule growth. After drying these granules are typically larger and less friable (Cutt et al., 1986).

Binder efficiency can be defined as the minimum binder level required to achieve a certain benchmark tablet strength. When comparing the efficiency of different binders, mechanical (plastic) and surface wetting properties tend to be of paramount importance. In terms of plastic properties, HPC and PVA–PVP tend to be most efficient. For hydrophobic surfaces, HPC is often most effective because of its better wetting ability. PGS is generally used at higher levels. These relationships will be discussed in more detail in later sections.

Formulators do not usually aim at producing the strongest tablets or granules possible; rather the minimum amount of binder needed to achieve acceptable tablet crushing strength or friability is usually desired. This results in optimal material cost, tablet size, strength (robustness), and avoids the risk of slowdown of drug release (which often is correlated with stronger, less porous tablets).

As a general rule tablet friability needs to be low enough to allow handling, packaging, and coating in commercial-scale coating pans (e.g., 48- to 72-in. coating pans). In general the friability of tablets that are 500 mg or less in weight should be lower than 0.8% w/w; and for larger tablets (e.g., 1000 mg or higher), friability below 0.3% should be targeted to allow for problem-free handling at commercial scale.

9.3.2 Stability and compatibility considerations for binders

Physical and chemical excipient compatibility with the API is a key consideration during preformulation to ensure that the chosen excipients do not compromise drug stability and safety (Waterman et al., 2004). During wet granulation binders are brought into intimate contact with APIs in the presence of water or other solvents. The presence of water and the elevated temperatures used during drying are common factors that can accelerate degradation kinetics if an incompatibility between the excipients and the API exists. Final drug product stability, therefore, is an important criterion for binder selection.

Excipients are generally designed to be inert. Direct chemical reactions between the excipient molecules and the drug molecules are relatively rare. A more frequently observed cause of drug-excipient interactions involves impurities that originate from the drug, the excipient, or the packaging materials (Waterman et al., 2004). Most such impurity-induced incompatibilities can be attributed to a small group of molecules, including water, electrophiles such as aldehydes, and their related carboxylic acids and peroxides.

9.3.3 Binder water content

Because of its omnipresence, environmental moisture is one of the major destabilizing factors for drug products. Water acts as a plasticizer, thus lowering glass transition temperatures and raising molecular mobility in solid-state systems. This can facilitate degradation reactions, such as oxidation or hydrolysis. Water can also introduce polymorphic transitions and recrystallization (Brady et al., 2009, chap. 9; Fitzpatrick et al., 2002; Kiekens et al., 2000; Uzunarslan & Akbuğa, 1991; Waterman et al., 2004), as well as tablet softening or hardening.

Water is associated with excipients and APIs in various states, including loosely held surface water, intermediate and tightly bound water (which is not freely available for reactions), and water of crystallization (which can be released only by transition in crystal form).

Equilibrium moisture sorption curves for various binders are shown in Fig. 9.2. In most cases the moisture content in a manufacturing environment controlled at or below 50% relative humidity is less than 10% and is in the intermediate or tightly bound nonmobile state.

An example where the hygroscopicity can be a problem involves PVP, which–at room temperature and 55% relative humidity–is in the glassy state. It undergoes transition to a rubbery state at higher humidities, resulting in higher molecular mobility and ultimately tablet hardening (Kiekens et al., 2000). In another example Fitzpatrick et al. (2002) reported a significant retardation in dissolution for the formulation of a new chemical entity wet granulated with PVP and stored at elevated humidity. Tablets stored at low humidity or made with HPC were not affected. The change in dissolution behavior was correlated with a lower glass transition temperature, caused by moisture sorption.

9.3.4 Stability considerations

9.3.4.1 Aldehydes and carboxylic acids

Trace amounts of low MW aldehydes and carboxylic acids are found in many excipients, including saccharides, polysaccharides, synthetic polymers, and unsaturated fats (Waterman et al., 2004). The most common reactive species in solid dosage forms are formaldehyde and its corresponding acidic degradation product, formic acid. Other reactive species of concern include acetaldehyde, glyoxal, furfural, glyoxalic acid, and acetic acid. Aldehydes and carboxylic acids are introduced mainly through auto-oxidation of the binder molecules, which, for example, leads to the formation of formaldehyde. Formaldehyde, because of its high volatility and reactivity, is oxidized rapidly to form formic acid.

The potential impact of aldehyde and carboxylic acid impurities should be considered for acid-labile drugs and those with nucleophilic functional groups, such as primary and secondary amines and hydroxyl groups (del Barrio et al., 2006; Waterman et al., 2004). Drug reactivity with these excipient impurities becomes particularly important for low-dose, highly potent drugs. Excipients used in wet granulation, in which the potential presence of formaldehyde and formic acid should be considered, include polysorbate, povidone, and polyethylene glycol (Nassar et al., 2004).

9.3.4.2 Peroxides

Peroxides are found in a number of excipients, including binders. The most common type of reactive peroxide moiety is organically bound peroxide (ROOH, where R is a carbon atom). Theoretically trace amounts of hydrogen peroxide (H_2O_2) are also possible, because hydrogen peroxide is sometimes used in the manufacturing of polymers, for instance, in MW reduction. H_2O_2, however, is very volatile and is therefore usually eliminated during unit operations such as drying, blending, and packaging. Peroxides can react directly with oxidation-sensitive drugs and generate free radicals, which can initiate oxidative chain reactions (Wasylaschuk et al., 2007; Waterman et al., 2004).

Excipients that historically were associated with peroxide impurities include polyethylene oxides, polysorbates, PVP, and PVP-related polymers. These excipients typically undergo auto-oxidation, meaning levels of impurities can grow upon aging. This phenomenon, however, is now well understood, and the major multinational manufacturers have taken steps to supply these products in a stabilized form

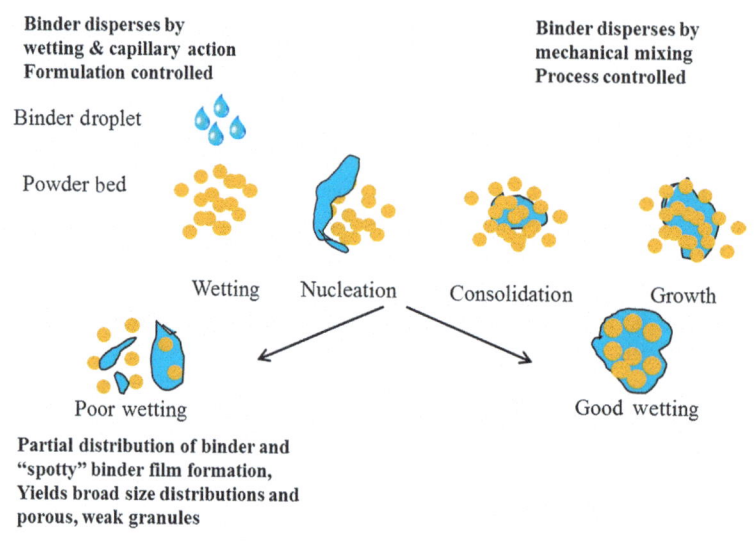

FIGURE 9.3

Schematic showing wetting, nucleation, consolidation, and granule growth process.

From Hapgood et al. (2002).

by inert packaging or through proprietary manufacturing steps. Scientists, however, should continue to treat less well-known, low-cost sources of these products with great caution.

9.3.5 Properties impacting binder mechanisms in wet granulation

9.3.5.1 Binder viscosity, wettability, and surface energetics

Wet granulation can be described as a combination of three rate processes (Fig. 9.3):

1. Powder bed wetting and binder distribution, leading to initial nuclei formation.
2. Consolidation and granule growth, because of collisions between granules and between granules and primary particles.
3. Attrition, as some of the granules deform and break under the shearing forces and affect the forces of the impeller and particle collisions (Iveson & Litster, 1998; Wells & Walker, 1983).

The ability of the binder solution to wet the porous powder substrate (by penetrating and spreading rapidly across host particles) and to distribute uniformly is a key factor in granulation process control. Poor wettability and binder spreading frequently lead to porous, weak granules with nonuniform binder distribution and wide particle size distribution.

During the early stages of nucleation and wetting the binder fluid disperses mainly by capillary action and depends on surface tension. The solid bridge strength of the initial granule nuclei and granule growth, however, is not only dependent on capillary forces, but also on viscous forces (i.e., the viscosity of the binder fluid). This pivotal stage is therefore strongly dependent on the powder formulation, substrate characteristics, and selection of a binder with favorable properties. The binder solution characteristics (surface tension and viscosity) and the substrate characteristics (surface free energy) determine the extent and the type of interaction between the binder fluid and the substrate, which can be characterized by substrate surface wetting (contact angle), spreading ability (spreading coefficient), work of adhesion, wet mass consistency, and the resultant granule characteristics (such as particle size distribution and granule strength) (Chitu et al., 2011; Litster & Ennis, 2004; Rowe, 1990). Granule strength also depends on the plastic properties of the binder.

After nucleus formation the mechanical shear forces of the mixer are responsible for the consolidation and growth phases of the granulation. Depending on the operating shear forces and wet strength of the granules, granule attrition can occur alongside growth and consolidation during the late phase. In Fig. 9.3, some of these processes are shown schematically.

Binder viscosity is an important factor in determining the granulation mechanism as well as granule strength. Generally a minimum viscosity is needed in order to successfully form granules. Moreover up to a certain critical value, increased binder viscosity will have a beneficial effect in terms of increased granule size and strength (Johansen & Schæfer, 2001; Keningley et al., 1997; Mills et al., 2000). Lower viscosity binders appear to induce granule growth mainly through a layering growth mechanism; however when using higher viscosity binders, coalescence appears to be the major mechanism.

As with viscosity, binder solution surface tension is an important factor in granulation (Krycer et al., 1983; Lusvardi et al., 2003; Zhang et al., 2002). Litster and Ennis (2004) studied the combined effect of viscosity and surface tension. They observed that the granulation mechanisms depended on capillary forces, modulated by surface tension, and viscous forces, as highlighted later:

A viscous capillary number, C_{avis}, was proposed as follows:

$$C_{avis} = \frac{\mu_L \times U}{\gamma_L} \tag{9.1}$$

where, U is the speed of particles, μ_L is the binder viscosity, and δ_L is the liquid binder surface tension. According to this relationship the viscous capillary number is proportional to viscosity and inversely proportional to surface tension.

A more practical and refined modified viscous capillary number, C^*_{avis}, was proposed by Benali et al. (2009):

$$C_{avis}^* = \frac{\mu_L \times \omega}{\gamma_L|(1 + \cos\theta)} \tag{9.2}$$

Here, impeller tip speed, ω, approximates particle velocity, and simple surface tension is replaced by the work of adhesion, $\gamma_L(1 + \cos\theta)$, which takes into account substrate polarity and surface energy.

These relationships were further examined by Chitu et al. (2011) using aqueous PVP and HPMC binder solutions at different concentrations and granulators at various scales. For low viscosity binder solutions, they found wet mass consistency and dry granule strength to be dependent on the work of adhesion. As viscosity increased, the amount of granulating liquid that was required decreased, indicating dominance of viscous forces and particle coalescence. Granule strength, however, tended to be weaker. Good adherence to the modified capillary viscous number was observed with a boundary of $C_{avis}^* < 0.8$, for which work of adhesion is the dominating parameter, and $C_{avis}^* \geq 00.8$, for which viscosity is the controlling parameter.

In practice most granulating fluids are of relatively low viscosity to allow pumping and spraying. These solutions, therefore, tend to be in the lower viscosity regime or close to the boundary condition.

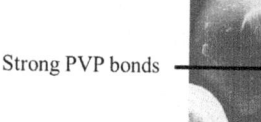 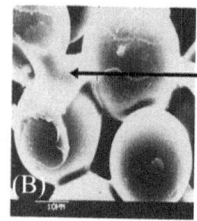

Strong PVP bonds — Weak PVP bonds resulting in adhesive failure

FIGURE 9.4

Fractured glass granules showing broken binder bonds. (A) Fracture within the PVP binder film for hydrophilic glass beads, and (B) adhesive failure of the PVP binder film on the hydrophobic glass beads because of weak bonds.

From Cutt et al. (1986).

Table 9.3 Glass bead granule properties treated with aqueous PVP binder.

Surface type	Failure strength (g)	Friability (%)
Hydrophilic	202	5.2
Hydrophobic	115	13.8

From Cutt et al. (1986)

Wettability and surface energetics can seldom be ignored in pharmaceutical granulations. These concepts are discussed in more detail later.

9.3.5.2 Surface properties and binder performance

The importance of substrate wetting was illustrated by Cutt et al. (1986) in their seminal study about hydrophilic and hydrophobic glass beads (Fig. 9.4 in Table 9.5). The water-based PVP solutions spread easily over the hydrophilic glass beads (high spreading coefficient), resulting in a high capillary adhesion force and continuous film between the particles. This led to strong, dense granules with binder film bonds at all contact points between substrate particles. When these strong granules were crushed, failure was evident within the binder film (cohesive failure) (Fig. 9.4A).

When hydrophobically modified glass beads were granulated with the same aqueous PVP solution, the binder film was deposited in a discontinuous manner (Table 9.3). The resultant granules were comparatively more porous and weaker than those granulations made with hydrophilic beads, and failure occurred at the surface of the glass beads, that is, between the substrate and the binder film (Fig. 9.4B).

9.3.5.3 Wetting fundamentals

Wettability of a solid by a liquid is a direct consequence of molecular interactions between the phases coming into contact (Lazghab et al., 2005). Adhesive forces required for wetting arise from different interatomic and intermolecular bonds that are established between atoms and molecules at the liquid/solid interface.

9.3 Important considerations for binder selection and use

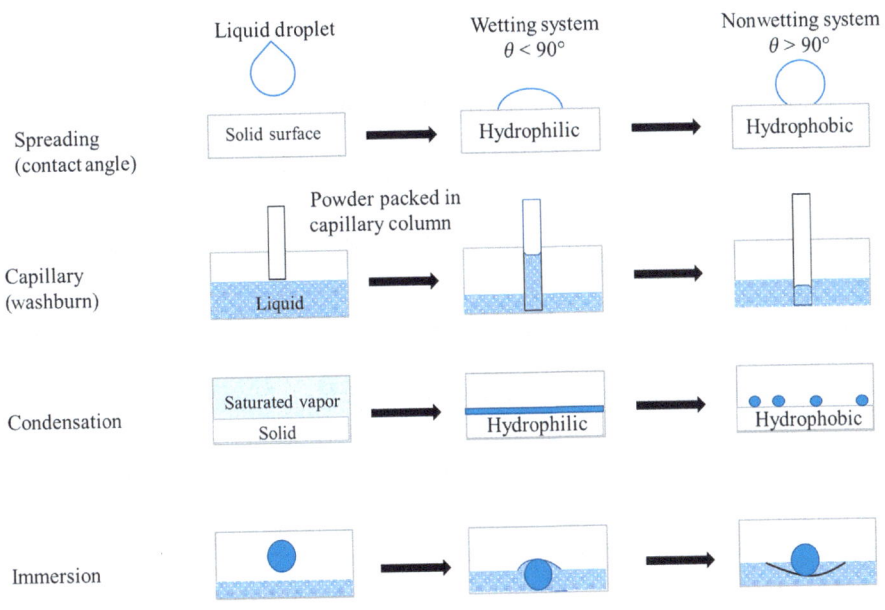

FIGURE 9.5

Different wetting measurement methods.

From Lazghab et al. (2005).

Various techniques can be used to measure the wetting and spreading abilities of binder solutions to aid in binder selection for a particular substrate formulation. Fig. 9.5 illustrates various approaches, of which the direct measurement of spreading via the contact angle and the capillary rise, or Washburn, methods tend to be the most common.

Wetting can be quantified by measuring the solution contact angle on the substrate and the liquid vapor surface energy of the wetting liquid (often referred to as surface tension). From these fundamental properties, one can calculate the surface free energies and the work of cohesion, adhesion, and spreading (commonly known as the spreading coefficient).

The contact angle, θ, is a measure of the affinity of the fluid for a solid, as described in Young's equation:

$$\gamma_{SV} - \gamma_{SL} = \gamma_{LV} \cos|\theta \quad (9.3)$$

where, γ_{SV} is the solid–vapor surface energy, γ_{SL} is the solid–liquid surface energy, and γ_{LV} is the liquid–vapor surface energy, which is more commonly referred to as the surface tension of the liquid.

For fluids to wet a solid surface, the contact angle must be lower than 90°. For this to occur, the solid–vapor surface energy must be greater than the solid–liquid surface energy. The degree of wetting is therefore determined by $\gamma_{LV} \cos \theta$, which sometimes is described as the adhesion tension (Litster & Ennis, 2004).

Surface energies are important fundamental properties for which the work of adhesion, cohesion, and spreading can be calculated.

The work of cohesion, W_{CS}, for a solid can be calculated as follows:

$$W_{CS} = 2|\gamma_{SV} \tag{9.4}$$

And for a liquid:

$$W_{CL} = 2|\gamma_{SV} \tag{9.5}$$

The work of adhesion for a solid–liquid interface can be written as:

$$W_a = \gamma_{SV} \mp \gamma_{LV} - \gamma_{SL}$$
$$\circ| = \gamma_{LV}(1 \mp \cos|\theta) \tag{9.6}$$

W_a represents the work that is done when a liquid adheres to a solid particle, in the process replacing air–particle and air–liquid interfaces with a particle-liquid interface.

The work of spreading, W_S, commonly described as the spreading coefficient, is the difference between the work of adhesion and the work of liquid cohesion:

$$W_S = W_a - W_{CL} = \gamma_{LV}(\cos|\theta - 1) \tag{9.7}$$

The work of spreading is the work done when a liquid spreads over a particle's surface, changing the particle–air and liquid–air interface. The spreading coefficient is key. It has been demonstrated experimentally that positive W_S values are correlated directly with lower granule friability. The larger the W_S, the better the binder distribution and adhesion, and the stronger the granules (Planinšek et al., 2000).

More recently Hapgood et al. (2002) have introduced the concept of liquid binder drop penetration time, t_p, which relates the key binder solution parameters important for granule formation, that is, solution viscosity, surface tension, and substrate contact angle, to powder bed penetration kinetics, as relevant in a high-shear granulator:

$$t_p = \frac{1.35|V_o^{2/3}}{\varepsilon^2 R |\gamma_{LV}| \cos\theta} \tag{9.8}$$

Or more simply, for a given droplet volume V_o, pore size, R, and packing density, [

$$t_p \infty \frac{|\eta}{|\gamma_{LV}|\cos\theta} \tag{9.9}$$

According to this relationship, t_p can be decreased by minimizing binder fluid viscosity and by maximizing the adhesion tension of the binder liquid.

From a formulator's perspective this can be achieved by choosing a low viscosity binder fluid with a surface tension such that the contact angle with the solid formulation substrate is as close to zero as possible (Hapgood et al., 2002). In practice this might imply that for a very polar, hydrophilic substrate such as metformin HCl, a binder solution with a higher surface tension such as PVP (mN/m) might be preferable; while for a very nonpolar, more hydrophobic substrate, a binder solution with a lower surface tension such as that provided by HPC (40 mN/m) might be more effective for rapid binder penetration and wetting kinetics (see later case study example).

The various experimental techniques to measure wettability and contact have been reviewed in detail by Lazghab et al. (2005). If the substrate can be rendered sufficiently nonporous, then the sessile drop technique and contact angle goniometer can be used to measure contact angles directly. As indicated the substrate needs to be rendered nonporous, which can be achieved by sintering, melting, or compressing powder into very hard and smooth compacts.

An alternative method to assess penetration kinetics of liquid into a powder bed is the Washburn method (Lusvardi et al., 2003; Washburn, 1921). This has the advantage of mimicking the binder penetration by capillary action and wetting processes that occur in wet granulation. In this technique, binder solution uptake by capillary action into a column of substrate powder is measured. The liquid uptake is described by the following equations:

$$t = Am^2 \tag{9.10}$$

where,

$$A = \frac{\eta}{c\rho^2 \gamma_{LV} \cos\theta} \tag{9.11}$$

where, t is the time after the solid and liquid are brought into contact; m is the mass of the liquid drawn into the solid; A is a constant dependent the liquid properties (viscosity, η; density, ρ; liquid–vapor interfacial tension, γ_{LV}; and solid–liquid contact angle, θ); and c is a material constant dependent on the porosity and tortuosity of the powder bed.

Additional tests include immersional and condensational tests and inverse gas chromatography (IGC). IGC allows surface energy determination by measuring preferential adsorption of various well-characterized vapor probes onto the substrate (Litster & Ennis, 2004; Thielmann et al., 2005).

9.3.5.4 Wetting Studies for Formulation Screening and Optimization

The work of Krycer et al. (1983) is among the earliest studies to highlight the importance of binder fluid wetting and spreadability in relation to granule friability and compressed tablet strength and capping tendency. Krycer et al. (1983) evaluated the relative binder efficiencies of HPMC, PVP, starch, acacia, and sugar in an acetaminophen granulation. They found that binder-substrate wetting, binder-substrate adhesion, and binder cohesion were important determinants of optimal granule and tablet properties. Film mechanical properties were also shown to be relevant. In this regard it is important to state that in addition to binder viscosity and binder-substrate interactions, the mechanical–thermoplastic properties of the binder also exert a significant effect on the granule and tablet strength. The thermomechanical properties of binders will be discussed later in this chapter. The results of Krycer et al. (1983) are summarized in Table 9.4. Strong correlations generally exist among spreading coefficient, surface tension, angle of contact, and granule and tablet strength.

Another example in which spreading coefficients were used to select and optimize binder fluid characteristics is illustrated in Fig. 9.6. This study demonstrates that in addition to selecting HPC, which has lower surface tension than PVP, replacing water with a hydroalcoholic solvent further increases the spreading coefficient, leading to lower granule friability (Zhang et al., 2002). A further example of the application of the classical Washburn capillary wetting approach to rational binder selection is provided by Lusvardi et al. (2003). These workers used the Washburn approach to determine the wetting kinetics of HPC, HPMC, and PVP on ibuprofen and naproxen (relatively low solubility drugs). Both these drugs have different degrees of hydrophobicity, Naproxen has a log P of 2.99; ibuprofen is more hydrophobic and has a log P of 3.84. Binder solution characteristics and spreading coefficients are given in Table 9.5. These show excellent correlation with final tablet characteristics, as shown in Fig. 9.7.

Table 9.4 Properties of 4% w/v binder solutions and resultant granule and tablet properties in an acetaminophen (APAP) model system.

Binder solution	Surface tension (dyn/cm)	Contact angle on APAP (°)	Work of spreading (dyn/cm)	Granule friability index	Tablet strength (N)[a]
HPMC	45.2	27.4	−5.07	14.8	180
Acacia	50.6	30.3	−6.92	19.8	162
Sucrose	50.4	32.8	−8.01	87.6	98
PVP	53.6	42.2	−13.9	26.5	57
Starch	58.7	47.3	−18.9	45.3	37
Water	70.3	59.6	−110	–	–

From Krycer et al. (1983).
[a] Diametrical crushing strength for tablets compressed at 120 MPa.

Table 9.5 Binder solution wetting characteristics on drugs with different degrees of hydrophobicity, ibuprofen, and naproxen.

Wetting solution	Surface tension (mN/m)	Viscosity (cps)	Contact angle on ibuprofen	Spreading coefficient for ibuprofen (mN/m)	Contact angle on naproxen	Spreading coefficient for naproxen (mN/m)
n-Hexane	18.4	0.3	0°	0	0°	0
HPC	40.0	2.3	68°	−25.0	0°	0
HPMC	48.4	1.9	81°	−40.8	37°	−9.7
PVP	53.6	1.5	88°	−51.7	63°	−29.3
Water	72.1	1.0	>90°	>72.1	85°	−65.0

From Lusvardi et al. (2003).

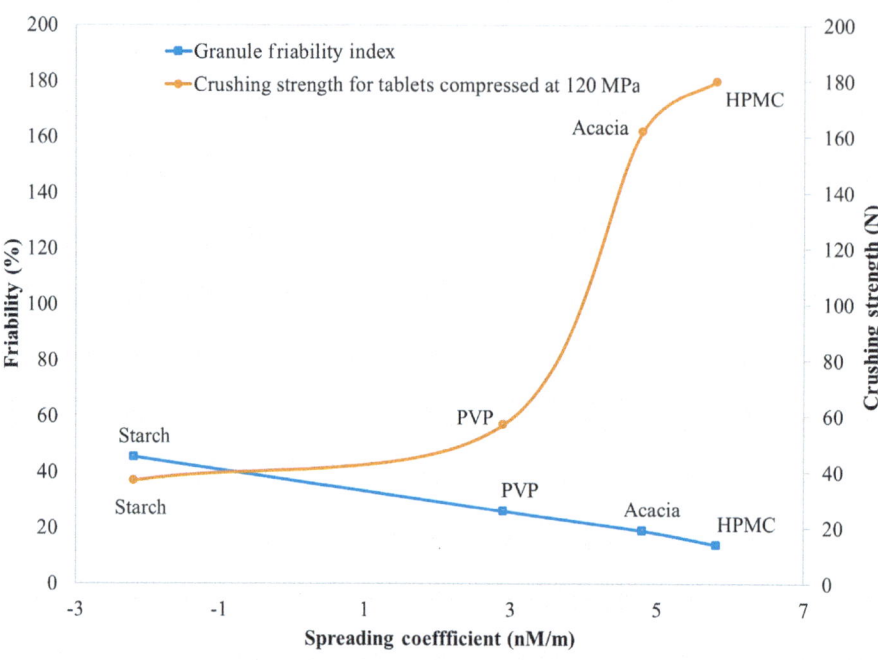

FIGURE 9.6

Relationships among spreading coefficients calculated by Rowe (1990) and tablet and granule strength data reported by Krycer et al. (1983).

The studies support the general conclusion that optimal wetting for a granulation can be achieved through rational binder selection. The polarity of a binder directly affects the surface tension of the binder solution, which in turn affects adhesion tension and spreadability on the substrate. The surface tensions for common aqueous binder solutions are summarized in Table 9.6.

Although wettability studies are useful tools to guide rational formulation and binder choices, one of their shortcomings relates to equipment and methodological complexity and time consumption. Such studies can require several days of work, while multiple typical granulation trials can be completed in a single day.

Taflioglu et al. (2014) have successfully tested an alternative, more practical approach based on approximating binder-substrate interaction potential based on binder solution surface tension and the log P of the compound (Taflioglu et al., 2014). The hypothesis is that for a high log P, hydrophobic API, a binder solution with low surface tension would be most effective to yield strong tablets and granules. Conversely, for low log P or negative log P, that is, very hydrophilic and polar APIs, a binder solution with a high surface tension would be optimal.

This approximation was tested on four APIs of different log P, as shown in Table 9.6. These APIs were granulated in a model formulation with binders of varying surface tension, as illustrated in Table 9.9. These studies showed that for APIs with high log P values, binders with lower surface tension, such as HPC, would be more effective. The opposite was true for very hydrophilic APIs (low log P), such as metformin HCl, where PVP and PVP–PVA were significantly better (Figs. 9.8 and 9.9).

Table 9.6 APIs of varying hydrophobicity (log P) and binders of various surface tension used to measure substrate binder interactions.

API	Water solubility	Lipophilicity (log P)	Flowability	Compactability	Plasticity
Metformin	>300 g/L	−0.50	Poor	Poor	Very poor
Albendazole	2.28 mg/L	2.70	Poor	Good	Good
Efavirenz	0.86 mg/L	4.60	Poor	Poor	Very poor
Simvastatin	1.22 mg/L	4.68	Moderate	Moderate	Poor
Property	Klucel EXF HPC	Benecel E15 HPMC	Plasdone S-630 PVP/VA	Plasdone K-29/32 PVP	
Surface tension (mN/m)	40.00	45.90	49.50	53.60	
Mean particle size (μm)	50.00	90.00	76.00	107.00	
Surface area (m²/cm³)	0.32	0.15	0.21	0.37	
Relative toughness	Very tough	Very low toughness	Tough	Low toughness	

From Tufttoglu et al. (2014) and Yu et al. (2016).

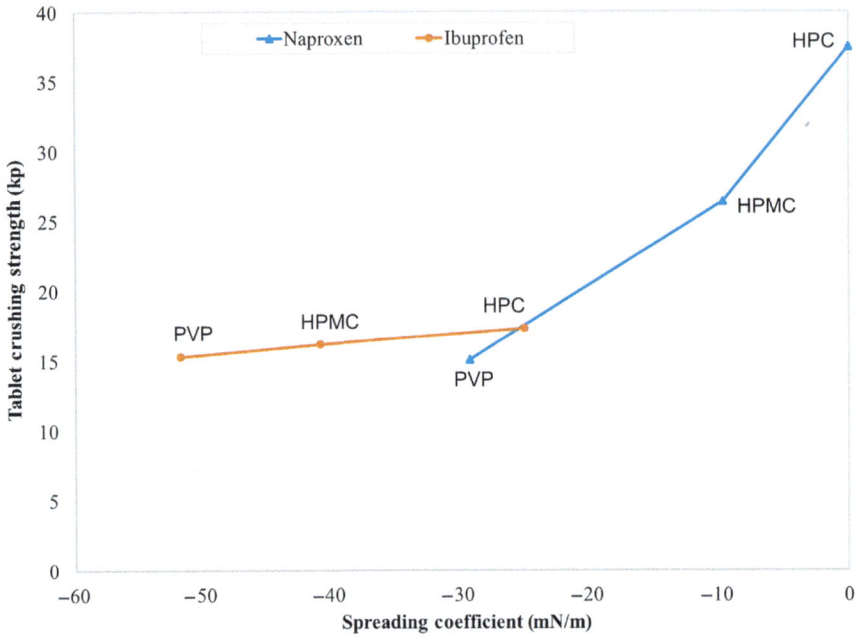

FIGURE 9.7

Relationships among spreading coefficients of binder solutions of HPC, HPMC, and PVP over ibuprofen and naproxen, and tablet crushing strength. Ibuprofen and naproxen were granulated with HPC, HPMC, and PVP solutions and compressed at a 15 kN compression force. Tablets weighed 600 mg and were compressed with 7/16″ standard concave tooling.

From Lusvardi et al. (2003).

For APIs with log P in the range of 2 to 3.5, binder choice and surface tension were less critical. These findings are summarized in Fig. 9.10. It should be emphasized that this approach provides a good and often sufficient first approximation, but it ignores some of the fundamentals inherent in the more rigorous contact angle measurement techniques or in IGC.

9.3.5.5 The role of wet granulation solvent

The solubility of a binder in any given solvent depends on the ability of the solvent to overcome the intermolecular forces of the material. It is maximized when the cohesive energy densities of both are equal and decreases as they become increasingly dissimilar (Barth & Mays, 1991). From previous studies it was noted that the greater the similarity in molecular structure, the higher the solubility of the binder in the solvent, that is, like dissolves like (Parikh et al., 2005).

As mentioned earlier a primary factor for good granulation is the wetting and spreading ability of the binder solution over the substrate. A simple way to modulate these properties is to modify the solvent composition to influence the solution polarity and wetting properties in the desired direction. In addition to choosing the binder with lower surface tension, adding surfactants to reduce the surface tension or replacing the aqueous solvent with a less polar organic solvent can also be explored.

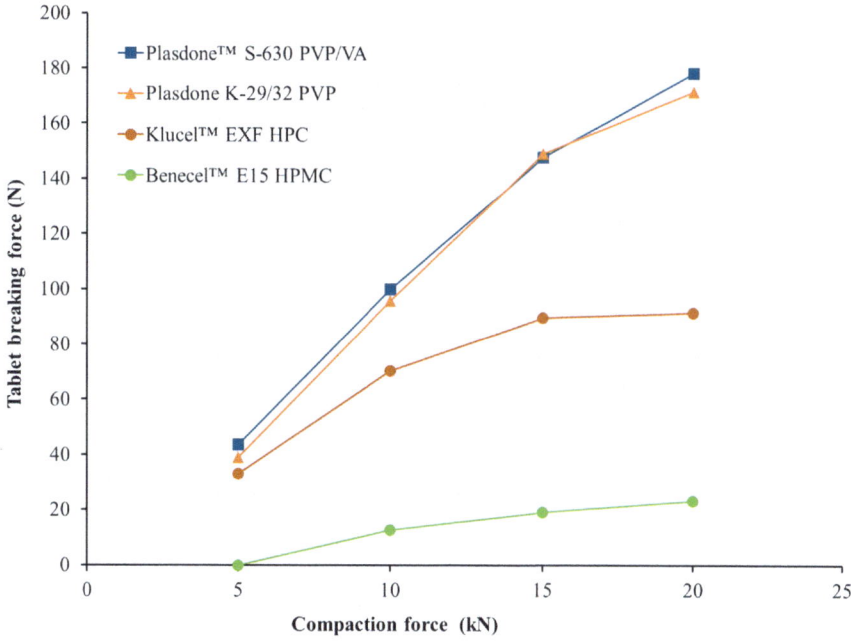

FIGURE 9.8

Compactibility of metformin formulations. More polar, high surface tension binders such as vinyl pyrrolidones are advantageous for this low log P API.

From Yu et al. (2016).

Krycer et al. (1983) were among the first to report improved granule properties by adding a surfactant to PVP. Their work showed that adding a surfactant, sodium lauryl sulfate (SLS), markedly improved the spreading coefficient of PVP solution, resulting in a significant decrease in granule friability. Tablet crushing strength, however, was only modestly influenced (Table 9.7). This work led to the occasional use of surfactants, such as SLS or Tween 80, in binder solutions as wetting agents. Because most surfactants have poor compressibility and binding characteristics, caution must be exercised in their use because excessive film of surfactant can lead to poor binding.

Use of less polar, organic solvents has been studied as an alternate approach to modify surface energetics. Commonly enlisted solvents for granulation are hydroalcoholic solutions of methanol, ethanol, IPA, and acetone.

Fig. 9.11 depicts the effect of a binary solvent system on the granule friability for two experimental drugs granulated using either PVP or HPC, and water or ethanol–water combination. Granule friability decreases substantially as the spreading coefficient increases by shifting from simply water to an ethanol–water solvent system.

An obvious limitation of this approach to enhance wetting is binder solubility in the solvent system. For instance starches and NaCMC generally are not soluble in organic solvents, and HPMC and MC require a minimum of 10% to 15% water to be soluble in polar organic solvents such as methanol, ethanol, IPA, and acetone.

Table 9.7 Properties of 4% w/v binder solutions, and resultant granule and tablet properties for a PVP–acetaminophen model system.

Binder solution	Surface tension (dyn/cm)	Contact angle on APAP (°)	Work of spreading (dyn/cm)	Granule friability index	Tablet strength (N)[a]
PVP + SLS (90:10)	44.1	44.1	−0.9	20.8	58
PVP + glycerol (90:10)	43.8	40.1	−10.7	25.5	80
PVP	53.6	42.2	−13.9	26.5	48

From Krycer et al. (1983).
[a] Diametrical crushing strength for tablets compressed at 120 MPa.

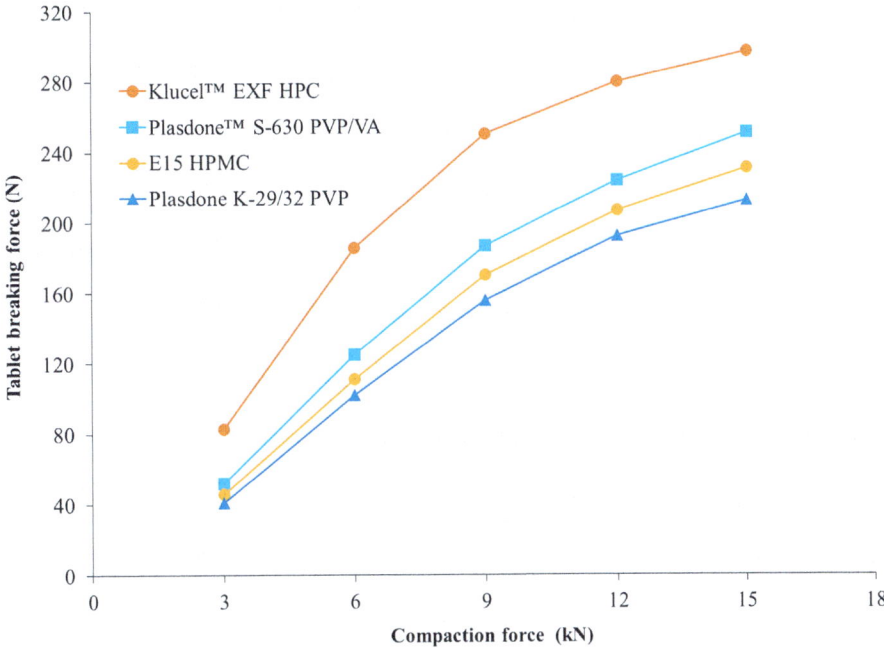

FIGURE 9.9

Compactibility of efavirenz formulations. Low surface tension, more nonpolar binders such as HPC provide better wetting for the high log P API.

From Taflioglu et al. (2014).

Additionally one needs to consider the complication that binder solutions' properties with the same binder concentration can vary considerably in different solvent systems. Table 9.8 shows the variation in viscosity of PVP solutions when the water-to-alcohol ratio is varied. Similar behavior occurs for most polymers.

Wells and Walker (1983) performed a detailed study on solvent effects. Their work accentuated the effect of solvent choice on PVP-acetylsalicylic acid granulations by varying the ratios of ethanol and water in the granulation fluid. Opposing the general rule, granule bulk density was found to decrease with decreased surface tension (increased ethanol). Tablet crushing strength and friability, however, were increased, being the highest with the lowest surface tension (100% ethanol). The authors pointed out that aspirin solubility was increased in this optimal solvent level, which possibly contributed to better bond formation through greater dissolution and recrystallization of aspirin.

9.3.5.6 Mechanical properties of binder

The mechanical and film-forming properties, along with surface energies associated with the binder and binder solution, dictate the characteristics of granules and their resultant tablets (Krycer et al., 1983). In their foundational study they concluded that in addition to having a favorable spreading coefficient

Table 9.8 Effect of solvent composition on PVP binder solution properties and PVP–acetylsalicylic acid granulations and tablet strength.

Binder solution	Viscosity (cP)	Surface tension (dynes/cm)	Wettability ($r \cdot \cos \theta \; 10^{-4}$)	Bulk density (g/mL)	Tablet strength (kP)	Tablet friability (%)
100% water	67	69.5	4.1	0.48	6.6	3.5
25% ethanol	194	42.6	12.16	0.44	6	4
50% ethanol	287	33.6	15.9	0.43	5.8	4.6
75% ethanol	240	30.1	18.68	0.425	8	3.4
100% ethanol	186	26.35	16.4	0.42	–	2.5

From Wells and Walker (1983).

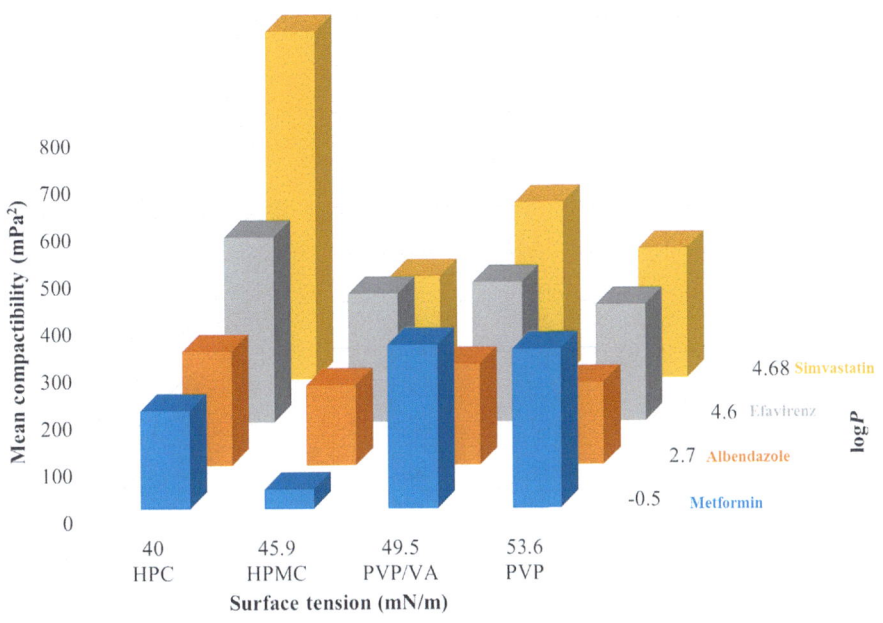

FIGURE 9.10

Effect of binder surface tension and API log P on tablet compactibility.

From Taflioglu et al. (2014) and Yu et al. (2016).

on acetaminophen, HPMC was also a superior film former when compared to PVP, acacia, and starch, in that it produced comparatively soft but tough films.

Reading and Spring (1984) further investigated these concepts using MC, PVP, starch, and gelatin as film-forming wet binders using sand as a hydrophilic inert substrate. As depicted in Table 9.9, MC showed significantly greater film strength and deformation ability than those of the other binders. Concurrently MC films were softer, showing the lowest Brinell (indentation) hardness. These properties correlated with significantly larger, less friable granules, resulting in stronger tablets.

The work of Krycer et al. (1983) and that of Reading and Spring (1984) suggest that binders that have good wettability and film forming ability with high toughness (i.e., high work of failure, as measured by area under the force–distance curve) and high percentage of elongation at break (which together are indicators of plastic flow), likely will produce more robust granules and tablets. These studies, however, do not directly relate to the compaction behavior of tablet binders when subjected to high-speed uniaxial compaction, which occurs in commercial tableting.

Fig. 9.12 shows a typical force-displacement chart obtained by subjecting pure polymer tablets to diametrical compression on a universal testing machine. These binder tablets were prepared by compressing the dry binder powders of similar fine particle size on a rotary press at high speed, replicating the strain rates typically encountered by the binder and other formulation ingredients in commercial tableting presses.

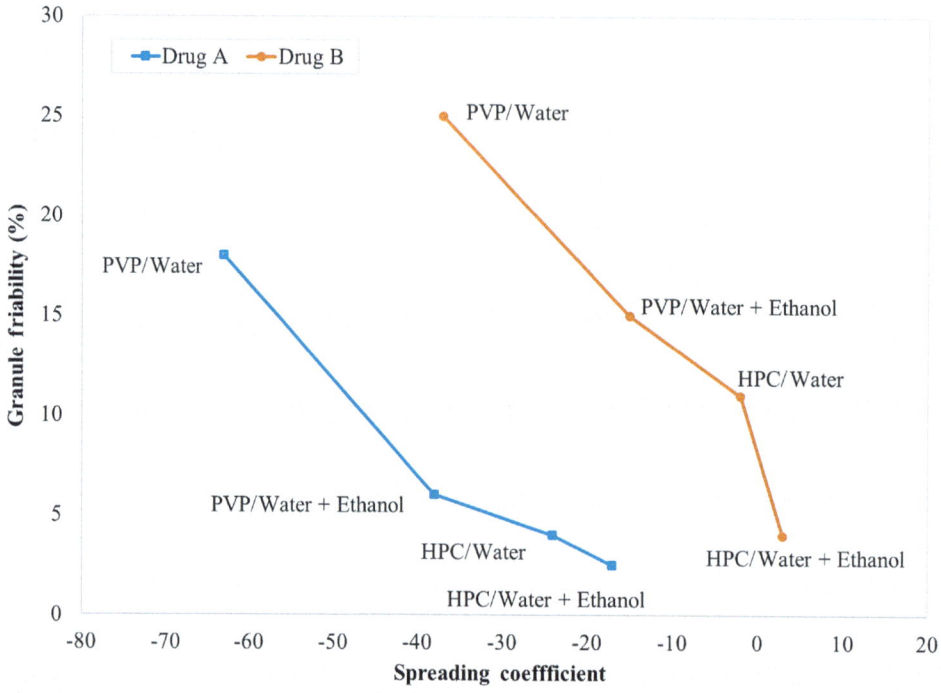

FIGURE 9.11

Effect of spreading coefficient on granule friability of two experimental drug formulations.

From Zhang et al. (2002).

Table 9.9 Selected free film, granule, and tablet properties for MC, PVP, gelatin, and starch binders. Sand was used as the binder substrate. Films were dried at 60°C.

Property	MC	PVP	Gelatin	Starch
Film tensile strength (MPa)	70	18	27	33
% elongation at break	37	5.3	3.1	3.2
Film toughness (J/m$^3 \cdot 10^5$)	192	8	9	10
Film Brinell hardness, 12% RH (MPa)	7.5	9.17	18.7	15.6
% granule friability	5.0	11.0	14.6	20.4
Mean granule size (μm)	680	445	365	200
Tablet crushing strength[a] (kPa)	345	240	200	70

From Reading and Spring (1984).
[a] *Tablets compressed at 120 MPa compression force.*

9.3 Important considerations for binder selection and use

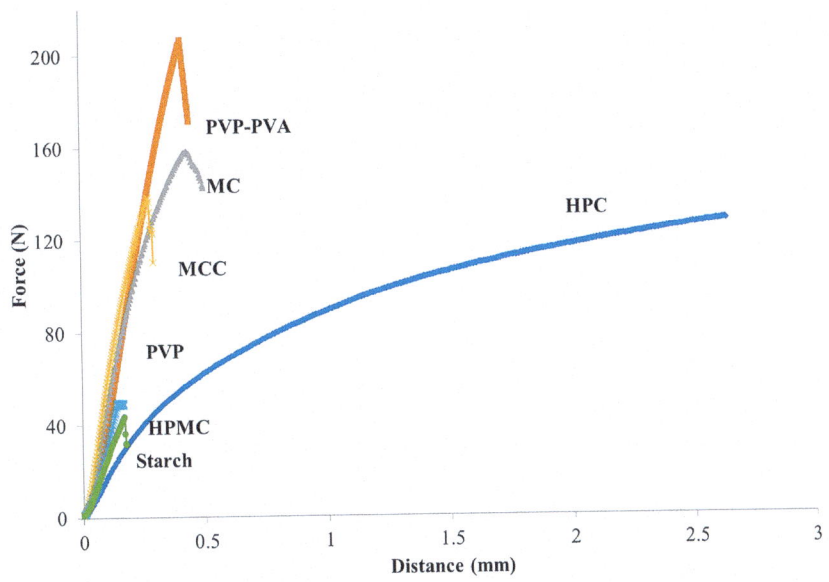

FIGURE 9.12

Load-deformation plots for pure polymer tablets subjected to diametrical compression on a universal testing machine (0.5 in./min cross-head speed). The 100% polymer tablets were made on a rotary tablet press.

From Skinner et al. (2003).

Table 9.10 Glass transition temperature at equilibrium moisture content as received, detected by modulated temperature differential scanning calorimetry.

Binder	Equilibrium moisture content (%)	Glass transition temperature (°C)
HPC	3.2	−2.6
MC	4.7	145
HPMC (type 2910)	3.1	160
MCC	4.9	~105
PVA-PVP	4.8	101
PVP	8.0	164

From Skinner et al. (2003).

HPC shows fundamentally different behavior, exhibiting significantly greater plasticity and toughness (area under the curve) as compared to the other binders, which uniformly show an increased tendency to undergo brittle fracture. PVA–PVP, MC, and MCC also show higher toughness. In these cases, very high loads are reached before the tablet fails, but deformability or distance to rupture is moderate (0.28–0.46 mm). Contrarily, HPC tablets do not show this brittle behavior. The HPC tablets

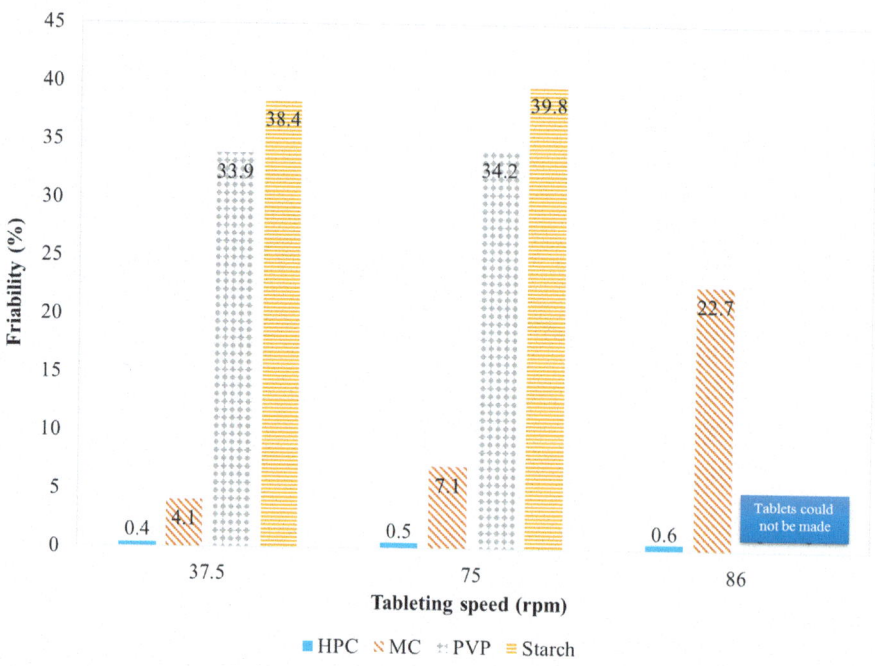

FIGURE 9.13

Friability for wet granulated acetaminophen tablets containing 6% HPC, MC, PVP, or starch as wet binders. Six hundred mg tablets (using a 7/16-in. standard round convex tooling) were compressed on a Manesty Beta press at 15 kN at three different turret speeds. HPC comprising tablets showed only a negligible increase in friability. PVP and starch comprising tablets could not be made at 86 RPM because of excessive capping on ejection from the press.

From Joneja et al. (1999).

were deformed beyond 2.6 mm without fracturing, while absorbing applied energy and providing much greater toughness. Relative to the other materials tested, the HPMC and PVP tablets in this idealized dry state clearly show lower toughness. In wet binder applications, however, apart from pure mechanical properties, one must take surface interactions and wettability into consideration.

Besides greater toughness and deformability, thermal analysis measurement confirms higher thermoplasticity of HPC. In contrast to other polymers, at a typical equilibrium moisture content (~3%), HPC exhibits a high intensity glass transition in the low temperature ranges (−3°C to 0°C) (Table 9.10) (Picker-Freyer & Dürig, 2007; Skinner et al., 2003). This highlights increased molecular mobility and plasticity generally associated with lower glass transition. Overall these results confirm a higher state of plasticity for HPC in relation to the other binders.

Joneja et al. (1999) found that key elements to ensure robust tablets were binder toughness and a high degree of plastic flow for wet granulated acetaminophen formulations using four binders, HPC, MC, PVP, and pregelatinized starch at 6% level. This work was consistent with the results for pure polymer tablets and earlier reports on binder film mechanical properties. Their study concluded that HPC yielded stronger, more deformable, and, therefore, tougher tablets. The differences in terms of tablet friability

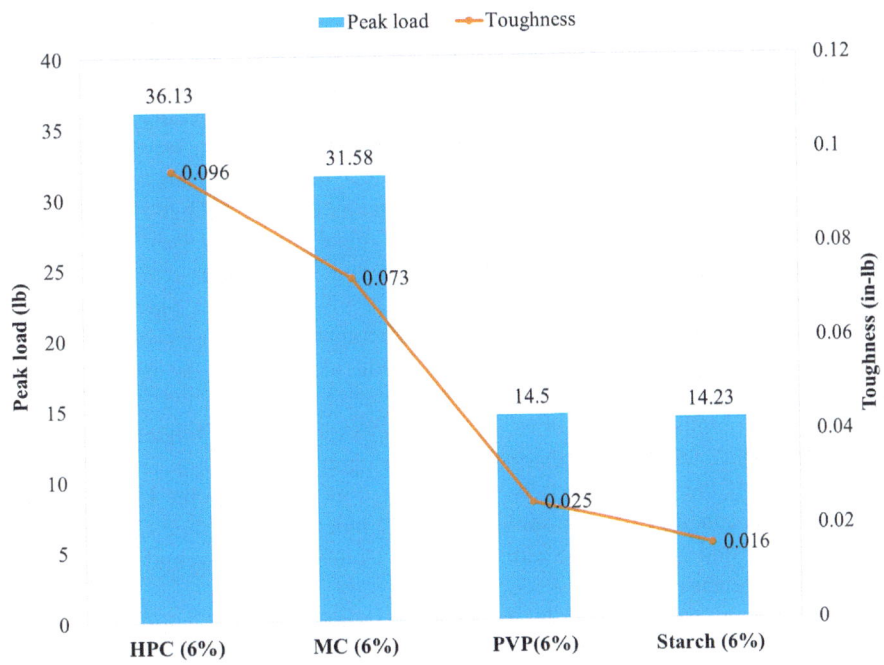

FIGURE 9.14

Tablet strength and toughness for wet granulated acetaminophen tablets comprising 6% HPC, MC, PVP, or pregelatinized starch binder. Tablets were compressed on a Manesty Beta press at 37.5 rpm turret speed at 15kN compression force. Tablet strength and toughness were assessed by diametral compression on a universal testing machine at 0.05 in./min crosshead speed.

From Joneja et al. (1999).

and strength were further heightened when the tablet press speed was increased to simulate typical speeds encountered in commercial production (Figs. 9.13 and 9.14). Further studies evaluated whether the use of various binder levels would result in comparable properties.

From the results of this study it is noteworthy how consistent the results are with the work performed by Krycer et al. (1983), Lusvardi et al. (2003), and Reading and Spring (1984), indicating that thermal and mechanical properties, such as binder plasticity and toughness, can be used to predict binder performance.

Additionally it is important to recognize that the results also correlate well with measures of binder wettability and spreading, such as aqueous surface tensions, binder solution contact angles on the substrate, and spreading coefficients. Surface energetics and mechanical properties are equally important. It is not practical to separate one from the other.

Based on evidence from multiple binder studies involving a variety of substrates, one can deduce that both binder solution-substrate wetting and mechanical properties of the binder and binder films are key determinants of granule and tablet properties such as density, strength, and overall robustness of the product.

9.4 Regulatory acceptance and supplier reliability

The technological aspects of formulation development projects are generally well understood by formulation scientists. Regulatory aspects and strategic material sourcing, for example, supplier reliability, however, are equally important to ensure product development success. Most excipients including binders are manufactured in large chemical plants, creating a large difference in scale as compared to typical pharmaceutical manufacturing. A typical binder lot size can be 10 to 30,000 kg. Similar polymer grades can be manufactured in a plant where other industrial uses occur. In fact pharmaceutical grades often represent only a small proportion of a plant's output. It is important, therefore, to ensure that the excipient manufacturer has the appropriate excipient good manufacturing practices and capabilities and has an audit history with the appropriate regulatory authorities.

From a regulatory compliance perspective, it is also important to choose binders that are established pharmaceutical excipients, ideally with pharmacopeia monographs and appropriate inactive ingredient database listings. Ideally the binder will have a harmonized monograph in the USP/NF, pH.Eur., and JP. For dietary supplements it is also advisable to choose binders listed in the FCC and recognized as GRAS ingredients, or listed as direct food additives by the United States Food and Drug Administration and other regulatory agencies.

Lastly in order to comply with quality-by-design (QbD) requirements it is important for the scientist to establish the relevant critical material attributes. This is best accomplished by working with the excipient supplier and testing lots near the CMA limits or process limits. When possible and if available it can be useful to obtain a QbD package of data that includes a multiyear lot history for the critical quality attributes and critical process parameters of excipient manufacture.

References

Ashland. (2001). Klucel[fi] hydroxypropyl cellulose. Physical and chemical properties. *Ashland aqualon functional ingredients*. Ashland.

Barth, H. G., & Mays, J. W. (1991). *Modern methods of polymer characterization*. Wiley-Interscience.

BASF. (2008). Soluble Kollidon[fi] grades povidone Ph. Eur., USP, JP. Technical information. *BASF SE*. Care Chemicals Division.

Benali, M., Gerbaud, V., & Hemati, M. (2009). Effect of operating conditions and physico-chemical properties on the wet granulation kinetics in high shear mixer. *Powder Technology, 190*(1), 160–169. https://doi.org/10.1016/j.powtec.2008.04.082.

Brady, J. E., Durig, T., & Shang, S. S. (2009). Polymer properties and characterization A2—Qiu, Yihong. In Y. Chen, G. G. Z. Zhang, L. Liu, & W. R. Porter (Eds.), *Developing solid oral dosage forms* (pp. 187–217). Academic Press.

Chitu, T. M., Oulahna, D., & Hemati, M. (2011). Wet granulation in laboratory scale high shear mixers: Effect of binder properties. *Powder Technology, 206*(1), 25–33. https://doi.org/10.1016/j.powtec.2010.07.012.

Colorcon. (1999). Starch 1500[fi] partially pregelatinized maize starch product brochure. Colorcon.

Cutt, T., Fell, J. T., Rue, P. J., & Spring, M. S. (1986). Granulation and compaction of a model system. I. Granule properties. *International Journal of Pharmaceutics, 33*(1), 81–87. https://doi.org/10.1016/0378-5173(86)90041-4.

del Barrio, M.-A., Hu, J., Zhou, P., & Cauchon, N. (2006). Simultaneous determination of formic acid and formaldehyde in pharmaceutical excipients using headspace GC/MS. *Journal of Pharmaceutical and Biomedical Analysis, 41*(3), 738–743. https://doi.org/10.1016/j.jpba.2005.12.033.

Fitzpatrick, S., McCabe, J. F., Petts, C. R., & Booth, S. W. (2002). Effect of moisture on polyvinylpyrrolidone in accelerated stability testing. *International Journal of Pharmaceutics, 246*(1–2), 143–151. https://doi.org/10.1016/S0378-5173(02)00375-7.

Hapgood, K. P., Litster, J. D., Biggs, S. R., & Howes, T. (2002). Drop penetration into porous powder beds. *Journal of Colloid and Interface Science, 253*(2), 353–366. https://doi.org/10.1006/jcis.2002.8527.

ISP. (2009). Plasdone^fi S-630 copovidone product guide. International Specialty Products.

Iveson, S. M., & Litster, J. D. (1998). Fundamental studies of granule consolidation. Part 2: Quantifying the effects of particle and binder properties. *Powder Technology, 99*(3), 243–250. https://doi.org/10.1016/S0032-5910(98).

Johansen, A., & Schæfer, T. (2001). Effects of interactions between powder particle size and binder viscosity on agglomerate growth mechanisms in a high shear mixer. *European Journal of Pharmaceutical Sciences, 12*(3), 297–309. https://doi.org/10.1016/S0928-0987(00)00182-2.

Joneja, S. K., Harcum, W. W., Skinner, G. W., Barnum, P. E., & Guo, J. H. (1999). Investigating the fundamental effects of binders on pharmaceutical tablet performance. *Drug Development and Industrial Pharmacy, 25*(10), 1129–1135. https://doi.org/10.1081/ddc-100102279.

Keningley, S. T., Knight, P. C., & Marson, A. D. (1997). An investigation into the effects of binder viscosity on agglomeration behaviour. *Powder Technology, 91*(2), 95–103. https://doi.org/10.1016/S0032-5910(96)03230-5.

Kiekens, F., Zelko, R., & Remon, J. P. (2000). Effect of the storage conditions on the tensile strength of tablets in relation to the enthalpy relaxation of the binder. *Pharmaceutical Research, 17*(4), 490–493. https://doi.org/10.1023/A:1007549625317.

Krycer, I., Pope, D. G., & Hersey, J. A. (1983). An evaluation of tablet binding agents part I. Solution binders. *Powder Technology, 34*(1), 39–51. https://doi.org/10.1016/0032-5910(83)87026-0.

Lazghab, M., Saleh, K., Pezron, I., Guigon, P., & Komunjer, L. (2005). Wettability assessment of finely divided solids. *Powder Technology, 157*(1), 79–91. https://doi.org/10.1016/j.powtec.2005.05.014.

Litster, J., & Ennis, B. (2004). Wetting, nucleation and binder distribution. *The science and engineering of granulation processes* (pp. 37–74). Springer. https://doi.org/10.1007/978-94-017-0546-2_3.

Lusvardi, K. M., Harcum, W. W., Skinner, G. W., & Durig, T. (2003). *Aqualon Pharmaceutical Technology Report PTR 26 (2003)*. AAPS Journal.

Mills, P. J. T., Seville, J. P. K., Knight, P. C., & Adams, M. J. (2000). The effect of binder viscosity on particle agglomeration in a low shear mixer/agglomerator. *Powder Technology, 113*(1), 140–147. https://doi.org/10.1016/S0032-5910(00)00224-2.

Nassar, M. N., Nesarikar, V. N., Lozano, R., Parker, W. L., Huang, Y., Palaniswamy, V., Xu, W., & Khaselev, N. (2004). Influence of formaldehyde impurity in polysorbate 80 and PEG-300 on the stability of a parenteral formulation of BMS- 204352: Identification and control of the degradation product. *Pharmaceutical Development and Technology, 9*(2), 189–195. https://doi.org/10.1081/pdt-120030249.

Parikh, D. E., Parikh, D., Swarbrick, J., Bertuzzi, G., Cameron, I., Celik, M., Ennis, B., Gokhale, R., Hamed, E., Heng, Y., Jambhekar, S., Kanneganti, P., Mehta, K., Miller, R., Schroeder, R., Rekhi, G., Watts, D., Wong, T., Yinghe, H., & Birudaraj, R. (2005). *Drug substance and excipient characterization* (2nd ed.). CRC Press.

Picker-Freyer, K. M., & Dürig, T. (2007). Physical mechanical and tablet formation properties of hydroxypropylcellulose: In pure form and in mixtures. *AAPS PharmSciTech, 8*(4), 82. https://doi.org/10.1208/pt0804092.

Planinšek, O., Pišek, R., Trojak, A., & Srčİcİ, S. (2000). The utilization of surface free-energy parameters for the selection of a suitable binder in fluidized bed granulation. *International Journal of Pharmaceutics, 207*(1), 77–88. https://doi.org/10.1016/S0378-5173(00)00535-4.

Reading, S. J., & Spring, M. S. (1984). The effects of binder film characteristics on granule and tablet properties. *Journal of Pharmacy and Pharmacology, 36*(7), 421–426. https://doi.org/10.1111/j.2042-7158.1984.tb04417.x.

Rowe, R. C. (1990). Correlation between predicted binder spreading coefficients and measured granule and tablet properties in the granulation of paracetamol. *International Journal of Pharmaceutics, 58*(3), 209–213. https://doi.org/10.1016/0378-5173(90)90197-C.

Rudnic, E., & Schwartz, J. B. (1995). *Oral solid dosage forms* (19th ed, p. 1629). Mack Publishing Company.

Skinner, G. W., et al. (2003). Aqualon Pharmaceutical Technology Report PTR. *Aqualon Pharmaceutical Technology Report PTR*: 25.

Skinner, G.W., & Harcum, W.W. (1998). Evaluation of low-viscosity polymers in a model high-dose, acetaminophen formulation. http://www.herc.com/aqualon/product/data/ptr/ptr011 (Accessed March 2009).

Taflioglu, B., Vago, T., Dere, G., Tuglu, T., Oren, Z., & Stoyanov, E. (2014). Evaluating role of the binders in model albendazole formulation done by high-shear wet granulation. In *Paper presented at the 9th PBP world meeting on pharmaceuticals, biopharmaceuticals, and pharmaceutical technology*.

Thielmann, F., Burnett, D., Lusvardi, K. M., & Durig, T. (2005). Correlating drug-binder adhesive strengths measured using inverse gas chromatography with tablet performance. *Journal of Pharmacy and Pharmacology, 57*, S91–S92.

Uzunarslan, K., & AkbugÆa, J. (1991). The effect of moisture on the physical characteristics of ranitidine hydrochloride tablets prepared by different binders and techniques. *Drug Development and Industrial Pharmacy, 17*(8), 1067–1081. https://doi.org/10.3109/03639049109043845.

Washburn, E. W. (1921). The dynamics of capillary flow. *Physical Review, 17*(3), 273–283. https://doi.org/10.1103/PhysRev.17.273.

Wasylaschuk, W. R., Harmon, P. A., Wagner, G., Harman, A. B., Templeton, A. C., Xu, H., & Reed, R. A. (2007). Evaluation of hydroperoxides in common pharmaceutical excipients. *Journal of Pharmaceutical Sciences, 96*(1), 106–116. https://doi.org/10.1002/jps.20726.

Waterman, K. C., Adami, R. C., & Hong, J. Y. (2004). Impurities in drug products. *Separation Science and Technology, 5*, 75–88. https://doi.org/10.1016/S0149-6395(03)80006-5.

Wells, J. I., & Walker, C. V. (1983). The influence of granulating fluids upon granule and tablet properties: The role of secondary binding. *International Journal of Pharmaceutics, 15*(1), 97–111. https://doi.org/10.1016/0378-5173(83)90070-90074.

Yu, S., Huang, Z., Gabbott, I., El Saleh, F., Muehlenfeld, C., & Warnke, G. et al. (2016).In *Experimental study of binder selection using twin-screw and batch high-shear wet granulation. Paper presented at the 10th PBP world meeting on pharmaceutics, biopharmaceuticals and pharmaceutical technology*.

Zhang, D., Flory, J. H., Panmai, S., Batra, U., & Kaufman, M. J. (2002). Wettability of pharmaceutical solids: Its measurement and influence on wet granulation. *Colloids and Surfaces A: Physicochemical and Engineering Aspects, 206*(1), 547–554. https://doi.org/10.1016/S0927-7757(02)00091-2.

CHAPTER 10

Effect of binder attributes on granule growth and densification

Ajit S. Narang[a], Li Tao[b], Junshu Zhao[b], Rekha Keluskar[b], Shruti Gour[b], Tim Stevens[b], Kevin Macias[c], Brenda Remy[b], Preetanshu Pandey[d], Richard D. LaRoche[e], Anna Sosnowska[e], Stephen Cole[e], Atul Dubey[f,g], Rohit Ramachandran[h], Jinjiang Li[i] and Dilbir Bindra[b]

[a]Pharmaceutical Sciences, ORIC Pharmaceuticals, Inc., South San Francisco, CA, United States, [b]Drug Product Science & Technology, Bristol-Myers Squibb Co., New Brunswick, NJ, United States, [c]Analytical and Bioanalytical Development, Bristol-Myers Squibb Co., New Brunswick, NJ, United States, [d]Drug Product Development, Kura Oncology, San Diego, CA, United States, [e]DEM Solutions, Denver, CO, United States, [f]Tridiagonal Solutions Pvt. Ltd., Pune, Maharashtra, India, [g]Department of Process Engineering, Aditya Birla Science and Technology Company Pvt. Ltd., Navi Mumbai, Maharashtra, India, [h]Department of Chemical & Biochemical Engineering, Rutgers University, Piscataway, NJ, United States, [i]Wolfe Labs, Woburn, MA, United States

10.1 Functional excipients and critical material attributes

The choice of excipients in the design of a drug product (DP) is based on several criteria, including salient elements of the quality target product profile, such as the rate and site of drug delivery; stability/target shelf life, packaging, and storage temperature; and appearance, target patient population, and palatability (Narang et al., 2017). These considerations, combined with input material properties, including the physicochemical and biopharmaceutical properties of the active pharmaceutical ingredient (API), and the manufacturing process, help determine the selection of excipients (Narang et al., 2017). Material attributes (MAs) of selected excipients that help achieve the critical quality attributes (CQAs) of the DP are enlisted as target material properties, while the MAs that are essential to achieving one or more DP CQAs and are potentially variable are enlisted as critical material properties (Narang et al., 2017). The identification and understanding of how the MAs of critical excipients affect their functionality is an important element of new product and process development. This chapter highlights literature and case studies that exemplify the methods of investigation of the effects of certain excipient (binder) attributes in a wet granulation (WG) process and product.

Excipients used in oral solid dosage forms have been classified based on their functionality into groups such as diluents, disintegrants, binders, glidants, lubricants, release-controlling polymers, stabilizers (such as antioxidants, chelators, and pH-modifiers), film-coating polymers, plasticizers, surfactants, colorants, sweeteners, and flavors (Narang et al., 2017). Among the intragranular excipients used in the wet granulation process, the most common functional excipients are:

- Fillers, typically a combination of microcrystalline cellulose (MCC) and lactose monohydrate;
- Binder, typically hydroxypropyl cellulose (HPC) or polyvinyl pyrrolidone (PVP);

- Disintegrant, typically croscarmellose sodium, sodium starch glycollate, or crospovidone.

The examples listed above and discussed in this chapter are representative of contemporary practices and are not intended to be comprehensive. They are selected to highlight the methodologies through which the effect of MAs of these functional excipients can be studied in a wet granulation unit operation. Some of these methodologies are expected to be generally applicable to other functional excipients as well.

10.2 Impact of binder attributes on granulation outcomes
10.2.1 Quality attributes of interest

Characteristics or properties of binders that can affect their functional performance can include surface tension; particle size, shape, and distribution; solubility; viscosity (which depends on the nature of the polymer structure and molecular weight and its distribution); polymeric structure (including monomer properties and sequence, functional groups, degree of substitution, and cross-linking); and variations in their sources (for natural polymers) (Narang et al., 2017). The functionality-related characteristics (FRCs) of some specific binders—for example, extracted from nonbinding parts of the monographs listed in the European Pharmacopeia (Ph. Eur.)—include:

- Molecular mass (viscosity, expressed as K-value) for PVP;
- Viscosity, degree of substitution, powder flow (as matrix polymer), and particle size (as matrix polymer) for HPC;
- Cold water-soluble matter, particle-size distribution, and powder flow for pregelatinized starch.

The QAs of pharmaceutical binders can be either the intrinsic material properties, such as molecular weight, viscosity as a function of concentration in solution, surface tension of the solution, particle size distribution (PSD), and the rate of hydration. In addition to these inherent characteristics of the binders, process options that define how a binder is used in a given product, such as concentration in the formulation and the mode (and rate) of addition, also have a bearing on the impact of binder attributes on granulation outcomes and product QAs. These variables can be termed the process-specific binder variables. Experimental studies that investigate the effect of binder attributes on granulation and product outcomes can include both intrinsic binder properties and process-specific binder variables.

The quality attributes of importance to a particular application have been termed as FRCs by the Ph. Eur. Excipients can be multifunctional, and the physical and chemical attributes that afford them their functionality will differ based on the application (Narang et al., 2017). Thus the functionality determining attributes of hypromellose (hydroxypropyl methyl cellulose [HPMC]) differ by the application: viscosity, degree of substitution, molecular mass distribution, particle-size distribution, and powder flow when applied as matrix polymer for extended release dosage forms; and only viscosity and degree of substitution when used as a binder, viscosity-increasing agent, or film-former (Narang et al., 2017).

Systematic studies have been reported on the effect of binder properties on granule growth and consolidation (Johansen & Schaefer, 2001; Kenningley et al., 1997; Mills et al., 2000). These studies tend to be multivariate in nature and specific to the bulk powder and particle surface properties of the formulation being studied.

10.2.2 Mechanistic basis of binder effects

Wet granulation is a size enlargement process that is accomplished through the use of a liquid binder, followed by evaporative loss of solvent to form dry granules that then possess superior properties in terms of powder flow, compaction, and distribution of minor components. Three principal mechanisms of the wet granulation process have been identified (Iveson et al., 2001):

1. Wetting and nucleation;
2. Consolidation and growth; and
3. Attrition and breakage.

Granulation outcomes of an HSWG process have been described as a balance of building forces and break-up forces, both of which are influenced by a combination of formulation and process design (Benali et al., 2009). The building forces are the particle surface phenomena that depend on the properties of the binder solution and the solid particles, while the break-up forces are process intensity dependent.

Physicochemical properties of the binder and the particle surface influence granule formation and growth through an interplay of two kinds of forces: capillary forces and viscous forces.

These forces are influenced by physicochemical properties, such as the surface tension and viscosity of the liquid; wettability and swellability of the solid particles; and the size, shape, and porosity of the primary solid particles. Wet particle agglomeration has been described as a phenomenon resulting from the formation of strong liquid bridges between solid particles, which can withstand dynamic breakage forces within the granulator. At the same time, capillary forces allow binder solution or granulating liquid penetration into agglomerates of primary particles, allowing for a stronger interparticle bond formation. Thus viscosity of the granulating liquid determines the strength of the dynamic liquid bridges on the surface of the forming agglomerate, while surface tension determines the strength of the static liquid bridges inside the forming agglomerate.

10.2.2.1 Granule formation and growth

Two key attributes of the binder liquid that play an important role in the granule growth rate and growth mechanism are its viscosity and surface tension.

Viscosity of the binder is an important attribute in governing the formation and growth of granules. Keningley et al. (1997) have shown, for example, that a minimum viscosity of the binder solution is necessary to achieve granulation. Other investigators, however, have shown that increasing the viscosity of the binder can have a beneficial effect on granulation growth but only up to a certain value (Johansen & Schaefer, 2001; Mills et al., 2000). Mills et al. (2000) postulated that granulation at lower viscosity is governed predominantly by layering mechanisms, while coalescence is the principal driving mechanism at higher viscosities.

Capes and Danckwerts (1965) reported that a minimal surface tension is necessary for particle agglomeration. The influence of viscosity and surface tension of the liquid on particle agglomeration outcomes has been described by Ennis et al. (1990) in terms of a capillary number, C_a, which is given by:

$$C_a = \frac{\mu}{\gamma} \times v$$

where μ represents the viscosity and γ represents the surface tension of the liquid, and v represents the impeller tip speed.

When the capillary viscous number is greater than a critical value, the cohesion of dynamic liquid bridges exceeds that of the static liquid bridges. This relationship was developed for spherical solid particles that do not swell in the presence of water. Common pharmaceutical systems, however, use swelling matrices such as cellulose-based diluents and disintegrants.

In working with hydrophilic and swellable matrix (MCC), Benali et al. (2009) studied the effect of physicochemical properties and wet granulation process parameters on granule growth kinetics in a high shear mixer. The authors carried out a series of experiments with water and with aqueous solutions of a variety of binders, namely, sodium carboxymethyl cellulose, PVP, HPMC, or a nonionic surfactant. The authors identified four regimes of granule growth (namely, wetting, nucleation, growth, and overwetting) and studied the effect of binder properties and process parameters in granulation transition through these regimes. The presence of these regimes was consistent across all experiments, with the influence of process parameters being predominantly on the transitions between the regimes.

The effect of physicochemical properties of the binder liquid and the particle surface was described in terms of a modified capillary viscous parameter, C_a', which is defined as a ratio between the viscous forces and the work of adhesion (Benali et al., 2009):

$$C_a' = \frac{\mu}{W_a} \times v$$

where μ represents the viscosity and v represents impeller tip speed. The work of adhesion, W_a, is defined as,

$$W_a = \gamma(1 + cos\theta)$$

where γ represents the surface tension of the liquid and θ represents the contact angle between the binder liquid and the particle surface.

The authors defined a value of the modified capillary viscous number ($C_a' < 1$), below which the viscosity of the solution does not significantly affect the granulation process, the interfacial forces dominate, and increasing the work of adhesion increases the granule growth rate. The granule growth rate is controlled by the viscous forces above a critical value of the modified capillary viscous number ($C_a' > 1.6$).

10.2.2.2 Granule breakage

Granule breakage was studied by van den Dries et al. (2003) using different particle size grades of lactose as the substrate, aqueous solutions of HPC as the binder, and a water soluble red colored dye, erythrosine, as a tracer. The authors investigated the effects of the particle size of lactose and the viscosity of HPC solutions on granule breakage during HSWG using a 10-L Gral high-shear mixer. They were able to predict granule breakage using the ratio of the impact energy and dynamic granule strength and found that an increase in viscosity of the binder solution and/or a decrease in the PSD of the substrate facilitated a transition from breakage to no-breakage behavior of the granules. Breakage, interestingly, also correlated with granule homogeneity. Granule breakage was associated with continuous exchange of particles, promoting homogeneous granules. Lack of granule breakage, however, promoted the layering mechanism of granule growth, leading to the formation of inhomogeneous granules (van den Dries and Vromans, 2002). The authors argued that granule breakage promotes the coalescence mechanism of granule growth, leading to the continuous mixing of various components in the granules. In the absence of granule breakage, the layering mechanism of granule growth dominates. With smaller primary particles having a greater propensity for layering than larger primary particles, this phenomenon can lead to an inhomogeneous distribution of ingredients with different PSDs among granules of different sizes.

To predict whether granule breakage will occur, the authors used the Stokes deformation number (St_{def}) proposed by Iveson et al. (2001) to predict granule breakage as a function of impact energy and granule strength:

$$St_{def} = \frac{\rho_g V_c^2}{2\rho_y}$$

where V_c is the impact velocity and ρ_g is the granule density. Granule breakage was predicted to occur if the St_{def} exceeded a critical value of 0.04. Lower primary particle size of the substrate and/or higher viscosity of the binder solution would result in lower St_{def}, and stronger granules.

At the level of physicochemical properties and interparticle forces governing granule breakage, Benali et al. (2009) also described this phenomenon as a balance between the impact pressure during processing and the static tensile strength of granules.

The impact pressure (σ_i) is given by Thornton et al. (1996), Vonk et al. (1997):

$$\sigma_i = \frac{2}{3}\rho U^2$$

where ρ is the particle density and U is the impeller tip speed. This equation, however, assumes the speed of particle impacts to be the same as the impeller tip speed, which is not necessarily true. The velocities of particles in the granulator and their relative collision frequency and velocities can be simulated by discrete element modeling ([DEM], as described in Section 10.4), which shows a regional and intensity distribution of impact velocities. Nonetheless, this equation emphasizes the role of particle density and particle velocity in governing the impaction pressure on the granules. The static tensile strength of the granules (σ_t) is given by Rumpf (1958),

$$\sigma_t = \frac{\gamma}{d} \times \frac{1-\varepsilon}{\varepsilon} \times CS\cos\theta$$

where γ is the surface tension of the liquid, d is the initial diameter of primary particles, [is granule porosity, θ is the contact angle, C is a material constant, and S is liquid pore saturation.

$$S = H\frac{\rho_s}{\rho_l} \times \frac{1-\varepsilon}{\varepsilon}$$

where H is the liquid-to-solid ratio and ρ_s and ρ_l are the true densities of the solid and the liquid, respectively.

10.2.3 Binder selection

Experimental studies to investigate the effect of binder type and grade, concentration, mode of addition, and particle size of the binder on the properties of the granulation are typically carried out early in DP development. Chemical compatibility studies carried out during preformulation stages help determine which binders might be suitable (Narang et al., 2017). Additional indication of binder suitability is obtained from an assessment of the compatibility profile of the drug substance, drug loading in the formulation, and the nature of other excipients projected for use in the DP formulation. For example, certain binders, such as starch, might not be suitable for high drug load formulation of a poorly compactible drug. The modality of use of binders, that is, addition as dry powder to the premix or as a solution in the binder fluid, can be determined empirically based on the current and historic practices in the organization. In terms of the particle size of the binder, a lower PSD is generally preferred to enable a rapid rate of dissolution (if the binder is to be dissolved in the granulating liquid) or a fast rate of hydration (if the binder is to be added dry into the premix). The selection of the molecular weight of the binder is typically based on the drug release profile desired, as higher molecular weight binders tend to create a gel matrix in contact with water and slow down the rate of drug release.

10.3 Impact of binders on granulation outcomes

Particle size and porosity of granules produced by WG processes are an outcome of the underlying mechanistic pathways such as nucleation, coalescence, layering, and consolidation. Of these, consolidation of the wet mass is the main mechanism that leads to granule densification, which affects the drug release from the granules and tablets. The rate of change in particle size and granule consolidation are indicators of how the binder MAs affect the wet granulation unit operation.

The dependence of granule growth kinetics on binder type, molecular weight, size, and mode of addition was studied using MCC as a model system and HPC or PVP (K29/32 grade) as the binder (Tao et al., 2013). The binder was added at 4% w/w concentration in a 10-L high-shear granulator using 75% w/w water as the granulation fluid. For HPC as a binder, the effect of binder particle size was studied by comparing granulations manufactured using the EF versus the EXF grade. To study the effect of binder molecular weight or viscosity, granulations manufactured with three different grades of HPC (of similar, fine, PSD) were compared–LXF, EXF, and GXF grades–by the dry addition method. The effect of the mode of addition of the binder was compared by dry powder versus aqueous solution addition. Granule size change during fluid addition and wet massing was monitored using an FBRM C35 probe, an inline monitoring process analytical technology. Several batches were manufactured in which the impeller was stopped at different times after the end of binder fluid addition. The wet massing times were selected based on the granule particle size growth observed using the FBRM probe. The granules were dried in an oven, and their size was measured by nested sieve analysis. The granules were also tested for pore size distribution by mercury intrusion porosimetry (Zhao et al., 2013). The rate of consolidation, represented by the porosity as a function of wet massing time for each system, was fitted to the exponential decay model.

The results of the granulation growth study indicated that:

- Dry versus wet addition.

    a. The EF grade of the binder HPC was added either as dry powder in the premix or as an aqueous solution. Binder added as a dry powder (Fig. 10.1A) led to faster granule growth than addition as an aqueous solution (Fig. 10.1C), even though the PSDs after wet massing overlapped regardless of the mode of addition.

10.3 Impact of binders on granulation outcomes

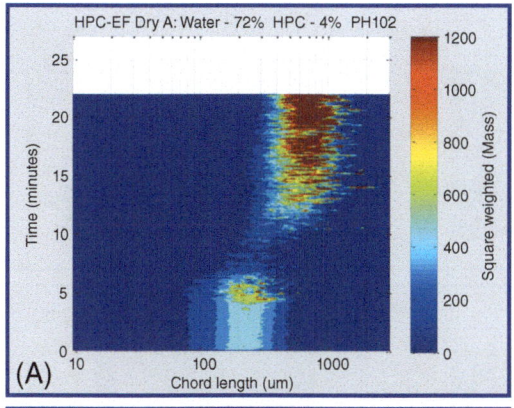
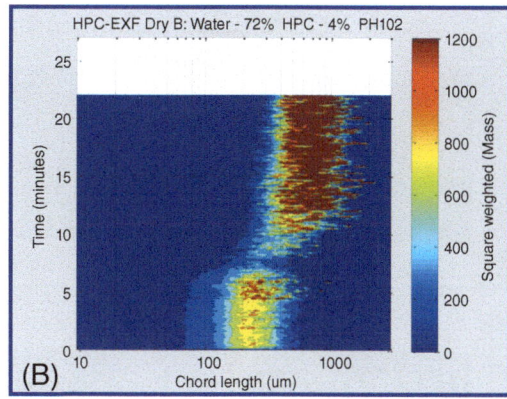
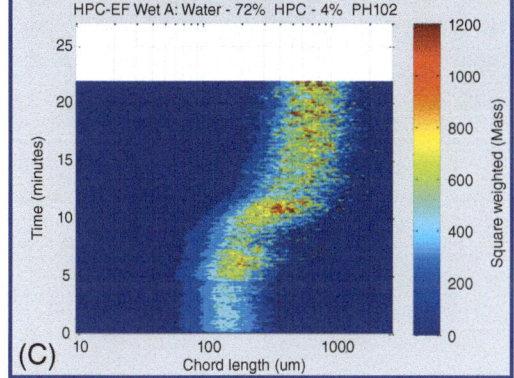

FIGURE 10.1

Effect of particle size and mode of addition of the binder on granule growth kinetics as measured by focused beam reflectance measurement (FBRM) C35 probe. Change in chord length distribution (CLD) of the particles in the granulator as a function of time is plotted in a three-dimensional plot comparing (A) dry addition of hydroxypropyl cellulose (HPC) EF grade, (B) dry addition of hydroxypropyl cellulose (HPC) EXF grade, and (C) wet addition of hydroxypropyl cellulose (HPC) EF grade as an aqueous solution.

  b. Comparing the dry versus the wet addition method for different molecular weight grades of HPC, the dry addition method consistently produced faster granule growth and narrower particle size distribution than the wet addition method (Fig. 10.2).

- Effect of binder particle size.

  a. In dry powder addition mode, faster granule size growth was observed when the particle size of the binder was smaller in the EXF grade (Fig. 10.1B), compared to the larger particle size EF grade (Fig. 10.1A).

- HPC versus PVP as a binder.

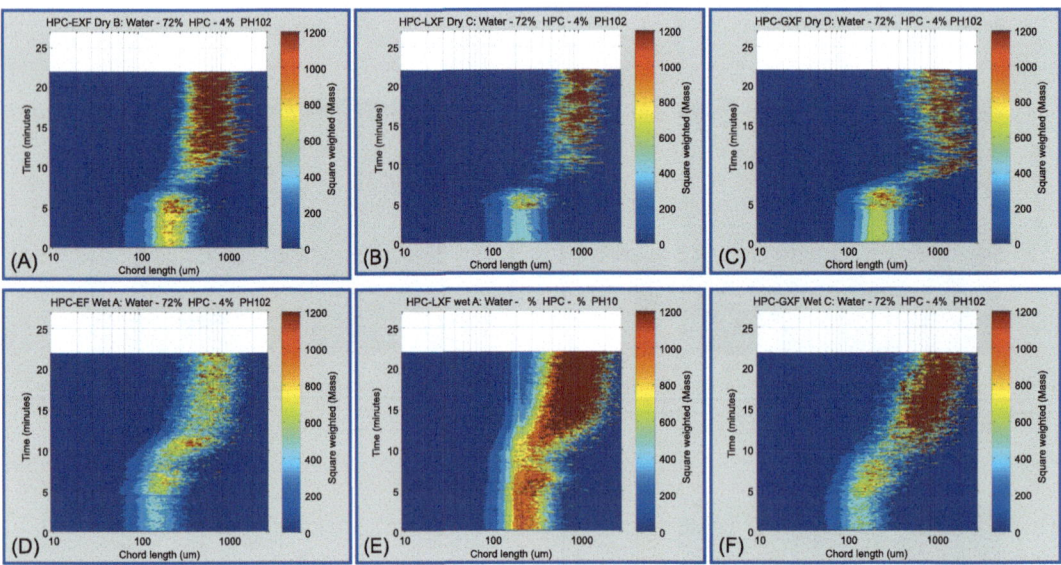

FIGURE 10.2

Effect of viscosity and mode of addition of binder, hydroxypropyl cellulose (HPC), on granule growth kinetics as measured by focused beam reflectance measurement (FBRM) C35 probe. Change in chord length distribution (CLD) of the particles in the granulator as a function of time is plotted in a three-dimensional plot comparing dry addition of (A) EXF, (B) LXF, and (C) GXF grades of HPC to their wet addition—(D) EXF, (E) LXF, and (F) GXF—as aqueous solutions.

   a. When added dry, HPC EF (Fig. 10.3A) produced larger granules and faster particle growth than PVP K30 (Fig. 10.3B). Nonetheless, the granules produced when using PVP as a binder were more uniform (i.e., narrow particle size distribution) and had higher proportions of fines.
   b. Similar results were obtained when the two binders were compared to the wet addition method. The addition of HPC EF grade (Fig. 10.3C) led to rapid particle size growth and larger granules compared to the wet addition of the PVP K30 grade (Fig. 10.3D).
- Effect of binder molecular weight. Three different molecular weight grades of HPC were investigated, represented by the commercially available grades EXF, LXF, and GXF.
   a. When added dry, a rapid rate of growth and larger-sized granules were obtained with increasing molecular weight of the binder (Fig. 10.2C).
   b. Similar trend of higher rate and extent of granule growth was obtained when the binders were added as aqueous solutions (Fig. 10.2D–F).

When comparing the change in the particle size of granules over a period of time, the PSD data obtained by nested sieve analysis (Fig. 10.4) were in good agreement with chord length distribution (CLD) data obtained using the inline FBRM C35 probe (Tao et al., 2013).

The rate-limiting steps in a granulation process should become evident in the outcome of experiments that look at different molecular weights, sizes, and modes of addition of the binder. Considering

10.3 Impact of binders on granulation outcomes

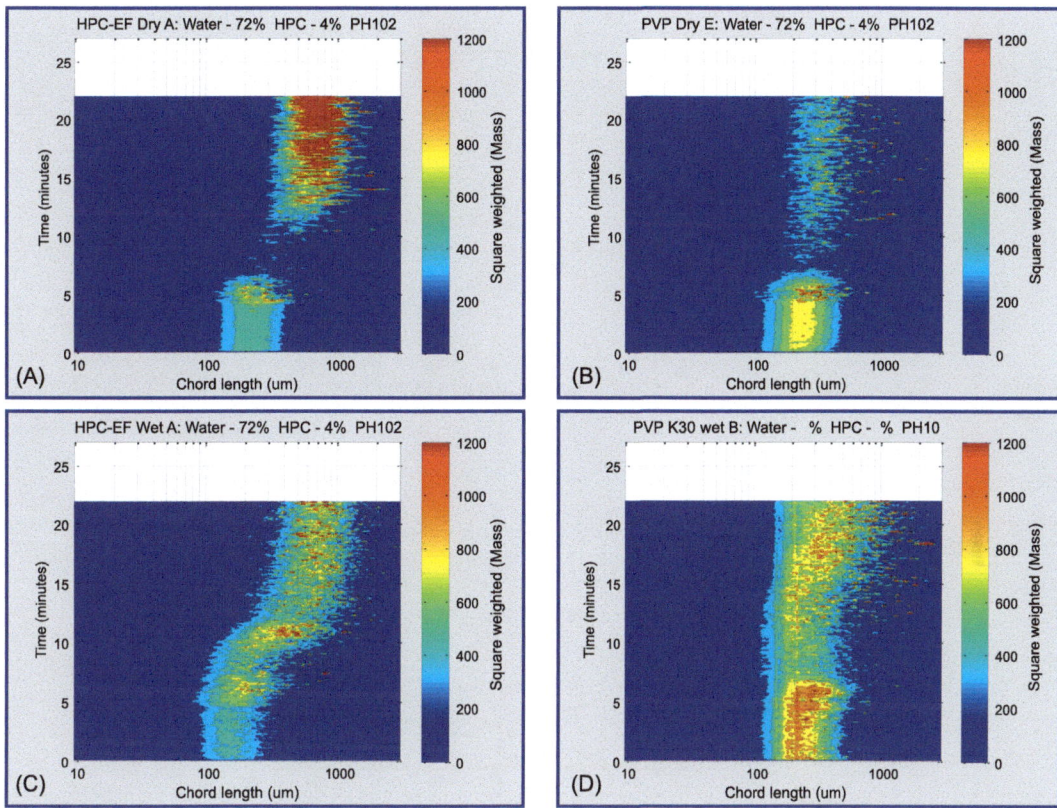

FIGURE 10.3

Effect of type and mode of addition of binder, hydroxypropyl cellulose (HPC) or poly(vinyl pyrrolidone) (PVP), on granule growth kinetics as measured by focused beam reflectance measurement (FBRM) C35 probe. Change in chord length distribution (CLD) of the particles in the granulator as a function of time is plotted in a three-dimensional plot comparing dry addition of (A) HPC with (B) PVP and wet addition of (C) HPC with (D) PVP.

this rate-limiting step to be either distribution of water/binder solution or activation of binder, for example,

- If the distribution of the binder fluid (such as water or the aqueous binder solution) was the limiting factor, the rate of binder fluid distribution would reduce further with increasing the molecular weight (and viscosity) of the binder in wet binder addition scenario. Therefore this rate-limiting factor should reflect in reduced rate of granule growth and/or densification with increasing binder molecular weight in wet addition mode. In addition, processes limited by the distribution of the binder fluid are likely to have a higher rate of granule growth with dry binder addition, which makes the binder fluid (water) the least viscous.

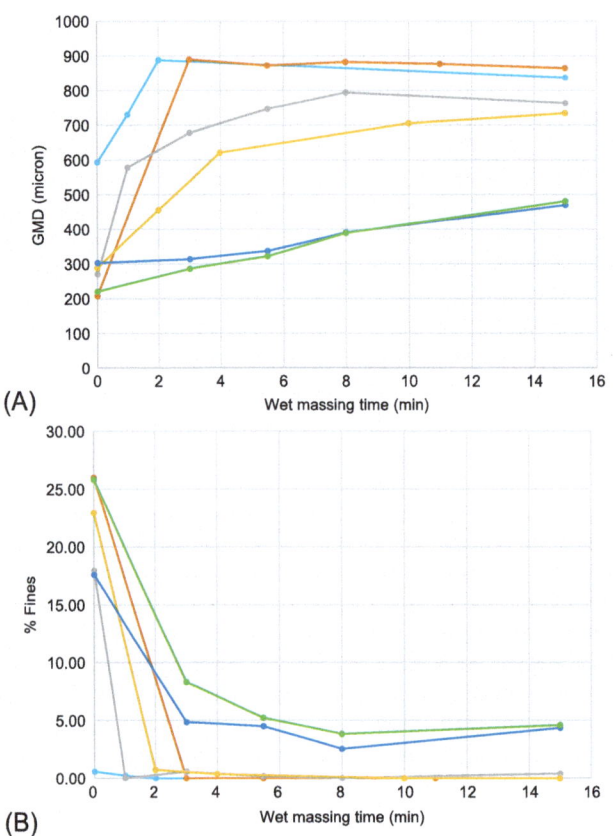

FIGURE 10.4

Particle size distribution (PSD) by sieve analyses for batches manufactured with different types, viscosity grades, and mode of addition of binders compared as (A) geometric mean diameter (GMD), or (B) % fines in the granulation.

- If the activation of the binder was the rate-limiting step in the granulation process, the effect of particle size of the binder during dry binder addition would show a lower rate of granule growth for larger size particles because the lower surface area of bigger particles limits the rate of hydration. Therefore, in systems where binder activation is the rate-limiting step, the granule growth rate and extent would be higher with the wet addition method of binder addition.

One must cautiously interpret the results of granulation studies because HSWG is inherently a multivariate phenomenon with several rate processes occurring concurrently. In addition, these phenomena likely would be specific to each granulating system, and specific changes in the process can alter the rate-limiting phenomenon of the granulation process. Thus the results of the MCC-HPC system reported here, for example, indicate that the distribution of the binder seems to be rate-limiting when comparing wet versus dry addition processes. When comparing the effect of particle size of the binder by the dry addition method, however, binder activation seems to be the rate-limiting step.

The role of binder properties on granulation outcomes was also studied by Chitu et al. (2011) using Mi-Pro high-shear mixer, fine particle size MCC, and povidone or HPMC as binders. The authors observed that interfacial forces are the predominant determinants of granulation outcomes for low viscosity binder, and that increasing binder solution viscosity by increasing the concentration of HPMC increases granulation growth rate. The accelerated growth for more viscous binders in reported studies could be an outcome of longer penetration time inside the granules, leading to the formation of a surface film that promotes coalescence (Chitu et al., 2011; Denesuk et al., 1993).

10.4 Simulation of particle velocities and impaction forces

In silico simulation is an essential modern tool in understanding the interaction of excipient attributes with process variables. Process simulation can allow for a better understanding of the impact of formulation and process variables on granule and product quality outcomes. It can also help understand the interdependence of the changes in formulation (e.g., excipient composition or variation in a functional excipient's FRCs) with changes in the process variables (e.g., within a design of experiment study) to support process operating ranges and design space.

To understand how the process parameters affect pressure and shear stress within a granulator, particle velocities and impaction forces were simulated at different locations within the granulator by the DEM of individual particles (Remy et al., 2015). This simulation modeled dry, spherical particles of ~1.8 mm mean particle diameter with a polydisperse size distribution. The input parameters to the DEM simulation for these particles were calibrated to match the experimental behavior of MCC, and, in particular, Avicel PH 101. These variables included the coefficient of rolling friction, coefficient of static friction, particle density, Young's modulus, coefficient of restitution, and Poisson's ratio. Material properties were used as is, if available from literature, and calibrated to match the angle of repose of MCC (35°). Calibration results from input parameters—coefficient of rolling friction (0.14), coefficient of static friction (0.9), and the coefficient of restitution (0.2)—were used to run the simulations.

A one-liter scale Diosna high-shear granulator was modeled, while ignoring the chopper. The results of this simulation indicated that fluctuations in shear stresses followed the frequency of blade rotation. The time-averaged stress profiles of shear stress and impaction pressure indicated that the pressure within the granulator was higher than the shear stress. In terms of spatial distribution, higher pressure and shear stresses were observed near the bottom plate by the wall (Fig. 10.5A and B). Particle velocities within the granulator followed the recirculating pattern of the roping motion, with the powder bed being lifted up and out by the impeller and the powder mass falling back toward the center (Fig. 10.5C). The proportion of total time in the granulator that particles spend in the high-pressure/shear zone is represented by the probability density function in Fig. 10.5D. These simulations helped identify the particle residence time, velocity, shear, and pressure distributions within the granulator (Remy et al., 2015).

Additional simulations were undertaken to understand the effect of certain process parameters on the pressure and shear stress profiles, particle velocity, and residence time of the particles in the high-pressure/shear zones. For example, Fig. 10.6 shows the effect of granulator fill; Fig. 10.7 shows the effect of changing the tip speed of the impeller; and Fig. 10.8 shows the effect of scaleup using constant

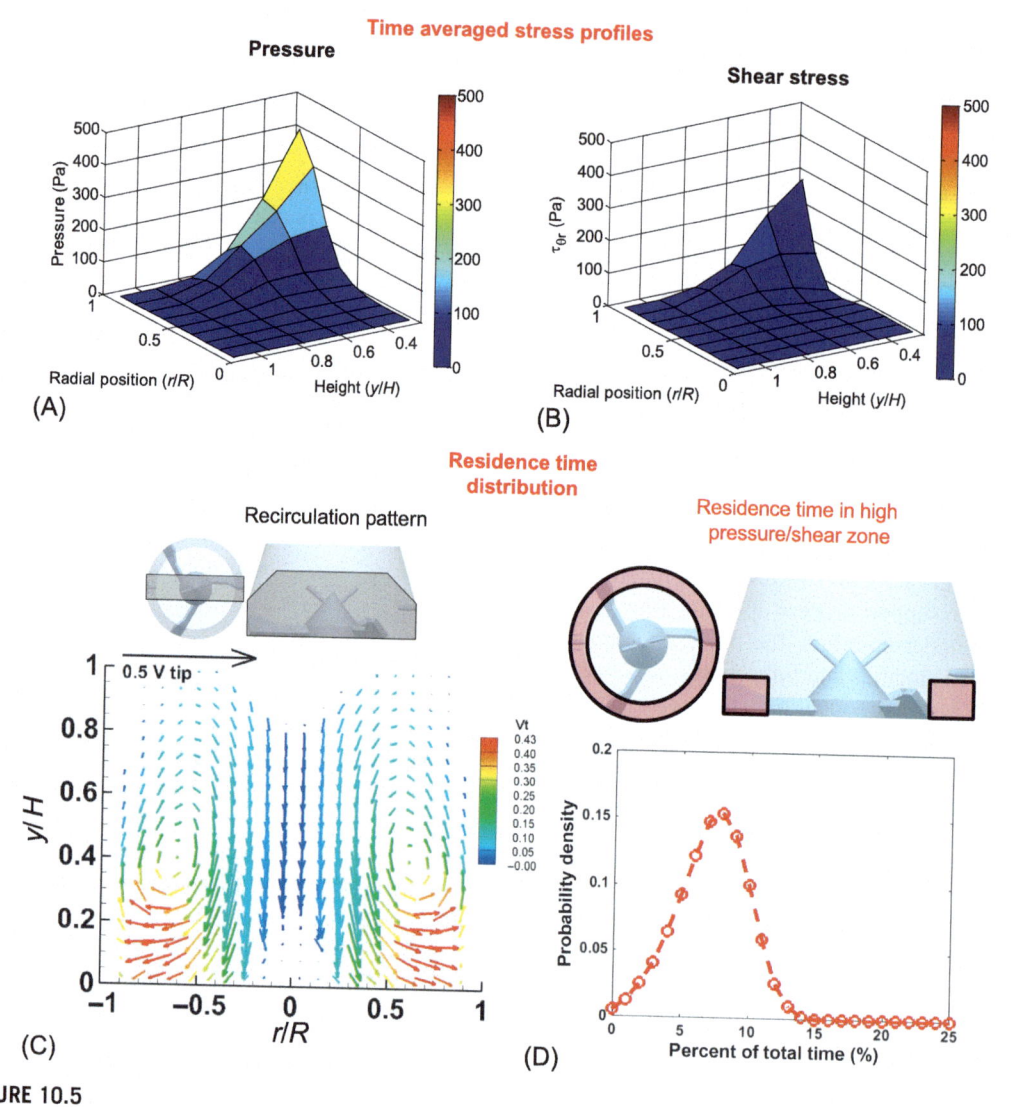

FIGURE 10.5

Time-averaged stress profiles of (A) shear stress and (B) impaction pressure within a 1 L Diosna high-shear granulator simulated by discrete element modeling (DEM). In these plots, the radial position is shown as the ratio of absolute distance from the shaft toward the outer wall "r" to the total radial distance "R"; height is plotted as the ratio of the absolute distance from the bottom of the granulator toward the top "y" to the total height "H"; and the pressure is plotted in Pascal with color coding of the plot by the observed pressures. As seen in the figure, higher pressure and shear stresses were observed near the bottom plate by the wall. (C) Particle velocities within the granulator follow the recirculating pattern of the roping motion, and (D) the proportion of total time in the granulator that particles spend in the high-pressure/shear zone is represented by the probability density function.

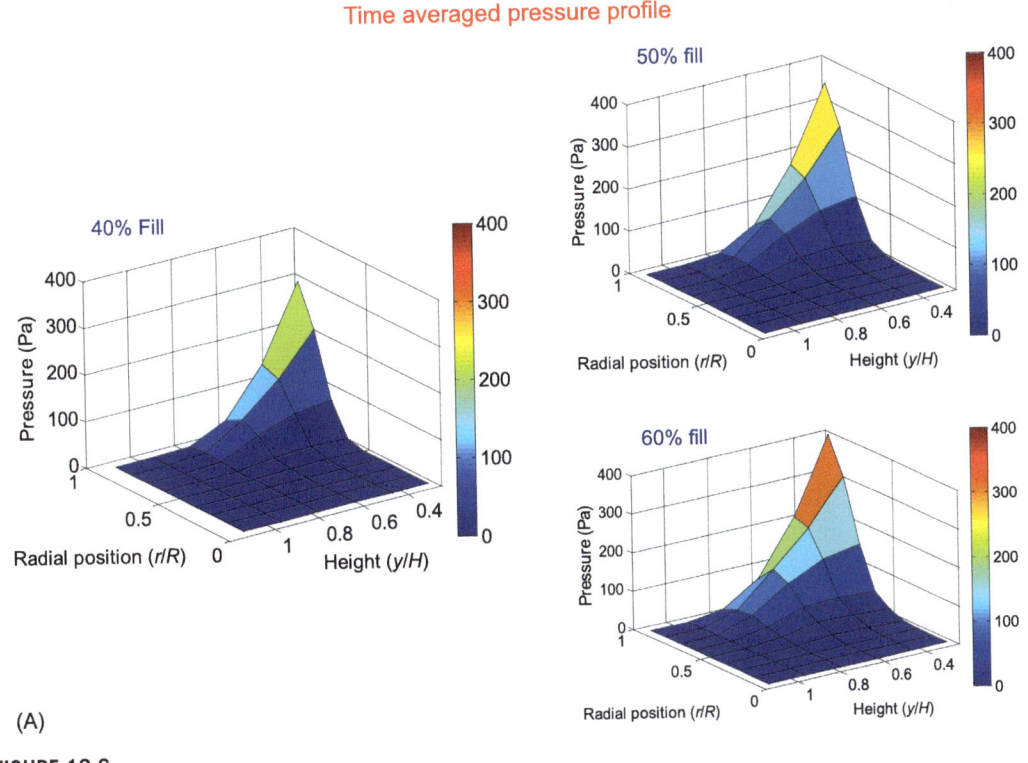

(A)

FIGURE 10.6

Effect of different volume proportions of powder fill in the granulator on the time (A) averaged pressure profile, (B) time averaged shear stress profile, (C) recirculation pattern, and (D) probability density of the residence time of the powder particles in the high-pressure/shear zone.

impeller tip speed and granulator fill. These simulations indicated that increasing powder fill level in the granulator, the tip speed, and the scale of manufacture all lead to higher pressure and shear stresses near the bottom plate and the wall. Because the pressure and the shear stress in the granulator are linked to the consolidation and attrition of the granules, these changes in process parameters can affect the rate and extent of granule growth and consolidation (Remy et al., 2015).

As useful as this simulation is to understanding the distribution of forces acting upon the granules, it does not take into account changes in particle size and density during the granulation process as primary powder particles agglomerate, consolidate, break, and undergo other changes with the progress of granulation.

10.5 Modeling binder mode of addition effects

Particle-level phenomena can help understand the mechanistic basis of differences in the rate of granule growth between the wet and the dry addition of binders. Granule formation starts with the formation

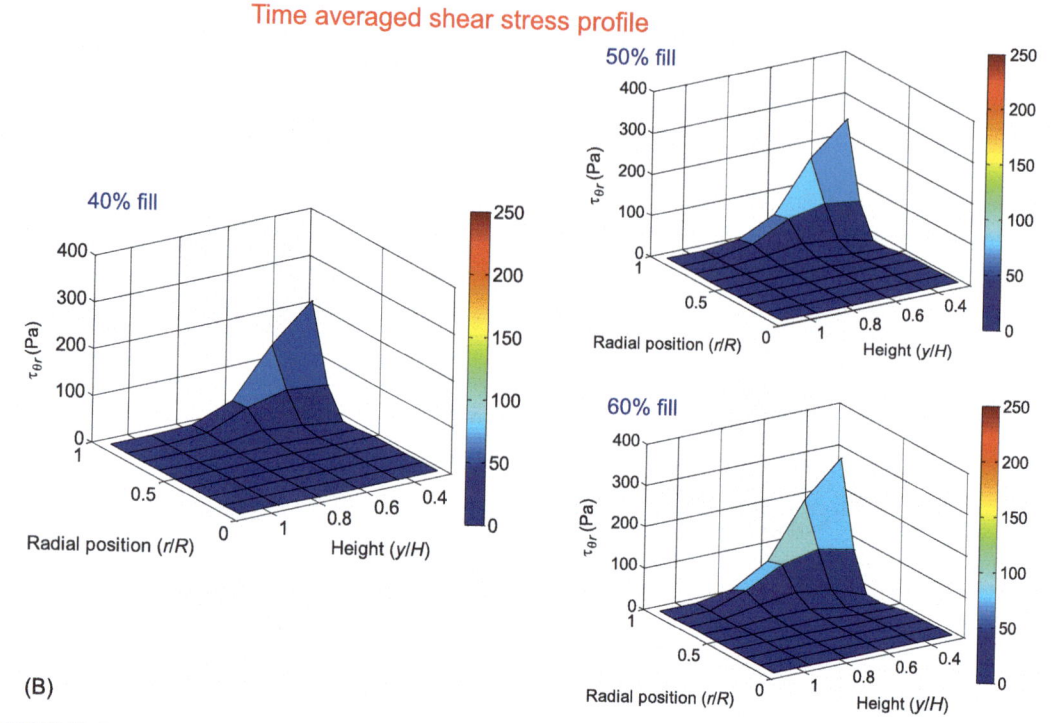

(B)

FIGURE 10.6, cont'd.

of cohesive/adhesive liquid bridges between particles. The balance of particle–particle cohesion forces with the compaction and attrition forces during the process then determine whether the particles will continue to stay together and compact to ultimately form granules or break apart into individual particles. Knowledge about the quantity of granulating liquid would be important for determining the number of primary powder particles that start to coalesce into granules. Therefore both the amount of water and the cohesive strength of the aqueous liquid bridges become important to the initiation of granule formation processes, persistence through process stresses, and the number of primary powder particles engaged through liquid bridges into structures that ultimately will form granules.

The cohesive strength of the liquid bridges is derived from the presence of the binder present in the dissolved–or at least surface hydrated (called activated)–state. Thus, binder activation could become a key rate-limiting step in processes in which the binder is added dry, and water is added subsequently as a binder fluid to effect granulation. For wet binder addition processes, however, the distribution of the binder solution within the powder bed could become a rate-limiting step.

To better understand the differences in process outcomes between the dry and the wet addition of the binder, these rate processes were modeled using a two-compartment population balance model (PBM) (Fig. 10.9) (Chaturbedi et al., 2017). This model is based on a volume-based three-dimensional PBM proposed by Verkoeijen et al. (2002), wherein the total volume of a particle is divided into three independent particle parameters (solid, liquid, and gas), which can keep track of different particle properties such as porosity, moisture content, and pore saturation. This model overcomes some

10.5 Modeling binder mode of addition effects

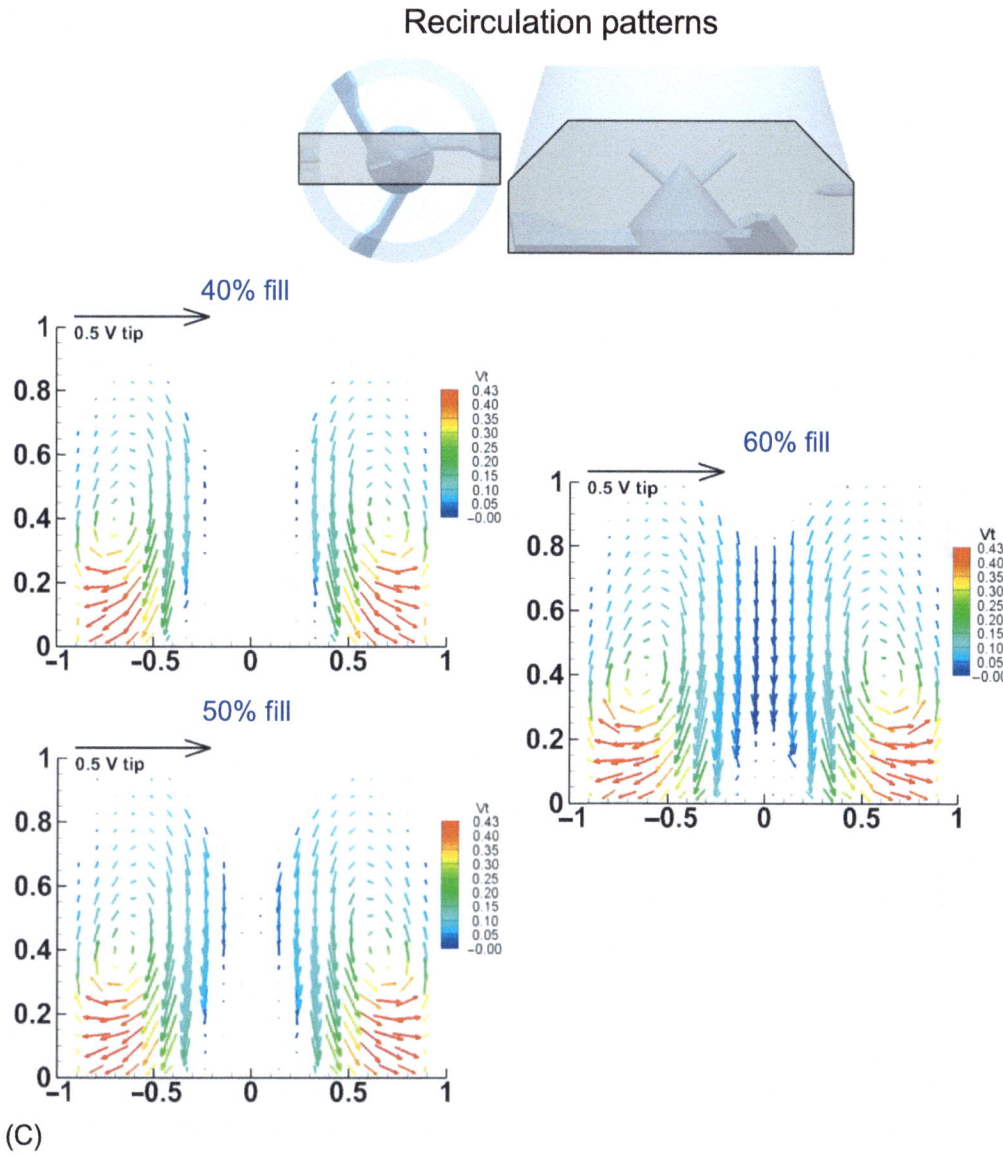

FIGURE 10.6, cont'd.

of the traditional limitations of one-dimensional PBM models (Iveson, 2002). This model includes mechanistic breakage and aggregation kernels, which allow more mechanistic incorporation of these mechanisms. This compartment-based model accounts for the heterogeneity in liquid content and viscosity between the spray and the bulk zone by solving the population balance separately for the two zones, while these zones or the compartments do exchange particles between themselves because

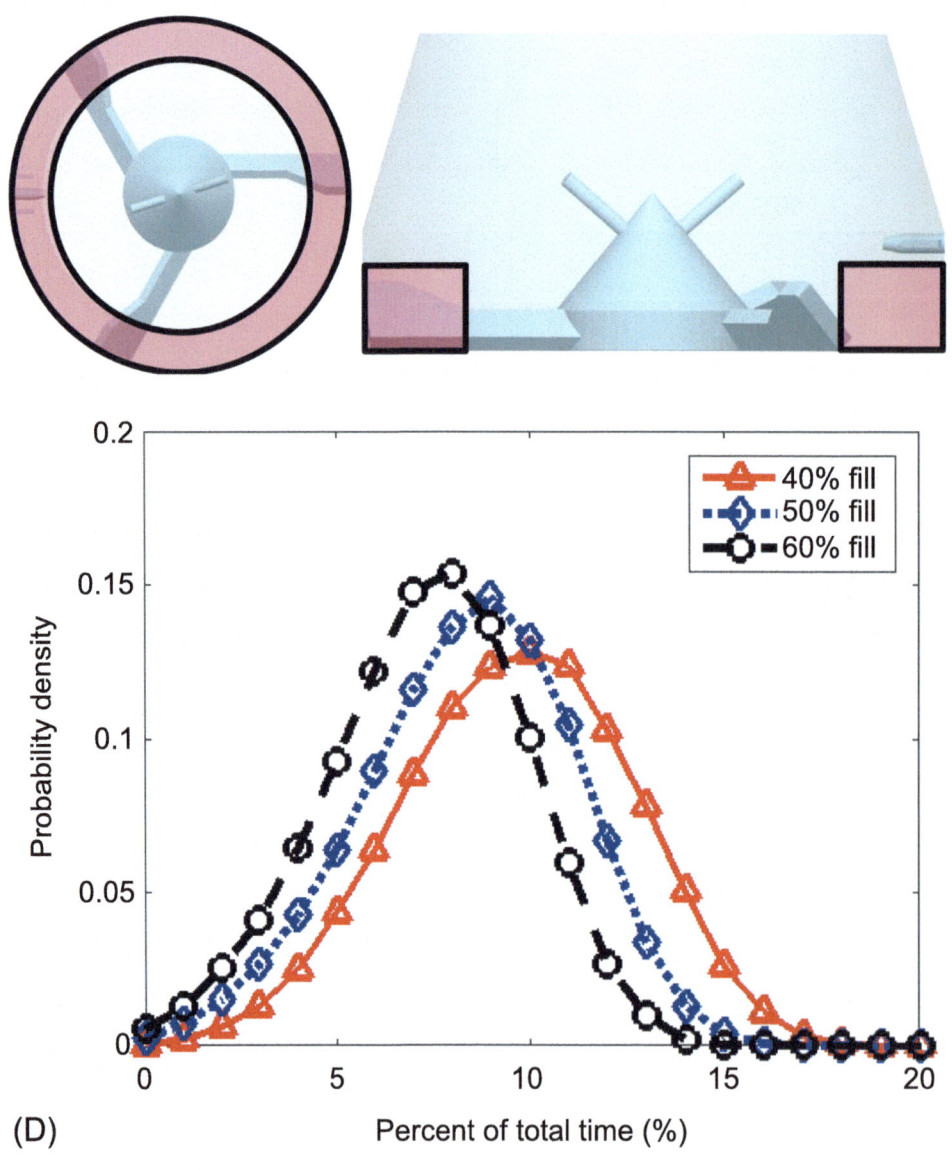

FIGURE 10.6, cont'd.

10.5 Modeling binder mode of addition effects

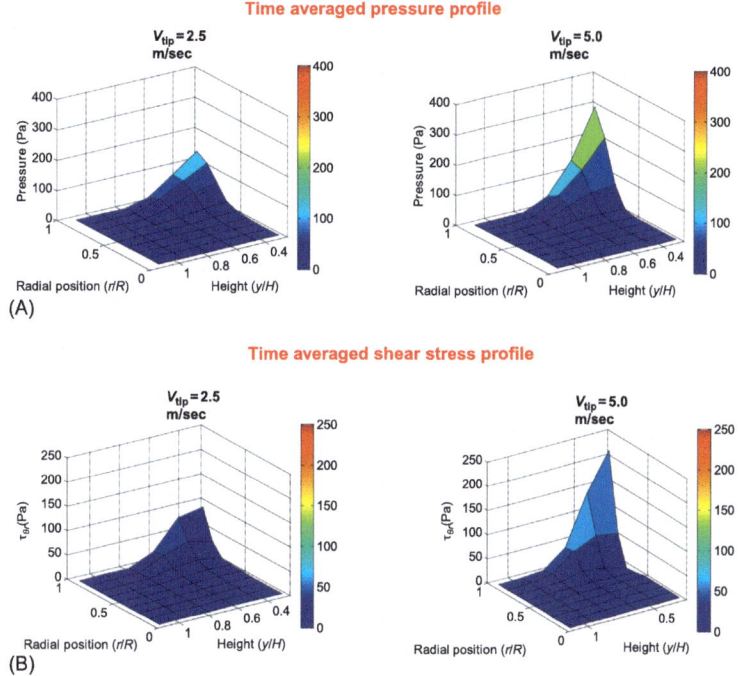

FIGURE 10.7

Effect of impeller speed on the (A) time-averaged pressure profile, (B) time-averaged shear stress profile, (C) recirculation pattern, and (D) probability density of the residence time of the powder particles in the high-pressure/shear zone.

of the rotation of the impeller. In addition, a binder dissolution submodel that computes the rate of dissolution of binder in water, which subsequently governs the change of viscosity in the system, has been developed and integrated into the compartment based PBM to account for the dynamic changes in viscosity with addition of water into the system with dry binder particles (Chaturbedi et al., 2017). This model was calibrated with the experimental data from wet granulation experiments with both dry and wet binder addition, shown in the previous section.

For the process of model development, the system is divided into two compartments based on the viscosity and liquid amount: the spray zone and bulk zone (Chaturbedi et al., 2017). The areas with more liquid and less binder come under the spray zone, while the bulk zone has areas with less liquid and more binder, as illustrated in Fig. 10.9. The compartments differ in viscosity, amount of liquid, and the number of particles—and are modeled separately. The liquid and gas volumes are lumped into the solid granule volume, and a lumped one-dimensional PBM model is developed for each compartment as follows (Chaturbedi et al., 2017).

PBM for compartment 1 (spray zone):

$$\frac{\partial F_1(s_1,t)}{\partial t} = R_{1,agg} + R_{1,break} - (R_{circulation}(F_1(s_1,t) - F_2(s_2,t))) \tag{10.1}$$

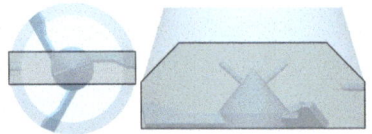
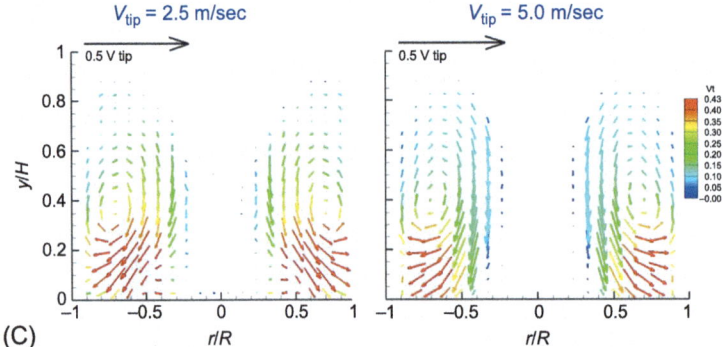
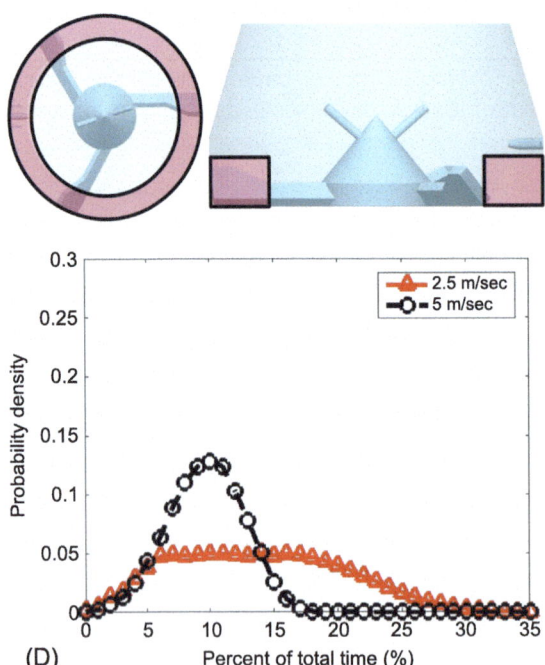

FIGURE 10.7, cont'd.

10.5 Modeling binder mode of addition effects

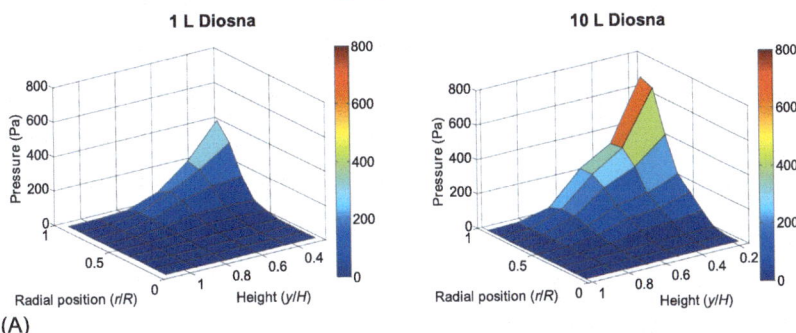

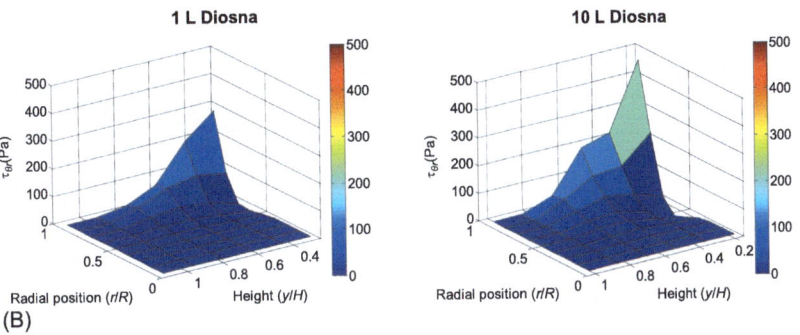

FIGURE 10.8

Effect of impeller speed on the (A) time-averaged pressure profile, (B) time-averaged shear stress profile, (C) recirculation pattern, and (D) probability density of the residence time of the powder particles in the high-pressure/shear zone.

$$\frac{\partial L_1(s_1,t)}{\partial t} = F_1(s_1,t)\frac{dl_1}{dt} + R_{1,agg,liq} + R_{1,break,liq} \tag{10.2}$$

$$\frac{\partial G_1(s_1,t)}{\partial t} = F_1(s_1,t)\frac{dg_1}{dt} + R_{1,agg,gass} + R_{1,break,gas} \tag{10.3}$$

PBM for compartment 2 (bulk zone):

$$\frac{\partial F_2(s_2,t)}{\partial t} = R_{2,agg} + R_{2,break} - \left(R_{circulation}\left(F_2(s_2,t) - F_1(s_1,t)\right)\right) \tag{10.4}$$

$$\frac{\partial L_2(s_2,t)}{\partial t} = F_2(s_2,t)\frac{dl_2}{dt} + R_{2,agg,liq} + R_{2,break,liq} \tag{10.5}$$

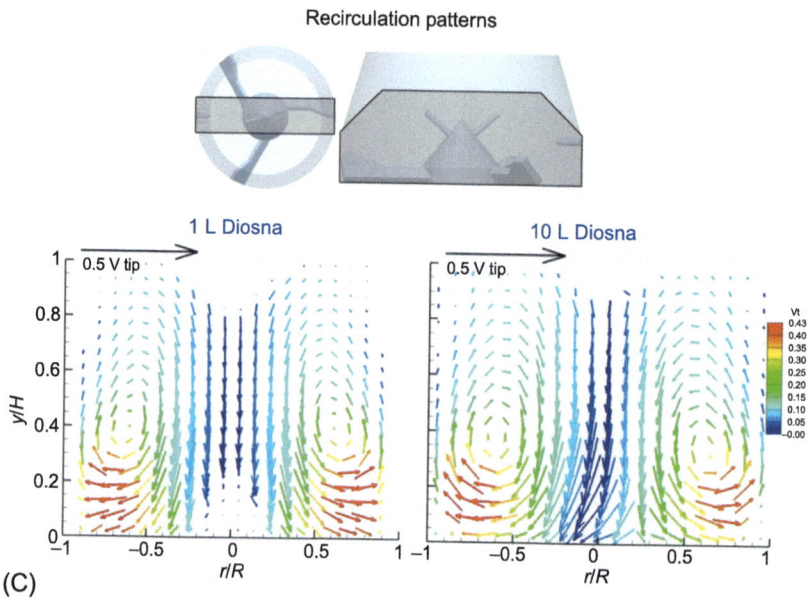

(C)

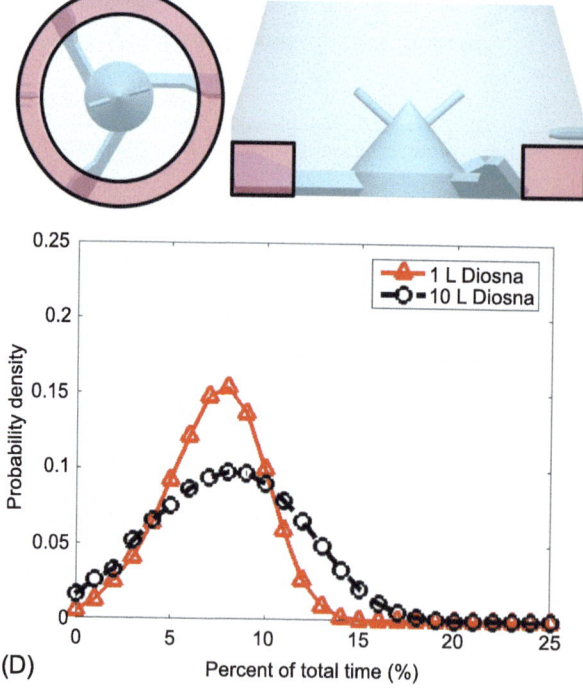

(D)

FIGURE 10.8, cont'd.

10.5 Modeling binder mode of addition effects

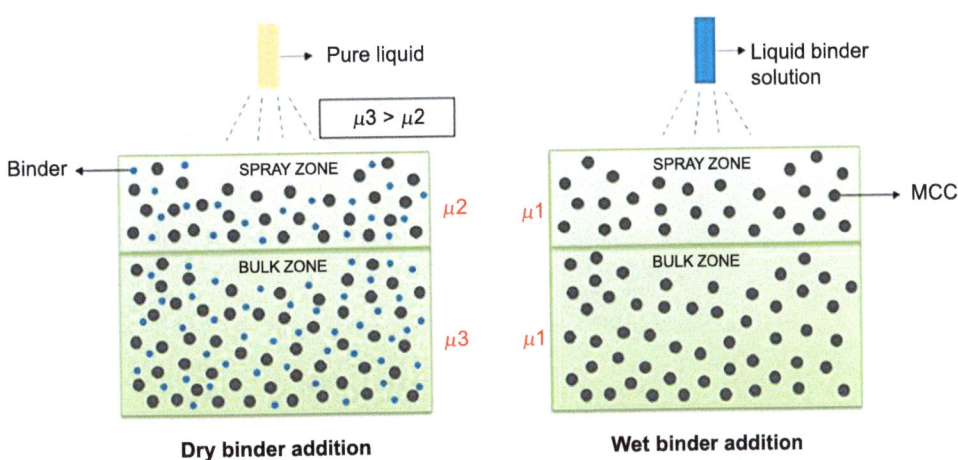

FIGURE 10.9

Schematic of a two-compartment population balance model (PBM) of high shear wet granulation showing the spray zone and the bulk zone within the granulator. The cases of dry versus wet binder addition are illustrated with the differences in the distribution of the binder, the composition of the binder fluid, and differences in viscosity (μ).

From Chaturbedi et al. (2017)

$$\frac{\partial G_2(s_2,t)}{\partial t} = F_2(s_2,t)\frac{dg_2}{dt} + R_{2,agg,gas} + R_{2,break,gas} \qquad (10.6)$$

where $F_x(s_x; t)$ represents the number of particles with solid volume s_x at time t in the compartment x, where $x = 1$ or 2 indicates the spray zone and bulk zone, accordingly. $L_x(s_x; t)$ and $G_x(s_x; t)$ are the total liquid volume and gas volume, respectively, present in the class of particles with solid volume of s_x in compartment x. $l_x(s_x; t)$, $g_x(s_x; t)$, respectively, represent the liquid and gas volumes of each particle having solid volume s_x in compartment x. R represents the rates of particle transfer between different size classes because of the various rate processes. The details of mathematical equations used in the various mechanistic kernels (namely, aggregation, breakage, liquid addition and consolidation, circulation, and binder dissolution) can be found in Chaturbedi et al. (2017).

This model incorporates several material properties and process parameters that feed into the rate processes (e.g., aggregation, breakage, consolidation, and binder dissolution) that determine granule growth. These parameters include binder material properties, such as binder density, size, diffusion coefficient, solubility, and contact angle; particle properties, such as granule density, minimum granule porosity, and granule saturation; and process parameters, such as impeller diameter, impeller speed, liquid addition rate, wet massing time, and circulation rate between compartments. Sensitivity of the average particle diameter with respect to these parameters was studied by offsetting their values in six different levels between -30% and $+30\%$ from their base value and measuring the percentage change in average diameter at the start and end of wet massing time. The results of these analyses are shown in Fig. 10.10.

368 Chapter 10 Effect of binder attributes on granule growth and densification

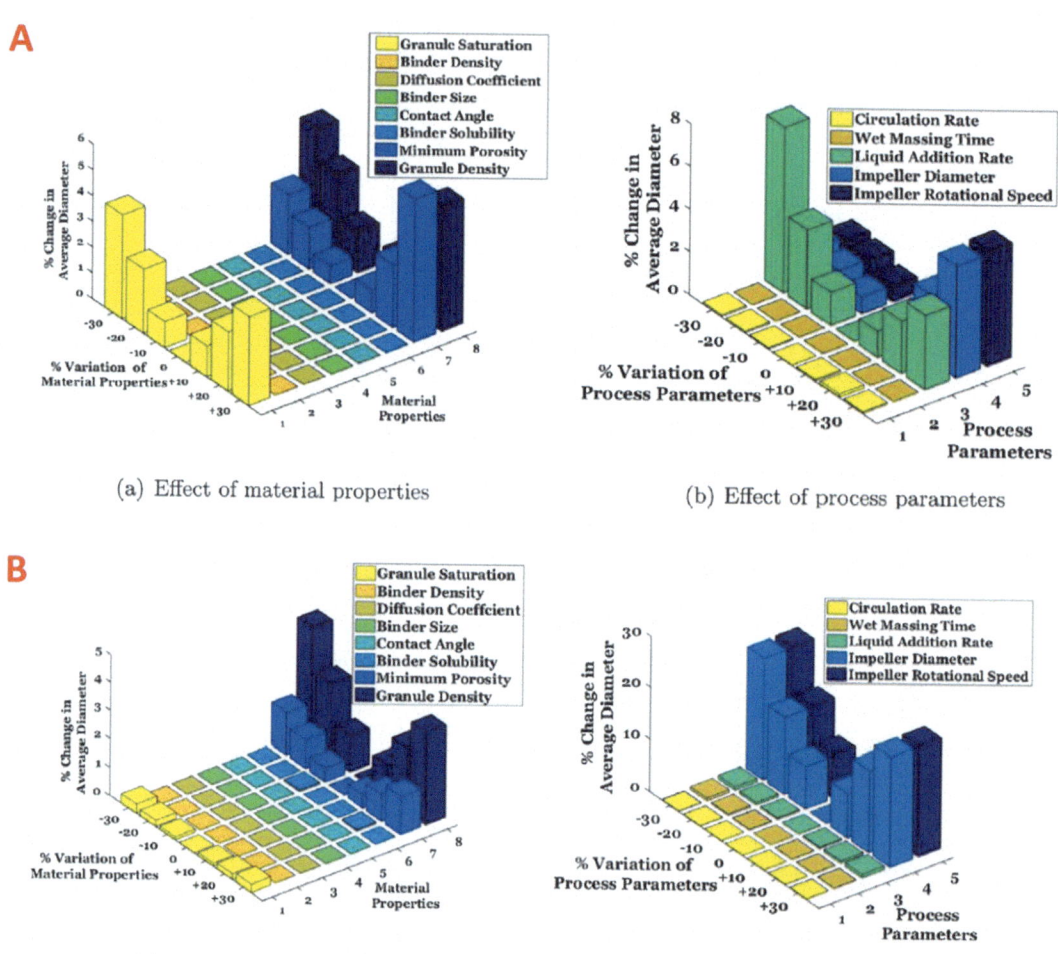

FIGURE 10.10

Sensitivity analysis of the average granule diameter at the beginning (A) or end (B) of the wet massing stage with (a) material properties and (b) process parameters.

From Chaturbedi et al. (2017)

Fig. 10.10A(a) illustrates the sensitivity of average particle diameter to the material properties before the start of the wet massing stage. It can be seen that granule saturation plays a major role in particle growth before the start of wet massing, probably because granule saturation dictates the demarcation between the internal and external liquid. Lower granule saturation leads to less internal and more external liquid and thus facilitates faster growth. Minimum porosity of particles also has a significant effect on particle growth because it affects the consolidation of the particles and, in turn, affects the extent of expulsion of internal liquid to the particle surface. Lower minimum porosity leads to higher consolidation of particles, leading to higher external liquid content and more growth. Granule density also plays an important role because denser, heavier granules have less mobility and thus result in fewer collisions between particles and therefore less growth (Chaturbedi et al., 2017).

Fig. 10.10A(b) shows the effect of process parameters on the particle diameter before the start of the wet massing stage. Because a faster and bigger impeller leads to more consolidation, leading to higher external liquid and also more collisions between particles, the diameter and the rotational speed of the impeller were seen to have a much more significant effect on the granule growth. Also, the liquid addition rate plays an important role because slower liquid addition helps in the uniform distribution of liquid on the powder bed, causing the viscosity and liquid content to be uniform throughout. This results in less growth compared to the situation with heterogeneity (pockets of higher viscosity at some places and higher liquid content at other places inside the powder bed) caused by the faster addition and thus nonuniform distribution of liquid. The wet massing time plays a significant role, although the sensitivity of average diameter to it is less compared to other parameters mentioned before (Chaturbedi et al., 2017).

Fig. 10.10B(a) shows the effect of material properties on average diameter of granules at the end of the wet massing stage. In comparison to Fig. 10.10A(a), it can be seen that the granule density and minimum porosity remain major factors even at the end of wet massing. Probably because, at the end of the wet massing, particles are consolidated and have fewer pores in them to hold internal liquid, the granule saturation that represents the saturation of the pores becomes less important. In Fig. 10.10B(b), which shows the effect of process parameters on the average diameter at the end of wet massing, the impeller speed and diameter still can be seen to play a significant role. The liquid addition rate, however, becomes less significant because at the end of wet massing, the liquid is relatively well distributed compared to at the start of wet massing (Chaturbedi et al., 2017).

Fig. 10.11A illustrates the experimentally observed and model-predicted average diameter of granules with time for wet granulation with HPC. Fig. 10.11A(a) and A(b), respectively, show the dynamic evolution of average diameter for wet and dry addition of HPC. Because the water added during dry binder addition is less viscous than the binder solution added during wet addition, the drop penetration rate is higher for the water. Therefore the water goes into the granule pores faster in dry addition than the binder solution in wet addition. As a result, particle growth starts at the onset of liquid addition stage in wet addition because of the higher amount of external liquid present, which helps the particles stick to each other after collision. In contrast, for dry addition, because the water goes into the pores quickly, no external liquid is present to facilitate particle growth. In the absence of external liquid, the binder particles also do not dissolve and create a viscous liquid to help the particles grow. After some time of continuous liquid addition, however, the pores get filled with water, and the additional liquid being added to the system stays on the particle surface and starts to dissolve the binder particles. Because the amount of water available for the binder to dissolve in at this point is much less, the resulting solution is saturated and significantly more viscous than the binder solution added during wet addition. Therefore the particles grow at a very fast rate that is higher than the growth rate during wet addition. Mainly because of the higher viscosity in the case of dry addition, for reasons explained above, the particles also have a higher average diameter at the end of the process (Chaturbedi et al., 2017).

Fig. 10.11B shows the experimentally observed and model-predicted average diameter with time for wet granulation with PVP. Fig. 10.11B(a) and B(b), respectively, represent wet and dry addition of PVP. PVP behaves a little differently from the HPC binder because of the difference in dissolution rate

370 Chapter 10 Effect of binder attributes on granule growth and densification

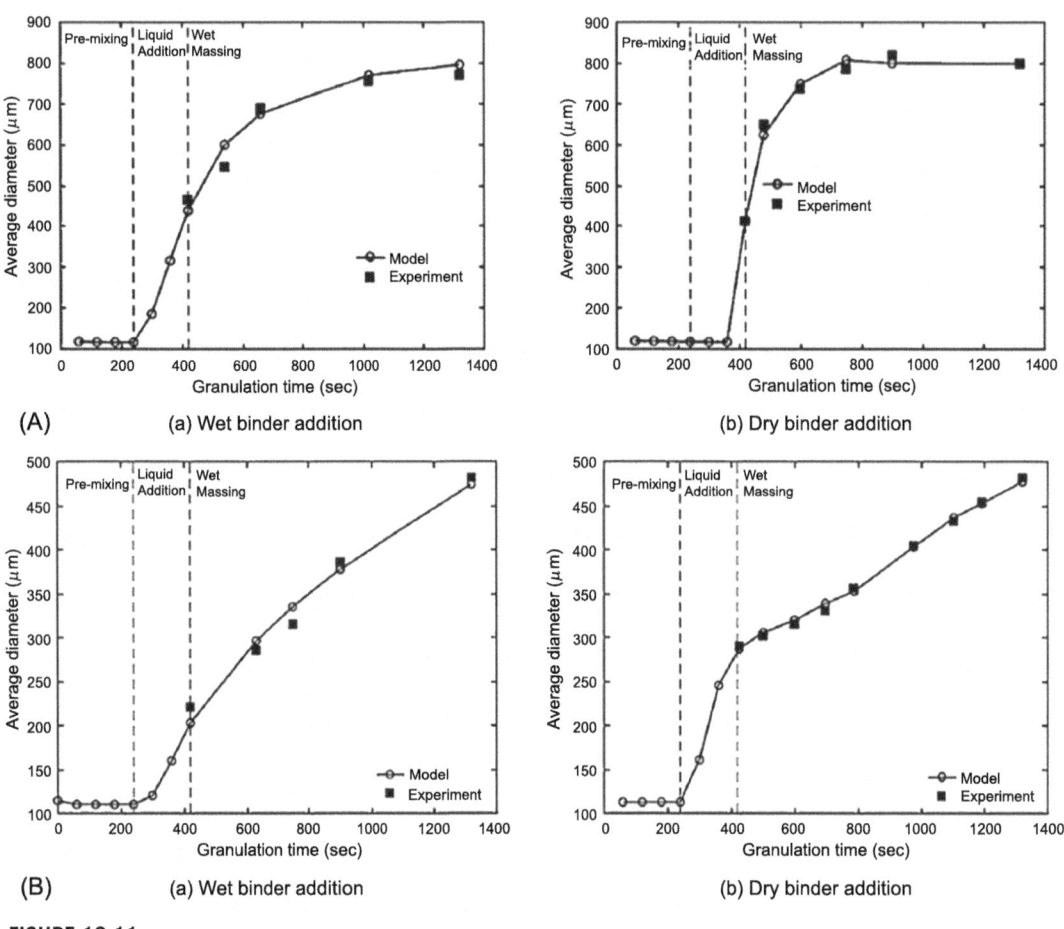

FIGURE 10.11

Experimental (data points shown in dots) and model-predicted (lines) average diameter over the course of granulation for (a) wet and (b) dry addition of HPC (A) and PVP (B) as binders.

From Chaturbedi et al. (2017)

between them. Because of the faster dissolution of PVP in water in case of dry addition, the binder solution becomes very viscous right at the start of the liquid addition stage. The binder solution for dry addition, being more viscous than the binder solution added in wet addition, penetrates slowly into the pores of the particles. This results in a higher amount of external liquid that is also more viscous, in the case of dry addition compared to wet addition. Therefore the particles grow very fast in dry addition from the onset of liquid addition. Because PVP dissolves in water quickly, however, the binder solution in dry addition becomes similar to the binder solution in wet addition in terms of concentration. This results in similar viscosities for both cases, resulting in almost identical growth rates. Similar to HPC, the particles attain a slightly higher diameter during dry addition. It is important to note that the difference between the final average diameter in wet and dry addition of HPC is higher than that of PVP because the rate of change of viscosity with binder concentration is considerably higher in case of HPC. As a result, the difference between the wet and dry addition systems, which is mainly because of their difference in viscosity, is higher with HPC than with PVP (Chaturbedi et al., 2017).

This two-compartment PBM model is able to account for the differences in granule growth dynamics for a dry binder addition system compared to wet addition. For example, binder added dry is well mixed, but the granulating liquid (water) is distributed unevenly–modeled by the presence of a spray zone and a bulk zone. As a result, the granulator has different viscosities in the two zones, which affects the overall granule growth dynamics. Simulations are also able to capture viscosity differences between dry and wet binder additions. A dissolution model is integrated with this PBM to account for binder dissolution in the granulating liquid (water). This allows incorporation of dynamic changes in viscosity in both zones during the granulation, and simulates its influence on the granule aggregation and breakage rates. Simulation of dry binder addition shows higher aggregation in the bulk zone, which has more particles/granules and a higher local viscosity.

Nonetheless, modeling a process incorporates certain assumptions and might not always be an accurate reflection of the physical state of matter, depending on the scale at which the processes are modeled. For example, uniformity of binder composition is an inherent assumption in the dry mixing models of wet granulation. This assumption might not hold true, however, at a microscopic, single-granule scale. For example, Reynolds et al. (2004) have shown that there is a broad variation in binder content between individual granules during early stages of the batch granulation process, and that this variation persists in granules of small size. Using calcium carbonate as a substrate and polyethylene glycol as a binder in high-shear melt granulation, these authors showed a distribution of binder concentration by the size class of granules, with smaller granules containing less than average proportions of binder and the larger ones containing more. Interestingly, the authors reported a sharp, step-change between the two regimes. In systems that support granule growth by layering, an increase in the size of larger granules happens over time in granulation, with the composition approaching the bulk average.

10.6 Granule consolidation kinetics

Granules are particle agglomerates that contain all three states of matter: solid, liquid, and gas. As particles start to bond together to form granules during HSWG, they entrain masses of water and binder solution. The impaction forces during mixing lead to the consolidation of granules with the ejection of the liquid and the gas content. These forces are generated by interparticle collisions, collisions of granules/particles with the wall of the granulator, and the weight of the powder bed as the impeller movement lifts the granules against gravity in each sweeping motion. The shear forces, however, lead to the breakage of fragile granules in turn, preventing the formation of large, relatively loose or lower density agglomerates. The compressive impaction forces lead to the formation of harder, denser granules with the squeezing out of the binder fluid to the surface.

10.6.1 Binder hydration

Granulation kinetics in high-shear wet granulation (HSWG) process can be limited by binder hydration (also called as binder activation) or distribution of binding solution. The hydration or activation of a

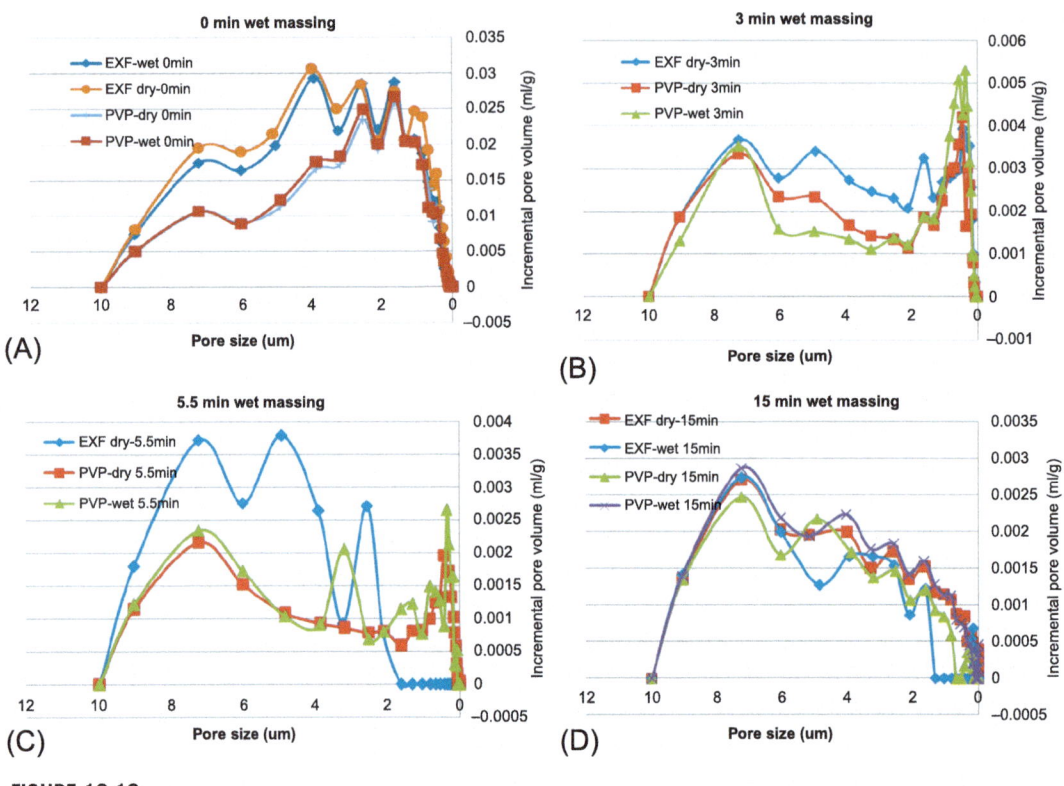

FIGURE 10.12

Granule porosity, measured by mercury intrusion porosimetry (MIP), as a function of wet massing time for different types and modes of addition of binder. Porosity of granulations made with dry and wet addition of HPC EXF grade is compared to that of PVP K30 grade at (A) 0 min, (B) 3 min, (C) 5.5 min, and (D) 15 min wet massing times.

binder can be described as a change in the physical state of the binder at room temperature as a function of its water content. This change involves a transition from a glassy state to a rubbery state, which can occur at different relative humidity thresholds and at different rates depending on the binder type and molecular weight. For example, Li et al. (2011) showed that PVP K12 (lower molecular weight grade povidone) undergoes phase transition from the glassy state to the rubbery state at a much lower water content than the higher molecular weight PVP K29/32. The authors also showed that this change in the physical state of the binder, and not necessarily complete dissolution of the binder in the liquid, is essential to initiate granulation.

10.6.2 Consolidation rate studies

The results of granulation consolidation rate investigation (reported in Section 10.3) as granule porosity at different wet massing times (Fig. 10.12) were fitted to the following exponential decay model of

10.6 Granule consolidation kinetics

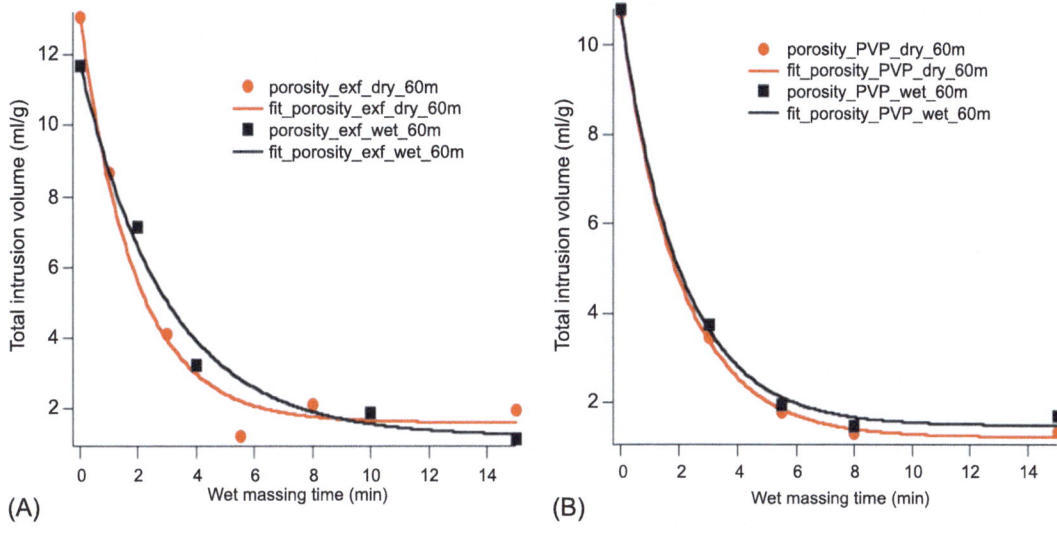

FIGURE 10.13

Granule consolidation kinetics as a function of wet massing time for different types and modes of addition of binder. Rate of granule consolidation of dry and wet addition of (A) HPC EXF grade is compared with dry and wet addition of (B) PVP.

granule consolidation (Fig. 10.13) (Kenningley et al., 1997; Zhao et al., 2013) and model parameters estimated for various types of binders and modes of binder addition (Table 10.1).

$$\frac{\varepsilon - \varepsilon_{min}}{\varepsilon_0 - \varepsilon_{min}} = e^{-kt}$$

where, ε_{min} is the minimum achievable porosity, k is the exponential consolidation rate constant, ε is granule porosity at time t, and ε_0 is initial granule porosity.

In studying the effect of binder viscosity and the size of constituent particles on agglomeration behavior in HSWG using calcium carbonate powders as the substrate and silicone fluids as binder liquids, Kenningley et al. (1997) showed that granules were formed readily with fine powders and binder liquids of high viscosity, but not with coarse powders and low viscosity binders. The authors hypothesized that the impact strength of granules was an important determinant in the survival of granules in the high-shear mixer. The use of a higher viscosity binder fluid, the authors concluded, led to forced viscous flow within the granule that gave higher impact resistance.

Our results of granule consolidation rate studies indicated that (Zhao et al., 2013):

- Systems with HPC EXF added dry have a higher porosity post binder addition, but a faster consolidation rate, suggesting hydration to be the rate-limiting step for granule consolidation.
- Comparing the two binders (HPC vs. PVP), at the end of water addition, granules made with HPC-EXF have higher porosity than those made with PVP. Upon wet massing, both reach a comparable minimum achievable porosity.

Table 10.1 Summary of fitting parameters for granule consolidation kinetics as a function of binder type, molecular weight, and mode of addition

Parameter	EXF-dry, 60 to 100 m	EXF-wet, 60 to 100 m	PVP-dry, 60 to 100 m	PVP-wet, 60 to 100 m	EXF-dry, 20 to 30 m	EXF-wet, 20 to 30 m	GXF-dry, 20 to 30 m	GXF-wet, 20 to 30 m
k	0.53167	0.34055	0.49323	0.48732	0.66317	0.28963	0.40652	0.61461
I_0	13.192	11.799	10.756	10.804	20.745	22.507	22.749	25.245
I_{min}	1.5869	1.1925	1.2057	1.4584	3.3946	2.7074	1.1366	1.5259

- Comparing the mode (dry powder vs. aqueous solution) of addition of PVP, the systems with PVP added dry and wet have similar granule consolidation kinetics. This is consistent with the higher water uptake and faster activation of PVP as a binder (Li et al., 2011).
- Comparing the mode (dry powder vs. aqueous solution) of addition of HPC low viscosity EXF grade, dry addition has a higher rate of consolidation than wet addition. This suggests that hydration is the rate-limiting step for granule consolidation for HPC.
- Comparing the effect of viscosity of the binder on the rate of consolidation during granulation, the systems with GXF grade of HPC added wet have a higher porosity post binder addition but a faster consolidation rate. This suggests that binder distribution is the rate-limiting step for this high viscosity grade.
- Comparing the mode (dry powder vs. aqueous solution) of addition of HPC high viscosity GXF grade, wet addition has a higher initial porosity but a faster consolidation rate. This suggests that binder distribution is the rate-limiting step for this high viscosity grade binder.

Granule porosity also varied as a function of granule size. Increasing the molecular weight of HPC increased pores in the larger size range (Zhao et al., 2013).

10.6.3 Granule consolidation

Compressive and shear forces during the granulation process tend to push the particles together, thus reducing the apparent density of the agglomerate, either when liquid bridges within an agglomerate are viscous and/or deformable or if the agglomerate is plastic and malleable (Simons & Fairbrother, 2000). Depending on the rate and efficiency of the binder in forming particle agglomerates, the agglomerates can have different porosity (differences in the structure and the amount of air entrapped within. The entrapped air gets released because of the external compressive and shear forces, consolidating the granule. The first stage after the removal of the air is liquid pore saturation, whereby all pores are occupied by the binder solution. Upon further consolidation, the liquid is squeezed out, resulting in the surface film of the binder and the agglomerates becoming more adhesive. These agglomerates then can start to rapidly coalesce or layer, with a significant increase in granule growth rate. Granulation processes typically are stopped at or before this stage, to achieve an optimum level of granule densification that will lead to a solid fraction (in the granules upon drying) somewhat lower than the solid fraction desired in the final compressed DP.

Because early stages of granule consolidation are associated with the exclusion of air and occupancy of intragranular pores with the granulating liquid, the degree of granule consolidation has also been linked to the extent of liquid pore saturation in the granules. Three distinct stages of granule pore saturation have been identified: pendular, funicular, and capillary.

These states are defined as when particles are held together by a liquid bridge at their contact points (pendular), all the voids are filled with the liquid and the surface liquid is drawn back into the pores under capillary action (capillary), and the transition state between the pendular and capillary states where the voids are not fully saturated with liquid (funicular) (Leuenberger et al., 2009). The amount of pore volume occupied by the liquid, that is, the moisture content of the agglomerate, therefore, can be defined quantitatively as the proportion of the volume of liquid to the total volume of the solid, liquid, and gas in the agglomerate (Flemmer, 1991). An agglomerate could be considered saturated at the transition from the pendular to the funicular state. Flemmer calculated this upper limit to be 13.6% for randomly packed equisized spheres.

10.7 Conclusion

Binders play an essential role in granule formation, growth, consolidation, breakage, and layering. Binder attributes and process parameters together play a critical role in determining not only the predominant granulation mechanisms and kinetics, but also the outcome of the granulation process in terms of quality attributes of dried granules and the finished DP. This chapter demonstrates the essential role of combining the fundamental principles-based particle-level mechanistic modeling and process modeling with experimental observations to fully understand and characterize the role of binder attributes on the process outcomes.

Acknowledgments

Authors would like to thank the following individuals, each for their contribution to some portion of the previously published work included in this chapter by reference: Megerle Escotet, Anik Chaturbedi, Chandra Kanth Bandia, Dheeraj Reddya; Rutgers University, New Brunswick, NJ.

Abbreviations

QbD quality-by-design

References

Benali, M., Gerbaud, V., & Hemati, M. (2009). Effect of operating conditions and physico-chemical properties on the wet granulation kinetics in high shear mixer. *Powder Technology, 190*(1–2), 160–169. https://doi.org/10.1016/j.powtec.2008.04.082.

Capes, C., & Danckwerts, P. (1965). Granule formation by the agglomeration of damp powders. Part I: The mechanism of granule growth. *Transactions Institute of Chemical Engineers, 43*, 116–124.

Chaturbedi, A., Bandi, C. K., Reddy, D., Pandey, P., Narang, A., Bindra, D., Tao, L., Zhao, J., Li, J., Hussain, M., & Ramachandran, R. (2017). Compartment based population balance model development of a high shear wet granulation process via dry and wet binder addition. *Chemical Engineering Research and Design., 123*, 187–200. https://doi.org/10.1016/j.cherd.2017.04.017.

Chitu, T. M., Oulahna, D., & Hemati, M. (2011). Wet granulation in laboratory scale high shear mixers: Effect of binder properties. *Powder Technology, 206*(1–2), 25–33. https://doi.org/10.1016/j.powtec.2010.07.012.

Denesuk, M., Smith, G. L., Zelinski, B. J. J., Kreidl, N. J., & Uhlmann, D. R. (1993). Capillary penetration of liquid droplets into porous materials. *Journal of Colloid & Interface Science, 158*(1), 114–120. https://doi.org/10.1006/jcis.1993.1235.

Ennis, B. J., Li, J., Tardos, G. I., & Pfeffer, R. (1990). The influence of viscosity on the strength of an axially strained pendular liquid bridge. *Chemical Engineering Science, 45*(10), 3071–3088. https://doi.org/10.1016/0009-2509(90)80054-I.

Flemmer, C. L. (1991). On the regime boundaries of moisture in granular materials. *Powder Technology, 66*(2), 191–194. https://doi.org/10.1016/0032-5910(91)80100-w.

Iveson, S. M. (2002). Limitations of one-dimensional population balance models of wet granulation processes. *Powder Technology, 124*(3), 219–229. https://doi.org/10.1016/S0032-5910(02)00026-8.

Iveson, S. M., Litster, J. D., Hapgood, K., & Ennis, B. J. (2001a). Nucleation, growth and breakage phenomena in agitated wet granulation processes: A review. *Powder Technology, 117*(1–2), 3–39. https://doi.org/10.1016/S0032-.5910(01)00313-8.

Iveson, S. M., Wauters, P. A. L., Forrest, S., Litster, J. D., Meesters, G. M. H., & Scarlett, B. (2001b). Growth regime map for liquid-bound granules: Further development and experimental validation. *Powder Technology, 117*(1–2), 83–97. https://doi.org/10.1016/S0032-5910(01)00317-5.

Johansen, A., & Schaefer, T. (2001). Effects of interactions between powder particle size and binder viscosity on agglomerate growth mechanisms in a high shear mixer. *European Journal of Pharmaceutical Sciences, 12*(3), 297–309. https://doi.org/10.1016/S0928-0987(00)00182-2.

Keningley, S. T., Knight, P. C., & Marson, A. D. (1997). An investigation into the effects of binder viscosity on agglomeration behavior. *Powder Technology, 91*(2), 95–103. https://doi.org/10.1016/S0032-5910(96)03230-5.

Leuenberger, H., Puchkov, M., Krausbauer, E., & Betz, G. (2009). Manufacturing pharmaceutical granules: Is the granulation end-point a myth? *Powder Technology, 189*(2), 141–148. https://doi.org/10.1016/j.powtec.2008.04.005.

Li, J., Tao, L., Dali, M., Buckley, D., Gao, J., & Hubert, M. (2011). The effect of the physical states of binders on high-shear wet granulation and granule properties: A mechanistic approach towards understanding high-shear wet granulation process. part i. physical characterization of binders. *Journal of Pharmaceutical Sciences, 100*(1), 164–173. https://doi.org/10.1002/jps.22260.

Mills, P. J. T., Seville, J. P. K., Knight, P. C., & Adams, M. J. (2000). The effect of binder viscosity on particle agglomeration in a low-shear mixer/agglomerator. *Powder Technology, 113*(1–2), 140–147. https://doi.org/10.1016/S0032-5910(00)00224-2.

Narang, A. S., Mantri, R. V., & Raghavan, K. (2017). Excipient compatibility and functionality. In Y. Qiu, Y. Chen, G. G. Z. Zhang, L. Lu, & R. V. Mantri (Eds.), *Developing solid oral dosage forms* (pp. 151–180). Elsevier. https://doi.org/10.1016/b978-0-12-802447-8.00006-6.

Remy, B., Narang Ajit, S., Pandey, P., Bindra, D., LaRoche, R., & Sosnowska, A. et al. (2015) In: DEM modeling of high shear wet granulation: Effect of process parameters and equipment scale on particle level mechanisms AIChE annual meeting, Salt Lake City, UT.

Reynolds, G. K., Biggs, C. A., Salman, A. D., & Hounslow, M. J. (2004). Non-uniformity of binder distribution in high-shear granulation. *Powder Technology, 140*(3), 203–208. https://doi.org/10.1016/j.powtec.2004.01.017.

Rumpf, H. (1958). Grundlagen und Methoden des Granulierens. *Chemie Ingenieur Technik - CIT, 30*(3), 144–158. https://doi.org/10.1002/cite.330300307.

Simons, S. J. R., & Fairbrother, R. J. (2000). Direct observations of liquid binder–particle interactions: The role of wetting behaviour in agglomerate growth. *Powder Technology, 110*(1—2), 44–58. https://doi.org/10.1016/s0032-5910(99)00267-3.

Tao, L., Zhao, J., Li, J., Keluskar, R., Gour, S., & Stevens, T. et al. (2013). *Identifying rate limiting processes determining granule growth kinetics during high shear wet granulation: Effect of type, molecular weight, particle size, and mode of addition of polymeric binders and correlation of mesh analysis with inline focused beam reflectance measurement (FBRM)AAPS annual meeting, San Antonio, TX.*

Thornton, C., Yin, K. K., & Adams, M. J. (1996). Numerical simulation of the impact fracture and fragmentation of agglomerates. *Journal of Physics D: Applied Physics, 29*(2), 424–435. https://doi.org/10.1088/0022-3727/29/2/021.

van den Dries, K., de Vegt, O. M., Girard, V., & Vromans, H. (2003). Granule breakage phenomena in a high shear mixer; influence of process and formulation variables and consequences on granule homogeneity. *Powder Technology, 133*(1–3), 228–236. https://doi.org/10.1016/S0032-5910(03)00106-2.

van den Dries, K., & Vromans, H. (2002). Relationship between inhomogeneity phenomena and granule growth mechanisms in a high-shear mixer. *International Journal of Pharmaceutics, 247*(1–2), 167–177. https://doi.org/10.1016/S0378-5173(02)00419-2.

Verkoeijen, D., Pouw, G. A., Meesters, G. M. H., & Scarlett, B. (2002). Population balances for particulate processes: A volume approach. *Chemical Engineering Science, 57*(12), 2287–2303.

Vonk, P., Guillaume, C. P. F., Ramaker, J. S., Vromans, H., & Kossen, N. W. F. (1997). Growth mechanisms of high-shear pelletization. *International Journal of Pharmaceutics, 157*(1), 93–102. https://doi.org/10.1016/S0378-.5173(97)00232-9.

Zhao, Z., Tao, L., Li, J., Bindra, D., & Narang Ajit, S. (2013). *Granules consolidation kinetics during high shear wet granulation: Effect of binder type, molecular weight, and method of addition*. American Association of Pharmaceutical Sciences (AAPS).

CHAPTER 11

Role of drug substance material properties in the processibility and performance of wet granulated products

Chandra Vemavarapu[a] and Sherif I.F. Badawy[b]
[a]Global Regulatory Sciences, Bristol-Myers Squibb, Hopewell, NJ, United States, [b]Drug Product Science & Technology, Bristol-Myers Squibb Co., New Brunswick, NJ, United States

11.1 Introduction

Various physical attributes of the excipients and drug substance govern the granulation process and, therefore, the final characteristics of the resultant drug product. Examples of physical properties of input materials (excipients and drug substance) that have been shown to have an effect on the granulation process include particle size and shape, surface area, solubility, porosity, and contact angle with the binder solution. The granulation phenomenon is also affected by the properties of the granules that are being formed during the process, such as the granule strength and granule porosity. This chapter discusses some of the main attributes.

11.2 Multivariate study of input material properties

The effect of material properties on their performance in wet granulation and on the attributes of produced granulation and final product has been the subject of a number of reported studies (Badawy & Hussain, 2004; Badawy et al., 2000; Iveson & Litster, 1998; Kristensen et al., 1985b; Vemavarapu et al., 2009). The study by Vemavarapu et al. examined the relationship between the physical properties of an active pharmaceutical ingredient (API) and excipients and their behavior during high shear wet granulation. Various properties of materials used in the study were chosen based on their perceived role in the different stages of high-shear wet granulation, including wetting and nucleation; coalescence, consolidation, and growth; and attrition of wet agglomerates. Material properties evaluated in the study included aqueous solubility, wettability (contact angle with water), water holding capacity (equilibrium wt. gain at 97% RH), particle size, and Brunauer-Emmett-Teller's (BET) surface area. The influence of these material properties on the processibility and performance of the granulation was evaluated by monitoring such responses as granule growth, compactability, and flow changes upon wet granulation.

Although previous evidence has been published about the effects of some of the material attributes, such studies were characteristically single-factor/univariate in design without adequate control over factors outside of the design. As a result, it had been difficult to delineate the main effects, often contributing to inconsistencies among studies. In contrast, the study of Vemavarapu et al. used a number of materials to probe the effects in a multifactorial design, allowing for delineation of not only the main effects but also the interactions among various factors. Twenty-five APIs and excipients were chosen to provide a large spread in the combinations of these properties. A primary criterion used in the selection of materials was to cover a large design space with variable interest. To serve this purpose, APIs and excipients with a wide range of molecular weights were included (Table 11.1). Because the primary intent was to investigate the effect of materials on the granulation process, care was taken to ensure that the physical form of the materials was unaltered during and after wet granulation. Accordingly salts and molecules known to undergo form conversions were mostly excluded from this study.

11.2.1 Input material properties

The different properties of input materials evaluated in the study were determined as follows: The aqueous solubility values of the various test materials were obtained from literature. Advancing contact angle, θ of test materials against water was determined using either the Washburn technique ($\theta < 90°$; Kruss Tensiometer, K12-MK5, Kruss GmbH, NC) or the goniometer ($\theta > 90°$; drop shape analysis system, DSA10-MK2, Kruss GmbH, NC). The values used from the goniometer determination were equilibrium contact angles at 30 s. The goniometer method employed samples prepared by gently pressing the powder bed between two glass slides without application of much shear or normal force because the compaction of materials at loads such as 4 to 5 kN was found to alter the surfaces and therefore the contact angles. Highly soluble materials that instantly dissolved upon contact with water disallowed measurement with either of the techniques and were arbitrarily assigned a value of 0°.

The water holding capacity of various test materials at room temperature was gravimetrically determined by placing 150 mg of the test material in a desiccator maintained at 97% relative humidity using saturated K_2SO_4 solution and recording the weight gain at the end of 5, 6, and 7 days.

Particle size analysis of dry powders was performed using a laser diffraction technique employing a laser light source ($\lambda = 632.8$ nm) and appropriate lenses (R2, R4, R5, and R7) on Sympatec (Helos and Vibri modules, Sympatec, NJ). The air pressure to disperse the powders without any significant attrition was determined following a pressure titration in the range 0.5 to 5 bar (6 points). After determination of the appropriate feed rate/obscurance, dispersion pressure, and the size of the lens, three independent measurements were made, and the reproducibility among measurements was tested. Typical variation between measurements expressed as mean %RSD in D10, D50, and D90 was 3.2%, 2.3%, and 4.1% respectively. The D[3,2] and D90/D10 statistics were used to represent the mean particle size and width of distribution, respectively.

The surface area was determined using the nitrogen adsorption technique by the Gemini surface area analyzer (2380 V-series Micromeritics, GA). The amount of nitrogen adsorbed was determined at partial nitrogen pressure (P/Po) ranging between 0.05 and 0.30. Surface area was determined by Gemini software using the BET calculation for the nitrogen data in the P/Po range from 0.05 to 0.30.

Compaction properties of the preblend, which has an identical composition to the final granulation but is not subjected to the wet-granulation process, were determined using the EHS compaction simulator and 3/8″ round standard flat-faced tooling. The die was manually filled with 300 mg of powder blend, and tablets were compressed at five different compression forces (range 5–30 kN). During this test, forces (compression, ejection) and displacement experienced by punches are recorded. Offline

Table 11.1 List of test materials and associated material properties.

	Model compound	H₂O solubility	Wettability	H₂O holding capacity	Particle size	Surface area	BET, m²/g
		mg/mL @ pH 5 to 7	Contact angle (°)	% wt gain @97%RH	d(3,2)μ	d90/d10	
1	Acetaminophen, USP	14.300	77	0.20	12	25.9	0.2698
2	Aspirin, USP	3.300	87	0.19	452	3.8	0.0189
3	Aspirin-REPEAT	3.300	87	0.19	452	3.8	0.0189
4	Avicel-PH101	0.100	73	16.59	41	6.4	1.1596
5	Avicel-PH102	0.100	73	16.57	70	6.0	1.0193
6	Avicel-PH200	0.100	73	10.97	124	4.6	1.1960
7	BMS-A	0.009	132	0.15	2	19.6	5.2573
8	BMS-B	0.016	145	2.08	5	12.2	0.4621
9	BMS-C	0.010	136	0.39	4	14.1	0.9872
10	BMS-D	0.003	128	0.33	4	10.7	0.3193
11	Caffeine anhydrous, USP	22.000	41	7.46	7	54.5	0.9104
12	Calcium carbonate USP, light powder	0.100	59	1.86	1	9.5	6.1330
13	Dibasic calcium phosphate anhydrous, USP	0.320	43	0.59	5	15.6	2.3976
14	Griseofulvin	0.009	136	0.17	11	14.2	1.9069
15	BMS-E	0.010	146	0.63	3	11.6	2.3296
16	BMS-E-REPEAT	0.010	146	0.63	3	11.6	2.3296
17	Isoniazid	140.300	0	0.18	72	12.1	0.0455
18	Lactose anhydrous DC NF	200.000	0	7.02	28	27.0	0.4550

(continued on next page)

Table 11.1 List of test materials and associated material properties—cont'd

		H$_2$O solubility	Wettability	H$_2$O holding capacity	Particle size	Surface area	
19	Lactose monohydrate SD Dry	200.000	0	1.97	85	3.4	0.3846
20	Mannitol 60	182.000	0	0.83	23	30.0	0.2626
21	Mannitol granular USP	182.000	0	3.98	252	3.8	0.5911
22	Pregelatinized starch	15.000	85	30.48	19	12.7	0.3442
23	Salicylic acid, USP	2.240	136	1.20	15	9.7	0.1965
24	Starch, maize	1.800	86	26.98	7	2.7	0.5082
25	Theophylline anhydrous, USP	7.300	70	10.50	15	28.3	0.6065

From Vemavarapu et al. (2009).

measurements on tablets include weight, thickness, and hardness. Using this data, a plot was generated between tablet tensile strength (y) and compaction pressure (x), and the slope of the fitted linear was reported as compactability. Changes as a result of wet granulation were evaluated in the study by comparing the properties of granulation (final) to those of the preblend (initial).

Flow of preblends was determined using a Flodex tester (Model-211, Hanson Research, CA), in which 30 g of test material was loaded onto a sample plate (with a fixed aperture) using a funnel. After 30 s, the aperture on this plate was opened, and the powder flow was examined. The minimum diameter of aperture in millimeters that allowed the flow of material from the plate (unassisted) was reported as the Flodex number. Accordingly, smaller Flodex numbers represent better flow and vice versa.

11.2.2 Statistical model for wet-granulated product attributes

Identical composition and process conditions were used to wet granulate the various test materials in Table 11.1 in order to be able to make direct comparisons between the responses. The template composition evaluated in this study contained 70% test compound + 26% Avicel-PH102 + 4% HPC-EXF (w/w). Materials were granulated at a batch size of 150 g in a 1 L Diosna bottom-driven mixer (P1-6, Diosna Dierks & Söhne, GmbH) at impeller speed = 600 rpm (tip speed = 4.8 m/s), 30% water for granulation (relative to dry weight), and wet-massing time = 30 s. The level of water for granulation was determined based on preliminary runs on materials at extreme ends of the solubility range. The wet granulation was tray dried in an oven at $55 \pm 5°C$ for 8 h.

The responses measured for the dried granulation included percent oversize (>1.4 mm), granule growth ratio, compactability ratio, and flow ratio. Percent oversize (>1.4 mm) was determined as the granulation mass fraction that was larger than 1.4 mm, which was separated from the bulk using a #14 mesh sieve, and the percentage of total mass was reported as oversize. Granulation that passed through the #14 mesh was recovered and used for the other tests.

The D[3,2] and D90/D10 of granulations were obtained using the same technique and equipment that was employed for starting materials. In case of granulations, dispersing air pressure (Sympatec, Gradis, and Rodos/Vibri modules used) was not used because granule flow/dispersion of granulation into the path of laser light was determined to be good in the majority of cases. Granule growth ratio was defined as the ratio of D[3,2] of the granulation to that of the test material.

Compactability of the granulations was determined using the same procedure as the preblend. Compactability ratio was defined as the ratio of compactability of the granulation to that of the preblend. Similarly, Flodex numbers for the granulations were determined, and the flow ratio was defined as the ratio of the Flodex number of the granulation to that of the preblend.

Response data from the study are shown in Table 11.2. The response data were analyzed by fitting statistical models to explain five responses based on the factors (input material properties) in Table 11.1. Mathematical transformation of the material properties and response data was performed as needed in order to attain a normal distribution of the population selected prior to model fitting. Model fitting was performed using stepwise regression involving the five factors (main effects and two-way interactions) for each of the five responses, and the significant relationships ($P < .05$) were identified. All the data analysis was performed using SAS-JMP software (SAS Institute Inc.). The output from the regression analysis using SAS-JMP is summarized in the leverage plots (Figs. 11.1–11.5) and summary of fit data (Tables 11.3–11.7). The leverage plot for the linear effect in a simple regression is the same as the

Table 11.2 Summary of responses from the study of the effect of material properties on wet granulation attributes.

	Model compound	% LOD	% oversize (>1.4 mm)	Flodex flow number, mm		Comparability, kPa/MPa		D(3,2) μm		D90/D10		Growth ratio	
				Preblend	Granulation	Preblend	Granulation	Preblend	Granulation	Preblend	Granulation	Preblend	Granulation
1	Acetaminophen, USP	0.85	20.2	26	5			1.73		3.01		272	3.6	23.1
2	Aspirin, USP	0.67	68.0	10	8			2.52		2.23		989	2.3	2.2
3	Aspirin-REPEAT	0.76	74.1	9	14			2.86		2.39		840	2.4	1.9
4	Avicel-PH101	2.71	0.0	14	5			8.70		7.47		62	5.5	1.5
5	Avicel-PH102	3.36	0.1	5	5			10.52		8.27		101	4.9	1.4
6	Avicel-PH200	2.98	0.3	4	4			7.03		6.48		144	4.2	1.2
7	BMS-A	0.85	2.7	32	24			2.40		5.67		21	51.5	13.5
8	BMS-B	0.70	12.9	30	6			1.85		3.14		158	5.3	33.2
9	BMS-C	0.90	11.0	30	6			2.76		5.77		135	7.6	33.9
10	BMS-D	0.90	13.4	24	4			3.64		5.28		165	4.8	44.8
11	Caffeine anhydrous, USP	0.89	16.8	24	8			6.41		7.68		191	8.1	26.2
12	Calcium carbonate USP, light powder	0.96	4.0	>32	26			5.67		5.39		161	3.5	114.7
13	Dibasic calcium phosphate anhydrous, USP	1.06	3.3	14	4			8.79		9.07		86	5.9	19.1
14	Griseofulvin	0.91	15.0	32	4			7.33		8.68		150	6.1	13.6
15	BMS-E	1.03	10.2	>32	30			2.81		6.24		17	36.4	5.8
16	BMS-E-REPEAT	0.99	5.2	>32	26			2.81		6.77		14	39.9	4.7
17	Isoniazid	0.70	96.8	14	–			2.35		3.13		1400	3.0	19.5
18	Lactose anhydrous DC NF	1.13	16.6	14	5			10.69		8.01		348	3.1	12.3

(continued on next page)

11.2 Multivariate study of input material properties

19	Lactose monohydrate SD dry	0.98	26.7	4	6	10.79	9.08	418	3.5	4.9
20	Mannitol 60	0.74	96.3	18	–	4.47	5.79	1400	3.0	60.4
21	Mannitol granular USP	0.79	30.4	8	6	4.57	5.17	665	2.9	2.6
22	Pregelatinized starch	4.63	35.3	16	8	3.89	2.76	283	7.1	14.7
23	Salicylic acid, USP	–	15.5	30	14	3.30	4.14	159	4.1	10.3
24	Starch, maize	5.92	7.5	26	12	3.85	0.86	111	10.7	15.2
25	Theophylline anhydrous, USP	1.12	10.2	30	8	6.00	7.78	176	9.6	12.0

From Vemavarapu et al. (2009).

Table 11.3 Statistical summary of fit-controlled granule growth.

Log (Gran Growth Ratio)

Summary of Fit	
R square	0.78
R-squared adj	0.75
Root mean square error	0.27
Mean of response	1.01
Observations (or sum wgts)	25

Parameter Estimates		
Term	Estimate	Prob > |t|
Intercept	3.32	<0.0001
ContactAngle	0.00	0.0008
Log(d3,2)	−0.97	<0.0001
Cuberoot (BET)	−0.79	0.0005

From Vemavarapu et al. (2009).

Table 11.4 Statistical summary of fit-uncontrolled granule growth.

Sqrt (% Oversize)				
Summary of Fit	**Parameter Estimates**			
		Term	Estimate	Prob > |t|
R square	0.79	Intercept	6.94	<0.0001
R square adj	0.76	Log(solubility)	0.82	0.0009
Root mean square error	1.34	Log(WHC)	−1.74	0.0001
Mean of response	4.08	Cuberoot(BET)	−2.74	0.0042
Observations (or sum wgts)	25			

From Vemavarapu et al. (2009).

Table 11.5 Statistical summary of the fit-width of the granule size distribution.

Log(d90/d10) of Gran Summary of Fit	Parameter Estimates			
		Term	Estimate	Prob > |t|
R square	0.94	Intercept	0.08	0.729
R square adj	0.91	Log(Solubility)	0.17	0.0006
Root mean square error	0.11	ContactAngle	0.01	0.0001
Mean of response	0.78	Log(d3,2)	−0.13	0.0298
Observations (or sum Wgts)	25	Cuberoot(BET)	0.71	< 0.0001
		(Log(Solubility) + 0.05883) * (ContactAngle − 78.36)	0.00	0.0009
		(Log(Solubility) + 0.05883) *(cuberoot(BET) − 0.9111)	0.94	< 0.0001
		(ContactAngle − 78.36) *(cuberoot(BET) − 0.9111)	0.03	< 0.0001
		(Log(d3,2) − 1.24536) *(cuberoot(BET) − 0.9111)	−0.43	0.0029

From Vemavarapu et al. (2009).

Table 11.6 Statistical summary of fit-compactability ratio.

Compactability Ratio Summary of Fit	Parameter Estimates		
	Term	Estimate	Prob > |t|
R square 0.94	Intercept	−0.39	0.3422
R square adj 0.91	log(WHC)	−0.14	0.0322
Root mean square error 0.17	ContactAngle	0.00	0.0012
Mean of response 1.26	log(d3,2)	0.11	0.2746
Observations (or sum wgts) 25	log(d90/d10)	0.85	<0.0001
	cuberoot(BET)	0.64	0.001
	log(preblend compactability)	−0.54	0.0304
	(log(WHC) − 0.18427)*(cuberoot(BET) − 0.9111)	0.78	0.0055
	(ContactAngle − 78.36)*(log(d3,2) − 1.24536)	−0.01	0.0005
	(cuberoot(BET) − 0.9111)*(log(preblend compactability) − 0.6414)	−1.73	0.0396

From Vemavarapu et al. (2009).

Table 11.7 Statistical summary of fit-flow ratio.

Log(Flow Ratio) Summary of Fit	Parameter Estimates		
	Term	Estimate	Prob > |t|
R square 0.89	Intercept	−0.37	0.2653
R-squared adj 0.82	log(Solubility)	0.26	0.0002
Root mean square error 0.13	ContactAngle	0.01	0.0014
Mean of response −0.33	log(d3,2)	−0.25	0.0441
Observations (or sum wgts) 23	log(d90/d10)	−0.36	0.024
	cuberoot(BET)	0.50	0.0066
	Preblend flow	−0.03	0.0021
	(ContactAngle − 85.1739)*(log(d90/d10) −1.00418)	−0.01	0.04
	(log(d3,2) − 1.21356)*(Preblend flow − 21.1739)	−0.04	0.0001

From Vemavarapu et al. (2009).

11.2 Multivariate study of input material properties

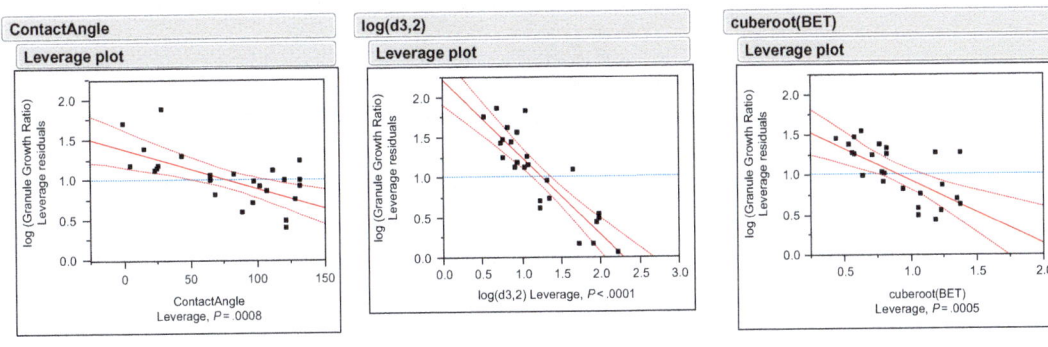

FIGURE 11.1

Significant material properties affecting controlled granule growth.

From Vemavarapu et al. (2009).

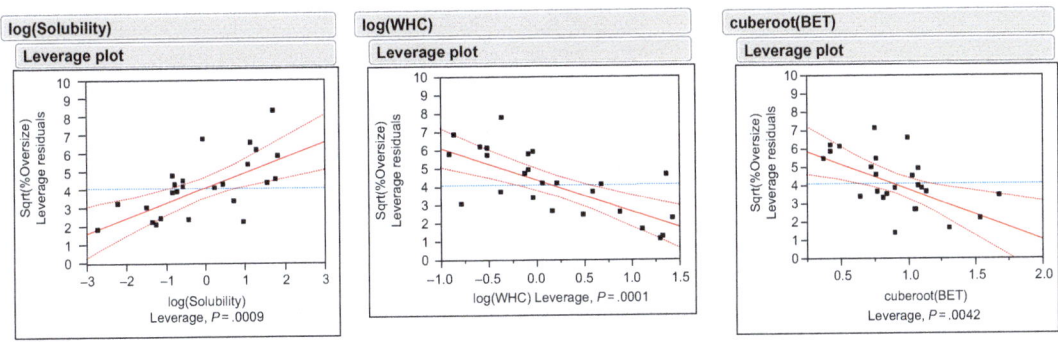

FIGURE 11.2

Significant material properties affecting uncontrolled granule growth.

From Vemavarapu et al. (2009).

traditional plot of actual response values (y) and the regressor (x). The distance from a point to the line of fit shows the actual residual. The distance from the point to the horizontal line of the mean shows what the residual error would be without the effect in the model. In other words, the mean line in this leverage plot represents the model where the hypothesized value of the parameter (x) is constrained to zero. Alternatively, adjusted r^2 values of the whole model, as well as the parameter estimates and their significance (P-values) are summarized in Tables 11.4–11.8. The relevant findings from the regression models are used in the discussions below to help illustrate the impact of various material properties on their performance in wet granulation.

Table 11.8 Granule size data for the granulations manufactured using the different lactose fractions.

Water Concentration (%)	Fraction I	Fraction II			60M Lactose				
	Geometric Mean Diameter (μm)	Normalized Granule Size[a]	Geometric Standard Deviation	Geometric Mean Diameter (μm)	Normalized Granule Size[a]	Geometric Standard Deviation	Geometric Mean Diameter (μm)	Normalized Granule Size[a]	Geometric Standard Deviation
0[b]	53.3	1	1.5	178.0	1	1.5	110.1	1	2.0
6	98.1	1.84	2.8	190.4	1.07	1.4	135.6	1.23	1.9
12	145.8	2.73	2.5	296.3	1.66	1.4	196.9	1.79	1.6

From Badawy and Hussain (2004).
[a] Normalized granule size is obtained by dividing the geometric mean diameter of the granulation by the geometric mean diameter of the starting material.
[b] The 0% water concentration refers to data for the starting material.

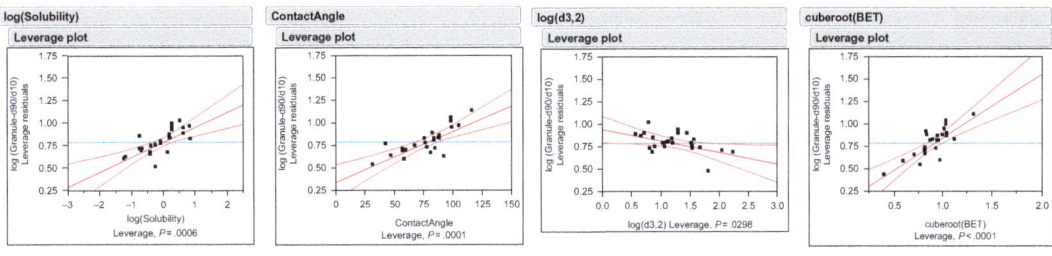

FIGURE 11.3

Significant material properties affecting the width of the granule size distribution.

From Vemavarapu et al. (2009).

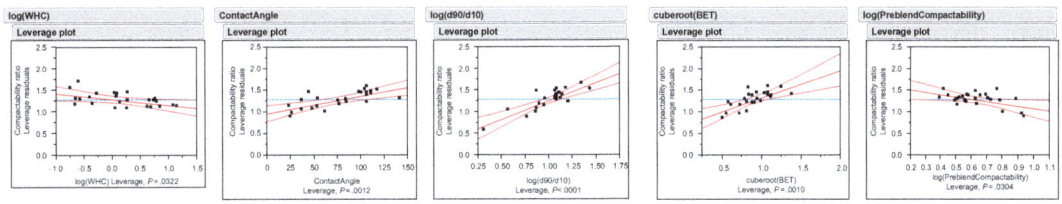

FIGURE 11.4

Significant material properties affecting compactability ratio.

From Vemavarapu et al. (2009).

11.3 Effect of particle size

Granule growth in high-shear wet granulation is a dynamic process in which granules are continuously forming and breaking down (Iveson et al., 2001). Granule size achieved at a given set of experimental conditions depends on the relative rates of granule formation and breakdown. Granule growth in a high-shear mixer proceeds initially by a nucleation mechanism. Liquid droplets are broken up and dispersed by the shear forces in the granulator and then wet the primary particle surface. Liquid bridge bondings are established between the surface wet particles, resulting in the formation of nuclei. Nucleation is the predominant granulation mechanism at low liquid concentration, and the rate of nucleation is proportional to the strength of the formed nuclei (Kristensen, 1996; Tardos et al., 1997). At the low liquid concentration characteristic of the nucleation stage, nuclei typically exhibit brittle behavior. Higher strength decreases the rate of nuclei breakdown, thus shifting the process toward larger nuclei sizes. As granulation continues, liquid saturation of the formed nuclei increases as a result of the continued addition of the binder liquid. After granules reach a certain limiting liquid saturation, they start to exhibit plastic behavior, and granule growth by coalescence becomes a predominant mechanism. Coalescence is the combination of two colliding granules to form a single larger granule, which results in rapid granule growth and a significant increase in granule growth rate. The probability of coalescence of two colliding granules increases as their ability to plastically deform under applied pressure increases. Plastic deformation of the colliding granules increases their area of contact, allowing for greater bonding

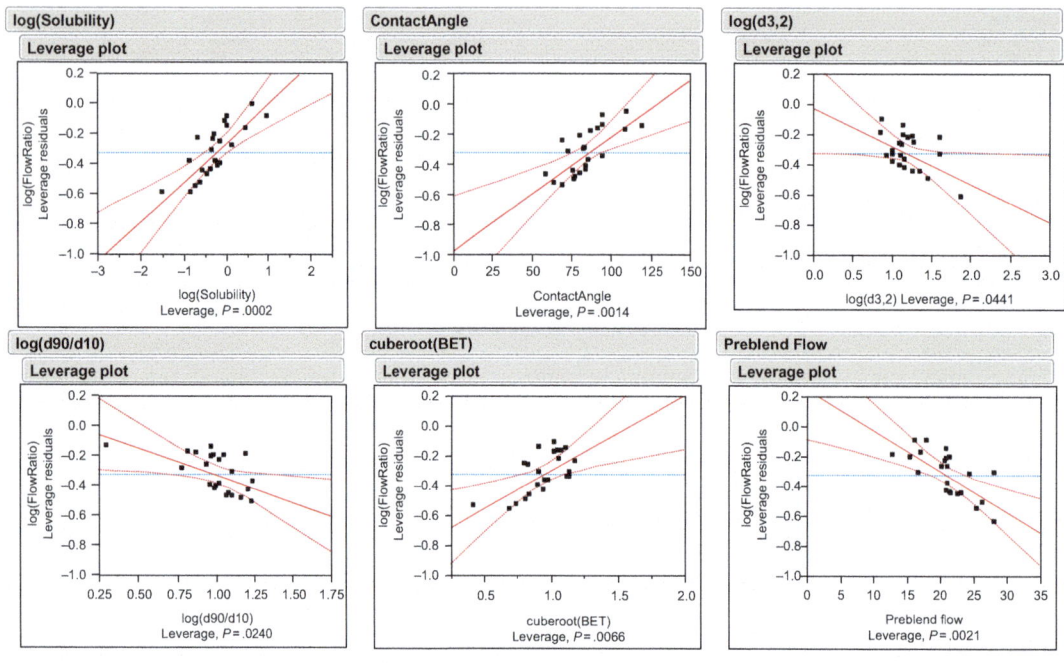

FIGURE 11.5

Significant material properties affecting flow ratio.

From Vemavarapu et al. (2009).

and successful coalescence. Growth by coalescence is enhanced by factors that result in a higher ability of granules to plastically deform upon application of stress and in a lower yield stress of the moist granules (Iveson et al., 2001; Iveson et al., 2002; Liu et al., 2000; Ouchiyama & Tanaka, 1975). Thus while a lower breakage strength of brittle granules is expected to have a negative effect on granulation behavior, a lower yield strength for plastically deformable granules would result in more pronounced granule growth. The mechanical properties of moist granules, therefore, play an important role in granule growth in wet granulation. Particle size of the starting material affects the strength and deformability of moist granules and their behavior in the high-shear granulator at both the nucleation and coalescence stages.

11.3.1 Particle Size and Wet Granule Mechanical Properties

The strength of wet agglomerates is directly impacted by the particle size distribution of the starting materials. According to the model proposed by Pietsch et al. (1969), the static strength of moist agglomerates is expressed as:

$$\sigma_T = SC\left[\frac{(1-\varepsilon)}{\varepsilon}\right]\left[\frac{\lambda_v \cos\theta}{d_p}\right] \qquad (11.1)$$

where σ_T is the tensile strength of a moist agglomerate, S is the liquid saturation of the agglomerate, [is agglomerate porosity, λ_v is liquid surface tension, θ is liquid–solid contact angle, d_p is the diameter of primary particles (assuming monosized spheres), and C is a material constant. The increase in granule strength at the smaller particle size is attributed to the increased volume density of interparticle contacts, which consequently increases frictional forces that contribute to the strength of wet granules (Iveson et al., 2001). Kristensen et al. proposed a model relating agglomerate tensile strength to the intrinsic interaction parameter of particles within the agglomerate and showed that this interaction parameter increased with the decrease in particle size and increase in the width of particle size distribution (Kristensen et al., 1985a).

Because of the practical difficulties of studying the mechanical properties of an actual granule formed in the high-shear granulator, stress-strain profiles were obtained for moist compacts as a surrogate for wet granules (Badawy & Hussain, 2004; Badawy et al., 2000). A uniaxial compression test was used to examine the impact of material particle size on the stress-strain behavior of moist compacts. The moist compact simulates a wet granule, and the study of the mechanical behavior of wet compacts was shown to correlate with the material performance in the high-shear granulator. In one study, the mechanical properties of moist agglomerates of three lactose samples with different particle size distributions were studied (Badawy & Hussain, 2004). An anhydrous lactose grade (60M lactose) was used in the study, in addition to a particle size fraction ranging between 40 and 75 μm (fraction I) and another particle size fraction between 212 and 250 μm (fraction II) obtained by screening of the 60M lactose through sieves with the appropriate mesh size (Table 11.8). The dry lactose sample was wetted to a specified moisture content by spraying with water and mixing very slowly with a glass rod to achieve uniform water distribution in the sample. A cylindrical mass was formed using a die and flat-faced punches, 1.27 cm in diameter, on the Instron mechanical tester. Wet compacts were compressed to either a target thickness (constant porosity experiment) or to a specified compression force (constant force experiment). For the constant porosity experiment, the Instron crosshead was programmed to travel down at a speed of 5 mm/min until the target thickness of the compact with 0.5 g of the solid was achieved. The thickness of the compact was chosen so that compacts with a porosity of 47% or 37% (on the dry basis) were obtained. For the constant compression force experiment, a weight of the moist sample equivalent to 1 g of the dry material was compressed using the target compression force of 700 N. In both experiments, the compact was removed from the die, and stress-strain profiles for the formed compact were determined by loading the compact between the flat-faced platens of the Instron driven at a constant rate of 5 mm/min. The applied force and displacement were obtained and converted to the corresponding stress and strain values, respectively.

Fig. 11.6 shows representative stress-strain profiles of wet lactose compacts. Stress-strain profiles for all compacts showed steady increase in strain as a function of the applied stress until a critical stress is reached. At the critical stress, a drop in the applied stress is observed indicating breakage of a brittle compact (Fig. 11.6, *top*). To the contrary, a plastic compact maintains the critical stress at continuing strain (Fig. 11.6, *bottom*).

11.3.1.1 *Constant porosity experiment*

For all three lactose fractions, compression force required to achieve the specified compact porosity decreased as the water concentration increased (Fig. 11.7). At the same water concentration, the force required to achieve the specified porosity was in the following order: Fraction I > Fraction II ≥ 60M lactose, indicating a higher resistance to densification for the smaller particle size. Stress-strain profiles

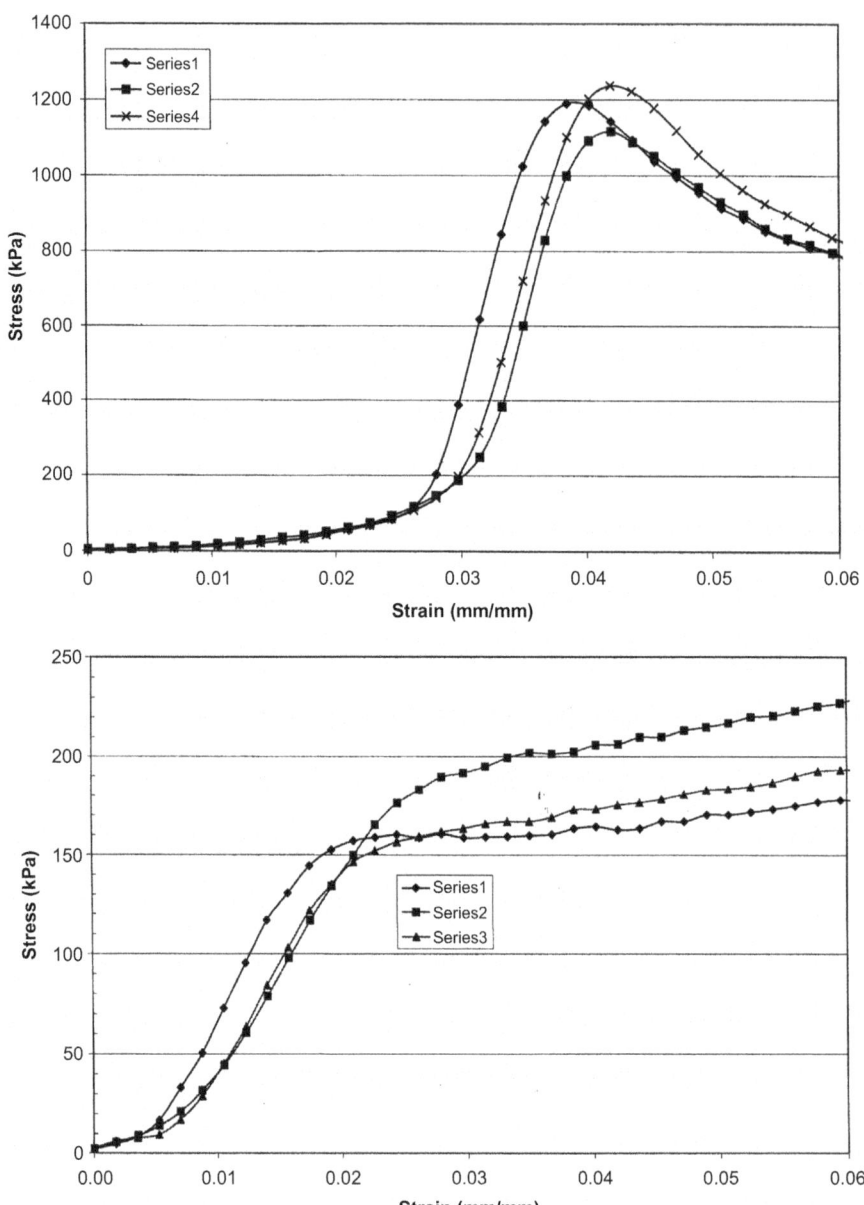

FIGURE 11.6

Stress-strain profiles (three replicate samples) for lactose moist compacts. *Top*, compact showing brittle behavior prepared using Fraction I and 10% water at 47% porosity; *bottom*, compact exhibiting plastic deformation prepared using 60M lactose and 20% water at 47% porosity.

From Badawy and Hussain (2004).

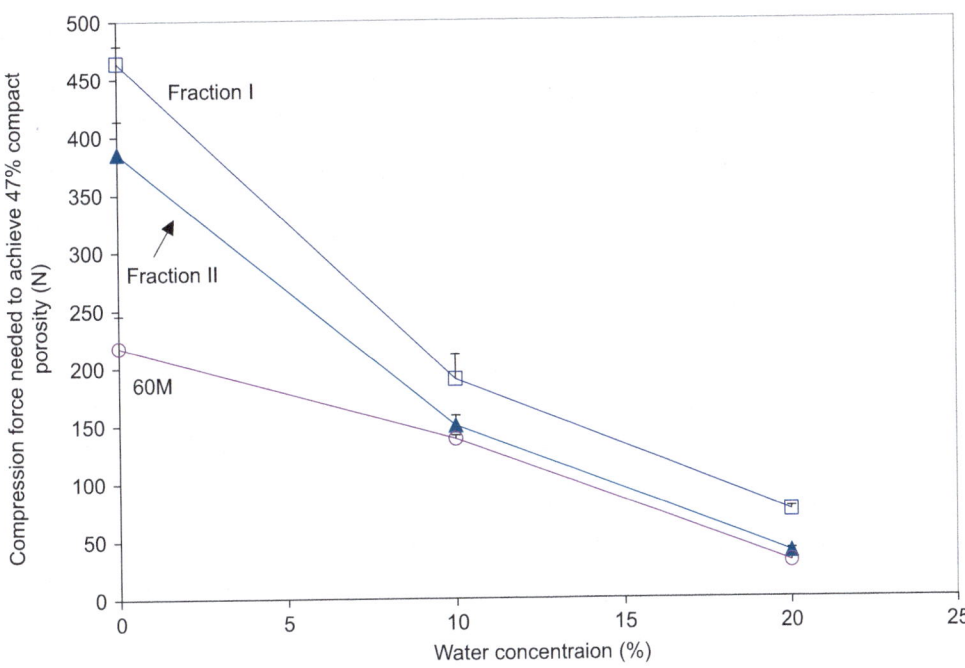

FIGURE 11.7

Effect of water content on compaction behavior of wet anhydrous lactose samples.

From Badawy and Hussain (2004).

of the wet compacts showed breakage rather than yield at the critical stress for all the compacts at the 10% water concentration, indicating that compacts were still brittle at this liquid saturation condition. The liquid saturation corresponding to the 10% water is probably more representative of the nucleation stage of granulation rather than the coalescence phase, in which the material usually exhibits plastic behavior. The breakage stress was higher for Fraction I than that for the other two samples at 10% water concentration and 47% porosity, suggesting that this material with the smaller particle size is capable of forming stronger nuclei and a more pronounced nucleation phase (Fig. 11.8).

Compacts prepared with the 60M lactose showed lower yield stress at the 20% water concentration than those prepared using Fraction II (Fig. 11.9). At 20% water concentration, compacts prepared using Fraction I were still brittle at the 37% and 47% porosity as indicated by compact breakage at critical stress. Compacts prepared using the 60M lactose and Fraction II, however, showed yield behavior rather than breakage at those liquid saturation conditions. The change from brittle to plastic behavior upon increase in water content is likely because of the lubricant effect of water, which disrupts interparticle frictional forces at points of particle contact. The smaller particle size material has a larger surface area available for interparticle contact and therefore requires more water to disrupt those contact points prior to the change of material behavior from brittle to plastic. Because granule coalescence in the high-shear granulator is usually enhanced significantly at the point where granule behavior starts to exhibit plastic

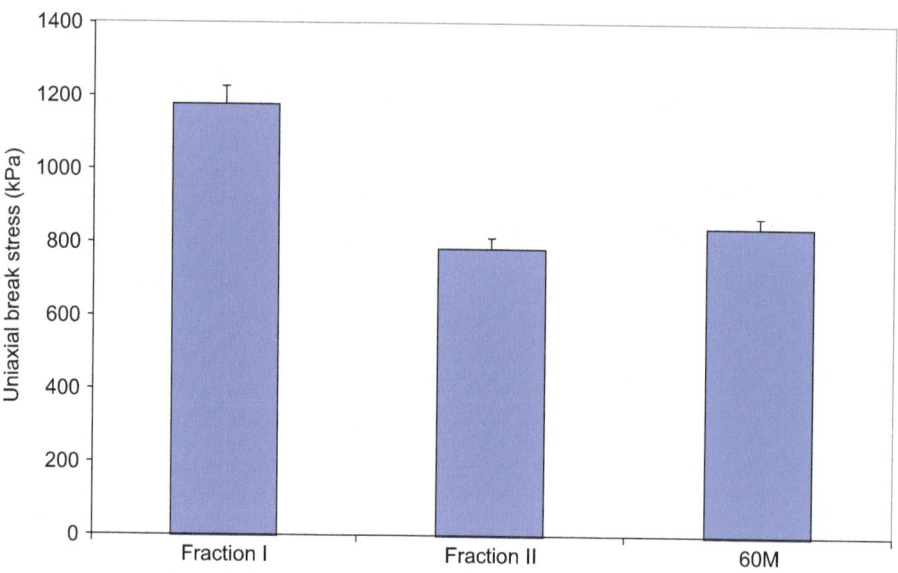

FIGURE 11.8

Strength of wet lactose compacts with 47% porosity and 10% water.

From Badawy and Hussain (2004).

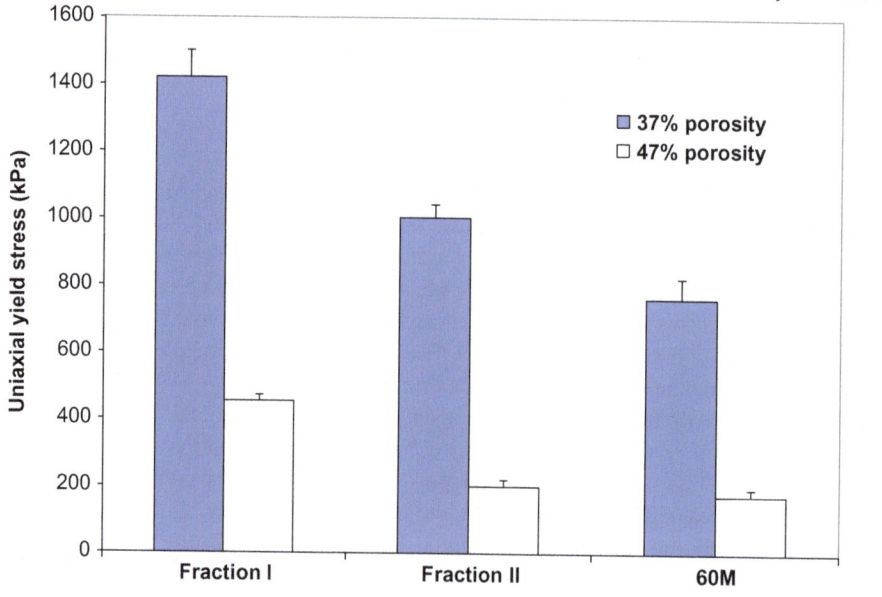

FIGURE 11.9

Strength of wet lactose compacts containing 20% water.

From Badawy and Hussain (2004).

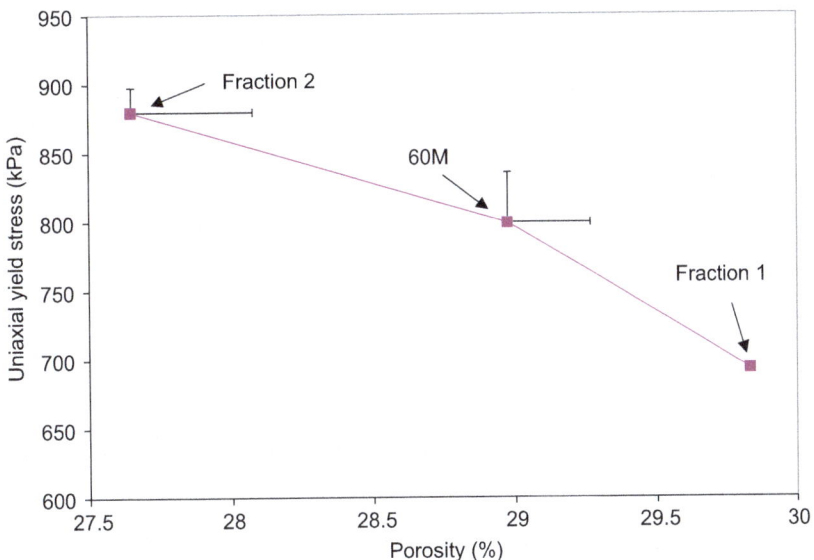

FIGURE 11.10

Yield stress and porosity of wet lactose compacts prepared using 20% water concentration and a compression force of 700 N.

From Badawy and Hussain (2004).

deformation, these data were considered an indication that the larger particle size material reaches the coalescence phase at a lower liquid saturation than the smaller particle size material.

11.3.1.2 Constant force experiments

Compacts prepared at the 700 N compression force showed the following order of porosity: Fraction I > 60M lactose > Fraction II (Fig. 11.10). Compact porosity ranged between 27% and 30%, which is lower than the above constant porosity experiments. As seen in the constant porosity experiments, smaller particle size material demonstrated again that it is more resistant to densification. All compacts showed a yield behavior at critical stress, indicating that all three lactose fractions exhibit plastic behavior at those experimental conditions. Experimental conditions for the constant force experiment were believed to more closely model material behavior during coalescence in a high-shear mixer than the constant porosity experiments, as liquid saturation was sufficiently high for the material to exhibit plastic deformation behavior. Coalescence usually occurs at a later stage in wet granulation, so the constant force rather than constant porosity approach for compact preparation was considered more appropriate because materials would have had the opportunity to densify according to their inherent densification propensity. Yield stress of compacts in the uniaxial compression test is believed to be inversely related to the probability of plastic deformation of granules upon collision in the high-shear mixer (Kristensen et al., 1985a; Holm et al., 1985, 1985b). As mentioned earlier, plastic deformation of colliding granules has paramount significance in determining whether the colliding granules would coalesce or bounce back, particularly when using low-viscosity granulating liquid (Liu et al., 2000). Yield stress of the

Table 11.9 Granule pore size data determined by mercury intrusion porosimetry for granulations manufactured using the different lactose fractions.

Granulation	Total pore volume (mL/g)	Average pore diameter (μm)	Porosity (%)[a]
Fraction I	0.411	1.18	38.9
Fraction II	0.123	0.28	16.0
60M	0.072	0.33	10.1

From Badawy and Hussain (2004).
[a] Calculated using the true density value of 1.552 g/mL.

compacts was in the following order: Fraction II > 60M lactose > Fraction I, suggesting that growth by coalescence would be more pronounced for the smaller particle size granules compared with the other two lactose samples. While granule strength generally is expected to be inversely related to particle size at constant porosity, differences in granule porosity could be a more important factor in determining granule strength in some instances. It is noteworthy that the yield stress of the compacts was in the reverse order of compact porosity, suggesting that material densification tendency could be an important factor in determining granule yield strength and, therefore, its deformability and coalescence behavior. The lower porosity of compacts prepared with a larger particle size resulted in a higher strength for those compacts compared with the more porous compacts obtained with the smaller particle size sample. Those results highlighted the importance of the effect of granule densification propensity on its mechanical properties and, therefore, its growth behavior.

11.3.2 Particle size and granule porosity

Consistent with the wet compact studies, granules manufactured in the high-shear granulator using material with a smaller particle size were found to be more porous than those with a larger starting material particle size under similar process conditions (Badawy & Pandey, 2016; Iveson et al., 1996). As previously described, the small particle size of the starting material results in moist agglomerates that are more resistant to densification in the uniaxial compaction test than the larger particle size (Badawy & Hussain, 2004; Kristensen et al., 1985b).

In the lactose study, granule porosity was found to increase with the decrease in starting material particle size. Granules prepared using Fraction I (small particle size) showed higher total pore volume by mercury intrusion porosimetry compared with granules prepared using the large particle size lactose sample (Fraction II) and the same manufacturing conditions (Table 11.9). The smaller particle size material thus showed a lower densification tendency compared with the larger particle size of the same material because it produced more porous granules under the same experimental conditions. The lower porosity of the 60M sample (as received lactose sample) compared with Fraction II might be attributed to its wider particle size distribution and the ability of the fine particles to fill in the voids between the larger particles.

Granule porosity for the development compound DPC 963 was also found to be higher when using the smaller drug substance particle size under the same process conditions (Badawy et al., 2000) (Fig. 11.11). Granulation manufactured using the SQ964-001 drug substance lot (small particle size,

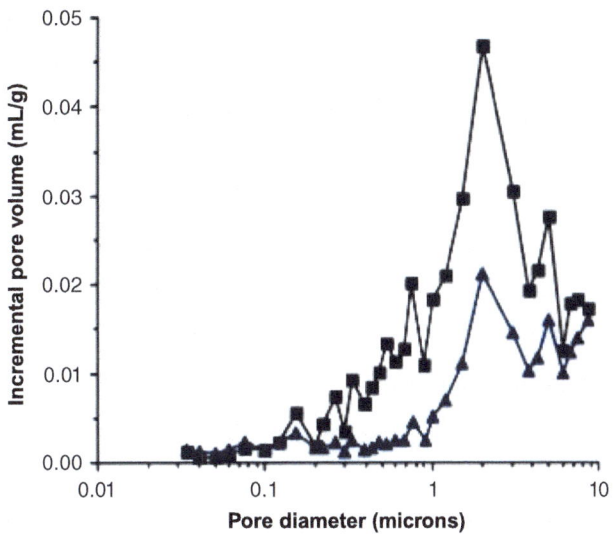

FIGURE 11.11

Porosity of DPC 963 granulations manufactured using small (■) and large (▲) drug substance particle size.

From Badawy et al. (2000a).

Table 11.10) showed higher intragranular pore volume by mercury intrusion porosimetry in the 1 to 10 μm diameter range than for the granulation manufactured using the drug substance with a larger particle size (SQ963-010).

11.3.3 Particle size and granule growth

In their study of 25 pharmaceutical compounds and excipients, Vemavarapu et al. (2009) found a statistically significant enhancement in granule growth resulting from the decrease in starting material particle size (Table 11.3 and Fig. 11.1). A similar effect of particle size on granule growth also was reported for lactose and DPC 963 (Badawy et al., 2000). Granule growth in the high-shear mixer was found to increase as the particle size of the DPC 963 drug substance decreased. A normalized granule size parameter was used as a measure of particle growth in the granulator and was obtained by dividing the geometric mean diameter of the granulation by the median particle size (D50) of the corresponding drug substance lot. Granulation manufactured with drug substance lot SQ964-001 with the smallest particle, showed a substantially higher normalized granule size compared with the granulation manufactured using drug substance lots with larger particle size (Table 11.10).

Similarly in the lactose study, normalized granule size was inversely related to the starting material geometric mean particle size at a given water concentration (Figs. 11.12 and 11.13). Small particle size lactose (Fraction I, geometric mean = 53.3 μm) showed more pronounced growth compared with the 60M lactose (as received sample; geometric mean = 110.1 μm) and the large particle size lactose fraction (Fraction II, geometric mean = 178.0 μm). Fraction II, however, showed a slightly smaller normalized granule size than the 60M lactose, suggesting that the effect of starting material particle size might

Table 11.10 Effect of drug substance particle size on granule growth of the DPC 963 granulation.

Drug substance lot	SQ964-001	SQ963-010	SQ963-010 (jet milled)	SQ963-011
Drug substance particle size (µm)	10%a—2.9% 50%b—5.1% 90%c—8.3%	10%—10.5% 50%—21.8% 90%—31.2%	10%—3.3% 50%—9.3% 90%—15.5%	10%—5.4% 50%—12.8% 90%—21.9%
Drug substance surface area (m^2/g)	1.36	0.46	ND	1.25
Drug substance bulk density (g/mL)	0.12	0.42	0.29	0.32
Normalized granule size[d]	36.9	6.1	19.1	9.6

From Badawy et al. (2000a).
[a] 10% of the particles are smaller than this number.
[b] 50% of the particles are smaller than this number.
[c] 90% of the particles are smaller than this number.
[d] Normalized granule size is obtained by dividing the geometric mean diameter of the granulation by the median particle size (D50) of the corresponding drug substance lot.

11.3 Effect of particle size   403

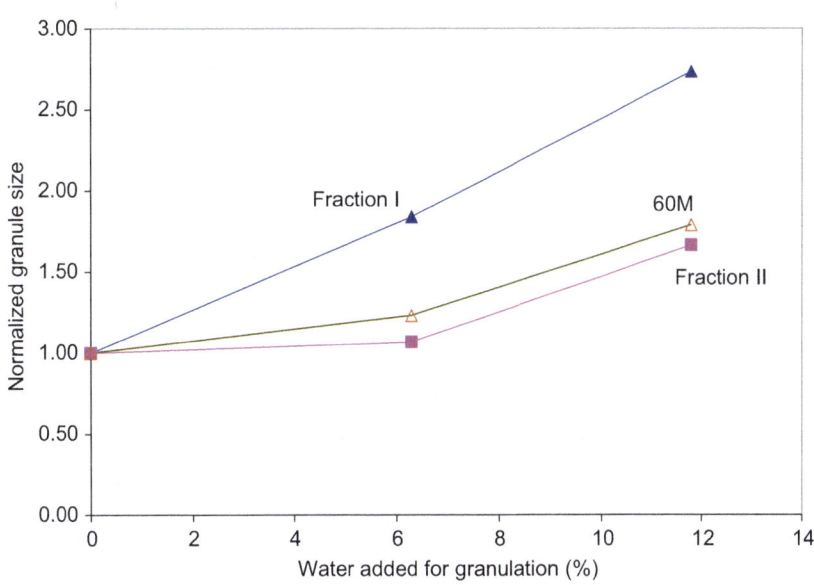

FIGURE 11.12

Effect of granulating water on granule growth of anhydrous lactose in the high shear granulator.

From Badawy and Hussain (2004).

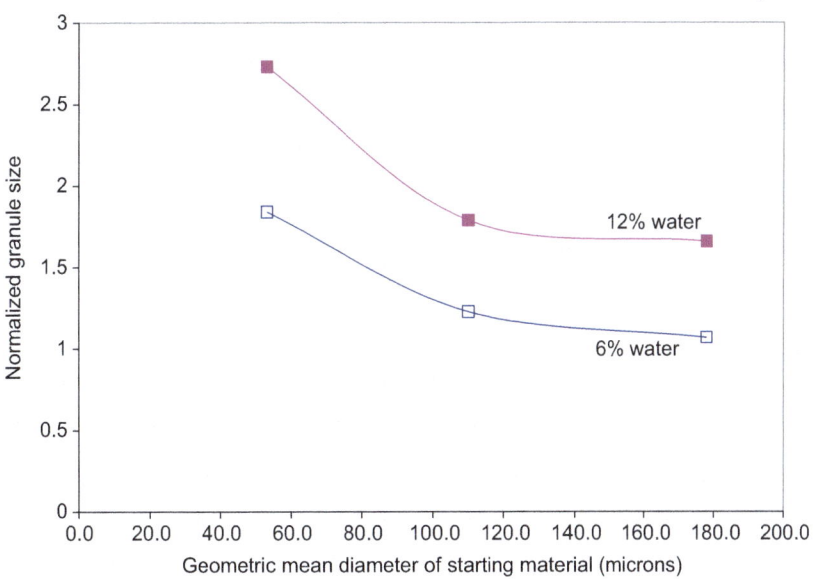

FIGURE 11.13

Effect of starting material particle size on granule growth of anhydrous lactose.

From Badawy and Hussain (2004).

be less pronounced in this range. It is noteworthy, however, that the 60M lactose has a wider particle size distribution than the other two particle size fractions, as indicated by its higher geometric standard deviation (Table 11.8). The growth behavior of the 60M lactose in comparison to Fractions I and II might be a reflection of its wider particle size distribution and not just the difference in mean particle size.

The higher strength of moist agglomerates consisting of small primary particles is conducive to more robust nucleation and reduced attrition at the liquid distribution stage. Increasing material particle size decreases the strength of formed granules and moves the systems up the growth regime map from induction to steady growth and eventually to the crumb region. A minimum particle size, therefore, is required for a given solid/liquid system above which stable granules cannot form (Iveson et al., 2001). At the coalescence phase, granule growth is enhanced by higher plasticity and lower yield stress of the granules (Liu et al., 2000). Therefore, at the same porosity and liquid saturation, granules formed from larger particle size material are expected to have lower yield stress and more rapid coalescence. The effect of particle size on coalescence, however, is confounded by its effect on consolidation tendency and, therefore, liquid saturation. Granules with a larger starting material particle size are more easily densified and possess lower porosity and higher liquid saturation at the same solid to liquid ratio. The lower porosity increases yield stress, while the higher liquid saturation tends to increase plasticity and reduce yield stress (Badawy & Hussain, 2004). The effect of starting material particle size on yield stress and granule coalescence, therefore, will depend on the system-dependent balance of these opposing mechanisms. For many pharmaceutical systems, increased porosity outweighs the impact of lower liquid saturation (Badawy & Hussain, 2004), and therefore, smaller starting material particle size was found to enhance granule growth in many cases (Badawy & Hussain, 2004; Badawy et al., 2000; Vemavarapu et al., 2009). An opposite effect of particle size on granule growth, however, was reported for dicalcium phosphate, where larger particle size showed faster growth (Kristensen et al., 1985b). Dicalcium phosphate demonstrated induction growth with brittle granule behavior and minimal calescence until very high liquid saturations. The impact of increased liquid saturation for low-porosity granules with larger starting material particle size appeared to be predominant in this case. For cohesive materials such as dicalcium phosphate, the effects of increased liquid saturation and diminished interparticle friction are significant, resulting in lower yield strength and rapid coalescence for denser agglomerates prepared using the larger particle size material (Badawy & Pandey, 2016).

11.4 Surface area

Surface area is inversely related to particle size, and therefore increased surface area resulting from the decrease in particle size is expected to have the same effects as previously described. Interestingly (Vemavarapu et al., 2009) did not find a correlation between the effect of inverse particle size and surface area on granule growth. Granule growth was found to decrease as both surface area and particle size increased (Fig. 11.1). Inclusion of both terms in the regression model was necessary to optimize model fit (Table 11.3). It appears that the two factors affect granule growth through independent mechanisms (Badawy & Pandey, 2016). The contribution of internal porosity to surface area was more significant than the impact of particle size on surface area. Surface area was construed primarily as the internal area (intraparticle porosity) with the additional contribution of surface irregularities. The penetration of water into the pores of the primary particles during granulation makes less of it available at the particle surface, which decreases liquid film thickness and granule pore saturation, reducing the probability of coalescence and growth. The width of granule distribution is also found to be high when the surface area of the test materials is high (Fig. 11.3, Table 11.5), which is consistent with the proposed mechanism of the surface area effect on granule growth.

Low surface area also was found to promote uncontrolled granule growth (Fig. 11.2, Table 11.4). Uncontrolled or rapid granule growth is undesirable because it makes the drug product sensitive to even moderate changes in formulation or process. High percent oversize was used as a measure of uncontrolled granule growth and overgranulation. Materials with low surface area exhibited high percent oversize values. At a given water level, the thickness of the liquid film around the particles is high when the surface area is low, resulting in high pore saturation and uncontrolled growth.

11.5 Contact angle

Wetting of the powder particles by the granulating liquid (low solid–liquid contact angle) is necessary for robust nucleation. High contact angle results in slow nucleation and low strength of wet granules. Slow nucleation is the result of decreased spreading coefficient and the less thermodynamically favorable wetting at the higher contact angle (Badawy & Pandey, 2016). Additionally, the static strength of the wet granule is lowered at high contact angles as predicted by the Rumpf equation (Eq. 11.1). After it is formed, the wet granule should withstand the high shear and impact forces generated in the mixer. Such a lack of strength because of poor wetting results in the breakage of wet granules, reducing granule growth. As an example, granulation of sulfur with different polymer solutions was not successful because of poor wetting of the sulfur particles by the granulating liquid (Ritala et al., 1988). The increase in solid–liquid contact angle by the addition of the hydrophobic salicylic acid to lactose resulted in the reduction of the resulting granule particle size (Iveson et al., 2001). Vemavarapu et al. (2009) reported a low growth ratio (defined as ratio of granulated blend particle size to particle size of starting material) at the same granulation conditions for materials with high contact angle with water (Fig. 11.1). In the same study, a high contact angle and inadequate wetting also was found to increase the width of the granule size distribution (Fig. 11.3). A similar negative influence of high contact angle and poor wettability on granule growth also was reported by Jaiyeoba and Spring (1980) (Iveson et al., 2001).

Poor wettability and high contact angle also result in decreased tendency for densification. Poor wetting by the granulating liquid reduces its effectiveness in lubricating and reducing frictional forces at interparticle contact points, which resists movements of particles (Iveson & Litster, 1998). Consequently poor wetting of the particles by the granulating liquid results in porous, weaker granules.

11.6 Solubility

During wet granulation, dissolution of soluble components in the granulating liquid takes place as liquid is added and mixed with the solids in the granulator. If the formulation contains a large fraction of a highly soluble API or excipient, a significant portion of the formulation can dissolve during wet granulation, resulting in a decrease in the solid-to-liquid ratio and an increase in granulating liquid viscosity. The decrease in the solid to liquid ratio shifts the system to the rapid growth and overwet mass regions of the growth regime map. The increased viscosity can also promote granule growth.

Consequently formulations with highly soluble components require less granulating liquid to achieve the desired growth and to avoid overgranulation and uncontrolled granule growth. Granule growth was reported to be uncontrolled in highly soluble compositions as reflected in the large oversize fraction of the resulting dried granules (Vemavarapu et al., 2009) (Fig. 11.2). In the same study, highly soluble compositions also demonstrated wider size distribution for the dried granules (Fig. 11.3), which was attributed to the higher tendency for uncontrolled growth. The increased viscosity of the granulating liquid upon dissolution of the highly soluble material, which hinders liquid distribution, was also suggested as a contributing mechanism to the formation of the wide granule size distribution. In addition, reprecipitation of the dissolved API at the granule surface after migration and subsequent evaporation of water could cause attrition of the API from dried granules. This, in turn, results in a wide distribution of granule sizes.

11.7 Compaction properties

Material physical properties affect compaction behavior directly, such as the known effect of particle size on compactibility of ungranulated material, and indirectly through their impact on material behavior in wet granulation. The former represents the baseline compaction behavior, while the latter is related to the change in material compaction properties upon wet granulation. The change in compaction properties caused by wet granulation is a combined effect of the loss in porosity and size enlargement of the constituent particles, with a concomitant increase in the bonding/plasticity of particles upon uniform application of the binder. While the former is known to decrease compactability following wet granulation, the latter enhances compactability through activation of the binder.

In the lactose study, granules manufactured using Fraction I (smaller lactose particle size) were more compactable than those manufactured using Fraction II (larger lactose particle size), as shown by the compression profiles for the two granulations (Fig. 11.14). The more porous granules manufactured using Fraction I have lower strength and, as a result, have higher fragmentation propensity under the applied compression force compared with the less porous granules manufactured from Fraction 2 (Badawy et al., 2000; Wikberg & Alderborn, 1991). Also under the applied compression force, granules usually fragment along failure planes created between primary particles. This together with the higher granule porosity, results in the formation of smaller fragments during compression of granules prepared using Fraction I. Consequently a higher surface area available for bonding and a smaller tablet pore volume are created during compression of those granules, resulting in a more compactible granulation. The smaller particle size of lactose, therefore, enhanced granule compaction properties because of its inherently higher compactability and also because of its resistance to densification in wet granulation.

Similarly, in the DPC 963 study, the more porous granules manufactured using the drug substance lot with a small particle size were found to have enhanced compaction properties. This was also attributed to increased fragmentation propensity and volume reduction behavior of the more porous granulation. In agreement with the proposed mechanism, tablets compressed using the more porous granules showed reduced pore volume in the 1 to 2 µm pore diameter range compared with tablets compressed using less porous granules under the same compression force (Fig. 11.15). This illustrated the higher tendency of the more porous granulation to densify upon application of the compression force, which resulted in closer packing of the particles.

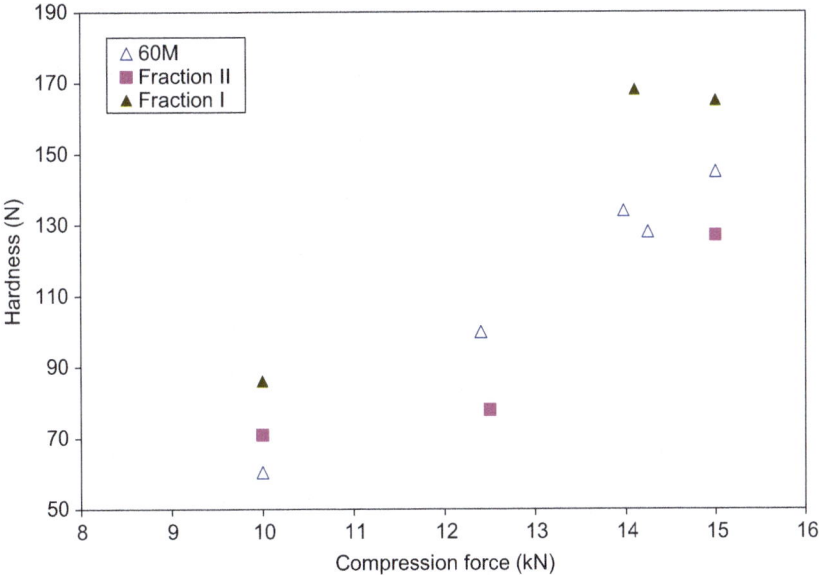

FIGURE 11.14

Compaction profiles of lactose granulation manufactured in the high granulator.

From Badawy and Hussain (2004).

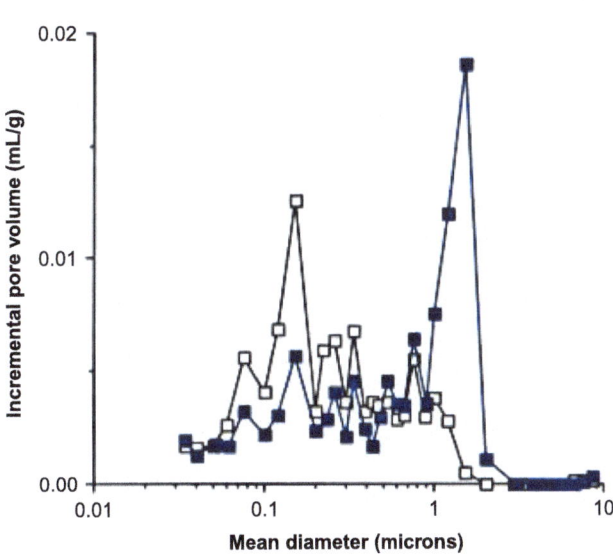

FIGURE 11.15

Porosity of DPC 963 tablets manufactured using small (□) and large (■) drug substance particle size.

From Badawy et al. (2000a).

The increase in bonding/plasticity of particles upon uniform application of the binder was found to be the dominant mechanism affecting compaction properties, outweighing the negative effect of densification and size enlargement in one wet granulation study (Vemavarapu et al., 2009). Thus the compactibility ratio was found to be greater than 1 (increased compactibility by wet granulation) for several materials (Table 11.2). Although it was difficult to delineate the mechanisms affecting compactability in the study, an attempt was made to connect the change in compaction properties upon wet granulation (compactability ratio) to the initial compactability of the starting material before granulation. Preblend compactability, therefore, was included as an independent variable in the regression model. Regression analysis of compactability ratio data resulted in several significant factors, including the preblend compactability, supporting its inclusion in the model (Table 11.6, Fig. 11.4).

The increase in compactability upon wet granulation was found to be highest in materials with low initial compactability. This result is consistent with the previously mentioned mechanism, whereby the application of a uniform binder layer enhances compactability of powders. Materials with high water uptake showed the least improvement in compactability upon wet granulation. This is believed to be a direct consequence of the loss of intraparticle porosity following granulation, with the underlying assumption that materials with high water uptake have primary particles that can be densified (e.g., celluloses, starches, large molecules) as a result of granulation (Badawy et al., 2006).

Contact angle is another material property that was found to have a significant influence on compactability ratio. As can be seen in Fig. 11.4, materials with high contact angles showed the largest improvement in compactability upon granulation. The effect of contact angle on compactability ratio is dual, in that the binder distribution is not optimal with difficult-to-wet materials. This should result in lower compactability ratios at high contact angles. To the contrary, the extent of consolidation is low at high contact angles because of the reduced interparticle lubrication by a nonwetting liquid (Iveson & Litster, 1998). The latter mechanism appeared to be more dominant in this study and, as a result, granules made from high contact angle materials had high porosity and thus higher compactability ratios. Further, in instances where granule size affects compactability (typical at low granule porosity), it can be argued that a high contact angle results in increased compactability ratio mediated through minimal size enlargement.

Compactability increase was found to be highest in materials with high surface area. Such an effect can be linked readily to low granule growth and wide granule size distributions, seen when the surface area of starting materials is high. In addition, a high surface area makes less water available at the particle surface. Because water acts as a lubricant in decreasing interparticle friction, the extent of consolidation is low in high surface materials, resulting in higher compactability ratios. Alternatively, the large internal area/porosity indicative of the high surface values of starting materials, if lost during wet granulation, should have resulted in an inverse relationship. It was found in this study that the former effects are more prominent than the latter.

11.8 Flow properties

Improvement in flow upon granulation is expected to be a combined result of size enlargement, densification, reduction in the width of powder size distribution, as well as the percentage of fines, and lastly, surface modification or rounding. Flow data in the Vemavarapu study was interpreted based on the inferred effect from the previously mentioned factors. Regression analysis of flow ratio data (change in Flodex flow index upon granulation) resulted in six significant factors, including the preblend flow, supporting its inclusion in the model as seen in Tables 11.7 and 11.5 in which the low flow ratio values indicate greater improvement in flow upon granulation. As seen with compactability, the change in flow of materials was also dependent on the initial flow of starting materials (preblends). Data indicated that greater improvement in flow can be expected of poorly flowing blends when wet granulated. Inclusion of this factor in the statistical model was necessary so that the influence of other terms could be assessed accurately.

An inverse relationship was observed between the aqueous solubility of materials and flow enhancement, likely because poorly soluble materials typically are finely milled and therefore have poor intrinsic flow properties. In addition, better flow improvement might be a manifestation of the tighter granule size distributions seen when poorly soluble materials are wet granulated. Similar argument holds for the effect of particle size, contact angle, and surface area, all of which have shown consistent effect on the width of granule size distribution and flow improvement. Larger improvement in flow when contact angle and surface area are low also can be attributed to the higher granule growth seen at these conditions. Consistent with the effect of contact angle on granule porosity, higher densification of low contact angle materials also can be implicated for higher flow enhancement. Lastly, flow improvement was found to be highest in materials with broader initial size distributions. By reducing the width of the size distribution, the tendency for powders to consolidate under their own weight is decreased, resulting in improved flow.

11.9 Summary and Conclusions

This chapter discussed the relationships among the material properties of a starting material and its behavior during high-shear wet granulation. The influence of particle size distribution, aqueous solubility, wettability, and water uptake properties was reviewed. The effect of these variables on the processibility and performance in wet granulation was shown to be of paramount importance because they affect granule quality attributes such as granule size, porosity, compactibility, and flow properties. Understanding the dependency of granule critical quality attributes on input material properties, therefore, is essential to ensure an adequate control strategy for the wet granulation process. Such critical input material attributes should be identified and controlled within a proven and acceptable range to ensure the quality of the final product.

References

Badawy, S., & Pandey, P. (2016). Design, development, and scale-up of the high-shear wet granulation process. *Developing solid oral dosage forms: Pharmaceutical theory and practice* (2nd ed, pp. 749–776). Elsevier.

Badawy, S. I., & Hussain, M. A. (2004). Effect of starting material particle size on its agglomeration behavior in high shear wet granulation. *Aaps Pharmscitech, 5*(3), e38. https://doi.org/10.1208/pt050338.

Badawy, S. I., Lee, T. J., & Menning, M. M. (2000a). Effect of drug substance particle size on the characteristics of granulation manufactured in a high-shear mixer. *Aaps Pharmscitech, 1*(4), E33. https://doi.org/10.1208/pt010433.

Badawy, S. I., Menning, M. M., Gorko, M. A., & Gilbert, D. L. (2000b). Effect of process parameters on compressibility of granulation manufactured in a high-shear mixer. *International Journal of Pharmaceutics, 198*(1), 51–61.

Badawy, S. I. F., Gray, D. B., & Hussain, M. A. (2006). A study on the effect of wet granulation on microcrystalline cellulose particle structure and performance. *Pharmaceutical Research, 23*(3), 634–640. https://doi.org/10.1007/s11095-005-9555-z.

Holm, P., Schaefer, T., & Kristensen, H. G. (1985). Granulation in high-speed mixers Part V. Power consumption and temperature changes during granulation. *Powder Technology, 43*(3), 213–223. https://doi.org/10.1016/0032-.5910(85)80002-4.

Iveson, S. M., Beathe, J. A., & Page, N. W. (2002). The dynamic strength of partially saturated powder compacts: The effect of liquid properties. *Powder Technology, 127*(2), 149–161. https://doi.org/10.1016/S0032-5910(02)00118-.3.

Iveson, S. M., & Litster, J. D. (1998a). Fundamental studies of granule consolidation part 2: Quantifying the effects of particle and binder properties. *Powder Technology, 99*(3), 243–250. https://doi.org/10.1016/s0032-5910(98)00116-8.

Iveson, S. M., & Litster, J. D. (1998b). Growth regime map for liquid-bound granules. *AIChE Journal, 44*(7), 1510–1518. https://doi.org/10.1002/aic.690440705.

Iveson, S. M., Litster, J. D., & Ennis, B. J. (1996). Fundamental studies of granule consolidation. Part 1: Effects of binder content and binder viscosity. *Powder Technology, 88*(1), 15–20. https://doi.org/10.1016/0032-5910(96)03096-3.

Iveson, S. M., Litster, J. D., Hapgood, K., & Ennis, B. J. (2001). Nucleation, growth and breakage phenomena in agitated wet granulation processes: A review. *Powder Technology, 117*(1–2), 3–39. https://doi.org/10.1016/S0032-.5910(01)00313-8.

Jaiyeoba, K. T., & Spring, M. S. (1980). The granulation of ternary mixtures: The effect of the solubility of the excipients. *The Journal of Pharmacy and Pharmacology, 32*(1), 1–5. https://doi.org/10.1111/j.2042-7158.1980.tb12833.x.

Kristensen, H. G. (1996). Particle agglomeration in high shear mixers. *Powder Technology, 88*(3), 197–202. https://doi.org/10.1016/S0032-5910(96)03123-3.

Kristensen, H. G., Holm, P., & Schaefer, T. (1985a). Mechanical properties of moist agglomerates in relation to granulation mechanisms Part I. Deformability of moist, densified agglomerates. *Powder Technology, 44*(3), 227–237. https://doi.org/10.1016/0032-5910(85)85004-x.

Kristensen, H. G., Holm, P., & Schaefer, T. (1985b). Mechanical properties of moist agglomerates in relation to granulation mechanisms. Part II. Effects of particle size distribution. *Powder Technology, 44*(3), 239–247. https://doi.org/10.1016/0032-5910(85)85005-1.

Liu, L. X., Litster, J. D., Iveson, S. M., & Ennis, B. J. (2000). Coalescence of deformable granules in wet granulation processes. *AIChE Journal, 46*(3), 529–539. https://doi.org/10.1002/aic.690460312.

Ouchiyama, N., & Tanaka, T. (1975). The probability of coalescence in granulation kinetics. *Industrial & Engineering Chemistry Process Design and Development, 14*(3), 286–289. https://doi.org/10.1021/i260055a016.

Pietsch, W., Hoffman, E., & Rumpf, H. (1969). Tensile strength of moist agglomerates. *Industrial & Engineering Chemistry Product Research and Development, 8*(1), 58–62.

Ritala, M., Holm, P., Schaefer, T., & Kristensen, H. G. (1988). Influence of liquid bonding strength on power consumption during granulation in a high shear mixer. *Drug Development and Industrial Pharmacy, 14*(8), 1041–1060. https://doi.org/10.3109/03639048809151919.

Tardos, G. I., Khan, M. I., & Mort, P. R. (1997). Critical parameters and limiting conditions in binder granulation of fine powders. *Powder Technology, 94*(3), 245–258. https://doi.org/10.1016/S0032-5910(97)03321-4.

Vemavarapu, C., Surapaneni, M., Hussain, M., & Badawy, S. (2009). Role of drug substance material properties in the processibility and performance of a wet granulated product. *International Journal of Pharmaceutics, 374*(1–2), 96–105. https://doi.org/10.1016/j.ijpharm.2009.03.014.

Wikberg, M., & Alderborn, G. (1991). Compression characteristics of granulated materials. IV. The effect of granule porosity on the fragmentation propensity and the compatibility of some granulations. *International Journal of Pharmaceutics, 69*(3), 239–253. https://doi.org/10.1016/0378-5173(91)90366-V.

CHAPTER 12

Critical material attributes in wet granulation

Arvind K. Bansal, Garima Balwani and Sneha Sheokand

Department of Pharmaceutics, National Institute of Pharmaceutical Education and Research (NIPER), Mohali, Punjab, India

12.1 Introduction

Granulation, derived from the Latin word *granulatum*, meaning grain, is a process of enlargement of powdered particles to form grain-like agglomerates. In the pharmaceutical industry, the granules formed from the particles of the active pharmaceutical ingredient (API) and excipient mix are further processed effectively into solid dosage forms, such as tablets and capsules, or multiparticulates, such as pellets, beads, or spheroids, to be filled into capsules or packed as sprinkle formulations. These granules offer the advantages of a favorable size distribution, improved bulk density, and flow properties that further prevent the problems of caking, segregation, and poor tableting performance in the downstream processing.

Granulation techniques are classified based on the type of binder and process used. Common granulation techniques include wet granulation, dry granulation, and hot melt granulation.

- In wet granulation, a liquid, generally a binder solution, is added to facilitate wetting and agglomeration of a powder blend (Kristensen & Schaefer, 1987). The wet mass is subsequently dried and screened to obtain granules with the desired size.
- Dry granulation involves granulation using dry binders without the aid of any liquid and is particularly useful for drugs that are sensitive to moisture or heat and, therefore, cannot be processed using wet granulation. In this technique, powder particles are agglomerated under high pressure either by using the process of slugging or roller compaction. The slugs or compacts are then milled to the desired size range granules (Bacher et al., 2008).
- Hot melt granulation is used as an alternative to wet granulation and employs binders that facilitate granulation in their molten form (Passerini et al., 2010). After granules are formed, they are cooled and screened to achieve the proper size. The process is carried out in a high-shear granulator fitted with heating device or using the relatively new technique of hot melt extrusion. The advantages and disadvantages offered by the three granulation techniques are captured in Table 12.1.

Despite some restrictions and disadvantages, wet granulation is the most preferred process for granulation of pharmaceutical solids. This method offers better binder–substrate interactions and yields homogenous granules with superior compaction properties. It is also applicable for a wide range of drug substances and excipients, and uses robust equipment with superior process controls that help in smooth scaleup.

Chapter 12 Critical material attributes in wet granulation

Table 12.1 Advantages and disadvantages of various granulation techniques.

	Wet granulation	Dry granulation	Hot melt granulation
Advantages	• More uniform mixing than dry granulation • More binding in less quantity of binder • Better content uniformity • Suitable for very low to very high drug content • Reduces sticking of the blend to compression tooling	• Suitable for moisture and temperature-sensitive APIs • Suitable for low to high drug contents • Time and cost-effective, unit operation of drying is not involved	• Intimate mixing, sometimes up to the molecular level, can produce solid dispersions and lead to improvement in solubility and bioavailability • Time and cost-effective, unit operation of drying is not involved
Disadvantages	• Not suitable for moisture-sensitive products • Time-consuming because of the involvement of additional unit operations of drying	• High force involved in compaction • Greater chances of the generation of dust and environmental contamination	• Not suitable for thermolabile drugs • Nontraditional process requiring greater effort in polymer screening and training on the instrument • Uses binders with low melting points, which can soften during storage and handling

With regulatory bodies promoting the modern philosophy of quality by design (QbD), there has been an increased understanding of the fundamental processes and unit operations involved in pharmaceutical manufacturing. Because wet granulation is one of the fundamental steps involved in the formulation of oral dosage forms such as tablets and multiparticulates, there has been a shift in the focus of the scientific community toward a better and in-depth understanding of the process. Applying QbD to wet granulation is being practiced globally and involves a comprehensive understanding of the parameters that have a significant effect on the quality of the granules produced. This chapter provides insights about one such parameter, that is, the material attributes that are critical to wet granulation and their impact on the quality of the resultant granules.

12.2 Basics of wet granulation: Understanding the process, the equipment, and the materials

In the wet granulation process, agglomeration of powder particles is aided with a liquid binder solution under agitation. As the binder solution is added, liquid bridges develop between the particles and granule formation commences with the application of capillary and viscous forces. Four states of liquid saturation contribute to the mechanical properties of the liquid bound granules (Iveson et al., 2001): pendular (liquid bridges at contact points only), funicular (liquid bridges expand and occupy more pore

12.2 Basics of wet granulation: Understanding the process, the equipment

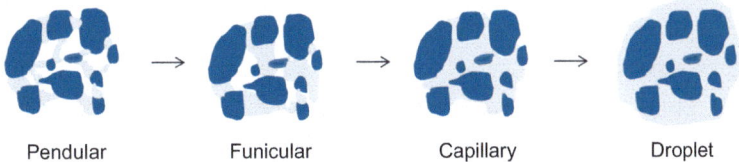

FIGURE 12.1

Different states of liquid saturation during wet granulation.

spaces), capillary (liquid filling all the voids), and droplet (particles held inside the liquid drop attain fluidity) states (Fig. 12.1).

Among these, the capillary state reflects the zone of maximum strength of the wet granules, and optimization of most of the wet granulation processes is done around this zone.

12.2.1 Rate processes

The complete behavior of the wet granulation process is presented in a Iveson et al. (2001). They discuss the three rate processes in the wet granulation process: wetting and nucleation, consolidation and growth, and attrition and breakage.

Wetting and nucleation are the processes in which a liquid binder is brought into contact with the powder, and initial agglomerates are formed. Simultaneously, granule growth from sticking and collisions between wet particles comes into effect and is referred to as coalescence. More numerous and greater force of collisions of the forming granules with each other and with equipment surfaces because of agitation leads to strengthening or consolidation of the granules. Breakage and attrition of wet granules occur from impact or compaction in the granulator. These simultaneous rate processes determine the final attributes of the granulated product. A thorough understanding of all these stages has helped in prediction of the effects of the material properties and process parameters that are critical to the wet granulation operation.

12.2.2 Types of wet granulation

On the basis of the processing conditions and the equipment used, types of wet granulation include low-shear, high-shear, and fluid-bed granulation (Fig. 12.2). In all the processes, the three major steps involved are dry blending, liquid binder addition, and distribution of liquid or wet massing (Hausman, 2004). The granulation techniques can affect the physical properties of the resulting granules, which then affect the tableting process.

- Low-shear granulation is carried out in mixers such as a planetary mixer, a sigma blade mixer, or a ribbon blender. The process involves agitation under low impact (blade speed up to 100 rpm), resulting in fluffy and porous granules having superior compressibility.
- High-shear granulation is the most commonly used process of wet granulation. In this process, mixing and agglomeration are carried out at under high speed with the aid of an impeller, with simultaneous cutting of the wet mass by an auxiliary chopper to form granules. Commercially available rapid mixer granulators, typically ranging from 25 to 1000 L in capacity, are equipped

416 Chapter 12 Critical material attributes in wet granulation

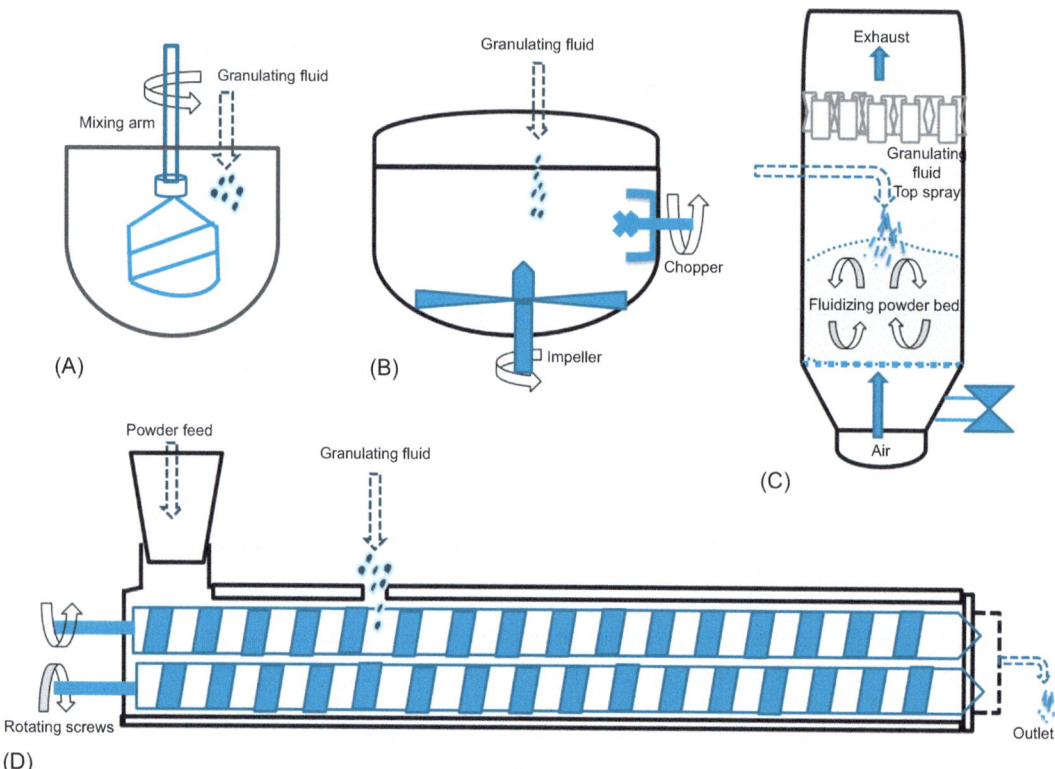

FIGURE 12.2

Equipment for wet granulation (A) low-shear mixer, (B) high-shear mixer, (C) fluid-bed processor, and (D) twin-screw granulator.

    with two dual-speed motors for the impeller to operate at 750 or 1500 rpm and the chopper to operate at 1500 or 3000 rpm. To operate the impeller and chopper at the same rpm, the power of the motors increases with an increase in the scale of the equipment. Because of higher shear forces encountered in the rapid mixer granulator, the resulting granules are less porous and denser, having less compressibility than those produced in low-shear granulators.
- Fluid-bed granulation involves spraying of the granulating fluid over a powder bed, which is first fluidized in a stream of air. It is a one-pot process in which blending, wet granulation, and drying take place in a single equipment, which reduces operator exposure to the material and results in overall shortening of manufacturing time and cost.

    By appropriately selecting the extra granular additives, tablets with similar physical properties and *in vitro* performance can be produced from granules obtained from all three processes. The choice of granulation process, thus, can be made based on the process cycle time, properties of the API, and accessibility to equipment.

12.2.3 Continuous wet granulation

With the implementation of QbD and process analytical techniques (PAT) in pharmaceutical manufacturing, continuous processing has been encouraged by the FDA. A relatively new, widely adopted continuous process of wet granulation is twin-screw granulation (Keleb et al., 2004; Seem et al., 2015). In this process, the powder blend is introduced into a chamber with two counter-rotating parallel screws and is conveyed along the screw length. The binder liquid is added at different zones along the length of the shaft, and granulation takes place as a result of the mechanical energy generated in the mixing zone by the rotating screws. The wet mass is transferred continuously from the outlet to the next processing equipment, generally a fluid-bed dryer. As with batch granulators, granules of desired properties can be produced in the twin-screw granulator by designing a control strategy for the material properties and process parameters involved. Continuous processing offers significant advantages over batch processing in terms of automation, reduction of batch-to-batch variation, labor cost, and processing time.

12.2.4 Endpoint

Endpoint determination in the wet granulation process is vital to produce granules of the desired physical properties that ultimately have an impact on final tablet properties. Traditionally, a hand-squeeze test has been used to determine the consistency of the wet mass formed on granulation. Because this technique is operator-dependent and subject to variation, it cannot be considered as a validated method (Sakr et al., 2012).

In the current scenario, where QbD is implemented in the entire process, endpoint determination is carried out by advanced monitoring techniques and has become an important part of the process control strategies. These techniques are often termed PAT tools and enable proactive and flexible control of unit operations. PAT tools implemented in the wet-granulation process include measurement of power consumption, capacitance, and torque, acoustic emissions, Raman and near infrared spectroscopy, focused beam reflectance measurements, and drag force flow sensors (Levin, 2006; Narang et al., 2016). The output of these techniques is correlated with the granules' properties. Granulation process can be stopped where the signal corresponds to the desired properties. This is particularly an important aspect to be considered during process scaleup. For example, when the process of high-shear wet granulation is transferred to high-intensity, large-scale equipment, granules with desired properties (such as density and compressibility) generally are formed at a lower liquid and binder levels than those required in lab-scale equipment. This behavior calls for reworking formula optimization for large-scale development, involving excessive time and cost. Endpoint determination at the lab-scale level, in terms of power, torque, or any other measurement tool, provides valuable input on where to stop granulation at the larger scale.

12.2.5 Excipients in wet granulation

Wet granulation is a multicomponent process, and the materials include the API and excipients. Excipients, also referred to as functional additives, are pharmacologically inactive and are present in virtually all dosage forms. Functional additives in wet granulation predominantly include diluents, binders, and disintegrants.

- Diluents are usually inert and added as fillers at levels up to 90% by weight, to form the structure of the dosage form. Typical diluents include lactose, microcrystalline cellulose (MCC), dicalcium phosphate (DCP), and mannitol.
- Binders provide cohesiveness to the powder mass and usually are effective in low quantities (up to 5%). Commonly used binders are starch, polyvinyl pyrolidone (PVP), hydroxypropyl cellulose (HPC), and hydroxypropyl methyl cellulose (HPMC).
- Disintegrants promote the breakup of the dosage form into smaller particles to facilitate drug release. Croscarmellose sodium (CCS), sodium starch glycolate (SSG), and crospovidone are the most commonly used disintegrants.

Excipients such as coloring agents, antioxidants, and dissolution modifiers can be included in the blend for granulation. Granulating fluid might be water or isopropyl alcohol or ethanol, used alone or in combination (Mills, 2010).

The choice of excipients for wet granulation mainly depends on the compatibility with the drug used, the purpose, and the process involved. For example, lactose interacts with primary and secondary amines through Maillard reaction. Therefore, it is not the excipient of choice with any drug having an amine moiety in its structure. Mannitol is a nonhygroscopic material, making it a good choice for moisture-sensitive drugs. It is also the preferred diluent for buccal tablets because its negative heat of solution causes a cooling effect. If the wet granulation is to be followed by extrusion and spheronization, MCC is the material of choice because of its ability to provide shape to the spheroids.

12.2.6 Functionality of excipients

In the traditional product development approach, excipients complying with compendial standards and from similar lots were used during development and commercial manufacturing to avoid variations. The QbD approach has enabled a profound understanding of the relationship between excipient variability and process parameters, and their effects on product quality attributes. Physical and chemical properties of the excipients, apart from those provided in their monographs, contribute toward the consistent performance of the product and robustness of the manufacturing process. This has led to the inclusion of functionality testing of excipients in pharmacopeias, and a chapter about excipient performance has been included in the US Pharmacopeia (USP30-NF25, 2007). The chapter states that "Not all critical material attributes of an excipient may be identified or evaluated by tests, procedures, and acceptance criteria in NF monographs. Excipient suppliers and users, therefore at times may wish to identify and control critical excipient attributes that go beyond monograph specifications." Therefore, the chapter provided an overview of typical material properties of excipients belonging to different functional categories and the additional tests that can be performed on the excipients. Such characterization of the material attributes would evaluate the suitability of the excipients for their intended use.

Similarly, the European Pharmacopeia has included certain functionality related characteristics sections to monographs of excipients (Ph.Eur. 7.0, 2012). These material attributes, also mentioned in Fig. 12.3, can affect different stages of wet granulation and influence the quality of the product. A critical assessment of these properties of the excipients, therefore, is desirable to identify the ones that have an impact on the critical quality attributes (CQAs) of the product. Such properties then would qualify to be the critical material attributes (CMAs) for the product and would demand establishment of a design space as a part of the control strategy for the development process.

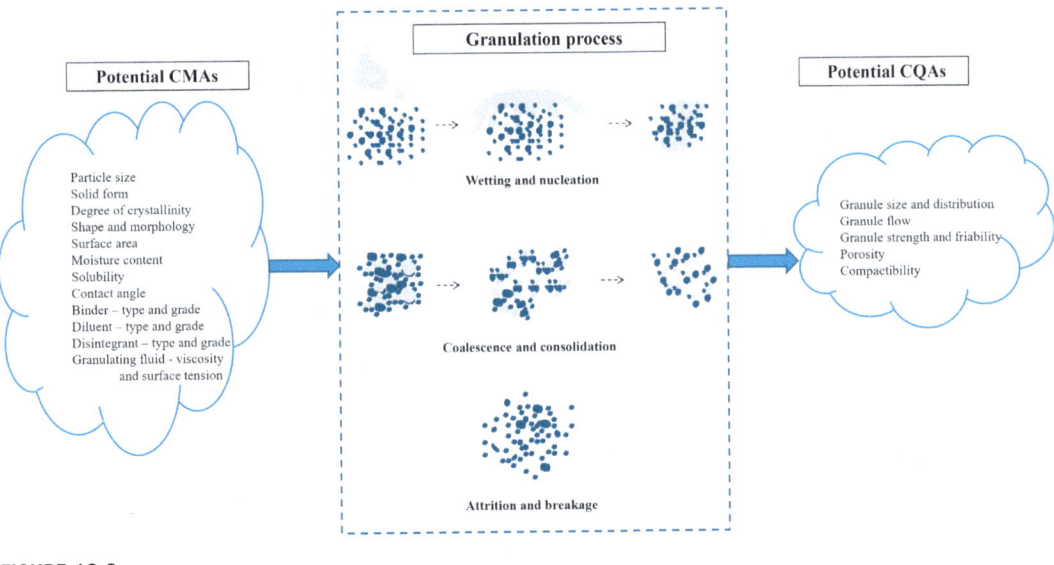

FIGURE 12.3

Material attributes affecting the granulation process and the resulting granules quality attributes.

12.3 Quality by design paradigm in wet granulation

The philosophy of QbD primarily involves the "voice of consumer," that is, patient-related aspects of safety, efficacy, and quality. This concept also conjointly involves the application of the scientific risk-based approach to formulation development, which minimizes consumers' risk. QbD has evolved with the release of three ICH guidelines; Q8(R2) (pharmaceutical development), Q9 (quality risk management), and Q10 (pharmaceutical quality system). These documents provide a direction with respect to the objective and definition of QbD as it applies to the pharmaceutical industry.

Over the years, QbD concepts have emerged with industry and regulatory agencies understanding that "quality cannot be tested into the products; quality should be built-in or should be by design" (Q8(R2), 2009). The central tenet of QbD is the quality target product profile (QTPP) of a product, which forms the basis for the development of CQAs of the drug product. Building quality into products is achieved by an exhaustive understanding of the product and process, and the development of relationships that link input material attributes and process parameters with CQAs of the product. This, in turn, enables manufacturing flexibility, process robustness, continuous improvement, and minimal postapproval changes.

Conventionally, the most important attribute related to the granulation process was the dissolution rate of the resultant tablet/capsule. If the dissolution of the resultant tablet/capsule showed the desired profile, scaleup was considered to be successful. This approach, therefore, considered granulation as more of an art than a science. Such statements, however, are obsolete because of the growing quantitative understanding of granulation processes and the mechanisms of granulation. Even though dissolution still remains a CQA of the final product, process, and product knowledge at the intermediate granule level is considered to be equally important.

With advancements in the tools and techniques for monitoring granulation processes, defining the criticality of material attributes and process parameters for granulation is possible within the QbD framework. Existing knowledge about granulation in the literature can be leveraged for a better process and product design and to save time.

The QbD continuum begins with defining QTPP, as it relates to quality that ensures the safety and efficacy of the product. The QTPP includes route of administration, dosage form, and stability [Q8(R2), 2009]. It provides the quality characteristics that the drug product should possess so as to deliver the therapeutic benefit promised in the label. The next step involves identifying CQAs, which is followed by the application of experimental and experience-based knowledge and performing risk analysis to define CMAs and critical process parameters (CPPs). The design of experiments (DoE) approach is implemented to establish the operating range of the CMAs and CPPs and the design space within which the quantitative parameters must be maintained. In the entire experimentation process, PAT tools to assess the wet mass properties are put into practice for continuous monitoring and improvement of the process.

12.3.1 Critical quality attributes of granules

A CQA has been defined in ICH Q8(R2) guidelines as "a physical, chemical, biological or microbiological property or characteristic that should be within an appropriate limit, range, or distribution to ensure the desired product quality" [Q8(R2), 2009]. Drug product CQAs are identified from QTPP, and intermediate CQAs are identified from the process map. Intermediate CQAs are the quality attributes of in-process materials that affect drug product CQAs. Because wet granulation is one of the processes in solid dosage form development, the intermediate CQAs for granules need to be defined. An intermediate CQA can become a CMA of that same intermediate for a downstream processing step. For the purpose of this chapter, CMAs are considered for input materials, including drug substance, excipients, and granulating fluid; and CQAs are for output materials, that is, granules. The potential quality attributes of granules are granule size distribution, porosity, granule strength, flowability of granules, solid state or polymorphic form of the API, compaction properties, granule shape and morphology, moisture content, surface area of granules, friability, and hygroscopicity (Lawrence et al., 2014; Pandey & Badawy, 2016).

12.3.2 Critical process parameters of wet granulation

The wet granulation process is considered to be well-understood when all sources of variability have been identified, explained, and managed, therefore allowing better prediction of CQAs. A process parameter is a measure that indicates the status of the process or unit operation (e.g., pressure and temperature) and can be changed to achieve the desired product quality. It is considered critical when its variability within the normal operating ranges can affect a CQA. Therefore, it should be monitored or controlled to ensure that the process produces the desired quality [Q8(R2), 2009].

In the case of wet granulation, the CPPs can differ with respect to the type of granulator (low shear, high shear, fluid bed, or twin screw). Table 12.2 lists potential process parameters of wet granulation. A better process understanding could be established by: identifying all potential process parameters

Table 12.2 A summary of potential process parameters related to wet granulation.

Type of granulation	Process parameters
High/low-shear granulation	Type of granulator (high/low shear, top/bottom drive)
	Fill level
	Pregranulation and postgranulation mix time
	Granulating liquid or solvent quantity
	Impeller speed, tip speed, configuration, location, power consumption/torque
	Chopper speed, configuration, location
	Spray nozzle type and location
	Method of binder excipient addition (dry/wet)
	Method of granulating liquid addition (spray or pump)
	Granulating liquid temperature, addition rate
	Wet massing time (postgranulation mix time)
	Bowl temperature (jacket temperature)
	Product temperature
	Pump type: Peristaltic, gear type
	Granulating liquid vessel (e.g., pressurized, heated)
Fluid-bed granulation	Type of fluid bed
	Inlet air distribution plate
	Spray nozzle (tip size/type/quantity/pattern/configuration/position)
	Filter type and orifice size
	Fill level
	Bottom screen size and type
	Preheating temperature/time
	Method of binder excipient addition (dry/wet)
	Granulating liquid temperature, quantity, holding time, delivery method, spray rate
	Inlet air, volume, temperature, dew point
	Atomization air pressure
	Product and filter pressure differentials
	Product temperature
	Exhaust air temperature, flow
	Filter shaking interval and duration
Twin-screw granulation	Powder feed rate
	Liquid/solid ratio
	Screw speed
	Throughput
	Screw design (number and angle of kneading elements)
	Binder addition rate
	Method of binder excipient addition (dry/wet)
	Barrel temperature
	Barrel length

From Lawrence et al. (2014) and Vercruysse et al. (2012).

that could affect the CQAs; screening out the high-risk process parameters based on prior knowledge and risk assessment; linking process parameters with CQAs by designing and conducting experiments; determining the criticality of process parameters based on experimental results; and defining acceptable ranges for CPPs.

Many studies that identified the CPPs for wet granulation and established their relationships with CQAs have been reported. For example, Wang et al. (2008) recorded an observation of the role of impeller speed and chopper speed on granule characteristics. They found that an optimum combination of impeller and chopper speeds provided a balance that established a good toroidal motion of the wet mass during the early stages of granulation. This toroidal flow was important for maintaining the homogeneity in the wet mass. By formation of granules during the kneading stage, the granule–chopper collision disturbed the toroidal motion and resulted in adhesion of the material to the walls. Loss of toroidal motion, therefore, caused material deposition and uncontrolled granule growth. By turning off the chopper toward the end of the kneading process, however, the toroidal motion can be maintained for a longer period and the granule size distribution can be narrowed, thereby producing a smoother granule surface (Wang et al., 2008). Further details about CPPs of wet granulation can be found elsewhere in the literature.

12.3.3 Critical material attributes for granules

Product design and understanding are considered equally important as the process design and understanding. Product design determines whether the product is of desired quality (to satisfy patient's needs) and maintains its performance throughout its shelf life. Considering granules as the product, the key elements of product design are similar to those of the process design: identifying material attributes of drug substances, excipients, and granulating fluid; establishing functional relationships between material attributes and CQAs; and developing control strategy for CMAs by defining their acceptable ranges.

The potential CMAs that can affect the different stages of the wet granulation process and the CQAs of the resultant granules are presented in Fig. 12.3. In addition, researchers have found that the material attributes also can interact with process parameters and have an impact on the output granulation CQAs. For instance, the effect of impeller speed and starting material particle size (API or excipient) has been found to be highly interactive on the rate of consolidation of granules. Larger starting material particle size can make granules more susceptible to consolidation at a higher impeller speed (Iveson & Litster, 1998b; Iveson et al., 1996).

Defining the criticality of a material attribute depends on the QTPP of a particular drug product and desired targets for the different CQAs. For example, if granule size distribution is a CQA, then starting material particle size of API and excipients might be a CMA (Badawy & Hussain, 2004). Typically, a wet granulated formulation includes one or more APIs (or drug substance), diluents, binders, disintegrants, and granulation fluid. The potential CMAs relevant to these materials are discussed in the following subsections.

12.3.3.1 Solid form
Knowledge about the solid form of API or excipients during processing and storage is important to ensure consistent stability and bioavailability, and therefore, the safety and efficacy of a drug product.

Solid state changes can influence the physicochemical properties, resulting in different processability (compressibility and powder flow), solubility, stability, and bioavailability. Therefore, it is important to evaluate this material attribute to avoid a safety, efficacy, or quality issue in the final product.

Wet granulation is appropriate for some materials, but it might not be appropriate for materials that are prone to solid-state transformations. Many studies have been reported that highlight the solid form changes in drugs as well as excipients (Fonteyne et al., 2013; Landin et al., 1994; Mirza et al., 2007; Yoshinari et al., 2002, 2003; Zhang et al., 2004). APIs can undergo a solvent-mediated phase transformation from the anhydrous to the monohydrate form during wet granulation, leading, in some cases, to a decrease in drug release rate, thereby affecting bioavailability. Shift from offline to inline monitoring of this phenomenon with the aid of PAT tools (such as Raman imaging and NIR spectroscopy) can help monitor the process closely and determine ways to prevent the transformation during early stages of product development (Wikstrom et al., 2005). Studies by several research groups have shown that it might be possible to inhibit this transformation by incorporating certain polymeric excipients in the formulation. For example, conversion of anhydrous theophylline to its hydrate form has been inhibited by low molecular weight methyl celluloses (Wikstrom et al., 2008). The most likely mechanism for this inhibition effect is the selective adsorption of methyl cellulose to fast-growing crystal surfaces of the hydrate crystal. It was hypothesized that the morphology of the crystal changed from needle-shaped in the absence of methyl cellulose to polygon-shaped in the presence of methylcellulose. The presence of methylcellulose was also found to delay the onset of the transformation and reduce the number of hydrate crystals produced, suggesting an effect on the nucleation stage of the conversion (ibid).

In another study, it was seen that hydrate formation of anhydrous nitrofurantoin during wet granulation was inhibited by excipients having predominant amorphous regions in their structure (low substituted HPC and pregelatinized starch), while the transformation was not prevented by crystalline excipients (lactose monohydrate) or hygroscopic partially crystalline excipient (silicified MCC) (Airaksinen et al., 2005). These excipients depict different water sorption properties. Crystalline excipients have a tendency to adsorb water and form hydrates, while amorphous materials have water absorption potential and prevent the hydrate formation of the anhydrous crystalline drug molecule (ibid).

In yet another study, cross-linked polyacrylic acid inhibited the transformation of caffeine, and HPMC inhibited the transformation of carbamazepine (Gift et al., 2009). It is noted that the inhibitory effects of the polymeric excipients are specific to drug–polymer combinations, necessitating the implementation of an excipient screening method to identify polymers that can be used to manipulate the hydrate formation of API during wet granulation.

The crystallinity of the starting material also affects the quality attributes of the granules produced upon wet granulation. Suzuki et al. (1994) studied the effect of crystallinity of MCC on granulation with cornstarch in a high-shear mixer (). They used intact MCC with 65% crystallinity and two types of milled MCCs, with almost equivalent (63%) or significantly lower (25%) crystallinity as that of the intact MCC. The authors found that the granule size and growth rate were highest in the case of MCC with 25% crystallinity. This was attributed to the presence of a large portion of amorphous form in the MCC, which has the tendency to absorb a large amount of water, therefore making the water less available for granule growth. Structurally, layered granules were formed with crystalline MCC, where the rigid crystal structure was difficult to break with the impeller, thereby leading to a broad particle size distribution and deposition of ungranulated material over MCC. Amorphous MCC, however, abraded easily under the impact of the impeller, leading to homogenous mixing with cornstarch during coalescence and formation of a more uniform structure. Drug content uniformity of granules was also better with the use of amorphous MCC because of the homogenous distribution of components during granulation.

A favorable solid-state transformation of mannitol is observed when it is used as the diluent during wet granulation. The δ-form of mannitol converted to its stable polymorphic form, that is, β-form, during wet granulation and was found to have better compaction properties (Yoshinari et al., 2002, 2003). These in-process phase transformations can be monitored by appropriate PAT tools (such as NIR or Raman) and, based on the requirements, the solid form of the starting material can be evaluated critically.

12.3.3.2 Particle size

The particle size of starting material (i.e., drug and excipients) is inversely related to granule growth, suggesting that a smaller particle size material would show rapid granule growth as compared to a material with a larger initial particle size (Badawy & Hussain, 2004; Vemavarapu et al., 2009). This can be explained based on the fact that nucleation and coalescence processes are more pronounced in the smaller particle fraction. The width of the granule size distribution is also high when the initial particle size of the material is low (Badawy & Hussain, 2004). Granule porosity and compactability are also affected by the changes in starting material particle size. Decreasing the initial particle size results in higher porosity and compactability of granules (Badawy & Hussain, 2004; Badawy et al., 2000; Fonteyne et al., 2014).

The influence of particle size of active substances on granules' properties was studied by Belohlav et al. (2007). Two batches of an active substance, one with a smaller mean diameter (63–125 μm) and more fines, and the other with a larger mean diameter (125–250 μm) and fewer fines, obtained from different manufacturers, were used in the study. The batch with the smaller particle size was wetted uniformly, whereas that with the larger particle size material was overwetted and had a paste-like consistency, with the same amount of granulating fluid. This was because of the lower surface area of the larger particle size material requiring less granulating fluid. In the same study, the effect of particle size distribution was also assessed by using the active substance from different levels of storage container. The samples taken from the bottom of the drum contained greater levels of finer particle size fraction (> 80% in 63–125 μm range) compared with the samples from the top layer of the storage drum (> 90% in 90–180 μm range). On granulation with an identical amount of wetting agent, it was observed that the material taken from bottom of the container was not sufficiently wet and the granulation period was longer.

The effect of the particle size of drug substance, DPC 963, on the granule properties prepared by high-shear wet granulation was evaluated by Badawy et al. (2000). Granule size was inversely proportional to the particle size of the drug substance, and the granules prepared with the smaller particle size had higher porosity and compressibility. Another study from the same author describes the effect of particle size of the commonly used excipient, anhydrous lactose, on the agglomeration behavior during wet granulation (Badawy & Hussain, 2004). Three grades of anhydrous lactose differing in their particle size, that is, 60 M grade, and two size fractions obtained by passing 60 M grade through screens of different mesh sizes were granulated with water, and the effect on granule growth, porosity, and compactibility were studied. The initial geometric mean diameter of 60 M grade, fraction I, and fraction II, as determined by mesh analysis, were 110.1, 53.3, and 178 μm, respectively. After granulation with water, fraction I showed pronounced granule growth with a 2.78 times increase in particle size from the initial as compared to 1.66 and 1.79 times in fraction II and 60 M grades, respectively. The granules prepared with fraction I also had a higher porosity of about 39% as compared to 16% for fraction II and

10% for the 60 M grade. As the more porous granules have lower strength and a tendency to fragment under applied compression force, fraction I granules depicted better compactability than the other two grades. The stress-strain profile of moist lactose compacts prepared at water levels of 10% and 20% that was compressed at a certain porosity also was determined. At a 10% liquid level, compacts of all the fractions exhibited brittle behavior, representing the nucleation stage of granulation at low water level. At a 20% water level, compacts of fraction I were still brittle, and compacts of fraction II and 60 M grade showed plastic deformation, representing dominance of coalescence stage behavior. These results suggested that lower particle size materials require higher liquid levels for transformation into a wet plastic mass suitable for compression or extrusion in the downstream processes.

Similar findings have been observed in studies involving different excipients such as lactose monohydrate (Mackaplow et al., 2000) and DCP (Kristensen et al., 1985), in which an increase in the primary particle size resulted in larger and less porous wet granules. This is consistent with the expectation that both the capillary and viscous interparticle forces decrease with increasing primary particle size, and the resulting granules become more deformable.

12.3.3.3 Particle shape and morphology

Shape and morphology of the starting material particles influence the strength of wet granules, which then affects their size distribution and compactability. Limited studies have reported that correlated the shape and morphology of particles with the quality attributes of granules. Granule growth kinetics and granule strength were shown to depend strongly on the primary particle shape and morphology (Iveson & Page, 2005; Rajniak et al., 2009; Simons & Pepin, 2003; Simons et al., 2003). Granule strength increases with an increase in a number of interparticle contacts, more interlocking, and higher coefficients of friction. Particles with smooth surfaces showed faster growth kinetics and resulted in larger granule size, while rough-surfaced particles showed an initial lag with no granulation, followed by slower growth and smaller granule size (Rajniak et al., 2009). Spherical particles produced granules with lower strength than nonspherical particles. Irregular shape (such as needle, angular, or dendritic) and rough surfaces exhibited higher stresses and, therefore, increased granule strength (Iveson & Page, 2005; Simons et al., 2003).

12.3.3.4 Surface area

The surface area of the starting material particles is an important parameter that can affect a number of granulation properties, such as granule size distribution, percentage of oversized particles or fines, compaction properties, and flowability. The extent of interaction between particles might not be appropriately accounted for by the size of the particles because their surface area also plays an important role. Particles of a similar size might not have a similar surface area if the surface roughness of the particles differs considerably. Vemavarapu et al. established a relationship between surface area and various granulation properties (Vemavarapu et al., 2009). An increase in the surface area of materials was found to decrease the granule size, percent oversize fraction, and flowability. Compactability of granules, however, was found to be higher in materials with greater surface area.

12.3.3.5 Moisture content

The moisture content of drugs and excipients, and/or processes involving water, can have a significant impact on the physical and chemical properties of the finished dosage form. Various processes require water (or another solvent) to accomplish their intended result. Wet granulation is one of the processes that requires a granulating fluid with or without a binder. Quality attributes of granules, such as flowability, compactability, polymorphic purity, or solid form, are affected by the moisture content of the input materials.

One of the most important issues is the stability of the drug molecule in the granules. The physical stability of an amorphous drug is affected by moisture because an amorphous solid absorbs more water into its bulk structure than a crystalline material does. Moisture acts as a plasticizer, which could increase the molecular mobility of the system, increase the rate of phase transformation, and hamper the physical stability of amorphous drugs (Shalaev & Zografi, 1996). Therefore, it has been a common practice to avoid excipients of high initial moisture content with moisture-sensitive drugs. Location and availability of moisture in materials, rather than the overall moisture content, however, are crucial for drug stability. For example, if the moisture is present in the bulk of the excipient rather than at the surface, it will be less available for interacting with the drug and therefore would cause less degradation.

Variations in initial moisture content were also found to affect the flowability and compactability of granules. Shi et al. showed that the initial moisture content can have a significant impact on the manufacturability of MCC granules by high-shear wet granulation process (Shi et al., 2011). MCC having different initial moisture content (0.9%, 2.6%, 4.6%, 7.2%, and 10.5%) was granulated with 65% water, keeping all parameters constant. The granule physical properties, flowability, and tabletability were studied. Morphological evaluation of the granules revealed the presence of loose agglomerates for low moisture level MCC (0.9% and 2.6%). These loose agglomerates were thought to be remnants of the nucleation stage and their absence in granules made from MCC with higher moisture levels (4.6%, 7.2%, and 10.5%) suggested completion of the granulation process. The higher granule size with increasing initial moisture content of MCC indicated the dominance of the coalescence phase, thus causing pronounced granule growth. This resembled an overgranulated state. Liquid saturation of MCC is reached at 70% w/w water level. It was observed that the porosity of the granules decreased with an increase in initial moisture content (up to 4.6%), until liquid saturation was reached, beyond which there was no significant effect on porosity. The specific surface area of granules prepared with MCC of 0.9% initial moisture content was more than twice that of granules prepared with MCC of 10.5% initial moisture content. Consistent with the increase in granule size, decrease in porosity, and specific surface area, the tabletability of granules was reduced, while granule flowability improved continuously with the increasing moisture content of MCC. This work highlighted the importance of controlling the initial moisture content of raw materials to ensure good batch-to-batch reproducibility in physical properties and manufacturing performance of the formed granules.

12.3.3.6 Solubility

The solubility of starting materials in granulating fluids is a key parameter affecting the granule characteristics. Highly soluble compounds favor uncontrolled granule coalescence and growth, leading to a wider granule size distribution. This is because of an increase in the liquid-to-solid ratio as the material dissolves and because of an increased viscosity of the liquid that compromises uniform liquid distribution and yields granules of variable sizes (Iveson et al., 2001; Vemavarapu et al., 2009). High solubility also leads to reprecipitation of the solid material on the granule surface after drying. The migration of liquid to the outer layers of granules and subsequent evaporation of water during drying could cause this reprecipitation, and attrition of granules could then remove the precipitated material from the granule surface, creating fines. This segregation of particles results in a wide granule size distribution (Iveson et al., 2001).

Jaiyeoba and Spring (1980) investigated the effect of solubility of excipients on granule properties after wet granulation using ternary mixtures. Granulation of lactose and boric acid blend was carried out with PVP binder solution, and the effects of third components of different solubilities on the granule features were studied. Addition of a highly soluble excipient, sucrose, to the powder blend led to a homogenous distribution of the binder fluid and formation of stronger liquid bridges, producing bigger granules of greater strength. In contrast, when the third component was a partially soluble and water-absorbing starch, Sta-Rx 1500, there was a reduction in binder volume availability for granulation, producing smaller granules with poor strength (Jaiyeoba & Spring, 1980).

An inverse relationship was observed between the solubility of materials and their flowability (Vemavarapu et al., 2009). This is because highly soluble materials cause wider granule size distribution and a higher percentage of fines, which results in poor flowability of granules. Therefore, care should be exercised during the selection or modification of excipients for wet granulation because a variation in solubility can impact the granule properties.

12.3.3.7 *Wettability (contact angle)*

As mentioned previously, granulation is a combination of three sets of rate processes: nucleation; coalescence and growth; and attrition and breakage. The nucleation zone is the area where the liquid binder and powder surface first come into contact and form the initial nuclei. The size distribution of these initial nuclei depends on the wettability of the powder with the binder solution. A higher wettability or lower contact angle of powder with binder decreases the width of granule size distribution because of enhanced nucleation (Ho et al., 2011; Nguyen et al., 2010; Thielmann et al., 2008).

Wettability of a material is linked simply to the type of functional groups exposed on its surface, that is, surface chemistry of particles. While spherical, monodisperse hydrophilic or hydrophobic particles are highly idealized systems, real pharmaceutical materials are polydisperse with surface heterogeneity. Therefore, the surface chemistry of particles has a considerable effect on the granulation behavior and the granule attributes. Ho et al. (2011) showed that the granule size distribution was affected by granulating different mass ratios of hydrophobic surface-silanized mannitol and untreated hydrophilic mannitol. As the hydrophobicity of the formulation increased (i.e., increase in amount of surface-treated mannitol), the proportion of smaller granules was higher, owing to poor distribution of granulating fluid because of the material's poor wettability.

The content uniformity or distribution of the drug in the granules is strongly dependent upon the wettability of the formulation. Nguyen et al. (2010) observed this effect, in which granulation of a blend composed of different ratios of salicylic acid (hydrophobic) and lactose (hydrophilic) was carried out with different granulating fluids, that is, water and aqueous PVP solution . The granulation batches with varying formulation hydrophobicity were sieved into different size fractions and checked for assay. For batches with water as the granulating fluid, the drug distribution was uneven; some sieve fractions were enriched, and some were deficient in drug content. When the PVP solution was used as the granulating fluid, the wettability of the formulation was improved, and a more uniform distribution of the drug was achieved across all sieve fractions. The average granule size was also found to decrease as the formulation hydrophobicity increased. This was attributed to the decreasing liquid bridge strength between the particles, which results in greater attrition and the generation of fines during drying.

Wettability also has a significant influence on compactability and flowability of granules. Materials with poor wettability or higher contact angles show largest improvement in compactability upon granulation. This is because binder distribution will not be perfect with difficult-to-wet materials, leading to higher porosity, reduced interparticle lubrication, and lower extent of consolidation. Consistent with the effect of contact angle on granule porosity, lower densification of high contact angle materials can lead to poor flowability (Vemavarapu et al., 2009). Krycer et al. (1983) studied the effect of wetting and film-forming ability of several binders (PVP, HPMC, PEG6000, starch, acacia, and sucrose) on friability of paracetamol granules and capping tendency in tablets (Table 12.3). It was observed that paracetamol granulated with HPMC and acacia solution had the least friable granules and depicted the least capping tendency. This was because of the low contact angle of HPMC and acacia on paracetamol surface, leading to uniform spreading of binder films and the formation of numerous liquid bridges at the particle contacts. In contrast, the granules bound with sucrose and PEG 6000 solution were the most friable and had the highest capping tendency. Sucrose and PEG 6000 formed only a semicoherent coat over paracetamol surface and revealed no film formation and only few interparticle bonds. Starch and PVP, being film-forming agents, were unable to produce granules of good strength because of their high contact angle and reduced wetting on the paracetamol surface. The effect of the addition of glycerol as a plasticizer or SLS as a wetting agent to the PVP binder solution was also studied. Addition of glycerol exerted only a minor improvement in paracetamol wetting and did not significantly alter granule strength, whereas SLS addition caused an increased wetting and bond formation and a significant improvement in granule strength (Krycer et al., 1983). Therefore, choosing a proper binder and/or wetting agent combination is essential for achieving the required granule and tablet strength.

12.3.3.8 Powder density

In the field of granulation, the term particle refers to fine aggregates/agglomerates. The bulk density of the starting material is an important attribute because it influences the mechanical properties and fluidization of powder. Particle density has a direct correlation with the maximum pore liquid saturation of the moist agglomerates formed during wet granulation, which thereby affects the granule growth behavior and intragranular porosity (Iveson et al., 2001). It is also desirable that the densities of the different components meant for wet granulation do not differ significantly. Differences in densities of the ingredients can lead to differential liquid saturation across the powder blend, resulting in granules with varying size distribution and nonuniformity in drug content. Such an effect is seen particularly when wet granulation is carried out in a fluid-bed granulator where segregation of nonsimilar density particles can take place on fluidization and lead to nonuniform granulation (Iveson et al., 2001).

12.3.3.9 Type of diluents

In most cases, diluents form the bulk of the powder blend meant for wet granulation and, therefore, play an important role by affecting properties such as content uniformity, compactability, and flowability of granules. Some of the commonly used diluents in wet granulation process include soluble fillers such as lactose and mannitol, and insoluble fillers such as MCC and DCP. These diluents differ in their material properties, such as solid form, crystallinity, particle size, and moisture content. The impact of these material properties on granule quality attributes have been described in the previous sections.

When wet granulation is to be followed by tableting, the type of diluent governs the mechanism of compactibility. For example, lactose and DCP show brittle fracture and have greater fragmentation tendency on application of shear forces. MCC, however, has a large water holding capacity and depicts a plastic deformation on shear. Wet granulation with MCC in a high-shear granulator leads to overgranulation at water level $\geq 55\%$, and results in loss of compressibility (Osei-Yeboah et al., 2014). Incorporation of brittle excipients (such as lactose and DCP) with MCC

Table 12.3 Characteristics and effects of different binders on friability and capping of paracetamol granulations.

Binders used for paracetamol granulation at 4% w/w concentration	Contact angle on paracetamol surface (θ)	Binder film formation and intragranular solid bridges (observed under SEM)	Granule friability index ($\times 10^{-3}$ min^{-1})	Tablet capping index (%/MPa)
HPMC	27.1	Yes	14.8	0.54
Acacia	30.3	Yes	19.8	0.81
Sucrose	32.8	No	87.6	4.4
PEG 6000	37.2	No	61.5	3.8
Starch	47.3	Yes	45.3	2.1
PVP	42.2	Yes	26.5	1.4
PVP + glycerol	40.1	Yes	25.5	0.72
PVP + SLS	11.5	Yes	20.8	0.87

From Krycer et al. (1983).

addresses the problem of overgranulation and yield tablets of desired tensile strength (Osei-Yeboah et al., 2014).

In contrast, when the process of wet granulation is to be followed by extrusion and spheronization, the large water holding capacity and plasticity of MCC render it an essential formulation component to form spherical pellets. MCC acts as a molecular sponge imbibing water and alters the rheological properties of the wet mass (Ek & Newton, 1998). It also possesses adhesion properties because of the interdiffusion of free cellulose polymer chains, leading to hard, noncompressible, and nondisintegrating pellets. Immediate release of drugs from these pellets requires the coaddition of a water soluble excipient (filler/disintegrant) into the formulation.

Diluents come in different grades, which differ in the particle size distribution, moisture content, particle shape, density, surface area, degree of polymerization, solubility, and polymorphic form. For example, α-lactose monohydrate and β-anhydrous lactose are among the most commonly used forms of lactose. The anhydrous β form has a higher solubility, less hygroscopicity, and higher compactability than the α form (Listiohadi et al., 2008). These differences in the grades of lactose affect the CQAs of granules. Similarly, out of the nine commercially available pharmaceutical grades of MCC, the Avicel PH 101 grade is used in wet granulation because of its low median diameter of $\sim$50 µm, which causes rapid and uniform water uptake during granulation. The coarser grades, Avicel PH 102 and Avicel PH 112, have a higher median diameter of $\sim$100 µm and are more suitable for direct compression applications because of their enhanced flow properties.

12.3.3.10 Type of disintegrants

Commonly used disintegrants in wet granulation are crospovidone, CCS, and SSG, which increase the hydrostatic pressure in the formulation when it comes in contact with water. The pressure is increased either by water wicking (for crospovidone) or swelling (CCS and SSG) phenomena. These water uptake behaviors of the disintegrants also occur during the wet granulation process and result in a reduced efficacy of disintegration in dosage forms produced thereafter, as compared to the dosage forms produced using dry granulation. The efficiency loss of disintegrants on wet granulation was studied by Zhao and Augsburger (2006). The authors reported that, when these disintegrants were wet granulated, a significant increase in disintegration time (i.e., reduction of disintegrant efficiency) was observed in tablets prepared with SSG and crospovidone, while CCS was the least affected by prewetting undergone during wet granulation. The efficiency of SSG and crospovidone showed dependence on their particle size, with a decrease in particle size tending to increase their efficiency (ibid).

In another study, the effects of CCS and crospovidone on process endpoint determination and drug release from the tablets were determined (Narang, 2015). Crospovidone showed a rapid and sudden granule growth with a sharp endpoint, whereas CCS-based formulations provided slower granule growth and a smooth endpoint. Both CCS- and crospovidone-based formulations showed higher initial dissolution, but storage under accelerated conditions resulted in a dissolution slowdown in the case of the crospovidone-based formulation, while no such significant change was observed in the CCS-based formulation. These studies indicated that CCS, as an intragranular disintegrant, is likely to provide greater process robustness as well as better stability and product performance.

Another material property related to disintegrants affecting their function is the degree of substitution and cross-linking (Rudnic et al., 1985). These two attributes greatly affect the water uptake capacity of the disintegrants. In addition, the ionizable functional groups in the backbone of SSG and CCS have potential for interaction with drug substances. If the interaction is significant to the extent that drug release or bioavailability is affected, these properties can become a CMA for the drug product (Narang et al., 2012).

12.3.3.11 *Type of binders*

Binder affects both the physical and mechanical properties of granules. The efficiency of a binder depends on its type (i.e., natural or synthetic polymers, sugars, etc.), grade, amount, and addition method (dry or wet mix). Commonly used binders in wet granulation are cellulose derivatives (e.g., HPC and HPMC), starches, povidone (of different sizes and molecular weight grades, such as, K12, K30, and K90), and gelatin.

Variations in the mechanical and film-forming properties of binders determine the strength and deformation behavior of binder films. Healey et al. (1974) found that acacia and PVP formed weak films with low tensile strength, while starch and gelatin films possessed higher tensile strength. Therefore, granules formed with gelatin or starch, would be expected to be less friable, while granules containing PVP or acacia would show high friability. In addition, PVP was also reported to exhibit high film deformability, which would improve consolidation during compaction.

Granule strength and friability have been linked with the wettability of powders with the binder fluid, which is affected by binder cohesion and binder–powder adhesion. Here, the term powder means the intragranular materials, which include primarily drug, diluents, and disintegrants. Nucleation commences when wetting is thermodynamically favorable as suggested by the contact angle and spreading coefficients of the binder solution and solid powder (Iveson et al., 2001). The spreading coefficient is defined as the propensity of a binder liquid to spread over solid powder and vice versa. It is the difference between the work of adhesion and work of cohesion. The work of adhesion separates a unit cross-sectional area of a solid–liquid interface from one another. The work of cohesion separates a unit area of a solid-liquid from itself. These surface-free energy parameters such as spreading coefficient, work of adhesion, and work of cohesion can be determined for different combinations of binders and substrates (solid powder), and the information can be used in the selection of suitable binders for various substrates. Work of adhesion W_a, work of cohesion W_c, and spreading coefficient, λ_{12} and λ_{21} can be calculated using the following equations:

$$Wa = 4\frac{y_1^d y_2^d}{y_1^d y_2^d} + 4\frac{y_1^p y_2^p}{y_1^p y_2^p} \tag{12.1}$$

where γ_1^d and γ_2^d are individual surface-free energies of the nonpolar components of the two phases (binder and substrate) and γ_1^p and γ_2^p are individual surface free energies of the polar components of the two phases (Planinsek et al., 2000; Rowe, 1989; Simons et al., 2005):

$$Wc = 2y \tag{12.2}$$

where γ is surface free energy of individual phase.

Surface free energy and polarity values can be determined using contact angle measurement by the Young's equation (Zografi & Tam, 1976).

The spreading coefficient is the difference between work of adhesion and work of cohesion. Positive values of spreading coefficient indicate granule formation, but there are two distinct modes. A positive spreading coefficient value of binder over substrate (λ_{12}) signifies that the tendency of binder liquid to spread on solid powder is high, thereby creating a surface film and forming strong, dense, and less friable granules. A positive substrate over binder spreading coefficient (λ_{21}) value, however, signifies that the tendency of substrate to spread on binder liquid is more. In this case, the binder does not form a film around the substrate particles, but the substrate adheres to the binder at isolated points, which causes the formation of more open porous structure of granules with increased friability.

Binder substrate interactions have been predicted by calculating the spreading coefficient values for several drug-binder combinations (Table 12.4). On the basis of these calculations, it was concluded that PVP or starch would be the best option as binders for low polarity powders (e.g., griseofulvin, β-sitosterol) because they generated positive spreading coefficient values, and cellulose derivatives or acacia would be pertinent binders for high polarity powders (e.g., aspirin, ethinamate) (Planinsek et al., 2000; Rowe, 1989; Simons et al., 2005).

12.3.3.12 Granulating fluid (or binder solution) viscosity and surface tension

Different types of binders are marketed in various grades that differ in molecular weight, degree of cross-linking, particle size, and viscosity. The granulating liquids prepared with these binders, therefore, would have different viscosities and surface tensions. The granule consolidation greatly depends on the granule porosity, which is related indirectly to binder solution viscosity and binder content. The effects of binder viscosity on physical properties of granules prepared by high-shear and fluid-bed granulation have been reported by Kazumi et al. (1992). An increase in viscosity of binder solution caused an increase in granule size in the high-shear granulator, while an initial rapid decrease and further increase in granule size was observed in the fluid-bed granulator. With the increase in viscosity of the binder solution, granule hardness decreased sharply followed by a slight increase. This result correlated with an increase followed by a reduction in granule porosity at the corresponding viscosity level.

Increasing the amount of low-viscosity binder decreases granule porosity (increases extent of consolidation), whereas increasing the amount of a high-viscosity binder produces highly porous granules (decreases extent of consolidation) (Kazumi et al., 1992). This higher granule porosity can be attributed to the higher resistance to consolidation of the granules because of the nonhomogenous distribution of the highly viscous binder.

Binder surface tension is inversely related to the granule porosity. Increasing the surface tension of binder solution decreases granule porosity and strength, which then increases the consolidation (Iveson et al., 1996; Iveson & Litster, 1998a; Kokubo et al., 1993, 1995; Zuurman et al., 1995). Iveson and Litster (1998a) observed the effect of surface tension of binder liquid on the granule porosity and rate of consolidation. They used glass beads of two different sizes (10 and 19 μm) and binder solutions of varying surface tension values between 31 and 72 mN/m. It was observed that the granule porosity always decreased with an increase in the binder surface tension. In contrast, decreasing binder surface tension generally increased the rate of consolidation. A minimum or leveling off in the rate of consolidation of granules, however, was also observed at surface tensions $\sim$ 60 mN/m. The reason for the increased consolidation rate of granules lies in the reduction of frictional resistance to particle rearrangement because of the lowering of binder surface tension. Moreover, the capillary forces resisting particle dilation and rearrangement were also reduced with a binder of low surface tension (Iveson & Litster, 1998a).

12.3.3.13 Granulating solvent

CQAs of granules can be affected significantly by changing the granulating solvent. Changing the granulating solvent has a profound effect on the granule formation and growth during the wetting

Table 12.4 Calculation of spreading coefficients of different binders and drug molecules of varying polarities.

Binder	Surface-free energy (mN/m)	Polarity fraction	W_c (mN/m)											
HPMC	48.4	0.62	96.8											
PVP	53.6	0.47	107.2											
Acacia	50.6	0.57	101.2											
Starch	58.7	0.51	117.4											

Substrate				Interaction with different binders										
				HPMC			PVP			Acacia			Starch	
				W_a	λ_{12}	λ_{21}	W_a	λ_{12}	λ_{21}	W_a	λ_{12}	λ_{21}	W_a	λ_{12}	λ_{21}
Griseofulvin	32.2	0.06	64.4	53.1	−43.7	−11.3	65.7	−41.5	+1.3	57.6	−43.6	−6.8	66.4	−51.0	+2.0
β-Sitosterol	34.9	0.11	69.8	59.3	−37.1	−10.2	72.4	−34.8	+2.6	64.2	−37.0	−5.6	73.3	−44.1	+3.4
Ethinamate	70.0	0.39	140.0	108.3	+11.5	−31.7	120.5	+13.3	−19.5	113.2	+12.0	−26.7	125.7	+8.3	−14.3
Aspirin	67.5	0.42	135.0	108.3	+11.5	−26.7	119.2	+12.0	−15.8	112.9	+11.7	−22.1	124.6	+7.2	−10.4

From Rowe (1989).

Table 12.5 Mean granule size and friability values obtained for aspirin-PVP granulations carried out using different hydro-alcoholic solutions.

Granulation fluid	Solubility of aspirin in 6% w/w PVP in binder solution (mg/mL), at 20°C	Mean Granule Size (μ) (Sieve Analysis)	% Granule Friability
Ungranulated	–	201.33	–
Water	4.66	454.91	29.51
25% ethanol in water	12.50	521.97	28.72
50% ethanol in water	79.00	531.60	18.20
75% ethanol in water	147.00	592.30	8.96
Ethanol	133.06	574.11	13.36

From Wells and Walker (1983).

stage and the structure of the granules. This effect stems from the change in the solubility and wettability of materials. In wet granulation, only aqueous and alcoholic (ethanol and isopropyl alcohol) solvent systems are used widely. A good solvent produces uniform granules and, therefore, narrow size distribution, because of increased wettability and solubility of materials (Wells & Walker, 1983). Increased solubility facilitates nucleation. In addition, the dissolution of intragranular components during granulation produces viscous films, which promote the adhesion of small particles onto larger granules. As a consequence, increased material solubility in the granulating solvent leads to uncontrolled growth and results in granules with reduced friability. It has been observed that granulation of lactose with PVP dissolved in water yielded granules with porosity of 24.5% and friability of 9.7%, whereas granulation with PVP dissolved in ethanol resulted in porosity of 33% and friability of 21% (Wikberg & Alderborn, 1993). Therefore by replacing water with ethanol for the granulation of water-soluble material, lactose resulted in higher porosity and friability of granules. When aspirin was granulated with PVP dissolved in different hydroalcoholic solutions, the largest mean granular size and reduced friability were observed in the binder preparation where aspirin's solubility was the highest—specifically in the 75% ethanol solution (Table 12.5) (Wells & Walker, 1983).

12.4 Execution of quality by design methodology in assessing criticality of material attributes

Identification of the criticality of material properties generates greater material understanding in terms of both patient-centric and formulator-centric attributes. Identifying the impact of these CMAs and setting a material control strategy ensures conformance and robustness of the drug product quality. The increase in knowledge of several elements and tools involved in QbD methodology has led to its effective implementation in modern product development.

12.4 Execution of quality by design methodology in assessing criticality of material attributes

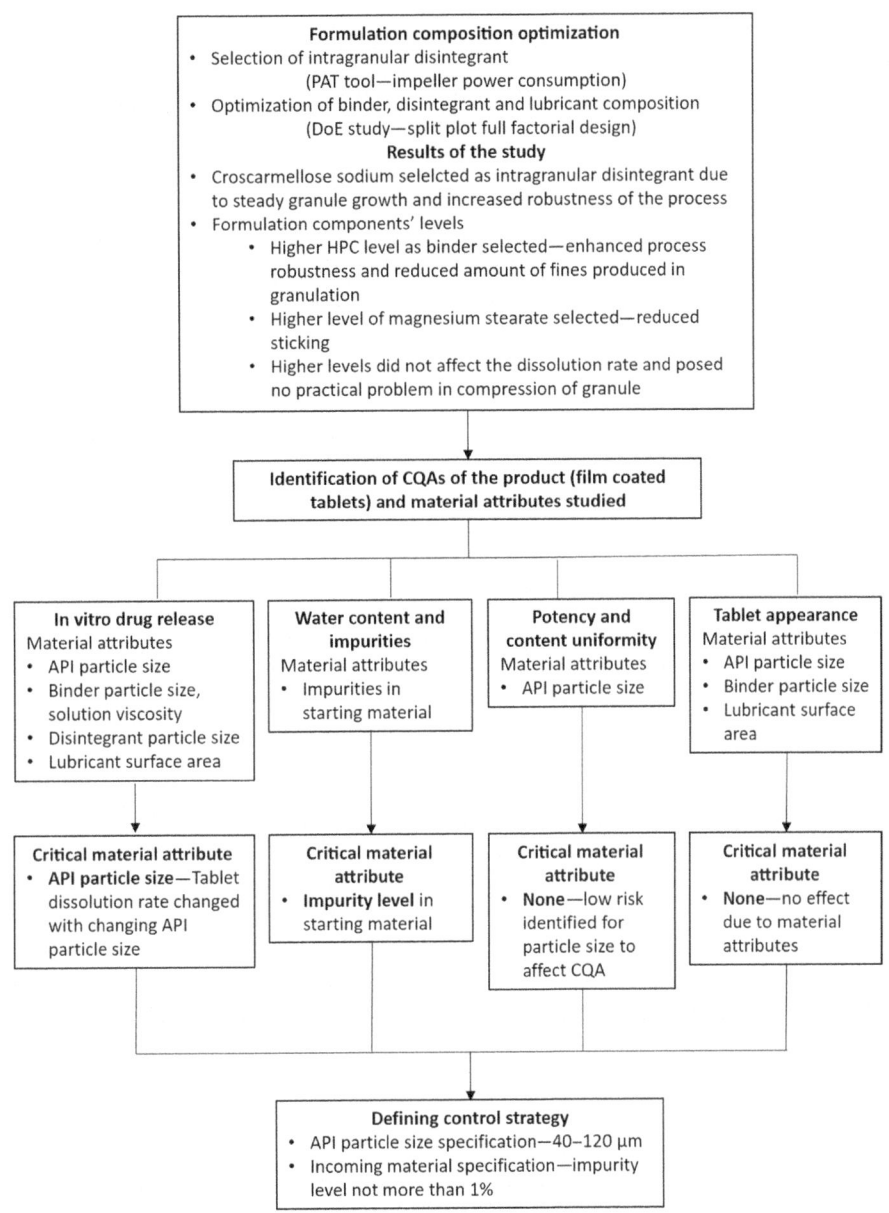

FIGURE 12.4

Case study exemplifying the impact of material attributes to CQAs in the formulation development process studied by QbD methodology.

436 Chapter 12 Critical material attributes in wet granulation

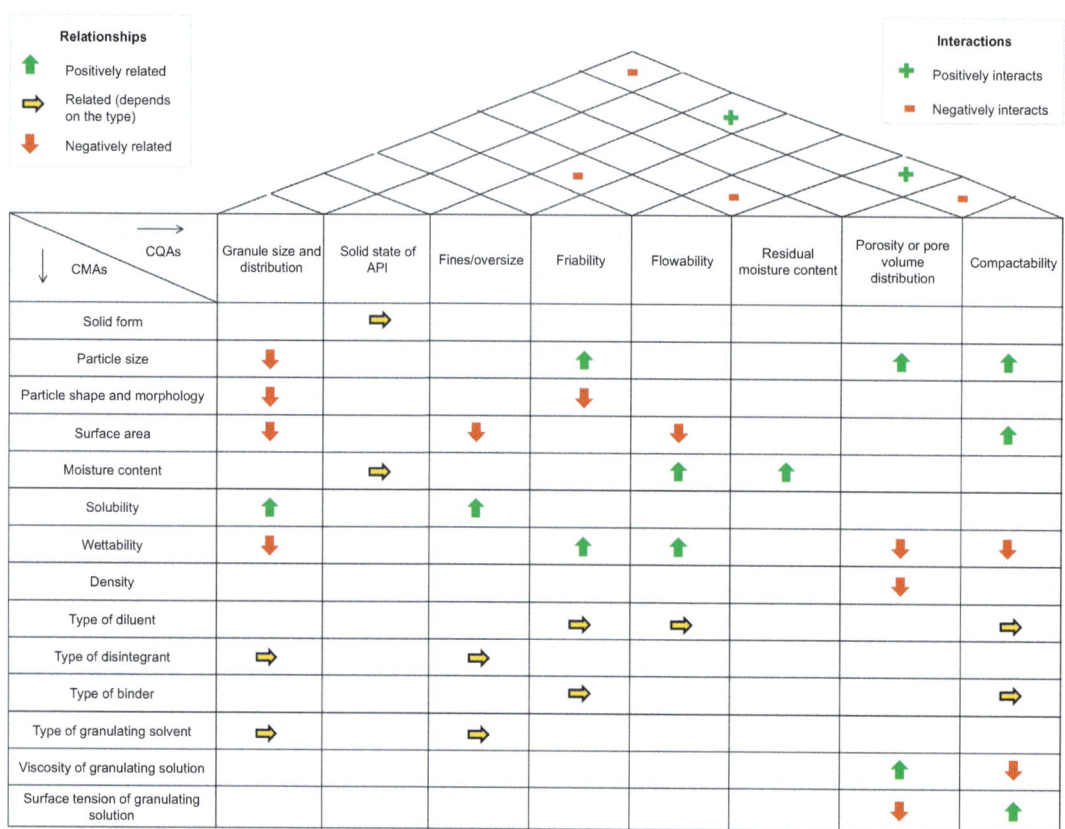

FIGURE 12.5

Relationships between CMAs and CQAs of granules prepared by wet granulation using QFD matrix.

A case study published by Badawy et al. provides a holistic demonstration of the application of QbD principles to product and process design for the development of film-coated tablet formulation (Badawy et al., 2016). Wet granulation process was selected as the suitable agglomeration process to reduce the risk of material sticking to compression tools and machinery. The systematic approach involved application of QbD tools, such as quality risk assessment and DoE, to study the parameters that pose risk to the achievement of the desired product attributes. This was followed by establishing acceptable ranges for material attributes and process parameters, and defining the control strategy for the commercial manufacture of the product. The methodology adopted in this case study for defining and controlling material attributes is captured in Fig. 12.4 and is representative of the currently followed practice of QbD.

12.4.1 Quality function deployment

On the basis of the knowledge assimilated from the literature, a summary of the effects of materials attributes on quality attributes of granules produced on wet granulation is provided in the form of a quality function deployment (QFD) matrix (Fig. 12.5). QFD is thought to be a consumer-oriented planning process to ensure the product quality during its development. In pharmaceuticals, the QFD translates "voice of consumer," that is, QTPP, into "voice of formulator," that is, CQAs, in the "house of quality" (Chan & Wu, 2002).

We have used the QFD matrix to translate the voice of formulator, that is, CQAs, with the "voice of process," that is, CPPs and CMAs. For the scope of this chapter, a modified QFD chart for CMAs and CQAs of wet granulation process has been framed, which can be used as a risk assessment tool to identify and prioritize the CMAs according to the CQAs. In Fig. 12.5, the part of the house of quality with green, yellow, and red arrows is the relationship matrix; while the one with plus and minus signs is the interactions matrix. For example, the CQA granule size is strongly or positively related to particle size, shape, morphology, and wettability of the material while negatively related to the solubility of the material in the granulating fluid. Likewise, the CQA compactibility is positively related to particle size and surface area of the material; and negatively related to the wettability of the material. Furthermore, the CQA compactibility is negatively correlated with another CQA, granule size. This QFD matrix for wet granulation process has been generated with qualitative as well as quantitative outcomes of the studies in the literature but there might be exceptions with respect to the relationship or interaction of CMAs and CQAs on a case-by-case basis.

12.5 Conclusions

The QbD approach has offered great opportunities to formulation scientists to recognize the importance of build-in quality in the process of product development. During the past two decades, there has been an increased emphasis on holistic understanding of the complete manufacturing process with emphasis on each unit operation involved. Being a multicomponent process, wet granulation involves assessment of numerous material attributes and determining their criticality in affecting the CQAs of the formed granules. Variability in the physicochemical properties of the starting materials affect the product CQAs. Findings from previous studies and experimental results prove beneficial in qualitatively predicting their performance in a wet granulation process. Identification of these CMAs and placing a material control strategy in practice, therefore, becomes a prerequisite in developing a quality-based drug product.

References

Pharmaceutical development Q8 (R2). (2009). Current step 4th International conference on harmonization. https://database.ich.org/sites/default/files/Q8_R2_Guideline.pdf.

Airaksinen, S., Karjalainen, M., Kivikero, N., Westermarck, S., Shevchenko, A., Rantanen, J., & Yliruusi, J. (2005). Excipient selection can significantly affect solid-state phase transformation in formulation during wet granulation. *AAPS PharmSciTech, 6*(2), E311–E322.

Bacher, C., Olsen, P., Bertelsen, P., & Sonnergaard, J. (2008). Compressibility and compactibility of granules produced by wet and dry granulation. *International Journal of Pharmaceutics, 358*(1), 69–74.

Badawy, S., & Hussain, M. A. (2004). Effect of starting material particle size on its agglomeration behavior in high shear wet granulation. *AAPS PharmSciTech, 5*(3), 16–22.

Badawy, S., Narang, A. S., LaMarche, K. R., Subramanian, G. A., Varia, S. A., Lin, J., Steven, T., & Shah, P. A. (2016). Integrated application of quality-by-design principles to drug product development: A case study of brivanib alaninate film-coated tablets. *Journal of Pharmaceutical Sciences, 105*(1), 168–181.

Badawy, S. I. F., Lee, T. J., & Menning, M. M. (2000). Effect of drug substance particle size on the characteristics of granulation manufactured in a high-shear mixer. *AAPS PharmSciTech, 1*(4), 55–61.

Belohlav, Z., Brenkova, L., Hanika, J., Durdil, P., Rapek, P., & Tomasek, V. (2007). Effect of drug active substance particles on wet granulation process. *Chemical Engineering Research and Design, 85*(7), 974–980.

Chan, L. K., & Wu, M. L. (2002). Quality function deployment: A literature review. *Journal of Operational Research, 143*(3), 463–497. https://doi.org/10.1016/S0377-2217(02)00178-9.

Ek, R., & Newton, J. M. (1998). Microcrystalline cellulose as a sponge as an alternative concept to the crystallite-gel model for extrusion and spheronization. *Pharmaceutical Research, 15*(4), 509–512.

Fonteyne, M., Vercruysse, J., Díaz, D. C., Gildemyn, D., Vervaet, C., Remon, J. P., & Beer, T. D. (2013). Real-time assessment of critical quality attributes of a continuous granulation process. *Pharmaceutical Development and Technology, 18*(1), 85–97. https://doi.org/10.3109/10837450.2011.627869.

Fonteyne, M., Wickstrom, H., Peeters, E., Vercruysse, J., Ehlers, H., Peters, B. H., Remon, J. P., Vervaet, C., Ketolainen, J., Sandler, N., Rantanen, J., Naelapää, K., & Beer, T. D. (2014). Influence of raw material properties upon critical quality attributes of continuously produced granules and tablets. *European Journal of Pharmaceutics and Biopharmaceutics, 87*(2), 252–263. https://doi.org/10.1016/j.ejpb.2014.02.011.

Gift, A. D., Luner, P. E., Luedeman, L., & Taylor, L. S. (2009). Manipulating hydrate formation during high shear wet granulation using polymeric excipients. *Journal of Pharmaceutical Sciences, 98*(12), 4670–4683.

Hausman, D. S. (2004). Comparison of low shear, high shear, and fluid bed granulation during low dose tablet process development. *Drug Development and Industrial Pharmacy, 30*(3), 259–266.

Healey, J., Rubinstein, M., & Walters, V. (1974). The mechanical properties of some binders used in tableting. *Journal of Pharmacy and Pharmacology, 26*(S1), 41P–46P.

Ho, R., Dilworth, S. E., Williams, D. R., & Heng, J. Y. (2011). Role of surface chemistry and energetics in high shear wet granulation. *Industrial & Engineering Chemistry Research, 50*(16), 9642–9649. https://doi.org/10.1021/ie2009263.

Iveson, S., & Litster, J. (1998a). Fundamental studies of granule consolidation Part 2: Quantifying the effects of particle and binder properties. *Powder Technology, 99*(3), 243–250.

Iveson, S., & Litster, J. (1998b). Growth regime map for liquid-bound granules. *AIChE Journal, 44*(7), 1510.

Iveson, S., Litster, J., & Ennis, B. (1996). Fundamental studies of granule consolidation Part 1: Effects of binder content and binder viscosity. *Powder Technology, 88*(1), 15–20.

Iveson, S., Litster, J., Hapgood, K., & Ennis, B. (2001). Nucleation, growth and breakage phenomena in agitated wet granulation processes: A review. *Powder Technology, 117*(1), 3–39.

Iveson, S., & Page, N. (2005). Dynamic strength of liquid-bound granular materials: The effect of particle size and shape. *Powder Technology, 152*(1), 79–89.

Jaiyeoba, K., & Spring, M. (1980). The granulation of ternary mixtures: The effect of the solubility of the excipients. *Journal of Pharmacy and Pharmacology, 32*(1), 1–5. https://doi.org/10.1111/j.2042-7158.1980.tb12833.x.

Kazumi, D., Akira, K., Eiji, I., Sakiko, T., Sunada, H., & Otsuka, A. (1992). Influence of granulating fluids in hydroxypropylcellulose binder solution on physical properties of lactose granules. *Chemical and Pharmaceutical Bulletin, 40*(9), 2505–2509.

Keleb, E., Vermeire, A., Vervaet, C., & Remon, J. P. (2004). Twin screw granulation as a simple and efficient tool for continuous wet granulation. *International Journal of Pharmaceutics, 273*(1), 183–194.

Kokubo, H., Nakamura, S., & Sunada, H. (1993). Effect of several cellulosic binders on particle size distribution of granules prepared by a high-speed mixer. *Chemical and Pharmaceutical Bulletin, 41*(12), 2151–2155.

Kokubo, H., Nakamura, S., & Sunada, H. (1995). Effect of several cellulosic binders on particle size distribution in fluidized bed granulation. *Chemical and Pharmaceutical Bulletin, 43*(8), 1402–1406.

Kristensen, H., Holm, P., & Schaefer, T. (1985). Mechanical properties of moist agglomerates in relation to granulation mechanisms. Part II. Effects of particle size distribution. *Powder Technology, 44*(3), 239–247.

Kristensen, H., & Schaefer, T. (1987). Granulation: A review on pharmaceutical wet-granulation. *Drug Development and Industrial Pharmacy, 13*(4–5), 803–872. https://doi.org/10.3109/03639048709105217.

Krycer, I., Pope, D. G., & Hersey, J. A. (1983). An evaluation of tablet binding agents. Part I. Solution binders. *Powder Technology, 34*(1), 39–51.

Landin, M., Rowe, R., & York, P. (1994). Structural changes during the dehydration of dicalcium phosphate dihydrate. *European Journal of Pharmaceutical Sciences, 2*(3), 245–252.

Lawrence, X. Y., Amidon, G., Khan, M. A., Hoag, S. W., Polli, J., Raju, G. K., & Woodcock, J. (2014). Understanding pharmaceutical quality by design. *The AAPS Journal, 16*(4), 771–783.

Levin, M. (2006). Wet granulation: End-point determination and scale-up. *Encyclopedia of pharmaceutical technology* (pp. 4078–4098). Marcel Dekker.

Listiohadi, Y., Hourigan, J., Sleigh, R., & Steele, R. (2008). Moisture sorption, compressibility and caking of lactose polymorphs. *International Journal of Pharmaceutics, 359*(1), 123–134.

Mackaplow, M. B., Rosen, L. A., & Michaels, J. N. (2000). Effect of primary particle size on granule growth and endpoint determination in high-shear wet granulation. *Powder Technology, 108*(1), 32–45.

Mills, S. (2010). Pharmaceutical excipients: An overview including considerations for pediatric dosing.. In *Paper presented at the training workshop: Pharmaceutical development with focus on paediatric formulations*. Beijing: World Health Organisation.

Mirza, S., Miroshnyk, I., Rantanen, J., Aaltonen, J., Harjula, P., Kiljunen, E., Heinämäki, J., & Yliruusi, J. (2007). Solid-state properties and relationship between anhydrate and monohydrate of baclofen. *Journal of Pharmaceutical Sciences, 96*(9), 2399–2408. https://doi.org/10.1002/jps.20894.

Narang, A. S. (2015). Addressing excipient variability in formulation design and drug development. In Narang, A.S., & S. H. S. Boddu (Eds.), *Excipient applications in formulation design and drug delivery* (pp. 541–567). Springer.

Narang, A. S., Sheverev, V., Freeman, T., Both, D., Stepaniuk, V., Delancy, M., Millington-Smith, D., Macias, K., & Subramanian, G. (2016). Process analytical technology for high shear wet granulation: Wet mass consistency reported by in-line drag flow force sensor is consistent with powder rheology measured by at-line FT4 powder Rheometer®. *Journal of Pharmaceutical Sciences, 105*(1), 182–187.

Narang, A. S., Yamniuk, A. P., Zhang, L., Comezoglu, S. N., Bindra, D. S., Varia, S., Doyle, M. L., & Badawy, S. (2012). Reversible and pH-dependent weak drug-excipient binding does not affect oral bioavailability of high dose drugs. *Journal of Pharmacy and Pharmacology, 64*(4), 553–565. https://doi.org/10.1111/j.2042-7158.2011.01435.x.

Nguyen, T. H., Shen, W., & Hapgood, K. (2010). Effect of formulation hydrophobicity on drug distribution in wet granulation. *Chemical Engineering Journal, 164*(2), 330–339. https://doi.org/10.1016/j.cej.2010.05.008.

Osei-Yeboah, F., Zhang, M., Feng, Y., & Sun, C. C. (2014). A formulation strategy for solving the overgranulation problem in high shear wet granulation. *Journal of Pharmaceutical Sciences, 103*(8), 2434–2440. https://doi.org/10.1002/jps.24066.

Pandey, P., & Badawy, S. (2016). A quality by design approach to scale-up of high-shear wet granulation process. *Drug Development and Industrial Pharmacy, 42*(2), 175–189. https://doi.org/10.3109/03639045.2015.1100199.

Passerini, N., Calogera, G., Albertini, B., & Rodriguez, L. (2010). Melt granulation of pharmaceutical powders: A comparison of high-shear mixer and fluidized bed processes. *International Journal of Pharmaceutics, 391*(1), 177–186.

Pharmacopoeia, E. (2012). Functionality-related characteristics of excipients. (7th Ed.), pp. 661–662).

Planinsek, O., Pisek, R., Trojak, A., & Srcic, S. (2000). The utilization of surface free-energy parameters for the selection of a suitable binder in fluidized bed granulation. *International Journal of Pharmaceutics, 207*(1), 77–88.

Rajniak, P., Stepanek, F., Dhanasekharan, K., Fan, R., Mancinelli, C., & Chern, R. T. (2009). A combined experimental and computational study of wet granulation in a wurster fluid bed granulator. *Powder Technology, 189*(2), 190–201. https://doi.org/10.1016/j.powtec.2008.04.027.

Rowe, R. (1989). Binder-substrate interactions in granulation: A theoretical approach based on surface free energy and polarity. *International Journal of Pharmaceutics, 52*(2), 149–154. https://doi.org/10.1016/0378-5173(89)90289-5.

Rudnic, E., Kanig, J., & Rhodes, C. (1985). Effect of molecular structure variation on the disintegrant action of sodium starch glycolate. *Journal of Pharmaceutical Sciences, 74*(6), 647–650.

Sakr, W. F., Ibrahim, M. A., Alanazi, F. K., & Sakr, A. A. (2012). Upgrading wet granulation monitoring from hand squeeze test to mixing torque rheometry. *Saudi Pharmaceutical Journal, 20*(1), 9–19.

Seem, T. C., Rowson, N. A., Ingram, A., Huang, Z., Shen, Y., de Matas, M. D., Gabbott, I., & Reynolds, G. K. (2015). Twin screw granulation: A literature review. *Powder Technology, 276*, 89–102. https://doi.org/10.1016/j.powtec.2015.01.075.

Shalaev, E. Y., & Zografi, G. (1996). How does residual water affect the solid-state degradation of drugs in the amorphous state? *Journal of Pharmaceutical Sciences, 85*(11), 1137–1141. https://doi.org/10.1021/js960257o.

Shi, L., Feng, Y., & Sun, C. C. (2011). Initial moisture content in raw material can profoundly influence high shear wet granulation process. *International Journal of Pharmaceutics, 416*(1), 43–48.

Simons, S., Pepin, X., & Rossetti, D. (2003). Predicting granule behavior through micro-mechanistic investigations. *International Journal of Mineral Processing, 72*(1), 463–475.

Simons, S., Rossetti, D., Pagliai, P., Ward, R., & Fitzpatrick, S. (2005). The relationship between surface properties and binder performance in granulation. *Chemical Engineering Science, 60*(14), 4055–4060. https://doi.org/10.1016/j.ces.2005.02.034.

Simons, S. J. R., & Pepin, X. (2003). Hardness of moist agglomerates in relation to interparticle friction, capillary and viscous forces. *Powder Technology, 138*(1), 57–62. https://doi.org/10.1016/j.powtec.2003.08.041.

Simons, S. J. R., Pepin, X., & Rossetti, D. (2003). Predicting granule behaviour through micro-mechanistic investigations. *International Journal of Mineral Processing, 72*(1—4), 463–475. https://doi.org/10.1016/s0301-7516(03)00120-0.

Suzuki, T., Watanbe, K., Kikkawa, S., & Nakagami, H. (1994). Effect of crystallinity of microcrystalline cellulose on granulation in high-shear mixer. *Chemical and Pharmaceutical Bulletin, 42*(11), 2315–2319.

Thielmann, F., Naderi, M., Ansari, M. A., & Stepanek, F. (2008). The effect of primary particle surface energy on agglomeration rate in fluidized bed wet granulation. *Powder Technology, 181*(2), 160–168.

USP 30-NF 25. (2007). (1059) Excipient performance. (pp. 1045–1049). USP.

Vemavarapu, C., Surapaneni, M., Hussain, M. A., & Badawy, S. (2009). Role of drug substance material properties in the processibility and performance of a wet granulated product. *International Journal of Pharmaceutics, 374*(1—2), 96–105. https://doi.org/10.1016/j.ijpharm.2009.03.014.

Vercruysse, J. C. D. D., Díaz, D. C., Peeters, E., Fonteyne, M., Delaet, U., Van Assche, I., De Beer, T., Remon, J. P., & Vervaet, C. (2012). Continuous twin screw granulation: Influence of process variables on granule and tablet quality. *European Journal of Pharmaceutics and Biopharmaceutics, 82*(1), 205–211.

Wang, S., Ye, G., Heng, P. W. S., & Ma, M. (2008). Investigation of high shear wet granulation processes using different parameters and formulations. *Chemical and Pharmaceutical Bulletin, 56*(1), 22–27.

Wells, J. I., & Walker, C. V. (1983). The influence of granulating fluids upon granule and tablet properties: The role of secondary binding. *International Journal of Pharmaceutics, 15*(1), 97–111. https://doi.org/10.1016/0378-5173(83)90070-4.

Wikberg, M., & Alderborn, G. (1993). Compression characteristics of granulated materials. VII. The effect of intragranular binder distribution on the compactibility of some lactose granulations. *Pharmaceutical Research, 10*(1), 88–94. https://doi.org/10.1023/A:1018929214629.

Wikstrom, H., Carroll, W. J., & Taylor, L. S. (2008). Manipulating theophylline monohydrate formation during high-shear wet granulation through improved understanding of the role of pharmaceutical excipients. *Pharmaceutical Research, 25*(4), 923–935.

Wikstrom, H., Marsac, P. J., & Taylor, L. S. (2005). In-line monitoring of hydrate formation during wet granulation using Raman spectroscopy. *Journal of Pharmaceutical Sciences, 94*(1), 209–219.

Yoshinari, T., Forbes, R. T., York, P., & Kawashima, Y. (2002). Moisture induced polymorphic transition of mannitol and its morphological transformation. *International Journal of Pharmaceutics, 247*(1), 69–77. https://doi.org/10.1016/S0378-5173(02)00380-0.

Yoshinari, T., Forbes, R. T., York, P., & Kawashima, Y. (2003). The improved compaction properties of mannitol after a moisture-induced polymorphic transition. *International Journal of Pharmaceutics, 258*(1), 121–131.

Zhang, G. G., Law, D., Schmitt, E. A., & Qiu, Y. (2004). Phase transformation considerations during process development and manufacture of solid oral dosage forms. *Advanced Drug Delivery Reviews, 56*(3), 371–390.

Zhao, N., & Augsburger, L. L. (2006). The influence of granulation on super disintegrant performance. *Pharmaceutical Development and Technology, 11*(1), 47–53. https://doi.org/10.1080/10837450500463828.

Zografi, G., & Tam, S. S. (1976). Wettability of pharmaceutical solids: Estimates of solid surface polarity. *Journal of Pharmaceutical Sciences, 65*(8), 1145–1149. https://doi.org/10.1002/jps.2600650805.

Zuurman, K., Bolhuis, G., & Vromans, H. (1995). Effect of binder on the relationship between bulk density and compactibility of lactose granulations. *International Journal of Pharmaceutics, 119*(1), 65–69.

CHAPTER 13

Critical material attributes during continuous twin-screw wet granulation

Valerie Vanhoorne, Phaedra Denduyver and Chris Vervaet
Laboratory of Pharmaceutical Technology, Ghent University, Ghent, Belgium

13.1 Continuous twin-screw wet granulation

The ongoing switch from batch-wise to continuous manufacturing in the pharmaceutical industry necessitates the introduction of novel techniques to support the continuous production of oral solid dosage forms. Wet granulation is a key intermediate step in tablet production, improving the physical properties of the blend, such as flowability, density, compressibility, and active pharmaceutical ingredient (API) uniformity (Bacher et al., 2008; Dun & Sun, 2018). Traditionally, wet granulation was performed through batch-wise fluidized bed granulation (FBG) or high shear granulation (HSG) which involves different single-step unit operations. In recent years, twin-screw wet granulation (TSWG) has emerged as a continuous wet granulation technique offering several advantages. Its continuous nature enables flexible production volumes, reduces the need for scale-up, allows for a smaller equipment footprint, and improves product quality through the implementation of control strategies and process analytical technology (PAT) (Lee et al., 2015; Seem et al., 2015). Additionally the twin-screw wet granulator typically features a modular screw design, flexible liquid addition ports, and a short residence time (2–40 s) (El Hagrasy et al., 2013; Kumar et al., 2016; Pradhan et al., 2019). TSWG can be performed on adapted twin screw extrusion equipment of manufacturers such as Thermo Fisher Scientific (Eurolab 16 TSG), Leistritz Extrusiontechnik (Leistritz Micro 27GL/28D), and APV Baker (MP 19 TC 25) (Dhenge et al., 2011; Djuric & Kleinebudde, 2008; Djuric et al., 2009; Keleb et al., 2004a; Van Melkebeke et al., 2008). However, other equipment manufacturers have developed twin screw granulation modules incorporated in fully continuous from powder-to-tablet lines, for example, GEA Pharma Systems (ConsiGma-system), Bohle (QbCon system), and Glatt (MODCOS system).

A twin-screw wet granulator consists of a temperature-controlled barrel that houses two co-rotating intermeshed screws (Fig. 13.1). The geometry of the system varies in terms of the diameter and length-to-diameter ratio of the barrel and screws, which, in combination with the screw configuration, determine the free volume (Djuric et al., 2009; Franke et al., 2023; Menth et al., 2020; Osorio et al., 2017). In contrast to batch-wise granulators, a twin-screw wet granulator operates as a regime-separated granulator, with the rate processes spatially separated along the length of the screw (Dhenge et al., 2012a; El Hagrasy & Litster, 2013).

The screw configuration is modular and generally consists of three types of elements: conveying, kneading, and distributive mixing elements. These elements created distinct compartments along the barrel, each with specific granule formation mechanisms (Fig. 13.2) (Dhenge et al., 2012a; Djuric & Kleinebudde, 2008; Verstraeten et al., 2017). The first compartment, which consists solely of conveying

Chapter 13 Critical material attributes during continuous twin-screw wet

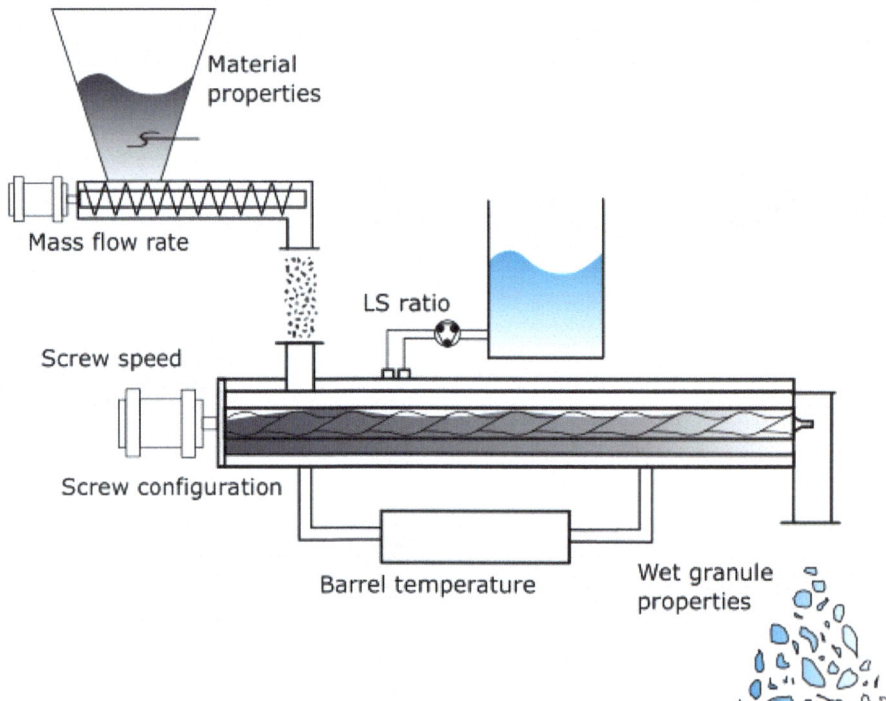

FIGURE 13.1

Schematic overview of a twin screw wet granulator.

From Vandeputte (2023).

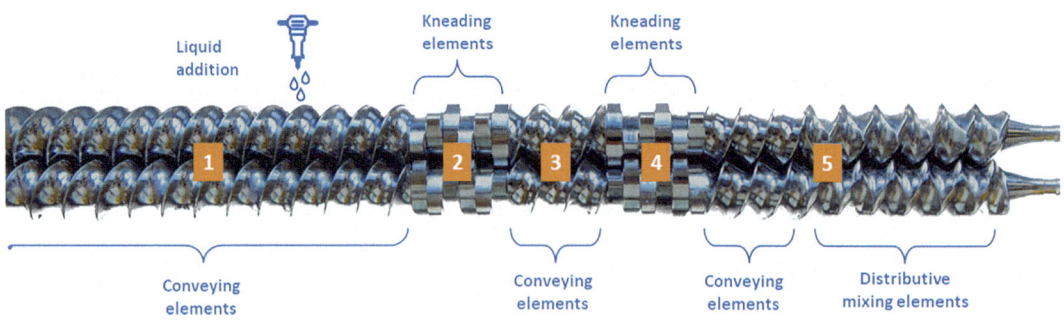

FIGURE 13.2

Example of a typical screw configuration. Compartments 1, 3, and 5 consist of conveying elements, while compartments 2 and 4 consist of kneading elements. Distributive mixing elements (e.g., size control elements) can be added at the end of the screws.

Modified from Denduyver et al. (2024).

elements, ensures easy transport of material along the barrel due to its low shear design (Sayin et al., 2015; Thompson & Sun, 2010). In this zone, the granulation liquid (demineralized water or binder solution) is typically added to wet the powder mass, resulting in a bimodal size distribution that includes nuclei or agglomerates and a fraction of ungranulated material (Hagrasy & Litster, 2013; Verstraeten et al., 2017). The second compartment is a kneading zone, composed of kneading discs at a specific stagger angle (30°, 60°, or 90°). Due to the design of the kneading discs and their configuration, the kneading zone exerts a significant amount of shear on the nuclei, pushing the granulation liquid towards the particle surface, causing coalescence, consolidation, and breakage of the granules (Dhenge et al., 2012a; Li et al., 2014). In the third compartment, the granules are transported by conveying elements to the second kneading zone, where further coalescence or breakage can occur (Dhenge et al., 2012a; Van Melkebeke et al., 2008; Verstraeten et al., 2017). In the second kneading zone (fourth compartment), further growth could occur with coalescence between particles dominating over consolidation and breakage. The final compartment transports the material to the outlet of the barrel using conveying elements, which may induce breakage of larger granules by exerting a limited amount of shear (Dhenge et al., 2012a). Additionally, distributive mixing elements can be incorporated to improve the size distribution and shape of the granules (Sayin et al., 2015). The wet granules are subsequently dried, milled, blended with external phase components (i.e., containing at least a lubricant, potentially a filler or disintegrant), and tableted.

13.2 Formulation adjustments: Batch-wise versus continuous manufacturing

While the initial studies about TSWG mainly focused on the impact of process parameters on granule quality, later studies also studied the effect of formulation parameters on this granulation process. As a result, a better understanding of the process has been achieved in recent years through the identification of critical process parameters (CPPs) and critical material attributes (CMAs) that impact the critical quality attributes (CQAs) according to the quality-by-design (QbD) principle (Yu et al., 2014a). This chapter focuses on the CMAs specific for TSWG. Switching from batch-wise to continuous granulation may require formulation adjustments due to the inherent differences in the feeding of raw materials, granule formation mechanisms, and downstream processing (Fig. 13.3).

Continuous feeding and blending of the raw materials into the granulator barrel is the initial step during continuous TSWG. While feeders must accurately dose the different ingredients of the formulation into the continuous blender, the blender will create a homogeneous blend by intensive mixing of the different components using a series of paddles. In batch manufacturing, formulations typically consist of several excipients with different functionalities. However, in continuous manufacturing, the number, as well as attributes of the excipients, are more critical due to the complexity of a fully integrated manufacturing line (Challener, 2020). First, the number of feeders—related to the number of ingredients—should be limited to ensure a robust process with good blend uniformity—it typically is restricted to 5 or 6 feeders positioned around the blender inlet (Meier et al., 2015; Portier et al., 2021). Furthermore, simplified formulations for continuous manufacturing would minimize traceability issues when different batches are mixed during continuous processing (Challener, 2020; Meier et al., 2017, 2015). Meier et al. (2017, 2015) studied the feasibility of simplified formulations and reported successful

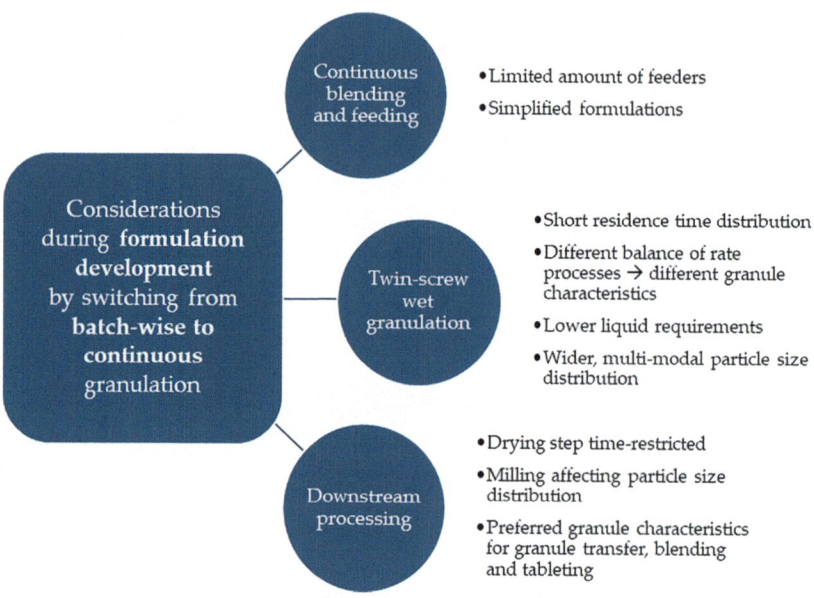

FIGURE 13.3

Formulation adjustments needed to be considered during formulation development when switching from batch-wise to continuous granulation.

granulation and tableting of high drug-loaded formulations in binary (drug and disintegrant) or ternary (drug, disintegrant, and binder) formulations. In addition, using coprocessed excipients would not only simplify formulations, but it could also improve the granule quality by combining functionalities into a single ingredient (Challener, 2020). As an example, Vanhoorne et al. (2020) reported that the inclusion of δ-mannitol as a filler allowed successful tableting of a highly dosed paracetamol formulation due to the polymorphic transition from δ- to β-mannitol after granulation making the addition of a binder unnecessary. Finally, excipient manufacturers are making efforts to adapt the attributes (flowability, density) of their excipients through particle engineering or development of new (coprocessed) excipients to ensure consistent feeding during continuous manufacturing (Challener, 2020; Erdemir et al., 2023). As an example, Merck developed a combination of mannitol with magnesium stearate (Parteck LM), functioning as filler, binder, and lubricant (Parteck®, 2024).

Continuous and batch-wise granulation processes differ in terms of the material residence time in the equipment, the design of the equipment, the liquid addition volume and method, and the balance between the rate processes. These differences determine the adjustments needed during formulation development in order to yield similar granule properties. Characteristic of TSWG is the short residence time (2–40 s) of material in the granulator barrel. This translates to a 50 to 100 times shorter duration for the granulation rate processes compared to batch-wise granulation, potentially limiting the use of conventional pharmaceutical excipients for TSWG (El Hagrasy et al., 2013; Kumar et al., 2014; Vanhoorne et al., 2020). Binders suitable for TSWG will require good wetting properties to be activated within the short residence time in the granulator barrel. Furthermore, the solubility and solubility rate

13.2 Formulation adjustments: Batch-wise versus continuous manufacturing

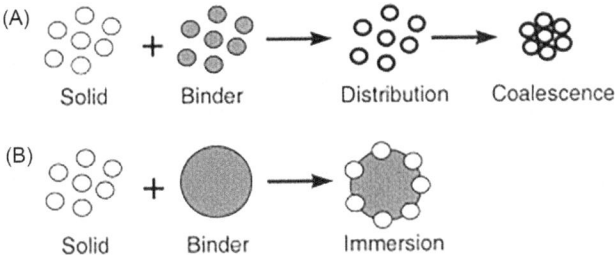

FIGURE 13.4

Two types of nucleation formation mechanisms depending on the relative size of the droplet to the primary particle size (A) Distribution mechanism, (B) Immersion mechanism.

From Iveson et al. (2001).

of the ingredients are important as the short residence time limits the fraction of material dissolved during granulation. This fraction crystallizes upon drying, forming strong solid bridges within the granules. Due to a different balance of the rate processes (wetting and nucleation, consolidation and coalescence, breaking and attrition), different granule properties are obtained during TSWG compared to the batch-wise granulation methods, HSG, and FBG. Granules produced via TSWG are typically less spherical and have a wider or even multimodal particle size distribution (PSD) compared to HSG and FBG (Arndt et al., 2018; Kyttä et al., 2020; Lee et al., 2013; Vercruysse et al., 2015). TSWG granules are more porous than HSG granules but less porous than FBG granules (Arndt et al., 2018). In HSG, consolidation is more significant due to the continuous impact of the granules with the granulation bowl, the impeller, and other granules, resulting in dense and spherical particles. In TSWG, the rate processes are spatially separated along the length of the granulation barrel, yielding irregular, porous granules (Lee et al., 2013; Seem et al., 2015). However, granules produced via TSWG are more consistent in mechanical property compared to batch-wise granulation as the granule strength remains similar across different granule size fractions and is relatively independent of process conditions (Lee et al., 2013). Moreover, less liquid needs to be added during TSWG compared to batch-wise granulation (HSG and FBG) to yield granules with similar properties (Beer et al., 2014; Keleb et al., 2002; Vercruysse, 2014). This can be attributed to a different interaction of the powder with the equipment in both techniques. Owing to the enclosed nature and relatively narrow barrel (channel depth and flight clearance) of a twin-screw wet granulator, the wet mass is exposed to higher shear forces due to the increased interaction of the wet mass with the equipment. In HSG, the frequency of wet particles colliding with the walls, impeller, and chopper will be less as the distance to the walls and moving elements is larger (Kyttä et al., 2020; Shah & Serajuddin, 2018). In FBG, less shear is exerted on the wet mass compared to TSWG (similar to HSG), and granulation liquid can evaporate. During TSWG, the granulation liquid is directly pumped onto the powder bed, the resulting relatively large droplets create an immersion type of nucleation (Verstraeten et al., 2017) (Fig. 13.4). In batch-wise granulation, the liquid is typically sprayed onto the powder bed, and the smaller droplets result in a distribution type of nucleation. Overall, the granulation liquid is more homogeneously distributed during batch-wise compared to continuous granulation, linked to the liquid addition method as well as to the longer mixing (residence) time of the material in the granulator (Beer et al., 2014). During TSWG, effective liquid distribution can only be obtained by mechanical mixing and the shear forces generated during processing

(El Hagrasy et al., 2013). El Hagrasy et al. (2013) hypothesized that the typical bimodal particle size distribution is caused by the insufficient mixing of powder and liquid during the short residence time. Vercruysse (2014) used NIR-chemical imaging to demonstrate that the bimodal size distributions are inherent to the TSWG mechanism, as improving the moisture content distribution over the granules (e.g., changing the orientation of the peristaltic pumps from in-phase to out-of-phase, changing the tubing configuration, using multiple injection ports) or increasing the number of kneading zones did not significantly improve the bimodality of PSD. Finally, TSWG is less sensitive to over-wetting due to the shorter residence time (Kyttä et al., 2020). Over-wetted granules are generally denser which can result in poor compactibility (Lee et al., 2013).

The downstream processing of TSWG involves granule transfer, drying, milling, blending with external phase, and tableting. It is crucial to take the preferred granule characteristics for downstream processing into account during formulation development. A wider granule size distribution can increase the segregation tendency during granule transfer, blending and tableting, and decrease the drying uniformity across granules (El Hagrasy et al., 2013). In addition, milling will affect the PSD of granules produced via TSWG more compared to batch-wise granulation as the TSWG granules typically exhibit a broader size distribution (Kyttä et al., 2020). Granules with a higher porosity and more irregular shape are preferred as they improve tabletability and interparticle bonding (due to particle entanglement), respectively (Arndt et al., 2018; Lee et al., 2013).

13.3 Solubility

While certain material characteristics are particularly important for an API, filler, or binder, the solubility of all ingredients is important. Given the short time for bond formation during TSWG, both the hydrophilicity or hydrophobicity (surface) and the solubility (entire particle) of the ingredients significantly influence the granule properties along the barrel length. In the initial rate process of wetting and nucleation, the interaction of the aqueous granulation liquid with the particle will be larger in case of a hydrophilic particle compared to a hydrophobic particle surface as the contact angle of the former will be smaller and the granulation liquid can easily penetrate into the capillary pores (Fig. 13.5). Consequently, more spherical and less dense granules will be created when granulating a hydrophilic formulation as particles can move closer to each other. For hydrophobic surfaces, achieving good wetting and nucleation becomes challenging due to poor liquid distribution (Mundozah et al., 2019). The amount of available liquid on the particle surface strongly determines the wetting of hydrophobic particles and their incorporation in granules. The poor liquid distribution results in a large fraction of ungranulated material and a wide PSD (Mundozah et al., 2019; Yu et al., 2014b). When the liquid-to-solid (L/S-ratio) is changed from a low to a high level, a significant increase in oversized material will be observed for hydrophobic material due to higher availability of liquid at the surface. In addition, hydrophobic particles will primarily be situated at the outer surface of the agglomerate due to the mechanism of layering, whereas hydrophilic particles are more concentrated in the inner core. During the consolidation and coalescence phase, hydrophilic particles will further aggregate as the applied force in the kneading zones expels the liquid from the inner core towards the surface, increasing the probability for bond formation. For hydrophobic particles, aggregation by layering due to higher liquid availability at the particle surface will occur initially, followed by breakage of brittle granules when additional shear is exerted (Verstraeten et al., 2017). Poorer wettability and aqueous solubility of the excipients

13.3 Solubility   449

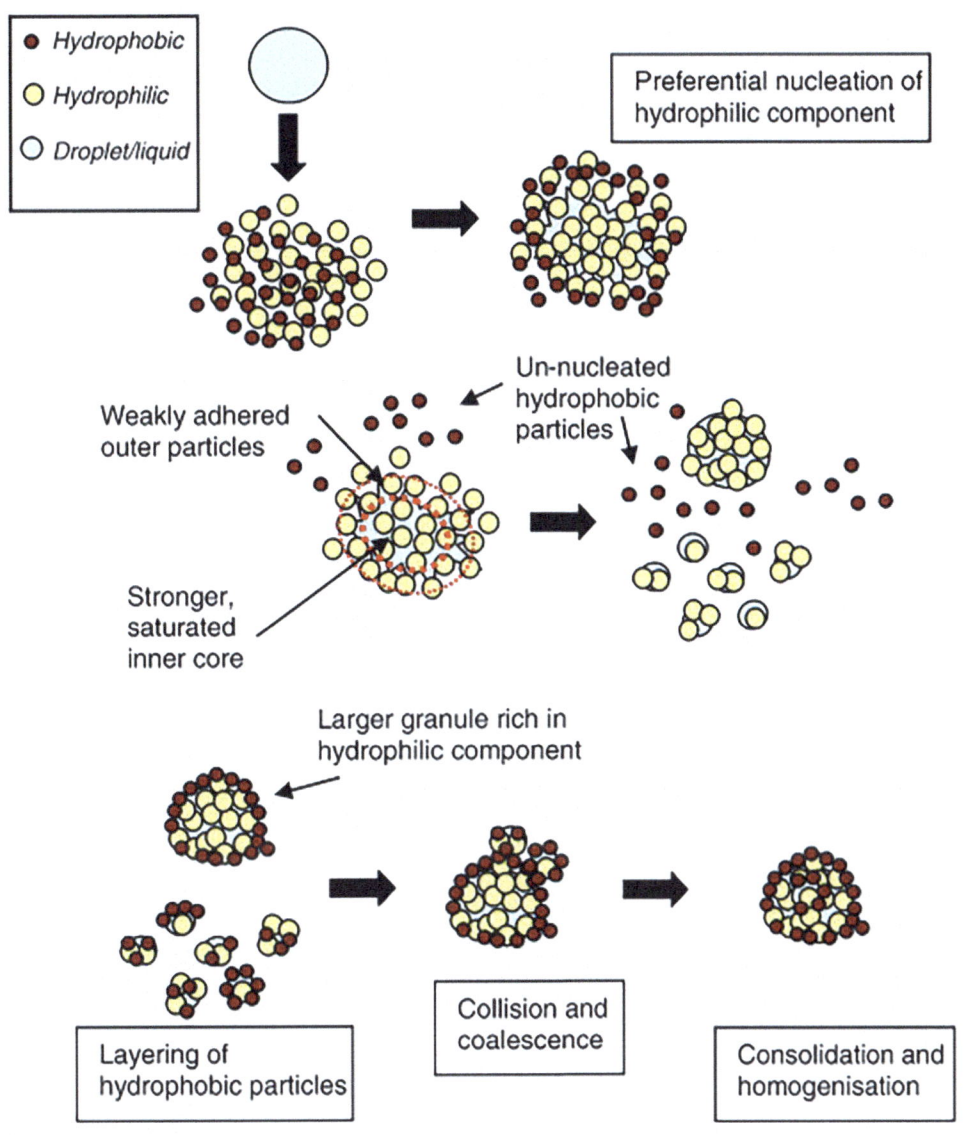

FIGURE 13.5

Proposed nucleation and granulation mechanism for granulation of mixtures containing hydrophilic and hydrophobic components.

From Charles-Williams et al. (2013).

can lead to smaller nuclei as well as more friable granules due to the lower amount of liquid and solid bridges between the particles (Portier et al., 2020a; Zupančič et al., 2023). In addition, the interplay between ingredients is of high importance. Portier et al. (2020b) reported that granulating a poorly soluble and wettable API (50% w/w mebendazole) with a highly wettable filler (lactose) did not yield granules with appropriate characteristics without a compound having sufficient amphiphilic properties (e.g., hydroxypropyl methylcellulose [HPMC]) included in the formulation. Performing surface energy measurements (contact angle method or inverse gas chromatography) upfront could be beneficial to calculate cohesion, adhesion, and spreading coefficients between ingredients and evaluate the suitability of ingredients for TSWG (Zhang et al., 2002). Poor solubility (rate) of ingredients can, to some extent be compensated by increasing the barrel temperature (e.g., up to 40°C). Higher barrel temperatures resulted in fewer fines and more oversized granules due to the increased solubility rate of the powder in the granulation liquid. This strengthened the solid bridges in the granules due to re-crystallization of the dissolved compounds upon drying (Ito & Kleinebudde, 2019; Vanhoorne et al., 2016a; Vercruysse et al., 2012).

13.4 Active pharmaceutical ingredient

API powders used in TSWG typically possess challenging properties such as a small particle size with a non-spherical shape, low density, high cohesivity, and poor flowability, solubility, and wettability (Cai et al., 2013; Cartwright et al., 2013; Meier et al., 2015). These APIs are generally difficult to formulate as high drug-loaded formulations suitable for processing via continuous direct compression (Van Snick et al., 2017). Although continuous TSWG allows for processing these high drug-loaded formulations, excipient selection and process optimization are crucial. The excipients need to address the deficiencies of the API, but the composition of the formulation is restricted by the high drug load (Sun et al., 2009; Willecke et al., 2018). Furthermore, APIs can exhibit batch-to-batch variability (e.g., due to sourcing from a different supplier, upscaling of the crystallization process). Therefore, improved understanding of the effect of the APIs' material attributes and their batch-to-batch variability on the granule CQAs facilitates product development and improves final product quality (Stauffer et al., 2018). Fonteyne et al. (2014) studied the effects of variability of the material properties of several theophylline grades on the critical quality attributes of granules produced through TSWG. A correlation was found between the particle size of the API and the granule size distribution, whereby a smaller starting material resulted in more fines. As a result, tablets made of those granules had the lowest porosity as the larger fines fraction resulted in tablets with a smaller pore structure. The studies by Stauffer et al. (2018, 2019) evaluated the impact of API batch-to-batch variability using multivariate data analysis. The API batches mainly differed in crystal length which is correlated with the agglomeration tendency of the API and flowability parameters. The span of the PSD also contributed to some extent to the variability. The APIs CMAs were linked to the granule CQAs, whereby API powder characterized by long crystals, high porosity, and high compressibility resulted in small granules with a narrow distribution. The high porosity of the powder was beneficial for liquid penetration into the powder bed, resulting in a limited fraction of ungranulated material and less loose, coarse nuclei which could easily be broken. APIs with a long crystal shape and a high porosity required more water to generate strong granules as the pore saturation during consolidation was limited. APIs with a large PSD span (i.e., a low powder bed porosity) had more pore saturation at the same liquid level which improved consolidation and reduced granule friability. The effect of 51 powder properties (evaluated for six different APIs) on granule porosity at the different compartments throughout the granulator barrel was studied by Peeters et al. (2023). At low L/S-ratio, size- and flow-related properties affected the granule porosity, whereas at high L/S-ratio, the water-related properties were most influential. Larger API particles increased the granule porosity as the number of interparticle contact points reduced, creating a larger intragranular void volume. As flow properties are related to particle size, powders characterized by better flow (i.e., larger particles) yielded granules with higher porosity. Stronger powder cohesion reduced the granule porosity as stronger interparticle interactions were formed, resulting in a lower granule void volume. At a high L/S-ratio, a higher API solubility and faster dissolution rate yielded granules with lower porosity as more powder particles dissolved, enhancing inter-particle contact due to the formation of more solid bridges. In addition, the API properties had a different impact on the granule porosity depending on the compartment in the granulator barrel. The particle size was important in the wetting zone and in the zone after the first kneading element. The water-related properties were most important in the zone after the second kneading element and at the granulator outlet due to the short residence time, limiting the ability to dissolve in the upstream compartments.

13.5 Fillers

Fillers are included to enhance the processability and robustness of TSWG and downstream processes and to improve the granule CQAs of difficult-to-process powders (e.g., those with small particle size, high hydrophobicity, and cohesive). Commonly used fillers include lactose, microcrystalline cellulose (MCC), mannitol, and mixtures thereof (Hwang et al., 2019; Portier et al., 2021; Willecke et al., 2017). While dicalcium phosphate (DCP) has been used in research studies, due to its insolubility which allowed to separate binder and filler effects, it is not a preferred filler for commercial formulations. As excipients are generally the major component in the formulation, a thorough understanding of the impact of different filler grades and the variability in their material properties on the granule CQAs is essential (Zarmpi et al., 2017). Willecke et al. (2017) defined a principal component analysis (PCA) plot of 9 preferred fillers and 9 preferred binders in TSWG to identify the similarities and differences in excipient characteristics of materials with varying chemical and physical properties. The original 23 variables could be condensed to four overarching properties (i.e., principal components) which describe moisture-related, flow-related, density/particle size-related, and charge/adhesion-related properties. Additionally, binary filler/binder formulations were processed via TSWG, followed by tablet compression to study the impact of the filler and binder properties on the granule and tablet quality attributes (Willecke et al., 2018). The filler properties had a significant effect on granule characteristics, such as particle size, friability, and specific surface area. Fillers with high values for the moisture-related properties and/or flow-related properties increased the fines fraction and reduced the oversized fraction of granules. Fillers with low values for the moisture-related and flow-related properties in combination with high values for the density/particle size-related properties yielded stronger granules (Willecke et al., 2018).

13.5.1 Lactose

Lactose is a disaccharide consisting of D-galactose and D-glucose units linked through a β-(1-4) glycosidic bond. It is used as diluent, but also possesses some binding properties due to its good water-solubility (Rowe et al., 2009). Two anomeric forms (α and β exist, differing in physical properties such as solubility, melting point, density, and specific optical rotation with the β-anhydrous form being more soluble than the α-monohydrate form (Zarmpi et al., 2017). In addition, various lactose grades exist, differing in particle size which affects the granulation and compression properties (Keleb et al., 2004b). El Hagrasy et al. (2013) studied the granule formation and growth behavior of three lactose grades with different PSD and observed similar growth behavior at different L/S-ratios. Similar results were found in a study by Lute et al. (2018), where comparable granules were produced at the lowest L/S-ratio with Pharmatose 200M and Supertab 21AN, which have small and large primary particle sizes, respectively. This behavior could be attributed to the enhanced dissolution of smaller particles compared to larger ones, which favors granule growth due to the formation of more liquid bridges. However, at a higher L/S-ratio granule growth for Supertab 21AN-based formulations was significant, yielding larger granules than Pharmatose 200M due to the presence of more granulation liquid and a larger fraction of dissolved lactose. The influence of the primary particle size of lactose on friability was investigated by Hwang et al. (2019) who found that a lactose grade with a larger surface area (i.e., smaller primary particle size) produced stronger granules because of the higher probability of interaction between liquid and powder at a low L/S-ratio. At a high L/S-ratio, the effect of primary particle size on friability disappeared as more liquid is available.

(Portier et al., 2020b, 2020c); Portier et al. (2020a) suggested that a combination of lactose and MCC (1:1 ratio) could serve as a robust filler combination, as similar granulation behavior was observed when processing APIs with different solubility and wettability. Additionally, the combination of lactose/MCC as filler and HPMC as binder could be considered as a versatile and robust platform formulation. These formulations were less influenced by diverse API characteristics, screw configurations, and process settings compared to identical formulations without MCC. This approach could accelerate early drug product development and reduce costs by limiting the API consumption.

13.5.2 Microcrystalline cellulose

MCC is a purified, partially depolymerized cellulose made from wood pulp. It is used as a diluent and also exhibits some binding, lubricant, and disintegration properties (Rowe et al., 2009). MCC is water-insoluble but has hydrophilic properties due to its high water-binding capacity and tendency to swell. The commercial grades primarily differ in particle size and moisture content (Ali et al., 2009). As mentioned in the previous section, MCC can contribute to the robustness of the formulation; however, some limitations must be considered when using MCC as the sole filler during TSWG. Its high water-binding capacity promotes the robustness of the granulation process, but also increases the required L/S-ratio to produce granules with good quality attributes. Since the twin-screw wet granulator unit is part of a continuous from powder-to-tablet line, the drying capacity of the subsequent drying unit is limited. Higher L/S-ratios could thus limit the total throughput capacity of the continuous line (Portier et al., 2020b); Portier et al. (2020a). Additionally, batch-to-batch variability of MCC is well known due to its natural origin, resulting in different degrees of crystallinity among batches (Landín et al., 1993). Fonteyne et al. (2015) investigated the effect of MCC variability on the granule properties and processability during TSWG. The granule size distribution (GSD) was significantly affected by the variability of the MCC grade (Avicel PH101). The batch with the lowest water-binding capacity yielded more oversized granules as more granulation liquid was available. This was even more pronounced when processing at high barrel filling degree (i.e., high throughput) and short residence time (i.e., high screw speed). Differences in water-binding capacity of MCC batches are attributed to variations in the degree of crystallinity, which may arise from producing MCC from different wood pulps.

13.5.3 Mannitol

Mannitol is a widely used excipient, favored due to its high water-solubility and solubility rate, nonhygroscopic nature, compatibility with amines, and high sweetness (Rowe et al., 2009). It exists in three polymeric forms (α-, β-, δ-mannitol), with β-mannitol being the thermodynamically stable form and the most abundantly used in commercial production (Burger et al., 2000). The different grades vary in size, shape, and structure, which determine the application depending on the preferred manufacturing process. Whereas most studies focused on using lactose or MCC as filler, some authors explored the use of mannitol during TSWG. (Vanhoorne et al., 2016a) investigated the moisture-induced polymorphic transition from δ-mannitol to β-mannitol during TSWG. This transition resulted in superior tabletability as granules with enhanced plastic deformation and an unique granule morphology (with a high specific surface) were manufactured using δ-mannitol instead of β-mannitol as starting material. The occurrence of a full transition depended on the L/S-ratio, screw speed, and number of kneading elements, although the improved tabletability could be obtained without full transition. Additionally, the granulation mechanism of mannitol was compared to lactose and MCC. Due to the high solubility and dissolution rate of mannitol, it was suggested that the granule growth of mannitol particles is primarily driven by the formation of liquid and solid bridges after crystallization of solubilized material rather than by compressive forces. To further assess its potential, the inclusion of a highly-dosed poorly compressible drug and influence of a binder on polymeric transition were investigated (Vanhoorne et al., 2020). The polymeric transition from δ-mannitol to β-mannitol still occurred at high drug loading. Granules and tablets with excellent quality attributes were obtained, whereas this was not achieved using β-mannitol as starting material. However, the transition was inhibited at low L/S-ratio and with the inclusion of HPMC or polyvinylpyrrolidone (PVP) as binder. This was not considered as an issue since high-quality granules could be produced without the need for binders. The mannitol grade as starting material has a significant effect on the granule morphology after TSWG. According to Megarry et al. (2020) Pearlitol 200SD, which consists of spherical particles, yielded needle-shaped granules, whereas this was not identified with Pearlitol 160C, which consisted of crystallized angular particles. This difference could be attributed to the higher dissolution rate of Pearlitol 200SD, resulting in a higher solubilized fraction that subsequently recrystallizes. This recrystallization was associated with a transition from a mixture of α-mannitol and β-mannitol to predominantly β-mannitol. While the mannitol grades behaved similarly during TSWG, they had a clear impact on the specific surface area and flowability of the granules due to differences in particle morphology.

13.6 Binders

Binders are crucial for enhancing the adhesion and cohesion between particle surfaces and strengthening the interparticulate bonds by creating a cohesive network during granulation. Their functionality depends on the intrinsic binder capacity and their distribution through the powder bed (Vandevivere et al., 2019). Due to the short residence time during TSWG, binders have to fulfill their binding potential in a significantly shorter time frame. Therefore, selecting a binder with appropriate attributes is essential to achieve fast activation (i.e., high degree of surface wetting and spreadability) and good compatibility (i.e., high degree of wet adhesion) with the other excipients (Dürig & Karan, 2018). Binders can also influence the tableting step following granulation. Köster and Kleinebudde (2024) studied the influence of different concentrations of two hydroxypropylcellulose (HPC) grades (HPC-SSL and HPC SLL-SFP) in granules produced via TSWG to achieve sufficient tabletability while maintaining a disintegration time that meets the pharmacopoeial requirements. When lactose was included as a filler, a higher binder concentration enhanced the tabletability, and the disintegration time was still within the limits. However, when a mixture of DCP and MCC was included as filler, the binder concentration in the granules did not affect tabletability, but tablets made from granules containing 5% binder and processed at high compression pressure did not meet the disintegration time limits. Commonly used binders in wet granulation are semi-synthetic cellulose derivatives (HPC, HPMC, methylcellulose, ethylcellulose, sodium carboxymethylcellulose), synthetic polymers (PVP, copovidone), natural polymers (modified starches), and sugars (maltodextrins) (Cantor et al., 2008; Dürig & Karan, 2018). Most binders are available in different grades with specific viscosities (molecular weight), which indirectly influence other properties like wetting properties and surface tension. Several studies have screened the potential of the commonly used binders and the binder addition method during TSWG, providing valuable insights into optimizing the granulation process. An overview is given in Table 13.1.

13.6.1 Binder addition method

Binders can be added in the powder blend (i.e., dry addition) or dissolved in the granulation liquid (i.e., wet addition). There is no consensus about the impact of wet or dry binder addition on binder efficiency, as evidenced by several studies. El Hagrasy et al. (2013) found that incorporating HPMC in the liquid phase yielded granules with a lower fines fraction and narrower PSD compared to addition in dry form for a formulation consisting of lactose and MCC. Similarly, the inclusion of PVP in the granulation liquid yielded granules with less fines and a higher oversized fraction for a formulation consisting of theophylline and lactose (Vercruysse et al., 2012) and reduced the required L/S-ratio (Portier et al., 2020d) for a formulation consisting of mebendazole, lactose, and MCC. Similar results were obtained for wet addition of HPMC, HPC, and PVP. A larger d50 value and span were obtained in a study by Ito and Kleinebudde (2019) (for a formulation consisting of acetaminophen and DCP or lactose), while a larger amount of oversized granules was observed in a study of Fonteyne et al. (for a formulation consisting of MCC and lactose) (Fonteyne et al., 2015). In contrast, Keleb et al. (2002) found no significant differences when comparing wet or dry addition of PVP for a formulation solely consisting of lactose. Vandevivere et al. (2020) studied the effect of wet or dry addition of 11 binders in a binary DCP-binder (95/5 \%w/w) formulation. Dry binder addition generally resulted in the production of high-quality granules. However, wet binder addition at the lowest L/S-ratios yielded stronger granules for most binders. Interestingly, at higher L/S-ratios, both addition methods showed similar friability values. Notably, binders HPMC E5, HPMC E15, and PVP K90 produced stronger granules via dry

Table 13.1 Overview of semisynthetic cellulose derivatives, sugars, synthetic polymers, and modified starches used as binders in dry and/or wet state during continuous twin-screw wet granulation[1].

Binder	Grade	Concentration	Formulation	Binder Addition	References
HPC	HPC-SSL	2% or 3%	Acetaminophen, sodium croscarmellose, DCP, and lactose (different compositions)	Dry and wet	Ito and Kleinebudde (2019)
HPC	HPC-SSL	2%, 3.5% or 5%	Lactose monohydrate or DCP:MCC (75:25)	Dry	Köster and Kleinebudde (2024)
HPC	HPC-SSL	5%	Lactose monohydrate (95%)	Dry	Köster et al. (2021)
HPC	Klucel EXF	3%	MCC (30%), lactose monohydrate (67%)	Dry and wet	Fonteyne et al. (2015)
HPC	HPC SSL-SFP	5%	Lactose monohydrate (95%)	Dry	Köster et al. (2021)
HPC	HPC SSL-SFP	2%, 3.5%, or 5%	Lactose monohydrate or DCP:MCC (75:25)	Dry	Köster and Kleinebudde (2024)
HPC	HPC SL-FP	5%	Lactose monohydrate (95%)	Dry	Köster et al. (2021)
HPC	Klucel-EF	3% or 6%	α-lactose monohydrate (80%), MCC (18.5%), crosscarmellose sodium (1.5%)	Wet	Dhenge et al. (2012a, 2012b, 2013)
HPC	Not specified	2% or 4%	Lactose, DCP, MCC, crosscarmellose sodium (different compositions)	Wet	(Yu et al., 2014b)
HPC	Nisso HPC-SL FP	5%	Lactose and disintegrants (different compositions)	Dry	Köster and Kleinebudde (2023)
HPC	Nisso HPC-SL	5%	Lactose and disintegrants (different compositions)	Dry	Köster and Kleinebudde (2023)
HPC	Nisso HPC-SSL SFP	5%	Lactose and disintegrants (different compositions)	Dry	Köster and Kleinebudde (2023)
HPMC	Hypromellose unspecified	5%	α-lactose monohydrate (73.5%), MCC (20%), sodium croscarmellose (1.5%)	Dry and wet	El Hagrasy et al. (2013)
HPMC	Pharmacoat 603	2% or 3%	Acetaminophen, sodium croscarmellose, DCP, and lactose (different compositions)	Dry and wet	Ito and Kleinebudde (2019)
HPMC	Pharmacoat 603	0%, 2.5%, or 5%	Lactose monohydrate, MCC, and sodium croscarmellose (different compositions)	Dry, wet, and mixture (1:1)	Saleh et al. (2015)
HPMC	Pharmacoat 603	1%–2% in dry granules	MCC/α-lactose monohydrate/mannitol (varying concentrations)	Wet	Willecke et al. (2018)
HPMC	Pharmacoat 606	5%	Lactose monohydrate (95%)	Dry	Köster et al. (2021)

(continued on next page)

[1] Table compiled by screening studies with keywords "binders," "twin-screw wet granulation," "formulation parameters," no time delimitation was used. The table is not exhaustive.

Table 13.1 Overview of semisynthetic cellulose derivatives, sugars, synthetic polymers, and modified starches used as binders in dry and/or wet state during continuous twin-screw wet granulation—cont'd

Binder	Grade	Concentration	Formulation	Binder Addition	References
HPMC	Methocel E5	3%	MCC (30%), lactose monohydrate (67%)	Dry and wet	Fonteyne et al. (2015)
HPMC	Methocel E5	5%	Anhydrous DCP (95%)	Dry and wet (only 95% DCP)	Vandevivere et al. (2020)
HPMC	Methocel E15	5%	Anhydrous DCP (95%)/ mannitol (95%)/APAP (50%) and mannitol (45%)	Dry and wet (only 95% DCP)	Vandevivere et al. (2020, 2021, 2022)
HPMC	Methocel E15	1%–5%	α-lactose monohydrate:MCC 1:1 (varying concentrations)	Dry and partially dispersed	Portier et al. (2020)
Maltodextrin	Glucidex 2	5%	Anhydrous DCP (95%)	Dry and wet (only 95% DCP)	Vandevivere et al. (2020)
Maltodextrin	Glucidex 6	5%	Anhydrous DCP (95%)/ mannitol (95%)	Dry and wet (only 95% DCP)	Vandevivere et al. (2020, 2021)
Maltodextrin	Lycatab DSH	5%	Anhydrous DCP (95%)	Dry and wet (only 95% DCP)	Vandevivere et al. (2020)
PVA	Parteck MXP	5%	Anhydrous DCP (95%)/ mannitol (95%)	Dry and wet (only 95% DCP)	Vandevivere et al. (2020, 2021)
PVP	Kollidon K12	5%	Anhydrous DCP (95%)/ mannitol (95%)/APAP (50%) and mannitol (45%)	Dry and wet (only 95% DCP)	Vandevivere et al. (2020, 2021, 2022)
PVP	Kollidon K30	2.5%	α-lactose monohydrate (97.5%)	Dry and wet	Keleb et al. (2002)
PVP	Kollidon K30	2.5%	Theophylline anhydrate (30%), α-lactose monohydrate (67.5%)	Dry and wet	Vercruysse et al. (2012)
PVP	Kollidon K30	3%	MCC (30%), lactose monohydrate (67%)	Dry and wet	Fonteyne et al. (2015)
PVP	Kollidon K30	2% or 3%	Acetaminophen, sodium croscarmellose, DCP, and lactose (different compositions)	Dry and wet	Ito and Kleinebudde (2019)
PVP	Kollidon K30	5%	α-lactose monohydrate (22.5%), MCC (22.5%), mebendazole (50%)	Dry and wet	Portier et al. (2020)
PVP	Kollidon K30	5%	Lactose monohydrate (95%)	Dry	Köster et al. (2021)
PVP	Kollidon K30	5%	Anhydrous DCP (95%)	Dry and wet (only 95% DCP)	Vandevivere et al. (2020)
PVP	Kollidon K30	1%–5%	α-lactose monohydrate:MCC 1:1 (varying concentrations)	Dry and partially dispersed	Portier et al. (2020)
PVP	Kollidon K30	1%–2% in dry granules	MCC/α-lactose monohydrate/mannitol (varying concentrations)	Wet	Willecke et al. (2018)

(continued on next page)

Table 13.1 Overview of semisynthetic cellulose derivatives, sugars, synthetic polymers, and modified starches used as binders in dry and/or wet state during continuous twin-screw wet granulation—cont'd

Binder	Grade	Concentration	Formulation	Binder Addition	References
PVP	Kollidon K30	1.25%, 2.5%, or 5%	α-lactose monohydrate	Dry and wet	Keleb et al. (2002)
PVP	Kollidon K30	1.25% or 2.5%	α-lactose monohydrate, hydrochlorothiazide (ratio not specified)	Wet	Keleb et al. (2004a)
PVP	Kollidon K90	5%	Anhydrous DCP (95%)/mannitol (95%)	Dry and wet (only 95% DCP)	Vandevivere et al. (2020, 2021)
PVP/VA	Kollidon VA64	5%	Lactose monohydrate (95%)	Dry	Köster et al. (2021)
Hydroxypropyl pea starch	Lycoat RS 720	5%	Anhydrous DCP (95%)/mannitol (95%)/APAP (50%) and mannitol (45%)	Dry and wet (only 95% DCP)	Vandevivere et al. (2020, 2021, 2022)
Native starches	Pea, maize, potato, and wheat starch	5%	Anhydrous DCP (95%)/mannitol (95%)	Dry	Vandevivere et al. (2019)
Starch octenyl succinate	Cleargum CO 01	5%	Anhydrous DCP (95%)/mannitol (95%)/APAP (50%) and mannitol (45%)	Dry and wet (only 95% DCP)	Vandevivere et al. (2020, 2021, 2022)

addition, irrespective of the L/S-ratio. This observation was linked to an uneven binder distribution caused by the high viscosity of these binder dispersions, resulting in a higher degree of bimodality in GSD in the case of the wet binder addition compared to the dry binder addition. The choice of whether to add the binder in dry form or dispersed in granulation liquid will depend on various factors. First, dissolving the binder in the granulation liquid is limited by the binder's viscosity, highly viscous binders could hinder consistent and stable liquid pumping (Willecke et al., 2018). Additionally, dispersing the binder in the granulation liquid adds an extra step to the manufacturing process. Therefore, dry binder addition might be the preferred method from an industrial standpoint. However, the cohesiveness and stickiness of the dry binder could potentially hinder accurate feeding (Vandevivere et al., 2020).

13.6.2 Binder attributes

Most studies initially focused on viscosity as the main binder attribute affecting the quality of granules produced via TSWG. However, subsequent research by various investigators identified additional binder attributes that also impact granule quality, thereby aiding in binder selection.

13.6.2.1 Viscosity

First of all, it should be acknowledged that binders added dry do not reach their maximal viscosity during TSWG due to the short residence times. Dhenge et al. (2012b) studied the influence of the binder viscosity on residence time, torque, and granule properties (e.g., size, shape, structure, strength, and friability) of a lactose/MCC-based formulation. A higher binder concentration in the granulation liquid (i.e., a higher viscosity) yielded larger granules with a PSD becoming mono-modal. In addition, stronger granules were produced with improved shape and flow (Dhenge et al., 2012b; Keleb et al., 2004b; Yu et al., 2014b). These improved granule properties can be attributed to an increased stickiness of the wet mass due to higher viscous forces in case of a highly viscous granulation liquid. More and stronger bridges between the particles are formed, and in combination with a longer residence time due to the higher binder viscosity, consolidation is facilitated, yielding denser and stronger granules. Hence, less granulation liquid is needed to yield high-quality granules if a high viscosity binder is included (Keleb et al., 2002). Nevertheless, sufficient shear must be applied to the viscous wet mass as the liquid penetration time increases due to slow liquid infiltration by a high viscosity solution. If a highly viscous granulation liquid was processed on a twin-screw granulator only equipped with conveying elements, smaller and more elongated granules with a bimodal PSD were manufactured due to poor distribution of binder liquid (Dhenge et al., 2013; Yu et al., 2014b).

13.6.2.2 Binder selection

As shown in Table 13.1, various studies have investigated the use of binders during TSWG. However, a thorough comparison between the binders is hindered by the use of different formulations (API, excipients, and binder concentration). Therefore, some researchers have aimed to facilitate formulation development by a science-based binder selection, taking the binder attributes into account. Willecke et al. (2017, 2018) characterized 9 binders in terms of their viscosity, glass transition temperature, surface tension, and molecular weight. PCA was performed, revealing that the first principal component (PC) described the viscosity variability (major variability), the second PC the surface tension variability, and the third PC the glass transition temperature variability. Binary formulations consisting of different percentages of filler (MCC, lactose, or mannitol) and binder (PVP or HPMC, wet addition) were granulated. Binder type and its concentration influenced the granule flowability, friability, and compactability. Higher binder concentrations improved the flowability and tabletability, and reduced the friability of the granules. PVP improved the tabletability and reduced the friability. Vandevivere et al. (2020, 2021) performed a more extensive binder characterization (both as dry particles and in the hydrated state), focusing on PSD, specific surface area, dissolution kinetics, wettability, surface tension, viscosity, and surface energy of 11 binders. Two model formulations, one water-soluble mannitol-based and one water-insoluble DCP-based, were used to assess the binder effectiveness (i.e., L/S-ratio needed to achieve a granule friability <30%) in function of formulation solubility. The intrinsic effect of the binders in the water-insoluble formulation was also assessed as DCP does not contribute to binding. Both studies identified that binder attributes affected the binder effectiveness and that different binder attributes were important depending on the formulation solubility (Fig. 13.6). To obtain strong granules at a low L/S ratio, a binder for a water-soluble formulation should exhibit fast activation (i.e., fast dissolution kinetics), low viscosity, and low surface tension. The binder should be wetted easily and, once activated, show good wetting properties on the surface of the mannitol particles. A suitable binder for a water-insoluble formulation should exhibit good wetting by water and good wetting of dispersive surfaces when activated, high viscosity, slow dissolution kinetics, and low surface tension (Vandevivere et al., 2021).

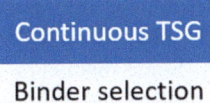

FIGURE 13.6

Overview of the critical binder attributes influencing the binder effectiveness for a highly and poorly water-soluble formulation.

From Vandevivere et al. (2021).

13.7 Surfactants

Only a few studies have investigated the effect of surfactant addition on granule properties. Dhenge et al. (2012b) studied the effect of the surface tension of the granulation liquid on granule properties by varying the concentration of sodium 1-decanesulonate (0.01%–0.03% w/w) in a placebo formulation (lactose/MCC-based). The surfactant had a negligible effect on the size, shape, surface, flowability, and strength of the granules. In contrast, Portier et al. (2020) reported that the addition of sodium lauryl sulphate (0.2% in the dry granules) in a highly-dosed poorly water-soluble formulation reduced the required L/S-ratio to achieve similar granule characteristics as larger, more spherical and stronger granules were obtained when including the surfactant. This improvement could be linked to more efficient wetting of the powder due to the amphiphilic surfactant properties.

13.8 Disintegrants

Disintegrants are added to the formulation to improve the rate and extent of tablet disintegration. Typically, they are mixed extragranularly after wet granulation, drying, or milling to prevent loss of their disintegrating power (Roy et al., 2011; Zhao & Augsburger, 2006). However, including the disintegrant intragranularly could simplify the continuous manufacturing process. Therefore, Meier et al. studied the addition of various disintegrants intragranularly during TSWG (Meier et al., 2015). In general, higher L/S-ratios were needed for intragranular addition of the disintegrant due to its high hygroscopicity and water-sorption capacity. Some disintegrants can be added intragranularly as they did not lose their disintegration power after wetting and subsequent drying. Croscarmellose sodium and sodium starch glycolate were the most promising disintegrants with the broadest workable range and the best granule and tablet characteristics. In a study by Köster and Kleinebudde (2023), the influence of the localization (intragranular, split, or extragranular) of three superdisintegrants (croscarmellose sodium, crospovidone, sodium starch glycolate) on granules and tablets produced via twin-screw wet granulation was investigated. Similar to the findings of Meier et al. (2015), intragranular croscarmellose sodium was beneficial, providing sufficient tensile strength in combination with a fast disintegration for the different HPC grades (included in a lactose-based formulation) compared to other disintegrants and localization. Extragranular crospovidone yielded similar results. While sodium starch glycolate showed promising results in the study by Meier et al. (2015), this disintegrant performed the worst in the study by Köster and Kleinebudde (2023), as the disintegration was not within the required limits.

13.9 Controlled release formulations

Controlled release (CR) formulations are of interest as they offer prolonged therapeutic effect, minimized toxicity, and improved patient compliance. Few studies investigated those formulations during TSWG. Thompson and O'Donnell (2015) reported the formation of long noodle-shaped granules when granulating two CR excipients (Methocel K4M and Kollidon SR) in a placebo formulation (Fig. 13.7). Changing the concentration (5%–20% w/w) of the CR excipient did not improve the shape of the granules. It was hypothesized that a rolling mechanism created the long twisted noodle-shaped granules, as increasing the screw speed had a different effect compared to immediate release formulations. Normally, higher screw speeds cause a higher rate of attrition and therefore more spherical particles. However, with the CR formulations, a higher screw speed produced less spherical particles. The powder accumulated at the flight tips of the screws, and due to the adhesive properties of the CR excipient, fragmentation is restricted. This formed long strands of powder, which kept growing due to the continuous nature of the flight. The long strands were also compressed into dense, twisted noodle-shaped granules by impact against the adjacent barrel wall. The most effective means to produce near-spherical particles was by changing the screw design, placing non-conveying screw elements at the end. In a study by (Vanhoorne et al., 2016b), granules with comparable aspect ratios to immediate release formulations were obtained using a formulation consisting of HPMC as a CR excipient, a very water-soluble drug, and a filler. To further investigate the discrepancy in findings about the shape of CR granules between different literature reports (Vanhoorne et al., 2016b) processed the same formulation (20% w/w HPMC, 16% w/w MCC, and 64% w/w lactose) used in the work of Thompson and O'Donnell (2015) on a different granulator, and reported that the twisted elongated granules were only formed at excessive L/S-ratio. This indicated that both granulator design and formulation composition are essential parameters during the continuous granulation of controlled-release formulations with HPMC, and that the combination of MCC and HPMC should be discouraged. In a follow-up study, the influence of HPMC grade and particle size of theophylline on the granule CQAs was investigated

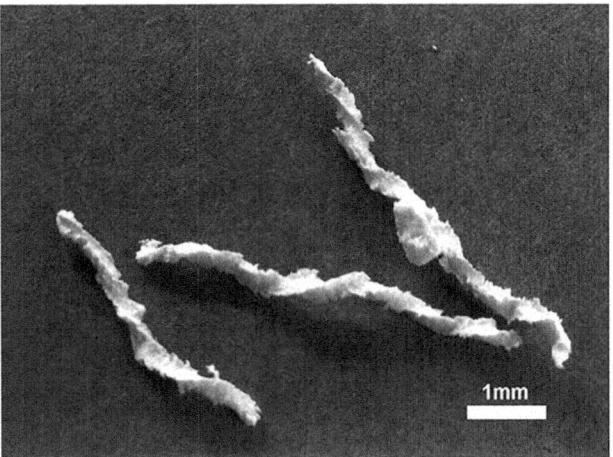

FIGURE 13.7

Example of "extended-form" granules with their twisted noodle-like appearance. Sample was prepared with 20\% METHOCEL K4M, 76\% lactose monohydrate, 19\% microcrystalline cellulose.

From Thompson and O'Donnell (2015).

by (Vanhoorne et al., 2016c). Similar to immediate release formulations, a broad PSD was obtained for all formulations. The viscous gel layer formed by HPMC limited granule breakage and uniform liquid distribution, making the formation of a monomodal PSD with CR excipients even more challenging. The substitution degree influenced the GSD as the ratio of hydrophilic hydroxypropyl and hydrophobic methoxyl groups differed between the HPMC grades. HPMC type 2208, with higher hydrophilicity, yielded granules with less fines and more oversized granules compared to HPMC type 2910 at a similar L/S-ratio. The molecular weight of the polymers, and thus the viscosity, did not influence the PSD. Including more HPMC in the formulation increased the amount of liquid needed to obtain a similar GSD, as HPMC has a high water-binding capacity. Additionally, the content uniformity of theophylline over different sieve fractions was analyzed, revealing theophylline underdosing in the fines fraction (<150 μm) and overdosing in the 150 to 250 μm and 250 to 500 μm fractions. This phenomenon was not correlated with the hydration rate or viscosity of HPMC. Denduyver et al. (2024) studied the root cause of the nonhomogeneous theophylline distribution of granules produced with HPMC as CR excipient. The influence of HPMC and filler type on the content uniformity was examined. It was found that HPMC causes nonhomogeneous API distribution over the different granule sieve fractions due to its rapid swelling upon contact with the aqueous granulation liquid, preventing granule breakage and subsequently particle exchange during granule growth. Different filler types yielded granules with various API concentrations over the sieve fractions, but this could not be correlated with their solubility. The most homogeneous API distribution over the size fractions was obtained using MCC as filler, possibly because HPMC and MCC both exhibit relatively strong water-binding properties. However, it was also hypothesized that differences in particle size of the filler and API could influence the uneven API distribution over the size fractions of these sustained release formulations.

13.10 Conclusion

TSWG has evolved in recent years into a mature, effective, and widely adopted technique. The increased understanding of the effect of the formulation parameters (CMAs) on the granule properties (CQAs) has partially contributed to this. More studies have focused on formulation development, industrially relevant drug-loaded formulations, and linking of raw material properties to the granulation process and granule quality. The complex interplay of formulation and process parameters complicates general conclusions based on literature data. Due to the continuous nature of TSWG, a well-considered choice of excipients (filler, binder, disintegrant) is essential to yield an efficient process and manufacture granules with properties suitable for further downstream processing (drying, milling, blending, and tableting). Additionally, formulation adaptations must be made when transitioning from batch-wise to continuous manufacturing due to different material feeding, the short residence time of material in the granulator barrel, and different granule growth mechanisms.

References

Ali, J., Saigal, N., Baboota, S., & Ahuja, A. (2009). Microcrystalline cellulose as a versatile excipient in drug research. *Journal of Young Pharmacists, 1*(1), 6. https://doi.org/10.4103/0975-1483.51868.

Arndt, O. R., Baggio, R., Adam, A. K., Harting, J., Franceschinis, E., & Kleinebudde, P. (2018). Impact of Different dry and wet granulation techniques on granule and tablet properties: a comparative study. *Journal of Pharmaceutical Sciences, 107*(12), 3143–3152. https://doi.org/10.1016/j.xphs.2018.09.006.

Bacher, C., Olsen, P. M., Bertelsen, P., & Sonnergaard, J. M. (2008). Compressibility and compactibility of granules produced by wet and dry granulation. *International Journal of Pharmaceutics, 358*(1-2), 69–74. https://doi.org/10.1016/j.ijpharm.2008.02.013.

Beer, P., Wilson, D., Huang, Z., & De Matas, M. (2014). Transfer from high-shear batch to continuous twin screw wet granulation: A case study in understanding the relationship between process parameters and product quality attributes. *Journal of Pharmaceutical Sciences, 103*(10), 3075–3082. https://doi.org/10.1002/jps.24078.

Burger, A., Henck, J. O., Hetz, S., Rollinger, J. M., Weissnicht, A. A., & Stöttner, H. (2000). Energy/temperature diagram and compression behavior of the polymorphs of D-mannitol. *Journal of Pharmaceutical Sciences, 89*(4), 457–468. https://doi.org/10.1002/(SICI)1520-6017(200004)89:4⟨457::AID-JPS3⟩3.0.CO;2-G.

Cai, L., Farber, L., Zhang, D., Li, F., & Farabaugh, J. (2013). A new methodology for high drug loading wet granulation formulation development. *International Journal of Pharmaceutics, 441*(1-2), 790–800. https://doi.org/10.1016/j.ijpharm.2012.09.052.

Cantor, S., Augsburger, L., & Gerhardt, A. (2008). Pharmaceutical granulation processes, mechanisms, and the use of binders. In *Pharmaceutical dosage forms—Tablets* (pp. 261–301). CRC Press. https://doi.org/10.1201/b15115-9.

Cartwright, J. J., Robertson, J., D'Haene, D., Burke, M. D., & Hennenkamp, J. R. (2013). Twin screw wet granulation: Loss in weight feeding of a poorly flowing active pharmaceutical ingredient. *Powder Technology., 238*, 116–121. https://doi.org/10.1016/j.powtec.2012.04.034.

Challener, C. A. (2020). Key ingredients needed to drive the success of contiuous manufacturing. *Pharmaceutical Technology, 44*(8), 21–24.

Charles-Williams, H., Wegeler, R., Flnore, K., Feise, H., Hounslow, M. J., & Salman, A. D. (2013). Granulation behaviour of increasingly hydrophobic mixtures. *Powder Technology., 238*, 64–76. https://doi.org/10.1016/j.powtec.2012.06.009.

Denduyver, P., Vervaet, C., & Vanhoorne, V. (2024). Studying the API distribution of controlled release formulations produced via continuous twin-screw wet granulation: Influence of matrix former, filler and process parameters. *Pharmaceutics, 16*(3). https://doi.org/10.3390/pharmaceutics16030341.

Dhenge, R. M., Cartwright, J. J., Doughty, D. G., Hounslow, M. J., & Salman, A. D. (2011). Twin screw wet granulation: Effect of powder feed rate. *Advanced Powder Technology., 22*(2), 162–166. https://doi.org/10.1016/j.apt.2010.09.004.

Dhenge, R. M., Cartwright, J. J., Hounslow, M. J., & Salman, A. D. (2012a). Twin screw wet granulation: Effects of properties of granulation liquid. *Powder Technology., 229*, 126–136. http://dx.doi.org/10.1016/j.powtec.2012.06.019.

Dhenge, R. M., Cartwright, J. J., Hounslow, M. J., & Salman, A. D. (2012b). Twin screw granulation: Steps in granule growth. *International Journal of Pharmaceutics, 438*(1-2), 20–32. https://doi.org/10.1016/j.ijpharm.2012.08.049.

Dhenge, R. M., Washino, K., Cartwright, J. J., Hounslow, M. J., & Salman, A. D. (2013). Twin screw granulation using conveying screws: Effects of viscosity of granulation liquids and flow of powders. *Powder Technology, 238*, 77–90. https://doi.org/10.1016/j.powtec.2012.05.045.

Djuric, D., & Kleinebudde, P. (2008). Impact of screw elements on continuous granulation with a twin-screw extruder. *Journal of Pharmaceutical Sciences, 97*(11), 4934–4942. https://doi.org/10.1002/jps21339.

Djuric, D., Van Melkebeke, B., Kleinebudde, P., Remon, J. P., & Vervaet, C. (2009). Comparison of two twin-screw extruders for continuous granulation. *European Journal of Pharmaceutics and Biopharmaceutics, 71*(1), 155–160. https://doi.org/10.1016/j.ejpb.2008.06.033.

Dun, J., & Sun, C. C. (2018). Structures and properties of granules prepared by high shear wet granulation. In *Handbook of pharmaceutical wet granulation: Theory and practice in a quality by design paradigm* (pp. 119–147). Elsevier Inc. https://doi.org/10.1016/B978-0-12-810460-6.00004-X.

Dürig, T., & Karan, K. (2018). Binders in wet granulation. In *Handbook of pharmaceutical wet granulation: Theory and practice in a quality by design paradigm* (pp. 317–349). https://doi.org/10.1016/B978-0-12-810460-6.00010-5.

El Hagrasy, A. S., Hennenkamp, J. R., Burke, M. D., Cartwright, J. J., & Litster, J. D. (2013). Twin screw wet granulation: Influence of formulation parameters on granule properties and growth behavior. *Powder Technology., 238*, 108–115. https://doi.org/10.1016/j.powtec.2012.04.035.

Erdemir, D., Gawel, J., Yohannes, B., Yates, P., Tang, D., Ha, K., Breza, B., DiMaso, E., Abebe, A., & Zombek, J. (2023). Continuous feeding and blending demonstration with co-processed drug substance. *Journal of Pharmaceutical Sciences, 112*(8), 2046–2056. https://doi.org/10.1016/j.xphs.2022.11.023.

Fonteyne, M., Correia, A., De Plecker, S., Vercruysse, J., Ilić, I., Zhou, Qi, Vervaet, C., Paul Remon, J., Onofre, F., Bulone, V., & De Beer, T. (2015). Impact of microcrystalline cellulose material attributes: A case study on continuous twin screw granulation. *International Journal of Pharmaceutics, 478*(2), 705–717. https://doi.org/10.1016/j.ijpharm.2014.11.070.

Fonteyne, M., Wickström, H., Peeters, E., Vercruysse, J., Ehlers, H., Peters, B. H., Remon, J. P., Vervaet, C., Ketolainen, J., Sandler, N., Rantanen, J., Naelapää, K., & De Beer, T. (2014). Influence of raw material properties upon critical quality attributes of continuously produced granules and tablets. *European Journal of Pharmaceutics and Biopharmaceutics, 87*(2), 252–263. https://doi.org/10.1016/j.ejpb.2014.02.011.

Franke, M., Riedel, T., Meier, R., Schmidt, C., & Kleinebudde, P. (2023). Comparison of scale-up strategies in twin-screw wet granulation. *International Journal of Pharmaceutics, 641*, 123052. https://doi.org/10.1016/j.ijpharm.2023.123052.

Hagrasy, A. E. l, & Litster, J. (2013). Granulation rate processes in the kneading elements of a twin screw granulator. *Aiche Journal, 59*. https://doi.org/10.1002/aic.14180.

Hwang, K.-M., Cho, C.-H., Yoo, S.-D., Cha, K.-I., & Park, E.-S. (2019). Continuous twin screw granulation: Impact of the starting material properties and various process parameters. *Powder Technology., 356*, 847–857. https://doi.org/10.1016/j.powtec.2019.08.062.

Ito, A., & Kleinebudde, P. (2019). Influence of granulation temperature on particle size distribution of granules in twin-screw granulation (TSG). *Pharmaceutical Development and Technology, 24*(7), 874–882. https://doi.org/10.1080/10837450.2019.1615089.

Iveson, S. M., Litster, J. D., Hapgood, K., & Ennis, B. J. (2001). Nucleation, growth and breakage phenomena in agitated wet granulation processes: A review. *Powder Technology., 117*(1-2), 3–39. https://doi.org/10.1016/S0032-5910(01)00313-8.

Keleb, E. I., Vermeire, A., Vervaet, C., & Paul Remon, J. (2002). Continuous twin screw extrusion for the wet granulation of lactose. *International Journal of Pharmaceutics, 239*(1-2), 69–80. https://doi.org/10.1016/S0378-5173(02)00052-2.

Keleb, E. I., Vermeire, A., Vervaet, C., & Remon, J. P. (2004a). Single-step granulation/tabletting of different grades of lactose: A comparison with high shear granulation and compression. *European Journal of Pharmaceutics and Biopharmaceutics, 58*(1), 77–82. https://doi.org/10.1016/j.ejpb.2004.03.007.

Keleb, E. I., Vermeire, A., Vervaet, C., & Remon, J. P. (2004b). Twin screw granulation as a simple and efficient tool for continuous wet granulation. *International Journal of Pharmaceutics, 273*(1-2), 183–194. https://doi.org/10.1016/j.ijpharm.2004.01.001.

Köster, C., & Kleinebudde, P. (2023). Evaluation of binders in twin-screw wet granulation—Optimal combination of binder and disintegrant. *European Journal of Pharmaceutics and Biopharmaceutics, 186*, 55–64. https://doi.org/10.1016/j.ejpb.2023.03.003.

Köster, C., & Kleinebudde, P. (2024). Evaluation of binders in twin-screw wet granulation—Optimization of tabletability. *International Journal of Pharmaceutics, 659*, 124290. https://doi.org/10.1016/j.ijpharm.2024.124290.

Köster, C., Pohl, S., & Kleinebudde, P. (2021). Evaluation of binders in twin-screw wet granulation. *Pharmaceutics, 13*(2), 1–17. https://doi.org/10.3390/pharmaceutics13020241.

Kumar, A., Alakarjula, M., Vanhoorne, V., Toiviainen, M., De Leersnyder, F., Vercruysse, J., Juuti, M., Ketolainen, J., Vervaet, C., Remon, J. P., Gernaey, K. V., De Beer, T., & Nopens, I. (2016). Linking granulation performance with residence time and granulation liquid distributions in twin-screw granulation: An experimental investigation. *European Journal of Pharmaceutical Sciences, 90*, 25–37. https://doi.org/10.1016/j.ejps.2015.12.021.

Kumar, A., Vercruysse, J., Toiviainen, M., Panouillot, P. E., Juuti, M., Vanhoorne, V., Vervaet, C., Remon, J. P., Gernaey, K. V., De Beer, T., & Nopens, I. (2014). Mixing and transport during pharmaceutical twin-screw wet granulation: Experimental analysis via chemical imaging. *European Journal of Pharmaceutics and Biopharmaceutics, 87*(2), 279–289. https://doi.org/10.1016/j.ejpb.2014.04.004.

Kyttä, K. M., Lakio, S., Wikström, H., Sulemanji, A., Fransson, M., Ketolainen, J., & Tajarobi, P. (2020). Comparison between twin-screw and high-shear granulation—The effect of filler and active pharmaceutical ingredient on the granule and tablet properties. *Powder Technology, 376*, 187–198. https://doi.org/10.1016/j.powtec.2020.08.030.

Landín, M., Martínez-Pacheco, R., Gómez-Amoza, J. L., Souto, C., Concheiro, A., & Rowe, R. C. (1993). Effect of batch variation and source of pulp on the properties of microcrystalline cellulose. *International Journal of Pharmaceutics, 91*(2-3), 133–141. https://doi.org/10.1016/0378-5173(93)90332-A.

Lee, K. T., Ingram, A., & Rowson, N. A. (2013). Comparison of granule properties produced using twin screw extruder and high shear mixer: A step towards understanding the mechanism of twin screw wet granulation. *Powder Technology, 238*, 91–98. https://doi.org/10.1016/j.powtec.2012.05.031.

Lee, S. L., O'Connor, T. F., Yang, X., Cruz, C. N., Chatterjee, S., Madurawe, R. D., Moore, C. M. V., Yu, L. X., & Woodcock, J. (2015). Modernizing pharmaceutical manufacturing: From batch to continuous production. *Journal of Pharmaceutical Innovation, 10*(3), 191–199. https://doi.org/10.1007/s12247-015-9215-8.

Li, H., Thompson, M. R., & O'Donnell, K. P. (2014). Understanding wet granulation in the kneading block of twin screw extruders. *Chemical Engineering Science, 113*, 11–21. https://doi.org/10.1016/j.ces.2014.03.007.

Lute, S. V., Dhenge, R. M., & Salman, A. D. (2018). Twin screw granulation: Effects of properties of primary powders. *Pharmaceutics, 10*(2). https://doi.org/10.3390/pharmaceutics10020068.

Megarry, A., Taylor, A., Gholami, A., Wikström, H., & Tajarobi, P. (2020). Twin-screw granulation and high-shear granulation: The influence of mannitol grade on granule and tablet properties. *International Journal of Pharmaceutics, 590*. https://doi.org/10.1016/j.ijpharm.2020.119890.

Meier, R., Moll, K. P., Krumme, M., & Kleinebudde, P. (2017). Simplified, high drug-loaded formulations containing hydrochlorothiazide for twin-screw granulation. *Chemie Ingenieur Technik, 89*(8), 1025–1033. https://doi.org/10.1002/cite.201600134.

Meier, R., Thommes, M., Rasenack, N., Krumme, M., Moll, K. P., & Kleinebudde, P. (2015). Simplified formulations with high drug loads for continuous twin-screw granulation. *International Journal of Pharmaceutics, 496*(1), 12–23. https://doi.org/10.1016/j.ijpharm.2015.05.060.

Menth, J., Maus, M., & Wagner, K. G. (2020). Continuous twin screw granulation and fluid bed drying: A mechanistic scaling approach focusing optimal tablet properties. *International Journal of Pharmaceutics, 586*, 119509. https://doi.org/10.1016/j.ijpharm.2020.119509.

Mundozah, A. L., Cartwright, J. J., Tridon, C. C., Hounslow, M. J., & Salman, A. D. (2019). Hydrophobic/hydrophilic powders: Practical implications of screw element type on the reduction of fines in twin screw granulation. *Powder Technology., 341*, 94–103. https://doi.org/10.1016/j.powtec.2018.03.018.

Osorio, J. G., Sayin, R., Kalbag, A. V., Litster, J. D., Martinez-Marcos, L., Lamprou, D. A., & Halbert, G. W. (2017). Scaling of continuous twin screw wet granulation. *Aiche Journal, 63*(3), 921–932. https://doi.org/10.1002/aic.15459.

Parteck® LM Excipient System. 2024,6–13. https://www.merckmillipore.com/BE/fr/products/small-molecule-pharmaceuticals/formulation/solid-dosage-form/parteck-excipients/Parteck-LM-Excipient/Szeb.hzPTrsAAAGGamJ_XUyF,nav?ReferrerURL=https%3A%2F%2Fwww.google.com%2F. (accessed July 2024).

Peeters, M., Jiménez, A. A. B., Matsunami, K., Stauffer, F., Nopens, I., & De Beer, T. (2023). Evaluation of the influence of material properties and process parameters on granule porosity in twin-screw wet granulation. *International Journal of Pharmaceutics, 641*. https://doi.org/10.1016/j.ijpharm.2023.123010.

Portier, C., De Vriendt, C., Vigh, T., Di, G., & De Beer, T. (2020a). Continuous twin screw granulation: Robustness of lactose /MCC-based formulations. *International Journal of Pharmaceutics, 588*, 119756. https://doi.org/10.1016/j.ijpharm.2020.119756.

Portier, C., Pandelaere, K., Delaet, U., Vigh, T., Di Pretoro, G., De Beer, T., Vervaet, C., & Vanhoorne, V. (2020b). Continuous twin screw granulation: A complex interplay between formulation properties, process settings, and screw design. *International Journal of Pharmaceutics, 576*, 119004. https://doi.org/10.1016/j.ijpharm.2019.119004.

Portier, C., Pandelaere, K., Delaet, U., Vigh, T., Kumar, A., Di Pretoro, G., De Beer, T., Vervaet, C., & Vanhoorne, V. (2020c). Continuous twin screw granulation: Influence of process and formulation variables on granule quality attributes of model formulations. *International Journal of Pharmaceutics, 576*, 118981. https://doi.org/10.1016/j.ijpharm.2019.118981.

Portier, C., Vervaet, C., & Vanhoorne, V. (2021). Continuous twin screw granulation: A review of recent progress and opportunities in formulation and equipment design. *Pharmaceutics, 13*(5). https://doi.org/10.3390/pharmaceutics13050668.

Portier, C., Vigh, T., Di Pretoro, G., De Beer, T., Vervaet, C., & Vanhoorne, V. (2020d). Continuous twin screw granulation: Impact of binder addition method and surfactants on granulation of a high-dosed, poorly soluble API. *International Journal of Pharmaceutics, 577*, 119068. https://doi.org/10.1016/j.ijpharm.2020.119068.

Pradhan, S. U., Zhang, Y., Li, J., Litster, J. D., & Wassgren, C. R. (2019). Tailored granule properties using 3D printed screw geometries in twin screw granulation. *Powder Technology., 341*, 75–84. https://doi.org/10.1016/j.powtec.2017.12.068.

Rowe, R. C., Sheskey, P. J., & Quinn, M. E. (2009). *Handbook of pharmaceutical excipients*. Pharmaceutical Press.

Roy, S., Hasan, S., & Kumar, M. (2011). Effect of mode of addition of disintegrants on dissolution of model drug from wet granulation tablets. *International Journal of Pharma Sciences and Research, 2*(2), 84–92.

Saleh, M. F., Dhenge, R. M., Cartwright, J. J., Hounslow, M. J., & Salman, A. D. (2015). Twin screw wet granulation: Binder delivery. *International Journal of Pharmaceutics, 487*(1-2), 124–134. https://doi.org/10.1016/j.ijpharm.2015.04.017.

Sayin, R., El Hagrasy, A. S., & Litster, J. D. (2015). Distributive mixing elements: Towards improved granule attributes from a twin screw granulation process. *Chemical Engineering Science, 125*, 165–175. https://doi.org/10.1016/j.ces.2014.06.040.

Shah, A. V., & Serajuddin, A. T. M. (2018). Twin screw continuous wet granulation. In *Handbook of pharmaceutical wet granulation: Theory and practice in a quality by design paradigm* (pp. 791–823). Elsevier Inc. https://doi.org/10.1016/B978-0-12-810460-6.00008-7.

Stauffer, F., Vanhoorne, V., Pilcer, G., Chavez, P. F., Rome, S., Schubert, M. A., Aerts, L., & De Beer, T. (2018). Raw material variability of an active pharmaceutical ingredient and its relevance for processability in secondary continuous pharmaceutical manufacturing. *European Journal of Pharmaceutics and Biopharmaceutics, 127*, 92–103. https://doi.org/10.1016/j.ejpb.2018.02.017.

Stauffer, F., Vanhoorne, V., Pilcer, G., François Chavez, P., Vervaet, C., & De Beer, T. (2019). Managing API raw material variability during continuous twin-screw wet granulation. *International Journal of Pharmaceutics, 561*, 265–273. https://doi.org/10.1016/j.ijpharm.2019.03.012.

Thompson, M. R., & O'Donnell, K. P. (2015). Rolling" phenomenon in twin screw granulation with controlled-release excipients. *Drug Development and Industrial Pharmacy, 41*(3), 482–492. https://doi.org/10.3109/03639045.2013.879723.

Thompson, M. R., & Sun, J. (2010). Wet Granulation in a twin-screw extruder: implications of screw design. *Journal of Pharmaceutical Sciences, 99*(4), 2090–2103. https://doi.org/10.1002/jps.21973.

Vandeputte, T (2023). *A formulation-generic pharmaceutical fluidized bed drying model including granule segregation as a vital component*. Ghent University. Faculty of Bioscience Engineering http://hdl.handle.net/1854/LU-01HA9WXNRA77PMY9CBZ9ZGC79B.

Vandevivere, L., Denduyver, P., Portier, C., Häusler, O., De Beer, T., Vervaet, C., & Vanhoorne, V. (2020). Influence of binder attributes on binder effectiveness in a continuous twin screw wet granulation process via wet and dry binder addition. *International Journal of Pharmaceutics, 585*, 119466. https://doi.org/10.1016/j.ijpharm.2020.119466.

Vandevivere, L., Denduyver, P., Portier, C., Häusler, O., De Beer, T., Vervaet, C., & Vanhoorne, V. (2022). The effect of binder types on the breakage and drying behavior of granules in a semi-continuous fluid bed dryer after twin screw wet granulation. *International Journal of Pharmaceutics, 614*, 121449. https://doi.org/10.1016/j.ijpharm.2022.121449.

Vandevivere, L., Portier, C., Vanhoorne, V., Häusler, O., Simon, D., De Beer, T., & Vervaet, C. (2019). Native starch as in situ binder for continuous twin screw wet granulation. *International Journal of Pharmaceutics, 571*, 118760. https://doi.org/10.1016/j.ijpharm.2019.118760.

Vandevivere, L., Vangampelaere, M., Portier, C., de Backere, C., Häusler, O., De Beer, T., Vervaet, C., & Vanhoorne, V. (2021). Identifying critical binder attributes to facilitate binder selection for efficient formulation development in a continuous twin screw wet granulation process. *Pharmaceutics, 13*(2), 1–19. https://doi.org/10.3390/pharmaceutics13020210.

Vanhoorne, V., Almey, R., De Beer, T., & Vervaet, C. (2020). Delta-mannitol to enable continuous twin-screw granulation of a highly dosed, poorly compactable formulation. *International Journal of Pharmaceutics, 583*, 119374. https://doi.org/10.1016/j.ijpharm.2020.119374.

Vanhoorne, V., Bekaert, B., Peeters, E., De Beer, T., Remon, J. P., & Vervaet, C. (2016a). Improved tabletability after a polymorphic transition of delta-mannitol during twin screw granulation. *International Journal of Pharmaceutics, 506*(1-2), 13–24. https://doi.org/10.1016/j.ijpharm.2016.04.025.

Vanhoorne, V., Janssens, L., Vercruysse, J., De Beer, T., Remon, J. P., & Vervaet, C. (2016b). Continuous twin screw granulation of controlled release formulations with various HPMC grades. *International Journal of Pharmaceutics, 511*(2), 1048–1057. https://doi.org/10.1016/j.ijpharm.2016.08.020.

Vanhoorne, V., Vanbillemont, B., Vercruysse, J., De Leersnyder, F., Gomes, P., De Beer, T., Remon, J. P., & Vervaet, C. (2016c). Development of a controlled release formulation by continuous twin screw granulation: Influence of process and formulation parameters. *International Journal of Pharmaceutics, 505*(1-2), 61–68. https://doi.org/10.1016/j.ijpharm.2016.03.058.

Van Melkebeke, B., Vervaet, C., & Remon, J. P. (2008). Validation of a continuous granulation process using a twin-screw extruder. *International Journal of Pharmaceutics, 356*(1-2), 224–230. https://doi.org/10.1016/j.ijpharm.2008.01.012.

Van Snick, B., Holman, J., Vanhoorne, V., Kumar, A., De Beer, T., Remon, J. P., & Vervaet, C. (2017). Development of a continuous direct compression platform for low-dose drug products. *International Journal of Pharmaceutics, 529*(1-2), 329–346. https://doi.org/10.1016/j.ijpharm.2017.07.003.

Vercruysse, J. (2014). *(Ph.D. thesis)*. Ghent University.

Vercruysse, J., Burggraeve, A., Fonteyne, M., Cappuyns, P., Delaet, U., Van Assche, I., De Beer, T., Remon, J. P., & Vervaet, C. (2015). Impact of screw configuration on the particle size distribution of granules produced by twin screw granulation. *International Journal of Pharmaceutics, 479*(1), 171–180. https://doi.org/10.1016/j.ijpharm.2014.12.071.

Vercruysse, J., Córdoba Díaz, D., Peeters, E., Fonteyne, M., Delaet, U., Van Assche, I., De Beer, T., Remon, J. P., & Vervaet, C. (2012). Continuous twin screw granulation: Influence of process variables on granule and tablet quality. *European Journal of Pharmaceutics and Biopharmaceutics, 82*(1), 205–211. https://doi.org/10.1016/j.ejpb.2012.05.010.

Verstraeten, M., Hauwermeiren, D. V., Lee, K., Turnbull, N., Wilsdon, D., Ende, M., Doshi, P., Vervaet, C., Brouckaert, D., Mortier, S. T. F. C., Nopens, I., & De Beer, T. (2017). In-depth experimental analysis of pharmaceutical twin-screw wet granulation in view of detailed process understanding. *International Journal of Pharmaceutics, 529*(1-2), 678–693. https://doi.org/10.1016/j.ijpharm.2017.07.045.

Willecke, N., Szepes, A., Wunderlich, M., Remon, J. P., Vervaet, C., & De Beer, T. (2017). Identifying overarching excipient properties towards an in-depth understanding of process and product performance for continuous twin-screw wet granulation. *International Journal of Pharmaceutics, 522*(1-2), 234–247. https://doi.org/10.1016/j.ijpharm.2017.02.028.

Willecke, N., Szepes, A., Wunderlich, M., Remon, J. P., Vervaet, C., & De Beer, T. (2018). A novel approach to support formulation design on twin screw wet granulation technology: Understanding the impact of overarching excipient properties on drug product quality attributes. *International Journal of Pharmaceutics, 545*(1-2), 128–143. https://doi.org/10.1016/j.ijpharm.2018.04.017.

Yu, L. X., Amidon, G., Khan, M. A., Hoag, S. W., Polli, J., Raju, G. K., & Woodcock, J. (2014a). Understanding pharmaceutical quality by design. *AAPS Journal, 16*(4), 771–783. https://doi.org/10.1208/s12248-014-9598-3.

Yu, S., Reynolds, G. K., Huang, Z., De Matas, M., & Salman, A. D. (2014b). Granulation of increasingly hydrophobic formulations using a twin screw granulator. *International Journal of Pharmaceutics, 475*(1-2), 82–96. https://doi.org/10.1016/j.ijpharm.2014.08.015.

Zarmpi, P., Flanagan, T., Meehan, E., Mann, J., & Fotaki, N. (2017). Biopharmaceutical aspects and implications of excipient variability in drug product performance. *European Journal of Pharmaceutics and Biopharmaceutics, 111*, 1–15. https://doi.org/10.1016/j.ejpb.2016.11.004.

Zhang, D., Flory, J. H., Panmai, S., Batra, U., & Kaufman, M. J. (2002). Wettability of pharmaceutical solids: Its measurement and influence on wet granulation. *Colloids and Surfaces A, 206*(1-3), 547–554. https://doi.org/10.1016/S0927-7757(02)00091-2.

Zhao, Na, & Augsburger, L. L. (2006). The influence of granulation on super disintegrant performance. *Pharmaceutical Development and Technology, 11*(1), 47–53. https://doi.org/10.1080/10837450500463828.

Zupančič, Ož, Doğan, A., Martins Fraga, R., Demiri, V., Paudel, A., Khinast, J., Spoerk, M., & Sacher, S. (2023). On the influence of raw material attributes on process behaviour and product quality in a continuous WET granulation tableting line. *International Journal of Pharmaceutics, 642*, 1–10. https://doi.org/10.1016/j.ijpharm.2023.123097.

SECTION III

PAT, scale-up, control strategy

14. Inline focused beam reflectance measurement during wet granulation 471
15. Principles and applications of drag force flow sensor 509
16. An introduction to powder characterization 557
17. A quality by design approach to scale-up of high-shear wet granulation process 601
18. Integrated application of quality-by-design principles to drug product and its control strategy development 633
19. Implementation of pharmaceutical quality by design in wet granulation 669

CHAPTER 14

Inline focused beam reflectance measurement during wet granulation

Ajit S. Narang[a], Tim Stevens[b], Srinivasa Paruchuri[c], Kevin Macias[d], Zhihui Gao[b], Sherif I.F. Badawy[b], Dilbir Bindra[b] and Mario Hubert[d]

[a]*Small Molecule Pharmaceutical Sciences, Genentech, Inc., South San Francisco, CA, United States,*
[b]*Drug Product Science & Technology, Bristol-Myers Squibb Co., New Brunswick, NJ, United States,*
[c]*Appco Pharma, Somerset, NJ, United States,* [d]*Analytical and Bioanalytical Development, Bristol-Myers Squibb Co., New Brunswick, NJ, United States*

14.1 Introduction

The critical quality attributes (CQAs) of the drug product (DP) made using high-shear wet granulation (HSWG) tend to depend predominantly on the quality of the granulation (Badawy et al., 2014, 2015; Narang et al., 2015; Tao et al., 2015). At the same time, the HSWG proceeds with concurrent multiple mechanistic pathways (e.g., particle aggregation, consolidation, layering, breakage, and attrition [Iveson et al., 2001]) in a short time (2–4 min), and does not lend itself easily to sampling or in-process controls. The kinetics of these competing processes are influenced by the same variables, which can cause an unpredictable drift in the process that can affect product quality.

A quality-by-design (QbD) DP development paradigm calls for attribute-based process robustness assessment and systematic study of the impact of incoming and in-process material attributes and process parameters (Badawy et al., 2015). Conventionally, such QbD studies typically are carried out as statistical design-of-experiment (DoE) investigations, in which the effects of changes in input variables is assessed on the attributes of dried granules or compressed tablets (Badawy et al., 2012). The resolution of the measured responses with respect to the range of input variables is often low. For example, the impact of changes in granulation process parameters on the flow characteristics of the final blend might be low, even though the particle size distribution (PSD) of the granules might be affected significantly.

Enabling QbD of HSWG requires a process analytical technology (PAT) tool that can provide rapid and reproducible inline assessment of a granule attribute of relevance–PSD–with a pharmaceutically relevant quantitative range and that correlates with conventional (nested sieve analyses of dried granules) and other PSD measurement techniques. Measurement of a granule attribute that is independent of equipment design and scale also can help during scale-up, technology transfer, equipment changes, and for process monitoring and control. Focused beam reflectance measurement (FBRM) has been proposed as one such tool (Arp et al., 2011; Kukec et al., 2014; Kumar et al., 2013).

Several PATs have been studied for HSWG, including power consumption, near-infrared spectroscopy, Raman spectroscopy, capacitance, microwave measurements, imaging, spatial filter velocimetry, stress and vibration measurements, acoustic emissions, and FBRM (Hansuld & Briens, 2014;

Huang et al., 2010). Of all these technologies, FBRM stands out as the one that directly reports the PSD of the granules in an independent equipment manner. While FBRM has found widespread application in crystallization (Saleemi et al., 2012), precipitation (Xu et al., 2013), and continuous manufacturing applications, HSWG has not had much success with this probe except for one reported study (Huang et al., 2010). The use of a wiper blade in front of the sapphire window in the Mettler Toledo (formerly Lasentec) C35 probe improves the quality and reproducibility of data collected through this probe.

In this chapter, systematic work highlighting the resolution and sensitivity of chord length distribution (CLD) reported by FBRM C35 probe is reported in pharmaceutically relevant size ranges using common excipients. The mechanistic basis of the effect of impeller tip speed and water on the measured CLD and its correlation to other particle sizing techniques is reported. In addition, the application of FBRM probe to measure the inline PSD during HSWG processing for a placebo and a high drug load active formulation are exemplified.

14.2 Focused beam reflectance measurement probe setup and operation

Setup of the FBRM C35 probe is critical to reproducible data collection. This section describes optimum positioning, orientation, and operational aspects of the FBRM C35 probe in an HSG (Fig. 14.1) that were found through several studies to be critical to ensuring reproducible data quality.

14.2.1 Position

The probe should be held firmly using a compression adapter to attach the probe to the granulator lid and the probe held firmly in the powder bed (Fig. 14.1). The probe should be placed away from the chopper because the chopper interrupts the flow of particles to the probe tip, creating a noisy signal. The probe also should be placed away from water addition port because the water droplets are measured by the probe and combined with the data collected from the granules. This results in data that is generally not reproducible and does not correlate with process changes. The probe should be placed well inside the resting powder bed, but no lower than 1 in. above the impeller. It is important that the probe wiper does not come into contact with the impeller and that powder always covers the tip of the probe. The probe should be placed as far to the edge of the bowl (away from the center) as possible based on the roping-type mixing behavior of the powder bed in an HSG. Placing the probe close to the edge of the bowl ensures that as the powder bed ropes through the probe. The probe should be inserted from the top of the bowl and perpendicular to the impeller (parallel to the rotor shaft). It is not necessary to angle the probe toward the flow of particles, as is sometimes done for measurement in slurries.

14.2.2 Orientation

When placing the probe into the granulation, it is important that the scraper is parallel to the impeller as it passes below the scraper. Fig. 14.1A shows the proper orientation of the scraper with respect to the impeller.

14.2 Focused beam reflectance measurement probe setup and operation

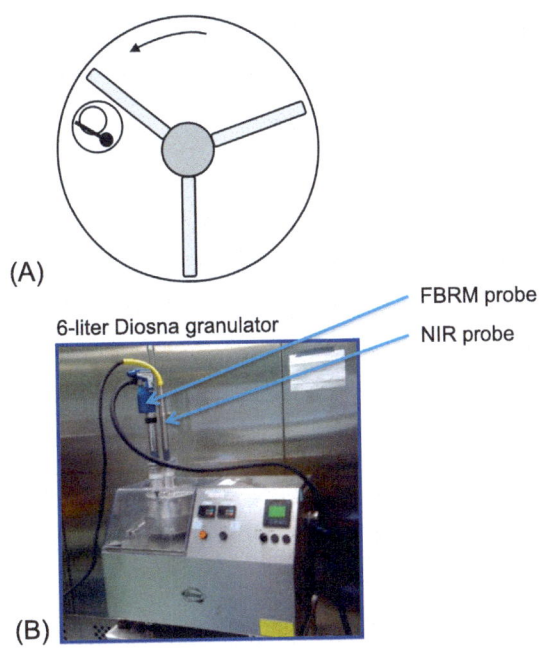

FIGURE 14.1

Focused beam reflectance measurement (FBRM) C35 probe setup with (A) recommended orientation of the probe wiper in a high-shear granulator and probe insertion from the lid in a (B) 6-L high-shear granulator (HSG). The high-shear granulator picture shows concurrent use of the FBRM probe and the near infrared (NIR) probe (data not discussed in this manuscript).

Reproduced with permission from Narang et al. (2016).

The scraper also has a resting or home position that should be chosen such that the measurement window of the probe is not blocked by the scraper. The proper scraper home position also is shown in Fig. 14.1A. This scraper orientation was chosen after experimenting with different positions in different sizes of granulators. The orientation shown in Fig. 14.1A provided the most consistent and lowest noise data between batches and granulators.

The optical cable connecting the probe to the control box should be positioned such that it does not rotate the probe and alter its orientation during the measurement process.

14.2.3 Operation

The FBRM C35 probe should be operated in the high-shear granulator only with the scraper working properly (cleaning the quartz window thoroughly) and be on the side in the resting position, such that the laser window is completely unobstructed. As shown in the equipment setup Fig. 14.1A, the wiper resting position should be radial with the impeller blades and away from the direction of flow of the powder bed. The location of water addition port should be after or downstream of the C35 probe–considering powder flow direction–to ensure that the incompletely mixed water does not affect C35 data. The position of

Table 14.1 Particle size distribution (PSD) of different grades of microcrystalline cellulose (MCC) by Malvern laser diffraction, measured prior to mixing for $n = 3$ samples.

Diameter in μm for $n = 3$	d (0.1)	d (0.5)	d (0.9)
Avicel PH 101	20.0 ± 0.2	52.0 ± 0.7	122.0 ± 7.6
Avicel PH 102	35.6 ± 1.1	114.6 ± 1.4	242.6 ± 3.4
Avicel PH 200	61.5 ± 50.5	205.5 ± 17.6	399.8 ± 37.2

The table lists mean and standard deviation of the measurements.

the probe within the granulator (in terms of angular distance and up/downstream from the chopper and the fluid addition port) and the orientation of the resting position of the scraper then should be fixed for all batches being compared. The probe position also should be consistent for the depth from the lid across all the batches being compared. For example, 9.0-in. depth from top surface of a custom-made plastic lid was found adequate with 40% v/v bowl filled with dry powder for a 6-L Diosna granulator. Similarly, 11.25-in. depth from top surface of a custom-made plastic lid was found adequate with 40% v/v bowl filled with dry powder for a 10-L PMA granulator. The powder bed height changes during the process of granulation, especially upon addition of water. The probe height should be selected such that the probe is always dipped within the powder bed. Good and reproducible results are obtained when the wiper is operated back and forth, twice, every 5 s with the data collected every 2 s.

14.3 Understanding chord length distribution

The FBRM operates by the principle of a rotating laser (Fig. 14.2B) that detects particle size as a chord length (Fig. 14.2C), depending on the location of laser beam reflection with the particle and the threshold of variation in signal input that is considered a particle by the software (Fig. 14.2D). The FBRM chord length is always a distribution even for uniform-size spherical particles (Fig. 14.2E) because the probe laser invariably scans multiple particles of the same size at different locations.

The concept of CLD is best understood using powders with known PSDs measured by different techniques. Particle sizes of pharmaceutical interest within the operating environment of a high-shear granulator (HSG) are exemplified by the different particle size grades of microcrystalline cellulose (MCC): Avicel PH 101, 102, and 200. The differences in the PSDs of these grades of MCC (Avicel PH 101 < Avicel PH 102 < Avicel PH 200) have been shown by offline FBRM D600 probe (no wiper) CLD data in acetone suspension (Fig. 14.3), laser diffraction (Table 14.1), Sympatec QicPic (Table 14.2), nested sieve analyses (Fig. 14.4), and Morphologi G3 (Gamble et al., 2011).

When these grades of MCC were placed separately in a Diosna 6-l HSG and CLD data was collected at dry mixing impeller tip speeds of 3, 4, 5, 6, and 7 m/s (corresponding to 230, 310, 380, 460, and 530 rpm, respectively) (Narang et al., 2016), the CLDs showed discernable differences between the three powder grades. A comprehensive 3-dimensional plot of all-encompassing data is shown in Fig. 14.5, while the data reported in Figs. 14.6 and 14.7 were collected by the FBRM probe at the end of the respective period of dry mixing for the noted blade speed. The data reported in Fig. 14.3 is from samples collected before and after high-shear mixing of the dry powder for 14 min following the above

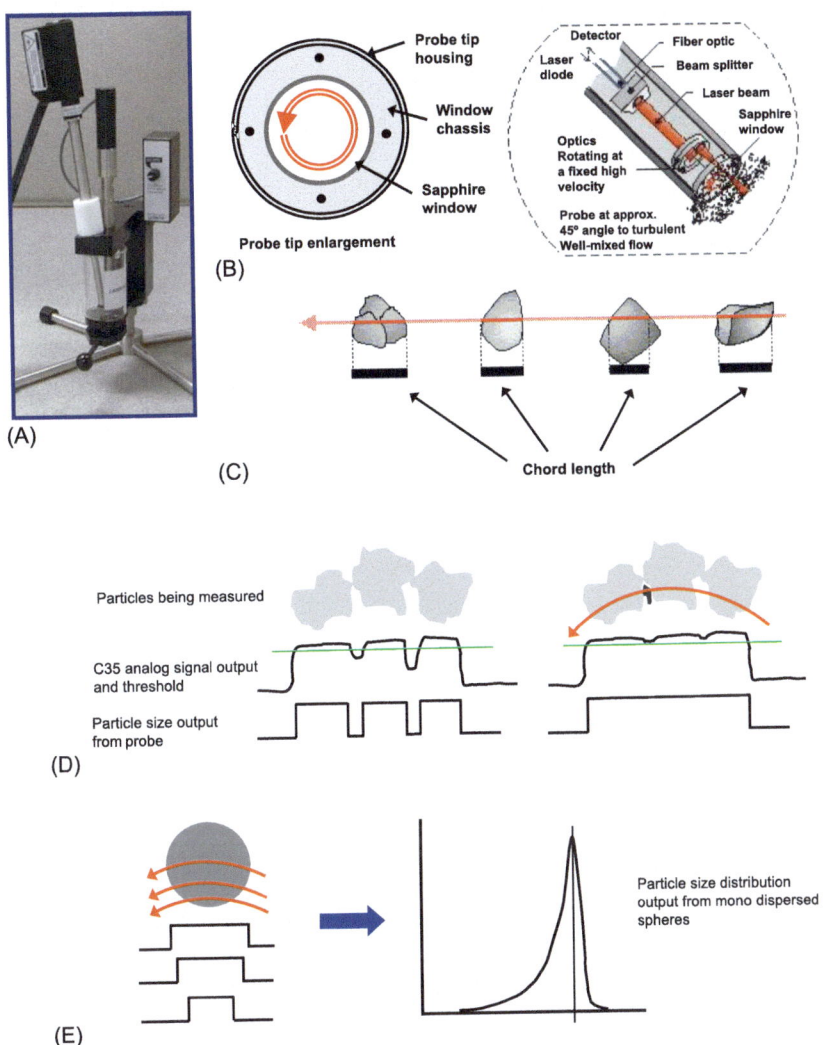

FIGURE 14.2

Focused beam reflectance measurement (FBRM) instrumentation (A) and principle of operation, demonstrating (B) a spinning laser beam that measures chord length (C) of particles at the focal point in front of a sapphire window and categorizes them into discrete particle sizes based on threshold of analog signal output (D) to provide a chord length distribution as an output. It is to be noted that even a mixture of homogeneous, noncohesive spherical particles would generate a distribution profile of chord lengths (E) because of different regions of focus of the flowing particles in front of the probe window. Thus, chord length distribution is a function both of measurement parameters and the particle size distribution of the bulk powder.

Reproduced with permission from Narang et al. (2016).

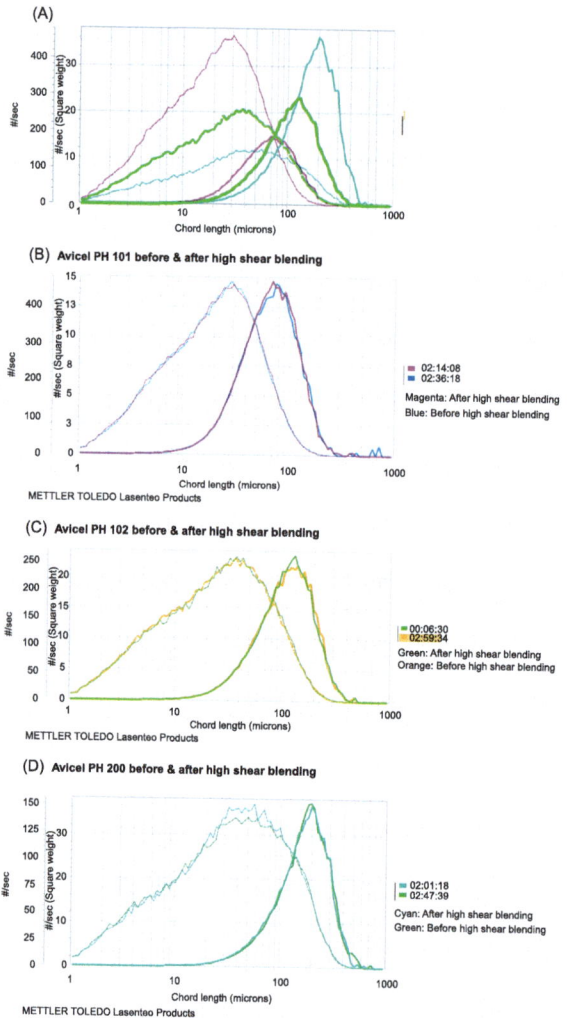

FIGURE 14.3

Resolution of particle size differences among different grades of microcrystalline cellulose (MCC)–Avicel PH 101, Avicel PH 102, and Avicel PH 200–by FBRM D600 probe during mixing as a dispersion in acetone at 3.3% w/v concentration (A). This figure demonstrates differences in the chord length distribution (CLD) of the measured samples as a function of number density of particles without or with square weighing. These data correlate with the findings of the FBRM C35 probe in the high-shear granulator (HSG) (Fig. 14.6), confirming the validity of FBRM as a useful tool for CLD analyses in high-shear wet granulation. In addition, offline Lasentec FBRM D600 probe measurements of samples before and after shearing in a high-shear mixer for Avicel PH 101 (B), Avicel PH 102 (C), and Avicel PH 200 (D) indicated that the extent of shear forces encountered in the high-shear granulator did not affect particle integrity.

Reproduced with permission from Narang et al. (2016).

Table 14.2 Particle size distribution (PSD) of different grades of microcrystalline cellulose (MCC) by Sympatec QicPic, measured prior to mixing for $n = 3$ samples.

Diameter in μm for $n = 3$	$d\,(0.1)$	$d\,(0.5)$	$d\,(0.9)$
Avicel PH 101	36.3 ± 0.6	74.8 ± 0.4	129 ± 1.7
Avicel PH 102	50 ± 0.2	125.1 ± 0.4	201.3 ± 54
Avicel PH 200	69.3 ± 1.8	193.0 ± 4.9	318.7 ± 15.7

The table lists mean and standard deviation of the measurements.

protocol for different blade speeds. The data reported in Fig. 14.4 is from samples collected before and after high-shear mixing of the dry powder for 14 min following the above protocol for different blade speeds; and mixing through the wet massing time with the same protocol after water addition.

The inline CLD data generated using the FBRM C35 probe in the Diosna HSG at 310 rpm impeller tip speed and dry MCC (Fig. 14.6) was able to clearly distinguish between the three different grades of MCC. These data are plotted as both unweighted (or numerically weighted) and square weighted number density of particles at recorded chord lengths. Empirically, square weighted PSD seems to correlate better with mass distribution of particles (e.g., by nested sieve analysis). The numerically weighted data for the three grades of MCC shows significantly different results than square weighted data because more volume is tied up in agglomerated particles.

Reported particle size of a sample could be weighed differently depending on the method of measurement and reporting. Particle sizes could be expressed as number weighted (by counting-based methods such as microscopy), length weighted (by reporting length of particles), square or surface area weighted (by reporting surface area of equivalent spheres, e.g., planar vision analysis systems), volume weighted (by expressing volumes of equivalent spheres, e.g., using laser diffraction analysis), or mass weighted (by reporting total mass of all particles in a given size fraction, e.g., by sedimentation and nested sieve analysis methods). Length and surface-weighted distributions are not common. Different weighted distributions are not comparable to one another unless they are transformed mathematically. For example, volume distribution can be transformed into mass distribution by incorporating density of particles.

FBRM generates number distribution of particles by the chord length. The number distribution can be transformed by square or cube weighting, which results in the shifting of the distribution to the right (as seen in Figs. 14.6, 14.7, and 14.3). Most conventional particle sizing methods, such as dynamic light scattering and nested sieve analysis, inherently report results that are weighted higher than number weighting. The weighting of larger size particles increases as the results are reported in square or cube weighted manner. Square weighting of the number distribution data generated by FBRM was adopted to provide a relative balance of emphasis of different size particles, while the data also show comparability to conventional particle size measurement and reporting methods.

The effect of process shear on the integrity MCC particles was investigated to verify the ability to use MCC with continuous mixing in an HSG to assess reproducibility of CLD measurements. The effect of high-shear mixing on the PSD of MCC particles was assessed by offline CLD measurement in a suspension in a nonaqueous solvent using the Lasentec D600 probe (Fig. 14.3) and offline nested sieve analyses (Fig. 14.4).

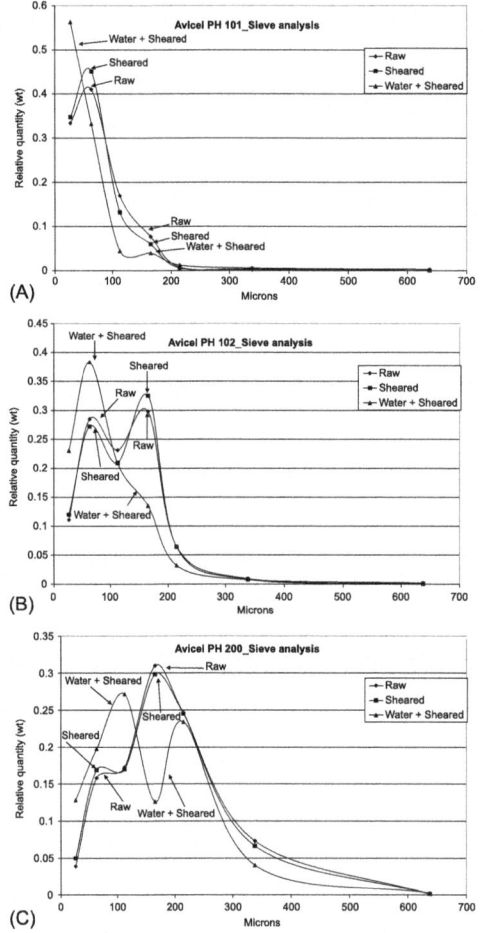

FIGURE 14.4

Effect of mixing in a high-shear granulator (HSG) with or without the addition of water on the particle size distribution (PSD) of different grades of microcrystalline cellulose (MCC): (A) Avicel PH 101, (B) Avicel PH 102, and (C) Avicel PH 200. The three grades of MCC were mixed in a high-shear granulator (HSG) as listed in the experimental section and the PSD determined by nested sieve analyses before and after mixing with or without the addition of water. Significant change in the PSD of MCCs was noted in smaller particle ranges.

Reproduced with permission from Narang et al. (2016).

In addition, PSD of MCC samples before and after sharing in an HSG by nested sieve analysis did not show any significant change (Fig. 14.4). These results indicate that even though HSWG process exhibits high enough shear to affect granule consolidation, attrition, and breakage, it did not break apart MCC particles.

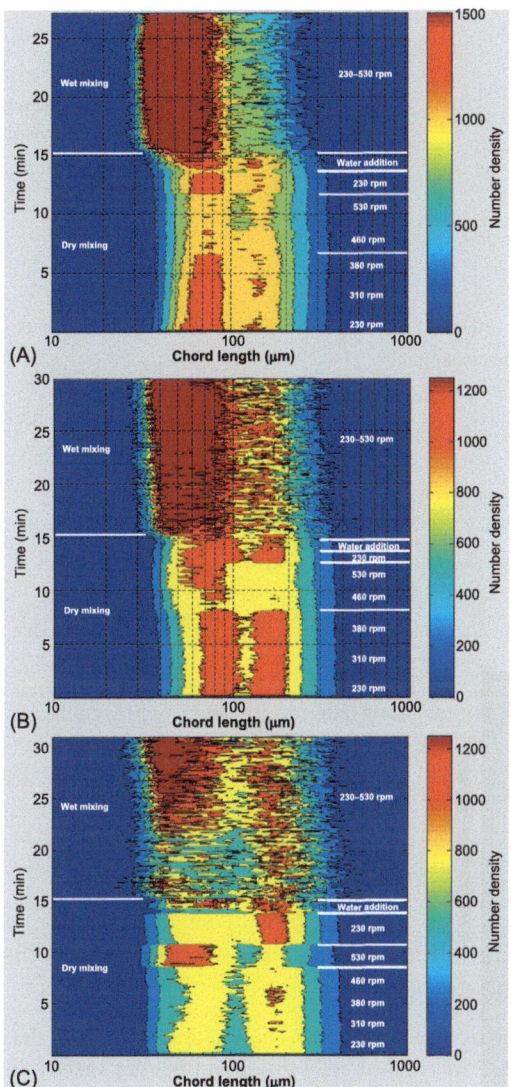

FIGURE 14.5

Effect of mixing in a high-shear granulator (HSG) with water addition on the chord length distribution (CLD) of different grades of microcrystalline cellulose (MCC): (A) Avicel PH 101, (B) Avicel PH 102, and (C) Avicel PH 200. The CLD of the measured samples was plotted as a function of number density of particles and time. The CLD profile of the particles reduced upon water addition.

Reproduced with permission from Narang et al. (2016).

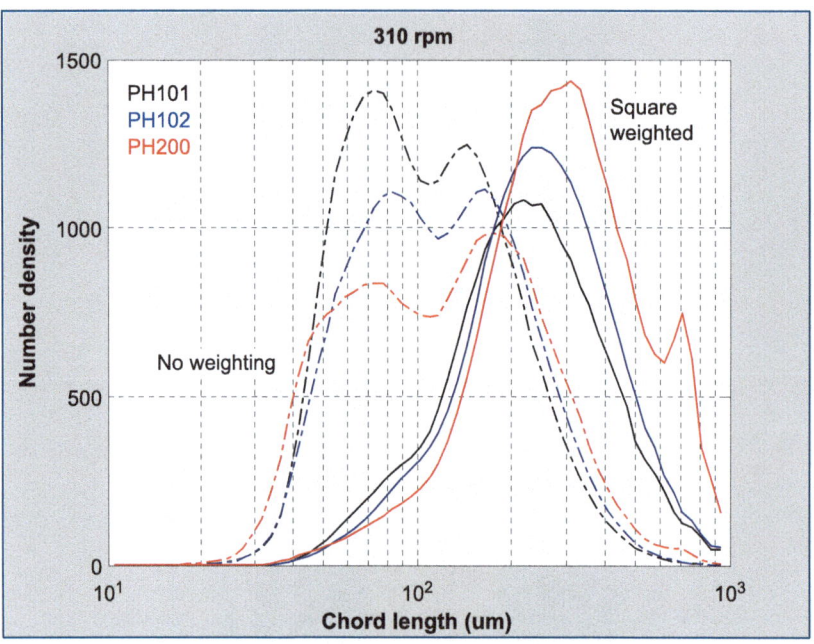

FIGURE 14.6

Resolution of particle size differences among different grades of microcrystalline cellulose (MCC)–Avicel PH 101, Avicel PH 102, and Avicel PH 200–by FBRM C35 probe during mixing in a high-shear granulator (HSG). This figure demonstrates differences in the chord length distribution (CLD) of the measured samples as a function of number density of particles without or with square weighing.

Reproduced with permission from Narang et al. (2016).

14.3.1 Effect of impeller speed on the inline chord length distribution of dry microcrystalline cellulose

To understand the effect of process parameters on analytical sensitivity of the measurement system, CLD of the same batch of dry MCC was measured at different impeller speeds. Figs. 14.7 and 14.5 show the number density plots of Avicel PH 101, PH 102, and PH 200 particles measured at different impeller speeds. Impeller speed is seen to influence the particle number density but not the distribution profile of the chord length (Figs. 14.7 and 14.5). The effect of impeller tip speed on the particle number density measurement is likely because of increased fluidization of particles at higher impeller speeds, leading to more particles being measured at the focal point of the laser in the FBRM probe. Changes in the tip speed also could lead to other potentially confounding changes in powder bed properties, such as changes in bed speed and bed height, which not only could affect particle density in the air but also could differentially affect the particle density for primary versus agglomerated particles. The fact that the distribution of chord lengths remained the same with increased impeller speed and particle fluidization indicated consistent and adequate sampling of the powder bed mass by the probe.

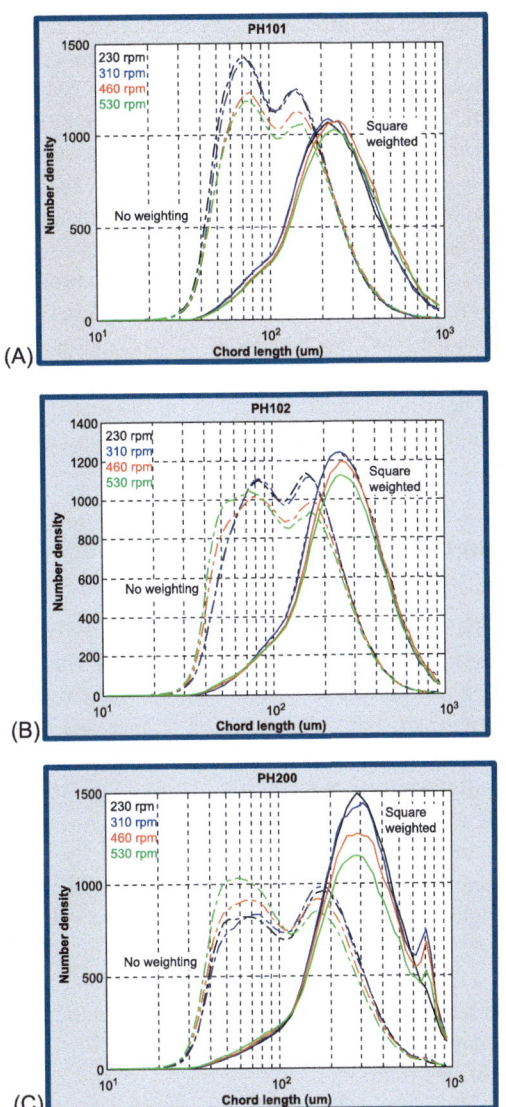

FIGURE 14.7

Effect of mixing in a high-shear granulator (HSG) at different impeller tip speeds on the chord length distribution (CLD) of different grades of microcrystalline cellulose (MCC): (A) Avicel PH 101, (B) Avicel PH 102, and (C) Avicel PH 200. The CLD of the measured samples was plotted as a function of number density of particles without or with square weighting. The CLD profile of the particles did not seem to change with the impeller tip speed. The powder mass seemed to increase in number density of particles, while maintaining the same CLD profile of the particles. This could be because of increased fluidization of particles in front of the probe window.

Reproduced with permission from Narang et al. (2016).

The effect of impeller tip speed on the number density of chord lengths was more for the unweighted (dotted lines) than for the square weighted (solid lines) particle number density data. Also, although the apparent CLD was affected by impeller tip speed for all grades of MCC, this effect was minimal up to 310 rpm. These results indicate that the FBRM CLD data are relatively robust at the impeller tip speed used conventionally in wet granulations. Nevertheless, these studies also indicated that greatest analytical method robustness is achieved when the batches used for the comparison of FBRM data are manufactured at the same impeller tip speed and comparisons are made using the square weighted data.

PSD of MCC was shown to be unaffected by high-shear processing conditions in the dry state. Using MCCs of different particle sizes, reproducibility of the probe was established. Impeller tip speed affected the number density of CLD, but not the shape of the CLD itself. This was likely because of change in the fluidization of the powder in the HSG. Such changes in fluidization also can be expected during the progress of the granulation, such as during the addition of water and consolidation, when granule size, density, and mass increases. Therefore, one should focus on the distribution of chord lengths, not the absolute number density, when interpreting monitoring the wet granulation processes using FBRM.

14.3.2 Effect of water on chord length distribution of dry microcrystalline cellulose

The ability to use inline FBRM for monitoring HSWG depends on the ability of the probe to reproducibly measure CLDs both in the presence and absence of water and being able to delineate the impact of water addition on powder particles. Addition of water to dry MCC particles, which can absorb a significant amount of water, could be expected to increase the CLD of the powder in the HSG. Addition of water to the MCC particles in the high-shear mixer with continuous mixing, however, showed reduction in the CLD after addition of water (Fig. 14.5). This data was consistent with the offline PSD measured by nested sieve analyses after drying (Fig.14.4).

To better understand the interaction of water with MCC in the HSG leading to reduction in CLD/PSD, with prior literature on the particles in the three grades of MCC indicating that the larger size MCC particles in higher particle size grades of Avicel PH 102 and PH 200 were likely agglomerated primary particles (Gamble et al., 2011), the samples were analyzed by nested sieve analyses. The nested sieve analyses data indicated a tendency toward bimodal distribution for Avicel PH 102 and 200 (Fig. 14.4). In the absence of a binder, the addition of 30% w/w water under high-shear mixing apparently deagglomerated MCCs to primary particles, thus shifting the CLD (Figs. 14.7 and 14.5)/PSD (Fig.14.4) profile toward the finer particle size fraction.

Addition of water to dry MCC generally would be expected to increase CLD because of water sorption and increase in primary particle size, as well as potential bridging and agglomeration of particles. The results of this study, however, showed reduction in the CLD of MCC upon addition to water. This reduction was more apparent for the larger particle size grades of MCC. These results were confirmed with offline measurement by orthogonal techniques such as nested sieve analyses. Although initially surprising, further microscopic investigation pointed these findings to potential deagglomeration of primary particles in the larger particle size grades of MCC in the presence of water. These findings underscore the need for a hydrophilic polymer (binder) in wet granulation processes to effect particle agglomeration and particle size growth, which is conventionally established in the art of wet granulation.

Table 14.3 Experimental design to study the impact of binder concentration and water concentration on a placebo formulation in a high-shear granulator.

C35 Experiment Matrix in Diosna 6-L Granulator

Water	HPC 2%	3%	4%	5%
65%		✓		
75%	✓	✓	✓	✓
85%		✓		

HPC, *hydroxypropyl cellulose*.

14.4 Granulations that predominantly increase the particle size

A hydrophilic polymer, called a binder, typically is used in pharmaceutical wet granulation, which causes the aggregation of fine powder mix to larger size granules. An inline CLD probe should be capable of delineating increase in particle size in the presence of a binder and should be responsive to the changes in the process trajectory depending on the concentration of the binder and the granulating fluid (water).

For the granulation of MCC with hydroxypropyl cellulose (HPC) as a binder, the pharmaceutical optimum high-shear granulation conditions were determined to be 75% w/w water concentration and 3% w/w HPC concentration. Formulation and process conditions that represent a pharmaceutical optimum desired process for the granulation of a single excipient, MCC, with HPC as a dry powder binder was determined based on conventional and empirical wet granulation process development methodologies. The parameters taken into consideration for these assessments included visual observation of completeness of granulation, absence of dry material in the bowl, brittle self-association of the wet mass when compressed in a fist, and change in the PSD (viz., reduction of fines) of the dried granules compared to the dry powder premix used for granulation. These conditions were consistent with previous studies that produced dense granules of larger particle size (Habib et al., 1999).

MCC (Avicel PH 101) was granulated with HPC (Klucel EXF grade) added as a dry powder, and water as the granulating fluid at different concentrations of water and HPC (Table 14.3). Batches were manufactured with different water and HPC concentrations to produce realistic changes in the granulation process. Three batches were manufactured at 65% w/w, 75% w/w, and 85% w/w water concentration and fixed 3% w/w HPC concentration. The other three batches were manufactured at 2% w/w, 4% w/w, and 5% w/w HPC concentration and fixed 75% w/w water concentration. These batches were manufactured in a 6-L (DIOSNA Dierks & Söhne GmbH, Osnabrück, Germany) HSG at an impeller speed of 4.8 m/s and low chopper speed (1750 rpm) at 40% v/v fill volume. The 6-L batch was manufactured with dry mixing time of 4 min, water addition time of 3 min, and wet massing time of ~20 min.

The CLD was measured by inline FBRM probe to capture differences between the placebo batches manufactured using different concentrations of the binder (HPC) at fixed water concentration (Fig. 14.8) and for the batches manufactured at different water concentrations at fixed concentration of the binder (Fig. 14.9). These results were compared with the results of conventional (nested sieve) analyses of

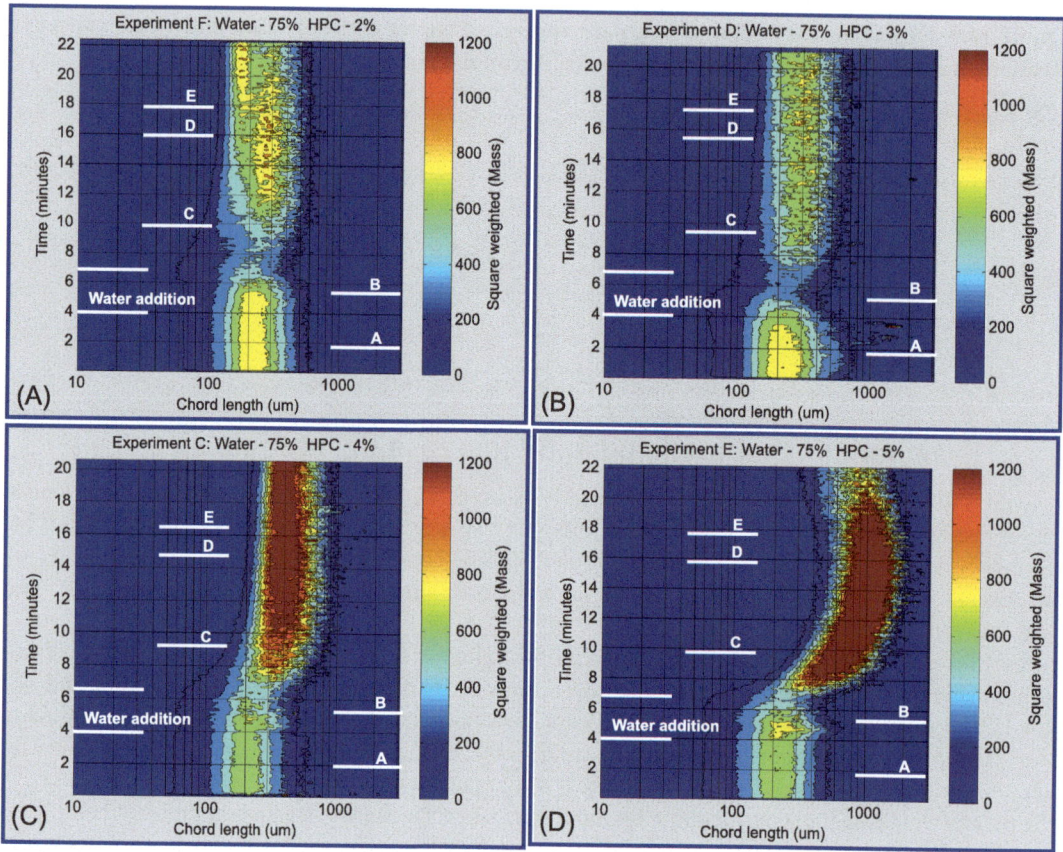

FIGURE 14.8

Sensitivity of chord length distribution (CLD) measured by FBRM C35 probe to different concentrations of the binder, hydroxypropyl cellulose (HPC) (A, 2%; B, 3%; C; 4%; D, 5%) in a placebo formulation. The results are presented as a three-dimensional plot of CLD as a function of time (I); CLDs at different times marked in I and color coded at 1.5, 5, 9, 15, and 21 min in black, blue, red, green, and purple, respectively (II); and change in the square weighted mass of particles in the selected size ranges as a function of time during granulation (III).

dried granules (Fig. 14.10). The reproducibility of the inline FBRM process monitoring capability was assessed at the center point of the placebo granulation (Fig. 14.11).

14.4.1 Effect of Binder Concentration on the Chord Length Distribution Profile

The inline CLD profiles for the wet granulation process were distinct for the batches manufactured using different concentrations of the binder, HPC, in the dry powder mix, while using a constant water concentration (75% *w*/w) (Fig. 14.8).

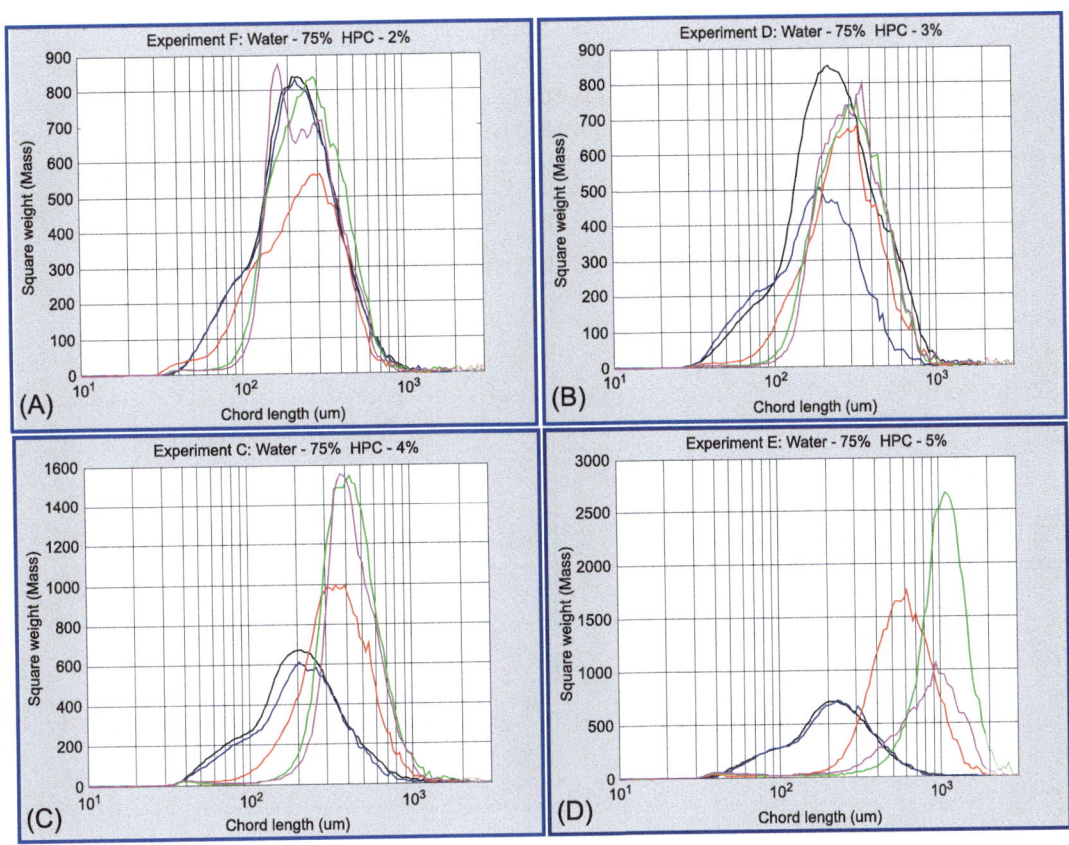

FIGURE 14.8, cont'd.

These binder concentrations were chosen based on a prior assessment of the optimum binder concentration (viz. 4% w/w) for this formulation, at a fixed water concentration (75% w/w). Manufacture of batches below (2% w/w and 3% w/w) and above (5% w/w) this optimum binder concentration was expected to lead to undergranulation and overgranulation, respectively. At lower binder concentration (2% w/w and 3% w/w), the PSD of the dried granules by nested sieve analyses showed greater proportion of the smallest size particles (Fig. 14.10A). At optimum binder concentration (4% w/w), the PSD of dried granules exhibited clear aggregation of powder mass to form larger granules–the peak of dried granule particle size is in the relatively larger size range (Fig.14.10A). At the highest binder concentration (5% w/w), the PSD of dried granules exhibited clear aggregation of powder mass to form large granules with bimodal PSD–with a peak of much larger particle size granules in addition to the one similar to that obtained with optimum particle size granulation (Fig. 14.10A). These results were consistent with the visual observations of the powder bed after granulation.

At the optimum binder concentration (4% w/w) (Fig. 14.8-I/II/IIIC), a three-dimensional plot of all CLD data collected during the granulation (Fig. 14.8-IC) shows a significant reduction in the proportion of fines (deep blue curve to the left) and a modest increase in the mean size of the granulation (shift of

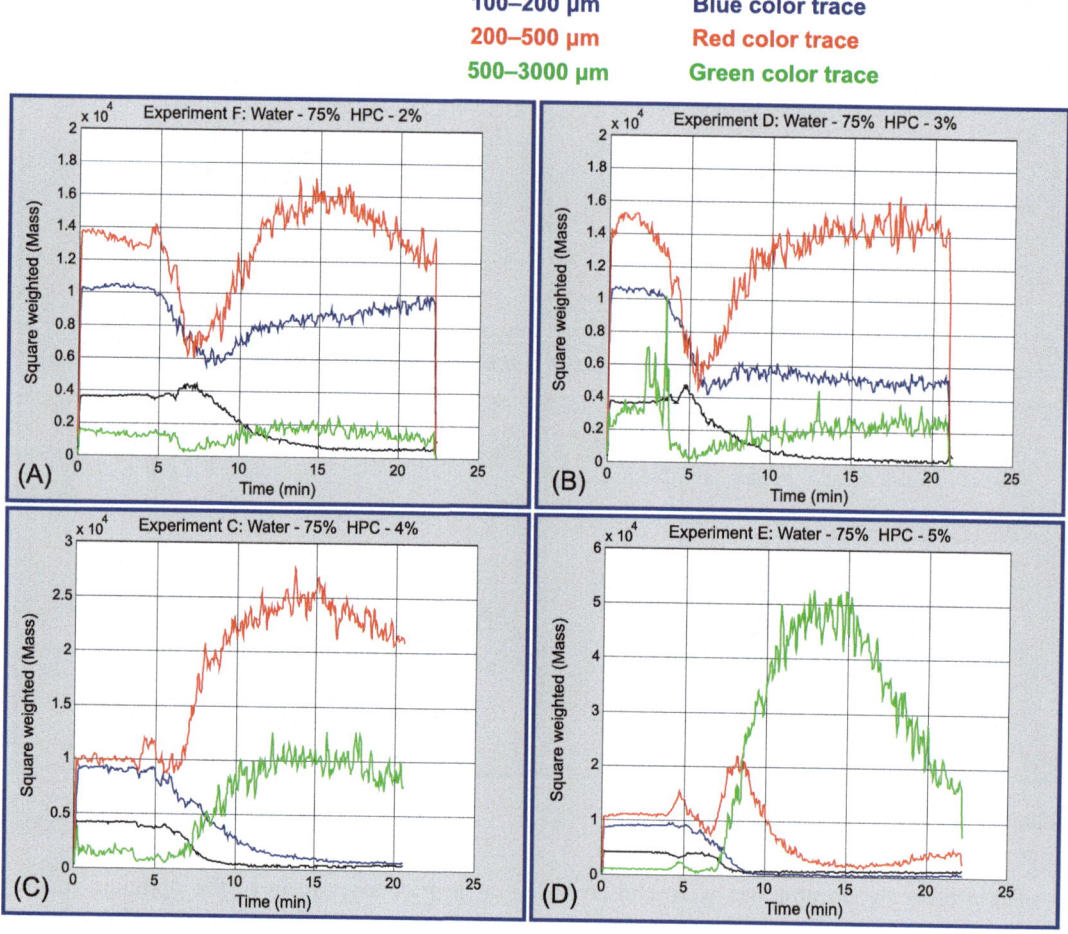

FIGURE 14.8, cont'd.

the peak region), after the addition of water (compared to before the addition of binder). The reduction of fines is more clearly evident in the CLDs plotted at different time points (Fig. 14.8-IIC) before, during, and after the binder addition phases, showing peak shift to higher chord lengths for the samples after water addition and reduction in the proportion of powder mass below the chord length of 200 μm. Similarly, tracing the changes in the square weighted mass of a given size fraction of the granules over the duration of granulation shows reduction in the black and the blue traces (Fig. 14.8-IIIC) after water addition, representing reduction of powder mass in the 0 to 100 μm and 100 to 200 μm chord length range, respectively; and increase in the size of particles in larger particle size range (red trace representing 200–500 μm and green trace representing 500–3000 μm).

14.4 Granulations that predominantly increase the particle size

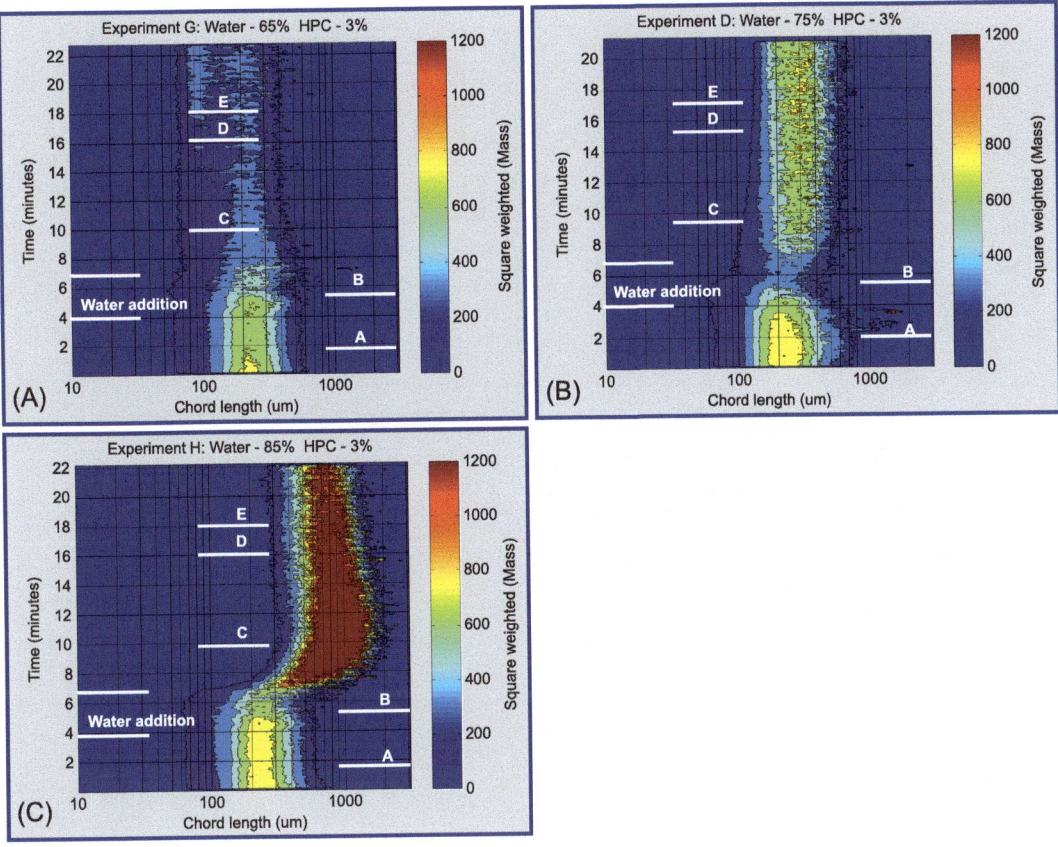

FIGURE 14.9

Sensitivity of chord length distribution (CLD) measured by FBRM C35 probe to different concentrations of the granulating fluid, water (A, 65%; B, 75%; C, 85%) in a placebo formulation. The results are presented as a three-dimensional plot of CLD as a function of time (I); CLDs at different times marked in I and color coded at 1.5, 5, 9, 15, and 21 min in black, blue, red, green, and purple, respectively (II); and change in the square weighted mass of particles in the selected size ranges as a function of time during granulation (III).

At 5% w/w binder concentration (Fig. 14.8-I/II/IIID), the CLD profile changes dramatically toward the end of binder addition with respect to reduction of fines (evident by the reduction of the lowest chord length regions of the size distribution) and increase in the overall CLD of particles (evident in the shift of the granulation toward larger chord lengths) (Fig. 14.8-ID). The reduction of fines is clearly distinguishable in the plot of CLDs at certain time points and before, during, and after the granulation (Fig. 14.8-IID), with the particles forming a distinct peak at about 200 μm chord length, which disappears and forms into larger peaks at about 600 μm or higher chord lengths. The line plot of different size fractions of particles in the powder mix (Fig. 14.8-IIID) through the process of water addition and into wet massing phase clearly shows both loss of fines (black and blue traces representing

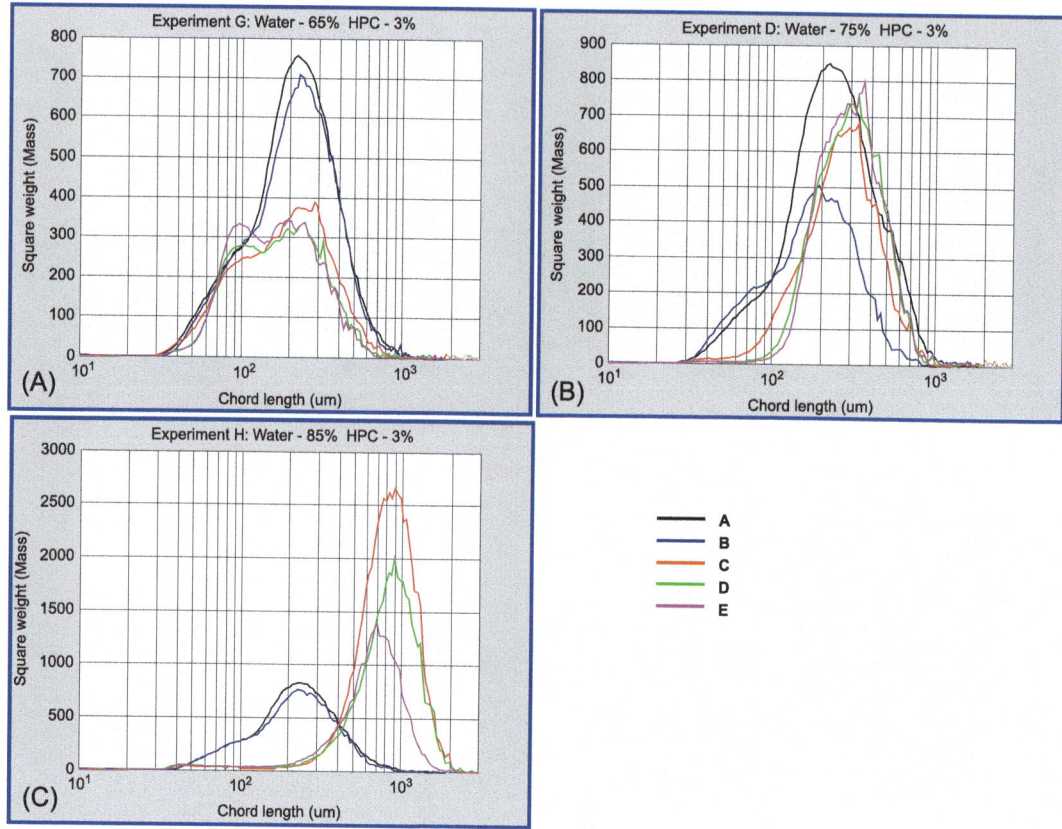

FIGURE 14.9, cont'd.

0–100 μm and 100–200 μm chord length particles, respectively) and increase in the proportion of large size granules, notable the green trace of 500 to 3000 μm chord length particles. The extent of granulation seen with 5% HPC concentration is increased significantly the particles of larger size range while being equivalently effective as the 4% HPC concentration in reducing the proportion of smaller particles after water addition.

In contrast, the use of lower binder concentrations (2% w/w HPC in Fig. 14.8-I/II/IIIA and 3% w/w HPC in Fig. 14.8-I/II/IIIB) during granulation showed progressively lower reduction of fines, while not showing any significant increase in the size of coarse particles. These data indicated that CLD was able to effectively delineate the expected impact of HPC concentration on both aspects of wet granulation: increase in the size of coarse particles or granules and decrease in the proportion of fines or smaller sized particles. Thus, inline FBRM not only is able to delineate changes in the powder mass through the course of granulation, but it also clearly distinguished the effect of binder concentration on the CLD of the wet mass in a placebo granulation.

14.4 Granulations that predominantly increase the particle size

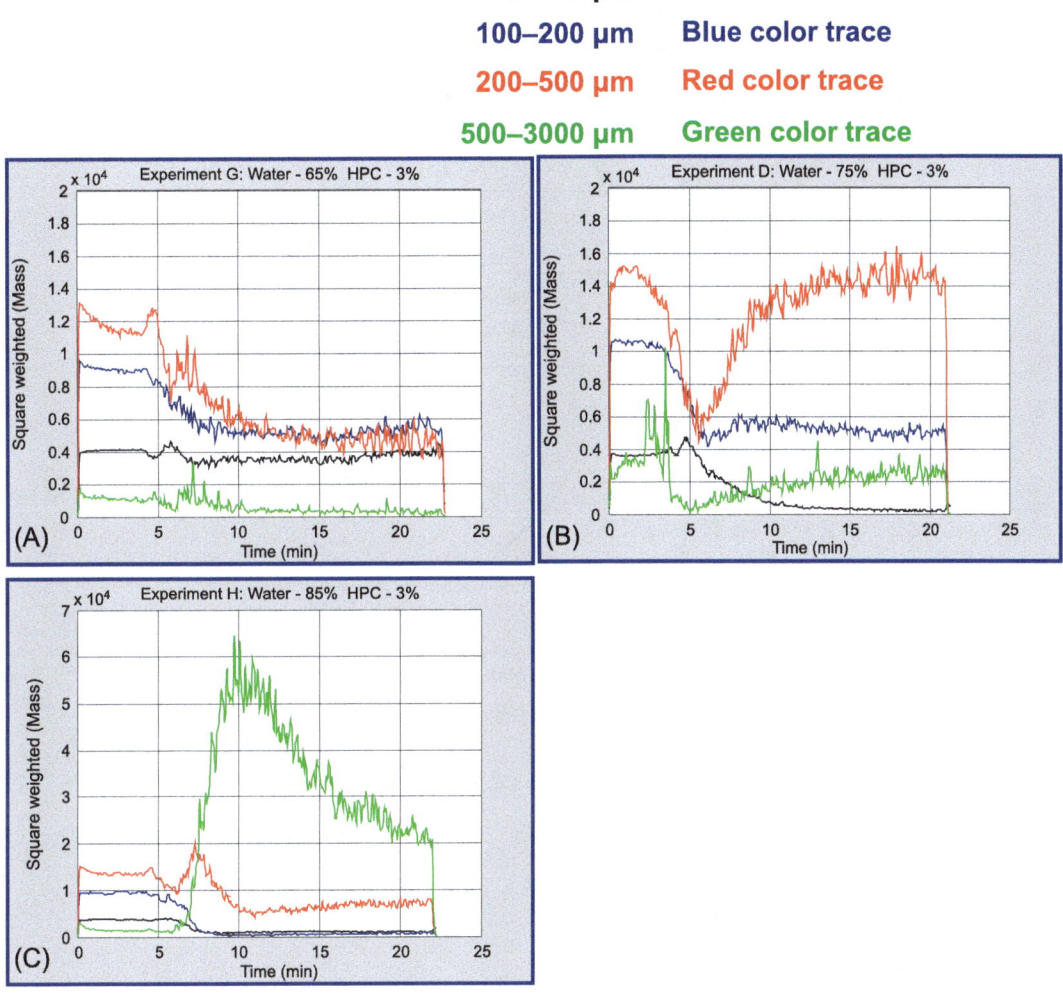

FIGURE 14.9, cont'd.

14.4.2 Effect of water concentration on the chord length distribution profile

The inline CLD profiles through the dry mixing, water addition, and wet massing phases of granulation were distinct for batches manufactured with low (65% w/w), medium (75% w/w), and high (85% w/w) water concentrations (Fig. 14.9). These water concentrations were chosen based on a prior assessment of the optimum water concentration for this formulation. The optimum water concentration for this placebo formulation was defined through visual observation of completeness of granulation, absence of dry material in the bowl, brittle self-association of the wet mass when compressed in a fist, and change in the PSD (viz., reduction of fines) of the dried granules compared to the dry powder premix

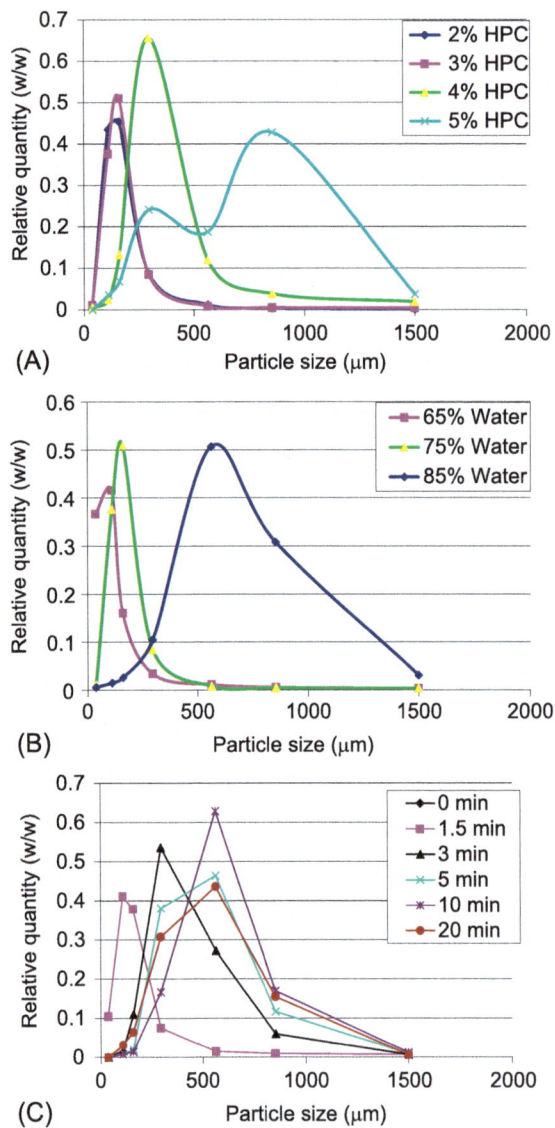

FIGURE 14.10

Particle size distribution (PSD) by nested sieve analyses of batches made with different concentrations of hydroxypropyl cellulose (HPC) (A), water (B), or wet massing time (C).

14.4 Granulations that predominantly increase the particle size

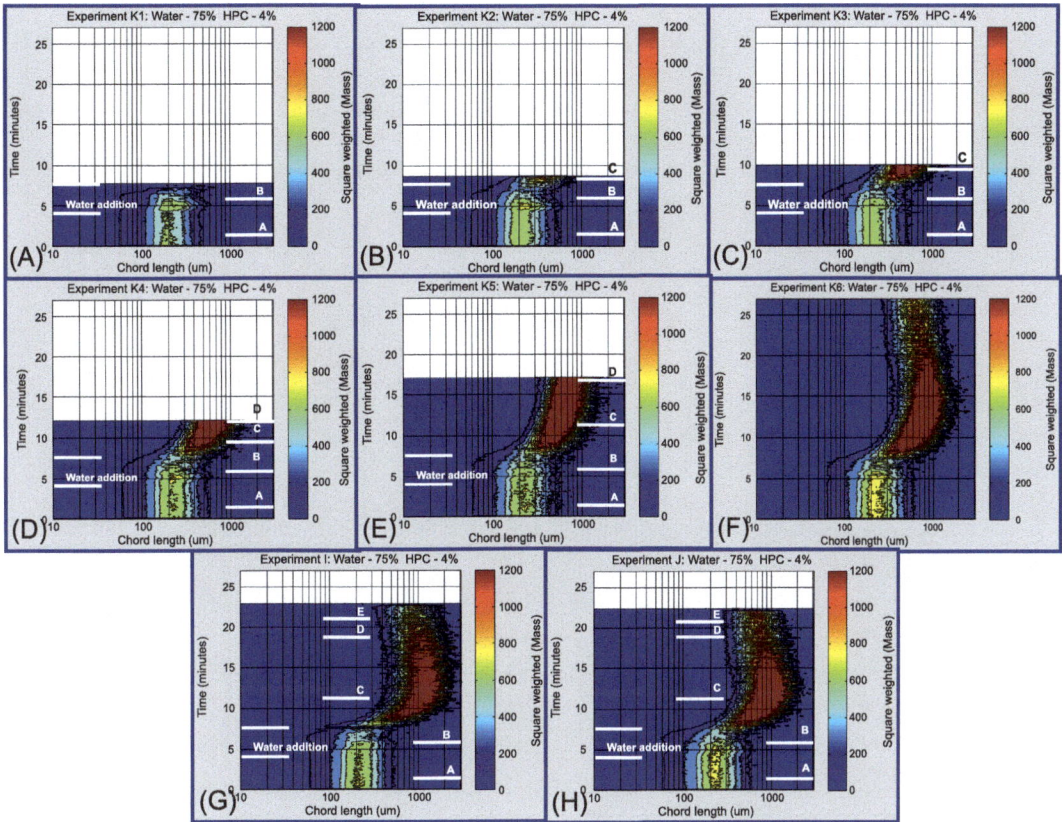

FIGURE 14.11

Reproducibility of FBRM measurement using C35 probe for eight different batches, A through H, that were manufactured until different wet massing times after the end of water addition. These eight batches represent the same formulation and process. The batches were dry mixed for 4 min followed by water addition, which was carried out over a 3-min period. The wet massing time for each batch was different (viz., 0, 1.5, 3, 5, 10, 20, 16, and 16 min for subbatches A, B, C, D, E, F, G, and H, respectively). The results are presented as a three-dimensional plot of CLD as a function of time (I); CLDs at different times marked in I and color coded at 1.5, 5, 9, 15, and 21 min in black, blue, red, green, and purple, respectively (II); and change in the square weighted mass of particles in the selected size ranges as a function of time during granulation (III).

used for granulation (Fig. 14.10B). Manufacture of batches below and above this optimum water concentration was expected to lead to undergranulation and overgranulation, respectively. At lower water concentration (65% w/w), the PSD of the dried granules by nested sieve analyses showed greater proportion of the smallest size particles (fines). At higher water concentration (85% w/w), the PSD of dried granules exhibited clear aggregation of powder mass to form large granules with the reduction of fines. These results were consistent with the visual observations of the powder bed after granulation.

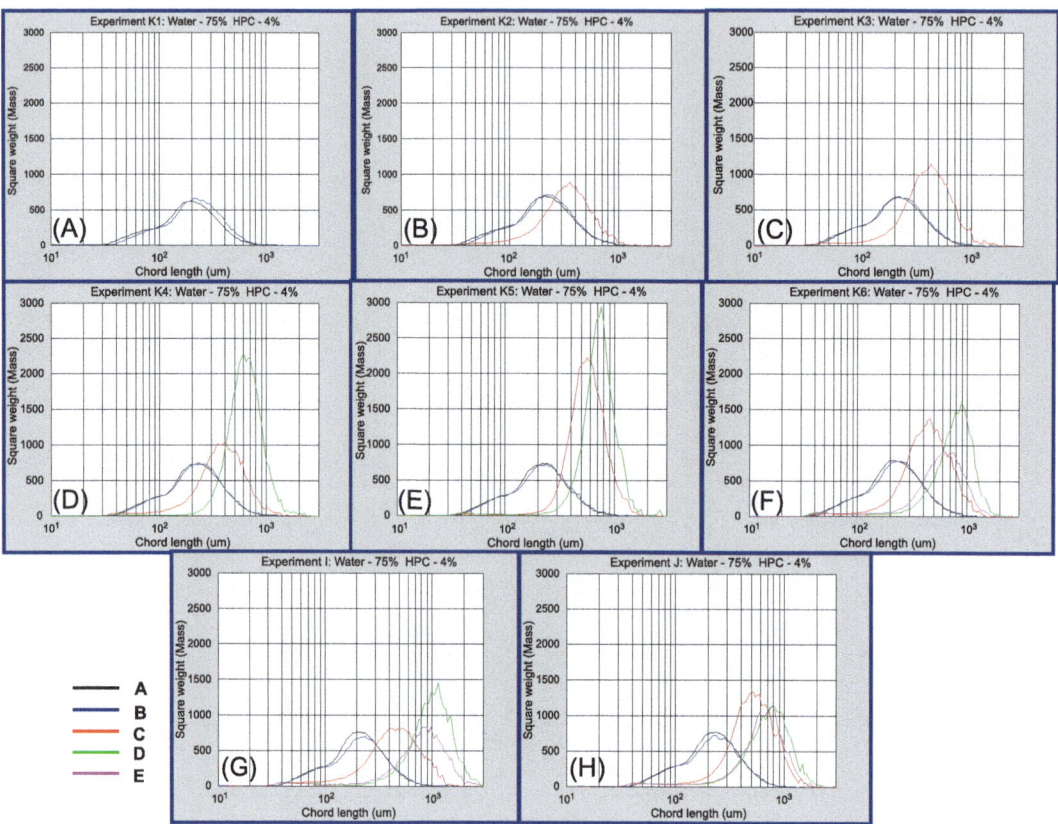

FIGURE 14.11, cont'd.

At the medium or optimum water concentration (75% w/w), a three-dimensional plot of all CLD data collected during the granulation (Fig. 14.9-IB) shows a significant reduction in the proportion of fines and a modest increase in the mean size of the granulation after water addition (compared to before the addition of water). The reduction of fines is more clearly evident in the CLDs plotted at different time points (Fig. 14.9-IIB) before, during, and after the water addition phases, showing progressive reduction in the proportion of powder mass below the chord length of 100 μm. Similarly, tracing the changes in the square weighted mass of a given size fraction of the granules over the duration of granulation shows reduction in the black and the blue traces (Fig. 14.9-IIIB) after water addition, representing reduction of powder mass in the 0 to 100 μm and 100 to 200 μm chord length range, respectively.

At 85% w/w water concentration, the CLD profile changes dramatically toward the end of water addition with respect to reduction of fines (evident by the reduction of the lowest chord length regions of the size distribution) and increase in the overall CLD of particles (evident in the shift of the granulation toward larger chord lengths). This is seen in the overall three-dimensional plot (Fig. 14.9-IC). The reduction of fines is clearly distinguishable in the plot of CLDs at certain times and before, during, and after the granulation (Fig. 14.9-IIC), while the increase in the proportion of coarse granules is seen

14.4 Granulations that predominantly increase the particle size 493

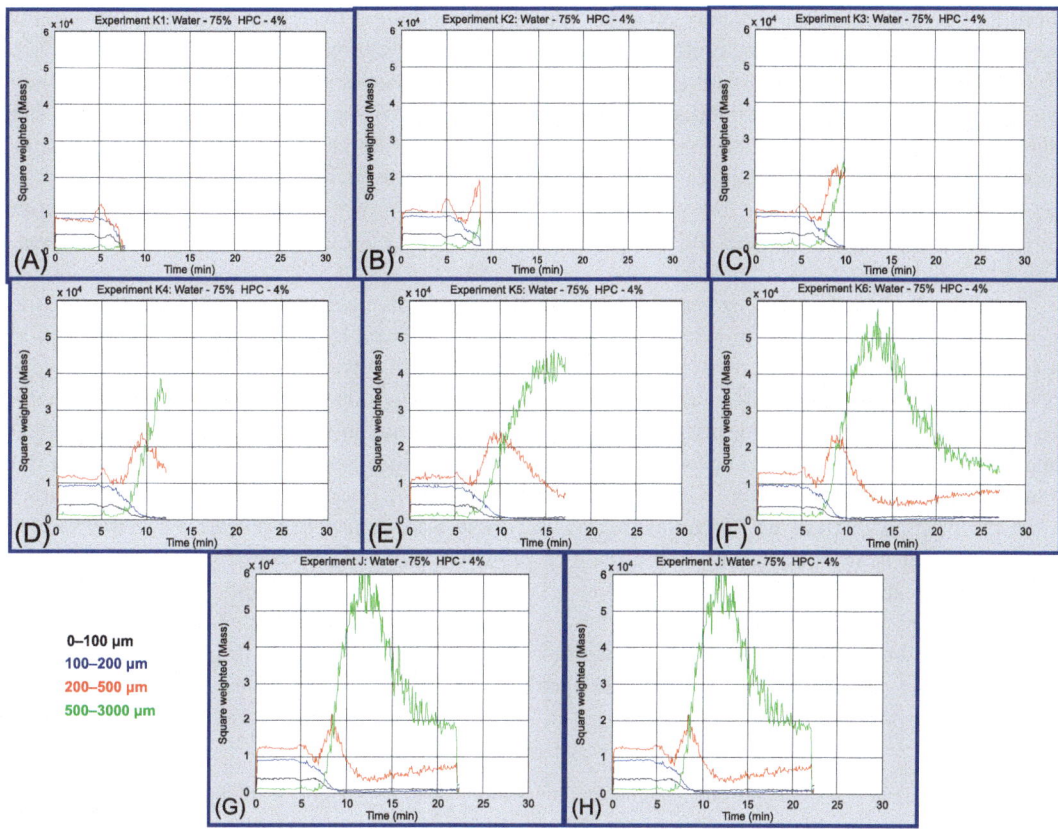

FIGURE 14.11, cont'd.

clearly when change in the proportion of different size fractions of granules are plotted as a function of time (e.g., the green trace of 500–3000 μm chord length granules in Fig. 14.9-IIIC).

In contrast, the use of low water concentration (65% w/w) during granulation does not show any appreciable reduction of fines or increase in the size of the coarse granules. Thus, a three-dimensional plot of all CLD data (Fig. 14.5-IA) shows only a modest decrease in the lowest size fraction of the powder mass. The plot of CLDs at certain times during dry mixing, water addition, and wet massing phases does not show significant difference in the proportion of particles below 100 μm chord length or increase in the highest size fraction or the overall CLD (Fig. 14.9-IIA). This is also evident in the trace of plots of the lowest size fraction of powder mass (Fig. 14.9-IIIA).

These data show that inline FBRM not only is able to clearly distinguish the effect of water on the CLD of the wet mass in a placebo granulation but also can delineate change in the powder mass through the course of granulation. The ability to inline FBRM to robustly and reproducibly discern differences in granule size as the granulation progresses, while maintaining low sensitivity to process changes (such as water) are critical to the successful application of this tool. The FBRM C35 probe used for this study displayed such robustness. For example, upon prolonged wet massing, the particles of the highest chord lengths generated immediately after the end of water addition disappear (e.g., CLD profile for the placebo granulation manufactured with 3% w/w HPC and 85% w/w water shown in Fig. 14.9-IC). This could be because of the predominance of one of several mechanistic granulation processes such as consolidation, attrition, and breakage. Thus, inline FBRM provides an extensive data set that can identify the phases of granulation in which one type of mechanistic granulation processes dominates over the others. In addition to being useful for mechanistic process understanding and modeling, such as population balance modeling, these data can help identify the growth phase of the granulation from the plateau or consolidation phase and help identify optimum range for granulation end-point selection, typically in the granule growth phase.

As much as the CLD data generated by the FBRM probe is designed to detect changes in the overall granule size distribution because of agglomeration of particles, the phase of water addition is likely to have surface adsorbed water that has not yet been absorbed fully by the dry powder mass or incorporated within the wet granules. This surface modification of the powder mass with the added water for granulation might affect the FBRM response in unpredictable ways. The CLD data collected during the phase of water addition sometimes does not harmoniously blend with the data trends before and after the water addition phase, while it does at other times. For example, when the placebo mixture was granulated with 75% w/w water, the CLD profile apparently shows apparent loss of peak of square weighted mass of particles during the phase of water addition (Fig. 14.9-IB). The transition of the dry powder mass CLD profiles from before to after wet addition seems much more congruent when the placebo mixture was granulated with high water concentration (85% w/w) (Fig. 14.9-IC). This phenomenon, nevertheless, might not affect the utility of this probe as a wet granulation monitoring tool because the changes in CLD that are of relevance for monitoring and control purposes are the temporal changes in the CLD of wet granules after the water addition phase.

14.4.3 Reproducibility of chord length distribution measurement

Reproducibility of inline CLD measurement by the FBRM C35 probe was investigated by repeat manufacturing of a fixed composition and process of the placebo granulation. Eight batches were manufactured for different wet massing times. These batches were dry mixed for 4 min followed by water addition, which was carried out over a 3-min period. The wet massing time for each batch was different (viz., 0, 1.5, 3, 5, 10, 20, 16, and 16 min for subbatches A, B, C, D, E, F, G, and H, respectively). At the end of each batch, the powder bed was dried in a tray oven overnight and the PSD analyzed by nested sieve analyses. The rank ordering of the CLD results obtained for these batches were compared with the PSD analyses, except the batches G and H—which were reproduced exactly.

The inline CLD profiles of the eight different batches manufactured at different wet massing times (Fig. 14.11-I/II/III) showed consistent profiles in terms of both the location and distribution of the CLD of the powder bed (Fig. 14.11-II) and the time profiles of changes in the proportion (square weighted mass) of specific particle size fractions (Fig. 14.11-III).

Offline PSD by nested sieve analyses (Fig. 14.10C) showed changes in the granule size over a period, including both increase in the large particle size fraction and decrease in the proportion of particles in the small particle size range. Overall PSD is seen to shift toward the right, larger particle size range, with increase in the granulation time.

The effect of wet massing time on the granulation progress, as monitored by the inline FBRM, was fairly reproducible when a single batch was manufactured for an extended wet massing time (comparing subfigures G and H in Fig. 14.11) or when different batches were manufactured for different wet massing times (comparing subfigures A through F, that were manufactured for 0, 1.5, 3, 5, 10, 20 min wet massing time, respectively, in Fig. 14.11). These FBRM data were also consistent with the results of offline PSD analyses by nested sieve analyses (Fig. 14.10C). Taken together, these results confirmed the ability of the inline FBRM probe to measure changes in the granulation that were real and reflected in the PSD differences in dried granules.

14.5 Granulations that predominantly reduce fines

Wet granulation unit operation seeks particle agglomeration as an outcome to increase the overall powder PSD with the intent to improve processability characteristics such as densification, flow, and adhesion or sticking during processing. The critical in-process quality attribute of granulated material that is desired to be altered through the HSWG unit operation can be one or more of densification, increase in the proportion of larger particles or granules, and decrease in the proportion of smaller particles, generally called fines. This usually depends on the characteristics of the active pharmaceutical ingredient (API), drug load, and formulation characteristics. For example, in placebo formulation, both increase in the proportion of larger granules and reduction in the fines were identified as quality attributes of interest that were affected by the HSWG unit operation. An in-line CLD probe should be capable of delineating decrease in the fine particle content of the powder blend during granulation in the presence of a binder and should be responsive to the changes in the process trajectory depending on the concentration of the binder and the granulating fluid (water).

The fines usually are understood to be the particles below 100 mesh size (150 μm) in nested sieve analyses. Fines contribute to sticking tendency (Shah et al., 2014a, 2014b, 2015, 2015; Wang et al., 2015) and increase the overall surface area of the powder mass. The fines can represent ungranulated powder mass or the dry powder blend or the particles whose agglomeration has not progressed beyond a few primary powder particles. In the case of brivanib alaninate (a model drug used in the study reported in this section), reduction of fines to mitigate observed sticking to stainless steel tooling during the compression unit operation was identified as a primary goal during development studies (Badawy et al., 2016). These studies indicated an impact of the water concentration used during granulation on the PSD of dried powder and the adhesion or sticking tendency of the final, lubricated powder blend.

A high drug load formulation consisting of 50% w/w brivanib alaninate as the API, 39.5% w/w MCC as the filler, 3% w/w croscarmellose sodium as a superdisintegrant, and 4% w/w HPC as a dry powder binder was granulated with 46.5% w/w water as a center point of a preestablished design space (Badawy et al., 2012, 2016; Narang et al., 2015). Formulation and process conditions that represent a pharmaceutical optimum desired process for a high drug load brivanib alaninate formulation were based on conventional and empirical wet granulation process development methodologies. The parameters for these assessments included visual observation of completeness of granulation, absence of dry material in the bowl, brittle self-association of the wet mass when compressed in a fist, and change in the PSD (viz., reduction of fines) of the dried granules compared to the dry powder premix used for granulation. These assessments were combined with the results of the impact of formulation and process parameters on DP CQAs through statistical DoE studies that defined the center point of a design space (Badawy et al., 2012, 2016; Narang et al., 2015).

Using the center point formulation and process, changes in the powder mass's CLD through various stages of the wet granulation process (viz., dry mixing, water addition, and wet massing) were monitored using an inline FBRM probe (Mettler Toledo's C35 probe). In addition, the sensitivity of the inline probe to changes in the formulation and process parameters of the brivanib alaninate granulation was investigated by manufacturing batches that differed systematically in the concentration of water used during granulation (Fig. 14.12). Several batches were manufactured in a 6-L and a 150-L HSG at an impeller speed of 4.8 m/s and low chopper speed (1750 rpm in the 6-L HSG) at 40% v/v fill volume. The 150-L batch was manufactured with 48% w/w water for granulation. The 6-L batch was manufactured with dry mixing time of 4 min, water addition time of 3 min, and wet massing time of ~20 min. The 150-L batch was manufactured with dry mixing time of 4 min, water addition time of 3 min, and wet massing time of 30 s. Four additional 6-L batches were manufactured at either side of the design space with respect to the amount of water used for granulation (at ± 2.5% w/w and ± 7.5% w/w water concentration). In addition to inline CLD measured using an FBRM C35 probe at the end of granulation, the granules were dried in a tray dryer and analyzed for PSD by nested sieve analysis.

The FBRM results in Fig.14.12 are presented in three different formats. Subfigure I presents the full spectrum of results of change in chord length as a function of time during processing as a three-dimensional figure with the square weighted mass of particles of each chord length being represented on the z-axis by the color scale shown on the right side of the figure. Subfigure II shows chord length distribution of the square weighted mass of particles at predefined time points identified in the subfigure I. Subfigure III shows time profile of changes in the square weighted mass of particles of a predefined selected size fraction during the process. The three subfigures represent the same batch and manufacturing process parameters represented in the batches A, B, C, D, and E, while providing different ways of looking at the same data set.

These results were compared with the results of conventional (nested sieve) analyses of dried granules (Fig. 14.13). The ability to use FBRM as an end-point detection tool (Fig.14.14) and to provide scale-independent assessment of granule attributes upon scale-up of the brivanib alaninate granulation (Fig. 14.15) was ascertained.

14.5.1 Effect of water concentration

The inline FBRM probe was used to study whether the probe was able to delineate the progress of the granulation through various stages of dry mixing, water addition, and wet massing at the target or center point of a pre-established design space (viz., 46.5% w/w water concentration in Fig. 14.12-I/II/IIIC). In addition, the ability of the inline FBRM measurement to distinguish batches manufactured using a range of water quantities within a relatively narrow range of the established design space (46.5 ± 2.5% w/w, viz. 44% w/w in Fig. 14.12-I/II/IIIB and 49% w/w in Fig. 14.12-I/II/IIID) and a wider range that was not studied earlier in the QbD DoE studies (46.5 ± 7.5% w/w, viz. 39% w/w in Fig.14.12-I/II/IIIA and 54% w/w in Fig. 14.12-I/II/IIIE) was investigated.

Manufacture of batches below and above the optimum water concentration (46.5% w/w) was expected to lead to undergranulation and overgranulation, respectively. At lower water concentration (44% w/w and 39% w/w), the PSD of the dried granules by nested sieve analyses showed progressively

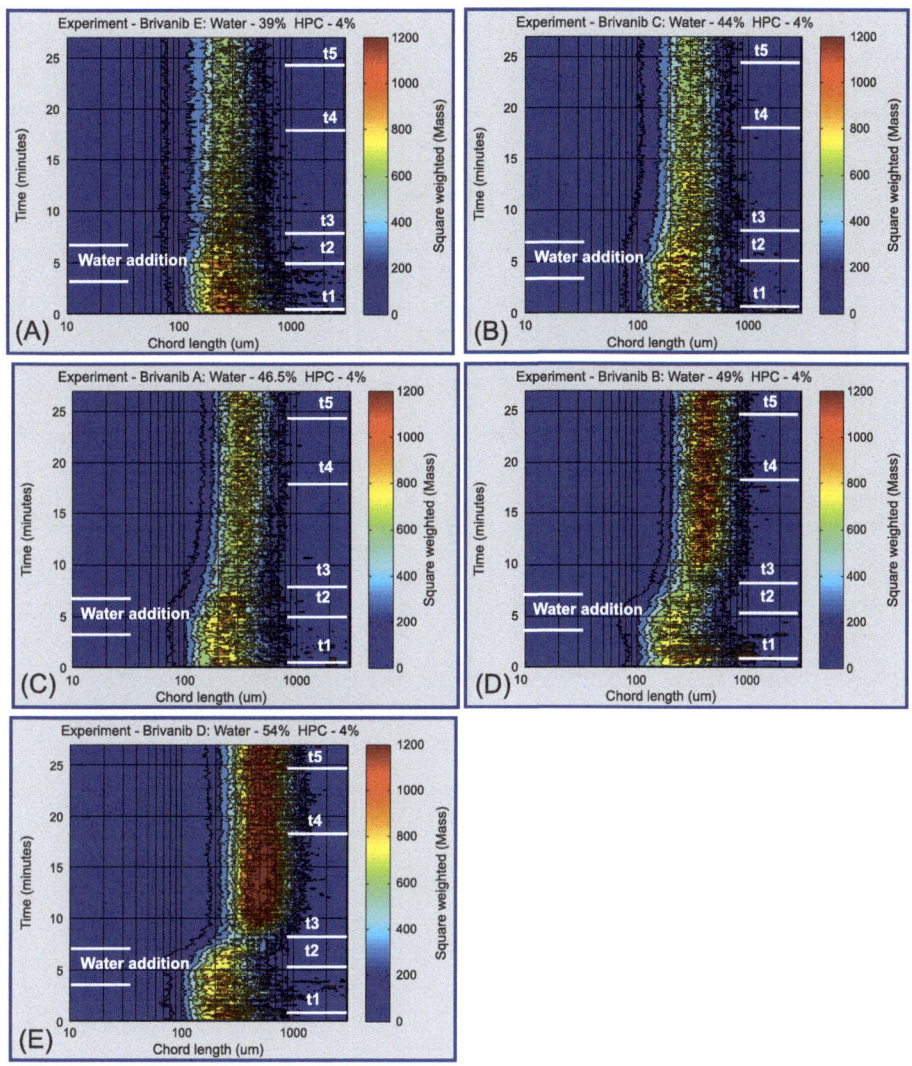

FIGURE 14.12

Effect of process parameters (wet massing time and the amount of water used during granulation) on granulation growth kinetics as measured by FBRM C35 probe during high-shear wet granulation (HSWG) of brivanib alaninate in a 6-L granulator. The results are presented as (I) three-dimensional plots of CLD as a function of time for granulations of brivanib alaninate manufactured with 39% w/w (A), 44% w/w (B), 46.5% w/w (C), 49% w/w (D), and 54% w/w (E) water concentration; (II) CLDs at different time points marked in I (time, t1, t2, t3, t4, and t5) and color coded at 1.5, 5, 9, 15, and 21 min in black, blue, red, green, and purple, respectively; and (III) change in the square weighted mass of particles in the selected size ranges as a function of time during granulation.

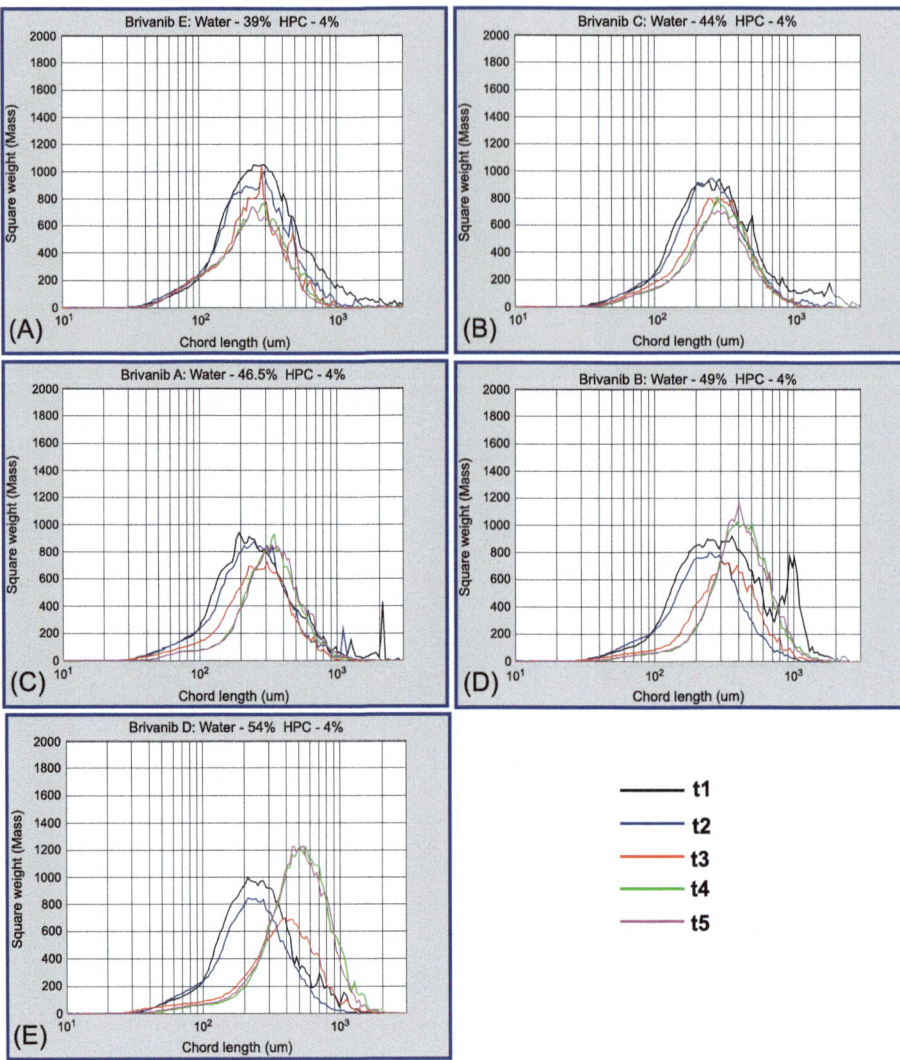

FIGURE 14.12, cont'd.

greater proportions of the smallest particles (fines) (Fig. 14.13). At higher water concentrations (49% w/w and 54% w/w), the PSD of dried granules showed the reduction of fines compared to the batch manufactured at target 46.5% w/w water concentration (Fig. 14.13). In the case of brivanib alaninate formulation, a significant increase in the size or formation of large size granules is not observed with increasing water amount. The nested sieve analyses data rather show narrowing of the size distribution with the reduction of fines as the water amount is increased (Fig. 14.13). These results were consistent with the visual observations of the powder bed after granulation.

14.5 Granulations that predominantly reduce fines

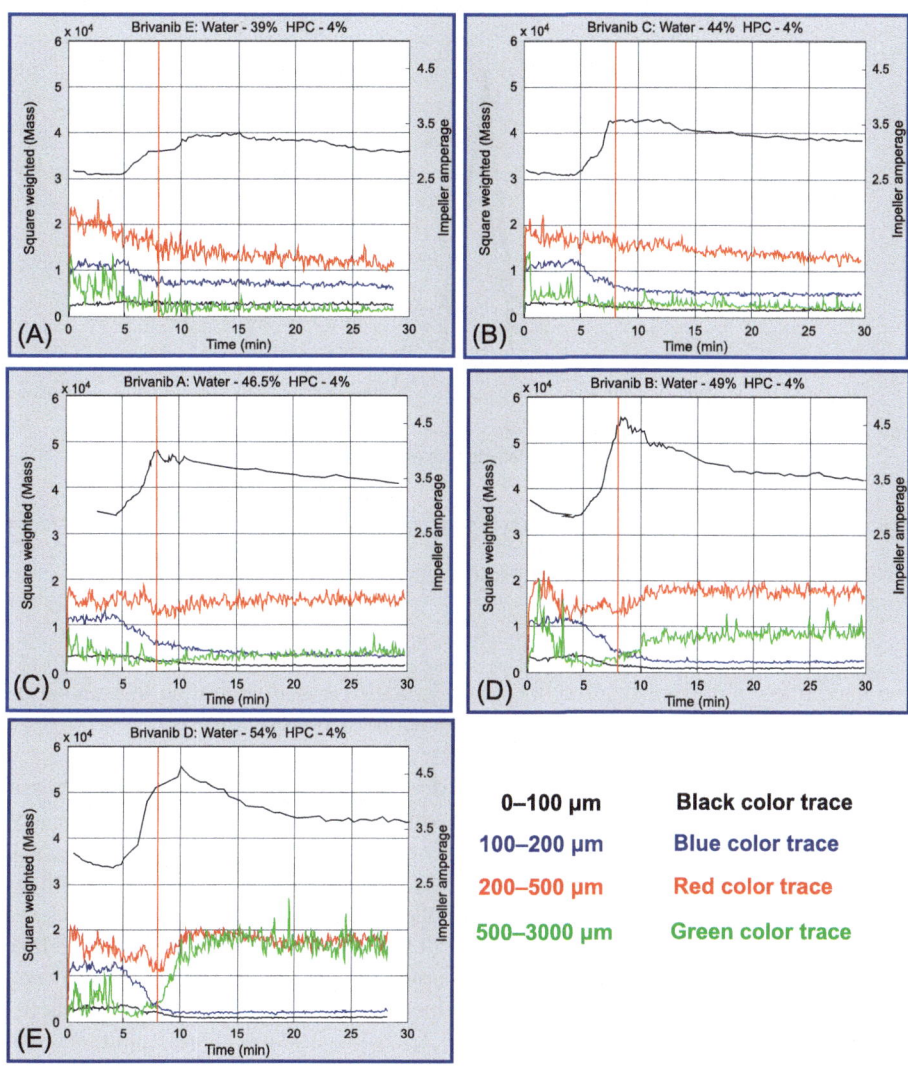

FIGURE 14.12, cont'd.

For the batch manufactured at target water concentration (46.5% w/w in Fig. 14.12-I/II/IIIC), the inline CLD profiles were able to distinguish the granulation phases of dry mixing, water addition, and wet massing. A consistent CLD of the powder mass is obtained during the dry mixing phase, which progressively narrows and transitions toward higher chord lengths during the water addition phase to result in a granule size distribution that is characterized mainly by the reduction of smaller particles or fines. The reduction of fines is more clearly evident in the CLDs plotted at different time points (Fig. 14.12-IIC) before, during, and after the water addition phases, showing progressive reduction in the proportion of powder mass below the chord length of 100 μm. Similarly, tracing the changes in the

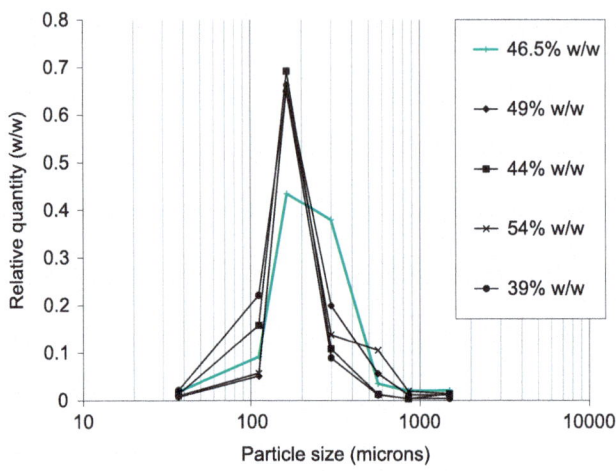

FIGURE 14.13

Particle size distribution (PSD) by nested sieve analysis of dried brivanib alaninate granulation of batches made with different concentrations of water with a wet massing time of 30 min in a 6-L granulator.

Reproduced with permission from Narang et al. (2017).

square weighted mass of a given size fraction of the granules over the duration of granulation shows reduction in the blue trace (Fig. 14.12-IIIC) after water addition, representing reduction of powder mass in the 100 to 200 μm chord length range.

The three phases of the granulation process (viz., dry mixing, water addition, and wet massing) also were distinctly identifiable for batches manufactured with high water concentrations of 49% w/w in Fig. 14.12-I/II/IIID and 54% w/w in Fig. 14.12-I/II/IIIE. The CLD profile changes toward the end of water addition with respect to reduction of fines (evident by the reduction of the lowest chord length regions of the size distribution) and increase in the overall CLD of particles (evident in the shift of the granulation toward larger chord lengths). This is seen in the overall three-dimensional plot (Fig. 14.12-ID and IE). The reduction of fines is clearly distinguishable in the plot of CLDs at certain times and before, during, and after the granulation (Fig. 14.12-IID and IIE), which show not only a progressive reduction in the relative proportion of smaller particles, such as particles below 100 μm chord length, but also a shift of the highest peak location to a longer chord. Increase in the proportion of coarse granules is seen clearly when change in the proportion of different size fractions of granules are plotted as a function of time (e.g., the green trace of 500–3000 μm chord length granules in Fig. 14.12-IIIE). The trace plot of different sizes also show significant reduction in small particles for the lower particle size range.

In contrast, the use of low water concentrations of 39% w/w in Fig. 14.12-I/II/IIIA and 44% w/w in Fig. 14.12-I/II/IIIB during granulation does not show any appreciable reduction of fines or increase in the size of the coarse granules. The CLD profile for the lowest water amount (39% w/w) does not show any appreciable change in the CLD of the powder bed mass with the progress of the granulation. This was consistent with the visual observations during manufacturing.

The granule size data generated using the inline FBRM probe (Fig. 14.12) correlated well with sieve analysis results (Fig. 14.13). The nested sieve analyses data indicated both reduction of fines and

14.5 Granulations that predominantly reduce fines   501

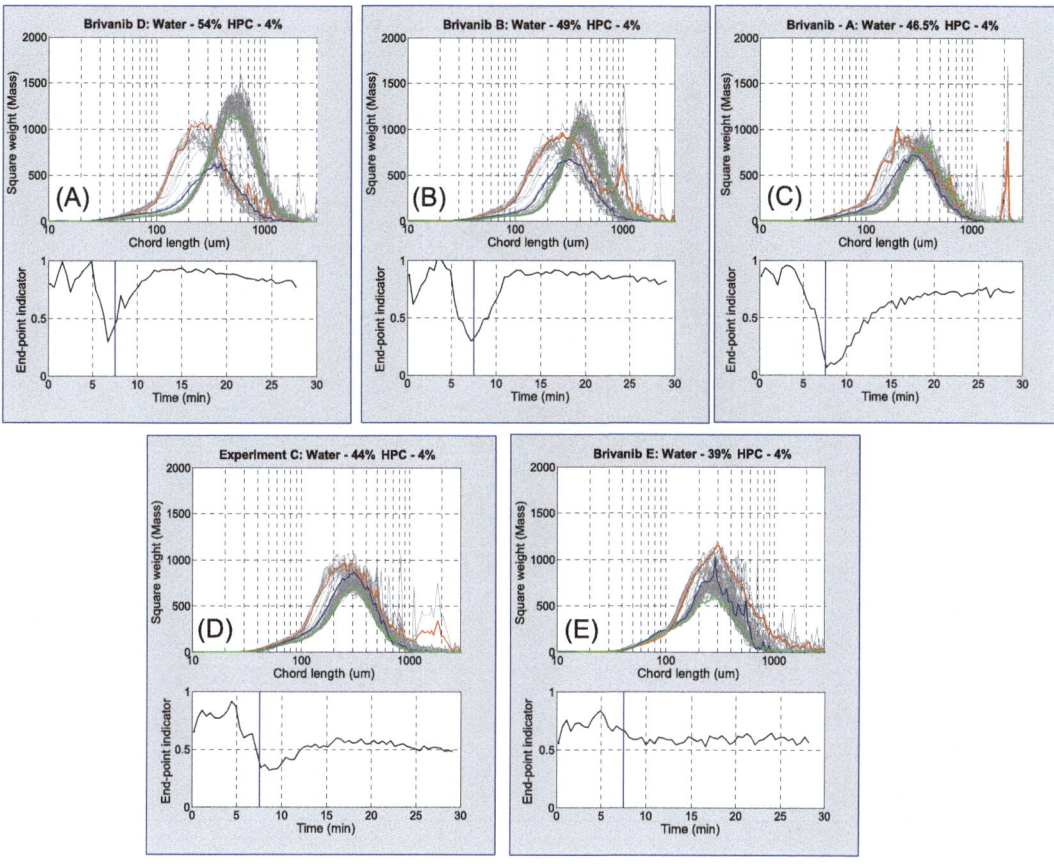

FIGURE 14.14

Demonstration of end-point detection algorithm based on difference minimization with target chord length distribution (CLD) marked in green for brivanib alaninate granulation batches manufactured in a 6-L granulator. Subbatch (C) represents granulation with the target 46.5% w/w water (D). The CLD profile is seen to achieve a minimum in the same time frame for water concentrations that are within ± 2.5% of the target water amount, viz. 49% (B) and 44% (D). At the same time, the CLD profile minimum is not achieved in this time frame for water concentrations that are within ± 7.5% of the target water amount, viz. 54% (A) and 39% (E). As discussed in the text, the ± 2.5% water concentration was identified within the process design space for brivanib alaninate, while ± 7.5% water concentration would be considered outside the qualified design space.

Reproduced with permission from Narang et al. (2017).

increase in the fraction of coarse granules by the rightward shift in the shape of the PSD of the dried, granulated powder, without any appreciable change in the peak location. These data show that CLD generated using inline FBRM can identify a granule attribute of interest and is not only able to clearly distinguish the effect of different concentrations of water on the CLD of the wet mass in a brivanib

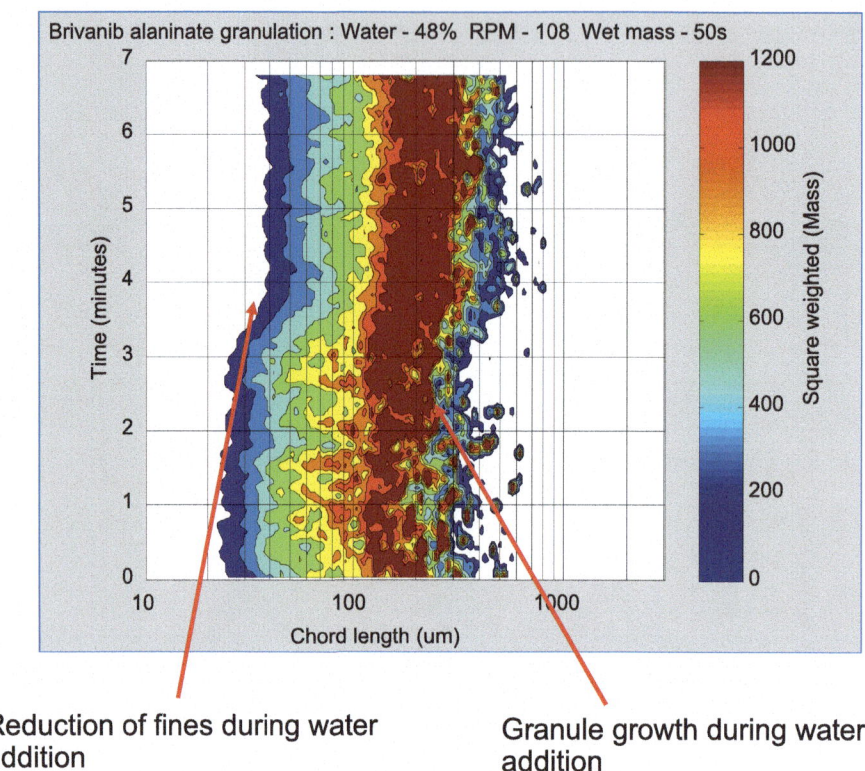

FIGURE 14.15

Application of FBRM C35 probe to the scale-up of brivanib alaninate granulation. This figure shows of FBRM C35 data of brivanib alaninate granulation made at the target water concentration of 46.5% w/w in a 150-L manufacturing scale granulator with the same tip speed as the 6-L pilot scale granulator used in the process design space studies. A characteristic feature of the particle CLD change during granulation, viz. reduction of fines, was reproducible at the manufacturing scale.

Reproduced with permission from Narang et al. (2017).

alaninate granulation but also is able to delineate change in the powder mass through the course of granulation.

14.5.2 End-point indicator algorithm

A CLD similarity algorithm was developed using the CLD at 30 s wet massing time for the center point batch (46.5% w/w water and 4% w/w HPC concentration) as the reference batch. This algorithm was used to assess the similarity of CLD at all points for all batches (different water concentration) manufactured in the 6-L HSG to provide a single-point measure of end point as a function of time.

Average root mean square error (RMSE) of the whole distribution of chord lengths compared against the reference CLD was used to identify a single numerical descriptor that compares two CLDs to identify their similarity. This similarity index was plotted over a period for each CLD data point when compared against a CLD profile observed at 30 s wet massing time, which was the identified end point of granulation through conventional DoE studies. This end-point indicator was plotted against time for all the brivanib alaninate granulations done using different water concentrations (bottom plots in Fig. 14.14). The vertical blue line plotted in the end-point indicator time profiles identifies a conventionally identified end point of granulation and the reference CLD profile that was used to compare the similarity of the test CLD (at any given point in time) with the reference CLD (at a predesignated end point) at all time points.

The end-point indicator showed a minimum in Fig. 14.14C, indicating a perfect match of the observed CLD profile with the reference CLD profile at the data point of the reference CLD profile. The similarity index is above this minimum at the data points before and after this data point, which is an indicator of evolving CLD profile evolves over time. When the same reference CLD profile (the minimum in Fig. 14.14C) was used to compare the CLD at all data points for all the other batches of brivanib alaninate (manufactured with different concentrations of water), different profiles of approach of the end-point indicator to the minimum are observed. In one case (that of 39% w/w water used for granulation), the end-point indicator never reaches the minimum.

Identification of the wet granulation end point relies primarily on empirical and experimental evidence of conformance of produced granules with in-process and end product desired quality attributes. The process is typically controlled by the combinations of process parameters that are able to reproducibly manufacture granules of such desired quality attributes in development and scale-up studies. The FBRM can replace wet massing time as a process parameter-based control of the granulation end point to granule CLD as a material attribute-based control.

A parametric approach is used to control the HSWG process to the desired, predefined end point. The primary application of FBRM for granulation end-point control is envisioned to be the replacement of parametric control (wet massing time, in this case) with an attribute-based control (target CLD during processing). We have demonstrated that FBRM can follow both the reduction in fines and the growth of granules in real time and in process during granulation. During development, the granulation end point was chosen at 30 s of wet massing after 4 min of dry mixing and 3 min of 46.5% w/w water addition using conventional DoE studies. To demonstrate the utility of FBRM as a potential attribute-based endpoint detection and control tool, we developed a simple algorithm that computes the RMSE between the current observed (in real time) CLD and the chosen end-point CLD, based prior experience at the preset 30 s of wet massing. The end point is determined as the RMSE between the two CLDs approaches zero or a minimum. Comparing the entire distribution, as an end-point indicator, provides a more accurate representation of the granules during the process than just trending fines or a granule size range, such at 500 to 1000 µm.

Fig. 14.14C illustrates this end-point indicator when the process is run at center point with 46.5% water addition. The top plots in Fig. 14.14 show the CLD plotted every 30 s; the red graph represents the dry blend CLD, the blue graph represents the 30-s wet massing end-point CLD, and the green graph represents the CLD after 20 min of wet massing. The bottom plots in Fig. 14.14 show the end-point indicator calculated every 30 s comparing the current CLD to the desired CLD at 30 s wet massing. In each bottom plot, a blue vertical line indicates the 30 s wet massing time point.

In Fig. 14.14C, the end-point indicator reaches a minimum close to zero at 30-s wet massing as the CLD distribution closely matches the desired CLD. As wet massing continues, the end-point indicator gradually increases to about 200 as the CLD continues to grow. In Fig. 14.14A and B, when 49% *w*/w and 54% w/w water is added, respectively, the end-point indicator approaches a minimum very quickly, and then just as quickly the granules grow beyond the optimum. In Fig. 14.14D, when 44% w/w water is added, the end-point approaches a minimum slowly, and then slowly the granules grow in size. These observations are indicators of the robustness of granulation process, with a process that shows relatively slower change in the granule attributes as a function of process parameters being considered more robust. In Fig. 14.14E, when 39% w/w water is added, the end-point never reaches a minimum, and very little granulation takes place.

14.5.3 Scale-up

Scale-up of HSWG processes has remained an empirical process, often used dimensionless numbers and general rules such as constant impeller tip speed across the scales (Tao et al., 2015). The scale-up challenges emanate from the short process time, usually 3 to 5 min, and relatively high intensity process conditions (in terms of shear, particle velocities, and particle collision frequency and energy) compared to other batch processes, such as low-shear wet granulation conducted in a planetary mixer (Tao et al., 2015). Historically, development of a high-intensity short-duration process was designed to overcome the key challenge of densification of highly bulky actives (Tao et al., 2015). The critical response variables for this goal were the manufacturability parameters of granule properties such as bulk density, PSD, flow, and sticking tendency (Tao et al., 2015). The emphasis on granule PSD, which affects density, flow, and sticking tendency, led to the development of inline PATs that monitor granulation PSD (Arp et al., 2011; Huang et al., 2010; Narang et al., 2010).

There is a significant unmet need for PAT tool that measures a granule attribute of interest in a scale-independence and can quantitatively track the progress of granulation. We explored the inline FBRM as one such tool for granule size distribution as an in-process quality attribute of interest. The setting of the FBRM probe in the granulator, insertion through the lid, makes it amenable to application across different scales and geometries of the granulator. Keeping the geometric location of the probe similar across the scales, measurement of powder CLD over the course of granulation produced similar response at the 150-L scale (Fig. 14.15) as that seen at the 6-L scale.

Brivanib alaninate granulation was scaled up 25 times from a 6-L HSG to a 150-L granulator using a constant tip speed method (Tao et al., 2015) and with a water content of 44% w/w and 48% w/w. A representative inline FBRM CLD data (for the 48% granulation batch) at the 150-L scale was plotted in Fig. 14.15. This data indicated a similar granule CLD time profile as the lab scale granulation (with the predominant response being the reduction of fines). The end-point indicator algorithm was plotted for both these batches (Fig. 14.16). The end-point indicator showed similarity to the batches manufactured at pilot scale with similar water concentration. These data identified the same process trajectory independent of the scale, thus validating the utility of this tool as a scale-independent granule attribute measurement.

In addition, CLD profiles and end-point detection algorithm were compared for representative granulation batches manufactured at the pilot scale (6-L granulator) and the commercial scale (150-L granulator). Batches shown in (B) and (D) in Fig. 14.16 were manufactured at 49% w/w and 44% w/w water concentration in the 6-L granulator, respectively. These are same as the batches (B) and (D) in Fig. 14.14. Batches (A) and (C) in Fig. 14.16 were manufactured with 48% w/w and 45% w/w water

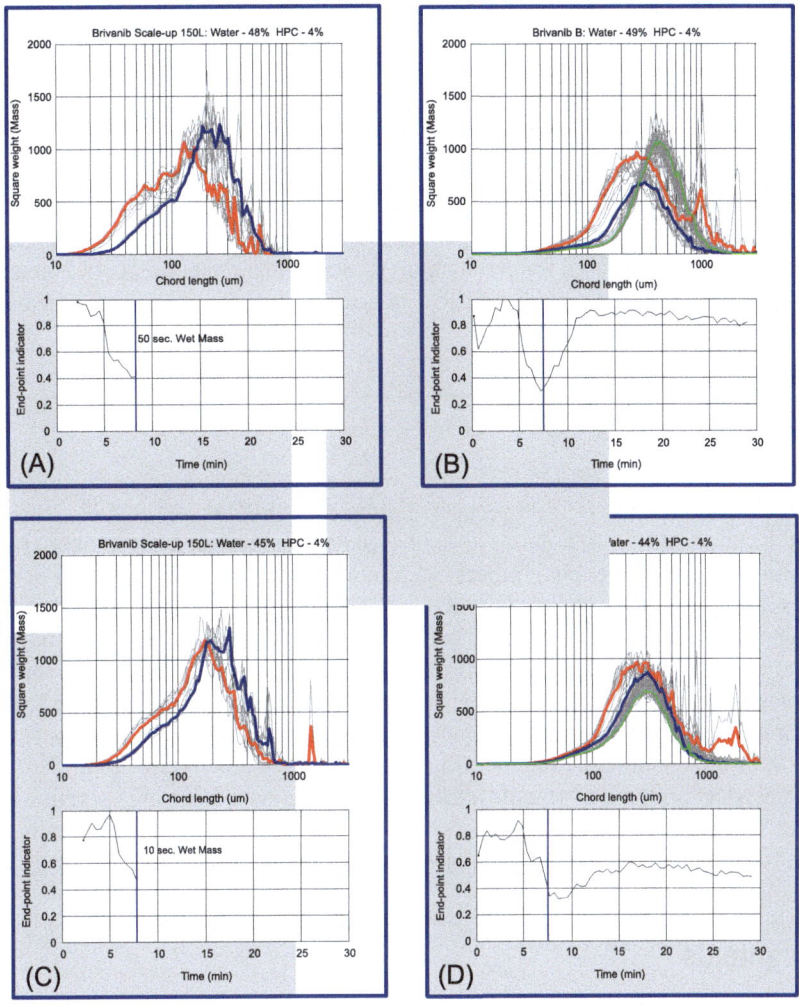

FIGURE 14.16

Comparison of chord length distribution (CLD) profiles and end-point detection algorithm for representative granulation batches manufactured at the pilot-scale (6-L granulator) and the commercial scale (150-L granulator). Batches (B) and (D) were manufactured at 49% w/w and 44% w/w water concentration in the 6-L granulator, respectively. These are same as the batches (B) and (D) in Fig. 14.6. Batches (A) and (C) were manufactured with 48% w/w and 45% w/w water concentration in the 150-L granulator, respectively. Although the scale-up batches (A) and (C) did not use the same process parameters as the pilot-scale batches (B) and (D), differing in the amount of water used by 1% w/w, they could be considered fairly similar for the purpose of comparison of the time-profile of the end-point indicator algorithm. The end-point algorithm seems to follow similar trajectory (i.e., the rate of change of the end-point indicator during granulation) across scales, batch (A) similar to batch (B) and batch (C) similar to batch (D), indicating the scale-independence of this measurement.

Reproduced with permission from Narang et al. (2017).

concentration in the 150-L granulator, respectively. While the scale-up batches (A) and (C) did not use the same process parameters as the pilot-scale batches (B) and (D), differing in the amount of water used by 1% w/w; these could be considered fairly similar for the purpose of comparing the time profile of the end-point indicator algorithm. As seen in Fig. 14.16, the end-point algorithm seems to follow similar trajectory (i.e., the rate of change of the end-point indicator during granulation) across scales, batch (A) similar to batch (B) and batch (C) similar to batch (D), indicating the scale independence of this measurement. Although it is true that the comparison across scales is not quantitative and the batches compared are not the same process parameters, the impact of process parameters (e.g., low versus high water concentration used for granulation) shows similar rank order correlation and trending across scales. These data do indicate that FBRM is a useful tool for quantitation of inline CLD of granules across scales.

14.6 Conclusion

Inline CLD measurement using FBRM C35 probe was a reproducible and sensitive measurement of particle size changes during the HSWG process. Using inline FBRM and RMSE between CLDs as an end-point indicator could be used during formulation development to design a more robust process with an end point that is more tolerant of process variation. For example, this technique could be used to design a process that reproducibly meets the end point and has a soft transition away from the minimum RMSE. This algorithm also could be used during scale-up to confirm that CLD evolution remains like development batches. It could be used as part of continuous process verification to routinely inform the operator of granulation consistency. This tool could be used as a single numerical monitoring and control parameter that translates a complex CLD profile into a single data point that is easily amenable to specification control and monitoring strategies. Furthermore, this tool offers a scale- and equipment-independent bridging of granule attributes from one manufacturing operation to another. In summary, inline FBRM offers a promising approach to real-time particle size monitoring for process understanding, scale-up, and control during high-shear wet granulation.

Acknowledgments

The authors thank Des O'Grady, Mettler-Toledo, Inc., for useful discussions during probe setup and use; their BMS colleagues John Gamble for discussions and David Trinkle for generating nested sieve analyses data. The chapter is an aggregate of the data and information previously published by the authors Narang et al. (2017) and is reproduced here with the consent of the publishers.

References

Arp, Z., Smith, B., Dycus, E., & O'Grady, D. (2011). Optimization of a high shear wet granulation process using focused beam reflectance measurement and particle vision microscope technologies. *Journal of Pharmaceutical Sciences, 100*(8), 3431–3440. https://doi.org/10.1002/jps.22556.

Badawy, S. I., Narang, A. S., LaMarche, K. R., Subramanian, G. A., Varia, S. A., Lin, J., Stevens, T., & Shah, P. A. (2016). Integrated application of quality-by-design principles to drug product development: A case study of brivanib alaninate film-coated tablets. *Journal of Pharmaceutical Sciences, 105*(1), 168–181. https://doi.org/10.1016/j.xphs.2015.11.023.

Badawy, S. I. F., Lin, J., Gokhale, M., Desai, S., Nesarikar, V. V., Lamarche, K. R., Subramanian, G. A., & Narang, A. S. (2014). Quality by design development of brivanib alaninate tablets: Degradant and moisture control strategy. *International Journal of Pharmaceutics, 469*(1), 111–120. https://doi.org/10.1016/j.ijpharm.2014.04.059.

Badawy, S. I. F., Narang, A. S., Lamarche, K., Subramanian, G., & Varia, S. A. (2012). Mechanistic basis for the effects of process parameters on quality attributes in high shear wet granulation. *International Journal of Pharmaceutics, 439*(1–2), 324–333. https://doi.org/10.1016/j.ijpharm.2012.09.011.

Badawy, S. I. F., Narang, A. S., LaMarche, K. R., Subramanian, G. A., Varia, S. A., Lin, J., Stevens, T., & Shah, P. A. (2015). Integrated application of quality-by-design principles to drug product development: Case study of brivanib alaninate film coated tablets. *Journal of Pharmaceutical Sciences, 105*(1), 168–181.

Gamble, J. F., Chiu, W. S., & Tobyn, M. (2011). Investigation into the impact of sub-populations of agglomerates on the particle size distribution and flow properties of conventional microcrystalline cellulose grades. *Pharmaceutical Development and Technology, 16*(5), 542–548. https://doi.org/10.3109/10837450.2010.495395.

Habib, Y. S., Abramowitz, R., Jerzewski, R. L., Jain, N. B., & Agharkar, S. N. (1999). Is silicified wet-granulated microcrystalline cellulose better than original wet-granulated microcrystalline cellulose? *Pharmaceutical Development and Technology, 4*(3), 431–437. https://doi.org/10.1081/PDT-100101379.

Hansuld, E. M., & Briens, L. (2014). A review of monitoring methods for pharmaceutical wet granulation. *International Journal of Pharmaceutics, 472*(1–2), 192–201. https://doi.org/10.1016/j.ijpharm.2014.06.027.

Huang, J., Kaul, G., Utz, J., Hernandez, P., Wong, V., Bradley, D., Nagi, A., & O'Grady, D. (2010). A PAT approach to improve process understanding of high shear wet granulation through in-line particle measurement using FBRM C35. *Journal of Pharmaceutical Sciences, 99*(7), 3205–3212. https://doi.org/10.1002/jps.22089.

Iveson, S. M., Litster, J. D., Hapgood, K., & Ennis, B. J. (2001). Nucleation, growth and breakage phenomena in agitated wet granulation processes: A review. *Powder Technology, 117*(1–2), 3–39.

Kukec, S., Hudovornik, G., Dreu, R., & Vrecer, F. (2014). Study of granule growth kinetics during in situ fluid bed melt granulation using inline FBRM and SFT probes. *Drug Development and Industrial Pharmacy, 40*(7), 952–959. https://doi.org/10.3109/03639045.2013.791832.

Kumar, V., Taylor, M. K., Mehrotra, A., & Stagner, W. C. (2013). Real-time particle size analysis using focused beam reflectance measurement as a process analytical technology tool for a continuous granulation-drying-milling process. *AAPS PharmSciTech, 14*(2), 523–530. https://doi.org/10.1208/s12249-013-9934-4.

Narang, A. S., Stevens, T., Hubert, M., Gambe, J., Paruchuri, S., Macias, K., Bindra, D. S., Gao, Z., & Badawy, S. F. (2010). *Resolution and sensitivity of lasentech FBRM C35 probe for real-time chord length distribution measurement during high shear wet granulation and correlation with other particle size distribution techniques*. New Orleans, LA: American Association of Pharmaceutical Sciences.

Narang, A. S., Stevens, T., Hubert, M., Paruchuri, S., Macias, K., Bindra, D., Gao, Z., & Badawy, S. (2016). Resolution and sensitivity of inline focused beam reflectance measurement during wet granulation in pharmaceutically relevant particle size ranges. *Journal of Pharmaceutical Sciences, 105*(12), 3594–3602. https://doi.org/10.1016/j.xphs.2016.09.001.

Narang, A. S., Stevens, T., Macias, K., Paruchuri, S., Gao, Z., & Badawy, S. (2017). Application of in-line focused beam reflectance measurement to brivanib alaninate wet granulation process to enable scale-up and attribute-based monitoring and control strategies. *Journal of Pharmaceutical Sciences, 106*(1), 224–233. https://doi.org/10.1016/j.xphs.2016.08.025.

Narang, V. A. S, Stepaniuk, V., Badawy, S., Stevens, T., Macias, K., Wolf, A., Pandey, P., Bindra, D., & Varia, S. (2015). Real-time assessment of granule densification in high shear wet granulation and application to scale-up of a placebo and a brivanib alaninate formulation. *Journal of Pharmaceutical Sciences, 104*(3), 1019–1034. https://doi.org/10.1002/jps.24233.

Saleemi, A. N., Steele, G., Pedge, N. I., Freeman, A., & Nagy, Z. K. (2012). Enhancing crystalline properties of a cardiovascular active pharmaceutical ingredient using a process analytical technology based crystallization feedback control strategy. *International Journal of Pharmaceutics, 430*(1–2), 56–64. https://doi.org/10.1016/j.ijpharm.2012.03.029.

Shah, U. V., Olusanmi, D., Narang, A. S., Hussain, M. A., Gamble, J. F., Tobyn, M. J., & Heng, J. Y. Y. (2014a). Effect of crystal habits on the surface energy and cohesion of crystalline powders. *International Journal of Pharmaceutics, 472*(1–2), 140–147. https://doi.org/10.1016/j.ijpharm.2014.06.014.

Shah, U. V., Olusanmi, D., Narang, A. S., Hussain, M. A., Tobyn, M. J., & Heng, J. Y. (2014b). Decoupling the contribution of dispersive and acid-base components of surface energy on the cohesion of pharmaceutical powders. *International Journal of Pharmaceutics, 475*(1–2), 592–596. https://doi.org/10.1016/j.ijpharm.2014.09.018.

Shah, U. V., Olusanmi, D., Narang, A. S., Hussain, M. A., Tobyn, M. J., Hinder, S. J, & Heng, J. Y. Y. (2015a). Decoupling the contribution of surface energy and surface area on the cohesion of pharmaceutical powders. *Pharmaceutical Research, 32*(1), 248–259. https://doi.org/10.1007/s11095-014-1459-3.

Shah, U. V., Wang, Z., Olusanmi, D., Narang, A. S., Hussain, M. A., Tobyn, M. J., & Heng, J. Y. (2015b). Effect of milling temperatures on surface area, surface energy and cohesion of pharmaceutical powders. *International Journal of Pharmaceutics, 495*(1), 234–240. https://doi.org/10.1016/j.ijpharm.2015.08.061.

Tao, J., Pandey, P., Bindra, D. S., Gao, J. Z., & Narang, A. S. (2015). Evaluating scale-up rules of a high-shear wet granulation process. *Journal of Pharmaceutical Sciences, 104*(7), 2323–2333. https://doi.org/10.1002/jps.24504.

Wang, Z., Shah, U. V., Olusanmi, D., Narang, A. S., Hussain, M. A., Gamble, J. F., Tobyn, M. J., & Heng, J. Y. Y. (2015). Measuring the sticking of mefenamic acid powders on stainless steel surface. *International Journal of Pharmaceutics, 496*(2), 407–413. https://doi.org/10.1016/j.ijpharm.2015.09.067.

Xu, X., Siddiqui, A., & Khan, M. A. (2013). Focused beam reflectance measurement to monitor nimodipine precipitation process. *International Journal of Pharmaceutics, 456*(2), 353–356. https://doi.org/10.1016/j.ijpharm.2013.08.083.

CHAPTER 15

Principles and applications of drag force flow sensor

Valery Sheverev[a], Vadim Stepaniuk[a] and Ajit S. Narang[b]

[a]Lenterra, Inc., Newark, NJ, United States, [b]Pharmaceutical Sciences, ORIC Pharmaceuticals, Inc., South San Francisco, CA, United States

15.1 Introduction

High-shear wet granulation (HSWG) is a complex, high energy, rapid process that uses granulating fluid (typically water) and a cohesive substance (binder) to effect the adhesion of primary powder particles. The primary powder particles are a mixture of different materials that agglomerate during HSWG into larger granules. The product formed at the end of HSWG unit operation is wet because it contains fluid used for granulation. This product is dehydrated to yield dry granules that can be used for downstream unit operations in pharmaceutical manufacturing. The wet and/or dry granules also are milled to remove any agglomerates and to produce more uniformly sized granules. The outcome of an HSWG drying-milling unit operations used in tandem is larger, denser, and more uniformly shaped particles (Table 15.1). While all three constituent unit operations (HSWG, drying, and milling) play significant role in the quality of the product of this process (granule density/porosity and size distribution), the HSWG is the most significant contributor because it drives interparticle adhesion and the formation of agglomerated granules.

Process analytical technology (PAT) tools are important to robust and reproducible manufacture of pharmaceutical dosage forms because they allow process monitoring, trending, and control. In addition, PATs can be used to guide development and scaleup of new drug product processes. Typical PAT tools used for the HSWG process in contemporary practice are outlined in Table 15.1. A key objective in monitoring and control of the HSWG unit operation is the need to identify optimum process parameters that result in consistent quality of the product. These process parameters could be, for example, the impeller speed mixing time, and the amount of water used for granulation. The quality attributes of the granules formed at the end of the HSWG unit operation and the HSWG process traditionally have been defined in terms of the attributes required for the robust, reproducible, and successful execution of the following unit operation and those that affect the quality of the finished drug product manufactured using the granules. The quality attributes that affect the following unit operation (wet milling and/or drying) include wet mass properties (adhesion or agglomerate formation, particle size distribution (PSD), and flowability). The quality attributes of the finished drug product that are affected by the HSWG unit operation include uniformity of drug content (affected by granule PSD affecting flow properties and manufacturability) and drug release or dissolution (affected by granule density or porosity). PAT tools help achieve these goals by defining, monitoring, and allowing the control of the end-point of the HSWG unit operation.

Table 15.1 Components of the high shear wet granulation (HSWG) process are described in terms of the process parameters and outcomes, material attributes, and typical process analytical technology (PAT) tools that have been used for process monitoring and control.

Descriptor	HSWG process		
	HSWG unit operation	Drying unit operation	Milling unit operation
Process description	Addition of granulating fluid, typically water or a binder solution, to primary particles to effect particle adhesion, leading to granule formation.	Temperature, air flow, and/or vacuum-assisted removal of the added granulating fluid from the wet granules.	Controlled breakage of wet or dry large particle agglomerates or coarse granules by mechanical force.
Input material	Powdered raw material and granulation fluid (usually water).	Wet mass from the HSWG.	Dried granules from the dryer.
Output material	Wet mass.	Dried granules.	Milled granules.
Quality attributes of input material	Solubility, wettability, particle size distribution, and density.	Particle size distribution, density, and cohesivity.	Particle size distribution, density, and fragility.
Quality attributes of output material	Wet material cohesivity, density, and size distribution.	Particle size distribution, density, and fragility.	Particle size distribution, density, and fragility.
Process variables	Bowl fill, amount and rate of addition of granulating fluid, impeller speed, and mixing time.	Drying temperature and time, dryer capacity utilization, air flow rate, and dew point.	Feed rate, speed of milling, and mill configuration.
Process analytical technology tools	Drag force flow sensor for densification, focused beam reflectance measurement (FBRM) for particle size distribution, impeller torque, or amperage.	FBRM for particle size distribution and near infra-red spectroscopy for water.	FBRM for particle size distribution.

15.2 In-line rheometry as process analytical technology

Modern description of granulation process includes three stages: wetting and nucleation; granule consolidation and growth; and attrition and breakage (Iveson et al., 2001). The desired granulation end-point typically lies somewhere between the second and third stages, where granule compaction and growth is balanced by the process of breakage because of impact or wear in the granulator. In pinpointing the end-point, size distribution is important, but PSD alone is not sufficient; granule densification is equally important. Visual observations, empirical process development methodologies, and PSD-focused PATs have been able to identify and control the approach of the desired state of granule agglomeration and PSD during the process, which has remained a challenge with granule densification. Nonetheless, the desired granulation end-point must take granule densification into consideration because the rate of drug release from most, if not all, pharmaceutical formulations are sensitive to granule porosity, which inversely correlates with granule densification. Mechanical force exerted by the wet mass on an obstacle (drag force) is related to the momentum of the flow, which is directly connected to the density of the mass. If a rate of the force measurement is comparable to the duration of the granule interaction with the obstacle, the measured instantaneous forces would characterize individual granule impacts thus providing information about the granule density.

HSWG was designed primarily to achieve higher granule densification than low-shear granulation processes. Traditionally HSWG process outcome focused on improving PSD and flow properties to enable manufacturability of drug substances that inherently had poor flow and low density, especially at high drug loading in the formulation.

Significant densification, however, was observed leading to disintegration and dissolution failures of the finished drug product even as the manufacturability attributes were well maintained. Thus, a narrower range of process parameters and in-process material attributes, and the associated end-point of the granulation process, were defined that would result in acceptable manufacturability as well as robust finished product critical quality attributes (CQAs). Accordingly the focus of the PAT applications in HSWG has evolved over the years to include granule densification (Narang et al., 2015a) in addition to PSD (Narang et al., 2010) as necessary readouts of critical in-process material attribute (iCMA) (Badawy et al., 2012, 2016). This chapter focuses on the PAT tools measuring mechanical forces during the HSWG unit operation, with particular emphasis on the drag force flow (DFF) sensor.

15.2.1 Impeller torque and power consumption

Among a variety of ways to monitor the HSWG unit operation online, those responding to forces exerted by the wet mass on the impeller blades are among the most popular (Levin, 2015; Parikh, 2010). Typically the first estimate of such forces is obtained by evaluating the load on the impeller by either measuring shaft torque or power consumption of the motor. The signal generated by measuring torque can be viewed as an integration of all the forces and moment arms along each of the mixing blades. A moment arm is the distance from the axis of rotation. As the granules grow in size and density, the forces exerted against the blades become greater, thereby increasing the torque. Torque and power consumption measurements correlate well with each other (Parikh, 2010). Measuring power consumption is an easier and cheaper method, and it is used most often in evaluation of the impeller mechanical load (Hansuld & Briens, 2014), along with measuring the impeller torque (Hansuld & Briens, 2014; Levin, 2015; Parikh, 2010). Typical power and torque time dependencies during a granulation cycle start with a flat low-level dry mixing stage, rise steeply when binder fluid or solution is added, then either level off into a plateau or slowly decrease, followed by flattening in an overgranulation stage. Early graduation theories (Leuenberger et al., 1979) suggested that useable granules can be obtained in the region that begins with the peak of the signal derivative. Therefore, real-time monitoring of torque or power consumption can provide an indication for the time when the process should be stopped (Holm et al., 2001; Laicher et al., 1997).

Although intuitively related to the physical parameter of the wet mass, the torque or power consumption is a volume-integrated measurement that did not prove sufficient for the process end-point determination and scaling for many formulations. Specifically a number of researchers found that a plateau of the torque versus time dependence did not correspond to the desired end-point for many formulations (Hansuld & Briens, 2014). The torque/power consumption monitoring is an attractive method of granulation monitoring because of its simplicity and direct physical connection between the measured value and the wet mass characteristics. The relationship between torque/power consumption measurements and granulation parameters, however, depends on a number of local variables, and it is not straightforward. Sensitivity of the measurement to changes in formulation properties also can be insufficient for successful monitoring and scaleup of the process.

15.2.2 Flow force measured locally inside granulator

Measuring impeller shaft torque provides information about the physical characteristics of the wet mass inside the granulator as a whole. Physical characteristics of the flow of particles alternatively can be monitored in a particular locality inside the wet mass during granulation. The local forces inside granulator vary strongly with time and position in space. For any given point within the granulator, the force changes along with the velocity of the flow at this position and the mass of the granule passing the point at a given instant. A measurement of the magnitude and direction of flow force in such a local position would provide information about individual granule momentum (product of mass and velocity) that is related directly to the density and size of the granule. Measuring instantaneous forces locally in different positions within the granulator would provide information about wet mass parameters, such as distribution of size and density of the particles, as well as about process robustness and day-to-day uniformity of the operation by showing force-time dependence in good temporal resolution.

15.2.2.1 *Diosna-boots flow sensor*

The measurement of local forces within the granulator was first attempted by the Boots Company in late 1970s, which developed an immersion probe for Diosna high-shear mixers (Kristensen & Schaefer, 1987). They used a device commonly known as target flow meter (Fig. 15.1). When immersed in the wet mass during the granulation process, flow particles, for example, granules, collide with the probe target so that a component of their momentum is transferred to the probe generating a flow force. The target typically is mounted to a stalk, and the stalk generally is affixed to a bendable beam configured to deflect/bend under the influence of the flow. Strain gauges affixed to the bendable beam, recessed within a chamber, measure the degree of deflection of the beam. The resistive strain gauges used in the Diosna-Boots probe generated an electrical signal proportional to the force exerted on the target and bendable beam. The force is proportional to the momentum of the particles. As granulation proceeds, the increasing density and size of the particles result in an increase in the momentum of the granule impact, which causes an increase in the output signal of this sensor. Therefore, the probe indirectly would characterize the granule size and density, which are critical characteristics of the granulation process.

The Diosna-Boots probe design, however, had several shortcomings that made it difficult to consistently measure forces in powder and multiphase flows. The strain gages chamber could trap particles and/or high viscosity components, altering the deflection of the stalk and skewing the measurements. Also, resistive strain gages need to be protected from the wet mass with a coating. The coating contributes to stiffness of the bendable beam and thus adversely affects the sensitivity of the probe. In addition, coating wears down by the particle flow, which can affect probe sensitivity and expose

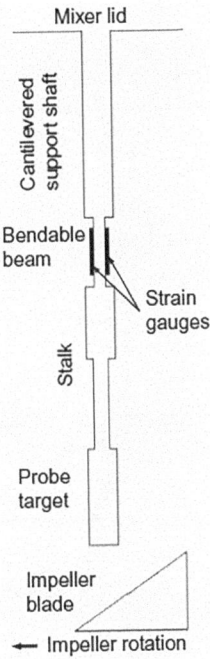

FIGURE 15.1

The Diosna-Boots flow sensor, an immersion probe for the measurement of local forces first developed by the Boots Company in late 1970s for Diosna high-shear mixers (Kristensen & Schaefer, 1987). When immersed in the wet mass during granulation process, this probe translates the momentum of colliding particles into flow force. The location of the strain gages is a sensitive location of the probe that could be susceptible to fouling during operation.

electrical parts of the device to the wet mass. Apparently because of these shortcomings, Diosna-Boots probe did not gain popularity in the HSWG community.

15.2.2.2 Drag force flow sensor

Development of fiber optic sensing enabled an improved design of a probe that is similar in its measurement principle to the Diosna-Boots probe but is durable and reliable (Narang et al., 2015a). The DFF probe is a hollow cylindrical pin, whose deflection is sensitively measured by an assembly of optical strain gages, or fiber Bragg gratings (FBGs). In this design, FBGs are affixed on the inner surface of the hollow pin and, therefore, not exposed to the flow (Figs. 15.2 and 15.3). Because of the FBG's high sensitivity to motion sensing, the pin can be thin (1–3 mm in diameter), rugged (stainless steel), and stiff. It, therefore, provides minimal intrusion to the flow, and its measurement sensitivity depends weakly on the amount of flow material that can stick to the sensor surface (shown in a photo of the installed probe taken after the completion of a measurement cycle in Fig. 15.4). The probe does not have any moving parts and/or traps (such as the recessed chamber for strain gages in the Diosna-Boots

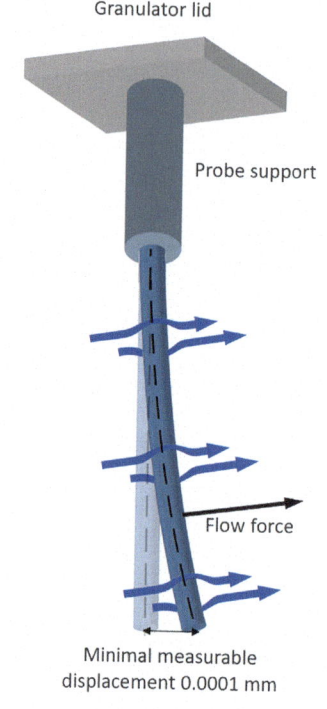

FIGURE 15.2

An illustration of the flow forces measured by the drag force flow (DFF) sensor. When placed inside a high-shear granulator hinged to the lid, the powder flow exerts a force upon the probe, leading to the physical bending of the probe in the direction of the flow.

probe shown in Fig. 15.1, which makes DFF sensor not paralyzable in particulate, bulky, and viscous materials.

The DFF sensor responds to a force component along the measurement axis (indicated with an arrow in Fig. 15.3), which typically is aligned with the direction of the flow at the probe tip. In addition to force, the optical assembly of the DFF sensor measures the temperature of the pin. Information is collected via optical fibers that connect the probe to an optical interrogator.

Optical sensing technology provides a number of advantages, such as not being susceptible to electromagnetic interference, presenting no ignition hazard (thus, having the potential to work with compounds that have low minimum ignition energy), and the ability to withstand higher temperatures. Being very thin (125 μm), several FBGs can fit into a millimeter-sized hollow pin.

15.3 Measurement principle of the drag force flow sensor

FBGs are periodic structures of varying refractive index embedded in the core of an optical fiber (Fig. 15.5A). FBG reflects light from a broad-spectrum interrogation optical source only at specific

15.3 Measurement principle of the drag force flow sensor

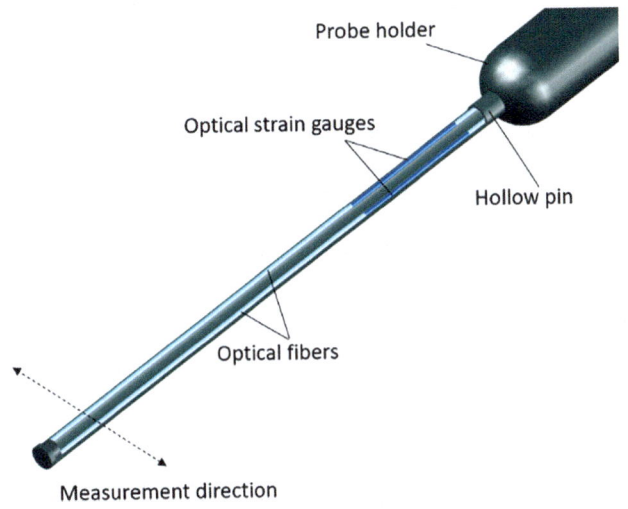

FIGURE 15.3
A schematic diagram of the drag force flow (DFF) sensor showing internal structure of the hollow pin design. The deflection of the fiber Bragg gratings (FBGs) affixed on the inner surface of the pin (labeled as optical strain gages) sensitively measures the force impact on the pin. Notice that the measurement direction of the flow force is along the imaginary straight line that is formed by the two optical strain gages.

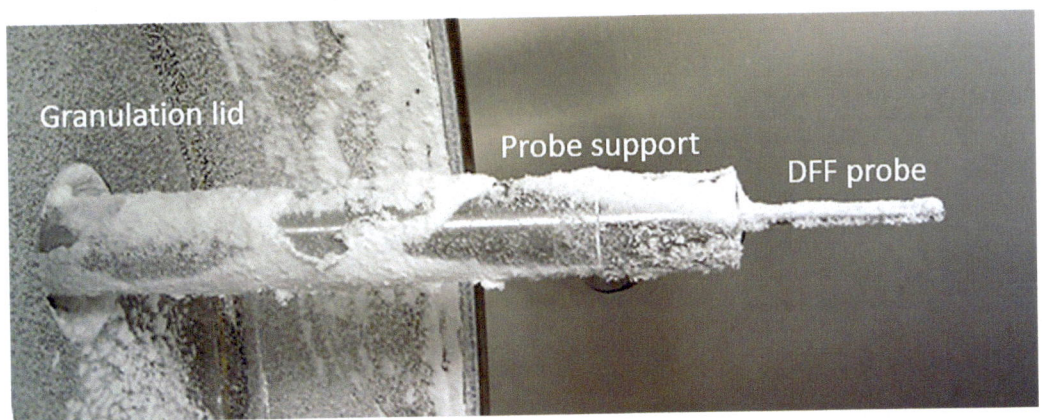

FIGURE 15.4
A representative photograph of the drag force flow (DFF) sensor taken after the completion of a measurement cycle in a high-shear granulator using a placebo formulation. This figure shows the granulation material sticking to the probe in the direction of the flow. The amount of material adhering would depend on the tackiness of the granulation powder mass. The measurement sensitivity of this probe depends weakly on the amount of flow material that can stick to the sensor surface.

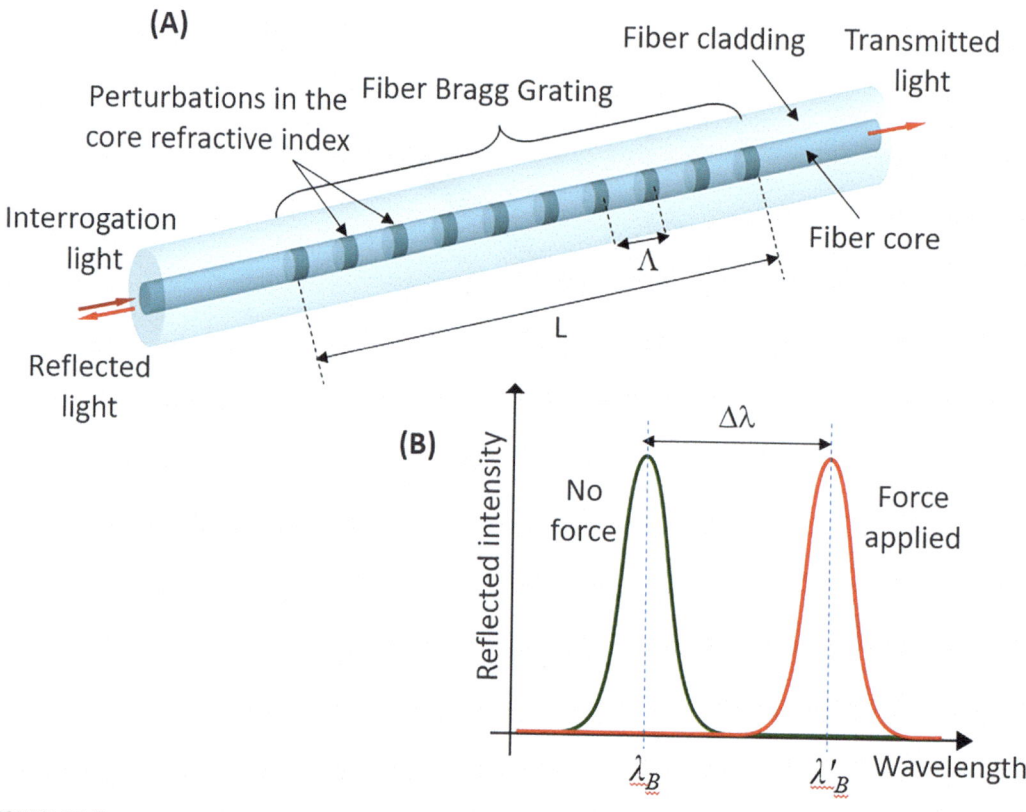

FIGURE 15.5

Detailed view of the fiber Bragg grating (FBG) as an optical strain gage (A) and its reflection spectra (B). Miniscule straining of the FBG leads to perturbations in the grating constant Λ that affects the characteristic wavelength of the reflected laser light. Measuring the shift in this wavelength, $\Delta\lambda$, is the core principle of DFF sensor's measurement of flow forces.

wavelengths (Fig. 15.5B). FBG wavelength, λ_B, the center wavelength that is reflected back, is calculated as $\lambda_B = 2n\Lambda$, where n is the effective index of refraction of the light propagation and Λ is the refractive index modulation period (see Fig. 15.5). This equation implies that the reflected wavelength λ_B is affected by any variation in the physical or mechanical properties of the grating region. For example, strain on the fiber increases Λ, and alters n via the stress-optic effect. Similarly changes in temperature lead to changes both in n via the thermo-optic effect, and in Λ because of thermal expansion or contraction of the grating material. As a result, the central wavelength λ_B shifts (Fig. 15.5). By monitoring the FBG spectrum shift, $\Delta\lambda$, therefore, it is possible to determine the displacement of the pin tip (Fig. 15.2) and the force acting on FBG and FBG's temperature.

Typical interrogation scheme involves a narrow band tunable laser source (Fig. 15.6). Light from the laser travels through the fiber and beam splitter to an FBG located on the inner surface of the pin of the sensor. Depending on the laser wavelength and FBG reflection spectrum, a part of the light is reflected

15.3 Measurement principle of the drag force flow sensor

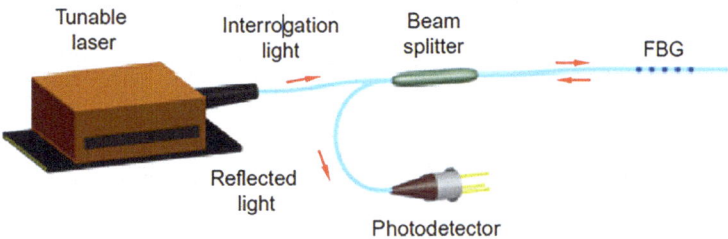

FIGURE 15.6

Instrumentation of the drag force flow (DFF) sensor demonstrating the principle of the interrogation: a tunable laser that passes light into the fiber Bragg grating (FBG) through an optical circuit, which also detects the intensity of the reflected light.

from FBG and travels back through the same fiber and the beam splitter to a photodiode, where its intensity is measured. The reflection spectrum of FBG is recorded by scanning laser wavelength through a predetermined range. The interrogator records the spectrum of each FBG repeatedly (with a typical frequency of 500 to 1000 spectra per second). When the probe pin is deflected because of force acting on it, one FBG becomes stretched and the other compressed, leading to FBG spectra to shifts along the wavelength axis in opposite directions (Fig. 15.7). On the contrary, the spectra for both FBGs shift in the same direction due to temperature change. The interrogator software determines relative spectra shifts and calculates force acting on the pillar as well as ambient temperature.

15.3.1 Sensing model

DFF sensor works by quantifying the force exerted by the flow against the pin. Depending on the nature of the flow, the sensor measures drag force exerted either by fluid or particles. A force, F, acting on the tip of a cylindrical hollow pin causes the elastic cylinder to bend, with pin tip deflection given by (Oberg et al., 2016)

$$\delta = \frac{4Fl^3}{3\pi E \left(r_0^4 - r_i^4\right)} \quad (15.1)$$

where E is the Young's modulus of the material, r_0 is the outer radius, r_i is the inner radius of the hollow pin, and l is its length. The FBG assembly (Fig. 15.3) provides an optical response based on the magnitude of the pin tip deflection, δ. By selecting pin material and size, it is possible to design and manufacture flow force sensor with desired sensitivity to δ and, therefore, to the applied force F.

Deflection of the pin tip causes FBGs attached to the inner surface of the hollow pin to change FBG length, L (see Fig. 15.5). Assuming that the force is acting along the sensor measurement direction (Fig. 15.3), the relative change in the FBG length, $\Delta L/L$, is

$$\frac{\Delta L}{L} = \frac{\delta(r_i - r_{FBG})}{Ll} \quad (15.2)$$

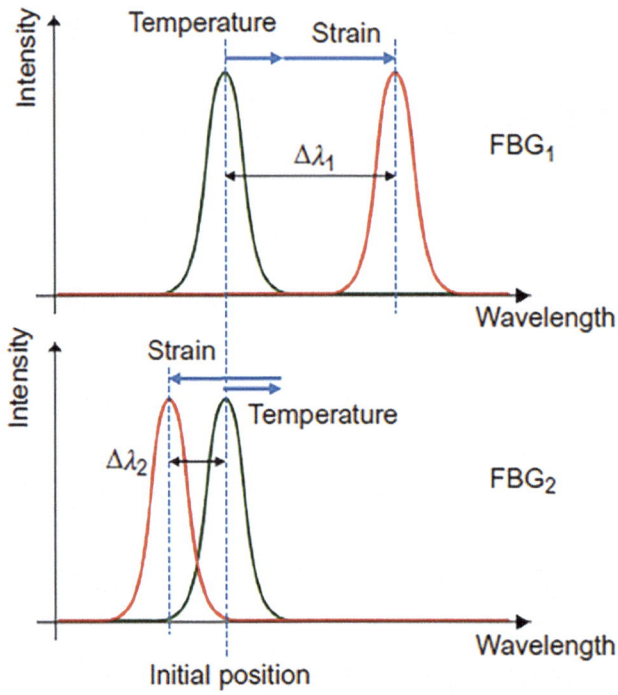

FIGURE 15.7

Measurement principle of temperature and strain by the drag force flow (DFF) sensor. Temperature shifts spectra of both FBGs in the same direction independently from strain. When force applied to the probe, first FBG is stretched and the second FBG is compressed causing the spectra to shift in opposite directions. Thus, interpretation of the optical wavelength shift responses from the two FBGs enables simultaneous detection of temperature and strain.

where r_{FBG} is the radius of the optical fiber (62 μm). Change in the FBG length, ΔL, leads to a shift of the FBG spectrum, $\Delta \lambda$ of

$$\Delta \lambda = \alpha \frac{\Delta L}{L} \tag{15.3}$$

where α is the is FBG strain sensitivity (typical value is 1.2 pm/$\mu \varepsilon$) (Campanella et al., 2018). Existing optical interrogators enable reliable detection of a 0.1 pm wavelength shifts of an FBG wavelength, at the measurement frequency of at least 500 spectra per second. Assuming the stainless-steel pin length of $l = 40$ mm, diameter of $r_0 = 1$ mm and wall thickness of 0.2 mm, and FBG length of $L = 1$ cm, the minimal detectable force calculated from is Eqs. (15.1)–(15.3) is 0.6 mN. The corresponding pin tip deflection is $\delta = 0.17$ μm.

Combining Eqs. (15.1)–(15.3) one can see that the FBG spectrum shift is proportional to force acting on the pin tip, or $F = \gamma \Delta \lambda$. Coefficient γ can be calculated from Eqs. (15.1)–(15.3) or determined experimentally by applying known force to the pin and measuring the corresponding spectrum shift.

15.3.2 Temperature compensation

FBG spectrum shifts because of both physical stress and change of ambient temperature. Temperature-related spectrum shift can be estimated from $\Delta \lambda = \beta \Delta T$, where β is the FBG thermal response coefficient (typical value is 9.9 pm/°C). To separate force and temperature action, a pair of FBGs are used in the DFF probe (Fig. 15.3). When the pin is deflected, one of the FBGs is stretched and another compressed by a comparable length. Their spectra, therefore, shift in opposite directions (Fig. 15.7). Assuming that temperature shifts are the same for both FBGs ($\Delta\lambda_{1temp} = \Delta\lambda_{2temp} \equiv \Delta\lambda_{temp}$), and the shift because of strain for FBG$_1$ is of the same magnitude as that for stress for FBG$_2$ ($\Delta\lambda_{1strain} = -\Delta\lambda_{2stress} \equiv \Delta\lambda_{force}$), temperature and force actions are separated as follows:

$$\Delta\lambda_{\text{force}} = \frac{|\Delta\lambda_1| - |\Delta\lambda_2|}{2}; \quad \Delta\lambda_{\text{temp}} = \frac{|\Delta\lambda_1| + |\Delta\lambda_2|}{2} \tag{15.4}$$

where $\Delta\lambda_1$ and $\Delta\lambda_2$ are total spectral shifts measured for FBG1 and FBG2, respectively, and $\Delta\lambda = \Delta\lambda_{\text{temp}} + \Delta\lambda_{\text{strain/stress}}$. As force, the temperature change is proportional to the combined spectral shift, $\Delta T = \frac{1}{\beta}\Delta\lambda_{\text{temp}}$. Values of the proportionality coefficients for the force and temperature measurements are usually determined by calibration.

15.3.3 Drag force flow probe in a uniform fluid flow

A quantitative connection between flow characteristics and force experienced by the cylindrical pin immersed into the flow can be established by considering interaction between a fluid traveling along the sensor surface. Multicomponent flows, such as those realized in a granulator, can be modeled as combination of a uniform fluid flow and a flow of particles. For a uniform fluid flow, the force acting on the cylindrical hollow pillar is (White, 1998):

$$F = \frac{C_d}{2} A \rho v^2 \tag{15.5}$$

where C_d is a drag coefficient for a cylinder, $A = 2r_0 l$ is cross-section of the cylinder, ρ is the fluid density, v is flow velocity. C_d is a function of the Reynolds number $Re = \frac{2r_0 v \rho}{\mu}$, where μ is the kinematic viscosity. Thus, if the flow velocity is known, the DFF probe output in a liquid flow can be connected to the viscosity of the fluid.

15.3.4 Drag force flow probe in a powder flow
15.3.4.1 Single granule impact

If the measurement rate is significantly greater than the frequency of the blades, an elementary interaction between a granule and the cylindrical probe would manifest itself as a pulse or peak in the force versus time measurement plot (Fig. 15.8A). An impact of a single particle of mass m and velocity v colliding head-on with the probe (Fig. 15.8) would cause the probe tip to start moving in the direction of the impact with certain velocity V. The deviation of the probe tip is measured by the sensor as a force that is determined by the relation between mass of the particle and the effective mass of the pin (Harris & Piersol, 2002) $M_{\text{eff}} = 0.23 m_{\text{pin}} = \frac{1}{4\pi^2}\frac{\kappa}{f^2}$, where m_{pin} is the probe pin mass, $\kappa = \frac{F}{\delta}$ is the

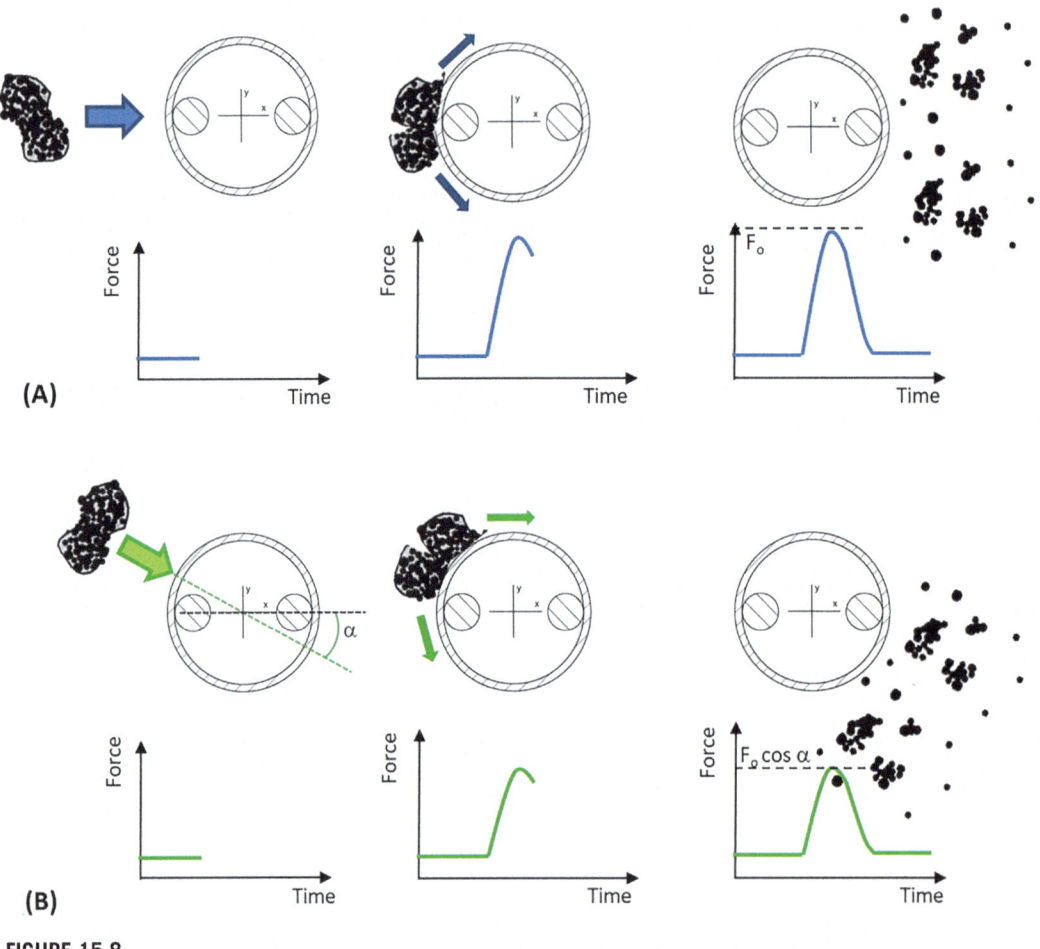

FIGURE 15.8

Mechanistic origin of flow forces on the drag force flow (DFF) sensor. A particle colliding with the sensor in the direction of the flow generates a force that is recorded by the sensor. (A) The impact is along the measurement axis of the DFF sensor (x-axis). (B) The particle impact is at an angle a to the measurement axis, the detected force pulse is reduced.

stiffness of the pin and f is its mechanical resonance frequency. For stainless steel pin of 40 mm length and 1 mm in radius with 0.2 mm walls, $M_{eff} \cong 0.08$ g, $\kappa \cong 3260$ N/m, $f \cong 1000$ Hz.

Assuming a perfectly inelastic collision, the momentum conservation yields and for $mv = (m + M_{eff})V$, which leads to $V = \frac{m}{m+M_{eff}}v$ or, if $m \ll M_{eff}$, to $V = \frac{m}{M_{eff}}v$. The energy of the impact is transferred to the potential energy of the probe at the maximum deviation δ_0: $\frac{M_{eff}V^2}{2} = \frac{\kappa \delta_0^2}{2} = \frac{F_0^2}{2\kappa}$. Therefore, assuming that the granule has diameter d, and that its density is ρ, for the force pulse magnitude (FPM) one obtains

$$F_0 = mv\sqrt{\frac{\kappa}{M_{\text{eff}}}} = \frac{\pi^2}{3}fv\rho d^3 \qquad (15.6)$$

The pulse magnitude is proportional to the mass of the granule or the third power of its diameter. Using Eq. 15.6, one can estimate the smallest particle impact that will be detected by the sensor. The minimum detectable force was estimated in Section 15.3.1 to be 0.6 mN, which approximately corresponds to the impact of the 0.5 mm diameter water droplet moving with the velocity of 1.5 m/s.

In this analysis, it is assumed that the granule impacts the probe along its measurements axis (see Figs. 15.3 and 15.8A). If the impact is at an angle α to the measurement axis, as shown in B, the sensor generates a smaller signal:

$$F_\alpha = F_0 \cos\alpha \qquad (15.7)$$

Since granules in a mixer move at most locations in multiple directions, the force measured by DFF sensor should be treated as a low estimate of the actual impact force on the pin.

15.3.4.2 Drag force flow probe in a thin powder flow

Another limiting case occurs when a continuous one-dimensional uniform flow of identical spherical small particles of mass m and density ρ ($m = \rho\frac{\pi d^3}{6}$ where d is particle diameter). Assuming that the particle diameter d is much smaller than the diameter of the probe, the equivalent (integrated over pin length) force on the tip of the pin can be evaluated in this case as follows (Gere & Goodno, 2013).

$$F_0 = \frac{3}{8}mv\frac{dN}{dt} = \frac{3}{8}mvnAv = \frac{\pi}{16}A\xi\rho v^2 \qquad (15.8)$$

Here N is the total number and n is the number density of the particles, A is the cross-sectional area of interaction, and v is the velocity of the flow. Particle number density is expressed as $n = \frac{\xi}{d^3}$, where $\xi < 1$ is the degree of packing. The maximum degree of packing. $\xi = 1$, is achieved when every particle is in touch with six neighboring particles so that the distance between the particle centers equals to its diameter. When the distance between the particle centers is, for example, twice the particle diameter, $\xi = \frac{1}{8}$. Assuming that the degree of packing in the wet mass does not change significantly, the measured force is the function of particle density and the interaction cross-section. The interaction cross-section depends strongly on the degree of binding between the particles. For example, dry powders particles can be considered not bound to each other and therefore its interaction cross-sectional area is close to the cross section of the pin (diameter times submerged length).

Eq. 15.8 also could be written in terms of bulk density, $\hat{\rho} = mn$:

$$F_0 = \frac{3}{8}A\hat{\rho}v^2 \qquad (15.9)$$

A flow of particles with bulk density $\hat{\rho} = 1$ g/cm³ moving along the measurement axis of the probe with a velocity of 2.5 m/s would deflect the pin equivalently to a force of 0.19 N applied to the pin tip. The lowest detectable force of 0.6 mN translates to minimal detectable bulk powder density of 0.003 g/cm³. In wet powders, where the interaction cross section is greater than the cross section of the pin, this estimate will reduce.

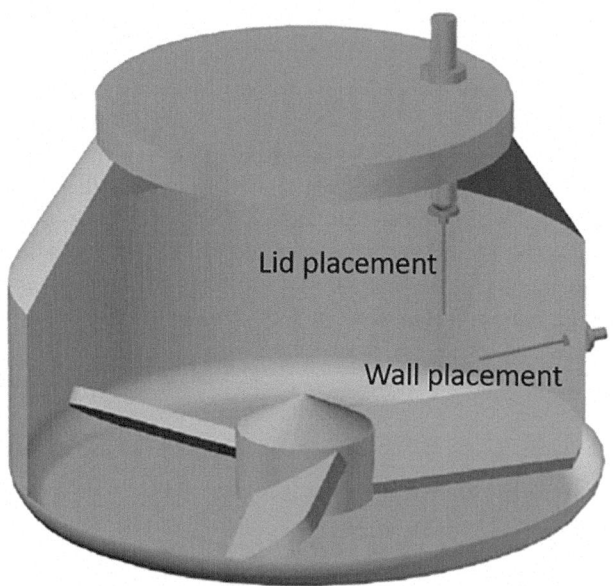

FIGURE 15.9

Placement of the drag force flow (DFF) sensor in the granulator from the lid or the side port. Consistent placement in the granulator across the batches that are being compared with respect to the height from the impeller and the radial distance from the shaft is important for consistent data generation and scaleup, because particle flow patterns vary in different regions of the granulator. In addition, the probe should be placed away from the port of granulating fluid addition and the chopper.

15.4 Drag force flow sensor data interpretation
15.4.1 Raw data

The DFF sensor response is a combined force of all collisions experienced by the sensor pin. Thus, the DFF sensor response represents a composite output of granule mass and size and rheological properties of the powder (such stickiness) that can be characterized as wet mass consistency.

Depending on the granulator type, DFF probe can be positioned through a port on the granulator lid or using a side port, if available (Fig. 15.9). Lid placement allows for flexible positioning of the probe using an adjustable shaft to survey the granulator volume.

Fig. 15.10A presents a typical measurement cycle taken with DFF sensor in a GEA PharmaConnect 10L granulator. These data were generated using a placebo formulation consisting of microcrystalline cellulose (MCC; 61% w/w), and lactose monohydrate (37% w/w) as diluents, croscarmellose sodium (1% w/w) as a disintegrant, and hydroxypropyl cellulose (HPC, 5% w/w) as a binder (Narang et al., 2016). Two kilograms of dry powder were granulated with 800 g water, added continuously over a period of 3 min. The impeller and chopper were ON during the dry mixing phase through the water addition and wet massing time. Impeller tip speed was kept at 4.8 m/s (300 rpm, blade frequency 15

15.4 Drag force flow sensor data interpretation

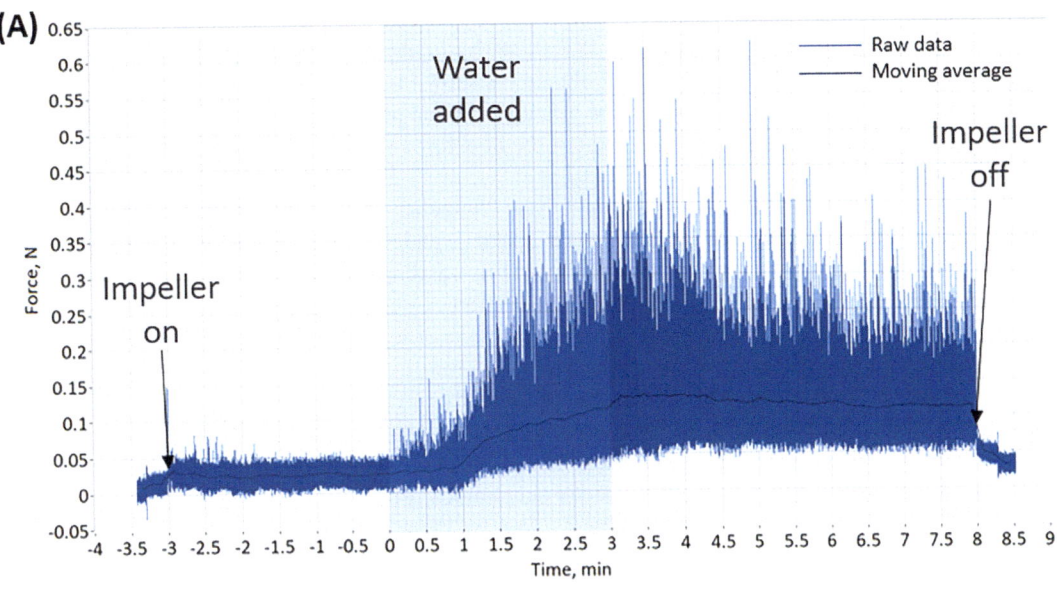

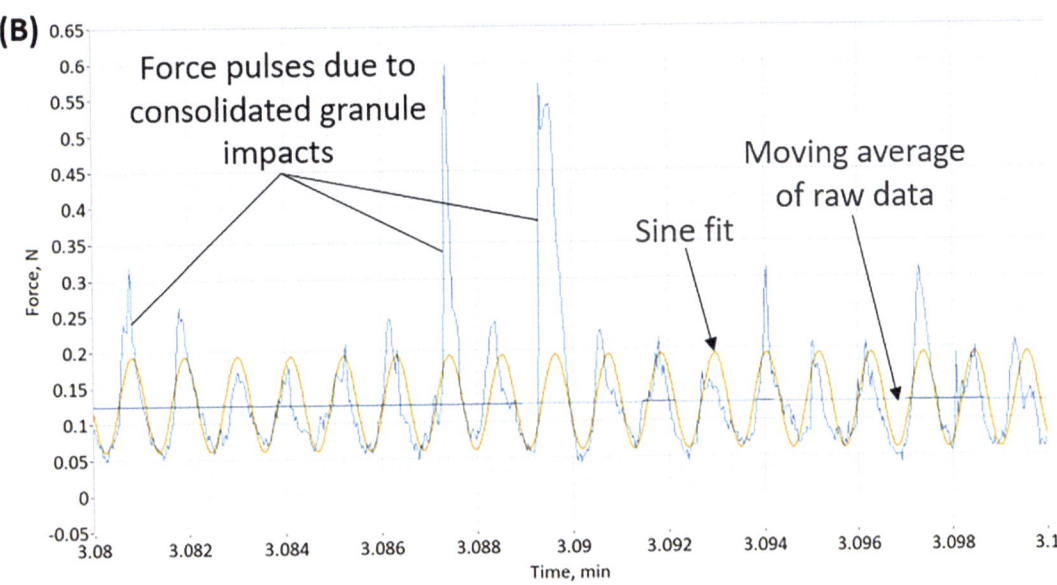

FIGURE 15.10

Raw data (A) and high temporal resolution portion of the plot (B) of the raw signal (blue) and the harmonic sine fit amplitude function (red) indicating the periodicity and the pulses of particle impacts on the DFF sensor in a high-shear granulator. High frequency data collection by DFF sensor enables high temporal resolution of granule impacts.

Hz) resulting at the impeller tip speed of 5 m/s. Chopper speed was kept at 1000 rpm. Granulation was monitored using an in-line DFF sensor type P-3000-40 by Lenterra Inc. The probe was installed from the granulator lid using an available ISO KF flange (as shown in Fig. 15.9) in such a way that the tip of the probe was 2.5 cm above the top of the granulator blade and 8.2 cm off the blade rotation axis. DFF sensor recorded the tangential component of the flow force with the measurement rate of 500 data points per second.

The unit operation consisted of a dry mixing phase ("impeller on" to "water added" in Fig. 15.10A), water addition phase (shaded area labeled "water added"), and wet massing phase ("water added" to "impeller off") while the contents were mixed. Plot in Fig. 15.10A contains 360,000 force measurement points taken during approximately 12 min, and the section of this plot shown in Fig. 15.10B includes 600 measurement points taken over 1.2 s. The signal consists of separate peaks, and the peak occurrence frequency matches the frequency of the blades passing below the sensor, as evident from Fig. 15.10B. When the blade is directly under the probe pin, the velocity of the wet mass is greatest, and the force measured by the probe is highest. Minima occur when flow velocity near the probe is lowest, which happens when the probe is between the blades. One also can observe a fine structure in each peak that includes a number of narrow pulses with various magnitudes, which are due to elementary impacts of granules of various sizes measured consecutively. Some of them overlap in time and some others have fairly large magnitudes. Orange line represents a harmonic fit to the data set shown in Fig. 15.10B (18 periods).

Overall, the raw signal can be understood as a periodic force versus time signal composed of pulses of various magnitudes. The width of these elementary pulses characterizes the time of contact between the granule and the probe, which is related to the granule size and density, and the magnitude of the pulse characterizes the mass of the granule (Fig. 15.8).

Peak forces exerted by dry powder (first 3 min in the Fig. 15.10A plot) are noticeably lower (about two orders of magnitude) than those observed during or after water addition, which is indicative of granulation dynamics. The greatest force is observed sometime after the water addition was stopped, with gradual decrease afterwards that is indicative of the gradual decrease in granule size This information can be used for identification of the granulation endpoint. As discussed in Section 15.2, granulation process starts with wetting and nucleation that roughly coincides in time with the water addition, and followed by the consolidation and growth stage that starts during water addition and continues afterwards. It is during this stage that the larges agglomerates, which manifest themselves as peaks of high magnitude, are formed. The attrition and breakage stage of the granulation process, where large agglomerates break down in collision with other granules or granulator walls, follows. Further, an equilibrium between breakage and coalescence is achieved that manifests itself as a plateau (after 4.5 min). It is typical to select granulation end-point when the consolidation of granules is in equilibrium with the breakage. For the example of Fig. 15.10A, this would be around 3.5 min.

Averaging raw data over a large number of blades would provide a simple integrated over time parameter that would be comparable to the impeller torque and will not have information about a single blade action. This force moving average for 150 number of periods (5000 number of data points) is shown with a dark blue line in A. Force moving average magnitude and its temporal evolution depends on the position of the sensor within the bowl, therefore it can be used in analyzing granulation process evolution along with other statistical metrics described in the following sections.

15.4 Drag force flow sensor data interpretation

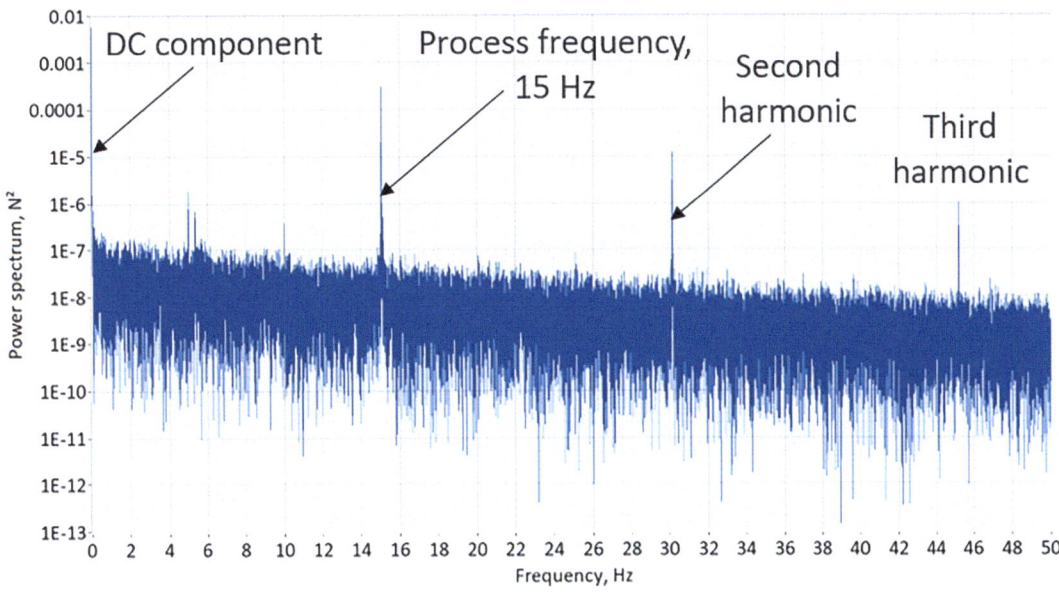

FIGURE 15.11

Frequency spectrum (fast Fourier transformation-FFT) of the signal of Fig. 15.10A showing the process frequency of the agitator blade rotation as well as its second and third harmonics.

15.4.2 Force pulse magnitude

More specific information about the granule interaction with the DFF probe can be obtained from statistical analysis of peak magnitudes observed in a force versus time dependence. Each such a peak represents action of a particular blade. Fast Fourier transformation (FFT) of the raw signal gives a value of the blade frequency. FFT for the raw data shown in Fig. 15.10A is given in Fig. 15.11. The blade frequency of 15 Hz (300 RPM for three-blade impeller) is referred to below as "process frequency." A useful parameter for characterization of the blade action is the FPM. It is introduced as a difference between the greatest and smallest values of force measured over one period. Each value of FPM, therefore, describes a passing of one blade in vicinity of the probe pin. Fig. 15.12A illustrates the computation of FPM for four consecutive blades and Fig. 15.12B provides a plot of FPM versus time for these four blades. Being a differential measurement, FPM provides a reliable characteristic of the wet mass densification and consistency because it is independent from possible zero drift in the raw signal.

15.4.2.1 Force pulse magnitude histograms

The FPMs vary from blade to blade randomly, reflecting the random distribution of granule size in the wet mass. Such a distribution could be observed by selecting a certain number of consecutive FPMs and constructing a histogram. Fig. 15.13A represents such an instantaneous histogram of FPMs calculated for the raw data presented in Fig. 15.10A, for the time interval between 3.22 and 3.77 min. This histogram contains 500 FPM values. Time interval of 0.55 min can be considered small compared

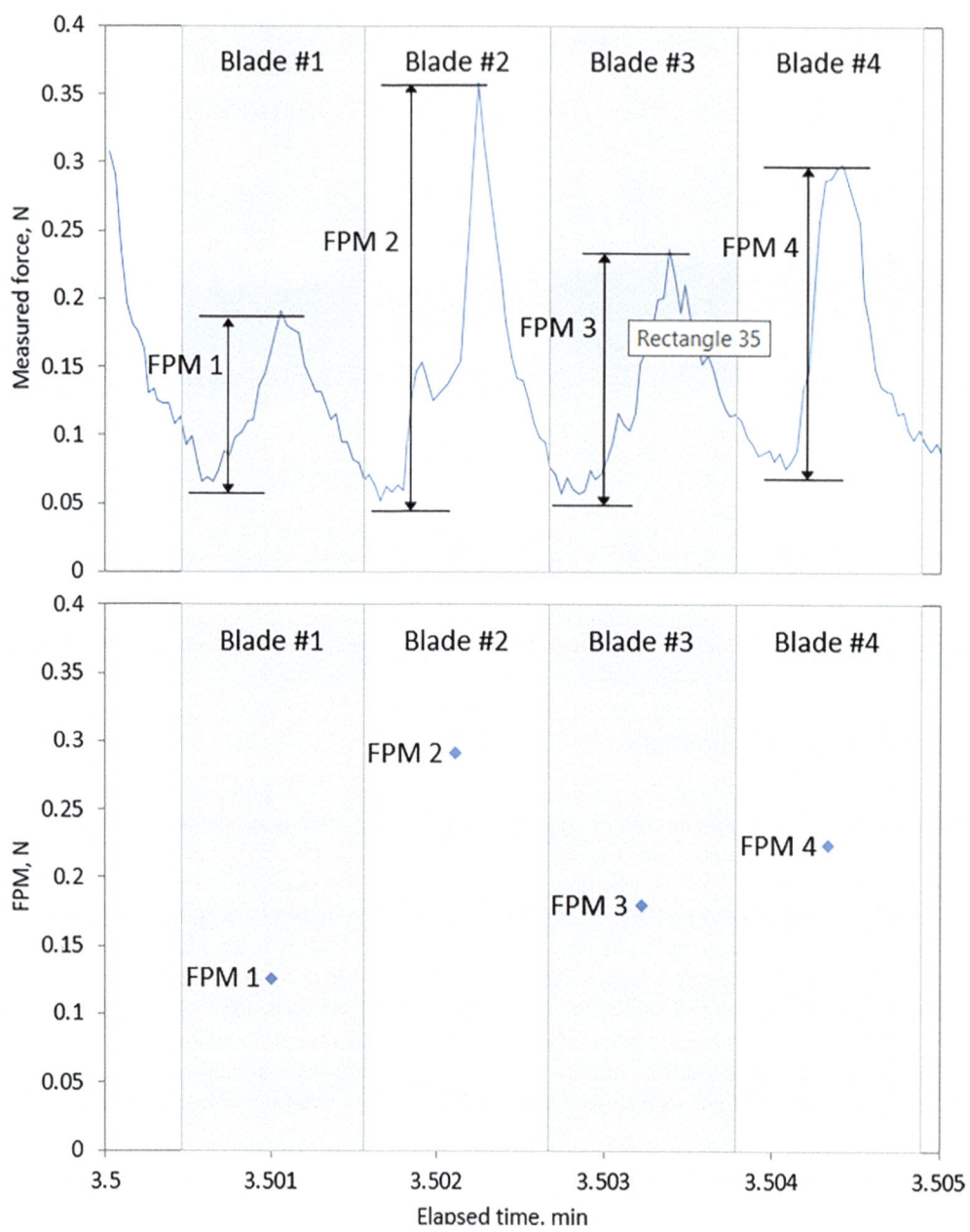

FIGURE 15.12
An illustration of calculation of force pulse magnitude (FPM) for each of the waves of particle impacts on the drag force flow (DFF) sensor as a blade passes below the sensor. The FPM reflects maximum measured force of impact during each blade pass.

15.4 Drag force flow sensor data interpretation

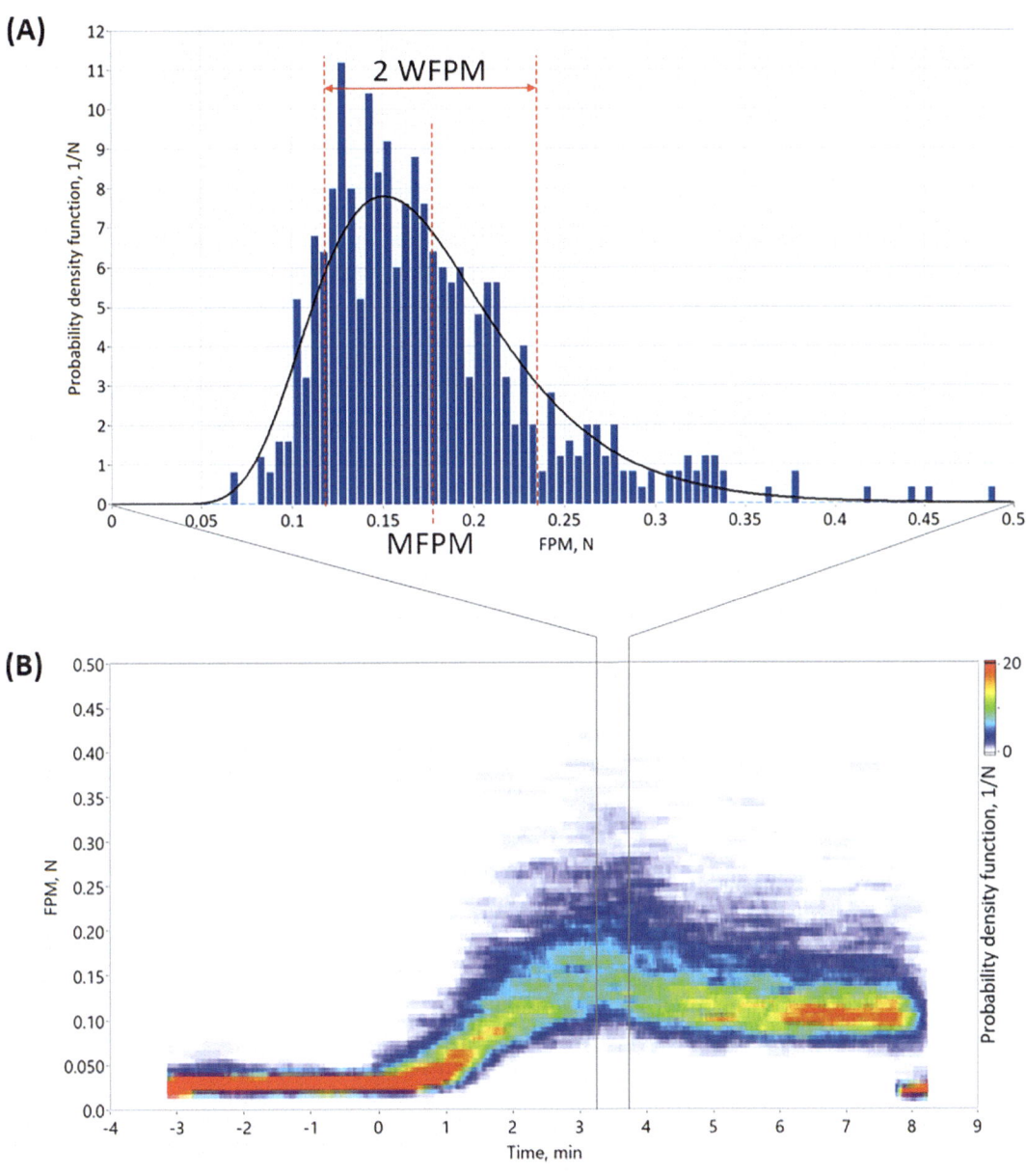

FIGURE 15.13

(A) A histogram of 500 force pulse magnitudes (FPMs) measured between 3.22 and 3.77 min of the test given in Fig. 15.10A, solid line represents a lognormal distribution restored from the subarray FPM values. (B) Three-dimensional plot of 500-blade moving histogram with probability density function (PDF) being represented on z-axis.

to total duration of granulation, therefore we will assume such a histogram to be "instantaneous." These instantaneous histograms carry information about wet mass densification at a particular time of the granulation process.

A reasonable approach to observing FPM distribution dynamics is to construct consecutive histograms for a fixed size array of FPMs, each shifted by one blade in time. Therefore, the first histogram is built for an array of N blades counted from blade number 1 to blade number N. The second histogram is built for blades from number 2 to number N + 1, and so on. In this way we obtain three dimensional plot (moving histogram), which is presented in Fig. 15.13B as a color map with the probability density function (PDF) being represented on the z-axis. The larger the array size, N, the more statistically significant the FPM distribution. But when N, measured in time units, significantly exceeds the characteristic time constant of physical and chemical processes occurring in the course of granulation, rapid changes of wet mass parameters might not be detected. But when plotting consecutive histograms with a small-time step of one blade, a reasonable time resolution is obtained even if the array size exceeds the characteristic time of the granulation process. For example, an array of 500 blades (33 s) for data shown in Fig. 15.10A might seem to exceed the characteristic times, but the following 500-blade histogram calculated with an increment of one blade (0.066 s) effectively increases the temporal resolution given by the moving histogram plot.

15.4.2.2 Basic metrics: Mean force pulse magnitude and CVFPM

The histogram in Fig. 15.13A is naturally described by the lognormal distribution, which is a typical distribution characterizing particle sizes of granulated powder (Masuda et al., 2006). The mean and standard deviation of the natural logarithm of FPM data array are used to restore a continuous lognormal distribution that fits the histogram (black line in Fig. 15.13A). Therefore, the instantaneous state of the FPM distribution can be characterized with two parameters: the lognormal distribution mean that is introduced here as MFPM indicated with a vertical line in Fig. 15.13A and calculated as

$$\text{MFPM} = e^{\left(\mu + \frac{\sigma^2}{2}\right)} \tag{15.10}$$

and the width of the lognormal distribution (square root of variance) which we call here width of force pulse magnitude distribution.

$$\text{WFPM} = \sqrt{\left(e^{\sigma^2} - 1\right)e^{(2\mu + \sigma^2)}} \tag{15.11}$$

where μ and σ are the mean and standard deviation of FPM's natural logarithm values, respectively. These values are referred to the time instant that is selected as a mid-range for the array. For example, time instant assigned to the distribution given in Fig. 15.13A is $\frac{3.33 + 3.67}{2} = 3.5$ min.

Generally the mean value and the width are independent from each other. MFPM is the mean force and WFPM characterizes uniformity of the FPM distribution in the array, that is how close the values are to each other. However, same value of WFPM can indicate a narrow distribution for a larger MFPM value and a wide distribution for a smaller MFPM. Therefore, to characterize the distribution width, a ratio of the width to the mean of distribution can be employed. For a normal distribution, this is known as the coefficient of variation, CV, the ratio of the standard deviation to the mean. The coefficient of variation measures the variability of a series of numbers independently of the unit of measurement used for these numbers, therefore it is a better metric for comparing variability of distributions measured in

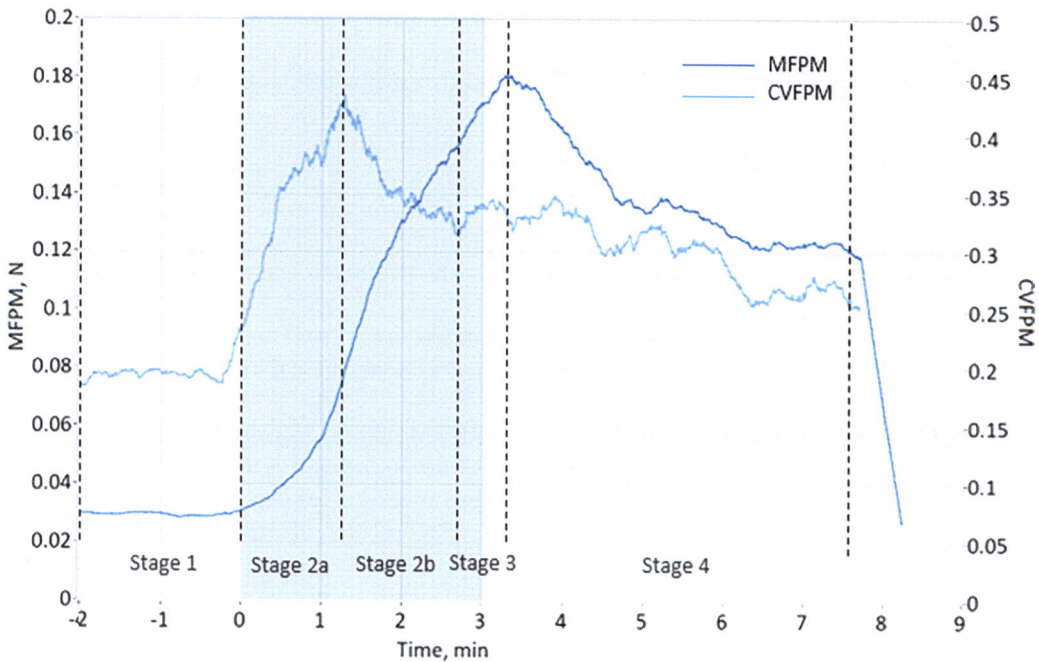

FIGURE 15.14

Process fingerprint: plots of moving mean (MFPM, blue) and coefficient of variation (CVFPM, light blue) of the lognormal distribution restored from consecutive arrays of 500 force pulse magnitudes (FPMs) measured for raw data shown in Fig. 15.10A. The shadowed area represents water addition cycle. Both MFPM and CVFPM change over the granulation cycle significantly. MFPM reaches maximum soon after the end of water addition, approximately at 3.3 min, and CVFPM—in the middle of water addition, at 1.3 min. Dynamics of MFPM and CVFPM provide convenient means for determination of granulation end point and differentiation between formulations.

different tests. For characterizing granulation processes, we introduce a coefficient of variation of FPM values, CVFPM, defined as:

$$\text{CVFPM} = \frac{\text{WFPM}}{\text{MFPM}} \qquad (15.12)$$

15.4.2.3 Process fingerprint

A combined plot of MFPM and CVFPM as functions of time represent the granulation process fingerprint. MFPM evolution indicates change in densification of the wet powder during granulation and the CVFPM characterizes uniformity of the granule sizes at various times. A process fingerprint for the measurement shown in Fig. 15.10A is given in Fig. 15.14. One can identify the following granulation stages:

- Stage 1 (−2 to 0 min) dry powder mixing—both MFPM and CVFPM are steady at a relatively low level of 0.04 N and 0.19, respectively, before the water addition starts;

- Stage 2 (0–2.7 min) extensive wetting and nucleation—consists of two sub-stages:
- Stage 2a (0–1.3 min) MFPM grows fast reflecting increasing number of heavier agglomerates being formed in the powder, CVFPM increases sharply as well, reaching maximum at 1.3 min; one would expect the distribution of masses being widest at this time in the wet mass, when many massive agglomerates formed, but a significant amount of dry powder still remains;
- Stage 2b (1.3–2.7 min) MFPM continues to grow since more and more heavier agglomerates are formed, but CVFPM starts to fall demonstrating the increasing uniformity of the powder when more water is added and less and less dry powder remains; the total wetting when all dry powder is linked into granules is reached at the end of this stage which is manifested with a local minimum of CVFPM;
- Stage 3 (2.7–3.3 min) granule consolidation and densification—MFPM continues to grow indicating continued formation of large and dense granules and agglomerates; CVFPM increases slightly, granule size/weight distribution widens;
- Stage 4 (3.3–7.8 min) breakage-dominating phase—MFPM starts to decrease reflecting breakage of large agglomerates into smaller granules; CVFPM continue to decrease as the distribution of masses becomes more uniform; most of physical and chemical processes leading to granule densification completed.

Following this granulation dynamics, it is reasonable to suggest that the granulation end-point should be at or soon after the end of granule consolidation and densification stage (Stage 3) conveyed by the MFPM maximum at 3.25 min. It is also possible to suggest that amount of water added for 3 min is appropriate since the amount required for total wetting was reached at 2.7 min.

For other formulations DFF sensor fingerprints may look differently and contain additional inflection points or extremums, such as the case for the brivanib alaninate formulations discussed in Section 15.5.2. However, the stages of granulation highlighted above are typically identifiable, which allows for identification of the granulation end-point.

15.4.3 Amplitude at process frequency

An alternative to FPM approach for the DFF sensor raw data analysis is retrieving an amplitude of harmonic function fitted into a subset of raw measurement data (see Fig. 15.10B). This amplitude at process frequency (APF) is another metric that is complimentary to FPM. Similarly to the process of calculating MFPM, a certain number of blades is selected (in example of Fig. 15.10B, 18 consecutive blades was processed to receive the sine fit), and a FFT algorithm is applied to the raw data within the subset. The resulting amplitude of the sine fit is categorized as a mean APF. The subset then shifts by one measurement point and another APF value is calculated, and so on. This way a moving plot of APF versus time is obtained. Fig. 15.15 shows a comparison between MFPM and APF for same data as in Fig. 15.14. One can see that the plots in are very similar reflecting similarity between MFPM and APF concepts.

This APF is a metric that is similar to FPM but they are not equivalent. The more raw data peaks deviate from harmonic function, for the same height, the more different APF will be from the "ideal" amplitude that would be realized when the signal is exactly harmonic. The FPM value, however, does not depend on the shape of the pulse, it depends only on maximum and minimum values in the pulse.

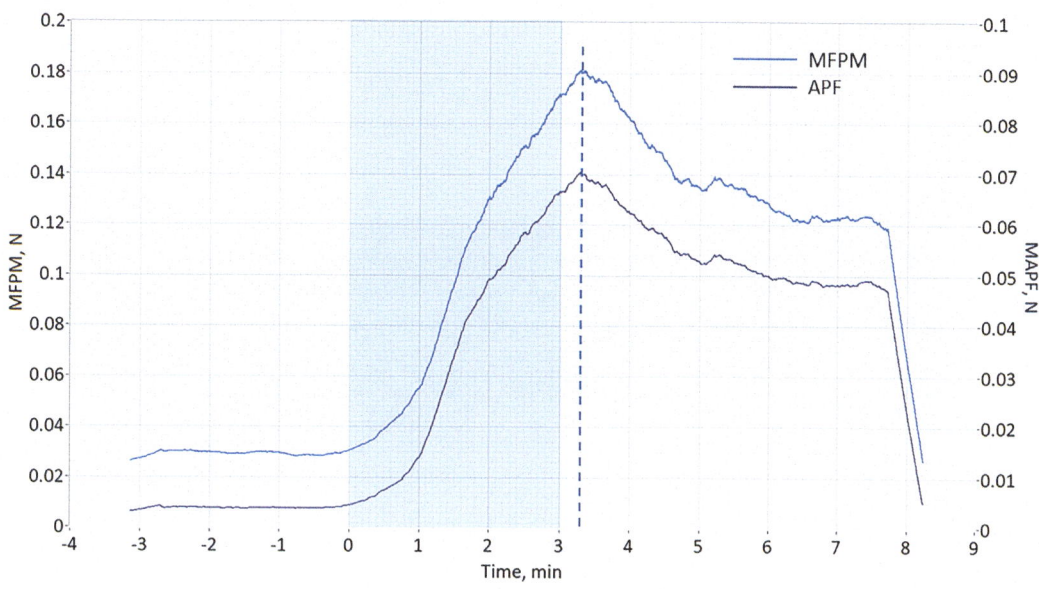

FIGURE 15.15

Plots of MFPM (*blue*) of the lognormal distribution restored from consecutive arrays of 500 force pulse magnitudes (FPMs) and mean amplitude at process frequency, APF, (*dark blue*) calculated for consecutive arrays of 20 periods, for raw data shown in Fig. 15.10A. The shadowed area represents water addition cycle.

For most application in HSWG, FPM analysis is sufficient, but for a signal where the signal-to-noise ratio of the measurement is low, MFPM plot may be too noisy to see any changes in the process dynamics, while APF approach may filter out a useful signal. An example of such a measurement is given in Fig. 15.16, where a force exerted on a sensor within in a 4.8 mm plastic tube by a peristaltic pump is not visually detectable from the raw data (Fig. 15.16A) or by the MFPM signal (lighter blue line in Fig. 15.16B), but the APF signal (darker blue in Fig. 15.16B) detects a force of 0.28 mN with a signal-to-noise ratio of 3.1. Pumping rate was 100 mL/min, process frequency 0.75 Hz, and the array size for both APF and MFPM calculations was 14 periods with moving average order of 6. The pump stated at 0.35 min.

15.5 Granule densification and scale-up in brivanib alaninate granulation

A high drug load formulation of brivanib alaninate (Diaz-Padilla & Siu, 2011) was granulated in two high-shear granulators at different binder liquid (water) concentrations and at two different scales of manufacture, while monitoring the granulations for particle size change and DFF sensor response

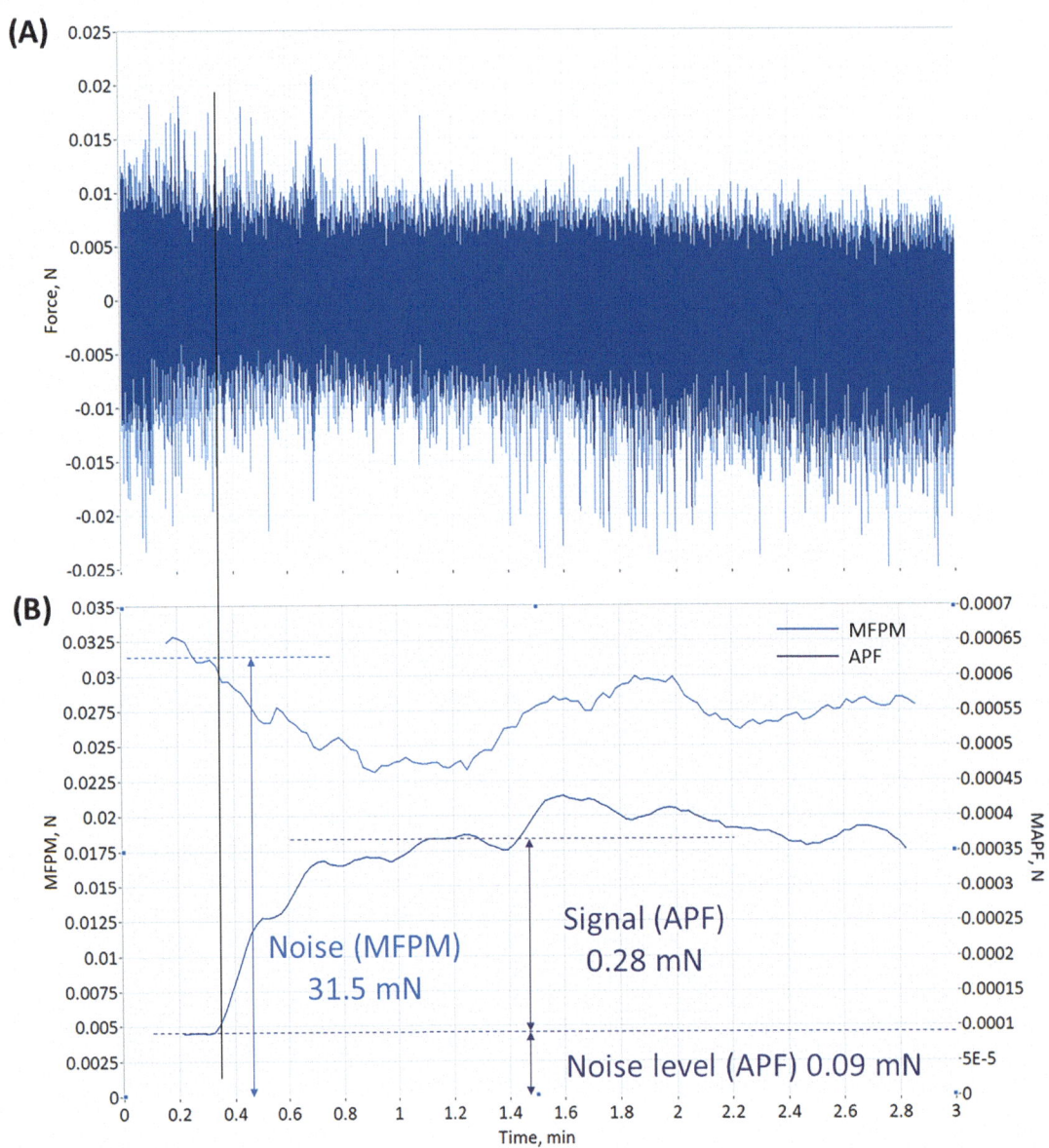

FIGURE 15.16

Comparison of MFPM and APF analysis of a noisy measurement of a force in a plastic tube exerted by a peristaltic pump. Raw data (A) and MFPM (B, *light blue*) do not indicate the presence of a force in the measurement, while APF (B, *dark blue*) detects a force of 0.28 mN with a signal-to-noise ratio of 3.1.

15.5 Granule densification and scale-up in brivanib alaninate granulation

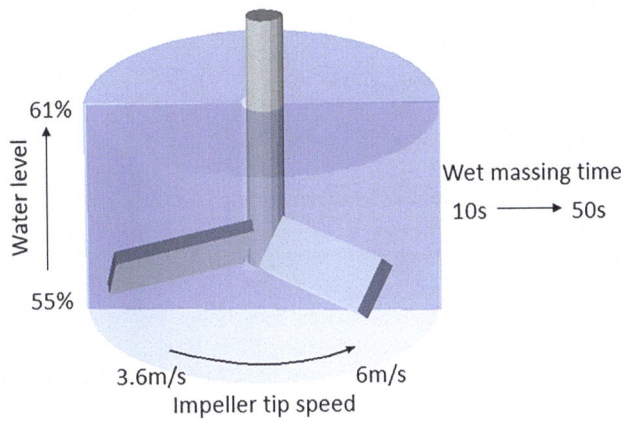

FIGURE 15.17

Process design space of brivanib alaninate granulation defined in terms of the range of parameters-water level, wet massing time, and impeller tip speed.

15.5.1 Design space of process parameters

Drug release or dissolution of drug from a tablet is defined as a critical drug product quality attribute (CQA) that can affect drug absorption (Badawy et al., 2012, 2016; Narang, Desai, & Badawy, 2012; Narang et al., 2015b; Panakanti & Narang, 2012). Dependence of tablet dissolution on the in-process material attributes of granules was investigated by studying the porosity and dissolution of granules prepared under different process conditions. Batches were tested for drug release using a discriminatory dissolution method, granule size distribution using sieve analysis, and granule pore distribution using mercury intrusion porosimetry (MIP). Statistical analysis of the results of this study identified optimal (center point) process parameters for granulation, as well as the range of process parameters within which the granule and tablet attributes were acceptable. The three input process parameters used in this DoE study included water level (range: 48%–67% w/w of the intragranular mass), wet massing time (range: 10–50 s after end of water addition), and impeller tip speed (range: 3.6–6.0 m/s). The center point for this DoE was water level, 58% w/w of the granulation; wet massing time, 30 s; and impeller tip speed, 4.8 m/s.

15.5.1.1 Correlation of tablet dissolution with granule porosity: Effect of process parameters

Experiments were carried out within and outside this design space (Fig. 15.17) to produce granules with different porosity profiles These experiments included batches that were manufactured at the lower or the higher end of design space with respect to the extremes of parameters that would be expected to produce least or greatest densification of the granules, respectively. Thus, low water concentration, low impeller tip speed, and low wet massing time was identified as the lower end of the design space; while high water concentration, high impeller tip speed, and high wet massing time was identified as the higher end of the design space. In addition, triplicate batches were manufactured at the center point of the design space with respect to these parameters. Conducting granule porosity measurements by MIP

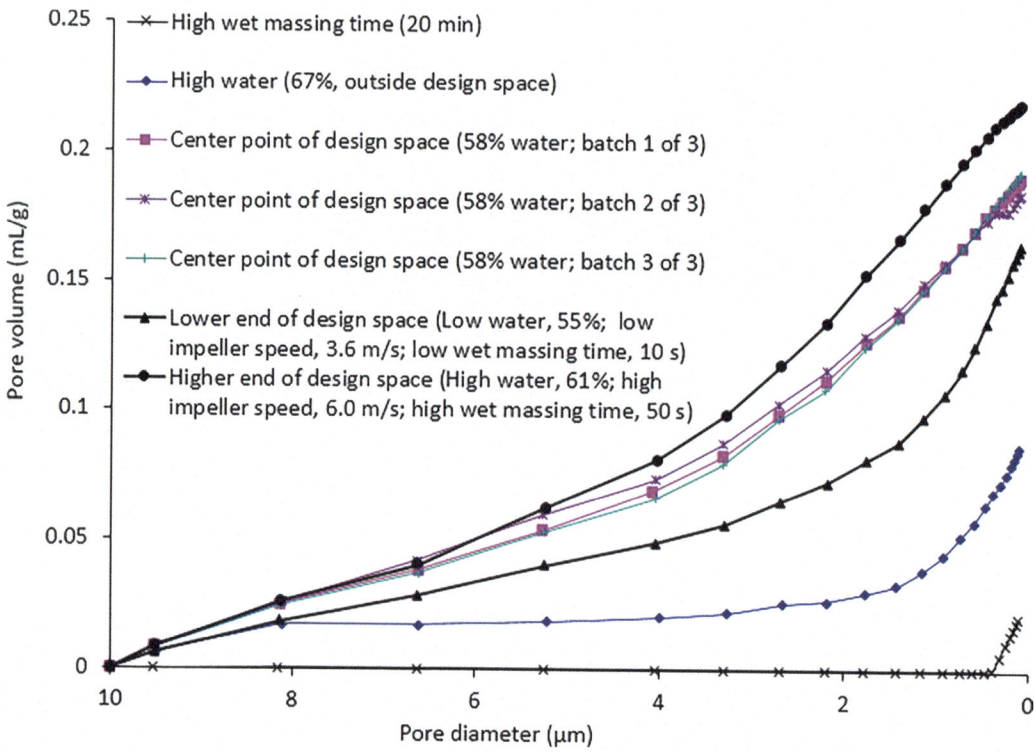

FIGURE 15.18

Porosity profiles of brivanib alaninate tablets manufactured with granulation batches within and outside process design space. Process design space for brivanib alaninate high-shear wet granulation unit operation established using a statistical design of experiment study that correlated the effect of three input parameters-water level, wet massing time, and impeller tip speed-on in-process and drug product quality attributes (Badawy et al., 2012). Subsequently, batches were manufactured at the center point (water level, 58% w/w of the granulation; wet massing time, 30 s; and impeller tip speed, 4.8 m/s) in triplicate. The porosity profile of these batches was compared to the batches made at the lower or higher end of the design space with respect to the influence of process parameters on drug release. These batches then were compared to the batches manufactured outside the design space (high wet massing time and high water amount). The dark black lines identify the limits of the design space with respect to granule porosity as a critical in-process material attribute, while lower porosity of the batches manufactured outside the design space highlights the impact and correlation of the granulation process parameters on porosity.

Reproduced with permission from (Narang et al., 2015a).

for these batches identified a range of granule porosity that can be expected within the design space, as highlighted by dark black lines in Fig. 15.18. The granule porosity profiles of the three batches manufactured at the center point was within this range, as expected. In addition, to identify achievable extremes of granule porosity profiles during WG, two batches were manufactured outside the design space using high wet massing time or high water, while keeping all other parameters same as the center point of the design space. As seen in Fig. 15.19, these batches showed lower granule porosity compared

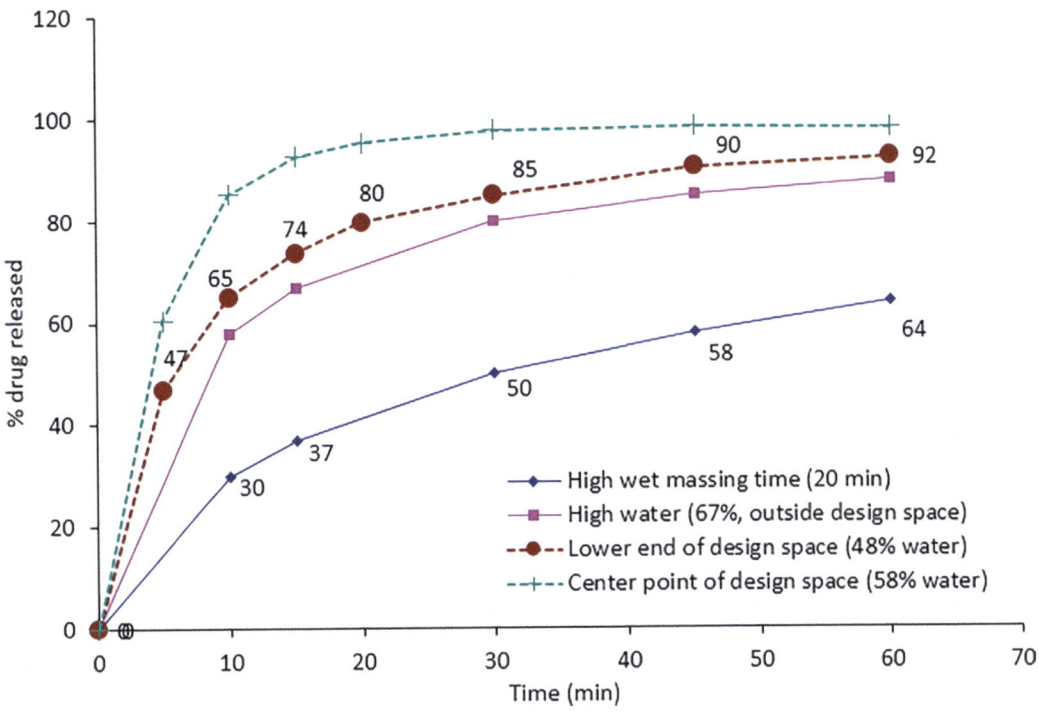

FIGURE 15.19

Tablet dissolution profiles of brivanib alaninate tablets manufactured with granulation batches within and outside process design space. Process design space for brivanib alaninate high-shear wet granulation unit operation established using a statistical design of experiment study that correlated the effect of three input parameters-water level, wet massing time, and impeller tip speed-on in-process and drug product quality attributes (Badawy et al., 2012). Subsequently, batches were manufactured at the center point (water level, 58% w/w of the granulation; wet massing time, 30 s; and impeller tip speed, 4.8 m/s) in triplicate. The porosity profile of these batches was compared to the batches made at the lower or higher end of the design space with respect to the influence of process parameters on drug release. These batches then were compared to the batches manufactured outside the design space (high wet massing time and high water amount). In this figure, dissolution from a batch manufactured at the center point of the design space was compared to batches manufactured at the lower end of the design space (with respect to dissolution impact of process parameters), and batches outside the design space (high wet massing time and high water amount). These data indicate affect of process parameters on the critical quality attribute of drug release profile from the tablets. Taken together, the dissolution and porosity data positively correlate with each other: the higher the porosity, the higher the dissolution.

Reproduced with permission from (Narang et al., 2015a).

to the batches manufactured within the design space. Interestingly reduction in granule porosity with high wet massing time was more than the reduction of granule porosity with high water content.

Dissolution studies on these batches (Fig. 15.19) indicated rank order correlation of drug release rate with granule densification (Fig. 15.18). In particular, the batch with high wet massing time showed least dissolution rate and extent, followed by the batch with high water amount outside the design space, and the batches at the lower end and the center point of the design space, respectively. This rank order correlation of drug release with granule porosity observed for batches outside the design space or a combination of process parameters that identified the higher end of the design space (worst-case scenario with respect to granule densification) was consistent with the results observed within the process design space DoE (Badawy et al., 2012).

15.5.1.2 Particle size change and granule densification on scale-up

Changes in the particle size and granule porosity for brivanib alaninate granulation were measured by sieve analysis and MIP, respectively, for batches manufactured at 10 and 65-L scale using process parameters that represented the center point of the process design space. Sieve analysis of granules collected after different wet massing times and dried in a tray oven yielded GMD and percentage of w/w fines content of the granulation (Fig. 15.20). These derived parameters of granule PSD are indicative of particle growth during granulation. These data indicated no appreciable change the granule GMD with wet massing time, while the proportion of fines in the powder mass reduced, with exponential reduction immediately after the end of water addition. These trends were consistent at both the 10 and the 65-L scales of manufacture. Focused Beam Reflectance Measurement (FBRM) (Arp et al., 2011; Gour et al., 2012; Huang et al., 2010; Macias et al., 2011; Narang et al., 2010, Narang et al., 2011) was used as an inline PSD assessment technique during the granulation of brivanib alaninate at two different scales of manufacture. The lack of a change in the CLD of brivanib alaninate during wet granulation by FBRM correlated well with the sieve analyses data indicating absence of any substantial change in the PSD of dried granules at the end of granulation.

Porosity of the granules was measured by MIP as the volume of mercury intruded into the granules, representing pore volume, as a function of pressure required for mercury intrusion into the granules, translated into pore diameter using the Washburn equation (Badawy et al., 2012). The granule consolidation data obtained with the brivanib alaninate formulation were fit to an exponential decay model of granule consolidation (Keningley et al., 1997),

$$\frac{\varepsilon - \varepsilon_{min}}{\varepsilon_0 - \varepsilon_{min}} = e^{-\kappa t} \tag{15.13}$$

where ε_{min} is the minimum achievable porosity, κ is the exponential consolidation rate constant, ε is granule porosity at time t, and ε_0 is initial granule porosity. Plots of cumulative distribution of granule pore volume as a function of pore diameter (Fig. 15.21A and B) at the two different scales of manufacture indicate progressive reduction in granule porosity as a function of wet massing time at both scales of manufacture, with apparent differences in the rate of densification at the two scales. Fitting of these data to an exponential decay model of granule densification (Eq. 15.13) revealed differences in the rate of granule densification at two different scales, while the minimum achievable porosity (εmin) was scale-independent. The exponential consolidation rate constant, κ, at 10-L scale was 0.41 min^{-1}, whereas it was 0.26 min^{-1} at 65-L scale, indicating lower densification rate at the larger scale (Fig. 15.21C). These data highlight the differences in granulation as a function of process scaleup.

15.5 Granule densification and scale-up in brivanib alaninate granulation

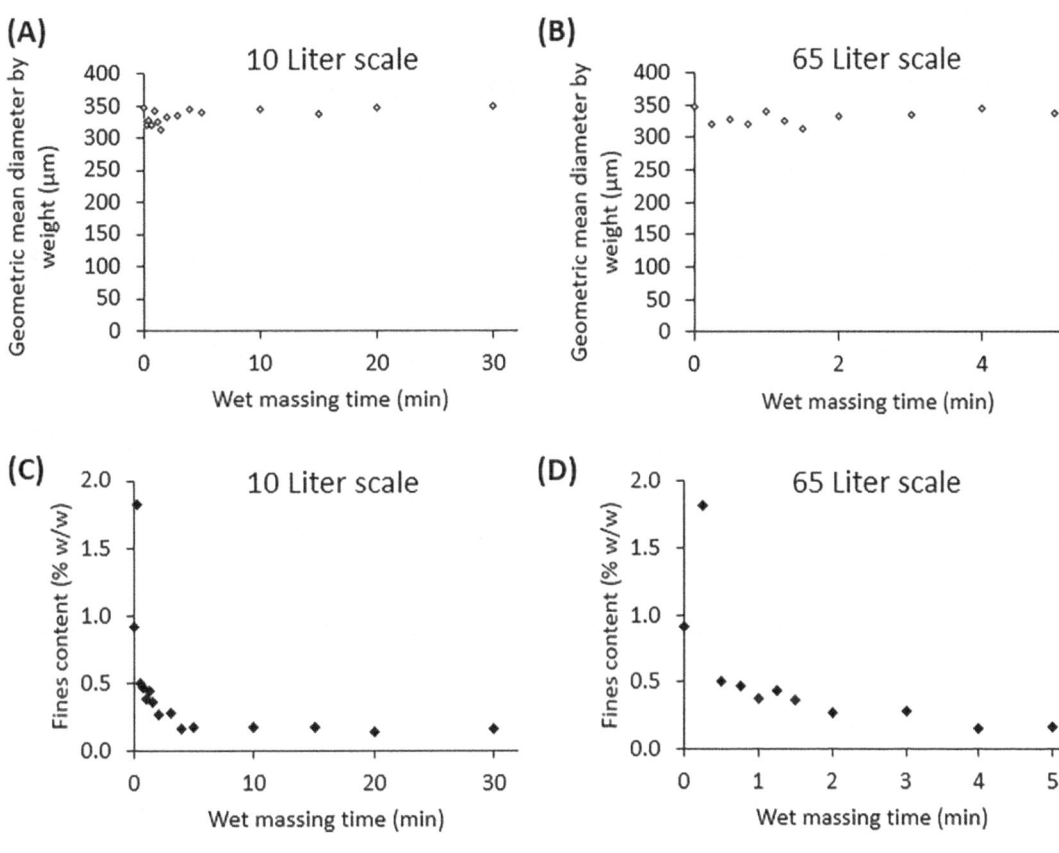

FIGURE 15.20

Granule particle size distribution (PSD), measured by nested sieve analysis and expressed in terms of geometric mean diameter by weight (A and B) and percent fines content (C and D), of brivanib alaninate granulation in 10 L (A and C) and 65L (B and D) high-shear granulator manufactured using 58% w/w water as a function of wet massing time.

15.5.2 Wet mass consistency by drag force flow sensor

Brivanib alaninate granules were prepared at high drug load (62% w/w) using 5% w/w HPC as a binder, 4% w/w croscarmellose sodium as a disintegrant, 29% w/w microcrystalline cellulose as a filler, and water as the granulating fluid. The batches at the center point of design space were monitored with the DFF sensor. Process design space for this formulation was defined at 40% v/v fill with the variables of impeller tip speed (3.6–6.0 m/s, target 4.8 m/s), amount of water used for granulation (55% to 61% w/w, target 58% w/w of granule composition), and wet massing time (10–50 s, target 30 s). To investigate the effect of substantial changes in process parameters outside the design space, two batches of brivanib

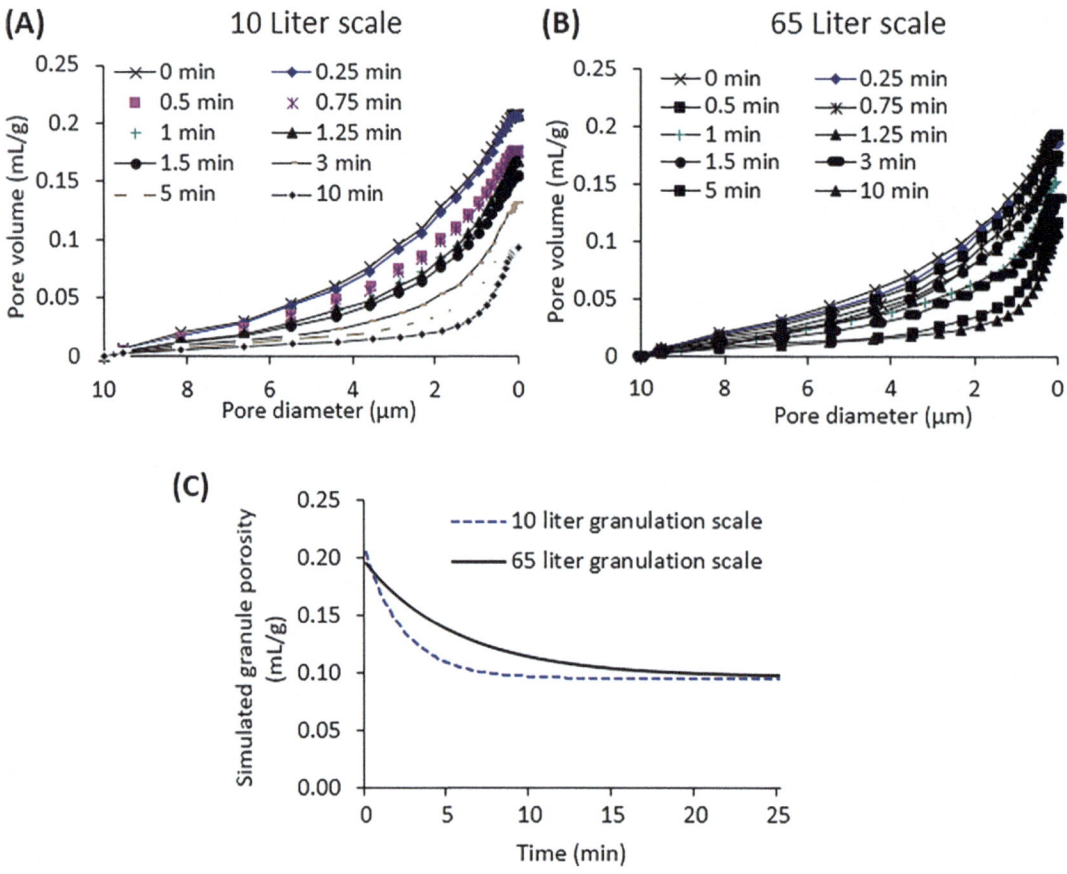

FIGURE 15.21

Granule porosity distribution of brivanib alaninate granulation in 10 L (A) and 65 L (B) high shear granulator manufactured using 58% w/w water as a function of wet massing time. The plot shows distribution of granulation porosity and total pore volume measured by mercury intrusion porosimetry (MIP). The data show difference in the evolution of granule porosity with wet massing time between the two scales of operation. Fitting the total pore volume distribution as a function of time to an exponential decay model allows simulation of granule porosity (C).

Reproduced with permission from (Narang et al., 2015a).

alaninate granulation were manufactured with water concentration outside the center of design space (at 48% and 67% w/w of granulation). All the other process conditions were at the center point of the design space except the wet massing time that was increased to 25 to 30 min to observe the evolution of the DFF sensor response during overmixing. Two DFF sensors were used simultaneously, one installed on the granulator lid in the same position as in the earlier described placebo studies, and the other from the side port as shown in Section 15.4.1, The side sensor was position on the wall, closer to the granulator blade, therefore the velocity of the wet mass interacting with the side probe and, thus, the force was higher than for the top sensor. Using GEA PharmaConnect granulators enabled a convenient side port for positioning a DFF probe. The formulation in the center point of design space (58% water) was scaled up from a 10 to a 60-L granulator maintaining volume fill of the granulator and blade tip speed at target parameters identified in the formulation design space.

These four batches (three in 10 L scale and one in 60 L scale) were monitored in-line with the DFF sensor. The batches were designed to investigate whether the DFF sensor signal of wet mass consistency is able to distinguish granules produced with changes in process parameters and whether the DFF sensor signal of wet mass consistency is able to identify any changes or consistency in the scaleup of the granulation from the 10 to the 60-L granulator.

15.5.2.1 *Effect of process parameters*

Wet mass consistency results obtained from DFF sensor (both top and side insertion) for four batches are illustrated in Fig. 15.22A–D, where evolutions of MFPM and CVFPM (fingerprints) calculated for the blade frequency of the granulators are presented. In these plots, the side sensor data is overlaid in darker color on the top sensor data. The duration of water addition is shown with shaded blue area, and the typical end-point of granulation at the center point of design space, 30s after the end of water addition, is shown with a vertical line. Fingerprints for the side and top sensor are similar indicating stability of the MFPM and CVFPM profiles over time for an arbitrary position of a sensor within a granulator. The MFPM peaks were higher for the side sensor, which was attributable to its proximity to the rotating blades. Expectedly the MFPM magnitude increased with increasing water concentration at the 10-L granulator scale. The DFF signal response also showed a pattern of increase in wet mass consistency (MFPM) during the phase of water addition and immediately after the end of water addition, and a decline during extended wet massing. As discussed in Section 15.4.1, such evolution is consistent with the modern model of granulation process (Iveson et al., 2001). The CVFPM patterns overlap for the side and top sensor after water addition ends, for the 10-L scale, for all three water contents, indicating that the sensor reflects the wet mass state independently of its placement during granulation stage. During water addition stage, one can observe a maximum in CVFPM dependencies. This maximum may indicate a transition between sub-stages 2a and 2b of the extensive wetting and nucleation stage (see Fig. 15.14 and explanation therein). The following minimum on CVFPM indicates the total wetting point, such as the minimum occurs earliest in a largest amount of water added (67%), and latest, after the end of water addition, for the least amount (48%).

The MFPM evolution for the center point batches (water 58% w/w concentration) manufactured in a 10 and a 60-L granulator showed similar shape profile (Fig. 15.22B and D). This profile was different from the MFPM signals obtained for batches with 48% and 67% w/w water concentration (Fig. 15.22A and C). This common shape features two salient points. One occurs at about 30 s of granulation time, exactly at the central point of the design space, after which the FPM signal decreases for about a minute, before picking up to reach another, greater maximum at about 5 min of granulation. The other two batches do not show this second delayed maximum. The fact that this delayed maximum feature is common for both 10 and 60-L granulators and is independent of the sensor placement indicates that DFF sensors pick up a chemical process that is specific for the central point of design space. DFF sensors, therefore, could provide a convenient way of identifying an optimal formulation by observing features on the MFPM and CVFPM evolutions over extended granulation time.

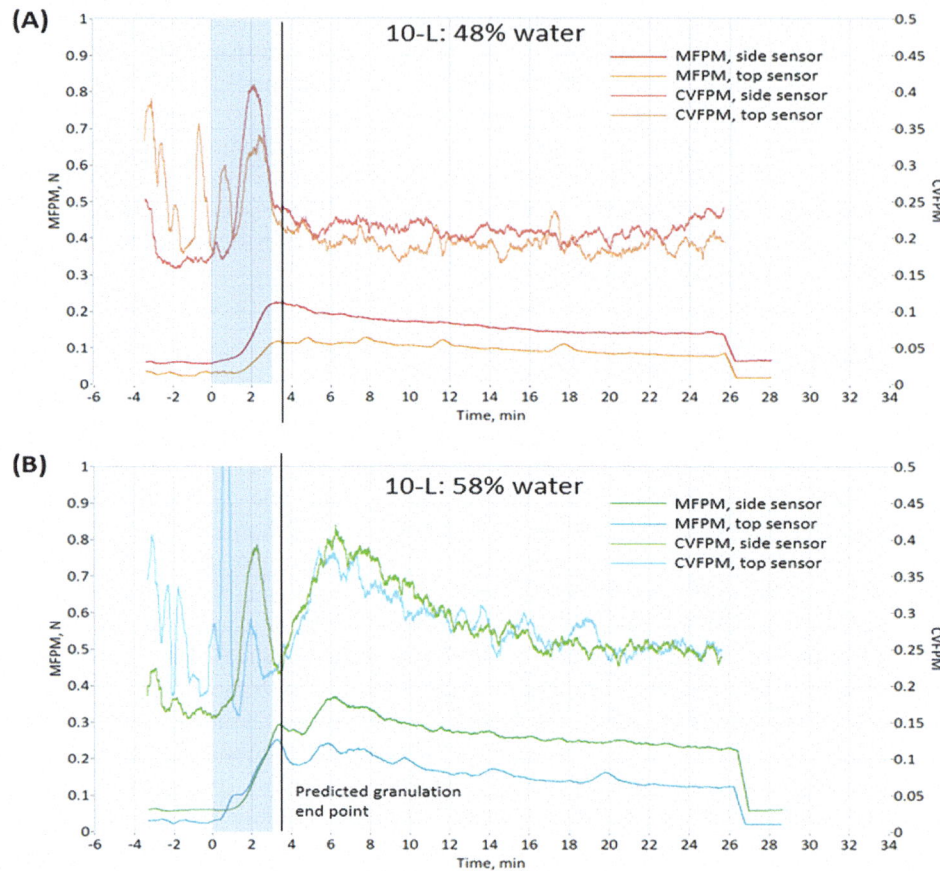

FIGURE 15.22

MFPM and CVFPM as a function of time for brivanib alaninate granulation batches manufactured at 10L scale with 67% w/w water (A), at 10-L scale with 58% w/w water (B), at 10-L scale with 48% w/w water (C), and at 60 L scale with 58% w/w water (D). The plot shows different stages of wet granulation demarked by the square shaded area, which represents the duration of water addition. Data from the side sensor is showing similar wet mass consistency profiles with different amplitudes. The vertical line indicates typical end of granulation at 30 s wet massing time. These data illustrate measurable variations in MFPM and CVFPM with differences in the amount of water, reflecting changes in wet mass consistency, in the 10 L scale, and a comparison of wet mass consistency profile between the 10 and the 60 L scale

15.5.2.2 Granulation end-point

The first maximum of the MFPM versus time dependence for brivanib alaninate formulation of 58% w/w water observed in both 10 and 60-L scale (30–50 s of WG) fits well into the center of the design space obtained via granule porosity and tablet dissolution studies. The delayed maximum in MFPM can indicate a second granulation wet mass consistency peak point at about 4 min of granulation time for

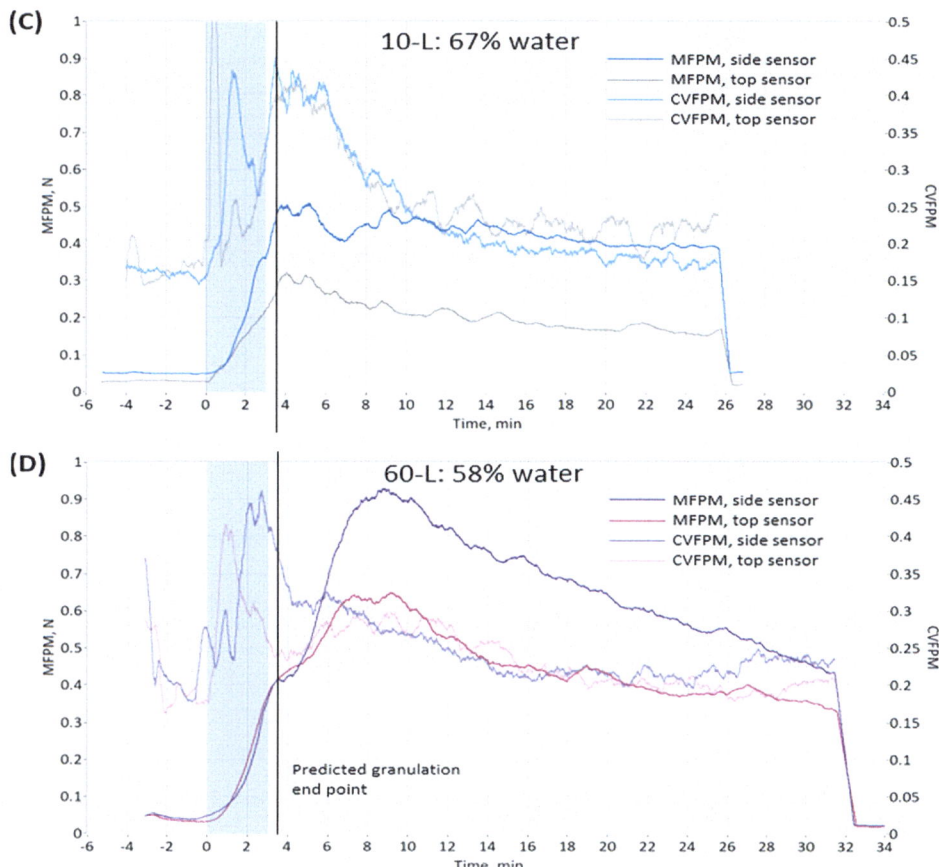

FIGURE 15.22, cont'd.

60-L granulator and 3 min for 10-L granulator that can be advantageous. The DoE studies described in Section 15.5.1 included only one batch that granulated for >50 s—the one granulated for 20 min (Fig. 15.18). This time is well past the second maximum discovered by the DFF sensor. It is possible that this delayed maximum indicates another state of wet mass where its properties are well suited for tablet preparation.

15.5.2.3 Scaleup

Comparison of the MFPM signal for the center point batches (water 58% w/w) manufactured in a 10 and a 60-L granulator (Fig. 15.22B and D) incorporate elements of both time domain and peak amplitude. The peak amplitude differences are expected, given the differences in the wet-mass pressure between the two granulators. The time domain differences vanished when the x-axis was plotted as the number of blades passing below the sensor (Fig. 15.23). The second maximum in the MFPM plots appears for both granulator sizes, and for both top and side sensors at approximately 3000 blades count, indicating that the time domain differences in sensor response

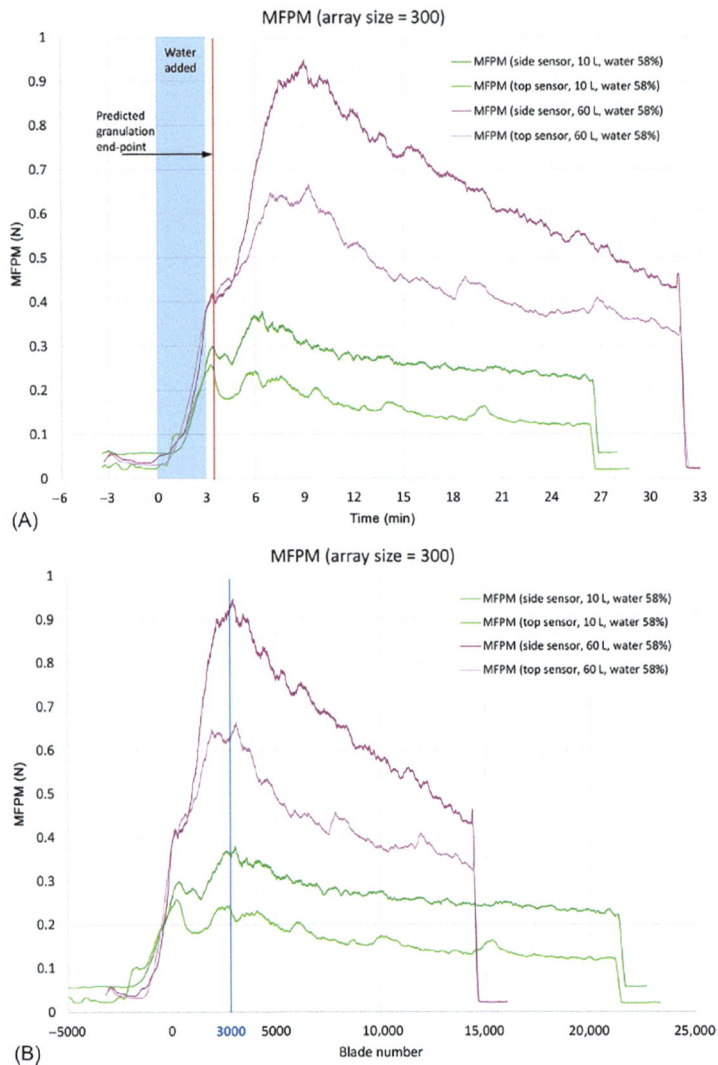

FIGURE 15.23

Comparison of MFPM between the 10-L and the 60-L scale using both top sensor and side sensor for brivanib alaninate granulation batches manufactured with 58% w/w water. Plot A is in time domain, while in plot B the signal is plotted against the number of blade passes under the DFF sensor, with the zero blade pass representing the end of water addition. Overlay of signal between the 10-L and the 60-L scale illustrates similarity in the shape and profile of DFF response, while highlighting the time-course and peak magnitude differences. In addition, normalization of data to the number of blade passes under the DFF sensor in B demonstrates scale-independence of granule wet mass consistency response measured by the DFF sensor between the two probes.

between the two scales was attributable to differences in number of impeller rotations per unit time because the impeller tip speed was kept constant. The consistent pattern of DFF sensor response of wet mass consistency between the two scales of manufacture, scaled to number of blade rotations under the sensor, indicated scale-independent attribute measurement by the sensor response. Data transformation to MFPM and CVFOM plots provides a single numeric data point for each time period plotted, allowing better design of process monitoring and control strategy using the DFF sensor.

15.5.2.4 Discussion

Time delay from the end of water addition to the peak of the MFPM signal response was 82, 160, and 57 s, respectively, at 48%, 58%, and 67% w/w water content for the granulations manufactured in a 10-L Granulator and monitored by the top sensor. For the granulations manufactured in the 10-L granulator and monitored by the side sensor, the time delay to the MFPM peak was 16, 173, and 86 s at 48%, 58%, and 67% w/w water content, respectively. The time delay to the MFPM peak 368 and 320s, respectively, for the granulations manufactured in the 60-L high-shear granulator at 58% water content and monitored by the top or the side sensor, respectively. Time delay to the peak of the MFPM signal correlated well with the expected granulation processes at different water levels and was consistent for the side sensor as well as the top sensor. WG mechanistically involves several simultaneous processes that proceed at different rates during different stages of granulation that include wetting, nucleation, granule growth, layering, agglomeration, breakage, attrition, and consolidation. For example, at the stop of water addition, if the water level is lower than optimum (48% w/w), the initial granule growth processes of nucleation and aggregation would be predominant. Similarly if the water level is higher than optimum (67% w/w), the final stage processes of attrition and breakage would be predominant. In both these cases, no further granule densification can be expected at the end of water addition. On the other hand, at optimum water concentration (58% w/w), various granulation mechanisms will be in an equilibrium and the granule densification would be sustained for a while. In the current set of experiments, this behavior is evident in the higher delay time to peak at 58% w/w water level, compared to 48% or 67% w/w water level.

Interestingly the time to peak was delayed even further when the formulation was scaled up to the 60-L granulator at the center point of water concentration (58% w/w). This delay in the time to peak DFF signal was consistent with reduced number of impeller rotations for a given period of time, when the process is scaled up with constant impeller tip speed. The greater time-to-peak lag at the larger scale indicated slower rate of granule densification at the larger scale. These findings were consistent with the observed changes in granule porosity as a function of time at 10 versus 65-L scale and the exponential fit to this porosity data (Fig. 15.18). These correlations indicated the value of the DFF sensor time-to-peak response as a parameter of interest that can inform the rate of granule densification and robustness of the HSWG process. Although the exact state of granulation and the responsible factor for the peak in the DFF response is not known, it is likely to be the culmination of dominance of granulation growth and consolidation mechanisms (over attrition and breakage) in the powder state. Thus, the peak of DFF sensor might not represent a desirable end-point of the granulation but could serve as an indicator of desirable ranges of wet massing times that can be acceptable as an end-point of the granulation. This time to peak can be used to derive parameters that can assist scaleup of granulations (Fig. 15.24).

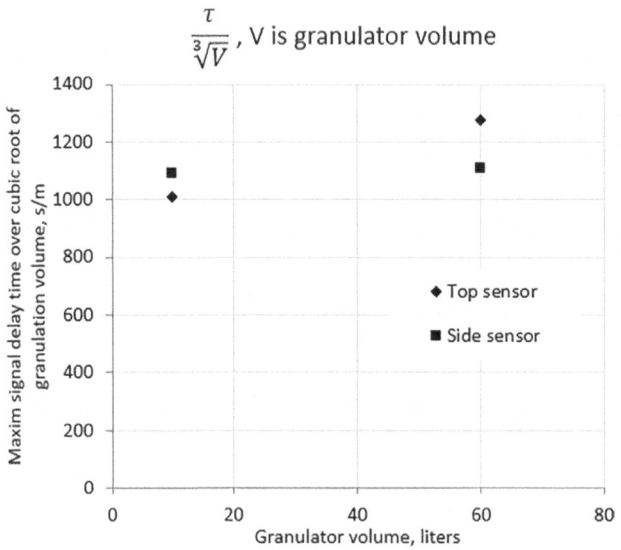

FIGURE 15.24

Development of a potential scale-independent measure of delay time for the maximum peak magnitude response of the drag force flow (DFF) sensor during the high-shear wet granulation (HSWG) of a brivanib alaninate formulation using granulator volume as parameters for defining an equation that provides scale-independent predictions. The plots show similarity of the observed time delay parameter for brivanib alaninate granulation batches manufactured at both 10 L (Test 2) and 60L (Test 4) scale with 58% w/w water using DFF side sensor measurement.

Reproduced with permission from (Narang et al., 2015a).

15.6 Mean force pulse magnitude and cvfpm compared with basic flowability energy measured by FT4 powder rheometer

In this section, we describe the correlation of the in-line response of DFF sensor with at-line measurement of basic flowability energy (BFE) measured by an FT4 Powder Rheometer by Freeman Technology Ltd. This correlation allows better understanding of the DFF sensor response as representing a fundamental rheological properties of the granules. Specifically MFPM is indicative of the resistance of a powder to flow and can also be quantitatively measured using an FT4 Powder Rheometer.

The FT4 measures the amount of energy required to move the powder in the test cylinder subsequent to the standard conditioning cycle (Osorio & Muzzio, 2013). The basic flow energy (BFE) is the specific energy needed to displace the powder and is related to the force and torque of the agitator submerged in the powder as the blade moves downwards in the cylinder with concurrent application of minor compressive stress. The BFE can quantify and differentiate among slightly cohesive powders with similar flow properties (Osorio & Muzzio, 2013).

15.6 Mean force pulse magnitude and CVFPM compared with basic flowability

Table 15.2 Time-points to which different batches were processed.

Run	Description	Time instant as marked on plots, min
A	5 min of wet mass mixing.	8
B	After 3 min of dry mixing, just before starting water addition.	0
C	3 min of wet mass mixing.	6
D	1 min of wet mass mixing.	4
E	Immediately after stopping water addition.	3
F	2 min after start of water addition.	2
G	1 min after start of water addition.	1

15.6.1 Comparison of drag force flow sensor and FT4 measurements during high-shear wet granulation

In-line measurements by DFF sensor were compared with at-line FT4 analysis of wet mass collected at different time points during processing of a placebo formulation. Tests were conducted in a 10-L GEA PharmaConnect granulator with formulation, process parameters and DFF sensor positioning described previously (see Section 15.4.1), but with different concentration of binder (HPC) of 1%, 3%, and 5% w/w. MCC concentration was adjusted to accommodate changes in HPC concentrations.

Six granulation batches with different end time points (1, 2, 3, 4, 6, and 8 min from start of water addition) plus one dry powder measurement were tested for each HPC concentration, separately. For every batch, granulation process was continuously monitored by the DFF sensor until it was stopped for at-line analysis of the wet mass by FT4 Powder Rheometer at end-time points indicated in Table 15.2. In every test, the whole granulation mass was retrieved from the granulator and three representative samples withdrawn. The samples were immediately tested in the FT4 rheometer. A new batch of the same formulation was then granulated and used for the next end-time point.

15.6.1.1 Reproducibility of drag force flow sensor measurements

Process fingerprints (evolution of MFPM and CVFPM) for an array size of 500 FPM values are displayed in Fig. 15.25A–C, respectively, for every end-point tested, for each formulation. One can see that MFPM profiles overlap quite well for different batches of each formulation indicating high batch-to-batch reproducibility of MFPM measurements. The CVFPM evolutions are also similar between the batches of the same formulation showing certain distinctions for the time period [−4, −3] where CVFPM picks up the variation of the sensor background noise (the impeller rotation started at -3 min), and for the period [−3, 0], the dry mixing stage. All salient points appear at approximately same time for all batches of same formulation.

15.6.1.2 Mean force pulse magnitude and CVFPM versus basic flowability energy

Fig. 15.26 compares the BFE measured by FT4 Powder Rheometer overlapped with MFPM and CVFPM measurements. MFPM and CVFPM dependencies are calculated for each formulation as the average of MFPMs for all batches processed that is averages of the curves given in Fig. 15.25A–C, respectively. The plotted value of BFE is the mean of measurements obtained from three samples of wet mass that

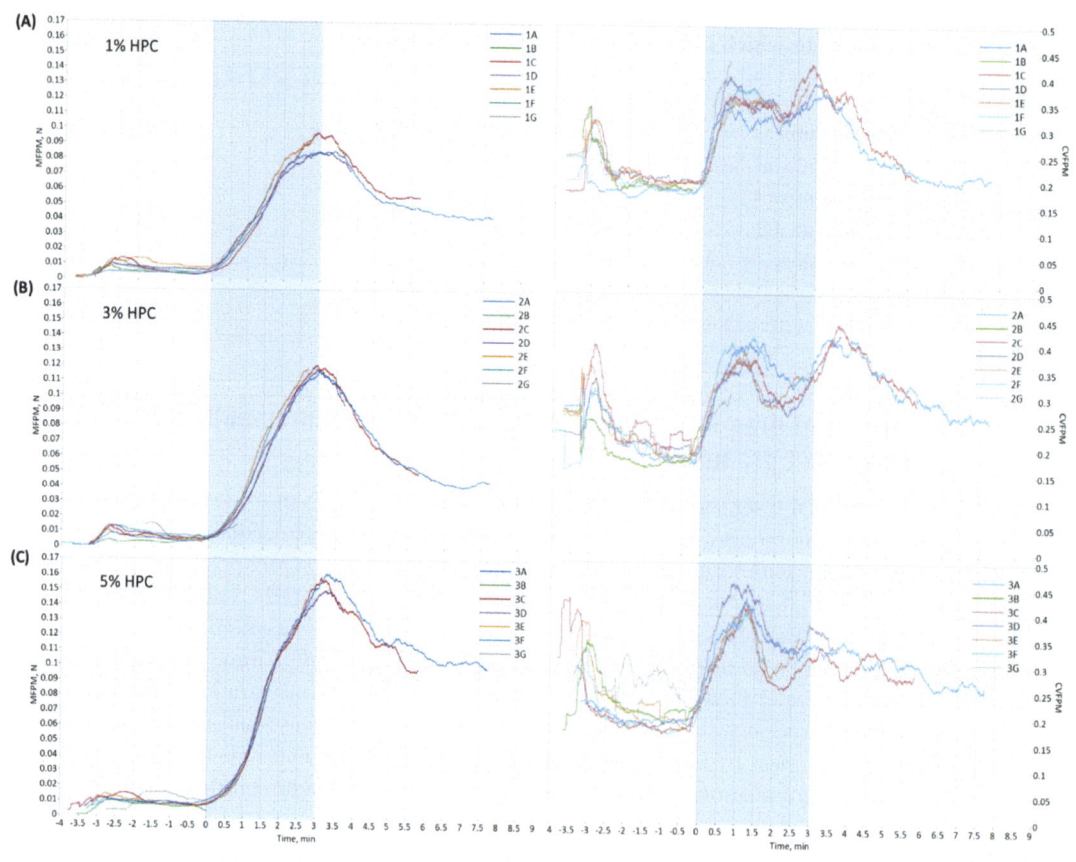

FIGURE 15.25

Reproducibility of drag force flow (DFF) sensor's force pulse magnitude (MFPM) and coefficient of variation (CVFPM). The DFF sensor response was recorded for several batches that were processed to a defined time point of the wet granulation cycle. Plots 1A through 1G in (A), 2A through 2G in (B), and 3A through 3G in (C) represent the identical batches processed to different end-point, as outlined in Table 15.2 and as can be inferred from the x-axis in the figure, respectively. Overlapping plots of MFPM and CVFPM are visually evident for all these samples, indicating high reproducibility of DFF measurements.

were collected after each run with the error bars being the standard deviations of mean over the three measurements.

Differences of granule wet mass consistency between formulations manufactured with different HPC concentrations were evident with both DFF sensor and FT4 measurements. The two measurement techniques showed similarities in the MFPM profiles, especially with respect to the effect of HPC concentration. There was a very sharp rise during water addition as measured by the DFF. The FT4 measures relatively small changes in BFE before (dry powder) and after water addition started (especially, 1% and 3% HPC). The DFF curves peak shortly after water addition is complete. The

15.6 Mean force pulse magnitude and CVFPM compared with basic flowability

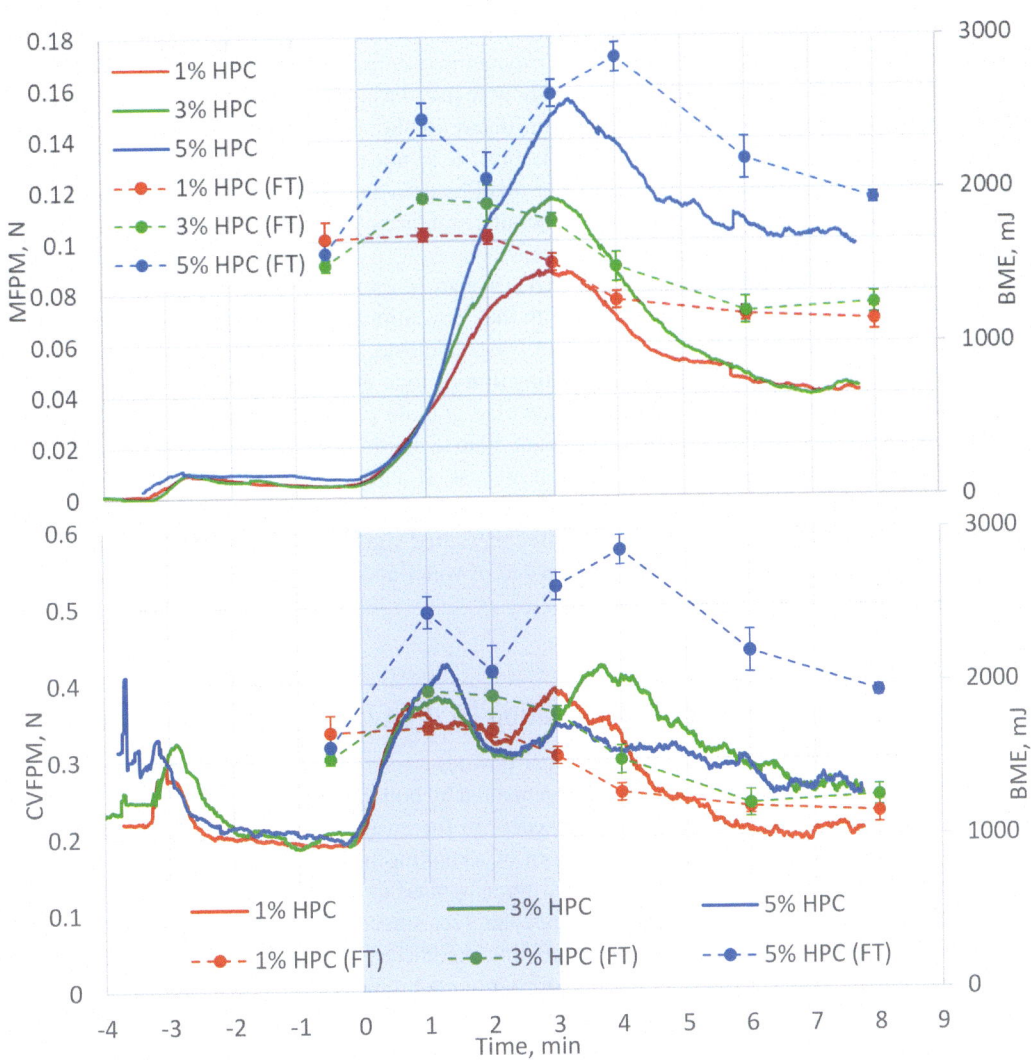

FIGURE 15.26

Comparison of DFF sensor's fingerprints (MFPM and CVFPM) collected in-line continuously during granulation process and the FT4 Powder Rheometer's basic flowability energy (BFE) of the wet powder sample collected at several time points. This figure shows data acquired for placebo formulations manufactured with 1%, 3%, or 5% hydroxypropyl cellulose (HPC) as a binder, respectively. The FT4 data points represent mean of three measurements conducted on wet granules at the end of each granulation cycle and the error bars show standard deviation of the mean. The FPM distributions from which MFPM and CVFPM values were incurred are FPM data arrays of 500 points or 33 s. Blue shadowed area shows time of water addition

FT4 measurements appear to peak during water addition, similarly to CVFPM signals. For the high binder content (5% w/w HPC) the granulation dynamics seems to be similarly described both by MFPM and FT4. Both temporal dependences show a rise during water addition, a maximum soon after the end of water addition, with subsequent decay of the signal afterwards. Because the BFE measurements have low temporal resolution (1 min), it is not possible to pinpoint the position of the maximum in the BFE dependence, however. MFPM demonstrates consistent temporal behavior for the lower binder content formulations, generally similar to the 5% HPC signal, but with less magnitude.

A salient feature of the FPM versus time signal is a characteristic maximum occurring soon after the end of water addition time. This is consistent with the conventional practice of granulation end-points being considered several seconds after the end of water addition. The FT4 measurements illustrate an increase and subsequent decrease in BFE as a function of time after water addition for the 5% HPC formulation, and as such can be a tool to characterize granule properties as a function of an offline variable (Freeman et al., 2015). From this standpoint, both technologies show the ability to track granule properties as they evolve through the granulation process. If a conventional definition of end-point is adopted, whereby the end-point occurs shortly after the end of the water addition phase, the DFF sensor identifies this occurrence as shown by a peak in the measured signal. In the FT4, however, for 1% and 3% HPC formulations, BFE decreases before the end of water addition. Thus, the DFF sensor seems a suitable in-line PAT for end-point determination.

15.6.2 Off-line dry powder characterization by drag force flow sensor and FT4

The rheological properties of seven common sugar powders were characterized using concurrently an FT4 Powder Rheometer and a DFF sensor. Data generated by both the FT4 and DFF sensor (specifically MFPM) reflect mechanical resistance of the powders to forced flow. During FT4 analysis, a precise volume of powder is subjected to the flow pattern of a rotating blade that moves downward through the powder bed. Measurements of the torque and force applied to the blade are combined to calculate basic flowability energy (BFE), quantifying the powder's resistance to the blade motion. For this study, an off-line DFF test station, a version of DFF off-line rheometer, was assembled, where a probe was installed into a laboratory scale mixer with a bottom-mounted impeller.

A 50 mm diameter test vessel containing 160 mL of powder was used for the FT4 analysis. Each sample was subjected to conditioning, where a precision blade was used to create low stress, homogeneous particle packing. Prior to testing, the vessel was also split to produce a consistent volume of materials for analysis. BFE measurements were conducted using an aggressive compaction mode, measuring the resistance to the motion of the rotating blade as powder was forced down the vessel.

A 2-L bottom-driven mixer with a DFF sensor installed vertically so that the probe tip was 8 mm above the blade was used as the test station for off-line DFF measurements. In every test, the mixer was loaded with 1.6 L of test powder so that the 4-cm probe was just covered with powder. The test protocol involved running the mixer for 60 s and taking measurements at a rate of 500 per second. The agitator rotational speed was kept at 600 RPM. The MFPM and CVFPM values were incurred for the data array value of 200 peaks (blades), and the MFPM and PCF values at a time instant of 30 s were collected in each test. Consistent with the FT4 measurements, the DFF tests were repeated in triplicate and the mean value used for comparison.

Fig. 15.27A compare the BFE and MFPM results for the powders, and Fig. 15.27B presents the CVFPM measurements by the DFF off-line rheometer. Error bars for both BFE and MFPM represent standard deviations over three measurements taken for each sugar.

Both devices demonstrated highly comparable trends in BFE (FT4) and MFPM (DFF sensor) for the seven different powders tested. Both methodologies categorized the powders into three distinct groups: brown sugar, four granulated sugars (Augason farm, domino, coconut palm, and caster sugar) and fine powders (confectioners sugar and corn starch). Brown sugar, characterized by its large tacky granules, displayed significantly higher BFE and MFPM values, approximately two times greater than the values generated by the four granulated sugars and around an order of magnitude greater than the fine powders. Corn starch was found to demonstrate the lowest value of BFE and MFPM, with Confectioners powdered sugar coming in close second place for both techniques. Differences between MFPMs measured for four granulated sugars exceeded the standard deviations by at least a factor of five, highlighting the excellent sensitivity of the DFF technique. The two finer powders, confectioners sugar and corn starch, displayed lower BFE and MFPM values that differ by a factor of two from one another for both techniques. The comparable trends observed between the BFE and MFPM demonstrate how both of these techniques can be used to quantify the material's resistance to forced, dynamic flow, with DFF sensor technology demonstrating a higher sensitivity in separation of powders of similar granule sizes (granulated sugars).

The DFF rheometer also reported the CVFPM, which is a measure of flow uniformity for each sugar (Fig. 15.27B).

This metric reflects cohesiveness or tackiness of the powder and it does not correlate to BFE or MFPM. The largest CVFPM values are realized for Confectioners Powder sugar, Coconut Palm sugar, and brown sugar indicating that these are tacky powders, what is consistent with "hand" tests for these powders. The two finer powders, confectioners sugar and corn Starch, displayed BFE and MFPM values that are similar and lowest for all sugars, but the CVFPM values proved to be the greatest among all powder for confectioners sugar and the lowest for corn starch, respectively. This confirms that the MFPM and CVFPM parameters measured simultaneously by DFF sensor reflect different properties of the powder.

15.7 Comparison of mean force pulse magnitude and CVFPM measured by drag force flow sensor with shaft amperage

Impeller parameters such as torque, amperage, and power consumption are the equipment responses that indicate the forces exerted by the wet mass on the impeller blades (Levin, 2015; Parikh, 2010). These can be viewed as an integration of all the forces and force moment arms along each of the mixing blade arms. According to granulation theory initially proposed by Leuenberger et al. (1979), usable granules can be obtained in the region that begins with the peak of the impeller torque derivative. Therefore, in real-time monitoring, observation of the amperage signal may provide an indication for the time when the process should be stopped (Holm et al., 2001; Laicher et al., 1997). While being intuitively related to the physical parameter of the wet mass, the torque or power consumption measurement was criticized as insufficient for end-point determination and scaling (Hansuld & Briens, 2014).

In contrast to the torque integrated over the surface of the impellers blades and shaft, the DFF sensor measures the impact force of granules in the locality of the probe tip. Measurements in a PharmaConnect

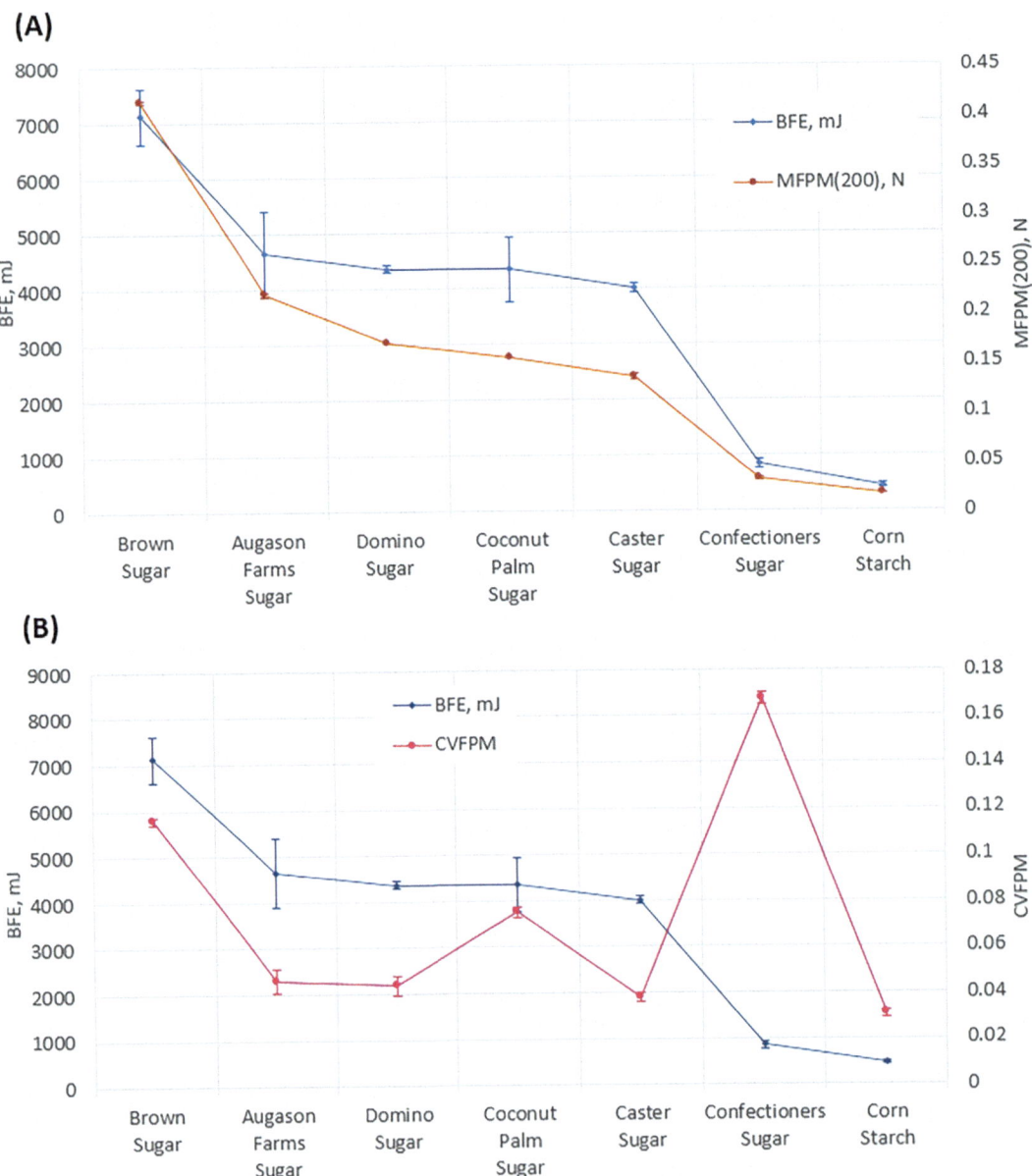

FIGURE 15.27

Comparison of BFE measured by FT4 and MFPM (A) measured with DFF test station. MFPM represent FPM averaging over 200 consecutive blades. Chart (B) presents corresponding CVFPM values.

10, 30, and 60 L granulators using a placebo and an active formulation indicated that the DFF sensor is readily capable of differentiating batches made with different formulation composition and process parameters (% w/w water used for granulation), as well as different stages of processing. In the presented study, the basic outputs of the DFF sensor, MFPM and CVFPM, were compared with simultaneous dynamic measurement of the motor current (amperage) during processing of a placebo formulations in a Bohle 4-L granulator at 375 RPM or 18.75 blades per second (Fig. 15.28).

The three formulations consisted of 37% anhydrous lactose, 1% croscarmellose sodium, and 1%, 3%, and 5% HPC with 61%, 59%, and 57% microcrystalline cellulose in respective batches. Granulation was performed with 40% water (320 g for the batch size of 800 g) added over 3 min. Granulator current was recorded once every 1.4 s along the DFF sensor force measurements of 500 per second. The probe was placed vertically inside the bowl so that its tip was at 63% radial position (5.6 cm from the blade axis) and 0.7 cm above the blade.

These data show that both impeller current or torque and DFF sensor (MFPM and CVFPM) provide differentiation between formulations with different binder concentration with DFF sensor demonstrating a higher sensitivity of MFPM for separating the formulations. But the variations in granule densification during water addition demonstrated by CVFPM are not picked up by the amperage.

Predicting downstream process parameters for tablet compression using Lenterra in-line rheometer (LIR) data in real-time was recently demonstrated by scientists from GSK and the University of Birmingham (Munu et al., 2024). by feeding the DFF sensor output to an artificial intelligence (AI)-based mathematical model.

In comparing the ability of torque measurement in the granulation bowl and the LIR sensor to predict the properties of granules and tablets, the authors reported that the granulation process was captured with greater sensitivity by in-line force measurements than impeller torque, demonstrating the evolution of granule formation. The LIR force profile identified the time instant of even distribution of the binder fluid and the end-point of granulation. Authors developed an AI-based closed form analytical equation model using the data from a face centered surface response design of experiment study. This model was able to provide a strong predictive indication of tablet tensile strength based only on the in-line data. The authors envision this model can be used for real-time decision making and process control, such as predicting the compaction force required to achieve the tablet tensile strength based on the upstream DFF sensor data.

The authors followed this work with the predictive ability of using Styl-One Evolution compaction simulator to produce tablets mimic rotary tablet presses used in routine production (Munu et al., 2025). Granule properties of basic flowability energy, specific energy, and conditioned bulk density, measured by FT4 rheometer, could be predicted using the integrated force and the magnitude of force pulses, measured by LIR during wet massing. Authors note that LIR is a reliable instrument for monitoring the granulation process, capable of determining the minimum volume of the binding fluid needed as well as the critical stages in the granulation process (wetting and nucleation, consolidation, and granule growth). Authors found that the model developed using a Gamlen tablet press had high predictability for the Styl'One Evolution compaction simulator, confirming that this model can be used in industrial scale applications. The integrated force during wet massing was found to be statistically significant and predictive of the behavior of granules during the die filling step.

Such an advancement of combining real-time DFF sensor data with AI mathematical modeling to predict downstream process parameters can revolutionize technologies such as continuous wet granulation with benefits of superior process control and efficiency.

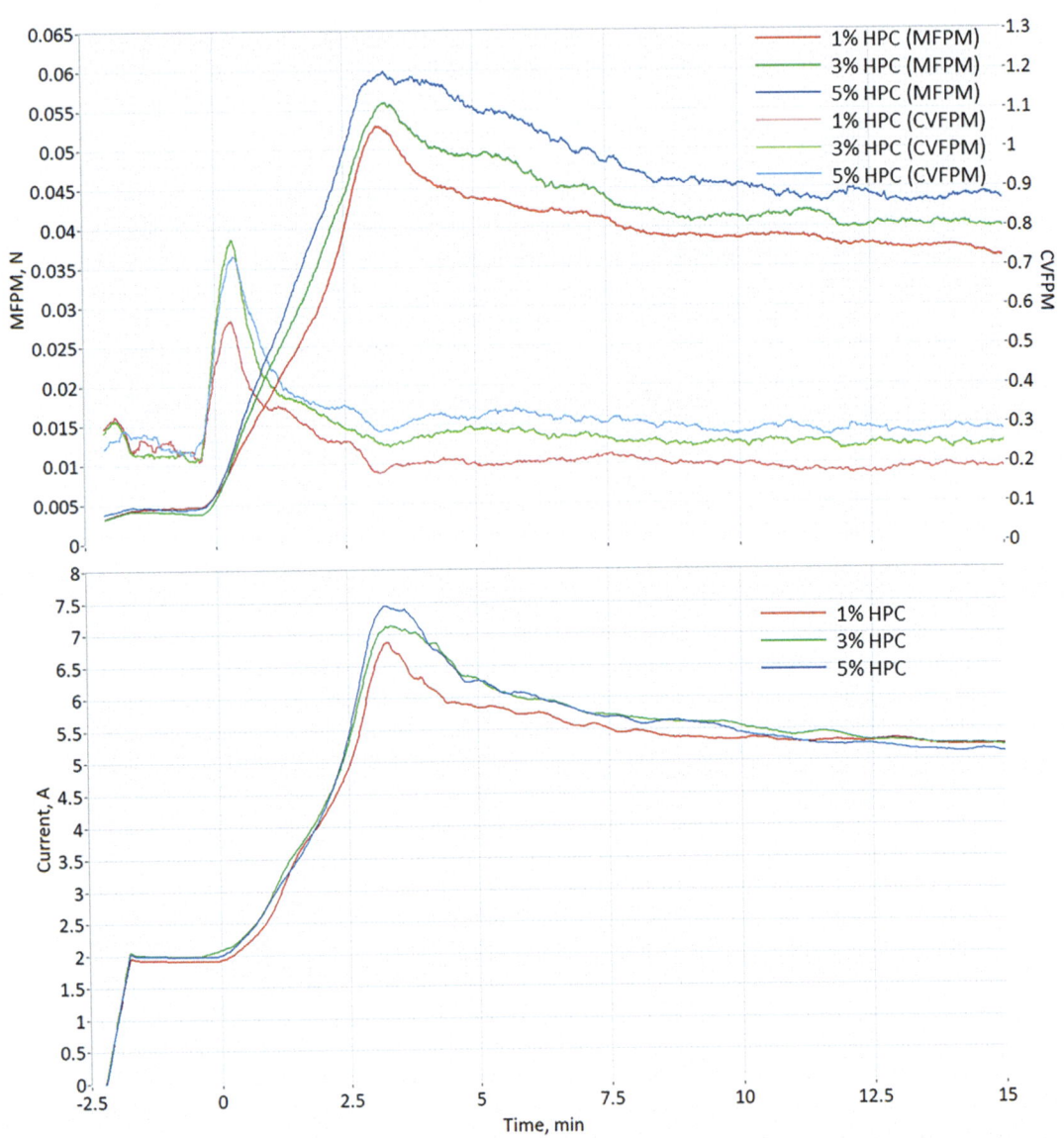

FIGURE 15.28

Time evolutions of MFPM and CVFPM (A) and shaft amperage (B). MFPM and CVFPM were computed over a subarray of 800 FPM points, corresponding to 42 s. The amperage was recorded using the output port of Bohle granulator power supply synchronously with DFF sensor for comparison. The plot shows moving averages over 27 s, or over 20 amperage data points. The HPC concentration was varied at 1%, 3% and 5% w/w to produce formulations that would granulate differently.

15.8 Conclusions

CQAs of drug products are defined by the properties and attributes of the in-process material during different stages of granulation. For HSWG processes, the WG and drying unit operations play a predominant role in determining the quality attributes of the granules that influence tablet properties. For example, the CQAs that influence the bioavailability of the drug product include drug release and dosage form integrity. These are influenced by the PSD and the porosity/density of the granules produced by the HSWG followed by drying unit operations. While the particle size change during WG processes has been well studied, granule densification during granulation is less understood. In addition, differences in the rate of granule densification across different scales of HSWG presents a significant challenge in the scaleup of WG process for formulations of new chemical entities and drug products. This challenge manifests in the unknown adjustment that might be needed in one or more process parameter (as the granulation is scaled up) to achieve similar granule attributes at the end of granulation (end-point).

The DFF sensor detects the drag flow of the particle mass in a high-shear granulator with tunable and very high sensitivity and temporal resolution. This sensor detects not only bulk flow properties but also individual force impacts on the sensor that can provide information about particle mass, size, density, and momentum. This sensing is based on FBG gratings that optically transmit information of probe bending with force impacts. Such a configuration allows electricity-free operations with minimal product contact surfaces and chances of fouling in the process, that are suitable for a highly regulated manufacturing environment that follows current good manufacturing practices for drug product manufacture.

WG process is associated with the formation and growth of powder agglomerates as granules. In addition, the presence of shear in the HSWG process can lead to significant densification of the granules with reduction in the size and volume of intragranular pores. Measurement of wet mass consistency using the DFF sensor appears to be a scale independent attribute. Output of the DFF sensor (MFPM and CVFPM) correlates with granule densification through an indirect measure of wet mass consistency, determined by the force impact peaks recorded by the sensor when immersed in the moving wet mass. For the placebo formulation, the DFF sensor response provides a measure of wet mass consistency that shows the impact of changes in the formulation composition (Figs. 15.25 and 15.26), which did not affect granule particle size during or at the end of granulation. For the brivanib alaninate granulation, the DFF sensor response is able to distinguish between a process variable (percentage w/w water used for granulation) (Figs. 15.22 and 15.23), which does not correlate with changes in granule PSD but does correlate with granule densification (Fig. 15.25). Granule densification is a critical material attribute (CMA) that correlates with tablet dissolution as a CQA of the drug product (Badawy et al., 2012). Taken together, these data demonstrate the utility of DFF sensor as a real-time inline PAT probe that responds to a property of the wet mass, which we call wet mass consistency that provides a metric of granulation progress different from the size distribution of the granules and correlates with granule densification. Therefore, DFF sensor can be used to study the effect of formulation and process parameters during formulation and process development, and to monitor reproducible manufacture of granulations. In addition, the DFF sensor response showed a time delay to the maximum after the end of water addition (Fig. 15.23). This time delay to the maximum is important since the HSWG processing typically is stopped a few seconds or minutes after the end of water addition. This conventional practice ensures complete the dispersion of water within the powder bed mass and it generally is understood that optimum granule quality is obtained only upon wet massing of the powder mass after completion of water addition. In the case of brivanib alaninate granulation, the wet massing time of 30 s was selected as the center point and 10 to 50 s as the range of the process DoE study. The optimum duration of wet massing time, however, remains a matter of tradition among different industries and practitioners, and can vary from a few seconds to several minutes. In this context, an identifiable peak of wet mass consistency after the end of water addition during WG by the DFF sensor enables identification of a potential inflection point after which the granule quality may deteriorate. In the case of brivanib alaninate granulation, this measure of granule quality is the porosity of granules, which showed positive correlation with tablet dissolution. The importance of identifying an appropriate wet massing time and water amount for the HSWG process also comes from the short processing time and irreversible nature of this unit operation. For example, a typical WG process could be as short as 3 min, and the quality attributes of the granules that are formed do not become evident until downstream processing is complete.

The MFPM response of the DFF sensor, reflecting wet mass consistency, is an indicator of powder rheology, as indicated by the concurrent measurements of BFE at different processing times for three different formulations using the FT4 Powder Rheometer (Fig. 15.27A). In addition to the wet mass consistency, DFF sensor is capable of simultaneously measuring another powder characteristic, its granule size/mass uniformity (Fig. 15.27B). In addition, MFPM and CVFPM measurements have higher sensitivity than impeller torque or amperage in detecting changes in granule attributes as a function of formulation variables. Overall, the DFF sensor provides data rich process fingerprinting in-line and in real-time with high sensitivity and data density (500 data points/s) to enable process design, scale-up, monitoring, and control applications, as well as for routine monitoring during manufacturing.

References

Arp, Z., Smith, B., Dycus, E., & O'grady, D. (2011). Optimization of a high shear wet granulation process using focused beam reflectance measurement and particle vision microscope technologies. *Journal of Pharmaceutical Sciences, 100*(8), 3431–3440. https://doi.org/10.1002/jps.22556.

Badawy, S. I. F., Narang, A. S., Lamarche, K., Subramanian, G., & Varia, S. A. (2012). Mechanistic basis for the effects of process parameters on quality attributes in high shear wet granulation. *International Journal of Pharmaceutics, 439*(1-2), 324–333. https://doi.org/10.1016/j.ijpharm.2012.09.011.

Badawy, S. I. F., Narang, A. S., Lamarche, K. R., Subramanian, G. A., Varia, S. A., Lin, J., Stevens, T., & Shah, P. A. (2016). Integrated application of quality-by-design principles to drug product development: A case study of Brivanib Alaninate film-coated tablets. *Journal of Pharmaceutical Sciences, 105*(1), 168–181. https://doi.org/10.1016/j.xphs.2015.11.023.

Campanella, C. E., Cuccovillo, A., Campanella, C., Yurt, A., & Passaro, V. M. N. (2018). Fibre Bragg grating based strain sensors: Review of technology and applications. *Sensors, 18*(9). https://doi.org/10.3390/s18093115.

Diaz-Padilla, I., & Siu, L. L. (2011). Brivanib alaninate for cancer. *Expert Opinion on Investigational Drugs, 20*(4), 577–586. https://doi.org/10.1517/13543784.2011.565329.

Freeman, T., Birkmire, A., & Armstrong, B. (2015). A QbD approach to continuous tablet manufacture. *Procedia Engineering, 102*, 443–449. https://doi.org/10.1016/j.proeng.2015.01.185.

Gere, J. M., & Goodno, B. J. (2013). *Mechanics of materials*. Cengage Learning.

Gour, S., Keluskar, R., Stevens, T., Gao, Z., Bindra, D., & Narang, A. (2012). Effect of type, particle size, and viscosity of binder used in high-shear wet granulation on granule growth as measured by chord length distribution using FBRM C35 probe, American Association of Pharmaceutical Sciences (AAPS) Annual Meeting. Chicago, Illinois

Hansuld, E. M., & Briens, L. (2014). A review of monitoring methods for pharmaceutical wet granulation. *International Journal of Pharmaceutics, 472*(1-2), 192–201. https://doi.org/10.1016/j.ijpharm.2014.06.027.

Harris, C. M., & Piersol, A. G. (2002). *Harris' shock and vibration handbook*. McGraw-Hill.

Holm, P., Schaefer, T., & Larsen, C. (2001). End-Point detection in a wet granulation process. *Pharmaceutical Development and Technology, 6*(2), 181–192. https://doi.org/10.1081/PDT-100000739.

Huang, J., Kaul, G., Utz, J., Hernandez, P., Wong, V., Bradley, D., Nagi, A., & O'Grady, D. (2010). A PAT approach to improve process understanding of high shear wet granulation through in-line particle measurement using FBRM C35. *Journal of Pharmaceutical Sciences, 99*(7), 3205–3212. https://doi.org/10.1002/jps.22089.

Iveson, S. M., Litster, J. D., Hapgood, K., & Ennis, B. J. (2001). Nucleation, growth and breakage phenomena in agitated wet granulation processes: A review. *Powder Technology, 117*(1-2), 3–39. https://doi.org/10.1016/S0032-5910(01)00313-8.

Keningley, S. T., Knight, P. C., & Marson, A. D. (1997). An investigation into the effects of binder viscosity on agglomeration behaviour. *Powder Technology, 91*(2), 95–103. https://doi.org/10.1016/s0032-5910(96)03230-5.

Kristensen, H. G., & Schaefer, T. (1987). Granulation: A review on pharmaceutical wet-granulation. *Drug Development and Industrial Pharmacy, 13*(4-5), 803–872. https://doi.org/10.3109/03639048709105217.

Laicher, A., Profitlich, T., Schwitzer, K., & Ahlert, D. (1997). A modified signal analysis system for end-point control during granulation. *European Journal of Pharmaceutical Sciences, 5*(1), 7–14. https://doi.org/10.1016/s0928-0987(96)00183-2.

Leuenberger, H., Bier, H. P., & Sucker, H. B. (1979). Theory of the granulating-liquid requirement in the conventional granulation process. *Pharmaceutical Technology, 6*, 61–68.

Levin, Michael (2015). *How to scaleup a wet granulation end point scientifically* (1st). Academic Press.

Macias, K., Narang, A.S., & Stevens, T. (2011). In Inline monitoring and control of high shear wet granulation. International Forum of Process Analysis and Control (IFPAC) annual meeting, Baltimore, MD. Academic press.

Masuda, H., Higashitani, K., & Yoshida, H. (2006). *Powder technology handbook*. CRC Press.

Munu, I., Nicusan, A. L., Crooks, J., Pitt, K., Windows-Yule, C., & Ingram, A. (2024). Predicting tablet properties using in-line measurements and evolutionary equation discovery: A high shear wet granulation study. *International Journal of Pharmaceutics, 661*, 124405. doi:10.1016/j.ijpharm.2024.124405.

Munu, I., Nicusan, AL., Crooks, J., Pitt, K., Windows-Yule, C., & Ingram, A. (2025). Using in-line measurement and statistical analyses to predict tablet properties compressed using a Styl'One compaction simulator: A high shear wet granulation study. *International Journal of Pharmaceutics, 669*, 125098. https://doi.org/10.1016/j.ijpharm.2024.125098.

Narang, A. S., Badawy, S., Ye, Q., Patel, D., Vincent, M., Raghavan, K., Huang, Y., Yamniuk, A., Vig, B., Crison, J., Derbin, G., Xu, Y., Ramirez, A., Galella, M., & Rinaldi, F. A. (2015b). Role of self-association and supersaturation in oral drug absorption modeling of a poorly soluble weakly basic drug. *Pharmaceutical Research, 32*, 2579–2594.

Narang, A. S., Desai, D., & Badawy, S. (2012). Physicochemical interactions in solid dosage forms. *Pharmaceutical Research, 29*(10), 2635–2638. https://doi.org/10.1007/s11095-012-0867-5.

Narang, A. S., Sheverev, V., Freeman, T., Both, D., Stepaniuk, V., Delancy, M., ... Subramanian, G. (2016). Process analytical technology for high shear wet granulation: Wet mass consistency reported by in-line drag flow force sensor is consistent with powder rheology measured by at-line FT4 Powder Rheometer. *Journal of Pharmaceutical Sciences, 105*(1), 182–187. https://doi.org/10.1016/j.xphs.2015.11.030.

Narang, V. A. S, Stepaniuk, V., Badawy, S., Stevens, T., Macias, K., Wolf, A., Pandey, P., Bindra, D., & Varia, S. (2015a). Real-time assessment of granule densification in high shear wet granulation and application to scale-up of a placebo and a brivanib alaninate formulation. *Journal of Pharmaceutical Sciences, 104*(3), 1019–1034. https://doi.org/10.1002/jps.24233.

Narang, A.S., LaMarche, K., Subramanian, G., Lin, J., Varia, S., & Badawy, S. (2011). Granule porosity is the overriding mechanism controlling dissolution rate of a model insoluble drug from an immediate release tablet. American Association of Pharmaceutical Sciences annual meeting, Washington, DC

Narang, A.S., Stevens, T., Hubert, M., Gambe, J., Paruchuri, S., Macias, K., Bindra, D.S., Gao, Z. & Badawy, S.F. (2010). Resolution and sensitivity of Lasentech FBRM C35 probe for real-time chord length distribution measurement during high-shear wet granulation and correlation with other particle size distribution techniques. American Association of Pharmaceutical Sciences annual meeting and exposition, New Orleans, LA.

Oberg, E., Jones, F. D., Horton, H. L., Ryffel, H. H., & Mccauley, C. J. (2016). *Machinery's handbook*. Industrial Press.

Osorio, J. G., & Muzzio, F. J. (2013). Effects of powder flow properties on capsule filling weight uniformity. *Drug Development and Industrial Pharmacy, 39*(9), 1464–1475. https://doi.org/10.3109/03639045.2012.728227.

Panakanti, R., & Narang, A. S. (2012). Impact of excipient interactions on drug bioavailability from solid dosage forms. *Pharmaceutical Research., 29*(10), 2639–2659. https://doi.org/10.1007/s11095-012-0767-8.

Parikh, D. M. (2010). *Handbook of pharmaceutical granulation technology*. CRC Press.

White, Frank (1998). *Fluid Mechanics* (4th). New York: McGraw-Hill Higher Education.

CHAPTER 16

An introduction to powder characterization

Jamie Clayton

Freeman Technology Ltd., Gloucestershire, United Kingdom

There is now widespread recognition that a good understanding of powder behavior is essential for the realization of many of the industry's long-term goals—more efficient production, continuous manufacture, and real-time release.

This chapter provides a straightforward, practical introduction to the challenges of powder characterization, the methods currently available (including various traditional techniques, uniaxial and biaxial shear testing, and dynamic flow testers), and what they measure. The fundamental properties of powders are also discussed to provide the necessary understanding to successfully apply quality-by-design (QbD) and process analytical technology (PAT) principles.

16.1 The relevance of powder flow for wet granulation

Though granulation confers a number of benefits, a primary motivation for its application is often to improve flowability. Fine, cohesive powders typically exhibit relatively poor flow properties that lead to suboptimal process efficiency or compromise product performance. Wet granulation, for example, is routinely used to improve the properties of tableting blends. Dense, free-flowing granules process more reliably through the tablet press at high rates, ensure uniform die filling, and compress well to form stable tablets of consistent quality.

One of the major challenges associated with wet granulation is endpoint detection. Identification of the point at which granules have attained optimal properties for their intended use is essential but especially difficult, as granules are typically an intermediate rather than the finished product, as in tableting. Here, the properties of the granules are only of interest with respect to their impact on tablet quality, making it highly beneficial to identify a measurable parameter that correlates directly with critical quality attributes (CQAs) of the tablet. Doing so eliminates the need to process granules through the press to assess their quality, thereby allowing the granulation to be treated as an independent unit operation.

The importance of flowability in defining process and product performance suggests that flowability measurements have considerable potential for endpoint detection, and this has proven to be the case. In experimental studies, the flow properties of granules have shown direct correlation with the hardness of tablets manufactured from them (Freeman et al., 2015). This is a significant finding. It suggests that a target specification for the wet granulation step of a tablet manufacturing process can be defined in terms of flowability. Granules meeting this specification will consistently produce tablets of desirable quality, regardless of manufacturing scale.

With such a specification in place, it is then possible to identify critical material attributes and critical process parameters for a wet granulation operation and define an associated design space. Scale-up is also eased by having a relevant granule specification that is independent of scale. As the size of processing equipment is increased, processing variables can simply be manipulated to meet the flowability specification. At the commercial scale, controlling to a specification that robustly correlates with finished product quality is entirely consistent with a quality-by-design (QbD) approach.

In summary, flowability measurements have the potential to support the development, optimization, monitoring, and control of wet granulation processes. This chapter provides an introduction to powder flowability, the mechanisms that impact it, and the methods that can be applied to measure it. It provides a valuable resource for those looking to better understand flowability measurements with respect to wet granulation applications.

16.2 Considerations and challenges in characterizing powder flow

16.2.1 Why do we test powders?

In an industrial setting powder testing is not an academic exercise, but rather a way to determine, for example, how to design a drug product, or whether a product is suitable for a given application. Evaluating the issues that powder testing is attempting to address is, therefore, a critical first step in assessing the techniques available and their potential value to a user and, more broadly, the industry. Typical questions might include:

- Will this blend result in high-quality tablets if manufactured using a direct compression process?
- Is this new raw material going to achieve blend uniformity in my existing blender?
- How closely do I need to control storage conditions for this material?
- What is the optimum water content required during granulation to ensure ideal final product attributes?
- Which process steps have the most influence on product quality?
- Will this new raw material discharge from an existing hopper?
- How can I develop an optimal process design for this new product or formulation?

16.2.2 A complex problem

Powders are bulk assemblies of solids, gases, and liquids. This explains why they are so complex, but it also underpins their industrial value. Powders are able to perform in many different ways, depending on their intrinsic properties, how they are formulated and manufactured, and the environment to which they are subjected. This is why they are so useful. For example, a powder can flow through a feeder or press but then exit as a stable, easily handled tablet.

Tables of physical properties and established mathematical theory often guide engineers when working with solids, liquids, or gases in a process. However, this toolbox is not defined for powders.

Efforts by US and European pharmacopeia organizations to provide direction and "approved" measurement techniques for powder testing have culminated in a harmonized approach, which was written into both the USP and EP publications in 2012 (United States Pharmacopeia, 2007). The USP and EP publications reflect the fact that "powder behavior is multifaceted" and suggest that "no single and simple test method can adequately characterize the flow properties experienced in the pharmaceutical industry." They include a number of powder flow measurement techniques and advise their combined application during pharmaceutical development.

These powder characterization methods undoubtedly provide a starting point for establishing a powder testing strategy, but are recognized as having certain limitations within the prevailing regulatory and economic environment. (See Section 16.3 for further discussion.)

Over the last decade, demands on drug producers have changed. Regulatory bodies have moved from simply requiring a well-defined set of mandatory data in order to approve a product to a scenario where they are looking for companies to demonstrate a truly rigorous understanding of a new drug and knowledge-based strategies for control of the risks associated with its manufacture. These principles are enshrined in QbD (ICH, 2006; U.S. Food and Drug Administration, 2007).

Commercial pressures have also intensified, increasing the premium associated with smarter manufacture and the realization of continuous processing (Pellek & Arnum, 2008). Continuous processing is associated with smaller, easier-to-manage plants, lower capital investment, and simplified scale-up (Clayton, 2012; Pellek & Arnum, 2008). However, it also demands a much deeper understanding of every aspect of formulation and process than has historically been required. Within this environment, answering the questions outlined above in sufficient detail can be challenging. Consequently new technology has been applied, and new ways of measuring powder flowability have become established in forward-thinking companies.

Discussions about powders and powder testing in particular tend to focus on flow from a hopper and hopper design. But implicit in the questions above is the fact that industrial interest in powder behavior actually extends far beyond this narrow definition. Considering a tableting process (Fig. 16.1), for example, the formulation scientists and process engineers would be interested in optimizing feeding, mixing/blending, granulation, drying, filling, tableting, aerosolization, pneumatic conveying, dosing, storage, and transportation, etc.

As with many complex problems, answers lie in understanding the fundamentals, in this case, the individual factors that influence powder behavior, while at the same time focusing on the importance of the overall goal—optimizing productivity and quality. Measuring properties such as bulk density, compressibility, and permeability are all useful, important, and potentially relevant by themselves. Perhaps more importantly, they all contribute to "flowability" and the behavior of a powder at every stage in a process—from bulk storage in a hopper to the various unit operations in production, for example.

16.2.3 Powder flowability

To understand how powders behave across the diversity of conditions encountered in pharmaceutical processing, further consideration must be given as to what influences their flowability.

The terms "particle" and "powder" are often used interchangeably, but this can be misleading. Powders are bulk assemblies, containing a subdivided solid mass (such as a crystalline drug) with gases (usually in the form of air), and liquid (typically water) on the surface of the particle or within its structure.

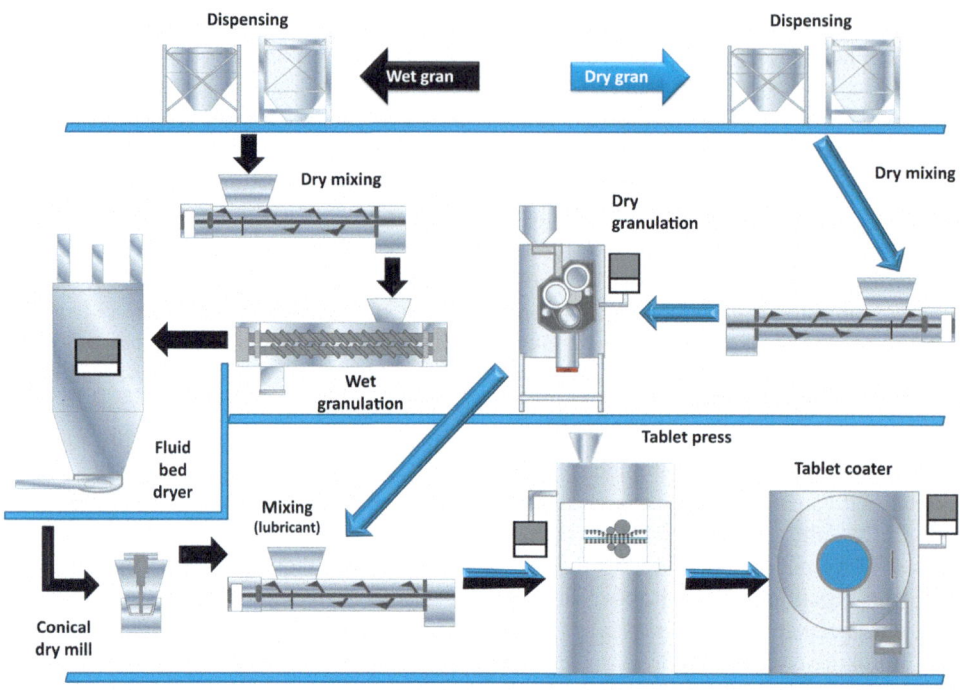

FIGURE 16.1

A stylized pharmaceutical tableting process, illustrating the multiple processes where powder behavior is important.

It is the properties of these three phases, and the interactions between them, that define bulk powder behavior. Behavior is, therefore, influenced by a large number of variables and an array of their potential interactions (Fig. 16.2). Many, but not all, of these variables relate to the physical properties of the particles, for example:

Note that many of these properties are distributions, rather than a single value.

Other variables relate to the process or "external" influences, such as the amount of air present, moisture levels, and the degree of consolidation. These variables are often uncontrolled and not measured.

Powder behavior is dependent on both the particle properties and external variables, which is why it is complex and why powder performance can't be accurately predicted from measurements of physical properties alone. From a practical perspective, too many variables impact powder behavior to make accurate mathematical modeling viable. Moreover, it is reasonable to assert that we don't yet fully understand all the possible interactions, nor do we have the capability to directly measure many of the influencing parameters.

Current understanding is limited to recognizing that:

$$\text{Feedrate} = 49.54\ \text{FRI} - 13.81\ \text{SE} + (163.8)\ \left(R^2 = 0.9466\right)$$

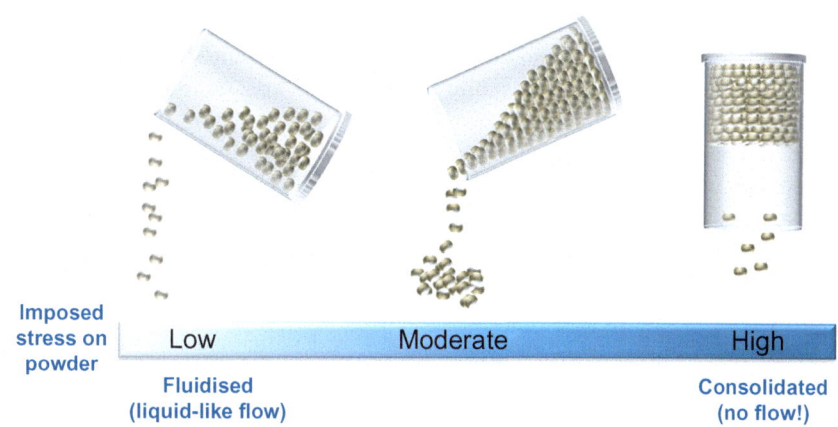

FIGURE 16.2

Particle properties are important when it comes to powder behavior, but so too are external variables such as the extent of aeration or consolidation.

and also, external variables, such as:

- Consolidation
- Aeration
- Humidity level
- Extent of shear/strain
- Equipment surface properties

To understand what this complex web of interactions means in practice, it is helpful to consider its impact with respect to powder flowability.

The concept of flowability can be appreciated by, for example, swirling a jar containing granulated sugar to produce fluid-like movement in the powder. If the jar is rolled on its side, the sugar will smoothly tumble, maintaining a steady flow pattern. Subjecting flour to the same movement induces a very different, intermittent flow. Small displacements of the jar may have no impact, but then an avalanching effect occurs once a critical point is passed. The flour is flowing—but in a very different way from the sugar, which makes it intuitively less appealing for those whose primary concern is steady, consistent flow.

In this loose packing state, these two powders lie at opposite ends of a flowability spectrum. In simple terms, the sugar would be described as "free-flowing" and the flour as "cohesive." Many powders lie at finely differentiated points between these two extremes. It is crucial to recognize that, because flowability is influenced by so many factors, the flowability of one powder relative to another is not fixed according to a linear scale of cohesion. Relative flowability may change dramatically depending on the conditions to which the powder is subjected (Freeman, 2007; Teunou & Fitzpatrick, 2000).

For example, if each powder is consolidated or aerated, then this will influence its flow properties in quite different ways. If the sugar particles are subjected to vibration, then they may lock together efficiently, resulting in a material that is very resistant to flow. Direct compression is likely to have relatively little effect. Conversely, compressing the flour excludes entrained air and is likely to further inhibit the limited flowability previously demonstrated. Compression will therefore detrimentally affect the flowability of the flour, compared to an unconsolidated baseline state. However, vibration is likely to have the opposite effect; that is, the flowability of the sugar may be more detrimentally affected compared to that of the flour.

In order to fully understand the range of flow characteristics that powders can exhibit, it is essential to assess their behavior under a wide range of stress and flow regimes, from aerated to consolidated, for example, under static and dynamic conditions or when subjected to high or low shear. Powders often exhibit flow rate sensitivity, meaning that the ease with which they flow is dependent on the rate of shear applied. A practical example that illustrates this point is a mixing process. Powders with high flow rate sensitivity will require more careful handling in a mixing process in order to achieve blend uniformity, with greater control required to achieve optimal results. Powders that are less flow rate sensitive can successfully be blended to uniformity under a wider range of conditions (Armstrong, 2011), providing the opportunity, for example, to utilize low-energy mixing processes, as opposed to high shear operations that may also damage particles and induce electrostatic charging.

Understanding how powders flow is critical for the efficient operation of many processes. Flow properties influence how easily powders will mix or blend, how they behave in a feeder, or how they flow into and fill a capsule or tablet die. For this reason, understanding the factors that influence flowability is important.

16.2.3.1 Cohesion

The term cohesion is commonly associated with a powder's flowability and indeed is often a highly influential property in regard to bulk powder behavior. Cohesion is the mechanism that acts between particles and has the tendency to "bond" a particle to neighboring particles. Tensile, cohesive forces are a combination of electrostatic charge, surface energy, and Van der Waals forces.

The magnitude of the cohesive force between adjacent particles is typically small and often insignificant when powders are in a consolidated state, where the forces of friction and mechanical interlocking are most dominant. However, when a powder is in a loose packing state, cohesive forces can be large relative to the other forces acting on the particle and may dominate the behavior of the bulk powder. As powders are often processed in low-stress conditions, the strength of the cohesive forces can make the difference between a successful process and failure.

16.2.3.2 Adhesion

While cohesion quantifies the strength of particle-particle interactions, adhesion is a measure of the propensity of particles to stick to a different surface or material—often the surface of the processing equipment. This is an important issue when it comes to manufacturing efficiency, since powder residue can result in blend inhomogeneity, require more thorough cleaning, increase the risk of cross-contamination between batches, and be a major initiator of blockages.

16.2.3.3 Compressibility

Compressibility is a measure of the volume change in a powder sample as a consequence of an applied consolidating stress. This stress forces particles closer together, packing the same number of particles into a smaller volume, and thereby increasing bulk density. Compressibility should not be confused with consolidation induced by vibration, an entirely different mechanism, dependent on particle reorganization as the bulk powder is jostled (Fig. 16.3).

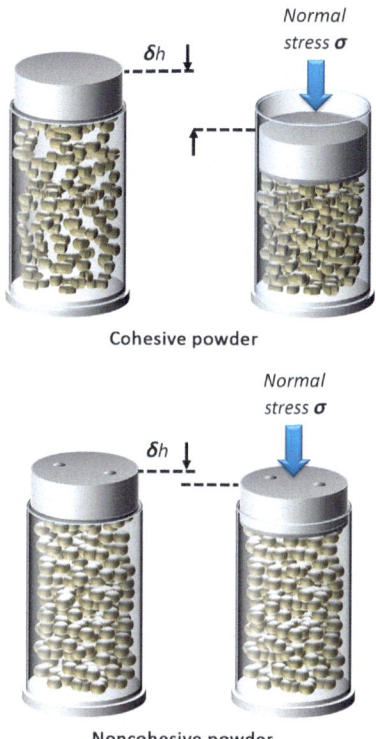

FIGURE 16.3

Cohesive powders, which tend to be inefficiently packed and, therefore, contain more air, are typically more compressible than less cohesive powders.

Compression may be intentional, during tableting, for example, or may occur in a relatively uncontrolled way, such as when a powder is stored under the pressure of its own weight.

Some powders are fairly incompressible, while others are readily reduced in volume. Returning to the previous examples, granulated sugar is relatively incompressible. In contrast, applying a consolidating stress to a sample of flour will cause a significant change in bed height, at the same time producing a compacted "plug" with increased structural strength.

There are links between compressibility and flow, although they are not as definitive as some traditional testing methods suggest. Powders exhibiting high compressibility tend to be characterized by an inefficient packing structure when in a low stress state, and this is mostly symptomatic of relatively cohesive materials. In contrast, noncohesive powders tend to be far less compressible.

It is this principle that underpins traditional bulk density methods and their use in predicting flowability. This is discussed further in Section 16.2; however, it should be noted that these methods rely on uniaxial deceleration (or tapping) to induce volume change. Whilst they are traditionally referred to as "compressibility" tests, this term should not be confused with compressibility induced by the application of an external force. Inferring flow from density measurements can be misleading and is far less sensitive than direct flow measurement (Freeman, 2007).

16.2.3.4 *Permeability*

The ease with which air can be transmitted through a powder bed is a function of the powder's bulk permeability. Powders with a large particle size tend to have high permeability due to the stable, uniform channels that exist between the particles, through which air can easily pass. Powders containing small particles may have a higher relative air content, or lower solids fraction or bulk density; however, due to smaller interstitial spaces, poorly connected "pockets" of air, and higher inter-particulate forces, the resistance of the bulk powder to air flow is typically greater.

Permeability is also influenced by other powder properties, such as the amount of fine particles present, the shape of the particles, and their surface properties. In addition, the permeability of a bulk powder is dependent on the extent of compression to which the powder is exposed. Powders that are highly compressible tend to exhibit larger changes in permeability upon consolidation. In contrast, powders whose packing structure changes to a lesser extent when subjected to a consolidating load exhibit more consistent permeability. Quantifying permeability across a range of consolidating stresses is, therefore, advisable.

Permeability measurements help to quantify how easily a powder retains or releases air. This aspect of powder behavior is influenced by the response of a powder to aeration, but is a separate property. Powders that are highly permeable transmit or release air very easily. If air is required to maintain their flow properties, then such powders have the potential to transition from liquid to solid-like behavior (or vice versa) very rapidly and become problematic in the process.

However, the ability to easily release air can, in fact, be a beneficial attribute in certain processes. For example, filling and compression processes are particularly sensitive to permeability, as the air inside the die or mold needs to exit rapidly to allow a complete fill. Here, low permeability restricts the back-flow of air between particles, as the particles fill the die, and may compromise fill rate. During tablet manufacture, the ability of the air to escape from the loose powder contained in the die during the compression step is strongly linked to the mechanical properties of the tablet, and can lead to undesirable characteristics such as lamination and capping (Seppälä et al., 2010).

It is relatively easy to see the relevance of permeability when it comes to fluidization processes and other unit operations, such as pneumatic conveying, where air (or another gas) plays a prominent role. However, the amount of air in a powder is typically uncontrolled and may change significantly during routine processing. Powders can easily pick up and lose air during discharge from a vessel, for example, or when left to settle under their own weight. With some powders, this results in a dramatic change in behavior (Fig. 16.4).

16.2.3.5 *Electrostatic charging*

Some particles are easily charged, changing the balance of forces within the powder and impacting bulk behavior. However, powders are generally electrical insulators, so simply earthing a charged powder does little to change its behavior. It is routine practice to "ground" processing equipment for safety reasons, but this action should not be assumed to eliminate charge evolution in a bulk powder. A far more effective method is to optimize the water content in the powder. Water is usually considered detrimental when it comes to powder flow, but in an electrostatically charged powder, water can act as a conductive

16.2 Considerations and challenges in characterizing powder flow

FIGURE 16.4

Powders can be intentionally aerated (left) or simply pick up or release air during routine processing. When flooding occurs, loss of process control can cause costly problems (right).

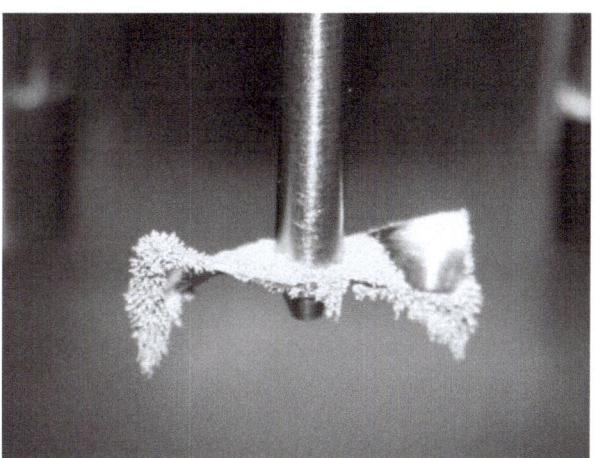

FIGURE 16.5

Electrostatic charge buildup in powders can transform behavior. Powders that are electrically charged may "fly away" and stick to nearby surfaces, such as the blades of an impeller in this example.

medium providing an effective route to ground, discharging the system, and potentially enabling the particles to flow more freely (Emery et al., 2009; Jones et al., 2003; Zeng et al., 2000) (Fig. 16.5).

16.2.3.6 Physical changes

When processing powders, it can be relatively easy to change some of the properties of the particles, intentionally or otherwise, thereby inducing a significant change in bulk powder behavior. For example,

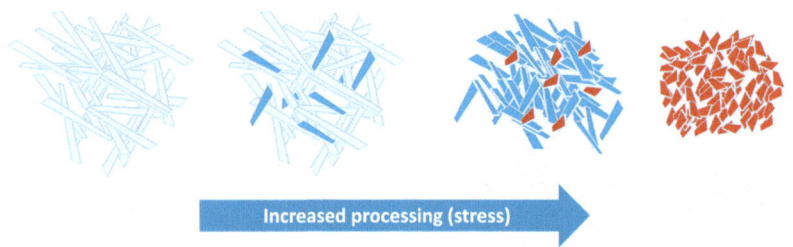

FIGURE 16.6

Particles may undergo attrition if subjected to stress, resulting in a reduction in particle size and a change in shape and surface area.

particles vary substantially in terms of their affinity for water. The impact of processing in a humid atmosphere will, therefore, vary depending on a powder's tendency to absorb or take on water, as will performance in certain unit operations such as wet granulation. Furthermore, particles that are not mechanically robust may undergo attrition when subjected to stress. Attrition can result in a reduction of particle size, changes in particle shape, an increase in surface area, and modification of surface properties—all of which will impact bulk powder behavior.

In many cases, a process step designed to produce one outcome may, at the same time, inadvertently and irreversibly change particle and powder properties. A common example of this is the agglomeration and densification phenomena that occur when feeding a cohesive powder. Screw feeders are utilized to control the flow rate of a feed material into the next step of a process, but they subject a powder to high levels of shear and compaction. As a result, certain powders will possess very different properties after being passed through the feeder, even though this is not the primary intention (Fig. 16.6).

16.2.4 Understanding the impact of the process environment

It is clear from the preceding analysis that powders exhibit a range of different behaviors and that different processes rely on these varying behaviors to work effectively. The logical conclusion from this is that, while intrinsic properties are important in defining the value of a powder, the key to optimizing efficiency is to match the bulk properties of a powder with the conditions in any individual process.

To illustrate this, consider two different unit operations in more detail. What stresses and conditions do they subject a powder to, and how does this influence which powder properties are most relevant?

First consider the environment and stress regimes acting on a powder discharging from a **hopper**. Powder flow is induced by gravity, and flow rates are, in general, relatively low. The powder is consolidated by the weight of material above it, potentially over a prolonged period, depending on its residence time, and this weight imposes a moderate to high stress environment on the powder just above the outlet. Frictional interaction between the powder and the hopper wall is an important additional consideration. At the time of discharge, a valve at the bottom of the hopper is opened, and the same consolidation stress that compacted the powder during storage is now relied upon to force the powder to begin to flow through the valve (Fig. 16.7).

In this environment, compressible powders are more susceptible to changes in their behavior as a consequence of the imposed weight of the powder bed (Peleg, 1977). A combination of powder and material of construction of the hopper that results in low friction at the hopper wall will be beneficial

16.2 Considerations and challenges in characterizing powder flow

Flow from a hopper
Low flow rate
High stress environment

FIGURE 16.7

To understand how a powder behaves in a process, it is best tested under conditions that simulate the processing environment.

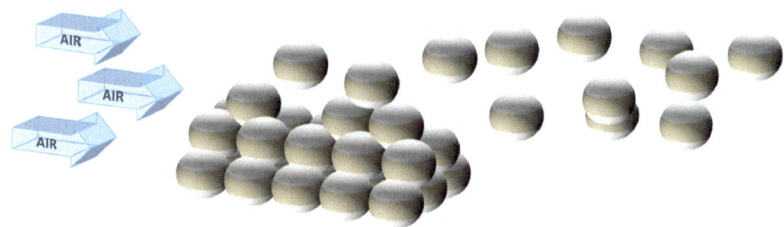

Aerosolization:
Low stress environment
Powder needs to dilate and fluidize

FIGURE 16.8

Understanding the conditions experienced during aerosolization.

in terms of enhancing flow, as will relatively low inter-particle forces. Under stress, particles are forced closer together, so frictional forces and mechanical interlocking are the mechanisms most likely to dominate flow behavior. Cohesive forces may exist, but they are negligible in contrast to these other mechanical interactions.

Therefore the information that is especially relevant when it comes to evaluating or predicting powder behavior in this environment is:

- Shear/flow properties of the powder under moderate to high stress
- Wall friction acting between the powder and the material from which the hopper is constructed
- Compressibility and density data
- Permeability, particularly if the outlet of the hopper is connected to a transfer chute or other hardware that mitigates backflow

Contrast the hopper conditions with what happens during **aerosolization** of a powder (Fig. 16.8), in dilute phase pneumatic conveying, or during drug delivery via a dry powder inhaler (DPI), for example. In these situations, there is little, if any, external compression stress acting on the powder. Air is being drawn through the sample with the potential for inter-particulate lubrication, aeration, and ultimately dilation of the powder into a stream of discrete particles.

Here, the more relevant questions to address are:

- What is the cohesive strength of the powder (as this will strongly influence its response to aeration)?
- How do the flow characteristics of the powder change when it is aerated?
- Does the powder fluidize?
- If so, what is the minimum fluidization velocity for this powder?
- Are the particles robust, or are they likely to suffer from attrition?

In this application, candidate powders must be assessed against very different criteria from those applied for the hopper environment. In this low-stress regime, cohesive forces will tend to define flow behavior, rather than mechanisms that rely on closer particle proximity. A highly cohesive powder is unlikely to flow easily or to exhibit uniform fluidization and progressive dispersion. However, cohesive powders often have low permeability and, therefore, their structure tends to fail catastrophically during an aeration process, giving rise to a high energy dispersion that may be of benefit to the application—producing a respirable cloud of particles in the DPI device (Shur et al., 2008). The optimum dispersion mechanism, therefore, needs to be carefully considered in relation to the specific application.

These two examples illustrate the difficulty of describing powders with just a single number, or even a single technique. A simple measurement, such as the Hausner ratio, for example, will give some insight into bulk powder behavior, however limited, but trying to develop the in-depth knowledge necessary for efficient processing from just a single, simple analysis is unrealistic. Each process, or unit operation, requires a different combination of powder properties to be understood. This is particularly significant in the context of QbD, which relies on understanding all the factors that impact a product or process.

The preceding analysis also emphasizes the point that it is unhelpful to think of powders as innately "good" or "bad." It is more appropriate to consider that they can be suited to different applications. Recognizing the impact of the processing environment is fundamental to translating an understanding of powder characteristics into better product or manufacturing performance. There exists a unique series of powder properties that define performance in any given process or unit operation. There will usually be more than one variable, and the combination will vary from application to application. The complexity of powder behavior and the need for sensitive, differentiating, and process-relevant information make single-number testing techniques both limited and relatively unhelpful. In contrast, multifaceted characterization, assessing a range of relevant power characteristics, provides the insight needed to ensure a good process-powder match, regardless of the industry, application, or powders in question.

This is ultimately what understanding and characterizing powders is about. Trying to process a powder that is ill-suited to the processing equipment and/or not conducive to a high-quality final product will be challenging. In contrast, a well-matched powder-process combination will contribute to long-term productivity, high performance, and profitability.

16.3 Defining a powder—fundamental principles of particles

When powders flow, the particles within them move relative to one another. The mechanisms that define the ease of that movement are:

16.3 Defining a powder—fundamental principles of particles

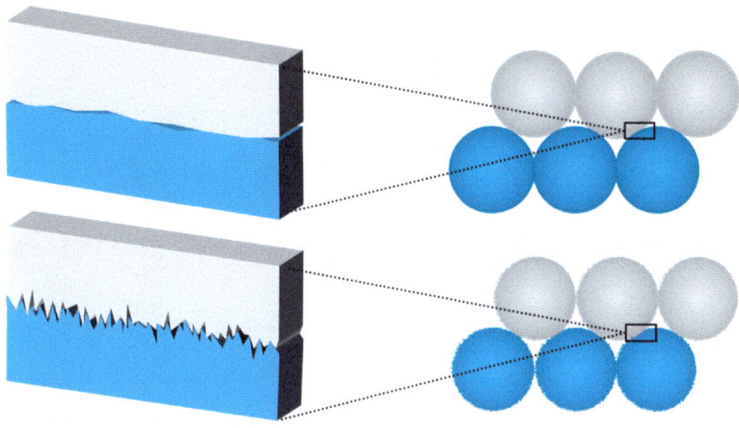

FIGURE 16.9

Particles with a smoother surface will generally have a lower frictional interaction and flow more easily than those that are rougher, assuming all other features are identical.

- Friction—between particles, and between particles and equipment surfaces
- Mechanical interlocking, which relates to the shape of the particles
- Liquid bridging—resulting in capillary forces
- Cohesion—inter-particle interactions due to electrostatics and Van der Waals forces
- Gravitational effect, which directly relates to the density and size of the particles

16.3.1 Friction

Smooth particles tend to slide more easily relative to each other as a consequence of the lower friction acting at the contact points. Likewise, the level of friction between particles and the walls of process equipment will be influenced by the surface properties of both the particle and the material of construction. Surface roughness, and also the chemical nature of the processing equipment surface, can influence frictional interactions (Fig. 16.9).

16.3.2 Mechanical interlocking

The particles shown in Fig. 16.10 have a shape that results in substantial mechanical interlocking. Once this occurs, the particles will strongly resist further movement, even if their surface friction is low. However, the particles shown in the image above left are less likely to be influenced by mechanical locking due to the more rounded nature of their morphology. In this case, the particles are more likely to move past one another, although their net interactions will still be influenced by other mechanisms, such as friction. It is generally the case that powders comprising particles of irregular shape tend to possess poorer flow properties.

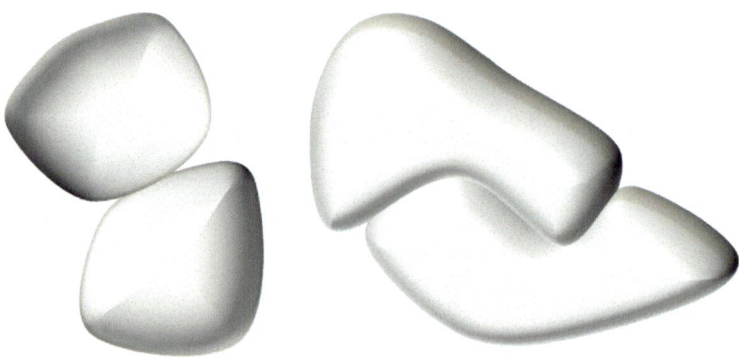

FIGURE 16.10

Particles of a certain shape can mechanically interlock and resist flow.

FIGURE 16.11

Liquid can bridge between particles, reducing particle independence.

16.3.3 Liquid bridging

Adsorbed moisture is widely perceived to have a negative impact on powder flowability. This is not always true, but liquid bridging can certainly be detrimental to flow. Liquids introduced into a powder have a tendency to coat the surface of the particles, forming bridges that inhibit the free movement of one particle relative to another (Fig. 16.11).

This effect is exploited in wet granulation, where water is used to bind individual particles to form more easily handled granules, often with improved compression properties. Elsewhere, the uncontrolled surface adsorption or ingress of water can significantly degrade material quality, compromising processability, promoting caking, and reducing productivity due to blockages, downtime, or material out of specification. Liquid bridging can also occur between powder particles and the walls of the processing equipment, creating further processing challenges.

However, as previously noted, water can help dissipate electrostatic charge, thereby improving flow properties. The challenge is to ensure the bulk powder contains sufficient moisture to minimize the effect

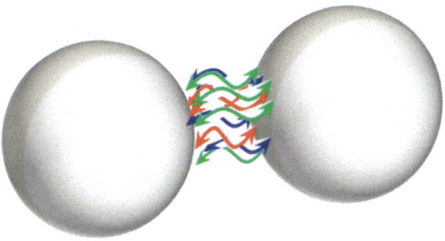

FIGURE 16.12

Inter-particulate forces act between contacting particles and those in close proximity.

of electrostatics, but not enough for capillary bonds to become established between particles, reducing flowability.

16.3.4 Inter-particulate forces of cohesion

The inherent strength of the inter-particulate forces of attraction (often called cohesive forces for similar particles) is derived from a combination of Van der Waals forces and electrostatic charge on the surface of the particles. These are typically tensile forces, holding particles close to their neighbors (Fig. 16.12) and inhibiting independent particle mobility. They result in the formation of agglomerates and can be the cause of many processing and quality issues.

The absolute values of inter-particulate forces can vary widely, depending on the chemical composition of the particles, their shape and surface texture, bulk water content, and processing history. They may also change with time, where powders can be seen to *relax* as the electrostatic charge dissipates (following a high shear blending process, for example).

Cohesive forces are perhaps the most challenging of all particle-particle interactions and are very difficult to quantify and model. However, their influence dominates powder behavior in many processes and applications, particularly where the powder is uncompressed and required to flow. Process operations such as filling, mixing, conveying, and applications such as DPIs are all dominated by the cohesive strength of the powder. It is imperative, therefore, that these forces are well understood and accurately quantified.

16.3.5 Gravity

This section has so far considered the forces that restrict particle-particle independence. Generally, the stronger their influence, the poorer the flow properties of the powder. However, many powders with high levels of cohesion, irregular morphology, and high surface friction still exhibit good flow properties, and this is due to the force imposed by the acceleration due to gravity.

Notwithstanding that particles in motion will have inertia, the primary force acting on a loosely packed stationary particle is that due to gravity—its weight (Fig. 16.13). Therefore the ability of a particle to begin to flow, particularly in powders that are unconsolidated, is largely dependent on the strength of the gravitational force acting on it. It is for this reason that powders containing large particles,

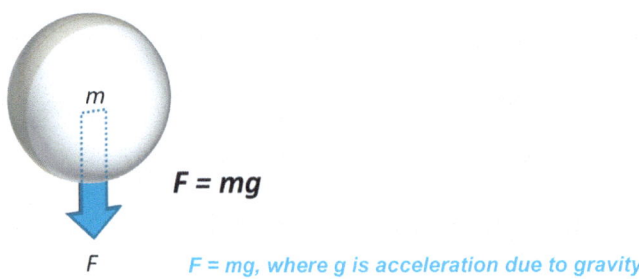

FIGURE 16.13

Gravitational forces are often the only motive force acting on the particle.

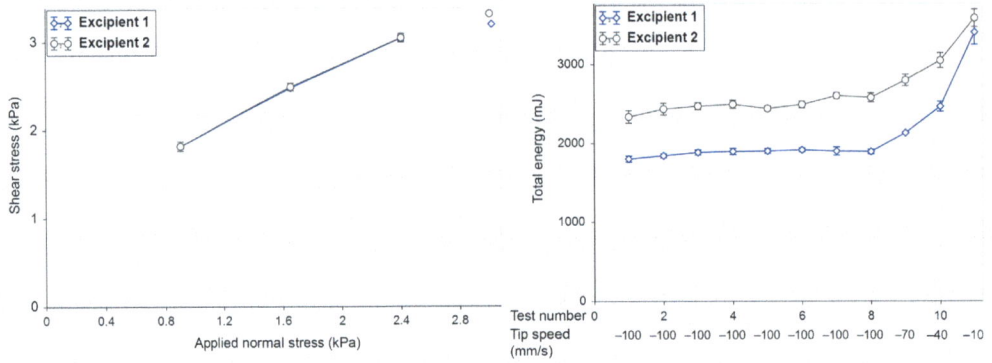

FIGURE 16.14

Data from shear cell and dynamic flow tests for two batches of an excipient.

or consisting of material that has high density, tend to flow better when loosely packed—each particle's individual mass and, therefore, the gravitational force acting on it, is high.

The relationship between all restrictive forces and the motivating force of gravity is what dictates whether particles can move independently or whether they will exist as part of a stronger agglomerate, in which case bulk flow is influenced by the mass of the agglomerate and its relationship with surrounding agglomerates. There is a common misconception that powders with a small particle size have stronger cohesive forces, but this is not necessarily true. The gravitational force acting on a small particle is low because the particle's mass is low, and hence, the cohesive force is relatively high compared to the gravitational force, even though the absolute value of the cohesive force could be low. A powder consisting of larger particles could have stronger cohesive forces acting between particles, but as particle mass is high, the gravitational force dominates, and the particles can flow independently.

Generally speaking, for loosely packed powders:

- If the gravitational force is **greater** than all restrictive forces, particles are mobile and the powder flows well. An extreme example would be a powder comprising large, spherical particles, with a low-friction surface and good electrical conductivity.

- If the gravitational force is **weaker** than all restrictive forces, particles are immobile, and the powder agglomerates and flows poorly. An extreme example would be a powder comprising small, irregularly shaped particles, with high surface energy, high surface friction, and some form of binder, such as excess water or fat.

There are an infinite number of cases between these two extremes, and this is further complicated by the fact that, at the particle scale, each force is variable and can change depending on the environment to which the powder is subjected.

16.4 Methods available to the modern scientist

There are many applications in the pharmaceutical industry where data on powder properties and powder flowability are crucial. In this section, we review some common powder testing methods.

A fundamental principle when considering any test method is the concept of relevance. To be relevant, and consequently useful, a test must measure a property that correlates directly with performance in the specific process of interest. It is not unusual to test a powder against a defined specification and find it acceptable, but then encounter problems when the material is introduced into the plant or released to the customer. This situation typically results from a failure to define a relevant specification, that is, one that relates to how the powder will actually perform during a particular processing stage.

In addition to relevance, other factors that influence the ability of a powder tester to address industrial requirements include sensitivity, reproducibility, and repeatability.

The sensitivity of a powder tester is crucial to its ability to identify and quantify differences. Certain techniques are inherently more sensitive than others, but repeatability and reproducibility are also critical in determining how successfully and precisely a tester will be able to differentiate samples. Where instruments utilize the same technique, the one with higher repeatability and reproducibility will more successfully detect smaller differences, thus offering greater insight to advance process understanding and/or supporting more exacting quality assurance/quality control requirements.

Features which enhance repeatability and reproducibility include well-defined measurement protocols, precision-engineered equipment, and a high level of automation. Such features also beneficially impact practicality—which is another important factor that influences the industrial suitability of a tester. Ease of use and measurement time both affect the ability of a powder tester to provide efficient support across the product lifecycle.

16.4.1 Traditional powder characterization techniques

Engineers have been assembling equipment to measure powder flowability for many decades in an attempt to understand powder behavior. Traditional tests with a long history of widespread use, and reference in USP/EP guidelines (United States Pharmacopeia, 2007), include angle of repose, flow through an orifice, and tapped density.

- Angle of repose—powder poured from a vessel forms a cone-like pile. The angle of repose—the angle between the slope of the pile and the horizontal—correlates with the strength of particle-particle interactions and, therefore, is measured to infer flowability.

- Flow through an orifice—measuring the rate at which powder flows through a hole of closely defined dimensions directly quantifies flowability. Instrumentation may be either manual or automated and may vary in terms of dimensions and configuration.
- Tapped density—tapping a powder sample causes the particles present to reposition and reorient with respect to each other, changing the bulk density. The extent of this change is related to the cohesivity of the powder and, hence, its flowability. Results are typically expressed in the form of ratios of tapped to untapped density—Hausner Ratio (Hausner, 1967) or Carr's compressibility index (CCI) (Carr, 1965).

16.4.1.1 Data generated

All of these techniques provide a flowability ranking based on the measured parameter. For example, a powder with an angle of repose of 66 degrees would be classified as having extremely poor flow properties compared with one with an angle of 25 to 30 degrees, which would be classified as "free-flowing." In tapped density testing, the designation "free-flowing" would be applied to powders with a CCI of <20%, which equates to a Hausner ratio of <1.25. While these techniques enable a simple ranking of the flow behavior, the results are not directly comparable from one technique to another, nor do they simulate the conditions in real-world processes (Hickey et al., 2007).

16.4.1.2 The practicalities

The apparatus employed to run these tests is typically manually operated and inexpensive. Measurement times are short, and simple principles of operation require minimal training.

16.4.1.3 Advantages and limitations

The practicalities of simple testing techniques make them an appealing choice for the manufacturing environment, and results tend to be intuitive. This is particularly true for flow through and orifice measurements, but less so for tapped density measurements, in which certain powders can behave counterintuitively.

Offsetting these practical advantages are a number of important limitations. In many cases, the test protocols for these traditional techniques can vary significantly, contributing to poor repeatability and reproducibility. This issue is compounded by the manual nature of the measurements, which introduces operator-to-operator variability, and by a lack of precision. For example, multiple angles can form in a pile of powder, making it difficult to ascertain the exact angle of repose (Lavoie et al., 2002; Taylor et al., 2000). Similarly, the accuracy of tapped density measurements may be limited by the scale on a measuring cylinder and/or an inability to achieve a level powder surface. This makes them insensitive to minor differences between powders and means that they are only likely to be useful when large differences are observed (Lavoie et al., 2002).

There is little flexibility with any of these tests to adapt the test conditions to reflect the processing environment. Furthermore, the results can be severely impacted by the lack of control over sample presentation, such as variability in the way that the sample is loaded into the instrument (Hancock et al., 2004; Taylor et al., 2000). With certain powders these simple techniques may fail completely—a prime example being the "no flow" result that can occur with a flow through an orifice or angle of repose tests on cohesive powders.

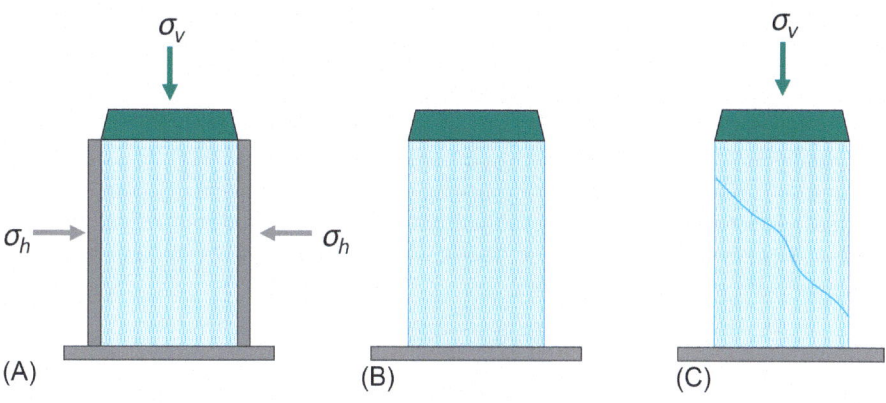

FIGURE 16.15

The stages of uniaxial powder testing: (A) consolidation, (B) removal of stress and confining wall, and (C) failure of the free-standing column.

16.4.2 Uniaxial testing

As previously discussed, the main parameters reported by biaxial shear cell testing are mathematically derived, rather than directly measured. Uniaxial testing provides a simpler, more direct method of quantifying the same parameters. In a uniaxial test, the force needed to break or fracture a free-standing column of powder that has been consolidated by a known amount is measured (Fig. 16.15).

This concept is much more straightforward than biaxial shear cell analysis, but the technique has not been as widely adopted in industry. The major challenge associated with uniaxial testing is ensuring uniform density and homogeneous stress distribution throughout the free-standing column of powder. Recent developments in both academia and industry have seen the introduction of devices that address these issues, thereby providing a robust assessment of the unconfined yield strength (UYS) of a diverse range of materials (Bell et al., 2007; Calvert et al., 2013; Freeman et al., 2016).

16.4.2.1 *How it works*

Uniaxial testing involves measuring the force required to fail a preconsolidated powder column. As illustrated in Fig. 16.15, measurement comprises: (A) construction of a uniformly consolidated column by the application of a vertical stress; (B) removal of the vertical stress and confining wall; and (C) failure of the resulting free-standing column through the application of a known vertically applied stress.

16.4.2.2 *Data generated*

The UYS of a powder as a function of consolidating stress is reported.

16.4.2.3 *The practicalities*

Uniaxial testing is a simple and quick methodology with measurements completed in a matter of minutes. As with biaxial shear testing, equipment costs can be high when compared to traditional techniques.

16.4.2.4 *Advantages and limitations*

Uniaxial testing offers high repeatability relative to other simple techniques, enhancing its ability to detect differences and making it an effective tool for ranking the flowability of powders. The practicalities of measurement are well-suited to production and process environments.

The main limitation of uniaxial testing is that, in common with biaxial shear cell testing, the sample is consolidated prior to measurement. The resulting data are, therefore, potentially more relevant to high-stress operations. There is little flexibility to modify test settings in order to reflect other conditions within the process environment. In addition, free-flowing materials, such as certain grades of lactose, may not be suitable for uniaxial testing.

16.4.3 Biaxial shear cell testing

The adoption of shear cell testing in the 1960s marked a major advance in powder testing and provided a foundation for a scientific approach to hopper design. It was developed specifically to generate the parameters required for understanding powder behavior in hoppers and for the application of hopper design methodologies. This methodology is still in use today, and many process engineers will be familiar with the principles of shear cell testing. The technique, therefore, provides a good baseline against which to compare alternatives.

16.4.3.1 *How it works*

In simple terms, shear cell testing involves measuring the forces required to shear one consolidated powder plane relative to another (Jenike, 1964; Roberts & Associates, 1993; Schulze, 1994). The resulting data are used to generate a yield locus, and a process of extrapolating this yield locus, followed by the application of Mohr's stress circles, allows the UYS to be derived. Combined with information on powder density, the data can also be used to generate flow function (FF) and flow factor (ff) plots. A similar technique can be applied, using a coupon (metallic surface insert) to represent an actual or potential material of construction, to produce information on wall friction

A wide range of shear cell designs is commercially available (Fig. 16.16), and these can be classified as translational, annular, or rotational. All measures are according to the same principle, but differ in terms of their level of automation, cost, ease of use, and reproducibility.

16.4.3.2 *Data generated*

Shear cell testing is used to derive parameters, including angle of internal friction, UYS, FF, and other flowability indices. These parameters are all used directly in hopper design calculations to determine the hopper half angle and outlet size that define acceptable discharge performance for a given powder/material of construction combination (Jenike, 1964). However, when considering the broader application of these parameters, and consequently the technique, it can be argued that they quantify the strength of frictional forces and the extent of mechanical interlocking in a sample, rather than being a universal measure of flowability.

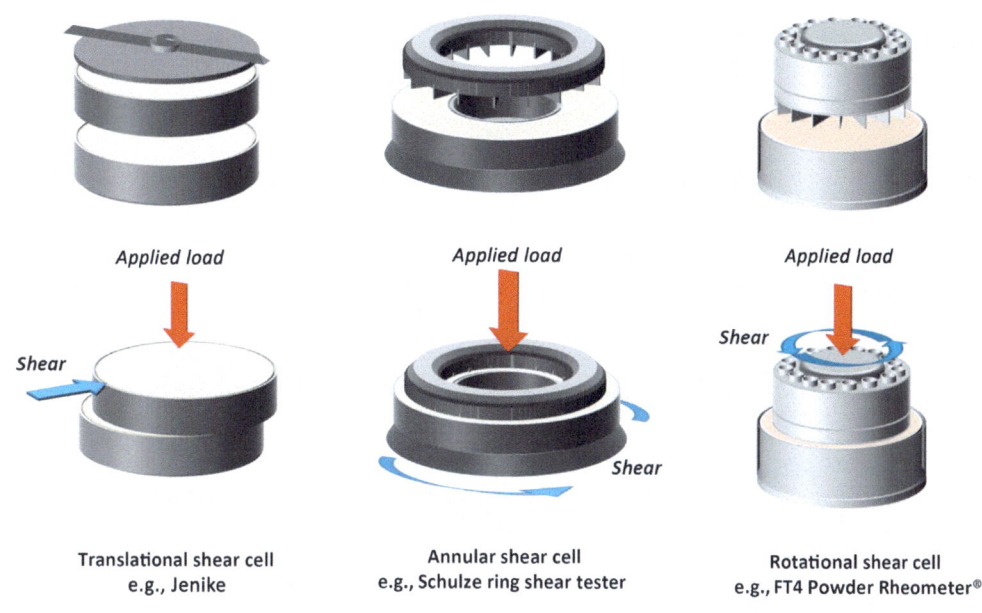

FIGURE 16.16

A range of commercial shear cell designs is now available. All operate on the same basic principle of measuring the shear forces required to move one consolidated powder plane relative to another.

16.4.3.3 *The practicalities*

Despite its longevity, shear cell analysis remains a relatively intensive and time-consuming procedure, depending on the device selected. It is often still considered an "expert task," and measurement can take a skilled operator 10 to 15 min. Equipment costs are high when compared to simpler techniques.

16.4.3.4 *Advantages and limitations*

Beyond the obvious link with hopper design, shear cell testing is useful for assessing the behavior of consolidated powders when subject to the moderate to high stress levels that are present when powder is stored under its own weight, or otherwise consolidated. It quantifies the ease with which a consolidated powder transitions from the static to dynamic state *and* can therefore be used to rank flowability. The technique offers some control over test conditions, most notably consolidation stress, and well-designed, automated instruments deliver high reproducibility.

Shear cell testing is less valuable when assessing behavior in a low-stress and/or dynamic environment, for example, in a mixer, or where a powder is in an aerated state (Bruni et al., 2007; Pingali et al., 2009). In addition, the conditions applied during shear cell testing can change powder properties through particle attrition or de-agglomeration, thereby compromising relevance. A key limitation is that results are inherently less differentiating for free-flowing powders because the forces measured are much smaller in magnitude (Aulton, 2002; Freeman, 2011).

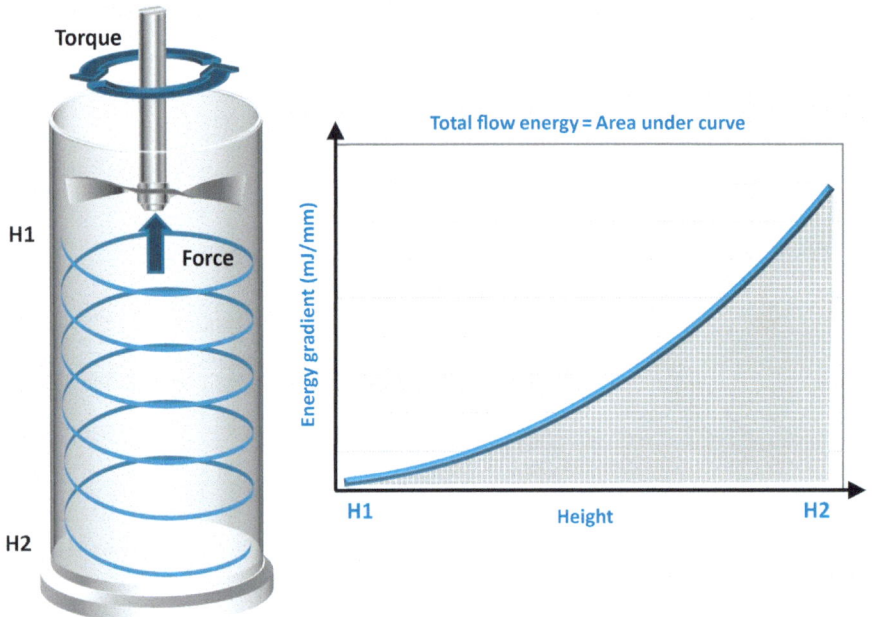

FIGURE 16.17

Dynamic powder testing measures the powder in motion and can be applied to samples in a wide range of different states to simulate numerous process environments.

16.4.4 Dynamic powder testing

Dynamic powder testing involves testing a powder in motion. The technique was developed in the 1990s in response to the need to directly measure powder flowability under conditions that represent the process environment. With dynamic testing, powders can be measured in a consolidated, moderate stress, aerated, or even fluidized state—representing the range of stress conditions that exist in routine powder handling operations and, thereby, delivering particular value to process-related investigations (Bharadwaj et al., 2010; Freeman, 2007; Hare et al., 2015).

16.4.4.1 How it works

Dynamic powder testing measures the axial and rotational forces acting on an impeller as it is rotated through a powder sample. The properties measured include flow energy, which quantifies how easily a powder flows under forcing conditions; and specific energy (SE), which defines unconfined flow properties (Fig. 16.17).

16.4.4.2 Data generated

Flow energy is measured as the blade descends, forcing the powder against the base of the test vessel. The flow energy reflects how a powder will flow under forced conditions, such as through an extruder or screw auger (Freeman & Fu, 2008; Freeman et al., 2012). SE, in contrast, is measured during an upward traverse of the blade and directly reflects behavior in low-stress, unconfined environments, for example, when a powder is flowing freely from a feed shoe in a tablet press (Freeman & Fu, 2008). A defining difference between dynamic powder testing and other techniques previously discussed is that powders can be tested under different conditions to directly quantify how they respond to varying stress and flow regimes.

The ability to measure a diverse range of powder flow properties makes dynamic powder testers particularly useful in a range of industrial applications. The sensitivity achieved allows subtle differentiation of raw materials, intermediate blends, and final products. Dynamic flow testing provides the ability to directly characterize the response of a powder to the introduction of air, up to and beyond the point of fluidization. This is a critical feature of this methodology and is important for the optimization of, for example, fluidized bed and pneumatic conveying processes (Kinnunen et al., 2014).

16.4.4.3 The practicalities

Dynamic test methodologies are well-defined, and instrumentation is precision-engineered and highly automated. As a result, dynamic powder testers, though relatively expensive, deliver unrivaled sensitivity and versatility. As indicated in the comparison presented above, it is possible to differentiate samples that would be classified as identical by other techniques but actually perform differently in a process.

Measurement times are in the region of 5 to 10 min for a basic assessment of flowability, but the range of testing available means that comprehensive investigations of the effect of variables such as strain rate, air and moisture content, storage time, and electrostatic charge can be carried out routinely.

16.4.4.4 Advantages and limitations

Dynamic powder testing provides intuitive results with direct relevance to process behavior. Small sample volumes are typically required, and samples can be tested in a consolidated, moderate stress, aerated, or fluidized state to simulate a range of process conditions.

Well-defined, substantially automated test methodologies ensure high repeatability and reproducibility.

However, dynamic testing is an empirical method, and the comprehensive data sets generated can require considered analysis. Equipment costs may be high compared to traditional techniques.

16.5 Practical process relevance—three case studies

16.5.1 Predicting feeder performance from powder flow measurements

16.5.1.1 Background

In powder handling industries, screw feeders are routinely used to control the flow of material from a hopper into the downstream stage of a process (Fig. 16.18). The properties of the powder directly impact the performance of the feeder, making it essential to tailor the design of the system to the product being handled. A poorly matched powder/feeder combination will typically result in low feed rates, high screw torques, and an accumulation of powder on the tube walls—all of which decrease both short and long-term operating efficiency. Optimizing this unit operation first requires powder properties to be identified that can be used to predict feeder performance.

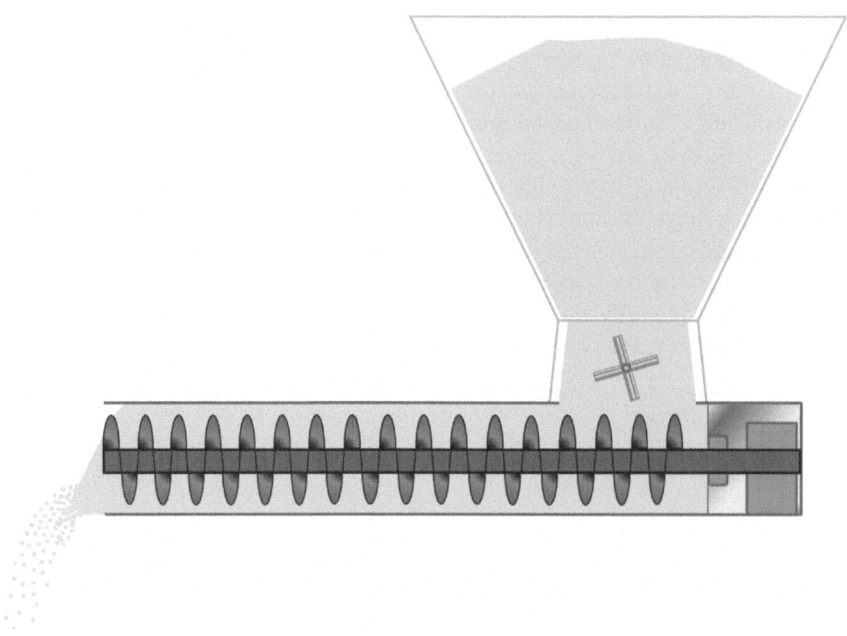

FIGURE 16.18

A screw feeder transfers powder using one or more augers enclosed in a tube.

16.5.1.2 Experimental

A study was carried out to investigate correlations between the properties of five different powders and their performance in two different screw feeders. The five powders tested were:

1. Calcium hydroxide
2. Maltodextrin
3. Milk protein
4. Cellulose
5. Calcium citrate

In the first stage of the study, samples of each powder were tested using an FT4 powder rheometer (Freeman Technology Ltd., Tewkesbury, UK) to measure a range of dynamic, bulk, and shear properties. Samples of each powder were then run through two screw feeders (Gericke AG, Zurich, Switzerland) to determine the volumetric feed rate (L/h) delivered at an auger rotation speed equivalent to 80 Hz. Volumetric feed rate was calculated from measurements of mass flow rate (in kg/h) and poured density.

The two screw feeders used were a DIWE-GLD-87 VR, a full-flight single-screw feeder using tube no. 3, and a DIWE-GZD flat-bottom double-screw feeder using a 12×13.5 mm tube with a conical core.

Table 16.1 shows the data, including average volumetric feed rate, for all five powders. A multiple linear regression (MLR) analysis was performed to identify correlations between feed rate and powder

16.5 Practical process relevance—three case studies

Table 16.1 Dynamic flow, bulk, and shear properties for five different powders alongside the volumetric feed rate each delivered when run through GLD and GZD screw feeders.

Material	Dynamic parameters						Bulk parameters							Shear parameters					Feeder parameters
	BFE (mJ)	SE m(J/g)	SI	FRI	AE40 (mJ)	AR40	NAS (S/mm)	CE50 tap (mJ)	BD50 tap (g/mL)	CBD (g/mL)	CPS (%)	PD1 kPa mbar	PD@15 kPa mbar	UYS (kPa)	MPS (kPa)	FF	Cohesion (kPa)	AIF (°)	WFA (°)	GLD (L/h)
GLD feeder																				
Calcium hydroxide	354	6.92	1.30	2.40	65	4.1	0.417	460	0.538	0.499	25.2	14.87	65.30	5.89	18.21	3.09	1.525	35.2	30.8	185.2
Maltodextrin	1282	5.19	1.11	1.16	13	107.3	0.156	1341	0.608	0.557	7.1	0.51	0.54	0.88	16.42	18.57	0.221	36.9	27.4	138.9
Milk protein	330	8.49	0.91	1.37	102	3.3	0.182	613	0.311	0.267	24.3	3.07	9.06	3.96	20.71	5.23	1.091	32.3	24.6	128.7
Cellulose	630	8.92	0.87	1.34	19	46.0	0.619	4124	0.376	0.327	22.2	2.75	3.97	6.68	22.06	3.30	1.579	39.4	17.5	115.8
Calcium citrate	680	12.50	1.07	1.40	225	3.0	0.077	914	0.261	0.234	41.6	1.02	37.82	8.14	22.59	2.78	1.877	40.5	41.0	50.13
GZD feeder																				
Cellulose	630	8.92	0.87	1.34	19	46.0	0.615	4124	0.376	1.327	22.2	2.75	3.97	6.68	22.06	3.30	1.579	39.4	17.5	34.98
Calcium hydroxide	354	6.92	1.30	2.40	65	4.1	0.417	460	0.538	0.499	25.2	14.87	65.30	5.85	18.21	3.09	1.525	35.2	30.8	33.02
Maltodextrin	1282	5.19	1.11	1.16	13	107.3	0.156	1341	0.608	0.557	7.1	0.51	0.54	0.88	16.42	18.57	0.221	36.9	27.4	29.88
Milk protein	330	8.49	0.91	1.37	102	3.3	0.182	613	0.311	0.267	24.3	3.07	9.06	3.96	20.71	5.23	1.091	32.3	24.6	18.67
Calcium citrate	680	12.50	1.07	1.40	225	3.0	0.077	914	0.261	0.234	41.6	1.02	37.82	8.14	22.55	2.78	1.877	40.5	41.0	10.39

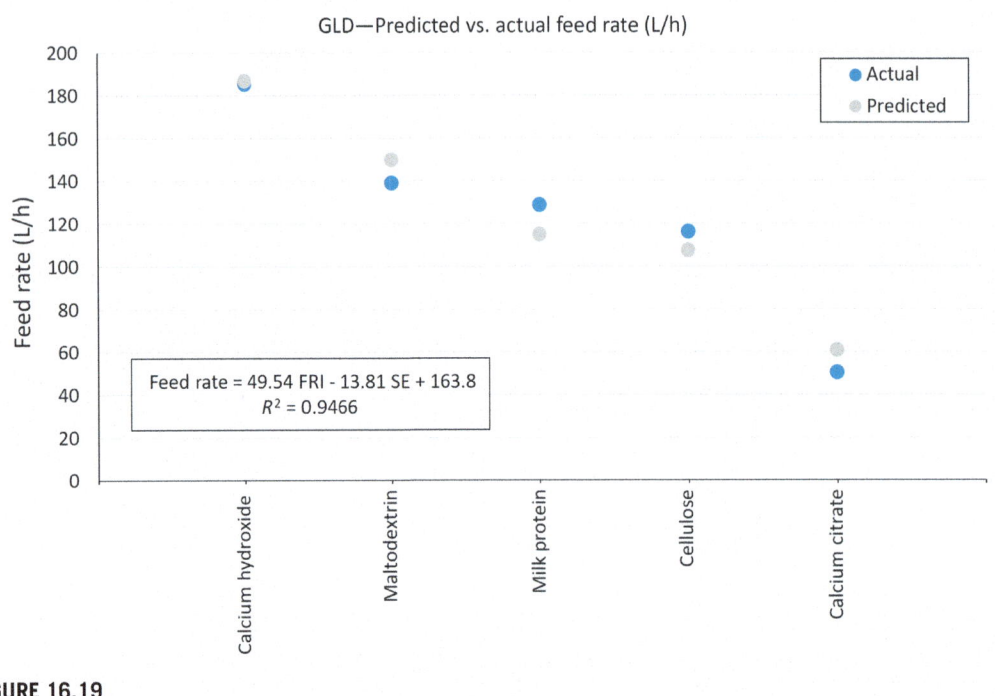

FIGURE 16.19

Predicted versus actual volumetric feed rate data for five powders.

flow properties. MLR generates a P value for each parameter, which indicates the probability that the parameter's contribution to the relationship is statistically insignificant. The lower the P value, the more likely that the parameter has an impact on the relationship. For the purposes of this study, a P value of 0.1 was considered the upper limit for relevance, and parameters with P values higher than this were eliminated to derive a robust relationship.

For the GLD feeder, MLR analysis produced the following relationship:

$$\text{Feedrate} = 49.54 \text{ FRI} - 13.81 \text{ SE} + (163.8) \; (R^2 = 0.9466)$$

R^2 is a measure of the "goodness of fit" of the model, with values closer to the upper limit of 1 indicating a close fit. This relationship suggests that feeder performance can be robustly predicted using two dynamic properties—SE and flow rate index (FRI). SE reflects how a powder behaves in an unconfined state. It is influenced by mechanical interlocking and friction between particles. FRI describes a powder's sensitivity to changes in flow rate. It is simulated by varying the speed of the helical blade during dynamic flow tests.

FRI values $>$ 1 indicate that resistance to flow is greater at lower flow rates. All the powders in this study generated an FRI $>$ 1, suggesting "shear thinning" behavior, that is, that they flow more easily at higher flow rates.

Fig. 16.19 shows the actual volumetric feed rates for the five powders along with the values predicted by the derived model. As suggested by the R^2 value, the predicted values accurately describe the observed performance of the powders in the GLD feeder.

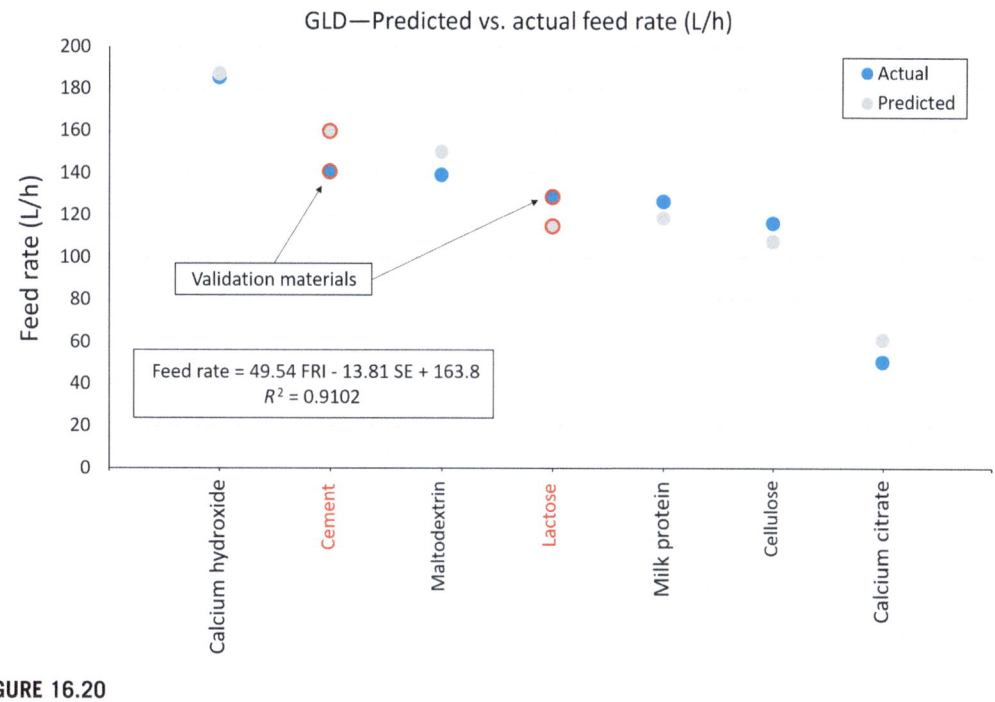

FIGURE 16.20

Predicted versus actual volumetric feed rate data for seven powders illustrate the ability to predict volumetric feed rate through the GLD feeder using dynamic flow properties.

To challenge the predictive ability of the derived relationship, two additional powders were tested—cement and lactose. Fig. 16.20 shows the actual flow rates for each of the original five powders, and the two additional materials (shown in red), along with the values predicted from measuring their flow properties. A revised R^2 confirms close agreement between the predicted and measured feed rates for the data sets incorporating all seven materials, and illustrates the feasibility of predicting volumetric feed rate from powder properties.

The same process was repeated for the alternative feeder. Here, a simpler relationship was observed with aerated energy (AE), the only parameter found to be highly relevant.

$$\text{Feedrate} = -0.1114 \, AE_{40} + (34.82) \; (R^2 = 0.8383)$$

AE is the flow energy measured when the sample is subjected to an upward airflow of defined linear velocity—in this case 40 mm/s, hence the descriptor AE_{40}. Cohesive powders tend to generate relatively high AE values, since aeration does little to reduce the resistance they present to flow, while for free-flowing powders, AE can approach zero as fluidization occurs.

Fig. 16.21 shows the actual volumetric feed rates for the five powders along with the values predicted by the derived model. As suggested by the R^2 value, the predicted values again accurately describe the performance of the powders in the feeder.

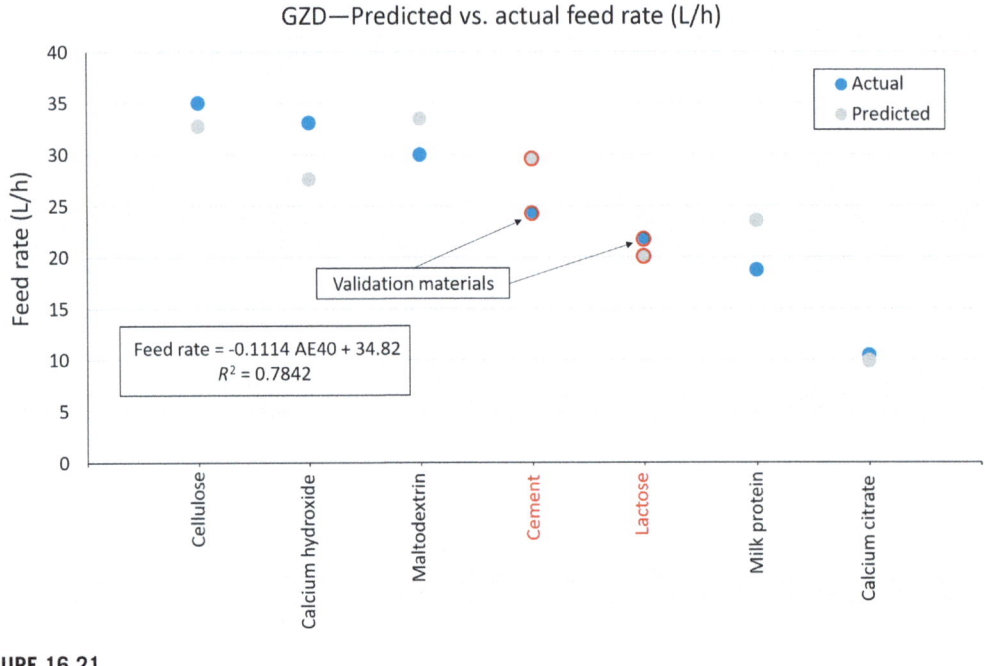

FIGURE 16.21

Predicted versus actual volumetric feed rate data for seven powders illustrate the ability to predict volumetric feed rate through the GZD feeder using dynamic flow properties.

As with the GLD feeder, the study was extended to verify the ability of the relationship to predict the volumetric flow rate achieved with cement and lactose. As before, the model performed robustly in this predictive mode (Fig. 16.21).

16.5.1.3 Conclusion

The results from this study demonstrate the ability to develop robust correlations between measurable powder properties and the volumetric feed rate delivered by different designs of screw feeders. Each screw feeder imposes different conditions on the powder, and this is reflected in the specific attributes of the powder that are found to be relevant for predicting performance. In both cases, dynamic flow properties were found to be most relevant, suggesting that dynamic tests more closely reflect the conditions in the process.

The approach taken in this study could equally be applied to a wide range of powder processing equipment. Multifaceted powder characterization provides an essential foundation for such an approach, supporting the identification of properties that are most relevant to the performance of a powder in any specific unit operation.

Table 16.2 Particle size and shape data for a series of four powders used in die filling tests.

Material	Material/Powder Description	D_{50} (mm)	Shape
(A)	GL Glass Beads	174	Spherical
(B)	GS Glass Beads	68	Spherical
(C)	Granular Aluminum Powder	134	Irregular
(D)	Tungsten Powder	4	Angular

16.5.2 Developing a design space for a die filling operation

16.5.2.1 Background

Filling is a common operation across many industries, although fill weights and tolerances vary widely. The pharmaceutical industry often requires milligram doses to be filled accurately and at high speed, in order to meet the stringent criteria and throughput requirements of tablet manufacture.

The factors that influence filling efficiency depend on the type of equipment being used. Some systems are purely gravity-driven, whilst others rely on force-feeding. In many applications, such as tablet manufacture on a rotary press, the powder fills the dies through a combination of gravity and forced flow. The influence of each of the two mechanisms will depend on the geometry of the feed frame, the flow rate through the press, and the characteristics of the powder. Given the wide variation in each of these variables, this is a complex process and remains challenging to model from knowledge of a limited number of powder properties and process parameters.

16.5.2.2 Experimental

A study was carried out to investigate correlations between the properties of four different powders and their performance in a die filling rig. The four powders were tungsten, large glass beads, small glass beads, and aluminum. Table 16.2 summarizes their physical properties.

Bulk and dynamic properties were measured for each powder using an FT4 powder rheometer (Freeman Technology Ltd., Tewkesbury, UK), specifically, AE_{30}, SE, and permeability. (Please refer to the preceding case study for descriptions of AE_{30} and SE.) Bulk permeability was quantified by measuring the pressure drop across a consolidated powder bed while air was introduced to the bottom of the powder column at a constant velocity.

The filling apparatus consisted of a stationary die and a motor-driven shoe capable of a range of steady-state velocities from 50 to 300 mm/s. The show and die were both cylindrical with volumes of 160 and 10 mL, and diameters of 50 and 25 mm, respectively. Filling ratio was calculated as the mass of powder in the die, following one pass of the shoe, with respect to the mass if the entire volume of the die had been filled with the same material. A value of 1.0 represents complete die fill, whilst a value of 0.2, for example, indicates that only 20% of the die has been filled. For each material, experiments were completed for a range of shoe speeds.

16.5.2.3 Results

The die filling performance of all the powders is summarized in Fig. 16.22. Higher shoe speeds increase the challenge of achieving maximum fill. This is reflected in the reduction in filling ratio at high speeds

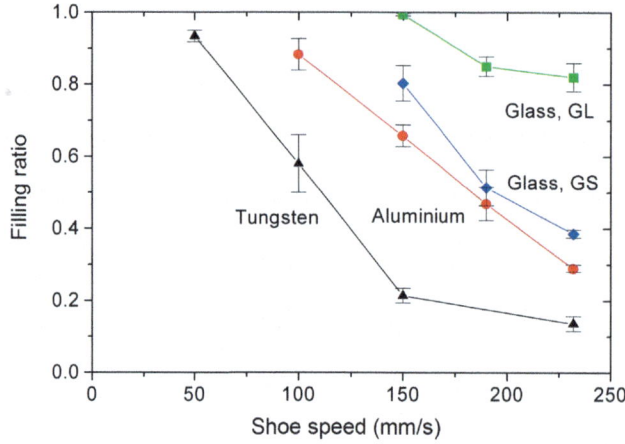

FIGURE 16.22

Filling performance for four materials.

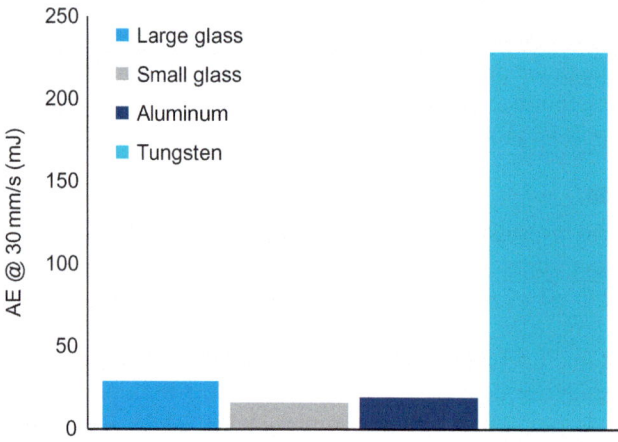

FIGURE 16.23

Aeration data for four materials.

for all four materials. However, there are clear differences in the behavior of the powders. Tungsten exhibited the worst filling performance at all shoe speeds. Conversely, large glass beads were the most effective at filling the die, achieving 100% fill at even moderate shoe speeds. small glass beads and Aluminum showed intermediate performance, achieving a >75% fill at moderate shoe speeds.

Comparing filling performance with measured powder properties (see Figs. 16.23, 16.24, and 16.25) highlights robust relationships. Tungsten generated the highest AE_{30} of the four samples, indicating it was the least sensitive to aeration. Low sensitivity to aeration is typically an indicator of cohesivity, as strong inter-particular forces prevent air from passing uniformly between the particles, leading to

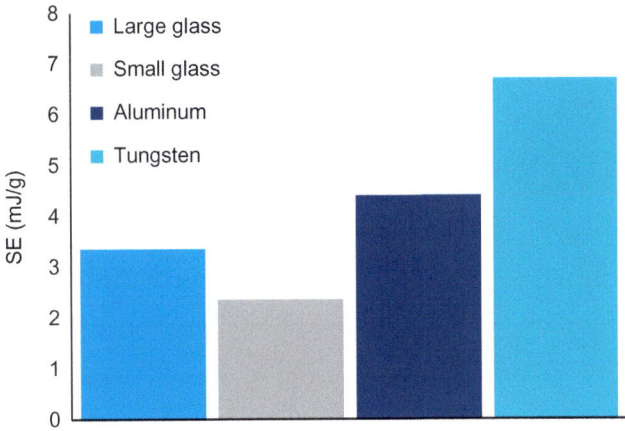

FIGURE 16.24

Specific energy for four materials.

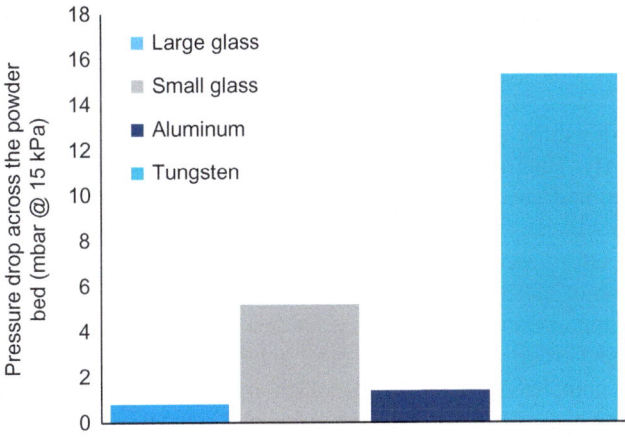

FIGURE 16.25

Permeability data for four materials.

minimal changes in the packing structure of the powder. Tungsten also generated the highest SE, indicating a high degree of mechanical interlocking and friction. The low permeability of the Tungsten powder indicated that air flow out of the die is restricted, negatively impacting both fill rate and fill consistency. In this process, therefore, the worst-performing powder is clearly identifiable from high values of all three of the powder properties measured.

For the other three samples, the results highlight the discrimination that is possible, having acquired the detailed data on each powder. Some aspects of the behavior are closely similar. For example, all three powders have a very low AE_{30}, indicating that they could be fully fluidized. When considering SE,

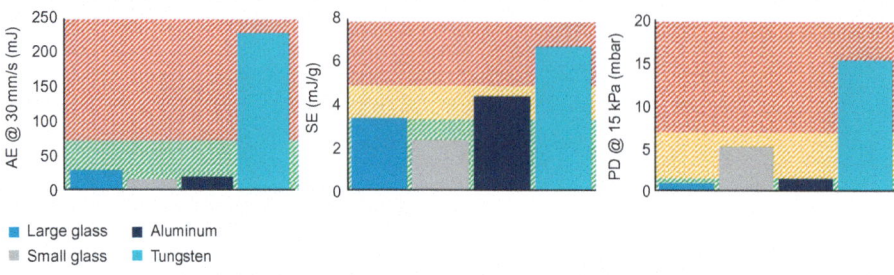

FIGURE 16.26

Developing a design space for the filling operation.

both glass bead samples generated a low value, but that of the small glass beads is lowest (suggesting lower inter-particular interlocking) despite the large glass beads filling the die the most effectively. Permeability data indicate that the small glass beads have considerably lower permeability than either the large glass beads or the aluminum. As with the Tungsten sample, this suggests that entrained air will take longer to escape from this material as the die fills, inhibiting effective filling.

In order to rationalize the competing effects of different properties, a model can be developed to estimate processing performance.

This information allows a "design space" to be constructed for the process, as exemplified by Fig. 16.26. Large glass beads exhibit good performance, so their properties can be used to define acceptable criteria for each property. Tungsten exhibits poor performance, so its properties define unacceptable criteria. The aluminum and small glass beads exhibit more complex behavior, with the aluminum meeting the criteria for AE_{30} and permeability, but not for SE. Small glass beads meet the AE_{30} and SE limits, but processing performance is likely to be compromised by the unacceptable permeability value.

If a new formulation or blend generates acceptable values for all properties, it can be expected to perform well in the filling operation. If the results are just outside the acceptable limits, there is a risk of poor performance, and process settings may need to be adjusted to achieve an acceptable fill ratio. Powders that generate completely unacceptable values are likely to perform poorly in the operation, regardless of process settings, and can, therefore, be screened out of the process—preventing blockages, downtime, and products out of specification.

16.5.2.4 Conclusion

In this study, it proved possible to rationalize differences in die filling performance with reference to the bulk and dynamic flow properties of powders. No single property alone successfully ranked the processing performance of the powders, but in combination, they could be used to construct a design space to identify powders that are likely to perform well.

The tests show that powders with a high sensitivity to aeration, a low degree of cohesion, mechanical interlocking and friction, and high permeability exhibit the best performance in this unit operation. The data also show that permeability is a highly influential parameter, meaning that a powder with less than optimal dynamic flow properties (such as large glass beads) can still perform effectively.

FIGURE 16.27

Continuous high shear wet granulation and drying system (GEA ConsiGma).

16.5.3 A QbD approach to continuous tablet manufacture

16.5.3.1 Background

The pharmaceutical industry is looking to continuous processing to enhance production efficiency and product quality, in line with guidance from the regulatory agencies. In many instances, wet granulation of a blend is an essential precursor to the efficient production of tablets, making the successful integration of wet granulation within a continuous manufacturing line an important goal. Establishing links between properties of a blend, processing parameters, and characteristics of the resulting granulate is essential for process development and optimization, particularly where such characteristics can be controlled to produce tablets with the required CQAs.

16.5.3.2 Experimental

In this study, experiments were carried out to explore relationships between the properties of granules produced by a continuous wet granulation process and variations in the formulations and processing parameters used in their manufacture. A continuous high shear wet granulation and drying system (ConsiGma, GEA, Maryland, United States, Fig. 16.27) was used to manufacture granules from two simple powder formulations using paracetamol (APAP) and dicalcium phosphate (DCP). A range of experiments was undertaken to evaluate the effect of varying water addition, input powder feed rate, and granulator screw speed. Granules were tested using an FT4 Powder Rheometer (Freeman Technology,

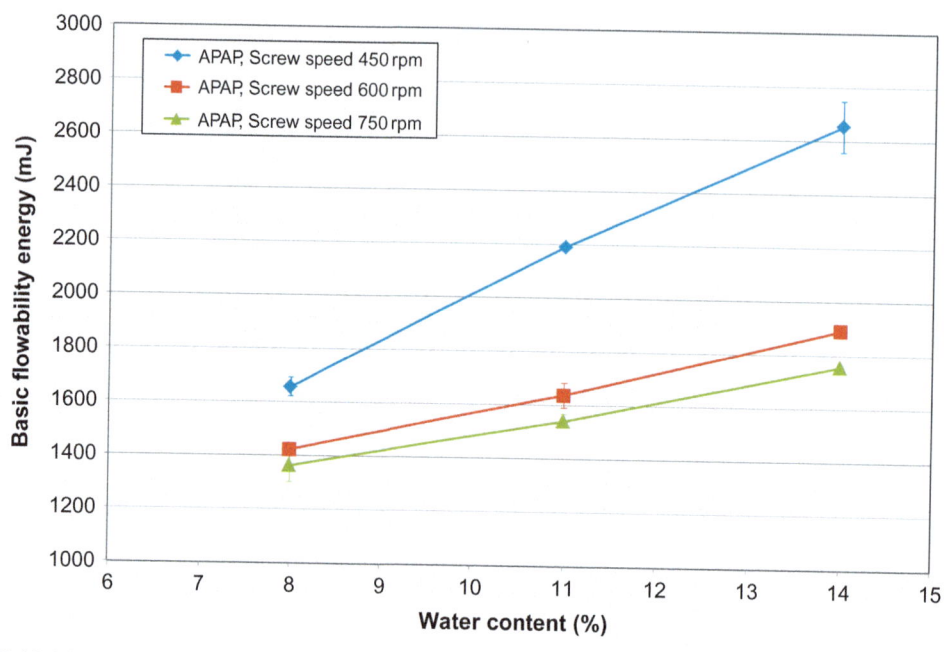

FIGURE 16.28

APAP formulation responds to changes in processing parameters.

Tewkesbury, UK) at four different stages in the manufacturing process: wet mass (ex-granulator); dry granules (post drying), milled granules (post milling), and lubricated granules (following the addition of lubricant as a precursor to tableting).

The work was subsequently extended to investigate correlations between granule properties and CQAs of finished tablets, such as hardness, manufactured from the granules produced.

16.5.3.3 Results

Data gathered for the APAP formulation (see Fig. 16.28) shows that increasing water content results in wet granules with a higher basic flowability energy (BFE) for all screw speeds (for a fixed powder feed rate). BFE quantifies the flow properties of the granulated mass by measuring resistance to flow. Reducing screw speed similarly results in a higher BFE.

Both of these trends were expected, as higher water content and lower screw speeds (which induce more shear) produce larger, denser, and more adhesive granules with a greater resistance to forced flow and, consequently, a higher BFE. However, the increase in BFE is linear with respect to water content, but not screw speed. In addition, the data shows that, for example, a screw speed of 600 rpm and a water content of 11% generate granules with a very similar BFE to those generated using a screw speed of 450 rpm and a water content of 8%.

Considering the DCP formulation (see Fig. 16.29), BFE substantially increases with reducing powder feed rate (at a water content of 15% and a fixed screw speed of 600 rpm). In addition, the data show it is possible to achieve comparable granule properties at higher water contents by increasing

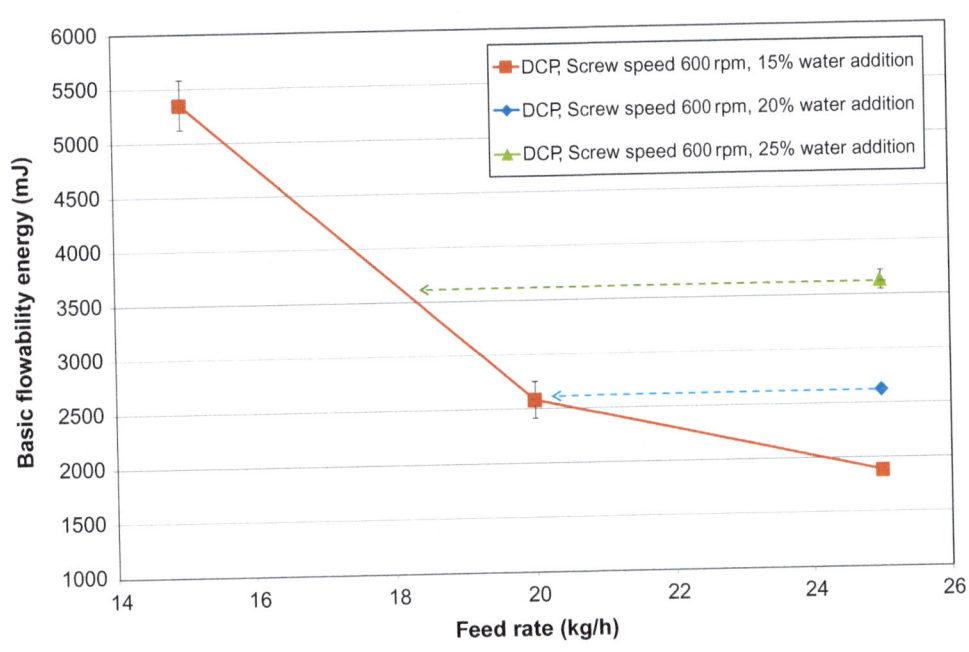

FIGURE 16.29

DCP formulation responds to changes in processing parameters.

the feed rate. For this formulation, granules with 25% water content and a feed rate of 25 kg/h have similar properties to granules containing 15% water, using a feed rate of 18 kg/h.

The evidence from these two formulations suggests that it is possible to generate granules with specific properties using different combinations of water content, screw speed, and feed rate. To test this hypothesis, granules were produced using the APAP formulation, targeting specific BFE values. As can be seen in Table 16.3, a BFE (wet mass) of around 2200 mJ was successfully produced using processing Conditions 1 and 2, while Conditions 3 and 4 produced wet mass BFE values of around 3200 mJ.

Table 16.3 and Fig. 16.30 also show how the bulk flow properties of the granules change during the different stages of manufacture. Conditions 3 and 4 show an increase in BFE following drying, in contrast to the trend observed for Conditions 1 and 2. This increase is attributed to the large relative size of these granules, which, combined with increased density and hardness, result in increased mechanical interlocking and, thus, higher resistance to forced flow. For the granules produced using Conditions 1 and 2, their weaker structure, lower density, and smaller relative size result in a comparable or lower BFE for the dried granules compared to their properties when in a wet form.

Following milling, the BFE values tend to converge as the particle size is normalized, but there is still a distinct difference in the properties of each pair of granules. These differences are retained following lubrication. Overall, as the granules progress through the manufacturing process, the relative BFE values remain consistently grouped—the BFEs of granules produced using Conditions 3 and 4 remain consistently higher than those produced under Conditions 1 and 2.

Table 16.3 Granules with a specific BFE can be produced using a different set of processing parameters.

Condition	Process Parameters				Granule Properties			
	Screw speed (rpm)	Powder feed rate (kg/h)	Liquid feed rate (g/min)	Moisture (%)	BFE – wet mass (mJ)	BFE – dry granules (mJ)	BFE – milled granules (mJ)	BFE – lubricated granules (mJ)
1	450	11.25	15.0	8.0	2217	1623	1283	1526
2	750	20.0	36.7	11.0	2133	1973	1463	1417
3	450	6.0	20.0	20.0	3172	4610	2268	1761
4	750	9.0	30.0	20.0	3342	4140	1951	1795

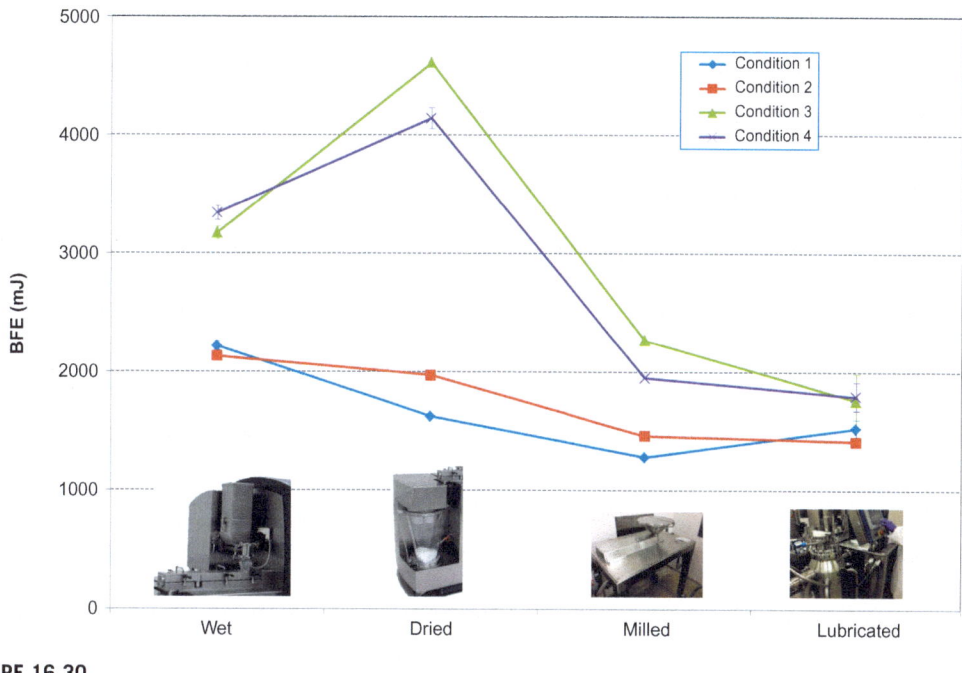

FIGURE 16.30

How bulk flow properties of the granules change with respect to the different stages of manufacture.

This observation suggests that granule properties are dependent on the process conditions and that it is possible to produce the desired granule quality using more than one manufacturing route. This is an extremely significant result for the formulation scientist, process engineer, and equipment designer, as it allows much greater scope in developing formulation/unit operation combinations to generate products of a specific quality. However, this conclusion has greater significance if there is also a strong relationship between the characteristics of the granules and the attributes of the final product—the manufactured tablet.

The four batches of the wet, dried, milled, and then lubricated granules were, therefore, tableted using a rotary tablet press (Modul S, GEA, Maryland, United States) and the strength of the subsequent tablets measured using a tablet hardness tester (Pharmatron 8M, Dr. Schleuniger Pharmatron, Solothurn, Switzerland).

Fig. 16.31 shows how the tablet hardness correlates with the flow properties of the granules at each stage.

The correlation is extremely strong between tablet hardness and the BFE for the dried and milled granules (R^2 values > 0.99). The slightly poorer (but still significant) correlation for the wet mass and lubricated granules can be attributed to the presence of the additional components—water or lubricant (magnesium stearate)—which are known to have an exaggerated influence on the bulk flow properties relative to their low concentrations.

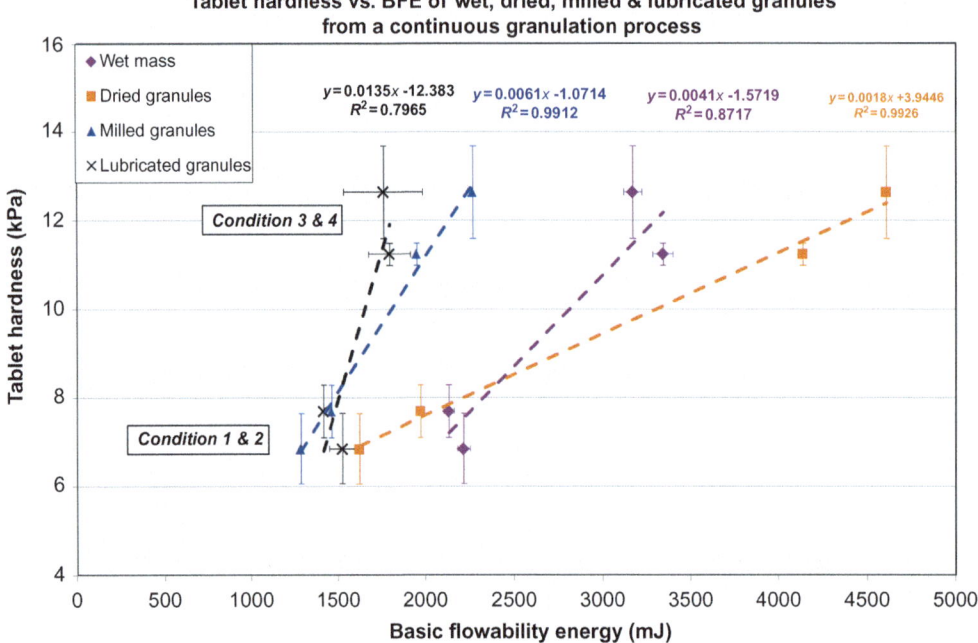

FIGURE 16.31

Tablet hardness versus BFE.

16.5.3.4 Conclusion

From the data presented here, it is possible to conclude that there is a direct relationship between the dynamic flow properties of the granules at all stages of manufacture and the CQA of the final tablet. The study shows that it is possible to generate specific tablet properties using different combinations of process conditions, which represents a significant step toward a full design space specification.

This approach also provides the opportunity to develop scaling criteria for batch granulation processes. Once a specific granule property has been identified as delivering the optimal CQA for the final product, manufacturing is no longer focused on equipment types, operational settings, or scale. As long as the wet granule attains the target property, tablet quality can be assured.

16.6 Conclusion

- Powders are multicomponent bulk assemblies consisting of particles, liquids, and gases.
- They exhibit complex behavior that can't be predicted from particle properties alone.
- This complexity makes powders challenging to process and characterize, but also delivers an array of industrially useful behavioral properties.

16.6 Conclusion

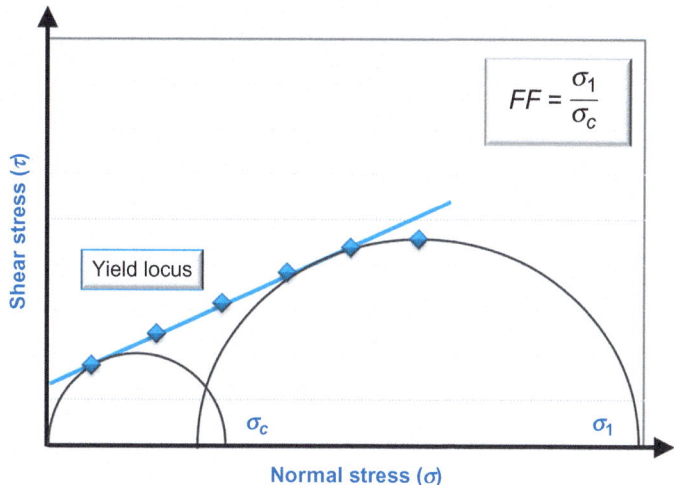

Unnumbered Fig. Schematic showing the application of a yield locus and Mohr's circles to data from a shear cell test.

- The flow behavior of a powder is a function of its inherent properties and the environment to which it is subjected. Therefore the processing environment is an additional variable which must be considered when determining powder behavior and the metrics that are most usefully measured to determine powder processability.
- Measuring a number of powder properties identifies those that are critical for a given application, helping to secure a good process-powder combination.
- Ensuring that a powder and process are compatible is essential for long-term, efficient manufacture, and high-quality product.
- When evaluating powder testing methods, relevance to the final application needs consideration. Sensitivity, repeatability, and reproducibility must be assessed, as well as how the methodology differentiates between powders in terms of their behavior in the processes of interest.

Unnumbered Table.

Size	Adhesion	Hardness/friability
Shape	Elasticity	Amorphous content
Surface texture	Plasticity	
Surface area	Porosity	
Density	Potential for electrostatic charge	
Cohesion	Hygroscopicity	

Key Learning Point

Understanding powder properties is important, but considering powder flowability as part of the complete process, not just storage and flow from a hopper, for example, is critical to success. Matching a powder to the process requires a set of relevant data to be generated, and, therefore, employing a multifaceted approach to powder testing is now recognized as being the most productive strategy to achieve this goal.

Key Learning Point

The fundamental forces acting on particles, and the mechanisms of interaction between them, define flow characteristics. Understanding how these translate to powder flowability in a particular environment and/or process is essential to effective powder processing.

Method Relevancy

Data is collected for samples of two excipients, one set using a biaxial shear cell tester (see Section 3.3), the other using a testing technique which produces a direct measure of dynamic flowability (see Section 3.4). The shear data suggest that the two samples have identical properties, but the dynamic flow data indicate that they don't (Fig. 16.14).

Both data sets are correct, but the techniques are measuring different characteristics of the powders. The two samples are likely to behave in the same way when in a consolidated environment, for example, during the onset of flow in a hopper, but they have significantly different flow properties when moving dynamically in a process.

If the analysis is to determine whether these powders will process identically, then it depends on what part of the process is being considered. Both powders are likely to behave in a similar way if discharging from the same hopper, but in more dynamic, moderate/low stress environments, such as mixing and filling processes, they are likely to exhibit different performance.

Definitions

- Repeatability is the variability associated with the same operator measuring the same sample, using the same instrument. It therefore assesses the variability introduced by the instrument itself.
- Reproducibility is the variability associated with different operators measuring different samples, potentially with different instruments. It therefore includes operator-to-operator variability as well as errors associated with sampling and sample preparation, and also due to the instrument.
- Sensitivity is the ability of a technique to detect variation. It is an inherent function of the technique, but also depends on repeatability and reproducibility.

> **Mohr's Circles Analysis**
>
> Yield Loci can be further analyzed by the application of Mohr's circles, constructed as follows:
> - A best-fit line is applied to the test data and extrapolated to intercept the y-axis
> - A semicircle is drawn where the circumference passes through the origin and is tangential to the best-fit line.
> - A second is drawn where the circumference passes through the preshear data point and is tangential to the best-fit line.
>
> The following parameters are then derived:
> - *Cohesion (C)*—the Shear Stress value where the best-fit line intercepts the y-axis
> - *Unconfined yield strength (UYS or σ_c)*—the value at which the smaller Mohr's Circle intercepts the x-axis
> - *Major principal stress (MPS or σ_1)*—the greater of the two values at which the larger Mohr's Circle intercepts the x-axis
> - *Angle of internal friction*—the angle created by the best-fit line with the horizontal axis.
> - *Flow function coefficient (ff_c)* can then be calculated as follows: $ff_c = $ MPS/UYS.

> **Key Learning Point**
>
> Test methods must be relevant to the process in question and, in particular, to the challenges being addressed. The methods employed must also provide the necessary sensitivity, repeatability, and reproducibility to be able to reliably identify and quantify subtle differences in powder flow properties that may influence in-process behavior.

References

Armstrong, B. (2011). The study of pharmaceutical powder mixing through improved flow property characterisation and tomographic imaging of blend content uniformity (EngD. Doctoral).

Aulton, M. E. (2002). *Pharmaceutics: The science of dosage form design*. Churchill Living stone.

Bell, T. A., Catalano, E. J., Zhong, Z., Ooi, J. Y., & Rotter, J. M. (2007). Evaluation of Edinburgh powder tester (EPT). In *Proceedings of the international congress on particle technology*.

Bharadwaj, R., Ketterhagen, W. R., & Hancock, B. C. (2010). Discrete element simulation study of a Freeman powder rheometer. *Chemical Engineering Science, 65*(21), 5747–5756. https://doi.org/10.1016/j.ces.2010.04.002.

Bruni, G., Lettieri, P., Newton, D., & Barletta, D. (2007). An investigation of the effect of the interparticle forces on the fluidization behaviour of fine powders linked with rheological studies. *Chemical Engineering Science, 62*(1–2), 387–396. https://doi.org/10.1016/j.ces.2006.08.059.

Calvert, G., Curcic, N., Redhead, C., Ahmadian, H., Owen, C., Ghadiri, M., & Beckett, D. (2013). A new environmental bulk powder caking tester. *Powder Technology, 249*, 323–329. https://doi.org/10.1016/j.powtec.2013.08.037.

Carr, R. E. (1965). Fundus flavimaculatus. *Archives of Ophthalmology, 74*(2), 163–168. https://doi.org/10.1001/archopht.1965.00970040165007.

Clayton, J. (2012). The attraction and challenge of continuous manufacturing. https://www.pharmaceuticalonline.com/doc/the-attraction-and-challenge-of-continuous-0001.

Emery, E., Oliver, J., Pugsley, T., Sharma, J., & Zhou, J. (2009). Flowability of moist pharmaceutical powders. *Powder Technology, 189*(3), 409–415. https://doi.org/10.1016/j.powtec.2008.06.017.

Freeman, R. (2007). Measuring the flow properties of consolidated, conditioned and aerated powders—A comparative study using a powder rheometer and a rotational shear cell. *Powder Technology, 174*(1–2), 25–33. https://doi.org/10.1016/j.powtec.2006.10.016.

Freeman, R. E., & Fu, X. (2008). Characterisation of powder bulk, dynamic flow and shear properties in relation to die filling. In *Paper presented at the world congress on powder metallurgy & particulate materials*.

Freeman, T. (2011). Powder characterization: Beyond shear testing. *Powder & Bulk Solids*. http://staging.powderbulksolids.com/article/powder-characterization-beyond-shear-testing.

Freeman, T., Birkmire, A., & Armstrong, B. (2015). A QbD approach to continuous tablet manufacture. *Procedia Engineering, 102*(Supplement C), 443–449. https://doi.org/10.1016/j.proeng.2015.01.185.

Freeman, T., Brockbank, K., & Yin, J. (2016). The development, operating, and measurement protocols for an advanced uniaxial powder tester. In *Paper presented at the AAPS 2016*.

Freeman, T., Moolchandani, V., Hoag, S. W., & Fu, X. (2012). Capsule filling performance of powdered formulations in relation to flow characteristics. *Particulate materials: Synthesis, characterisation, processing and modelling* (pp. 131–136). The Royal Society of Chemistry.

Hancock, B. C., Vukovinsky, K. E., Brolley, B., Grimsey, I., Hedden, D., Olsofsky, A., Doherty, R. A., et al. (2004). Development of a robust procedure for assessing powder flow using a commercial avalanche testing instrument. *Journal of Pharmaceutical and Biomedical Analysis, 35*(5), 979–990. https://doi.org/10.1016/j.jpba.2004.02.035.

Hare, C., Zafar, U., Ghadiri, M., Freeman, T., Clayton, J., & Murtagh, M. J. (2015). Analysis of the dynamics of the FT4 powder rheometer. *Powder Technology, 285*, 123–127. https://doi.org/10.1016/j.powtec.2015.04.039.

Hausner, H. (1967). Friction conditions in a mass of metal powder. *International Journal of Powder Metallurgy*, (7), 1967.

Hickey, A. J., Mansour, H. M., Telko, M. J., Xu, Z., Smyth, H. D. C., Mulder, T., et al. (2007). Physical characterization of component particles included in dry powder inhalers. II. Dynamic characteristics. *Journal of Pharmaceutical Sciences, 96*(5), 1302–1319. https://doi.org/10.1002/jps.20943.

ICH. (2006). International conference on harmonisation of technical requirements for registration of pharmaceuticals for human use, quality risk management, Q9. *Federal Register, 71*(106), 32105–32106.

Jenike, A. W. (1964). *Storage and flow of solids*. UT: University of Utah.

Jones, R., Pollock, H. M., Geldart, D., & Verlinden, A. (2003). Inter-particle forces in cohesive powders studied by AFM: Effects of relative humidity, particle size and wall adhesion. *Powder Technology, 132*(2), 196–210. https://doi.org/10.1016/S0032-5910(03)00072-X.

Kinnunen, H., Hebbink, G., Peters, H., Shur, J., & Price, R. (2014). An investigation into the effect of fine lactose particles on the fluidization behaviour and aerosolization performance of carrier-based dry powder inhaler formulations. *AAPS PharmSciTech, 15*(4), 898–909. https://doi.org/10.1208/s12249-014-0119-6.

Lavoie, F., Cartilier, L., & Thibert, R. (2002). New methods characterizing avalanche behavior to determine powder flow. *Pharmaceutical Research, 19*(6), 887–893. https://doi.org/10.1023/A:1016125420577.

Peleg, M. (1977). Flowability of food powders and methods for its evaluation—A review. *Journal of Food Process Engineering, 1*(4), 303–328. https://doi.org/10.1111/j.1745-4530.1977.tb00188.x.

Pellek, A., & Arnum, V. (2008). Continuous processing: Moving with or against the manufacturing flow. *Pharmaceutical Technology, 32*.

Pingali, K. C., Shinbrot, T., Hammond, S. V., & Muzzio, F. J. (2009). An observed correlation between flow and electrical properties of pharmaceutical blends. *Powder Technology, 192*(2), 157–165. https://doi.org/10.1016/j.powtec.2008.12.012.

Roberts, A. W., & Associates, T.B.S.H.R. (1993). Basic principles of bulk solids storage, flow and handling. The University of Newcastle.

Schulze, D. (1994). Entwicklung und Anwendung eines neuartigen Ringscherätes (development and application of a novel ring shear tester). *Aufbereitungstechnik, 35*(10), 524–535.

Seppälä, K., Heinämäki, J., Hatara, J., Seppälä, L., & Yliruusi, J. (2010). Development of a new method to get a reliable powder flow characteristics using only 1 to 2 g of powder. *AAPS PharmSciTech, 11*(1), 402–408. https://doi.org/10.1208/s12249-010-9397-9.

Shur, J., Harris, H., Jones, M. D., Kaerger, J. S., & Price, R. (2008). The role of fines in the modification of the fluidization and dispersion mechanism within dry powder inhaler formulations. *Pharmaceutical Research, 25*(7), 1631–1640. https://doi.org/10.1007/s11095-008-9538-y.

Taylor, M. K., Ginsburg, J., Hickey, A., & Gheyas, F. (2000). Composite method to quantify powder flow as a screening method in early tablet or capsule formulation development. *AAPS PharmSciTech, 1*(3), 20–30. https://doi.org/10.1208/pt010318.

Teunou, E., & Fitzpatrick, J. J. (2000). Effect of storage time and consolidation on food powder flowability. *Journal of Food Engineering, 43*(2), 97–101. https://doi.org/10.1016/S0260-8774(99)00137-5.

U.S. Food and Drug Administration. (2007). Pharmaceutical quality for the 21st century: A risk-based approach progress report. Retrieved from USA.

United States Pharmacopeia. (2007). Powder flow general chapters—Physical analysis. vol. *1174*. Pharmacopeial Convection, United States

Zeng, X. M., Martin, G. P., & Marriott, C. (2000). *Particulate interactions in dry powder formulation for inhalation.* Taylor & Francis.

CHAPTER 17

A quality by design approach to scale-up of high-shear wet granulation process

Preetanshu Pandey[a] and Sherif I.F. Badawy[b]

[a]*Drug Product Development, Kura Oncology, San Diego, CA, United States,* [b]*Drug Product Science & Technology, Bristol-Myers Squibb Co., New Brunswick, NJ, United States*

17.1 Introduction

Granulation is a commonly used technique to improve the bulk density and flow characteristics of a formulation. It also helps to minimize segregation risk and offers better content uniformity (Rahmanian et al., 2008). In certain instances, it can also be used to improve the bioavailability of a formulation by providing intimate contact between a poorly water-soluble drug and a surfactant (Pandey et al., 2014). The two main methods of granulation employed in the pharmaceutical industry for preparing solid dosage form include wet granulation and roller compaction. The roller compaction has the advantage of being a completely dry process thereby eliminating the need for an additional unit operation of drying. Additionally no use of water leads to a better drug product stability profile and less risk of drug substance form conversion. However roller compaction results in a loss in compaction of the powder. This is due to the "work" done on the powder as it passes through the rolls and gets compacted, thereby losing some inherent compactability (He et al., 2007). Wet granulation, on the other hand, uses a binder in the formulation that helps to enhance the compaction characteristics of the granulation significantly (Nguyen et al., 2013). Wet granulation is the preferred method for formulations with high drug loading, as most pharmaceutical actives have small particle size and have challenges associated with flow and sticking to surfaces such as the rolls of a roller compactor. Additionally, the wet granulation process is relatively more forgiving to changes in active pharmaceutical ingredient (API) powder properties, which becomes important because, routinely, the API manufacturing process development has significant overlap in timing with the drug product development programs.

Wet granulation can be conducted in three main ways: high-shear, low-shear, and fluid bed granulation. There are other enabling technologies such as extrusion, rotor granulation, and melt granulation (Palzer, 2011). Both high- and low-shear wet granulation utilize a mixer with a bowl and an impeller that agitates the powder while adding the granulating liquid. The low-shear granulation is performed using low-shear mixers and was a popular method for wet granulation before high-shear mixers became available for the pharmaceutical industry. Low-shear granulators, although have limited use in the modern pharmaceutical industry, are still employed in some cases for processing drug substance that are sensitive to shear (Knight et al., 2001). A comparison between the high-shear and the fluid bed granulation processes is shown in Table 17.1 (Pandey & Badawy, 2016). The main difference between the two techniques is the method of solid agitation and the mechanism of granule growth

Table 17.1 A comparison between high-shear and fluid-bed granulation techniques.

Granulation process	Pros	Cons
High-shear	Better and more intimate mixing. Good content uniformity for low drug loading formulations. Higher densification could be an advantage for high dose compounds to reduce size of dosage unit	Separate equipment for granulation and drying. Water contact time longer. Higher densification can reduce dissolution rate and compactability
Fluid-bed	More porous granules, better dissolution and compactability. One-step granulation and drying process. Minimal (time) exposure to water (due to continual drying). Easier scale-up	Segregation or sticking potential (vessel wall and filter bags). Low density granulation can be a disadvantage for high dose compounds

(Faure et al., 2001). Fluid bed has the advantage of providing a short exposure of the granulation to the water (granulating liquid) compared to high-shear granulation. It also imparts less shear and densification of the granulation and results in more porous granules that can have a higher dissolution rate and better compaction properties. A choice between the two can therefore be made using prior drug product formulation knowledge, such as sensitivity to moisture or temperature. The current chapter will focus on the high-shear wet granulation process.

17.2 High-shear wet granulation process

A high-shear wet granulation process involves a granulation bowl equipped with an impeller blade that maintains the powder in agitation and imparts shear to the powder during granulation. The bowl contains a chopper that helps to break-up large agglomerates formed during granulation. A binder is usually added (sprayed through a nozzle or pumped through a tube) as part of a binder solution from the top of the bowl. The binder may also be added dry by including it as part of the powder formulation, with water alone used as a granulating liquid. This mode of dry binder addition is gaining popularity as it gets rid of the additional processing steps of ensuring binder solution mixing, dispersion, and appropriate storage (Ax et al., 2008; D'alonzo et al., 1990). However, one may argue that dry binder addition doesn't give enough time for the binder to be fully hydrated, and thereby renders it slightly less effective.

A schematic of a high-shear wet granulator is shown in Fig. 17.1. The overall process can be subdivided into various steps. The first step ("preblending") involves charging the intragranular portion of the formulation into the granulator bowl and mixing all the components to ensure a uniform mix. The second step is when the granulating liquid is added at a certain rate to the moving powder bed, and is called the "water-addition" step. In this stage, the binder is distributed on the powder surfaces as the powder is moved by the impeller, promoting particle-particle collisions and thereby causing particle

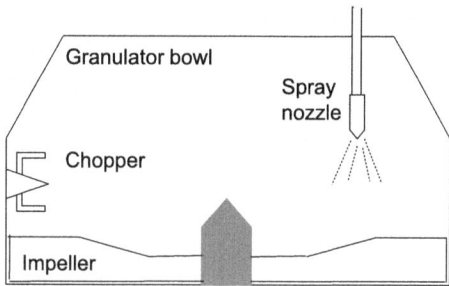

FIGURE 17.1

Schematic of a high-shear wet granulation process.

agglomeration. The chopper helps to break up any large agglomerates formed during this step. In the third and final step of granulation, the wet mass is mixed in the granulator for some additional time, and is termed as the "wet-massing" step. This step ensures further uniform distribution of water and is mainly dominated by granule consolidation and breakage. After the granulation step, the wet granules typically go through a wet milling step through a coarse mesh to break up over-sized agglomerates followed by a drying step either in a fluid bed dryer or a tray oven. Tray oven is often used for early-stage development work (e.g., formulation screening studies) where batch sizes are small (<500 g). Fluid bed drying is the method of choice for pilot-scale and commercial-scale batches. The fluid bed drying method is likely to result in more fines than the tray drying method due to granule attrition in the fluid bed, and a comparison of particle size distribution before and after the fluid bed drying process can give an indication of the granule strength. The granules coming out of the dryer are then passed through a screen (typically using a conical mill) that helps to break-up any large granules and result in a more uniform granule size distribution. The dry milled granules are blended with the extragranular excipients and lubricant in the final blending step(s) to form the "final blend," which is then ready to be compressed into tablets (Pandey & Badawy, 2016).

During the granulation process, there are several competing granulation and breakage mechanisms ongoing at the same time. A comprehensive overview of the main mechanisms is provided by Iveson et al. (2001). There are three main rate processes governing the granulation process: wetting and nucleation, consolidation and growth, and breakage and attrition.

Wetting and nucleation occur when the liquid binder get into contact with the powder (nucleation zone) and forms the initial nuclei. The extent and rate of wetting depend on the contact angle between the binder droplet and powder particle, and the spreading coefficients between the liquid and the solid phase. The nucleation kinetics also depend on the mixing dynamics in the granulator, given the intensive shear force environment inside the bowl. Hapgood et al. (2003) have proposed nucleation regime maps based on dimensionless spray flux (a measure of the liquid density on the powder surface), and the ratio of drop penetration time to particle circulation time (Hapgood et al., 2003). The map shows domains where the nucleation process would be droplet-controlled versus mechanical dispersion-controlled, and can aid in explaining and predicting granulation process changes. It was observed that for low viscosity liquid, optimization of the liquid delivery system is important, and therefore a better understanding of spray characteristics (e.g., droplet size distribution) is required. For this case, spray flux should be maintained in the droplet-controller regime (spray flux ~ 0.1) to ensure reproducible nuclei distribution (Hapgood et al., 2003). On the other hand, for high viscosity liquid (binder), the granule properties are controlled by mechanical dispersion and requires more focus toward parameters such as impeller speed, granulation time, etc.

The consolidation and growth (coalescence) rate processes start simultaneously with the wetting and nucleation rate processes but also carry into the wet massing step. This coalescence rate process involves the collision of two granules or nuclei in the presence of a certain amount of liquid binder, whereas consolidation process refers to the reduction in porosity of a granule due to the consolidating forces experienced by the granules. There are several physical models discussed in literature for coalescence phenomena that take into account the kinetic energy of the colliding particles relative to energy dissipation due to viscous forces for surface wet granule or granule deformation in case of deformable granules. This ratio is expressed in the form of a Stokes number and determines if that collision will lead to successful coalescence or rebound of the colliding granules (Iveson et al., 2001; Tardos et al., 1997). Iveson and Litster (1998b) proposed a granule growth regime map where the growth behavior was described by two dimensionless groups: Stokes deformation number, which accounts for the granule deformation during collision, and the maximum pore saturation (Iveson & Litster, 1998b). The widespread industrial utility of such models is generally limited due to lack of knowledge of the dynamic mechanical properties of the granules. Additionally, during scale-up of the process, another missing piece of information is the particle collision frequency across the granulators. Granule consolidation behavior is often modeled empirically by an exponential decay model (Hapgood & Litster, 2010; Iveson et al., 2001) and is generally the dominant rate process during the wet massing stage. Such empirical models, however, don't include effects of capillary, viscous, and frictional forces.

The breakage and attrition rate process occurs when the wet granules break due to impact, wear, or compaction in the granulator bowl. There are a few models that attempt to model this rate process, which essentially calculate if the kinetic energy is large enough to cause the granule to break, and their approach is the opposite of the coalescence model. In order to improve the predictions from these models, better estimates of granule mechanical properties such as dynamic yield stress or fracture mechanics are required. All three rate processes occur simultaneously for most of the granulation, and the formulation and process conditions would dictate which of them would dominate the granulation process at any particular point in time. The success of the models in making such predictions would depend on the estimates of the input variables of the model, all of which change due to the dynamic nature of the granulation process and are further complicated by the complexity of a multicomponent pharmaceutical formulation.

The multicomponent composition of a pharmaceutical formulation manufactured by wet granulation generally consists of API, fillers, disintegrants, binders, and lubricants. The API and fillers often dictate the amount of water needed for granulation in order to achieve the desired granule growth and consolidation, although both disintegrant and binder also influence the water requirements. Components with high water solubility tend to require less water for granulation, and are also more prone to "overgranulation" (uncontrolled granule growth leading to an excessive fraction of oversized granules). For example, a lactose-based formulation will require less water than a microcrystalline-based formulation if the filler is the major component in the formulation. When granulating powders with high water-solubility, care must be taken to avoid local over-wetting (localized regions of dissolved material), and in such cases, water is added using a spray nozzle to reduce the spray flux.

17.3 Quality by design in high-shear wet granulation

Conventionally, for pharmaceutical applications, the most important response associated with the granulation process would be the dissolution rate of the resultant tablet. There are scenarios where other attributes, such as content uniformity, flow, and/or resultant tablet hardness, may be equally important. When scaling up the granulation process, if the dissolution of the resultant tablet showed the desired dissolution profile, a successful scale-up was considered to be achieved. This approach therefore considered the granulation process like a "black-box." In the quality by design (QbD) paradigm, where the FDA has emphasized the importance of detailed process understanding, such an approach is not fit. Even though dissolution still remains a critical quality attribute (CQA), it is important to understand the process in more detail and at a granule level rather than at the tablet scale.

QbD is a systematic approach to pharmaceutical drug development that begins with predefined objectives and emphasizes product and process understanding and process control, based on sound science and quality risk management (ICH guideline: pharmaceutical development annex to Q8). QbD necessitates an understanding of how formulation and process variables influence final drug product quality, and is a way to design and develop formulations and processes to ensure a predefined product quality (or target product profile [TPP]). In general, a QbD development process involves defining a TPP and CQAs. CQAs are product attributes that affect product safety and efficacy and typically include potency, content uniformity, purity, drug release, and microbiological properties of the oral solid dosage form manufactured by wet granulation. Using prior knowledge, risk analysis, and development studies, critical material attributes (CMAs) and critical process parameters (CPPs) that affect CQAs are identified, and then a control and monitoring strategy for the overall process is devised to ensure the desired product quality. A summary of potential CMAs, CPPs, and CQAs is listed in Table 17.2 (Pandey & Badawy, 2016).

17.3.1 Critical material attributes

Material attributes are critical when a realistic change in that attribute can impact the final product quality. CMAs are often related to the input (raw or starting) materials or intermediate (in-process) materials. In the case of wet granulation, few potential input material CMAs were reported in the literature, such as API or excipient particle size, contact angle with the granulating liquid, surface area, and porosity. Critical attributes of the output granulation can be viewed as intermediate (or in-process) CMAs as they affect processing in subsequent unit operations such as compression or final product performance. Examples of in-process CMAs in wet granulation would include wet granule bulk density and rheology, and dried granule size distribution, porosity, moisture content, strength, morphology, flowability, and compaction characteristics (Table 17.2). Input material attributes and process parameters dictate quality attributes of the resulting granulation. It is therefore essential to identify the critical input material properties for a given formulation and ensure that these input material powder properties are controlled within an acceptable range. In addition, it is important to understand how variation of the input material within this range interacts with process parameters for their impact on output granulation properties. For example, larger starting material particle size (API or excipient) can make granules more sensitive to consolidation at higher impeller speed (Iveson & Litster, 1998b; Iveson et al., 1996). Lactose and DPC 963 input materials with small particle size was reported to be

Table 17.2 A summary of potential CMAs (critical material attributes), CPPs (critical process parameters) and potential critical granule properties as they relate to high-shear wet granulation.

Potential CMAs (Input Material Properties)	Potential CPPs (Process Parameters)	Potential Critical Granule Properties
Particle size distribution	Impeller speed	Particle size distribution
Surface area	Wet massing time	Pore volume distribution
Particle shape and morphology	Water addition time	Granule morphology
Solubility	Spray rate	Granule mechanical properties and strength
Binder viscosity	Nozzle properties	Bulk and tap density
Surface tension of granulating liquid	Water amount (water to solids ratio)	Flowability
Bulk density	Bowl fill	Compaction properties
Water uptake capacity	Granulator geometry	Content uniformity
Contact angle (wettability)	Impeller design/configuration	–
Particle surface properties	–	–

more resistant to densification compared to the larger particle size of the same material under similar process conditions, which was attributed to the larger area of inter-particle contact in a given volume. (Badawy et al., 2000; Badawy & Hussain, 2004) The frictional forces at these contact points impede the consolidation of particles during granulation. Input material with high water holding capacity also tends to produce porous granules as water uptake by the material makes it less available on the particle surface and hence less effective in disrupting frictional forces between particles (Vemavarapu et al., 2009). Poor wettability of the powder by the granulating liquid (high liquid-solid contact angle) was also reported to produce porous and weaker granules (Iveson & Litster, 1998b).

A case study published by Vemavarapu et al. (2009) evaluates the various input material properties that can dictate granulation outcome for about 25 APIs and excipients. These included solubility, wettability (contact angle), water holding capacity, particle size, surface area, pore volume, and preblend compactability and flow. The responses that were looked at after granulation were granule particle size distribution, compaction, and flow. It was shown that all these input material properties have a role to play in the final granulation outcome. Some of the main conclusions were that high solubility materials led to a greater tendency toward uncontrolled growth, surface area and particle size both dictated several granulation properties, high contact angle led to a more polydisperse granulation, large particle sizes showed lower granule growth, and compounds with higher water uptake showed reduced compactability. Whether all or any of these input material properties can be defined as CMAs for a particular drug product depends largely on the TPP of that drug product and the desired targets for the different CQAs. For example, if granule compaction is a critical property to a particular drug product, then water uptake capacity of the formulation components may be a CMA for that drug product. This study didn't look at the interplay between the potential CMAs and process conditions, but as pointed out previously, there's a significant interaction between formulation and process in a high-shear granulation process. In general, water amount has the biggest impact on granulation outcomes (e.g., granule porosity, particle size, etc.), and the magnitude of effects of other variables was reported to be dependent on the water amount used for granulation. For example, impeller speed will have a less dramatic effect when less water is being used (under-granulated), whereas small changes in impeller speed and wet massing times have a bigger impact when a high water amount is being used (over-granulated) (Pandey et al., 2013; Tao et al., 2015).

The granule properties that get the most attention are particle size (including number of fines), flowability (generally lower risk for wet-granulated products if optimal water amount is used), and compaction. Recent studies have shown the importance of granule porosity and have shown links between granule porosity and dissolution (Badawy et al., 2012) (Fig. 17.2) and compaction properties (Pandey et al., 2013,2013). Interesting to note in those case studies is that there was no clear relationship observed between granule particle size and compaction or dissolution properties, even though granule particle size generally gets the most attention during characterization. The distribution of the pores within a granule has an effect on the dissolution and release characteristics and can impact the bioavailability of a drug (Rahmanian et al., 2009; Štěpánek, 2004). There is a strong link between granule particle size (or amount of fines) and granule flow (Badawy et al., 2000; Pandey et al., 2013, 2013), with higher content of fines leading to poor flow properties. This relationship is opposite between number of fines and compaction with higher fines showing better compaction properties; therefore, a good balance between flow and compaction properties is targeted. The key to achieving successful scale-up would be to keep the granule properties (in-process CMAs) constant across scales by controlling the process variables (Pandey & Badawy, 2016).

17.3.2 Critical process parameters

Process parameters are considered critical when a realistic change in the parameter can impact the CQAs. A CPP has a direct impact on product CQA and therefore should be monitored and controlled to ensure product quality. CPPs are operating process parameters that should be combined with CMAs to describe the relationship between unit operation inputs and outputs.

A list of potential CPPs are first identified based on prior knowledge and risk analysis (Table 17.2). Small or pilot-scale studies are conducted on potential CPPs to identify the CPPs pertinent to that particular drug product. The typical process parameters that are evaluated as part of such a study include water amount, impeller speed, water addition rate, granulation (or water addition) time, mode of water addition, and wet massing time. Several of these factors are known to have significant interactions, and therefore such a study often requires some sort of a factorial design of experiments. A few case studies are presented in the literature (Badawy et al., 2000; Badawy et al., 2012; Pandey et al., 2013; Rahmanian et al., 2009; Wang et al., 2008). Badawy et al. (2000) showed that an increase in the moisture content of the granulation and a decrease wet massing time or impeller speed led to increased granulation compressibility (Badawy et al., 2000). Increasing impeller speed or wet massing time was found to decrease granule porosity, which led to decreased granulation compressibility. In another study, Badawy et al. (2012) showed that water amount was the most significant factor for granule growth, across three different model drug compounds with varying physical properties. They showed that impeller speed effect depends on the granule mechanical properties and also the liquid distribution in the granulator. Water addition rate did not have a significant effect on the density/porosity of any of the three formulations. Pandey et al. conducted a full factorial design of experiments on water amount,

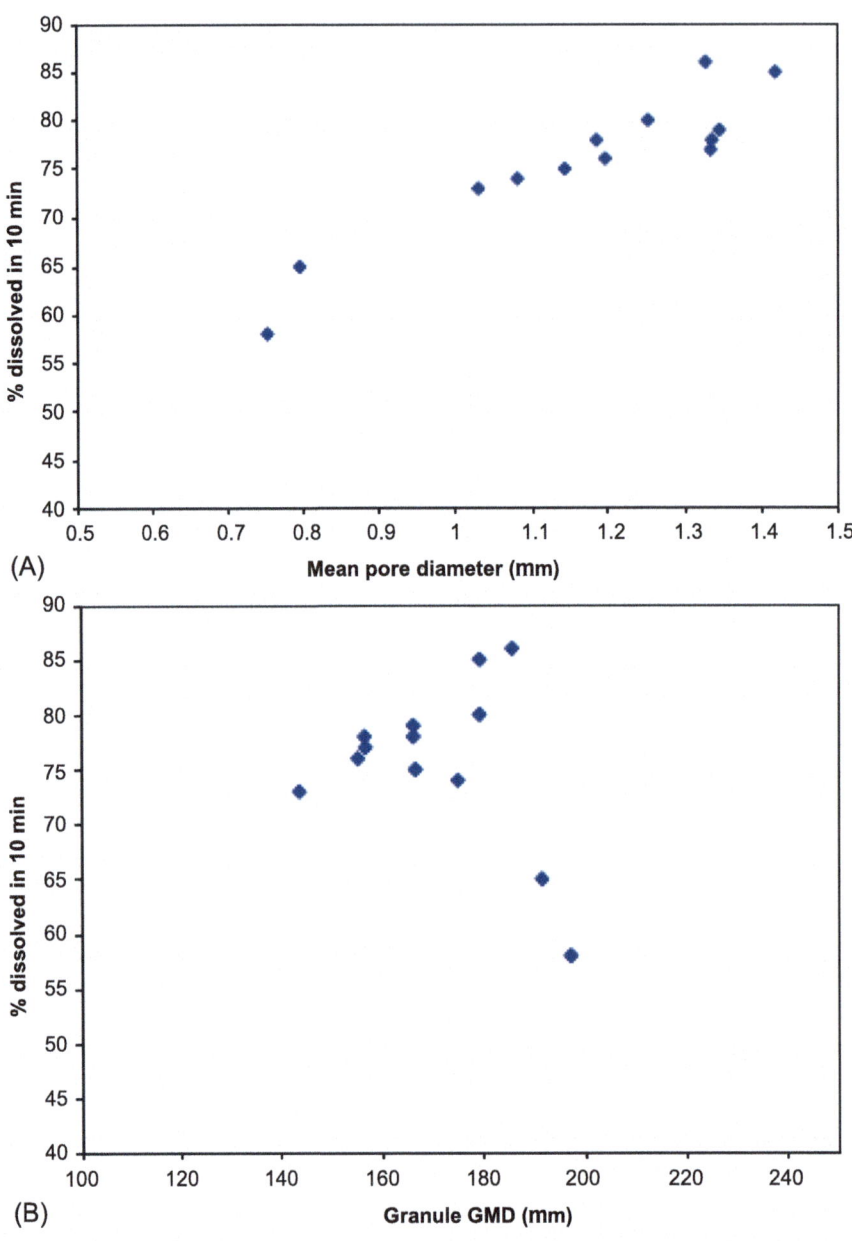

FIGURE 17.2

Correlation between dissolution (at 10min) and (A) granule porosity (pore diameter) and (B) granule GMD.

From Badawy et al. (2012)

impeller speed and wet massing time and showed that the water amount has the largest impact on the granule characteristics (Pandey et al., 2013). They also observed that the effect of other process variables was more pronounced at higher water amount. At high water amounts, an increase in impeller speed and/or wet massing time showed a decrease in granule porosity and compactability (Figs. 17.3A and B). A strong correlation between granule porosity and compactability was observed (Fig. 17.4).

Studies have also established the role of the chopper during granulation, where it was shown that a chopper can be used to dictate the shape of the resulting granules. It was shown that by turning the chopper off for the last 120 s of the process, a narrow granule size distribution was obtained, revealing that granule homogeneity was improved and uncontrolled granule growth was prevented effectively (Wang et al., 2008). It was observed that granules prepared with the chopper off toward the end of the process were more spherical and had a smoother surface. Wang et al. also concluded that the effect of impeller speed is dependent on the starting material formulation. Another variable often studied separately is the mode of water addition. Kayrak-Talay & Litster (2011) showed that the addition of the liquid in the dripping mode gave the widest granule particle size distribution and the highest percentage of "lumps" compared to the spraying mode. On the other hand, the addition of the liquid by spraying gave the narrowest granule particle size distribution with the lowest number of lumps. Such an effect is likely to be formulation-dependent, with more water-soluble formulations showing a more pronounced effect of mode of water addition. It is also likely to be scale-dependent, where the higher liquid addition rates required for the larger scales often cause the nucleation to operate in the mechanical dispersion regime (reference) if even a spray method is used for granulating liquid addition.

In general, it is clear that CPPs will be a strong function of the physical properties of the formulation. In order for the process to reproducibly result in acceptable CQAs, it is essential the CPPs are identified and controlled within an acceptable multivariate space (design space) that is established from the process development studies. In most cases, water amount used for granulation is a CPP and should be studied and optimized carefully. An optimal water amount will also reduce process sensitivity to other process variables and ultimately result in a more robust process. The design space of water amount and other CPPs is typically established at a lab or pilot scale and scaled up to the commercial scale using the approaches described later in this chapter. Verification of the scaled-up design space is then performed and is essential to ensure a successful commercial manufacturing process (Pandey & Badawy, 2016).

17.4 Scale-up principles

Scale-up of high-shear wet granulation has been a subject of study for several years. Given that the wet granulation process can have a significant impact on the granule properties and thereby influence some of the critical drug product properties such as dissolution, content uniformity, flow, and tablet hardness, it is important to be able to scale the process successfully (Badawy et al., 2012; Campbell et al., 2011). A QbD approach necessitates the need to establish a process design space for a drug product. Conducting experiments at the commercial scale to establish design space can be expensive. Therefore, it is desirable to conduct more experiments at small-scale and have the ability to scale-up successfully. This makes scale-up even more important. Additionally, the drug substance supply for investigational drugs is often limited during early stages of the drug product, and it is important to establish an experimental system that utilizes minimal drug substance quantities and can be extrapolated to commercial scales as a means of realizing the concept of ICH Q8 (Aikawa et al., 2008). If a one-to-one link between small-scale and

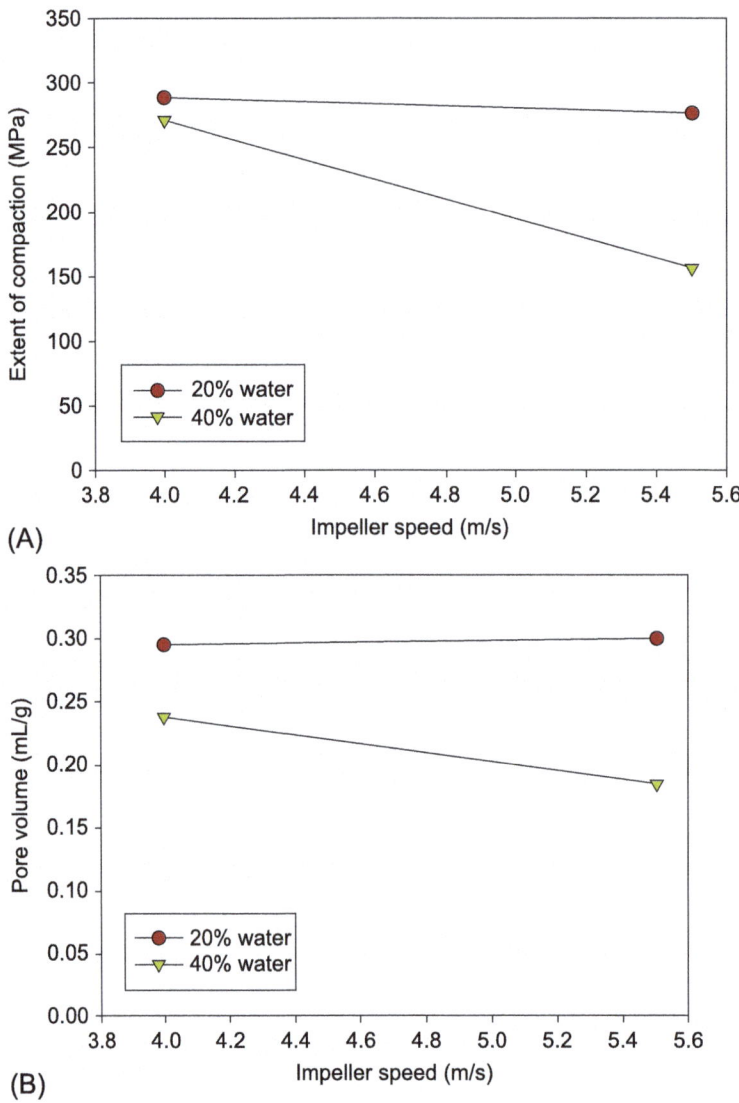

FIGURE 17.3

Effect of impeller speed on (A) extent of compaction, and (B) porosity (pore volume) as a function of water amount.

From Pandey et al. (2013b)

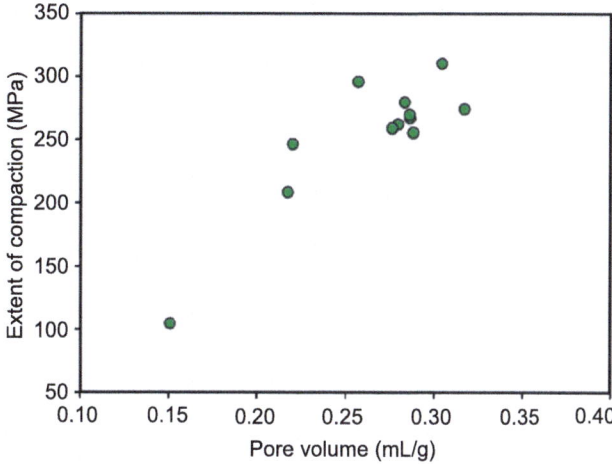

FIGURE 17.4

Correlation between compaction properties and granule porosity.

From Pandey et al. (2013b)

pilot and commercial-scale granulators can be established, this approach is an acceptable one. Some studies have focused on the scalability of small-scale granulators that utilize minimal resources (e.g., 30 g batch size). Aikawa et al. (2008) showed the feasibility of scaling up from a mini-scale (0.2 L, 30 g batch size) to pilot-scale (11 L, 1 kg batch size) to commercial-scale granulator (200 L, 46 kg batch size). Similarly, Ameye et al. (2002) down-scaled a Gral 10 granulator process to Mi-Pro small or mini-scale granulators (5 L, 1900, 900, and 250 mL).

There is significant existing literature addressing the scale-up of the high-shear wet granulation process, with a wide variety of scale-up rules used. The ultimate objective of all scale-up rules in general is to ensure that particles experience the same conditions across scales so that the same granule quality can be achieved (Tao et al., 2015). The key to successful scale-up is maintaining similarity across scales. The three main types of similarities are geometric, dynamic, and kinematic similarity (Agrawal & Pandey, 2015; Pandey et al., 2006). In relation to high-shear wet granulation, the geometric similarity refers to the granulator bowl geometry (including impeller design, etc.). Two systems are geometrically similar if the ratio of linear dimensions across scales is maintained. Ideally, if the same equipment manufacturer is maintained across scales then geometric similarity is attained more easily. However, in real practice, the lab-scale manufacturers are different from the commercial-scale ones. Additionally, over time the lab-scale units get upgraded with newer machines (including newer manufacturers) with better capabilities (e.g., torque profiles) but the commercial scale machines remain the same for a longer period of time. The dynamic and kinematic similarities refer to the forces experienced by the particles inside the granulator and the particle velocities. The forces or shear experienced by the particles inside the granulator result mainly from particle-particle, particle-wall, and particle-impeller collisions. The effect of the impeller (design and speed) generally dictates the shear experienced by the particles. This shear, in turn, governs the interplay between the various granulation mechanisms discussed in Section 2, such as aggregation, consolidation, and breakage.

Maintaining geometric, dynamic, and kinematic similarities across different scales of high-shear granulators is very challenging. While geometric similarity may appear to be the most straightforward among the three, nevertheless, granulators with the same design and dimensions' ratio across scales were shown to have different mixing and material flow patterns. This results in different distribution profiles of particle velocities and forces at the different scales, which makes the goal of attaining dynamic and kinematic similarities more complicated. In addition, it is also challenging to simultaneously obtain similar peak particle velocity and force across scales. As described below, the different approaches commonly used for scale-up of impeller speed target to achieve one of them (i.e., similar particle velocity or force) which does not usually achieve similarity for the other. As a result of these challenges in scaling up the high-shear wet granulation parameters, the concept of granulation "end point" is very common in high-shear wet granulation. According to this concept, binder (water) amount and/or wet massing time are based on an in-process measurement, which is used to determine when granulation reaches desired properties or "end point" and consequently water addition or wet massing are stopped when this end point is achieved.

Scale-up strategies for high-shear granulation can therefore be categorized as (1) attribute-based scale-up strategy in which parameters are adjusted at the different scales to achieve target in-process attributes (or granule properties) or end-point as determined by an in-process measurement, or (2) parametric-based scale-up strategy in which target parameters identified at the small scale and shown to provide the desired final product CQAs are scaled up using certain calculations. A combination of the two approaches has also been utilized. For example, impeller speed and granulating water amount can be scaled up using a parametric approach, while wet massing time is based on an attribute control strategy. A more detailed discussion on each of these categories is provided in the subsequent subsections (Pandey & Badawy, 2016).

17.4.1 Attribute-based scale-up strategy

An attribute-based scale-up strategy focuses on granule attributes that are targeted to be kept constant during scale-up. The term granule attributes refers to the in-process granule properties such as particle size distribution, pore size distribution, granule hardness, compaction properties, bulk and tap density, etc. In this approach, the equipment process parameters are adjusted such that equivalent granule attributes are obtained across scales. An in-process in-line or at line-process measurement using a process analytical technology (PAT) tool is typically used to monitor the progress of the granulation and to guide process parameters mainly water (binder) amount and/or wet massing time. In other words, an in-process granulation attribute is used to provide feedback control on process parameters and eventually dictate the granulation end point. While the in-process attribute may not be a CMA itself, the assumption is that it correlates with desired product CQAs. Given that there are several in-process properties that can be potentially monitored (e.g., particle size, density, porosity), the key to using this scale-up strategy successfully is to first identify the appropriate end-point measurement which ensures desired product quality for that particular formulation/system. This can be formulation-specific and is governed by the desired CQAs for a particular drug product. It is also important to not only look at the granule properties at the end of the granulation but also to monitor the kinetics (progression) of the granulation. Consider a case where a certain particle size distribution of granules is set as a desired property. It may be possible to reach the same particle size distribution via different granulation kinetics (e.g., slow water addition versus fast water addition) by manipulating other process parameters. So, even though the particle size distribution of those granules may be the same, the granule formation process was different, which would result in other granule properties being different. If the kinetics of the granulation (granulation fingerprint) can be maintained similarly across scales, there is a good chance that the resultant granules will have similar properties.

One example of challenges associated with the attribute-based scale-up approach would be if, let's say, a certain granule particle size distribution is targeted as the end point and is controlled by the wet massing time. This means that the wet massing time is adjusted to reach the desired granule particle size distribution. If some property of the input excipient or API changes in this case, then the wet massing time may have to be prolonged or reduced to reach the same granule particle size. If no other control is put in place, this may result in an extended period of wet massing where the granules would experience an increased amount of shear and possibly result in a significant loss of compaction. Additionally, an increased shear may result in other issues such as a potential physical form change of the drug substance, which may have not been previously identified as critical within a certain operating process space. Such changes in drug substance physical form caused by manufacturing process are called process-induced transformations (Debnath & Suryanarayanan, 2004; Wikstrom et al., 2005; Wikström et al., 2008). Wikstrom et al. (2005) showed the transformation of theophylline anhydrous to theophylline monohydrate during high-shear wet granulation process after 3 min of binder solution addition. Mixing speed was shown to have the greatest effect on the transformation kinetics.

End point measurements have been studied extensively, but the findings vary greatly on the best way to determine granulation end point. Interestingly, to date, one of the most prevalent methods in the pharmaceutical industry to determine granulation end point is still the "hand-squeeze" test, where the operator evaluates the granules visually and by hand and makes an end point decision based on prior experience. Some of the end point methods published in the literature include near-infrared, particle size by focused beam reflectance microscopy (FBRM), power or torque, acoustics, wet mass rheology, imaging, etc. (Campbell et al., 2011; Corvari et al., 1992; Frake et al., 1997; Hancock et al., 1982, 1994; Kristensen et al., 1985a, 1985b; Kristensen, 1996; Landin et al., 1996; Lindberg et al., 1982; Ohike et al., 1999; Sakr et al., 2012; Sato et al., 2005; Satoru, n.d.; Watano, 2001; Watano et al., 2005; Whitaker et al., 2000).

One of the most commonly used granule properties when using this type of scale-up strategy is particle size distribution. In-process measurement of granule size distribution using FBRM has recently been gaining popularity (Arp et al., 2011). FBRM probe measures the chord length of particles that pass through a spinning laser beam, and therefore, tracks real-time changes in particle size and distribution in the granulation process. FBRM (Lasentec, C35 probe, and Mettler-Toledo) technology gained attention over the last few years with the development of a modified probe with a mechanical scraper on the sapphire window that prevents probe fouling. Huang et al. (2010) showed the applicability of the FBRM probe to scale-up in monitoring process changes and also in capturing batch-to-batch variability and reproducibility. They did not observe a correlation of the chord length data with downstream quality attributes, possibly due to the limited data set and incomplete characterization of the various granule properties. However, such real-time monitoring of the process provides a great opportunity for scale-up purposes and for matching granulation kinetics across scales to ensure successful scale-up, and also to enhance process and equipment understanding.

Even though particle size is important, there is existing literature that clearly indicates that granules with the same particle size distributions can have differences in pore structures, which can in turn affect compaction characteristics (Badawy et al., 2000; Pandey et al., 2013). Some researchers have pointed out this limitation and used a different property, such as granule strength (Hassanpour et al., 2009;

Rahmanian et al., 2008), granule porosity and bulk density (He et al., 2008), etc., when evaluating the success of process scale-up. It should be pointed out that monitoring just one of the granule attributes is unlikely to be sufficient in completely characterizing a granule. Some of the granule attributes can be counteracting; for example, a reduced percentage of fines is good for flow but not for compaction, and therefore, a good balance would be required, and both of these properties would need to be constantly monitored.

There are other existing PAT techniques to monitor the granulation kinetics that have gained more attention recently, such capacitive sensor, strain-gauge torque sensor, near infrared (NIR), Raman, acoustic monitoring, etc. (Briens et al., 2007; Corvari et al., 1992; Findlay et al., 2005; Li et al., 2005; Luukkonen et al., 2008; Rantanen et al., 2004; Tok et al., 2008; Whitaker et al., 2000; Wikström et al., 2005). Rantanen et al. (2004) and Luukkonen et al. (2008) both demonstrated the utility of an in-line NIR spectroscopy to high-shear wet granulation process. They showed that NIR can be used to identify the different phases of granulation and determine the end-point of the process. They also found a good correlation of NIR data to granule and tablet physical properties. Rantanen et al. (2004) concluded that this technique can be valuable for scale-up but should be used in combination with other measurements, such as power consumption and torque, to provide full characterization of the process because NIR does not measure directly the properties of the wet mass. Another limitation with NIR is that it can only measure the moisture that is present at the surface of the granules. Acoustic monitoring is another monitoring technique that is shown to provide a signature of the granulation process. It is noninvasive in nature and can be correlated to dynamic changes in granule particle size, flow, and compaction properties (BRIENS et al., 2007; Whitaker et al., 2000). Whitaker et al. (2000) showed that a reasonable correlation was obtained between the acoustic signal and unmilled granule physical properties, such as particle size and flow. These PAT techniques provide valuable real-time information about the progression of the granulation process and in most cases, are used for process end point determination and control. These techniques usually work better (less noisy data) at larger granulator scales due to less interference from water addition.

Another approach for end point determination of the wet granulation process is related to the rheological or "consistency" changes of the wet mass during granulation. There are several ways in which this is accomplished. In some cases, this was done "at-line" by taking a sample of the wet mass from the granulator and using a mixer torque rheometer to determine sample "viscosity" (Hancock et al., 1991; Landín et al., 1995; Luukkonen et al., 1999; Parker et al., 1990). In-line measurements related to wet mass consistency are also common by using an instrumented impeller to measure impeller torque, impeller power, or other derived parameters based on impeller power (e.g., work done). There's significant existing literature on this topic with researchers using parameters such as impeller torque, impeller power, impeller amperage, work done, normalized impeller work etc. to monitor the process, define the granulation end-point, and also during scale-up (Campbell et al., 2011; Faure et al., 2001; Ghanta et al., 1984; Holm et al., 1985a; Kristensen et al., 1985b; Landin et al., 1996; Leuenberger, 1982; Sato et al., 2005; Sirois & Craig, 2000; Watano et al., 2005). Similar to the other above-mentioned PAT tools, this method captures the kinetics of the granulation process, and the data have often enabled researchers to identify the different phases of granulation. Additionally, it is not a one-point (location) measurement such as FBRM, and presents a cumulative effort that the impeller makes to move the entire mass of the material. There is also no noise associated with the data during the water-addition phase (often a drawback of FBRM), and is nonintrusive in nature.

While impeller power consumption was shown to correlate well to granule densification as well as to granule saturation (ratio of liquid-filled pore volume to total pore volume) (Kristensen & Schaefer, 1987; Lindberg et al., 1982; Ritala et al., 1988), a major limitation of using power-based approach is that the motor power varies between different equipment and can even change for a given equipment over time. Therefore, some normalization techniques, such as normalized impeller work, were proposed by researchers to account for differences in scale and impeller blade design, which can be different between granulators during scale-up (Campbell et al., 2011; Sirois & Craig, 2000). Dimensional analysis based on impeller power has also been applied to the scale-up of the wet granulation process. One dimensionless number derived from such analysis is the Newton power number shown in Eq. (17.1).

$$N_p = \frac{\Delta P}{\rho \omega^3 R^5} \tag{17.1}$$

where ΔP is the power consumption (corrected by the baseline power during dry mixing stage), ρ is the wet mass density, R is the impeller radius, and ω is the rotational speed.

Newton power number (N_p) can be expressed as a function of other dimensionless numbers such as Froude number (ratio of angular to gravitational acceleration, Eq. 17.2), pseudo Reynolds number, specific amount of binder, volume fraction of powder, and fill or geometric ratio (Am Ende et al., 2010; Faure et al., 2001) as shown in Eq. (17.3).

$$Fr = \frac{\omega^2 (2R)}{g} \tag{17.2}$$

$$\log_{10}(N_p) = a \cdot \log_{10}(\varphi Re \cdot Fr \cdot f) + b \tag{17.3}$$

where a and b are scale-independent regression constants, φRe is the pseudo Reynolds number (defined by Eq. 17.4), Fr is the Froude number, and f is the fill ratio (granulator volume to wet mass ratio), and g is the acceleration due to gravity.

$$\phi Re = \frac{\rho R^2 \omega}{\mu} \tag{17.4}$$

where μ is the wet mass viscosity.

The validity of Eq. (17.3) has been shown across several scales and granulators (Faure et al., 1999; Landin et al., 1996; Landin et al., 1999). In order to use this scale-up approach, data need to be collected for establishing Eq. (17.3). The unknowns in Eq. (17.3) are constants "a" and "b." Thus, experiments need to be conducted at different values of variables such as Froude number (vary impeller speed), fill ratio (batch size), and pseudo Reynolds number (vary impeller speed, granulator size, granulation time). Once the relationship in Eq. (17.3) is verified using that data, values of a and b are established for a particular formulation. This dimensionless relationship can be then used to predict the power end point at a larger scale for a desired wet mass viscosity to match granulation properties across scales. In the absence of wet mass density and viscosity data, impeller torque measurements can be used to estimate the pseudo Reynolds number, as they are directly correlated. Froude number during scale-up can either be kept constant or estimated if the impeller speed for the larger is determined by an impeller speed-related scale-up rule (e.g., constant tip speed) as described in Section 4.2.3.

17.4.2 Parametric-based scale-up strategy

Due to the limitations of end-point determination methods described above, some researchers believe that parametric-based scale-up is a more robust approach. A parametric-based scale-up strategy is where the process parameters at the larger scale are determined from the small-scale parameters using certain scale-up principles/factors. The product is then manufactured using the scaled-up parameters and tested to assess its attributes. If desired attributes do not match the target established at the small scale, large-scale process parameters can be modified to more closely match target product attributes. If this is the case, process knowledge established at the small scale is very valuable in deciding what parameters to vary (and which direction to vary them) in order to achieve the goal.

Scale-up factors are based on aspects related to the equipment, and some of them can be based on the manufacturer's recommendations for a particular equipment/scale. These factors are derived from prior experiences with the equipment at that scale, taking into account any equipment design constraints. It is important to understand the fundamentals of scale-up in more detail so that experimentation at commercial scale can be minimized. In relation to high-shear wet granulation, these scale-up factors are most commonly related to the impeller speed, wet massing time, water amount, and sometimes to the water or binder addition (e.g., spray rate, water addition time, droplet size, dimensionless spray flux, mode of addition). By definition, this approach doesn't take into account the potential effects of downstream processes such as fluid bed drying, milling, etc. (Sprockel & Stamato, 2010). however, in real practice, this approach is combined with the first attribute-based approach, and some of the important attributes of the granules produced downstream are evaluated.

The first step in a parametric-based approach would be to conduct a process parameter ranging study. This small or pilot-scale factorial study, including parameters such as impeller speed, water amount, wet massing time, and water addition rate, helps to identify CPPs related to the particular formulation that is to be scaled-up (He et al., 2008; Iskandarani et al., 2001; Wehrlé et al., 1993). The potential CPPs identified can be broadly classified into the following categories, and the subsequent subsections discuss the scale-up principles associated with them:

- Powder bed height/batch size
- Water amount
- Impeller speed
- Wet massing time
- Spray-related factors

It is noteworthy that there is no universal scale-up rule for the chopper. The primary purpose of the chopper is to break-up any large granules formed during granulation and studies have shown that it does not affect any potential critical granule properties. In many cases, granulators have fixed chopper speed (or high and low settings only) which provides little room for applying scale rules even if desired (Pandey & Badawy, 2016).

17.4.2.1 Powder bed height or batch size

Batch size during scale-up often does not receive the necessary attention. First, if batch size is not scaled appropriately, geometric similarity is not achieved. Additionally, the forces and stress experienced by the particles during granulation can be linked to the mass of particles above it. When scaling up between granulators from the same manufacturer, a simple volume proportionality rule for deciding the batch size works well and ensures geometric similarity. When the designs of the granulators are not exactly the same, a ratio of bed height (h) to impeller or bowl diameter (D) Eq. (17.5) is a commonly used scale-up rule to determine the batch size in order to achieve a certain level of geometric similarity.

$$\frac{h}{D} = constant \qquad (17.5)$$

17.4.2.2 Water amount

Water amount refers to the amount of water that is used for granulation and is the most important variable of the wet granulation process. The amount of water is directly dependent on the formulation and is almost impossible to predict this a priori based solely on powder properties. There is a wide variety of powder properties that dictate the amount of water required for granulation. Vemavarapu et al. (2009) conducted an extensive study using APIs and excipients with a wide variety of physical properties to establish a link between the physical properties and granulation behavior (discussed earlier). They showed that several physical properties, such as API solubility, water uptake capacity, contact angle, particle size, and surface area all have a significant role in the granulation process, which makes the determination of the water amount requirement for a new formulation solely from physical property data challenging. The amount of water determines granule liquid saturation (ratio of granule pore volume filled with water to total granule pore volume) which was reported by researchers to be a critical factor with respect to granule growth and consolidation (Hapgood et al., 2002; Litster, 2003; Marston et al., 2013; Oulahna et al., 2003).

The general strategy for scale-up of water amount would be to conduct a water ranging study at the small-scale to identify the optimal amount of water required for a given formulation. The amount of water is then increased linearly with batch size at the large sale, in order to keep % water (% of intragranular portion) constant at the different scales (Rekhi et al., 1996). This approach makes sense as the water amount required should be governed by the formulation components (physical properties), which do not change during scale-up. If other parameters have been scaled appropriately, there should be no need to make further adjustments in the water amount at the larger scale. However, amount of water is sometimes adjusted to compensate for other changes during granulation (He et al., 2008). Because it is usually not possible to exactly match particle velocity and force distributions at the different scales, the amount of water may need to be adjusted at the large scale in some cases in order to achieve similar growth and consolidation across scales. For example, if the % fines are observed to be higher at the larger scale, a slight increase in water amount can compensate for the observed change. This may not be an ideal way to scale-up, as this requires adjustments to be made at the larger or commercial scale, which can have cost implications, and in some ways becomes a trial-and-error method.

17.4.2.3 Impeller speed

Impeller speed scale-up rules get the most attention in regard to the scale-up of high-shear wet granulation process. Impeller speed is directly related to the shear that is applied to the granules. There are numerous publications that discuss impeller speed related scale-up rules (Faure et al., 2001; Hassanpour et al., 2007; Horsthuis et al., 1993; Rekhi et al., 1996; Rahmanian et al., 2008; Tao et al., 2015; Tardos et al., 2004). In general, one of the shortcomings of some such literature articles has been the incomplete characterization of the resultant granules across scales, as most studies focused on a single granule property and ignored others, thus providing an incomplete picture.

The most commonly used methods of scaling impeller speed use a power law correlation as shown in Eq. (17.6).

$$\frac{\omega_2}{\omega_1} = \left(\frac{D_1}{D_2}\right)^n \tag{17.6}$$

where ω_2 and ω_1 are the impeller speeds in revolutions per minute, and D_1 and D_2 are the impeller diameters in the two granulators, and n is the power law number. The three most commonly used values of n are 0.5, 0.8, and 1. A value of $n = 0.5$ corresponds to the case when the Froude number across the scales is kept constant. Froude number is a fundamental dimensionless number widely used in fluid mechanics and is defined as the ratio of inertial to gravitational forces. A constant Froude number across scales would aid in maintaining dynamic similarity across scales. A value of $n = 1$ corresponds to keeping the tip speed constant across scales and is the more commonly used value of n which aims to keep similar particle velocity (kinematic similarity) across scales. An experimental study was conducted by Tardos et al. (2004) using tracer pellets to measure the stress experienced by the granules across scales. Findings from this study suggested that a value of $n = 0.8$ results in similar shear stress across different scales. The value of n varied between 0.8 and 0.85 depending on fill level and impeller design (Hassanpour et al., 2007; Tardos et al., 2004). This empirically derived value of n (constant empirical shear stress) was used successfully to scale-up the wet granulation process (Horsthuis et al., 1993; Tardos et al., 1997, 2004). There are several examples of conflicting results when it comes to choosing between these three values of n, with case studies that show success with either one of them.

Rahmanian et al. (2008) studied the effects of impeller speed at different scales of vertical high-shear granulators (Cyclomix—1, 5, 50, and 250 L) using constant tip speed, constant shear stress, and constant Froude number rules on the mechanical strength of granules (Rahmanian et al., 2008). They concluded that the constant tip speed rule produces granules of comparable strength, whereas the differences between scales were larger at the $n = 0.8$ and 0.5 rules. They also confirmed by discrete element modeling (DEM) modeling and positron emission particle tracking (PEPT) experimental techniques that the granules at smaller scales (1 L) experience different flow fields and shear stress (when using $n = 0.8$ rule) and yield more elongated granules with a lower packing fraction and lower strength. Rekhi et al. (1996) successfully conducted scale-up experiments on Niro-Fielder PMA granulators (geometrically similar—10, 65, 150, and 300 L) using a constant tip speed rule, linearly scaling up the water amount by batch size (constant % water amount), but also recommending scale-up of wet massing time (discussed in Section 4.2.4). They observed good scalability of granules properties such as bulk and tap density, blend uniformity, particle size distribution, and tablet dissolution. Horsthuis et al. (1993) showed that the tip speed rule didn't result in the same particle size distribution using Gral granulators (10, 75, and 300 L), and that the Froude number rule resulted in similar granule particle size distributions for lactose granulations. They also showed that power consumption profiles were different across scales, implying possible differences in velocities and shear experienced by the granules.

Another scale-up parameter found in literature that is related to impeller speed is known as "relative swept volume" (Holm, 1987; Holm et al., 1985; Horsthuis et al., 1993; Kristensen & Schaefer, 1987; Schaefer et al., 1986, 1985, 1987). It is defined as the volume swept by the impeller per second divided by the volume of the granulator. This is associated with the work input on the material which provides densification of the consolidated mass (Horsthuis et al., 1993). Horsthuis et al. (1993) conducted a study where it was observed that the relative swept volume parameter did not provide equivalent scale-up for a lactose-PVP formulation between Gral 10, 75, and 300 L, despite the strong relation between relative swept volume and the specific energy consumption observed in the study. They conducted their experiment slightly differently in that they added the binder all at once (for ease and reproducibility reasons) before starting the mixer. They also observed significant changes in temperature during the granulation process.

Existing literature varies in opinion on which impeller speed scale-up rule works best. One of the limitations when using the Froude number, which is more fundamental in nature, is the constraints with the equipment itself. Depending on the scale difference between the scales, this rule may predict impeller speed values that may not be achievable with certain granulators. Therefore, a practical limitation with this rule often limits the use of the Froude number. The most commonly used rule is the one where tip speed is kept constant across scales. However, there is existing literature that suggests that a constant tip speed will not ensure similar mixing patterns inside the bowl, which means that the granulation process will be different. In fact, it has been shown that at the same tip speeds, there is more efficient mixing in lab-scale granulators, with the powder moving in a "roping regime," and that the powder surface velocity decreases with increase in scale (Litster et al., 2001; Plank et al., 2003; Tardos et al., 2004).

It is well-established that if the water amount chosen is higher than the optimal amount for that formulation, then sensitivity to other parameters, such impeller speed is more pronounced (Pandey et al., 2013), as shown in Fig. 17.3A and B. Therefore, if the process is well-optimized and understood at small-scale as dictated by QbD principles, an operating design space can be established where the process is less sensitive to small changes in impeller speed. Once such a process space is defined, then scale-up is less challenging and not as sensitive to impeller speed, and any of the typically used impeller speed power law rules (Eq. 17.2) may be used successfully (Tao et al., 2015).

17.4.2.4 Wet massing time

Wet massing time refers to the time for which the granulator mixer is operated after all the required water/binder has been added. This granulation phase helps to provide even distribution of water along with some further agglomeration, reduction of fines, consolidation, and attrition of granules. During this phase, the granules that were formed during water addition get further consolidated which results in reduced porosity and compactability of granules (Shi et al., 2011). Typically, an exponential decay of porosity is observed during this phase (Am Ende et al., 2010; Hapgood & Litster, 2010) as shown in Eq. (17.7). Consolidation rate during wet massing can also be dependent on other process conditions including the amount of water used during granulation. Because higher porosity is generally a desired property, the wet massing time is generally kept short.

$$\frac{\varepsilon - \varepsilon_{min}}{\varepsilon_0 - \varepsilon_{min}} = exp(-k_c t) \qquad (17.7)$$

where [is the average granule porosity, [$_{min}$ is the minimum achievable porosity at given process conditions, [$_0$ is the initial porosity, k_c is the granulation consolidation rate constant, and t is the wet massing time.

There is no universal rule of how wet massing time should be scaled-up. Quite often, the scientists use the recommended ranges at the scale of their operation. Rekhi et al. (1996) recommended a rule on how wet massing time should be scale-up. They postulated that the wet massing time should be adjusted based on a ratio of the impeller speeds from one scale to the other, in order to keep the same total number of impeller rotations across scales. Because they recommended that impeller speeds should be scaled up by constant tip speed, the wet massing time rule will be proportional to the granulator bowl diameters. This rule serves as a good rule of thumb, but limitations may exist when scaling between granulators that vary greatly in size. For example, if the starting scale is a mini-granulator using a 30-s wet massing time, this rule may estimate high wet massing times for commercial scale units, which can have a negative impact on the granule porosity and compactability. This approach implicitly assumes that granule growth and consolidation are predominantly limited to the regions of peak force in the granulator, and hence, the same number of rotations across scales aims to maintain a similar number of "passes" in these peak force regions. An opposite extreme would be to keep the wet massing time constant at the different scales with the underlying assumption of a narrow distribution of forces within the granulator.

Sometimes, wet massing time is used as a means to reach a desired end-point during scale-up as described above under attribute-based approach. This means that the wet massing phase is continued until a desired set of granule properties is reached. This method may ensure that some of the granule properties are matched between scales, but this may not be a desirable approach, given that an extended wet massing time is generally detrimental to most of the important granule properties (discussed previously in Section 4.1). Additionally, if the water amount is not chosen appropriately (e.g., undergranulated), an extended wet massing may not alter the granule properties to the desired end point. The wet massing times should be kept relatively short, in the range of 30 s to 2 min for small-scale and pilot-scale granulators and in the range of 2 to 5 min for commercial-scale granulators (Pandey & Badawy, 2016).

17.4.2.5 Spray-related Parameters

Liquid is added to the granulator bowl either through a tube or sprayed through a nozzle. There are several factors to be considered in terms of water addition namely:

- Liquid addition time
- Liquid addition rate or spray rate
- Nozzle properties (dictate droplet size, spray area, droplet velocity, etc.)

Some of these spray-related factors can be combined into a single parameter called dimensionless spray flux (Hapgood & Litster, 2010; Hapgood et al., 2004; Hapgood et al., 2009; Hapgood et al., 2010; Kariuki et al., 2013; Litster et al., 2001; Michaels et al., 2009; Nguyen et al., 2013). Dimensionless spray flux is a measure of the density of droplets falling on the powder bed surface, and is defined by Eq. (17.8):

$$\Psi_a = \frac{\dot{a}}{\dot{A}} = \frac{3|\dot{V}}{2|\dot{A}|d_d} \tag{17.8}$$

where Ψ_a is the dimensionless spray flux, $\dot{A}$ is the powder flux or spray area in the nucleation zone, $\dot{V}$ is the volumetric spray rate flowrate, d_d is the average droplet size, $\dot{a}$ total projected area of drops per unit time.

A very low value of Ψ_a (<<1) means low droplet density, and that the droplets will not coalesce with one another before hitting the powder surface. When the droplet comes in contact with the powder, each individual droplet will form a separate nucleus granule, and the nucleation process is considered to be in the "droplet-controlled" regime (also a function of droplet penetration time) (Hapgood et al., 2003). In this case, the initial nuclei size will be directly related to the droplet size, and therefore, can be controlled by spray nozzle, etc. On the other hand, a very high value of Ψ_a (>1) will result in the entire powder bed surface (size of the spray area) being covered by droplets that have coalesced, and the individual initial droplet size information bears little meaning. In that case, the water distribution is governed by the mechanics of the granulator (e.g., impeller speed and design, chopper speed and design) and is called the "mechanical dispersion" regime. Given that dimensionless spray flux is not a scale-dependent property, it has been a subject of several scale-up studies as a parameter to be kept constant during scale-up (Litster et al., 2004; Plank et al., 2003). It can serve as a good mechanistic basis for scale-up of spray zone conditions in order to achieve equivalent liquid distribution and nucleation across scales. It has also been shown that at low dimensionless spray flux a narrow nuclei size distribution is obtained (Litster et al., 2001). One of the main challenges with this approach would be the need to have full characterization of the spray (e.g., spray area, droplet size) (Mort, 2009).

In practice pharmaceutical high-shear wet granulation usually operates in the mechanical dispersion regime in production-scale equipment. While it may be possible to operate in the droplet-controlled regime at the small scale using an appropriately selected nozzle and spray rate, scale-up of such a process is usually not feasible from a practical perspective. If the liquid addition time is kept constant across scales, then the liquid spray rate has to increase significantly when scaling up. This means that the dimensionless spray flux would increase significantly, which would shift the nucleation regime. On the other hand, if the spray rate is kept constant, then it would lead to a significant increase in liquid addition time, which can affect the granule agglomeration and consolidation process, in addition to increasing the overall process time. A way to handle this would be to increase the number of spray nozzles used (Rambali et al., 2003), which may not always be a practical solution and would require re-engineering of the granulator bowl. Additionally, during scale-up, the batch size generally increases by volume, and it would be difficult to proportionally increase the number of spray nozzles.

An overall summary of the scale-up strategies discussed in Section 4 is provided in Fig. 17.5 (Pandey & Badawy, 2016).

17.5 Modeling techniques

The scale-up methodologies discussed so far can be considered experimental and mostly empirical in nature, with relatively weak physical relevance to the granulation process itself (Niklasson Björn et al., 2005). In a QbD paradigm there's a need to move from an empirical approach to more mechanistic models. If such a mechanistic model can be established, the scale-up of the process would be simpler, with increased fundamental and mechanistic understanding of the process and reduced experimentation. Some of the commonly used modeling approaches include regime-map based modeling, population balance modeling (PBM), DEM, PBM coupled with DEM, PBM coupled with computational fluid dynamics (CFD), PBM with compartmental model and DEM, and PBM coupled with Volume of fluid (VoF) models (Barrasso & Ramachandran, 2012; Chaudhury et al., 2014; Chaudhury et al., 2014; Darelius et al., 2005; Gantt & Gatzke, 2005; Immanuel & Doyle Iii, 2005; Kumar et al., 2013; Nakamura et al., 2013; Pandey et al., 2013; Poon et al., 2009; Rajniak et al., 2009; Ramachandran et al., 2009; Štěpánek et al., 2009). A recent review of regime-map based models for a priori design and scaling

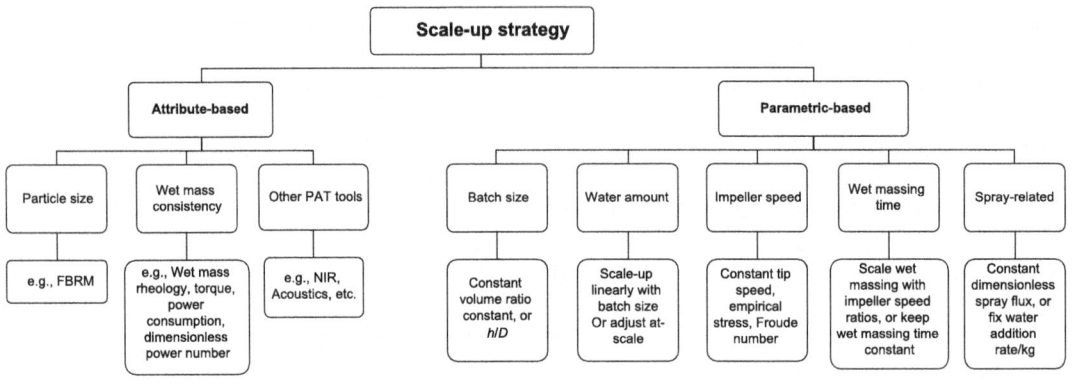

FIGURE 17.5

Scale-up strategy for high-shear wet granulation process. In parametric-based scale-up, all the parameters must be scaled-up simultaneously based on certain rules (as shown) in order to achieve success during scale-up.

was provided by Kayrak-Talay et al. (2013). These regime-map-based models are based on the first principle that models have a sound mechanistic basis. They are not fully predictive in nature but can offer mechanistic understanding and useful insights during scale-up and scale-down.

Population balance models use a number balance method around each granule size fraction where the change in the number of granules in a given size is equal to the number of granules formed minus the number of granule leaving the size fraction. Population balances are mechanistically-derived and can include terms that account for the different granulation mechanisms described in Section 2, such as nucleation, coalescence, consolidation, and attrition. Earlier models developed for wet granulation were "one-dimensional" as the coalescence kernels (frequency of collisions between granules that result in successful coalescence) were only dependent on granule size (Adetayo & Ennis, 1997; Adetayo et al., 1995). More recent population balance models are "multidimensional" with kernels which acknowledge that successful coalescences between granules is also a function of other granule properties, such as binder content and porosity, and not only size (Iveson, 2002; Poon et al., 2009). Traditionally, the value of these kernels was estimated by fitting data from granulation studies to the model, an approach which limits the utility of these models. The limitations in such cases would be that predictions can be made only for the same scale and within a certain process space spanned by the experimental design. Recently, the development of physical-model-based kernels is becoming more common, where physical models of granulation are used to define the kernels, hence increasing the scope and applicability of PBM models. However, almost always, there are some fitting parameters involved with those kernels, which makes PBM not entirely mechanistic in nature. When coupled with other techniques, such as VoF, DEM, and/or a compartment model, PBM can be more efficient, mechanistic, and useful. However, given the complex nature of granulation process, PBM still offers several advantages and has been widely used (Adetayo et al., 1995; Chaudhury et al., 2014; Darelius et al., 2006; Hounslow et al., 2001; Iveson, 2002; Liu & Litster, 2002; Pandey, et al., 2013; Poon et al., 2009; Ramachandran et al., 2009; Sanders et al., 2003; Song et al., 1997; Verkoeijen et al., 2002; Vreman et al., 2009). One of the utilities of PBM model is the ability to conduct a sensitivity analysis around some of the formulation and process parameters (e.g., viscosity, impeller speed, contact angle, and spray rate) using minimal experiments (Chaudhury et al., 2014). Such a sensitivity analysis can be useful in establishing a process and formulation design space of a drug product. PBM predictions can also be used to reduce experimentation as well once a good model can be established using some initial experiments (Chaudhury et al., 2014; Darelius et al., 2005; Pandey et al., 2013; Poon et al., 2009).

DEM is a numerical simulation technique which tracks the position of every particle within the defined geometry of the granulator and estimates the velocity and force exerted on every particle. DEM models have some advantages over PBM as simulation results contain a lot of detailed information about the particle dynamics, such as time evolution of individual particle velocities and shear experienced by the particles (Gantt & Gatzke, 2005; Kuo et al., 2004; Nakamura et al., 2013; Remy et al., 2009). However, DEM models face their own set of challenges; they are time-intensive, especially when dealing with multicomponent pharmaceutical powders that are not "hard spheres." Additionally, incorporation of water into the system adds another level of sophistication for DEM. As a result, some literature reports used DEM to study a limited aspect of wet granulation, rather than applying it to for complete prediction of the wet granulation process. Gantt et al. (2006) used a dynamic DEM approach to develop kernels for a PBM model of wet granulation. They estimated the frequency and size dependence of granule collisions in the granulator using DEM simulation. They also used DEM to determine the probability of successful coalescence between granules as a function of granule size, binder content, and porosity, which was used to build the PBM coalescence kernels (Gantt et al., 2006). In another study, DEM was used to understand differences in flow patterns and particle velocity distribution between different granulators and scales, which are useful from scale-up perspective (Nakamura et al., 2013). They also developed a scale-up method using DEM with an aim to maintain dynamic and kinematic similarities in particle behavior across scales (2–112 L). They confirmed by DEM simulations that when the tip speed is kept constant across scales, kinematic similarity is achieved. However, to achieve dynamic similarity, the wet massing time needed to be increased upon scale-up, and this increase was suggested based on DEM simulations. This scale-up rule is similar in approach as proposed by Rekhi et al. (1996), where the wet massing time was suggested to be scaled-up by the ratio of impeller speeds (or impeller diameter). DEM simulations provide a mechanistic reasoning behind such a rule and also offer a methodology for wet massing scale-up (matching cumulative particle collision energy per unit time).

A model's ability to make accurate and useful predictions relies heavily on some experimentally determined parameters. For high-shear granulation, examples of such parameters include dynamic viscosity, rheology, particle porosity, particle shape, particle Young's modulus, coefficient of restitution, particle-wall friction, particle-particle friction, and many others. For a multicomponent and multidimensional system such as a pharmaceutical granulation, these parameters are not easily determined, especially in a dynamic manner, and are often responsible for the shortcomings of such models. However, modeling techniques offer invaluable insight into the process and, as discussed, can provide good fundamental basis behind scale-up principles (Pandey & Badawy, 2016).

17.6 Summary

High-shear wet granulation is a complex process which involves multiple and simultaneous rate processes: nucleation, coalescence, consolidation, and attrition. Hence, outcome of wet granulation is determined by multiple CMAs and CPPs. In addition, flow patterns in the batch granulator are also complex and vary significantly across scales. Successful scale-up of the high-shear wet granulation process can therefore be a daunting task. Two main approaches to wet granulation scale-up have reported in the literature and utilized in the pharmaceutical industry. The first is an attribute-based approach where parameter values are adjusted at the larger scale based on an in-process measurement such as particle size or granulation viscosity. The other is a parametric approach where scale-up calculations are used to determine parameter values at the larger scale. Both approaches have their challenges, and a hybrid of the two approaches is sometimes used to maximize the probability of successful scale-up. Mathematical modeling of the wet granulation process is becoming more popular and has demonstrated its value in overcoming the challenges associated with wet granulation process development and scale-up.

References

Adetayo, A. A., & Ennis, B. J. (1997). Unifying approach to modeling granule coalescence mechanisms. *AICHE Journal, 43*(4), 927–934. https://doi.org/10.1002/aic.690430408.

Adetayo, A. A., Litster, J. D., Pratsinis, S. E., & Ennis, B. J. (1995). Population balance modelling of drum granulation of materials with wide size distribution. *Powder Technology, 82*(1), 37–49. https://doi.org/10.1016/0032-5910(94)02896-V.

Agrawal, A. M., & Pandey, P. (2015). Scale up of pan coating process using quality by design principles. *Journal of Pharmaceutical Sciences, 104*(11), 3589–3611. https://doi.org/10.1002/jps.24582.

Aikawa, S., Fujita, N., Myojo, H., Hayashi, T., & Tanino, T. (2008). Scale-up studies on high shear wet granulation process from mini-scale to commercial scale. *Chemical and Pharmaceutical Bulletin, 56*(10), 1431–1435. https://doi.org/10.1248/cpb.56.1431.

Am Ende, M.T., Bharadwaj, R., García-Muñoz, S., Ketterhagen, W., Prpich, A., & Doshi, P. (2010). Process modeling techniques and applications for solid oral drug products. In *Chemical engineering in the pharmaceutical industry*. (pp. 633–662). John Wiley & Sons, Inc.

Ameye, D., Keleb, E., Vervaet, C., Remon, J. P., Adams, E., & Massart, D. L. (2002). Scaling-up of a lactose wet granulation process in mi-pro high shear mixers. *European Journal of Pharmaceutical Sciences, 17*(4-5), 247–251. https://doi.org/10.1016/S0928-0987(02)00218-X.

Arp, Z., Smith, B., Dycus, E., & O'Grady, D. (2011). Optimization of a high shear wet granulation process using focused beam reflectance measurement and particle vision microscope technologies. *Journal of Pharmaceutical Sciences, 100*(8), 3431–3440. https://doi.org/10.1002/jps.22556.

Ax, K., Feise, H., Sochon, R., Hounslow, M., & Salman, A. (2008). Influence of liquid binder dispersion on agglomeration in an intensive mixer. *Powder Technology, 179*(3), 190–194. https://doi.org/10.1016/j.powtec.2007.06.010.

Badawy, S. I., & Hussain, M. A. (2004). Effect of starting material particle size on its agglomeration behavior in high shear wet granulation. *AAPS PharmSciTech, 5*(3). https://doi.org/10.1208/pt050338.

Badawy, S. I., Lee, T. J., & Menning, M. M. (2000). Effect of drug substance particle size on the characteristics of granulation manufactured in a high-shear mixer. *AAPS PharmSciTech, 1*(4). https://doi.org/10.1208/pt010433.

Badawy, S. I., Menning, M. M., Gorko, M. A., & Gilbert, D. L. (2000). Effect of process parameters on compressibility of granulation manufactured in a high-shear mixer. *International Journal of Pharmaceutics, 198*(1), 51–61. http://www.ncbi.nlm.nih.gov/pubmed/10722950.

References

Badawy, S. I., Narang, A. S., LaMarche, K., Subramanian, G., & Varia, S. A. (2012). Mechanistic basis for the effects of process parameters on quality attributes in high shear wet granulation. *International Journal of Pharmaceutics, 439*(1–2), 324–333. https://doi.org/10.1016/j.ijpharm.2012.09.011.

Barrasso, D., & Ramachandran, R. (2012). A comparison of model order reduction techniques for a four-dimensional population balance model describing multi-component wet granulation processes. *Chemical Engineering Science, 80*, 380–392. https://doi.org/10.1016/j.ces.2012.06.039.

Briens, L., Daniher, D., & Tallevi, A. (2007). Monitoring high-shear granulation using sound and vibration measurements. *International Journal of Pharmaceutics, 331*(1), 54–60. https://doi.org/10.1016/j.ijpharm.2006.09.012.

Campbell, G. A., Clancy, D. J., Zhang, J. X., Gupta, M. K., & Oh, C. K. (2011). Closing the gap in series scale up of high shear wet granulation process using impeller power and blade design. *Powder Technology, 205*(1–3), 184–192. https://doi.org/10.1016/j.powtec.2010.09.009.

Chaudhury, A., Barrasso, D., Pandey, P., Wu, H., & Ramachandran, R. (2014a). Population balance model development, validation, and prediction of CQAs of a high-shear wet granulation process: Towards QbD in drug product pharmaceutical manufacturing. *Journal of Pharmaceutical Innovation, 9*(1), 53–64. https://doi.org/10.1007/s12247-014-9172-7.

Chaudhury, A., Wu, H., Khan, M., & Ramachandran, R. (2014b). A mechanistic population balance model for granulation processes: Effect of process and formulation parameters. *Chemical Engineering Science, 107*(0), 76–92. https://doi.org/10.1016/j.ces.2013.11.031.

Corvari, V., Fry, W., Seibert, W., & Augsburger, L. (1992). Instrumentation of a high-shear mixer: Evaluation and comparison of a new capacitive sensor, a watt meter, and a strain-gage torque sensor for wet granulation monitoring. *Pharmaceutical Research, 9*(12), 1525–1533. https://doi.org/10.1023/A:1015843820526.

D'alonzo, G. D., O'connor, R. E., & Schwartz, J. B. (1990). Effect of binder concentration and method of addition on granule growth in a high intensity mixer. *Drug Development and Industrial Pharmacy, 16*(12), 1931–1944. https://doi.org/10.3109/03639049009028348.

Darelius, A., Brage, H., Rasmuson, A., Niklasson Björn, I., & Folestad, S. (2006). A volume-based multi-dimensional population balance approach for modelling high shear granulation. *Chemical Engineering Science, 61*(8), 2482–2493. https://doi.org/10.1016/j.ces.2005.11.016.

Darelius, A., Rasmuson, A., Björn, I. N., & Folestad, S. (2005). High shear wet granulation modelling—A mechanistic approach using population balances. *Powder Technology, 160*(3), 209–218. https://doi.org/10.1016/j.powtec.2005.08.036.

Debnath, S., & Suryanarayanan, R. (2004). Influence of processing-induced phase transformations on the dissolution of theophylline tablets. *AAPS PharmSciTech, 5*(1), 39–49. https://doi.org/10.1007/bf02830576.

Faure, A., Grimsey, I. M., Rowe, R. C., York, P., & Cliff, M. J. (1999). Applicability of a scale-up methodology for wet granulation processes in Collette Gral high shear mixer-granulators. *European Journal of Pharmaceutical Sciences, 8*(2), 85–93.

Faure, A., York, P., & Rowe, R. C. (2001). Process control and scale-up of pharmaceutical wet granulation processes: A review. *European Journal of Pharmaceutics and Biopharmaceutics, 52*(3), 269–277. https://doi.org/10.1016/s0939-6411(01)00184-9.

Findlay, W. P., Peck, G. R., & Morris, K. R. (2005). Determination of fluidized bed granulation end point using near-infrared spectroscopy and phenomenological analysis. *Journal of Pharmaceutical Sciences, 94*(3), 604–612. https://doi.org/10.1002/jps.20276.

Frake, P., Greenhalgh, D., Grierson, S. M., Hempenstall, J. M., & Rudd, D. R. (1997). Process control and end-point determination of a fluid bed granulation by application of near infra-red spectroscopy. *International Journal of Pharmaceutics, 151*(1), 75–80. https://doi.org/10.1016/S0378-5173(97)04894-1.

Gantt, J. A., Cameron, I. T., Litster, J. D., & Gatzke, E. P. (2006). Determination of coalescence kernels for high-shear granulation using DEM simulations. *Powder Technology, 170*(2), 53–63. https://doi.org/10.1016/j.powtec.2006.08.002.

Gantt, J. A., & Gatzke, E. P. (2005). High-shear granulation modeling using a discrete element simulation approach. *Powder Technology, 156*(2–3), 195–212. https://doi.org/10.1016/j.powtec.2005.04.012.

Ghanta, S. R., Srinivas, R., & Rhodes, C. T. (1984). Use of mixer-torque measurements as an aid to optimizing wet granulation process. *Drug Development and Industrial Pharmacy, 10*(2), 305–311. https://doi.org/10.3109/03639048409064652.

Hancock, B. C., York, P., & Rowe, R. C. (1982). Characterization of wet masses using a mixer torque rheometer. 2. Mixing kinetics. *International Journal of Pharmaceutics, 83*(1–3), 147–153. https://doi.org/10.1016/0378-5173(82)90017-5.

Hancock, B. C., York, P., & Rowe, R. C. (1994). An assessment of substrate-binder interactions in model wet masses. 1. Mixer torque rheometry. *International Journal of Pharmaceutics, 102*(1–3), 167–176. https://doi.org/10.1016/0378-5173(94)90052-3.

Hancock, B. C., York, P., Rowe, R. C., & Parker, M. D. (1991). Characterization of wet masses using a mixer torque rheometer. 1. Effect of instrument geometry. *International Journal of Pharmaceutics, 76*(3), 239–245. https://doi.org/10.1016/0378-5173(91)90276-T.

Hapgood, K. P., Amelia, R., Zaman, M. B., Merrett, B. K., & Leslie, P. (2010). Improving liquid distribution by reducing dimensionless spray flux in wet granulation—A pharmaceutical manufacturing case study. *Chemical Engineering Journal, 164*(2–3), 340–349. https://doi.org/10.1016/j.cej.2010.05.007.

Hapgood, K. P., & Litster, J. D. (2010). Wet granulation processes. *Chemical engineering in the pharmaceutical industry* (pp. 757–780). Wiley & Sons, Inc. https://doi.org/10.1002/9780470882221.ch39.

Hapgood, K. P., Litster, J. D., Biggs, S. R., & Howes, T. (2002). Drop penetration into porous powder beds. *Journal of Colloid and Interface Science, 253*(2), 353–366. https://doi.org/10.1006/jcis.2002.8527.

Hapgood, K. P., Litster, J. D., & Smith, R. (2003). Nucleation regime map for liquid bound granules. *AIChE Journal, 49*(2), 350–361. https://doi.org/10.1002/aic.690490207.

Hapgood, K. P., Litster, J. D., White, E. T., Mort, P. R., & Jones, D. G. (2004). Dimensionless spray flux in wet granulation: Monte-Carlo simulations and experimental validation. *Powder Technology, 141*(1–2), 20–30. https://doi.org/10.1016/j.powtec.2004.02.005.

Hapgood, K. P., Tan, M. X. L., & Chow, D. W. Y. (2009). A method to predict nuclei size distributions for use in models of wet granulation. *Advanced Powder Technology, 20*(4), 293–297. https://doi.org/10.1016/j.apt.2008.09.004.

Hassanpour, A., Antony, S. J., & Ghadiri, M. (2007). Modeling of agglomerate behavior under shear deformation: Effect of velocity field of a high shear mixer granulator on the structure of agglomerates. *Advanced Powder Technology, 18*(6), 803–812.

Hassanpour, A., Kwan, C. C., Ng, B. H., Rahmanian, N., Ding, Y. L., Antony, S. J., Jia, X. D., & Ghadiri, M. (2009). Effect of granulation scale-up on the strength of granules. *Powder Technology, 189*(2), 304–312. https://doi.org/10.1016/j.powtec.2008.04.023.

He, X., Lunday, K. A., Li, L. C., & Sacchetti, M. J. (2008). Formulation development and process scale up of a high shear wet granulation formulation containing a poorly wettable drug. *Journal of Pharmaceutical Sciences, 97*(12), 5274–5289. https://doi.org/10.1002/jps.21410.

He, X., Secreast, P. J., & Amidon, G. E. (2007). Mechanistic study of the effect of roller compaction and lubricant on tablet mechanical strength. *Journal of Pharmaceutical Sciences, 96*(5), 1342–1355. https://doi.org/10.1002/jps.20938.

Holm, P. (1987). Effect of impeller and chopper design on granulation in a high speed mixer. *Drug Development and Industrial Pharmacy, 13*(9–11), 1675–1701. https://doi.org/10.3109/03639048709068687.

Holm, P., Schaefer, T., & Kristensen, H. (1985a). Granulation in high-speed mixers. Part V. Power consumption and temperature changes during granulation. *Powder Technology, 43*(3), 213–223.

Holm, P., Schaefer, T., & Kristensen, H. G. (1985b). Granulation in high-speed mixers. Part VI. Effects of process conditions on power consumption and granule growth. *Powder Technology, 43*(3), 225–233. https://doi.org/10.1016/0032-5910(85)80003-6.

Horsthuis, G. J. B., van Laarhoven, J. A. H., van Rooij, R. C. B. M., & Vromans, H. (1993). Studies on upscaling parameters of the Gral high shear granulation process. *International Journal of Pharmaceutics, 92*(1–3), 143–150. https://doi.org/10.1016/0378-5173(93)90273-I.

Hounslow, M. J., Pearson, J. M. K., & Instone, T. (2001). Tracer studies of high-shear granulation. II. Population balance modeling. *AICHE Journal, 47*(9), 1984–1999. https://doi.org/10.1002/aic.690470910.

Huang, J., Kaul, G., Utz, J., Hernandez, P., Wong, V., Bradley, D., Nagi, A., & O'Grady, D. (2010). A PAT approach to improve process understanding of high shear wet granulation through in-line particle measurement using FBRM C35. *Journal of Pharmaceutical Sciences, 99*(7), 3205–3212. https://doi.org/10.1002/jps.22089.

Immanuel, C. D., & Doyle Iii, F. J. (2005). Solution technique for a multi-dimensional population balance model describing granulation processes. *Powder Technology, 156*(2–3), 213–225. https://doi.org/10.1016/j.powtec.2005.04.013.

Iskandarani, B., Shiromani, P. K., & Clair, J. H. (2001). Scale-up feasibility in high-shear mixers: Determination through statistical procedures. *Drug Development and Industrial Pharmacy, 27*(7), 651–657.

Iveson, S. M. (2002). Limitations of one-dimensional population balance models of wet granulation processes. *Powder Technology, 124*(3), 219–229. https://doi.org/10.1016/S0032-5910(02)00026-8.

Iveson, S. M., & Litster, J. D. (1998a). Fundamental studies of granule consolidation. Part 2. Quantifying the effects of particle and binder properties. *Powder Technology, 99*(3), 243–250. https://doi.org/10.1016/S0032-5910(98)00116–00118.

Iveson, S. M., & Litster, J. D. (1998b). Growth regime map for liquid-bound granules. *AICHE Journal, 44*(7), 1510–1518. https://doi.org/10.1002/aic.690440705.

Iveson, S. M., Litster, J. D., & Ennis, B. J. (1996). Fundamental studies of granule consolidation. Part 1. Effects of binder content and binder viscosity. *Powder Technology, 88*(1), 15–20. https://doi.org/10.1016/0032-5910(96)03096–03093.

Iveson, S. M., Litster, J. D., Hapgood, K., & Ennis, B. J. (2001). Nucleation, growth and breakage phenomena in agitated wet granulation processes: A review. *Powder Technology, 117*(1–2), 3–39. https://doi.org/10.1016/S0032-5910(01)00313-8.

Kariuki, W. I. J., Freireich, B., Smith, R. M., Rhodes, M., & Hapgood, K. P. (2013). Distribution nucleation: Quantifying liquid distribution on the particle surface using the dimensionless particle coating number. *Chemical Engineering Science, 92*(0), 134–145. https://doi.org/10.1016/j.ces.2013.01.010.

Kayrak-Talay, D., Dale, S., Wassgren, C., & Litster, J. (2013). Quality by design for wet granulation in pharmaceutical processing: Assessing models for a priori design and scaling. *Powder Technology, 240*(0), 7–18. https://doi.org/10.1016/j.powtec.2012.07.013.

Kayrak-Talay, D., & Litster, J. D. (2011). A priori performance prediction in pharmaceutical wet granulation: Testing the applicability of the nucleation regime map to a formulation with a broad size distribution and dry binder addition. *International Journal of Pharmaceutics, 418*(2), 254–264. https://doi.org/10.1016/j.ijpharm.2011.04.019.

Knight, P. C., Seville, J. P. K., Wellm, A. B., & Instone, T. (2001). Prediction of impeller torque in high shear powder mixers. *Chemical Engineering Science, 56*(15), 4457–4471. https://doi.org/10.1016/S0009-2509(01)00114-2.

Kristensen, H. G. (1996). Particle agglomeration in high shear mixers. *Powder Technology, 88*(3), 197–202. https://doi.org/10.1016/S0032-5910(96)03123-3.

Kristensen, H. G., Holm, P., & Schaefer, T. (1985a). Mechanical properties of moist agglomerates in relation to granulation mechanisms. Part I. Deformability of moist, densified agglomerates. *Powder Technology, 44*(3), 227–237. https://doi.org/10.1016/0032-5910(85)85004-X.

Kristensen, H. G., Holm, P., & Schaefer, T. (1985b). Mechanical properties of moist agglomerates in relation to granulation mechanisms. Part II. Effects of particle size distribution. *Powder Technology, 44*(3), 239–247. https://doi.org/10.1016/0032-5910(85)85005-1.

Kristensen, H. G., & Schaefer, T. (1987). Granulation: A review on pharmaceutical wet-granulation. *Drug Development and Industrial Pharmacy, 13*(4–5), 803–872. https://doi.org/10.3109/03639048709105217.

Kumar, A., Gernaey, K. V., Beer, T. D., & Nopens, I. (2013). Model-based analysis of high shear wet granulation from batch to continuous processes in pharmaceutical production—A critical review. *European Journal of Pharmaceutics and Biopharmaceutics, 85*(3, Part B), 814–832. https://doi.org/10.1016/j.ejpb.2013.09.013.

Kuo, H. P., Knight, P. C., Parker, D. J., Adams, M. J., & Seville, J. P. K. (2004). Discrete element simulations of a high-shear mixer. *Advanced Powder Technology, 15*(3), 297–309. https://doi.org/10.1163/156855204774150109.

Landín, M., Rowe, R. C., & York, P. (1995). Characterization of wet powder masses with a mixer torque rheometer. 3. Nonlinear effects of shaft speed and sample weight. *Journal of Pharmaceutical Sciences, 84*(5), 557–560. https://doi.org/10.1002/jps.2600840508.

Landin, M., York, P., Cliff, M. J., & Rowe, R. C. (1999). Scaleup of a pharmaceutical granulation in planetary mixers. *Pharmaceutical Development and Technology, 4*(2), 145–150. https://doi.org/10.1081/pdt-100101349.

Landin, M., York, P., Cliff, M. J., Rowe, R. C., & Wigmore, A. J. (1996). The effect of batch size on scale-up of a pharmaceutical granulation in a fixed bowl mixer granulator. *International Journal of Pharmaceutics, 134*(1–2), 243–246. https://doi.org/10.1016/0378-5173(96)04461-4.

Leuenberger, H. (1982). Granulation, new techniques. *Pharmaceutica Acta Helvetiae, 57*(3), 72–82.

Li, W., Worosila, G. D., Wang, W., & Mascaro, T. (2005). Determination of polymorph conversion of an active pharmaceutical ingredient in wet granulation using NIR calibration models generated from the premix blends. *Journal of Pharmaceutical Sciences, 94*(12), 2800–2806. https://doi.org/10.1002/jps.20501.

Lindberg, N.-O., Leander, L., & Reenstierna, B. (1982). Instrumentation of a Kenwood major domestic-type mixer for studies of granulation. *Drug Development and Industrial Pharmacy, 8*(5), 775–782. https://doi.org/10.3109/03639048209042702.

Litster, J., Ennis, B., & Lian, L. (2004). *The science and engineering of granulation processes*. Springer Science Business Media.

Litster, J. D. (2003). Scaleup of wet granulation processes: Science not art. *Powder Technology, 130*(1–3), 35–40. https://doi.org/10.1016/S0032-5910(02)00222-X.

Litster, J. D., Hapgood, K. P., Michaels, J. N., Sims, A., Roberts, M., Kameneni, S. K., & Hsu, T. (2001). Liquid distribution in wet granulation: Dimensionless spray flux. *Powder Technology, 114*(1–3), 32–39. https://doi.org/10.1016/S0032-5910(00)00259-X.

Liu, L. X., & Litster, J. D. (2002). Population balance modelling of granulation with a physically based coalescence kernel. *Chemical Engineering Science, 57*(12), 2183–2191. https://doi.org/10.1016/S0009-2509(02)00110-0.

Luukkonen, P., Fransson, M., Björn, I. N., Hautala, J., Lagerholm, B., & Folestad, S. (2008). Real-time assessment of granule and tablet properties using in-line data from a high-shear granulation process. *Journal of Pharmaceutical Sciences, 97*(2), 950–959. https://doi.org/10.1002/jps.20998.

Luukkonen, P., Schæfer, T., Hellen, L., Juppo, A. M., & Yliruusi, J. (1999). Rheological characterization of microcrystalline cellulose and silicified microcrystalline cellulose wet masses using a mixer torque rheometer. *International Journal of Pharmaceutics, 188*(2), 181–192. https://doi.org/10.1016/S0378-5173(99)00219-7.

Marston, J. O., Sprittles, J. E., Zhu, Y., Li, E. Q., Vakarelski, I. U., & Thoroddsen, S. T. (2013). Drop spreading and penetration into pre-wetted powders. *Powder Technology, 239*, 128–136. https://doi.org/10.1016/j.powtec.2013.01.062.

Michaels, J. N., Farber, L., Wong, G. S., Hapgood, K., Heidel, S. J., Farabaugh, J., Chou, J. H., & Tardos, G. I. (2009). Steady states in granulation of pharmaceutical powders with application to scale-up. *Powder Technology, 189*(2), 295–303. https://doi.org/10.1016/j.powtec.2008.04.028.

Mort, P. R. (2009). Scale-up and control of binder agglomeration processes—Flow and stress fields. *Powder Technology, 189*(2), 313–317. https://doi.org/10.1016/j.powtec.2008.04.022.

Nakamura, H., Fujii, H., & Watano, S. (2013). Scale-up of high shear mixer-granulator based on discrete element analysis. *Powder Technology, 236*(0), 149–156. https://doi.org/10.1016/j.powtec.2012.03.009.

Nguyen, T. H., Morton, D. A. V., & Hapgood, K. P. (2013). Application of the unified compaction curve to link wet granulation and tablet compaction behaviour. *Powder Technology, 240*(0), 103–115. https://doi.org/10.1016/j.powtec.2012.07.001.

Niklasson Björn, I., Jansson, A., Karlsson, M., Folestad, S., & Rasmuson, A. (2005). Empirical to mechanistic modelling in high shear granulation. *Chemical Engineering Science, 60*(14), 3795–3803. https://doi.org/10.1016/j.ces.2005.02.012.

Ohike, A., Ashihara, K., & Ibuki, R. (1999). Granulation monitoring by fast Fourier transform technique. *Chemical & Pharmaceutical Bulletin (Tokyo), 47*(12), 1734–1739.

Oulahna, D., Cordier, F., Galet, L., & Dodds, J. A. (2003). Wet granulation: The effect of shear on granule properties. *Powder Technology, 130*(1–3), 238–246. https://doi.org/10.1016/S0032-5910(02)00272-3.

Palzer, S. (2011). Agglomeration of pharmaceutical, detergent, chemical and food powders—Similarities and differences of materials and processes. *Powder Technology, 206*(1–2), 2–17. https://doi.org/10.1016/j.powtec.2010.05.006.

Pandey, P., & Badawy, S. (2016). A quality by design approach to scale-up of high-shear wet granulation process. *Drug Development and Industrial Pharmacy, 42*(2), 175–189. https://doi.org/10.3109/03639045.2015.1100199.

Pandey, P., Hamey, R., Bindra, D. S., Huang, Z., Mathias, N., Eley, T., et al. (2014). From bench to humans: Formulation development of a poorly water soluble drug to mitigate food effect. *AAPS PharmSciTech, 15*(2), 407–416. https://doi.org/10.1208/s12249-013-0069-4.

Pandey, P., Sinko, P. D., Bindra, D. S., Hamey, R., Gour, S., & Vema-Varapu, C. (2013a). Processing challenges with solid dosage formulations containing vitamin E TPGS. *Pharmaceutical Development and Technology, 18*(1), 296–304. https://doi.org/10.3109/10837450.2012.737807.

Pandey, P., Tao, J., Chaudhury, A., Ramachandran, R., Gao, J. Z., & Bindra, D. S. (2013b). A combined experimental and modeling approach to study the effects of high-shear wet granulation process parameters on granule characteristics. *Pharmaceutical Development and Technology, 18*(1), 210–224. https://doi.org/10.3109/10837450.2012.700933.

Pandey, P., Turton, R., Joshi, N., Hammerman, E., & Ergun, J. (2006). Scale-up of a pan-coating process. *AAPS PharmSciTech, 7*(4), 102. http://www.ncbi.nlm.nih.gov/pubmed/17285748.

Parker, M., Rowe, R., & Upjohn, N. (1990). Mixer torque rheometry: A method for quantifying the consistency of wet granulations. *Pharmaceutical Technology International, 2*(8), 50–64.

Plank, R., Diehl, B., Grinstead, H., & Zega, J. (2003). Quantifying liquid coverage and powder flux in high-shear granulators. *Powder Technology, 134*(3), 223–234. https://doi.org/10.1016/s0032-5910(03)00171-2.

Poon, J. M. H., Ramachandran, R., Sanders, C. F. W., Glaser, T., Immanuel, C. D., Doyle, F. J., Litster, J. D., Stepanek, F., Wang, F. Y., & Cameron, I. T. (2009). Experimental validation studies on a multi-dimensional and multi-scale population balance model of batch granulation. *Chemical Engineering Science, 64*(4), 775–786. https://doi.org/10.1016/j.ces.2008.08.037.

Rahmanian, N., Ghadiri, M., & Ding, Y. (2008a). Effect of scale of operation on granule strength in high shear granulators. *Chemical Engineering Science, 63*(4), 915–923. https://doi.org/10.1016/j.ces.2007.10.027.

Rahmanian, N., Ghadiri, M., Jia, X., & Stepanek, F. (2009). Characterisation of granule structure and strength made in a high shear granulator. *Powder Technology, 192*(2), 184–194. https://doi.org/10.1016/j.powtec.2008.12.016.

Rahmanian, N., Ng, B., Hassanpour, A., Ghadiri, M., Ding, Y., Jia, X., et al. (2008b). Scale-up of high shear mixer granulators. *Kona: Powder Science and Technology in Japan, 26*, 190–204.

Rajniak, P., Stepanek, F., Dhanasekharan, K., Fan, R., Mancinelli, C., & Chern, R. T. (2009). A combined experimental and computational study of wet granulation in a Wurster fluid bed granulator. *Powder Technology, 189*(2), 190–201. https://doi.org/10.1016/j.powtec.2008.04.027.

Ramachandran, R., Immanuel, C. D., Stepanek, F., Litster, J. D., & Doyle Iii, F. J. (2009). A mechanistic model for breakage in population balances of granulation: Theoretical kernel development and experimental validation. *Chemical Engineering Research and Design, 87*(4), 598–614. https://doi.org/10.1016/j.cherd.2008.11.007.

Rambali, B., Baert, L., & Massart, D. L. (2003). Scaling up of the fluidized bed granulation process. *International Journal of Pharmaceutics, 252*(1–2), 197–206.

Rantanen, J., Wikström, H., Turner, R., & Taylor, L. S. (2004). Use of in-line near-infrared spectroscopy in combination with chemometrics for improved understanding of pharmaceutical processes. *Analytical Chemistry, 77*(2), 556–563. https://doi.org/10.1021/ac048842u.

Rekhi, G., Caricofe, R., Parikh, D., & Augsburger, L. (1996). A new approach to scale-up of a high-shear granulation process. *Pharmaceutical Technology, 20*(10), 1–10.

Remy, B., Khinast, J. G., & Glasser, B. J. (2009). Discrete element simulation of free flowing grains in a four-bladed mixer. *AICHE Journal, 55*(8), 2035–2048. https://doi.org/10.1002/aic.11876.

Ritala, M., Holm, P., Schaefer, T., & Kristensen, H. G. (1988). Influence of liquid bonding strength on power consumption during granulation in a high shear mixer. *Drug Development and Industrial Pharmacy, 14*(8), 1041–1060. https://doi.org/10.3109/03639048809151919.

Sakr, W. F., Ibrahim, M. A., Alanazi, F. K., & Sakr, A. A. (2012). Upgrading wet granulation monitoring from hand squeeze test to mixing torque rheometry. *Saudi Pharmaceutical Journal, 20*(1), 9–19. https://doi.org/10.1016/j.jsps.2011.04.007.

Sanders, C. F. W., Willemse, A. W., Salman, A. D., & Hounslow, M. J. (2003). Development of a predictive high-shear granulation model. *Powder Technology, 138*(1), 18–24. https://doi.org/10.1016/j.powtec.2003.08.046.

Sato, Y., Okamoto, T., & Watano, S. (2005). Scale-up of high shear granulation based on agitation power. *Chemical & Pharmaceutical Bulletin (Tokyo), 53*(12), 1547–1550.

Schaefer, T., Bak, H., Jaegerskou, A., Kristensen, A., Svensson, J., Holm, P., et al. (1986). Granulation in different types of high speed mixers. I. Effects of process variables and up-scaling. *Pharmazeutische Industrie, 48*(9), 1083–1089.

Schaefer, T., Bak, H., Jaegerskou, A., Kristensen, A., Svensson, J., Holm, P., et al. (1987). Granulation in different types of high speed mixers. II. Comparisons between mixers. *Pharmazeutische Industrie, 49*(3), 297–304.

Shi, L., Feng, Y., & Sun, C. C. (2011). Massing in high shear wet granulation can simultaneously improve powder flow and deteriorate powder compaction: A double-edged sword. *European Journal of Pharmaceutical Sciences, 43*(1–2), 50–56. https://doi.org/10.1016/j.ejps.2011.03.009.

Sirois, P. J., & Craig, G. D. (2000). Scaleup of a high-shear granulation process using a normalized impeller work parameter. *Pharmaceutical Development and Technology, 5*(3), 365–374. https://doi.org/10.1081/pdt-100100552.

Song, M., Steiff, A., & Weinspach, P. M. (1997). A very effective new method to solve the population balance equation with particle-size growth. *Chemical Engineering Science, 52*(20), 3493–3498. https://doi.org/10.1016/S0009-2509(97)00152-8.

Sprockel, O. L., & Stamato, H. J. (2010). Design and scale-up of dry granulation processes. *Chemical engineering in the pharmaceutical industry* (pp. 727–755). John Wiley & Sons, Inc. https://doi.org/10.1002/9780470882221ch38.

Štěpánek, F., Rajniak, P., Mancinelli, C., Chern, R. T., & Ramachandran, R. (2009). Distribution and accessibility of binder in wet granules. *Powder Technology, 189*(2), 376–384. https://doi.org/10.1016/j.powtec.2008.04.015.

Štěpánek, F. (2004). Computer-aided product design: Granule dissolution. *Chemical Engineering Research and Design, 82*(11), 1458–1466. https://doi.org/10.1205/cerd.82.11.1458.52035.

Tao, J., Pandey, P., Bindra, D. S., Gao, J. Z., & Narang, A. S. (2015). Evaluating scale-up rules of a high-shear wet granulation process. *Journal of Pharmaceutical Sciences, 104*(7), 2323–2333. https://doi.org/10.1002/jps.24504.

Tardos, G. I., Hapgood, K. P., Ipadeola, O. O., & Michaels, J. N. (2004). Stress measurements in high-shear granulators using calibrated "test" particles: Application to scale-up. *Powder Technology, 140*(3), 217–227. https://doi.org/10.1016/j.powtec.2004.01.015.

Tardos, G. I., Khan, M. I., & Mort, P. R. (1997). Critical parameters and limiting conditions in binder granulation of fine powders. *Powder Technology, 94*(3), 245–258. https://doi.org/10.1016/S0032-5910(97)03321-4.

Tok, A., Goh, X., Ng, W., & Tan, R. H. (2008). Monitoring granulation rate processes using three PAT tools in a pilot-scale fluidized bed. *AAPS PharmSciTech, 9*(4), 1083–1091. https://doi.org/10.1208/s12249-008-9145-6.

Vemavarapu, C., Surapaneni, M., Hussain, M., & Badawy, S. (2009). Role of drug substance material properties in the processibility and performance of a wet granulated product. *International Journal of Pharmaceutics, 374*(1–2), 96–105. https://doi.org/10.1016/j.ijpharm.2009.03.014.

Verkoeijen, D., Pouw, G. A., Meesters, G. M. H., & Scarlett, B. (2002). Population balances for particulate processes—A volume approach. *Chemical Engineering Science, 57*(12), 2287–2303. https://doi.org/10.1016/S0009-2509(02)00118-5.

Vreman, A. W., van Lare, C. E., & Hounslow, M. J. (2009). A basic population balance model for fluid bed spray granulation. *Chemical Engineering Science, 64*(21), 4389–4398. https://doi.org/10.1016/j.ces.2009.07.010.

Wang, S., Ye, G., Heng, P. W., & Ma, M. (2008). Investigation of high shear wet granulation processes using different parameters and formulations. *Chemical & Pharmaceutical Bulletin (Tokyo), 56*(1), 22–27.

Watano, S. (2001). Direct control of wet granulation processes by image processing system. *Powder Technology, 117*(1–2), 163–172. https://doi.org/10.1016/S0032-5910(01)00322-9.

Watano, S., Okamoto, T., Sato, Y., & Osako, Y. (2005). Scale-up of high shear granulation based on the internal stress measurement. *Chemical & Pharmaceutical Bulletin (Tokyo), 53*(4), 351–354.

Wehrle, P., Nobelis, P., Cuine, A., & Stamm, A. (1993). Scaling-up of wet granulation a statistical methodology. *Drug Development and Industrial Pharmacy, 19*(16), 1983–1997. https://doi.org/10.3109/03639049309069336.

Whitaker, M., Baker, G. R., Westrup, J., Goulding, P. A., Rudd, D. R., Belchamber, R. M., & Collins, M. P. (2000). Application of acoustic emission to the monitoring and end point determination of a high shear granulation process. *International Journal of Pharmaceutics, 205*(1–2), 79–91. https://doi.org/10.1016/S0378-5173(00)00479-8.

Wikström, H., Carroll, W., & Taylor, L. (2008). Manipulating theophylline monohydrate formation during high-shear wet granulation through improved understanding of the role of pharmaceutical excipients. *Pharmaceutical Research, 25*(4), 923–935. https://doi.org/10.1007/s11095-007-9450-x.

Wikstrom, H., Marsac, P. J., & Taylor, L. S. (2005). In-line monitoring of hydrate formation during wet granulation using Raman spectroscopy. *Journal of Pharmaceutical Sciences, 94*(1), 209–219. https://doi.org/10.1002/jps.20241.

CHAPTER 18

Integrated application of quality-by-design principles to drug product and its control strategy development

Sherif I.F. Badawy[e], Ajit S. Narang[d], Keirnan R LaMarche[c], Ganeshkumar A. Subramanian[b], Sailesh A. Varia[e], Judy Lin[a], Tim Stevens[e] and Pankaj A. Shah[e]

[a]*Global Drug Development, Regulatory Affair – CMC, Novartis Pharmaceutical Corp., East Hanover, NJ, United States,* [b]*Manufacturing Science and Technology, Bristol-Myers Squibb Co., New Brunswick, NJ, United States,* [c]*Patheon Part of Thermo Fisher Scientific, Greenville, NC, United States,* [d]*Pharmaceutical Sciences, ORIC Pharmaceuticals, Inc., South San Francisco, CA, United States,* [e]*Drug Product Science & Technology, Bristol-Myers Squibb Co., New Brunswick, NJ, United States*

18.1 Introduction

The principles of quality-by-design (QbD) emphasize the need for understanding sources of variability in the manufacturing process and how they link to product quality as a prerequisite for robustness. A risk-based approach is used to define the goals of QbD development and to design appropriate development studies to achieve these goals. Product quality attributes based on patient needs and clinical relevance are identified, and the risks to these attributes due to input material variability and manufacturing process parameters are assessed. The output of QbD development takes the form of a process and product quality control strategy which ensures that the final product consistently meets preestablished acceptance criteria for the critical quality attributes (CQAs) (Yu, 2008; Yu et al., 2014). Since QbD principles have been introduced more than a decade ago by the International Conference on Harmonization guidelines, numerous reports were published invoking QbD principles in the study of a specific issue (Awotwe-Otoo et al., 2012; Bohlin et al., 2009; Kayrak-Talay et al., 2013; Kona et al., 2014; Kushner et al., 2014; Merritt et al., 2013; Prpich et al., 2008; Saurí et al., 2014; Visser et al., 2015). Nevertheless, few of them provide a wide view of the application of QbD to a comprehensive development program. This chapter provides a case study of the integrated application of QbD to the different aspects of development of a drug product manufactured by a high-shear wet granulation process (Badawy et al., 2016).

Brivanib alaninate (BMS-582664) is a potent, orally active inhibitor of vascular endothelial growth factor (VEGF) receptors (Huynh et al., 2008). A film-coated tablet formulation for brivanib alaninate was developed using a wet granulation manufacturing process. The principles of QbD were applied during the formulation and process development of brivanib alaninate film-coated tablets. Using a risk-based approach, the strategy for the formulation design and understanding were developed by (1) identifying desired product attributes, (2) identifying quality risks to the attributes during development, and (3) adopting the appropriate strategy and performing experiments to mitigate or reduce risks. Quality risk assessments and statistically designed experiments were performed to understand the quality of the input raw materials required for a robust formulation and the impact of manufacturing process parameters on the CQAs of the drug product. The data obtained were used to establish acceptable material attributes and process ranges and to define the control strategy for the commercial manufacture of brivanib alaninate film-coated tablets. The proposed commercial manufacturing process uses process analytical technology (PAT), as well as conventional analytical tests, to control the in-process material attributes. The product quality is controlled and assured through a holistic approach of input material, in-process attribute, and parametric controls combined with appropriate finished product tests. In this chapter, an integrated case study on the systematic application of QbD to product development is described, and the implementation of QbD concepts in the different aspects of product and process design is demonstrated. The chapter illustrates how QbD concepts, such as product robustness and scale-independent design space, are reduced to practice in the development of brivanib alaninate tablets.

18.2 QbD methodology

Formulation selection and optimization studies were conducted to establish formulation composition. After formulation composition was established as described below, studies were conducted to evaluate effect of material attributes and process parameters on product attributes of the selected formulation. CQAs for brivanib alaninate tablets were identified as tablet potency, dosage uniformity, impurities, dissolution rate, and appearance. Appearance was considered a CQA as it is perceived to affect patient compliance and, hence, outcome of therapy.

Subsequent to the identification of the CQAs, failure mode effect analysis (FMEA) was conducted to identify material attributes and process parameters to be studied for their impact on CQAs. While all material attributes and process parameters that can potentially affect CQAs were considered as part of the risk analysis, only a subset of these attributes and parameters were selected for development studies as warranted by the risk analysis. Studies were conducted to generate knowledge regarding the effect of the selected input material attributes and process parameters on CQAs and identify critical material attributes (CMAs) and parameters based on these studies. Critical attributes and parameters are defined as such if their variation within a reasonable range results in a measurable impact on any of the CQAs. Proven acceptable ranges (PARs) were established for the CMAs and critical process parameters (CPPs) based on these development studies. Controlling CMAs and CPPs within their PARs, together with in-process tests and controls of intermediate materials, constituted the control strategy for the brivanib alaninate tablet manufacturing process (Fig. 18.1).

All non-CPPs which were either not identified for evaluation as per the risk analysis or were found to be noncritical from development studies were not included in the control strategy. These parameters were set at a fixed value at batch record level, and no PARs were established for these parameters.

The following two sections describe methods used for the characterization of granulation in tablets, and the PAT methods used throughout development and implemented in the control strategy.

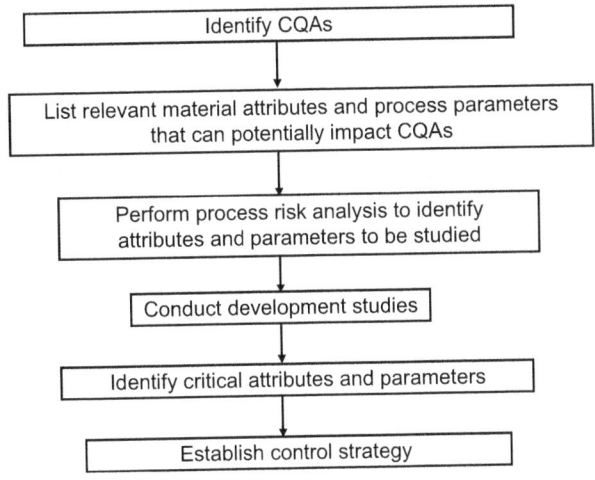

FIGURE 18.1

QbD methodology applied to the development of brivanib alaninate tablets.

From Badawy et al. (2016).

18.3 Physical characterization

The particle size distribution of the final granulation was determined by mesh analysis using an Allen Bradley Sonic Sifter (Allan Bradley, Milwaukee, WI) equipped with a series of six screens and a pan. The geometric mean diameter (GMD) of the granulation was calculated from the granulation weight fraction retained on each screen. The percent fines (<75 μm) was calculated by the summation of the percent material retained over the 325-mesh screen and dust collected in the pan of the sonic sifter.

Flow properties of the lubricated blend were tested using an Erweka GTB powder flow tester (Erweka GmbH, Heusenstamm, Germany). The time required for 100 g of the blend to flow through a 10 mm aperture was determined and converted to a flow rate in g/s.

A Presster compaction simulator (Metropolitan Computing Corporation, East Hanover, NJ) was used to generate the compaction profile data at 14 ms dwell time using oval-shaped B tooling and 800 mg tablet weight. Different compaction forces were used to compress the granulation, and tablet hardness obtained at each compression force was plotted against the corresponding compaction force to obtain the compaction profile. The Presster data was used to determine compactibility parameter (initial slope of the compaction profile), hardness to force ratio at target tablet hardness, and maximum hardness achieved for each batch.

Pore size distribution of granulation and tablets was determined by mercury intrusion porosimetry (MIP, AutoPore III, Micromeritics, Norcross, GA). A fraction of the tablet or a granulation sample of the particle size fraction passing through 60-mesh screen and retained on a 100-mesh screen was used for the porosity determination. Incremental pore volume was determined at different pressures ranging from 1 to 61,000 psi. The pore diameter corresponding to a given pressure was calculated by the software of the instrument according to the Washburn equation using mercury surface tension value of 485 dyn/cm and mercury contact angle of 130°. Pores in the 0.1 to 10 μm range were used for the assessment and comparison of the different batches.

Compression runs for all batches were monitored for sticking to the tablet compression tooling and defects on the tablets. Tooling was removed after each run to assess for sticking, filming, and filling of the embossing. A scoring system was devised to help quantify the extent of sticking on punches, and extent and frequency of defects on the tablets. The criterion for tablet picking was based on the severity and frequency of picking. A score of "0" indicated no more than minor picking on <30% of tablets, while a score of "5" indicated severe picking in >30% of tablets. A scoring system was also established to rank order the formulation for tendency to stick to the punches, wherein no sticking was assigned a score of "0" and severe sticking that obliterated punch face emboss was assigned a score of "5."

18.4 Process analytical technology

18.4.1 NIR moisture methods

An on-line near infrared (NIR) spectroscopic method was also developed as an in-process control to measure and control moisture in Brivanib granulation during fluid bed drying using a Brimrose 4030-EX explosion-proof NIR spectrometer. The spectrometer is attached to a window on the side of the dryer bowl and is used to measure granule moisture through the window in real-time during the drying process. A chemometric model was developed and calibrated to water content (as % w/w) in the granulation using results from at-line loss on drying (LOD) method determined by a moisture balance. The on-line NIR measurements determine when the material has reached the desired moisture level and provide the endpoint for the drying step. Calibration standards were made of granules equilibrated to varying moisture levels. During method development, additional samples were created to test the robustness of the method with respect to granule density, granule size, excipient, and drug substance concentration, and drug substance particle size.

18.4.2 NIR tablet potency method

The at-line NIR method was developed and validated in the reflectance mode using a Thermo Antaris NIR spectrometer in conjunction with a chemometric model calibrated to brivanib alaninate concentration (% w/w). Using prepared calibration tablet standards, the NIR measurements were calibrated with results from the primary HPLC tablet potency method computed from the average of two preparations of five tablets each. An appropriate partial least squares (PLS) chemometric model was selected to ensure that the spectral variation was covariant with the brivanib concentration. The NIR method computes the percentage concentration of brivanib alaninate from 30 individual tablet measurements, which is then adjusted to percentage label claim using the individual tablet weights. The batch potency is computed from the average of the 30 tablet results. Because the NIR method measures only the surface of the tablet, a statistical analysis was conducted to determine the appropriate number of tablets that would provide a similar assurance of batch potency as measuring potency of two sets of five tablets by HPLC analysis.

Tablet standards were prepared to build the calibration model for the NIR method. The composition of the tablets was varied using a structured design so that obtained spectra would encompass expected variations that may be encountered during routine use. The calibration standards were developed using a development granulation batch and then, through design of experiments (DOEs), extra-granular excipient levels were uniformly varied to achieve a drug substance calibration independent of extra-granular excipient level. All calibration tablets were manufactured using pilot-scale processes that mimicked the commercial manufacturing process. Robustness was further built into the model by identifying sources of method variation and either testing these variations directly or incorporating them into the calibration set of prepared tablets. Sources of method variation that were identified and assessed included drug substance particle size, granule size, tablet moisture content, tablet hardness, laboratory humidity, and instrument-to-instrument variability.

18.5 Formulation selection and optimization

The wet granulation process shown in Fig. 18.2 was selected to reduce the risk of drug substance adhesion to manufacturing equipment and compression tooling. The wet granulation reduces sticking of the blend to tablet compression tooling compared to a roller compaction process (Narang et al., 2012). The lower sticking tendency of wet granulation over roller compaction is attributed to the differences in the ability of the granulation process to modify exposed particle surface through the agglomeration of particles. The presence of a greater proportion of primary drug particles in the formulation is expected to exacerbate sticking, whereas the presence of agglomerated drug particles reduces sticking. Fines in the wet granulated formulation are expected to be agglomerates of primary drug particles, in contrast to fines in roller compacted samples which are largely unagglomerated primary drug particles (Narang et al., 2012).

18.5.1 Formulation selection

Formulation and process screening studies were conducted to select a prototype formulation and manufacturing process for brivanib alaninate tablets. Key formulation selection studies focused on the selection of tablet disintegrant, in which performance of both croscarmellose sodium (CCS) and crospovidone (CPVP) was evaluated. Formulation selection batches were manufactured at a 600 g batch size by a wet granulation process in a 4 L high-shear Diosna granulator (Diosna Dierks & Söhne GmbH, Osnabrück, Germany). All batches were granulated using 45% to 50% w/w water, relative to total batch size, at an impeller speed of 5.2 m/s. Wet granulation was screened through an 8 mesh screen, and granules were subsequently dried in a Glatt GPCG-1 fluid bed dryer (Glatt Air Techniques, Ramsey, NJ) until the LOD was ≤2% w/w. Dried granulations were milled using a conical mill (Comil 197S, Quadro, Waterloo, Ontario, Canada) and then blended with the extragranular excipients in a bin blender. The final blend was compressed into 400 mg strength tablets (800 mg tablet weight) on a 6-station Korsch PH 100 tablet press (Korsch AG, Berlin, Germany) using embossed tooling.

Fig. 18.3 shows impeller power consumption versus amount of water added for formulations with CCS or CPVP as the intragranular disintegrant. As water is added, impeller power increases as granule growth takes place. A rapid increase in impeller power around granulation endpoint is generally undesirable as it indicates rapid uncontrolled granule growth which makes the process more difficult to control. The use of CCS as the intragranular disintegrant showed steady controlled granule growth, which increases the robustness of the manufacturing process. In contrast, the use of

638 Chapter 18 Integrated application of quality-by-design

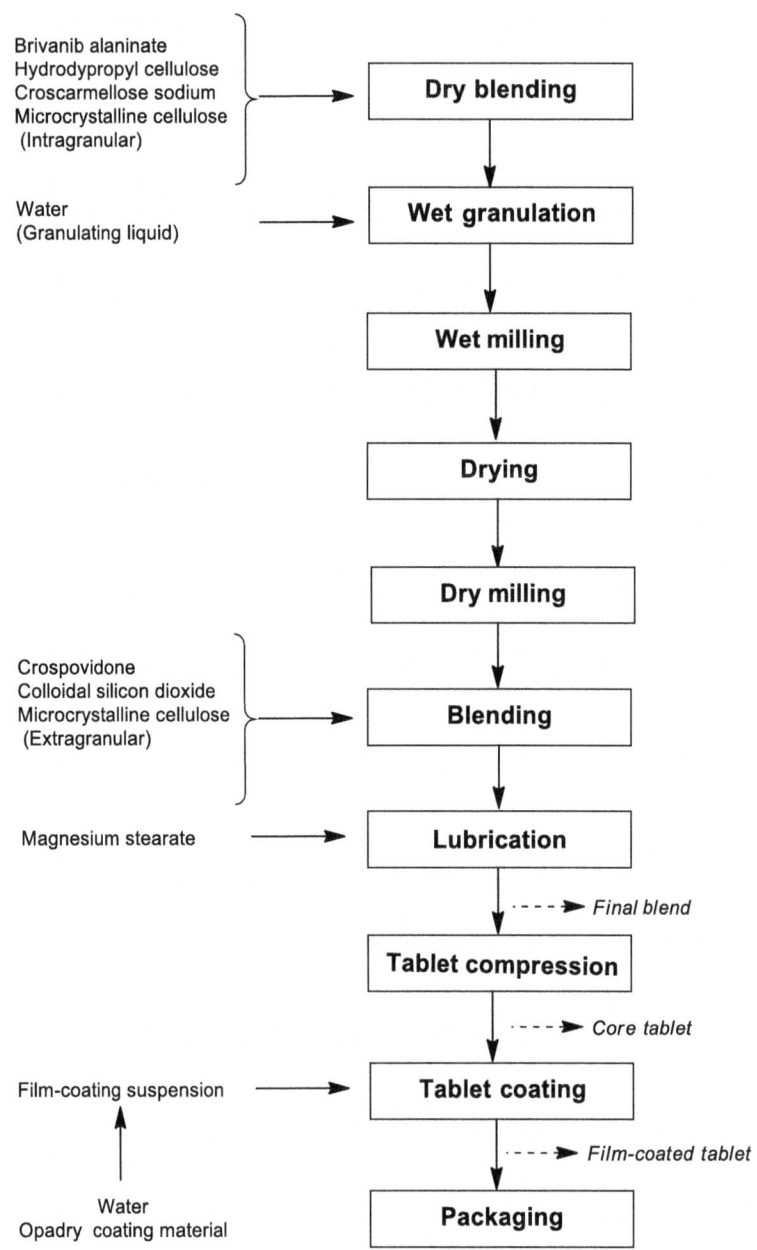

FIGURE 18.2

Flow diagram of the manufacturing process for brivanib alaninate film-coated tablet.

From Badawy et al. (2016).

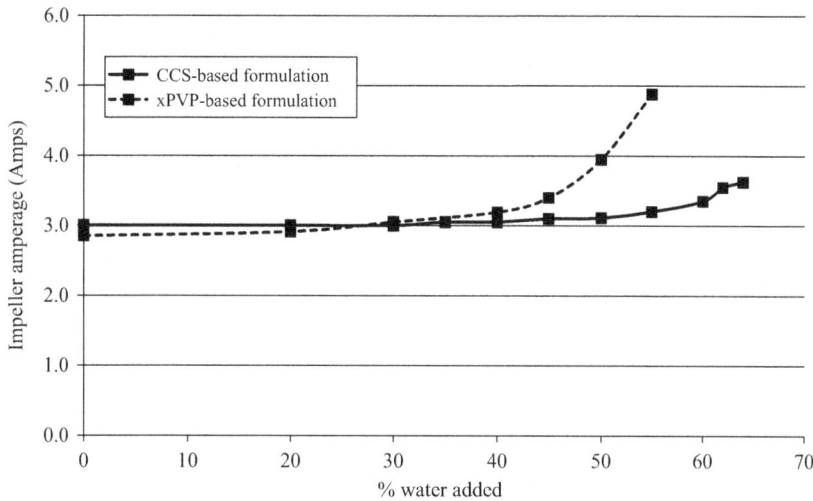

FIGURE 18.3

Impeller current profile during granulation of formulations with different disintegrants.

From Badawy et al. (2016).

CPVP as the intragranular disintegrant showed rapid granule growth and "sharp" endpoint, which increases sensitivity to changes in process parameters. This was shown by impeller power consumption measurement (Fig. 18.3) and visual observations during granulation.

On the other hand, the use of CCS as a disintegrant was associated with incomplete in vitro drug release. This is due to pH-dependent and reversible ionic binding of brivanib alaninate by CCS. A pharmacokinetic study in monkeys indicated that bioavailability of brivanib alaninate is not negatively impacted by the presence of CCS (Narang et al., 2012). This in vitro binding phenomenon was not observed with the use of CPVP as a disintegrant. Although binding of brivanib alaninate by CCS is not expected to be biorelevant based on this study, bioavailability risk posed by CCS still remains. Based on the findings from the wet granulation and CCS binding studies, CCS was selected as an intra-granular disintegrant to enhance robustness of the wet granulation process. However, CPVP was selected as the extra granular disintegrant to reduce the total amount of CCS in the tablet and minimize the risk of potential in vivo brivanib alaninate-CCS binding, which could compromise bioavailability. Composition of the prototype formulation is shown in Table 18.1

18.5.2 Formulation optimization

After the prototype formulation was identified, a formulation DOE study was conducted to assess formulation ruggedness and optimize composition with respect to the binder (hydroxypropyl cellulose, HPC), disintegrant (intra- and extragranular), and lubricant (magnesium stearate). A split-plot full factorial experimental design was used to complete the DOE. The design entails the manufacture of a larger-scale granulation batch and splitting the batch into two parts in order to study the two levels of magnesium stearate, instead of manufacturing a separate batch for each of the combinations. This

Table 18.1 Composition of brivanib alaninate film-coated tablets. From Badawy et al. (2016).

Ingredient	Function	Prototype formulation	Final formulation
Concentration (% w/w)			
Brivanib alaninate (BMS-582664-02)	Active	50.00	50.00
Microcrystalline cellulose (PH102)	Filler/Binder	39.50	38.25
Croscarmellose sodium	Disintegrant	3.00	3.00
Hydroxypropyl cellulose	Binder	3.00	4.00
Crospovidone	Disintegrant	3.00	3.00
Colloidal silicon dioxide	Glidant/Compaction aid	0.50	0.50
Magnesium stearate	Lubricant	1.00	1.25
Total, Core Tablet[a]		100.0	100.0
Opadry (weight gain)	Film coat	3.00	3.00

[a] Total core tablet weight = 800 m.

design reduces the number of granulation batches that need to be manufactured by half, while still providing sufficient information to understand the main effects and interactions of the various excipients. Experimental design batches are shown in Table 18.2.

DOE batches were manufactured in a Fuji high-shear granulator with a 25 L bowl at a 5 kg total batch size. The impeller speed was maintained at 240 rpm (equivalent tip speed of 4.8 m/s), and the chopper was run at low speed (1000 rpm). A peristaltic pump was used to add water at a rate of 100 g/min/kg over a period of 4 min and 30 s. The wet massing time was kept constant at 15 s for all batches while maintaining same impeller and chopper speeds. The wet granulation was passed through a conical mill (Comil 197S) with 0.125-in. (3.2 mm) round opening screen at an impeller speed of ~1350 rpm. The granulation was then dried using the GPCG-1 fluid bed dryer at 70°C to a moisture content of ≤1.50%. The dried granulation was milled using a conical mill (Comil 197S) with a 0.045-in. (1.1 mm) screen opening at an impeller speed of ~1350 rpm. The milled granulation was split into two equal sublots was then placed in a bin blender (8.3 L) and mixed with the extra granular excipients and appropriate amount of magnesium stearate per the experimental design. The final blend was compressed into 400 mg strength tablets (800 mg tablet weight) on the 6-station Korsch PH 100 tablet press using embossed tooling. Statistical analysis of response data was done by regression analysis in SAS JMP (SAS Institute Inc., Cary, NC).

The formulation DOE study showed that a higher level of HPC improves the formulation/process robustness by reducing the number of fines in the granulation, which lowers its sticking potential to tablet tooling as well as picking of tablets in compression. Also, a higher amount of the lubricant (magnesium stearate) reduced the risk of sticking and did not affect the dissolution rate. The negative effect of magnesium stearate on compaction properties was minimal and of little practical significance. In addition, a two-way interaction was observed between HPC and magnesium stearate (Fig. 18.4),

18.5 Formulation selection and optimization

Table 18.2 Experimental design for the formulation optimization study. From Badawy et al. (2016).

Batch number	Croscarmellose sodium (% w/w)	Hydroxypropyl cellulose (% w/w)	Crospovidone (% w/w)	Magnesium stearate (% w/w)
1	3.0	3.0	3.0	1.0
2	3.0	3.0	3.0	0.5
3	4.5	1.5	1.5	1.5
4	4.5	1.5	1.5	0.5
5	1.5	1.5	4.5	1.5
6	1.5	1.5	4.5	0.5
7	4.5	4.5	4.5	1.5
8	4.5	4.5	4.5	0.5
9	4.5	1.5	4.5	1.5
10	4.5	1.5	4.5	0.5
11	1.5	4.5	1.5	1.5
12	1.5	4.5	1.5	0.5
13	1.5	4.5	4.5	1.5
14	1.5	4.5	4.5	0.5
15	4.5	4.5	1.5	1.5
16	4.5	4.5	1.5	0.5
17	3.0	3.0	3.0	1.0
18	3.0	3.0	3.0	1.0
19	1.5	1.5	1.5	1.5
20	1.5	1.5	1.5	0.5

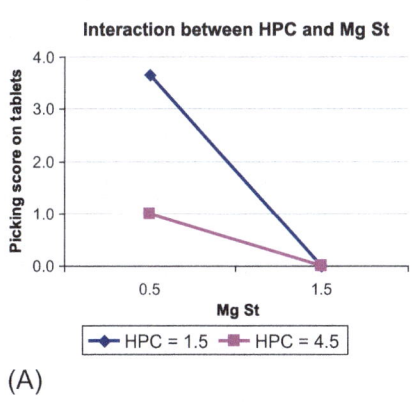
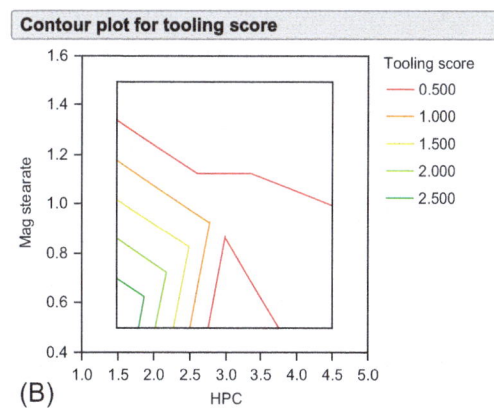

FIGURE 18.4

(A) Interaction plot of magnesium stearate and hydroxypropyl cellulose on tablet picking. (B) Contour plot of the effect of magnesium stearate and hydroxypropyl cellulose on sticking to compression tooling.

From Badawy et al. (2016).

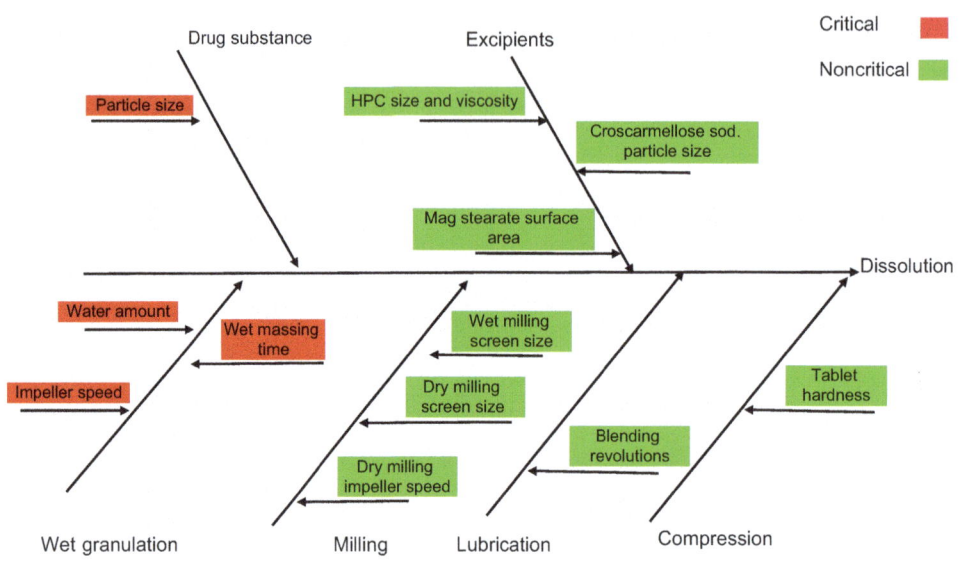

FIGURE 18.5

Material attributes and process parameters impacting tablet dissolution.

From Badawy et al. (2016).

which showed that formulations with lower binder content have higher lubricant requirement. At the higher HPC level, the powder is better granulated resulting in magnesium stearate being less critical to reducing sticking. However, at the lower level of HPC, the powder is less granulated with more fines, and the amount of magnesium stearate is critical to reduce sticking of powder to the punch surface, hence resulting in the two-way interaction between HPC and magnesium stearate. The outcome of the study indicated that the formulation needs to be well-granulated in order to minimize sticking risk. Accordingly, concentrations of HPC and magnesium stearate were increased in the optimized formulation selected for the commercial product (Table 18.1).

18.6 Critical quality attributes, process parameters, and material attributes

The film-coated tablet CQAs are presented below with respect to material attributes and process parameters assessed in development studies and those identified as critical based on development studies. Figs. 18.5, 20.10, 20.14, and 20.15 show fishbone diagrams, one for each CQA, showing the different parameters and attributes evaluated for their impact on the CQA. Those identified as critical based on development studies are colored red, while noncritical attributes and parameters are colored green.

18.6 Critical quality attributes, process parameters, and material attributes

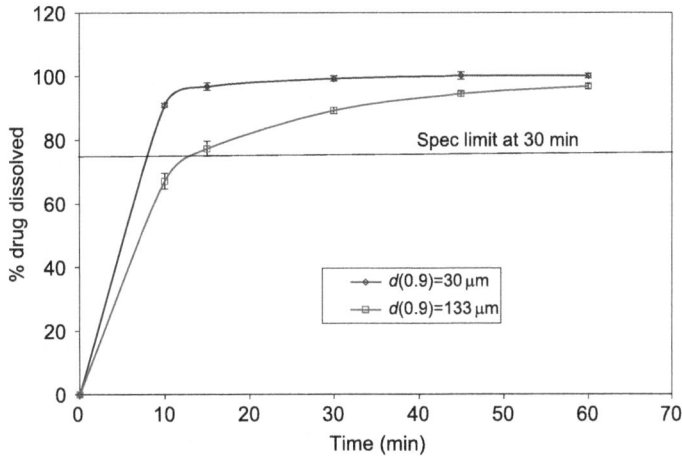

ᵃDissolution in 50 mM phosphate buffer at pH 6.8 with 1% Triton-X; USP Apparatus II; paddle speed 75 rpm.
Error bar represents one standard deviation

FIGURE 18.6

ᵃ Dissolution profiles of brivanib alaninate tablets manufactured using drug substance with different particle size. Badawy et al. (2016).

18.6.1 In vitro drug release (tablet dissolution)

Brivanib alaninate is a BCS class II compound with pH-dependent solubility and high permeability (Narang et al., 2015). Dissolution test of brivanib alaninate tablets was conducted in 1000 mL of 50 mM phosphate buffer, pH 6.8, containing 1.0% Triton X-100 at 37°C using USP apparatus II at 75 rpm. The target dissolution limit was NLT 75% dissolved after 30 min, which was based on the dissolution profiles of formulations used in the different stages of clinical development.

Process parameters of individual unit operations and material attributes studied based on risk assessment of dissolution are shown in Fig. 18.5.

18.6.1.1 Material attributes and dissolution

As shown in Fig. 18.5, drug substance particle size was evaluated for its effect on the dissolution rate for this BCS II compound. Tablet batches were manufactured at target process parameters using drug substance lots with different particle sizes and evaluated for their dissolution behavior. Tablet dissolution rate was found to be dependent on drug substance particle size. Tablets manufactured with drug substance particle size around the upper limit of the range typically produced by the drug substance manufacturing process ($D_{90} = 133$ μm) still met dissolution limit (Fig. 18.6). However, the dissolution of tablets manufactured with this large drug substance particle size was lower compared to the smaller drug substance particle size (% dissolved at 30 min was 89% and 99% for drug substance $D_{90} = 133$ μm and 30 μm, respectively). Hence, drug substance particle size was considered a CMA as variation in particle size was reflected in a change in the tablet dissolution rate.

Binder (HPC) particle size and solution viscosity, disintegrant (CCS) particle size, and lubricant (magnesium stearate) surface area were also evaluated for their potential impact on tablet dissolution

Table 18.3 Experimental design[a] for the wet granulation unit operation study of the brivanib formulation. From Badawy et al. (2016)[b].

Water level (w/w%)	Impeller speed (m/s)	Wet massing time (s)	Water addition time (min)
49	6	50	3
44	3.6	10	5
49	6	50	5
44	3.6	10	3
49	6	10	3
46.5	4.8	30	3
44	3.6	50	3
46.5	4.8	30	3
49	3.6	50	3
44	6	50	3
46.5	4.8	30	3
49	3.6	10	3
44	6	10	3

[a] 2^3 factorial design with three center point replicates for the water level, impeller speed, and wet massing time; effect of liquid addition time was tested at the extreme combinations of the other three parameters.
[b] Intragranular components: microcrystalline cellulose, Brivanib (50% w/w), hydroxypropyl cellulose, croscarmellose sodium; extragranular: crospovidone, colloidal silicon dioxide, microcrystalline cellulose, magnesium stearate —2.0 kg batch size.

rate. This was achieved by using alternate vendors for these excipients exhibiting differences in the above attributes of interest. Variation in these excipient attributes resulted in minimal change in dissolution rate, and hence, they were considered noncritical (Fig. 18.5).

18.6.1.2 Process parameters and dissolution

Based on risk assessment, development studies were conducted to assess the impact of process parameters in the following unit operations on tablet dissolution rate: wet granulation, wet milling, dry milling, and lubrication.

18.6.1.2.1 Wet granulation unit operation

A preliminary screening study was conducted to select wet granulation parameters to be studied in a full factorial DOE study. The results of the screening study identified the following process parameters for the granulation process DOE study; (1) water amount, (2) impeller tip speed, and (3) wet massing time. A 2^3 full factorial design with three replicate center points was used to study the effect of these parameters on tablet dissolution at a 2 kg scale in a 10-L high-shear granulator using a single batch of drug substance. Experimental design batches are shown in Table 18.3 (Badawy et al., 2012). Regression analysis showed that all main effects of the three granulation parameters and two-way interactions were statistically significant ($P < .05$) for the amount of drug dissolved at 30 min. All three wet granulation process parameters affected tablet dissolution in the same direction. Higher levels of water, impeller speed, or wet massing time resulted in a decrease in dissolution rate. Tablet dissolution rate increased with the increase in granule pore diameter (determined by mercury intrusion porosimetry) (Fig. 18.7).

18.6 Critical quality attributes, process parameters, and material attributes

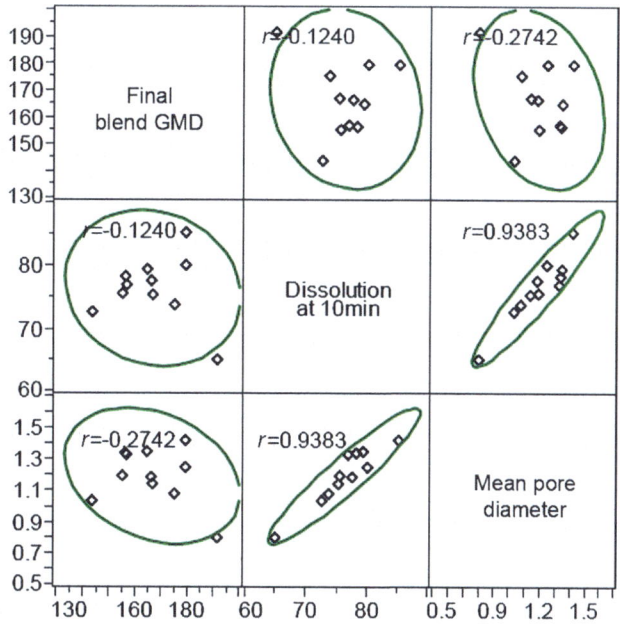

FIGURE 18.7

Correlation of granulation particle size (geometric mean diameter, GMD) and granule mean pore diameter with tablet dissolution.

From Badawy et al. (2016).

In addition, granule pore diameter showed similar dependence on granulation parameters as tablet dissolution rate (Fig. 18.8), suggesting that the effect of granulation parameters on tablet dissolution is achieved through their impact on granule porosity. In contrast, dissolution showed poor correlation with granule size (mean granule diameter) (Fig. 18.7). Also, tablet disintegration did not correlate with granulation parameters or tablet dissolution rate. This indicated that granule disintegration (rather than tablet disintegration) is the rate-controlling step for dissolution. Hence, the three wet granulation parameters highlighted in Fig. 18.5 were considered critical for dissolution through their impact on granule porosity. Other studies also highlighted that granule porosity is the overriding mechanism impacting drug product patient-centric CQAs such as drug release (dissolution) (Wolfe et al., 2013) and bioavailability. Efforts have been made to study the kinetics of granule consolidation during wet granulation and to make real-time measurements of granule densification during wet granulation processes (Narang et al., 2015). These reports collectively indicate that particle size alone may not be an adequate response parameter to assess granulation outcomes (Pandey et al., 2013a, 2013b).

18.6.1.2.2 Milling unit operation

A milling study was conducted to evaluate three milling parameters: (1) screen size for wet milling, (2) screen size for dry milling, and (3) impeller speed for dry milling. A split-plot DOE design was used for the study by making a single wet granulation batch, which was split into two sublots for wet

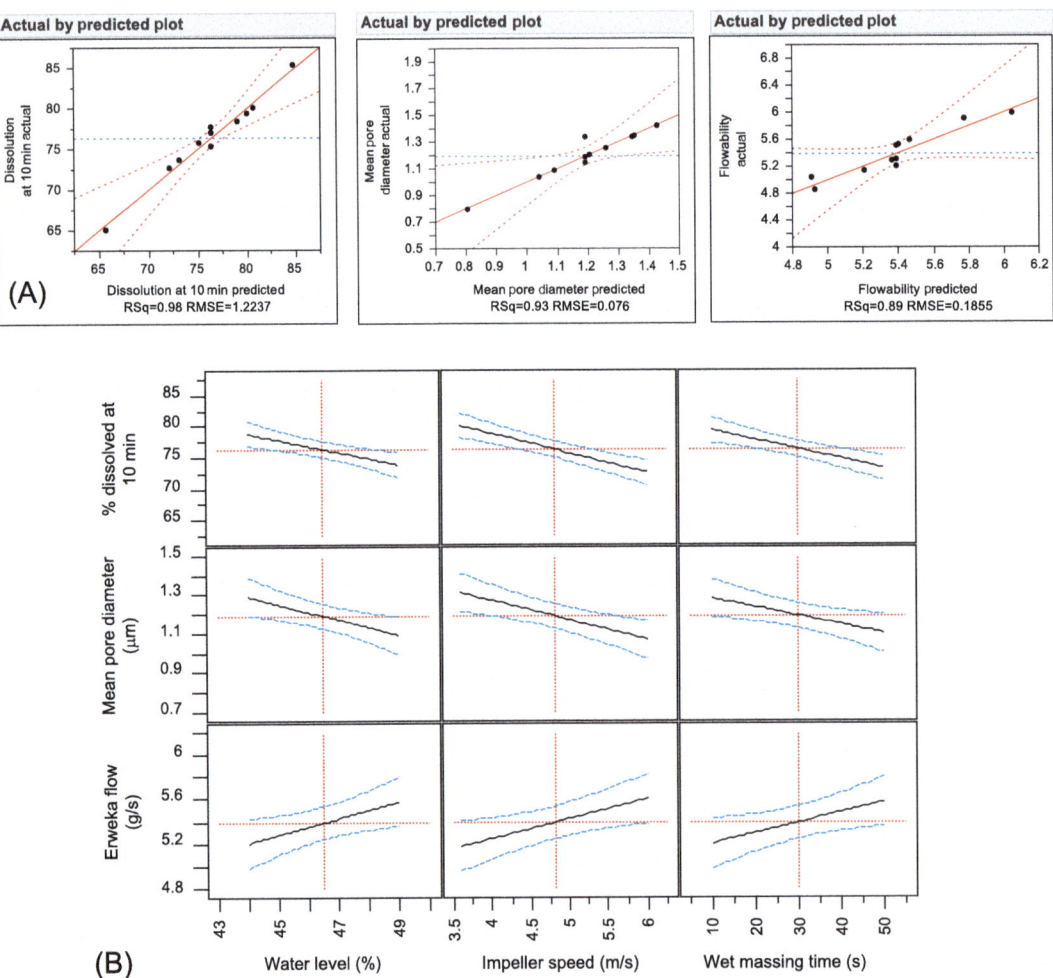

Figure represents main effects plots with 95% confidence intervals, showing the change in response as the variable is changed from the low to the high levels while holding other variables at their center point value.

FIGURE 18.8

Regression model for the effect of wet granulation process parameters on in-process material and tablet attributes. (A) Model fit to DOE data. (B) Model prediction.

From Badawy et al. (2016).

milling; each was then used for a full factorial study of the two dry milling parameters. There was no impact of the change in milling parameters on tablet dissolution despite the small changes observed in the granule particle size distribution resulting from the different milling conditions. This is consistent with the findings from the wet granulation study, which concluded that granule size does not impact tablet dissolution rate. Milling parameters were, therefore, designated as noncritical with respect to tablet dissolution as shown in Fig. 18.5.

18.6.1.2.3 Blending/lubrication unit operation

A magnesium stearate blending (lubrication) study was carried out to assess the risk of over-blending of magnesium stearate on tablet dissolution. Increasing blending time of magnesium stearate is known to reduce tablet dissolution rate in some cases due to the increased shearing of magnesium stearate layers, which subsequently coat and impart more hydrophobic characteristics to granule surfaces (a phenomenon commonly referred to as over-lubrication) (Lerk et al., 1982; Pingali & Mendez, 2014; Sheskey et al., 1995). In the manufacturing process for Brivanib alaninate tablets, magnesium stearate is blended for 250 revolutions in a bin blender. Increasing number of revolutions to 1000 did not have any impact on tablet dissolution. To further assess the risk of over-lubrication, the concentration of magnesium stearate was also increased to 1.5% w/w (from 1.25% w/w in the selected formulation) and also blended for 1000 revolutions. These changes did not impact tablet dissolution which confirmed low risk of over-lubrication. Magnesium stearate blending time (number of blender revolutions) was, hence, designated as noncritical.

Tablet dissolution was also found to be independent of tablet hardness, which was not unexpected considering that the rate-limiting step of tablet dissolution is granule disintegration rather than tablet disintegration.

18.6.2 Water content and impurities

Brivanib alaninate is an ester pro-drug which undergoes a hydrolysis reaction, resulting in the formation of parent compound BMS-540215 (brivanib) (Zhao et al., 2012). BMS-540215 is the only degradant observed in the drug product on accelerated and long-term stability studies. Hence, this degradant is the major determinant of product shelf-life. The hydrolysis reaction of brivanib alaninate involves water as a reactant. Thus, the rate of formation of BMS-540215 was found to be a function of relative humidity (water activity) and the storage temperature (Badawy et al., 2014). At constant temperature, apparent zero-order rate of formation of BMS-540215 showed exponential increase as a function of relative humidity (Fig. 18.9).

Process parameters and material attributes studied based on risk assessment for water content and impurities are shown in Fig. 18.10. Relative humidity inside the package on stability is a function of initial tablet water content and package design (moisture permeation rate through the pack and the amount of desiccant in the package, if present).

18.6.2.1 Degradation kinetics model

A mathematical model was developed which quantitatively predicts the effect of tablet initial water content and package parameters on the rate of formation of BMS-540215 on stability (Badawy et al., 2014). In order to accomplish this goal, two models were initially developed. First, the concentration of BMS-540215 versus time profile was simulated using a kinetic model for the formation of BMS-540215 as a function of relative humidity in the environment. A modified Arrhenius approach was used to model the formation of BMS-540215 in the tablets. A model was developed which assumes reaction to take place in a noncrystalline reactive phase in the tablet. This is consistent with the literature which suggests that majority of hydrolysis reactions in predominantly crystalline materials occur in amorphous regions

648 Chapter 18 Integrated application of quality-by-design

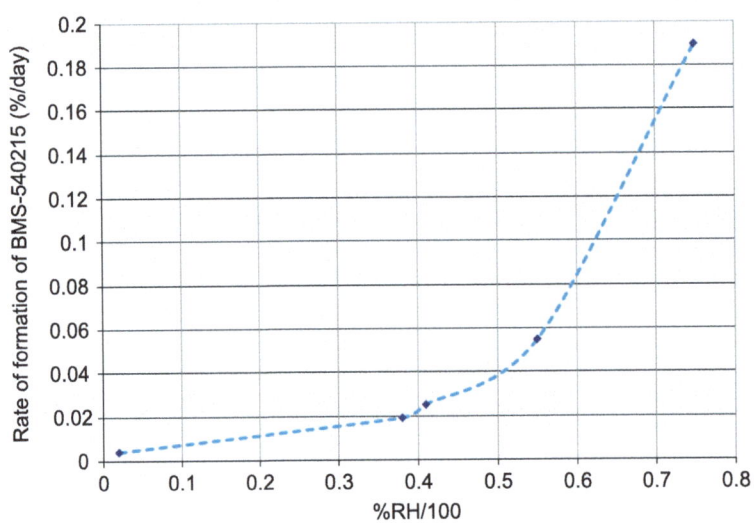

FIGURE 18.9

Effect of relative humidity on the rate of formation of BMS-540215 in film-coated tablets at 40°C.

From Badawy et al. (2014).

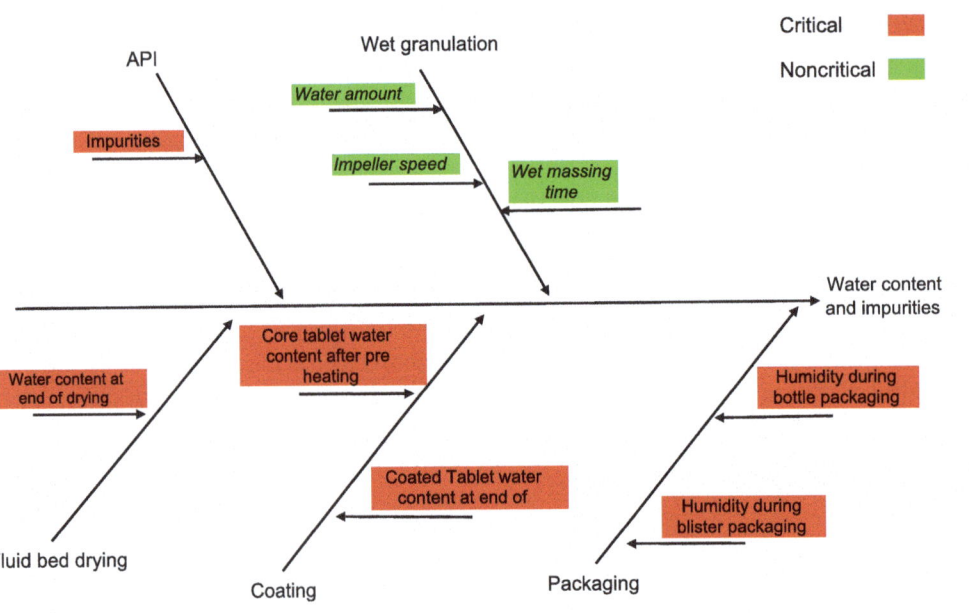

FIGURE 18.10

Material attributes and process parameters impacting water content and impurities.

From Badawy et al. (2016).

or crystal defects (Waterman et al., 2002). Brivanib alaninate concentration is assumed to be constant (C_s) in the noncrystalline phase as the extent of brivanib alaninate degradation in the reported studies is sufficiently low. A pseudo-zero-order reaction model is hence assumed as follows:

$$dX/dt = kC_sV_t \qquad (18.1)$$

where V_t is the volume of the noncrystalline phase where reaction takes place, and k is the first-order degradation rate constant. In order to account for the falling reaction rate with time observed during the course of the stability study, V_t is assumed to decrease with time in a first-order manner from an initial value of V_0 with a rate constant k_2 as follows:

$$V_t = V_0 e^{-k_2 t} \qquad (18.2)$$

the rate of BMS-540215 formation can therefore be expressed as:

$$dX/dt = kC_s V_0 e^{-k_2 t} \qquad (18.3)$$

k is assumed to follow modified Arrhenius dependence on temperature and humidity (Waterman et al., 2007), so Eq. (18.3) can be written as follows:

$$dX/dt = Ae^{-E_a/RT} e^{B_1 RH} C_s V_0 e^{-k_2 t} \qquad (18.4)$$

where A is the frequency factor, E_a is the activation energy, R is the gas constant, B_1 is the moisture sensitivity parameter, and RH is the relative humidity.

Substituting A_1 for the constants $AC_s V_0$ and C_1 for $-E_a/R$ yields:

$$dX/dt = A_1 e^{C_1/T} e^{B_1 RH} e^{-k_2 t} \qquad (18.5)$$

Similarly, k_2 is assumed to follow modified Arrhenius dependence on temperature and humidity:

$$k_2 = A_2 e^{C_2/T} e^{B_2 RH} \qquad (18.6)$$

Eqs. (18.5) and (18.6) are used together to calculate the amount of BMS-540215 in the tablet at different times, relative humidities, and temperatures.

A sorption-desorption moisture transfer (SDMT) model, which calculates humidity inside the package as a function of time, was also developed for two packages, HDPE bottles with silica gel desiccant and aluminum foil blisters. The SDMT model uses the moisture sorption isotherm of the tablets to link tablet moisture content to headspace relative humidity. According to the SDMT model, relative humidity in the package is initially determined by tablet water content and desiccant amount/water content (if present) which is added to lower humidity in the package (Badawy et al., 2001). In most cases, the initial amount of moisture in the headspace has a negligible contribution to the total moisture and humidity in the package. During storage, water permeates into the package, resulting in increase in humidity (Kontny et al., 1992). Humidity profile in the package can therefore be optimized by control of initial tablet moisture, package permeability, and, in case of HDPE bottles, by the use of desiccant. In the case of aluminum blisters where no permeation of moisture into the package takes place, humidity is determined exclusively by the initial tablet water content throughout shelf-life.

The SDMT model was combined with the degradation kinetics model (Eqs. 18.5 and 18.6), and the combined model was used to simulate the BMS-540215-time profile for the different packages and storage conditions. Figs. 18.11 and 18.12 show the model-predicted profiles which are in good

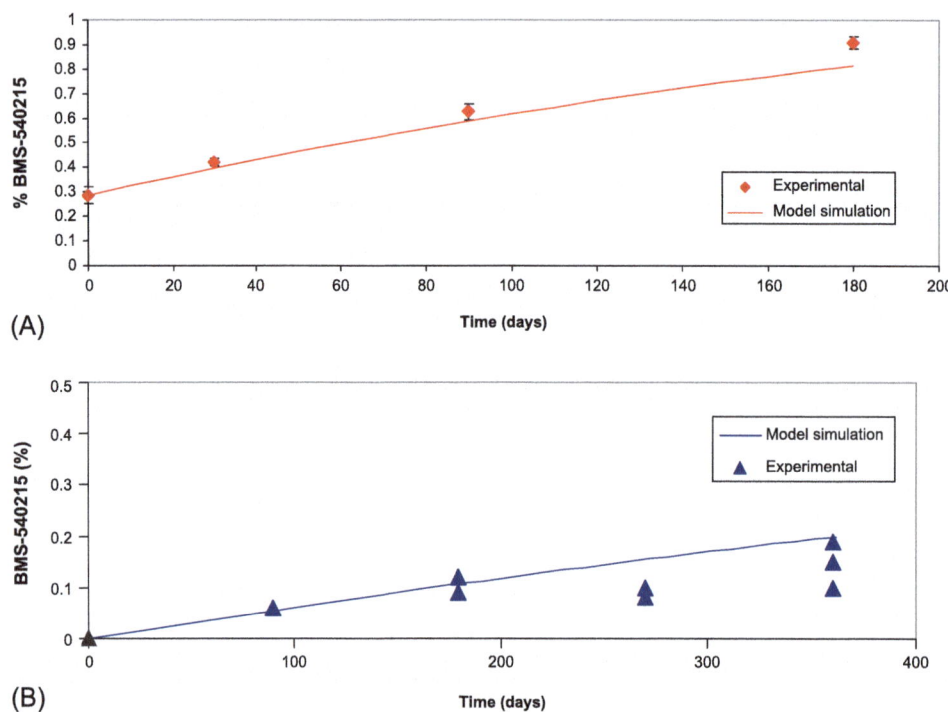

Data represent mean ± standard deviation (*n*=3) in (A). Each point in (B) represents a single determination; data is shown for three different batches tested according to a matrix design.

FIGURE 18.11

Stability of brivanib alaninate tablets in HDPE bottles. (A) 40°C/75% RH and (B) 25°C/60% RH. Points represent experimental data and solid lines represent model simulation for the same package and storage conditions.

From Badawy et al. (2014).

agreement with the experimental data. The combined model is then used to predict the effect of initial tablet water content on the humidity profile, which can be varied as input into the model. Two initial tablet water content levels were simulated, 1.7% w/w and 2.4% w/w (Fig. 18.13). Simulation of the BMS-540215 profile using initial tablet water content of 2.4% w/w indicated acceptable stability up to 3 years, based on a target of NMT 2.0% increase in BMS-540215 over its shelf-life. Bottles can more easily achieve this target considering the presence of desiccant and the 25°C/60% RH storage condition. The aluminum blisters show a higher degradant level due to the 30°C storage temperature and the lack of desiccant which lowers initial humidity in the package.

Since model simulations indicated that BMS-540215 formation on stability is dependent on initial tablet water content, particularly for the aluminum blister package, the initial water content of the tablets was, therefore, designated as a CQA.

Control of water in the final product is achieved by (1) drying of the granulation and tablets in the fluid bed and coating unit operations, respectively, (2) bulk storage of the tablets in sealed aluminum

18.6 Critical quality attributes, process parameters, and material attributes

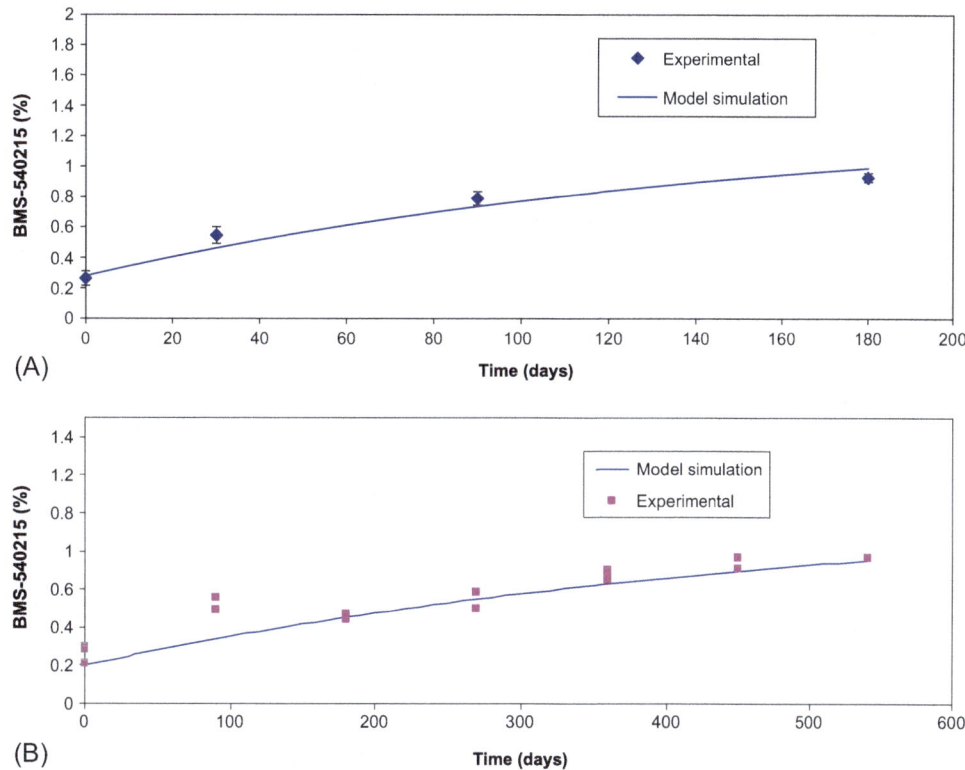

FIGURE 18.12

Stability of brivanib alaninate tablets in aluminum blisters. (A) 40°C/75% RH and (B) 30°C/75% RH. Points represent experimental data and solid lines represent model simulation for the same package and storage conditions.

Data represents mean ± standard deviation ($n=3$) in (A). Each point in (B) represents a single determination; data is shown for three different batches tested according to a matrix design.

From Badawy et al. (2014).

bags with desiccants, (3) final packaging of the tablets under controlled humidity conditions, and (4) package design to control water ingress into the package on shelf life. Water added during wet granulation is removed by the fluid bed drying unit operation to an in-process limit aligned with the target water limit of the tablets at release. An in-line NIR method was developed for the determination of granulation water content in the fluid bed dryer. Subsequent to wet granulation and fluid bed drying, moisture pick up by the product can take place during milling, blending, compression, and coating unit operations. Moisture pick-up could be avoided by humidity control of the manufacturing environment. Alternatively, tablet drying in the coating unit operation can be utilized to remove any water absorbed during process steps subsequent to fluid bed drying. The latter strategy was implemented to obviate the need for strict humidity control of the manufacturing environment. An initial drying step of the core tablets was utilized to ensure acceptable core tablet moisture control prior to initiation of the

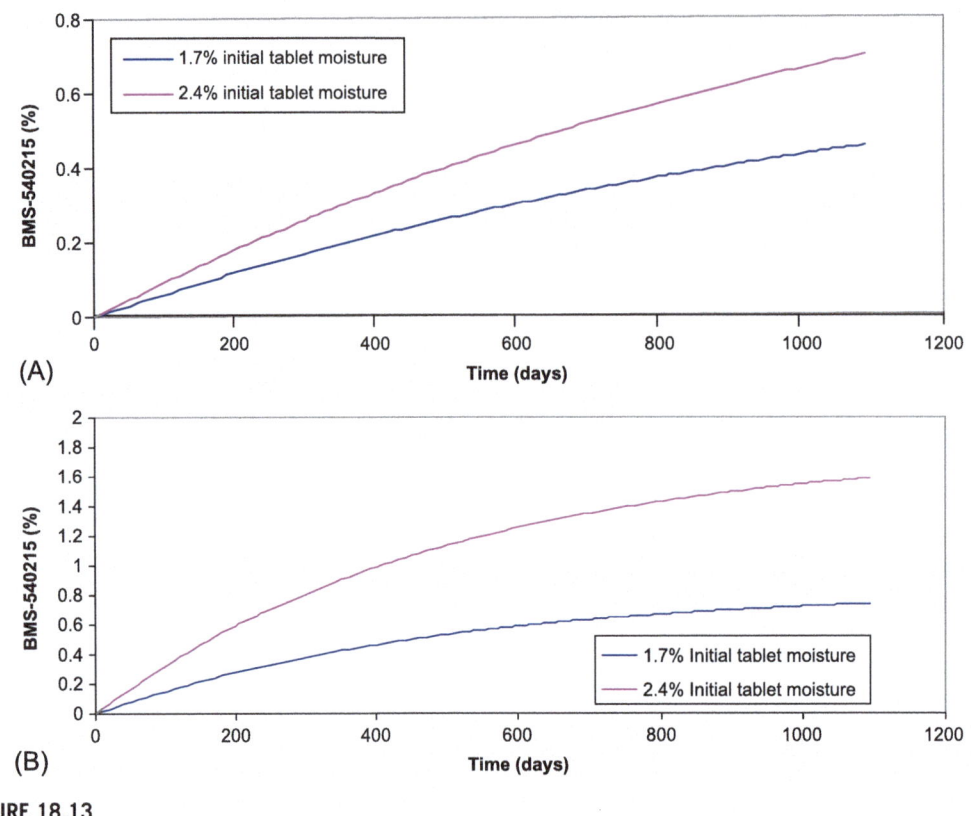

FIGURE 18.13

Simulated degradant profile in the package as function of initial tablet moisture content. (A) HDPE bottle with desiccant at 25°C/60% RH and (B) aluminum blister at 30°C/75% RH.

From Badawy et al. (2014).

coating process. This was considered necessary as drying of the tablet core after coating is applied becomes kinetically more difficult. In addition, a drying step after application of the film coat was also implemented to remove any additional moisture pick up during coating and ensure moisture control of the final coated tablet. Both drying steps were controlled by a tablet LOD method. Subsequent to film coating, tablet water content was controlled by the use of a protective (moisture impermeable) bulk package and control of humidity in the packaging environment.

18.6.2.2 Degradation during processing

Since BMS-540215 is formed by a hydrolytic reaction, the risk presented by aqueous wet granulation to degradation was extensively assessed. Studies using target wet granulation parameters showed no increase in the level of BMS-540215 (or any other impurity) during wet granulation, even when the wet mass was stored for 25 h at room temperature prior to drying. In addition, the effect of wet granulation process parameters on the formation of BMS-540215 during wet granulation was also

18.6 Critical quality attributes, process parameters, and material attributes

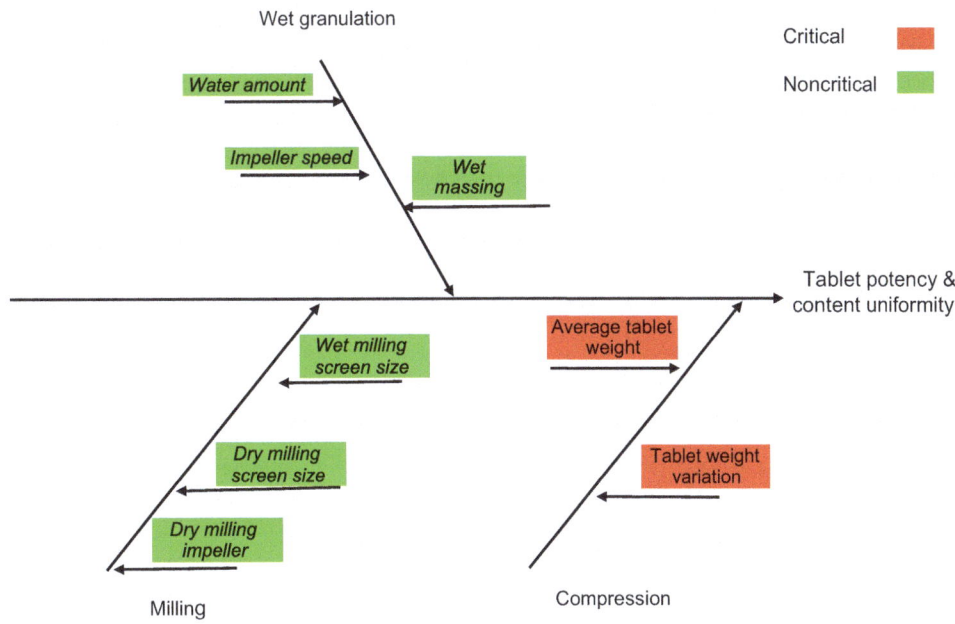

FIGURE 18.14

Material attributes and process parameters impacting potency and content uniformity.

From Badawy et al. (2016).

assessed. Granulation conditions imparting a high level of water and shear forces were considered potential risk factors for brivanib alaninate hydrolysis. Therefore, batches manufactured using the high levels of water, impeller speed, and wet massing time (in the DOE study described above) were tested for their impurity content. No increase in any impurity level was observed in these batches compared to the drug substance indicating that wet granulation process parameters are not critical for BMS-540215 formation in the tablet. The low risk presented by wet granulation to tablet impurity content is attributed to the low solubility of brivanib alaninate in the granulating liquid (water). Initial impurity level in the tablet is hence determined solely by its level in the drug substance used in manufacturing. Impurity level in the drug substance, including BMS-540215, is therefore considered CMA and is controlled by the drug substance specifications.

18.6.3 Potency and content uniformity

Process parameters and material attributes studied based on risk assessment for potency and content uniformity are shown in Fig. 18.14.

Blending parameters and drug substance particle size presented low risk due to the high drug loading of brivanib alaninate in the tablets (50% *w*/w). Therefore, they were not studied for their impact on tablet potency and content uniformity. For high drug load formulations, potency and content uniformity are typically controlled primarily by average tablet weight and weight variation. These were, hence, considered critical for the control of these CQAs.

In-process material attributes known to impact tablet weight variation during compression include granulation particle size distribution and flow properties. Wet granulation and milling process parameters were evaluated for their effect on granulation particle size distribution, flow properties, and tablet weight variation in compression in the abovementioned DOE studies. Although a change in wet granulation and milling parameters caused some variation in final blend particle size distribution and flow rate (measured by an Erweka flow tester) (Fig. 18.8), this variation in granule size distribution and flow rate had no impact on tablet weight variation in the compression studies. Wet granulation and milling parameters are hence considered noncritical with respect to their impact on potency and content uniformity.

18.6.4 Tablet appearance

Picking/sticking tendency during tablet compression caused by inherent properties of the brivanib alaninate drug substance was identified early in development as a risk to tablet appearance and to process yield and cycle time. The wet granulation process was primarily selected to manufacture brivanib alaninate tablets in order to mitigate sticking risk. Incorporation of the drug substance particles in granules manufactured by wet granulation was shown to eliminate sticking (Narang et al., 2012). Moreover, the concentration of the binder (HPC) was found to be statistically significant with respect to the sticking response in the formulation DOE study and hence was optimized in the selected formulation to eliminate sticking as described above. HPC reduces sticking by reducing fines and un-granulated drug substance particles in the final blend. In addition to HPC, the concentration of the lubricant (magnesium stearate) was found to be statistically significant in the formulation DOE and was similarly optimized to mitigate the sticking risk.

Based on the understanding of the sticking risk, drug substance particle size, HPC, and magnesium stearate material attributes, and wet granulation, milling, and lubrication process parameters were studied for their effect on sticking (Fig. 18.15).

18.6.4.1 *Material attributes and sticking*

The effect of drug substance particle size on sticking was evaluated. Small drug substance particle size at the lower specification limit did not show sticking when used for drug product manufacturing. Variation in HPC particle size and viscosity and magnesium stearate surface area were also studied for potential effect on tablet sticking during compression. As mentioned above, this was accomplished by using alternate vendors for these excipients exhibiting differences in the attributes of interest. No effect was observed upon variation in these excipient properties, and all batches made with the different excipient properties showed no sticking during compression.

18.6.4.2 *Process parameters and sticking*

None of the wet granulation or milling DOE batches described above showed any sign of sticking to the compression tooling. Similarly, the shorter blending time of magnesium stearate did not result in any sticking in the lubrication study.

Based on these results, no parameters or attributes were found to be critical for sticking. The selection of the manufacturing process (wet granulation) and the optimization of HPC and magnesium stearate

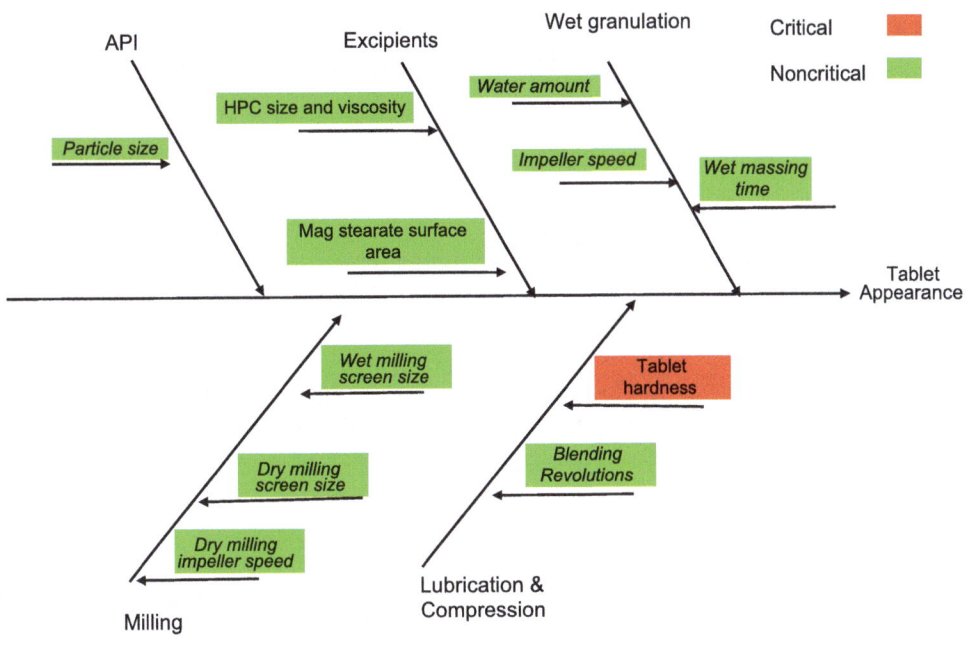

FIGURE 18.15

Material attributes and process parameters impacting tablet appearance.

From Badawy et al. (2016).

levels in the selected formulation resulted in a robust formulation that showed no sensitivity to process parameters or input material attributes with respect to blend sticking tendency during compression.

Tablet erosion during coating is a typical risk to the appearance of a film-coated product and hence presents a risk to the appearance of brivanib alaninate tablets. Tablet hardness was therefore controlled to minimize erosion risk. A lower limit of tablet hardness was identified which showed low friability in the USP test (<0.2%) and resulted in no erosion during tablet coating. Analysis of compaction data from the wet granulation DOE study revealed no significant main effects of granulation parameters with respect to compaction responses. All batches in the study had comparable compaction profiles indicating that variation in granulation parameters had no impact on the compaction properties of the final blend. Moreover, the maximum hardness achieved in the study for all batches was well above the target tablet hardness.

18.7 Control strategy

Based on development studies and knowledge of CPPs and attributes impacting CQAs, a combination of control elements is implemented to ensure that the drug product meets the required quality standards. The drug product manufacturing control strategy includes (1) process parameter controls (process design space), (2) input material controls, (3) in-process attribute controls during manufacturing, and

Table 18.4 Control strategy: process parameters design space. From Badawy et al. (2016).

Process parameter	Unit	Target (design space range)
Water for granulation	% w/w	46.5 (44–49)
Impeller tip speed	m/s	4.8 (3.6–6.0)
Wet massing time	seconds	30 (10–50)

Table 18.5 Control strategy: raw materials, in-process controls, and release tests. From Badawy et al. (2016).

Tablet quality attribute	Raw material	In-process control	At-line release test	Off-line release test
Drug release	- Drug substance particle size			Dissolution
Impurities	- Drug substance impurities			
Water content		- Water content at end of fluid bed drying of granules (NIR, NMT 1.3%) - Water content of preheated core tablets in coating pan (LOD, NMT 1.6%) - Water content of tablets at end of coating (LOD, NMT 1.8%)		
Potency		- Tablet weight during compression	Core tablet potency by NIR	
Content uniformity		- Tablet weight during compression	- Uniformity of dosage units by weight variation on core tablets	
Appearance		Tablet hardness		Physical imperfections AQL[a]

[a] Acceptable quality limit of tablet defects.

(4) end product testing. The control strategy is essentially a combination of limits on all critical starting material attributes, in-process material attributes, and process parameters identified for each CQA (Figs. 18.5, 18.10, 18.14, and 18.15). If a parameter or attribute is critical for more than one CQA, the limit in the control strategy is based on the CQA with the tightest requirement.

In addition, the control strategy also includes at-line release tests and finished product release tests. The at-line release tests include testing of core tablets for potency and uniformity of dosage units. Release tests of the finished product (film-coated tablets) include dissolution, water content, appearance, and identification. The control strategy is listed in Tables 18.4 and 18.5 and described in the following sections.

18.7.1 Process parameters design space

Wet granulation parameters (water amount, impeller speed, and wet massing time) are the only process parameters identified as critical based on the process studies. Wet milling, dry milling, and lubrication parameters are noncritical as all CQAs showed little sensitivity to these parameters.

The results of the wet granulation DOE study described above formed the basis for assigning the three wet granulation parameters as critical parameters due to their impact on tablet dissolution rate. Since all batches from the DOE study met the acceptance criteria for dissolution (NLT 75% dissolved in 30 min) and all other CQAs, parameter ranges in the DOE are a multivariate process design space (Table 18.3). Parameter values for the center point batches are established as the target parameters.

The design space for the wet granulation parameters is established based on the DOE study at a 2.5 kg scale. According to the ICH guideline, a design space can be developed at any scale and be applicable to the commercial scale if the design space is shown to be scale-independent (*International Conference on Harmonization of Technical Requirements for Registration of Pharmaceuticals for Human Use*, 2009). As a result, a strategy to justify the extrapolation of the design space from the small scale to the commercial scale was established. The strategy described below provides the necessary justification, while minimizing the number of batches manufactured at the commercial scale.

A prerequisite for the extrapolation among scales is to have the design space expressed in scale-independent terms. The critical wet granulation parameters of the design space are expressed in scale-independent terms as shown in Table 18.4. The suitability of the scale-independent terms used for the design space was verified by comparing product quality attributes at three scales, representing laboratory scale, pilot scale, and commercial scale. Key attributes of these batches manufactured at the three scales using target parameter values are comparable as shown in Fig. 18.16.

In addition, the design space was also verified by manufacturing two batches at two extreme regions in the design space at the commercial scale (off-target batches). The strategy for selecting the parameters for those batches was based on selection of worst-case combination of the three critical parameters. Since the three critical granulation parameters affect dissolution in the same direction (Fig. 18.8), a combination of the high levels of all three parameters represents the highest risk for slow dissolution rate. Conversely, a combination of the low levels of all parameters represents the highest risk of "under-granulation" and associated risks of poor flow properties and sticking to tablet compression tooling. Results from the two off-target batches met all acceptance criteria, hence providing additional justification for the applicability of the design space to the commercial scale.

18.7.2 Input (Raw) material controls

Drug substance particle size and impurity content are the only input material attributes identified as critical to product performance or CQAs. None of the excipient-related attributes are critical as concluded from the development studies. A correlation between the particle size of the drug substance and dissolution rate was established as previously discussed. Control of the drug substance particle size was therefore necessary to ensure consistent and acceptable dissolution of the tablets. A particle size specification of 40 to 120 μm (D_{90}) is established to ensure an acceptable dissolution rate (NLT 75% dissolved at 30 min). While the upper limit is established to ensure acceptable dissolution, the lower limit of particle size was set to minimize the risk of sticking to compression tooling resulting from fine drug substance particles.

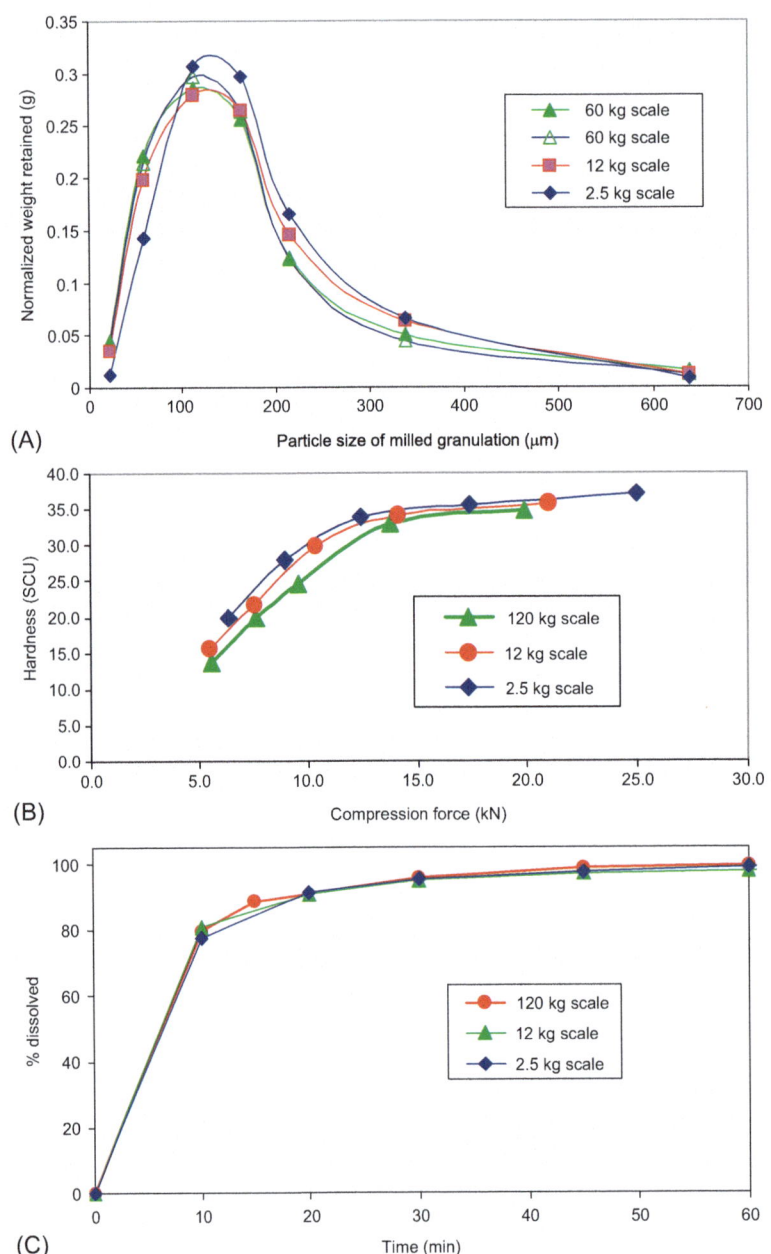

FIGURE 18.16

Comparison of brivanib alaninate in-process material and tablet attributes for product manufactured at three different scales. (A) Particle size distribution of final blend, (B) compaction profile of final blend, and (C) tablet dissolution profile.

From Badawy et al. (2016).

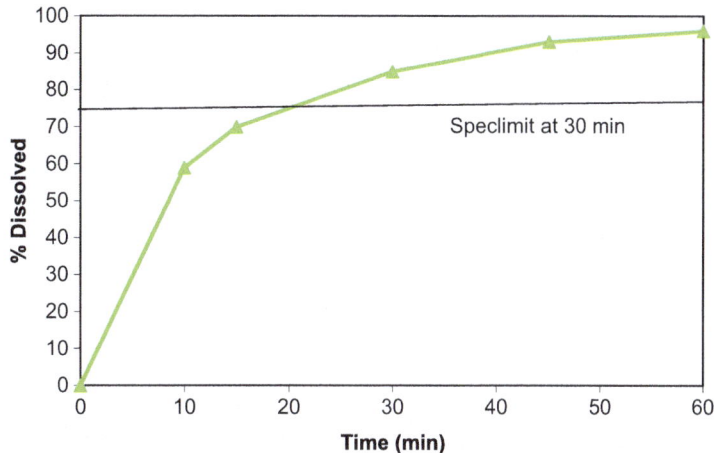

FIGURE 18.17

^aDissolution profiles of tablets manufactured with large drug substance particle size and high level of wet granulation parameters (water amount, impeller speed, and wet massing time).

From Badawy et al. (2016).

Since both wet granulation parameters and drug substance particle size are critical to dissolution rate, it was necessary to establish that the upper limit of the drug substance particle size range shows acceptable tablet dissolution even at the edge of the process design space. Since the wet granulation DOE was conducted with drug substance having D_{90} of 68 μm, additional study was necessary to demonstrate performance of the larger particle size at the edge of the process design space. A risk-based approach was used to achieve this purpose, leveraging the knowledge established from the wet granulation study which illustrated that a combination of the high levels of all three parameter values results in the lowest dissolution rate. Hence, the worst-case combination of granulation parameters which was identified from the DOE study (high water level, high impeller speed and high wet massing time) was evaluated in conjunction with large drug substance particle size ($D_{90} = 106$ μm) to ensure that the worst-case combination of process parameters and drug substance particle size would result in an acceptable dissolution rate. As shown in Fig. 18.17, dissolution of tablets manufactured with this worst-case combination of process parameters and particle size showed an acceptable dissolution profile which met the acceptance criteria.

The other critical input material attribute (drug substance impurity content) was controlled by incoming material specifications, with a limit of NMT 1.0% on the parent compound (BMS-540215).

18.7.3 In-process attributes control

18.7.3.1 Fluid bed drying

Water content of the granules during drying was tested and controlled by the NIR spectroscopy method to a limit of NMT 1.3% w/w. The limit was implemented to ensure that water added during wet

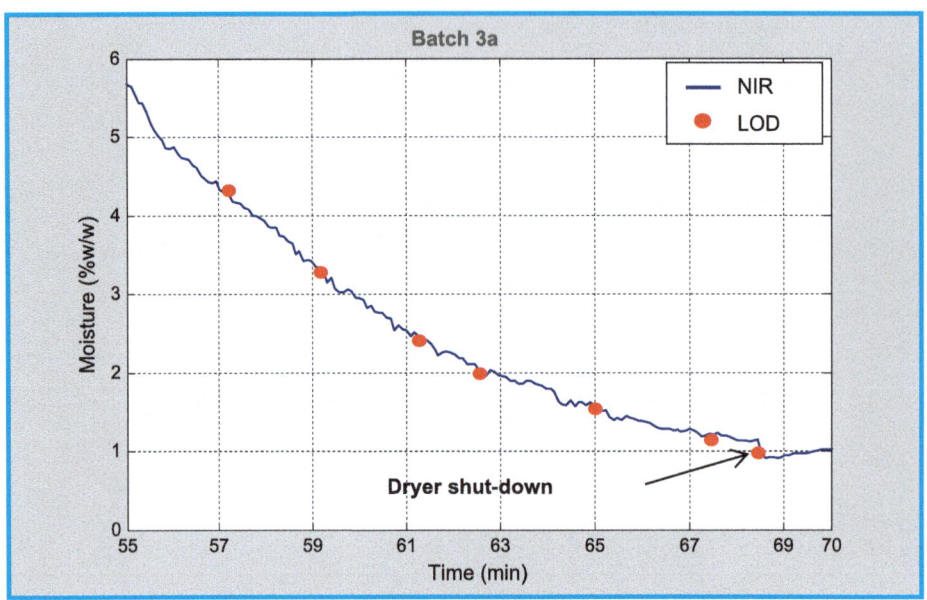

FIGURE 18.18

Comparison of online NIR and at-line LOD for moisture monitoring during fluid-bed drying of brivanib alaninate granulation.

From Badawy et al. (2016).

granulation is removed by drying in the fluid bed dryer and to account for moisture associated with the extragranular excipients added downstream. In addition, the limit was set to enable the coated tablet to meet the water content acceptance limit in the final product.

The established NIR method was used to monitor granulation moisture during fluid-bed drying of several development and commercial-scale batches. Comparison between the NIR and LOD results showed close agreement (Fig. 18.18), and it was determined that moisture levels could be accurately measured between 0% and 3%, with approximately a 0.25% resolution. The NIR method suitably monitored the drying process of commercial-scale batches at the manufacturing site, enabling determination of the drying endpoint.

18.7.3.2 Compression

As typically implemented in tablet compression, upper and lower acceptance limits for core tablet weight (both individual and mean tablet weight) were established to ensure acceptable assay and content uniformity of the core tablets. Similarly, control of tablet hardness within the acceptance limits was implemented to ensure that tablets could withstand the coating process, packaging, and shipping.

18.7.3.3 Film coating

As mentioned above, tablet coating process was designed to control the water level in the core tablets and to remove any additional water picked up by the drug product during film coating of the tablets.

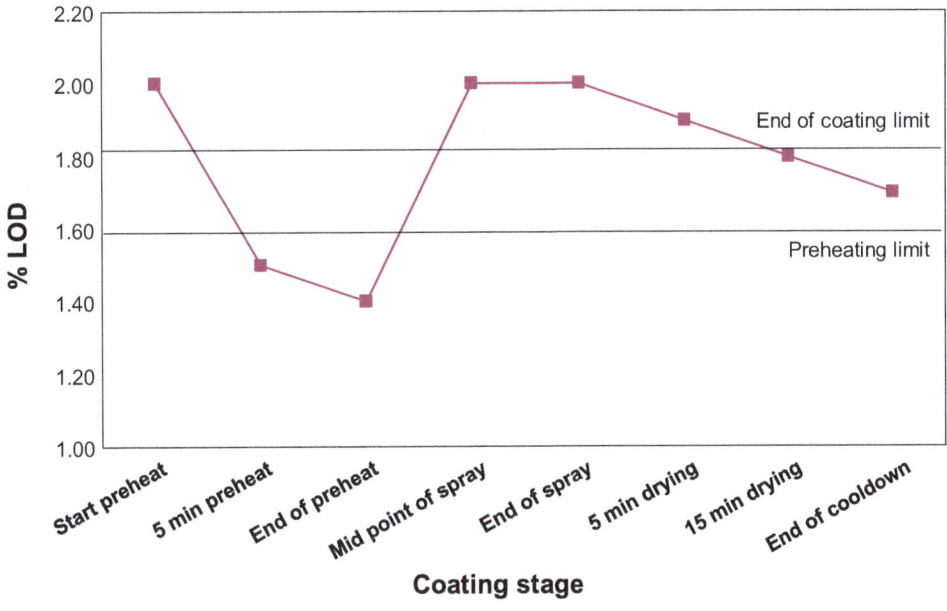

FIGURE 18.19

Change of water content of tablets during the film-coating process.

From Badawy et al. (2016).

To achieve these objectives, the coating process was designed to include three stages: (1) drying of core tablets, (2) application of film coat (spray), and (3) drying of the coated tablets. Water content of uncoated tablets (before the start of coating) and coated tablets (after the end of coating) was determined by the LOD method and controlled by an in-process specification limit. Water content limit in the core tablet was established at NMT 1.6% w/w. The control on the water level in the core tablets ensured that any water picked up by the product during the stages of manufacture after fluid bed drying was removed before coating.

A limit on the water level for the coated tablet was also set at NMT 1.8% w/w to ensure that any water picked up by the tablets during coating was removed. Effectiveness of this coating approach to achieve an acceptably low water level in the coated tablets is shown in Fig. 18.19 (Badawy et al., 2014). Water limit of the coated tablets was based on the simulation of the effect of initial water content on the rate of hydrolysis of brivanib alaninate. Using the predictive model mentioned above, the formation of BMS-540215 as a function of the initial tablet water content was examined, and results identified an acceptable water content release limit of NMT than 2.4% w/w for final coated tablets (Fig. 18.13). The 1.8% in-process limit of the coated tablets is tighter than release limit in order to account for the variability of the in-process LOD method and the Karl Fischer release test method, and ensures tablets meet the specification limit for water content upon release testing (Badawy et al., 2014).

18.7.4 Product release testing

18.7.4.1 At-line release tests

At-line release testing for tablet potency is performed on the core tablets by the NIR method. Tablet samples are collected throughout compression and used for the potency analysis. Tablet potency by the NIR method must meet the potency limit on the release specification. No potency release testing is done on the final coated tablets as the potency does not change during film coating. Data from development and technology transfer batches confirmed this requirement.

Fig. 18.20A shows the brivanib alaninate tablet calibration curve for the NIR method with seven concentration levels of brivanib alaninate and 17 combinations of excipient concentration levels. The calibration root mean squared error (RMSEC) of approximately 1.6% w/w and a correlation coefficient of 0.99 demonstrated strong correlations between the HPLC measurements and NIR predictions. A comparison between the HPLC composite potency results and the NIR potency results was performed for production full-scale batches. The results, plotted in Fig. 18.20B, show excellent agreement between NIR and HPLC results, demonstrating the capability of the NIR method to quantitate brivanib alaninate level in the core tablets.

Tablet uniformity was also tested on the core tablets using a weight variation method as allowed by the compendia at the high drug loading (50% w/w).

18.7.4.2 Off-line release tests

Off-line release tests are implemented for dissolution and tablet identity (Raman spectroscopy method) and must conform to the acceptance criteria of the release specifications. In addition, a water content release test by a Karl Fischer method is implemented with a limit of NMT 2.4% w/w as described above.

No final product release test is implemented for impurities. This is based on product understanding and input material controls in place. BMS-540215 is the only degradant of the drug product. However, BMS-540215 increases only in shelf-life and does not change during drug product manufacturing, even at the worst-case extremes of the granulation parameter design space. Therefore, controlling the level of BMS-540215 in the drug substance ensures an acceptable degradant level in the drug product at release and obviates the need for testing of impurities for tablet release. Impurities are tested on market life stability as BMS-540215 level increases over-shelf-life.

18.8 Key considerations

An integrated and overarching application of QbD principles to modern drug product development requires a holistic understanding of the effect of input variables which include (1) raw material properties and their potential variability, (2) formulation composition (qualitative and quantitative), and (3) manufacturing process parameters. In this case study, a drug product control strategy was developed which encompasses both material property and process parameter controls, with appropriate in-process and finished product analytical tests and specification limits (Tables 18.4 and 18.5). These analytical tests encompass both conventional analytical tests and PAT. In the section below, some of the key elements from the brivanib alaninate QbD development are discussed further.

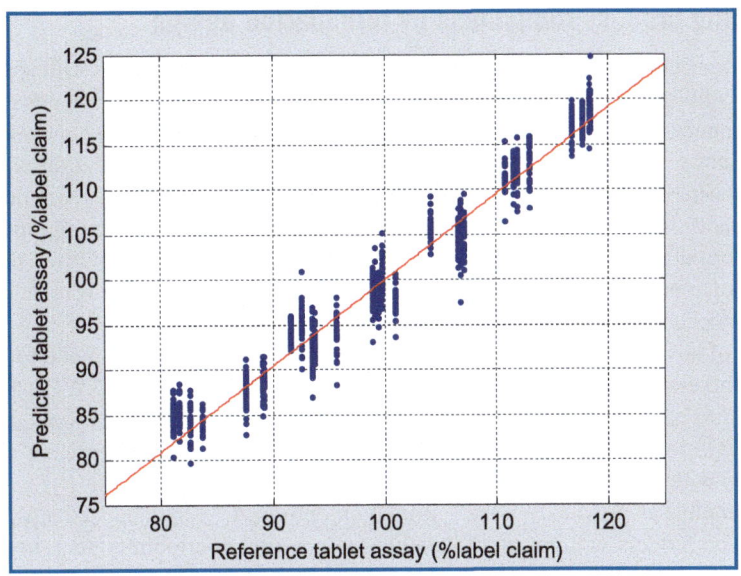

(A) Predicted tablet assay by NIR method, reference assay by HPLC

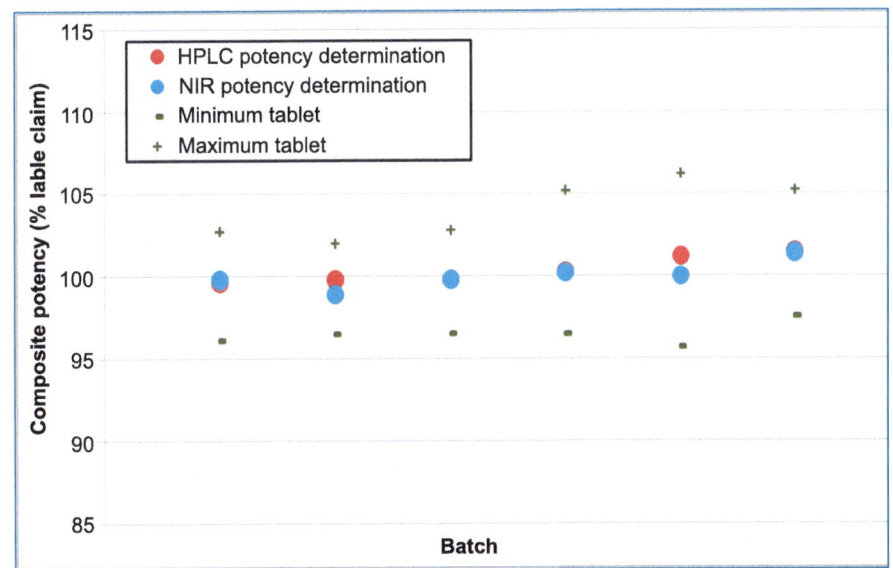

(B) "Minimum Tablet" and "Maximum Tablet" are for the NIR potency determination

FIGURE 18.20

(A) NIR calibration curve for at-line measurement of core tablet potency. (B) Potency results for full-scale brivanib tablet batches comparing composite HPLC (for coated tablets) versus at-line NIR (for core tablets).

From Badawy et al. (2016).

18.8.1 Building process robustness by formulation design

It should be recognized that building manufacturing process robustness starts with appropriate formulation design. Traditional product development efforts focus on sequential development of formulation composition, manufacturing process, and drug product control strategy. Because excipients affect process performance, selection of formulation composition without consideration of the process can limit process robustness. Process development efforts aimed at establishing a parameter space where process performance is minimally impacted by sources of variability would be significantly hampered by suboptimal formulation composition. In the integrated development exemplified in this case study, the assessment of robustness was studied across the unit operations at the formulation design stage through fewer, mechanistically driven experiments.

An example of how process robustness can be built into excipient selection and formulation design is provided by the evaluation of the effect of the disintegrant, CCS versus CPVP, on the endpoint of the wet granulation process. Performance of the wet granulation process with both disintegrants was evaluated, and CCS was selected as the intragranular disintegrant due to the greater robustness of the CCS formulation with respect to the granulation endpoint.

Another example was provided by the formulation optimization to enhance its performance in the tablet compression process with respect to sticking to compression tooling. As a key risk to process robustness, the effect of formulation composition on sticking potential was studied in the formulation ruggedness DOE. Findings related to the role of binder and lubricant and their interaction on sticking were the basis for the optimization of the concentration of HPC and magnesium stearate in the final formulation. The optimized formulation demonstrated excellent robustness and showed no sticking in the process development studies across the full range of parameters tested.

18.8.2 Design space strategy

An integrated design space that incorporates a range of permissible or expected variability in material properties and process variables presents a significant logistical challenge. These challenges stem from the wide range of input variables and responses that must be evaluated. Two strategies can be adopted to address these challenges: (1) generate an extended experimental space using a multifactorial statistical DOE study or (2) adopt a selective and sequential study design in which parameters and material attributes are evaluated based on a mechanistic understanding of their effect on product attributes.

A single DOE that incorporates variables across material properties and process parameters collectively, although feasible, becomes prohibitive in material, time, and effort requirements (Narang, 2015). In order to overcome the limitations of a single comprehensive study, this case study highlights the alternative strategy of using a sequential and selective DOEs. Using this approach, studies do not necessarily utilize exhaustive statistical designs but are based on sound scientific and mechanistic understanding of underlying causes of product responses. For example, the effects of wet granulation process parameters and material properties on the dissolution rate of brivanib alaninate tablets were evaluated in separate studies. CPPs, their design space and the mechanism of their effect on tablet dissolution were established in the process DOE. Material properties study identified drug substance particle size as the critical property for dissolution and established an acceptable particle size range. Since the wet granulation study used target drug substance particle size, while the particle size study was conducted at target wet granulation parameters, combination of the two acceptable ranges from the two studies required further justification. Product and process understanding established in the respective studies were leveraged to generate an experimental space that captures the worst-case combination of wet granulation parameters and drug substance particle size. Additional experimentation of this worst-case space was conducted to confirm the multivariate acceptability of the combination of acceptable ranges from the process and material property studies. The application of such strategy, as demonstrated in this case study, rationally reduces the extent of experimentation required by leveraging the scientific/mechanistic understanding of variables that impact product quality and enabling holistic approach to the development of a control strategy for assuring product quality.

The manufacturing scale used to establish design space presents another challenge. Ideally, design space would be established at the commercial production scale. However, this strategy presents significant challenges associated with cost and availability of drug substance. Limited availability of drug substance during drug product development stage makes this approach impractical in many cases. ICH Q8 permits development of design space at pilot scale and extrapolation to the production scale if properly justified. However, the ICH guideline does not prescribe an approach for justifying extrapolation of design space from pilot scale. In this case study, an approach for the extrapolation of wet granulation design space is utilized which consists of three components: (1) wet granulation parameters at the pilot scale are expressed in scale-independent terms (impeller tip speed, ratio of water to solids and wet massing time); (2) scale independence of these parameters was demonstrated by manufacturing product at three different scales (form pilot to production scales) using target parameters. Key product attributes were shown to be similar across the three scales; (3) production scale batches were manufactured at two points in the process parameter space which represent worst-case parameter combinations with respect to CQAs. Both batches showed acceptable attributes. The approach used was hence based on process understanding established from the development studies and was successful in minimizing number of batches manufactured at the production scale.

18.8.3 Model-based specification limits

Final product and in-process specification limits are key components of the control strategy. Specifications need to be linked to one or more CQA with limits established within which product performance is acceptable, and should be established based on understanding of the factors affecting the CQA. Traditionally, specification limits are based on data from key batches (e.g., long-term stability and clinical batches used in pivotal trials) and perceived process capability. Such approach is not very effective because there are usually other factors affecting a given CQA, in addition to the attribute for which the specification limit is being set. Established limits using the traditional approach do not account for the multivariate effect of these factors on the CQA. Mathematical modeling is a useful tool to quantify the effect of these factors on CQAs and provides the basis for product and in-process specifications. For brivanib alaninate tablets, degradant formation depends on initial water content of tablets and on packaging factors such as package permeability and presence of desiccant. A model-based approach was used to predict the effect of these different factors on rate of degradant formation and to rationally establish a specification limit on initial tablet water content. The model quantitatively predicted the effect of the different variables on degradation rate. Hence, the model allowed for the initial tablet water content limit to be set based on an in silico study of the variable space instead of relying on limited set of batch data.

18.9 Concluding remarks

Identification of CQAs, risk assessment and focused development studies are instrumental to the identification of an effective control strategy based on QbD principles. Controls are implemented on input and in-process materials and on manufacturing process parameters, which ensure final product consistently meets criteria established for CQAs. This demonstration of QbD principles to integrated and holistic drug product development highlights the value of mechanistically driven experiments in building product and process robustness while addressing key risks to the patient, manufacturing plant, and product development.

References

Awotwe-Otoo, D., Agarabi, C., Wu, G. K., Casey, E., Read, E., Lute, S., Brorson, K. A., Khan, M. A., & Shah, R. B. (2012). Quality by design: Impact of formulation variables and their interactions on quality attributes of a lyophilized monoclonal antibody. *International Journal of Pharmaceutics, 438*(1-2), 167–175. https://doi.org/10.1016/j.ijpharm.2012.08.033.

Badawy, S. I., Narang, A. S., Lamarche, K., Subramanian, G., & Varia, S. A. (2012). Mechanistic basis for the effects of process parameters on quality attributes in high shear wet granulation. *International Journal of Pharmaceutics, 439*(1–2), 324–333. https://doi.org/10.1016/j.ijpharm.2012.09.011.S0378-5173(12)00876-9.

Badawy, S. I. F., Gawronski, A. J., & Alvarez, F. J. (2001). Application of sorption-desorption moisture transfer modeling to the study of chemical stability of a moisture sensitive drug product in different packaging configurations. *International Journal of Pharmaceutics, 223*(1–2), 1–13. https://doi.org/10.1016/S0378-5173(01)00693-7.

Badawy, S. I. F., Lin, J., Gokhale, M., Desai, S., Nesarikar, V. V., Lamarche, K. R., Subramanian, G. A., & Narang, A. S. (2014). Quality by design development of brivanib alaninate tablets: Degradant and moisture control strategy. *International Journal of Pharmaceutics, 469*(1), 111–120. https://doi.org/10.1016/j.ijpharm.2014.04.059.

Badawy, S. I. F., Narang, A. S., Lamarche, K. R., Subramanian, G. A., Varia, S. A., Lin, J., Stevens, T., & Shah, P. A. (2016). Integrated application of quality-by-design principles to drug product development: A case study of brivanib alaninate film-coated tablets. *Journal of Pharmaceutical Sciences, 105*(1), 168–181. https://doi.org/10.1016/j.xphs.2015.11.023.

Bohlin, M., Jones, H., & Black, S. (2009). New developments in scale-up and QbD to ensure control over product quality. *American Pharmaceutical Review, 12*(7), 52–57.

Huynh, H., Ngo, V. C., Fargnoli, J., Ayers, M., Khee, C. S., Heng, N. K., Choon, H. T., Hock, S. O., Chung, A., Chow, P., Pollock, P., Byron, S., & Tran, E. (2008). Brivanib alaninate, a dual inhibitor of vascular endothelial growth factor receptor and fibroblast growth factor receptor tyrosine kinases, induces growth inhibition in mouse models of human hepatocellular carcinoma. *Clinical Cancer Research, 14*(19), 6146–6153. https://doi.org/10.1158/1078-0432.CCR-08-0509.

International Conference on Harmonization of Technical Requirements for Registration of Pharmaceuticals for Human Use. *8* (2009a).

International Conference on Harmonization of Technical Requirements for Registration of Pharmaceuticals for Human Use. (2009b). *ICH harmonized tripartite guideline, pharmaceutical development, Q8(R2), Current Step 4 version*.

Kayrak-Talay, D., Dale, S., Wassgren, C., & Litster, J. (2013). Quality by design for wet granulation in pharmaceutical processing: Assessing models for a priori design and scaling. *Powder Technology, 240*, 7–18. https://doi.org/10.1016/j.powtec.2012.07.013.

Kona, R., Fahmy, R. M., Claycamp, G., Polli, J. E., Martinez, M., & Hoag, S. W. (2014). Quality-by-design. III. Application of near-infrared spectroscopy to monitor roller compaction in-process and product quality attributes of immediate release tablets. *AAPS PharmSciTech, 16*(1), 202–216. https://doi.org/10.1208/s12249-014-0180-1.

Kontny, M. J., Koppenol, S., & Graham, E. T. (1992). Use of the sorption-desorption moisture transfer model to assess the utility of a desiccant in a solid product. *International Journal of Pharmaceutics, 84*(3), 261–271. https://doi.org/10.1016/0378-5173(92)90164-W.

Kushner, J., Langdon, B. A., Hicks, I., Song, D., Li, F., Kathiria, L., Kane, A., Ranade, G., & Agarwal, K. (2014). A quality-by-design study for an immediate-release tablet platform: Examining the relative impact of active pharmaceutical ingredient properties, processing methods, and excipient variability on drug product quality attributes. *Journal of Pharmaceutical Sciences, 103*(2), 527–538. https://doi.org/10.1002/jps.23810.

Lerk, C. F., Bolhuis, G. K., Smallenbroek, A. J., & Zuurman, K. (1982). Interaction of tablet disintegrants and magnesium stearate during mixing. II. Effect on dissolution rate. *Pharmaceutica Acta Helvetiae, 57*(1011), 282–286.

Merritt, J. M., Viswanath, S. K., & Stephenson, G. A. (2013). Implementing quality by design in pharmaceutical salt selection: A modeling approach to understanding disproportionation. *Pharmaceutical Research, 30*(1), 203–217.

Narang, A. (2015). Addressing excipient variability in formulation design and drug development. In A. Narang, & S. Boddu (Eds.), *Excipient applications in formulation design and drug delivery*. Springer https://doi.org/10.1007/978-3-319-20206-8_18.

Narang, A. S., Badawy, S., Ye, Q., Patel, D., Vincent, M., Raghavan, K., Huang, Y., Yamniuk, A., Vig, B., Crison, J., Derbin, G., Xu, Y., Ramirez, A., Galella, M., & Rinaldi, F. A. (2015a). Role of self-association and supersaturation in oral absorption of a poorly soluble weakly basic drug. *Pharmaceutical Research*. https://doi.org/10.1007/s11095-015-1645-y.

Narang, A.S., Badawy, S.I.F., Subramanian, G.A., LaMarche, K.R., Bindra, D.S., & Rao, V.M. (2012a). *High drug load tablet formulation of brivanib alaninate.* WO/2012/097222.

Narang, A. S., Sheverev, V. A., Stepaniuk, V., Badawy, S., Stevens, T., Macias, K., Wolf, A., Pandey, P., Bindra, D., & Varia, S. (2015b). Real-time assessment of granule densification in high shear wet granulation and application to scale-up of a placebo and a brivanib alaninate formulation. *Journal of Pharmaceutical Sciences, 104*(3), 1019–1034. https://doi.org/10.1002/jps.24233.

Narang, A. S., Yamniuk, A. P., Zhang, L., Comezoglu, S. N., Bindra, D. S., Varia, S., Doyle, M. L., & Badawy, S. (2012b). Reversible and pH-dependent weak drug-excipient binding does not affect oral bioavailability of high dose drugs. *Journal of Pharmacy and Pharmacology, 64*(4), 553–565. https://doi.org/10.1111/j.2042-7158.2011.01435.x.

Pandey, P., Sinko, P. D., Bindra, D. S., Hamey, R., Gour, S., & Vema-Varapu, C. (2013a). Processing challenges with solid dosage formulations containing vitamin E TPGS. *Pharmaceutical Development and Technology, 18*(1), 296–304. https://doi.org/10.3109/10837450.2012.737807.

Pandey, P., Tao, J., Chaudhury, A., Ramachandran, R., Gao, J. Z., & Bindra, D. S. (2013b). A combined experimental and modeling approach to study the effects of high-shear wet granulation process parameters on granule characteristics. *Pharmaceutical Development and Technology, 18*(1), 210–224. https://doi.org/10.3109/10837450.2012.700933.

Pingali, K. C., & Mendez, R. (2014). Nanosmearing due to process shear influence on powder and tablet properties. *Advanced Powder Technology, 25*(3), 952–959. https://doi.org/10.1016/j.apt.2014.01.016.

Prpich, A., Am Ende, M.T., Katzschner, T., Lubczyk, V., Weyhers, H., & Bernhard, G. (2008). Application of quality by design principles in the model predictions for scale-up of tablet film coating. Paper presented at the AIChE annual meeting, conference proceedings United States.

Saurí, J., Millán, D., Suñé-Negre, J. M., Colom, H., Ticó, J. R., Miñarro, M., Pérez-Lozano, P., & García-Montoya, E. (2014). Quality by design approach to understand the physicochemical phenomena involved in controlled release of captopril SR matrix tablets. *International Journal of Pharmaceutics, 477*(1–2), 431–441. https://doi.org/10.1016/j.ijpharm.2014.10.050.

Sheskey, P. J., Robb, R. T., Moore, R. D., & Boyce, B. M. (1995). Effects of lubricant level, method of mixing, and duration of mixing on a controlled-release matrix tablet containing hydroxypropyl methyl cellulose. *Drug Development and Industrial Pharmacy, 21*(19), 2151–2165. https://doi.org/10.3109/03639049509065898.

Visser, J. C., Dohmen, W. M. C., Hinrichs, W. L. J., Breitkreutz, J., Frijlink, H. W., & Woerdenbag, H. J. (2015). Quality by design approach for optimizing the formulation and physical properties of extemporaneously prepared orodispersible films. *International Journal of Pharmaceutics, 485*(1–2), 70–76. https://doi.org/10.1016/j.ijpharm.2015.03.005.

Waterman, K. C., Adami, R. C., Alsante, K. M., Antipas, A. S., Arenson, D. R., Carrier, R., Hong, J., Landis, M. S., Lombardo, F., Shah, J. C., Shalaev, E., Smith, S. W., & Wang, H. (2002). Hydrolysis in pharmaceutical formulations. *Pharmaceutical Development and Technology, 7*(2), 113–146. https://doi.org/10.1081/PDT-120003494.

Waterman, K. C., Carella, A. J., Gumkowski, M. J., Lukulay, P., MacDonald, B. C., Roy, M. C., & Shamblin, S. L. (2007). Improved protocol and data analysis for accelerated shelf-life estimation of solid dosage forms. *Pharmaceutical Research, 24*(4), 780–790. https://doi.org/10.1007/s11095-006-9201-4.

Wolfe, S., Pafiakis, S., Tang, D., Jennings, S., Abebe, A., Gao, Z., Varia, S.A., & Narang, A.S. (2013). Effect of tablet pore size distribution on dissolution upon accelerated storage conditions. Paper presented at the AAPS annual meeting, San Antonio, TX.

Yu, L. X. (2008). Pharmaceutical quality by design: Product and process development, understanding, and control. *Pharmaceutical Research, 25*(4), 781–791. https://doi.org/10.1007/s11095-007-9511-1.

Yu, L. X., Amidon, G., Khan, M. A., Hoag, S. W., Polli, J., Raju, G. K., & Woodcock, J. (2014). Understanding pharmaceutical quality by design. *The AAPS Journal, 16*(4), 771–783. https://doi.org/10.1208/s12248-014-9598-3.

Zhao, F., Derbin, G., Miller, S., Badawy, S., & Hussain, M. (2012). Acid-catalyzed hydrolysis of BMS-582664: Degradation product identification and mechanism elucidation. *Journal of Pharmaceutical Sciences, 101*(9), 3526–3530. https://doi.org/10.1002/jps.23028.

CHAPTER 19

Implementation of pharmaceutical quality by design in wet granulation

Xiang Yu, Lawrence X. Yu, Yue Teng, Dhaval K. Gaglani, Bhagwant D. Rege and Susan Rosencrance

Center for Drug Evaluation and Research, Office of Pharmaceutical Quality, Food and Drug Administration, Silver Spring, MD, United States

19.1 Introduction

Decades ago most pharmaceutical companies followed the "pharmaceutical quality by testing" (QbT) regulatory framework for drug products. Under that system, the formulation was designed empirically and iteratively, the manufacturing processes were fixed, and product quality was achieved through a series of extensive and inefficient testing on raw materials, in-process intermediates, and final products. This situation remained unchanged until the publication of the FDAfor the 21st century Initiative (FDA, 2004b) and process analytical technology (PAT) innovative framework (FDA, 2004a).

Pharmaceutical quality by design (QbD) is a systematic, scientific, risk-based, and holistic approach to pharmaceutical development in which quality is built into the product and process understanding and control. The principles of QbD were systemically described in the ICH Q8 (R2) guidance (pharmaceutical development). The Q8, Q9, and Q10 questions and answers and their appendix, known as the points to consider document (FDA, 2011, 2012), also provided high-level directions for QbD implementation. In addition to this guidance and these documents, recent publications have further detailed the pharmaceutical QbD concept and objectives, and explained the QbD elements and implementation tools (Yu, 2008; Yu et al., 2014; Lionberger et al., 2016). These regulatory efforts have achieved great success, with improved drug quality and process capability.

QbD can be applied to all types of drug products because the quality requirements of drug products, the management of manufacturing facilities, and the operation of the manufacturing process are all on the same regulatory basis. This is especially true for drug products in solid dosage forms. No matter whether the manufacturing process is batch manufacturing or continuous manufacturing, the whole manufacturing process can be seamlessly assembled by individual but common unit operations such as blending, wet and dry granulation, compression, coating, etc. Therefore successful implementation of QbD in the manufacturing process means a full understanding of each unit operation involved in the process, which also means using QbD tools to identify all critical quality attributes (CQAs), critical material attributes (CMAs), and critical process parameters (CPPs), and then developing a corresponding control strategy.

The FDA is the bridge between the United States' public and the pharmaceutical industry. It is also the information warehouse for the pharmaceutical industry. The FDA approved and documented about 388 new drug applications (NDAs) and 2851 abbreviated new drug applications (ANDAs) in tablet and capsule dosage forms between 2006 and 2015 (FDA, 2016). Reviewing these approved drug product applications enables us to better evaluate the status of QbD implementation in drug product development by the pharmaceutical industry. Also, systematically utilizing the process data in these applications can also establish a knowledge base, which can help improve regulatory review quality and efficiency.

This chapter documents our effort to develop a unit operation profile for systemically describing the general research pattern of a single unit operation under the QbD paradigm. On the basis of process information available from the pharmaceutical industry, we created two-unit operation profiles: one for high/low-shear (HS/LS) wet granulation and the other for FB granulation. These unit operation profiles help quantify the importance of material attributes, process parameters, and process outputs in the profile, and summarize the behavior of wet granulation in the whole manufacturing process development and the utilization of QbD tools in wet granulation development. As an extension of "understanding pharmaceutical QbD" (Yu et al., 2014), we hope this book chapter can enhance the prior knowledge of wet granulation development, and help regulatory agencies further enhance review efficiency and review quality.

19.2 Function and popularity of wet granulation

Granulation is a typical unit operation of particle enlargement carried out by the pharmaceutical industry to produce pharmaceutical dosage forms like tablets and capsules for over 70 years (Parikh, 2009). The purpose of wet granulation can be, but is not limited to:

- Enhance the uniformity of the drug substance in the final dosage form.
- Increase the density of the blend.
- Increase the flowability and compressibility of the powder mixture.
- Reduce the dust during the manufacturing process.
- Alter the physical appearance or surface properties.
- Improve the wettability of a poorly soluble drug substance.

Generally, the wet granulation process commences after fully mixing the drug substance and other necessary intragranular excipients to ensure the uniform distribution of each ingredient. The granule size used in the pharmaceutical industry is usually controlled between 0.2 and 4 mm, with a narrow particle size distribution. Drying and milling are required before tablet compression or capsule filling (Parikh, 2009; Shanmugam, 2015).

The pharmaceutical industry prefers direct compression/encapsulation over granulation. Wet granulation is to be considered next if the physical properties of the final powder mixture cannot satisfy the requirements for direct compression or encapsulation. In order to understand how often pharmaceutical companies employ wet granulation in their manufacturing process to produce drug products in solid oral dosage forms, we conducted a survey by counting any type of wet granulation employed in the manufacturing processes of 98 approved NDAs and 128 approved ANDAs. Among these 226 applications, 73.9% of drug products are in tablet dosage form, and the remaining 26.1% are in capsule dosage form. Among these, 46.7% of the tablets and 18.6% of the capsules employed at least one wet granulation step in the manufacturing process.

The wet granulation process can be classified into three major types according to the shear force applied to the granulation materials: HS granulation, LS granulation, and fluidized bed (FB) granulation.

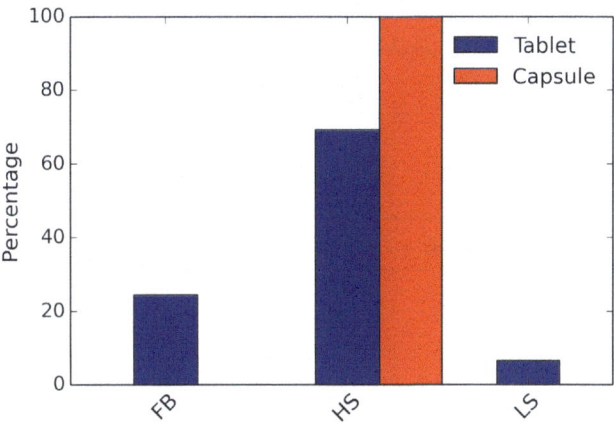

FIGURE 19.1

Popularity of FB granulation, HS granulation, and LS granulation in manufacturing processes to produce drug products in tablet and capsule dosage forms.

The granulation mechanism of FB granulation is completely different from that of HS/LS granulation, while the strength of its shear force on the granule is between HS granulation and LS granulation.

Fig. 19.1 shows a comparison of the popularity of three types of wet granulation used for tablet and capsule manufacturing. Clearly, HS granulation is the most popular wet granulation method in the pharmaceutical industry. About 69.2% of all tablet manufacturing processes with wet granulation and all capsule manufacturing processes with wet granulation employed HS granulation for wet granulation. The second most popular wet granulation technique is FB granulation, which was used among 24.4% of tablet manufacturing processes. Because of its time-consuming nature and relatively higher difficulty in controlling the granule particle size, the LS granulation technique is much less popular than HS granulation and FB granulation, and only about 6.4% of tablet manufacturing processes still apply this technique.

The HS granulation offers several advantages over FB and LS granulation, which make it the most popular wet granulation method in the pharmaceutical industry. These advantages include short granulation time, use of less granulation fluid, and easy determination of granulation endpoint (Liu et al., 2009). Of course, FB granulation would be of benefit if time/cost factors are not major considerations, or if a formulation has potential stability issues due to moisture exposure (Parikh & Jones, 2009). Although comparison of these three wet granulation methods has been well discussed in the literature, we notice that the selection of a specific wet granulation technique for the manufacturing process may not be fully science-based. In fact, in many situations, these three wet granulation methods can all produce granules with acceptable quality for the following downstream operations. At that time, previous R&D experience, facility availability from lab scale to final commercial scale, cost of wet granulation, and time required for wet granulation all became considerable factors for a pharmaceutical company to design and develop the manufacturing process.

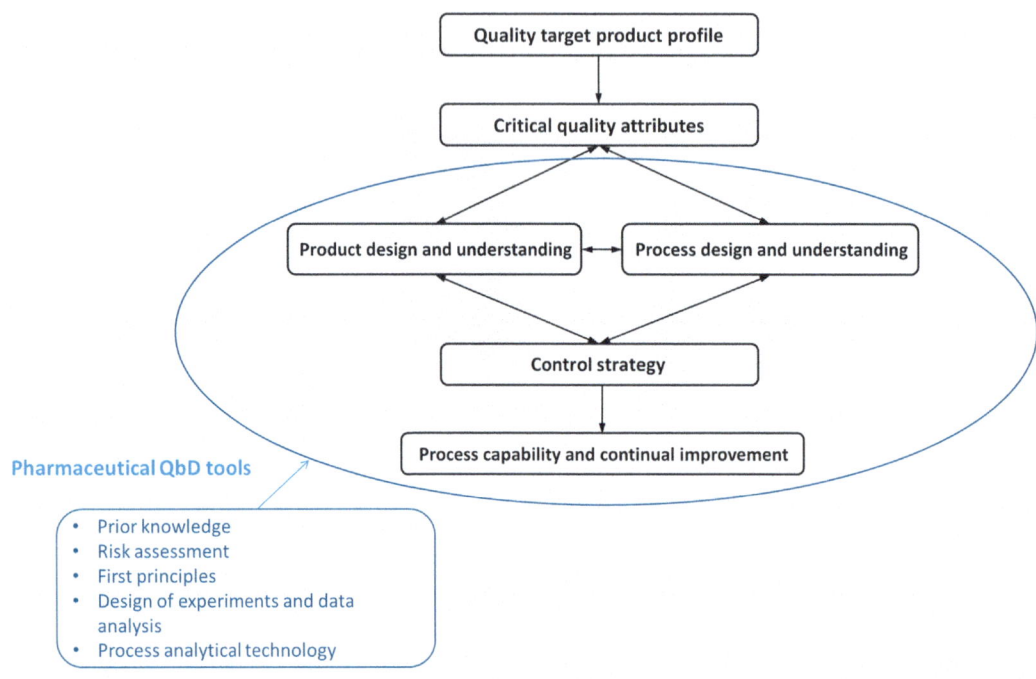

FIGURE 19.2

Typical elements of pharmaceutical QbD for product development.

During the review of these applications, we did observe that some pharmaceutical companies switched from FB granulation to HS granulation even after completing the initial formulation development in order to save process cost. Meanwhile, we also observed that some companies persisted in using LS granulation because of their prior knowledge of similar drug products and the availability of LS granulation facilities for research and manufacturing.

19.3 Quality by design and unit operation profiles for wet granulation
19.3.1 From product to unit operation

As shown in Fig. 19.2, the entirety of drug product development based on pharmaceutical QbD is composed of six typical elements (Yu et al., 2014):

1. A quality target profile (QTPP) that identifies CQAs of the drug product.
2. Product design and understanding, including the identification of CMAs based on risk assessment, prior knowledge, first principles, and design of experiments (DOEs).
3. Process design and understanding, including the identification of CPPs and a thorough understanding of scale-up principles, which link CMAs and CPPs to CQAs.

4. A control strategy that includes specifications for the drug substance(s), excipient(s), and drug product, as well as controls for each step of the manufacturing process.
5. Process capability and product lifecycle management, including continual improvement.

These QbD elements are designed to help pharmaceutical companies identify characteristics that are critical to consumers, translate them into drug product CQAs, and establish the relationship between formulation/manufacturing variables and CQAs in order to reproducibly deliver the therapeutic promis in the label to the patients (Yu, 2008).

When the development target is narrowed down to a specific step of the manufacturing process, such as wet granulation, the elements of QbD can be rationally simplified to:

- Identification of CQAs of wet granulation from the list of granulation outputs.
- Product and process design and understanding of wet granulation; and
- Control strategy of wet granulation.

In order to better understand the research pattern of single-unit operation, we propose to create a profile for each of these common pharmaceutical unit operations. The unit operation profile can be understood as:

- A data structure in-line with QbD to store material and process information during unit operation development, which includes drug substance(s) and the corresponding physical, chemical, and biological properties, excipients and their physical/chemical properties, process parameters, process outputs, equipment type, and specific equipment model, etc.
- A data set of unit operations with information extracted from scientific documents and standardized, following the same data structure; and
- A combination of statistical analysis and a summary of the collected data set.

The feasibility of creating profiles to describe the general wet granulation development is based on three considerations:

1. The majority of recently approved drug product applications have followed the QbD paradigm.
2. Drug products in solid dosage forms have similar QTPP and product CQAs. Therefore, the CQAs of wet granulation are also similar.
3. For the same type of wet granulation, the equipment made by different manufacturers can be operated in a similar manner and can execute the same function.

Due to different granulation mechanisms and operation procedures, we created two independent profiles for HS/LS wet granulation and FB wet granulation. We manually collected all the information on wet granulation from previously approved applications to prepare the data set to create an HS/LS granulation profile and an FB granulation profile. The information includes the granulation procedure, process outputs, material attributes involved in process development, all reported process parameters, and QbD tools (PAT, risk assessment, and DOE). Limited by our time and labor resources, 70 applications in total were carefully reviewed and processed to prepare the data sets for HS/LS granulation and FB granulation. After aligning the operation procedures of wet granulation from different drug product applications, associated process parameters were further grouped, standardized, and analyzed. In this section, we only describe the operation procedures of HS/LS granulation and FB granulation. Details of quality characterization of granules, material attributes, process parameters, and application of QbD

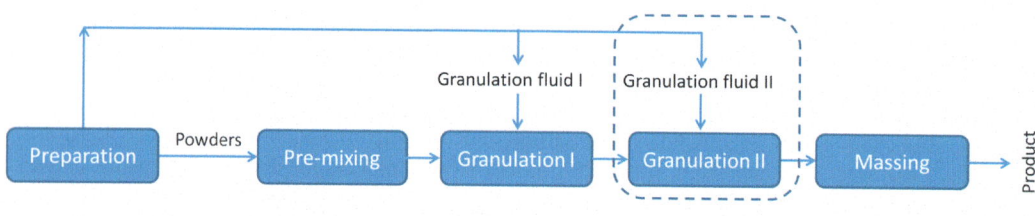

FIGURE 19.3

The unit operation profile of HS/LS granulation.

tools in wet granulation development (risk assessment, DOEs, and PAT), summarized from these unit operation profiles, will be introduced in the following sections.

19.3.2 Process operation procedure in high-shear/low-shear granulation profile

The created HS/LS granulation profile consists of five operation procedures, which are preparation, premixing, granulation I, granulation II (optional), and massing (Fig. 19.3).

1. *Preparation*: The granulator is cleaned and examined before any granulation operation. Granulation fluid is prepared by following the established protocol. All other intragranular material should be preprocessed (e.g., micronization of drug substance, screening of intragranular components) and accurately weighed.
2. *Premixing*: In this stage, the intragranular components are transferred into the HS/LS granulator. During this stage, the impeller is turned on, and the chopper is normally turned off.
3. *Granulation I*: After premixing, the granulation fluid is pumped or sprayed into the granulator. At the same time, both impeller and chopper (if available) are turned on, and the appropriate impeller speed and chopper speed need to be adjusted.
4. *Granulation II (optional)*: After completing the spraying of granulation fluid, extra granulation fluid or water may be further sprayed onto the top of the granule mixture. To decrease the number of process parameters to be studied during granulation, the process parameters (including impeller speed, chopper speed, and spray speed) of Granulation II are normally kept the same as those of Granulation I. The step of Granulation II is not necessary for most of the HS/LS granulation operations. It may be employed for one of the following considerations:
    - Use a small amount of water to rinse the container after completing the spray of granulation fluid I.
    - Spray the granulation fluid with the new granulation fluid of granulation II.
    - Continue the granulation so that the reading of the impeller torque or power consumption matches the predetermined granulation profile.
5. *Wet massing*: The alternative name for wet massing is kneading. The purpose of wet massing is to further unify the liquid distribution and densify the granules. Both the impeller and chopper are normally set to "on," and the setting of these parameters is usually kept the same as the previous granulation step.

19.3 Quality by design and unit operation profiles for wet granulation

Table 19.1 The common input variables of HS/LS granulation.

Preparation	Premixing	Granulation I	Granulation II (Optional)	Massing
Type of granulator	Impeller (tip) speed	Impeller (tip) speed		
Batch size and volume fill ratio		Chopper speed		
Spray system	Premixing time	Amount of liquid I	Amount of liquid II	
Drug substance properties		Spray rate/time		Massing time
Method of binder addition	Jacket/product temperature			
Type, amount, and grade of intragranular excipient	Power consumption/torque			

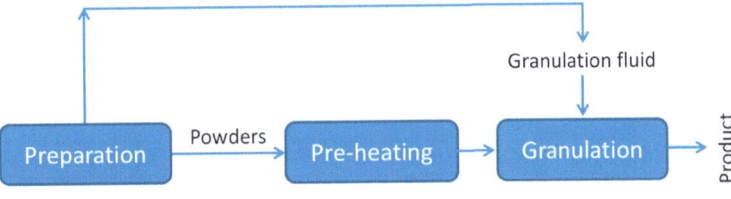

FIGURE 19.4

The unit operation profile of FB granulation.

The common input variables, including material attributes and process parameters of these five steps of HS/LS granulation that are frequently studied and reported in drug product applications, are listed in Table 19.1.

19.3.3 Process operation procedure in fluidized bed granulation profile

One advantage of FB granulation is that wet granules can be dried in the same equipment directly following the granulation operation. Many process scientists consider FB granulation and FB drying as a completely integrated unit operation. But keeping in line with HS/LS granulation, here the drying procedure is not included in our current profile of FB granulation. The general procedure of FB granulation is given in Fig. 19.4, which includes preparation, preheating, and granulation, three steps.

1. *Preparation*: Similar to HS/LS granulation, this step includes preparing and checking fluidized bed equipment, preparing granulation fluid, and preparing the dry powder mixture.
2. *Preheating*: The powder mixture is transferred into the fluidized bed. By blowing hot air through the bottom of the fluidized bed, the mixture is further mixed and fully fluidized, and the whole bed is heated to the intended temperature (normally the product temperature during granulation).
3. *Granulation*: Within this stage, granulation fluid is continuously sprayed onto the top of the powder mixture to form wet granules.

Table 19.2 The list of common input variables of FB granulation, aligned with preparation, preheating, and granulation (Fig. 4).

Preparation	Preheating	Granulation
Type of fluidized bed	Preheat inlet air temperature and preheat time	Amount of granulation fluid
Batch size and volume fill ratio		Spray rate/time
Spray system		Atomization air pressure
Process filter system		Atomization air flow
Drug substance properties		Inlet air temperature
Method of binder addition		Inlet air volume
Type, amount, and grade of intragranular excipient		Inlet dew point
		Product (bed) temperature
		Exhaust air temperature
	Shake time/interval	

The common input variables, including material attributes and process parameters of these three steps of FB granulation that are frequently studied and reported in drug product applications, are listed in Table 19.2.

19.4 Quality characterizations of wet granulation product

The main purpose of wet granulation is to increase the flowability and compactibility of powder mixture for tablet compression or encapsulating, or to increase the blend uniformity of the powder mixture due to the low dose of drug substance. However, because of the natural properties of wet granules (such as massy, sticky, and easy to agglomerate), it is difficult to directly and fully characterize the wet granules and then choose appropriate outputs of wet granules to correlate with the CQAs of the drug product. In such a case, process scientists prefer to consider the following downstream unit operations, such as drying, dry milling, and/or blending/lubricating, as an extension of wet granulation by fixing the process parameters of these unit operations during the study of wet granulation. Alternatively process scientists consider wet granulation plus these downstream unit operations to be a tightly integrated functional assembly to be studied together.

Table 19.3 lists the outputs of wet granulation and the reported frequency of use of each output for tablet products and capsule products summarized from our HS/LS profile and FB profile. The report frequency indicates how often a variable (an output or a material attribute, or a process parameter) of wet granulation was described in the drug product applications, which can be calculated by the number of applications that described the variable divided by the number of reviewed applications. This result shows that the majority of process scientists put the focus of granulation quality evaluation on the characterization of dried granules (usually after the step of dry milling).

Table 19.3 The reported frequency of wet granulation outputs for both capsule products and tablet products.

Step	Output	Capsule (%)	Tablet (%)
Wet granulation	Premixing blend uniformity	22.22	13.04
	In-process LOD and final LOD of granules	0	10.7
	PSD of wet granules	0	2.17
After drying[a]	Compressibility	22.22	8.7
	Strength and content uniformity	11.11	15.22
	Sphericity	11.11	0
	Bulk density	55.56	34.78
	Tapped density	44.44	32.61
	Granule flow	22.22	13.04
	Appearance	0	4.35
	LOD/humidity	55.56	78.26
	Yield	11.11	4.35
	PSD of dry granules	100	60.87
	Impurity	22.22	17.39
After lubrication and/or final blending	Compressibility	0	17.39
	Strength and content uniformity	11.11	30.43
	Bulk density	0	21.74
	Tapped density	0	19.57
	Granule flow	0	28.26
	Appearance	0	4.35
	LOD/humidity	0	4.35
	PSD of blend	33.33	34.78
After capsuling	Capsule strength and content uniformity	33.33	0
	Capsule dissolution	87.78	0
After compression	Tablet strength and content uniformity	0	65.65
	Tablet dissolution	0	71.74
	Tablet disintegration time	0	45.65
	Tablet hardness and tensile strength	0	47.83
	Tablet compression defect	0	8.7
	Tablet friability	0	15.22
	Tablet thickness	0	13.04
	Tablet weight	0	21.74
API	API polymorphic form	11.11	0
	API impurity and degradation	22.22	10.87
	Residual solvents of dried granules	0	2.17

[a] Usually the dry granules are sampled after both drying and milling.

The top four outputs of granulation for capsule products are:

1. Particle size and size distribution (PSD) of dried granules (100%),
2. Capsule dissolution (87.78%),
3. Loss on drying (LOD) of dried granules (55.56%), and
4. Bulk density (55.56%).

The top three outputs of granulation for tablet products include:

1. LOD of dried granules (78.26%),
2. Tablet dissolution (71.74%), and
3. Particle size and the size distribution of dried granules (60.87%).

Clearly, directly measuring the impact of granule properties on at least one of the product CQAs is a standard approach to evaluate the quality of granulation.

It is important to point out that this table should not be directly considered a weight matrix of intermediate CQAs for wet granulation for two reasons. First, this table is summarized from two general profiles, and the importance of specific quality requirements of some drug substance/product cannot be fully reflected in this table. For example, if the component in drug product formulation tends to degrade or transform morphology under high temperature/humidity environment, substance-related physical/chemical property should always be continuously monitored throughout the whole development process and should be considered a CQA if that variable qualifies the definition of CQA in ICH Q8(R2) guidance (FDA, 2009). Second, the risk evaluation of these outputs should always be iteratively updated throughout the process development and the lifecycle of the drug product. Consequently the importance of these outputs evaluated by process scientists may also change.

19.5 Product design and understanding of wet granulation

As addressed in a recent publication (Yu et al., 2014), the key elements of product design and understanding for wet granulation under a QbD quality system include the following:

- Identify physical, chemical, and biological attributes of drug substance(s) that could impact the performance of the final product.
- Identify and determine intragranular excipient type, grade, and amount.
- Use risk assessment, prior knowledge, and experiments to understand drug-excipient interactions and identify potentially high-risk material attributes.
- Identify CMAs of both excipients and drug substance, and optimize the formulation.
- Develop a control strategy for CMAs of both drug substance and excipients that can have an effect on product CQAs.

A sound understanding of properties of the drug substance(s) can be helpful in (1) rationally selecting excipients whose properties can compensate for the properties of the drug substance, and (2) developing a robust granulation process. Before any formulation and process development, the physical properties, chemical properties, and biological properties of the drug substance and their potential influence on granulation performance and final product quality should be considered (Table 19.4). Understanding the properties of drug substance(s) is the key to selecting appropriate intragranular excipients. The input variables of the drug substance usually include:

Table 19.4 Physical properties, chemical properties, and biological properties of drug substance that should be considered during initial product design.

Physical Properties	Chemical Properties	Biological Properties
Bulk/tapped density	pKa	Partition coefficient
Particle size distribution	Chemical stability in different environments	Membrane permeability
Morphology (crystalline or amorphous)	Photolytic stability	Bioavailability
Solubility in different solvents	Oxidative stability	
Intrinsic dissolution rate		
Hygroscopicity		
Thermostability		
Melt point(s)		

- Drug load
- Drug particle size
- Drug addition method (dry addition or wet addition)

Pharmaceutical excipients, which are defined as substances other than the active pharmaceutical ingredient, are intentionally included in all drug products in solid dosage forms and are essential to product performance. As in the same situation to quality control of drug products, both the manufacturing of excipients and the use of the excipients in drug products are controlled under regulatory oversight. The USP and NF general chapter on excipients listed 48 functional categories of excipients used in pharmaceutical products (The United States Pharmacopeial Convention, 2016). Intragranular formulation usually contains one to five excipients. These mainly include diluent, disintegrant, wet binder, and sometimes sweetener, coloring agent, glidant, and antioxidant (if necessary).

During process development and process validation, the major input variables of intragranular excipients can be summarized as:

- Type, level, and grade of wet binders,
- Type, level, and grade of disintegrants, if needed,
- Type, level, and grade of diluents, if needed.
- Type, level, and grade of other intragranular excipients, if needed.
- Binder addition method (dry addition, wet addition, or the combination of two methods).

Because the formulation development is such a large topic, it deserves to be an independent chapter. Here we only summarize some considerations of the formulation that may influence the granulation performance and, therefore the quality of the drug product. Obviously the binder is the pivotal excipient in intragranular formulation. The USP-NF lists 72 different wet binders and other binders. Binders that are frequently used in tablets/capsules include hydroxypropyl cellulose, hypromellose, polyvinylpyrrolidone (PVP), pregelatinized starch, and corn starch.

These wet binders normally have multiple functions in the formulation. For example, corn starch and microcrystalline cellulose are more frequently considered as disintegrants and diluents, and lactose monohydrate is mainly considered as a diluent in tablets and capsules (The United States Pharmacopeial Convention, 2016).

Throughout our review study, over 70% of pharmaceutical companies still follow the classical "trial and error" strategy for formulation development. As a consequence, the whole formulation was considered a single variable in the manufacturing process development. Such a development practice does not really help evaluate the importance of formulation variables. It unintentionally undermines the interactions between material attributes and process parameters, and leads to a local optimization of the drug product. For example, the majority of applicants didn't pay attention to the potential interaction of the particle size of the drug substance, the grade/supplier of binder, and the ratio of intragranular binder versus extragranular binder with other process parameters on granulation quality.

Another objective of the formulation development of wet granulation is to determine the binder addition method. The binder addition method affects the granule properties and thus the dissolution profile of the drug product (Holm et al., 2001). The addition methods include dry addition (binder is mixed with other intragranular components), wet addition (binder is dissolved in granulation fluid), and the combination method (binder is included in both intragranular mixture and granulation fluid). Usually, the method of binder addition is determined by the formulation scientist at the early stage of formulation development and will not be changed throughout the manufacturing process development. The reviewing result shows that about 49% of the applications chose the dry addition method, 33% chose the wet addition method, and the remaining 18% chose the combination of dry addition and wet addition.

Finally pharmaceutical developers should consider the following at the early stage of formulation development:

- Unless sufficient toxicity data of the new excipient can be provided to the agency to demonstrate its safety, the amount of excipient in the final formulation is not encouraged to exceed the safety limit listed in the FDA inactive ingredients database (https://www.accessdata.fda.gov/scripts/cder/iig/index.cfm), which is based on the amount of excipient used in prior approved drug products.
- Systematic drug-excipient compatibility studies should be performed prior to formulation optimization. Excipients can be a major source of variability in drug quality. Performing drug-excipient compatibility studies at an early stage of formulation development can help develop a robust formulation by maximizing the stability of a formulation and, hence, the shelf life of the final drug product. Also identification of potential interactions between drug substance and excipients helps root cause analysis, if and when a stability problem occurs.
- Risk assessment and DOEs as powerful research tools are highly recommended to be employed in formulation development, even if the QbD paradigm is not planned in the product development.

19.6 Process design and understanding of wet granulation

Process parameters are referred as to the input operating parameters (e.g., speed and flow rate) or process state variables (e.g., temperature and pressure) of a process step or unit operation. Lawrence et al. (n.d.) listed the typical material attributes, process parameters, and outputs of HS/LS granulation and FB granulation in their 2014 QbD paper (Yu et al., 2014), but the importance of these process parameters

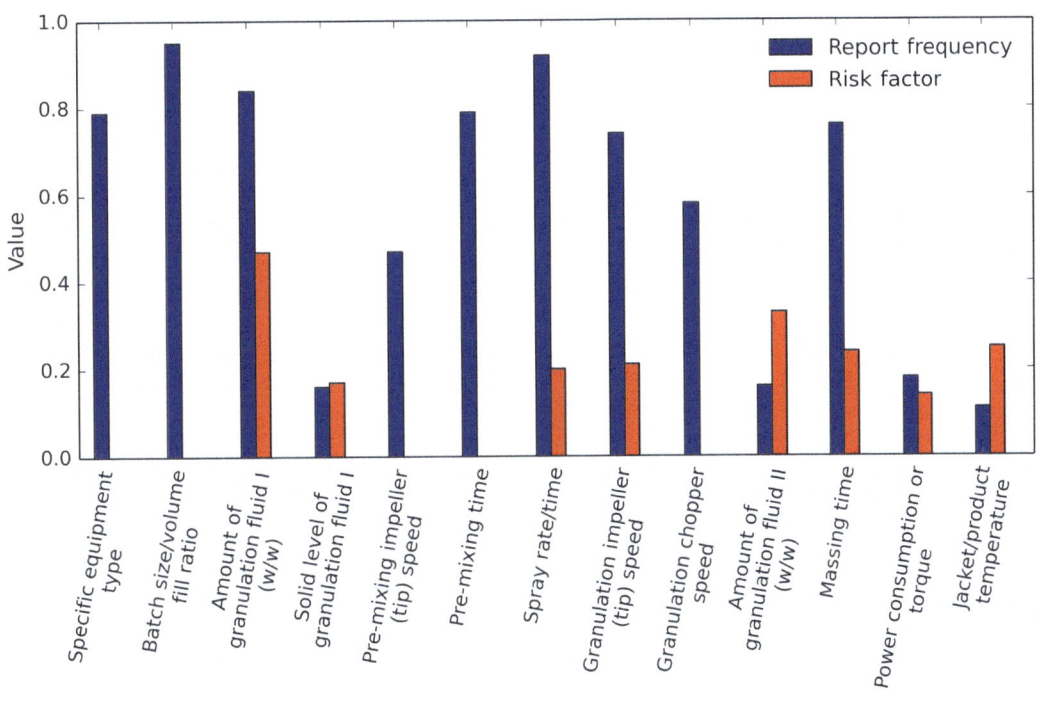

FIGURE 19.5

Process parameters with report frequency and the ratio of high risk of the listed process parameters from HS/LS-granulation profile.

was not further evaluated and reported by the authors later. This is the initial driving force compelling us to prepare this book chapter. Here, based on the HS/LS granulation profile and FB granulation profile generated from recently approved drug product applications, we calculated the reported frequencies and risk factors of common process parameters, and characterized the research behavior of each process parameter at laboratory scale and pilot/submission batch scale. The submission batch is a drug product batch manufactured under a GMP environment with its stability data to be submitted to the regulatory agency for quality review. There is a mandatory requirement of batch size for the submission batches submitted by both the NDA applicant and the ANDA applicant.

19.6.1 Process parameters for high-shear/low-shear granulation

The 13 process parameters of HS/LS granulations that are commonly studied by process scientists are listed in Fig. 19.5. Because process scientists usually choose to provide an agency with details about the process information, they think that it is important for process understanding and development; the reported frequency of these process parameters may reflect the process scientists' opinion about the importance of process parameters in HS/LS granulation. However, the current HS/LS granulation profile is a general unit operation profile without considering some special situations (e.g., temperature sensitivity of drug substance, dry addition method vs wet addition method). The reported frequency for

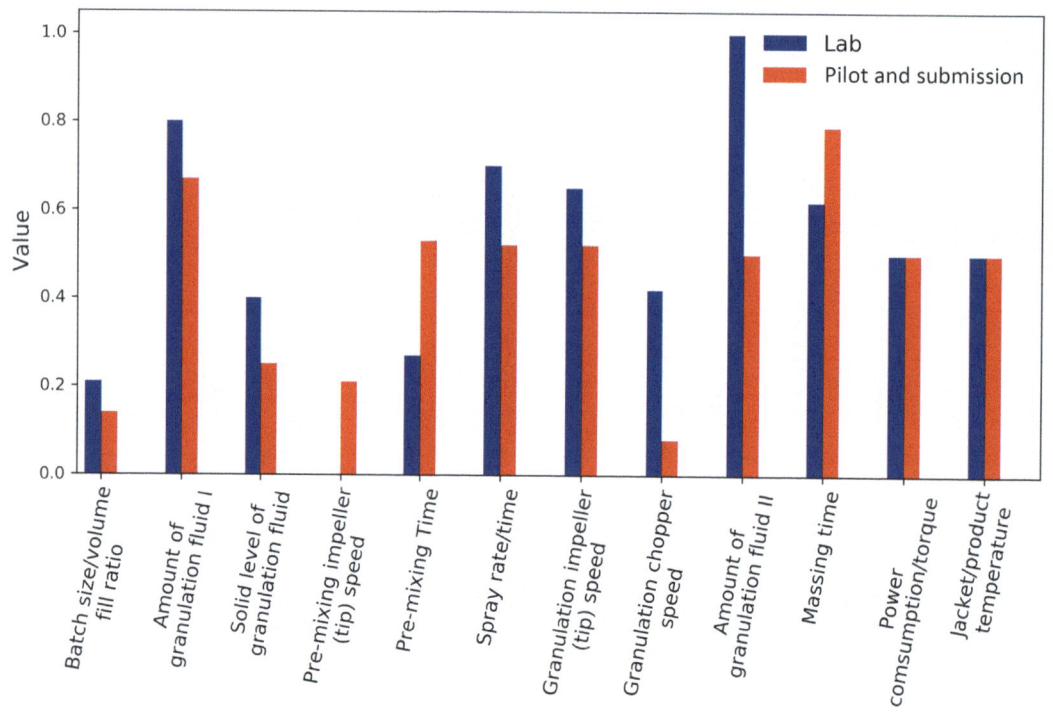

FIGURE 19.6

Study frequencies of process parameters that were reported (with investigation range) at laboratory scale, pilot scale, and submission batches from HS/LS-granulation profile.

some process parameters (e.g., jacket or product temperature, amount of granulation fluid II) may be potentially biased.

Calculating the frequency of a process parameter considered to be a CPP is an ideal way to represent the risk of that process parameter to the granulation process in general. But the judgment of a process parameter to be a CPP or not is greatly influenced by the setting of the final operation range of that parameter in the proven operating space (Lionberger et al., 2008). Therefore, the risk factor of each process parameter was calculated and provided in Fig. 19.5 in order to help process scientist better perform risk assessment in their own process development.

The risk factor of a process parameter is defined as the reported number of high risk divided by the number of our reviewed applications, while the reported number of high risk is represented as the number of times that a process parameter was determined to be CPP in the application or that process parameter was statistically significant in DOE studies at any scale. To each of the listed process parameters in Fig. 19.5, process scientists may choose to either fix the process parameter to a specific value or obtain an operation range of that parameter through a series of trials. In order to obtain the parameter research pattern from these process studies, we also calculate the rate at which the process parameter was studied and the range of investigation (Fig. 19.6). Any process parameter that was not reported in the application was considered to be a fixed variable and was not to be studied in the trials

Table 19.5 The batch size ranges and representative high-shear granulator models for each range of batch size.

Batch Size (kg)	Example
0.2–2	Diosna P1-6
3–6	PMA-1
9–40	Gral 150
40–125	VG400
>125	VG600

The common process parameters for HS/LS granulation are discussed below, and the behavior of these process parameters at laboratory scale and pilot scale is summarized in Table 19.5.

19.6.1.1 Equipment type

About 79% of companies reported the type of HS/LS granulators used for process development and scale-up. For LS granulation, the planetary mixer is the only type of LS granulator employed for LS granulation. For HS granulation, the HS granulators can be classified into vertical HS granulators and horizontal HS granulators per the geometric position and orientation of the primary impeller. The impeller shaft of a vertical HS granulator rotates in the vertical plane, and the impeller is bladelike, while the impeller shaft of a horizontal HS granulator rotates in the horizontal plane. In our study, we observe that the vertical (either top-driven or bottom-driven) HS granulator is more popular than the horizontal HS granulator, and the two types of vertical HS granulators (top-driven and bottom-driven) are both frequently employed in HS granulation. FDA SUPAC manufacturing equipment addendum (FDA, 2014) lists the major manufacturers of both LS and HS granulators. Different granulator types, or even granulators of the same granulator type but produced by different manufacturers, can definitely impact granulation quality attributes, such as the average particle size of granules (Bouwman et al., 2004; Liu et al., 2009). Nonetheless, this equipment impact can be well compensated by adjusting the process parameters (e.g., the amount of granulation fluid, granulation impeller speed, massing time). Therefore no applicant had even considered the granulator type to be critical.

19.6.1.2 Batch size and volume fill ratio

The FDA highly encourages the pharmaceutical industry to both provide batch sizes with batch identification numbers and to list specific equipment models used during process development in the application to facilitate the review and evaluation. Except for a few applications with historical issues, most of the pharmaceutical companies provided at least the size of submission batches to the agency. The normal ranges of batch size for HS granulation development are summarized in Table 19.5.

During process development, the batch size of experimental trials at various scale levels is normally fixed, as to the volume fill ratio. Unlike batch size, which is always reported in the drug product application explicitly, the volume fill ratios can able to be captured from only a few applications. Even though we consider the volume fill ratio is provided by pharmaceutical companies as long as batch size and equipment capacity are given in the application, the overall reported volume fill ratio only reached around 0.63. Normally the volume fill ratio is controlled between 0.4 and 0.6, but this number can be extended from 0.2 up to 0.8 in some special cases. Fig. 19.6 indicates that pharmaceutical companies prefer to fix the batch size and volume fill ratio for regulatory considerations.

19.6.1.3 Amount of granulation fluid I (w/w) and amount of granulation fluid II (w/w)

The amount of granulation fluid should always be controlled within a narrow range because it is widely acknowledged to dictate the properties of the final granules. The amount of granulation fluid I and the amount of granulation fluid II are the top 2 process parameters with the highest risk factor in Fig. 19.5. The amount of granulation fluid can be expressed as the weight of granulation fluid divided by the weight of dry intragranular powder mixture, by which this process parameter is converted to a scale-independent variable. Because the required amount of granulation fluid is also affected by other variables (e.g., formulation change during scale-up, geometry difference of two granulators, and change of impeller speed and massing time), process scientists may need to verify or optimize the amount of granulation fluid at each scale level. The initial investigation value of granulation fluid can be within the range of 10% to 55% (w/w).

19.6.1.4 Solid level of granulation fluid I

The granulation fluid can be as simple as purified water (normally seen in wet granulation with dry binder addition method), but can also be as complex as a mixture of solvent, drug substance, binders, and other excipients (frequently seen in wet granulation with wet binder addition method or the combined binder addition method). In the latter case, a solid level of granulation fluid becomes a process parameter and may need to be evaluated during process development. The overall report frequency is 0.16 as shown in Fig. 19.5.

As discussed in the previous section, the formulation was considered as a fixed input variable by the majority of process scientists. Consequently, the solid level in granulation fluid can not only impact the viscosity of granulation fluid and the spray performance, but can also affect the amount of granulation fluid for wet granulation. Process scientists may optimize this parameter at a small scale and fix the value after scaling up. The initial investigation value of the solid level is normally selected between 2% and 20%, and theoretically, it is a scale-independent variable.

19.6.1.5 Premixing impeller speed and premixing time

The premixing stage is usually considered low risk. The premixing impeller speed can be the intended impeller speed for granulation or a lower speed, and the chopper of the HS granulator is normally turned off. The typical premixing time for initial investigation is chosen between 2 and 20 min at various scales.

19.6.1.6 Spray rate and spray time

Generally, the spray rate is chosen such that local overwetting of the powder mass is not a concern. At the same time, the spray rate should be fast enough to accommodate the whole granulation process time. Therefore, the process scientists prefer to study and control the spray rate within a range rather than at a fixed value. The spray rate of granulation fluid is associated with the theory of granule growth. It can be as low as 2 to 10 g/[min·kg] and as high as >100 g/[min·kg]. The spray time is typically controlled between 2 and 15 min at various scales.

19.6.1.7 Granulation impeller speed and chopper speed

Because the mechanical energy is mainly delivered by rotating impeller, the impeller speed/tip speed is always studied by scientists (Bardin et al., 2004; Chitu et al., 2011; Oulahna et al., 2003). The general trend is that the faster the impeller speed, the larger the granules. Similar to the analysis result of the spray rate, it is difficult to suggest a general investigation range for impeller speed. The initial setting of impeller speed in our observation can be as low as 40 rpm but as high as 800 rpm, taking into account considerable factors such as the intended mechanical and dissolution properties of granules, specific granulator configuration, batch size, etc.

The function of the chopper is to break down the wet lumps into granules. The rotation speed of the chopper may range from 0 to 3600 rpm. Because the chopper speed of many HS granulators with large bowl capacity is not adjustable, it is usually a fixed value at pilot scale or commercial scale.

19.6.1.8 Wet massing time

Wet massing affects the density of the granules and the granule size. Normally the impeller speed and chopper speed are kept the same as the granulation stage; the research focuses on exploring an appropriate wet massing time. A typical investigation range of wet massing time is selected within 6 min at various scales.

19.6.1.9 Power consumption or torque

Power consumption and torque are electrical and mechanical characteristics of the impeller motor. Traditionally, monitoring power consumption or torque is an indirect measurement to determine the endpoint of wet granulation. The correlation between power consumption or impeller torque and granule/product properties has been extensively studied in the past 20 years (Achanta et al., 1997; Holm et al., 1985; Holm et al., 2001;Kopcha et al., 1992; Kornchankul et al., 2001). However, in current practice, the majority of the process scientists choose to use massing time or total processing time as the endpoint of granulation, but they may still monitor power consumption or torque as a reference to prevent the potential risk of overgranulation during process development.

19.6.1.10 Jacket or Product Temperature

Only a few applications reported that the granulation temperature was controlled or monitored, but one-fourth of the reports suggested the importance of this parameter. Generally speaking the higher the granulation temperature, the larger or faster the granules grow. The impact of temperature should be considered if:

- The drug substance or excipient is temperature sensitive.
- Site transfer occurs between different climate zones during and after process development.
- Multiple subbatches may need to be executed continuously by using the same granulator.

It should be noted that many process parameters, such as spray nozzle position, nozzle type and configuration, nozzle angle, and impeller diameter, are not included in Fig. 19.5. It is not because these parameters are not important, but because these parameters are generally considered fixed variables during process development. Based on our analysis of the reported frequency, risk factor, and parameter research behavior listed in Figs. 19.5 and 19.6, we suggest that the spray rate/time, the amount of granulation fluid I, massing time, and granulation impeller (tip) speed are the top four process

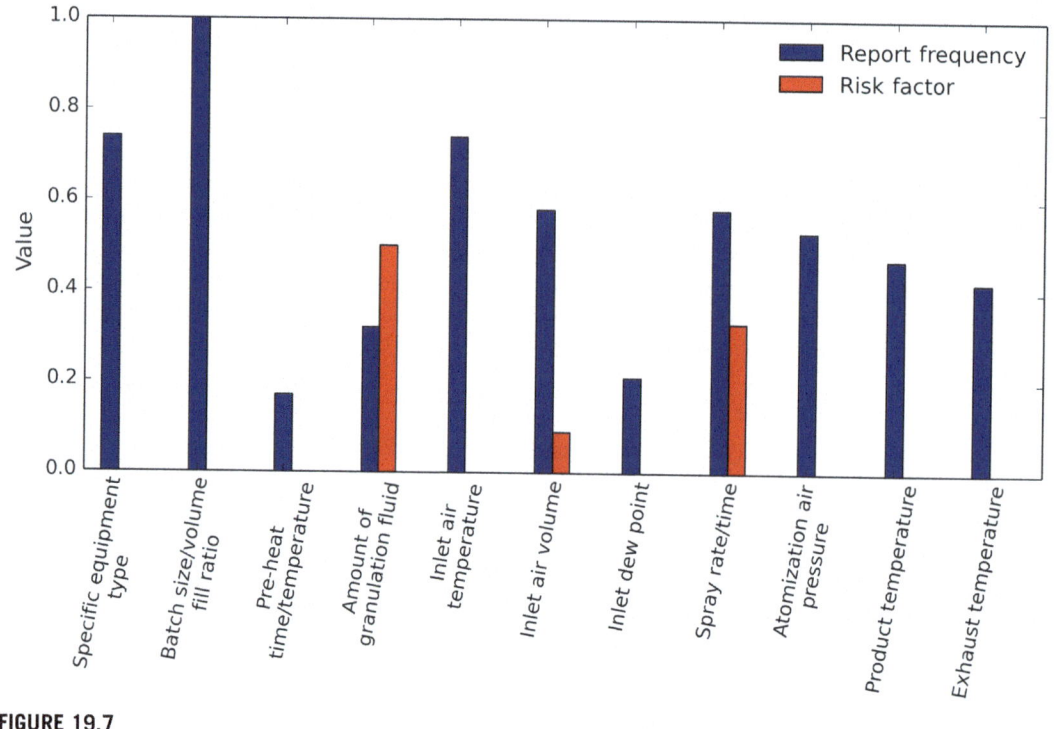

FIGURE 19.7

Process parameters with report rate and the importance ratio of the reported parameter, summarized from FB-granulation profile.

parameters that process scientists should pay attention to throughout wet granulation development. We are considering developing a scoring matrix that balances the reported frequency, risk factor, parameter research behavior, expert opinion, and other considerations to objectively evaluate the risk of process parameters on HS/LS in the future.

19.6.2 Process parameters for fluidized bed granulation

Based on our unit operation profile of FB granulation, the process parameters of FB granulations that were frequently studied and reported by applicants are listed in Fig. 19.7. Also, the study frequency of process parameters listed in Fig. 19.7 is counted and given in Fig. 19.8.

19.6.2.1 Equipment type

The information on equipment type is clearly addressed in about 74% of reviewed applications. Theoretically, top spraying FB, bottom spraying FB, and tangential spraying FB are all capable of FB granulation. In practice, top-spraying FB is the default choice for FB granulation due to equipment availability and prior knowledge of FB granulation with top-spraying FB.

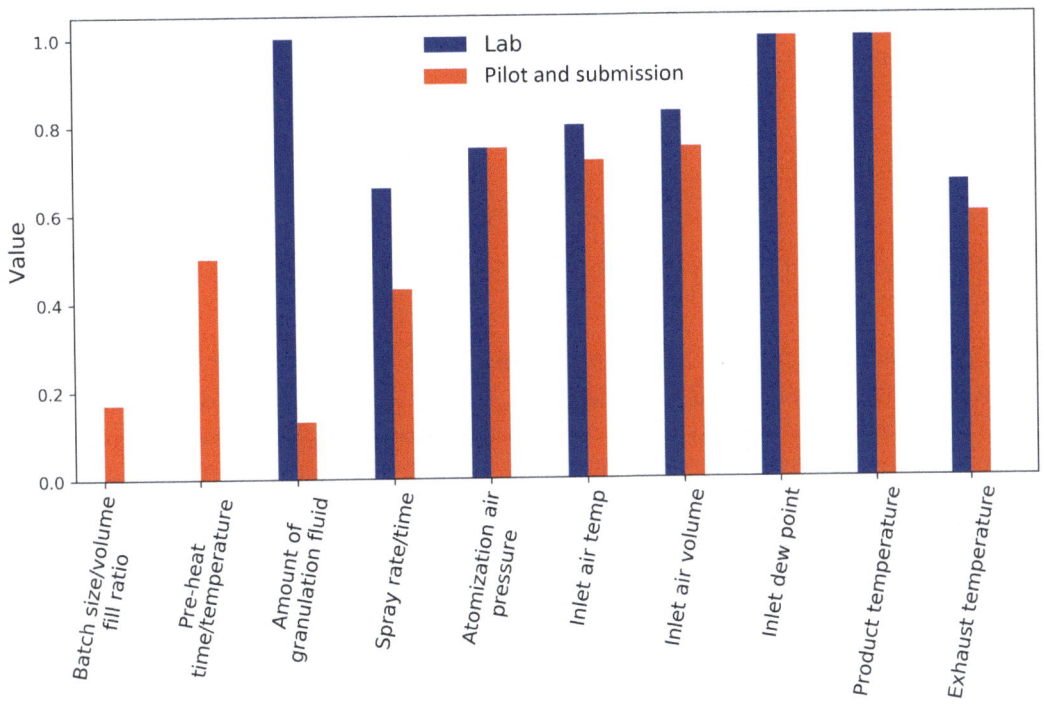

FIGURE 19.8

Study frequencies of process parameters that were reported with the investigation range at laboratory scale, pilot scale, and submission batches from FB-granulation profile.

Table 19.6 The batch size ranges and representative FB granulator model for each range.

Batch Size (kg)	Example
0.4–1	GPCG 1.1
1–5	Aeromatic S2
10–20	GPCG 5
30–80	GPCG 60
>100	CPGC 120

19.6.2.2 Batch size and volume fill ratio

We can retrieve the batch size information from most of our reviewed applications and the ranges of batch size are summarized in Table 19.6. Similar to HS/LS granulation, we cannot retrieve the volume fill ratio information from the majority of applications. But we may estimate the volume fill ratio as long as the batch size and equipment model are available in the application. Therefore, the adjusted overall reported frequency of volume fill ratio can reach one. Like the unit operation profile of HS/LS granulation, process scientists prefer to fix the batch size to minimize the impact of volume fill ratio.

Yamamoto and Shao (2009) provided two general equations to estimate the minimum and maximum batch size for top spraying FB granulation:

$$S_{min} = V \times 0.5 \times BD$$

$$S_{max} = V \times 0.8 \times BD$$

where S is the batch size in kilograms, V is the product bowl capacity in liters, and BD is the bulk density of finished granules in g/cm^3.

19.6.2.3 Preheat temperature and preheat time

The FB should be heated to the target temperature before the granulation step. Normally the preheat inlet air temperature is the intended inlet air temperature for granulation. The preheat time is scale and equipment-dependent, which may need to be monitored and recorded again after scale-up.

19.6.2.4 Amount of granulation fluid (w/w)

Similar to HS/LS granulation, water is the first choice as a solvent to prepare granulation fluid, and binders can be added by dry addition method, wet addition method, or a combined method. It is interesting to find out that process scientists had different research behaviors on HS/LS granulation development and FB granulation development regarding this process parameter. According to the FB granulation profile we prepared from our reviewed applications, the effect of granulation fluid on FB granulation was not frequently reported by process scientists (Fig. 19.7). At the same time, process scientists preferred to systematically study this process parameter at small scale and considered it a fixed variable at larger scales (Fig. 19.8). This is significantly different from the research behavior of this variable in HS/LS granulation profile, in which its reported frequency is over 0.8 and its study frequencies at both laboratory scale and pilot scale are all above 0.6. Moreover, because the granulation fluid for FB granulation needs to be atomized into tiny droplets, the viscosity or the solid content of the granulation liquid is limited.

FB granulation may require more granulation fluid than HS/LS granulation. The initial investigation value of among of granulation fluid can be any value between 30% and 90% (w/w), and, in some applications, this value may be extended to over 100%.

19.6.2.5 Inlet air condition

The inlet air condition includes inlet air temperature, volume, and dew point. These three variables, plus the spray rate are acknowledged as four key factors. Their interactions have a direct impact on nucleus formation and granule agglomeration, and, therefore, the properties of the granulation product (Loh et al., 2011). Inlet air volume also controls the status of FB. Proper air volume should fully fluidize the bed without clogging the filters.

Although inlet air temperature and inlet air dew point are scale-independent variables and values of risk factor for these three process parameters are below 0.1, process scientists are used to continuously optimizing them after scaling up to pilot scale and to commercial scale. The initial investigation range for inlet air temperature is usually selected within 40°C to 90°C, and the range for inlet air dew point is normally selected within 5°C to 15°C. The setting of inlet air volume is influenced by equipment manufacture, specific configuration of equipment, physical properties, and batch size. Because the reported frequency for inlet air volume is only 0.58, the ranges of inlet air volume for different ranges of batch size can only be roughly estimated, which is 10 to 300 cubic feet per minute (CFM) for batch sizes smaller than 20 kg and >300 CFM for batch sizes larger than 20 kg.

19.6.2.6 Spray condition

The importance of droplet size of granulation fluid at the spray zone and the amount of granulation fluid in controlling the granule formation and growth has been well acknowledged (Lipps & Sakr, 1994; Panda et al., 2001; Rambali et al., 2003). Among many variables such as spray rate, solution viscosity, or granulation fluid concentration, atomization air pressure, atomization air flow, nozzle position, nozzle orifice size, and nozzle type, spray rate and atomization air pressure are more frequently reported by the process scientists. The common investigation range of spray rate could be 1 to 80 g/[min • kg • nozzle] for batch size $\leq$4.5 kg, and 0.8 to 5 g/[min • kg • nozzle] for batch size $\geq$10 kg. The initial investigation range of atomization air pressure could start from 0.5 to 4 bars at various scales.

19.6.2.7 Product temperature and exhaust temperature

Both product temperature and exhaust temperature are scale-independent process state variables mainly controlled by inlet air temperature, inlet air volume, and spray rate. Pharmaceutical companies may have different and unique process development/control strategies, such that either product temperature or exhaust temperature can be selected as the target for process control. The investigation ranges for these temperature parameters are usually selected within 25°C to 60°C.

Other important but not listed process parameters include spray nozzle system (number of nozzles, position of nozzles, nozzle type, and orifice diameter) and process filter system (filter material, shaking time, and shaking interval). These variables are always preselected variables and would not be changed unless a granulation failure happened during process development. Compared with HS/LS granulation, the process development pattern of FB granulation is much clearer. Amount of granulation fluid, spray rate, time, and inlet air volume are the top three process parameters having a high impact on CQAs of granules and final product. Inlet air condition is a set of process parameters that process scientists like to validate and optimize after scale-up from laboratory scale.

19.6.3 Scale-up consideration

By definition, process scale-up is the transfer of a controlled process from small scale to large scale. To a process scientist, successful scale-up means maintaining the properties of the granules at a similar level and producing a final drug product of consistent quality. To the regulatory agency, scale-up is an important component of risk management and is of interest to reviewers for evaluating whether a pharmaceutical company can still produce drug products of consistent quality after technique transfer or after scaling up to the proposed commercial scale. In this review study, only a very limited number of applicants fully reported their scale-up procedure in the documents. Instead, many applicants chose to provide experimental information at the intended commercial scale level to alleviate the regulatory concerns.

The process control and scale-up strategies of the HS/LS granulation process have been well developed (Faure et al., 2001). Under the premise that the spraying time was kept constant by increasing the spraying rate of granulation fluid, HS/LS granulation can be scaled up linearly by following:

- Keep the relative swept volume constant, or
- Keep the tip speed of the impeller constant, or
- Keep certain dimensionless numbers (i.e., Froude number, spray flux) constant.

Theoretically, keeping certain dimensionless numbers constant should be a better approach for HS/LS scale-up, and many granulator manufacturers are likely developing equipment based on a certain similarity of dimensionless numbers (e.g., Froude number) rather than geometrical similarity (Horsthuis et al., 1993). In practice, our review of applications indicates that keeping the impeller tip speed constant is currently the most popular approach for HS/LS process scale-up. We also observed a scale-up case by maintaining a self-defined dimensionless number from small scale to large scale, but no scale-up case using the approach by keeping the swept volume constant. An ideal scenario of linear scale-up of the HS/LS granulation process is that the granulators used for wet granulation at various scales are geometrically, kinematically, and dynamically similar. This is extremely difficult to achieve simultaneously. Additional experimental work may be required for some compensatory changes in the amount of granulation fluid, impeller speed, or wet massing time to produce equivalent granules. DOE is highly encouraged for the purpose of process optimization and validation of design space after scale-up (Yu et al., 2014).

On the other hand, keeping the droplet size and FB humidity consistent across different scales is widely acknowledged to be two universal and essential keys of FB granulation scale-up (Rambali et al., 2003). There are many theories and models that help guide the scale-up of FB granulation (Glicksman et al., 1993; Horio et al., 1986; Sanderson & Rhodes, 2003). In practice the commonly used method is the linear scale-up approach, which is quite straightforward.

In an ideal situation, the following process parameters should be kept constant:

- Fluidization velocity of the inlet air through the granulation system.
- The ratio of granulation spray rate to drying capacity of the inlet air volume.
- Droplet size of the granulation fluid.

The fluidization velocity can be estimated as the inlet air volume divided by the cross-sectional area of the equipment. The drying capacity of the inlet air volume is determined by the inlet air volume, inlet air temperature, and inlet air dew point (humidity) together. By keeping the inlet air temperature and the inlet air dew point constant, the drying capacity is in a linear relationship with the inlet air volume. Droplet size of the granulation fluid is controlled by spray rate, specific nozzle type and nozzle configuration, atomization air pressure and air volume, and, of course, the formulation—especially if a formulation change can alter the physical properties of the granulation fluid. All process parameters related to the spray system need to be preadjusted at each scale. Atomization air pressure may need to be further optimized at the pilot scale and final commercial scale. Prior knowledge and experience still play significant roles in adjusting appropriate parameters for scale-up, but DOE is highly encouraged for process optimization and design space validation of FB granulation (Yu et al., 2014).

19.7 Quality by design tools and wet granulation development

As a tightly integrated part of the QbD approach, the QbD tools include, but are not limited to, risk assessment, DOE, and PAT as shown in Fig. 19.2.

Table 19.7 Example of early risk assessment across unit operations of the drug product CQAs for a modified release (MR) tablet.

Unit operation	Drug product CQAs				
	Appearance	Impurity	Assay	Content Uniformity	Dissolution
Dry mixing	Low	Low	Medium	Medium	Medium
Granulation	Low	Low	High	High	High
Drying	Low	Low	Low	Low	High
Milling	Low	Low	Low	Low	Low
Blending	Low	Low	Low	Medium	Low
Lubrication	Low	Low	Low	Low	Low
Compression	High	Low	High	High	High
Film coating	High	Low	Low	Low	High

19.7.1 Risk assessment

The key objective of risk assessment in pharmaceutical development is to identify potentially high-risk API properties, formulation variables, and process variables that could impact the quality of the final drug product, so that an appropriate control strategy can be implemented to ensure CQAs are within expectations. Risk assessment is also an iterative process and occurs throughout the development. During the initial phase of risk assessment, prior knowledge serves as the primary basis for the development designation. Then, the outcome of experimental investigations provides the basis to further re-evaluate and correct the risk score of each variable. ICH Q9 (FDA, 2006) provides a nonexhaustive list of common risk assessment tools, which includes basic risk management facilitation methods; failure mode effects analysis (FMEA); failure mode, effects, and criticality analysis fault tree analysis; hazard analysis and critical control points; hazard operability analysis; risk ranking and filtering; and supporting statistical tools.

It might be appropriate to adapt these tools for risk assessment in specific area. The subject of risk assessment can be as large as a whole manufacturing process (Table 19.7), but can also be as small as single unit operation development (Fig. 19.9). Currently, simple and qualitative risk assessment approaches are still the preference by process scientists, but advanced tools are also frequently used by experienced scientists to gain a more comprehensive understanding of the risks involved. We observed that, compared with process scientists who only followed minimum requirement of risk assessment in ICH guidance, process scientists who used advanced tools (e.g., FMEA) were more likely to identify the significant interactions between material attributes and process parameters in wet granulation process development.

19.7.2 Design of experiment

DOEs is a structured and organized method to determine the relationship among factors that influence outputs of a process. The meaning of DOEs in QbD implementation has been well addressed in ICH Q8 guidance and other scientific publications (Huang et al., 2009; Lionberger et al., 2008; Yu, 2008;

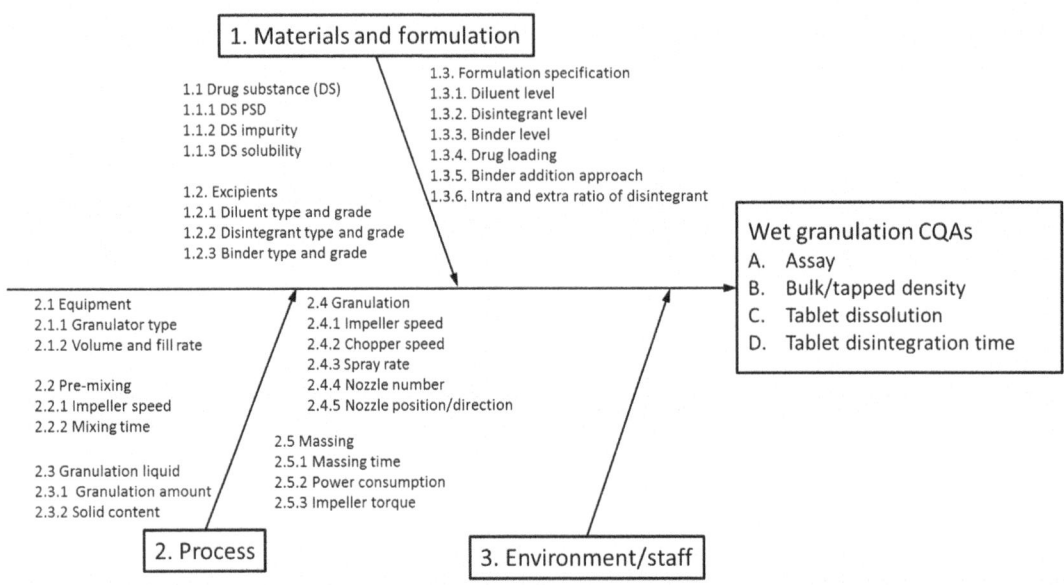

FIGURE 19.9

Example of fishbone diagram for HS/LS granulation.

Yu et al., 2014). In a QbD quality system DOE should be used in both product and process development for product/process understanding and optimization.

With respect to DOE application in process development using wet granulation, currently our review results show that DOE was applied in approximately 50% of our reviewed HS granulation research and about 20% of LS granulation research. Because of the special status of wet granulation in the whole manufacturing process, about 30% of these DOE studies cover input variables and process parameters from more than one unit operation. The amount of granulation fluid, massing time, and impeller (tip) speed are three process parameters most commonly selected for DOE study. Among these DOE studies, 18% included material attributes such as the ratio of intragranular binder against extragranular binder, and the solid content of granulation fluid.

On the other hand, DOE was used only for approximately 20% of FB granulation research. Compared with HS/LS granulation, the implementation of DOE in FB granulation research is more challenging because inappropriate selection of process parameters and their operation ranges may result in process interruption and failure. The amount of granulation fluid and the spray rate are two process parameters most frequently selected in a DOE study. Because inlet air temperature, inlet air volume, product temperature, and exhaust temperature are several parameters that directly control or indicate the physical status and humidity of the FB, it is reasonable to choose just one of these process parameters in the experimental design to avoid potential control failure.

In order to maximally benefit from DOE, an appropriate type of experiment design should be selected at different stages of process development. At the early stage of process development, depending on the amount of knowledge and prior experience of wet granulation, *screening DOE* is necessary to evaluate the risks and gain knowledge from the experiment. Screening DOE enables us to focus on the key variables that may significantly impact a CQA when there are a large number of variables to evaluate. The representative types of screening DOE include full/fractional factorial design. Once sufficient knowledge is obtained from these studies, *optimization DOE* can be very useful to further understand the complex interactions and even quadratic effects of process variables and develop the design space on CPPs. The representative types of optimization DOE include central composite design and Box-Behnken design. Finally, according to the QbD paradigm, the design space of the process can be directly derived from DOE studies. The established design space at small scale should be further verified after scaling up to pilot scale and to commercial scale.

19.7.3 Process analytical technology

PAT is defined by the FDA as a system for designing, analyzing, and controlling manufacturing through timely measurements of critical quality and performance attributes of raw and in-process materials and processes, with the goal of ensuring final product quality (FDA, 2004a). As an essential part of the QbD paradigm, PAT provides continuous monitoring of CPPs, CMAs, and/or CQAs in commercial manufacturing to examine the process condition and to ensure the process remains within an established design space. In order to successfully apply PAT to achieve advanced quality control of the manufacturing process and final drug product, PAT should be initiated at the early stage of process development on every unit operation of high risk identified by risk assessment.

However, our review of NDA shows that only a very limited number of applicants have ever applied PAT in their process design and understanding. Several possible reasons that keep PAT absent from the wet granulation process include, but are not limited to:

- The acceptable quality range of granules is wide, and the downstream process that includes one or two milling operations significantly decreases the risk of wet granulation.
- Applicants prefer to adopt the quality control strategy of similar legacy products approved by the FDA previously to avoid regulatory concerns caused by the application of new technology.
- The majority of applicants are still used to traditional quality control strategies by tightly constraining both formulation variables and process parameters rather than following the QbD paradigm. However, such a quality strategy does not necessarily eliminate the failures within the manufacturing facilities that may result in poor drug quality.

There is a lot of literature reporting the successful implementation of PAT in both HS/LS granulation process and FB granulation process (Alcala et al., 2010; Awotwe-Otoo et al., 2014; Halstensen et al., 2006; Huang et al., 2010; Luukkonen et al., 2008; Matero et al., 2009; Rantanen et al., 1998; Rantanen et al., 2005; Sandler, 2011; Watano et al., 2001; Wikstrom et al., 2005; Whitaker et al., 2000) at various scale levels. The common process analytical chemistry tools used for multivariate data collection include image analysis (Sandler, 2011; Watano et al., 2001), acoustic emission (Halstensen et al., 2006; Matero et al., 2009; Whitaker et al., 2000), near infrared spectroscopy (Alcala et al., 2010; 2005; Luukkonen et al., 2008; Rantanen et al., 1998), Raman spectroscopy (Wikstrom et al., 2005), and focused beam reflectance (Huang et al., 2010; Matero et al., 2010). The overall purpose of this PAT application research is to quickly measure one or multiple outputs of granulation such as in-process granule moisture, particle size distribution, and/or drug substance polymorphic transformation during processing so that the process scientist can better master the status of the granulation process.

The advantage of PAT application in wet granulation is obvious:

1. The end-point determined by PAT is more flexible than massing time, and also more accurate than power consumption or impeller torque (only for HS granulation).
2. As discussed by Yu et al. (2014) in their recent publication, PAT helps lift the control strategy from level 3 (relying on extensive end-product testing and tightly contained material attributes and process parameters) to level 2 (controlling with reduced end-product testing, flexible material attributes, and process parameters within the established design space) and even level 1 (automatic engineering control in real time).
3. It improves the process capability and makes the improvement possible. For example, the particle size distribution of granules and the content deviation can be further narrowed down, the process flow "granulation → wet milling → drying → dry milling" may be simplified to "granulation → drying → dry milling" so that processing time and manufacturing expense can be saved with less GMP concerns at the same time.

The current pharmaceutical industry is under transition from batch manufacturing practice to continuous manufacturing practice (Lee et al., 2015). Continuous manufacturing is faster, more reliable, and efficient than traditional batch manufacturing. It also requires the PAT system to be used throughout the whole manufacturing process for quality control. In 2015 and 2016 both Vetex and Janssen had drug products using continuous manufacturing approved by the FDA. Over the past decade, the FDA has expended great effort on the regulation side to encourage the application of new technology in the manufacturing process. It is continuously making its own effort to ensure that the application of regulatory policies reflects state-of-the-art manufacturing technology and ensures a continuous supply of drugs for the US public (FDA, 2015).

19.8 Summary

Unit operation is the basic unit of the pharmaceutical manufacturing process. QbD is a systematic and risk-based approach to pharmaceutical development, which has been widely accepted by the pharmaceutical industry for many years. By systematically collecting quality- and process-related information on wet granulation from recently approved drug product applications, we created a general profile for HS/LS granulation and a general profile for FB granulation. Based on these unit operation profiles, we are able to quantify the reported frequencies of quality attributes of granulation products, list the consideration points of product design and product understanding related to wet granulation development, quantify the reported frequencies and risk factors of process parameters, and also pattern the research behaviors on these process parameters. Based on general profiles for HS/LS granulation and FB granulation, we suggest the spray rate/time, the amount of granulation fluid I, massing time, and granulation impeller (tip) speed for HS/LS granulation; and the amount of granulation fluid, spray rate and time, and inlet air volume for FB granulation are important process parameters that process scientists should pay attention to during wet granulation development. We hope this review work can help the pharmaceutical industry better prepare for the development of wet granulation under the QbD paradigm, and can help regulatory agencies improve the review quality and consistency.

This is our first effort to develop a profile for a specific unit operation. Actually, many unit operations, such as bin blending, milling, roll compaction, Wurster coating, and pan coating, have been well studied and have been employed in the manufacturing process for many years. Enriched process data accumulated at the agency enables us to gradually develop general profiles for these common unit operations. Consequently, we can also derive a more dedicated unit operation profile targeting a specific dosage form, a time period, drug substance, a similar formulation, or an excipient involved in the unit operation development, etc.

Developing an in-house integrated unit operation profile system with a quality scoring matrix in the future can be a powerful knowledge base for regulatory review, and help the regulatory agency gradually transition from qualitative CMC review to quantitative CMC review in the future.

References

Achanta, A. S., Adusumilli, P. S., & James, K. W. (1997). Endpoint determination and its relevance to physico-chemical characteristics of solid dosage forms. *Drug Development and Industrial Pharmacy, 23*(6), 539–546.

Alcala, M., Blanco, M., Bautista, M., & Gonzalez, J. M. (2010). On-line monitoring of a granulation process by NIR spectroscopy. *Journal of Pharmaceutical Sciences, 99*(1), 336–345.

Awotwe-Otoo, D., Agarabi, C., & Khan, M. A. (2014). An integrated process analytical technology (PAT) approach to monitoring the effect of supercooling on lyophilization product and process parameters of model monoclonal antibody formulations. *Journal of Pharmaceutical Sciences, 103*(7), 2042–2052.

Bardin, M., Knight, P. C., & Seville, J. P. K. (2004). On control of particle size distribution in granulation using high-shear mixers. *Powder Technology, 140*(3), 169–175.

Bouwman, A. M., Visser, M. R., Eissens, A. C., Wesselingh, J. A., & Frijlink, H. W. (2004). The effect of vessel material on granules produced in a high-shear mixer. *European Journal of Pharmaceutical Sciences, 23*(2), 169–179.

Chitu, T. M., Oulahna, D., & Hemati, M. (2011). Wet granulation in laboratory-scale high shear mixers: Effect of chopper presence, design and impeller speed. *Powder Technology, 206*(1–2), 34–43.

Faure, A., York, P., & Rowe, R. C. (2001). Process control and scale-up of pharmaceutical wet granulation processes: A review. *European Journal of Pharmaceutics and Biopharmaceutics, 52*(3), 269–277.

Glicksman, L. R., Hyre, M., & Woloshun, K. (1993). Simplified scaling relationships for fluidized-beds. *Powder Technology, 77*(2), 177–199.

Halstensen, M., de Bakker, P., & Esbensen, K. H. (2006). Acoustic chemometric monitoring of an industrial granulation production process-a PAT feasibility study. *Chemometrics and Intelligent Laboratory Systems, 84*(1–2), 88–97. https://doi.org/10.1016/j.chemolab.2006.05.012.

Holm, P., Schaefer, T., & Kristensen, H. G. (1985). Granulation in high-speed mixers part VI. Effects of process conditions on power consumption and granule growth. *Powder Technology, 43*, 225–233.

Holm, P., Schaefer, T., & Larsen, C. (2001). End-point detection in a wet granulation process. *Pharmaceutical Development and Technology, 6*(2), 181–192.

Horio, M., Nonaka, A., Sawa, Y., & Muchi, I. (1986). A new similarity rule for fluidized-bed scale-up. *AICHE Journal, 32*(9), 1466–1482.

Horsthuis, G. J. B., Vanlaarhoven, J. A. H., Vanrooij, R. C. B. M., & Vromans, H. (1993). Studies on upscaling parameters of the Gral high shear granulation process. *International Journal of Pharmaceutics, 92*(1–3), 143–150.

Huang, J., Kaul, G., Cai, C. S., Chatlapalli, R., Hernandez-Abad, P., Ghosh, K., & Nagi, A. (2009). Quality by design case study: An integrated multivariate approach to drug product and process development. *International Journal of Pharmaceutics, 382*(1–2), 23–32.

Huang, J., Kaul, G., Utz, J., Hernandez, P., Wong, V., Bradley, D., Nagi, A., & O'Grady, D. (2010). A PAT approach to improve process understanding of high shear wet granulation through in-line particle measurement using FBRM C35. *Journal of Pharmaceutical Sciences, 99*(7), 3205–3212.

Kopcha, M., Roland, E., Bubb, G., & Vadino, W. A. (1992). Monitoring the granulation process in a high shear mixer granulator—An evaluation of 3 approaches to instrumentation. *Drug Development and Industrial Pharmacy, 18*(18), 1945–1968.

Kornchankul, W., Parikh, N. H., & Sakr, A. (2001). Correlation between wet granulation kinetic parameters and tablet characteristics. *Pharmazeutische Industrie, 63*(7), 764–774.

Lee, S. L., O'Connor, T. F., Yang, X., Cruz, C. N., Chatterjee, S., Madurawe, R. D., Moore, C. M. V., Yu, L. X., & Woodcock, J. (2015). Modernizing Pharmaceutical Manufacturing: From batch to continuous production. *Journal of Pharmaceutical Innovation, 10*(3), 191–199. https://doi.org/10.1007/s12247-015-9215-8.

Lionberger, R. A., Lee, S. L., Lee, L. M., Raw, A., & Yu, L. X. (2008). Quality by design: Concepts for ANDAs. *AAPS Journal, 10*(2), 268–276. https://doi.org/10.1208/s12248-008-9026-7.

Lipps, D. M., & Sakr, A. M. (1994). Characterization of wet granulation process parameters using response surface methodology. 1. Top-spray fluidized bed. *Journal of Pharmaceutical Sciences, 83*(7), 937–947. https://doi.org/10.1002/jps.2600830705.

Liu, L., Levin, M., & Sheskey, P. (2009). Process development and scale-up of wet granulation by the high shear process. In Qiu, Y., Chen, Y., & Zhang, G. (Eds.), *Developing solid oral dosage forms-pharmaceutical theroy and practice* (pp. 667–699). Academic Press. https://doi.org/10.1016/B978-0-444-53242-8.00029-1

Loh, Z. H., Er, D. Z. L., Chan, L. W., Liew, C. V., & Heng, P. W. S. (2011). Spray granulation for drug formulation. *Expert Opinion on Drug Delivery, 8*(12), 1645–1661.

Luukkonen, P., Fransson, M., Bjorn, I. N., Hautala, J., Lagerholm, B., & Folestad, S. (2008). Real-time assessment of granule and tablet properties using in-line data from a high-shear granulation process. *Journal of Pharmaceutical Sciences, 97*(2), 950–959. https://doi.org/10.1002/jps.20998.

Matero, S., Poutiainen, S., Leskinen, J., Jarvinen, K., Ketolainen, J., Poso, A., & Reinikainen, S. P. (2010). Estimation of granule size distribution for batch fluidized bed granulation process using acoustic emission and N-way PLS. *Journal of Chemometrics, 24*(7–8), 464–471. https://doi.org/10.1002/cem.1269.

Matero, S., Poutiainen, S., Leskinen, J., Jarvinen, K., Ketolainen, J., Reinikainen, S. P., Hakulinen, M., Lappalainen, R., & Poso, A. (2009). The feasibility of using acoustic emissions for monitoring of fluidized bed granulation. *Chemometrics and Intelligent Laboratory Systems, 97*(1), 75–81. https://doi.org/10.1016/j.chemolab.2008.11.001.

Oulahna, D., Cordier, F., Galet, L., & Dodds, J. A. (2003). Wet granulation: The effect of shear on granule properties. *Powder Technology, 130*(1–3), 238–246.

Panda, R. C., Zank, J., & Martin, H. (2001). Modeling the droplet deposition behavior on a single particle in fluidized bed spray granulation process. *Powder Technology, 115*(1), 51–57.

Parikh, D.M. (2009). In Parikh, D. M. (Ed.), *Handbook of pharmaceutical granulation technology*. 3rd ed. CRC Press.

Parikh, D. M., & Jones, D. M. (2009). Batch fluid bed granulation. In D. M. Parikh (Ed.), *Handbook of pharmaceutical granulation technology* (3rd ed.) (pp. 204–260). CRC Press.

Rambali, B., Baert, L., & Massart, D. L. (2003). Scaling up of the fluidized bed granulation process. *International Journal of Pharmaceutics, 252*(1–2), 197–206. https://doi.org/10.1016/S0378-5173(02)00646-4.

Rantanen, J., Lehtola, S., Rämet, P., Mannermaa, J. P., & Yliruusi, J. (1998). On-line monitoring of moisture content in an instrumented fluidized bed granulator with a multi-channel NIR moisture sensor. *Powder Technology, 99*(2), 163–170. https://doi.org/10.1016/S0032-5910(98)00100-4.

Rantanen, J., Wikstrom, H., Turner, R., & Taylor, L. S. (2005). Use of in-line near-infrared spectroscopy in combination with chemometrics for improved understanding of pharmaceutical processes. *Analytical Chemistry, 77*(2), 556–563.

Sanderson, J., & Rhodes, MJ. (2003). Hydrodynamic similarity of solids motion and mixing in bubbling fluidized beds. *AIChE Journal, 49*(9), 2317–2327. https://doi.org/10.1002/aic.690490908.

Sandler, N. (2011). Photometric imaging in particle size measurement and surface visualization. *International Journal of Pharmaceutics, 417*(1-2), 227–234. https://doi.org/10.1016/j.ijpharm.2010.11.007.

Shanmugam, S. (2015). Granulation techniques and technologies: Recent progresses. *BioImpacts, 5*(1), 55–63. https://doi.org/10.15171/bi.2015.04.

The United States Pharmacopeial Convention. (2016). *First supplement to USP 39-NF 34 USP and NF excipients, listed by functional category* (pp. 7949–7958). The United States Pharmacopeial Convention.

U.S. Food and Drug Administration. (2004a). *Guidance for industry: PAT—a framework for innovative pharmaceutical development, manufacturing, and quality assurance.*

U.S. Food and Drug Administration. (2004b). *Innovation and continuous improvement in pharmaceutical manufacturing: Pharmaceutical CGMPs for the 21st century.*

U.S. Food and Drug Administration. (2006). *Guidance for industry: Q9 quality risk management.*

U.S. Food and Drug Administration. (2009). *Guidance for industry: Q8(R2) pharmaceutical development.*

U.S. Food and Drug Administration. (2011). *Guidance for industry: Q8, Q9, and Q10 questions and answers (R4).*

U.S. Food and Drug Administration. (2012). *Guidance for industry: Q8, Q9, and Q10 questions and answers-appendix Q&As form training sessions.*

U.S. Food and Drug Administration. (2014). *Guidance for industry. SUPAC: Immediate release and modified release solid oral dosage forms. Manufacturing equipment addendum.*

U.S. Food and Drug Administration. (2015). *Guidance for industry: Advancement of emerging technology applications to modernize the pharmaceutical manufacturing base.*

U.S. Food and Drug Administration. (2016). *Orange book: Approved drug products with therapeutic equivalence evaluations.*

Watano, S., Numa, T., Miyanami, K., & Osako, Y. (2001). A fuzzy control system of high shear granulation using image processing. *Powder Technology, 115*(2), 124–130. https://doi.org/10.1016/s0032-5910(00)00332-6.

Whitaker, M., Baker, G. R., Westrup, J., Goulding, P. A., Rudd, D. R., Belchamber, R. M., & Collins, M. P. (2000). Application of acoustic emission to the monitoring and end point determination of a high shear granulation process. *International Journal of Pharmaceutics, 205*(1-2), 79–91. https://doi.org/10.1016/S0378-5173(00)00479-8.

Wikstrom, H., Marsac, P. J., & Taylor, L. S. (2005). In-line monitoring of hydrate formation during wet granulation using Raman spectroscopy. *Journal of Pharmaceutical Sciences, 94*(1), 209–219.

Yamamoto, K., & Shao, Z. J. (2009). Process development, optimization, and scale-up: Fluid-bed granulation. In Y. Qiu, Y. Chen, G. Zhang, L. Liu, & W. Porter (Eds.), *Developing solid oral dosage forms: Pharmaceutical theory and practice* (pp. 701–714). Academic Press. https://doi.org/10.1016/B978-0-444-53242-8.00030-8.

Yu, L. X. (2008). Pharmaceutical quality by design: Product and process development, understanding, and control. *Pharmaceutical Research, 25*(4), 781–791. 1573904X https://doi.org/10.1007/s11095-007-9511-1 .

Yu, L. X., Amidon, G., Khan, M. A., Hoag, S. W., Polli, J., Raju, G. K., & Woodcock, J. (2014). Understanding pharmaceutical quality by design. *AAPS Journal, 16*(4), 771–783. https://doi.org/10.1208/s12248-014-9598-3.

Further reading

Yu, L. X., Akseli, I., Allen, B., Amidon, G., Bizjak, T. G., Boam, A., Caulk, M., Doleski, D., Famulare, J., Fisher, A. C., Furness, S., Hasselbalch, B., Havel, H., Hoag, S. W., Iser, R., Johnson, B. D., Ju, R., Katz, P., Lacana, E., … Zezza, D. (2016). Advancing product quality: A summary of the second FDA/PQRI conference. *AAPS Journal, 18*(2), 528–543.

SECTION IV

Process modeling and emerging trends

20. Numerical modeling for wet granulation processes 701
21. Application of the discrete element method to scale-up of high-shear granulation ...725
22. Advances in computational modelling and simulation of wet granulation processes 747
23. Twin-screw continuous wet granulation 801
24. Melt granulation: Granulation mechanisms, formulation and process design for batch and twin-screw systems 837
25. The application of the state-of-art material library/material database approach to the process understanding and process modeling of wet granulation 867
26. Emerging paradigms in pharmaceutical wet granulation 911

CHAPTER 20

Numerical modeling for wet granulation processes

Satoru Watano and Hideya Nakamura
Department of Chemical Engineering, Osaka Prefecture University, Osaka, Japan

In this chapter, various numerical models and computer simulation methods which are applicable to wet granulation processes will be discussed. A basic overview of the recent trends in modeling and simulation will also be reviewed.

20.1 Introduction

Granulation, defined as a size enlargement in which small particles stick together by having a liquid bind onto their surface, plays an extremely important role in the powder-handling process. It is widely applied in particle manufacturing processes of the pharmaceuticals, food, fertilizer, forage, agricultural chemicals, ceramics, iron ore, and other chemical industries. It is conducted mainly to produce particulate materials of desired size, shape, density, and so on. Depending on the application, requirements for the granulated product and the feed to be granulated can be quite diverse. For example, the average particle size of the product could be as small as 100 μm or as large as 10 mm, and the feed could consist of a solution, slurry, or dry fine powder. One common requirement for all applications is that the granulated products are easy to process, handle, store, transport, and use. Another purpose is to prepare uniform mixes to prevent the segregation that can occur among blends containing two or more solids of different densities or sizes. In the pharmaceutical and food industries, granulation is especially useful in controlling dissolution and disintegration speed of granulated products including active pharmaceutical ingredient (API). Adding many functions to the original particle by layering granulation and improving product appearance and properties are also important aspects of granulation.

Despite its wide application in industry, however, many industrial plants still operate the granulation process based on their experts' knowledge and experience because of the lack of science and engineering expertise pertaining to granulation. Recently, numerous research projects have been devised to control and optimize the granulation process by using online measurement of important process parameters and granule properties (e.g., size, moisture content), and using numerical models that accurately predict granule growth behavior. Modeling and numerical simulation, in particular, have gathered much attention because, through a rapid advancement in computer science and technology, it has become possible to treat actual granulation phenomena precisely, taking into account the powder's properties, size distribution, irregular shape, and large number of particles.

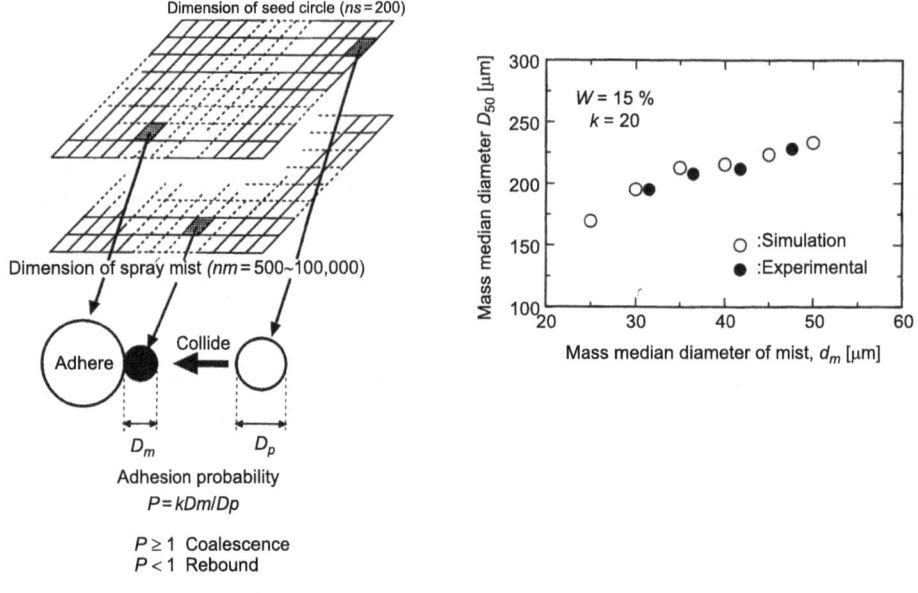

FIGURE 20.1

Random coalescence model for fluidized bed granulation.

From Watano et al. (1995a, 1995b).

This chapter discusses the numerical models that can be used to understand the granule growth mechanism and its behavior in the granulation process.

20.2 Random models

Various mathematical models and computer simulation methods have been studied for elucidating the mechanism of granulation. In the early stages of numerical simulation for granulation, simple models were preferred because computer performance was worse than it is currently.

In computer simulation studies of granulation, there are two basic considerations: (1) a model that allows particles to collide with one another, and (2) allows particles to adhere to one another in collision. In the first case, when two particles collide, they simply adhere or separate. These models, referred to as "random models" or "Monte Carlo methods," were mainly used for random packing (Jodrey & Tory, 1981) or particle mixing (Cahn & Fuerstenau, 1967; Strek et al., 1978; Too et al., 1979). Fukumori et al. (1992) used a random coalescence model to investigate the relationship between the droplet size distribution of the binder liquid and the size of agglomerates in the Wurster coating process. Watano et al. (1995) also used a random model to simulate the effect of spray mist size on the size of agglomerates in fluidized bed granulation. In their model, seed particles and spray mist particles were set up, respectively, and two particles were randomly selected from the same dimension (Fig. 20.1).

A spray mist particle was also selected randomly from another dimension. Assuming that a single spray mist particle on the surface of a core particle can stick to one particle smaller than the core particle, and the size of the smaller particle can be determined by the size of the spray mist, this model could mimic the phenomenon in which the adhesion force (evidenced by a liquid bridge between particles) was mainly determined by liquid volume. By using this simple model, the effect of the spray mist size on the granule size was roughly elucidated.

In the second case, whether two colliding particles adhered or not was determined by adhesion probability, which was presumed to correlate with the operational variables in actual granulation. Vold (1959, 1963) formulated closely packed granules (floc) using a random addition model, in which the particles were allowed one by one to join the central core with the adhesion probability $P=1$. Sutherland (1967) refined the probability function of this model and obtained additional insights on the agglomerate formation. Kawahima et al. (1986, 1989) studied the agglomeration kinetics and micromeritics properties of agglomerates of binary mixtures. Watano et al. (1995) also studied the relationship between granule structure and operating conditions using the random addition model.

Most of the above studies on computer simulation of granulation in the early stages, however, lacked sufficient correlation between simulated and actual results, because they could not introduce the granulation kernel which accurately expressed the coalescence phenomena.

20.3 Population balance models

In the granulation process, granule size distribution and other granule properties change along with elapsed time. In order to predict granule growth and analyze its mechanism, it is important to understand these temporal changes in granule properties. A population balance model (PBM) has been used to analyze granule growth because it can track temporal changes in size distribution and other power properties.

Population balance is one of the material balances for particle assembly systems. It is a number balance around each size fraction of granule size distribution based on a law of "conservation of numbers." The simplest population balance that considered coalescence alone in a batch system can be expressed using the rate of formation (R_{in}) and disappearance (R_{out}) of granules of size u at time t due to the coalescence of granules (Fig. 20.2).

$$\frac{\partial n(u,t)}{\partial t} \& = R_{in} - R_{out}$$
$$= -\frac{1}{N(t)} \int_{v=0}^{\infty} \beta^*(u,v,t) n(u,t) n(v,t) dv \qquad (20.1)$$
$$+ \frac{1}{2N(t)} \int_{v=0}^{u} \beta^*(v, u-v, t|) n(v,t) n(u-v,t) dv$$

In this equation, $\frac{\partial n(u,t)}{\partial t}$ expresses the accumulation of granules of size u with time. $N(t)$ is the total number of granules per unit volume in the system (granulator), and $\beta^*(u, v, t)$ is the coalescence "kernel" describing the rate of aggregation expressed by the frequency of collisions between granules having sizes u and v. In the complicated system, terms indicating the appearance and disappearance of granules due to nucleation and breakage can be added to the equation. In the population balance equation, the coalescence kernel is the most important parameter.

The coalescence kernel can be divided into two parts:

$$\beta^*(u,v,t) = \beta_0^*(t)\beta(u,v) \qquad (20.2)$$

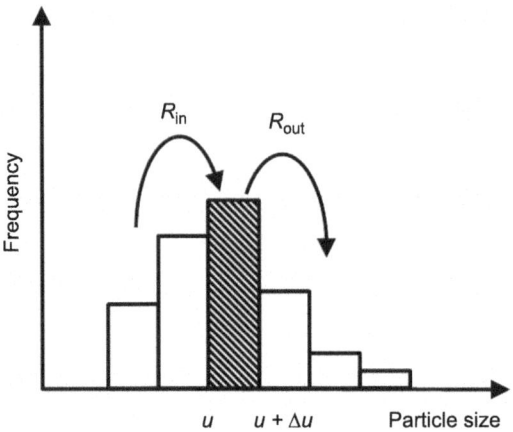

FIGURE 20.2

Concept of population balance model.

Table 20.1 Different types of kernel found in the literature.

Kernel	References
$\beta(u, v) = 1$	Kapur and Fuerstenau (1969)
$\beta(u, v) = u + v$	Golovin (1963)
$\beta(u,v) = \frac{(u+v)^a}{(uv)^b}$	Kapur (1972), Ouchiyama and Tanaka (1975)
$\beta(u,v) = \frac{(u^{2/3}+v^{2/3})}{1/u+1/v}$	Sastry and Fuerstenau (1970)
$\beta(u, v) = 1$ for $t \leq t_1$ $\beta(u, v) = u + v$ for $t > t_1$	Adetayo et al. (1995)
$\beta(u, v) = 1$ for $W \leq W^*$ $\beta(u, v) = u + v$ for $W > W^*$ $W^* =$ critical granule volume	Adetayo and Ennis (1997)
$\beta(u, v) = \beta_1$ for Type I or Type II coalescence with no permanent deformation $\beta(u, v) = \beta_1$ for Type II coalescence with permanent deformation $\beta(u, v) = 0$ rebound Type I, II, and rebound regions were classified by Stokes number and Stokes deformation number	Liu et al. (2000).

The first term, $\beta_0^*(t)$, indicates the granulation rate constant, which is determined by the operating conditions of granulator and powder properties. The second term, $\beta(u, v)$, expresses the coalescence probability when two granules of sizes u and v meet. The expression $\beta(u, v)$ is also known as the coalescence kernel. Different types of coalescence kernels have been proposed, which are summarized in Table 20.1.

20.3 Population balance models

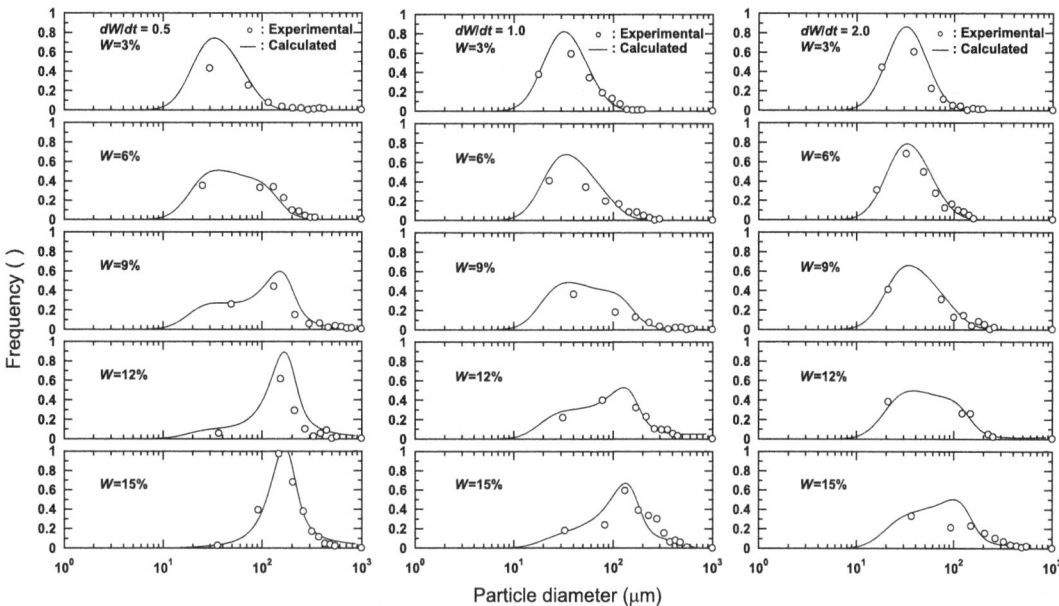

FIGURE 20.3

Temporal change in granule size distribution.

From Watano et al. (1996).

The first mathematical PBM in batch granulation was developed by Kapur and Fuerstenau (1969). They used a simple coalescence kernel shown as $\beta(u,v) = 1$. Kapur (1972) and Ouchiyama and Tanaka (1975) proposed a simple kernel considering colliding two particle sizes. Other coalescence kernels were proposed by Adetayo and Ennis (1997), Adetayo et al. (1995), Liu et al. (2000), and Sastry and Fuerstenau (1970).

There are many studies for PBMs, calculating particle size distribution in granulation, coating, and agglomeration in fluidized beds (Adetayo & Ennis, 1997; Adetayo et al., 1995; Cryer, 1999; Ennis et al., 1990, 1991; Heinrich et al., 2002; Hounslow, 1990; Hounslow et al., 1988; Iveson, 2001; Iveson et al., 1996, 2001, 2001; Kumar et al., 2009; Lister et al., 1995; Ramkrishna, 2000).

Here, temporal change in granule size distribution using a kernel proposed by Kapur (1972) and Ouchiyama and Tanaka (1975) is exemplified in Fig. 20.3. In this study, binder moisture content and its damping speed (moisture increase speed) were taken into account (Watano et al., 1996). Temporal changes in granule size distribution of experimental data indicated good agreement with the numerical results. The figure also implies that the probability of coalescence in large granule size is restricted, while one in small size is promoted, leading to the sharp difference in granule size distribution. Also, it indicates that the slower damping speed (moisture increased speed) promotes narrower granule size distribution. In this way, the PBM can track granule size change over elapsed time.

As mentioned above, the simple PBM can track granule size change in wet granulation. Now, multidimensional (high-order) population balance equations have been proposed to track other granule properties in addition to granule size distribution (Barrasso & Ramachandran, 2012; Barrasso et al., 2015; Poon et al., 2008).

For example, a three-dimensional population balance equation can track temporal change in granule size, liquid binder content, and porosity.

$$\frac{\partial}{\partial t}F(s,l,g,t) + \frac{\partial}{\partial g}\left(F(s,l,g,t)\frac{dg}{dt}\right) + \frac{\partial}{\partial s}\left(F(s,l,g,t)\frac{ds}{dt}\right) + \frac{\partial}{\partial l}\left(F(s,l,g,t)\frac{dl}{dt}\right) = R_{nuc} + R_{aggre} + R_{break} \quad (20.3)$$

where F, s, l, and g indicate the number of particles, particle size, liquid volume, and gas volume, respectively. Also, R_{nuc}, R_{aggre}, and R_{break} represent nucleation, aggregation, and breakage rates, respectively. More detailed granule formation mechanisms such as nucleation, aggregation, and breakage can be taken into account.

Multidimensional PBMs provide detailed information about coalescence, but they still cannot provide information related to particle physical collision such as the number of collisions, relative speed of two colliding particles, impact energy, etc. In fact, coalescence and breakage in granulation are greatly affected by particle collision inside the granulation vessel. So, hybrid type models which couple PBM with discrete element method (DEM), computational fluid dynamics (CFDs), and finite element method have been proposed. By using the hybrid type models, information that with PBM alone is impossible to obtain can be compensated. Details of the hybrid type population models will be discussed in the following section.

20.4 Fundamentals of the discrete element method

A powder is an assembly of single solid particles. Thus, a granular flow can be predicted if the individual particle motion can be modeled. The DEM, originally proposed by (Cundall & Strack, 1979), was developed based on this concept. In the DEM, unsteady translational motion and rotational motion of an individual solid particle are computed based on the following Newton's second law:

Translational motion:

$$m\frac{d^2x}{dt^2} = \sum F_c + mg \quad (20.4)$$

Rotational motion:

$$I\frac{d\omega}{dt} = \sum T_c \quad (20.5)$$

where m, x, t, g, ω, and I are the mass of a particle, the position vector of a particle, time, gravity, angular velocity of a particle, and the inertia moment of a particle, respectively. F_c and T_c are the contact force and contact torque acting on a particle, respectively. d^2x/dt^2 and $d\omega/dt$ denote the translational and angular acceleration of a particle, respectively. The DEM describes the motion of each particle using Newton's second law, allowing for the external forces acting on the particle.

Among the external forces, the particle-to-particle and particle-to-wall contact forces are the most dominant forces in granular flows. The contact force acting on a particle is computed using a soft sphere model (Cundall & Strack, 1979). In the soft sphere model, three kinds of mechanical contact forces, including the elastic rebound force, viscous damping force, and the friction force between contacting particles, are calculated. To calculate these contact forces, the Voigt model is employed; that is, the

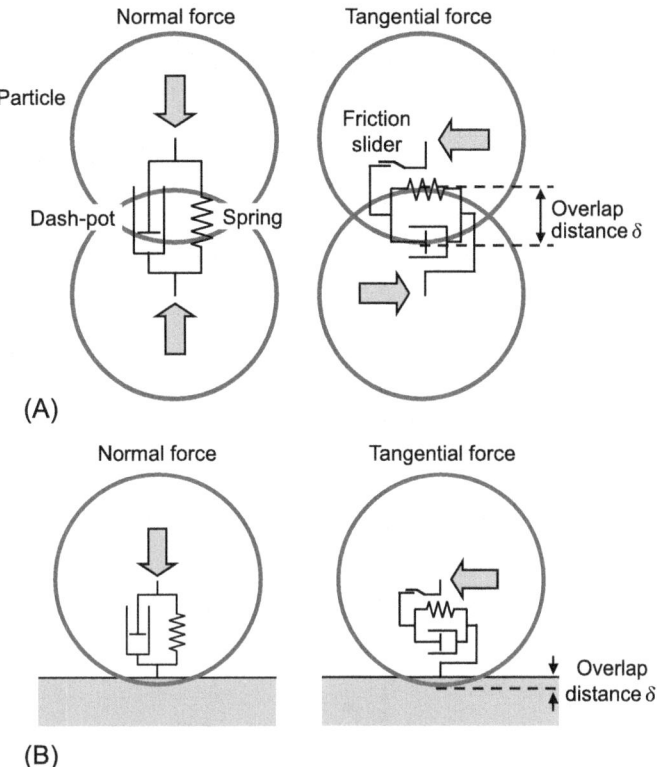

FIGURE 20.4

Model of contact forces in DEM. (A) Particle-to-particle. (B) Particle-to-wall.

elastic rebound force, viscous damping force, and the friction force are modeled using mechanical elements composed of a spring, dashpot, and friction slider, respectively (Fig. 20.4A). The particle deformation in the contact of particles is expressed as the overlap distance between the particles (δ in Fig. 20.4). The particle-to-wall contact is dealt with in the same manner as the particle-to-particle contact (Fig. 20.4B). In the DEM framework, not only can the mechanical contact forces be taken into account, but other external forces such as the fluid drag force and cohesive interparticle forces (i.e., van der Waals force, liquid bridge force, and electrostatic force) may also be included, if required.

Fig. 20.5 shows a typical calculation algorithm of the DEM simulation. After setting the initial conditions, the contact forces acting on a single particle are calculated. This step is repeated N times (N = number of particles), resulting in the calculation of the contact forces and contact torques [$\Sigma\, F_c$ and $\Sigma\, T_c$ in Eqs. (20.1) and (20.2)] on all individual particles. Subsequently, velocity and position of a single particle for the next tiny time step (Δt) are calculated using Newton's second law: the velocity and position of a particle are calculated by numerically solving the Eqs. (20.1) and (20.2) with respect to time from t (= current time) to $t + \Delta t$. This step is also repeated N times, resulting in the calculation of translational and rotational motion at $t + \Delta t$ for all individual particles. Finally, required information

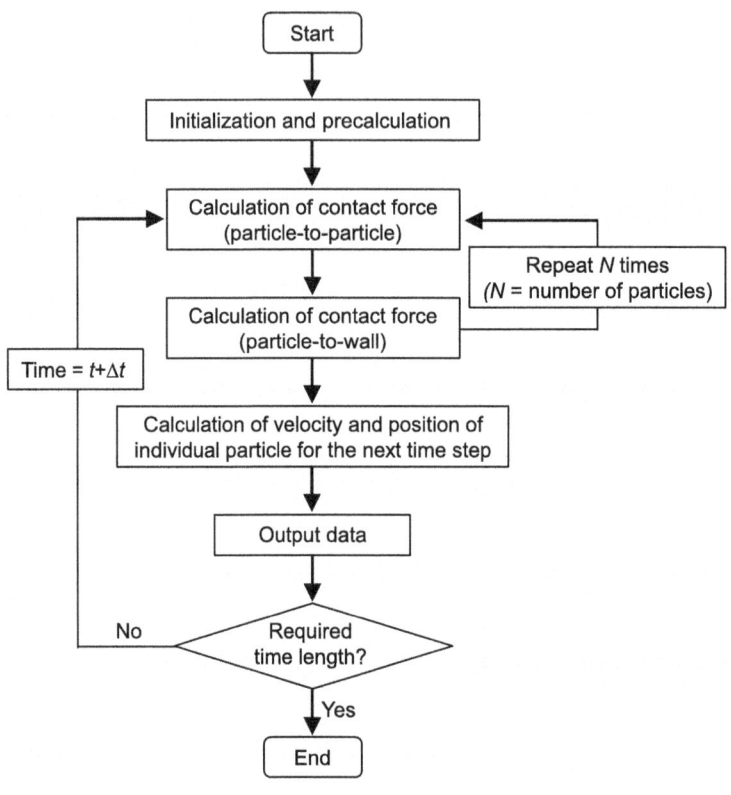

FIGURE 20.5

Typical calculation algorithm of the DEM simulation.

(e.g., positions, velocities, forces, etc.) is output. These steps are repeated for the required time length, resulting in the simulation of the granular flow for a certain time length.

Understanding of the granular flow properties and powder mixing behavior inside granulators can significantly contribute to the rationale design of a granulation process, which is the basic concept of the Quality by Design. For investigation of the internal granular flow and powder mixing, DEM is a powerful tool, because the calculation of the granular flow properties at a single particle scale is very straightforward. Hence, DEM appears to be one of the most promising approaches for modeling and analysis of the granulation processes. However, there are some critical issues for the application of DEM. The most critical issue is a limitation of the number of particles; the number of particles treated in the current DEM simulations is still much less than that in actual granulation processes. Moreover, the current DEM does not cover the actual wet granulation phenomenon, as agglomeration transformation from original fine particles to granule products with binder liquids cannot be simulated yet. Therefore, the current DEM should be regarded as a tool that can provide important insights into the design of granulation processes, while the actual wet granulation phenomenon itself cannot be predicted.

The following sections present the application of the DEM to granulation processes. First, examples of the application of DEM to the analysis and design of typical granulators (such as high-shear mixer-granulators and fluidized bed spray granulators) are reviewed. Finally, advanced modeling of the particle transformation in wet granulation by DEM-related methods is presented.

20.5 Analysis and design of granulation processes using the discrete element method

DEM can provide instantaneous velocity and trajectories of all individual particles. In addition, the forces acting on individual particles can also be calculated. These granular flow properties at a single particle scale cannot be obtained by experimental approaches. Thus, DEM has been applied for modeling, design, and analysis of various powder handling processes (Zhu et al., 2008). This section focuses on the DEM simulations of high-shear mixer granulators and fluidized bed spray granulators, which are the typical granulation processes in the pharmaceutical industry. In particular, to demonstrate the capability of DEM, studies on the analysis and design of the granulation processes utilizing DEM simulation are presented in the following subsections.

20.5.1 Discrete Element Method Simulation of High-shear Mixer Granulators

A DEM simulation of a shear-type mixer granulator has been reported for the first time by Muguruma et al. (2000). They conducted a DEM simulation of a centrifugal tumbling granulator and analyzed its internal granular flow. They showed a particle velocity contour map and a velocity vector field on a vertical cross-section of the powder bed. From the simulation result, a shear flow field induced by the rotor at the bottom of the vessel was confirmed. Moreover, a vertical circulating flow pattern consisting of upward flow along the vessel wall and downward flow near the center of the vessel, that is called "roping flow" (Litster et al., 2002), was clearly visualized by the DEM simulation. Their result demonstrated that DEM can be exploited for analyzing a complicated granular flow in a mixer granulator. Hassanpour et al. (2009) performed a DEM simulation of a commercial high-shear mixer granulator. An experimental investigation of the internal particle flow was also conducted by means of a positron emission particle tracking (PEPT) technique. They reported that the DEM simulation result showed qualitative agreement with the experimental PEPT result. They also analyzed the induced shear flow at a gap between the impeller blade and the vessel wall. The induced shear flow was quantitatively analyzed, and a simple formulation describing the velocity gradient of the shear flow was derived.

Both granular flow and the force acting on a particle are key properties needed to characterize a high-shear mixer granulator. Our research group analyzed the force acting on a single particle in a high-shear mixer granulator using DEM simulation (Sato et al., 2008). Fig. 20.6 shows temporal change in the force acting on a single particle in the mixer. The force acting on a particle instantaneously increased when the particle was located near the bottom of the vessel due to the rotating impeller blade at the bottom of the vessel. This result also showed that the force exerted by the rotating impeller blade on the particle at the bottom of the vessel was much higher than the forces acting on the particle when the particle was at the middle or upper height in the vessel, suggesting that the force acting on a single particle in a high-shear mixer granulator was mainly generated from the rotating impeller blade.

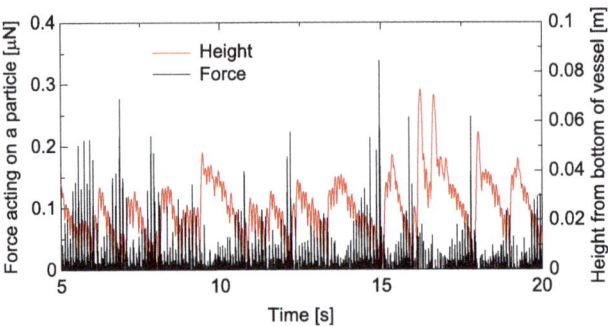

FIGURE 20.6

Temporal changes in forces acting on single particle in high-shear mixer granulator.

From Sato et al. (2008), Copyright: (2008) Elsevier B. V.

The torque acting on the impeller blade in a high-shear mixer granulator, the agitation torque, has often been measured in experiments as well as operations in the industrial sector (Ghanta et al., 1984; Lindberg et al., 1982). The agitation torque has been utilized to monitor internal states in the high-shear mixer granulator. Although agitation torque is empirically considered to relate with the internal particulate flow in the mixer granulator, its physical meaning has not been well understood. Our research group calculated the agitation torque using DEM simulation for the first time (Sato et al., 2008). We then investigated how the agitation torque could be correlated with the internal granular flow. Fig. 20.7A shows typical agitation torques calculated from a DEM simulation at different impeller rotating speeds. The calculated agitation torque showed a periodic fluctuation as a function of time. This behavior is quite similar to the agitation torque observed in the actual experiments (Corvari et al., 1992; Kopcha et al., 1992). We then investigated a correlation between the agitation torque and the internal granular flow. In terms of energy balance, it could be assumed that the energy input provided from the rotating impeller blade results in a change in the kinetic energy of the internal granular flow. Thus, we explored a correlation with the particle kinetic energy. Fig. 20.7B shows a relationship between the agitation torque and the particle kinetic energy. We found that the agitation torque was linearly correlated with the particle kinetic energy, suggesting that, physically, monitoring the agitation torque in a high-shear mixer granulator could monitor the kinetic energy of the granular flow.

Understanding of the powder mixing behavior in a high-shear mixer granulator is a very important subject for design and operation of the high-shear mixer granulator. Terashita et al. (2002) investigated the powder mixing in a commercial high-shear mixer granulator by means of a DEM simulation. They focused on mixing/segregation of binary mixture of particles with different particle densities. The mixing/segregation kinetics inside the vessel were clearly visualized and quantitatively analyzed. Through DEM simulation, they identified key process parameters for obtaining a suitable mixing state with less segregation.

DEM simulation has been utilized for the design and optimization of high-shear mixer granulators. Sinnott and Cleary (2003) investigated the effect of geometry of the impeller blade on the particle mixing using DEM simulation. Two types of impeller blades, namely a horizontal bottom disc and a vertical rectangular blade, were examined. The degree of the particle mixing (i.e., uniformity of the

20.5 Analysis and design of granulation processes using the discrete element method

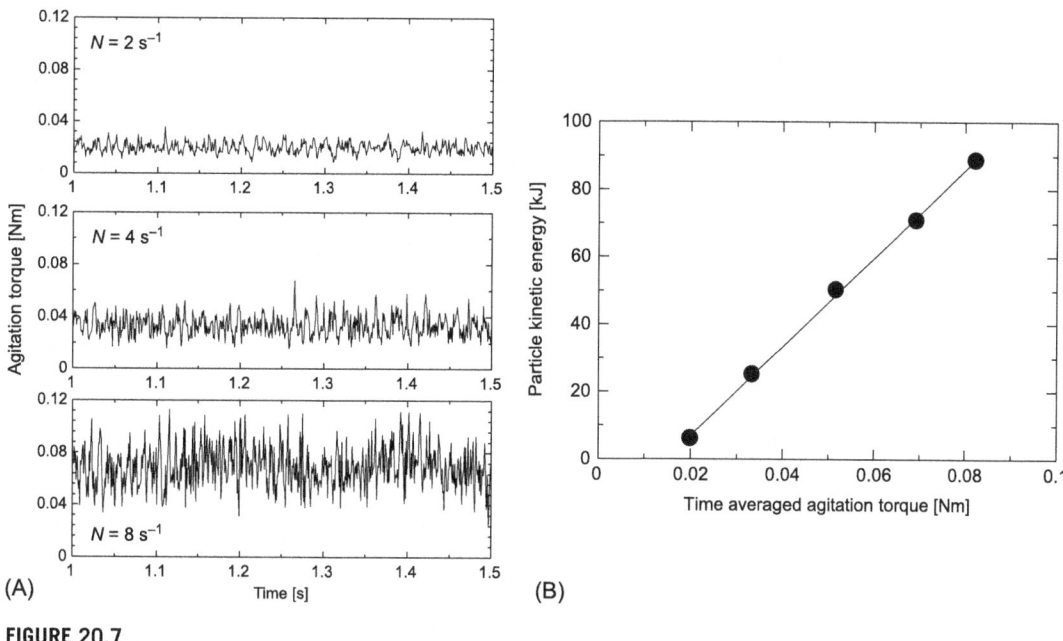

FIGURE 20.7

(A) Agitation torque at different impeller rotating speeds (N) calculated from DEM simulation. (B) Correlation between agitation torque and internal particle kinetic energy.

From Sato et al. (2008), Copyright: (2008) Elsevier B. V.

particle mixture) was evaluated at the two types of impeller blades. They revealed that the blade impeller resulted in better mixing performance than the disc impeller. Börner et al. (2016) employed a DEM simulation to design a commercial high-shear mixer granulator with new impeller geometry. The shear forces acting on particles and the particle kinetic energy were calculated from the DEM simulation, confirming the performance of the new impeller geometry. Terashita et al. (2002) performed a DEM simulation to investigate the effect of a powder loading (granulator fill volume?) on the particle kinetic energy in a high-shear mixer granulator. The calculated kinetic energy showed a maximum value at a middle range of powder loading. Granulation experiments were then conducted at various particle loadings. The experimental results showed that the yield of granules with targeted size fraction showed the maximum value at the middle range of powder loading where the particle kinetic energy reached the maximum in the DEM simulation. This result indicates that powder loading in a high-shear mixer granulator can be optimized by means of a DEM simulation based on particle kinetic energy. These studies demonstrate that DEM simulation can be a useful tool to design new granulator equipment and to find optimum process parameters. Recently, a DEM simulation has been utilized for scale-up of a high-shear mixer granulation, and this topic is presented in another chapter in this book (Nakamura et al., 2013). Moreover, a DEM simulation has been applied to modeling continuous processes. Kulju et al. (2016) performed a DEM simulation of a continuous high-shear granulator. They have successfully investigated the influence of a process operating parameter (such as the rotating speed of the impeller conveyor) on the critical internal process conditions (such as the particle residence time distribution (RTD) and the particle flux).

20.5.2 Discrete element method simulation of fluidized bed spray granulators/coaters

DEM simulation has been utilized for modeling fluidized bed spray granulators and coaters. In a fluidized bed, the solid particle motion is caused by fluid motion, and vice versa. Thus, when the fluidized bed is simulated, both the solid particle motion and the fluid motion should be solved simultaneously. This approach is referred to as the DEM-CFD method (Tsuji et al., 1993). In the DEM-CFD method, the individual particle motion is solved by the DEM, while the fluid (air) motion is solved by CFDs. The DEM and CFD are coupled with each other by taking into account Newton's third law, the momentum exchange between the solid particle phase and the fluid phase. Details on the simulation methods can be found in the literature (Tsuji, 2007; Tsuji et al., 1993). The DEM-CFD method has been a major simulation method for gas-solid fluidized beds including the fluidized bed spray granulators/coaters.

We employed the DEM-CFD method for the first time to investigate the performance of a fluidized bed spray coater. Nakamura et al. (2006) conducted a numerical simulation of the spray coating process in a rotating fluidized bed coater using a DEM-CFD method. The rotating fluidized bed coater, as illustrated in Fig. 20.8, has been developed in our research group (Watano et al., 2003, 2004). In this new type of fluidized bed granulator/coater, the particles are fluidized under a centrifugal force field, leading to smooth granulation and coating of fine particles, but this feature has not been well characterized. Through a DEM-CFD simulation, Nakamura et al. (2006) investigated uniformity of the coating amount (i.e., amount of the sprayed material on the core particles) at various process parameters. For the calculation of the coating amount on an individual particle, it was assumed that the coating amount is proportional to the residence time of the particle in the stationary spray zone as shown in Fig. 20.8.

The coating amount (C_m) of an individual particle was then determined by a ratio of residence time (T_R) of the individual particle in the stationary spray zone to the simulation time step (Δt), $C_m = T_R/\Delta t$. Fig. 20.9 shows a typical simulation result of the coating mass distribution at different processing times. Color on each particle shows the coating mass C_m. By using the DEM-CFD simulation, the progress of the particle coating as the coating time elapsed was clearly visualized. It should be noted that the DEM-CFD model is the only simulation method to visualize such coating amount distribution, which is a great advantage in the design of the spray coating process.

Fig. 20.10 shows the temporal change in the calculation and experimental results of coefficients of variation (CV) of the coating amount at different fluidizing air velocities. The calculation results showed qualitative agreement with the experimental results. With an increase in the coating time, the CV decreased at each gas velocity. This indicates that the coating amount of the particles became more uniform as the coating time elapsed. The CV also decreased with an increase in the gas velocity.

The rotating fluidized bed coater was then compared to a conventional fluidized bed coater (as illustrated in Fig. 20.11A) to clarify the characteristics of the rotating fluidized bed coater. Fig. 20.11B shows the temporal change in the calculation results of CV of the coating amount in the rotating FB and

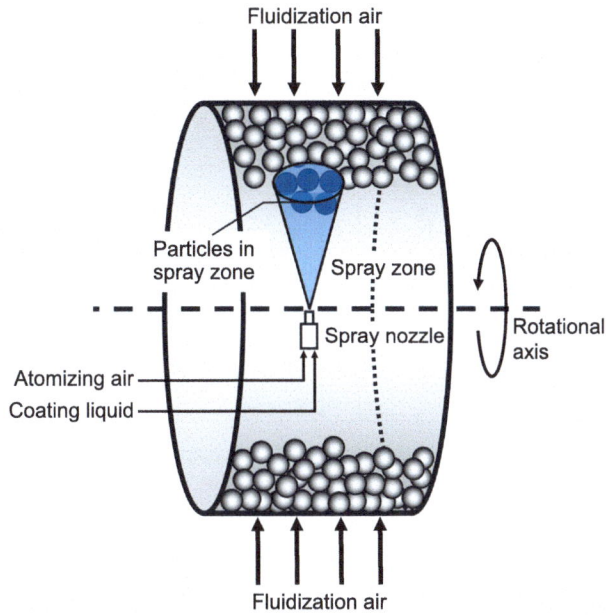

FIGURE 20.8

Schematic illustration of rotating fluidized bed coater (RFBC).

conventional FB. It was found that the CV in the rotating FB was smaller than that in the conventional FB, suggesting that the uniformity of the coating amount can be significantly improved in the rotating FB as compared to that in the conventional FB. The DEM-CFD simulation also provided the reason that for a smaller CV in the rotating FB, it is due to considerably smaller particle circulation period in the rotating FB ($= 0.68$ s), as compared to that in the conventional FB ($= 2.43$ s).

In recent studies, Fries et al. (2011, 2013) employed DEM-CFD simulation to characterize different types of fluidized bed granulators including a top-spray fluidized bed, a Wurster-type fluidized bed (a fluidized bed with a draft tube), and a spouted fluidized bed. From raw data of the DEM-CFD simulation, Fries et al. (2011) extracted the RTD of the particles inside a spray zone. RTD analysis allows assessment of the uniformity of the binder liquid content among the particles. The RTD in the Wurster-type fluidized bed exhibited a narrower distribution than that in the top-spray fluidized bed, indicating that the binder liquid distribution can be more uniform in the Wurster-type fluidized bed than the top-spray fluidized bed. This insight from the DEM-CFD simulation was confirmed by an actual wet granulation experiment with water-soluble food particles. Fries et al. (2013) also analyzed particle collision dynamics. They calculated a collision frequency and relative collision velocity between particles inside the spray zone, allowing for evaluation of agglomeration rate and liquid spreading in each type of fluidized bed granulator. Moreover, they have calculated the angular velocity of a particle (ω in Eq. 20.2) in the spray zone, providing an insight into the uniformity of the applied coating layer on a single particle. Based on these outputs from the DEM-CFD simulation, the three different types

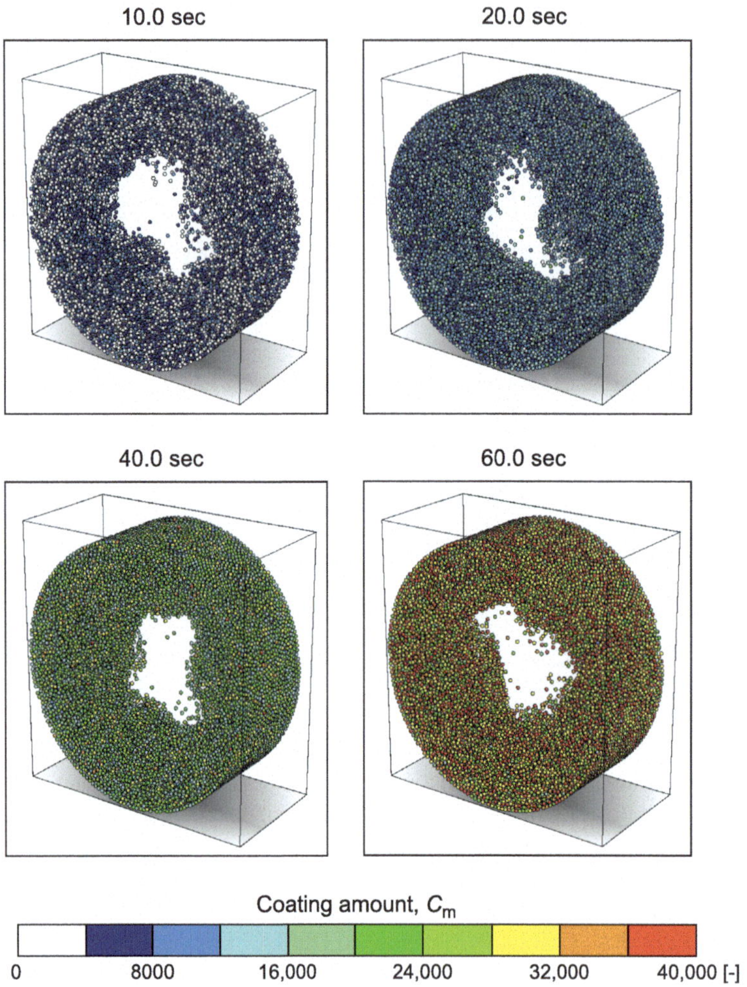

FIGURE 20.9

Simulated coating mass distribution in rotating fluidized bed coater.

From Nakamura et al. (2006).

of fluidized bed granulators (a top-spray fluidized bed, the Wurster-type fluidized bed, and a spouted fluidized bed) were characterized. The insight obtained from the DEM simulation was confirmed by wet granulation experiments.

These previous works demonstrated the high capabilities of the DEM-CFD simulation for characterizing the performance of fluidized bed granulators/coaters. Although the current DEM-CFD cannot simulate the agglomeration transformation phenomenon, the RTD and particle collision dynamics, which are extracted from the current DEM-CFD simulation, can be valuable information for understanding, designing, and optimizing the process.

20.5 Analysis and design of granulation processes using the discrete element method

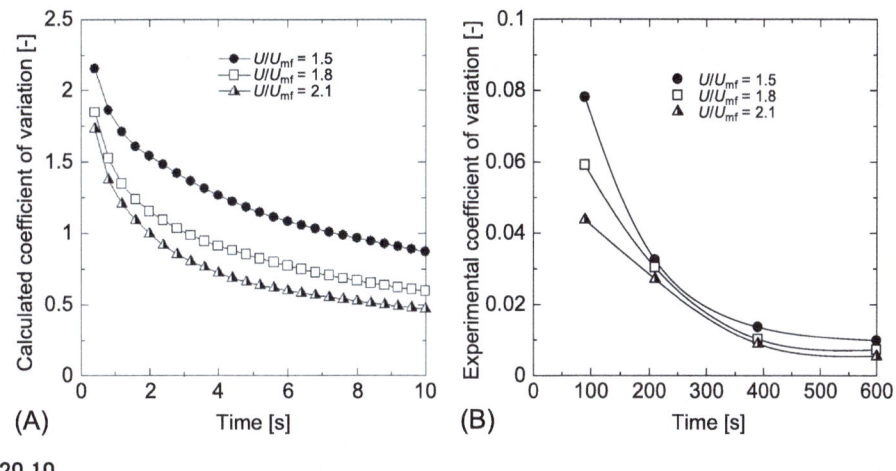

FIGURE 20.10

Temporal change in coefficient of variation of coating mass at various fluidizing air velocities. (A) Calculated. (B) Experimental.

From Nakamura et al. (2006).

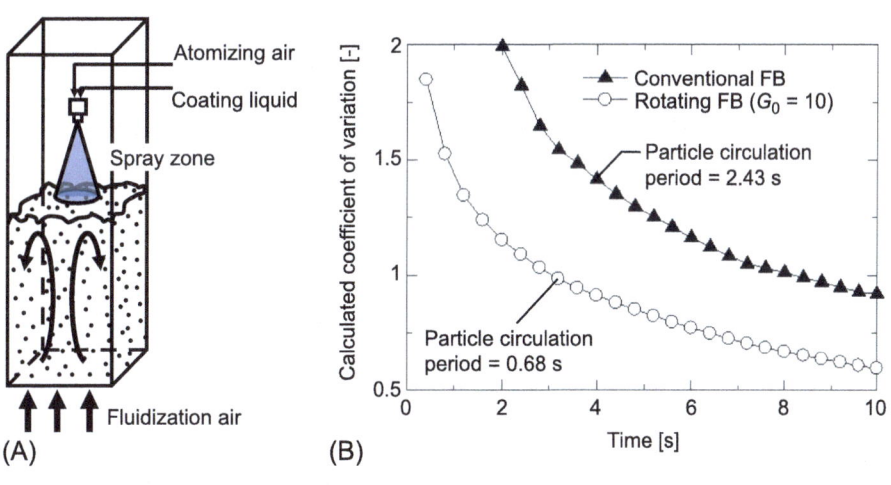

FIGURE 20.11

(A) Schematic of conventional fluidized bed coater. (B) Comparison of coefficients of variation of coating mass in rotating FB with that in conventional FB.

From Nakamura et al. (2006).

20.6 Advanced modeling of particle transformation in wet granulation by discrete element method-related methods

Although DEM is an outstanding approach for understanding the granulation processes, the current DEM cannot predict the granulation phenomenon itself; the particle transformation phenomena, consisting of wetting and nucleation, particle coalescence and growth, and attrition and breakage of granules, cannot be simulated (Ivesonet al., 2001). Toward predicting granulation phenomena, some advanced modeling studies using DEM-related methods have been reported. These studies fall into three categories: (1) DEM simulation coupled with a particle adhesion/coalescence model; (2) DEM simulation coupled with a PBM; (3) direct numerical simulation of fundamental processes for the particle transformation. In the following subsections, these advanced modeling studies are reviewed.

20.6.1 Coupling with particle adhesion/coalescence model

Numerical simulation of wet granulation based on a DEM approach was first reported by Talu et al. (2000). They simulated agglomeration of wet particles under a two-dimensional simple shear flow. In order to simulate formation of the agglomeration from primary small wet particles, the viscous force and capillary liquid bridge force were taken into account as additional external forces acting on a particle. The wet particles were assumed to be covered by uniform binder liquid layer, and the viscous force between the particles interacting via the binder liquid layer was estimated from a lubrication theory (Ennis et al., 1991). As a result, formation and growth of agglomeration from wet primary particles was successfully simulated. The simulation results also revealed that the agglomeration growth rate can be correlated by the Stokes number and the capillary number. Gantt and Gatzke (2005) proposed a simulation method for predicting the dynamic changes in particle size distribution. In their simulation method, DEM is coupled with a particle coalescence model, proposed by Liu et al. (2000). They present an analytical model for coalescence of deformable surface-wet particles. Their coalescence model allows estimating coalescence criteria when two wet particles collide with each other. The coalescence criteria is estimated based on key factors including particle properties (size, density, and mechanical properties), binder properties (amount and viscosity), and collision velocity. Gantt and Gatzke (2005) incorporated the Liu et al.'s model into the DEM simulation; if a particle-to-particle contact in the DEM simulation meets the coalescence criteria estimated from the Liu et al.'s model, the colliding particles coalesce with each other and transform into a larger particle. Gantt and Gatzke (2005) demonstrated that the dynamic changes in granule size distribution can be predicted at any process operating conditions (e.g., impeller rotating speed) and formulation conditions (e.g., amount of binder liquid).

Goldschmidt et al. (2003) reported for the first time a DEM simulation of wet granulation in a fluidized bed spray granulation. In their simulation model, both individual particle motion and individual sprayed droplet motion were computed by the DEM-CFD coupling model. When a particle contacts a droplet, the particle is treated as a surface-wet particle. When the surface-wet particle collides with another particle, the two particles are computed to be agglomerating. This scheme allows for directly simulating fundamental granulation processes such as wetting of particles and subsequent particle agglomeration. The simulation results demonstrated that the model can predict the influence of key process operating conditions such as fluidizing air velocity, spray rate, and spray pattern on the

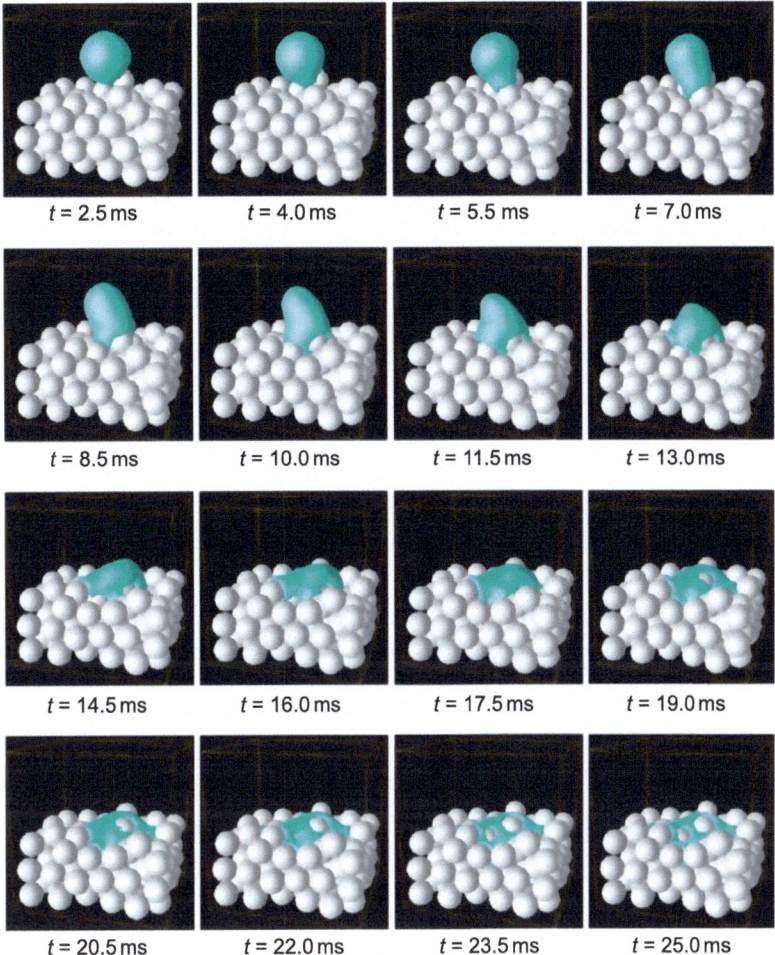

FIGURE 20.12

Direct numerical simulation of droplet penetration into a particle bed.

From Washino et al. (2013), Copyright: (2013) Elsevier B. V.

growth of granules. Kafui and Thornton (2008) proposed another simulation method for a fluidized bed spray granulation. They incorporated a surface energy-driven spray zone concept into a DEM-CFD simulation. In their simulation method, wetting of a particle in a spray zone is modeled as picking-up a surface energy on the particle. The picking-up and accumulation of the surface energy is calculated as a function of a particle's location in the spray zone and a particle's residence time in the spray zone. The interparticle adhesive force causing the particle agglomeration is then computed according to the accumulated surface energy. This model does not require computation of motion of individual sprayed droplets, leading to less computing load compared to the model of Goldschmidt et al. (2003).

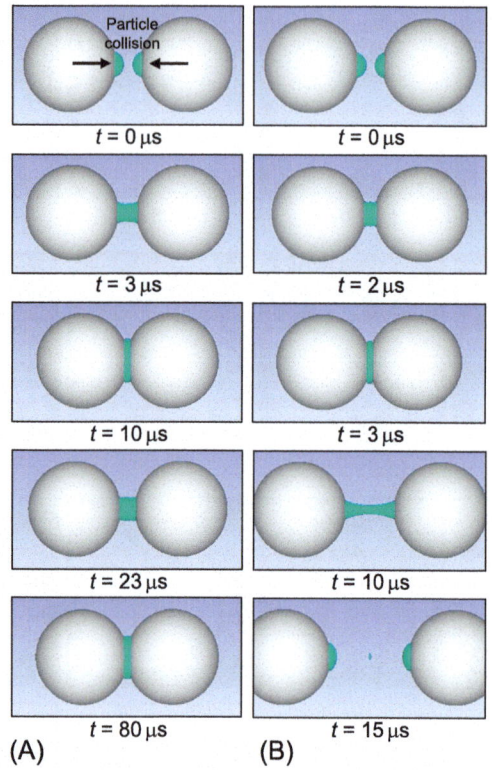

FIGURE 20.13

Direct numerical simulation of adhesion of two colliding particle though dynamic liquid bridge. (A) $v = 1.0$ m/s. (B) $v = 5.0$ m/s.

From Kan et al. (2015), Copyright: (2015) Elsevier B. V.

Kafui and Thornton (2008) demonstrated that formation of agglomerates can be adequately captured by their simulation method.

20.6.2 Coupling with population balance modeling

As mentioned above, DEM simulation coupled with a particle adhesion/coalescence model is a potential approach for simulating the wet granulation phenomena. However, even by the coupling approach, DEM is still computationally expensive; the number of the particles and time available for the simulation are limited and far from experimental scale. Meanwhile, PBM is suitable for large spatial scale and long-time scale due to its lighter computational load. However, PBM is an intrinsically empirical approach; influences of material properties and process operating parameters are lumped into empirical constants included in aggregation/breakage kernels (probability functions). Thus, determination of suitable coalescence kernels and fitting of the empirical constants are necessary for the PBM, limiting the predictive capability of the model. To address these advantages and limitations in DEM and PBM, some attempts have been made to develop a DEM-PBM coupled model.

The first category of DEM-PBM models is one in which DEM is utilized to derive coalescence kernels and determine their empirical parameters (Gantt et al., 2006; Lee et al., 2017). Gantt et al. (2006) calculated particle-to-particle collision rates in a high-shear mixer granulator from raw data of the DEM simulation. By using this result, they determined a suitable collision rate function for the high-shear mixer granulation among various forms proposed. They also derived a mathematical function expressing the particle coalescence probability from data of the DEM simulation. By means of these analytical functions derived from the DEM simulation data, a PBM using a Monte Carlo method was then conducted (Gantt & Gatzke, 2006). The simulation results were found to be in good agreement with experimental results.

The second category of the DEM-PBM model is one in which DEM is exploited for developing a multicompartment PBM (Chaudhury et al., 2015; Kulju et al., 2016; Lee et al., 2017). The PBMs generally assume that internal granular flow in granulation processes is homogeneous and in a perfectly mixed state, while the reality is far from these assumptions. To take this inhomogeneity into account in the PBM model, a multicompartment approach has been proposed. In the multicompartment approach, the simulation domain (i.e., whole granulation process) is divided into multiple compartments (e.g., sprayed compartment, mixing compartment, drying compartment, etc.) according to their characteristics. Suitable aggregation/breakage kernels are then individually defined in each compartment, while each compartment is considered a perfectly mixed state. Major obstacles to the multicompartment approach are the determination of the number of compartments and the powder flux (exchange) rates between the compartments. DEM can be greatly helpful in solving these issues. Based on this motivation, multicompartment approaches coupled with DEM simulation were conducted for modeling batch (Chaudhury et al., 2015; Lee et al., 2017) and continuous high-shear mixer granulators (Kulju et al., 2016).

The last category of the DEM-PBM model is a two-way coupling of DEM with PBM (Barrasso & Ramachandran, 2015; Barrasso et al., 2015). In the two-way coupling, the calculation result is transferred bidirectionally between DEM and PBM; once the PBM calculation is performed using data collected from the DEM simulation for a specified time period, the PBM result is applied to update the particle properties in the DEM. In actual wet granulation, both the coalescence/breakage kernels and the internal granular flow must be changed due to the growth of granules. Thus, the two-way coupling of DEM with PBM can be the most accurate of the DEM-PBM models, but its computational load is also the highest among them all.

20.6.3 Direct numerical simulation of fundamental processes of the particle transformation

Fundamental processes of the particle transformation in wet granulation, such as wetting, particle adhesion, and drying, are regarded as a solid-liquid-gas three-phase flow. Therefore, if the solid-liquid-gas three-phase flow can be solved, the particle transformation in wet granulation can be predicted in a deterministic way without any probabilistic model like the PBM. Although accurate direct numerical simulation of such a three-phase flow is still difficult, and the development of the simulation method is ongoing, some attempts at direct numerical simulation of fundamental processes of the particle transformation have been made.

Nishiura et al. (2010) have conducted a direct numerical simulation of drying a droplet of particulate suspension. The particle motion was solved by the DEM-based method, while the gas-liquid two-phase flow was solved by a CFD method for a multiphase flow. This is a critical process of particle transformation in spray-drying granulation. The simulation results exhibited the dynamic change in the granule structure as the drying proceeds. Washino et al. (2013) developed a new capillary force model and implemented it in a direct numerical simulation of droplet penetration into a powder bed, the wetting and nucleation phenomena (Fig. 20.12).

Our research group has conducted a direct numerical simulation of the particle–particle adhesion through a binder liquid droplet (Kan et al., 2015, 2016). In wet granulation processes, a liquid bridge formed between particles is not static but dynamic due to particle motion. Thus, we solved both the particle motion and the liquid bridge deformation by taking into account the interaction between the particle and the liquid bridge. This is considered the most fundamental phenomenon of particle agglomeration in wet granulation. Fig. 20.13 shows the results of simulation of the adhesion of two colliding particles through a binder droplet on a particle surface, corresponding to a particle-scale agglomeration phenomenon in a fluidized bed spray granulation. The simulation results represent well particle adhesion via the dynamic liquid bridge: that is, formation of the liquid bridge; particle–particle collision with compressing the liquid bridge; elongation of the liquid bridge; particle adhesion at lower collision velocity, or rupture of the liquid bridge at higher collision velocity. These direct numerical simulations are very informative for understanding the physical mechanism of particle transformation in wet granulation.

References

Adetayo, A. A., & Ennis, B. J. (1997). Unifying approach to modeling granule coalescence mechanisms. *AICHE Journal, 43*(4), 927–934. https://doi.org/10.1002/aic.690430408.

Adetayo, A. A., Litster, J. D., Pratsinis, S. E., & Ennis, B. J. (1995). Population balance modelling of drum granulation of materials with wide size distribution. *Powder Technology, 82*(1), 37–49. https://doi.org/10.1016/0032-5910(94)02896-V.

Barrasso, D., El Hagrasy, A., Litster, J. D., & Ramachandran, R. (2015a). Multi-dimensional population balance model development and validation for a twin screw granulation process. *Powder Technology, 270*, 612–621. https://doi.org/10.1016/j.powtec.2014.06.035.

Barrasso, D., Eppinger, T., Pereira, F. E., Aglave, R., Debus, K., Bermingham, S. K., & Ramachandran, R. (2015b). A multi-scale, mechanistic model of a wet granulation process using a novel bi-directional PBM–DEM coupling algorithm. *Chemical Engineering Science, 123*, 500–513. https://doi.org/10.1016/j.ces.2014.11.011.

Barrasso, D., & Ramachandran, R. (2012). A comparison of model order reduction techniques for a four-dimensional population balance model describing multi-component wet granulation processes. *Chemical Engineering Science, 80*, 380–392. https://doi.org/10.1016/j.ces.2012.06.039.

Barrasso, D., & Ramachandran, R. (2015). Multi-scale modeling of granulation processes: Bi-directional coupling of PBM with DEM via collision frequencies. *Chemical Engineering Research and Design, 93*, 304–317. https://doi.org/10.1016/j.cherd.2014.04.016.

Börner, M., Michaelis, M., Siegmann, E., Radeke, C., & Schmidt, U. (2016). Impact of impeller design on high-shear wet granulation. *Powder Technology, 295*, 261–271. https://doi.org/10.1016/j.powtec.2016.03.023.

Cahn, D. S., & Fuerstenau, D. W. (1967). Simulation of diffusional mixing of particulate solids by Monte Carlo techniques. *Powder Technology, 1*(3), 174–182. https://doi.org/10.1016/0032-5910(67)80029-9.

Chaudhury, A., Armenante, M. E., & Ramachandran, R. (2015). Compartment based population balance modeling of a high shear wet granulation process using data analytics. *Chemical Engineering Research and Design, 95*, 211–228. https://doi.org/10.1016/j.cherd.2014.10.024.

Corvari, V., Fry, W. C., Seibert, W. L., & Augsburger, L. (1992). Instrumentation of a high-shear mixer: Evaluation and comparison of a new capacitive sensor, a watt meter, and a strain-gage torque sensor for wet granulation monitoring. *Pharmaceutical Research, 9*(12), 1525–1533. https://doi.org/10.1023/A:1015843820526.

Cryer, S. A. (1999). Modeling agglomeration processes in fluid-bed granulation. *AICHE Journal, 45*(10), 2069–2078. https://doi.org/10.1002/aic.690451005.

Cundall, P. A., & Strack, O. D. L. (1979). A discrete numerical model for granular assemblies. *Geotechnique, 29*(1), 47–65. https://doi.org/10.1680/geot.1979.29.1.47.

Ennis, B. J., Li, J., Tardos, G. I., & Robert, P. (1990). The influence of viscosity on the strength of an axially strained pendular liquid bridge. *Chemical Engineering Science, 45*(10), 3071–3088. https://doi.org/10.1016/0009-2509(90)80054-I.

Ennis, B. J., Tardos, G., & Pfeffer, R. (1991). A microlevel-based characterization of granulation phenomena. *Powder Technology, 65*(1–3), 257–272. https://doi.org/10.1016/0032-5910(91)80189-P.

Fries, L., Antonyuk, S., Heinrich, S., Dopfer, D., & Palzer, S. (2013). Collision dynamics in fluidised bed granulators: A DEM-CFD study. *Chemical Engineering Science, 86*, 108–123. https://doi.org/10.1016/j.ces.2012.06.026.

Fries, L., Antonyuk, S., Heinrich, S., & Palzer, S. (2011). DEM–CFD modeling of a fluidized bed spray granulator. *Chemical Engineering Science, 66*(11), 2340–2355. https://doi.org/10.1016/j.ces.2011.02.038.

Fukumori, Y., Ichikawa, H., Jono, K., Takeuchi, Y., & Fukuda, T. (1992). Computer simulation of agglomeration in the Wurster process. *Chemical & Pharmaceutical Bulletin, 40*(8), 2159–2163. https://doi.org/10.1248/cpb.40.2159.

Gantt, J. A., Cameron, I. T., Litster, J. D., & Gatzke, E. P. (2006). Determination of coalescence kernels for high- shear granulation using DEM simulations. *Powder Technology, 170*(2), 53–63. https://doi.org/10.1016/j.powtec.2006.08.002.

Gantt, J. A., & Gatzke, E. P. (2005). High-shear granulation modeling using a discrete element simulation approach. *Powder Technology, 156*(2–3), 195–212. https://doi.org/10.1016/j.powtec.2005.04.012.

Gantt, J. A., & Gatzke, E. P. (2006). A stochastic technique for multidimensional granulation modeling. *AICHE Journal, 52*(9), 3067–3077. https://doi.org/10.1002/aic.

Ghanta, S. R., Srinivas, R., & Rhodes, C. T. (1984). Use of mixer-torque measurements as an aid to optimizing wet granulation process. *Drug Development and Industrial Pharmacy, 10*(2), 305–311. https://doi.org/10.3109/03639048409064652.

Goldschmidt, M., Weijers, G., Boerefijn, R., & Kuipers, J. (2003). Discrete element modelling of fluidised bed spray granulation. *Powder Technology, 138*(1), 39–45. https://doi.org/10.1016/j.powtec.2003.08.045.

Golovin, A. M. (1963). The solution of the coagulation equation for raindrops. Taking Condensation into account. *Soviet Physics Doklady, 8*, 191.

Hassanpour, A., Kwan, C. C., Ng, B. H., Rahmanian, N., Ding, Y. L., Antony, S. J., Jia, X. D., & Ghadiri, M. (2009). Effect of granulation scale-up on the strength of granules. *Powder Technology, 189*(2), 304–312. https://doi.org/10.1016/j.powtec.2008.04.023.

Heinrich, S., Peglow, M., & Mörl, L. (2002). Unsteady and steady-state particle size distributions in batch and continuous fluidized bed granulation systems. *Chemical Engineering Journal, 86*(1–2), 223–231. https://doi.org/10.1016/S1385-8947(01)00293-5.

Hounslow, M. J. (1990). A discretized population balance for continuous systems at steady state. *AICHE Journal, 36*(1), 106–116. https://doi.org/10.1002/aic.690360113.

Hounslow, M. J., Ryall, R. L., & Marshall, V. R. (1988). A discretized population balance for nucleation, growth, and aggregation. *AICHE Journal, 34*(11), 1821–1832. https://doi.org/10.1002/aic.690341108.

Iveson, S. (2001). Granule coalescence modelling: Including the effects of bond strengthening and distributed impact separation forces. *Chemical Engineering Science, 56*(6), 2215–2220. https://doi.org/10.1016/S0009-2509(00)00506-6.

Iveson, S. M., Litster, J. D., & Ennis, B. J. (1996). Fundamental studies of granule consolidation part 1: Effects of binder content and binder viscosity. *Powder Technology, 88*(1), 15–20. https://doi.org/10.1016/0032-5910(96)03096-3.

Iveson, S. M., Litster, J. D., Hapgood, K., & Ennis, B. J. (2001a). Nucleation, growth and breakage phenomena in agitated wet granulation processes: A review. *Powder Technology, 117*(1–2), 3–39. https://doi.org/10.1016/S0032-5910(01)00313-8.

Iveson, S. M., Wauters, P. A. L., Forrest, S., Litster, J. D., Meesters, G. M. H., & Scarlett, B. (2001b). Growth regime map for liquid-bound granules: Further development and experimental validation. *Powder Technology, 117*(1–2), 83–97. https://doi.org/10.1016/S0032-5910(01)00317-5.

Jodrey, W. S., & Tory, E. M. (1981). Computer simulation of isotropic, homogeneous, dense random packing of equal spheres. *Powder Technology, 30*(2), 111–118. https://doi.org/10.1016/0032-5910(81)80003-4.

Kafui, D., & Thornton, C. (2008). Fully-3D DEM simulation of fluidised bed spray granulation using an exploratory surface energy-based spray zone concept. *Powder Technology, 184*(2), 177–188. https://doi.org/10.1016/j.powtec.2007.11.038.

Kan, H., Nakamura, H., & Watano, S. (2015). Numerical simulation of particle–particle adhesion by dynamic liquid bridge. *Chemical Engineering Science, 138*, 607–615. https://doi.org/10.1016/j.ces.2015.08.043.

Kan, H., Nakamura, H., & Watano, S. (2016). Effect of particle wettability on particle-particle adhesion of colliding particles through droplet. *Powder Technology, 302*, 406–413. https://doi.org/10.1016/j.powtec.2016.08.066.

Kapur, P. C. (1972). Kinetics of granulation by non-random coalescence mechanism. *Chemical Engineering Science, 27*(10), 1863–1869. https://doi.org/10.1016/0009-2509(72)85048-6.

Kapur, P. C., & Fuerstenau, D. W. (1969). Coalescence model for granulation. *Industrial & Engineering Chemistry Process Design and Development, 8*(1), 56–62. https://doi.org/10.1021/i260029a010.

Kawahima, Y., Handa, T., Takeuchi, H., Niwa, K., Sunada, H., & Otsuka, A. (1986). Computer simulation of agglomeration by a two-dimensional random addition model. IV. Agglomeration kinetics and micromeritic properties of agglomerate of binary mixtures of adhesive circles. *Chemical & Pharmaceutical Bulletin, 34*(2), 833–837. https://doi.org/10.1248/cpb.34.833.

Kawashima, Y., Handa, T., Takeuchi, H., & Niwa, T. (1989). Computer simulation of agglomeration by a two-dimensional random addition model—Agglomeration kinetics and micromeritic properties of agglomerate accompanied by compaction process. *Powder Technology, 57*(3), 157–163. https://doi.org/10.1016/0032-5910(89)80071-3.

Kopcha, M., Roland, E., Bubb, G., & Vadino, W. A. (1992). Monitoring the granulation process in a high shear mixer/granulator: An evaluation of three approaches to instrumentation. *Drug Development and Industrial Pharmacy, 18*(18), 1945–1968. https://doi.org/10.3109/03639049209052411.

Kulju, T., Paavola, M., Spittka, H., Keiski, R. L., Juuso, E., Leiviskä, K., & Muurinen, E. (2016). Modeling continuous high-shear wet granulation with DEM-PB. *Chemical Engineering Science, 142*, 190–200. https://doi.org/10.1016/j.ces.2015.11.032.

Kumar, J., Warnecke, G., Peglow, M., & Heinrich, S. (2009). Comparison of numerical methods for solving population balance equations incorporating aggregation and breakage. *Powder Technology, 189*(2), 218–229. https://doi.org/10.1016/j.powtec.2008.04.014.

Lee, K. F., Dosta, M., McGuire, A. D., Mosbach, S., Wagner, W., Heinrich, S., & Kraft, M. (2017). Development of a multi-compartment population balance model for high-shear wet granulation with discrete element method. *Computers and Chemical Engineering, 99*, 171–184. https://doi.org/10.1016/j.compchemeng.2017.01.022.

Lindberg, N.-O., Leander, L., & Reenstierna, B. (1982). Instrumentation of a Kenwood major domestic-type mixer for studies of granulation. *Drug Development and Industrial Pharmacy, 8*(5), 775–782. https://doi.org/10.3109/03639048209042702.

Lister, J. D., Smit, D. J., & Hounslow, M. J. (1995). Adjustable discretized population balance for growth and aggregation. *AICHE Journal, 41*(3), 591–603. https://doi.org/10.1002/aic.690410317.

Litster, J., Hapgood, K., Michaels, J., Sims, A., Roberts, M., & Kameneni, S. (2002). Scale-up of mixer granulators for effective liquid distribution. *Powder Technology, 124*(3), 272–280. https://doi.org/10.1016/S0032-5910(02)00023-2.

Liu, L. X., Litster, J. D., Iveson, S. M., & Ennis, B. J. (2000). Coalescence of deformable granules in wet granulation processes. *AICHE Journal, 46*(3), 529–539. https://doi.org/10.1002/aic.690460312.

Muguruma, Y., Tanaka, T., & Tsuji, Y. (2000). Numerical simulation of particulate flow with liquid bridge between particles (simulation of centrifugal tumbling granulator). *Powder Technology, 109*(1–3), 49–57. https://doi.org/10.1016/S0032-5910(99)00226-0.

Nakamura, H., Fujii, H., & Watano, S. (2013). Scale-up of high shear mixer-granulator based on discrete element analysis. *Powder Technology, 236*, 149–156. https://doi.org/10.1016/j.powtec.2012.03.009.

Nakamura, H., Iwasaki, T., & Watano, S. (2006). Numerical simulation of film coating process in a novel rotating fluidized bed. *Chemical & Pharmaceutical Bulletin, 54*(6), 839–846. https://doi.org/10.1021/ie50643a003.

Nishiura, D., Shimosaka, A., Shirakawa, Y., & Hidaka, J. (2010). Simulation of drying of particulate suspensions in spray-drying granulation process. *Journal of Chemical Engineering of Japan, 43*(8), 641–649. https://doi.org/10.1252/jcej.10we040.

Ouchiyama, N., & Tanaka, T. (1975). The probability of coalescence in granulation kinetics. *Industrial & Engineering Chemistry Process Design and Development, 14*(3), 286–289. https://doi.org/10.1021/i260055a016.

Poon, J. M.-H., Immanuel, C. D., Doyle, F. J., III, & Litster, J. D. (2008). A three-dimensional population balance model of granulation with a mechanistic representation of the nucleation and aggregation phenomena. *Chemical Engineering Science, 63*(5), 1315–1329. https://doi.org/10.1016/j.ces.2007.07.048.

Ramkrishna, D. (2000). *Population balances: Theory and applications to particulate systems in engineering.* Academic Press.

Sastry, K. V. S., & Fuerstenau, D. W. (1970). Size distribution of agglomerates in coalescing dispersed phase systems. *Industrial & Engineering Chemistry Fundamentals, 9*(1), 145–149. https://doi.org/10.1021/i160033a023.

Sato, Y., Nakamura, H., & Watano, S. (2008). Numerical analysis of agitation torque and particle motion in a high shear mixer. *Powder Technology, 186*(2), 130–136. https://doi.org/10.1016/j.powtec.2007.11.028.

Sinnott, M., & Cleary, P. (2003). 3D DEM simulations of a high shear mixer. In *Third International Conference on CFFD in the Minerals and Processing Industries* (pp. 217–222). Melbourne, VIC, Australia: CSIRO Materials. Retrieved from. http://www.cfd.com.au/cfd_conf03/papers/110Sin.pdf .

Strek, F., Rochowiecki, A., & Karcz, J. (1978). A mathematical model of mixing of particulate solids. *Powder Technology, 20*(2), 243–248. https://doi.org/10.1016/0032-5910(78)80055-2.

Sutherland, D. N. (1967). A theoretical model of floc structure. *Journal of Colloid and Interface Science, 25*(3), 373–380. https://doi.org/10.1016/0021-9797(67)90043-4.

Talu, I., Tardos, G. I., & Khan, M. I. (2000). Computer simulation of wet granulation. *Powder Technology, 110*(1–2), 59–75. https://doi.org/10.1016/S0032-5910(99)00268-5.

Terashita, K., Nishimura, T., & Natsuyama, S. (2002). Optimization of operating conditions in a high-shear mixer using DEM model: Determination of optimal fill level. *Chemical & Pharmaceutical Bulletin, 50*(12), 1550–1557. https://doi.org/10.1248/cpb.50.1550.

Terashita, K., Nishimura, T., Natsuyama, S., & Satoh, M. (2002). DEM simulation of mixing and segregation in high-shear mixer. *Journal of the Japan Society of Powder and Powder Metallurgy, 49*(7), 638–645. https://doi.org/10.2497/jjspm.49.638.

Too, J. R., Rubison, R. M., Fan, L. T., & Lai, F. S. (1979). Studies on multicomponent solids mixing and mixtures II. Estimation of mixing index and contact number by spot sampling of a multicomponent mixture in an incompletely mixed state. *Powder Technology, 23*(1), 99–113. https://doi.org/10.1016/0032-5910(79)85029-9.

Tsuji, Y. (2007). Multi-scale modeling of dense phase gas–particle flow. *Chemical Engineering Science, 62*(13), 3410–3418. https://doi.org/10.1016/j.ces.2006.12.090.

Tsuji, Y., Kawaguchi, T., & Tanaka, T. (1993). Discrete particle simulation of two-dimensional fluidized bed. *Powder Technology, 77*(1), 79–87. https://doi.org/10.1016/0032-5910(93)85010-7.

Vold, M. J. (1959). A numerical approach to the problem of sediment volume. *Journal of Colloid Science, 14*(2), 168–174. https://doi.org/10.1016/0095-8522(59)90041-8.

Vold, M. J. (1963). Computer simulation of floc formation in a colloidal suspension. *Journal of Colloid Science, 18*(7), 684–695. https://doi.org/10.1016/0095-8522(63)90061-8.

Washino, K., Tan, H., Hounslow, M. J., & Salman, A. D. (2013). A new capillary force model implemented in micro-scale CFD–DEM coupling for wet granulation. *Chemical Engineering Science, 93*, 197–205. https://doi.org/10.1016/j.ces.2013.02.006.

Watano, S., Fukushima, T., & Miyanami, K. (1995a). Computer simulation of fluidized bed granulation by a two-dimensional random coalescence model. *Journal of Chemical Engineering of Japan, 28*(1), 8–13. https://doi.org/10.1252/jcej.28.8.

Watano, S., Imada, Y., Hamada, K., Wakamatsu, Y., Tanabe, Y., Dave, R. N., & Pfeffer, R. (2003). Microgranulation of fine powders by a novel rotating fluidized bed granulator. *Powder Technology, 131*(2–3), 250–255. https://doi.org/10.1016/S0032-5910(03)00007-X.

Watano, S., Morikawa, T., & Miyanami, K. (1995b). Computer simulation of agitation fluidized bed granulation by a two-dimensional random addition model. *Journal of Chemical Engineering of Japan, 28*(2), 171–178. https://doi.org/10.1252/jcej.28.171.

Watano, S., Morikawa, T., & Miyanami, K. (1996). Mathematical model in the kinetics of agitation fluidized bed granulation. Effects of moisture content, damping speed and operation time on granule growth rate. *Chemical & Pharmaceutical Bulletin, 44*(2), 409–415. https://doi.org/10.1248/cpb.44.409.

Watano, S., Nakamura, H., Hamada, K., Wakamatsu, Y., Tanabe, Y., Dave, R. N., & Pfeffer, R. (2004). Fine particle coating by a novel rotating fluidized bed coater. *Powder Technology, 141*(3), 172–176. https://doi.org/10.1016/j.powtec.2003.03.001.

Zhu, H., Zhou, Z., Yang, R., & Yu, A. (2008). Discrete particle simulation of particulate systems: A review of major applications and findings. *Chemical Engineering Science, 63*(23), 5728–5770. https://doi.org/10.1016/j.ces.2008.08.006.

CHAPTER 21

Application of the discrete element method to scale-up of high-shear granulation

Hideya Nakamura
Department of Chemical Engineering, Osaka Prefecture University, Osaka, Japan

21.1 Introduction

Scale-up of high-shear granulation has been an important yet challenging issue in the pharmaceutical industries. The goal of the scale-up is to maintain the targeted attributes of the granule products, such as size distribution, density, hardness, flowability, compactibility, etc., on scale-up from a smaller granulator used for lab experiment to a larger granulator used for manufacturing. Scale-up of high-shear granulation has often been carried out by relying on trial-and-error and empirical knowledge. However, this approach requires great cost and time, and this is far from the quality by design concept. Thus, there is a great need for developing a scientific based methodology providing rational guidelines for the scale-up.

A computer simulation is a promising tool for the rational design of powder handling processes. In particular, a discrete element method (DEM) appears to be a standard computational method for modeling and analysis of powder handling processes, including high-shear mixer-granulators. Recently scale-up of high-shear mixing and granulation has been attempted by utilizing DEM.

This chapter describes the application of DEM to the scale-up of high-shear granulation. First, DEM is briefly introduced. Physical properties, which we should investigate for the scale-up of high-shear granulation, are then discussed, and the advantages of DEM to the scale-up issue are described. Second, calculation outputs, which we can obtain from DEM simulation for the scale-up of high-shear granulation, are provided. Third some application examples reported to date are reviewed. Finally a summary and forward look are presented.

21.1.1 A brief introduction of discrete element method

The DEM, originally proposed by Cundall and Strack (1979), was developed based on the idea that a granular flow (i.e., flow of an assembly of solid particles) could be simulated when individual particle behavior could be calculated accurately. In fact, the validity of this idea has been confirmed, and DEM has become one of the most successful numerical simulation methods for granular flows. In DEM, the unsteady motion of the individual particles is computed based on Newton's second law of translational and rotational motion, allowing for contact force between particles. The contact force acting on a particle is calculated using the Voigt model, by which the elastic rebound force, the viscous damping force, and the friction force between contacting particles are calculated. Other external forces, such as cohesive forces and a fluid drug force, can also be implemented very easily. Through the discrete element modeling, various properties of granular flows inside the powder handling processes can be obtained. However, it should be noted that the current DEM cannot simulate the actual wet granulation, that is, the transformation from original fine particles to agglomerates with binder liquids cannot be predicted using the current DEM. Thus, we should interpret the current DEM as a tool that can provide important insights into the scale-up issue, while the wet granulation phenomena itself cannot be predicted.

21.1.2 What should we investigate for scale-up of high-shear granulation?

Agglomeration transformation in high-shear granulation can be considered a function of material properties and process parameters (Mort, 2005). When scaling up pharmaceutical granulation, the formulation of the starting material is rarely changed. Thus, the scale-up is typically carried out by adjusting the process parameters. So far, various dimensionless or dimensional parameter groups have been proposed as scale-up criteria for determining process parameters of the larger granulator. These parameter groups include power number and Reynolds number (Faure et al., 2001); Froude number (Horsthuis et al., 1993, Litster et al., 2002); dimensionless spray flux (Litster et al., 2002); relative swept volume (Ramaker et al., 1998); specific agitation power (Bardin et al., 2004; Campbell et al., 2011); and constant tip speed (Ameye et al., 2002; Bock & Kraas, 2001; Rahmanian et al., 2008; Watano et al., 2005). Details of these scale-up criteria are reviewed in several pieces of literature (e.g., Mort, 2005).

In general, as pointed out by Leuenberger (1983), the following three types of similarity should be maintained across granulator sizes for the successful scale-up: geometric (i.e., structural parameters of the granulator and particle bed), kinematic (i.e., the particle flow), and dynamic (i.e., forces or energy acting on the particles) similarities. However, in most of the previous scale-up studies where the dimensionless or dimensional parameters were employed, this principle of the similarities has not been assured due to difficulties in measuring the granular flow properties inside the granulators. In order to establish a rational scale-up methodology, it is necessary to investigate the granular flow properties inside the granulator (e.g., granular flow pattern and velocity, stress acting on the powder, and so on), and thereby the geometric, kinematic, and dynamic similarities should be discussed. Although some experimental studies attempted to measure the internal particle flow (e.g., Saito et al., 2011) and stresses (Tardos et al., 2004; Watano et al., 2005) in high-shear granulators, it is still hard to obtain the internal properties by means of experimental techniques. Determination of a granulation end point is also important for a successful scale-up. In the high-shear mixer granulator, the end-point determination has often been conducted by monitoring the power consumption and torque acting on the impeller blade. However, these monitored values do not directly represent the internal granular flow properties. Thus, it is also desirable to determine the granulation end-point based on the internal granular flow properties.

21.1.3 Advantages of discrete element modeling in the scale-up issue

In terms of investigation of the internal properties, DEM is powerful, because calculations of the particle flow and stresses at a single particle scale are very straightforward within the framework of DEM. Alternative modeling strategies for the scale-up of a wet granulation, such as population balance

Table 21.1 Summary of geometric properties calculated from DEM simulations of high-shear mixers.

Name	Comments/definition	References
Local solid fraction, void fraction, bulk density	Characterizing compression and dilation of particle bed	Boonkanokwong et al. (2016), Chan et al. (2015), (2016), Remy et al. (2010), Remy et al. (2009), and Sarkar and Wassgren (2015)
Local bed height	Characterizing geometry of free surface of powder bed	Chan et al. (2016)
Relative swept volume	Ratio of powder volume swept away by impeller blade to whole powder volume or vessel volume	Chan et al. (2015)
Coordination number	Characterizing the particle packing structure of particle bed	Sarkar and Wassgren (2015)

modeling (PBM) cannot provide information on what is going on inside of the granulator. Recent advances in computer hardware and simulation methods have enabled us to perform DEM simulation of the powder handling processes with a large scale (Radeke et al., 2010, Sakai et al., 2014)). Thus DEM is a potential way to investigate the internal properties that are inaccessible by experimental approaches and to solve the scale-up issue. Moreover, DEM simulation can greatly contribute to the reduction of the cost of the scale-up investigations in industrial sectors, because experimental investigations in larger granulator are very costly and time-consuming.

21.2 What can we extract from the discrete element modeling?

As mentioned in the above section, an important concept for the scale-up is the principle of similarity. When scaling-up high-shear granulation, the three types of similarities should be maintained across the different granulator sizes: the geometric similarity, the kinematic similarity, and the dynamic similarity. Thus, when we apply DEM to the scale-up, the physical properties that correspond to the geometric, kinematic, and dynamic properties should be characterized from raw data of DEM simulations. In the following sections, various geometric, kinematic, and dynamic properties in high-shear granulators, that are available from DEM simulations, are reviewed. This will help the readers to consider what we should/can extract from DEM simulations when the readers utilize DEM for scale-up of high-shear granulation.

21.2.1 Geometric properties

Table 21.1 lists geometric properties calculated from DEM simulations of high-shear mixers. As a geometric property, a local solid fraction, void fraction, and bulk density of the powder bed are easily available from DEM simulation (Boonkanokwong et al., 2016; Chan et al., 2015; Remy et al., 2010; Sarkar & Wassgren, 2015), because positions of the individual particles at any time are calculated. By using these properties, compression and dilation of the particle bed can be characterized. The local

Table 21.2 Summary of kinematic properties calculated from DEM simulations of high-shear mixers.

Name	Definition	Comments	References
Particle velocity field	—	Time- and spatial-average velocity	Boonkanokwong et al. (2016), Nakamura et al. (2013), Remy et al. (2010), Remy et al. (2009), and Sarkar and Wassgren (2015)
Normalized particle velocity	$\frac{\text{Particle velocity}}{\text{Impeller velocity}}$	Normalized by impeller velocity	Chan et al. (2015), Nakamura et al. (2013), and Radl et al. (2010)
Normalized shear rate	$\dot{\gamma}_{ij}^* = \frac{\partial U_j^*}{\partial x_i^*}$	Particle velocity gradient	Nakamura et al. (2013) and Radl et al. (2010)
Granular temperature	$T = \frac{1}{2} u' u'$	Degree of random motion of particles at macroscopic scale	Boonkanokwong et al. (2016), Radl et al. (2010), Remy et al. (2010), and Remy et al. (2009)
Particle diffusion coefficient	$D_{ij} = \frac{(\Delta x_i - \overline{\Delta x_i})(\Delta x_j - \overline{\Delta x_j})}{2t}$	Degree of random motion of particles at microscopic scale	Boonkanokwong et al. (2016), Campbell (1997), Remy et al. (2010), and Sarkar and Wassgren (2015)
Peclet number	$P_{e_{ij}} = \frac{U_i R}{D_{ij}}$	Convective and diffusive contributions to particle motion	Boonkanokwong et al. (2016) and Remy et al. (2010)

bed height is also easily calculated from DEM simulations (Chan et al., 2016). This is useful for characterizing the geometry of the free surface of the powder bed flowing in high-shear granulators. As a dimensionless property, the relative swept volume is calculated (Chan et al., 2015). This dimensionless quantity is defined as a ratio of the powder volume swept away by the main impeller blade to the whole powder volume or vessel volume per unit time or per impeller rotation. As a microscopic property, the coordination number is also analyzed (Sarkar & Wassgren, 2015). The coordination number characterizes particle packing structure in the particle bed, providing an internal geometric property at a microscale.

21.2.2 Kinematic properties

Kinematic properties available from DEM simulations of high-shear mixers are summarized in Table 21.2. Various dimensional or dimensionless physical quantities have been used for characterizing the internal particle flow. The particle velocity field is a commonly used in DEM simulations (Boonkanokwong et al., 2016; Remy et al., 2010; Sarkar & Wassgren, 2015) (e.g., Nakamura et al., 2013). The particle velocity field is visualized as a vector and/or contour plot, providing an intuitive understanding of the internal particle flow. In DEM simulations of high-shear granulators, the time- and spatial-average particle velocity is often calculated. The particle velocity has often been normalized by the impeller velocity (Chan et al., 2015; Nakamura et al., 2013; Radl et al., 2010). This normalized particle velocity is useful to investigate the relative particle velocity against the impeller and to discuss the similarity of the particle velocities between the different impeller velocities.

21.2 What can we extract from the discrete element modeling?

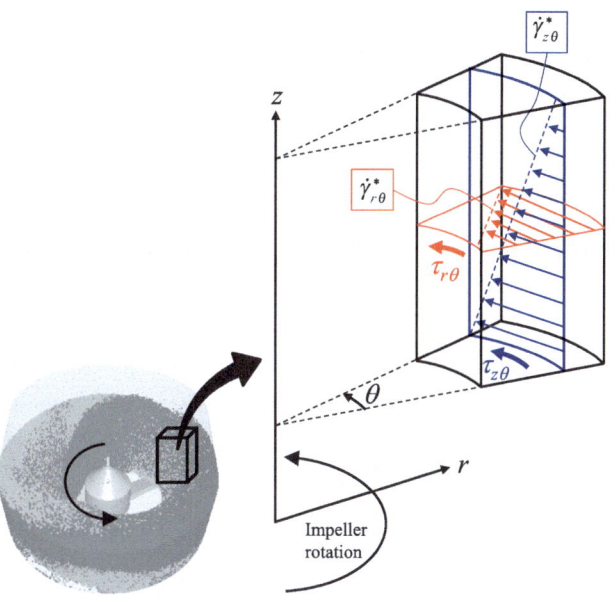

FIGURE 21.1

Dominant components of shear rates and shear stresses in high shear granulator.

The particle flow in high-shear granulators can be regarded as a shear flow. Thus, the shear rates (velocity gradients) of the particle flow can be a representative kinematic property in high-shear granulators. In some previous works, the following normalized shear rate ($\dot{\gamma}_{ij}^*$) is calculated:

$$\dot{\gamma}_{ij}^* = \frac{\partial U_j^*}{\partial x_i^*} \tag{21.1}$$

where U_j^* is the particle velocity normalized by the impeller velocity, and x_i^* is the coordinate normalized by a representative length. As a representative length a vessel diameter (Radl et al., 2010) and bed height (Nakamura et al., 2013) were selected. The subscripts i and j are the components in the coordinate system. In high-shear granulators, as illustrated in Fig. 21.1, $\dot{\gamma}_{r\theta}^*$ (a shear rate on a horizontal $r\theta$-plane) and $\dot{\gamma}_{z\theta}^*$ (a shear rate on a vertical $z\theta$-plane) are dominant components among the nine components of the shear rate tensor (Nakamura et al., 2013). Thus, to characterize the shear rates in high-shear granulators, these two components should be focused and calculated. Radl et al. (2010) and Nakamura et al. (2013) calculated the $\dot{\gamma}_{r\theta}^*$ and $\dot{\gamma}_{z\theta}^*$, respectively.

Another kinematic property is the diffusivity of the particle motion, that is, the degree of the random motion of the particles. Granular temperature is one of the commonly used measures of the particle diffusivity in granular flows (Table 21.2). It has also been used in DEM simulations of high-shear granulators (Boonkanokwong et al., 2016; Radl et al., 2010; Remy et al., 2009; 2010). The granular temperature (T) is defined as:

$$T = \frac{1}{2} u' u' \tag{21.2}$$

where u' is the ensemble average of the fluctuation velocity of particles in a control volume. The fluctuation velocity is the instantaneous deviation from the mean particle velocity in the control volume around the point being examined. The mean particle velocity is calculated by spatially averaging the velocity of the particles existing in the control volume at each time step. The angular bracket $\langle\rangle$ denotes time-averaging of the quantity $u'u'$ within the control volume. Thus, the granular temperature corresponds to the degree of random motion of the particles at a macroscopic scale (i.e., within the control volume size for spatial averaging). A higher granular temperature (i.e., "hot" granular flow) means vigorous particle flow with high velocity fluctuation, while a lower granular temperature (i.e., "cold" granular flow) means uniform particle flow with little velocity fluctuation. Details about the calculation procedures for the granular temperature within the DEM framework can be found in the literature (Boonkanokwong et al., 2016; Radl et al., 2010; Remy et al., 2009; 2010).

Another measure gauging the degree of random motion of particles is the particle diffusion coefficient (Table 21.2). The particle diffusion coefficient corresponds to the particle diffusivity at a microscopic scale (individual particle scale). The particle diffusion coefficient is expressed by a tensor quantity. The particle diffusion coefficient in granular flows is defined as (Campbell, 1997):

$$D_{ij} = \frac{\left(\Delta x_i - \overline{\Delta x_i}\right)\left(\Delta x_j - \overline{\Delta x_j}\right)}{2t} \tag{21.3}$$

where Δx_i is the particle displacement in the i-direction relative to the particle's initial position, and $\overline{\Delta x_i}$ is the mean particle displacement. D_{ij} is the corresponding diffusion coefficient in the i-direction due to a gradient in the j-direction. The angular bracket $\langle\rangle$ means time-averaging of the expression within the time interval t for the averaging. Details about the calculation of the particle diffusion coefficients in high-shear granulators can be found in some literatures (Boonkanokwong et al., 2016; Remy et al., 2010; Sarkar & Wassgren, 2015). As pointed out by Choi et al. (2004) and Boonkanokwong et al. (2016), the particle diffusion coefficients can depend on system geometry and size in granular flows. Therefore, Boonkanokwong et al. (2016) proposed to use the following normalized diffusion coefficient $D_{ij}*$ for high-shear granulators:

$$D^*_{ij} = \frac{D_{ij}}{DV_{tip}} \tag{21.4}$$

where D and V_{tip} are the vessel diameter and the impeller tip speed, respectively. These normalized diffusion coefficients can also be useful for investigating the similarity of the particle diffusivity between different granulator sizes. Once the particle diffusion coefficients are obtained, one can quantify convective and diffusive contributions to the particle motion using the Peclet number (Pe). The Pe for high-shear granulation is defined as (Boonkanokwong et al., 2016; Remy et al., 2010):

$$Pe_{ij} = \frac{U_i R}{D_{ij}} \tag{21.5}$$

where U_i is the averaged particle velocity of the i-component, and R is the vessel radius. The higher Pe indicates that convection is the dominant mechanism for the particle motion rather than diffusion, while the lower Pe indicates that diffusion is dominant.

Table 21.3 Summary of dynamic properties calculated from DEM simulations of high-shear mixers.

Name	Definition	Comments	References
Normal contact force network		Particle contact force distribution	Boonkanokwong et al. (2016)
Collisional stress tensor	$\tau_{ij} = \frac{d_p F_j k_i}{V_c}$	Stress tensor derived from particle-to-particle collision	Campbell (2002), Remy et al. (2009), and Remy et al. (2010)
Internal pressure	$P = \frac{\tau_{rr} + \tau_{\theta\theta} + \tau_{zz}}{3}$	Pressure (normal stress) inside particle bed	Remy et al. (2009) and Remy et al. (2010)
Blade-particle contact force or stress		Contact forces or stress between particles and impeller blade	Chan et al. (2016) and Chandratilleke et al. (2012)
Impeller torque	$T = (\mathbf{r}_i \times \mathbf{F}_{ci})_z$	Torque acting on rotating impeller	Remy et al. (2010), Sarkar and Wassgren (2015), and Sato et al. (2008)
Particle collision energy	$E_c = \frac{1}{2} \frac{m_{p,i} m_{p,j}}{m_{p,i} + m_{p,j}} U_{c,i-j}^2$	Energy between colliding particles	Nakamura et al. (2013)

21.2.3 Dynamic properties

Table 21.3 lists dynamic properties calculated from DEM simulations of high-shear mixers. The internal dynamic properties in high-shear mixers have been characterized by various forces, stresses, and energy.

One of the dynamic properties commonly used in DEM simulations is the normal contact force network. This is depicted by lines representing the normal contact vector connecting the center of the particles contacting with each other. Spatial distribution of the lines is visualized, providing intuitive information that is helpful to understand the internal force distribution within the particle bed. The contact force network represents locations where the momentum transfer causing the particle flow occurs. (Boonkanokwong et al., 2016) employed the contact force network to investigate force propagation from the rotating impeller blade to particles.

As a measure of the internal stress, the collisional stress tensor has been employed in DEM simulations of high-shear mixers (Remy et al., 2009;2010). The collisional stress tensor was proposed by Campbell (2002) and defined as follows:

$$\tau_{ij} = \frac{d_p F_j k_i}{V_c} \tag{21.6}$$

where d_p is the particle diameter, V_c is the control volume, F_j is the total contact force in the j-direction, and k_i is the i-component of the unit vector connecting the centers of the contacting particles. The collisional stress can be affected by the size of the control volume. Thus, a suitable size of the control volume, where the calculated stress is independent of the size of the control volume, should be investigated in advance. More details of the calculation procedure can be found in Remy et al. (2009, 2010). As described before, in high-shear granulators, shear stresses on a horizontal $r\theta$-plane ($\tau r\theta$) and on a vertical $z\theta$-plane ($\tau z\theta$) can be dominant components (Fig. 21.1). Thus these two components should also be focused on when calculating the collisional stress tensor. Once the collisional stresses are obtained, the internal pressure P is calculated from the normal components of the collisional stresses

(Remy et al., 2009;2010):

$$P = \frac{\tau_{rr} + \tau_{\theta\theta} + \tau_{zz}}{3} \qquad (21.7)$$

where τ_{rr}, $\tau_{\theta\theta}$, τ_{zz} are the normal collisional stresses in all three components. For investigating the similarity of the stresses and pressure between different vessel sizes or different particle bed heights, the stresses and pressure have been scaled by the hydrostatic pressure $\rho_p g H$ (where ρ_p, g, H are the particle density, gravity acceleration, particle bed height, respectively) (Remy et al., 2010).

In high-shear granulation, the internal forces and stresses within the particle bed arise from the external forces exerted by the rotating impeller blades. Thus, the stress between the blade and particles, that is, the blade-particle bed stress, can be an important dynamic property. Chan et al. (2016) and Chandratilleke et al. (2012) investigated the normal component of the blade-particle bed stress at various conditions, including different blade angles, blade speeds, and vessel sizes using DEM simulations. They utilized the simulated blade-particle bed stress for deriving semiempirical equations to predict the blade-particle bed stress. These equations can explain a dynamic similarity based on the blade-particle bed stress. Torque acting on the rotating impeller has also been investigated by means of DEM simulations. Remy et al. (2010) and Sato et al. (2008) found that the impeller torque is linearly correlated with the kinetic energy of particles and the particle fill level. Sarkar and Wassgren (2015) reported that a dimensionless impeller torque can be correlated as a single master curve by a simple Froude number scaling.

Although the aforementioned properties can be utilized to characterize internal dynamic properties of high-shear mixers, there is no physical explanation which assures that the agglomeration transformation in wet granulation can be controlled by the aforementioned properties. As a dimensionless number which can describe granule coalescence and growth in high-shear granulation, the Stokes deformation number St_{def} was proposed as (Iveson & Litster, 1998; Tardos et al., 1997):

$$St_{def} = \frac{0.5 m_p U_c^2}{V_p Y_p} \qquad (21.8)$$

where m_p and V_p are the mass and volume of a particle, U_c is the particle collision velocity, and Y_p is the yield stress of the particle. The Stokes deformation number describes the particle collision energy scaled by the energy required to break the particle. The Stokes deformation number suggests that the particle collision energy can be a dynamic property correlating with the agglomeration transformation in high-shear granulation. Based on this consideration, Nakamura et al. (2013) calculated the particle collision energy, which is the numerator of St_{def}, and used it as a measure of the dynamic property. They also investigated the dynamic similarity based on the collision energy across different sizes. Details will be presented in the following Section 21.3.4.

21.3 Application examples

In the following sections, application examples of DEM simulation to the scale-up of high-shear mixer-granulators are described. First, some similarity analyses between high-shear mixers with different sizes are reviewed (Sections 21.3.1–21.3.3). Finally, a combined kinematic-dynamic scale-up method proposed by our research group (Nakamura et al., 2013) is explained, and its application to high-shear granulation is presented.

21.3.1 Similarities between different sizes of vertical mixers at bumping flow regime

Remy et al. (2010) investigated the effects of D/d (ratio of the vessel diameter to the particle diameter) on geometric, kinematic, and dynamic properties in a four-bladed vertical mixer by means of a DEM simulation. A granular flow of monodisperse and cohesionless spherical glass beads with 10 mm diameter was simulated. Dimensions of the mixer were linearly scaled-up with satisfying geometric similarity, while the particle diameter was kept constant. Simulations were performed under a constant blade rotation speed of 20 rpm. Under this relatively small rotation speed, the particle flow patterns appear to be a bumping flow regime (Litster et al., 2002), where surface of the powder bed remains horizontal and the powder bed bumps up and down with little vertical circulation as the impeller blade passes underneath. Thus, this literature (Remy et al., 2010) can be regarded as a work in which similarities of vertical mixers under the bumping flow regime were investigated using DEM simulations.

Fig. 21.2 shows the effects of D/d on internal properties in the mixers (Remy et al., 2010). They calculated the following internal properties: solids fraction as a geometric property; normalized particle velocities as kinematic properties; normalized pressure, shear stress on the horizontal plane, and bulk friction coefficients as dynamic properties. The particle velocities were normalized by the impeller tip speed, and the pressure and shear stress were normalized by the hydrostatic pressure $\rho_p g H$. Interestingly, the calculated internal properties were found to be scaled according to the mixer size at $D/d \geq 63.0$, while the internal properties were not be scaled at the smallest scale $D/d = 31.5$. The differences in properties observed at $D/d = 31.5$ were considered to be attributed to the wall friction effect, which is more significant for smaller vessel size. This result suggests that the geometric, kinematic, and dynamic similarities can be simultaneously achieved when the mixer scale is larger than a critical size such that the wall effect is minimized. However, it should be noted that this scaling law may be limited within a lower blade rotation speed, that is, within the bumping flow regime.

21.3.2 Kinematic similarity based on relative swept volume and impeller speed-mixer diameter scaling relationship

Chan et al. (2015) performed a DEM simulation of high-shear mixers with different vessel sizes, followed by analyzing the kinematic similarity. They simulated a horizontal high-shear mixer. They also used monodisperse and cohesionless particles with different sizes within 1.25 to 10 mm. DEM simulations were performed under various operating conditions including different impeller tip speeds and fill levels. The simulations were also conducted at different ratios of the blade-to-wall gap to the vessel radius ($= G/R$).

Through the systematic investigation, they found a similarity rule for the particle velocity: that is, the normalized particle velocity can be linearly correlated with the relative swept volume per impeller rotation ($RSVP_{par}$) regardless of the vessel volume (Fig. 21.3). This result provided new insight that the kinematic similarity can be satisfied by keeping the $RSVP_{par}$ constant when scaling-up the horizontal high-shear mixer. This also implies that a constant rule of geometric property (i.e., $RSVP_{par}$) can result in a kinematic similarity when scaling up high-shear mixer.

Chan et al. (2015) also determined a scaling exponent (n) for the impeller speed-mixer diameter scaling relationship (Tardos et al., 2004) from the simulation results. The scaling relationship between

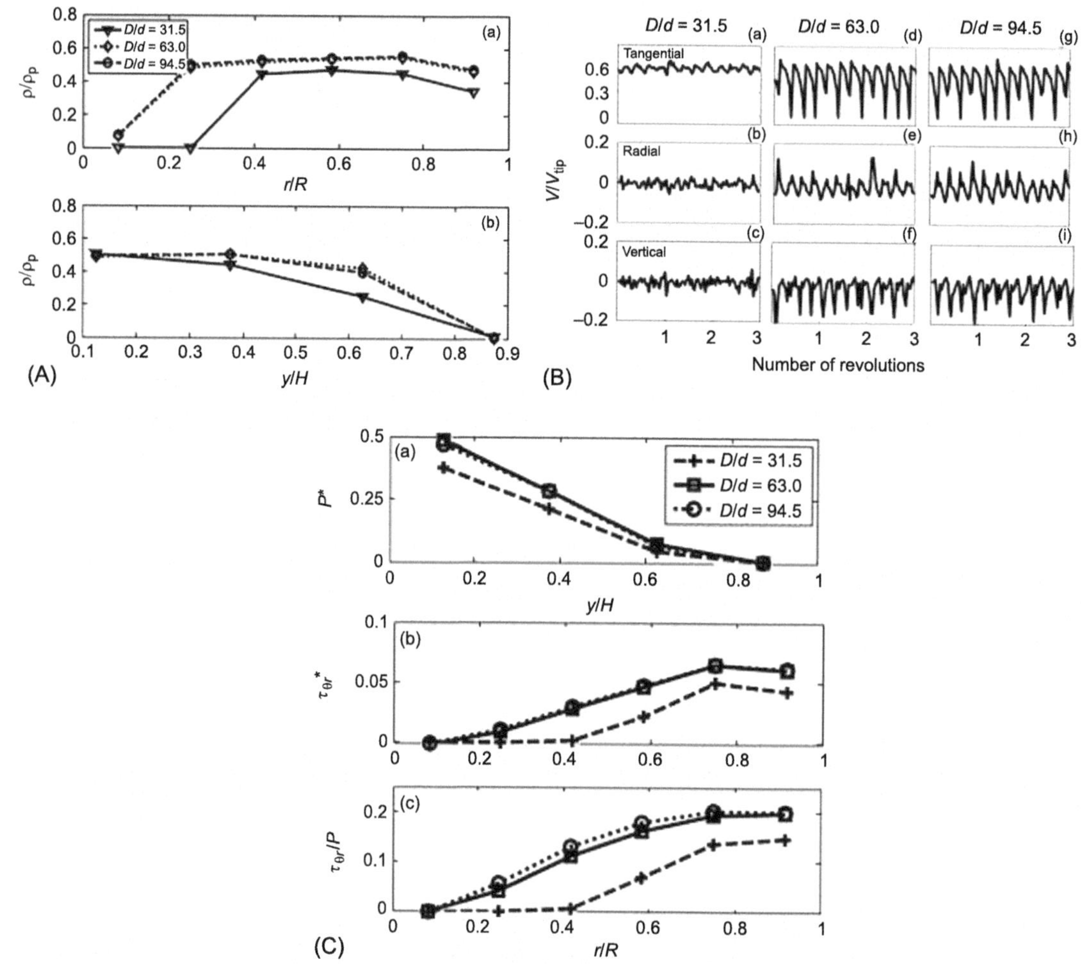

FIGURE 21.2

Effects of D/d on geometric, kinematic, and dynamic properties in four-bladed vertical mixer. (A) Effects of D/d on solid fraction along (a) radial and (b) vertical directions. (B) Normalized particle velocities in each component at different D/d. (C) normalized pressure, stress and bulk friction at different D/d.

From Remy et al. (2010).

different vessel sizes (subscripts 1 and 2) is expressed as (Tardos et al., 2004):

$$\frac{\omega_2}{\omega_1} = \left(\frac{D_1}{D_2}\right)^n \tag{21.9}$$

where D is a vessel diameter. ω is the impeller rotational speed to maintain the normalized particle velocity inside the mixer. Table 21.4 lists the scaling exponent n at different G/R ratios and particle sizes (Chan et al., 2015). They found that the scaling exponent n was sensitive to the G/R ratio. At a

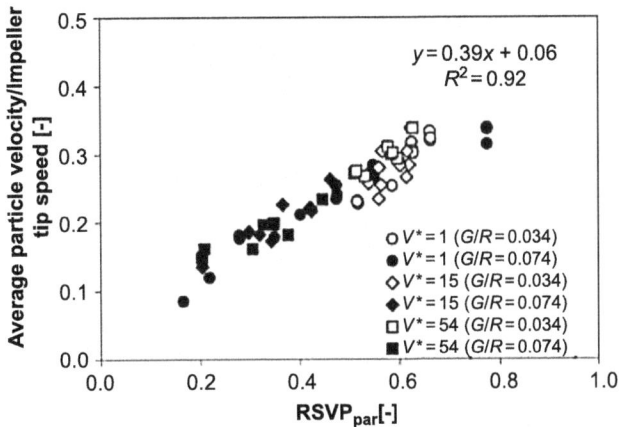

FIGURE 21.3

ormalized particle velocity as a function of relative swept volume (particle volume-based) per impeller rotation (RSVP$_{par}$) at different vessel sizes. V_* is a normalized vessel volume. G/R is the ratio of blade-wall gap to vessel radius.

From Chan et al. (2015).

Table 21.4 Scaling exponent (n) obtained from DEM simulations. G/R is the ratio of blade-wall gap to vessel radius.

d_p (mm)	N	
	$G/R = 0.074$	$G/R = 0.034$
10	0.7–0.8	0.8–0.9
5	0.7–0.8	0.9–1.0
2.5	0.75–0.8	0.9–1.0

From Chan et al. (2015).

smaller relative gap size ($G/R = 0.034$), the scaling exponent n was close to 1, which follows the constant tip speed, while at a larger relative gap size ($G/R = 0.074$), the scaling exponent n decreased to 0.7 to 0.8.

21.3.3 Dynamic similarity based on the blade-particle bed stress

Chan et al. (2016) investigated a dynamic similarity between vertical high-shear mixers with different vessel sizes by means of a DEM simulation. They focused on the blade-particle bed stress as a measure of the dynamic properties. Monodisperse and cohesionless particles from 2 to 10 mm sizes were used. DEM simulations were systematically performed under various properties for the particle and particle bed (including particle diameter, bed load, particle density, and friction coefficient between particles) as well as various impeller geometries (including blade angle, number of blades, and blade width).

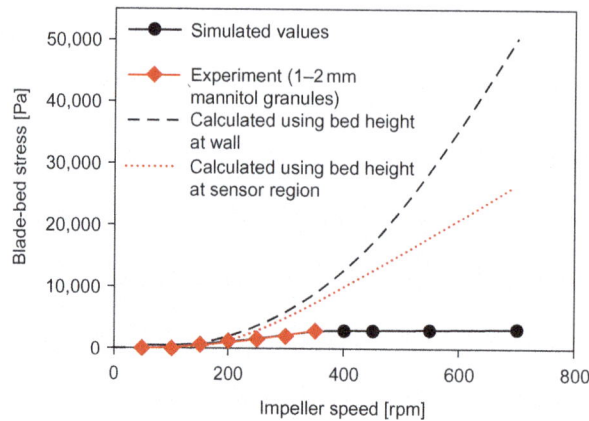

FIGURE 21.4

Comparison of the simulated blade-bed stress by DEM simulation with the experimental results.

From Chan et al. (2016).

In their work, a comparison between simulation results of the blade-particle bed stress with the experimental results was conducted, as shown in Fig. 21.4 (Chan et al., 2016). The blade-particle bed stress was experimentally measured by means of a custom-built telemetric impeller pressure sensor system. The simulation result obtained from the DEM simulation showed good agreement with the experimental results. They concluded that DEM simulation can reasonably predict the blade-particle bed stress within impeller speeds up to 350 rpm (upper limit of the experimental measurement system).

For analyses of the blade-particle bed stress at higher impeller speeds than 350 rpm, DEM simulation was employed. They proposed a dynamic similarity scaling law based on the blade-particle bed stress calculated from DEM simulation. The following semiempirical correlation to estimate the blade-particle bed stress across different vessel sizes was derived:

$$\sigma_I = B_2 \frac{\rho_b H_{g,r} v_{rel}^2 (1 - \cos\theta_I)}{h_I} \tag{21.10}$$

where σ_I, ρ_b, $H_{g,r}$, v_{rel}, θ_I, and h_I are the blade-particle normal stress at a certain radial position, bulk density of the particle bed, particle bed height at a certain radial position, relative velocity between the blade and particles, impeller blade angle, and impeller blade height, respectively. B_2 is a correction factor. This formulation was originally derived by Knight et al. (2001). Chan et al. (2016) attempted to modify the original formulation with the correction factor B_2, which is applicable across different vessel sizes. DEM simulation results were utilized to obtain the following correlation regarding the correction factor B_2 in the Eq. (10):

$$B_2 = 0.32 \frac{1}{Fr_{rel}^{0.5}} \left(\frac{2h_w}{D_I}\right)^{0.3} \theta_I^{-0.5} \tag{21.11}$$

where Fr_{rel}, h_w, D_I are the relative Froude number, impeller bed width, and impeller diameter, respectively. Validity of the Eq. (11) was confirmed within mixer vessel sizes up to 90 L.

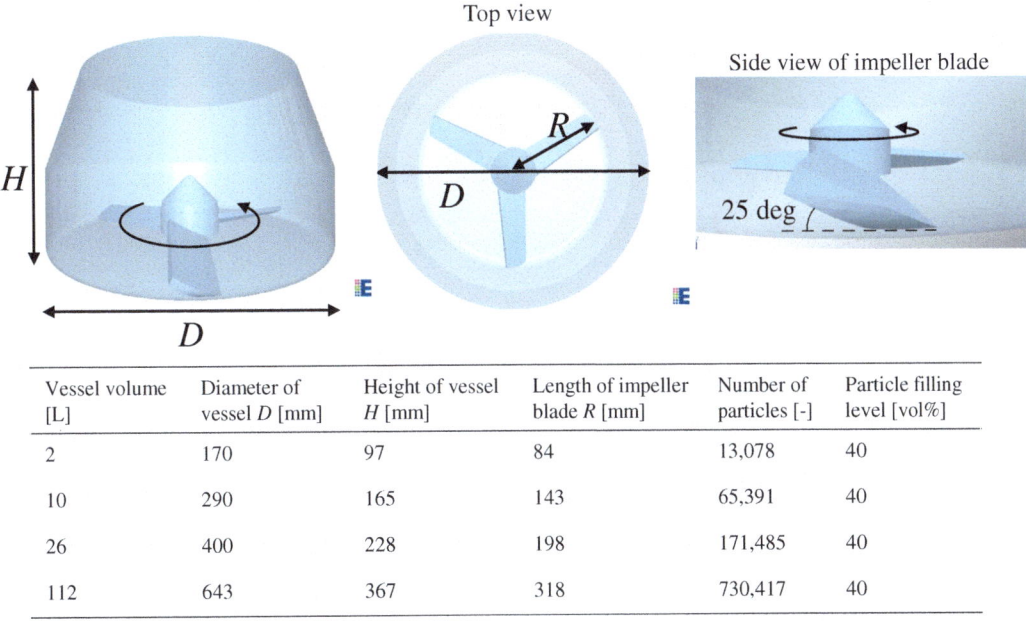

FIGURE 21.5

Geometry and dimensions of high shear mixer-granulator simulated.

From Nakamura et al. (2013).

Vessel volume [L]	Diameter of vessel D [mm]	Height of vessel H [mm]	Length of impeller blade R [mm]	Number of particles [-]	Particle filling level [vol%]
2	170	97	84	13,078	40
10	290	165	143	65,391	40
26	400	228	198	171,485	40
112	643	367	318	730,417	40

21.3.4 Combined kinematic-dynamic scale-up method

Our research group investigated both of kinematic and dynamic similarities between vertical high-shear mixers with different sizes using a DEM simulation (Nakamura et al., 2013). Based on the simulation result, we proposed a combined kinematic-dynamic scale-up method for high-shear granulation. Finally, the validity of the proposed method was confirmed by conducting an experiment of wet granulation. In the following section, detail of this work is described (Nakamura et al., 2013).

In the DEM simulation by Nakamura et al. (2013), the Hertz–Mindlin contact model was employed for the calculations of the contact forces. The rolling friction torque was also taken into account in the contact forces. A model particle with 4 mm diameter, spherical shape, monodispersity, nonbreakable (rigid) property, and cohesionless property was used in this study. This model particle obviously differed from particles used in real wet granulations. However, the objective of this study was to investigate influences of the vessel size on the kinematic and dynamic similarities but not to simulate the actual agglomeration transformation phenomena. Thus, this simplified model particle was used based on an assumption that the kinematic and dynamic similarities can be analyzed even by the simplified model particle. A vertical high-shear mixer-granulator shown in Fig. 21.5 was used. A three-dimensional representation of this mixer-granulator was created using a computer aided design (CAD) software. The three-dimensional CAD drawing was imported into the simulator so that its complex geometries were accurately taken into account. Four sets of mixer-granulators with different vessel volumes

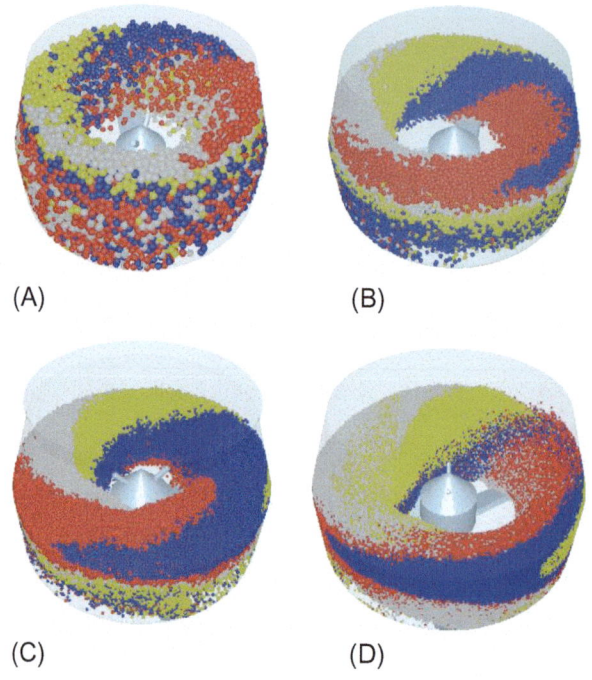

FIGURE 21.6

Snapshots of simulated particle bed at different vessel sizes with a constant impeller tip speed (= 6.66 m/s). (A) 2 L; (B) 10 L; (C) 26 L; (D) 112 L.

(2–112 L) were simulated, and the geometric similarity was satisfied across the mixer-granulators. The DEM simulations were performed at various impeller rotating speeds. EDEM software (ver. 2.2; DEM Solutions Ltd.) was employed for the simulations.

Fig. 21.6 shows typical snapshots of the simulated particles at different vessel sizes. Each particle was dyed depending on its initial position. With an increase in the vessel sizes, the particle size appears smaller due to the increase in the vessel size. Based on these DEM simulations, an internal particle flow property (i.e., kinematic property) was first analyzed. Fig. 21.7A shows a typical snapshot of particles on a vertical cross-section above the impeller blade. Each particle was dyed depending on the magnitude of its velocity. The particle velocity gradient along the vertical direction was clearly observed. Fig. 21.7B shows tangential, vertical, and radial components of the averaged particle velocity as a function of height from the bottom of the vessel. The magnitude of the tangential particle velocity was more than twenty times as much as that of the vertical and radial particle velocities. This means that in the high-shear mixer-granulator, the particle motion in the tangential direction is largely dominant over the other directions. This result indicates that the internal particle flow can be characterized by the tangential component of the particle velocity and its gradient along the vertical direction. Based on this investigation, Nakamura et al. (2013) used $\dot{\gamma}^*_{z\theta}$ (a normalized shear rate in the vertical direction, shown in Fig. 21.1) as a measure of the kinematic properties.

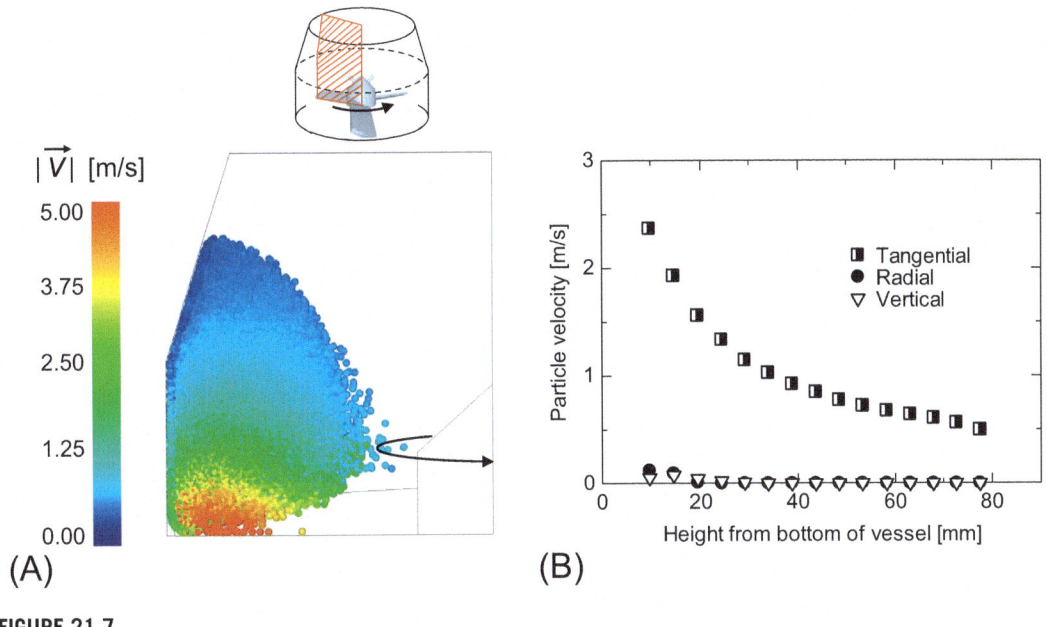

FIGURE 21.7

(A) Example of particle velocity distribution on vertical cross-section (impeller tip speed = 6.50 m/s). Each particle was dyed depending on magnitude of velocity. (B) Example of particle velocities in each component as a function of height from bottom of vessel (impeller tip speed = 6.50 m/s).

From Nakamura et al. (2013).

Fig. 21.8 shows the normalized shear rate $\dot{\gamma}_{z\theta}^*$ as a function of the impeller tip speed at various vessel sizes. The normalized shear rate was well correlated by the impeller tip speed regardless of the vessel size, suggesting that the kinematic similarity can be achieved by maintaining the impeller tip speed across the different vessel sizes. With an increase in the impeller tip speed, the normalized shear rate increased within the range of the impeller tip speed = 1.5 to 5 m/s, followed by an almost constant value regardless of the vessel size. It was found that this change in the normalized shear rate was correlated with a change in the particle flow pattern: that is, in the increasing range of the normalized shear rate the particle flow pattern changed from the bumping flow to roping flow, while within the constant range of the normalized shear rate the internal particle flow pattern remained the roping flow (Nakamura et al., 2013). The shear rate in Fig. 21.8 was normalized by the impeller tip speed. Thus, the result in Fig. 21.8 suggests that when scaling-up the high-shear mixer-granulator, the impeller tip speed should be kept constant in order to maintain the magnitude of the shear rate. The enclosed plots in Fig. 21.8 denote the normalized shear rates at a constant impeller tip speed (= 6.66 m/s). The normalized shear rate was collapsed into almost same value regardless of the vessel size, supporting the aforementioned suggestion. Consequently, the DEM analysis by (Nakamura et al., 2013) provided insight into the kinematic similarity: that is, the rule of constant impeller tip speed should be employed to achieve the kinematic similarity. This result is consistent with many experimental and empirical findings where

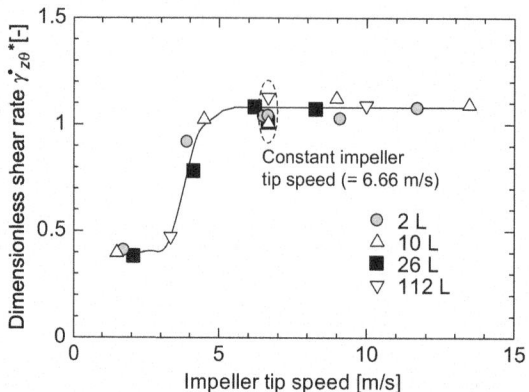

FIGURE 21.8

Normalized shear rate as a function impeller tip speed at various vessel sizes.

From Nakamura et al. (2013).

the rule of constant impeller tip speed was employed for the scale-up of high-shear granulations (Ameye et al., 2002; Bock & Kraas, 2001; Rahmanian et al., 2008; Watano et al., 2005).

Subsequently Nakamura et al. (2013) analyzed a dynamic similarity. The particle collision energy was used as a measure of the dynamic property. The maximum particle collision energy generated during the collision of the particles was calculated, because it can be the most effective. Nakamura et al. (2013) calculated the maximum collision energy within the DEM framework as shown in Fig. 21.9A; that is, the collision energy just after collision of a pair of particles was only calculated, but the collision energy during overlapping was not counted. Fig. 21.9B shows the calculation procedure of the collision energy, proposed by (Nakamura et al., 2013). At each calculation time, the particle-to-particle collisions which newly occurred were detected. The collision energy E_c was then calculated for only the newly occurred particle-to-particle collisions. The collision energy E_c was defined as:

$$E_{c,i-j} = \frac{1}{2} \frac{m_{p,i} m_{p,j}}{m_{p,i} + m_{p,j}} U_{c,i-j}^2 \tag{21.12}$$

where subscripts i and j indicate the two colliding particles. m_p and U_c are the mass of the particle and the particle collision velocity. In high-shear granulation, it is also important to understand how much collision energy is imposed on a single particle. The collision energy acting on a single particle can be determined from both of the collision energy per particle-to-particle collision (i.e., E_c) and the number of particle-to-particle collisions per single particle. The collision energy acting on individual particles E_p was defined as:

$$E_p = \sum_{}^{N_c} E_c \tag{21.13}$$

where N_c is the number of particle-to-particle collisions on a single particle. E_p corresponds to the instantaneous collision energy acting on a single particle. The cumulative particle collision energy per unit time (E_{cum}) was then calculated by summing up E_p for a certain sampling duration. E_{cum} was defined

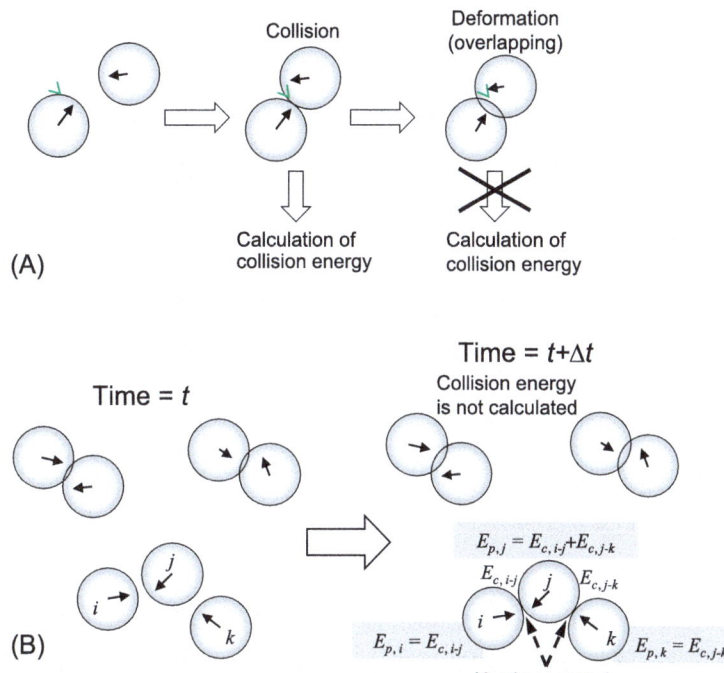

FIGURE 21.9

(A) Calculation of collision energy between colliding particles. (B) Calculation procedure of collision energy at each time step.

From Nakamura et al. (2013).

as:

$$E_{cum} = \frac{1}{T_s} \int_0^{T_s} E_p dt \tag{21.14}$$

where T_s is the sampling duration. Finally, the number averaged cumulative particle collision energy per unit time (E_t) was calculated as:

$$E_t = \frac{1}{N_p} \sum^{N_p} E_{cum} \tag{21.15}$$

where N_p is the total number of particles.

Fig. 21.10A shows distributions of E_p (the instantaneous collision energy acting on the individual particles) at different vessel sizes with a constant impeller tip speed. Although slight variations were observed with a change in the vessel size, it can be considered that the instantaneous collision energy acting on the individual particles was nearly maintained. This implies that the instantaneous collision energy can be maintained by keeping the impeller tip speed constant. Fig. 21.10B shows E_t (the

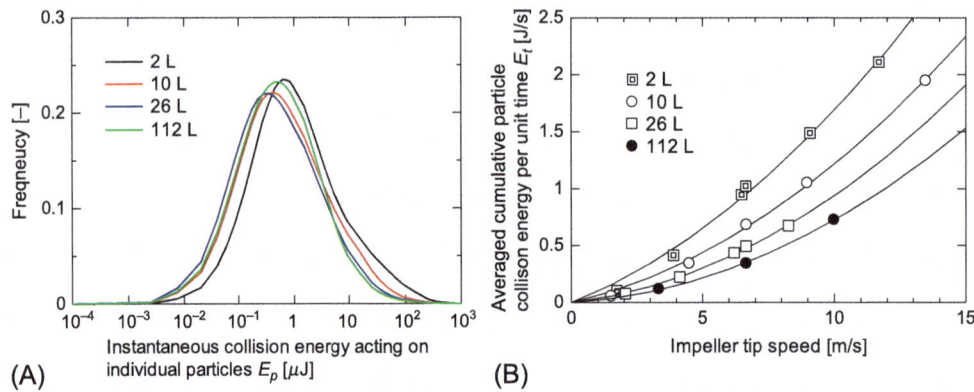

FIGURE 21.10

(A) Distributions of instantaneous collision energy acting on individual particle at different vessel sizes under constant impeller tip speed (= 6.66 m/s). (B) Cumulative particle collision energy per unit time at different vessel sizes as a function of impeller tip speed.

From Nakamura et al. (2013).

cumulative particle collision energy per unit time) as a function of the impeller tip speed at different vessel sizes. Although the instantaneous collision energy (E_p) was nearly maintained, the E_t was not maintained but reduced with the vessel size under a constant impeller tip speed. Nakamura et al. (2013) considered that this reduction in the cumulative collision energy at larger size was derived from less particle circulation in the vertical direction. The cumulative particle collision energy is determined by both the magnitude of the instantaneous collision energy and its accumulation rate. In general, scale-up of the high-shear mixer-granulator under the constant impeller tip speed leads to less vertical circulation of particles in larger scale, resulting in less accumulation rate of the instantaneous collision energy. Consequently, DEM simulation Nakamura et al. (2013) provided an insight into the dynamic similarity: that is, the cumulative particle collision energy per unit time cannot be maintained but reduced if a high-shear mixer-granulator is scaled up under a constant impeller tip speed, although the instantaneous particle collision energy can be maintained.

Based on the aforementioned results obtained from DEM simulation, Nakamura et al. (2013) proposed a combined kinematic-dynamic scale-up method. In this method, the constant rule of the impeller tip speed is adopted for satisfying the kinematic similarity when scaling-up a granulator. In order to maintain the cumulative collision energy, Nakamura et al. (2013) proposed to scale the processing time. The proposed constant rule is expressed as:

$$E_{t1}t_{g1} = E_{t2}t_{g2} \tag{21.16}$$

where t_g is the processing time, and the subscripts 1 and 2 represent smaller and larger mixer-granulators, respectively. Once the optimal impeller tip speed and processing time (t_{g1}) is experimentally determined in the small scale mixer-granulator and the cumulative particle collision energy per unit time in the smaller and larger mixer-granulators (E_{t1} and E_{t2} under the constant impeller tip speed) is calculated from DEM simulation. It is possible to determine the processing time in the larger mixer-granulator (t_{g2}) according to the scaling law of Eq. (16). Consequently, this method can simultaneously achieve

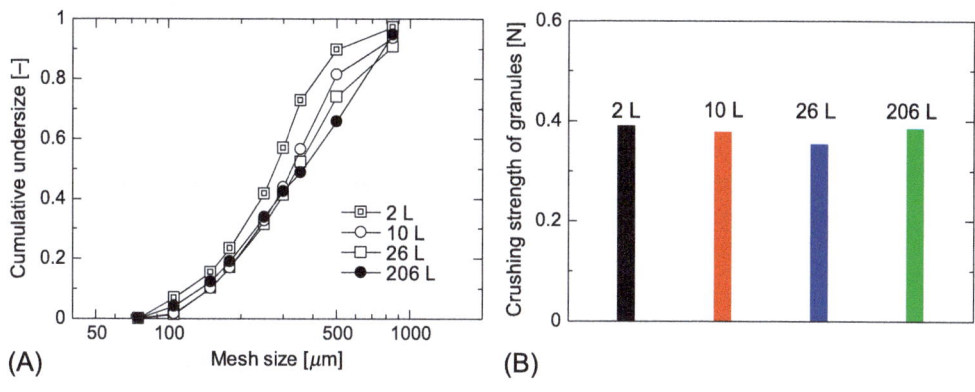

FIGURE 21.11

(A) Size distribution and (B) crushing strength of granules prepared in different vessel sizes under operating conditions determined by the combined kinematic-dynamic scale-up method.

From Nakamura et al. (2013).

Table 21.5 Operating conditions determined by the combined kinematic-dynamic scale-up method.

Vessel volume (L)	Impeller rotational speed (rpm)	Impeller tip speed (m/s)	Averaged cumulative particle collision energy per unit time E_t (J/s)	Processing time t_g (s)
2	721	6.26	0.873[b]	193
10	418	6.26	0.589[b]	286
26	301	6.26	0.418[b]	403
206[a]	150	6.26	0.281[c]	600

From Nakamura et al. (2013).
[a] Reference size.
[b] Calculated from DEM simulation.
[c] Extrapolated from the correlation of E_t within 2 to 26 L.

the kinematic and dynamic similarities between different vessel sizes, that is, the combined kinematic-dynamic scale-up method.

Validity of the proposed method was evaluated by an experiment of wet granulation (Nakamura et al., 2013). In the experiment, a ternary mixture of pharmaceutical excipients, which composed of lactose, corn starch, and microcrystalline cellulose, was used as the starting material. Hydroxypropyl cellulose was used as a binder and added as a form of dry powder. Purified water was used as a binder solution. Scaling-down of the 206 L mixer-granulator to the 26 L, 10 L, and 2 L mixer-granulators was conducted, because the experiment at the 206 L had been already conducted by our research group (Watano et al., 2005). Table 21.5 listed the operating conditions in each vessel size, which were determined by the combined kinematic-dynamic scale-up method. The averaged cumulative particle collision energy per unit time E_t was preliminarily obtained by DEM simulation at each vessel size under the constant impeller tip speed (= 6.26 m/s in this case). Fig. 21.11 shows the experimental results. The size distribution and crushing strength of granules prepared at the reference size (206 L) were well maintained when the scaling-down. These experimental results support the validity of the proposed scale-up method.

21.4 Summary and outlook

Application of the DEM to the scale-up of high-shear granulation was presented in this chapter. First of all, a brief introduction of DEM was described (Section 21.1.1). Physical properties, which we should investigate for scale-up of high-shear granulation, were then discussed (Section 21.1.2). When scaling-up high-shear granulation, the following three types of similarities should be maintained across the different sizes; the geometric, kinematic, and dynamic similarities. Thus, it is necessary to analyze the physical properties corresponding to these similarities. Advantages of DEM were then described (Section 21.1.3). Within the framework of DEM, calculations of the individual particle motion and forces or energy on the individual particle are very straightforward. This feature makes DEM a powerful tool to investigate the internal properties. In Section 21.2, geometric, kinematic, and dynamic properties that are available from DEM simulation of high-shear mixers were summarized. Finally, application examples of DEM to similarity analyses and scale-up of high-shear mixer-granulators were reviewed (Section 21.3). In particular, a combined kinematic-dynamic scale-up method proposed by Nakamura et al. (2013) was explained in detail.

Many questions remain unanswered, and many issues are still unsolved for application of DEM to the scale-up of high-shear granulation. In the application examples, which have been performed so far, large (mm-size) and cohesionless particles were used in the simulations for simplicity and reducing the computing cost. However, in actual wet granulation, fine particles (several hundred micrometer size) are used. Thus, a new method which enables us to perform large-scale DEM simulation is strongly required. Recently, many efforts, including high-performance computing methods (e.g., Radeke et al., 2010) and coarse-grained methods (e.g., Sakai et al., 2014) have been proposed to realize the large scale DEM simulation for powder handling processes with an industrial large scale. Moreover, in wet granulations, cohesive particles including binder liquid are usually used as a starting material, and its transformation (such as wetting, agglomerate growth, consolidation, and breakage) occur in the real wet granulation process. Thus these fundamental steps constituting an agglomerate transformation should be modeled. Recently, some related phenomena such as binder penetration and spreading into the moving particle bed (Washino et al., 2013) and particle adhesion by the dynamic liquid bridge (Kan et al., 2015, 2016) have been modeled. To predict the particle growth kinetics, a new coupling model of DEM with a PBM has been proposed (e.g., Barrasso & Ramachandran, 2015). A combination of these efforts can realize a direct and a priori modeling methodology for scale-up of high-shear wet granulation.

References

Ameye, D., Keleb, E., Vervaet, C., Remon, J. P., Adams, E., & Massart, D. L. (2002). Scaling-up of a lactose wet granulation process in mi-pro high shear mixers. *European Journal of Pharmaceutical Sciences, 17*(4–5), 247–251. https://doi.org/10.1016/S0928-0987(02)00218-X.

Bardin, M., Knight, P., & Seville, J. (2004). On control of particle size distribution in granulation using high-shear mixers. *Powder Technology, 140*(3), 169–175. https://doi.org/10.1016/j.powtec.2004.03.003.

Barrasso, D., & Ramachandran, R. (2015). Multi-scale modeling of granulation processes: Bi-directional coupling of PBM with DEM via collision frequencies. *Chemical Engineering Research and Design, 93*, 304–317. https://doi.org/10.1016/j.cherd.2014.04.016.

Bock, T. K., & Kraas, U. (2001). Experience with the Diosna mini-granulator and assessment of process scalability. *European Journal of Pharmaceutics and Biopharmaceutics, 52*(3), 297–303. https://doi.org/10.1016/S0939-6411(01).00197-7.

Boonkanokwong, V., Remy, B., Khinast, J. G., & Glasser, B. J. (2016). The effect of the number of impeller blades on granular flow in a bladed mixer. *Powder Technology, 302*, 333–349. https://doi.org/10.1016/j.powtec.2016.08.064.

Campbell, C. S. (1997). Self-diffusion in granular shear flows. *Journal of Fluid Mechanics, 348*. May https://doi.org/10.1017/S0022112097006496.

Campbell, C. S. (2002). Granular shear flows at the elastic limit. *Journal of Fluid Mechanics, 465*(2002), 261–291. https://doi.org/10.1017/S002211200200109X.

Campbell, G. A., Clancy, D. J., Zhang, J. X., Gupta, M. K., & Oh, C. K. (2011). Closing the gap in series scale up of high shear wet granulation process using impeller power and blade design. *Powder Technology, 205*(1–3), 184–192. https://doi.org/10.1016/j.powtec.2010.09.009.

Chan, E. L., Washino, K., Ahmadian, H., Bayly, A., Alam, Z., Hounslow, M. J., et al. (2015). Dem investigation of horizontal high shear mixer flow behaviour and implications for scale-up. *Powder Technology, 270*(PB), 561–568. https://doi.org/10.1016/j.powtec.2014.09.017.

Chan, E. L., Washino, K., Reynolds, G. K., Gururajan, B., Hounslow, M. J., & Salman, A. D. (2016). Blade-granule bed stress in a cylindrical high shear granulator: Further characterisation using DEM. *Powder Technology, 300*, 92–106. https://doi.org/10.1016/j.powtec.2016.02.010.

Chandratilleke, G. R., Yu, A. B., & Bridgwater, J. (2012). A DEM study of the mixing of particles induced by a flat blade. *Chemical Engineering Science, 79*, 54–74. https://doi.org/10.1016/j.ces.2012.05.010.

Choi, J., Kudrolli, A., Rosales, R. R., & Bazant, M. Z. (2004). Diffusion and mixing in gravity-driven dense granular flows. *Physical Review Letter, 92*(17), 174301. https://doi.org/10.1103/PhysRevLett.92.174301.

Cundall, P. A., & Strack, O. D. L. (1979). A discrete numerical model for granular assemblies. *Geotechnique, 29*(1), 47–65. https://doi.org/10.1680/geot.1979.29.1.47.

Faure, A., York, P., & Rowe, R. (2001). Process control and scale-up of pharmaceutical wet granulation processes: A review. *European Journal of Pharmaceutics and Biopharmaceutics, 52*(3), 269–277. https://doi.org/10.1016/S0939-.6411(01)00184-9.

Horsthuis, G., Vanlaarhoven, J., Vanrooij, R., & Vromans, H. (1993). Studies on upscaling parameters of the Gral high shear granulation process. *International Journal of Pharmaceutics, 92*(1–3), 143–150. https://doi.org/10.1016/0378-5173(93)90273-I.

Iveson, S. M., & Litster, J. D. (1998). Growth regime map for liquid-bound granules. *AIChE Journal, 44*(7), 1510–1518. https://doi.org/10.1002/aic.690440705.

Kan, H., Nakamura, H., & Watano, S. (2015). Numerical simulation of particle–particle adhesion by dynamic liquid bridge. *Chemical Engineering Science, 138*, 607–615. https://doi.org/10.1016/j.ces.2015.08.043.

Kan, H., Nakamura, H., & Watano, S. (2016). Effect of particle wettability on particle-particle adhesion of colliding particles through droplet. *Powder Technology, 302*, 406–413. https://doi.org/10.1016/j.powtec.2016.08.066.

Knight, P., Seville, J., Wellm, A., & Instone, T. (2001). Prediction of impeller torque in high shear powder mixers. *Chemical Engineering Science, 56*(15), 4457–4471. https://doi.org/10.1016/S0009-2509(01)00114-2.

Leuenberger, H. (1983). Scale-up of granulation processes with reference to process monitoring. *Acta Pharmaceutica Technologica, 29*(4), 274–280.

Litster, J., Hapgood, K., Michaels, J., Sims, A., Roberts, M., & Kameneni, S. (2002). Scale-up of mixer granulators for effective liquid distribution. *Powder Technology, 124*(3), 272–280. https://doi.org/10.1016/S0032-5910(02)00023-2.

Mort, P. R. (2005). Scale-up of binder agglomeration processes. *Powder Technology, 150*(2), 86–103. https://doi.org/10.1016/j.powtec.2004.11.025.

Nakamura, H., Fujii, H., & Watano, S. (2013). Scale-up of high shear mixer-granulator based on discrete element analysis. *Powder Technology, 236*, 149–156. https://doi.org/10.1016/j.powtec.2012.03.009.

Radeke, C. A., Glasser, B. J., & Khinast, J. G. (2010). Large-scale powder mixer simulations using massively parallel GPU architectures. *Chemical Engineering Science, 65*(24), 6435–6442. https://doi.org/10.1016/j.ces.2010.09.035.

Radl, S., Kalvoda, E., Glasser, B. J., & Khinast, J. G. (2010). Mixing characteristics of wet granular matter in a bladed mixer. *Powder Technology, 200*(3), 171–189. https://doi.org/10.1016/j.powtec.2010.02.022.

Rahmanian, N., Ghadiri, M., & Ding, Y. (2008). Effect of scale of operation on granule strength in high shear granulators. *Chemical Engineering Science, 63*(4), 915–923. https://doi.org/10.1016/j.ces.2007.10.027.

Ramaker, J., Jelgersma, M. A., Vonk, P., & Kossen, N. (1998). Scale-down of a high-shear pelletisation process: Flow profile and growth kinetics. *International Journal of Pharmaceutics, 166*(1), 89–97. https://doi.org/10.1016/S0378-5173(98)00030-1.

Remy, B., Glasser, B. J., & Khinast, J. G. (2010). The effect of mixer properties and fill level on granular flow in a bladed mixer. *AIChE Journal, 56*(2), 336–353. https://doi.org/10.1002/aic.11979.

Remy, B., Khinast, J. G., & Glasser, B. J. (2009). Discrete element simulation of free flowing grains in a four-bladed mixer. *AIChE Journal, 55*(8), 2035–2048. https://doi.org/10.1002/aic.11876.

Saito, Y., Fan, X., Ingram, A., & Seville, J. P. K. (2011). A new approach to high-shear mixer granulation using positron emission particle tracking. *Chemical Engineering Science, 66*(4), 563–569. https://doi.org/10.1016/j.ces.2010.09.028.

Sakai, M., Abe, M., Shigeto, Y., Mizutani, S., Takahashi, H., Viré, A., Percival, J. R., Xiang, J., & Pain, C. C. (2014). Verification and validation of a coarse grain model of the DEM in a bubbling fluidized bed. *Chemical Engineering Science, 244*, 33–43. doi:10.1016/j.cej.2014.01.029.

Sarkar, A., & Wassgren, C. R. (2015). Effect of particle size on flow and mixing in a bladed granular mixer. *AIChE Journal, 61*(1), 46–57. https://doi.org/10.1002/aic.14629.

Sato, Y., Nakamura, H., & Watano, S. (2008). Numerical analysis of agitation torque and particle motion in a high shear mixer. *Powder Technology, 186*(2), 130–136. https://doi.org/10.1016/j.powtec.2007.11.028.

Tardos, G. I., Hapgood, K. P., Ipadeola, O. O., & Michaels, J. N. (2004). Stress measurements in high-shear granulators using calibrated "test" particles: Application to scale-up. *Powder Technology, 140*(3), 217–227. https://doi.org/10.1016/j.powtec.2004.01.015.

Tardos, G. I., Khan, M. I., & Mort, P. R. (1997). Critical parameters and limiting conditions in binder granulation of fine powders. *Powder Technology, 94*(3), 245–258. https://doi.org/10.1016/S0032-5910(97)03321-4.

Washino, K., Tan, H. S., Hounslow, M. J., & Salman, A. D. (2013). Meso-scale coupling model of DEM and CIP for nucleation processes in wet granulation. *Chemical Engineering Science, 86*, 25–37. https://doi.org/10.1016/j.ces.2012.04.020.

Watano, S., Okamoto, T., Sato, Y., & Osako, Y. (2005). Scale-up of high shear granulation based on the internal stress measurement. *Chemical and Pharmaceutical Bulletin, 53*(4), 351–354. https://doi.org/10.1248/cpb.53.351.

CHAPTER 22

Advances in computational modeling and simulation of wet granulation processes

Nejat Rahmanian and Tony Bediako Arthur

Chemical Engineering Program, School of Engineering, Faculty of Engineering and Technologies, University of Bradford, Bradford, United Kingdom

22.1 Introduction

Granulation is a crucial process in the pharmaceutical industries that has become increasingly important due to the demand for innovative solutions to new medical problems (Bandari et al., 2020). Pharmaceutical industries are proficient in creating new products but frequently depend on obsolete and ineffective techniques for manufacturing therapeutic components (Suresh et al., 2017). Granulation, the process of combining small powder particles to create bigger granules, provides numerous benefits compared to working with fine powders. These advantages include improved flowability, enhanced dispersion, and increased dissolving rates. Granulation can be classified into two primary categories: dry granulation, which involves mechanical compression without the use of a liquid medium. However, this chapter focuses on wet granulation, which utilizes a liquid binder to enhance particle adhesion (Iveson et al., 2001b).

Wet granulation is an essential process in multiple industries, especially in the pharmaceutical sector, where it has a vital function in manufacturing solid dosage forms. The intricate interaction among process parameters, equipment design, and material qualities has resulted in a growing use of computer modeling and simulation tools to comprehend, enhance, and manage granulation operations (Litster & Ennis, 2004).

This chapter offers an in-depth examination of the latest progress in modeling approaches employed for simulating wet granulation in different types of equipment, such as high-shear granulators, fluidized bed granulators, twin-screw granulators, and other specialized systems. This chapter primarily examines three modeling techniques: the discrete element method (DEM), population balance modeling (PBM) and computational fluid dynamics (CFD).

DEM has proven to be an effective method for understanding the intricate interactions between particles, as well as the processes of mixing and granule formation in granulation. DEM is a computational technique that models the movement and interactions of individual particles. It yields important data on particle velocities, collision frequencies, and force distributions. This information may be utilized to enhance the design of equipment and improve operating conditions (Börner et al., 2016; Gantt et al., 2006).

PBM, in contrast, is a mathematical framework that precisely characterizes the progression of particle size distribution (PSD) as it changes during the granulation process. PBM utilizes rate equations for different processes involved in granulation, including nucleation, aggregation, breakage, and consolidation, to forecast the dynamic evolution of PSD (Barrasso & Ramachandran, 2015; Ramachandran et al., 2008).

The combination of the DEM and PBM has demonstrated significant potential in precisely forecasting the distribution of granule sizes and other important quality characteristics. This is achieved by integrating mechanistic kernels produced from DEM data into the PBM framework. This chapter also discusses the integration of the DEM with CFD and experimental characterization techniques to create multiscale models that accurately represent the interaction between fluid dynamics and particle behavior in granulation processes. The increasing utilization of these sophisticated modeling methods has a substantial influence on enhancing product quality, minimizing development duration and expenses, and augmenting process adaptability and resilience, as firms adopt continuous manufacturing and quality-by-design principles. This chapter is a significant resource for scholars and industry practitioners. It provides a complete overview of the current state-of-the-art and future possibilities of computer modeling and simulation in granulation processes.

22.1.1 Wet granulation

The wet granulation, which makes use of liquid as a medium for binding the fine powder particles to form granules of larger size has become the widely used granulation method in the pharmaceutical industry (Singh et al., 2022). In wet granulation, the resultant product is a wet mass, and it is dried to form the granules. The granulation liquid serves as a binder between the powder particles through forces of capillary and viscous. In a pharmaceutical continuous setup where wet granulation is used, there is a need for another setup for granule drying for advanced processing (Rogers, Hashemi, & Ierapetritou, 2013). The properties of the granules produced after granulation are dependent on the respective particle sizes of the granulation ingredients, the amount, or the concentration levels of the materials, the binding medium, the kind of granulation technique, and the operating conditions.

The process, although is being adopted in the pharmaceutical industry due it its obvious merits, has its own demerits. These merits of wet granulation include the ability to transform powder particle surfaces, which are hydrophobic, into hydrophilic; it prevents powder segregation, promotes the flowability of powders, and improves the compressibility properties of powders, it also promotes uniformity of constituent mixture and reduces dustiness of powders. However, there are drawbacks to consider, such as the potential for compatibility issues between the different components of the formulation and the risk of losing granulating material during the processing phases. In addition, there is the issue of powders that are not suited due to their sensitivity to moisture. Wet granulation is a complex process that involves multiple processing phases and demands space, power, and labor, which makes the entire system costly.

Different equipment and granulation techniques employ the wet granulation principles. These include fluidized bed granulators, high-shear granulators and twin-screw granulators. The most known fluidized granulators are the column-based fluidized bed granulators, which consist of an overhead sprayer of binder solution on the powder bed and an underneath sprayer of fluidized gas. The fluidized airflow dries the granules. The fluidized bed is ideally suitable for the pharmaceutical process due to its very high throughput, usually above 20 kg/h (Vervaet & Remon, 2005). The high-shear granulators are employed in batch-mode industrial setups; although, the newer version of the high-shear granulator

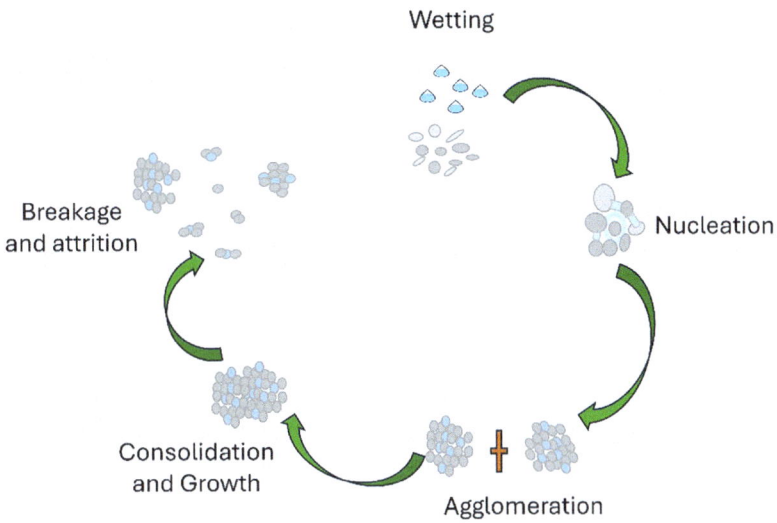

FIGURE 22.1

Schematic diagram of wet granulation rate process.

called the "instant agglomerator" can operate in continuous processing. The downside of this instant agglomerator is that it has a very small throughput, that is, a very small amount of granulating material can be processed, and its residence time is very short (Faure, York, & Rowe, 2001). The twin-screw granulation (TSG) is an extrusion equipment that has two screws lying side by side in a barrel. The corotating twin-screw granulator has become the preferred choice of granulating equipment in the pharmaceutical industry. This is a result of its continuous operation capabilities (Tu et al., 2013).

22.1.2 Wet granulation rate processes

Wet granulation is an intricate process that comprises various physical contesting occurrences taking place in the granulating equipment which result in granule production. As shown in Fig. 22.1, these mechanisms are classified into three main rate processes, that is:

1 the wetting and nucleation;
2 consolidation and coalescence; and
3 breakage and attrition.

These groupings of the mechanism could expand to include abrasion transfer and layering, but these inclusions are dependent on the researcher's interest and ability to measure particle size. These phenomena are classified under coalescence and breakage. Irrespective of the wet granulation method employed, the physical phenomena that guide the mechanisms will continue to remain the same. The final granule characteristics, including granule size distribution, pore spaces, and granule shape and size are governed by the control of the rate processes taking place in the granulator (Iveson et al., 2001a).

Wetting and nucleation: Wetting and nucleation is the first stage of granulation where the powder particles are brought into contact with the granulating liquid (Liu et al., 2000). The liquid is dispersed through the powder bed to produce a cluster of granule nuclei. The rate at which nucleation occurs is determined by the thermodynamics of wetting, the kinetics of drop penetration, and the dispersion of the binder. The size distribution of nuclei is influenced by the respective rates of nuclei creation and binder dispersion (Iveson et al., 2001b).

Consolidation and agglomeration: Following the creation of nuclei, granule development takes place through two primary mechanisms:

- Coalescence refers to the process in which two granules, granules and fine particles, or a granule and equipment come into contact and adhere to each other. Coalescence is contingent upon achieving equilibrium between the cohesive forces and the disruptive forces (Iveson et al., 2001b).
- Layering refers to the process in which small particles stick to the surface of bigger granules. This is also seen as an integral component of the consolidation process (Iveson et al., 2001b).

The consolidation and agglomeration stage of the granulation is also regarded as the growth stage of granules. Consolidation is the process by which the granules become more tightly packed, resulting in a decrease in their size and porosity. This compression forces out any trapped air and may also cause liquid binder to rise to the surface, which might facilitate more growth (Iveson et al., 2001a). The consolidation process is affected by the impact forces that the granules experience.

The loosely packed granules, as a result of nucleation are distributed through the equipment. This results in collisions involving the granules-granules, dry powder particles-granules, and/or with the involvement of the granulating equipment (Liu et al., 2000). This leads to granule growth and compaction. This collision causes the granule size and voidage reduction and drives out any liquid (binder) or air entrapped inside the granule out to the surface. The liquid presence on the surface of consolidated granules is a vital component for coalescence. The final voidage of the granule is influenced by the actions of the granule consolidation. There are two major factors influencing growth by consolidation (Iveson et al., 2001b; Liu et al., 2000). These factors are the ability of colliding granules to absorb the energy resulting from the impact so that the granules do not rebound, and the bonding forces at the contact of collision between the granules should be very strong (Iveson et al., 2001).

Agglomeration refers to the process of granule expansion that occurs when nucleation, coalescence, and layering combine together. The growth kinetics can exhibit either a steady growth regime, characterized by a linear increase in granule size over time, or an induction growth regime, which involves an initial slow growth phase followed by rapid expansion. The growth regime is dictated by the granule stokes deformation number, which is contingent upon the mechanical properties of the granule and the impact forces it undergoes(Iveson et al., 2001a).

Agglomeration refers to the process of granules growing larger through the combination of nucleation, coalescence, and stacking. The growth can occur through two mechanisms: steady growth, where the size of the granules rises linearly over time, or induction growth, where there is minimal growth initially followed by a rapid increase in size (Iveson et al., 2001b).

Breakage and attrition: The breakage and attrition stage of the granulation is the stage where wet granules break due to the impact of compression and wear in the granulator. The attrition and breaking of wet grains result from the impact and shear forces within the granulator (Iveson et al., 2001b). The maximum granule size might be limited by the strength of the granules.

The breakage of wet granules controls and has an impact on the final granule size distribution (GSD). Breakage phenomena in granulation can be employed to limit the granule maximum size or help in distributing viscous fluids (Liu et al., 2000). The level of breakage is regulated by the strength of the granules, which is dictated by the content and qualities of the binder, as well as the consolidation process. In addition, the magnitude of the destructive forces also plays a role (Liu et al., 2000; Ramachandran et al., 2008; Vonk et al., 1997). Attrition is the process of minute fragments being lost from the surface of a granule, which can then lead to layering (Vonk et al., 1997).

This section is a discussion about the fundamental physics behind the respective rate process and defines each of the rate process parameters and the controlling formulating properties. The overall granulation rate and development of granule characteristics are contingent upon the equilibrium and interplay among the fundamental rate processes of nucleation, coalescence, layering, consolidation, and breakage. Gaining comprehension and mastery over these minute-scale procedures is essential for maximizing the efficiency of the granulation process.

22.2 Approaches to modeling of wet granulation processes

There are two main patterns for the mathematical modeling of wet granulation processes. These are the physically based models and the data-driven models (Sampat & Ramachandran, 2021).

22.2.1 The data-driven models (Empirical)

These models use time series data of real plants based on the input and output date at a specified time range. The model is developed by choosing an appropriate model structure and applying the model parameters to the data to achieve the best fit for the model. There are diverse ways to achieve the best fit which includes altering the parameters of the structure and the parameters of the models too.

This type of modeling is significantly limited and only applicable to systems within the interval set. It is also useful when a model is required to control the application but not necessarily looking for insight into the model.

22.2.2 The physical (Mechanistic)-based models

These models integrate into the model's primary concept of the chemistry and physics behind the processes. The conservation and constitutive facets are the two main aspects covered by physical-based modeling and they include:

1 Application of thermodynamic conservation principles for mass, energy, and momentum;
2 development of appropriate constitutive relations that define intensive properties, mass, and heat transfer mechanisms as well as particle growth and breakage mechanisms; and
3 application of population balances that track PSDs as various particulate phenomena take place.

The mechanistic models are a very laborious and complicated process of developing a model as compared to the development of empirical models. The model is useful when a full insight understanding of the relevant constitutive relations is accessible, and it is useful for a larger variation of applications.

22.2.3 Modeling through population balance model

The PBM monitors the transformation of particles through the granulation mechanism. The PBM, in either its one-dimensional or multidimensional form, has been utilized to forecast the progressive alterations in granule characteristics, including granule size distribution, moisture content, density, and porosity (Ismail et al., 2020). Eq. (22.1) represents the general equation for the PBM, Eq. (22.2) illustrates the concept of particles in a specific region of the particle phase, and Eq. (22.3) provides a generic depiction of the super-structure of the PBM.

$$\frac{\delta n}{\delta t} + \nabla \cdot (vn) - B + D = 0 \tag{22.1}$$

Where v is the particle velocity, n is the number of density particles, particle's birth rate, and death rate are B and D, respectively.

$$\begin{aligned}&\{\text{densityfunctionchangeinclass, locationandtime}\} = \{\text{disperseinthroughboundary}\}\\ &-\{\text{disperseoutthroughboundary}\} + \{\text{flowinboundary}\} - \{\text{flowoutofboundary}\} +\\ &\{\text{growinfromlowerclasses}\} - \{\text{growoutfromcurrentclass}\} +\\ &\{\text{birthduetocoalescene}\} - \{\text{deathduetocoalescence}\} +\\ &\{\text{breakupinfromupperclases}\}\end{aligned} \tag{22.2}$$

22.2.4 Population balance modeling governing equations: 3D population balance models

The PBM has traditionally been the predominant method for simulating granulation processes. This has the potential to incorporate the different processes involved in granulation rates, such as nucleation, consolidation, aggregation, and breakage, as described by Poon et al. (2008). Accounting for the distribution of particle population is crucial in a granulation process, taking into consideration factors such as granule voidage (porosity), binder concentration, and particle size. The distribution of the particle population is a result of the actions of the three particle characteristics discussed before. When creating a 3D population balance, the characteristics of particles are categorized into corresponding qualities, such as liquid volume, solid volume, and gas volume (Verkoeijen et al., 2002).

Eq. (22.3) is a typical multidimensional PBM that has been developed for granule solid, liquid, and volume of gas tracking (Wang et al., 2020).

$$\frac{\partial}{\partial t}n(s,l,g,t) + \frac{\partial}{\partial s}\left[n(s,l,g,t)\frac{ds}{dt}\right] + \frac{\partial}{\partial l}\left[n(s,l,g,t)\frac{dl}{dt}\right] + \frac{\partial}{\partial g}\left[n(s,l,g,t)\frac{dg}{dt}\right]$$
$$= B_{nuc}(s,l,g,t) + B_{lay}(s,l,g,t) + \text{Break}_{break}(s,l,g,t) - D_{break}(s,l,g,t) \tag{22.3}$$

The vector denoting the solid particles, liquid, and the volume of gas of granule are (s, l, g), and these parameters' granule population density over time is represented as $n(s, l, g, t)$. The three terms on the left-hand side of the equation in the form of partial differential are expressions for change of state due to layering, addition of liquid and consolidation of gas, respectively. The nucleation and layering net rates are denoted by the first two terms on the right side of the equation as $B_{nuc}(s,l,g,t)$, and $B_{lay}(s,l,g,t)$, respectively. While the last two terms represent the birth and death rates due to breakage.

The 3D population balance model proposed by Wang et al. (2020) in Eq. (22.3) does not have the aggregation kernel factored into it (Poon et al., 2008). Eq. (22.4) proposed a 3D PBM where the net rate of aggregation of granules was factored into the equation.

$$\frac{\partial}{\partial t}n(s,l,g,t) + \frac{\partial}{\partial s}\left[n(s,l,g,t)\frac{ds}{dt}\right] + \frac{\partial}{\partial l}\left[n(s,l,g,t)\frac{dl}{dt}\right] + \frac{\partial}{\partial g}\left[n(s,l,g,t)\frac{dg}{dt}\right] = \Re_{nuc} + \Re_{aggre} + \Re_{break} \quad (22.4)$$

The layering process where fines lie on the surface of granules is expressed as the partial derivative with respect to s. The rewetting of granules with new binder and binder drying are both expressed as the partial derivative with respect to l whilst the consolidation process where gas is taken out and the continuous decrease of voidage but with an unchanged liquid and solid volume is the partial differential with respect to g. The granule porosity is decreased whilst the saturation of binder liquid is increased because of the granule compaction process of consolidations; thus the consolidation rate process is given a negative growth outlook with respect to the volume of gas.

Nucleation kernel for use in PBM: The nucleation process describes the diffusion process of the binder drop into the powder bed, and this vital rate process leads to the creation of granule nuclei. In determining the nuclei size, the size of the drop is incredibly significant in that, larger binder droplet size compared to the size of powder particles is very crucial. This phenomenon is very dominant in twin-screw and high-shear granulation, and it is referred to as penetration or immersion nucleation. Whereas the nucleation whereby the powder particle size is far larger than the drop is called distributive nucleation (Kariuki et al., 2013).

Hounslow et al. (2009) proposed two methods to model the formation of nuclei through penetration/immersion. The approaches are the surface tension drive flow and deformation-driven diffusive flow. Hounslow et al. (2009) surface tension approach for modeling the nucleation rate is grouped into two the use of planar geometry and spherical geometry of a nucleus shape to track the granule growth. The velocities and fluxes of the liquid and powder particles at a particular point in time were modeled using (Bird, 2002). The local superficial velocity was equated to the pressure gradient for slow flow by applying the equation from Blake-Kozeny (Bird, 2002) under Darcy's law. Combining these models with the kinetics of capillary flow, Hounslow developed Eq. (22.5) for the thickness of the nucleus as a function of time using the planar geometry method and Eq. (22.6) for the spherical geometry method as:

$$\frac{\upsilon - \upsilon_0}{\upsilon_{max} - \upsilon_0} = \frac{\upsilon' - 1}{\upsilon'_{max} - 1} = \frac{\hat{h} - 1}{\frac{1}{\phi_{cp}} - 1} = \sqrt{\hat{t}} \quad (22.5)$$

where υ is the nucleus volume, h is the radius of the nucleus, t is the time, and ϕ_{cp} is the critical volume packing.

$$v = v_L\left(1 + \frac{1 - \phi_{cp}}{\phi_{cp}}\sqrt{\frac{t}{t_{max}}}\right) \quad (22.6)$$

Poon et al. (2008) modeled the nucleation phenomena by focusing on the droplet-control regime. The vital process variables that were accounted for in this model were the droplet size and the penetration rate. These affect the nucleation phenomena and the wetting of powders in an unequivocal manner, thus helping to control the effect of lumping. The assumptions made for the formation of the Poon et al.'s (2008) nucleation kernel include:

- The assumption relaxed form Wildeboer et al. (2005) assumes that there is no overlapping between the droplets and that they are well distributed.

- Two scenarios were assumed for the presence of fine particles and that the particles of the fine powder will only receive binder under the condition that, (1) the droplets are not fast enough to wet the bed floor hence, left on the surface of the powder at time of bed turnover or (2) the droplets are not fast enough to penetrate the bed and that the surface of the powder bed still hold the binder at the time of its turnover. It is assumed that surface binders of such are distributed uniformly to all granules through mixing in the granulator.

Using the Hapgood et al.'s (2002, 2003) equations on surface free energies (Hancock et al., 1994), for computing the spreading coefficient, the Washburn equation to develop the drop penetration model and using Hapgood et al.'s (2003) model for the penetration time, the nucleation rate of nuclei particle formulation and nucleation rate of primary particle depletion was derived by Poon et al. (2008) in Eqs. (22.7) and (22.8), respectively.

$$\mathcal{R}_{nuc}(s_{nuc}, l_{nuc}, g_{nuc}) = k_{nuc} \tag{22.7}$$

$$\mathcal{R}_{nuc}(s, l, g) = -k_{nuc}(s, l, g) \tag{22.8}$$

22.2.5 Discrete element method simulation

DEM modeling and simulation of the granulation process has shown great progress in recent years. The development of free open platforms such as Open FOAM and LAMMPS improved for general granular and granular heat transfer simulation (LIGGGHTS) and commercial software packages such as Rocky, Altair EDEM, Berker 3D, Three particle/CAE, and CADFEM have increased research into the granular flow and, for that matter granulation. DEM has been used to investigate process optimization, process modeling, and material formulation in fluidized bed, high-shear, drum, and twin-screw granulators

Modeling a process always allows for a more in-depth investigation of the system dynamics, followed by the development of an efficient control plan to improve operational efficiency and scale up more effectively if necessary. When investigating the various aspects of a dynamic process such as granulation would be labor and capital-intensive at the plant level, a model-based approach is always useful. In order to gain a deeper comprehension of the granulation process, simulations were conducted at three different scales: macro (examining the overall flow behavior), meso (analyzing ensembles of granules and particles), and micro (studying individual granules/particles). Equations were solved using both continuum and volume-averaged approaches, considering both macroscopic and microscopic scales. Several of studies have been conducted on the simulation and modeling of wet granulation in order to gain a deeper understanding of the intricacies involved in the process and its mechanisms, leading to improved process control and product quality.

The DEM also known as the distinct element method was developed in the 1970s by Cundall and Strack (1979) after using Alder, Berni, & Thomas Everett (1959) model for discrete elements meant for molecular dynamics studies. The DEM has been widely employed in modeling rock engineering problems (Boon et al., 2012) and in recent times for addressing both granular and discontinuous material challenges (Govender et al., 2015).

In modeling granular (particulate systems) matter, the continuum (Eulerian) and discrete (Lagrangian) are the two major approaches. The Eulerian approach describes the granular material constitutive behavior through differential equations to represent the constitutive laws that relate the stress and strain field variables (Edem, 2023). Whiles in a discrete method, the granular material is made up of a combination of individual idealized single particles which are discrete entities. Thus, the discrete approach is the most preferred approach as the continuum form has issues relating to a lack of understanding and overly complicated issues of the strain-stress laws for granular material. Whereas the particle level behavior directly dictates the behavior and phenomenon of the particulate system. It is therefore more realistic to model granular material using the discrete approaches (Edem, 2023).

The discrete approach can be performed through two main methods, that is, the soft-sphere and hard-sphere methods. The soft ball/sphere assumes the particles to be rigid, but the deformation is represented by a little overlap between the particles during contact. While the hard-sphere method is also rigid in this case, it assumes that the collision of particles forms the only medium of momentum exchange and also the forces of interaction are also impulsive (Cundall & Strack, 1979).

22.2.6 Base models

Eqs. (22.9) and (22.10) describe both the translational and rotational motions of each particle, as previously stated. The equations are numerically integrated using explicit time integration strategies, such as the forward Euler or velocity Verlet algorithms (Scherer, 2017).

$$\frac{m_i(dv_i)}{dt} = F_{C,i} + m_i g V \qquad (22.9)$$

$$I_i \frac{d\omega_i}{dt} = T_i \qquad (22.10)$$

Let m_i represent the mass of the particle i, and I_i represent the moment of inertia of particle i V_i represents the velocity of translation, g represents the acceleration due to gravity, ω_i represents the angular velocity, $F_{C,i}$ represents the combined force of contact between particle i and nearby particles or walls, and T_i represents the total torque applied to particle i.

The contact forces between particles are determined by the utilization of several contact models. The Hertz-Mindlin model, the linear spring-dashpot (LSD) model, and the Johnson-Kendall-Roberts (JKR) model (Johnson, 1982) are the most often utilized models. The Hertz-Mindlin model combines Hertz theory to calculate the normal force (F_n) and Mindlin theory to calculate the tangential force (F_t).

$$F_n = -k_n x_n - \eta_n \frac{dx_n}{dt} \qquad (22.11)$$

$$F_t = -k_t x_t - \eta_t \frac{dx_t}{dt} \qquad (22.12)$$

k_n and k_t represent the spring stiffness in the normal and tangential directions, respectively. η_n and η_t represent the damping coefficients in the normal and tangential directions, respectively. x_n and x_t represent the overlaps in the normal and tangential directions, respectively.

The Hertz-Mindlin model combines the Hertz theory, which deals with normal contact, and the Mindlin theory, which deals with tangential contact. This model considers the elastic deformation of the particles and is appropriate for simulating rigid, nonsticky particles. Eqs. (22.11) and (22.12) provide the values for the normal and tangential forces, respectively. The spring stiffnesses (k_n and k_t) and damping coefficients (η_n and η_t) depend on the particle properties, including Young's modulus, Poisson's ratio, and the coefficient of restitution (Di Renzo & Di Maio, 2004).

The Linear spring-dashpot LSD model is a simplified variant of the Hertz-Mindlin model, in which the spring stiffnesses remain constant and are not affected by the overlap (Tsuji et al., 1993). This model exhibits higher computational efficiency, but worse accuracy compared to the Hertz-Mindlin model.

The JKR model expands upon the Hertz-Mindlin model by incorporating adhesive forces between particles (Johnson, 1982). It considers the surface energy of the particles and the adhesion work. The JKR model is appropriate for simulating cohesive particles, namely, moist granular materials (Nase et al., 2001). DEM simulations necessitate a meticulous selection of simulation parameters, including contact models, time steps, particle characteristics, and boundary conditions (Cleary & Sawley, 2002). The time step should be sufficiently tiny to accurately capture the contact dynamics, usually a fraction of the Rayleigh time step (O'Sullivan & Bray, 2004).

DEM has been extensively employed to investigate many particulate systems, including powder flow, granular mixing, and particle segregation (Coetzee, 2017). The information it offers is highly extensive and encompasses the dynamics at the particle level, including particle velocities, contact forces, and coordination numbers. Nevertheless, DEM simulations are highly resource-intensive, particularly for systems containing a substantial number of particles (Matuttis et al., 2000). In order to address the constraints imposed by computation, several strategies have been suggested, including the utilization of reduced particle geometries (Coetzee, 2017), coarse-graining approaches (Kobayashi et al., 2013), and parallel computing. Advancements in GPU processing have greatly sped up DEM simulations, allowing for the modeling of millions of particles (Govender et al., 2016). The DEM is a flexible tool used to simulate particulate systems, offering important insights into the intricate behavior of granular materials. Due to the growing computer power and the advancement of algorithms, DEM is anticipated to have a vital impact on the model design and optimization of particle processes across many industries.

22.2.7 CFD simulation

CFDs is a method that uses numerical solutions to analyze the movement of fluids and the transportation of substances. CFD, or CFDs, is widely utilized in engineering research and industries to address and evaluate issues pertaining to heat transfer, biological engineering, aerodynamics, flight testing, weather simulation, and system design and analysis across diverse sectors (Dadvand & Motlagh, 2021).

The origins of modern fluid dynamics can be traced back to Leonardo Da Vinci (1452–1510). Subsequently, diligent scientific scholars have devoted significant effort to developing a mathematical description of the phenomenon. Claude Louis Marie Henry Navier and George Gabriel Stokes proposed the Navier-Stokes equation in the early 1900s. They modified the Euler equations by including viscous transport (Dadvand & Motlagh, 2021).

22.2.8 Equations of conservation

The conservation equations are the primary governing equations of fluid dynamics. The equations encompass the continuity equation (which accounts for mass conservation), the momentum equation, and the energy conservation equation (Wendt et al., 2009).

Mass conservation (Continuity equation): The continuity equation is the most familiar term for mass conservation. The mass conservation equation, commonly referred to as the continuity equation, is derived from the fundamental premise that mass is conserved and cannot be generated or lost during a chemical reaction. Put simply, the mass of a system remains unchanged as time progresses. The mass conservation equation is expressed in a general form as follows (Ferziger et al., 2019).

$$\frac{\partial \rho}{\partial t} + \nabla \cdot (\rho V) = 0 \tag{22.13}$$

Where ρ and V are the density and volume of the material.

Momentum conservation: The principle of conservation of momentum states that the total momentum of a given system remains constant. Momentum cannot be created or destroyed; rather, it can only be modified by external forces under Newton's laws of motion (Ferziger et al., 2019). The principle of momentum conservation asserts that an object or a collection of objects in motion will preserve its overall momentum, defined as the product of mass and vector velocity, unless influenced by an external force. In an isolated system, such as the cosmos, where external effects are absent, momentum remains conserved. Momentum conservation ensures that its components in all directions remain conserved. The law of conservation of momentum is crucial in solving collision situations (Pope, 2001). The equation of momentum conservation is given as

$$\rho \frac{\partial v}{\partial t} = \rho + \nabla \cdot \sigma \tag{22.14}$$

where stress tensor $\sigma = -p + \varepsilon$ and ε is the viscous stress and p is pressure, hence the momentum conservation equation becomes

$$\rho \frac{\partial v}{\partial t} = -\nabla p + \nabla \cdot \varepsilon + \rho \tag{22.15}$$

In the case of Newtonian fluids, the stress tensor is directly proportional to the strain rate tensor, with the proportionality constant being the fluid's viscosity (Kundu et al., 2015).

Energy conservation: Energy conservation is a principle that is derived from the first rule of thermodynamics. It takes into consideration the movement and transfer of energy inside a fluid. The energy conservation equation can be expressed in a general form, as stated in reference (Patankar, 2018);

$$\rho \left(\frac{\partial E}{\partial t} + V \cdot \nabla E \right) = -\nabla \cdot q + \nabla \cdot (\tau \cdot V) + \rho g \cdot V + Q \tag{22.16}$$

The symbol E indicates the total energy per unit mass, q denotes the heat flux vector, and Q represents any additional heat sources or sinks. The total energy is comprised of internal energy, kinetic energy, and potential energy. The energy conservation equation delineates the equilibrium of energy transfer resulting from convection, conduction (heat flux), stress-induced work, and external heat sources or sinks (Chung, 2002).

CFD models commonly use supplementary equations to represent turbulence, chemical reactions, multiphase flows, and other intricate phenomena, alongside the conservation equations. Turbulence models, such as the k-ε or k-ω models, are employed to finalize the system of equations by offering equations for the turbulent stresses and fluxes (Uddin, 2008).

CFD has become an essential tool in multiple engineering disciplines, such as aerospace, automotive, chemical, and biomedical engineering (Blazek, 2015). Engineers and researchers can use it to examine intricate fluid flow issues, enhance designs, and forecast the performance of systems in various operating situations. Due to the progress in processing resources and numerical algorithms, CFD is constantly improving and addressing increasingly complex fluid dynamics issues.

22.2.9 Approaches for computational fluid dynamics multiphase simulations

CFD is a highly effective method for simulating the behavior of granular flows including multiphases in different systems, such as fluidized bed granulators. The CFD methodology relies on solving the fundamental equations of mass, momentum, and energy conservation. There are two primary methods for modeling gas–solid multiphase flows:

The Eulerian-Eulerian (Two-fluid) approach: This approach considers both the gas and solid phases as interpenetrating continua. The governing equations are obtained by calculating the average of the conservation equations over a control volume. The continuity equations for the gas and solid phases are provided by Gidaspow (1994) and Ding and Gidaspow (1990).

$$\frac{\partial(\varepsilon_g\rho_g)}{\partial t} + \nabla \cdot (\varepsilon_g\rho_g u_g) = 0 \tag{22.17}$$

$$\frac{\partial(\varepsilon_s\rho_s)}{\partial t} + \nabla \cdot (\varepsilon_s\rho_s u_s) = 0 \tag{22.18a}$$

The variables ε_g and ε_s represent the volume fractions, while ρ_g and ρ_s represent the densities. The velocities of the gas and solid phases are denoted as u_g and u_s, respectively. The momentum equations can be expressed as:

$$\frac{\partial(\varepsilon_g\rho_g u_g)}{\partial t} + \nabla \cdot (\varepsilon_g\rho_g u_g) = -\varepsilon_g\nabla p + \nabla \cdot \tau_g + \varepsilon_g\rho_g g + K_{gs}(u_s - u_g) \tag{22.18b}$$

$$\frac{\partial(\varepsilon_s\rho_s u_s)}{\partial t} + \nabla \cdot (\varepsilon_s\rho_s u_s) = -\varepsilon_s\nabla p - \nabla p_s + \nabla \cdot \tau_s + \varepsilon_s\rho_s g + K_{gs}(u_g - u_s) \tag{22.19}$$

In this context, p represents the gas pressure, τ_g and τ_s denote the viscous stress tensors, g represents the acceleration due to gravity, ps represents the solid phase pressure, and *Kgs* represents the gas–solid momentum exchange coefficient. In order to complete this set of equations, it is necessary to establish constitutive relations for the stress tensor and pressure of the solid phase. The stress in the solid phase is represented by the kinetic theory of granular flow (KTGF) by Gidaspow (1994), in addition to frictional stress models by Schaeffer (1987). The KTGF is a modified version of the kinetic theory of dense gases that takes into consideration inelastic particle collisions. It establishes a connection between the tension experienced by a solid phase and the temperature of the granular material. The interphase momentum exchange is represented by drag laws, such as the Gidaspow drag model (Gidaspow, 1994) which combines the Wen-Yu equation for high void fractions with the Ergun equation for low void fractions.

The Eulerian-Lagrangian (CFD-DEM) approach: This method considers the gas phase as a continuous medium, whereas the solid phase is represented by individual particles that are monitored using a Lagrangian approach. The governing equations for the gas phase are identical to those used in the

Eulerian-Eulerian method, namely, Eqs. (Eq. 22.18a) and (22.19). Nevertheless, the discrete particles are individually monitored by solving Newton's equations of motion (Deen et al., 2007).

The gas-particle interaction is resolved by averaging the gas equations over suitable control volumes and subsequently employing drag laws to couple the two phases. The interactions between particles are simulated using contact mechanics and are identified by a collision technique. The CFD-DEM approach is characterized by higher processing costs compared to the two-fluid approach. However, it offers more comprehensive insights into particle dynamics and is not constrained by the KTGF assumptions. Nevertheless, it typically has a restricted scope, including a lesser quantity of particles (Deen et al., 2007).

Examples of their application in granular flows: Ding and Gidaspow (1990) created a model for bubbling fluidization utilizing the Eulerian-Eulerian approach with KTGF closures. They successfully documented crucial fluidization phenomena such as the generation of bubbles and the expansion of the bed. Hoomans et al. (1996) employed a CFD-DEM technique to model a 2D gas-fluidized bed using hard-sphere particles. They then compared the outcomes with the two-fluid model, demonstrating the benefits of the discrete approach. Deen et al. (2007) conducted a comprehensive analysis of the Eulerian-Eulerian and Eulerian-Lagrangian methods used in fluidized bed modeling.

Fries et al. (2013) employed a coupled CFD-DEM methodology to simulate particle collisions in a fluidized bed granulator. They extracted valuable information on the dynamics of these collisions. Sen et al. (2014) employed a multiscale computational fluid dynamics-discrete element method-population balance modeling (CFD-DEM-PBM) technique to simulate a fluidized bed granulation (FBG) process.

CFD modeling, employing either the Eulerian-Eulerian or Eulerian-Lagrangian technique, has demonstrated its use in comprehending the intricate gas-solid hydrodynamics within fluidized bed systems. The selection of methodology relies on the desired level of specificity and the computing cost involved. The ongoing research in granular and multiphase flows focuses on the advancement of CFD models and their validation through experiments.

22.3 Simulation of high granulators

High shear granulation depicted in Fig. 22.2 is a commonly employed method in the pharmaceutical sector for creating granules with specific characteristics (Litster & Ennis, 2004; Liu et al., 2021). Optimizing and controlling high shear granulation processes pose challenges because of the intricate interaction among process parameters, equipment design, and material qualities (Iveson et al., 2001a). Computational modeling and simulation tools, such as the DEM and PBM, are being used more frequently to gain insight into the fundamental processes and forecast the characteristics of granules in high shear granulators (Barrasso & Ramachandran, 2015; Sen et al., 2014).

This section presents a comprehensive review of the most advanced DEM, CFD, and PBM simulation techniques used in high shear granulation. The focus will be on how these techniques have been applied to analyze the impact of process parameters, equipment design, and material qualities on the performance of granulation. The studies mentioned encompass the precise forecasting of agglomerate breakage through the utilization of the Timoshenko beam bond model (Chen et al., 2022), the impact of fill level on the characteristics of granules and tablets (Matsushita et al., 2022), and the exploration of mixing uniformity in binary particle systems using DEM (Fan et al., 2023).

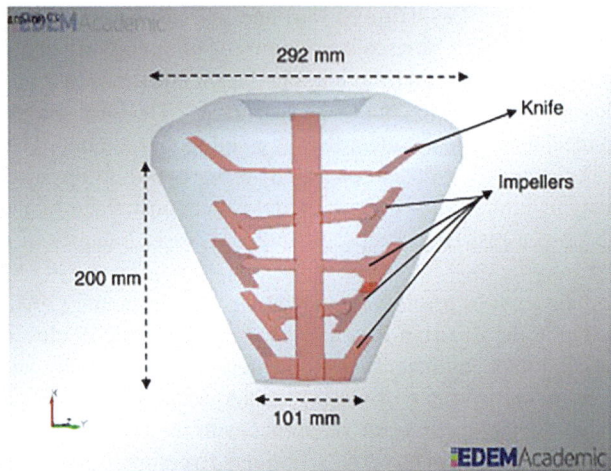

FIGURE 22.2

Design modeled of a high shear granulator.

From Hassanpour et al. (2013).

22.3.1 Discrete element method simulation of high shear granulation

The process of high-shear wet granulation (HSWG) is intricate and entails the clustering of particles by introducing a liquid binder. The procedure is extensively employed in diverse sectors, such as medicine, food, and metallurgy, to enhance the characteristics of powders, such as their capacity to flow, compress, and maintain uniformity. The utilization of the DEM has become increasingly prominent in the study of the underlying mechanics of HSWG in recent times. Differential equation modeling (DEM) simulations offer a comprehensive understanding of particle-level interactions, force distributions, and granule growth dynamics, which are challenging to acquire just through experimental approaches. In addition, the progress made in the coarse-grained discrete element method (CG-DEM) has facilitated the realistic representation of extensive systems, such as vertical high-shear mixers, with enhanced computing efficacy. This chapter centers on the utilization of DEM and computational geodesic models (CG-DEM) for investigating different facets of HSWG. These aspects encompass the determination of coalescence kernels, the impact of impeller design on shear forces and energy input, the influence of binder addition techniques, the granulation of iron ore fines, and the dynamics of particle flow and mixing in vertical high-shear mixers.

The innovative approach to determining coalescence kernels for population balance equation (PBE) models of high-shear granulation processes was developed by Gantt et al. (2006). This approach involved the integration of DEM simulations with a physically based coalescence model for deformable surface wet granules, as described by Liu et al. (2000). In order to mimic particle movement and collisions within a high-shear mixer, the DEM model utilized a soft-sphere technique and incorporated periodic boundary constraints. During the granulation process, the simulations monitored many parameters including collision frequency, aggregation frequency, kinetic energy dissipation, coalescence efficiency, and compaction rates. The study examined the viscosities of two binders: a silicon oil with high viscosity and a water binder with low viscosity, using calcium carbonate as the solid material. The model utilized a log-normal size distribution to mimic 5000 particles within a 32 wedge-shaped periodic volume, which accurately represents the high-shear mixer.

According to Gantt et al. (2006), the assumptions of the DEM model encompassed the following: (1) the exclusive consideration of binary, instantaneous particle collisions; (2) the utilization of a flat plate impeller to circumvent the representation of a dynamic geometry; and (3) the exclusion of actual particle coalescence or breakage to uphold a consistent number and size distribution of particles. The coalescence model proposed by Liu et al. (2000) encompasses two distinct modes: type I coalescence, characterized by the complete dissipation of kinetic energy through the viscous forces present in the binder layer before particle contact; and type II coalescence, wherein particles undergo collision and deformation, yet can still coalesce if the elastic energy is dissipated by the binder layer. The aforementioned model was utilized to analyze every collision event within the DEM simulation, to ascertain the likelihood of coalescence by considering the collision velocity and forces.

The examination of the DEM data unveiled a compaction rate constant that varies with size, around $10^0 s^{-1}$. This finding aligns with the anticipated behavior of a high-shear process, as opposed to the $10^{-2} s^{-1}$ rate observed in low-shear drum granulators. According to Gantt et al. (2006), the compaction rate was observed to be predominantly influenced by the collision frequency, with smaller particles exhibiting a substantially greater frequency. Gantt et al. (2006) separated the coalescence kernel β (u, v) into two separate functions: a collision rate function C (u, v) and a coalescence efficiency function Ψ (u, v). The DEM collision frequency data are most accurately represented by an induced shear kernel. The coalescence efficiency was assessed by utilizing Liu et al.'s (2000) model to map the collision data of the DEM (including particle size, pore saturation, porosity, and speed) to the coalescence results. The efficiency function Ψ was modeled in a multidimensional manner, considering factors such as particle size, pore saturation, porosity, and velocity. The study conducted by Gantt et al. (2006) revealed that smaller particles with more porosity and saturation, along with a binder with higher viscosity, exhibited higher efficiency.

The study conducted by Gantt et al. (2006) revealed that smaller particles with more porosity and saturation, along with a binder with higher viscosity, exhibited higher efficiency. The study conducted by Gantt et al. (2006) introduces an innovative method for obtaining physically based coalescence kernels through the integration of micro-scale DEM models with mesoscale population balance models. The utilization of size-dependent compaction rates and multidimensional coalescence kernels in the determination of high-shear granulation processes can enhance the mechanistic and predictive capabilities of PBE models. Additional investigation is required to authenticate these kernels using empirical data and integrate them into comprehensive PBE simulations in order to evaluate their influence on model efficacy and predictive capacity.

Börner et al. (2016) utilized the DEM simulations to examine the distribution of shear forces and energy input at the individual particle level in HSWG. The investigation focused on impeller designs with two and three blades. This study aimed to replicate the ideal operating speeds of a 600 L production-scale mixer for each impeller. Specifically, the two-blade impeller and the impeller with three blades were simulated at a lower and high speed of 30 and 90 RPM, respectively. In order to guarantee the precision of the simulations conducted at a large scale, the DEM model was initially verified through experimental mixing trials in a 10 L laboratory mixer. During these trials, the velocity vectors of particles on the surface of the bed were compared to the data obtained using particle image velocimetry (PIV) (Börner et al., 2016).

The DEM simulations revealed that the shear forces exhibited similar characteristics in both impeller designs, as no significant high-shear regions were detected close to the impellers. According to Börner et al. (2016), the bulk material exhibited the greatest shear forces, mostly due to the presence of significant velocity gradients resulting from the alteration of axial flow directions among particles. This observation presents a counterargument to the prevailing belief that the impeller is the predominant contributor to the generation of significant shear forces during the granulation process. The study conducted by Börner et al. (2016) revealed that the shear forces exerted on particles in close interaction with the rotor blades of the impeller were roughly 30% of the maximum shear forces measured in the bulk. This can be explained by the comparatively low velocity of the particles in relation to the impeller.

The DEM simulations demonstrated a notable disparity in the kinetic energy of the particles between the two impeller systems, despite the similarity in shear forces. The particles were subjected to significantly greater kinetic energy when the impeller with three blades, operating at a higher speed, was compared to the impeller with two blades running at a lower speed. In the three-blade system, the particles located in the lower portion of the mixer exhibited the most kinetic energy due to their displacement caused by the impeller's rapid motion. Börner et al. (2016) discovered that the particle kinetic energy in the impeller with two-blade system was notably reduced. The highest values were seen for particles that were transported to the upper part of the mixer by the outer impeller wings. According to the authors, the elevated kinetic energy within the three-blade system may result in heightened collision frequency and rates of granule breakage.

The DEM simulations performed by Börner et al. (2016) provided significant insights into the distribution of shear forces and the features of energy input in impeller designs with two and three blades, specifically in the context of HSWG. The findings demonstrated that the impeller design had a restricted impact on the particles' maximum shear forces, with the most significant forces observed in the bulk material rather than near the impeller blades. Previous studies may have overstated the role of the impeller in generating large shear forces, as indicated by this finding. Nevertheless, the particle kinetic energy was greatly influenced by the impeller design and rotation speed. Specifically, the three-blade impeller, when running at higher speeds, led to a considerably greater energy input in comparison to the two-blade impeller (Börner et al., 2016). The proposition put forth by the authors suggests that the elevated kinetic energy within the three-blade system may result in heightened rates of granule breakage. This underscores the significance of considering both shear forces and energy input when optimizing impeller design and operating conditions for HSWG processes.

The effect of binder addition, both wet and dry, in a higher shear granulation process has been investigated using the DEM model simulation method. Tamrakar et al. (2019) explored two different types of binder-adding processes. The objective of this work is to investigate the disparities between wet binder addition (WBA) and dry binder addition (DBA) methodologies in HSWG processes using a DEM model. In order to accommodate capillary and viscous liquid bridge forces, binder dissolution, and liquid penetration mechanisms, a special contact model was constructed for the DEM. The experimental setup was utilized to mimic the granulation of WBA and DBA in a 1-L HSG using the model. The simulations effectively monitored the movements of individual particles and their interactions, so offering valuable insights at the particle level that are challenging to acquire through experimental means (Tamrakar et al., 2019).

The DEM model used in this simulation incorporated a liquid penetration model which is derived from the Washburn equation, capillary forces, viscous forces which are based on the liquid bridge, and the binder dissolution model.

The liquid penetration equation is given in Eqs. (22.20) and (22.21) as stated in Tamrakar et al. (2019):

$$m_{liq^2} = w_c \left(\frac{\rho^2 \gamma \cos\theta}{2\mu} \right) t \tag{22.20}$$

m_{liq} is the penetrating liquid mass, w_c is the Washburn constant, ρ is the density, and γ is the difference in concentration between the binder's bulk concentration and its maximum solubility in liquid determines the dissolving rate is given as:

$$\frac{dm_d}{dt} = \frac{D\rho_l A}{L}(C_s - C_b) \tag{22.21}$$

where D represents the binder diffusion coefficient ρ_l denotes the density of the liquid, L signifies the thickness of the diffusion layer, C_s signifies maximal solubility, and C_b signifies the bulk binder concentration in the volume of the liquid m_d is the mass of the dissolved binder.

The capillary force acting between particles ($F^{p-p}{}_{cap}$) and between particles and the wall ($F^{p-w}{}_{cap}$) are given in Eq. (22.22) and (22.23) as.

$$F^{p-p}{}_{cap} = \frac{4\pi \bar{r} \gamma \cos\theta}{1 + \frac{1}{\sqrt{1 + \frac{V_{liq}}{\pi \bar{r} S^2}} - 1}} \tag{22.22}$$

$$F^{p-w}{}_{cap} = \frac{4\pi r \gamma \cos\theta}{1 + \frac{S}{\sqrt{\frac{\pi r_j}{V_{liq}}}}} \tag{22.23}$$

The symbols $\bar{r}$ denote the equivalent radii of the particles in contact, the liquid bridge contacting angle of the fluid and (θ and γ, respectively), the liquid volume utilized in the bridge's creation (v_{liq}), and the separation between the wall and particle.

The liquid bridge's viscous forces (F_{vis}) are composed of both normal F^n_{vis} and tangential components F^t_{vis} are shown in Eqs. (22.24), (22.24a), and (22.24b) as. Where:

$$F^t_{vis} = \left(\frac{8}{15} \ln\frac{\bar{r}}{S} + 0.9588 \right) 6\pi \mu \bar{r} v'_r \tag{22.24}$$

and:

$$F^n_{vis} = \frac{6\pi \mu \bar{r} v^r_r r}{S} \tag{22.24a}$$

thus:

$$F_{vis} = F^t_{vis} + F^n_{vis} \tag{22.24b}$$

The viscosity of the fluid is denoted as μ, $\bar{r}$ represents the equivalent radii of the particles in contact, v^r_r and v^t_r denotes the normal and tangential relative velocities between the two particles, tangential relative velocity as, and the particle distance separation as S. When particle-to-geometry interactions occur, the radius r of the particle is utilized in place of $\bar{r}$.

In WBA, the binder is introduced as a pre-dissolved solution, whereas in DBA, it is introduced as solid particles that dissolve during granulation. As a result, the two systems exhibit different viscosity distributions. In WBA, regions of high viscosity are located at the upper part near the spray nozzle, whereas regions of low viscosity are found along the vessel walls in areas with poor mixing. The phenomenon of high viscosity is observed at lower levels within the bed, where the binder particles effectively blend and dissolve. Conversely, regions of low viscosity are comparatively smaller, owing to the presence of stronger liquid bridges resulting from the greater viscosity. In general, the regions of high viscosity are contingent upon the dispersion of binder and liquid, which is influenced by the process of mechanical mixing (Barrasso et al., 2014; Tamrakar et al., 2019).

In brief, this study presents a unique DEM that offers quantitative and mechanistic insights into the variations observed in binder addition techniques during wet granulation. The analysis of viscosity distributions, liquid bridge forces, velocities, and agglomeration indicators at the particle level provides insights into the mechanism underlying the delayed yet accelerated growth of granules in DBA as compared to WBA (Chaturbedi et al., 2017). The interplay between binder dissolution, liquid addition, and shear mixing plays a critical role in determining the high-viscosity zones (Litster et al., 2002). The comprehension of these intricate multiphasic processes is improved by the model and its conclusions, which can assist in the logical design and optimization of wet granulation operations (Tamrakar et al., 2019).

DEM method has been used to explore the granulation of iron ore fines in HSG. This investigation by You et al. (2021) intended to comprehend the inter-particle force, particle motion, and adhesion that occur during the process of granulating. The granulation of coke, fluxes, and iron ore with water to produce a sinter is a necessary stage of developing a blast furnace as this enhances the sinter bed's air permeability with the right granule size and strength. The research focused on horizontal high-shear granulators which are appropriate for continuous granulation. The studies were done to investigate the efficacy of iron ore granulation under varying water contents and different impeller rotation speeds. The liquid bridge model was used to model the influence of water in the DEM simulation. As the water content increased, so did the liquid bridge force and the quantity of bonded small particles, which led to the formation of aggregates with broader size distributions and greater dimensions. As a result of increased particle collisions, granulation efficacy was enhanced as impeller speed was increased. The results obtained from the simulation illustrating the granulation of particles with various size ratios indicated that as the size ratios of nucleating and adhesive particles increased, so did the granulation efficiency (You et al., 2021). You et al. (2021), however, noted that granule size distribution is not properly accounted for in the DEM modeling and simulation and further recommended the need for other software to be coupled with DEM for proper GSD analysis. It is also noted that this simulation did not consider the right operating and material characteristics in developing the DEM model.

In their study, Kishida et al. (2021) introduce a simulation of particle flow and mixing in a vertical high-shear mixer using a coarse-grained DEM. The authors employ a newly developed coarse-grained approach for granular shear flow (CGSF). The concept underlying the CGSF involves the alignment of four distinct types of energy (friction damping, kinetic, elastic, and viscous damping) between the original particles and the coarse-grained (CG) particles in the context of shear flow. By applying this energy-matching technique, scaling rules for the DEM's contact force parameters can be defined, including the stiffness (k), friction (μ), and restitution (e) coefficients (Table 22.1) (Kishida et al., 2021). The fundamental principles governing the scaling of particle-to-particle interactions are as follows: The

22.3 Simulation of high granulators

Table 22.1 Contact parameters and coefficients for different interaction types.

Parameter	Particle-to-wall	Particle-to-Particle
Sliding friction coefficient, μ	$\mu_{CPW} = \mu_{0,PW} + \mu_{0,PP} \frac{|(\alpha-1)\beta(\alpha)d_o|\omega_i^{CG}*\hat{n}_{pw}|}{|(v_t^{wall}-v_t^{CG})+\beta(\alpha)(r_c-r_o)\omega_i^{CG}*\hat{n}_{pw}|}$	$\mu_{CPP} = \mu_{0,PP} + \mu_{0,PP} \frac{(\alpha-1)\beta(\alpha)d_o\left(|\omega_i^{CG}*\hat{n}_j|+|\omega_j^{CG}*\hat{n}_{ij}|\right)}{|(v_{t,i}^{CG}-v_{t,j}^{CG})+\beta(\alpha)(r_c-r_o)(\omega_i^{CG}*\omega_j^{CG})*\hat{n}_{ij}|}$
Stiffness, k	$k_{CPW} = \alpha^2 \left(\frac{1}{k_{0,PW}} + \frac{\alpha-1}{k_{0,PP}}\right)^{-1}$	$k_{C,PP} = \frac{\alpha^2}{2\alpha-1} k_{0,PP}$
Restitution coefficient, e	$e_{CPW} = \sqrt{1 - \frac{1}{\alpha}(1 - e_{0,PW}^2)}$	$e_{C,PP} = \sqrt{1 - \frac{1}{\alpha}(1 - e_{0,PP}^2)}$
Where $\beta(\alpha) = \left\{\left(\frac{1}{\alpha}\right)^2 + \left(1 - \frac{1}{\alpha}\right)^2\right\}^{-\frac{1}{2}}$ and $\alpha = d_c/d_o$		

equation can be expressed as:

$$\mu_{C,pp} = \mu_{0,PP} + \mu_{0,PP} \frac{(\alpha-1)\beta(\alpha)d_0\left(|\omega_I^{CG}+\hat{n}_{If}|+|\omega_J^{CG}+\hat{n}_{If}|\right)}{\left|(v_{t,I}^{CG}-v_{t,f}^{CG})+\beta(\alpha)(r_c-r_o)(\omega_I^{CG}+\omega_J^{CG})*\hat{n}_{If}\right|}$$

Where $\beta(\alpha)$

$$= \left\{\left(\frac{1}{\alpha}\right)^2 + \left(1 - \frac{1}{\alpha}\right)^2\right\}^{-\frac{1}{2}} \tag{22.25}$$

The simulation of the CG-DEM is conducted using different coarse-graining ratios and rotational speeds of the impeller. The equations that govern the translational motions of the individual cosmic ray particles are provided in Eqs. (22.25a), (22.25b), and (22.25c).

$$k_{C,PP} = \frac{\alpha^2}{2\alpha-1} K_{O,PP} \tag{22.25a}$$

$$m^{CG}\frac{dV^{CG}}{dt} = F_{cn}^{CG} + F_{ct}^{CG} + m^{CG} \tag{22.25b}$$

$$gI^{CG}\frac{d\omega^{CG}}{dt} = r_c\hat{n} * F_{Ct}^{CG} \tag{22.25c}$$

The variables m^{CG}, I^{CG}, V^{CG}, and ω^{CG} represent the mass, inertia, translational velocity, and angular velocity of a CG particle, respectively. The variables r_C, $\hat{n}$, g and t represent the radius of the CG particle, the unit normal vector connecting the center of the contacting CG particles, gravity, and time, respectively. The variables F_{cn}^{CG} and F_{ct}^{CG} represent the normal and tangential contact forces acting on a CG particle, respectively. The contact forces are determined through the utilization of a soft sphere model incorporating linear springs. $F_{cn}^{CG} = -k_C\delta_n^{CG} - \eta_C V_{rn}^{CG}$

$$F_{ct}^{CG} = \begin{cases} -k_C\delta_t^{CG} - \eta_C V_{rt}^{CG} \\ -\mu_C |F_{cn}^{CG}|\hat{t} \end{cases} \quad damping coefficient (\eta_C) \; the restitution coefficient of the CG particle (e_c) \tag{22.26}$$

The determination of the damping coefficient is based on restitution coefficient of the CG-particle the Eq. (22.26a)

$$\eta_C = 2\sqrt{m^{CG}K_C}\left\{1 + \left(\frac{\pi}{\ln e_C}\right)^2\right\}^{-\frac{1}{2}} \tag{22.26a}$$

The evaluation of the CGSF's performance is conducted by considering factors such as particle velocity and kinetic energy, the geometry of the powder bed in motion, and the depth of powder mixing, which is measured using Lacey's mixing index (Lacey, 1954). The findings indicate that the CG-DEM simulations utilizing CGSF display particle velocity distributions, moving powder bed geometries, and particle mixing behaviors that closely resemble those seen in the original scenario, even when considering high coarse-graining ratios ($\alpha \geq 10$). The CG-DEM simulations, in contrast, exhibit notable departures from the original situation when the contact force parameters are not scaled.

In addition, the efficacy of the CGSF is examined at different rotational velocities of the impeller (ranging from 90 to 640 rpm). Under fully developed granular shear flows, such as the roping flow regime, the CGSF demonstrates a high level of accuracy. However, its accuracy diminishes when dealing with weak granular shear flows, such as the bumping flow regime. The study showcases the suitability of the CGSF for CG-DEM simulations of vertical high-shear mixers characterized by dense granular shear flows of greater complexity and intensity, in contrast to rotating drum mixers (Kishida et al., 2021).

Insights into the intricate mechanisms driving particle interactions, granule formation, and the impacts of different process parameters have been gained by the utilization of DEM and CG-DEM in the examination of HSWG processes. This work emphasizes the significance of taking into account several parameters, including coalescence kernels, impeller design, binder addition methods, material qualities, and particle flow and mixing, to optimize HSWG operations.

The study conducted by Gantt et al. (2006) showcases the capability of combining DEM simulations with physically based coalescence models to create predictive PBE models that are both mechanistic and predictive. Incorporating size-dependent compaction rates and multidimensional coalescence kernels into PBE models for high-shear granulation processes can greatly improve their accuracy and predictive capabilities. The study conducted by Börner et al. (2016) offers a novel viewpoint on the influence of impeller design on the production of shear forces and energy input in the context of HSWG. The results of their study indicate that the main contributor to large shear forces, as a result of substantial velocity gradients, is the bulk material rather than the impeller. Furthermore, the design of the impeller and the speed at which it rotates significantly affect the kinetic energy of particles, hence influencing the frequency of collisions and the rates at which granules shatter.

Tamrakar and Ramachandran (2019) provide significant contributions to the understanding of the distinctions between WBA and DBA techniques in the context of HSWG. Their study uses a DEM that incorporates specialized contact models. The creation of high-viscosity zones, which are crucial for granule growth kinetics, is determined by the interaction among binder dissolution, liquid addition, and shear mixing.

The study conducted by You et al. (2021) showcases the practicality of employing DEM to examine the process of granulation in a horizontal high-shear granulator for iron ore fines. Their research emphasizes the significance of water content, impeller rotation speed, and particle size ratios in determining the effectiveness of granulation. Nevertheless, they acknowledge the constraints of DEM in precisely depicting granule size distributions and the necessity of integrating with additional software for thorough research.

The coarse-grained discrete element technique (CG-DEM) proposed by Kishida et al. (2021) is utilized to simulate the dynamics of particle movement and mixing within a vertical high-shear mixer. In this study, a novel coarse-grained technique for granular shear flow (CGSF) is utilized, which effectively aligns the energy between the original particles and the coarse-grained particles. The CFD model has a notable level of precision for simulating dense granular shear flows, notably within the rope flow domain. Nevertheless, the precision of the system decreases when confronted with weak granular shear flows, such as the regimen of bumping flow.

The research examined in this study emphasizes the importance of DEM and computer-generated discrete element method (CG-DEM) as effective instruments for comprehending and enhancing HSWG processes. Future study should prioritize the enhancement of these models, their integration with other modeling methodologies, and the validation of the results through empirical investigations. Researchers and practitioners can enhance the efficiency, consistency, and predictability of HSWG processes in different industries by utilizing the knowledge acquired from DEM and CG-DEM simulations.

Seeded granulation was first introduced by Rahmanian et al. (2011) to describe the granulation process which intends to produce granules with a core structure that is made up of coarse particles and covered by fine particles. This granulation process has been simulated by Hassanpour et al. (2013) in batch Cyclomix granulators using DEM. In this work, an attempt is made to explore the process conditions that lead to the production and breaking of seeded granules. In this instance, simulation, the Hertz-Mindlin contact model is utilized to model the bonding forces between particles and capillary and viscous forces as a result, liquid binder was not taken into consideration in the modeling process. The result from the simulation indicates that, during the granulation process, seeded granules increase and shrink continuously. The simulation findings demonstrate that a high percentage of seeded granules are covered by fine particles to about 60% coverage. At high rotational velocities, seeded granules with extensive surface coverage are frequently observed, based on additional quantitative analyses. This implies that there is a robust connection between the rate of rotation and the production of seeded granules. Further research is needed to determine the most important conditions, such as the Stokes number, which are required to improve seeded granulation; however, the work has shown the potential of DEM for modeling granulation processes and exploring the fundamental processes of granulation at large and seeded granulation in particular (Hassanpour et al., 2013).

Fan et al. (2023) offer a thorough examination of the uniformity of mixing in binary particle systems within high-shear wet granulators. This is a critical factor in the processing of particles in the pharmaceutical business. The authors utilize DEM simulations to examine the impact of particle size, density, and volume fraction on mixing behavior and homogeneity. This level of analysis is difficult to attain just through experimental approaches (Fan et al., 2023). The study's findings emphasize the intricate relationship between particle characteristics and the dynamics of mixing. Observing phenomena akin to the Brazil nut effect and Reverse-Brazil nut effect at various points of the mixing process highlights the significance of taking into account the time-dependent nature of particle interactions. Moreover, the determination of a specific size limit for a particular binary particle system indicates that the mixing characteristics can vary significantly based on the relative sizes of the particles present (Fan et al., 2023). Fan et al.'s (2023) examination of vertical driving forces and granular temperature offers a further understanding of the factors that contribute to particle mixing in high-shear wet granulators. The substantial fluctuation in vertical driving forces when particle size or density is altered emphasizes the criticality of meticulously choosing particle characteristics to enhance mixing efficiency. The granular temperature is distributed spatially, with the highest values found near the wall of the pelletizer and the tip of the impeller. This emphasizes the important influence of equipment design on the efficiency of mixing (Fan et al., 2023).

The findings obtained from this study have significant ramifications for the advancement and enhancement of HSWG procedures in the pharmaceutical sector. By comprehending the aspects that affect the evenness of mixing, scientists and professionals can develop particle systems and processing settings that encourage consistent mixing, ultimately resulting in enhanced product quality and uniformity. Furthermore, the utilization of DEM simulations showcased in this research provides a potent instrument for investigating the behavior and interactions of particles in granulation procedures, facilitating the creation of enhanced and more productive processing approaches (Fan et al., 2023).

Overall, Fan et al. (2023) study on the uniformity of mixing in binary particle systems offers useful insights into the intricate dynamics of HSWG. The study's findings highlight the significance of considering particle characteristics, time-varying phenomena, and equipment configuration when optimizing mixing efficiency. This research enhances our comprehension of these elements, hence aiding the advancement of more effective particle processing techniques in the pharmaceutical business.

The use of DEM and CG-DEM in studying HSWG processes has provided valuable insights into the complex mechanisms driving particle interactions, granule formation, and the effects of various process parameters. The studies highlighted in this chapter emphasize the importance of considering multiple factors, such as coalescence kernels, impeller design, binder addition methods, material properties, and particle flow and mixing, to optimize HSWG processes.

Recommendation is made for future research (Vanarase et al., 2010) to focus on validating these models with experimental data, integrating them with multiscale PBE simulations, and exploring their applicability to a wider range of materials and operating conditions. In addition, efforts should be made to address the limitations identified in these studies, such as improving the representation of granule size distributions in DEM models and considering actual operating and material characteristics in the simulations.

As DEM and CG-DEM techniques continue to advance, they will play an increasingly crucial role in the rational design and optimization of HSWG processes across various industries. By providing a deeper understanding of the fundamental mechanisms at the particle level, these simulation tools will enable the development of more efficient, consistent, and high-quality granulation processes, ultimately benefiting the pharmaceutical, food, and metallurgical sectors, among others.

22.3.2 CFD simulation of high shear granulation

High-shear mixers and granulators are extensively utilized in diverse industries, such as pharmaceutical, chemical, and food processing, to manufacture granular products with specific qualities. Comprehending the intricate flow characteristics of dense granular substances in these devices is essential for enhancing their design and performance. CFD is a highly effective method for studying the hydrodynamics and efficiency of high-shear mixers and granulators. It offers vital knowledge about the processes of mixing and granulation. Nevertheless, precisely simulating the dense granular flow in these systems continues to be a difficult task because of the intricate interactions between particles and the walls of the equipment.

The progress in CFD modeling of dense granular flow in high-shear mixers and granulators has been significantly influenced by two noteworthy studies conducted by Darelius et al. (2008) and Ng et al. (2009). In their study, Darelius et al. (2008) utilized the Eulerian-Eulerian multiphase flow method, using the KTGF and frictional stress models, to represent the dense gas-solid flow occurring in a high shear mixer. The researchers examined the impact of different solid phase boundary conditions on the vessel wall and verified their computer simulations by comparing them to experimental velocity profiles obtained using high-speed video and image analysis. In contrast, Ng et al. (2009) conducted a comparison between CFD simulations based on a continuum model of dense-gas kinetic theory and experimental data obtained by the positron emission particle tracking (PEPT) technique. Their study specifically focused on modeling granular flow in a high shear mixer granulator.

In their study, Darelius et al. (2008) discovered that the partial slip model, which enhanced the dissipation of energy at the wall, yielded a more accurate forecast of the powder bed height when compared to the free slide boundary condition. Nevertheless, the forecasts about the directions of velocity were insufficient, since the simulations indicated a notable transformation of particle tangential momentum into axial momentum at the vessel wall, a phenomenon that was not observed in the experimental results. The disparities between the simulated and measured velocities were especially noticeable in the concentrated particle area near the impeller. This indicates that there is a requirement for additional improvement of the frictional stress models in order to accurately represent the cohesive nature of dense particle flow. Surprisingly, the simulated outcomes were not considerably affected by the selection of restitution coefficients in the partial slip model. This emphasizes the necessity of developing partial slip models that are specifically designed for dense systems with continuous particle-wall contact (Darelius et al., 2008).

Ng et al. (2009) showed that the Eulerian-based continuous model well represented the key characteristics of solids movement in the HSG, including the height of the bed and the primary tangential velocity. Nevertheless, the model's failure to accurately represent the intricate vertical swirl pattern observed in the PEPT trials emphasized the necessity for more enhancements in the modeling methodology. The disparities between the CFD forecasts and empirical observations, specifically the exaggerated calculation of tangential velocity and abrupt declines near the wall area, indicated a failure in the transmission of forces inside the granular bed. The work highlighted the importance of creating enhanced constitutive relations that more accurately represent the frictional interactions and momentum transmission in dense granular flows (Ng et al., 2009).

Both works showcase the capability of utilizing the Eulerian-Eulerian CFD method, in conjunction with the KTGF and frictional stress models, to replicate intricate dense multiphase flow in high shear mixers and granulators. Nevertheless, they also emphasize the difficulties in precisely forecasting the speed patterns and movement of the sediment on the seabed, particularly in areas with a large number of particles and substantial interactions between the particles and the walls. In order to provide reliable quantitative predictions, it is required to make further advancements in frictional stress models and construct suitable partial slip boundary conditions for dense particle systems (Darelius et al., 2008; Ng et al., 2009).

The research conducted by Ng et al. (2009) and Darelius et al. (2008) has greatly enhanced our comprehension and modeling of dense granular flow in high-shear mixers and granulators by the application of CFD. Although there are limitations and problems mentioned in these studies, validated CFD models can offer useful insights into the processes of mixing and granulation. This can help optimize the design and operation of high-shear mixers and granulators.

The results of these investigations have significant ramifications for the future advancement and use of CFD models in simulating the movement of granular materials in high-shear mixers and granulators. These findings highlight the necessity for additional studies to improve the precision and predictive capabilities of these models. Researchers can make progress in developing more accurate and predictive CFD tools for designing and optimizing high-shear mixers and granulators by tackling specific challenges. These challenges include enhancing frictional stress models and creating suitable partial slip boundary conditions for dense particle systems. Ultimately, this advancement will lead to more efficient and robust processes in industries like pharmaceuticals.

22.3.3 Coupled simulation of high shear granulation

In the pharmaceutical sector, the process of HSWG holds significant importance as it serves as a crucial step in the manufacturing of tablets and other solid dosage forms. The procedure entails the consolidation of tiny powder particles into bigger granules by introducing a liquid binder and exerting shear pressures. The granulation process is intricate and diverse, encompassing several physicochemical occurrences that transpire at distinct intervals and time scales, despite its extensive utilization. Comprehending and enhancing the granulation process is of utmost significance in guaranteeing the quality of the product, the efficiency of the process, and the ability to reproduce results. In recent times, the utilization of computational modeling and simulation has been increasingly prominent in the examination of the granulation process and the anticipation of granule characteristics. The utilization of multiscale modeling methodologies has garnered considerable interest owing to its capacity to effectively reflect the intricate relationship between processes at the particle level and dynamics at the equipment scale. The primary objective of this chapter is to examine three prominent studies that have made noteworthy advancements in the domain of high-shear granulation modeling. These studies include the DEM-PBM framework developed by Sampat et al. (2018), the multiscale modeling approach proposed by Kulju et al. (2016), and the compartmental PBM with spatial dependence developed by Yu et al. (2017). Through the analysis of these studies, our objective is to present a comprehensive summary of the most advanced techniques in granulation modeling and emphasize the potential of these methods for comprehending, enhancing, and managing processes.

The DEM is a computational approach utilized for the observation and analysis of the motion and interactions of minuscule particles. The software employs chemical principles to calculate the occurrences of particle collisions, adhesion, and aggregation, enabling a thorough simulation of the granule formation process. The PBM is a theoretical framework that characterizes the progression of PSD throughout the granulation process. Mathematical frameworks are employed to represent the rates at which particles aggregate, fracture, and grow, thus providing a macroscopic viewpoint on the phenomenon.

The proposed approach combines the benefits of both DEM and PBM by integrating them concurrently. The DEM accurately represents the complex interactions among particles at a microscopic scale, while the particle beam model (PBM) provides a more comprehensive representation of the distribution of particle sizes at a macroscopic scale. The application of this integrated technique offers a comprehensive perspective on the granulation process, enabling a deep understanding of the involved processes.

This work aims to efficiently manage the computational needs associated with executing the connected DEM-PBM model by utilizing high-performance computing (HPC) resources. According to Sampat et al. (2018), a specialized scheduler is employed to supervise the distribution of resources and ensure the efficient simultaneous execution of the two models. The use of this computational technology enables faster and more accurate representations of the granulation phenomena, hence facilitating the study of particle interactions and size distributions.

The study by Sampat et al. (2018) utilizes a quality by design (QbD) approach, which emphasizes achieving a full understanding of the granulation procedure to ensure consistent product quality. The objective of this study is to get quantitative knowledge about the granulation process and improve the management and optimization of process parameters by employing the coupled DEM-PBM model.

According to Sampat et al. (2018), this proactive approach aims to obtain the necessary attributes of the product during the initial stages of production, rather than solely relying on testing the finished product. The proposed methodology for multiscale modeling has the potential to greatly enhance the understanding and control of the granulation process in the field of pharmaceutical manufacturing. The model provides a thorough understanding of the granulation process by combining the detailed information from the DEM with the larger representations from the PBM. The acquisition of this knowledge holds promise for the enhancement of process parameters, the elevation of product quality, and the general optimization of tablet manufacturing efficiency. The present study provides a significant contribution to the overarching goal of ensuring consistent production of pharmaceutical products that are both safe and effective (Sampat et al., 2018).

Kulju et al. (2016) present a novel multiscale modeling approach in their research, which seeks to replicate the complex and scalable challenges related to continuous HSWG. This process holds significant importance in the pharmaceutical manufacturing industry, as highlighted by Leuenberger (2001) and Vervaet and Remon (2005). The methodology employed by Kulju et al. (2016) effectively integrates the utilization of DEM simulations to precisely depict the dynamics of particle flow, while also incorporating a population balance (PB) model to clarify the mechanisms at the particle level, such as nucleation, aggregation, and breaking. Kulju et al. (2016) found that the DEM-PB model closely matches experimental data on PSD and residence time distribution (RTD), demonstrating its ability to make accurate predictions.

In their study, Kulju et al. (2016) employ the DEM simulations to gain a thorough comprehension of the intricate flow patterns and spatial configurations of particles within a continuous high-shear granulator at various impeller velocities. The results indicate notable phenomena, such as the concentration of particles in certain regions at low velocities and the presence of low-velocity areas that promote clumping (Kulju et al., 2016). The findings from the study align with previous research that highlights the importance of understanding the hydrodynamics of granulators (Litster & Ennis, 2004). The decomposition of particles into forward and backward fluxes using the DEM is then strategically utilized to inform the development of the particle blending (PB) model, ensuring a physically precise representation of the system.

The compartmental methodology of the PB model, as described by Freireich et al. (2011) and Li et al. (2012), involves conceptualizing the granulator as a network of continuously stirred tank reactors (CSTRs). The integration of expedited routes and recycling channels among the CSTRs is influenced by the particle flow patterns observed in the DEM. The objective of this approach is to accurately represent the processes of mixing and transport (Kulju et al., 2016). A comprehensive 3D PBE is employed to track the changes in particle size, binder content, and porosity distributions inside each

CSTR. According to Poon et al. (2008), this equation incorporates significant rate processes, including nucleation, aggregation, breaking, and consolidation. The PB model has a high level of agreement with experimental PSDs, particularly at higher impeller speeds, by precisely modifying parameters. This highlights its ability to accurately predict granule features. A significant finding of this study is the considerable influence of impeller speed on the characteristics of granules, mostly ascribed to its effect on flow patterns and accumulation. According to Kulju et al. (2016), when the velocities are reduced, there is an increased accumulation of material within the spray zone, which has a significant effect on the creation and expansion of granules. The observation made by Knight (2004) underscores the importance of carefully selecting operating conditions and design parameters to achieve the desired characteristics of granules.

The integration of the DEM-PB model proves to be a valuable instrument in the optimization of process parameters and equipment geometry. It enables the systematic planning and management of uninterrupted wet granulation operations with high shear. Kulju et al.'s (2016) research shows how using multiscale modeling tools may effectively explain the complex connection between equipment dynamics and microscale phenomena in challenging particle processes like HSWG. The integration of DEM and PB modeling approaches enables the development of a comprehensive framework for comprehending and predicting granule qualities in continuous granulators. This technique has the potential to enhance model-driven design and control of continuous granulation processes by integrating particle-level impacts into the DEM and improving the structure and parameters of the particle boundary (PB) model (Kulju et al., 2016). The present study provides a significant contribution to the ongoing efforts in continuous manufacturing within the pharmaceutical industry, to improve operational efficiency, adaptability, and the overall quality of products (Schaber et al., 2011; Vanarase et al., 2010).

Yu et al. (2017) have developed a novel compartmental PBM that incorporates spatial dependence in the context of HSWG processes. The objective of the model is to address the limitations of traditional PBMs that depend on the assumption of perfectly mixed conditions and neglect to account for the varied spatial distribution of granulation rates (Litster & Ennis, 2004). Yu et al. (2017) employed a heuristic approach to construct the compartmental structure, utilizing CFD analysis to investigate the flow pattern, velocity profile, and solids concentration in two laboratory-scale granulators: a 1.9L MiPro granulator and a 4L DIOSNA granulator.

The study conducted by Darelius et al. (2008) utilized CFD analysis to identify several zones within the granulator. These regions encompassed the spray zone located near the liquid inlet, the high-shear impeller zone, and the zone characterized by slower movement in the bulk. According to Yu et al. (2017), the granulators were separated using multiple compartments, and the transport rates between these compartments were evaluated using CFD simulations. The multicompartment PBM demonstrated the presence of crucial granulation mechanisms, including nucleation, rewetting, layering, coalescence, and breakage. The decision to incorporate these methods was based on the consideration of the specific characteristics of granule velocity, size, and liquid content within each compartment, which are influenced by spatial parameters (Hounslow et al., 2009; Yu et al., 2017). The constant volume Monte Carlo (CVMC) approach was employed to solve the multidimensional PBEs in each compartment. The primary focus of the study was to investigate the processes of nucleation and stacking in the 1.9L MiPro case, employing a low liquid content. In order to coincide with the observed granule size distribution in the experimental data, the parameters, including the layering rate constant and critical packing fraction, were varied (Yu et al., 2017). The research incorporated the mechanics of coalescence and fracture in the 4L DIOSNA scenario, which involved a significant liquid component. The evaluation encompassed the measures of interest, namely, the coalescence kernel and breakage rate.

According to Yu et al. (2017), the compartmental PBM has shown the ability to effectively predict the advancement of granule size distribution, liquid distribution, and primary particle depletion in both low and high liquid content situations. The model effectively represented the progressive changes in the mean granule size over time, while also providing forecasts for the overall size distribution and its associated standard deviation. Furthermore, it is feasible to obtain localized granule data, including the spatial arrangement of liquid constituents. By modifying parameters such as the layering rate constant, critical packing percentage, coalescence kernel, and breakage rate, the model successfully achieved a high level of accuracy in fitting the experimental data.

The model developed by Yu et al. (2017) presents a pragmatic framework for the integration of granulation kinetics and spatial heterogeneity at a mesoscopic scale. This integration is crucial for a comprehensive understanding and improvement of various processes. The technique presented by Yu et al. (2017) known as compartmental computational fluid dynamics (CFD-PBM, enables the simulation of complex wet granulation behavior in the presence of high shear conditions. The integration of particle-level kinetics with hydrodynamics at the equipment scale enables the achievement of this objective. The multiscale model, as proposed by Yu et al. (2017), incorporates the granulation mechanisms of nucleation, growth, consolidation, and breaking. Furthermore, it accounts for spatial variations in granule properties. This methodology provides a reliable tool for examining the interplay between operational parameters, equipment configuration, and granule properties. The model has the potential to aid the understanding, improvement, control, and enlargement of various processes. However, there remain opportunities to improve the precision of the CFD submodels, validate the individual granulation kernel techniques, and broaden the framework to include more complex systems. Yu et al. (2017) argue that the incorporation of the compartmental approach is a crucial foundation for the advancement of comprehensive universal models of granulation processes.

The thorough comprehension of the intricate interaction among process parameters, equipment design, and granule properties is facilitated by the combination of particle-level mechanisms, as captured by DEM, with equipment-scale dynamics, as represented by particle beam modeling (PBM) or CFD. The aforementioned multiscale modeling methodologies offer significant contributions to understanding the impact of operational parameters, such as impeller speed, on the distribution of granule sizes, binder content, and porosity. In addition, these approaches shed light on the spatial variability of granulation rates within the granulator.

The potential for revolutionizing the design, operation, and management of HSWG processes in the pharmaceutical industry exists with the implementation of these modeling tools. Researchers and practitioners can achieve optimization of process parameters, improvement of product quality, and enhancement of overall manufacturing efficiency through the utilization of these models. Furthermore, these methodologies make a valuable contribution to the continuous manufacturing endeavors within the pharmaceutical sector, facilitating the advancement of production processes that are characterized by enhanced flexibility, efficiency, and resilience.

Nevertheless, despite the notable advancements achieved, there are still prospects for enhancing and broadening these modeling frameworks. Subsequent investigations ought to prioritize the augmentation of sub-model correctness, the validation of individual granulation kernel approaches, and the expansion of frameworks to include more intricate systems. The practical value of these multiscale models in

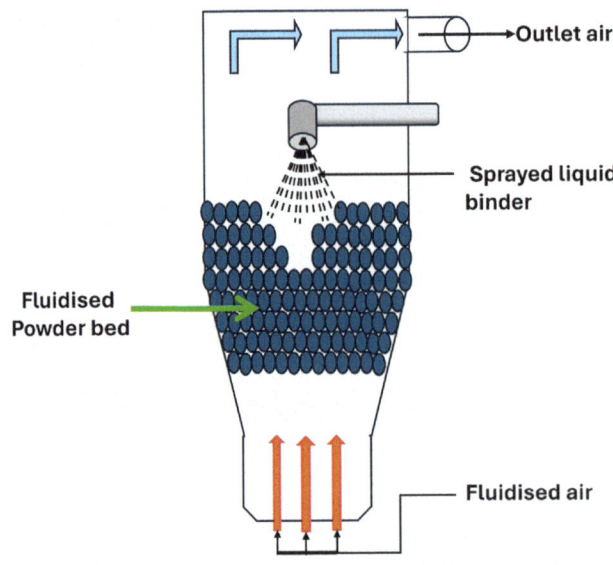

FIGURE 22.3

Schematic of fluidized bed granulator.

A diagram depicting a typical spray top fluidized bed.

industrial settings could be further enhanced through the incorporation of modern process control strategies and real-time monitoring systems.

The utilization of multiscale modeling techniques to simulate HSG holds significant potential for enhancing our comprehension and regulation of this pivotal pharmaceutical manufacturing procedure. The ongoing advancement and use of these methodologies will undeniably make a significant contribution to the manufacturing of pharmaceutical items that are of superior quality, safe, and efficacious.

22.4 Fluidized bed granulation simulation

Fluidized bed granulation (FBG), a typical schematic shown in Fig. 22.3, is a versatile process used in numerous industries, such as medicines, food, and agriculture, to produce granular particles with desired qualities (Diez et al., 2018). The intricate interplay between the fluid and solid phases in FBG systems provides considerable obstacles in understanding and managing the process (Darelius et al., 2008). Computational modeling and simulation tools, such as CFD, DEM, and population balance modeling (PBM), are increasingly being employed to gain insights into the underlying mechanisms and optimize FBG processes (Fries et al., 2013; Heinrich et al., 2002; Tamrakar & Ramachandran, 2019). This section will concentrate on the implementation of these modeling methodologies in FBG, with a specific emphasis on the creation of combined CFD-DEM-PBM frameworks to capture the interaction between fluid dynamics and particle behavior. The studies to be covered include the CFD-DEM simulation of a coating process in a fluidized bed rotor granulator by Grohn et al. (2020), the development of a model to improve granular temperature prediction in CFD-DEM simulations by Yu et al. (2020), and the CFD-DEM modeling of the FBG of food powders by Kim and Chung (2019).

22.4.1 Discrete element method simulation of fluidized bed granulation

To prove the concept of particles gaining higher surface energy as they pass through the spray zones of FBG, Kafui and Thornton (2008) developed a DEM simulation of a fluidized bed where the formation and breakage of the granules are due to the theoretical mechanical contacts. A notion based on surface energy was introduced and utilized in a 3D DEM simulation of FBG for a Geldart group A particle beds. The idea utilizes basic mathematical concepts to represent the absorption of surface energy by particles in the spray zone.

The granulation simulation based on the DEM model efficiently achieves a homogeneous dispersion of spray-layered particles in the system. The development of net granules that were identified was linked to patterns in bond formation, binding strength, and maximal granule size. The process of granule formation and shattering was accurately represented, together with the changes in granule mass and linear dimension, which demonstrated both growth and reduction in mass/size. The fractal analysis of the initial granules indicates that the results do not exhibit fractal characteristics. The modeling technique employed in this simulation shows great potential. It, however, recommended the utilization of pendular liquid bridges where spray droplets of the binder solution would represent a distinct element as a feasible expansion of the methodology.

Researchers sought to identify the location within the granulation apparatus where this compaction takes place in order to study the compression process that happens during the interaction of particle-particles. As shown by Abu Bakar et al. (2013), the DEM was used to examine the impact of collisions in a particular kind of FBG known as pressure swing. Using a customized code for 3D simulation, the researchers carried out DEM simulation, enabling a thorough analysis of particle interactions and compaction dynamics. This study calculated the collision frequency and force between particles. The DEM simulation coefficient of restitution was 0.9. MPIA was used to calculate the Hookean spring constant k from the experimental force-displacement curve. Deformation values obtained and were varied between 300 and 790 N/m. According to the DEM modeling and experiment, researchers noted a significant occurrence of collisions and a remarkably low magnitude of collision force. The results of this study suggest that the primary factor contributing to granule densification in PSG is reverse flow, which results in compaction. Nevertheless, it has been shown that the process of granule mixing during fluidization should also exert a substantial influence on the sphericity (Abu Bakar et al., 2013).

22.4.2 Coupled granulation simulations and modeling of fluidized bed granulation

Most granulation phenomena cannot be simulated using just one simulation method. The most common coupled method is the DEM-PBM, where the DEM simulates the particle dynamics in the granulation, and the PBM is used to determine the GSD of the particles simulated in DEM (Suresh et al., 2017). This section will discuss some coupling methods employed in the granulation simulation process, give a brief overview of how the model was developed, and discuss the outcomes from such models.

Arthur et al. (2022) utilized gPROMS Formulated Product 2.0 (gFP) software to perform process simulations of FBG. The objective was to examine how process factors impact the distribution of granule sizes. The gFP software utilizes the PBM technique to construct different kernels for the granulation rate process. The study examined the impact of air flow rate, binder concentration, and binder spray rate on the granule size distribution of the final product. The simulations were conducted both with and without the inclusion of agglomeration and consolidation kernels.

The findings indicated that the application spray rate of the binder had the greatest influence on the distribution of granule sizes, as higher rates of spraying resulted in a greater proportion of larger granules in the final product. The granule size distribution was hardly affected by changes in air flow rate and binder concentration in the absence of agglomeration and consolidation kernels. Nevertheless, the inclusion of these kernels in the simulation enhanced the accuracy of predicting the impact of binder concentration and air flow rate on granule size distribution. The comparison between the simulation findings and experimental data from the literature by Suresh et al. (2020) emphasized the significance of including agglomeration and consolidation kernels in the simulation design to achieve precise outcomes. While the granule size distribution patterns were comparable between the simulations and actual data, there were discrepancies in the specific values. This suggests that additional research is necessary to refine the agglomeration and consolidation models used in the simulation design. The study highlights the significance of agglomeration and consolidation rate processes in FBG. It suggests that these processes should be considered as essential factors when simulating the effects of formulation and process parameters on granule size distribution and other granule characteristics. The gFP software's use of the PBM approach to implementing these kernels demonstrates the effectiveness of PBM simulations in modeling FBG processes.

FBG is a commonly employed method in diverse manufacturing sectors, including medicines, agriculture, food, and detergents, for the production of granular particles with specific qualities (Uhlemann & Mörl, 2000). The process entails fluidizing and granulating particles by spraying liquid binder droplets onto a fluidized powder bed. This results in repeated wetting and collisions among particles, leading to an increase in size. Nevertheless, the intricate interplay between fluids and solids in FBG systems presents a formidable obstacle to comprehending and managing the quality of the result (Boerefijn & Hounslow, 2005; Burggraeve et al., 2013).

Various modeling approaches have been used to acquire insights into the underlying mechanisms and optimize FBG processes. CFD is frequently employed to model the fluid flow and pressure variations in FBG systems (Börner et al., 2017; Darelius et al., 2008; Fries et al., 2011). Nevertheless, CFD alone may not sufficiently reflect the intricate dynamics of particles in situations with high concentrations of solids. The combination of CFD with the DEM, which monitors the movement of individual particles using a Lagrangian framework, has shown great potential in modeling the behaviors of both fluids and particles (Bokkers et al., 2004; Fries et al., 2013).

Furthermore, population balance models (PBM) have been widely employed, alongside CFD-DEM, to simulate FBG processes (Heinrich et al., 2002; Vreman et al., 2009). PBM utilizes rate expressions for aggregation, breaking, consolidation, and other processes to explain the phenomena of size enlargement and offer insights into the dynamic alteration of PSD (Heinrich et al., 2002). However, conventional PBM assumes complete homogeneity, which may not be accurate for heterogeneous FBG systems. Compartmental physiologically-based models (PBMs) have been suggested as a solution to overcome this constraint (Liu & Li, 2014).

One effective method involves extracting data on particle kinetics, such as velocities and collisions, from simulations that combine CFD with the DEM. These data may then be used in growth rate kernels of population balance models (PBM) to predict how particle size changes over time (Gantt et al., 2006; Tamrakar & Ramachandran, 2019). This integrated computational framework, including CFD, DEM, and PBM, enables the connection between microscopic particle-level events and the overall behavior of granulation processes. Previous studies have shown the effectiveness of high-shear granulation (Sen et al., 2014), but further research is required to adapt it for FBG and consider the presence of the fluid phase (Tamrakar & Ramachandran, 2019).

Rajniak et al. (2009) have demonstrated the use of CFD-PBM and CFD-DEM-PBM methodologies, respectively, in the context of FBG systems. Dosta et al. (2012) developed a multiscale CFD-DEM-PBM framework. This framework involves moving data from the micro-scale CFD-DEM simulations to the macroscopic PBM simulations.

The PBM is constructed using a modeling framework that establishes the rates at which granules increase and decrease, considering mass balance. It can accurately replicate changes over time in granule size distribution (GSD) and other characteristics of granules (Bandari et al., 2020). The most important component of PBMs for granulation is the agglomeration kernel, which represents the particle agglomeration rate. Conversely, the agglomeration kernels utilized in PBMs do not reflect the granule characteristics inside a granulator, and so remain empirical, with undetermined parameters (Barrasso et al., 2015a).

The study introduces an innovative methodology for simulating the granulation behavior of food powders utilizing a coupled CFD-DEM approach. This work by Kim and Chung (2019) utilizes the ANSYS Fluent CFD code coupled with the EDEM DEM code to simulate the granulation behavior of food powders in a fluidized bed.

Food powders are categorized into two groups: self-agglomerating and nonagglomerating powders. The simulations encompass the analysis of rheological changes and variations in water content that occur throughout the granulation process. The study focuses on modeling multiple aspects of the granulation process, including nucleation, granulation based on powder properties, and changes in particle water content due to evaporation, which have not been previously modeled in this manner. For self-agglomerating powders, the study develops a model that considers the glass transition temperature of the powder component, specifically dextrin, and incorporates a newly derived evaporation model (Kim & Chung, 2019). The model aims to simulate the granulation process of a mixture consisting of cricket and dextrin powder.

The primary objective of the model is to predict the temporal variation in the moisture content of particles, a crucial factor influencing the granulation process. This is because the stickiness of particles, and subsequently their granulation behavior, is impacted by their moisture content. The evaporation model includes a set of equations that describe the rate of water loss from the surface of powder particles during the FBG process. These equations consider variables such as the temperature of the granulation environment, humidity levels, and the physical properties of the powder, including the glass transition temperature (Kim & Chung, 2019).

The study reports that the simulation model accurately describes the granulation behavior of food powders by incorporating rheological changes and water content variations, accurately representing nucleation, granulation, and changes in particle water content (Kim & Chung, 2019). However, the research does not consider simulations conducted using the DEM-PBM.

Simulation modeling of the process of production of pharmaceutical tablets using wet granulation has been conducted by Wang et al. (2019). The study introduces a comprehensive theoretical framework for the simulation modeling of the pharmaceutical tablet manufacturing process (PTMP) using spray fluidized bed granulation (SFBG). The framework integrates the utilization of the DEM and PBM methodologies to facilitate a full simulation of the granulation step.

The DEM model records the movement and interactions of individual particles during the process of granulation, whereas the PBM model monitors the changes in the distribution of particle sizes over time. The integration of various methodologies facilitates the anticipation of crucial granule characteristics, such as the average particle size (APS). The APS model proposed by Wang et al. (2019) incorporates many operating variables, such as spray rate and inlet air rate, as well as material parameters, including particle-particle interactions, which have an impact on the kinetics of granulation.

To provide a comprehensive comprehension, the developed framework integrates a drying model in conjunction with the APS model. The present drying model is utilized to assess the variations in particle moisture content during the process of granulation. These variations have a substantial influence on several aspects such as granule development, densification, and bulk properties, including density, flowability, and compatibility.

The Heckel equation is utilized in the second step of tablet manufacture to represent the correlation between the applied compression pressure and the resulting porosity of the tablet during the compaction process. The microstructural characteristic of porosity plays a crucial role in determining the performance of tablets. Mathematical models established were subsequently employed to compute crucial quality indicators such as tensile strength, hardness, disintegration time, and dissolving rate, which exhibit a correlation with the porosity of the tablet. It is possible to utilize these models to forecast the influence of granule characteristics and compression circumstances on the essential quality parameters of the end tablets (Wang et al., 2019). This approach combines granulation, drying, and porosity models to accurately represent the intricacies of the PTMP using SFBG. The study aims to provide a comprehensive understanding of the impact of formulation factors and process parameters on many aspects of granulation mechanisms, particle characteristics, tablet microstructure, and finally, product performance metrics such as strength, disintegration, and drug release kinetics.

The comprehensive simulation framework improves comprehension of processes, enabling the development, expansion, and optimization of processes to meet desired quality objectives. This approach decreases the need for considerable testing and facilitates the implementation of quality-by-design principles inside the complex pharmaceutical production process (Wang et al., 2019).

Yu et al. (2020) proposed a model that aims to enhance the accuracy of granular temperature prediction in computational fluid dynamic-discrete element method (CFD-DEM) simulations of gas-solid flows in fluidized bed granulation. The underprediction of granular temperature in CFD-DEM simulations is commonly attributed to the utilization of average drag force models. The drag force fluctuations on individual particles are augmented in the proposed model to align with the amplitude of variations seen in PR-DNS (particle-resolved direct numerical simulation) experiments (Tang et al., 2015; Tenneti et al., 2011). The model alters the fluctuation of the drag force $F'_{d,i}$ on a particle in the following manner in Eq. (22.27):

$$F''_{d,i} = \left(e_i\sqrt{F_{d,j} * F_{d,j}} - \overline{e_i\sqrt{F_{d,j} * F_{d,j}}}\right)h + F'_{d,i} \qquad (22.27)$$

The unit vector in the direction i is denoted as e_i, the average drag force in a computational cell is represented by F_d and h is a parameter that guarantees the mean relative deviation of the modified drag force fluctuation $F''_{d,i}$ approaches the expected value $\sigma^{exp}_{F_d}$, as determined by PR-DNS data. The ultimate adjusted drag force exerted on the particles is:

$$F^{modify}_{d,i} = \overline{F_{d,i}} + F''_{d,j} \tag{22.27a}$$

The predicted value for the expected drag deviation, denoted as $\sigma^{exp}_{F_d}$, is:

$$\sigma^{exp}_{F_d} = cor * \sigma^{dns}_{F_d} \tag{22.27b}$$

When considering the solid-fluid density ratio, solid volume fraction, and Reynolds number, the correction coefficient, denoted as *cor*, is fitted as a function. The variable "$\sigma^{dns}_{F_d}$" represents the average relative deviation of the drag force derived from the PR-DNS method. The model was implemented in MFIX-DEM software to conduct posteriori tests (Garg, Galvin-Carney, Li, & Pannala, 2012). The updated CFD-DEM accurately corresponds to the steady-state and transient granular temperatures obtained from PR-DNS for gas-solid fluxes in a tri-periodic domain. The improved CFD-DEM greatly enhances the accuracy of predicting granular temperature in liquid-solid fluidized beds, surpassing the normal CFD-DEM. However, it is important to note that inhomogeneities at high Reynolds numbers still result in some degree of underprediction. The modified CFD-DEM exhibits comparable performance to the normal CFD-DEM for gas-solid fluidized beds characterized by substantial inhomogeneities.

The model under consideration offers significant insights into the significance of incorporating unresolved fluctuations in drag force within computational fluid dynamics-definition model simulations. Prior research has emphasized the notable disparities between the drag forces exerted by individual particles and the average drag force (Akiki et al., 2016). However, this study represents one of the initial endeavours to explicitly incorporate these variations into the CFD-DEM model. The findings indicate that incorporating these variations can greatly enhance the accuracy of granular temperature forecasts, which is a crucial factor in characterizing the behavior of gas-solid fluxes (Hrenya & Sinclair, 1997). Nevertheless, the model does possess constraints, especially in extremely heterogeneous systems where unresolved structures impact the oscillations of drag force. Additional research is required to expand the model's scope to encompass these impacts and to evaluate its performance in extensive, industrially significant scenarios. However, this study signifies a significant advancement in the development of more accurate CFD-DEM simulations for gas-solid fluxes.

The process of FBG is a multifaceted phenomenon that encompasses the dynamics of multiphase flow, interactions between particles and fluid, as well as many mechanisms governing granule growth and breakup. The development of predictive models is of utmost importance in comprehending, managing, and enhancing processes. However, this task remains arduous due to the complex and multifaceted character of the process. In recent years, there has been a growing recognition of the efficacy of integrating CFD and DEM as a robust methodology for simulating gas-solid flow and particle dynamics within fluidized beds (Deen et al., 2007). CFD and DEM models have yielded significant findings about the impact of operating conditions and equipment geometry on fluidization behavior (Börner et al., 2017). Nevertheless, these models in isolation are incapable of forecasting the progression of particle size throughout the granulation process.

Fries et al. (2013) conducted a comprehensive CFD-DEM analysis to compare the behavior of particles and collisions in three different industrial FBG setups: top-spray, Wurster-coater, and spouted bed. An investigation was conducted to examine the impact of geometry on particle translational and rotational velocities, as well as collision frequencies and energy while keeping process parameters constant. The work offers vital insights into the micro-scale mechanisms that influence the formation and structure of granules in each arrangement. The top spray exhibited the lowest average particle velocity, wetting intensity, and collision frequency in the spray zone, suggesting minimal growth rates. The Wurster-coater exhibited a consistent flow of particles with high velocities of collision, but a low frequency of collisions in the spray zone. This indicates that the rate of particle clumping is low, as well as the rate of particle breakdown, resulting in rapid overall growth. The spouted bed exhibited a high level of gas-liquid-solid interaction, characterized by elevated average particle rotation, collision frequency, and breakage rates. This environment promoted the development of tightly packed, dense granules.

The experimental results confirmed the predictions made by the CFD-DEM model. The Wurster coater exhibited the highest growth rate, followed by the spouted bed, while the top spray had a significantly lower growth rate. The work showcases the capabilities of CFD-DEM models to establish a connection between granule size and structure with equipment design and collision dynamics, hence assisting in the optimization of processes (Fries et al., 2013).

Population balance modeling (PBM) is a widely recognized and proven methodology utilized to describe the dynamics of particle size in granulation systems (Immanuel & Doyle, 2003). Powder bed granulation (PBM) techniques have demonstrated successful application in both batch and continuous FBG processes (Chaudhury et al., 2014). The primary difficulty lies in creating kernels that are based on physical principles for different granulation rate processes such as aggregation, breaking, and consolidation. The combination of CFD-DEM and PBM presents a favorable solution by capitalizing on the advantages of each modeling approach. CFD-DEM offers comprehensive data regarding particle velocities, collisions, and residence times, hence enabling the utilization of mechanistic PBM kernels. The potential of this multiscale modeling methodology for FBG has been proven in several research (Sen et al., 2014).

A crucial factor in FBG modeling is the precise depiction of capillary forces and wetting behavior at the level of individual particles. et al. created a novel and effective continuum capillary force (CCF) model for determining capillary forces on particles within the color function framework for interface tracking in micro-scale CFD-DEM coupling. The CCF model enables the assessment of capillary forces, even in cases where the three-phase contact line is spread out due to numerical diffusion. The model was verified using theoretical values and demonstrated high accuracy, even when using relatively coarse meshes.

In their study, Washino et al. (2013) utilized the CFD-DEM coupling technique along with the CCF model to simulate the process of droplet penetration into stationary particle beds. This phenomenon plays a crucial role in the wetting and nucleation stage of FBG. The simulations accurately depicted the initial process of the droplet being drawn into the bed as a result of surface wetting, followed by its penetration driven by capillary pressure. The wetting dynamics were characterized by quantifying the total wet surface area of particles. In addition, the CFD-DEM model was employed to replicate the impact of droplets on a moving particle bed within a high-shear granulator (Washino et al., 2013). The simulations yielded droplet deformation and velocity decay characteristics that closely matched the experimental results documented in the literature. This indicates that the CFD-DEM coupling approach devised is dependable.

Tamrakar and Ramachandran (2019) propose a comprehensive modeling framework that combines CFD, digital elevation model (DEM), and particle belt method (PBM) to simulate a top-spray FBG process. Although there has been recent research on capturing particle movement and aggregation mechanisms in FBG, there has been little focus on connecting important process parameters to quality features. The primary aims of the study conducted by Tamrakar and Ramachandran (2019):

1. Develop a coupled CFD, DEM, and PBM framework for FBG that allows for the transfer of mechanistic data from CFD-DEM simulations to a compartmental PBM.
2. Incorporate mechanistic kernels into the process-based model (PBM) to represent the impact of key process parameters (CPPs) such as air flow rate, temperature, spray rate, etc., on critical quality attributes (CQAs) such as granule size and liquid content.
3. Verify the framework by conducting a quantitative comparison between simulation results and experimental data. The objective of this study is to establish a correlation between key process factors, such as inlet air flow rate and temperature, and the size and moisture content of granules.

The PBM mechanistic kernels receive inputs from CFD-DEM simulations conducted at various air flow rates. The integrated model demonstrates the capability to accurately forecast dynamic granule size distributions that exhibit strong agreement with empirical observations, effectively capturing the influence of crucial operational parameters. In the realm of rational design and optimization of FBG processes, the multiscale CFD-DEM-PBM approach exhibits considerable potential. Future research should consider the dynamics of spray and verify the accuracy of the model across a broader spectrum of operational circumstances.

The recent study conducted by Tamrakar and Ramachandran (2019) constitutes a noteworthy advancement in the field. The CFD-DEM-PBM framework utilized in this study integrates essential granulation mechanisms and intercompartment particle transport, enabling the prediction of the impact of process parameters on granule size and moisture. The strong concurrence seen with empirical data serves to validate the modeling methodology and underscores its efficacy in the realm of process design and optimization. Nevertheless, it is important to acknowledge several limits and potential avenues for future research. The existing model has a simplistic approach to the spray zone and does not explicitly consider the dynamics of droplets, which can influence the process of granule nucleation and growth (Fries et al., 2011). Enhancing the forecasting skills could be achieved by incorporating models that account for droplet trajectories, spreading, and absorption. Furthermore, conducting validation experiments using a broader spectrum of formulations and operating conditions would enhance the reliability and applicability of the modeling approach. Ultimately, the CFD-DEM-PBM methodology introduced by Tamrakar and Ramachandran (2019) represents a significant advancement in the field of computational modeling for FBG. Through additional improvements and verification, these multiscale models can function as potent instruments for the advancement, expansion, and regulation of processes, finally facilitating the creation of more effective and dependable granulation processes.

Kim and Chung (2019) proposed a new model for FBG that takes into account the glass transition temperature of the powder as a determining factor for particle stickiness and agglomeration. In addition, a model for evaporation was included. The cricket powder, which lacks inherent agglomeration properties, was combined with dextrin, a substance that demonstrates a glass transition temperature. The glass transition temperature of dextrin was utilized as the determining factor for agglomeration, with moisture content being the variable. The models and experiments showed similar patterns of agglomeration, indicating the capability to forecast the behavior of nucleation and the agglomeration process, which is influenced by the liquid.

Grohn et al. (2020) examined the movement of particles during the coating process in a rotor granulator using CFD and DEM. The simulation treated droplets as individual entities and a model of liquid bridges was used to account for capillary and viscous forces. The coating model encompassed the processes of droplet deposition, liquid transfer, and drying. When comparing dry conditions to liquid injection, the particle velocities and contact times were higher because the momentum transfer from the rotor plate to wet particles was improved. Raising the viscosity of the binder initially led to an increase in contact forces and particle velocities, but later caused a decrease in both factors as a result of higher energy dissipation.

The CFD-DEM model offered valuable insight into the particle's motion, their collisions, and the process of wetting under various operational circumstances. Increased air flow rates resulted in higher levels of bed expansion, particle velocities, and transfer between compartments. The frequency of collisions decreased, but the energy of collisions rose as the air velocity increased. The particles achieved a somewhat uniform wet state within approximately 10 s as a result of circulation. The hybrid CFD-DEM-PBM framework accurately predicted the dynamic evolution of granule size distribution, closely matching the experimental trends under various process circumstances. The main effects that were recorded were:

Increasing the spray rate of the binder results in accelerated development of the granules. Decreasing the temperature of the incoming air results in an increase in the size of the granules. An increased quantity of binder leads to the formation of smaller granules. The framework demonstrated strong predictive power, with an average error in anticipated granule size of less than 10% compared to experimental results. The model can accurately represent the variations in local wetting and drying circumstances when it is used on a larger-scale and can forecast the effects on the size of the granules.

This chapter shows how linked multiscale CFD-DEM-PBM frameworks are employed to describe the intricate relationship between fluid dynamics and particle interactions in FBG processes. The primary benefits include the capacity to:

1 Acquire essential knowledge about the behavior of particles in motion, their interactions during collisions, and the dynamics of wetting.
2 Establish a connection between important process parameters and granule quality features using mechanistic kernels.
3 Use quantitative methods to accurately forecast and confirm the changes in the distribution of granule sizes.

The inclusion of particle characteristics such as the glass transition temperature, the viscosity of the coating solution, and the precise depiction of capillary forces and wetting behavior through models like CCF enhances the capabilities of mechanistic modeling. These frameworks are useful tools for designing, optimizing, and scaling up science-based processes. Subsequent research should prioritize the development of models for particle breakage, the inclusion of the impact of nozzle spray dynamics, and the expansion to other FBG systems to enhance the reliability of the process.

22.5 Simulation of twin-screw granulation

TSG is a critical technology for the continuous production of solid pharmaceutical dosage forms, offering several advantages over batch granulation processes. However, the successful development and scale-up of TSG processes require a deep understanding of the complex interplay between equipment design, operating conditions, and material properties. Computational modeling tools, such as the DEM

22.5 Simulation of twin-screw granulation

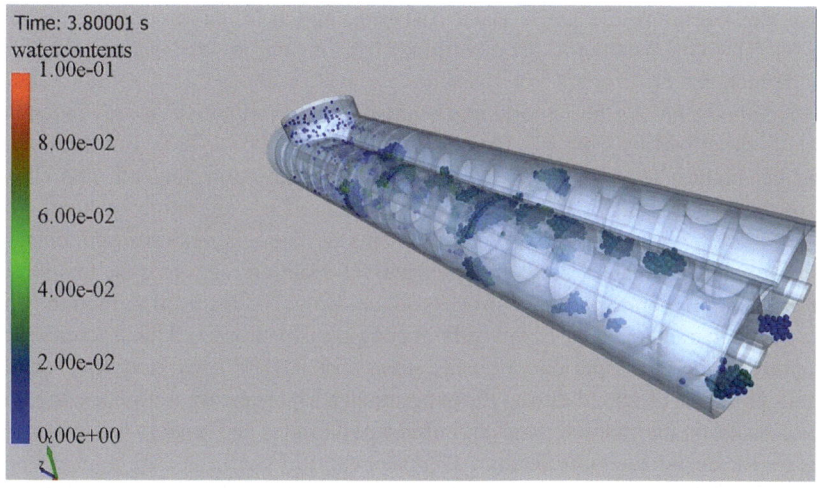

FIGURE 22.4

Typical DEM simulation of TSG wet granulation.

DEM simulation of a twin-screw granulation process.

are increasingly being used alongside experimental studies to gain insights into the TSG process and optimize its performance.

This chapter provides a comprehensive review of the state-of-the-art computational modeling of TSG, with a focus on DEM and PBM approaches. The chapter discusses the application of these models to analyze the impact of process parameters, equipment design, and material properties on TSG performance, as well as their integration with experimental characterization techniques for the development of predictive process models.

DEM simulations offer detailed information about the microscopic mechanisms involved in mixing, shearing, and wetting processes that control the formation and growth of granules. By integrating DEM with PBM, it is possible to accurately forecast the distribution of granule sizes and other characteristics across the TSG barrel.

22.5.1 DEM simulation twin-screw granulation

DEM simulations, along with experimental characterization techniques such as near-infrared (NIR) and Raman spectroscopy, facilitate the creation of predictive process models. These models can guide for making decisions on equipment selection, process optimization, and control system design. Promising areas for future research involve the incorporation of particle wetting and liquid distribution processes, as well as the utilization of high-performance computing for comprehensive simulations at a larger scale.

The DEM simulation of the twin-screw wet granulation example shown in Fig. 22.4 has been gaining prominence in recent years. Chen et al. (2022) offer valuable insights into the correct prediction of agglomerate breakup in DEM simulations by utilizing the Timoshenko beam bond model (TBBM).

The TBBM model considers the axial, shear, twisting, and bending properties of the bonds that connect particles. The bond parameters are determined by using empirical measurements of the material properties of the binder.

The TBBM method employs the Timoshenko beam theory to compute the forces and moments acting on each bond, as described by Eqs. (22.11)–(22.17) in Chen et al. (2022). Bond failure is assessed by comparing the highest levels of stress to the bond's compression, tension, and shear strengths, as indicated by Eqs. (22.25)–(22.27) in Chen et al. (2022). The calibration of bond characteristics is accomplished by utilizing experimental measurements of Young's modulus and tensile strength of the binder. The simulation results offer comprehensive information regarding the temporal and spatial development of the agglomerate breakage process. Chen et al. (2022) unveil a fracture zone that takes the shape of a cone, characterized by high levels of compressive stresses. This fracture zone is flanked by an arch-shaped region that experiences tensile stress. The TBBM accurately forecasted the precise sequence of failure modes observed during the experiments. The capacity to forecast the fragmentation of agglomerates based on the characteristics of individual particles and binders has substantial practical consequences for the design and optimization of processes.

Matsushita et al. (2022) investigate the influence of fill level in a twin-screw granulator on the characteristics of granules and tablets through a combination of experiments and DEM simulations. Comprehending the mechanisms that dictate the impact of fill level on product attributes is essential for optimizing continuous TSG. Experimental measurements were performed to evaluate the fill level and its influence on the size, flowability, and strength of the granules. Analyzing particle motion in the granulator, DEM simulations were conducted using the Johnson-Kendall-Roberts (JKR) theory and the Hertz-Mindlin contact model. The Rumpf equation was employed to approximate the interparticle adhesion force based on tablet hardness data.

The results of the DEM simulations were consistent with the experimental findings about the mixed diffusion coefficient at various screw speeds. The simulations demonstrated that the compressive force exerted on particles increased as the fill level grew, while it remained consistent at a fixed fill level, regardless of the screw speed. According to the hypothesized mechanism, a higher fill level increases the amount of force applied to the particles, resulting in greater strength of the granules and hardness of the tablets. Toson and Khinast (2023) conducted a study on refill techniques for a twin-screw feeder using a DEM model. They highlighted the need to consider the blending of past and current material batches to ensure precise material monitoring in continuous manufacturing lines. The residence time distribution (RTD) and mixing behavior of particles within a twin-screw feeder during a whole discharge process were analyzed using DEM simulations. Three predictive models were presented for analyzing behavior during many refill events: (1) a basic exponential RTD assuming complete mixing, (2) an RTD model derived from DEM results, and (3) particle-level material tracking utilizing a "relay race" methodology.

DEM analysis uncovered intricate flow patterns in the feeder. The analysis showed that material is discharged early from the area above the agitator, whereas material near the rotational axis of the agitator tends to become trapped and is discharged at lower fill levels. The simple perfect mixing model was determined to be a suitable approximation for refills up to 30% fill level, however, the RTD and relay race models yielded more precise outcomes for greater refill levels.

Mateo-Ortiz et al. (2021) examine the practicality of combining dry powders in the pre-melting areas of a twin-screw extruder using DEM modeling and NIR spectroscopy. The objective is to merge blending and extrusion/granulation procedures into a single device for uninterrupted manufacturing.

The utilization of DEM simulations with the Hertz-Mindlin contact model allowed for a detailed understanding of the mixing patterns occurring within the extruder at the individual particle level. An investigation was carried out utilizing a Coperion ZSK-18 extruder that was fitted with NIR spectroscopy to assess the uniformity of the blend. The mass of the sample analyzed by the NIR probe was determined using Eq. (22.28).

$$M = \rho \left[\pi \left(\frac{d}{2} \right)^2 + d \left(t_{acq} * V_{pow} \right) \right] * H \qquad (22.28)$$

where ρ represents bulk density of the sample, d represents the diameter of the NIR beam, t_{acq} represents the acquisition time, V_{pow} represents the linear velocity of the powders, and H represents the experimental depth of penetration of the NIR beam (Mateo-Ortiz et al., 2021). The twin-screw extruder successfully combined two powder feed streams at a small-scale dosage for the tested formulation including a cohesive active pharmaceutical ingredient (API). The RTD profiles of screw topologies consisting solely of conveying components exhibited characteristics similar to those of a plug flow reactor (PFR), whereas configurations with kneading devices displayed a closer resemblance to an ideal continuous stirred tank reactor (CSTR).

Cheng et al. (2022) examined the impact of twin-screw pulping technology on the performance of straw pulping. Cheng et al. (2022) utilized DEM and experimentally validated their findings using the Tavares mathematical model. A DEM model was created to simulate the breakage of straw particles. The model is based on the Tavares breakage model, which posits that the likelihood of breakage follows an upper-truncated log-normal distribution. The model considers the reduction in energy of breakage ratio caused by the accumulation of damage during particle impact. The Voronoi fracture equation is utilized to estimate the shape of the fragments, which are subsequently dispersed using the Gaudin-Schumann function. Through various simulations, it was shown that the casing combination of negative-positive-negative-positive (NPNP) and tooth groove angle arrangement of 45°–30°–15° led to the largest occurrence of broken straw particles. A Box-Behnken experimental methodology was employed to establish a mathematical model (Eq. 22.18a and 22.18b) that links pulp yield with tooth groove angle (A), screw speed (B), and straw moisture content (C). Validation experiments have verified that employing DEM and the Tavares model is appropriate for analyzing straw breakage in twin-screw pulping.

Karkala and Ramachandran (2022) created a detailed DEM model to investigate the mixing behavior of cohesive particles in a corotating twin-screw mixer (TSM). They employed the JKR contact model to simulate the cohesive forces between the particles. The JKR model utilizes fundamental equations to establish the relationship between normal contact force and the properties of particles and contacts. The Mindlin-Deresiewicz model is employed to calculate tangential forces. The characteristics of DEM particles were calibrated by angle of repose simulations for particles with good flow (GF) and poor flow (PF), and dynamic yield strength simulations for particles with very very poor flow (VVPF). Findings: The simulations yielded valuable information about how particles with varying flowabilities mix. GF, PF, and VVPF particles were analyzed to determine their mixing zones and mechanisms. The flowability of particles had a substantial impact on the steady-state holdup, mean residence time, and mixing efficiency. The authors proposed modifying screw designs according to the flowability of particles to optimize mixing efficiency and reduce demixing.

Zheng et al. (2022) conducted a study to examine the impact of particle shape on conveying properties in a full-scale twin-screw granulator. They utilized a GPU-enhanced DEM approach to simulate the TSG process and investigated the effects of different particle shapes, including sphere, cube, bilunabirotunda, and hexagonal prism. The Blaze DEM-GPU methodology was utilized, employing an explicit forward Eulerian time integration method on GPU to solve the particle motion equations. The precise approach of volume contact detection was utilized to determine the contact forces between particles. The Kelvin-Voigt linear viscoelastic spring dashpot model was used to determine normal forces, whereas the Cundall-Stark model (Eqs. [22.29] and [22.30]) was used to calculate tangential forces.

$$F_n = \left(K_n \Delta V^{\frac{1}{3}}\right)n - C_n(V_R * n)n \tag{22.29}$$

$$F_T = -K_T(V_T dt) - C_T V_T + F'_T \tag{22.30}$$

K_n represents the stiffness of the spring, n refers to the normal direction of the force, C_n represents the damping coefficient, and V_R represents the relative velocity between two particles in contact (Zheng et al., 2022). F'_T represents the component of the force that is parallel to the current plane. K_T refers to the stiffness of the spring in the same direction. C_T represents the damping coefficient in the same direction. V_T represents the relative velocity in the same direction (Zheng et al., 2022). Spherical particles demonstrated the fastest movement and shortest duration of stay, whereas polyhedral particles exhibited more elaborate movement patterns and longer durations of stay due to complex collisions between particles and between particles and walls. The characterization of residence time distribution was conducted by utilizing the average residence time, residence time variance, and cumulative exit age distribution function. The conveying qualities throughout the TSG process were considerably affected by the geometry of the particles. Spherical particles showed flow patterns that were closer to optimal plug flow, while cubic particles resembled perfect mixing flow. Polyhedral particles necessitated greater power consumption during the TSG process as a result of their more frequent and powerful collisions with the walls.

The use of computational modeling, particularly DEM has significantly advanced our understanding of the complex dynamics driving TSG processes. These modeling approaches have provided valuable insights into powder behaviors, granule formation and growth, and the impact of various factors on TSG performance. The integration of these models with experimental characterization techniques has facilitated the development of predictive process models, which can inform equipment selection, process optimization, and control system design.

However, there are still significant opportunities for further research and development in this field, such as coupling DEM with other modeling tools, utilizing high-performance computing platforms for full-scale equipment modeling, and integrating these modeling approaches into the product development and manufacturing workflow. As the pharmaceutical industry increasingly adopts continuous manufacturing and QbD principles, the continued advancement of TSG modeling will be crucial for improving product quality, reducing development time and cost, and enhancing process flexibility.

22.5.2 Couple simulation of the twin-screw granulation

TSG is a highly favorable method of continuous granulation due to its several benefits, including a brief time in which the material remains, effective distribution of liquid, and enhanced regulation of granule characteristics in comparison to batch granulation (Dhenge et al., 2010; Seem et al., 2015).

In order to implement QbD in the operation of TSG, it is necessary to create mechanistic models that can accurately explain the impact of process parameters and material qualities on the important quality attributes of the granules (Patil & Pethe, 2013). Population balance modeling (PBM) and DEM are the primary frameworks employed for modeling TSG, as discussed by Kumar et al. (2013) and Litster and Ennis (2004). The PBM method monitors the evolution of particle characteristics, such as size, liquid content, and porosity, by considering rate processes such as nucleation, aggregation, and breaking (Chaudhury et al., 2014). Nevertheless, PBM is dependent on semiempirical kernels that necessitate the use of experimental data for estimating unknown parameters (Barrasso et al., 2015b; Chaudhury et al., 2014). On the other hand, DEM is characterized by a mechanical approach that monitors the movement and pressures exerted on every single particle (Kumar et al., 2013). DEM is capable of providing comprehensive collision and velocity data (Gantt & Gatzke, 2005; Hassanpour et al., 2011; Yang et al., 2003), but it is unable to consider variations in particle size and is computationally demanding (Ketterhagen et al., 2009).

Recent research has concentrated on combining PBM and DEM methods to take advantage of the benefits of both approaches in TSG modeling (Barrasso et al., 2015b; Wang et al., 2012). In their study, Barrasso and Ramachandran (2015) introduced a bi-directional linkage between PBM and DEM, where DEM is used to determine collision frequencies and liquid distribution data for a 2-D PBM model that considers both particle size and liquid content. The PBM subsequently computes the net variations in the quantity of particles within each size/liquid category during a specific time period. These variations are then included into the DEM by generating or eliminating particles. The coupled model accurately represented the temporal rise in average particle size and the formation of a bimodal size distribution, in line with experimental findings (Barrasso & Ramachandran, 2015). Nevertheless, the complete PBM-DEM model remains computationally burdensome mostly because of the DEM component (Barrasso & Ramachandran, 2015).

In order to enhance efficiency, Barrasso et al. (2014) devised a reduced-order PBM-ANN model as an alternative. A trained artificial neural network (ANN) utilized data from DEM simulations to forecast collision rates by considering factors such as particle size, size distribution, and impeller speed. The collision frequencies predicted by the ANN were subsequently utilized in the aggregation kernel of a one-dimensional population balance model (PBM). The coupled PBM-ANN model demonstrated excellent concurrence with the whole PBM-DEM predictions while requiring significantly fewer computer resources. This allows for its application in dynamic simulations and optimization, as stated by Barrasso et al. (2015). Furthermore, Ismail et al. (2020) constructed a compartmental PBM (CPBM) of TSG that takes into consideration the spatial variability of the screw components. The screws were separated into transporting and kneading zones, with each zone characterized by its unique aggregation and breakage kernels. The PBM was solved using both finite volume and cell average approaches, and it was found that the finite volume method achieved superior mass conservation compared to the cell average method (Ismail et al., 2020).

The study by Ismail et al. (2020) and Kapur (1972) indicated that Kapur's kernel offers greater flexibility in terms of the aggregation rate compared to the sum kernel. The empirical kernel parameters were obtained by fitting the model to experimental size distribution data. These parameters were then correlated with TSG process parameters (liquid-to-solid ratio, screw speed) using kriging interpolation and polynomial regression techniques (Ismail et al., 2020). The CPBM accurately predicted the impact of liquid-to-solid ratio and screw speed on the distribution of granule sizes, in line with experimental results. It was found that the liquid-to-solid ratio had the most significant influence, which is consistent with the findings of Dhenge et al. (2010) and Meng et al. (2017). Combining mechanistic DEM information with semiempirical population balance model (PBM) allows for the creation of predictive models for the thermally stimulated growth (TSG) process (Barrasso & Ramachandran, 2015).

To enhance computational efficiency, one can train surrogate models such as ANNs using the DEM data (Barrasso et al., 2014). The PBM is able to capture the granulation mechanisms that vary in space by considering the screw configuration zones and establishing a relationship between kernel parameters and process inputs (Ismail et al., 2020). Coupled multiscale models enhance the mechanistic comprehension and model-based optimization of TSG. Additional research is required to experimentally validate the findings, including the inclusion of other mechanisms such as consolidation and distributive mixing (Kumar et al., 2014), and expanding the scope to include downstream processes such as drying and tableting.

22.6 DEM simulation of other granulation equipment

Seeded granulation was first introduced by Rahmanian et al. (2011) to describe the granulation process which intends to produce granules with a core structure that is made up of coarse particles and covered by fine particles. This granulation process has been simulated by Hassanpour et al. (2013) in batch Cyclomix granulators using DEM. In this work, an attempt is made to explore the process conditions that lead to the production and breaking of seeded granules. In this instance simulation, the Hertz-Mindlin contact model is utilized to model the bonding forces between particles and capillary and viscous forces as a result liquid binder was not taken into consideration in the modeling process. The result from the simulation indicates that, during the granulation process, seeded granules increase and shrink continuously. The simulation findings demonstrate that a high percentage of seeded granules are covered by fine particles to about 60% coverage. At high rotational velocities, seeded granules with extensive surface coverage are frequently observed, based on additional quantitative analyses. This implies that there is a robust connection between the rate of rotation and the production of seeded granules. Further research is needed to determine the most important conditions, such as the Stokes number, which are required to improve seeded granulation; however, the work has shown the potential of DEM for modeling granulation processes and exploring the fundamental processes of granulation at large and seeded granulation in particular (Hassanpour et al., 2013).

The work by Hassanpour et al. (2013) has been continued by Behjani et al. (2017) but in this case in a drum granulator by including MATLAB code the calculate the surface coverage of the seeded granules. Following the trend of Hassanpour et al. (2013), the Hertz-Mindlin contact model was used as the base model for computing the contact between the particles. A dimensionless number called Cohesion number based on "work of cohesion" which is the required energy needed to break two bonding particles was developed as

$$Cohession number (Coh) = \frac{work of cohession}{gravitation potential energy} \tag{22.31}$$

$$Coh = \frac{1}{\rho g}\left(\frac{\Gamma^5}{E^{*2} R^{*8}}\right)^{\frac{1}{3}} \tag{22.31a}$$

Where R^* is the equivalent radius, g is the gravitational potential energy E^* is the equivalent young modulus and Γ is the particle's surface energy. The simulation suggested that a higher seed-fine size ratio of about 8:1 increases the possibility of producing seeded granules. It was evident that surface energies of the particles play a crucial role in the formation of seeded granules, as low sow surface energy of 0.5 J/m² did not produce seeded granules whereas surface energies of 1.5 and 3.0 J/m² showed promising signs of seeding and above these values of surface energies produced coagulations of granules. These values are case-specific to these simulations and cannot serve as guide values for other works as there is no calibration of the particles used in this simulation and no form of experimental validation provided to back these values. Also, increasing drum rotation speed had a positive correlation with the seed coverage. It was found that continuous granulators with baffles have a reasonable potential for application in the seeded granulation process. It is also observed that particles tend to separate within the basic drum and stick together after segregation, but the scooped granulator successfully uniforms the granulation (Behjani et al., 2017).

Modeled simulation of the single screw extruder has been performed by Arthur et al. (2024) for granulation purposes using DEM. The DEM model employed the liquid bridge model to investigate the effect of liquid in single screw granulation. The results from the DEM simulation were compared with experimental results. The research suggests that increasing binder content results in a more homogenous product with fewer fine particles (710–1400 μm). Lower binder concentrations provide uneven products with more particulates. In formulations with low liquid-to-solid (L/S) ratios, fine granules with fragility and a wide GSD are produced more often, according to experimental results. In limiting the creation of smaller granules and increasing granule size, liquid content is more important than concentration. Similarly, granule strength depends more on liquid content than concentration (Arthur et al., 2024).

More DEM simulations show that contact forces increase L/S ratios. In the experimental study, a high number of particles coupled with bound particles indicates fine development. Higher L/S ratios correlate with worse bonds and more contacts. This link can be explained by the chance of fewer particles at the parameter measurement point. Granule strength is directly proportional to the contact force in the simulation, indicating that contact force magnitude affects granule strength (Arthur et al., 2024).

Another iron ore granulation in the drum granulator has been modeled using DEM numerical simulation to investigate the kinetics of iron ore granulation by Kano et al. (2006). The spring and dashpot methods were considered for the DEM model as opposed to JKR. Also, liquid (water) properties were not factored into this model. The DEM was used to calculate the impact energy and rotational kinetic energy of the granules. It established a correlation for what they termed granulation energy to be a result of the impact energy and the rotational kinetic energy.

The granulation energy is defined as a function of the granulation index G', the granulation index is given as

$$G' = \frac{E_r}{E_i} \quad (22.32)$$

where E_r and E_i are the granule rotation and impact energies given as

$$E_r = \frac{1}{2}I\omega^2 \quad (22.32a)$$

and

$$E_i = \sum \frac{1}{2}mv_r^2 \quad (22.32b)$$

where I is the inertia, m is the mass of the granule, v and ω are the relative velocity and angular velocity of the granule. The granulation energy G is there given as

$$G = \frac{E_r}{E_i} E_r \tag{22.33}$$

It was observed from the DEM simulation that, an increase in drum rotation speed increases the granulation energy which intends to decrease the charge ratio of the granules. Also, increasing the drum diameter decreases the granulation due to an increase in the impact energy.

A numerical simulation of the continuous drum granulation using DEM has been performed by (Soda et al., 2009) which discusses the validity of the numerical simulation method for particulate, concerning the occupation ratio and retention time. In addition, the influence of drum angle and length on the behavior of granules is investigated. The simulation method where the granulated particle is treated as one particle is employed to investigate the granulation mechanism with the following assumptions adopted for modeling the simulation model.

1. Granules are considered a single particle.
2. Spherical shaped particle.
3. Uniform particle radius.
4. In the simulation, particle diameter and other parameters do not change.
5. The impact of moisture is disregarded.

The influence of different equipment lengths and orientations was experimented with and compared with simulation results. The quantity of granules increases significantly during the initial phase and subsequently reaches a state of saturation. This condition is known as a stable state. As the inclination of the mixer increases, the quantity of particulates within the mixer diminishes. The duration required to attain a steady state in a mixer increases as the gradient angle decreases. It is reported in both simulation and experiment a decrease in retention time with a lowing gradient angle and vice versa. The model's validity was assessed through a comparison with the experimental outcomes, specifically about the granules' retention time and occupation ratio. By utilizing the simulation model, one can analyze the granulation behavior of the raw materials, devise effective granulation equipment, and optimize the granulation parameters (Soda et al., 2009).

22.7 Conclusion

This chapter has examined the recent advancements in computational modeling and simulation of different granulation processes, specifically emphasizing the DEM and population balance modeling (PBM) techniques. The evaluation encompasses a diverse array of granulation equipment, such as high-shear granulators, fluidized bed granulators, twin-screw granulators, and other specialized granulation systems.

The utilization of the DEM has demonstrated its efficacy as a potent instrument for acquiring a profound understanding of intricate interactions between particles at a microscopic level, as well as the processes of mixing and production of granules in granulation processes. DEM simulations offer important insights into particle velocities, collision frequencies, and force distributions. This information may be utilized to enhance the design of equipment and optimize operating conditions. The combination of the DEM with the population balance model (PBM) has become a highly promising method for accurately predicting the distribution of granule sizes and other important quality characteristics. This is achieved by integrating mechanistic kernels, which are produced from DEM data, into the PBM framework. Multiple significant research has been emphasized, showcasing the effective use of the DEM and coupled DEM-PBM models in different granulation systems. The studies mentioned encompass the analysis of coalescence kernels in high shear granulation (Gantt et al., 2006), the examination of the influence of impeller design on shear forces and energy input (Börner et al., 2016), the investigation of the impact of binder addition methods (Tamrakar et al., 2019), and the modeling of particle flow and mixing in vertical high-shear mixers using coarse-grained DEM (Kishida et al., 2021).

This chapter examines the use of CFD-DEM-PBM frameworks in FBG to accurately represent the intricate relationship between fluid dynamics and particle interactions (Tamrakar & Ramachandran, 2019). Studies have demonstrated that including particle-level phenomena, such as the glass transition temperature and capillary forces, improves the ability of these models to make accurate predictions (Kim & Chung, 2019; Washino et al., 2013).

This chapter emphasize the significance of DEM simulations in comprehending the influence of process factors, equipment design, and material qualities on the performance of TSG. The combination of DEM with population balance model (PBM) and experimental characterization approaches has enabled the creation of accurate process models that may be used to assist in equipment selection, process optimization, and control system design (Barrasso & Ramachandran, 2015). This chapter also discusses the utilization of DEM in various specialized granulation systems, including seeded granulation in both batch and continuous granulators (Behjani et al., 2017; Hassanpour et al., 2013) and single screw granulation (Arthur et al., 2024). These studies showcase the adaptability and capacity of DEM in simulating different granulation processes and investigating fundamental granulation mechanisms. This chapter highlights the notable progress made in computer modeling and simulation of granulation processes through the utilization of the DEM and coupled DEM-PBM techniques. These modeling techniques are extremely important for comprehending the intricate relationship between process parameters, equipment design, and material qualities. This understanding eventually enables the optimization and control of granulation processes. The ongoing advancement and implementation of advanced modeling techniques will have a vital impact on enhancing product quality, reducing development time and cost, and improving process flexibility and robustness. This is particularly significant as industries such as pharmaceuticals adopt continuous manufacturing and QbD principles.

Although there has been considerable progress in using computational modeling and simulation to study granulation processes, there is a noticeable lack of research on the use of CFD and coupled CFD-DEM simulations for twin-screw wet granulation. The pharmaceutical industry has shown growing interest in TSG because of its ability to enable continuous manufacturing and enhance control over granule qualities (Seem et al., 2015). Nevertheless, the intricate interaction among the screw shape, operating parameters, and multiphase flow in twin-screw granulators poses considerable difficulties for modeling and simulation.

The limited presence of CFD and CFD-DEM investigations in the field of twin-screw wet granulation can be ascribed to various factors:

- *Challenging geometry*: The intricate shape of the screws and the dynamic behavior of the granulation process in twin-screw granulators make it difficult to create a mesh for CFD and define boundary conditions.

- *Multiphase flow*: TSG involve the presence of solid, liquid, and gas phases. To fully understand the interactions between these phases and the granulation mechanisms, advanced multiphase flow models are necessary.
- The computational expense of twin-screw granulator simulations using CFD and CFD-DEM is high because of the intricate geometry, multiphase flow, and the requirement to accurately model particle-level interactions.
- Insufficient experimental validation: The absence of extensive experimental data on twin-screw wet granulation impedes the verification and improvement of CFD and CFD-DEM models.
- *Multiple rotation zones*: TSG have two distinct rotation zones, each with its own unique rotational axis. This adds complexity to the modeling process, as the CFD and CFD-DEM models need to accurately capture the rotation and interaction of the screws in these zones. The presence of multiple rotation zones requires specialized numerical techniques and boundary conditions to ensure the stability and accuracy of the simulations.

Notwithstanding these difficulties, the advancement of CFD and CFD-DEM models for twin-screw wet granulation is essential for acquiring a more profound comprehension of the procedure and optimizing granule characteristics. CFD simulations can offer useful information about the flow patterns, mixing, and distribution of liquid binder in twin-screw granulators. On the other hand, CFD-DEM models can accurately represent the interactions between particles and the mechanisms of granule growth. The progress in computational resources, multiphase flow models, and experimental characterization methodologies is anticipated to expedite the creation of dependable CFD and CFD-DEM models for twin-screw wet granulation in the foreseeable future.

This chapter showcases the notable advancements achieved in the field of computer modeling and simulation of granulation processes. It underscores the importance of doing more research specifically focused on CFD and CFD-DEM modeling of twin-screw wet granulation. With the pharmaceutical sector increasingly adopting continuous manufacturing and quality-by-design principles, the creation of predictive models for TSG will be vital for optimizing processes and ensuring product quality. The modeling process of TSgs becomes more complex due to the existence of several rotation zones. To provide accurate and reliable simulations, specialized numerical approaches and boundary conditions are necessary.

Acknowledgment

The authors would like to express their gratitude to the Ghana Scholarship Secretariat for their indispensable support, which has enabled this research to be conducted.

References

Abu Bakar, N. F., Anzai, R., & Horio, M. (2013). Microscopic evaluation of binderless granulation in a pressure swing granulation fluidized bed. *Chemical Engineering Science, 98*, 51–58. https://doi.org/10.1016/j.ces.2013.04.048.

Akiki, G., Jackson, T. L., & Balachandar, S. (2016). Force variation within arrays of monodisperse spherical particles. *Physical Review Fluids, 1*(4). https://doi.org/10.1103/PhysRevFluids.1.044202.

Alder, Berni, J., & Thomas Everett, Wainwright. (1959). Studies in molecular dynamics. I. General method. *The Journal of Chemical Physics, 31*(2), 459–466.

Arthur, T. B., Chauhan, J., & Rahmanian, N. (2022). Process simulation of fluidized bed granulation: Effect of process parameters on granule size distribution. *Chemical Engineering Transactions, 95*, 241–246. https://doi.org/10.3303/CET2295041.

Arthur, T. B., Sekyi, N. K. G., & Rahmanian, N. (2024). DEM simulation of a single screw granulator: The effect of liquid binder on granule properties. *Chemical Engineering Research and Design, 203*, 233–242. https://doi.org/10.1016/j.cherd.2024.01.028.

Bandari, S., Nyavanandi, D., Kallakunta, V. R., Janga, K. Y., Sarabu, S., Butreddy, A., & Repka, M. A. (2020). Continuous twin screw granulation—An advanced alternative granulation technology for use in the pharmaceutical industry. *International Journal of Pharmaceutics, 580*. https://doi.org/10.1016/j.ijpharm.2020.119215.

Barrasso, D., El Hagrasy, A., Litster, J. D., & Ramachandran, R. (2015a). Multi-dimensional population balance model development and validation for a twin screw granulation process. *Powder Technology, 270*, 612–621. https://doi.org/10.1016/j.powtec.2014.06.035.

Barrasso, D., Eppinger, T., Pereira, F. E., Aglave, R., Debus, K., Bermingham, S. K., & Ramachandran, R. (2015b). A multi-scale, mechanistic model of a wet granulation process using a novel bi-directional PBM-DEM coupling algorithm. *Chemical Engineering Science, 123*, 500–513. https://doi.org/10.1016/j.ces.2014.11.011.

Barrasso, D., & Ramachandran, R. (2015). Multi-scale modeling of granulation processes: Bi-directional coupling of PBM with DEM via collision frequencies. *Chemical Engineering Research and Design, 93*, 304–317. https://doi.org/10.1016/j.cherd.2014.04.016.

Barrasso, D., Tamrakar, A., & Ramachandran, R. (2014). A reduced order PBM-ANN model of a multi-scale PBM-DEM description of a wet granulation process. *Chemical Engineering Science, 119*, 319–329. https://doi.org/10.1016/j.ces.2014.08.005.

Behjani, M. A., Rahmanian, N., Fardina bt Abdul Ghani, N., & Hassanpour, A. (2017). An investigation on process of seeded granulation in a continuous drum granulator using DEM. *Advanced Powder Technology, 28*(10), 2456–2464. https://doi.org/10.1016/j.apt.2017.02.011.

Bird, R. B. (2002). Transport phenomena. *Applied Mechanics Reviews, 55*(1), RR1–RR4.

Blazek, J. (2015). *Computational fluid dynamics: Principles and applications*. Butterworth-Heinemann.

Boerefijn, R., & Hounslow, M. J. (2005). Studies of fluid bed granulation in an industrial R&D context. *Chemical Engineering Science, 60*(14), 3879–3890. https://doi.org/10.1016/j.ces.2005.02.021.

Bokkers, G. A., van Sint Annaland, M., & Kuipers, J. A. M. (2004). Mixing and segregation in a bidisperse gas–solid fluidised bed: A numerical and experimental study. *Powder Technology, 140*(3), 176–186. https://doi.org/10.1016/j.powtec.2004.01.018.

Boon, C. W., Houlsby, G. T., & Utili, S. (2012). A new algorithm for contact detection between convex polygonal and polyhedral particles in the discrete element method. *Computers and Geotechnics, 44*, 73–82. https://doi.org/10.1016/j.compgeo.2012.03.012.

Börner, M., Bück, A., & Tsotsas, E. (2017). DEM-CFD investigation of particle residence time distribution in top-spray fluidised bed granulation. *Chemical Engineering Science, 161*, 187–197. https://doi.org/10.1016/j.ces.2016.12.020.

Börner, M., Michaelis, M., Siegmann, E., Radeke, C., & Schmidt, U. (2016). Impact of impeller design on high-shear wet granulation. *Powder Technology, 295*, 261–271. https://doi.org/10.1016/j.powtec.2016.03.023.

Burggraeve, A., Monteyne, T., Vervaet, C., Remon, J. P., & Beer, T. D. (2013). Process analytical tools for monitoring, understanding, and control of pharmaceutical fluidized bed granulation: A review. *European Journal of Pharmaceutics and Biopharmaceutics, 83*(1), 2–15. https://doi.org/10.1016/j.ejpb.2012.09.008.

Chaturbedi, A., Bandi, C. K., Reddy, D., Pandey, P., Narang, A., Bindra, D., Tao, L., Zhao, J., Li, J., Hussain, M., & Ramachandran, R. (2017). Compartment based population balance model development of a high shear wet granulation process via dry and wet binder addition. *Chemical Engineering Research and Design, 123*, 187–200. https://doi.org/10.1016/j.cherd.2017.04.017.

Chaudhury, A., Wu, H., Khan, M., & Ramachandran, R. (2014). A mechanistic population balance model for granulation processes: Effect of process and formulation parameters. *Chemical Engineering Science, 107*, 76–92. https://doi.org/10.1016/j.ces.2013.11.031.

Chen, X., Wang, L. G., Morrissey, J. P., & Ooi, J. Y. (2022). DEM simulations of agglomerates impact breakage using Timoshenko beam bond model. *Granular Matter, 24*(3). https://doi.org/10.1007/s10035-022-01231-9.

Cheng, H., Gong, Y., Zhao, N., Zhang, L., Lv, D., & Ren, D. (2022). Simulation and experimental validation on the effect of twin-screw pulping technology upon straw pulping performance based on tavares mathematical model. *Processes, 10*(11), 2336. https://doi.org/10.3390/pr10112336.

Chung, T. J. (2002). *Computational fluid dynamics*. Cambridge University Press.

Cleary, P. W., & Sawley, M. L. (2002). DEM modelling of industrial granular flows: 3D case studies and the effect of particle shape on hopper discharge. *Applied Mathematical Modelling, 26*(2), 89–111. https://doi.org/10.1016/S0307-904X.(01)00050-6.

Coetzee, C. J. (2017). Review: Calibration of the discrete element method. *Powder Technology, 310*, 104–142. https://doi.org/10.1016/j.powtec.2017.01.015.

Cundall, P. A., & Strack, O. D. L. (1979). A discrete numerical model for granular assemblies. *Géotechnique, 29*(1), 47–65. https://doi.org/10.1680/geot.1979.29.1.47.

Dadvand, A., & Motlagh, S. Y. (2021). Introduction on principle of computational fluid dynamics. In *Current trends and future developments on (bio-) membranes: Techniques of computational fluid dynamic (CFD) for development of membrane technology* (pp. 1–26). Elsevier. https://doi.org/10.1016/B978-0-12-822294-2.00006-0.

Darelius, A., Rasmuson, A., van Wachem, B., Niklasson Björn, I., & Folestad, S. (2008). CFD simulation of the high shear mixing process using kinetic theory of granular flow and frictional stress models. *Chemical Engineering Science, 63*(8), 2188–2197. https://doi.org/10.1016/j.ces.2008.01.018.

Deen, N. G., Van Sint Annaland, M., Van der Hoef, M. A., & Kuipers, J. A. M. (2007). Review of discrete particle modeling of fluidized beds. *Chemical Engineering Science, 62*(1–2), 28–44. https://doi.org/10.1016/j.ces.2006.08.014.

Dhenge, R. M., Fyles, R. S., Cartwright, J. J., Doughty, D. G., Hounslow, M. J., & Salman, A. D. (2010). Twin screw wet granulation: Granule properties. *Chemical Engineering Journal, 164*(2–3), 322–329. https://doi.org/10.1016/j.cej.2010.05.023.

Diez, E., Meyer, K., Bück, A., Tsotsas, E., & Heinrich, S. (2018). Influence of process conditions on the product properties in a continuous fluidized bed spray granulation process. *Chemical engineering research and design, 139*, 104–115.

Ding, J., & Gidaspow, D. (1990). A bubbling fluidization model using kinetic theory of granular flow. *AIChE Journal, 36*(4), 523–538. https://doi.org/10.1002/aic.690360404.

Di Renzo, A., & Di Maio, F. P. (2004). Comparison of contact-force models for the simulation of collisions in DEM-based granular flow codes. *Chemical Engineering Science, 59*(3), 525–541. https://doi.org/10.1016/j.ces.2003.09.037.

Dosta, M., Antonyuk, S., & Heinrich, S. (2012). Multiscale simulation of the fluidized bed granulation process. *Chemical Engineering and Technology, 35*(8), 1373–1380. https://doi.org/10.1002/ceat.201200075.

Fan, R., Zhao, M., Luo, L., Wang, Y., Zhou, K., Liu, Z., Zhou, Y., Guan, T., Sun, H., & Dai, C. (2023). Investigation of mixing homogeneity of binary particle systems in high-shear wet granulator by DEM. *Drug Development and Industrial Pharmacy, 49*(2), 179–188. https://doi.org/10.1080/03639045.2023.2194993.

Faure, A., York, P., & Rowe, R. C. (2001). Process control and scale-up of pharmaceutical wet granulation processes: a review. *European Journal of Pharmaceutics and Biopharmaceutics, 52*(3), 269–277.

Ferziger, J. H., Perić, M., & Street, R. L. (2019). *Computational methods for fluid dynamics* (pp. 1–596). Springer. https://doi.org/10.1007/978-3-319-99693-6.

Freireich, B., Li, J., Litster, J., & Wassgren, C. (2011). Incorporating particle flow information from discrete element simulations in population balance models of mixer-coaters. *Chemical Engineering Science, 66*(16), 3592–3604. https://doi.org/10.1016/j.ces.2011.04.015.

Fries, L., Antonyuk, S., Heinrich, S., Dopfer, D., & Palzer, S. (2013). Collision dynamics in fluidised bed granulators: A DEM-CFD study. *Chemical Engineering Science, 86*, 108–123. https://doi.org/10.1016/j.ces.2012.06.026.

Fries, L., Antonyuk, S., Heinrich, S., & Palzer, S. (2011). DEM-CFD modeling of a fluidized bed spray granulator. *Chemical Engineering Science, 66*(11), 2340–2355. https://doi.org/10.1016/j.ces.2011.02.038.

Gantt, J. A., Cameron, I. T., Litster, J. D., & Gatzke, E. P. (2006). Determination of coalescence kernels for high-shear granulation using DEM simulations. *Powder Technology, 170*(2), 53–63. https://doi.org/10.1016/j.powtec.2006.08.002.

Gantt, J. A., & Gatzke, E. P. (2005). High-shear granulation modeling using a discrete element simulation approach. *Powder Technology, 156*(2–3), 195–212. doi:10.1016/j.powtec.2005.04.012.

Edem, A., The Hertz-Mindlin with JKR Model. (2023). Altair Engineering Inc. ⟨16:italic ⟩https://help.altair.com/edem/Creator/Physics/Base_Models/Hertz-Mindlin_with_JKR.htm⟨/16:italic⟩ Accessed 26/03/2024.

Garg, Rahul, Galvin-Carney, Janine, Li, Tingwen, & Pannala, Sreekanth. *Documentation of open-source MFIX–DEM software for gas-solids flows. Tingwen Li Dr.* https://doi.org/10.1016/j.powtec.2011.09.019.

Gidaspow, D. (1994). *Multiphase flow and fluidization: Continuum and kinetic theory descriptions.* Academic Press.

Govender, N., Wilke, D. N., & Kok, S. (2015). Blaze-DEMGPU: Modular high performance DEM framework for the GPU architecture. *SoftwareX, 5*, 62–66. https://doi.org/10.1016/j.softx.2016.04.004.

Govender, N., Wilke, D. N., & Kok, S. (2016). Blaze-DEMGPU: Modular high performance DEM framework for the GPU architecture. *SoftwareX, 5*, 62–66.

Grohn, P., Lawall, M., Oesau, T., Heinrich, S., & Antonyuk, S. (2020). CFD-DEM simulation of a coating process in a fluidized bed rotor granulator. *Processes, 8*(9). https://doi.org/10.3390/pr8091090.

Hancock, B. C., York, P., & Rowe, R. C. (1994). An assessment of substrate-binder interactions in model wet masses. 1: Mixer torque rheometry. *International Journal of Pharmaceutics, 102*(1), 167–176.

Hapgood, K. P., Litster, J. D., Biggs, S. R., & Howes, T. (2002). Drop penetration into porous powder beds. *Journal of Colloid and Interface Science, 253*(2), 353–366. https://doi.org/10.1006/jcis.2002.8527.

Hapgood, K. P., Litster, J. D., & Smith, R. (2003). Nucleation regime map for liquid bound granules. *AIChE Journal, 49*(2), 350–361. https://doi.org/10.1002/aic.690490207.

Hassanpour, A., Pasha, M., Susana, L., Rahmanian, N., Santomaso, A. C., & Ghadiri, M. (2013). Analysis of seeded granulation in high shear granulators by discrete element method. *Powder Technology, 238*, 50–55. https://doi.org/10.1016/j.powtec.2012.06.028.

Hassanpour, A., Tan, H., Bayly, A., Gopalkrishnan, P., Ng, B., & Ghadiri, M. (2011). Analysis of particle motion in a paddle mixer using discrete element method (DEM). *Powder Technology, 206*(1–2), 189–194. https://doi.org/10.1016/j.powtec.2010.07.025.

Heinrich, S., Peglow, M., Ihlow, M., Henneberg, M., & Mörl, L. (2002). Analysis of the start-up process in continuous fluidized bed spray granulation by population balance modelling. *Chemical Engineering Science, 57*(20), 4369–4390. https://doi.org/10.1016/S0009-2509.(02)00352-4.

Hoomans, B. P. B., Kuipers, J. A. M., Briels, W. J., & van Swaaij, W. P. M. (1996). Discrete particle simulation of bubble and slug formation in a two-dimensional gas-fluidised bed: A hard-sphere approach. *Chemical Engineering Science, 51*(1), 99–118. https://doi.org/10.1016/0009-2509.(95)00271-5.

Hounslow, M. J., Oullion, M., & Reynolds, G. K. (2009). Kinetic models for granule nucleation by the immersion mechanism. *Powder Technology, 189*(2), 177–189. https://doi.org/10.1016/j.powtec.2008.04.008.

Hrenya, C. M., & Sinclair, J. L. (1997). Effects of particle-phase turbulence in gas-solid flows. *AIChE Journal, 43*(4), 853–869. https://doi.org/10.1002/aic.690430402.

Immanuel, C. D., & Doyle, F. J. (2003). Computationally efficient solution of population balance models incorporating nucleation, growth and coagulation: Application to emulsion polymerization. *Chemical Engineering Science, 58*(16), 3681–3698. https://doi.org/10.1016/S0009-2509.(03)00216-1.

Ismail, H. Y., Shirazian, S., Singh, M., Whitaker, D., Albadarin, A. B., & Walker, G. M. (2020). Compartmental approach for modelling twin-screw granulation using population balances. *International Journal of Pharmaceutics, 576*. https://doi.org/10.1016/j.ijpharm.2019.118737.

Iveson, S. M., Litster, J. D., Hapgood, K., & Ennis, B. J. (2001a). Nucleation, growth and breakage phenomena in agitated wet granulation processes: A review. *Powder Technology, 117*(1–2), 3–39. https://doi.org/10.1016/S0032-5910.(01)00313-8.

Iveson, S. M., Wauters, P. A. L., Forrest, S., Litster, J. D., Meesters, G. M. H., & Scarlett, B. (2001b). Growth regime map for liquid-bound granules: Further development and experimental validation. *Powder Technology, 117*(1–2), 83–97. https://doi.org/10.1016/S0032-5910.(01)00317-5.

Johnson, K. L. (1982). One hundred years of Hertz contact. *Proceedings of the Institution of Mechanical Engineers, 196*(1), 363–378. https://doi.org/10.1243/pime_proc_1982_196_039_02.

Kafui, D. K., & Thornton, C. (2008). Fully-3D DEM simulation of fluidised bed spray granulation using an exploratory surface energy-based spray zone concept. *Powder Technology, 184*(2), 177–188. https://doi.org/10.1016/j.powtec.2007.11.038.

Kapur, P. C. (1972). Kinetics of granulation by non-random coalescence mechanism. *Chemical Engineering Science, 27*(10), 1863–1869. https://doi.org/10.1016/0009-2509.(72)85048-6.

Kariuki, W. I. J., Freireich, B., Smith, R. M., Rhodes, M., & Hapgood, K. P. (2013). Distribution nucleation: Quantifying liquid distribution on the particle surface using the dimensionless particle coating number. *Chemical Engineering Science, 92*, 134–145. https://doi.org/10.1016/j.ces.2013.01.010.

Karkala, S., & Ramachandran, R. (2022). Investigating the effects of material properties on the mixing dynamics of cohesive particles in a twin screw mixer using a discrete element method approach. *Powder Technology, 409*, 117762. https://doi.org/10.1016/j.powtec.2022.117762.

Ketterhagen, W. R., Am Ende, M. T., & Hancock, B. C. (2009). Process modeling in the pharmaceutical industry using the discrete element method. *Journal of Pharmaceutical Sciences, 98*(2), 442–470. https://doi.org/10.1002/jps.21466.

Kim, J. E., & Chung, Y. M. (2019). CFD-DEM Simulation of the fluidized-bed granulation of food powders. *Biotechnology and Bioprocess Engineering, 24*(1), 191–205. https://doi.org/10.1007/s12257-018-0382-6.

Kishida, N., Nakamura, H., Takimoto, H., Ohsaki, S., & Watano, S. (2021). Coarse-grained discrete element simulation of particle flow and mixing in a vertical high-shear mixer. *Powder Technology, 390*, 1–10. https://doi.org/10.1016/j.powtec.2021.05.028.

Knight, P. (2004). Challenges in granulation technology. *Powder technology, 140*(3), 156–162.

Kobayashi, T., Tanaka, T., Shimada, N., & Kawaguchi, T. (2013). DEM-CFD analysis of fluidization behavior of Geldart Group A particles using a dynamic adhesion force model. *Powder Technology, 248*, 143–152. https://doi.org/10.1016/j.powtec.2013.02.028.

Kulju, T., Paavola, M., Spittka, H., Keiski, R. L., Juuso, E., Leiviskä, K., & Muurinen, E. (2016). Modeling continuous high-shear wet granulation with DEM-PB. *Chemical Engineering Science, 142*, 190–200. https://doi.org/10.1016/j.ces.2015.11.032.

Kumar, A., Gernaey, K. V., Beer, T. D., & Nopens, I. (2013). Model-based analysis of high shear wet granulation from batch to continuous processes in pharmaceutical production – A critical review. *Journal of Pharmaceutics and Biopharmaceutics, 85*(3), 814–832. https://doi.org/10.1016/j.ejpb.2013.09.013.

Kumar, A., Vercruysse, J., Toiviainen, M., Panouillot, P. E., Juuti, M., Vanhoorne, V., Vervaet, C., Remon, J. P., Gernaey, K. V., De Beer, T., & Nopens, I. (2014). Mixing and transport during pharmaceutical twin-screw wet granulation: Experimental analysis via chemical imaging. *European Journal of Pharmaceutics and Biopharmaceutics, 87*(2), 279–289. https://doi.org/10.1016/j.ejpb.2014.04.004.

Kundu, P. K., Cohen, I. M., & Dowling, D. R. (2015). *Fluid mechanics*. Academic Press.

Lacey, P. M. C. (1954). Developments in the theory of particle mixing. *Journal of Applied Chemistry, 4*(5), 257–268. https://doi.org/10.1002/jctb.5010040504.

Leuenberger, H. (2001). New trends in the production of pharmaceutical granules: The classical batch concept and the problem of scale-up. *European Journal of Pharmaceutics and Biopharmaceutics, 52*(3), 279–288. https://doi.org/10.1016/S0939-6411.(01)00200-4.

Li, J., Freireich, B., Wassgren, C., & Litster, J. D. (2012). A general compartment-based population balance model for particle coating and layered granulation. *AIChE Journal, 58*(5), 1397–1408.

Litster, J., & Ennis, B. J. (2004). *The science and engineering of granulation processes*: 15. Springer.

Litster, J. D., Hapgood, K. P., Michaels, J. N., Sims, A., Roberts, M., & Kameneni, S. K. (2002). Scale-up of mixer granulators for effective liquid distribution. *Powder Technology, 124*(3), 272–280. doi:10.1016/S0032-5910 (02)00023-2 00325910.

Liu, B., Wang, J., Zeng, J., Zhao, L., Wang, Y., Feng, Y., & Du, R. (2021). A review of high shear wet granulation for better process understanding, control and product development. *Powder Technology, 381*, 204–223. https://doi.org/10.1016/j.powtec.2020.11.051.

Liu, H., & Li, M. (2014). Two-compartmental population balance modeling of a pulsed spray fluidized bed granulation based on computational fluid dynamics (CFD) analysis. *International Journal of Pharmaceutics, 475*(1–2), 256–269. https://doi.org/10.1016/j.ijpharm.2014.08.057.

Liu, L. X., Litster, J. D., Iveson, S. M., & Ennis, B. J. (2000). Coalescence of deformable granules in wet granulation processes. *AIChE Journal, 46*(3), 529–539. https://doi.org/10.1002/aic.690460312.

Mateo-Ortiz, D., Villanueva-Lopez, V., Muddu, S. V., Doddridge, G. D., Alhasson, D., & Dennis, M. C. (2021). Dry powder mixing is feasible in continuous twin screw extruder: Towards lean extrusion process for oral solid dosage manufacturing. *AAPS PharmSciTech, 22*(7). https://doi.org/10.1208/s12249-021-02148-x.

Matsushita, M., Ohsaki, S., Nara, S., Nakamura, H., & Watano, S. (2022). Effect of fill level in continuous twin-screw granulator: A combined experimental and simulation study. *Advanced Powder Technology, 33*(11), 103822. https://doi.org/10.1016/j.apt.2022.103822.

Matuttis, H. G., Luding, S., & Herrmann, H. J. (2000). Discrete element simulations of dense packings and heaps made of spherical and non-spherical particles. *Powder Technology, 109*(1–3), 278–292. https://doi.org/10.1016/S0032-5910.(99)00243-0.

Meng, W., Oka, S., Liu, X., Omer, T., Ramachandran, R., & Muzzio, F. J. (2017). Effects of process and design parameters on granule size distribution in a continuous high shear granulation process. *Journal of Pharmaceutical Innovation, 12*(4), 283–295. https://doi.org/10.1007/s12247-017-9288-7.

Nase, S. T., Vargas, W. L., Abatan, A. A., & McCarthy, J. J. (2001). Discrete characterization tools for cohesive granular material. *Powder Technology, 116*(2–3), 214–223. https://doi.org/10.1016/S0032-5910.(00)00398-3.

Ng, B. H., Ding, Y. L., & Ghadiri, M. (2009). Modelling of dense and complex granular flow in high shear mixer granulator—A CFD approach. *Chemical Engineering Science, 64*(16), 3622–3632. https://doi.org/10.1016/j.ces.2009.05.011.

O'Sullivan, C., & Bray, J. D. (2004). Selecting a suitable time step for discrete element simulations that use the central difference time integration scheme. *Engineering Computations, 21*(2/3/4), 278–303. https://doi.org/10.1108/02644400410519794.

Patankar, S. (2018). *Numerical heat transfer and fluid flow*. CRC Press.

Patil, A. S., & Pethe, A. M. (2013). Quality by design (QbD): A new concept for development of quality pharmaceuticals. *International Journal of Pharmaceutical Quality Assurance, 4*(2), 13–19.

Poon, J. M. H., Immanuel, C. D., Doyle, F. J., & Litster, J. D. (2008). A three-dimensional population balance model of granulation with a mechanistic representation of the nucleation and aggregation phenomena. *Chemical Engineering Science, 63*(5), 1315–1329. https://doi.org/10.1016/j.ces.2007.07.048.

Pope, S. B. (2001). Turbulent flows. *Measurement Science and Technology, 12*(11), 2020–2021. https://doi.org/10.1088/0957-0233/12/11/705.

Rahmanian, N., Ghadiri, M., & Jia, X. (2011). Seeded granulation. *Powder Technology, 206*(1–2), 53–62. https://doi.org/10.1016/j.powtec.2010.07.011.

Rajniak, P., Stepanek, F., Dhanasekharan, K., Fan, R., Mancinelli, C., & Chern, R. T. (2009). A combined experimental and computational study of wet granulation in a Wurster fluid bed granulator. *Powder Technology, 189*(2), 190–201. https://doi.org/10.1016/j.powtec.2008.04.027.

Ramachandran, R., Poon, J. M. H., Sanders, C. F. W., Glaser, T., Immanuel, C. D., Doyle, F. J., Litster, J. D., Stepanek, F., Wang, F. Y., & Cameron, I. T. (2008). Experimental studies on distributions of granule size, binder content and porosity in batch drum granulation: Inferences on process modelling requirements and process sensitivities. *Powder Technology, 188*(2), 89–101. https://doi.org/10.1016/j.powtec.2008.04.013.

Rogers, A. J., Hashemi, A., & Ierapetritou, M. G. (2013). Modeling of particulate processes for the continuous manufacture of solid-based pharmaceutical dosage forms. *Processes, 1*(2), 67–127.

Sampat, C., Baranwal, Y., Paraskevakos, I., Jha, S., Ierapetritou, M., & Ramachandran, R. (2018). HPC enabled parallel, multi-scale & mechanistic model for high shear granulation using a coupled DEM-PBM framework. *Computer Aided Chemical Engineering, 44*, 1459–1464. https://doi.org/10.1016/B978-0-444-64241-7.50238-X.

Sampat, C., & Ramachandran, R. (2021). Identification of granule growth regimes in high shear wet granulation processes using a physics-constrained neural network. *Processes, 9*(5), 737. https://doi.org/10.3390/pr9050737.

Schaber, S. D., Gerogiorgis, D. I., Ramachandran, R., Evans, J. M. B., Barton, P. I., & Trout, B. L. (2011). Economic analysis of integrated continuous and batch pharmaceutical manufacturing: A case study. *Industrial and Engineering Chemistry Research, 50*(17), 10083–10092. https://doi.org/10.1021/ie2006752.

Schaeffer, D. G. (1987). Instability in the evolution equations describing incompressible granular flow. *Journal of Differential Equations, 66*(1), 19–50. https://doi.org/10.1016/0022-0396.(87)90038-6.

Scherer, Philipp OJ. (2017). Equations of motion. In Computational Physics: Simulation of Classical and Quantum Systems (pp. 289–321). Cham: Springer International Publishing.

Seem, T. C., Rowson, N. A., Ingram, A., Huang, Z., Yu, S., de Matas, M., Gabbott, I., & Reynolds, G. K. (2015). Twin screw granulation—A literature review. *Powder Technology, 276*, 89–102. https://doi.org/10.1016/j.powtec.2015.01.075.

Sen, M., Barrasso, D., Singh, R., & Ramachandran, R. (2014). A multi-scale hybrid CFD-DEM-PBM description of a fluid-bed granulation process. *Processes, 2*(1), 89–111. https://doi.org/10.3390/pr2010089.

Singh, M., Shirazian, S., Ranade, V., Walker, G. M., & Kumar, A. (2022). Challenges and opportunities in modelling wet granulation in pharmaceutical industry—A critical review. *Powder Technology, 403*, 117380.

Soda, R., Sato, A., Kano, J., Kasai, E., Saito, F., Hara, M., & Kawaguchi, T. (2009). Analysis of granules behavior in continuous drum mixer by DEM. *ISIJ International, 49*(5), 645–649. https://doi.org/10.2355/isijinternational.49.645.

Suresh, P., Saketharam Reddy, N., Hariharan, R., & Sreedhar, I. (2020). Studies on fluid bed granulation of lactose-MCC mixture. *Materials Today: Proceedings, 24*, 519–530. https://doi.org/10.1016/j.matpr.2020.04.305.

Suresh, P., Sreedhar, I., Vaidhiswaran, R., & Venugopal, A. (2017). A comprehensive review on process and engineering aspects of pharmaceutical wet granulation. *Chemical Engineering Journal, 328*, 785–815. https://doi.org/10.1016/j.cej.2017.07.091.

Tamrakar, A., Chen, S. W., & Ramachandran, R. (2019). A DEM model-based study to quantitatively compare the effect of wet and dry binder addition in high-shear wet granulation processes. *Chemical Engineering Research and Design, 142*, 307–326. https://doi.org/10.1016/j.cherd.2018.12.016.

Tamrakar, A., & Ramachandran, R. (2019). CFD–DEM–PBM coupled model development and validation of a 3D top-spray fluidized bed wet granulation process. *Computers and Chemical Engineering, 125*, 249–270. https://doi.org/10.1016/j.compchemeng.2019.01.023.

Tang, Y. Y, Frank Peters, E. A. J. F., Hans Kuipers, J. A. M., Sebastian Kriebitzsch, S. H. L., & Martin van der Hoef, M. A. (2015). A new drag correlation from fully resolved simulations of flow past monodisperse static arrays of spheres. *AIChE Journal, 61*(2), 688–698. https://doi.org/10.1002/aic.14645.

Tenneti, S., Garg, R., & Subramaniam, S. (2011). Drag law for monodisperse gas-solid systems using particle-resolved direct numerical simulation of flow past fixed assemblies of spheres.

International Journal of Multiphase Flow, 37(9), 1072–1092. https://doi.org/10.1016/j.ijmultiphaseflow.2011.05.010.

Toson, P., & Khinast, J. G. (2023). A DEM model to evaluate refill strategies of a twin-screw feeder. *International Journal of Pharmaceutics, 641*. https://doi.org/10.1016/j.ijpharm.2023.122915.

Tsuji, Y., Kawaguchi, T., & Tanaka, T. (1993). Discrete particle simulation of two-dimensional fluidized bed. *Powder technology, 77*(1), 79–87.

Tu, W. D., Ingram, A., & Seville, J. (2013). Regime map development for continuous twin screw granulation. *Chemical Engineering Science, 87*, 315–326. https://doi.org/10.1016/j.ces.2012.08.015.

Uddin, N. (2008). *Turbulence modeling of complex flows in CFD*. Stuttgart: Institute of Aerospace Thermodynamics Universität Stuttgart.

Uhlemann, H., & Mörl, L. (2000). Anwendungen der Wirbelschicht-Sprühgranulation. In *Wirbelschicht-Sprühgranulation* (pp. 491–564). Springer. https://doi.org/10.1007/978-3-642-57004-9_16.

Vanarase, A. U., Alcalà, M., Jerez Rozo, J. I., Muzzio, F. J., & Romañach, R. J. (2010). Real-time monitoring of drug concentration in a continuous powder mixing process using NIR spectroscopy. *Chemical Engineering Science, 65*(21), 5728–5733. https://doi.org/10.1016/j.ces.2010.01.036.

Verkoeijen, D., Pouw, G. A., Meesters, G. M. H., & Scarlett, B. (2002). Population balances for particulate processes—A volume approach. *Chemical Engineering Science, 57*(12), 2287–2303. https://doi.org/10.1016/S0009-2509.(02)00118-5.

Vervaet, C., & Remon, J. P. (2005). Continuous granulation in the pharmaceutical industry. *Chemical Engineering Science, 60*(14), 3949–3957. https://doi.org/10.1016/j.ces.2005.02.028.

Vonk, P., Guillaume, C. P. F, Ramaker, J. S., Vromans, H., & Kossen, N. W. F (1997). Growth mechanisms of high-shear pelletisation. *International Journal of Pharmaceutics, 157*(1), 93–102. https://doi.org/10.1016/s0378-5173.(97)00232-9.

Vreman, A. W., van Lare, C. E., & Hounslow, M. J. (2009). A basic population balance model for fluid bed spray granulation. *Chemical Engineering Science, 64*(21), 4389–4398. https://doi.org/10.1016/j.ces.2009.07.010.

Wang, L. G., Pradhan, S. U., Wassgren, C., Barrasso, D., Slade, D., & Litster, J. D. (2020). A breakage kernel for use in population balance modelling of twin screw granulation. *Powder Technology, 363*, 525–540.

Wang, M. H., Yang, R. Y., & Yu, A. B. (2012). DEM investigation of energy distribution and particle breakage in tumbling ball mills. *Powder Technology, 223*, 83–91. https://doi.org/10.1016/j.powtec.2011.07.024.

Wang, Z., Pan, Z., He, D., Shi, J., Sun, S., & Hou, Y. (2019). Simulation modeling of a pharmaceutical tablet manufacturing process via wet granulation. *Complexity, 2019*, 1–16. https://doi.org/10.1155/2019/3659309.

Washino, K., Tan, H. S., Hounslow, M. J., & Salman, A. D. (2013). A new capillary force model implemented in micro-scale CFD–DEM coupling for wet granulation. *Chemical Engineering Science, 93*, 197–205. https://doi.org/10.1016/j.ces.2013.02.006.

Wendt, J. F., Anderson, J. D., Degroote, J., Degrez, G., Dick, E., Grundmann, R., & Vierendeels, J. (2009). *Computational fluid dynamics: An introduction computational fluid dynamics* (pp. 1–332). Springer. https://doi.org/10.1007/978-3-540-85056-4.

Wildeboer, W. J., Litster, J. D., & Cameron, I. T. (2005). Modelling nucleation in wet granulation. *Chemical Engineering Science, 60*(14), 3751–3761. doi:10.1016/j.ces.2005.02.005.

Yang, R. Y., Zou, R. P., & Yu, A. B. (2003). Microdynamic analysis of particle flow in a horizontal rotating drum. *Powder Technology, 130*(1–3), 138–146. doi:10.1016/S0032-5910 (02)00257-7.

You, Y., Guo, J., Li, G., Lv, X., Wu, S., Li, Y., & Yang, R. (2021). Investigation the iron ore fine granulation effects and particle adhesion behavior in a horizontal high-shear granulator. *Powder Technology, 394*, 162–170. https://doi.org/10.1016/j.powtec.2021.08.047.

Yu, X., Hounslow, M. J., Reynolds, G. K., Rasmuson, A., Niklasson Björn, I., & Abrahamsson, P. J. (2017). A compartmental CFD-PBM model of high shear wet granulation. *AIChE Journal, 63*(2), 438–458. https://doi.org/10.1002/aic.15401.

Yu, Y., Zhao, L., Li, Y., & Zhou, Q. (2020). A model to improve granular temperature in CFD-DEM simulations. *Energies, 13*(18), 4730. https://doi.org/10.3390/en13184730.

Zheng, C., Govender, N., Zhang, L., & Wu, C. Y. (2022). GPU-enhanced DEM analysis of flow behaviour of irregularly shaped particles in a full-scale twin screw granulator. *Particuology, 61*, 30–40. https://doi.org/10.1016/j.partic.2021.03.007.

Kano, J., Kasai, E., Saito, F. and Kawaguchi, T. (2006) Numerical Simulation Model for Granulation Kinetics of Iron Ores Based on Discrete Element Method. *Tetsu to hagane.*

CHAPTER 23

Twin-screw continuous wet granulation

Niyati Niranjan Kodange[a], Tongzhou Liu[b], Adwait Pradhan[a], Ankita V. Shah[c], Abu T. Serajuddin[d] and Feng Zhang[a]

[a]*Division of Molecular Pharmaceutics and Drug Delivery, College of Pharmacy, University of Texas at Austin, Austin, TX, United States,* [b]*Department of Product Development, Science and Technology Operations, AbbVie Inc., North Chicago, IL, United States,* [c]*Department of Deerfield Discovery and Development, Deerfield Management, New York, NY, United States,* [d]*Department of Pharmaceutical Sciences, College of Pharmacy, St. John's University, Queens, NY, United States*

23.1 Introduction

Granulation is a fundamental process in pharmaceutical manufacturing, wherein fine powders are agglomerated to form larger, multiparticle structures known as granules. This unit operation plays a vital role in the production of solid oral dosage forms such as tablets and capsules. Compared to unprocessed powders, granules offer several distinct advantages, including improved flowability, increased bulk density, reduced dust formation, and enhanced uniformity in content distribution. Moreover, granulation contributes to the mechanical robustness of the final dosage form, thereby ensuring better manufacturability and product quality (Thapa et al., 2019).

There are mainly three types of granulation processes used in pharmaceutical manufacturing, namely, dry granulation, melt granulation, and wet granulation (WG). As the name suggests, dry granulation is a dry process where no granulating liquid is added to aid granulation. In this process, mixtures of active pharmaceutical ingredient (API) and excipients are roller compacted to prepare dense ribbons, which are then milled and compressed into tablets (Parrott, 1981). The primary requirement for this process is that the raw materials must have adequate inherent binding and cohesive properties for compression (Zhou et al., 2017). The melt granulation is another dry process that attracted much attention in recent years (Batra et al., 2017; Lakshman et al., 2011; Vasanthavada et al., 2011). In this process, mixtures of APIs and excipients are passed through melt extruders at relatively high temperatures, where polymeric binders melt but not the drug substances, and granules are formed by the action of the molten binders that solidify when the temperatures are reduced.

WG is the third granulation process that is widely used in the manufacture of tablets due to its versatility and reliability in achieving the desired granule properties. It involves the addition of a liquid binder to a powder blend under agitation to form a moist mass. This mass is subsequently dried and further processed. Cohesion within the granules is primarily facilitated through adhesive interactions between the binder and the particles, supported by capillary and viscous forces (Suresh et al., 2017). However, traditionally, WG has been performed as a batch. Despite its widespread use, the batch WG is often associated with several limitations, including difficulties in scale-up, long processing times, and significant batch-to-batch variability (Järvinen et al., 2015).

In response to these challenges, the pharmaceutical industry is increasingly adopting continuous manufacturing strategies. One of the most promising advancements in this domain is the implementation of twin-screw extrusion technology for continuous WG. This approach offers improved process control, enhanced efficiency, and reduced variability, making it a compelling alternative to traditional batch processing (Suresh et al., 2017; Thapa et al., 2019; Thompson, 2015).

This chapter provides a comprehensive overview of the continuous twin-screw wet granulation (TSWG) process. It explores the influence of key formulation and process parameters on granule quality and performance. Particular attention is given to the use of process analytical technology (PAT) tools for real-time monitoring, control, and optimization of the TSWG process. The granulation mechanism is also discussed in detail, with emphasis on how it governs granule formation and properties. Additionally the chapter addresses the current limitations of TSWG, such as the need for separate drying and milling steps, and highlights recent innovations aimed at achieving a fully continuous, integrated process. Considerations related to process scale-up and industrial implementation are also discussed.

23.2 Continuous wet granulation

Continuous WG has gained recognition as a viable alternative to traditional batch methods, primarily due to its enhanced process control, improved scalability, and consistent product quality. In this method, materials are continuously introduced, blended, granulated, and further processed in an integrated and efficient workflow. Research by Schaber et al. (2011) has also demonstrated the economic advantages of continuous manufacturing over batch processing. To maintain rigorous control and ensure product quality, the approach necessitates the use of advanced in-line or real-time analytical technologies (Schaber et al., 2011; Suresh et al., 2017). Various continuous WG methods have been developed, including TSWG, continuous fluidized bed granulation, ring-layer granulation, and foam-based continuous granulation (Järvinen et al., 2015; Thompson et al., 2012).

Twin-screw extruders have long been utilized in the food and plastics industries for the continuous processing of materials due to their ability to provide high shear forces for effective mixing, melting, and extrusion. In the pharmaceutical field, their application to WG was first reported by Gamlen and Eardley (1986), who prepared rod-shaped extrudates using a blend of paracetamol, lactose, microcrystalline cellulose (MCC), and hydroxypropylmethyl cellulose (HPMC) in the presence of moisture. Subsequently Lindberg et al. (1987) employed the twin-screw extrusion technique to produce effervescent granules. Throughout the early 1990s, multiple research groups further explored and refined the TSWG process, contributing to its development and optimization (Baert et al., 1993; Elbers et al., 1992; Kleinebudde & Lindner, 1993). The TSWG process has more recently emerged as an advanced and efficient approach to WG. A detailed description of this process is provided in Section 23.3 of this chapter.

Continuous fluid bed WG represents a modern approach to pharmaceutical manufacturing by combining granulation and drying within a single, continuously operating system. In this process, powders are steadily introduced into a fluidized bed chamber, where a granulation liquid is sprayed to promote particle agglomeration, followed by either simultaneous or sequential drying of the resulting wet granules within the same apparatus. Compared to traditional batch fluid bed granulation (FBG), this continuous method offers improved control over key process variables, enhanced product uniformity, and greater operational efficiency in terms of both throughput and energy consumption. Additionally granules produced through FBG have been reported to exhibit higher porosity, lower density, and greater compressibility than those generated by high-shear wet granulation (HSWG), highlighting the potential impact of granulation method on final product properties (Hausman, 2004; Suresh et al., 2017).

Continuous foam granulation is a WG technique wherein the binder is introduced as a pregenerated foam rather than in liquid form, enabling more uniform binder distribution and reduced liquid requirement. Initially explored in batch high-shear granulators (Keary & Sheskey, 2004), this approach was later adapted for twin-screw extrusion to facilitate continuous processing (Thompson et al., 2012). The use of foam allows for improved binder dispersion and minimizes the risk of oversaturation, as its low soak-to-spread ratio promotes efficient surface coverage within the powder bed. Suitable binders for this method typically exhibit surfactant-like properties in aqueous systems, such as hydroxypropyl cellulose (HPC), hypromellose, and other cellulose ethers. In a comparative study by Thompson et al. (2012), the use of foamed binder in twin-screw extrusion demonstrated advantages over conventional liquid injection, including enhanced powder wetting, improved lubricity, and reduced frictional heat generation. These effects collectively contribute to lower equipment wear and help maintain a lower granule outlet temperature, supporting both product quality and process efficiency (Rocca et al., 2015; Thompson et al., 2012).

23.3 Twin-screw wet granulation

TSWG has emerged as a more advanced and robust alternative to traditional WG techniques, offering several operational advantages (Haser et al., 2023). These include flexible batch sizes, simplified scale-up, reduced liquid requirements to achieve comparable granule porosity, enhanced process tunability, shorter processing times, and minimized material waste (Beer et al., 2014; Thompson, 2015).

In the TSWG process, liquid is introduced into the extruder through a pump and nozzle system connected to an injection port located on the barrel. The binder may be added in either a dissolved form within the liquid phase or as a dry component blended with the powder mixture. Once the liquid is injected, the binder solution is uniformly distributed throughout the powder blend. The granule formation primarily takes place in the mixing zone of the extruder, where mechanical energy is imparted through kneading or combing elements. These mixing components facilitate compaction and agglomeration of the material, leading to granule formation (Sayin et al., 2015a; Vandevivere et al., 2020). Due to the high mixing efficiency of the twin-screw system, less liquid is required to achieve the same level of granule porosity as seen in HSWG or FBG, thereby potentially lowering the risk of moisture-related degradation (Beer et al., 2014).

Fig. 23.1 illustrates a typical twin-screw extruder used in the granulation process (Process 11, Thermo Fisher Scientific, Darmstadt, Germany). An extruder generally consists of one or two screws, either co-rotating or counter-rotating, housed within a closed barrel (Breitenbach, 2002). The barrel is typically divided into four functional sections: the feeding zone, two transition zones comprising conveying and mixing (kneading) elements, and the final metering zone, which includes a die plate that can be replaced to modify the diameter and shape of the extrudate. However, during WG, this die plate is usually removed to facilitate the collection of granules. The powder blend is introduced into the system via a feeder and then transported through the barrel, where it encounters various conveying and kneading sections. Granulation is initiated when water or a granulating fluid is added through a designated

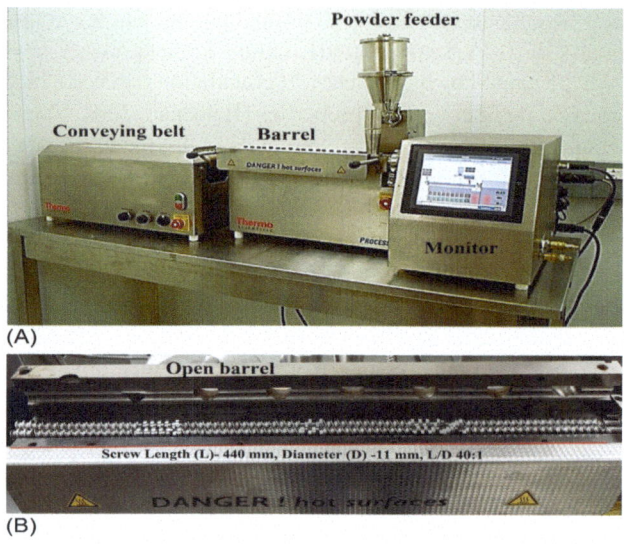

FIGURE 23.1

(A) Twin-screw extruder (Process 11, Thermo Scientific, Darmstadt, Germany); (B) open barrel view illustrating distinct processing zones and twin-screw configuration. The extruder features a length-to-diameter (L/D) ratio of 40:1, with a screw diameter of 11 mm and a barrel process section length of 440 mm (ref: image reused from the 1st ed).

injection port, allowing the binder to interact with the formulation as it moves through the extruder. This interaction, assisted by the mechanical action of the screws, leads to the formation of granules, which are ultimately discharged from the system. Twin-screw extruders are also characterized by their geometric design, specifically the screw diameter and the length-to-diameter (L/D) ratio (Thompson, 2015). A higher L/D ratio generally corresponds to greater processing capacity, allowing more material to be granulated in a given period.

The design of a twin-screw extruder can significantly influence its performance, and the quality of the granules produced. In a study conducted by Djuric et al. (2009), two geometrically distinct twin-screw extruders from different manufacturers were evaluated using comparable screw configurations. The formulations used for granulation included both a water-insoluble material (dicalcium phosphate) and a water-soluble material (α-lactose monohydrate). The researchers assessed the impact of process parameters (input feed rate and screw speed [SS]) on the characteristics of the resulting granules and the properties of the final tablets. Although both extruders successfully produced granules, the results revealed that the type of extruder had a substantial influence on granule quality. The study emphasized that substituting one extruder with another is not straightforward and requires thorough process validation. For instance, even though both extruders had the same L/D ratio of 24, their performance varied. As shown in Fig. 23.2, the Leistritz extruder demonstrated a noticeable increase in median granule size (d50) with an increasing feed rate. In contrast, the APV extruder showed minimal variation in granule size across different input rates. This difference was attributed to the internal

23.4 Effects of formulation parameters on twin-screw wet granulation

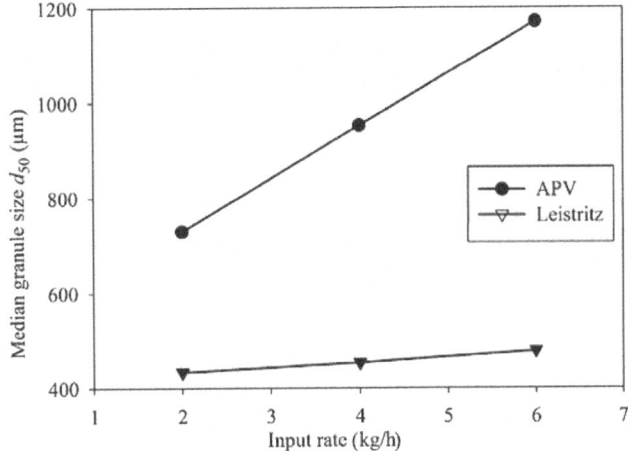

FIGURE 23.2

Interaction plot illustrating the median particle size of dicalcium phosphate granules as influenced by the type of twin-screw extruder. The study compared the APV Baker extruder and the Leistritz micro extruder, which have screw diameters of 19 and 27 mm, respectively, but identical length-to-diameter (L/D) ratios of 24.

From Djuric et al. (2009) (ref: image reused from the 1st ed).

free chamber volume of each machine. The APV extruder, which had a larger free chamber volume, accommodated higher feed rates with limited effect on granule size (Djuric et al., 2009).

In addition to equipment selection, formulation composition, and process parameters play a critical role in determining the properties of the resulting granules. In fact, it is the interaction between these two factors that ultimately governs the critical quality attributes (CQAs) of the granules. These attributes include particle or granule size distribution (PSD or GSD), shape, strength, friability, porosity, flowability, moisture content, bulk and tapped densities, and compressibility. These granule characteristics, in turn, significantly affect the final tablet properties, such as tensile strength, disintegration time, friability, and dissolution rate (Dhenge et al., 2010, 2012a; Fonteyne et al., 2014a; Meena et al., 2017; Tan et al., 2011).

23.4 Effects of formulation parameters on twin-screw wet granulation

23.4.1 Active pharmaceutical ingredient properties

The physicochemical properties of active pharmaceutical ingredient (API) have a significant influence on the performance of TSWG. These properties can impact key aspects such as the optimal liquid-to-solid (L/S) ratio, granule size, porosity, and mechanical strength. Variations in API properties such as particle size, morphology, wettability, flowability, and solubility can significantly alter wetting behavior, liquid distribution, and the extent of granule nucleation and growth. Understanding these API characteristics early in development helps in choosing the right formulation strategy and optimizing the granulation process.

The granulation behavior of three different APIs (acetaminophen, allopurinol, and metformin HCl) in combination with fillers was studied by Kyttä et al. (2020). Water absorption capacity and thus, the optimal amount of water in the WG were influenced by the solubility and particle size of API when combined with fillers. However, the tensile strength of the tablets was mostly unaffected by the compaction properties of pure APIs (Kyttä et al., 2020).

Peeters et al. (2023) studied the impact of API's physicochemical properties on the granule porosity. Particle size, density, flow properties, and water-related properties such as solubility and dissolution rate of the API have been reported to have major effects on the granule porosity. The particle size and flow properties were more significant at a low L/S ratio, while water-related properties were predominant at high L/S ratios. With an increase in particle size of API, there are fewer interparticle contact points, which resulted in an increase in intragranular void volume and granule porosity. Generally larger particle size powders have improved flow properties, that is, higher flow-function coefficient (ffc) and lower Hausner ratio (HR). Therefore, ffc and HR showed positive and negative correlation with granule porosity, respectively. However, the granule porosity decreased with an increase in powder cohesion due to greater interparticle interactions. The solubility and dissolution rate of the API were found to be negatively correlated with granule porosity. This is because more powder particles that dissolve form solid bridges that bring the individual powder particles closer together, reducing the intragranular void volume (Peeters et al., 2023).

Peeters et al. (2024) also examined how the physicochemical properties and drug load of APIs affected the granule formation. The granule growth and its corresponding size and shape at each of the twin-screw granulation (TSG) compartments was significantly influenced by flow, size, density, and water-related properties of API. It was observed that the solubility of APIs plays important role in determining the formulation compositions and screw configurations. When granulating APIs with very low water solubility, a significant amount of fines were generated for formulations with high dose. This observation was based on ineffective interactions of API particles with binder liquid. Hence, the drug load should be decreased, or formulation compositions should be optimized.

According to a study by Fonteyne et al. (2014b), the coarser grades of theophylline anhydrous produced bigger granules and a lower percentage of fines. The differences in granule bulk and tapped density were much lower, even though premixes with these coarser grades had higher bulk and tapped density than their finer grades. Li et al. (2015) investigated the impact of API properties on a TSG system using foamed binder delivery for formulations with a 15% drug load. They found that formulations with more hydrophobic APIs (ibuprofen or griseofulvin) required larger L/S ratios for the granules to resemble those in formulations with more hydrophilic APIs (acetaminophen or caffeine).

23.4.2 Binder type, concentration, and its properties

The granulation process (i.e., nucleation and granule growth) is often facilitated by the addition of a binder. The quality by design (QbD) approach requires that the key binder attributes are identified as they affect the final quality attributes of the granules and tablets, such as granule friability and tablet tensile strength (Bansal et al., 2019). The most frequently used pharmaceutical binders in TSWG are sugars, synthetic polymers such as polyvinylpyrrolidone (PVP), and cellulose-based polymers like HPC, HPMC (Portier et al., 2020a, 2021a; Rahimi et al., 2020).

The binder type affects the granule quality; therefore, selection of the binder should be based on scientific approach (Vandevivere et al., 2020, 2021). The average residence time in the barrel is only 5 to 20 s, making it essential for binders to get activated quickly and reach their full binding potential. This process is not critical in the case of batch granulation process as the typical residence time is around 10 to 30 min (De Leersnyder et al., 2018; Portier et al., 2021a). Cellulosic based binders are not always the most suitable choice for TSWG. A study by Portier et al. (2020b) on granulating a high-dose, poorly soluble, and poorly wettable API demonstrated that switching the binder from HPMC (Methocel E15 LV) to PVP (Kollidon K30) produced granules with greater strength at similar or lower L/S ratios. The same grade of PVP was used as a binder for granulating carvedilol (Fülöp et al., 2021) and theophylline (Fonteyne et al., 2014b). Although PVP offers potential benefits, formulators must consider reactive impurities such as peroxide residues from chemical synthesis, which can degrade oxidation-sensitive APIs (Wu et al., 2011).

Binder properties, such as binder viscosity and wettability, also have an impact on granule properties. In the case of HPC with varying concentrations, it was observed that an increase in binder viscosity and amount of HPC resulted in unimodal granules with greater strength and improved flow properties (Dhenge et al., 2012a, 2012b). In addition to high binder viscosity and low surface tension, good wettability of the formulation by the binder is also important. For a binder to be very effective at low L/S ratios, it must wet quickly and be able to adequately wet the powder bed. Binders that adequately wet the powder bed required less granulation liquid to obtain low granule friability (Vandevivere et al., 2020). A study reported by Vandevivere et al. (2021) recommended that for highly soluble formulations, binders with low viscosity, low surface tension, effective wettability, and rapid dissolution kinetics should be used.

In addition to the conventional immediate release binders, recent studies involve the use of sustained or controlled release binders for various applications. Several HPMC grades are employed as matrix formers for sustained release. A study by Vanhoorne et al. (2016c) showed the sustained release of 20% metoprolol tartrate for 16 h when 20% HPMC 90SH-4000 was used as a binder. The formulation and process settings had no influence on the sustained release of this API. MCC as a filler in this formulation resulted in elongated granules at high L/S ratios, in contrast to granules with regular shape when lactose or lactose and native maize starch were used as filler. Hence, it has been concluded that the HPMC and MCC should not be combined in case of sustained or controlled release formulations. Thompson and O'Donnell (2015) evaluated Methocel K4M and Kollidon SR as controlled release agents at concentrations ranging from 5% to 20% and reported that their incorporation led to the formation of elongated granules. This effect was attributed to the ability of these excipients to enhance bridging forces within the wet agglomerated mass, thereby promoting the development of extended-form granules. Vanhoorne et al. (2016b) conducted a second study comparing the efficacy of three grades of HPMC (Metolose 90SH-4000-SR, 90SH-100000-SR, and 60SH-4000) as controlled-release binders within a concentration range of 20% to 40%. They found that the release of theophylline could be sustained for about 24 h by adjusting the viscosity and degree of substitution of HPMC.

23.4.3 Binder form (powder versus solution)

In addition to selecting an appropriate binder, the form of the binder (powder versus solution) can also influence the properties of the granules produced. Binders can be incorporated either as a dry powder (dry binder addition) or by preparing a solution beforehand (wet binder dispersion) (Vandevivere et al., 2020). In a study reported by El Hagrasy et al. (2013b), it was observed that the amount of fines could be reduced significantly by adding HPMC (binder) in liquid form in comparison to adding as a dry powder. Granule porosity decreased as the concentration (%) of binder in the liquid phase and the L/S ratio increased, suggesting greater consolidation. A similar reduction in the amount of fines was observed when PVP and HPC were added as wet binder dispersions. Since the residence time is low in TSWG, the binder was more effective when premixed in a granulating liquid (Vercruysse et al., 2012). By ensuring enhanced binder distribution throughout the granulation process, dissolving the binder in the granulating liquid prior to addition improved its binding efficiency (El Hagrasy et al., 2013b). However, the amount of binder that can be added to make wet dispersion is restricted by the viscosity of the resulting binder dispersion. The high viscosity of the binder solutions can limit the distribution of the granulating liquid throughout the powder bed. Hence, low viscosity binders should be used to achieve higher binder content in the dispersion (Vandevivere et al., 2020).

In another study involving granulation of extra fine acetaminophen powder, particle size distribution (PSD) of granules became broad when binder was added in the solution form (Ito & Kleinebudde, 2019). Vandevivere et al. (2020) studied the impact of binder addition method on granule friability. Wet binder addition produced granules with lower friability than dry binder addition for the majority of binders at lower L/S ratios. However, for some specific binders like PVP K90, HPMC E15, and HPMC E5, adding dry binder led to decreased granule friability. At higher L/S ratios, there was no difference in friability of granules obtained by both the methods.

23.4.4 Fillers and filler combinations

MCC, lactose, mannitol, and dibasic calcium phosphate (DCP) are commonly used as fillers in the TSWG process. Since each excipient has both pros and cons, often two different fillers are combined in formulations to achieve the desired material properties (Kyttä et al., 2020; Vanhoorne et al., 2020; Willecke et al., 2018).

MCC exhibits distinctive compaction behavior, attributed to its porous structure comprising both crystalline and amorphous domains. Although MCC is insoluble in water, it can engage in hydrogen bonding with water molecules. Upon exposure to water, MCC particles swell, and subsequent drying induces particle shrinkage, leading to reduced porosity. Additionally since MCC is manufactured through partial hydrolysis of cellulose obtained from natural sources, batch-to-batch variability in its properties may arise (Kyttä et al., 2020). According to a study by Fonteyne et al. (2015) differences in the water binding capacities of MCC batches were caused by variable crystallinity levels. It was determined that variations in MCC batches should be considered during manufacturing because the addition of binders was unable to reduce these discrepancies. Portier et al. (2021b) analyzed the impact of batch-to-batch variability in MCC on the TSWG process and found that granules with the most desired quality attributes were formed by MCC batches that had low moisture content, high bulk density, and low water binding capacity. Usually MCC is combined with a water-soluble filler, such as lactose and mannitol, in commercial formulations. To avoid adding excessive water required by MCC, the ratio of the water-soluble filler to MCC is often greater than 1 (Portier et al., 2020a).

Hwang et al. (2019) developed a multivariate method to explain the various granulation behaviors of different lactose grades (Granulac 200, Granulac 70, Prismalac 40, and Flowlac 90). The lactose grade influenced the characteristics of the granules. Due to its wide surface area and stronger contact between the granulation liquid and powder, Granulac 200 exhibited lower granule friability compared to the other lactose grades at low L/S ratios. However, this effect was eliminated at higher L/S ratios. (El Hagrasy et al., 2013b) compared three different lactose grades (Pharmatose 200M, Supertab 30GR, and Lactose Impalpable) and found only minor changes in PSD. However, at higher L/S ratios, the formulation incorporating granular lactose (Supertab 30GR) exhibited much lower porosity.

Mannitol, a polyol, is occasionally substituted for lactose as a water-soluble filler in formulations due to its faster dissolution rate. The solubility of mannitol in water facilitates the formation of new pores upon recrystallization, and WG further enhances its tabletability (Kyttä et al., 2020). During TSG, Vanhoorne et al. (2016a) noted a polymorphic transformation from δ- to β- mannitol, enhancing the plastic deformability and tabletability of the granules. Megarry et al. (2020) investigated the effects of various grades of mannitol (Pearlitol 160C and 200SD) on granules and tablet characteristics. They observed that during granulation, 200SD underwent a polymorphic transformation from a mixture of α- and β- forms to predominantly β form. This transformation was accompanied by a change in morphology, resulting in the formation of mannitol needles and producing granules with increased porosity and specific surface area. These granules exhibited poorer flow properties but higher tablet tensile strength.

DCP is a practically insoluble and brittle excipient that complements plastic fillers like MCC in tableting due to its high yield pressure and low strain-rate sensitivity (Djuric & Kleinebudde, 2010). Khorsheed et al. (2019) attributed the consistent tabletability behavior of DCP granules to the insoluble and nondeformable nature of the powder particles. Consequently, in formulations containing DCP, the bonding ability of granules is typically influenced by other ingredients rather than by DCP itself.

23.4.5 Surfactants

There are limited reports regarding the use of surfactants in TSWG. The existing literature primarily addresses the effects of surfactant additions on the necessary L/S ratios. Portier et al. (2020b) reported that incorporating 0.2% sodium lauryl sulfate (SLS) into a formulation containing 50% mebendazole, an API with low solubility, significantly improved the required L/S ratio for granulation. This adjustment allowed for a 20% reduction in the L/S ratio without compromising the PSD, density, or friability of the granules. This effect could prove advantageous in formulations where the required L/S ratio constrains achievable throughput due to the drying unit's capacity limitations (Portier et al., 2021a). On the contrary, Dhenge et al. (2012b) found that higher concentrations of SLS had no impact on the water requirement for granulation. Their study revealed that variations in SLS amount, and thus the surface tension of the granulation liquids, did not significantly alter PSD, flow characteristics, or granule strength. Schmidt et al. (2018) investigated the influence of incorporating polysorbate 80 on the necessary L/S ratio. They found that higher concentrations of polysorbate 80 increased the solubility of the API (ibuprofen), thereby delaying the onset of paste formation. However, this study did not include data on how polysorbate 80 affected the physical properties of the granules.

Table 23.1 Properties of acetaminophen granules produced at 70°C and 4.5 g/min feed rate, 300 RPM screw speed with 5% polyvinylpyrrolidone (PVP) or 5% pregelatinized starch (PGS) using 7% water.

Granule Property	Binders			
	PVP[a]	PGS[a]	PVP[a]	PGS[a]
	Powder Feed Rate			
	4.5 g/min		5.5 g/min	
Oversized granules (%)[b]	3.0 ± 1.1	2.0 ± 0.9	9.9 ± 0.2	3.0 ± 0.5
Fines (%)[b]	4.1 ± 0.1	6.2 ± 0.5	3.3 ± 0.3	5.6 ± 0.3
d_{90} (μm)[b]	875 ± 63	833 ± 28	1000 ± 0	947 ± 6
d_{50} (μm)[b]	443 ± 6	417 ± 25	510 ± 3	522 ± 20
d_{10} (μm)[b]	175 ± 5	153 ± 7	180 ± 10	177 ± 20
$d_{90/10}$ (μm)[b]	5.0	5.4	5.6	5.4
Tensile strength (MPa)[c]	1.7 ± 0.2	2.1 ± 0.1	1.4 ± 0.1	1.1 ± 0.1

Reproduced with permission from Meena et al. (2017) (ref: table reused from the 1st ed).
[a] *Torque generated during granulation was 9%.*
[b] *Mean ± SD (n = 3).*
[c] *Mean ± SD (n = 6), 250-mg granules compressed at 177 MPa using flat-faced punches of 8 mm diameter.*

23.5 Effects of process parameters on twin-screw wet granulation

23.5.1 Powder feed rate

Powder feed rate (PFR) is a critical process parameter that significantly influences granule attributes during TSWG (Dhenge et al., 2011; Kumar et al., 2016a). Dhenge et al. (2010) observed that operating at a low feed rate leads to an extended residence time within the barrel, primarily due to the reduced powder fill, which limits the material throughput. This lower feed rate also resulted in smaller granule sizes, attributed to insufficient compaction and particle packing. In contrast, higher feed rates reduced residence time as the barrel became densely filled, causing the material to move forward in a plug-like manner, with its velocity governed by the SS. This plug flow increased the torque requirement, as greater mechanical force was necessary to move the material through the extruder. Additionally higher feed rates enhanced granule sphericity and strength due to improved powder compaction and interlocking under elevated shear forces. Similar outcomes were noted by Fonteyne et al. (2013).

Beyond pharmaceutical applications, several studies in the food and polymer industries using co-rotating twin-screw extruders have demonstrated that increased feed rates or throughput generally reduce residence time (Kao and Allison, 1984; Yeh et al., 1992). However, not all studies have reported a direct correlation between feed rate and granule size. For instance, Meena et al. (2017) and Yu et al. (2014) reported only marginal changes in granule size with varying feed rates. In a study by Meena et al. (2017), reducing the feed rate from 5.5 to 4.5 g/min improved granule quality by decreasing the proportion of oversized particles (Table 23.1). Similarly Vercruysse et al. (2012) found that increasing the feed rate had no notable effect on granule size distribution during TSWG.

These discrepancies across studies are likely due to differences in extruder scale, screw configuration, and formulation properties (Gorringe et al., 2017). Alongside SS, feed rate also determines the specific feed load (SFL), a parameter critical to defining the fill level of the extruder barrel. SFL, introduced by Kolter et al. (2012), refers to the mass of material transported per screw revolution and is calculated by dividing the PFR by the SS.

The fill level, which is the ratio of occupied volume to the total barrel capacity, directly impacts material mixing and granule formation. Meier et al. (2017) emphasized that optimizing SFL is essential for ensuring consistent product quality. They found that increasing SFL led to a higher incidence of oversized granules, likely due to enhanced material consolidation and greater contact among wetted particles. Specifically as SFL increased from 0.2 to 0.275 g and then to 0.45 g, the median granule size (X_{50}) grew from 668 to 750 μm and 1167 μm, respectively. Interestingly even at the highest SFL of 0.45 g, the barrel fill level was only around 30%, indicating substantial unused capacity within the extruder.

23.5.2 Screw speed

In general, increasing the SS during TSWG results in a reduced residence time, as the material is transported more quickly through the barrel. A higher SS also typically leads to a decrease in torque because the powder mass load on the screws becomes lighter (Tan et al., 2011). However, this acceleration in screw rotation introduces more intense shear forces, which can fragment developing granules and generate particles with irregular shapes and rough surfaces. These higher shear forces may cause erosion of granules, leading to less efficient granulation and a higher proportion of fines (Dhenge et al., 2010). Conversely reducing the SS increases the degree of barrel fill and prolongs the residence time. This extended residence provides more opportunity for effective granule growth, often resulting in larger granules and a reduction in the number of fines.

Kumar et al. (2016a) further highlighted that both the mean residence time (MRT) and the mixing efficiency are governed by the combined influence of material throughput, SS, and the screw configuration, particularly the number of kneading elements (KEs). While a moderate increase in SS can improve the yield of granules within the desired size range, excessive speed may reduce mixing efficiency if the residence time becomes too short for uniform wetting and granule growth. Therefore, optimizing SS is crucial to balance shear input, granule integrity, and process yield.

23.5.3 Screw elements

The screw elements in a twin-screw extruder are essential for enabling continuous material processing and can be tailored to regulate the level of shear applied within the barrel. These elements are generally classified into three categories based on their function in granulation: conveying elements, kneading or mixing elements, and combing mixer elements (Fig. 23.3A–C, respectively). Mixing elements are further distinguished as either dispersive or distributive. Dispersive mixing elements, as illustrated in Fig. 23.3B, are designed to break down structural units such as droplets, phase domains, and aggregates. They operate by capturing material within localized pressure zones, subjecting it to squeezing, shearing, and elongation. KEs, a type of dispersive mixer, can be configured with different angles and geometries to perform both dispersive and distributive mixing functions.

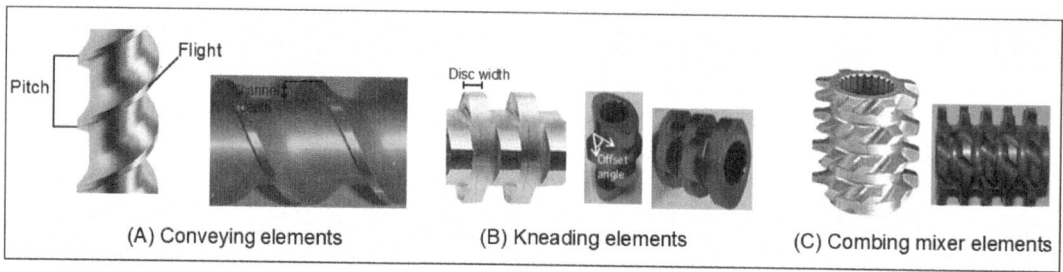

FIGURE 23.3

Common screw elements employed in Leistritz twin-screw extruders, including (A) conveying elements for material transport, (B) Kneading elements for shear-intensive mixing, and (C) combing mixer elements mainly for distributive mixing.

Distributive mixers, in contrast, continuously divide and recombine the material without applying excessive localized pressure. Unlike dispersive mixers, they facilitate material movement through open pathways, promoting mixing at lower energy and stress levels. Narrow kneading discs and combing mixer elements (Fig. 23.3C) are typical examples of distributive mixing components (Thiele, 2018). The design parameters of these screw elements, including the pitch and length of conveying sections, the thickness and offset angles of kneading discs, the number of KEs, and their cross-sectional area, significantly influence the process. A shorter pitch increases residence time by requiring the material to make more revolutions to travel a given distance, whereas a longer pitch enhances conveying capacity (Giles et al., 2005).

Additionally increasing the number of KEs promotes more intense interaction between the powder and granulating liquid, which can result in a higher proportion of oversized agglomerates and fewer fine particles (Vercruysse et al., 2012). Screw design is achieved by arranging various types of screw elements in sequence within the barrel (Fig. 23.1B). This configuration is a crucial process parameter that must be optimized to ensure the production of high-quality granules. Multiple studies have evaluated the effects of different screw elements and configurations on granulation performance (Djuric & Kleinebudde, 2008, 2010; El Hagrasy & Litster, 2013; Keleb et al., 2004c; Li et al., 2014; Meier et al., 2016; Shah, 2005; Thompson & Sun, 2010). An increase in the offset angle of KEs has been associated with a corresponding rise in shear intensity applied to the material during processing (Batra et al., 2017).

The influence of feed rate, SS, and screw configuration on the fill fraction within a TSG was investigated by Lee et al. (2012). Fig. 23.4 presents the overall fill fraction observed under various processing conditions. An increase in feed rate from 10 to 20 g/min at a fixed SS of 300 rpm led to a rise in fill fraction, regardless of the screw configuration. Conversely increasing the SS from 150 to 300 rpm at a constant feed rate of 10 g/min resulted in a reduction in fill fraction, which was attributed to the enhanced conveying capacity of the granulator at higher SSs. Regarding screw configuration, increasing the offset angle of KEs from 30 to 90 degrees, while maintaining constant feed rate and SS, also led to a higher fill fraction. This trend was associated with the reduced conveying efficiency of the screws as the offset angle increased. As illustrated in Fig. 23.4, the granulator operated under starved conditions throughout the study, with fill fractions remaining below 0.3. This suboptimal fill level likely influenced

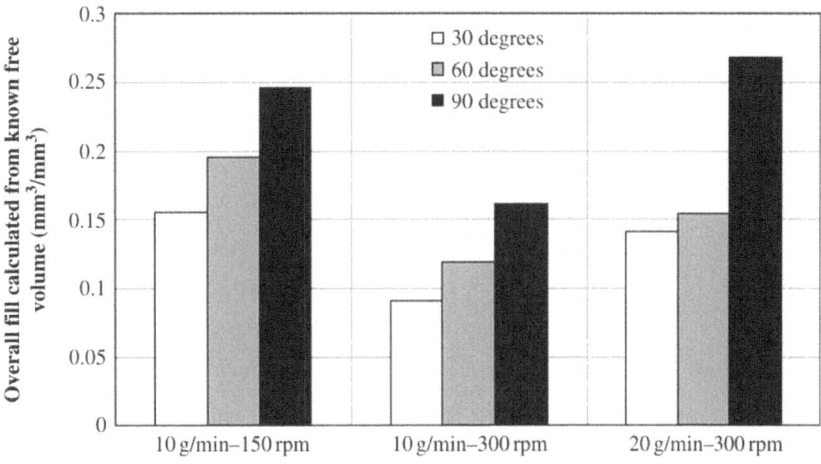

FIGURE 23.4

Overall occupancy of the granulator for various screw geometry at different process parameters. The mixture of α-lactose monohydrate (73.5%), MCC (20%), HPC (5%), and superdisintegrant (1.5%) was used for all the experiments at the constant L/S ratio of 0.35.

Reproduced with permission from Lee et al. (2012) (ref: image reused from the 1st ed).

the compression behavior during granulation and, consequently, the physical attributes of the resulting granules, such as porosity (Lee et al., 2012).

Djuric and Kleinebudde (2008) investigated the influence of various screw elements, specifically conveying elements, combing mixers, and KEs, on the WG of lactose using a twin-screw extruder. Their findings indicated that incorporating kneading blocks into the screw configuration led to complete agglomeration of lactose. In contrast, combining kneading blocks with combing mixers resulted in smaller granule sizes. Additionally granule properties such as porosity, friability, flowability, and tabletability were found to be dependent on the specific screw elements employed. The study concluded that the interplay of different screw geometries enables the production of granules with diverse quality attributes, highlighting the versatility of TSWG. As illustrated in Fig. 23.5, screw designs comprising only conveying elements (GFA 15 to GFA 40) yielded granules with higher porosity compared to those incorporating kneading blocks. For instance, GFA 15 produced granules with a porosity of 50.6%, whereas the inclusion of kneading blocks set at 60 to 90 degrees resulted in significantly lower porosity (17.4%). Furthermore, when compared at identical pitch values, combing mixers generated granules with reduced porosity relative to conveying elements. This reduction is likely due to the enhanced fill capacity provided by the longitudinal slots on the combing mixer elements (Fig. 23.3C).

In a separate study, Lute et al. (2016) also confirmed that the type and number of conveying and KEs, along with their configuration, significantly influence granule characteristics. They found that increasing the number of KEs in conjunction with a higher L/S ratio led to the formation of larger, denser, smoother, and more elongated granules. These effects were attributed to the elevated shear and compaction forces generated by the screws, combined with the enhanced availability of granulating liquid.

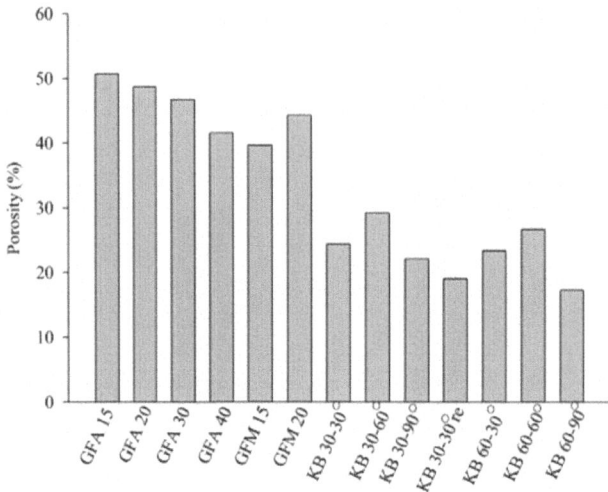

FIGURE 23.5

Granule porosities calculated by using helium and mercury densities; n = 2, mean ± min/max. In this figure, GFA 15, GFA 20, GFA 30, and GFA 40 represent conveying elements with pitch length of 15, 20, 30, and 40 mm, respectively. GFM 15 and GFM 20 represent combing mixer elements with pitch length of 15 and 20 mm, respectively. KB 30-30 degrees represents 30 mm kneading block with offset angle of 30 degrees. Similar labeling can be applied to KB 30-60, KB 30-90, KB 60-60, KB 60-60, and KB 60-90 degrees. KB 30-30 degrees represents a reverse-flighted 30 degrees kneading block with a length of 30 mm.

Reproduced with permission from Djuric and Kleinebudde (2008) (ref: image reused from the 1st ed).

23.5.4 Residence time

TSGs are characterized by significantly shorter residence times, typically measured in seconds, compared to conventional batch granulators, where material often remains in the system for several minutes (Kumar et al., 2013; Martin, 2013). As a result, it is essential to carefully optimize key process parameters, such as screw configuration, SS, and PFR, to ensure effective mixing between the solid material and the granulating liquid. This optimization is crucial for consistently producing high-quality granules.

Several researchers have reported that twin screw WG often results in a bimodal PSD. This means that the granule population commonly shows two distinct peaks, one representing oversized agglomerates or lumps, and the other corresponding to fine, ungranulated particles (El Hagrasy et al., 2013b). In an effort to understand the origin of this phenomenon, Fonteyne et al. (2014a) investigated the distribution of binder within granules using hyperspectral coherent anti-Stokes Raman scattering (CARS) microscopy. They conducted TSG on a formulation composed of theophylline, lactose, and PVP in a weight ratio of 30:67.5:2.5. The analysis revealed that PVP was homogeneously distributed throughout the granules, despite the short residence time. This finding suggested that nonuniform binder distribution was not responsible for the observed bimodal size distribution. Therefore, the authors proposed that other factors might be contributing to this outcome.

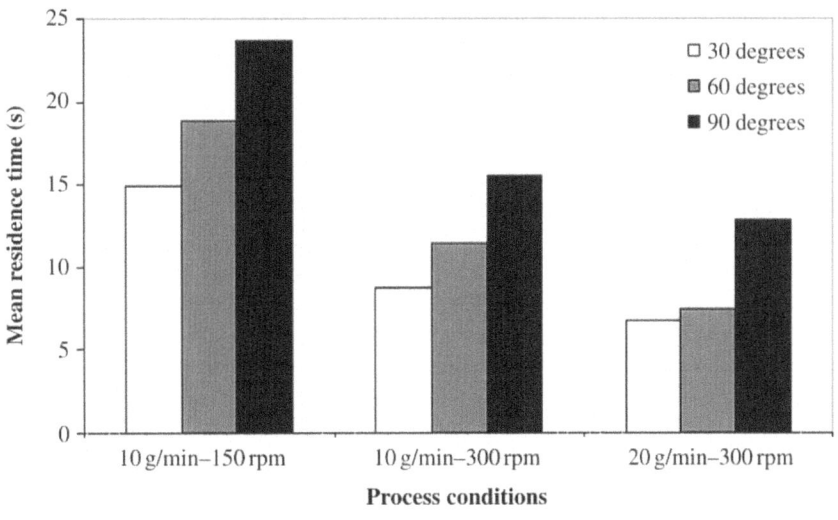

FIGURE 23.6

Mean residence time (MRT) for various screw configuration at different process conditions. α-lactose monohydrate (73.5%), MCC (20%), HPC (5%), and superdisintegrant (1.5%) were used for all the experiments at constant L/S ratio of 0.35.

Reproduced with permission from Lee et al. (2012) (ref: image reused from the 1st ed).

Lee et al. (2012) conducted a comprehensive investigation on the influence of kneading block configuration, specifically the offset angle between kneading discs (30, 60, and 90 degrees), on the MRT under varying feed rates (10 and 20 g/min) and SSs (150 and 300 rpm). The combined effects of screw configuration in the mixing zones and processing parameters on MRT are illustrated in Fig. 23.6. The impact of feed rate and SS is clearly depicted. At a fixed feed rate of 10 g/min and constant screw configuration, an increase in SS from 150 to 300 rpm led to a reduction in MRT, attributed to the enhanced conveying efficiency at higher SSs. Additionally the influence of kneading disc offset angle was evident under identical feed and speed conditions. Across all tested conditions, kneading blocks with a 30 degree offset yielded the shortest MRT, followed by those with a 60 degree offset, while the 90 degree configuration resulted in the longest MRT. This trend reflects the progressive reduction in conveying capacity as the offset angle increases from 30 to 90 degree.

23.5.5 Liquid-to-solid ratio

The concentration of water or granulating liquid used during the TSWG process plays a pivotal role in determining the characteristics of the resulting granules (Dhenge et al., 2012b; Kleinebudde & Lindner, 1993; Meena et al., 2017). This parameter is commonly represented as the L/S ratio, and identifying an optimal L/S value is essential to achieving granules with the desired size distribution and mechanical strength.

Typically increasing the L/S ratio results in a reduction in the proportion of fine particles and a corresponding increase in the fraction of oversized granules. In contrast, reducing the water content below a certain threshold tends to generate a higher proportion of fines (El Hagrasy et al., 2013b; Kumar et al., 2016a, 2016b). According to Dhenge et al. (2010), as the L/S ratio rises, the material within the extruder barrel becomes wetter and more cohesive, which increases its resistance to movement. This results in longer residence times and a noticeable rise in the torque required for processing.

Furthermore, a higher L/S promotes the formation of liquid bridges between particles, contributing to the development of more spherical granules. Conversely lower L/S values tend to produce granules with elongated shapes and rougher surfaces. In the study conducted by Kumar et al. (2016a), a direct correlation was observed between increasing L/S ratios and the formation of oversized granules. This effect was further amplified when the material throughput was also increased.

Optimizing the concentration of water during TSG is essential and must be tailored to the specific characteristics of the formulation. Meena et al. (2017) investigated the granulation behavior of two model compounds: water-soluble lactose and relatively water-insoluble acetaminophen, both formulated with 5% PVP as a binder. Their study revealed that the optimal water content required to achieve high granule yield varied significantly between the two materials, primarily due to differences in their solubility and chemical nature. Lactose, being highly soluble in water, achieved effective surface wetting and optimal granulation with just 6% water. In contrast, when acetaminophen was granulated at the same water concentration, it produced an unacceptably high level of fines, approximately 23%. To achieve an optimal granule size distribution, acetaminophen required a minimum of 7% water.

Further research by Dhenge et al. (2012b) explored the effects of water and aqueous granulating solutions containing HPC at concentrations of 3% and 6% w/v on the granulation of commonly used excipients, including α-lactose monohydrate, MCC, and croscarmellose sodium. Their findings emphasized that the viscosity of the granulating fluid, which is influenced by the binder concentration, has a significant impact on the granulation process. Specifically higher viscosity fluids led to increased residence time and torque within the extruder, along with notable changes in granule characteristics such as size distribution, flowability, and mechanical strength. Under high-viscosity conditions, the granule size distribution tended to become monomodal, and granules exhibited improved strength and flow properties.

Since the residence time of materials in the extruder is relatively short, even slight variations in wetting time can significantly influence granule development (El Hagrasy et al., 2013b; Saleh et al., 2015; Vercruysse et al., 2012). Wetting time, also referred to as wet massing time, describes the duration of mixing between the powder and granulating liquid following the completion of liquid addition (Badawy et al., 2019).

Oka et al. (2015) explored the influence of three critical process parameters, L/S ratio, impeller speed, and wet massing time, on the quality attributes of acetaminophen and MCC granules produced via high-shear WG. Their findings highlighted that both the L/S ratio and the duration of wet massing significantly affected granule porosity. Specifically, a reduction in the L/S ratio from 0.77 to 0.73, coupled with a decrease in wet massing time from seven minutes to three minutes, led to a marked increase in granule porosity. These observations were consistent with earlier work by Badawy et al. (2000), who reported a decrease in tablet hardness as wet massing time increased from zero to three minutes, attributed to reduced granule porosity. Similar mechanisms are believed to operate during continuous WG processes, particularly across various zones within the extruder barrel, as suggested by Dhenge et al. (2012a).

23.5.6 Barrel temperature

While twin screw extruders are capable of operating at elevated temperatures, the majority of reported studies on TSWG have been carried out at ambient conditions, typically around 25°C. Consequently, the granules produced at the end of the extrusion process remain wet and require an additional drying step prior to further processing. Although some temperature rise can occur inside the barrel due to friction between the material, screws, and barrel walls, this increase is generally insufficient to facilitate in-line drying of the granules. As a result, drying must be performed separately using conventional batch techniques such as oven or fluidized bed drying, which often take several hours. This additional drying step disrupts the continuity of the manufacturing process, making downstream operations such as tableting noncontinuous (Dhenge et al., 2012b; Djuric & Kleinebudde, 2008; El Hagrasy et al., 2013b). To address this limitation, recent research efforts have begun exploring the feasibility of in-barrel drying techniques, as discussed in Section 23.9.4.

Liu et al. (2016) investigated the effect of processing temperature on TSWG by evaluating three distinct temperature conditions: 30°C, 55°C, and 80°C. The study focused on formulations composed of HPMC, lactose, or MCC, both with and without acetaminophen, which is known for its hydrophobic nature and poor compactibility. The researchers observed that as the barrel temperature increased, the granules produced became more spherical in shape, accompanied by a reduction in the amount of coarse particles. One of the key findings was that elevated temperatures appeared to decrease the water absorption capacity of certain excipients, particularly HPMC and MCC. This reduction in water uptake left a greater proportion of the added water available for the formation of liquid bridges between particles, which in turn facilitated granule formation. However, the presence of a hydrophobic compound such as acetaminophen introduced complexity into the distribution of water within the granules, potentially affecting the uniformity of granule formation.

In another related study by Fonteyne et al. (2013) employed in-line particle size monitoring using Raman and near-infrared (NIR) spectroscopy to determine the key process parameters affecting granule characteristics during continuous TSG. The study involved a formulation consisting of anhydrous theophylline, lactose monohydrate, and PVP, and examined barrel temperatures ranging from 25°C to 40°C. Results indicated that among the parameters evaluated, PFR, water content, and barrel temperature had the most significant impact on granule size. Specifically an increase in barrel temperature to 40°C resulted in larger median particle sizes (D50) compared to those produced at 25°C under constant feed rate and SS. This was attributed to enhanced solubility of the powder components in water at elevated temperatures, which likely facilitated greater liquid bridge formation and, consequently, the generation of larger granules. These findings were consistent with observations reported by Vercruysse et al. (2012).

23.6 Process monitoring of twin-screw wet granulation

To generate granules of the desired quality, PAT tools are employed in the TSWG process to monitor, control, and optimize the process parameters. Real-time in-process CQA measurement can help detect and remove off-spec products from the stream, thereby maintaining safety and quality. Additionally real-time monitoring of the manufacturing process can aid in visualizing and providing product information throughout the process (Zhao et al., 2023). Since most twin-screw extruders have built-in capabilities for continuous torque monitoring during WG, torque analysis is the most widely utilized in-line monitoring method. This analysis allows for the evaluation of energy consumption during granulation, offering insights into process performance and, ultimately, product quality. Once the optimal torque is established, it is typically maintained within a very narrow range (Meena et al., 2017). Granules produced by the TSWG process are qualitatively and quantitatively analyzed using PAT tools to assess API content, PSD, particle density, moisture content, and residence time distribution (RTD), etc. (Zhao et al., 2023). Numerous studies have investigated the use of PAT tools, including Raman spectroscopy, NIR spectroscopy, and spatial filter velocimetry (SFV), for process monitoring in TSWG. By enhancing process understanding, these methods can improve the quality of the final product. Furthermore, PAT tools are essential for the implementation of process control strategies (Kim et al., 2021).

Fonteyne et al. (2013) measurements for the TSWG process using complementary PAT tools, including an SFV probe, a Raman spectrometer, and an NIR spectrometer. Raman spectroscopy is a potential PAT technique for API determination and process monitoring in the TSWG process. This is evidenced by the development of an in-line approach using Raman spectroscopy to quantify APIs and ascertain API concentration during the process. It should be noted that since light affects Raman spectra, measurements must be conducted in complete darkness. To allow measurements in illuminated environments, the Raman on-line analytical approach must be adjusted (Harting & Kleinebudde, 2018, 2019).

NIR spectroscopy is another widely used PAT technique due to its flexibility and nondestructive nature. It is specific and sensitive to particle properties that affect path length and light penetration (Rosas et al., 2012). To replace the laborious loss-on-drying (LOD) measurements previously used, Haser et al. (2023) investigated NIR spectroscopy as a real-time method for tracking the moisture content of granules as they exit the extruder. In a study by Román-Ospino et al. (2020), particle dynamics and API content of the granules were determined through multivariate analysis of NIR spectra recorded during the process. This analysis involved defining NIR interfaces at various positions within the TSWG process. The mixing and material flow in TSWG have been characterized using NIR chemical imaging (NIR-CI). This qualitative and quantitative analytical technique combines conventional optical imaging with NIR spectroscopy (Kumar et al., 2014). Kumar et al. (2016a) also qualitatively and quantitatively characterized the flow, axial mixing of the tracer material, and mixing of powder and granulating liquid using an NIR-CI tool. This characterization was conducted as a function of SS, material throughput, and L/S ratio. Vercruysse et al. (2014) employed NIR-CI technology to understand the liquid mixing and distribution process during TSG. To maintain granule quality, NIR-CI also enables process optimization and provides enhanced quantification and visualization of particle segregation along the TSG barrel (Mundozah et al., 2019).

Imaging techniques were used to record the temperature of granules. Stauffer et al. (2020) employed an FLIR A655sc infrared camera with a 45 degree lens and a detector to monitor the granule's temperature. It improved the overall understanding of the wetting mechanisms during TSWG when combined with the characteristics of the granules produced by TSWG. Eyecon^{TM} 3D imaging equipment (Innopharma Labs, Dublin, Ireland), a direct imaging particle analyzer, is typically used for TSWG process detection. It provides high-quality in-line images and PSD information for granules with a dense flow (Kumar et al., 2015). When the L/S ratio changes, it can detect size enlargement and a reduction in particle count. The system demonstrates sensitivity to changes and disturbances in the TSWG process (El Hagrasy et al., 2013a; Sayin et al., 2015b). An image processing technique (IPT) for evaluating the two-dimensional (2D) property distributions of granules formed in TSG was introduced by

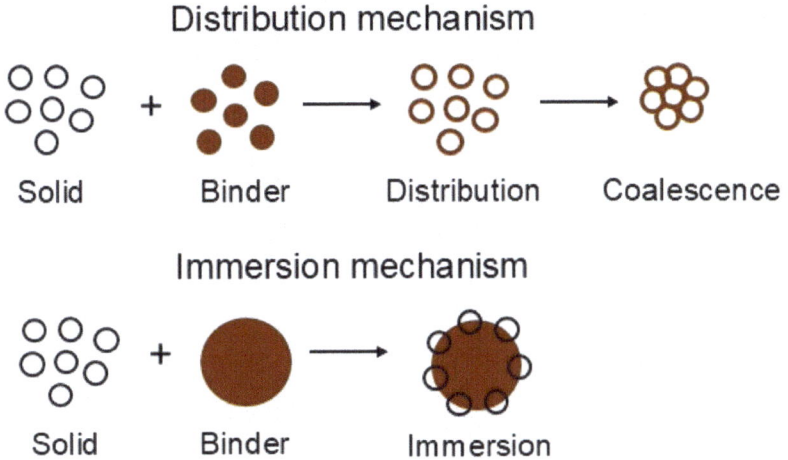

FIGURE 23.7

Nuclei formation mechanism.

From Schæfer and Mathiesen (1996).

Ismail et al. (2021). The entire 2D particle size and liquid content distributions of the granules generated in TSG were measured using this designed IPT. Abdulhussain and Thompson (2021) used ultrasonic acoustic emissions as a nondestructive PAT tool for monitoring the complex PSD of granules produced by TSWG. Lee et al. (2012) utilized the positron emission particle tracking (PEPT) technology to trace the path of individual particles through the conveying and mixing zones of a TSG.

Meng et al. (2019) monitored the TSWG process using three PAT techniques: imaging, NIR, and Raman spectroscopy. Raman spectroscopy measured the content of drug components and solid-state transformations, while Eyecon monitored granule size and shape variations. NIR, combined with chemometrics, predicted the physical properties of granules, including size, porosity, bulk/tapped densities, and flowability. This study illustrated the use of various PAT technologies for continuous in-line and online monitoring of granule production, thereby enhancing the design and monitoring of the TSWG process.

23.7 Granulation mechanism in twin-screw wet granulation and its effect on granule properties

Granulation involves mechanisms such as nucleation, coalescence, consolidation (growth), and breakage. In WG, nucleation refers to the initial stage where powder particles are loosely joined together. These initial agglomerates are wet and rich in liquid, as they form early in the process with minimal agitation when liquid droplets come into contact with powder particles. The nucleation process can follow one of two primary mechanisms, immersion or distribution, which are largely influenced by the manner of liquid addition (Fig. 23.7) (Lute et al., 2016; Schæfer & Mathiesen, 1996). In the immersion mechanism, the liquid droplet size is larger than the particle size. In contrast, the distribution mechanism involves spraying liquid onto the powder, where each small droplet is surrounded by powder particles to form wet masses or nuclei. These larger masses or small nuclei then undergo consolidation, coalescence, and breakage (Iveson et al., 2001). In TSG, the binder is pumped onto the powder inside the barrel, causing the liquid to drip and form immersion-type nuclei (Dhenge et al., 2012a).

The type of mixing components, the length of the mixing zone, and process parameters all significantly influence the mixing characteristics of TSWG. KEs are the most commonly utilized due to their effective distributive and dispersive mixing capabilities. KEs vary in their design based on several factors, including the stagger angle (e.g., forward 30 degree, forward 60 degree, neutral, or reverse), disc width, and the number of discs. Notably the conveying capacity of forward angled KEs diminishes with increasing stagger angle (Andersen, 1998; Thompson, 2015). In addition to KEs, turbine mixing elements (TMEs) are also frequently incorporated into TSWG processes. TMEs consist of multiple parallel tooth rings, generated by radial slotting, which form planes oriented perpendicular to the shaft axis. Variations among TMEs can include differences in tooth orientation, slot dimensions, and the number of teeth per ring. Based on the orientation of the teeth relative to the flow direction, TMEs can be categorized as forward, neutral, or reverse. Unlike KEs, TMEs do not contribute to material conveyance regardless of their tooth orientation (Brouwer et al., 2002; Thompson & Sun, 2010; Zhang et al., 2021).

The granulation mechanism varies along the barrel and is primarily influenced by the screw profiles (Lute et al., 2016). The upstream conveying zone mainly facilitates nucleation, while the mixing zone focuses on compression and packing, leading to granule growth. Granule breakage occurs in both the mixing zone and the downstream conveying zone due to granule-screw interactions (Zhang et al., 2021). Granule growth in the upstream conveying zone is minimal and primarily results from layering, where unwetted powder adheres to wet granules (Verstraeten et al., 2017). The granulation mechanism in the mixing zone is highly complex due to the simultaneous occurrence of additional nucleation, granule growth, and its breakage. In KEs, granule growth primarily occurs through compaction as material is forced through high-shear zones. The formation of ribbon-like granules supports the compaction action of KEs (Rahimi et al., 2020). This compaction forces granules into close contact, causing them to consolidate within a small area. Additionally the compaction produced by the KEs further consolidates the granules, accelerating their growth. Different levels of compaction produce granules with varying porosity. Effective compaction, achieved through a higher staggering angle of KEs, results in denser granules with lower porosity (Li et al., 2019). In contrast to KEs, TMEs do not actively compact the material. Instead, TMEs use packing to promote granule growth. During the TSWG operation, TMEs tend to accumulate wet mass within their structure due to their lack of conveying capability. This accumulation, or packing, may contribute to a reduction in granule porosity. However, the extent of consolidation achieved through packing is generally lower than that obtained via active compaction mechanisms. Comparative studies have shown that granules produced using neutral KEs exhibit greater density and a higher degree of consolidation than those generated by TMEs (Rahimi et al., 2020; Thompson & Sun, 2010).

The fragmentation of granules in the mixing zone is a critical aspect of the TSWG mechanism. There are two possible mechanisms of granule breakage in KEs: First, as granules traverse the adjacent kneading discs, they may be severed by the disc edges. Given that most granules accumulate along the kneading disc edges during operation, this process is anticipated to be predominant in systems with narrow kneading discs. Second, the breakage of granules may result from the radial motion of the kneading discs, which imparts shear forces to the granules (Kumar et al., 2016a; Zhang et al., 2021). In a TME, granule breakage occurs by crushing and chipping mechanism. Crushing is characterized by complete granule disintegration upon entering the intermesh zone between two co-rotating TMEs. During the chipping process, granules are compressed between the barrel wall and the TME teeth. The maximum granule size that can withstand a TME is determined by the volume enclosed by the TME teeth and the barrel wall. Larger granules are more susceptible to chipping (Pradhan et al., 2017; Wang et al., 2020). Further breakdown of oversized granules occurs in the downstream conveying zone before they exit the barrel. In some instances, limited granule growth can also occur through a layering mechanism (Dhenge et al., 2012a).

The more uniform and effective distribution of binders on the granule surface may account for the enhanced properties and improved compactibility of granules produced by TSWG (Fonteyne et al., 2014a; Saleh et al., 2015). The research by Meena et al. (2017) provides further support for this theory. The tensile strength of tablets containing acetaminophen granules with 5% PVP or 5% pregelatinized starch (PGS) as binders, produced by TSWG, increased when the granulation temperature was raised from 50°C to 70°C. The improved and more uniform distribution of binders on drug particles was attributed to the enhanced dissolution of PVP in water or the gelatinization of PGS in water, both of which result from the temperature increase. Additionally they found that acetaminophen-binder (5% PVP or PGS) granules produced by TSWG at 70°C exhibited significantly higher tensile strength compared to granules produced by high-shear granulation at the same temperature. The tablets produced from granules generated by these two techniques had tensile strengths of 1.7–2.1 and 0.4–1.0 MPa, respectively. The higher tensile strength observed in tablets made from TSWG may be attributed to the high shear forces delivered during the process, resulting in a more even mixing and distribution of the drug and binder on the granule surface.

Additional studies by Osei-Yeboah and Sun (2015) and Sun (2017) emphasized the importance of surface coating or particle coating in enhancing the compactibility of weakly compactible drug particles. Osei-Yeboah and Sun (2015) demonstrated that tablets made from physical mixtures of 40% HPC and 60% acetaminophen were frail and did not hold together. In such physical mixtures, there was no coating of the drug particles by the polymer. On the other hand, tablets made from acetaminophen particles coated with 1% to 10% HPC exhibited high tensile strength ranging from 1.9 to 7.0 MPa. Such surface coating without and with the addition of water work similarly. Without the addition of water, a twin-screw extruder can be utilized to perform melt granulation, producing a surface coating of binder on drug particles. Batra et al. (2017) reported an improvement in the tablet tensile strengths (>2 MPa) of poorly compactible drugs, such as acetaminophen and metformin HCl, in the presence of a very small amount of binder (10%). This enhancement was achieved by performing granulation at an elevated temperature sufficient to melt the binder but below the drug's melting point. The high temperature and shear applied in the extruder during melt granulation cause the polymeric binder to transition into a rubbery state, which subsequently coats the surface of the drug particles. This type of coating facilitates the bridging of particles to form granules. The rapid and uniform heat transfer in a twin-screw extruder, as opposed to a traditional high-shear mixer granulator, can result in a consistent distribution of the melted binder throughout the powder bed. These findings clearly indicate that, unlike high-shear granulators, the twin-screw extruder's unique design and enhanced process parameter control enable homogeneous surface coating in twin screw melt granulation (Meena et al., 2017).

23.8 Comparison of twin-screw granulation and conventional batch high-shear granulation

TSWG functions as a continuous and integrated manufacturing process, in contrast to HSWG, which is traditionally executed in batch mode. This fundamental difference leads to significant variations in scalability, processing efficiency, and granule characteristics. TSWG enables scale-up by extending production duration or using parallel equipment lines, without substantially affecting granule properties. In contrast, increasing batch size in HSWG often leads to densification and coalescence, making it challenging to maintain consistent porosity across batches (Beer et al., 2014; Tao et al., 2015).

TSWG typically requires a lower quantity of granulation liquid to achieve granules of the desired size, owing to its efficient mixing dynamics. Additionally TSWG appears to be better at handling high viscosity granulating liquids compared to HSWG (Dhenge et al., 2012b). The granules produced through TSWG are generally less spherical, more porous, and possess intermediate strength and density when compared to those obtained via HSWG. These structural features promote greater fragmentation during tableting, potentially enhancing the tensile strength of the final dosage form (Kyttä et al., 2020; Lee et al., 2013; Portier et al., 2021a). Granules produced through HSWG typically exhibit smoother and denser surfaces due to the intensive consolidation forces applied during processing. Additionally these granules tend to have a narrower PSD compared to those obtained via TSWG (Lee et al., 2013; Macho et al., 2021).

Keleb et al. (2004a) conducted a comparative study evaluating co-rotating TSWG and HSWG for processing various grades of lactose, as well as formulations containing acetaminophen and cimetidine. Across all tested formulations, TSWG consistently produced granules and tablets with higher yield and lower friability than those obtained through HSWG. Furthermore, the twin-screw extrusion process required significantly less water to achieve effective granulation. In another study, Keleb et al. (2004b) demonstrated that tablets manufactured using TSWG exhibited superior quality characteristics, including enhanced tensile strength and reduced disintegration time, compared to those produced via HSWG.

Tan et al. (2011) demonstrated that when a formulation containing 90% acetaminophen and 10% Eudragit EPO was granulated, the high-shear granulator failed to produce tablets with adequate tensile strength, achieving only 0.55 MPa, in contrast to 2.0 MPa obtained with TSWG. This disparity was attributed to insufficient wetting and mixing efficiency in the HSWG process. Additionally the HSWG method required a substantially greater amount of granulation liquid (11.7%–16.5%) compared to the lower quantity needed for TSWG (7.5%), further highlighting the superior efficiency of the twin-screw approach.

23.9 Challenges in twin-screw wet granulation and emerging solutions

23.9.1 Abrasion

In TSG, friction is a critical inherent factor. While it facilitates effective dispersive and distributive mixing, excessive friction can lead to undesirable outcomes such as abrasion of the screw and barrel surfaces (Stauffer et al., 2019). In a study conducted by Menth et al. (2022), tablets produced via

TSWG exhibited visual defects, specifically gray spots on their surfaces, attributed to abrasion-related discoloration. These spots were linked to a gray film formed by powder accumulation on the inner barrel wall during processing, likely due to abrasive interactions. The researchers evaluated this phenomenon using the wall friction angle (WFA), a parameter that quantifies the friction between the powder and equipment surfaces. Measurements were performed using the Schulze ring shear tester (Menth et al., 2022). Their findings indicated that a lower WFA correlated with a reduction in tablet surface defects. As a result, wall friction measurements were proposed as a valuable tool for both equipment selection and defining optimal process conditions before initiating TSWG trials. Furthermore, WFA assessment proved useful not only for early-stage process development but also as a diagnostic aid for identifying and mitigating abrasion-related issues (Hancock, 2019; Menth et al., 2022; Pillai et al., 2007).

23.9.2 Post-granulation milling

TSWG often results in the formation of elongated, rod-like, or oversized agglomerates, especially under suboptimal processing conditions. These oversized granules often necessitate a subsequent milling step to achieve the desired PSD. For instance, operating at low L/S ratios can produce long, rough granules (Arthur & Rahmanian, 2024), while certain screw configurations have been associated with a high proportion (36%–78%) of oversized particles (Kiricenko et al., 2024). These irregularly shaped granules, particularly needle-like ones, typically exhibit poor flowability and may hinder consistent feeding during downstream operations (Megarry et al., 2020). Furthermore, a broad size distribution can prolong drying times and negatively affect tablet quality by causing surface defects or weakening mechanical strength. To address these challenges, a postgranulation comminution step, often using a high-shear conical mill, is commonly employed (Kotamarthy et al., 2020). However, integrating such a milling unit into a continuous manufacturing line presents significant drawbacks. It increases equipment requirements, energy usage, and material handling complexity while disrupting the seamless nature of continuous processing. Although modifications to the formulation such as increasing binder content or to the process settings can improve granule morphology and reduce the formation of oversized particles, many formulations still require additional size reduction (Megarry et al., 2020). As a result, post-granulation milling remains a well-recognized limitation in TSWG, often cited as a barrier to achieving truly uninterrupted, efficient continuous granulation.

23.9.3 Post-granulation drying

TSWG typically requires an additional drying step, as the process is generally conducted at room temperature (Dhenge et al., 2012b; Djuric & Kleinebudde, 2008; El Hagrasy et al., 2013b). Even when modestly elevated temperatures (up to 40°C) are applied during extrusion, the resulting granules remain wet and must be dried before they can undergo further processing (Vercruysse et al., 2012). These moist granules are usually transferred to an external drying unit, such as a continuous fluid bed or vibratory belt dryer. In many cases, conventional drying methods like oven or fluid bed drying are employed, with durations ranging from 4 to 48 hours depending on the drying temperature used (Dhenge et al., 2012b; Djuric & Kleinebudde, 2008; El Hagrasy et al., 2013b; Lute et al., 2016; Vanhoorne et al., 2016b).

This post-granulation drying introduces significant challenges to the process. It adds complexity by requiring additional equipment, such as pneumatic or vibratory transfer systems, to move the wet granules from the extruder to the dryer. This increases both capital investment and operational costs. Moreover, the throughput of continuous dryers is limited by their fixed airflow and residence volume, making it difficult to handle high L/S ratios (Portier et al., 2021a; Ryckaert et al., 2021). Excessive liquid loads can reduce drying efficiency, necessitating longer drying times and potentially creating bottlenecks in production (Barriga et al., 2023).

Beyond logistical and economic drawbacks, these additional steps can negatively impact granule quality. For instance, pneumatic transport of wet granules has been associated with high levels of breakage and attrition, leading to increased fines formation (Ryckaert et al., 2021). The presence of excess fines not only broadens the PSD but also increases the risk of segregation during downstream blending and tablet compression. This can ultimately compromise content uniformity and tablet integrity (Portier et al., 2021a; Ryckaert et al., 2021). Therefore, the need for a separate drying step in TSWG remains a critical limitation from both process and product quality perspectives.

23.9.4 Recent advancements to eliminate multiple unit operations

The address the limitations in TSWG highlighted in Sections 23.9.2 and 23.9.3, specifically the need for separate drying and milling steps, recent developments aim to eliminate these multiple unit operations by integrating them into semi-continuous or fully continuous process. Fully integrated powder-to-tablet manufacturing lines, such as Consigma developed by GEA Pharma Systems and QbCon developed by L.B. Bohle, exemplify this transition. In these systems, raw materials including APIs and excipients are individually introduced into a continuous inline blender through the use of gravimetric feeders. The resulting blend is directed into a TSG, where wet granules are produced and subsequently transferred either pneumatically or by gravimetric means into a continuous or semicontinuous drying unit. After drying, the granules are milled to achieve the target PSD. These granules are then blended with extragranular excipients such as magnesium stearate, which are also introduced via gravimetric feeders. The final blend is processed using a rotary tablet press. Following compression, the tablets undergo dedusting and, if required, coating before being directed to the packaging unit. These integrated systems are often equipped with real-time PAT tools, such as NIR moisture probes, which facilitate recipe-driven control and enable quality to be designed into the process through QbD principles (Portier et al., 2021a).

Further advancements in continuous manufacturing were demonstrated in a study by Fülöp et al. (2021), where low-dose (50 µg) tablets with excellent uniformity were produced using a fully integrated system that included a TSWG, a continuous fluid bed dryer, and a continuous sieving unit. Wet granules were dried on a vibratory belt in the CFBD and then passed directly through a continuous sieving unit. This streamlined sequence minimized manual intervention, reduced fines, and ensured a consistent granule size distribution.

In another study by Meena et al. (2017), it was demonstrated that twin-screw extruders can be used to optimize processing temperatures to achieve *in-situ* drying of granules, thereby removing the need for a subsequent drying step. By adjusting the screw configuration, the researchers were also able to obtain granules within an acceptable size range for direct tableting. Their findings indicated that at an elevated temperature of approximately 70°C, granules could be sufficiently dried within the extruder, eliminating the necessity for additional drying equipment such as a fluid bed dryer. Furthermore, there was no requirement for post-drying milling. The study also proposed that the tableting process could continue in a fully integrated manner, with lubricants being applied in-line during transfer to the tablet press or externally through spray lubrication of the punches and dies during compression (Meena et al., 2017; Wang et al., 2010). The TSWG process developed by Meena et al. (2017) is illustrated in Fig. 23.8, which shows that the wet granules produced can not only be dried to <1% moisture content by conducting extrusion at an elevated temperature of 70°C, the granule size can also be optimized by selecting appropriate processing parameters. When fully optimized, this process can lead to fully continuous manufacturing of tablets, starting from the blending of powders to the production of dry granules ready for compression into tablets, without the need for intermediate drying and granule size reduction steps.

In another study, in-barrel drying was implemented, allowing WG and drying to occur within a single extruder by applying heated zones and vacuum. Haser et al. (2023) demonstrated this method using an 18 mm twin-screw extruder (40:1 L/D), where the initial zones were used for liquid addition and granulation, while the later zones were dedicated to drying at 100°C to 110°C under vacuum. This setup successfully reduced the residual moisture content to below 2% in a single continuous pass. The use of inline NIR sensors enabled real-time moisture monitoring of wet granules, facilitating dynamic adjustments in barrel temperature and vacuum to meet targeted LOD values. This integrated, single-unit approach significantly reduces processing time and equipment footprint by eliminating the need for separate drying steps.

23.10 Scale-up in twin-screw wet granulation

Djuric et al. (2009) compared two TSGs with screw diameters of 19 and 27 mm, featuring markedly different screw geometries, for an understanding of the scalability of the TSWG process. A 2^3 full-factorial design of experiments (DOE) was employed to examine the effects of PFR and SS at three levels, while maintaining a constant L/S ratio. The PFR levels of 2, 4, and 6 kg/h were consistent across both scales. To assess the dynamic similarity between the two systems, the Froude number (Fr) (Eq. 23.1) was utilized. The findings revealed significant differences in product properties based on the granulator configurations. The authors suggested that further research should be undertaken with granulators that have more geometrically comparable setups. Osorio et al. (2017) formulated scaling rules through dimensional analysis and evaluated three distinct scales (11, 16, and 24 mm). These scaling rules were derived to maintain consistent L/S ratios, Fr, which are utilized in high-shear granulation to characterize the ratio of angular to gravitational acceleration, and powder feed number (PFN) (Eq. 23.2) to estimate fill levels. The results showed that the particle sizes of the unmilled granules were primarily influenced by the L/S ratio and the scale of the granulator. An increase in the L/S ratio led to larger granule sizes. The d_{90} parameter was significantly influenced by the TSG scale, with lump sizes increasing roughly linearly with the barrel diameter. Conversely the Fr and PFN had no notable impact on d_{10} or d_{50} and only a minimal effect on d_{90}. Menth et al. (2020) conducted a scale-up using a screw configuration consisting solely of conveying elements. The three control tests were performed across three different scales (11, 16, and 24 mm) with PFRs of 1.5, 5, and 20 kg/h, respectively. The experiments aimed to examine the effect of the Fr while keeping the SS, PFN, and L/S ratio constant. Due to the drying process being a limiting factor in this investigation, MCC was excluded to maintain low L/S ratios. They achieved a successful scale-up with consistent tablet properties by maintaining constant fill level inside the barrel, SS, and the moisture levels in the granules during the initial scale-up approach. Consequently they adjusted the PFR and liquid feed rate according to the scale.

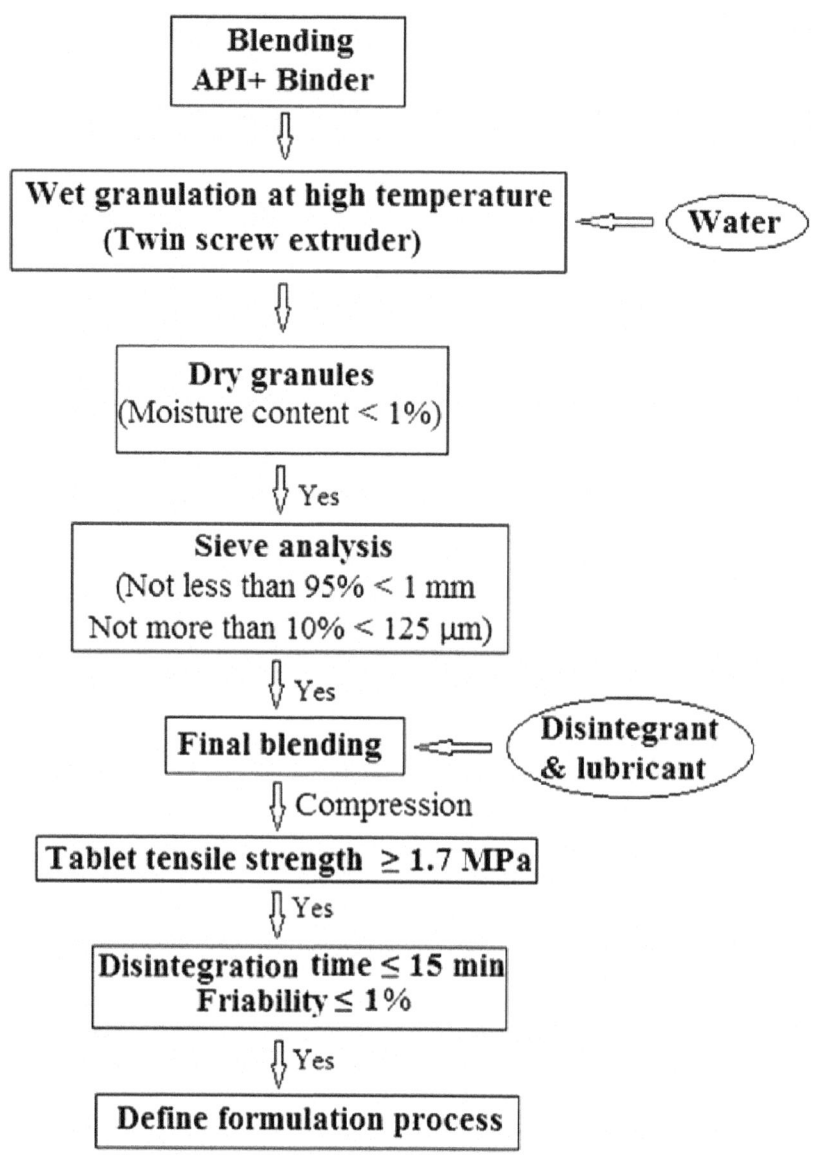

FIGURE 23.8

Schematic representation (flowchart) of granulation and tableting processes with the acceptance criteria used by Meena et al. (2017) in the investigations of a twin-screw granulation process.

Reproduced with permission from Meena et al. (2017) (ref: image reused from the 1st ed).

Franke et al. (2023a) conducted the process transfer from a QbCon 1 line (16 mm screw diameter) to a QbCon 25 line (25 mm screw diameter) by employing three distinct scale-up procedures while maintaining a similar PFN. These approaches were based on RTD, as well as circumferential speed (CS) and SS. They found that the chosen scale-up technique did not affect the granule and tablet properties for the selected formulation, despite significant variations in the adjusted process parameters. As long as the L/S ratio and PFN remained constant, there were no appreciable differences in product attributes across methods and scales. Overall, it was demonstrated that the impact of the L/S ratio on the characteristics of granules and tablets was significantly greater than the effects of variations in lines or scale-up techniques. This study highlighted the need to test their scale-up strategies on different formulations to check for the robustness of TSWG process. Thus, Franke et al. (2023b) utilized three different formulations within a design space (screw configurations and scale-up strategies) similar to that developed earlier to investigate the impact of various formulations and APIs on the scale-up process (Franke et al., 2023a). The results of their investigation demonstrated that TSWG can be successfully scaled up with different formulations, exhibiting a high degree of robustness. This is achievable provided that the L/S ratio and screw configurations remain consistent across both scales, and the flowability of the granulation blend is sufficient to ensure continuous refilling of the large granulation unit via the automatic feeding and blending system. The study further emphasized that TSWG is a robust and versatile process. It allows for most process optimizations to be performed on a small scale before being successfully transferred to a larger production line.

$$Fr = \frac{Screw\,diameter \times (Screw\,speed)^2}{acceleration\,due\,to\,gravity\,(g)} \tag{23.1}$$

$$PFN = \frac{Powder\,feed\,rate}{Screw\,speed \times Powder\,bulk\,density \times (Screw\,diameter)^3} \tag{23.2}$$

23.11 Summary

This chapter presents an extensive review of the continuous TSWG process and its emerging significance as a modern alternative to conventional batch-based WG in pharmaceutical manufacturing. It begins with a general overview of granulation and emphasizes the functional advantages of granules over powders in the formulation of solid oral dosage forms. These include improved powder flow, enhanced compressibility, and better content uniformity (Thapa et al., 2019). TSWG is introduced as a continuous processing technology that offers notable benefits, particularly in terms of operational efficiency, scalability, and consistent product quality (Schaber et al., 2011).

Key sections of the chapter focus on the influence of various process parameters, such as SS, feed rate, screw element configuration, MRT, processing temperature, and the L/S ratio, on granule formation and quality. In addition, the review explores the role of formulation-related factors including the physicochemical properties of APIs, binder selection and concentration, the type of excipients used, and the method of binder incorporation (Portier et al., 2021a). Mechanistic insights into the granulation process are also included, detailing stages such as wetting, nucleation, coalescence, consolidation, growth, and breakage, and how these phenomena are distributed across different regions of the twin screw extruder barrel (Lute et al., 2016; Schæfer & Mathiesen, 1996)

The application of PAT tools, such as NIR and Raman spectroscopy, is discussed in the context of real-time process monitoring and control. It also highlights comparisons between TSWG and HSWG, focusing on distinctions in liquid usage, granule morphology, and process scalability. Additionally it discusses the limitations of TSWG, including the need for postgranulation drying and milling, as well as challenges related to equipment abrasion. Recent research efforts aimed at addressing these limitations, such as in-barrel drying and fully integrated powder-to-tablet systems, are also presented (Haser et al., 2023; Meena et al., 2017). The section concludes with insights into reported scale-up strategies, reflecting the growing industrial interest in implementing TSWG as part of continuous pharmaceutical manufacturing platforms.

References

Abdulhussain, H., & Thompson, M. (2021). Predicting the particle size distribution in twin screw granulation through acoustic emissions. *Powder Technology, 394*, 757–766.

Andersen, P. (1998). The Werner and Pfleiderer twin-screw corotating extruder system. In David B. Todd (Ed.). In *Plastics compounding: equipment and processing: 133* (pp. 71–123). Hanser Publishers, Munich.

Arthur, T. B., & Rahmanian, N. (2024). Process Simulation of twin-screw granulation: A review. *Pharmaceutics, 16*, 706.

Badawy, S. I., Narang, A. S., LaMarche, K. R., Subramanian, G. A., & Varia, S. A. (2019). Mechanistic basis for the effects of process parameters on quality attributes in high shear wet granulation. In Ajit S. Narang, & Sherif I. F. Badawy (Eds.), *Handbook of pharmaceutical wet granulation* (pp. 89–118). Elsevier.

Badawy, S. I. F., Menning, M. M., Gorko, M. A., & Gilbert, D. L. (2000). Effect of process parameters on compressibility of granulation manufactured in a high-shear mixer. *International Journal of Pharmaceutics, 198*, 51–61.

Baert, L., Remon, J. P., Elbers, J., & Van Bommel, E. (1993). Comparison between a gravity feed extruder and a twin screw extruder. *International Journal of Pharmaceutics, 99*, 7–12.

Bansal, A. K., Balwani, G., & Sheokand, S. (2019). Critical material attributes in wet granulation. In Ajit S. Narang, & Sherif I. F. Badawy (Eds.), *Handbook of pharmaceutical wet granulation* (pp. 421–453). Elsevier.

Barriga, R., Romero, M., Hassan, H., & Nettleton, D. F. (2023). Energy consumption optimization of a fluid bed dryer in pharmaceutical manufacturing using EDA (exploratory data analysis). *Sensors, 23*, 3994.

Batra, A., Desai, D., & Serajuddin, A. T. (2017). Investigating the use of polymeric binders in twin screw melt granulation process for improving compactibility of drugs. *Journal of Pharmaceutical Sciences, 106*, 140–150.

Beer, P., Wilson, D., Huang, Z., & De Matas, M. (2014). Transfer from high-shear batch to continuous twin screw wet granulation: A case study in understanding the relationship between process parameters and product quality attributes. *Journal of Pharmaceutical Sciences, 103*, 3075–3082.

Breitenbach, J. (2002). Melt extrusion: From process to drug delivery technology. *European Journal of Pharmaceutics and Biopharmaceutics, 54*, 107–117.

Brouwer, T., Todd, D., & Janssen, L. (2002). Flow characteristics of screws and special mixing enhancers in a co-rotating twin screw extruder. *International Polymer Processing, 17*, 26–32.

De Leersnyder, F., Vanhoorne, V., Bekaert, H., Vercruysse, J., Ghijs, M., Bostijn, N., Verstraeten, M., Cappuyns, P., Van Assche, I., & Vander Heyden, Y. (2018). Breakage and drying behaviour of granules in a continuous fluid bed dryer: Influence of process parameters and wet granule transfer. *European Journal of Pharmaceutical Sciences, 115*, 223–232.

Dhenge, R. M., Cartwright, J. J., Doughty, D. G., Hounslow, M. J., & Salman, A. D. (2011). Twin screw wet granulation: Effect of powder feed rate. *Advanced Powder Technology, 22*, 162–166.

Dhenge, R. M., Cartwright, J. J., Hounslow, M. J., & Salman, A. D. (2012a). Twin screw granulation: Steps in granule growth. *International Journal of Pharmaceutics, 438*, 20–32.

Dhenge, R. M., Cartwright, J. J., Hounslow, M. J., & Salman, A. D. (2012b). Twin screw wet granulation: Effects of properties of granulation liquid. *Powder Technology, 229*, 126–136.

Dhenge, R. M., Fyles, R. S., Cartwright, J. J., Doughty, D. G., Hounslow, M. J., & Salman, A. D. (2010). Twin screw wet granulation: Granule properties. *Chemical Engineering Journal, 164*, 322–329.

Djuric, D., & Kleinebudde, P. (2008). Impact of screw elements on continuous granulation with a twin-screw extruder. *Journal of Pharmaceutical Sciences, 97*, 4934–4942.

Djuric, D., & Kleinebudde, P. (2010). Continuous granulation with a twin-screw extruder: Impact of material throughput. *Pharmaceutical Development and Technology, 15*, 518–525.

Djuric, D., Van Melkebeke, B., Kleinebudde, P., Remon, J. P., & Vervaet, C. (2009). Comparison of two twin-screw extruders for continuous granulation. *European Journal of Pharmaceutics and Biopharmaceutics, 71*, 155–160.

Elbers, J., Bakkenes, H., & Fokkens, J. (1992). Effect of amount and composition of granulation liquid on mixing, extrusion and spheronization. *Drug Development and Industrial Pharmacy, 18*, 501–517.

El Hagrasy, A., Cruise, P., Jones, I., & Litster, J. (2013a). In-line size monitoring of a twin screw granulation process using high-speed imaging. *Journal of Pharmaceutical Innovation, 8*, 90–98.

El Hagrasy, A., Hennenkamp, J., Burke, M., Cartwright, J., & Litster, J. (2013b). Twin screw wet granulation: Influence of formulation parameters on granule properties and growth behavior. *Powder Technology, 238*, 108–115.

El Hagrasy, A., & Litster, J. (2013). Granulation rate processes in the kneading elements of a twin screw granulator. *AIChE Journal, 59*, 4100–4115.

Fonteyne, M., Correia, A., De Plecker, S., Vercruysse, J., Ilić, I., Zhou, Q., Vervaet, C., Remon, J. P., Onofre, F., & Bulone, V. (2015). Impact of microcrystalline cellulose material attributes: A case study on continuous twin screw granulation. *International Journal of Pharmaceutics, 478*, 705–717.

Fonteyne, M., Fussell, A. L., Vercruysse, J., Vervaet, C., Remon, J. P., Strachan, C., Rades, T., & De Beer, T. (2014a). Distribution of binder in granules produced by means of twin screw granulation. *International Journal of Pharmaceutics, 462*, 8–10.

Fonteyne, M., Vercruysse, J., Díaz, D. C., Gildemyn, D., Vervaet, C., Remon, J. P., & Beer, T. D. (2013). Real-time assessment of critical quality attributes of a continuous granulation process. *Pharmaceutical Development and Technology, 18*, 85–97.

Fonteyne, M., Wickström, H., Peeters, E., Vercruysse, J., Ehlers, H., Peters, B.-H., Remon, J. P., Vervaet, C., Ketolainen, J., & Sandler, N. (2014b). Influence of raw material properties upon critical quality attributes of continuously produced granules and tablets. *European Journal of Pharmaceutics and Biopharmaceutics, 87*, 252–263.

Franke, M., Riedel, T., Meier, R., Schmidt, C., & Kleinebudde, P. (2023a). Comparison of scale-up strategies in twin-screw wet granulation. *International Journal of Pharmaceutics, 641*, 123052.

Franke, M., Riedel, T., Meier, R., Schmidt, C., & Kleinebudde, P. (2023b). Scale-up in twin-screw wet granulation: Impact of formulation properties. *Pharmaceutical Development and Technology, 28*, 948–961.

Fülöp, G., Domokos, A., Galata, D., Szabó, E., Gyürkés, M., Szabó, B., Farkas, A., Madarász, L., Démuth, B., & Lendér, T. (2021). Integrated twin-screw wet granulation, continuous vibrational fluid drying and milling: A fully continuous powder to granule line. *International Journal of Pharmaceutics, 594*, 120126.

Gamlen, M., & Eardley, C. (1986). Continuous extrusion using a raker perkins MP50 (multipurpose) extruder. *Drug Development and Industrial Pharmacy, 12*, 1701–1713.

Giles, H. F., Wagner, J. R., & Mount, E. M. (2005). Twin Screw Extruder Equipment. In H. F. Giles, J. R. Wagner, & E. M. Mount (Eds.), *Extrusion: the definitive processing guide and handbook* (pp. 95–113). New York, SAD: William Andrew Inc.

Gorringe, L., Kee, G., Saleh, M., Fa, N., & Elkes, R. (2017). Use of the channel fill level in defining a design space for twin screw wet granulation. *International Journal of Pharmaceutics, 519*, 165–177.

Hancock, B. C. (2019). The wall friction properties of pharmaceutical powders, blends, and granulations. *Journal of Pharmaceutical Sciences, 108*, 457–463.

Harting, J., & Kleinebudde, P. (2018). Development of an in-line Raman spectroscopic method for continuous API quantification during twin-screw wet granulation. *European Journal of Pharmaceutics and Biopharmaceutics, 125*, 169–181.

Harting, J., & Kleinebudde, P. (2019). Optimisation of an in-line Raman spectroscopic method for continuous API quantification during twin-screw wet granulation and its application for process characterisation. *European Journal of Pharmaceutics and Biopharmaceutics, 137*, 77–85.

Haser, A., Kittikunakorn, N., Dippold, E., DiNunzio, J. C., & Blincoe, W. (2023). Continuous twin-screw wet granulation process with in-barrel drying and NIR setup for real-time moisture monitoring. *International Journal of Pharmaceutics, 630*, 122377.

Hausman, D. S. (2004). Comparison of low shear, high shear, and fluid bed granulation during low dose tablet process development. *Drug Development And Industrial Pharmacy, 30*, 259–266.

Hwang, K.-M., Cho, C.-H., Yoo, S.-D., Cha, K.-I., & Park, E.-S. (2019). Continuous twin screw granulation: Impact of the starting material properties and various process parameters. *Powder Technology, 356*, 847–857.

Ismail, H. Y., Albadarin, A. B., Iqbal, J., & Walker, G. M. (2021). Image processing for detecting complete two dimensional properties' distribution of granules produced in twin screw granulation. *International Journal of Pharmaceutics, 600*, 120472.

Ito, A., & Kleinebudde, P. (2019). Influence of granulation temperature on particle size distribution of granules in twin-screw granulation (TSG). *Pharmaceutical Development and Technology, 24*, 874–882.

Iveson, S. M., Litster, J. D., Hapgood, K., & Ennis, B. J. (2001). Nucleation, growth and breakage phenomena in agitated wet granulation processes: A review. *Powder Technology, 117*, 3–39.

Järvinen, M. A., Paavola, M., Poutiainen, S., Itkonen, P., Pasanen, V., Uljas, K., Leiviskä, K., Juuti, M., Ketolainen, J., & Järvinen, K. (2015). Comparison of a continuous ring layer wet granulation process with batch high shear and fluidized bed granulation processes. *Powder Technology, 275*, 113–120.

Kao, S., & Allison, G. (1984). Residence time distribution in a twin screw extruder. *Polymer Engineering & Science, 24*, 645–651.

Keary, C. M., & Sheskey, P. J. (2004). Preliminary report of the discovery of a new pharmaceutical granulation process using foamed aqueous binders. *Drug Development and Industrial Pharmacy, 30*, 831–845.

Keleb, E., Vermeire, A., Vervaet, C., & Remon, J. P. (2004a). Extrusion granulation and high shear granulation of different grades of lactose and highly dosed drugs: A comparative study. *Drug Development and Industrial Pharmacy, 30*, 679–691.

Keleb, E., Vermeire, A., Vervaet, C., & Remon, J. P. (2004b). Single-step granulation/tabletting of different grades of lactose: A comparison with high shear granulation and compression. *European Journal of Pharmaceutics and Biopharmaceutics, 58*, 77–82.

Keleb, E., Vermeire, A., Vervaet, C., & Remon, J. P. (2004c). Twin screw granulation as a simple and efficient tool for continuous wet granulation. *International Journal of Pharmaceutics, 273*, 183–194.

Khorsheed, B., Gabbott, I., Reynolds, G. K., Taylor, S. C., Roberts, R. J., & Salman, A. D. (2019). Twin-screw granulation: Understanding the mechanical properties from powder to tablets. *Powder Technology, 341*, 104–115.

Kim, E. J., Kim, J. H., Kim, M.-S., Jeong, S. H., & Choi, D. H. (2021). Process analytical technology tools for monitoring pharmaceutical unit operations: A control strategy for continuous process verification. *Pharmaceutics, 13*, 919.

Kiricenko, K., Meier, R., & Kleinebudde, P. (2024). Systematic investigation of the impact of screw elements in continuous wet granulation. *International Journal of Pharmaceutics: X, 8*, 100273.

Kleinebudde, P., & Lindner, H. (1993). Experiments with an instrumented twin-screw extruder using a single-step granulation/extrusion process. *International Journal of Pharmaceutics, 94*, 49–58.

Kolter, K., Karl, M., Gryczke, A., & Ludwigshafen am Rhein, B. (2012). Hot-melt extrusion with BASF pharma polymers: Extrusion compendium. In *Process Variables of Twin-Screw Extruders* (2nd edition) (pp. 35–53).

Kotamarthy, L., Metta, N., & Ramachandran, R. (2020). Understanding the effect of granulation and milling process parameters on the quality attributes of milled granules. *Processes, 8*, 683.

Kumar, A., Alakarjula, M., Vanhoorne, V., Toiviainen, M., De Leersnyder, F., Vercruysse, J., Juuti, M., Ketolainen, J., Vervaet, C., & Remon, J. P. (2016a). Linking granulation performance with residence time and granulation liquid distributions in twin-screw granulation: An experimental investigation. *European Journal of Pharmaceutical Sciences, 90*, 25–37.

Kumar, A., Dhondt, J., De Leersnyder, F., Vercruysse, J., Vanhoorne, V., Vervaet, C., Remon, J. P., Gernaey, K. V., De Beer, T., & Nopens, I. (2015). Evaluation of an in-line particle imaging tool for monitoring twin-screw granulation performance. *Powder Technology, 285*, 80–87.

Kumar, A., Dhondt, J., Vercruysse, J., De Leersnyder, F., Vanhoorne, V., Vervaet, C., Remon, J. P., Gernaey, K. V., De Beer, T., & Nopens, I. (2016b). Development of a process map: A step towards a regime map for steady-state high shear wet twin screw granulation. *Powder Technology, 300*, 73–82.

Kumar, A., Gernaey, K. V., De Beer, T., & Nopens, I. (2013). Model-based analysis of high shear wet granulation from batch to continuous processes in pharmaceutical production–a critical review. *European Journal of Pharmaceutics and Biopharmaceutics, 85*, 814–832.

Kumar, A., Vercruysse, J., Toiviainen, M., Panouillot, P.-E., Juuti, M., Vanhoorne, V., Vervaet, C., Remon, J. P., Gernaey, K. V., & De Beer, T. (2014). Mixing and transport during pharmaceutical twin-screw wet granulation: Experimental analysis via chemical imaging. *European Journal of Pharmaceutics and Biopharmaceutics, 87*, 279–289.

Kyttä, K. M., Lakio, S., Wikström, H., Sulemanji, A., Fransson, M., Ketolainen, J., & Tajarobi, P. (2020). Comparison between twin-screw and high-shear granulation—The effect of filler and active pharmaceutical ingredient on the granule and tablet properties. *Powder Technology, 376*, 187–198.

Lakshman, J. P., Kowalski, J., Vasanthavada, M., Tong, W. Q., Joshi, Y. M., & Serajuddin, A. T. (2011). Application of melt granulation technology to enhance tabletting properties of poorly compactible high-dose drugs. *Journal of Pharmaceutical Sciences, 100*, 1553–1565.

Lee, K. T., Ingram, A., & Rowson, N. A. (2012). Twin screw wet granulation: The study of a continuous twin screw granulator using positron emission particle tracking (PEPT) technique. *European Journal of Pharmaceutics and Biopharmaceutics, 81*, 666–673.

Lee, K. T., Ingram, A., & Rowson, N. A. (2013). Comparison of granule properties produced using twin screw extruder and high shear mixer: A step towards understanding the mechanism of twin screw wet granulation. *Powder Technology, 238*, 91–98.

Li, H., Thompson, M., & O'donnell, K. (2014). Understanding wet granulation in the kneading block of twin screw extruders. *Chemical Engineering Science, 113*, 11–21.

Li, H., Thompson, M., & O'donnell, K. (2015). Examining drug hydrophobicity in continuous wet granulation within a twin screw extruder. *International Journal of Pharmaceutics, 496*, 3–11.

Li, J., Pradhan, S. U., & Wassgren, C. R. (2019). Granule transformation in a twin screw granulator: Effects of conveying, kneading, and distributive mixing elements. *Powder Technology, 346*, 363–372.

Lindberg, N.-O., Tufvesson, C., & Olbjer, L. (1987). Extrusion of an effervescent granulation with a twin screw extruder, Baker Perkins MPF 50 D. *Drug Development and Industrial Pharmacy, 13*, 1891–1913.

Liu, Y., Thompson, M., O'donnell, K., & Grasman, N. (2016). Effect of temperature on the wetting behavior of hydroxypropyl methylcellulose in a twin-screw granulator. *Powder Technology, 302*, 63–74.

Lute, S. V., Dhenge, R. M., Hounslow, M. J., & Salman, A. D. (2016). Twin screw granulation: Understanding the mechanism of granule formation along the barrel length. *Chemical Engineering Research and Design, 110*, 43–53.

Macho, O., Gabrišová, Ľ., Peciar, P., Juriga, M., Kubinec, R., Rajniak, P., Svačinová, P., Vařilová, T., & Šklubalová, Z. (2021). Systematic study of the effects of high shear granulation parameters on process yield, granule size, and shape by dynamic image analysis. *Pharmaceutics, 13*, 1894.

Martin, C. (2013). Twin screw extrusion for pharmaceutical processes. In Michael A. Repka, Nigel Langley, & James DiNunzio (Eds.), *Melt extrusion: Materials, technology and drug product design* (pp. 47–79). Springer.

Meena, A. K., Desai, D., & Serajuddin, A. T. (2017). Development and optimization of a wet granulation process at elevated temperature for a poorly compactible drug using twin screw extruder for continuous manufacturing. *Journal of Pharmaceutical Sciences, 106*, 589–600.

Megarry, A., Taylor, A., Gholami, A., Wikström, H., & Tajarobi, P. (2020). Twin-screw granulation and high-shear granulation: The influence of mannitol grade on granule and tablet properties. *International Journal of Pharmaceutics, 590*, 119890.

Meier, R., Moll, K.-P., Krumme, M., & Kleinebudde, P. (2017). Impact of fill-level in twin-screw granulation on critical quality attributes of granules and tablets. *European Journal of Pharmaceutics and Biopharmaceutics, 115*, 102–112.

Meier, R., Thommes, M., Rasenack, N., Moll, K.-P., Krumme, M., & Kleinebudde, P. (2016). Granule size distributions after twin-screw granulation—Do not forget the feeding systems. *European Journal of Pharmaceutics and Biopharmaceutics, 106*, 59–69.

Meng, W., Román-Ospino, A. D., Panikar, S. S., O'Callaghan, C., Gilliam, S. J., Ramachandran, R., & Muzzio, F. J. (2019). Advanced process design and understanding of continuous twin-screw granulation via implementation of in-line process analytical technologies. *Advanced Powder Technology, 30*, 879–894.

Menth, J., Maus, M., & Wagner, K. G. (2020). Continuous twin screw granulation and fluid bed drying: A mechanistic scaling approach focusing optimal tablet properties. *International Journal of Pharmaceutics, 586*, 119509.

Menth, J., Maus, M., & Wagner, K. G. (2022). Assessment of abrasion-induced visual defects in twin screw wet granulation using wall friction measurements. *AAPS PharmSciTech, 23*, 47.

Mundozah, A. L., Yang, J., Tridon, C. C., Cartwright, J. J., Omar, C. S., & Salman, A. D. (2019). Assessing particle segregation using near-infrared chemical imaging in twin screw granulation. *International Journal of Pharmaceutics, 568*, 118541.

Oka, S., Emady, H., Kašpar, O., Tokárová, V., Muzzio, F., Štěpánek, F., & Ramachandran, R. (2015). The effects of improper mixing and preferential wetting of active and excipient ingredients on content uniformity in high shear wet granulation. *Powder Technology, 278*, 266–277.

Osei-Yeboah, F., & Sun, C. C. (2015). Tabletability modulation through surface engineering. *Journal of Pharmaceutical Sciences, 104*, 2645–2648.

Osorio, J. G., Sayin, R., Kalbag, A. V., Litster, J. D., Martinez-Marcos, L., Lamprou, D. A., & Halbert, G. W. (2017). Scaling of continuous twin screw wet granulation. *AIChE Journal, 63*, 921–932.

Parrott, E. L. (1981). Densification of powders by concavo-convex roller compactor. *Journal of Pharmaceutical Sciences, 70*, 288–291.

Peeters, M., Jiménez, A. A. B., Matsunami, K., Ghijs, M., dos Santos Schultz, E., Roudgar, M., Vigh, T., Stauffer, F., Nopens, I., & De Beer, T. (2024). Analysis of the effect of formulation properties and process parameters on granule formation in twin-screw wet granulation. *International Journal of Pharmaceutics, 650*, 123671.

Peeters, M., Jiménez, A. A. B., Matsunami, K., Stauffer, F., Nopens, I., & De Beer, T. (2023). Evaluation of the influence of material properties and process parameters on granule porosity in twin-screw wet granulation. *International Journal of Pharmaceutics, 641*, 123010.

Pillai, J. R., Bradley, M., & Berry, R. (2007). Comparison between the angles of wall friction measured on an on-line wall friction tester and the Jenike wall friction tester. *Powder Technology, 174*, 64–70.

Portier, C., Pandelaere, K., Delaet, U., Vigh, T., Di Pretoro, G., De Beer, T., Vervaet, C., & Vanhoorne, V. (2020a). Continuous twin screw granulation: A complex interplay between formulation properties, process settings and screw design. *International Journal of Pharmaceutics, 576*, 119004.

Portier, C., Vervaet, C., & Vanhoorne, V. (2021a). Continuous twin screw granulation: A review of recent progress and opportunities in formulation and equipment design. *Pharmaceutics, 13*, 668.

Portier, C., Vigh, T., Di Pretoro, G., De Beer, T., Vervaet, C., & Vanhoorne, V. (2020b). Continuous twin screw granulation: Impact of binder addition method and surfactants on granulation of a high-dosed, poorly soluble API. *International Journal of Pharmaceutics, 577*, 119068.

Portier, C., Vigh, T., Di Pretoro, G., Leys, J., Klingeleers, D., De Beer, T., Vervaet, C., & Vanhoorne, V. (2021b). Continuous twin screw granulation: Impact of microcrystalline cellulose batch-to-batch variability during granulation and drying—A QbD approach. *International Journal of Pharmaceutics: X, 3*, 100077.

Pradhan, S. U., Sen, M., Li, J., Litster, J. D., & Wassgren, C. R. (2017). Granule breakage in twin screw granulation: Effect of material properties and screw element geometry. *Powder Technology, 315*, 290–299.

Rahimi, S. K., Paul, S., Sun, C. C., & Zhang, F. (2020). The role of the screw profile on granular structure and mixing efficiency of a high-dose hydrophobic drug formulation during twin screw wet granulation. *International Journal of Pharmaceutics, 575*, 118958.

Rocca, K., Weatherley, S., Sheskey, P., & Thompson, M. (2015). Influence of filler selection on twin screw foam granulation. *Drug Development and Industrial Pharmacy, 41*, 35–42.

Román-Ospino, A. D., Tamrakar, A., Igne, B., Dimaso, E. T., Airiau, C., Clancy, D. J., Pereira, G., Muzzio, F. J., Singh, R., & Ramachandran, R. (2020). Characterization of NIR interfaces for the feeding and in-line monitoring of a continuous granulation process. *International Journal of Pharmaceutics, 574*, 118848.

Rosas, J. G., Blanco, M., Gonzalez, J. M., & Alcalà, M. (2012). Real-time determination of critical quality attributes using near-infrared spectroscopy: A contribution for process analytical technology (PAT). *Talanta, 97*, 163–170.

Ryckaert, A., Ghijs, M., Portier, C., Djuric, D., Funke, A., Vervaet, C., & De Beer, T. (2021). The influence of equipment design and process parameters on granule breakage in a semi-continuous fluid bed dryer after continuous twin-screw wet granulation. *Pharmaceutics, 13*, 293.

Saleh, M. F., Dhenge, R. M., Cartwright, J. J., Hounslow, M. J., & Salman, A. D. (2015). Twin screw wet granulation: Binder delivery. *International Journal of Pharmaceutics, 487*, 124–134.

Sayin, R., El Hagrasy, A., & Litster, J. (2015a). Distributive mixing elements: Towards improved granule attributes from a twin screw granulation process. *Chemical Engineering Science, 125*, 165–175.

Sayin, R., Martinez-Marcos, L., Osorio, J. G., Cruise, P., Jones, I., Halbert, G. W., Lamprou, D. A., & Litster, J. D. (2015b). Investigation of an 11 mm diameter twin screw granulator: Screw element performance and in-line monitoring via image analysis. *International Journal of Pharmaceutics, 496*, 24–32.

Schaber, S. D., Gerogiorgis, D. I., Ramachandran, R., Evans, J. M., Barton, P. I., & Trout, B. L. (2011). Economic analysis of integrated continuous and batch pharmaceutical manufacturing: a case study. *Industrial & Engineering Chemistry Research, 50*, 10083–10092.

Schæfer, T., & Mathiesen, C. (1996). Melt pelletization in a high shear mixer. IX. Effects of binder particle size. *International Journal of Pharmaceutics, 139*, 139–148.

Schmidt, A., de Waard, H., Moll, K.-P., Kleinebudde, P., & Krumme, M. (2018). Simplified end-to-end continuous manufacturing by feeding API suspensions in twin-screw wet granulation. *European Journal of Pharmaceutics and Biopharmaceutics, 133*, 224–231.

Shah, U. (2005). Use of a modified twin-screw extruder to develop a high-strength tablet dosage form. *Pharmaceutical Technology, 29*, 52–66.

Stauffer, F., Ryckaert, A., Van Hauwermeiren, D., Funke, A., Djuric, D., Nopens, I., & De Beer, T. (2019). Heat transfer evaluation during twin-screw wet granulation in view of detailed process understanding. *Aaps Pharmscitech, 20*, 1–13.

Stauffer, F., Ryckaert, A., Vanhoorne, V., Van Hauwermeiren, D., Funke, A., Djuric, D., Vervaet, C., Nopens, I., & De Beer, T. (2020). In-line temperature measurement to improve the understanding of the wetting phase in twin-screw wet granulation and its use in process development. *International Journal of Pharmaceutics, 584*, 119451.

Sun, C. C. (2017). Microstructure of tablet—pharmaceutical significance, assessment, and engineering. *Pharmaceutical Research, 34*, 918–928.

Suresh, P., Sreedhar, I., Vaidhiswaran, R., & Venugopal, A. (2017). A comprehensive review on process and engineering aspects of pharmaceutical wet granulation. *Chemical Engineering Journal, 328*, 785–815.

Tan, L., Carella, A. J., Ren, Y., & Lo, J. B. (2011). Process optimization for continuous extrusion wet granulation. *Pharmaceutical Development and Technology, 16*, 302–315.

Tao, J., Pandey, P., Bindra, D. S., Gao, J. Z., & Narang, A. S. (2015). Evaluating scale-up rules of a high-shear wet granulation process. *Journal of Pharmaceutical Sciences, 104*, 2323–2333.

Thapa, P., Tripathi, J., & Jeong, S. H. (2019). Recent trends and future perspective of pharmaceutical wet granulation for better process understanding and product development. *Powder Technology, 344*, 864–882.

Thiele, W. (2018). *Twin-screw extrusion and screw design, Pharmaceutical extrusion technology* (pp. 71–94). CRC Press.

Thompson, M. (2015). Twin screw granulation–review of current progress. *Drug Development and Industrial Pharmacy, 41*, 1223–1231.

Thompson, M., & O'Donnell, K. (2015). Rolling" phenomenon in twin screw granulation with controlled-release excipients. *Drug Development and Industrial Pharmacy, 41*, 482–492.

Thompson, M., & Sun, J. (2010). Wet granulation in a twin-screw extruder: Implications of screw design. *Journal of Pharmaceutical Sciences, 99*, 2090–2103.

Thompson, M., Weatherley, S., Pukadyil, R., & Sheskey, P. (2012). Foam granulation: New developments in pharmaceutical solid oral dosage forms using twin screw extrusion machinery. *Drug Development and Industrial Pharmacy, 38*, 771–784.

Vandevivere, L., Denduyver, P., Portier, C., Häusler, O., De Beer, T., Vervaet, C., & Vanhoorne, V. (2020). Influence of binder attributes on binder effectiveness in a continuous twin screw wet granulation process via wet and dry binder addition. *International Journal of Pharmaceutics, 585*, 119466.

Vandevivere, L., Vangampelaere, M., Portier, C., de Backere, C., Häusler, O., De Beer, T., Vervaet, C., & Vanhoorne, V. (2021). Identifying critical binder attributes to facilitate binder selection for efficient formulation development in a continuous twin screw wet granulation process. *Pharmaceutics, 13*, 210.

Vanhoorne, V., Almey, R., De Beer, T., & Vervaet, C. (2020). Delta-mannitol to enable continuous twin-screw granulation of a highly dosed, poorly compactable formulation. *International Journal of Pharmaceutics, 583*, 119374.

Vanhoorne, V., Bekaert, B., Peeters, E., De Beer, T., Remon, J. P., & Vervaet, C. (2016a). Improved tabletability after a polymorphic transition of delta-mannitol during twin screw granulation. *International Journal of Pharmaceutics, 506*, 13–24.

Vanhoorne, V., Janssens, L., Vercruysse, J., De Beer, T., Remon, J. P., & Vervaet, C. (2016b). Continuous twin screw granulation of controlled release formulations with various HPMC grades. *International Journal of Pharmaceutics, 511*, 1048–1057.

Vanhoorne, V., Vanbillemont, B., Vercruysse, J., De Leersnyder, F., Gomes, P., De Beer, T., Remon, J. P., & Vervaet, C. (2016c). Development of a controlled release formulation by continuous twin screw granulation: Influence of process and formulation parameters. *International Journal of Pharmaceutics, 505*, 61–68.

Vasanthavada, M., Wang, Y., Haefele, T., Lakshman, J. P., Mone, M., Tong, W., Joshi, Y. M., & Serajuddin, A. T. (2011). Application of melt granulation technology using twin-screw extruder in development of high-dose modified-release tablet formulation. *Journal of Pharmaceutical Sciences, 100*, 1923–1934.

Vercruysse, J., Díaz, D. C., Peeters, E., Fonteyne, M., Delaet, U., Van Assche, I., De Beer, T., Remon, J. P., & Vervaet, C. (2012). Continuous twin screw granulation: Influence of process variables on granule and tablet quality. *European Journal of Pharmaceutics and Biopharmaceutics, 82*, 205–211.

Vercruysse, J., Toiviainen, M., Fonteyne, M., Helkimo, N., Ketolainen, J., Juuti, M., Delaet, U., Van Assche, I., Remon, J. P., & Vervaet, C. (2014). Visualization and understanding of the granulation liquid mixing and distribution during continuous twin screw granulation using NIR chemical imaging. *European Journal of Pharmaceutics and Biopharmaceutics, 86*, 383–392.

Verstraeten, M., Van Hauwermeiren, D., Lee, K., Turnbull, N., Wilsdon, D., Am Ende, M., Doshi, P., Vervaet, C., Brouckaert, D., & Mortier, S. T. (2017). In-depth experimental analysis of pharmaceutical twin-screw wet granulation in view of detailed process understanding. *International Journal of Pharmaceutics, 529*, 678–693.

Wang, J., Wen, H., & Desai, D. (2010). Lubrication in tablet formulations. *European Journal of Pharmaceutics and Biopharmaceutics, 75*, 1–15.

Wang, L. G., Pradhan, S. U., Wassgren, C., Barrasso, D., Slade, D., & Litster, J. D. (2020). A breakage kernel for use in population balance modelling of twin screw granulation. *Powder Technology, 363*, 525–540.

Willecke, N., Szepes, A., Wunderlich, M., Remon, J. P., Vervaet, C., & De Beer, T. (2018). A novel approach to support formulation design on twin screw wet granulation technology: Understanding the impact of overarching excipient properties on drug product quality attributes. *International Journal of Pharmaceutics, 545*, 128–143.

Wu, Y., Levons, J., Narang, A. S., Raghavan, K., & Rao, V. M. (2011). Reactive impurities in excipients: Profiling, identification and mitigation of drug–excipient incompatibility. *Aaps Pharmscitech, 12*, 1248–1263.

Yeh, A.-I., Hwang, S.-J., & Guo, J.-J. (1992). Effects of screw speed and feed rate on residence time distribution and axial mixing of wheat flour in a twin-screw extruder. *Journal of Food Engineering, 17*, 1–13.

Yu, S., Reynolds, G. K., Huang, Z., de Matas, M., & Salman, A. D. (2014). Granulation of increasingly hydrophobic formulations using a twin screw granulator. *International Journal of Pharmaceutics, 475*, 82–96.

Zhang, Y., Liu, T., Kashani-Rahimi, S., & Zhang, F. (2021). A review of twin screw wet granulation mechanisms in relation to granule attributes. *Drug Development and Industrial Pharmacy, 47*, 349–360.

Zhao, J., Tian, G., & Qu, H. (2023). Pharmaceutical application of process understanding and optimization techniques: A review on the continuous twin-screw wet granulation. *Biomedicines, 11*, 1923.

Zhou, D., Porter, W. R., & Zhang, G. G. (2017). Drug stability and degradation studies. In Yihong Qiu, Yisheng Chen, Geoff G. z. Zhang, Lawrence Yu, & Rao V. Mantri (Eds.), *Developing solid oral dosage forms* (pp. 113–149). Elsevier.

CHAPTER 24

Melt granulation: Granulation mechanisms, formulation and process design for batch and twin-screw systems

Adwait Pradhan[a], Niyati Niranjan Kodange[a], Fengyuan Yang[b], Kapish Karan[b], Thomas Durig[c] and Feng Zhang[a]

[a]*Division of Molecular Pharmaceutics and Drug Delivery, The University of Texas at Austin, Austin, TX, United States,* [b]*Pharmaceutical Technology, Ashland Specialty Ingredients, Wilmington, DE, United States,* [c]*Barentz North America HQ, Avon, OH, United States*

24.1 Introduction

Granulation is one of the oldest and most widely used unit operations in the pharmaceutical industry particularly for the manufacture of oral solid dosage forms like tablets and capsules. It is an agglomeration process that involves the particle enlargement of powders through binding, typically using a combination of polymers and common solvents like water, alcohol (Shanmugam, 2015). Granulation is frequently adopted in the pharmaceutical industry to enhance the flowability, processability and tabletability of powders, along with reducing the risk of segregation and improving uniformity in development of oral solid dosage forms (Tovey, 2018). Over the past few decades, granulation has evolved through two primary methodologies: dry and wet granulation followed by technological advancements in melt granulation (Liu et al., 2021), steam granulation (Rodriguez et al., 2002), foam granulation (Keary & Sheskey, 2004), moisture-assisted dry granulation (MADG) (Ullah, 1987).

Among these, melt granulation has gained prominence as a solvent-free alternative to the conventional wet granulation process. It is perfectly suitable for processing moisture-sensitive API's, and the absence of a drying step after granulation significantly minimizes the energy consumption, cost, and time (Shanmugam, 2015; Walker et al., 2006). For instance, metformin hydrochloride tablets formulated via melt granulation were stronger and less friable than dry and wet granulation, showcasing its promise as a processing technique (Lakshman et al., 2011). This chapter presents a comprehensive review of melt granulation from both formulation and processing, with a focus on continuous melt granulation/Twin-screw melt granulation (TSMG).

Melt granulation is a thermal particle enlargement and densification process where granule growth is achieved through molten or a meltable binder(Kittikunakorn, Liu, & Zhang, 2020; Passerini, Calogerà, Albertini, & Rodriguez, 2010). As a result, melt granulation is also referred to as melt agglomeration, wax granulation, thermoplastic granulation, hot melt granulation, etc.(Royce, Suryawanshi, Shah, & Vishnupad, 1996; Schaefer, Holm, & Kristensen, 1990; Shanmugam, 2015). McTaggart et al. first introduced melt granulation in 1984 for the granulation of isoxepac using a Henschel mixer, where the impact of wax type, drug-wax ratio, and batch size was investigated (McTaggart et al., 1984). Later, Royce et al. (1996) demonstrated the feasibility of using melt granulation as an alternative for processing moisture-sensitive drugs in a high-shear Collete Gral mixer—traditionally used for wet-granulation—successfully developing both immediate and sustained release dosage forms of clemastine fumarate (Royce, Suryawanshi, Shah, & Vishnupad, 1996). However, the foundational framework for melt granulation was established through the systematic studies by Schæfer, Kristensen, and Mathiesen et al. investigating the impact of process variables, binder particle size, binder viscosity for granulation of binary mixtures and cohesive powders (Schæfer, 1996a, 1996b; Schaefer, Holm, & Kristensen, 1990; Schæfer & Mathiesen, 1996a, 1996b; Schæfer, Taagegaard, Juul Thomsen, & Gjelstrup Kristensen, 1993). Since then, extensive research has been conducted using both batch and continuous processing approaches.

Today, melt granulation finds applications in a wide range of pharmaceutical contexts, including for processing high dose and moisture-sensitive API (Batra et al., 2017), fixed dose combination products (Kelleher et al., 2020), solubility enhancement (Kipping, 2024), taste masking (Forster & Lebo, 2021), controlled and sustained release dosage forms (Royce, Suryawanshi, Shah, & Vishnupad, 1996), osmotic release tablets (Panda & Tiwary, 2012) and even manufacturing directly compressible excipients (Pradhan et al., 2022). These versatile applications are made possible through a careful selection of the thermal binder with hydrophilic binders like PEG, hydroxypropyl cellulose used for immediate release, while waxes, fatty acids are the binder of choice for sustained/controlled release formulations (Wong et al., 2005). A summary of applications of melt granulation with the binder used and equipment utilized is provided in Table 24.1. Despite the comprehensive research over the past 30 years, commercial success has been limited to a few products, including Eucreas, Orlissa, Oriahnn, Certican, Fenoglide, and Comtess (Forster et al., 2021; Parikh, 2005; Qiu, 2024; Vahervuo & Orion Oyj, 2008).

24.2 Similarities in the granulation mechanism between wet and melt granulation processes

Shanmugam et al.(2015) describe melt granulation as a type of wet granulation, wherein a meltable binder replaces the granulating fluid typically used in a conventional wet-granulation process (Shanmugam, 2015). The underlying granulation mechanisms in both processes share notable similarities. The mechanisms for wet granulation through high shear mixer or fluid bed processor proposed by Ennis and Litster consisted of three main steps (Iveson, Litster, Hapgood, & Ennis, 2001a; Litster & Ennis, 2004)

1. Wetting and nucleation
2. Granule growth and consolidation
3. Granule attrition

Despite the absence of a solvent drying step and its role in bridge formation of granules, the general mechanisms for melt granulation follow the same steps as wet granulation: nucleation, granule growth, and granule attrition (Liu, Kittikunakorn, Zhang, & Zhang, 2021; Parikh, 2005)

Table 24.1 Applications of melt granulation with summary of drug, binder, and the processing equipment.

Sr no.	Application	Drug	Optimal Binder	Processing Equipment	References
1	Immediate release	Fevipiprant	Hydroxypropylcellulose (Klucel)	Twin-screw extruder	Vicente Martin et al. (2022)
2		Metformin hydrochloride	Hydroxypropylcellulose (Klucel)	Twin-screw extruder	Lakshman et al. (2011)
3		Carbamazepine	PEG 4000	High shear mixer	Perissutti et al. (2003)
4	Sustained release	Ondansetron hydrochloride	Stearic acid / Ethyl cellulose	Twin-screw extruder	Patil et al. (2015)
5		Theophylline	PEG 6000 / Glyceryl monostearate	High shear mixer	Ochoa et al. (2010)
6	Extended-release	Lovastatin	PEG 6000	High shear mixer	Ochoa et al. (2011)
7	Controlled release	Tramadol hydrochloride	Glyceryl behenate	Twin-screw extruder	Keen et al. (2015)
8		Ibuprofen	Glycerylpalmitostearate (Precirol ATO 5)	Fluid bed processor	Prado et al. (2014)

(continued on next page)

Table 24.1 Applications of melt granulation with summary of drug, binder, and the processing equipment—cont'd

Sr no.	Application	Drug	Optimal Binder	Processing Equipment	References
9	Taste masking	Ibuprofen	Glycerylpalmitostearate (Precirol ATO 5)	Twin-screw extruder	Forster and Lebo (2021)
10	Fixed dose combination	Acetaminophen, Ibuprofen	Kollicoat IR	Twin-screw extruder	Munnangi et al. (2024)
11	Osmotic Release	Glipizide	PEG 6000	Twin-screw extruder	Panda and Tiwary (2012)
12	Dissolution/Solubility enhancement	Indomethacin	PEG 6000	Fluid bed processor	Andrade et al. (2015)
13		Simvastatin	Soluplus + Kollicoat IR	Twin-screw extruder	Elkanayati et al. (2024)
14		Fenofibrate	PEG 3350	Twin-screw extruder	Mamidi et al. (2021)
15		Ibuprofen	Poloxamer 188	High shear mixer	Passerini et al. (2002)
16	Directly compressible excipient	Mannitol	Hydroxypropylcellulose (Klucel)	Twin-screw extruder	Pradhan et al. (2022)
17	Effervescent tablets	Citric acid / sodium bicarbonate	PEG 6000	Fluid bed processor	Yanze et al. (2000)

24.2.1 Nucleation mechanism

The nucleation phase of the melt granulation process involves the formation of small agglomerates as the binder melts and interacts with powders. Two primary mechanisms have been identified: distribution and immersion. The distribution mechanism resembles the wet granulation process, where the binder coats the drug particles owing to coalescence forces, forming a thin film around them (Iveson, Litster, Hapgood, & Ennis, 2001a; Schæfer & Mathiesen, 1996a). The immersion mechanism involves drug particles adhering to the large binder droplets, resulting in the formation of nuclei through agglomeration (Iveson, Litster, Hapgood, & Ennis, 2001a; Schæfer & Mathiesen, 1996a). These two mechanisms are influenced by several factors like the size ratio of binder to drug particles, rheological properties of binder, and processing temperature (Schæfer & Mathiesen, 1996a, 1996b). Studies by Johansen & Schæfer et al. and Mu & Thompson et al. have investigated the occurrence of these mechanisms with batch (high-shear and fluid bed processors) and continuous processes (TSMG), respectively (Johansen & Schæfer, 2001; Mu & Thompson, 2012; Zhai et al., 2010). Generally, a lower binder particle size or binder viscosity favors the distribution mechanism, whereas a higher binder particle size or binder viscosity favors an immersion mechanism (Johansen & Schæfer, 2001; Mu & Thompson, 2012; Steffens & Wagner, 2019). The impact of nucleation mechanisms on granule porosity and binder distribution has been investigated, and they are observed to influence critical quality attributes (CQA) like tensile strength and disintegration/dissolution (Schæfer et al., 2004).

24.2.2 Granule growth

Following nucleation, granule growth occurs primarily through a coalescence/layering-based mechanism, depending on the binder liquid saturation, similar to a wet granulation process (Iveson et al., 2001b). This coalescence/fusion of nuclei must be carefully controlled to achieve a uniform particle size distribution. Schæfer et al (1990) reported that a binder saturation of 80-100% was necessary for effective agglomerate growth in batch-melt granulation (Schaefer et al., 1990). However, the impact of liquid saturation on granule growth can also be enhanced by molecular mixing or capillary forces between the drug and binder, promoting better fusion of nuclei (Liu et al., 2021). These findings underscore the importance of binder level and drug-binder interactions in controlling granule growth/consolidation in melt granulation. The modular nature of screws in a continuous process like TSMG facilitates efficient molecular mixing in a confined space, enhancing granule consolidation with lower binder levels compared to batch processes (Liu et al., 2021).

24.2.3 Granule attrition

The final stage of granule attrition and breakage is primarily governed by the processing equipment and mechanical strength of the granules. Granule strength is a result of the interparticle forces (capillary, frictional, and viscous) and mechanical forces at play during the granule growth process, and its relationship with porosity, binder viscosity, and nucleation mechanism is well investigated in literature (Liu et al., 2020; Monteyne, Vancoillie, Remon, Vervaet, & De Beer, 2016d; Zhai, Li, Jones, Walker, & Andrews, 2010). Granules formed via the immersion mechanism tend to have binder at the core, resulting in non-uniform binder distribution, lower porosity, but higher granule strength, contrary to granules formed via the distribution mechanism (Zhai et al., 2010). Ultimately, the granule properties

Table 24.2 Critical quality attributes for a melt granulation process and which processing equipment is beneficial for achieving the desired properties. (1 means difficult to achieve, 2 means a narrow design space to achieve and 3 means easy to achieve).

Critical Quality Attribute	High-shear Mixer	Fluid Bed Processor	Twin-screw Extruder
Improvement in tabletability	2	2	3
Spherical shape	3	1	1
Uniform particle size distribution	1	2	3
Porosity	2	3	1
Density	3	2	2
Flow	3	2	3

Modified and compiled from Parikh et al. (2021); Lakshman et al. (2011); Passerini et al. (2010) and Steffens et al. (2020).

are shaped by the interplay of these three stages and are critical in defining CQA of the final dosage form.

24.3 Critical quality attributes for melt granulation

Granule strength, bulk density, flowability, drug release, and particle size distribution are routinely evaluated during process optimization of granulation processes. The selection of a suitable design space is guided by the desired granule attributes, such as improved tabletability/flowability or accelerated dissolution rate. Melt granulation has demonstrated the ability to enhance the dissolution rate of different drugs, like griseofulvin (Yang et al., 2007), carbamazepine (Perissutti, Rubessa, Moneghini, & Voinovich, 2003), and ibuprofen (Passerini, Albertini, González-Rodrı́guez, Cavallari, & Rodriguez, 2002), primarily due to the incorporation of hydrophilic binders. However, the extent of dissolution enhancement is influenced by the binder type and nature of drug-binder interactions. These aspects are further elaborated in section 24.4. Other granule properties referred to as CQA are influenced by processing equipment, as shown in Table 24.2. Depending on the mechanism through which granule growth occurs, some granule properties are favored through specific processes. For instance, TSMG has been shown to significantly improve tabletability, as supported by multiple studies (Kittikunakorn, Koleng, Listro, Sun, & Zhang, 2019b; Lakshman et al., 2011; Liu et al., 2020; Pradhan et al., 2022). This understanding can be leveraged to build a decision tree for selecting the most appropriate processing equipment based on the desired granule properties.

24.4 Formulation design for melt granulation

Despite the use of various processing equipment in melt granulation, a consistent feature across all the processing techniques is the simplicity of formulation composition. The granule growth process is initiated by the melting and softening of only the binder; the other components, like drug, diluent, and flow aid, do not melt during processing. Nevertheless, the drug does undergo in-process physicochemical changes, not limited to particle size reduction, amorphization, polymorphism, and even degradation due to the high-energy environment during processing (Liu et al., 2021). Such effects have been documented during the melt granulation of gabapentin (Kittikunakorn et al., 2019), gliclazide (Liu et al., 2022), and acetaminophen (Liu et al., 2020) using the twin-screw extruder. Therefore, successful formulation optimization requires a thorough understanding of the physicochemical properties of the drug, binder, and their potential interactions. This section attempts to provide a brief overview of the same.

24.4.1 Binder as the most important formulation variable

Since densification is achieved through the melting or softening of the binder, the choice of binder is paramount not only for achieving the drug release profile but also for ensuring optimal granule attributes. The desired formulation properties are directly influenced by the choice of binder (hydrophobic or hydrophilic), binder molecular weight, and binder level, and their impact has been thoroughly investigated in the literature by various researchers. This section summarizes the impact of binder properties on the melt granulation process.

24.4.1.1 Binder type

Depending on the desired release, a hydrophilic or hydrophobic binder is chosen, with hydrophilic binders preferred for immediate release formulations and hydrophobic binders preferred for sustained release formulations. Table 24.3 provides a detailed summary of the two classes of binders, their melting/softening temperatures, and their operating ranges available in literature. Depending on the targeted release profile, a suitable chemistry binder can be selected.

The hydrophilic binders can also be split into two main classes based on their melting points: wax-based binders or polymeric binders. Wax-based binders are typically melted due to their low melting point and injected through a liquid nozzle or sprayed, while polymeric binders are thermoplastic polymers added as solids, and they undergo softening/melting upon exposure to elevated temperatures (Dürig & Karan, 2021). Due to their low melting temperatures, waxy binders are primarily employed in batch melt granulation processes (high-shear mixer and fluid-bed) (Andrade, Martins, & Freitas, 2015; Ochoa et al., 2010, 2011; Perissutti, Rubessa, Moneghini, & Voinovich, 2003)

24.4.1.2 Binder viscosity

Binder chemistry plays a pivotal role in determining binder viscosity, and its influence is modulated by the type of processing equipment used. In batch processes, such as high-shear mixers or fluid bed processors, low-viscosity, meltable binders are preferred to ensure uniform coating of drug particles. On the contrary, the twin-screw extruder relies on frictional heat dissipation, facilitating the use of high-viscosity binders like thermoplastic polymers in TSMG.

The effect of binder viscosity on granule properties in the batch melt granulation process is complex and dependent on the nucleation mechanism, granule strength, and the mechanical forces responsible for agglomerate breakage. Schæfer & Mathiesen et al.(1996b) and Walker et al.(2006) reported slower initial granule growth with high viscosity binders during melt granulation of lactose in high-shear and fluid-bed processes, respectively (Schæfer & Mathiesen, 1996b; Walker, Andrews, & Jones, 2006). However, over the course of the process, contradicting trends were observed by the researchers, highlighting the dynamic interplay between granule growth and mechanical stresses. Zhai et al. demonstrated that a more viscous binder, Poloxamer 407, led to a faster onset of granulation with larger binder particles, whereas the opposite was observed for lower sizes (Zhai et al., 2010). This behavior was attributed to

Table 24.3 List of binders used in melt granulation with their transition/melting temperature and binder levels.

Chemical Name	Glass Transition Temperature (°C)	Melting Temperature (°C)	Use Range (%)	References
		Hydrophilic Polymers		
Hydroxypropylcellulose (Klucel)	25, 120	N/A	5–30	Lakshman et al. (2011), Pradhan et al. (2022),Liu et al., (2020)(Kittikunakorn et al., 2019b), Vasanthavada et al. (2011)
Poloxamer 188	N/A	52–57	23–37	Batra et al. (2017), Passerini, Albertini, González-Rodrǵuez, Cavallari, & Rodriguez, (2002); Passerini, Albertini, Perissutti, & Rodriguez, 2006)
Poloxamer 407 / Kolliphor P407	N/A	56	5–20	Batra et al. (2017), Guimarães et al. (2017), Steffens and Wagner (2019, 2020)
Kollidon VA64	108	N/A	15–35	Batra et al. (2017), Grymonpré et al. (2018), Nyavanandi et al. (2021)
Kollidon-12PF	72	N/A	15–35	Batra et al. (2017), Nyavanandi et al. (2021)

(continued on next page)

Table 24.3 List of binders used in melt granulation with their transition/melting temperature and binder levels—cont'd

Chemical Name	Glass Transition Temperature (°C)	Melting Temperature (°C)	Use Range (%)	References
Kollicoat IR	209	N/A	25–35	Munnangi et al., 2024
Hydroxypropyl methylcellulose (HPMC K4M)	96	N/A	5–25	Batra et al. (2017, 2022), Grymonpré et al. (2018), Tan et al. (2014)
Hydroxypropyl methylcellulose (HPMC 15LV)	97	N/A	10–25	Batra et al. (2017), Grymonpré et al. (2018)
Eudragit EPO	56	N/A	10–15	(Batra, Desai, & Serajuddin, 2017; Batra, Thongsukmak, Desai, & Serajuddin, 2021), Grymonpré et al. (2018)
Soluplus	70	N/A	5–30	Batra et al. (2017), Elkanayati et al. (2024), Grymonpré et al. (2018), (Monteyne et al., 2016d)
Polyethylene glycol (PEG 6000)	N/A	61	5–20	Panda and Tiwary (2012), Andrade et al. (2015), Ochoa et al. (2010, 2011)
Polyethylene glycol (PEG 4000)	N/A	65	5–25	Andrade et al. (2015), (Monteyne et al., 2016d) (Van Melkebeke et al., 2006)

(continued on next page)

Table 24.3 List of binders used in melt granulation with their transition/melting temperature and binder levels—cont'd

Chemical Name	Glass Transition Temperature (°C)	Melting Temperature (°C)	Use Range (%)	References
Polyethylene glycol (PEG 2000)	N/A	53	5–20	Mašić et al. (2012)
Hydrophobic Polymers				
Glyceryl palmitostearate (Precirol ATO5)	N/A	52	10–30	Evrard et al. (1999), Forster and Lebo (2021), Mašić et al. (2012), Prado et al. (2014)
Glyceryl monostearate	N/A	57	10–20	Ochoa et al. (2010), Thomsen et al. (1994)
Glyceryl behenate (Compritol 888)	N/A	70	15–30	Evrard et al. (1999), Keen et al. (2015), (Kittikunakorn et al., 2019b)
Stearic acid	N/A	69	12–85	(Monteyne et al., 2016a), Patil et al. (2015)
Carnauba Wax	N/A	60	25–45	Nart et al. (2017)
Ethyl cellulose	133	N/A	10–40	Patil et al. (2015), Vasanthavada et al. (2011)

the complex interplay between binder viscosity, particle size, and prevailing nucleation mechanisms, immersion versus distribution, which varied with binder particle size.

Monteyne et al. (2016d) investigated tabletability and the amount of fines generated during TSMG of a low viscosity binder, PEG 4000, against a high viscosity binder, Soluplus (Monteyne et al., 2016d) Thermoplastic polymer Soluplus produced more tabletable granules with fewer fines, supporting the widespread use of thermoplastic binders for enhancing tabletability in melt granulation. Similar findings were reported by Kittikunakorn et al. (2019b), who observed a significant improvement in tabletability with Klucel™ (Hydroxypropylcellulose) over Compritol and PEG during TSMG of gabapentin (Kittikunakorn et al., 2019b). Interestingly, when comparing binder viscosity across the same chemistry class, e.g., PEG 2000 vs PEG 10000, lower viscosity binders yielded stronger granules due to the uniform binder distribution during TSMG with PEG and HPC alike (Kittikunakorn, Koleng, Listro, Sun, & Zhang, 2019b; Mu & Thompson, 2012). These findings underscore that, irrespective of processing method, final granule attributes are governed by the intricate interplay between binder properties and growth mechanisms.

24.4.1.3 Binder level

Binder level plays a critical role in determining granule properties in melt granulation, as it directly influences liquid saturation and granule strength, particularly important in the absence of solvent-mediated bridge formation (Liu et al., 2021). At higher binder levels, saturation is more easily achieved, providing sufficient binder for effective layering/agglomeration, resulting in bigger and stronger granules. Conversely, lower binder level often leads to broader particle size distributions and increased generation of fines.

Mašić et al. (2012) demonstrated that the binder level had a significant effect on the granule growth during fluid-bed melt granulation of paracetamol using PEG 2000 and Precirol as the binders (Mašić et al., 2012). For both binders, a threshold binder level was necessary to achieve acceptable quality, as 5% binder level resulted in increased fines with a broader particle size distribution, whereas a 15% yielded well-formed granules with a narrow particle size distribution (Mašić et al., 2012).

In the context of TSMG, Liu et al (2020) investigated the effect of Klucel (Hydroxypropylcellulose (HPC)) on the granulation of acetaminophen (Liu et al., 2020). They found that while 5% HPC could generate acceptable granules at a high degree of fill, higher HPC levels of 10% to 20% yielded robust granules at all degrees of fill (Liu et al., 2020). However, at a 20% HPC level, a decline in tabletability was observed due to 'over-granulation'(Liu et al., 2020). Similar findings were reported by Pradhan et al. (2022) and Batra et al. (2021), who evaluated different binders like HPC, Eudragit (Batra, Thongsukmak, Desai, & Serajuddin, 2021; Pradhan et al., 2022). Their studies concluded that a minimum of 10% HPC level was generally required for a robust TSMG process. While a 5% binder level can improve tabletability, it often results in a narrow design space or a significant amount of fines, whereas 2% binder level was insufficient to attain any meaningful improvement in tabletability (Batra, Thongsukmak, Desai, & Serajuddin, 2021; Pradhan et al., 2022). Recent advances in binder chemistry have shown promise in overcoming these limitations. Kodange et al. (2025) reported good granules having minimal fines with significant improvement in tabletability with merely 5% of Klucel Fusion X (Kodange et al., 2025). This finding highlights the promise of novel binders like Klucel Fusion X and underscores the need for future research to evaluate their applicability across a broader range of drugs in TSMG.

24.4.2 Physicochemical properties of the drug

24.4.2.1 Drug stability

Melt granulation is a thermally intensive process, and prolonged exposure to elevated temperatures can pose significant risks, including degradation or unintended physicochemical changes. (Forster et al., 2021). The rheological property of the binder plays a critical role in the frictional heat generation, directly influencing the product temperature during processing. This may be unsuitable for pharmaceutical drugs. Several drugs, such as gabapentin, meloxicam, fenbendazole, and gliclazide, degrade upon melting, and utmost care must be taken during processing to ensure that the local granule temperature is significantly below the melting point of the drug. Owing to this, TGA (Thermogravimetric analysis) is a part of the preformulation toolbox to assess the thermal stability of the drug within the operating window. Haser et al. (2018) successfully enhanced the thermal stability of meloxicam by formulating a salt with meglumine to facilitate thermal processing (Haser et al., 2018). Similarly, Liu et al. (2022) emphasized the importance of degradation kinetics of gliclazide in maintaining drug stability during thermal processing (Liu et al., 2022). In another study, Pradhan et al. (2025) demonstrated a material sparing approach using Differential Scanning Calorimetry in conjunction with HPLC to screen drug-excipient mixtures for elucidating drug stability at elevated temperatures prior to melt granulation (Pradhan et al., 2025). These findings highlight the necessity of a comprehensive understanding of drug degradation pathways, particularly in the presence of excipients at elevated temperatures prior to melt granulation.

24.4.2.2 Drug-binder interactions and miscibility

Beyond thermal stability, drug solubility in the polymer melt or the formation of hydrogen bonds can significantly influence binder behavior during melt granulation. These interactions may plasticize the binder, enhancing its spreadability over drug particle surfaces and thereby affecting granule properties. For instance, Kidokoro et al. (2001) reported improved processability in fluid bed melt granulation by lowering the binder's glass transition temperature through hydrogen bonding between ibuprofen and Eudragit PO (Kidokoro et al., 2001). Such drug-binder interactions have been investigated using various analytical techniques like melting point depression using DSC (Weatherley et al., 2013), Hansen solubility parameters (Greenhalgh et al., 1999), changes in glass transition temperature (Monteyne et al., 2016b), and changes in tan δ (Loss Modulus/Storage Modulus) through rheological characterization (Monteyne et al., 2016c).

The impact of drug-binder miscibility on binder plasticization and distribution in the granules was explored in a rheological study comparing an immiscible and miscible system with the same binder, Soluplus (Monteyne et al., 2016c). Monteyne et al. (2016c) observed higher tan δ values (i.e., a measure of material flow) for a miscible system compared to an immiscible system due to plasticization of Soluplus by the drug. However, despite the reduction in glass transition temperature due to plasticization, hydrogen bonding in the miscible system interfered with binder distribution, resulting in discontinuous patches of binder and reduced tabletability (Monteyne et al., 2016c).

Due to miscibility between the drug-binder, wettability is enhanced, promoting binder spreading over the drug surface and more agglomeration (Liu et al., 2021). This was confirmed by Weatherley et al. (2013), who observed larger granules in a miscible ibuprofen-PEG system than in an immiscible caffeine-PEG one (Weatherley et al., 2013). The miscibility between ibuprofen and PEG imparted a lubricating effect, improving processability and potentially reducing local heat generation during TSMG (Weatherley et al., 2013).

In addition to influencing granule properties, the drug-binder interactions can enhance product performance. For instance, the interaction between ibuprofen and Poloxamer 188 resulted in the formation of an eutectic mixture upon fluid bed melt granulation, enhancing ibuprofen dissolution (Passerini et al., 2002). Drug solubility in the polymer melt can serve as an early indicator for any potential incompatibilities prior to melt granulation. While the melting point depression of glipizide in the presence of PEG suggested good miscibility, impurity profiling of physical mixtures revealed a new unidentified peak (Panda & Tiwary, 2012). Although total impurity was within ICH limits, this highlights the importance of early screening for potential incompatibilities in drug-binder systems.

Based on the drug stability of drug and drug-binder interactions, to guide formulation design, Liu et al. classified drug-binders into four classes based on their thermal stability and miscibility (Liu et al., 2020). When integrated with process design considerations, this framework offers a systematic approach for developing the melt granulation of new drug-polymer combinations.

24.5 Process design for melt granulation

Melt granulation process, also referred to as melt agglomeration, was originally developed in the 1990s as a batch process using the high-shear mixer (Royce, Suryawanshi, Shah, & Vishnupad, 1996; Schaefer, Holm, & Kristensen, 1990) . In the early 2000s, research conducted by Abberger (2001) along with Kojima and Nakagami (2001) pioneered the fluid-bed melt granulation process (Abberger, 2001; Kojima & Nakagami, 2001). With the advent of twin-screw processing in the pharmaceutical industry, TSMG emerged as a new, cost-effective, and robust manufacturing platform for processing drugs. Since then, numerous studies have been conducted in the literature about TSMG along with the commercialization of three products: Eucreas, Orlissa, and Oriahnn processed through TSMG. The following section reviews critical aspects during the design of a batch and continuous melt granulation process.

24.5.1 Batch melt granulation (*In-situ* melt granulation)

Batch melt-granulation can be classified into two types: based on the mode of binder addition: spray-on melt granulation and *in-situ* melt granulation (Passerini et al., 2010). Additionally, this chapter primarily focuses on the *in-situ* melt granulation process due to challenges with hot polymer melt flow for spray-on melt granulation.

24.5.1.1 *High-shear mixer*
Some of the earliest studies on the batch melt granulation process were conducted using a high-shear mixer such as the Collete Gral mixer, which is a jacketed mixing bowl equipped with a chopper and impeller blades (Royce, Suryawanshi, Shah, & Vishnupad, 1996). The main process variables that are typically investigated are the jacket temperature, mixing time, and impeller speed. The thermal energy from the jacket or the friction from impeller blades causes the melting of the binder, followed by agglomeration via particle collisions. Due to frictional heat generation, jacket temperature is not a reliable parameter to optimize the melt granulation process.

Influence of process parameters on granule properties for high-shear melt granulation has been extensively studied (Schæfer, 1996b; Schaefer, Holm, & Kristensen, 1990; Schæfer, Taagegaard, Juul Thomsen, & Gjelstrup Kristensen, 1993). The mechanism of nucleation is primarily controlled by the binder size, irrespective of the different processing parameters. Granules produced via this method are more spherical and less porous, as demonstrated by ibuprofen, ketoprofen, and lactose melt granulation processes through a high-shear mixer (Passerini, Calogerà, Albertini, & Rodriguez, 2010). Among the tested variables, impeller speed has shown good correlation with granule yield and particle size distribution. Higher impeller speeds led to increased agglomerate breakage and further densification, often resulting in a broader particle size distribution (Schæfer et al., 1993; Steffens et al., 2020). Advanced in-line tools like Spatial Filter Velocimetry (SFV) probes have been employed for in-line particle analysis to monitor granule size and shape in real time (Kukec et al., 2012). Additionally, Voinovich (1999) investigated the effect of impeller blade geometry, finding no significant impact on the % fines produced, though different shapes did affect granule size distribution and mean diameter (Voinovich et al., 1999).

A robust granulation process of lansoprazole using a high viscosity binder was difficult unless a longer mixing time was used (Bukovec et al., 2009). Longer time allowed for complete melting of the binder, leading to agglomeration. However, prolonged exposure to elevated temperatures (Typical processing time is 20–40 min for high-shear mixer) is detrimental to drug stability, highlighting two main drawbacks of high-shear melt granulation: poor temperature control and limited choice of binders (Kittikunakorn, Liu, & Zhang, 2020; Passerini, Calogerà, Albertini, & Rodriguez, 2010).

24.5.1.2 Fluid bed processor

Fluid-bed melt granulation process emerged as a viable alternative to high-shear melt granulation, offering improved temperature control, lower shear forces, and a more controlled granulation environment (Van de Steene, Vanhoorne, Vervaet, & De Beer, 2021). Unlike high-shear mixers, which typically require a processing time of 20 to 40 min, fluid bed melt granulation can be completed in as little as 15 min, making it suitable from a drug stability perspective (Guimarães et al., 2017). The solid binder undergoes softening/melting because of the hot air flowing through the powder bed, leading to agglomeration (Abberger, 2001). Similar to a high-shear mixer, the nucleation mechanism is primarily controlled by the binder size.

Key process variables that are investigated during optimization include granulation time, fluidized air temperature, and fluidized air flow rate. Mašić et al. (2012) reported that granule size distribution during paracetamol melt granulation was largely unaffected by granulation time and fluidized air flow rate, suggesting binder properties are the dominant factor in granule growth (Mašić et al., 2012). However, Ponga et al. (2023) observed that excessively high flow rates can cause granule breakage, impacting particle size distribution for weaker lactose granules (Ponga et al., 2023). This can be attributed to the binder-dependent granule strength for the two cases— lactose and paracetamol. Granulation time has been shown to significantly impact the flow properties, like angle of repose, compressibility index, during paracetamol melt granulation (Aleksić et al., 2014; Mašić et al., 2012). Fluidizing air temperature affects binder melting and viscosity, ultimately impacting granule strength (Ponga et al., 2023).

Recent advancements include the application of machine learning algorithms, artificial neural networks, and process analytical technology (PAT) tools like Focused Beam Reflectance measurement and Spatial Filtering techniques for better elucidation of granule growth kinetics and temperature inside the fluid bed. These tools reflect the evolving landscape of fluid-bed melt granulation processes (Korteby, Kristó, Sovány, & Regdon, 2018; Kukec, Hudovornik, Dreu, & Vrečer, 2014)

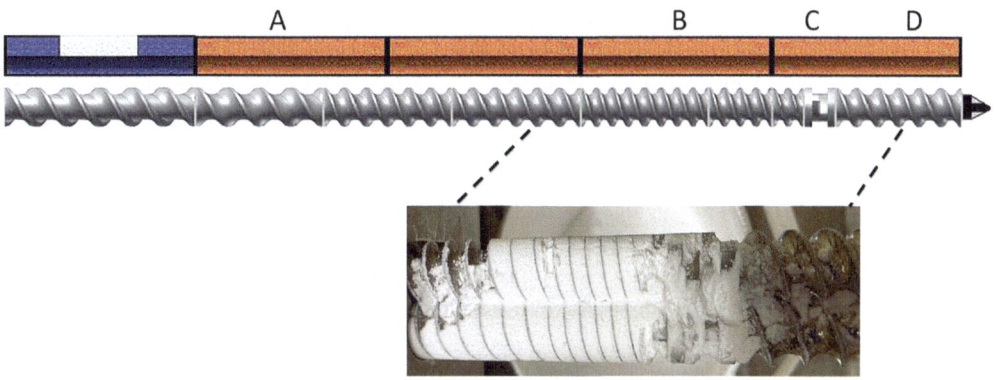

FIGURE 24.1

Typical screw configuration for TSMG.

Top: Section A is conveying elements (wide pitch); Section B is conveying elements (narrow pitch), Section C - kneading elements, Section D - Conveying elements (Wide pitch). Bottom: Pre-compaction and increased degree of fill upstream of kneading elements.

24.5.2 Continuous melt granulation (twin-screw melt granulation)

TSMG is typically performed using a twin-screw extruder, a technology long established in the plastics industry. For pharmaceutical processing, fully intermeshing, co-rotating twin-screw extruders are used to minimize the stagnation of material (DiNunzio et al., 2011). TSMG has been shown to successfully produce more compressible granules in comparison to a fluid-bed or high-shear melt granulation process (Lakshman et al., 2011). This is attributed to the unique mixing mechanism of the extruder. The material experiences both dispersive and distributive mixing as it travels across a set of heated zones from feed to open-end discharge (without a die). Therefore, granule properties are primarily influenced by the extruder screw configuration, choice of mixing elements, and their position, followed by the process variables. The below section reviews the role of screw design, process variables while highlighting the importance of system parameters and PAT tools in TSMG.

24.5.2.1 Extruder screw configuration

The screw elements in a twin-screw extruder are broadly classified as: Conveying elements, combining elements, and kneading elements. A combination of these elements makes up the extruder screw configuration (Fig. 24.1), with conveying elements responsible for material transport, while combining and kneading elements are responsible for mixing (Ghebre-Sellassie & Martin, 2003). Conveying elements of different pitches are used during processing as they can be used not only to transport materials but also for pre-compaction and oversized agglomerate breakage (Kittikunakorn et al., 2020). Large pitch elements have better conveying capacity and are typically placed near the feed zone (Fig. 24.1A), whereas small pitch elements, due to their poor conveying efficiency, are placed just upstream of the kneading zone to promote pre-compaction and increased degree of fill (Fig. 24.1B), (Van de Steene, 2023a; Todd & Andersen, 1998). This enhances drug-binder interactions, inducing nucleation as confirmed by the screw-pullout studies (Pradhan et al., 2025, Fig. 24.1). Conveying elements downstream of kneading (Fig. 24.1D) are used for granule breakage to reduce oversized agglomerates and improve yield as confirmed by (Van Melkebeke et al., 2008). Combining elements facilitates distributive mixing with lower mechanical stress and is commonly used in the wet granulation process to distribute liquid binders (Kittikunakorn et al., 2020). Their application in TSMG is limited to cases involving binder injection through liquid pumps.

Kneading elements are critical in a screw configuration as they are responsible for densification and granule growth through dispersive and distributive mixing. Kneading elements are defined by three parameters: disc width, offset angle, and overflight clearance (gap between screw tip and barrel wall). Dispersive mixing involves higher shear stress to overcome cohesion forces, causing morphological changes, whereas distributive mixing involves repeated division and recombination of material without morphological changes (Ghebre-Sellassie & Martin, 2003). Depending on the extent of mixing required, kneading elements are chosen. Increasing the offset angles from 30 degrees to 90 degrees raises shear intensity while reducing conveying capacity, promoting increased granule growth and densification (Ghebre-Sellassie & Martin, 2003; Kittikunakorn et al., 2021; Kittikunakorn, Sun, & Zhang, 2019a). However, this is accompanied by an increase in torque and frictional heat generation, causing in-process physicochemical changes leading to degradation (Kittikunakorn et al., 2021; Kittikunakorn, Sun, & Zhang, 2019a; Pradhan et al., 2025). For instance, Kittikunakorn et al. (2019a) reported higher gabapentin degradation with increased offset angles, linked to polymorphic transition and crystal defects (Kittikunakorn et al., 2019a).

Disc width directly influences dispersive mixing, responsible for granule densification. Hence, the wide disc width (5 mm) and narrow disc width (2.5 mm) follow different granulation mechanisms. Kittikunakorn et al. (2021) observed that for a narrow disc width, an increased degree of fill and higher barrel temperature are required to compensate for the lost shear (Kittikunakorn et al., 2021). Increased degree of fill leads to better precompaction, facilitating faster consolidation of agglomerates (Liu et al., 2021). While for a wider disc width element, granule densification occurs via the mechanical energy and shear stress due to more intensive mixing (Liu et al., 2021). Pradhan et al. observed similar trends with narrow disc width elements and increased overflight clearance (Pradhan et al., 2024). Additionally, increased overflight clearance results in lower drug crystal size reduction after processing, which impacts dissolution and drug stability.

Screw pullout studies by Pradhan et al. (2025) and Kittikunakorn et al. (2021) confirmed that the kneading zone is the primary site of granule growth and shear (Kittikunakorn et al., 2021; Pradhan et al., 2025). However, this zone also has peak shear, causing local heat generation, posing a risk to sensitive drugs. Therefore, careful selection of kneading element geometry, offset angle, and their position is essential for balancing shear intensity, residence time, and degree of fill to maintain drug stability in TSMG.

24.5.2.2 Process variables

Apart from screw configuration, the three primary adjustable process variables are barrel temperature, screw speed, and feed rate. The effects of screw speed and feed rate are interdependent and are best interpreted through the system parameters such as residence time and degree of fill, which are discussed in the following section.

Barrel temperature is a critical process parameter that must exceed a minimum threshold, as a cold barrel could trigger machine overtorquing or significant local heat generation. For example, Kittikunakorn observed granule temperatures of 90°C to 125°C during TSMG of gabapentin despite a fixed barrel temperature of 60°C (Kittikunakorn et al., 2021). The elevated granule temperatures are due to the frictional heat and are unsuitable for good process control. To mitigate this, barrel temperatures are typically set above the glass transition temperature/melting point of the binder for a smooth process. Pradhan et al. demonstrated the need for preliminary runs for setting barrel-temperature limits during TSMG of thermally sensitive drugs like gabapentin (Pradhan et al., 2025). Barrel temperature can be optimized by either a flat temperature profile across all the zones or a gradient profile with increasing barrel temperature till the kneading zone, followed by a reduced barrel temperature to initiate solidification of the binder. Despite several studies utilizing either approach, no systematic study has explored the benefit of a gradient temperature profile or a flat temperature profile. For thermoplastic binders like hydroxypropylcellulose, Soluplus, a higher barrel temperature reduces melt viscosity, promoting uniform binder distribution and increased granule growth (Kittikunakorn et al., 2020). This effect was observed in the case of caffeine-Soluplus granules or mannitol-HPC granules, where larger and stronger granules were observed at higher barrel temperatures (Monteyne, Vancoillie, Remon, Vervaet, & De Beer, 2016d; Pradhan et al., 2022). However, high barrel temperatures are counterproductive as they may trigger drug stability issues due to elevated temperatures. Additionally, since frictional heat exceeds the thermal energy input, granule temperature is a superior indicator of thermal history during granule growth as demonstrated (Pradhan et al., 2025).

24.5.2.3 System parameters

Granule properties in TSMG are governed by the complex interplay among various process variables. Hence, system parameters were introduced to capture the effect of multiple process variables into a single parameter. Examples of key system parameters include residence time, degree of fill, and specific mechanical energy. Previous studies have revealed that granule attributes correlated well with system parameters, highlighting their importance (Kittikunakorn et al., 2021; Kodange et al., 2025; Pradhan et al., 2024).

Degree of fill is the proportion of extruder free volume occupied by the material and is determined by the feed rate (Q)/screw speed (N) ratio while being dependent on other constants like extruder free volume, specific gravity, and screw efficiency (Liu et al., 2021). A higher degree of fill achieved through increasing feed rate or reducing screw speed enhances drug-binder interactions, powder bed compaction, promoting granule growth and densification, ultimately leading to stronger granules (Monteyne, Vancoillie, Remon, Vervaet, & De Beer, 2016d; Pradhan et al., 2024). This effect is further amplified at lower binder levels or with narrow disc width kneading (Kodange et al., 2025; Pradhan et al., 2024). With the novel narrow disc width kneading elements, Pradhan et al. reported that the lost dispersive shear was compensated through an increased degree of fill, leading to densification. This led to increased tabletability and local heat generation as observed by the strong correlation of Q/N ratio with tabletability and gabapentin degradation (Pradhan et al., 2024).

Residence time is defined as the time material is exposed to the high-energy conditions inside the extruder. Residence time is typically measured using a dye tracer study where the color intensity of dye is measured and analyzed using MATLAB to determine the residence time distribution (Fig. 24.2) (Pradhan et al., 2025). Typical residence time for TSMG is <60 sec with exposure to peak shear at the kneading zone reported as low as 6 to 10 sec (Pradhan et al., 2025). A long tail or a secondary peak in residence time distribution (Fig. 24.2b) may indicate prolonged exposure to peak shear, raising drug stability concerns. While longer residence time may result in a more uniform particle size distribution (due to reduced feed rate/ higher screw speeds), it also increases the risk of thermal degradation (Monteyne et al., 2016d).

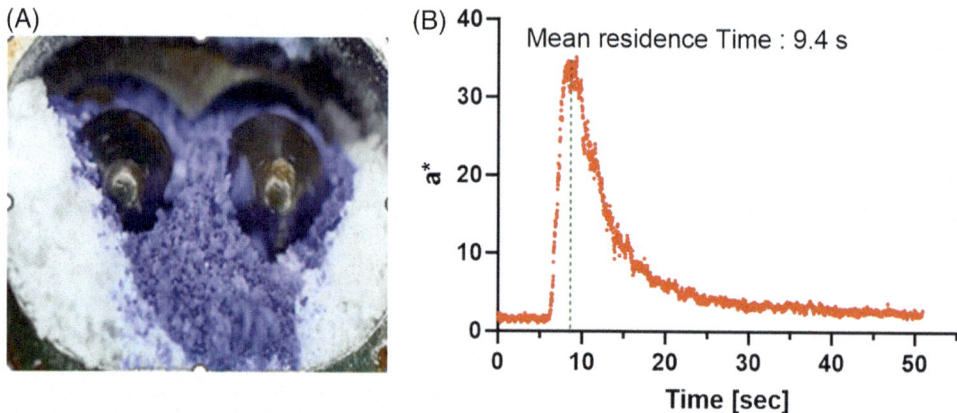

FIGURE 24.2

Residence time using a dye tracer study

Residence time measurement (A) Color intensity of dye measured at discharge. (B) Typical residence time distribution for a TSMG run at high screw speed.

Specific mechanical energy (SME) is the mechanical energy imparted by the extruder on the material and can be calculated as

$$\text{SME} = \frac{\text{Motorrating(kW)} * \% \text{Torque} * \text{Screwspeed} * 0.97}{\text{Feedrate} * \text{Maxscrewspeed}}$$

Numerous studies have demonstrated SME as a critical system parameter for TSMG, especially on larger extruders, and it is frequently used to scale-up efficiently. SME encompasses both frictional and mechanical energy inputs provided by the screws to soften the binder. Hence, stronger granules have been reported at higher SME (faster screw speed, lower feed rate) through numerous studies (Kittikunakorn et al., 2021; Kodange et al., 2025). For instance, Kittikunakorn et al. (2021) and Kodange et al. (2025) reported positive correlations between SME and tensile strength in both immiscible and miscible drug-binder systems (Kittikunakorn et al., 2021; Kodange et al., 2025). Despite shorter residence time, granule growth is influenced by the significantly increased frictional energy input, enhancing drug-binder mixing and densification (Liu et al., 2021). However, SME was found to correlate well with gabapentin degradation as increased energy input potentially promotes crystal defects, amorphization leading to high local heat generation and elevated granule temperature (Kittikunakorn et al., 2021). Thus, for TSMG, relying on the degree of fill for granulation is recognized as better process control than SME, especially for thermally or shear-sensitive drugs.

24.5.2.4 Process analytical technology for twin-screw melt granulation

As the pharmaceutical industry advances towards continuous pharmaceutical manufacturing, real-time process monitoring and development of PAT are critical for ensuring optimal product quality and enabling efficient process control. PAT tools are designed to monitor, measure, and analyze real-time properties of raw material and granules for supporting process robustness and consistency (Fonteyne et al., 2015). A successful integration of the PAT framework in the product development cycle can significantly save costs and time. Given that granulation is a particle agglomeration process, knowledge of the granule characteristics after TSMG is essential for not only quality but also downstream processing. Fonteyne et al. (2013) demonstrated this by the use of an in-line particle analyzer during the wet granulation process of theophylline using the ConsiGma 25. The detection of oversized agglomerates indicated the need for milling prior to drying (Fonteyne et al., 2013). Further process optimization trials may aid in reducing the formation of these agglomerates, making the downstream operation of milling irrelevant. Additionally, advanced imaging techniques based on acoustic emission analysis, photometric stereo imaging, and high-speed imaging to capture particle movement have been used to characterize granule size, shape, and surface roughness (Abdulhussain & Thompson, 2021; Fonteyne et al., 2012, Fonteyne et al., 2015). The high-speed imaging technique, in particular, has been widely used in twin-screw wet granulation via the Eyecon 3D Particle Characterizer (Kumar et al., 2015; Meng et al., 2019).

Similarly, the solid state of the drug needs to be monitored due to concerns with form changes. To address this, in-line NIR probes, Raman spectroscopy probes, and UV probes were developed and incorporated in conjunction with design of experiments (DoE) and principal component analysis (PCA) for optimal product quality (Fonteyne et al., 2013; Nandi, Trivedi, Ross, & Douroumis, 2021; Wahl et al., 2013). In twin-screw wet granulation, NIR probes have been utilized for evaluating form changes, drug-binder interactions, quantifying drug concentration, and even moisture content (Harting & Kleinebudde, 2019; Haser et al., 2023; Nagy et al., 2017). These tools can be applied to TSMG processes as demonstrated in the case of fevipiprant, where a multipurpose in-line NIR analyzer was used to monitor assay and content uniformity during scale-up (Vicente Martin et al., 2022).

Due to the reliance on frictional heat for granule growth, local heat is generated inside the extruder as observed through the granule temperature of 120°C for a barrel temperature of 60°C (Kittikunakorn et al., 2021). Infrared cameras can be employed for temperature mapping at discharge to monitor granule temperature in real time. Van de Steene et al. (2023b) and Pradhan et al. (2025) successfully utilized Teledyne FLIR Infrared Imaging cameras to investigate granulation mechanisms and local heat generation during TSMG (Pradhan et al., 2025; Van de Steene et al., 2023b). These cameras can be valuable additions to the PAT toolbox for long-duration TSMG runs, enabling continuous temperature monitoring and improved process control.

24.6 Advantages of twin-screw melt granulation

Continuous melt granulation/TSMG is a solvent-free process that offers a sustainable manufacturing alternative for oral solid dosage forms. With the potential to operate entirely on green energy, TSMG presents a pathway towards near-zero carbon emissions (Kittikunakorn et al., 2020; Sampat & Ramachandran, 2024). Owing to its long-standing applications in the plastics industry, twin-screw processing benefits from established scale-up methodologies that require minimal adjustment in processing parameters, whereas batch melt granulation frequently requires re-optimization during scale-up, complicating the development process (Ghebre-Sellassie & Martin, 2003).

One of the major drawbacks of batch melt granulation is inefficient heating and non-uniform temperature control, adversely impacting granule growth. For example, high-shear mixers are typically limited to a processing temperature of 100°C, narrowing the formulation design space as well as restricting the choice of binders (Vasanthavada et al., 2011). In contrast, TSMG offers precise process control and superior mixing mechanisms, enabling a wide range of formulation and processing conditions. The uniform heating and reliance on frictional energy dissipation during TSMG allow for the use of both waxy as well as thermoplastic binders, the latter being more popular for improving tabletability (Forster et al., 2021). Compared to the 10 to 40 min residence time in the batch melt granulation process, the residence time for a TSMG is typically under 1 min, significantly limiting the exposure to elevated temperatures (Kittikunakorn et al., 2020). This short residence time enables the processing of thermolabile compounds like gabapentin despite the high mechanical stresses during TSMG. Pradhan et al. reported that the residence time where the material experiences peak mechanical stresses could be as low as 6 to 10 sec (Pradhan et al., 2025). The use of modular mixing elements and small mixing volumes facilitates efficient mass and heat transfer, leading to improved content uniformity and better product quality (Kittikunakorn et al., 2020; Lodaya and Thompson, 2018). Another significant advantage of TSMG over batch processing is the ability to run at high drug loads. Numerous studies have reported that drug loading as high as 90% to 95% can be achieved with improvement in tabletability (Kodange et al., 2025; Lakshman et al., 2011; Liu, Kittikunakorn, Zhang, & Zhang, 2021; Pradhan et al., 2024). Due to recent technological advancements in PAT, real-time monitoring of product temperatures, feed rates, and content uniformity through NIR, particle size measurements are possible, allowing for enhanced process control as compared to a batch process. Despite the long list of advantages, there are a few challenges that need to be addressed to fully utilize the potential of TSMG for product development.

24.7 Challenges of twin-screw melt granulation

Due to the thermal and shear-intensive nature of melt granulation, exposure to elevated temperatures can induce unexpected in-process changes compromising drug stability. In-process physicochemical changes like amorphization, particle size reduction, and polymorphism are frequently attributed to the local heat generation at the kneading zone during the granule growth process (Kittikunakorn et al., 2021; Kittikunakorn, Sun, & Zhang, 2019a; Liu et al., 2022). Numerous studies on TSMG of gabapentin, acetaminophen have shown significant particle size reduction under different formulation, processing conditions, and extruder configurations, highlighting the influence of mechanical stresses on the material behavior (Kittikunakorn et al., 2019, 2019, 2021; Pradhan et al., 2024). In addition to particle size reduction, crystal defects have been observed in gabapentin during TSMG (Kittikunakorn et al., 2019; Pradhan et al., 2025). These crystal defects will then act as hotspots due to their high free energy and increase gabapentin degradation (Kittikunakorn, Sun, & Zhang, 2019a; Newman & Zografi, 2014; Zong, Qiu, Tinmanee, & Kirsch, 2012).

Kittikunakorn et al. reported increased degradation of gabapentin due to the polymorphic transition of gabapentin from form II to metastable form IV during TSMG with neutral kneading elements (Kittikunakorn et al., 2019a). Similarly, Xu et al. reported polymorphic transformation of indomethacin during TSMG with PEG 3350 as the binder, causing the drug to convert to metastable α form (Xu, Nahar, Bates, Morris, & Dave, 2018). In some other cases, partial amorphization or conversion may occur due to the high-energy environment inside the extruder. This amorphization/conversion increases the molecular mobility and will not only accelerate drug degradation but also adversely affect granule properties like tabletability and flowability due to molecular changes (Kittikunakorn et al., 2020). These findings underscore the importance of drug-binder miscibility and the mechanistic understanding of the granulation process to minimize in-process physicochemical changes.

Despite numerous articles and considerable research on TSMG, the widespread adoption remains limited primarily due to the absence of a rational material sparing approach for feasibility and process optimization. Traditional process optimization typically requires kilograms of material and multiple trials to evaluate drug stability under TSMG conditions. To address this, Pradhan et al. recently published an article on the development of a material-sparing empirical model for stability assessment of thermolabile drugs like gabapentin; however, other efforts in that space are scarce (Pradhan et al., 2025). Additionally, the limited availability of thermoplastic binders with low glass transition temperatures continues to pose challenges for the processing of thermal/shear-sensitive drugs.

24.8 Expert's opinion on the future of twin-screw melt granulation

Despite the potential for zero carbon emissions with twin-screw processing, the application of machine learning algorithms to maximize the energy efficiency of continuous melt granulation remains unexplored. For instance, Sampat et al. achieved a 27% higher efficiency using the long-term memory model algorithm for a continuous (twin-screw) wet granulation process (Sampat & Ramachandran, 2024). Applying similar machine learning algorithms in continuous melt granulation could significantly accelerate its industrial adoption.

In alignment with the FDA (Food and Drug Administration) guidance, recent efforts have been focused on the use of empirical modelling techniques, including statistical analysis, DoE, and artificial neural networks, to ensure optimal quality of granules. Korteby et al. successfully demonstrated the potential of artificial neural networks in enhancing product quality through process optimization for the in-situ fluid bed melt granulation process of lactose with PEG as the binder (Korteby et al., 2018). Additionally, regime maps were also developed for batch melt granulation processes, particularly the fluid-bed processor, using dimensionless numbers to support predictive modelling in process design (Ponga et al., 2023). Despite extensive efforts on process optimization of TSMG, there remains a notable gap in the development of predictive models, especially regime maps for TSMG. This can be attributed to the complex interplay between local heat generation, binder rheological properties and their impact on granule properties Given that TSMG is both a material and time-consuming effort, the creation of a robust regime map could significantly alter the paradigm of TSMG by reducing the experimental trials needed to effectively optimize the design space and achieve desired granule properties.

24.9 Summary

The book chapter critically examines the evolving domain of melt granulation as an alternative to the conventional wet-granulation technique. Being a solvent-free process and lacking of drying step, melt granulation offers a distinct advantage over wet-granulation in terms of unit operation and energy consumption. Both processes share similar steps of nucleation, granule consolidation, and granule breakage during granule growth, with binder properties influencing the nucleation mechanism. Various CQAs of the melt granulation process are emphasized while summarizing the preferred granulation technique to attain that property. For instance, an improvement in tabletability can be more readily achieved using TSMG. Binder type was observed to influence the choice of melt granulation equipment, with batch melt granulation used only with low-viscosity binders. Furthermore, a detailed understanding of binder particle size, rheological properties, and drug-binder miscibility holds the key to product development using a melt granulation process. Despite the differing equipment, the granule growth in a batch process is primarily controlled by the binder properties as observed through numerous studies. Contrary to the batch process, TSMG not only is a continuous process but also offers superior mixing capabilities, ability to process high drug loads and easier to scale up making it the preferred option for melt granulation. The choice of an optimal screw configuration holds the key to balance the shear stress generated by kneading elements with residence time to maintain drug stability. The good correlation between system parameters (residence time, degree of fill and specific mechanical energy) with granule attributes make the process design and optimization easier. The integration of in-line IR, Raman and particle size probes have allowed real-time monitoring of granule properties to enhance process control. The addition of machine learning algorithms and predictive modelling will further aid in the broader adoption of TSMG.

References

Abberger, T. (2001). Influence of binder properties, method of addition, powder type and operating conditions on fluid-bed melt granulation and resulting tablet properties. *Die Pharmazie, 56*(12), 949–952.

Abdulhussain, H. A., & Thompson, M. R. (2021). Predicting the particle size distribution in twin screw granulation through acoustic emissions. *Powder Technology, 394*, 757–766. doi:10.1016/j.powtec.2021.08.089.

Aleksić, I., Đuriš, J., Ilić, I., Ibrić, S., Parojčić, J., & Srčič, S. (2014). In silico modeling of in situ fluidized bed melt granulation. *International Journal of Pharmaceutics, 466*(1—2), 21–30. doi:10.1016/j.ijpharm.2014.02.045.

Andrade, T. C., Martins, R. M., & Freitas, L. A. P. (2015). Granulation of indomethacin and a hydrophilic carrier by fluidized hot melt method: The drug solubility enhancement. *Powder Technology, 270*, 453–460. doi:10.1016/j.powtec.2014.07.030.

Batra, A., Desai, D., & Serajuddin, A. T. M. (2017). Investigating the use of polymeric binders in twin screw melt granulation process for improving compactibility of drugs. *Journal Pharmaceutics Science, 106*(1), 140–150. doi:10.1016/j.xphs.2016.07.014.

Batra, A., Thongsukmak, A., Desai, D., & Serajuddin, A. T. M. (2021). The effect of process variables and binder concentration on tabletability of metformin hydrochloride and acetaminophen granules produced by twin screw melt granulation with different polymeric binders. *AAPS PharmSciTech, 22*(4). doi:10.1208/s12249-021-02018-6.

Batra, A., Yang, F., Kogan, M., Sosnowik, A., Usher, C., Oldham, E. W., Chen, N., Lawal, K., Bi, Y., & Dürig, T. (2022). Comparison of hydroxypropylcellulose and hot-melt extrudable hypromellose in twin-screw melt granulation of metformin hydrochloride: Effect of rheological properties of polymer on melt granulation and granule properties. *Macromol, 2*(1), 1–19. doi:10.3390/macromol2010001.

Bukovec, P., Krošelj, V., Turk, S., & Vrečer, F. (2009). Optimization of melt pelletization in a high shear mixer. *International Journal of Pharmaceutics, 381*(2), 192–198. doi:10.1016/j.ijpharm.2009.06.036.

DiNunzio, J. C., Zhang, F., Martin, C., & McGinity, J. W. (2011). Melt Extrusion. In *Formulating poorly water soluble drugs* (pp. 311–362). Springer Science and Business Media LLC.

Dürig, T., & Karan, K. (2021). Binders in pharmaceutical granulation. In Dilip Parikh (Ed.), *Handbook of Pharmaceutical Granulation Technology* (pp. 78–97). CRC Press (Taylor and Francis group).

Elkanayati, R. M., Karnik, I., Uttreja, P., Narala, N., Vemula, S. K., Karry, K., & Repka, M. A. (2024). Twin screw melt granulation of simvastatin: Drug solubility and dissolution rate enhancement using polymer blends. *Pharmaceutics, 16*(12). doi:10.3390/pharmaceutics16121630.

Evrard, B., Amighi, K., Beten, D., Delattre, L., & Moës, A. J. (1999). Influence of melting and rheological properties of fatty binders on the melt granulation process in a high-shear mixer. *Drug Development and Industrial Pharmacy, 25*(11), 1177–1184. doi:10.1081/DDC-100102285.

Fonteyne, M., Soares, S., Vercruysse, J., Peeters, E., Burggraeve, A., Vervaet, C., Paul Remon, J., Sandler, N., & De Beer, T. (2012). Prediction of quality attributes of continuously produced granules using complementary pat tools. *European Journal of Pharmaceutics and Biopharmaceutics, 82*(2), 429–436. doi:10.1016/j.ejpb.2012.07.017.

Fonteyne, M., Vercruysse, J., De Leersnyder, F., Van Snick, B., Vervaet, C., Paul Remon, J., De Beer, T., et al. (2015). Process analytical technology for continuous manufacturing of solid-dosage forms. *TrAC Trends in Analytical Chemistry, 67*, 159–166. doi:10.1016/j.trac.2015.01.011.

Fonteyne, M., Vercruysse, J., Díaz, D. C., Gildemyn, D., Vervaet, C., Remon, J. P., & Beer, T. D. (2013). Real-time assessment of critical quality attributes of a continuous granulation process. *Pharmaceutical Development and Technology, 18*(1), 85–97. doi:10.3109/10837450.2011.627869.

Forster, S. P., Dippold, E., & Chiang, T. (2021). Twin-screw melt granulation for oral solid pharmaceutical products. *Pharmaceutics, 13*(5). doi:10.3390/pharmaceutics13050665.

Forster, S. P., & Lebo, D. B. (2021). Continuous melt granulation for taste-masking of ibuprofen. *Pharmaceutics, 13*(6). doi:10.3390/pharmaceutics13060863.

Ghebre-Sellassie, I., & Martin, C. (2003). *Pharmaceutical extrusion technology* (pp. 1–401). CRC Press, Eritrea CRC Press. Retrieved date.https://www.taylorfrancis.com/books/e/9780203911532.

Greenhalgh, D. J., Williams, A. C., Timmins, P., & York, P. (1999). Solubility parameters as predictors of miscibility in solid dispersions. *Journal of Pharmaceutical Sciences, 88*(11), 1182–1190. doi:10.1021/js9900856.

Grymonpré, W., Verstraete, G., Vanhoorne, V., Remon, J. P., De Beer, T., & Vervaet, C. (2018). Downstream processing from melt granulation towards tablets: In-depth analysis of a continuous twin-screw melt granulation process using polymeric binders. *European Journal of Pharmaceutics and Biopharmaceutics, 124*, 43–54. doi:10.1016/j.ejpb.2017.12.005.

Guimarães, T. F., Comelli, A. C. C., Tacón, L. A., Cunha, T. A., Marreto, R. N., & Freitas, L. A. P. (2017). Fluidized bed hot melt granulation with hydrophilic materials improves enalapril maleate stability. *Aaps Pharmscitech [Electronic Resource], 18*(4), 1302–1310. doi:10.1208/s12249-016-0593-0.

Harting, J., & Kleinebudde, P. (2019). Optimisation of an in-line Raman spectroscopic method for continuous API quantification during twin-screw wet granulation and its application for process characterisation. *European Journal of Pharmaceutics and Biopharmaceutics, 137*, 77–85. doi:10.1016/j.ejpb.2019.02.015.

Haser, A., Cao, T., Lubach, J. W., & Zhang, F. (2018). *In-situ* salt formation during melt extrusion for improved chemical stability and dissolution performance of a meloxicam–copovidone amorphous solid dispersion. *Molecular Pharmaceutics, 15*(3), 1226–1237. doi:10.1021/acs.molpharmaceut.7b01057.

Haser, A., Kittikunakorn, N., Dippold, E., DiNunzio, J. C., & Blincoe, W. (2023). Continuous Twin-Screw wet granulation process with In-Barrel drying and NIR setup for Real-Time moisture monitoring. *International Journal of Pharmaceutics, 630*, 122377. doi:10.1016/j.ijpharm.2022.122377.

Iveson, S. M., Litster, J. D., Hapgood, K., & Ennis, B. J. (2001a). Nucleation, growth and breakage phenomena in agitated wet granulation processes: A review. *Powder Technology, 117*(1—2), 3–39. doi:10.1016/S0032-5910(01)00313-8.

Iveson, S. M., Wauters, P. A. L., Forrest, S., Litster, J. D., Meesters, G. M. H., & Scarlett, B. (2001b). Growth regime map for liquid-bound granules: Further development and experimental validation. *Powder Technology, 117*(1-2), 83–97. doi:10.1016/S0032-5910(01)00317-5.

Johansen, A., & Schæfer, T. (2001). Effects of interactions between powder particle size and binder viscosity on agglomerate growth mechanisms in a high shear mixer. *European Journal of Pharmaceutical Sciences, 12*(3), 297–309. doi:10.1016/s0928-0987(00)00182-2.

Keary, C. M., & Sheskey, P. J. (2004). Preliminary report of the discovery of a new pharmaceutical granulation process using foamed aqueous binders. *Drug Development and Industrial Pharmacy, 30*(8), 831–845. doi:10.1081/DDC-200030504.

Keen, J. M., Foley, C. J., Hughey, J. R., Bennett, R. C., Jannin, V., Rosiaux, Y., Marchaud, D., & McGinity, J. W. (2015). Continuous twin screw melt granulation of glyceryl behenate: Development of controlled release tramadol hydrochloride tablets for improved safety. *International Journal of Pharmaceutics, 487*(1—2), 72–80. doi:10.1016/j.ijpharm.2015.03.058.

Kelleher, J. F., Madi, A. M., Gilvary, G. C., Tian, Y. W., Li, S., Almajaan, A., Loys, Z. S., Jones, D. S., Andrews, G. P., & Healy, A. M. (2020). Metformin hydrochloride and sitagliptin phosphate fixed-dose combination product prepared using melt granulation continuous processing technology. *Aaps Pharmscitech [Electronic Resource], 21*(1). doi:10.1208/s12249-019-1553-2.

Kidokoro, M., Shah, N. H., Malick, A. W., Infeld, M. H., & McGinity, J. W. (2001). Properties of tablets containing granulations of ibuprofen and an acrylic copolymer prepared by thermal processes. *Pharmaceutical Development and Technology, 6*(2), 263–275. doi:10.1081/PDT-100002203.

Kipping, T. (2024). *Process for continuous hot melt granulation of low soluble pharmaceuticals.* Google Patents.

Kittikunakorn, N., Sun, C. C., & Zhang, F. (2019a). Effect of screw profile and processing conditions on physical transformation and chemical degradation of gabapentin during twin-screw melt granulation. *European Journal of Pharmaceutical Sciences, 131*, 243–253. doi:10.1016/j.ejps.2019.02.024.

Kittikunakorn, N., Koleng, J. J., Listro, T., Sun, C. C., & Zhang, F. (2019b). Effects of thermal binders on chemical stabilities and tabletability of gabapentin granules prepared by twin-screw melt granulation. *International Journal of Pharmaceutics, 559*, 37–47. doi:10.1016/j.ijpharm.2019.01.014.

Kittikunakorn, N., Liu, T., & Zhang, F. (2020). Twin-screw melt granulation: Current progress and challenges. *International Journal of Pharmaceutics, 588*, 119670. doi:10.1016/j.ijpharm.2020.119670.

Kittikunakorn, N., Paul, S., Koleng, J. J., Liu, T., Cook, R., Yang, F., Bi, V., Durig, T., Sun, C. C., Kumar, A., & Zhang, F. (2021). How does the dissimilarity of screw geometry impact twin-screw melt granulation? *European Journal of Pharmaceutical Sciences, 157*, 105645. doi:10.1016/j.ejps.2020.105645.

Kodange, N. N., Pradhan, A., Yang, F., Karan, K., Schwing, Q., Durig, T., & Zhang, F. (2025). Unraveling local heat generation in twin-screw melt granulation. *International Journal of Pharmaceutics, 675*, 125563. doi:10.1016/j.ijpharm.2025.125563.

Kojima, M., & Nakagami, H. (2001). Preparation of the controlled-release matrix tablets of theophylline with micronized low-substituted hydroxypropyl cellulose by a fluidized hot-melt granulation method. *S.T.P. Pharma Sciences, 11*(2), 145–150.

Korteby, Y., Kristó, K., Sovány, T., & Regdon, G. (2018). Use of machine learning tool to elucidate and characterize the growth mechanism of an in-situ fluid bed melt granulation. *Powder Technology, 331*, 286–295. doi:10.1016/j.powtec.2018.03.052.

Kukec, S., Dreu, R., Vrbanec, T., Srčič, S., & Vrečer, F. (2012). Characterization of agglomerated carvedilol by hot-melt processes in a fluid bed and high shear granulator. *International Journal of Pharmaceutics, 430*, 74–85. doi:10.1016/j.ijpharm.2012.03.041.

Kukec, S., Hudovornik, G., Dreu, R., & Vrečer, F. (2014). Study of granule growth kinetics during in situ fluid bed melt granulation using in-line FBRM and SFT probes. *Informa Healthcare, Slovenia Drug Development and Industrial Pharmacy, 40*(7), 952–959. doi:10.3109/03639045.2013.791832.

Kumar, A., Dhondt, J., De Leersnyder, F., Vercruysse, J., Vanhoorne, V., Vervaet, C., Paul Remon, J., Gernaey, K. V., De Beer, T., & Nopens, I. (2015). Evaluation of an in-line particle imaging tool for monitoring twin-screw granulation performance. *Powder Technology, 285*, 80–87. doi:10.1016/j.powtec.2015.05.031.

Lakshman, J. P., Kowalski, J., Vasanthavada, M., Tong, W. Q., Joshi, Y. M., & Serajuddin, A. T. M. (2011). Application of melt granulation technology to enhance tabletting properties of poorly compactible high-dose drugs. *Journal of Pharmaceutical Sciences, 100*(4), 1553–1565. doi:10.1002/jps.22369.

Litster, J., & Ennis, B. (2004). *The science and engineering of granulation processes*: 15. Springer Science & Business Media.

Liu, T., Kaur, N., Chen, B., Phillips, B., Chang, S.-Y., Yang, F., Bi, V., Durig, T., & Zhang, F. (2022). Physicochemical changes and chemical degradation of gliclazide during twin-screw melt granulation. *International Journal of Pharmaceutics, 619*, 121702. doi:10.1016/j.ijpharm.2022.121702.

Liu, T., Kittikunakorn, N., Zhang, Y., & Zhang, F. (2021). Mechanisms of twin screw melt granulation. *Journal of Drug Delivery Science and Technology, 61*, 102150. doi:10.1016/j.jddst.2020.102150.

Liu, Tongzhou, Paul, Shubhajit, Beeson, Brian T, Alexander, Johnny, Yang, Fengyuan, Bi, Vivian, ... Zhang, Feng (2020). Effect of hydroxypropyl cellulose level on twin-screw melt granulation of acetaminophen. *AAPS PharmSciTech, 21*, 1–14.

Lodaya, M., & Thompson, M. (2018). Continuous oral solid dose manufacture. In Isaac Ghebre-Sellasie, Charlie Martin, Feng Zhang, & James DiNunzio (Eds.), *Pharmaceutical extrusion technology* (pp. 337–361). CRC Press, CRC Press.

Mamidi, H. K., Palekar, S., Nukala, P. K., Mishra, S. M., Patki, M., Fu, Y., Supner, P., Chauhan, G., & Patel, K. (2021). Process optimization of twin-screw melt granulation of fenofibrate using design of experiment (DoE). *International Journal of Pharmaceutics, 593*, 120101. doi:10.1016/j.ijpharm.2020.120101.

Mašić, I., Ilić, I., Dreu, R., Ibrić, S., Parojčić, J., & Đurić, Z. (2012). An investigation into the effect of formulation variables and process parameters on characteristics of granules obtained by in situ fluidized hot melt granulation. *International Journal of Pharmaceutics, 423*(2), 202–212. doi:10.1016/j.ijpharm.2011.12.013.

McTaggart, C. M., Ganley, J. A., Sickmueller, A., & Walker, S. E. (1984). The evaluation of formulation and processing conditions of a melt granulation process. *International Journal of Pharmaceutics, 19*(2), 139–148. doi:10.1016/0378-5173(84)90156-X.

Meng, W., Román-Ospino, A. D., Panikar, S. S., O'Callaghan, C., Gilliam, S. J., Ramachandran, R., & Muzzio, F. J. (2019). Advanced process design and understanding of continuous twin-screw granulation via implementation of in-line process analytical technologies. *Advanced Powder Technology, 30*(4), 879–894. doi:10.1016/j.apt.2019.01.017.

Monteyne, T., Adriaensens, P., Brouckaert, D., Remon, J.-P., Vervaet, C., & De Beer, T. (2016a). Stearic acid and high molecular weight PEO as matrix for the highly water soluble metoprolol tartrate in continuous twin-screw melt granulation. *International Journal of Pharmaceutics, 512*(1), 158–167. doi:10.1016/j.ijpharm.2016.07.035.

Monteyne, T., Heeze, L., Mortier, S. T. F. C., Oldörp, K., Cardinaels, R., Nopens, I., Vervaet, C., Remon, J.-P., & Beer, T. D. (2016b). The use of rheology combined with differential scanning calorimetry to elucidate the granulation mechanism of an immiscible formulation during continuous twin-screw melt granulation. *Pharmaceutical Research, 33*(10), 2481–2494. doi:10.1007/s11095-016-1973-6.

Monteyne, T., Heeze, L., Mortier, S. T. F. C., Oldörp, K., Nopens, I., Remon, J.-P., Vervaet, C., & De Beer, T. (2016c). The use of rheology to elucidate the granulation mechanisms of a miscible and immiscible system during continuous twin-screw melt granulation. *International Journal of Pharmaceutics, 510*(1), 271–284. doi:10.1016/j.ijpharm.2016.06.055.

Monteyne, T., Vancoillie, J., Remon, J.-P., Vervaet, C., & De Beer, T. (2016d). Continuous melt granulation: Influence of process and formulation parameters upon granule and tablet properties. *European Journal of Pharmaceutics and Biopharmaceutics, 107*, 249–262. doi:10.1016/j.ejpb.2016.07.021.

Mu, B., Thompson, M. R., et al. (2012). Examining the mechanics of granulation with a hot melt binder in a twin-screw extruder. *Chem Eng Sci, 81*, 46–56. doi:10.1016/j.ces.2012.06.057.

Munnangi, S. R., Narala, Nagarjuna, Lakkala, Preethi, Vemula, Sateesh Kumar, Narala, Sagar, Johnson, Lindsay, ... Repka, Michael, et al. (2024). Optimization of a twin screw melt granulation process for fixed dose combination immediate release tablets: Differential amorphization of one drug and crystalline continuance in the other. *International Journal of Pharmaceutics, 665*(124717), 1–12. doi:10.1016/j.ijpharm.2024.124717.

Nagy, B., Farkas, A., Gyürkés, M., Komaromy-Hiller, S., Démuth, B., Szabó, B., Nusser, D., Borbás, E., Marosi, G., & Nagy, Z. K. (2017). In-line raman spectroscopic monitoring and feedback control of a continuous twin-screw pharmaceutical powder blending and tableting process. *International Journal of Pharmaceutics, 530*(1—2), 21–29. doi:10.1016/j.ijpharm.2017.07.041.

Nandi, U., Trivedi, V., Ross, S. A., & Douroumis, D. (2021). Advances in twin-screw granulation processing. *Pharmaceutics, 13*(5), 624. doi:10.3390/pharmaceutics13050624.

Nart, V., Beringhs, A. O., França, M. T., de Espíndola, B., Pezzini, B. R., & Stulzer, H. K. (2017). Carnauba wax as a promising excipient in melt granulation targeting the preparation of mini-tablets for sustained release of highly soluble drugs. *Materials Science and Engineering: C, 70*(Pt 1), 250–257. doi:10.1016/j.msec.2016.07.070.

Newman, A., & Zografi, G. (2014). Critical considerations for the qualitative and quantitative determination of process-induced disorder in crystalline solids. *Journal of Pharmaceutical Sciences, 103*(9), 2595–2604. doi:10.1002/jps.23930.

Nyavanandi, D., Kallakunta, V. R., Sarabu, S., Butreddy, A., Narala, S., Bandari, S., & Repka, M. A. (2021). Impact of hydrophilic binders on stability of lipid-based sustained release matrices of quetiapine fumarate by the continuous twin screw melt granulation technique. *Advanced Powder Technology, 32*(7), 2591–2604. doi:10.1016/j.apt.2021.05.040.

Ochoa, L., Igartua, M., Hernández, R. M., Gascón, A. R., Solinis, M. A., & Pedraz, J. L. (2011). Novel extended-release formulation of lovastatin by one-step melt granulation: In vitro and in vivo evaluation. *European Journal of Pharmaceutics and Biopharmaceutics, 77*(2), 306–312. doi:10.1016/j.ejpb.2010.11.024.

Ochoa, L., Igartua, M., Hernández, R. M., Solinís, M. Á., Gascón, A. R., & Luis Pedraz, J. (2010). *In vivo* evaluation of two new sustained release formulations elaborated by one-step melt granulation: Level A *in vitro–in vivo* correlation. *European Journal of Pharmaceutics and Biopharmaceutics, 75*(2), 232–237. doi:10.1016/j.ejpb.2010.02.008.

Panda, R. R., & Tiwary, A. K. (2012). Hot melt granulation: A facile approach for monolithic osmotic release tablets. *Drug Development and Industrial Pharmacy, 38*(4), 447–461. doi:10.3109/03639045.2011.609562.

Parikh, D. M. (2005). *Handbook of pharmaceutical granulation technology* (pp. 1–625). CRC Press. Retrieved date.https://www.taylorfrancis.com/books/e/9780849354953.

Passerini, N., Albertini, B., González-Rodrıǵuez, M. L., Cavallari, C., Rodriguez, L., et al. (2002). Preparation and characterisation of ibuprofen–poloxamer 188 granules obtained by melt granulation. *European Journal of Pharmaceutical Sciences, 15*(1), 71–78. doi:10.1016/s0928-0987(01)00210-x.

Passerini, N., Albertini, B., Perissutti, B., & Rodriguez, L. (2006). Evaluation of melt granulation and ultrasonic spray congealing as techniques to enhance the dissolution of praziquantel. *International Journal of Pharmaceutics, 318*(1—2), 92–102. doi:10.1016/j.ijpharm.2006.03.028.

Passerini, Nadia, Calogerà, Giacomo, Albertini, Beatrice, Rodriguez, Lorenzo, et al. (2010). Melt granulation of pharmaceutical powders: A comparison of high-shear mixer and fluidised bed processes. *International Journal of Pharmaceutics, 391*(1—2), 177–186. doi:10.1016/j.ijpharm.2010.03.013.

Patil, H., Tiwari, R. V., Upadhye, S. B., Vladyka, R. S., & Repka, M. A. (2015). Formulation and development of pH-independent/dependent sustained release matrix tablets of ondansetron HCl by a continuous twin-screw melt granulation process. *International Journal of Pharmaceutics, 496*(1), 33–41. doi:10.1016/j.ijpharm.2015.04.009.

Perissutti, B., Rubessa, F., Moneghini, M., Voinovich, D., et al. (2003). Formulation design of carbamazepine fast-release tablets prepared by melt granulation technique. *International Journal of Pharmaceutics, 256*(1—2), 53–63. doi:10.1016/s0378-5173(03)00062-0.

Ponga, J. C. L., Piña, J., & Cotabarren, I. M. (2023). Fluidized bed co-melt granulation: New insights in the influence of process variables and validation of regime map theory. *Powders, 2*(3), 639–658. doi:10.3390/powders2030040.

Pradhan, A., Costello, M., Yang, F., Bi, V., Durig, T., & Zhang, F. (2022). Using twin-screw melt granulation to co-process mannitol and hydroxypropylcellulose. *Journal of Drug Delivery Science and Technology, 77*, 103880. doi:10.1016/j.jddst.2022.103880.

Pradhan, A., Phillips, B., Yang, F., Karan, K., Durig, T., Haight, B., Martin, C., & Zhang, F. (2024). Optimizing twin-screw melt granulation: The role of overflight clearance on granulation behavior. *International Journal of Pharmaceutics, 653*, 123900. doi:10.1016/j.ijpharm.2024.123900.

Pradhan, A., Yang, F., Karan, K., Durig, T., Schwing, Q., Haight, B., Costello, M., Anderson, M., & Zhang, F. (2025). Material-sparing degradation-kinetics model for thermolabile drug stability assessment during twin-screw melt granulation—Insights with gabapentin. *International Journal of Pharmaceutics, 674*, 125421. doi:10.1016/j.ijpharm.2025.125421.

Prado, H. J., Bonelli, P. R., & Cukierman, A. L. (2014). In situ fluidized hot melt granulation using a novel meltable binder: Effect of formulation variables on granule characteristics and controlled release tablets. *Powder Technology, 264*, 498–506. doi:10.1016/j.powtec.2014.05.058.

Qiu, Y. (2024). *Solid pharmaceutical formulations for treating endometriosis, uterine fibroids, polycystic ovary syndrome or adenomyosis*. Google Patents.

Rodriguez, L., Cavallari, C., Passerini, N., Albertini, B., González-Rodrıguez, M. L., & Fini, A. (2002). Preparation and characterization by morphological analysis of diclofenac/PEG 4000 granules obtained using three different techniques. *International Journal of Pharmaceutics, 242*(1—2), 285–289. doi:10.1016/s0378-5173(02)00189-8.

Royce, A., Suryawanshi, J., Shah, U., Vishnupad, K., et al. (1996). Alternative granulation technique: Melt granulation. *Drug Development and Industrial Pharmacy, 22*(9—10), 917–924. doi:10.3109/03639049609065921.

Sampat, C., & Ramachandran, R. (2024). Optimizing energy efficiency of a twin-screw granulation process in real-time using a long short-term memory (LSTM) network. *ACS Engineering Au, 4*(2), 278–289. doi:10.1021/acsengineeringau.3c00038.

Schæfer, T. (1996a). Melt pelletization in a high shear mixer. X. Agglomeration of binary mixtures. *International Journal of Pharmaceutics, 139*(1—2), 149–159. doi:10.1016/0378-5173(96)04615-7.

Schæfer, T. (1996b). Melt pelletization in a high shear mixer VI. Agglomeration of a cohesive powder. *International Journal of Pharmaceutics, 132*(1—2), 221–230. doi:10.1016/0378-5173(95)04374-8.

Schaefer, T., Holm, P., & Kristensen, H. G. (1990). Melt granulation in a laboratory scale high shear mixer. *Drug Development and Industrial Pharmacy, 16*(8), 1249–1277. doi:10.3109/03639049009115960.

Schæfer, T., Johnsen, D., & Johansen, A. (2004). Effects of powder particle size and binder viscosity on intergranular and intragranular particle size heterogeneity during high shear granulation. *European Journal of Pharmaceutical Sciences, 21*(4), 525–531. doi:10.1016/j.ejps.2003.12.002.

Schæfer, T., & Mathiesen, C. (1996a). Melt pelletization in a high shear mixer. VIII. Effects of binder viscosity. *International Journal of Pharmaceutics, 139*(1—2), 125–138. doi:10.1016/0378-5173(96)04549-8.

Schæfer, T., & Mathiesen, C. (1996b). Melt pelletization in a high shear mixer. IX. Effects of binder particle size. *International Journal of Pharmaceutics, 139*(1—2), 139–148. doi:10.1016/0378-5173(96)04548-6.

Schæfer, T., Taagegaard, B., Juul Thomsen, L., & Gjelstrup Kristensen, H. (1993). Melt pelletization in a high shear mixer. IV. Effects of process variables in a laboratory scale mixer. *European Journal of Pharmaceutical Sciences, 1*(3), 125–131. doi:10.1016/0928-0987(93)90002-r.

Shanmugam, S. (2015). Granulation techniques and technologies: Recent progresses. *BioImpacts, 5*(1), 55–63. doi:10.15171/bi.2015.04.

Van de Steene, S., Van Renterghem, J., Vanhoorne, V., Vervaet, C., Kumar, A., & De Beer, T. (2023a). Elucidation of granulation mechanisms along the length of the barrel in continuous twin-screw melt granulation. *International Journal of Pharmaceutics, 639*.

Steffens, K. E., Brenner, M. B., Hartig, M. U., Monschke, M., & Wagner, K. G. (2020). Melt granulation: A comparison of granules produced via high-shear mixing and twin-screw granulation. *International Journal of Pharmaceutics, 591*, 119941. doi:10.1016/j.ijpharm.2020.119941.

Steffens, K. E., & Wagner, K. G. (2019). Improvement of tabletability via twin-screw melt granulation: Focus on binder distribution. *International Journal of Pharmaceutics, 570*, 118649. doi:10.1016/j.ijpharm.2019.118649.

Steffens, K. E., & Wagner, K. G. (2020). Dissolution enhancement of carbamazepine using twin-screw melt granulation. *European Journal of Pharmaceutics and Biopharmaceutics, 148*, 77–87. doi:10.1016/j.ejpb.2020.01.006.

Tan, D. C. T., Chin, W. W. L., Tan, E. H., Hong, S., Gu, W., & Gokhale, R. (2014). Effect of binders on the release rates of direct molded verapamil tablets using twin-screw extruder in melt granulation. *International Journal of Pharmaceutics, 463*(1), 89–97. doi:10.1016/j.ijpharm.2013.12.053.

Thomsen, L. J., Schaefer, T., & Kristensen, H. G. (1994). Prolonged release matrix pellets prepared by melt pelletization II. Hydrophobic substances as meltable binders. *Drug Development and Industrial Pharmacy, 20*(7), 1179–1197. doi:10.3109/03639049409038360.

Todd, D. B., Andersen, Paul, et al. (1998). The Werner and Pfleiderer Twin-screw co-rotating extruder system. In *Plastics compounding: Equipment and processing* (pp. 71–124). Hansen Publishers.

Tovey, G. D. (2018). *Pharmaceutical formulation: The science and technology of dosage forms drug discovery*. Royal Society of Chemistry. doi:10.1039/9781782620402.

Ullah, I. (1987). Moisture activated dry granulation: A general process. *Pharmaceutical Technology, 11*(9), 48–54.

Van de Steene, S., Van Renterghem, J., Vanhoorne, V., Vervaet, C., Kumar, A., & De Beer, T. (2023b). Visualization of the granule temperature using thermal imaging to improve understanding of the granulation mechanism in continuous twin-screw melt granulation. *International Journal of Pharmaceutics, 645*, 123423.

Van de Steene, Shana, Vanhoorne, Valerie, Vervaet, Chris, & De Beer, Thomas (2021). Melt Granulation. In Parikh, Dilip (Ed.), Handbook of Pharmaceutical Granulation Technology (4th ed.). CRC Press (Taylor and Francis group).

Van Melkebeke, B., Vermeulen, B., Vervaet, C., & Paul Remon, J. (2006). Melt granulation using a twin-screw extruder: A case study. *International Journal of Pharmaceutics, 326*(1–2), 89–93. doi:10.1016/j.ijpharm.2006.07.005.

Van Melkebeke, B., Vervaet, C., & Remon, J. P. (2008). Validation of a continuous granulation process using a twin-screw extruder. *International Journal of Pharmaceutics, 356*(1–2), 224–230. doi:10.1016/j.ijpharm.2008.01.012.

Vasanthavada, M., Wang, Y., Haefele, T., Lakshman, J. P., Mone, M., Tong, W., Joshi, Y. M., & Serajuddin, A. T. M. (2011). Application of melt granulation technology using twin-screw extruder in development of high-dose modified-release tablet formulation. *Journal of Pharmaceutical Sciences, 100*(5), 1923–1934. doi:10.1002/jps.22411.

Vicente Martin, C., Stocker, S., Bautista, M., Rogue, V., Steib-Lauer, C., Häcker, H.-G., Spickermann, D., Hirsch, S., & Dhareshwar, S. S. (2022). Commercial scale transfer of a twin-screw melt granulation process for high drug load fevipiprant tablets. *Drug Development and Industrial Pharmacy, 48*(5), 211–225. doi:10.1080/03639045.2022.2104307.

Voinovich, D., Campisi, B., Moneghini, M., Vincenzi, C., & Phan-Tan-Luu, R. (1999). Screening of high shear mixer melt granulation process variables using an asymmetrical factorial design. *International Journal of Pharmaceutics, 190*(1), 73–81. doi:10.1016/s0378-5173(99)00278-1.

Wahl, P. R., Treffer, D., Mohr, S., Roblegg, E., Koscher, G., & Khinast, J. G. (2013). Inline monitoring and a PAT strategy for pharmaceutical hot melt extrusion. *International Journal of Pharmaceutics, 455*(1–2), 159–168. doi:10.1016/j.ijpharm.2013.07.044.

Walker, G. M., Andrews, G., & Jones, D. (2006). Effect of process parameters on the melt granulation of pharmaceutical powders. *Powder Technology, 165*(3), 161–166. doi:10.1016/j.powtec.2006.03.024.

Weatherley, S., MU, B. O., Thompson, M. R., Sheskey, P. J., & O'Donnell, K. P. (2013). Hot-melt granulation in a twin screw extruder: Effects of processing on formulations with caffeine and ibuprofen. *Journal of Pharmaceutical Sciences, 102*(12), 4330–4336. doi:10.1002/jps.23739.

Wong, T. W., Cheong, W. S., & Heng, P. W. S. (2005). Melt granulation and pelletization. In *Handbook of pharmaceutical granulation technology* (pp. 385–406). CRC Press. Retrieved from. https://www.taylorfrancis.com/books/e/9780849354953.

Xu, Ting, Nahar, Kajaljit, Bates, Simon, Morris, Kenneth, & Dave, Rutesh (2018). Polymorphic transformation of indomethacin during hot melt extrusion granulation: Process and dissolution control. *Pharmaceutical Research, 35*(7). doi:10.1007/s11095-017-2325-x.

Yang, D., Kulkarni, R., Behme, R., & Kotiyan, P. (2007). Effect of the melt granulation technique on the dissolution characteristics of griseofulvin. *International Journal of Pharmaceutics, 329*(1—2), 72–80. https://doi.org/10.1016/j.ijpharm.2006.08.029.

Yanze, F. M., Duru, C., & Jacob, M. (2000). A process to produce effervescent tablets: Fluidized bed dryer melt granulation. *Drug Development and Industrial Pharmacy, 26*(11), 1167–1176. https://doi.org/10.1081/DDC-100100988.

Zhai, H., Li, S., Jones, D. S., Walker, G. M., & Andrews, G. P. (2010). The effect of the binder size and viscosity on agglomerate growth in fluidised hot melt granulation. *Chemical Engineering Journal, 164*(2—3), 275–284. https://doi.org/10.1016/j.cej.2010.08.056.

Zong, Z., Qiu, J., Tinmanee, R., & Kirsch, L. E. (2012). Kinetic model for solid-state degradation of gabapentin. *Journal of Pharmaceutical Sciences, 101*(6), 2123–2133. https://doi.org/10.1002/jps.23115.

Vahervuo, K., Orion Oyj, 2008. *Oral Dosage Form*. U.S. Patent Application 11/916,460. (https://patents.google.com/patent/US20080187590A1/en).

Non-Print Items

Abstract

Melt granulation is a promising alternative to wet-granulation for manufacturing oral solid dosage forms, especially for moisture-sensitive drugs, as evidenced by Eucreas. Despite similarities in mechanisms, it relies on thermal and frictional energy to melt/soften binders. This chapter explores melt granulation from formulation and process design perspectives, emphasizing critical quality attributes. Binder characteristics like type, viscosity, level, and drug interactions are pivotal as the binder undergoes thermal activation. Although batch melts granulation using high-shear mixers/fluid bed processors is studied, adoption is limited due to binder viscosity constraints and poor temperature control. Twin-screw melt granulation (TSMG) has emerged as a preferred technology, offering better mixing, scalability, and higher drug loading. Process analytical technology enhances real-time monitoring, improving robustness. However, TSMG's thermal and shear intensity induces in-process physicochemical changes, and the lack of material-sparing approaches for assessment makes it challenging. The chapter briefly touches on integrating modeling and machine learning to support broader adoption.

Keywords

Binder properties; Continuous manufacturing; Granulation; Molten binder; Process design; Twin-screw melt granulation

CHAPTER 25

The application of the state-of-the-art material library/material database approach to the process understanding and process modeling of wet granulation

Zichen Liang[a,b], Gan Luo[a,b] and Bing Xu[a,b]

[a]Department of Chinese Medicine Informatics, Beijing University of Chinese Medicine, Beijing, China, [b]Beijing Key Laboratory of Chinese Medicine Manufacturing Process Control and Quality Evaluation, Beijing, China

25.1 The material library/material database concept and its application in pharmaceutical formulation development

In pharmaceutical design and development, the material library or material database is an emerging, new, and efficient approach for organizing physical property data of materials such as active pharmaceutical ingredients (APIs), excipients, or intermediates. The material library method aims to develop a standard material characterization framework to collect and store the physiochemical properties and related information of pharmaceutical materials (Escotet-Espinoza et al., 2018). By using a material library, both new and generic drug development activities could be enhanced in different ways, such as by understanding the relationship among different material quality attributes and simplifying the material characterization workload, finding the surrogate or equivalent materials for costly APIs during initial process development, and supporting the development of process models by linking the materials' physical properties to unit operations (Alwosheel et al., 2018; Dai et al., 2019b; Drašković et al., 2018; Kassem et al., 2020; Katz et al., 2013; Yu et al., 2019).

In practice, the material library was presented in the form of tabular data, in which the rows were different materials, and the columns were material quality attributes. Table 25.1 shows the typical sizes of reported pharmaceutical material libraries after retrieving articles in Web of Science using "material library" or "material database" as keywords (Bekaert et al., 2021; Benedetti et al., 2019; Bostijn et al., 2019; Dai et al., 2019; Dhondt et al., 2022a, Dhondt et al., 2022b; Escotet-Espinoza et al., 2018; Hayashi et al., 2021; Kassem et al., 2020; Ryckaert et al., 2021; Sousa, Serra, Estevens, Costa, & Ribeiro, 2023; Su et al., 2023; Van Snick et al., 2018; Y. Wang et al., 2022; Y. Wang, O'Connor, Li, Ashraf, & Cruz, 2019; Z. Wang et al., 2021; Yu et al., 2019; Bekaert et al., 2022; Escotet-Espinoza et al., 2019; Zhang et al., 2019b). From the aspect of material attributes, it can be seen that the numbers of material

Table 25.1 The sizes and applications of material libraries reported in 2018–2023.

No.	Number of Samples	Number of Material Attributes	Year	Application Area	References
1	20	30	2018	Find surrogate materials for pharmaceutical process development.	Escotet-Espinoza et al. (2018)
2	55	Over 100	2018	Build predictive models for *in silico* process.	Van Snick et al. (2018)
3	41	8	2019	Develop a direct compression decision-making tool to accelerate materials' screening.	Benedetti et al. (2019)
4	15	25	2019	Predict the volumetric and gravimetric feeding behavior of a low-feed-rate feeder.	Bostijn et al. (2019)
5	130	18	2019	Develop a compression behavior classification system for direct compression.	Dai et al., 2019b
6	20	32	2019	Study the effect of tracer material properties on the residence time distribution of continuous powder-blending operations.	Sebastian Escotet-Espinoza et al. (2019)
7	20	44	2019	Evaluate material performance on a loss-in-weight feeder.	Wang et al. (2019)
8	111	22	2019	Develop a compression behavior classification system for roll compaction.	Yu et al. (2019)
9	12	18	2019	Analyze the effect of material attributes on the dissolution profile of the matrix tablet.	Zhang et al. (2019b)
10	10	30	2020	Analyze the impact of material attributes on the performance of an auger dosing process.	Kassem et al., 2020b

(*continued on next page*)

11	13	44	2021	Predict feeding performance based on material properties.	Bekaert et al. (2021)
12	81	28	2021	Develop machine learning models by linking material properties and direct compression tablet properties.	Hayashi et al., 2021)
13	27	48	2021	Develop a TPLS model for the twin-screw wet granulation process and formulation development.	Ryckaert et al., 2021)
14	56	18	2021	Develop a formulation process quality model for high-shear wet granulation.	Wang et al., 2021)
15	12	44	2022	Analyze the impact of material attributes on the gravimetric feeding process.	Bekaert et al., 2022a)
16	14	55	2022	Develop a TPLS model for the direct compression process and formulation design.	Dhondt et al., 2022ab)
17	32	19	2022	Develop a PCA model to recognize the highest amount of variability in physical powder properties.	(Dhondt et al., 2022b)
18	30	19	2022	Develop a tabletability change classification system for high-shear wet granulation and tableting.	Wang et al., 2022)
19	15	14	2023	Analyze the impact of material attributes on direct compression extended-release formulations.	Sousa et al., 2023)
20	31	18	2023	Develop a tabletability change classification system for roll compaction, dry granulation, and tableting.	Su et al., 2023)

From Cao et al. (2024).

property descriptors are mainly spread in the range of 8 to 55. In particular, Van et al.(2018) built a material library involving over 100 raw material descriptors. The investigated material properties are usually function-related or process-oriented. For instance, Wang et al. (2019) tried to predict feeder performance based on material flow properties. Generally speaking, a thorough evaluation of material properties is the prerequisite for building predictive models for *in silico* processes and formulation development. But this does not mean that more material property descriptors are better, since there may be interrelationships among them and more descriptors mean a higher cost of characterization. Van Snick et al. (2018) proved that correlated descriptors in the raw material property database could be simplified using a multivariate data analysis (e.g., principal component analysis).

In addition, the requirements for sample size and sample diversity of a material library also need to be considered to obtain a high-quality dataset. As shown in Table 25.1, the sample sizes of most material libraries range from 10 to 130. Material types are often determined empirically by considering the following aspects: (1) selecting materials with different deformation behavior (e.g., plastic or brittle) (Katz et al., 2013; Lamešić et al., 2018; Wang et al., 2020); (2) choosing materials with different chemical compositions such as APIs, cellulose, lactose, starch, or calcium hydrogen phosphate (Hildebrandt et al., 2019; Ono & Yonemochi, 2020; Rojas et al., 2015); (3) using different pharmaceutical excipients such as fillers, binders, lubricants, or disintegrants (Hildebrandt et al., 2019; Van Snick et al., 2018); (4) enriching material variability by incorporating materials from different suppliers or with different grades (Hernández et al., 2021; Nofrerias et al., 2019; Paul et al., 2018).

Conventionally, sufficiently more observations serving as a training set are considered favorable before modeling in machine learning. For instance, in the field of artificial neural networks, there is a set of rules of thumb regarding sample size requirements: (1) The sample size needs to be at least a factor 50 to 1000 times the number of prediction classes; (2) the sample size needs to be at least a factor 10 to 100 times the number of the features; and (3) the sample size needs to be at least a factor 10 times the number of parameters (i.e., synaptic weights and biases) in the network (Alwosheel et al., 2018; Haykin, 2009). In fact, collecting and annotating high-volume data meeting these above requirements is time-consuming and expensive. To overcome this obstacle, some research has discussed the feasibility of minimum dataset size and few-shot learning, which does not result in significant model performance loss in the presence of limited data. Bongiorno et al.(2022) constructed sample sets ranging from 10 to 50,000 to study the effect of dataset size on model training performance and found that approximately 200 examples were generally sufficient to train a machine learning algorithm, and increasing the number of training samples did not significantly improve the accuracy of the results. Li et al. (2020) proposed an indicator g^2, which was used to assess the model structure to analyze the minimum size of data to construct a valid model. The verification found that with the increase in the number of samples of the modeling dataset, the model became stable, as the g^2 index converged to zero. Althnian et al. (2021) found that the overall performance of classifiers depended on how well a dataset represented the original distribution rather than its size. These studies demonstrated that it was possible to find the suitable sample size for modeling purposes. It was found through our study that small-size material library consisting of representative and diverse materials are feasible for developing models. We will describe how to construct a material library in the follow content.

The self-constructed database of intelligent TCMs (iTCMs) with a wide range of naturally produced powders (NPPs) and pharmaceutical excipients contains 27 quality attribute descriptions (18 physical powder parameters and 9 compression descriptors), as outlined in Table 25.2.

Table 25.2 Overview of powder characterization techniques, corresponding descriptors, and abbreviations.

Property	Characterization Technique	Descriptor	Abbreviation
Powder properties	Powder Density Test	Bulk and tapped density	ρ_b and ρ_t
		Ture density and porosity	D_t and ε
		Carr's index	IC
		Interparticle porosity	Ie
		Hausner ratio	IH
	Diametrical crushing test under maximum compression pressure	Cohesion index	Icd
	Flow through an orifice	Angle of repose	AOR/α
		Flow time	t''
	Rapid moisture test	Moisture content	HR%
	Moisture sorption	Water uptake at 76% (±2%) of the relative humidity	H%
	Laser diffraction	10%, 50%, and 90% cumulative undersize of volumetric particle size distribution	D_{10}, D_{50}, D_{90}
		Width and span of volumetric particle size distribution	Span
		Percentage of particles measuring below 50 μm	%Pf
		Homogeneity index	Iθ
Compression descriptor	The *Kawakita* model	$\frac{P}{C} = \frac{1}{ab} + \frac{P}{a}$	a, ab, b^{-1}
	The *Heckel* model	$\mathrm{Ln}(\frac{1}{\varepsilon}) = kP + A$, $P_y = \frac{1}{k}$	P_y
	The *Gurnham* model	$\varepsilon = -\frac{1}{K}\mathrm{Ln}(\frac{P}{P_0})$	K
	The *Ryshkewitch–Duckworth* model	$TS = TS_0 \exp(-k_b \varepsilon)$	k_b
	The *Shapiro* model	$\mathrm{Ln}(\varepsilon) = \mathrm{ln}\varepsilon_0 - kP - fP^{0.5}$	f
	The *Power* model	$TS = dP^g$	d, g
Tablet mechanical property	Diametrical crushing test	Tensile strength	TS

Among the 27 powder properties, 12 parameters were measured or calculated by standard testing procedures of the SeDeM expert system methodology (Dai et al., 2019a; Perez et al., 2006). These parameters include bulk density (ρ_b, g•cm^{-3}), tapped density (ρ_t, g•cm^{-3}), interparticle porosity (*Ie*), Carr's index (*IC*), Hausner ratio (*IH*), angle of repose (*AOR*, °), flow time (*t''*, s), cohesion index (*Icd*, N), loss on drying (*HR%*), hygroscopicity (*H%*), proportion of particles smaller than 50 μm (*%Pf*), and homogeneity index (*Iθ*). The dimensions of powder can be expressed by ρ_b and ρ_t. The parameters *IC, Ie,* and *Icd* characterized the compressibility of powders. Descriptors *AOR, t''*, and *IH* reflect the flowability of powder. The stability of powder can be described by the parameters *HR%* and *H%*. Physical properties *%Pf* and *Iθ* represent the uniformity of the powder. The remaining six physical properties include true density (D_t, g·cm^{-3}), particle sizes (i.e., D_{10}, D_{50}, and D_{90}, μm), particle size distribution width (Span), and solid fraction of powder (SFp). The compression curve (*TS* vs. pressure *P*) data for each material were also stored in the iTCM database. As for compression descriptors, different compression equations are used, respectively, to interpret the compressibility, compactability, and tabletability. The compressibility of a powder is the powder's ability to deform under pressure, and it is described by the indexes of *Kawakita a, ab,* and b^{-1}, *Heckel* P_y, *Shapiro f,* and *Gurnham K* (Leuenberger, 1982; Sonnergaard, 2006; Šantl et al., 2011). The compactability of a powder is the ability to form mechanically strong compacts and is expressed by *Ryshkewitch–Duckworth* k_b (Šantl et al., 2011; Sonnergaard, 2006; Sun, 2008). The tabletability of powders, defined as the capacity of a powdered material to be transformed into a tablet of sufficient strength under the prescribed pressures, can be indicated by the index of *Power d* (Dai et al., 2019b; Sun, 2008; Sun & Kleinebudde, 2016).

25.2 Wet granulation data fusion modeling based on material library

In recent decades, wet granulation (WG) has been widely used as a crucial process for manufacturing of oral solid dosage (OSD) forms. A survey by Leane et al. (2018) showed that WG was the most popular process choice compared with direct compression (DC), dry granulation (DG), and other technologies during the pharmaceutical development of over 80% of early-stage compounds. From manufacturing point of view, high shear wet granulation (HSWG), not only improves raw materials' flowability and content uniformity but also reduces dustiness and minimizes segregation, by which particles get ideal properties that adjust to subsequent processing such as compaction or coating (Thapa et al., 2019). It is the fact that desired granule attributes can be obtained by optimizing both the formulation and process conditions. To achieve this goal, lots of effort have been made on understanding multidimensional relationships among formulation properties, process parameters, and granule properties (Thapa et al., 2019). Nevertheless, wet granulation in practice has partly remained "more of an art than a science"(Iveson et al., 2001; Thapa et al., 2019), due to the lack of knowledge in the multiple mechanisms (de Koster et al., 2019).

HSWG is an intricate particle size enlargement technique, which could be typically separated into three simultaneous rate processes, namely, wetting and nucleation, consolidation and coalescence, and attrition and breakage (Wade et al., 2020). In order to understand the complex mechanisms of the HSWG process, the nucleation regime map and the growth regime map were derived (de Koster et al., 2019; Hapgood et al., 2003; Iveson et al., 2001; Pohlman & Litster, 2015; Realpe & Velázquez, 2008). Some dimensionless parameters, such as the drop penetration time and the spray flux, summarizing the key features of multiple process parameters, were brought forward (Hapgood et al., 2003; Wade et al., 2020). Besides, population balance model was developed to describe the evolution of granule properties over time, from single dimensional to multidimensional (Yu et al., 2017). The complexity of HSWG further embraced advanced computational approaches such as discrete element model, computational fluid dynamics, and hybrid models, seeking for more accurate prediction of granulation behaviors (Barrasso et al., 2015; Braumann et al., 2010; Kulju et al., 2016; Yu et al., 2017).

Although there was a large amount of effort focusing on modeling the HSWG process from first principles as described above, a rational approach for modeling or optimizing the wet granulation process was often realized by quality by design (QbD) experiments and statistical analysis in practice (Thapa et al., 2019). Until now, QbD principles have been applied to a considerable number of researches in HSWG process development (Ferreira & Tobyn, 2015; Han et al., 2019; Pandey & Badawy, 2016; Silva et al., 2017; Singh, 2019; Thoorens et al., 2015; Zhang et al., 2019a). For instance, Lee et al. (2017) optimized a high shear granulation process of a bilayer tablet based on QbD approaches, in order to achieve better physical stability via economical and simpler manufacturing processes. In another case, Fayed et al. (2017a) investigated the critical influence of water amount and wet mixing time on both granules produced by high shear granulation and final tablets using the toolkit of QbD. From each QbD study, it was observed that the designed experiments often generated a solution to a particular product formulation at given granulation conditions. However, from various literature, information about the HSWG process can be acquired in multiple product formulations or experiments, at different granulation scales and conditions. An aggregation of available literature data through data fusion provides a possibility of obtaining more complete knowledge of the HSWG process.

Data fusion is a technique that integrates data and knowledge from multiple sources to produce a single model. There are three levels of data fusion, namely, low-level (data level), mid-level (feature level) and high-level (decision level) (Borràs et al., 2015). So far, data fusion has been extensively applied in multisensor or multimodality environments (Borràs et al., 2016; Mu et al., 2020). The research of data fusion in pharmaceutical process development is at the preliminary stage. A data fusion model combining near infrared features and process parameters for the prediction of drug dissolution from controlled release multiparticulate beads was developed by Ibrahim et al. (2019). Casian et al. (2019) first applied a data fusion strategy to increase the quantitative ability of a process analytical technology platform consisting of four sensors. Han et al. (2018) extracted 145 direct compressed oral disintegrating tablets (ODT) formulation data from 1218 articles in Web of Science database, and built a deep neural network model to predict the disintegrating time of ODT formulations. To the best of our knowledge, there has been no study using the data fusion method to develop an HSWG process model.

In this chapter, we aim to establish a novel comprehensive dataset of HSWG from historical articles and to derive a statistical model for predicting the critical quality attribute (CQA) of granules. The diversity across the datasets from different HSWG-related literature not only laid the foundation of data fusion but also brought forward challenges in transforming multiple datasets into unified representations. With regard to data alignment, the current chapter provides the following contributions. From the aspect of formulation, the formulation materials and their mass fractions were recorded, but the material characterization would be inadequate or even missing. With the help of an established material database, several material matching rules were self-developed to give each formulation material a properties vector. Then, using the ideal mixing rule, the formulation properties were estimated. From the aspect of HSWG process, the process variables were described in many scales from the lab to the commercial. The dimensionless parameters, regardless of the granulator scale, were used to tackle this problem. To be specific, the Froude number (Fr) was used to mitigate variations of batch scales, and the theoretical maximum pore saturation (S'_{max}) was used to explain the influence of binder addition on granule size (Badawy et al., 2019). Consequently, by using the formatted dataset with adequate information, a multivariate calibration model is more likely to be derived. In addition, the literature datasets may not have the same level of information quality or reliability. In order to enhance the prediction accuracy of HSWG process model, a few laboratory experiments designed by the Monte

$$\begin{array}{c}\text{Ingredients}\\ \text{Formulation No.}\begin{bmatrix} a_{11} & a_{12} & a_{13} & \cdots & a_{1j} \\ a_{21} & a_{22} & a_{23} & \cdots & a_{2j} \\ & & \vdots & & \\ a_{i1} & a_{i2} & a_{i3} & \cdots & a_{ij} \end{bmatrix}\end{array} \times \begin{array}{c}\text{Material properties}\\ \text{Ingredients}\begin{bmatrix} b_{11} & b_{12} & b_{13} & \cdots & b_{1k} \\ b_{21} & b_{22} & b_{23} & \cdots & b_{2k} \\ & & \vdots & & \\ b_{j1} & b_{j2} & b_{j3} & \cdots & b_{jk} \end{bmatrix}\end{array} = \begin{array}{c}\text{Mixture properties}\\ \text{Formulation No.}\begin{bmatrix} c_{11} & c_{12} & c_{13} & \cdots & c_{1k} \\ c_{21} & c_{22} & c_{23} & \cdots & c_{2k} \\ & & \vdots & & \\ c_{i1} & c_{i2} & c_{i3} & \cdots & c_{ik} \end{bmatrix}\end{array}$$

FIGURE 25.1

Prediction of formulation properties based on the ideal mixing rule.

Carlo simulation and the random sampling were carried out. Two types of data, that is, laboratory measurements and literature observations, were further integrated to build a more robust statistical model.

25.2.1 Data sources

The multivariate relationships of the HSWG process were assumed to be preserved in the data from the designed experiments of literature. To ensure the reliability, data were extracted from historical articles searched in the Web of Science and China National Knowledge Infrastructure database. Using "high shear wet granulation" as keywords, the articles that employed water as binder agent of HSWG were collected. The criterions of selecting articles are described as follows. First, the articles should give the mass fraction and properties (if mentioned) of each formulation material. Some articles employed single material as the research object, and the corresponding mass fraction was regarded as 100%. Then, at least three process parameters such as the liquid-to-solid (L/S) ratio, the impeller speed and the granulator scale were provided. Next, the median granule size G_{50} as the CQA must be measured and given, since the granule size had a significant impact on the subsequent processes (e.g., tableting) as well as the final product quality (Grdešič et al., 2018). As a result, 10 articles were found to fit for the above requirements, and 143 pieces of experimental data were successfully extracted.

25.2.2 Formulation properties estimation

Before the HSWG process, the ingredients of the formulation must be mixed uniformly. The mixture properties can be estimated using the ideal mixing rule (Grassmann et al., 1971), as shown in Fig. 25.1. Linear combinations of physical properties of formulation materials were realized by multiplying the material properties matrix ($j \times k$) by the formulation matrix ($i \times j$) using MATLAB R2016b software (The MathWorks Inc., United States). In the collected 143 pieces of experimental data, 22 formulation materials including 19 pharmaceutical excipients and 3 APIs are involved in 17 formulations. Hence, the formulation matrix was organized to contain 22 columns and 17 rows.

In the collected articles, there was a lack of characterization of input materials. The material properties matrix was built by taking advantage of an open-access material database named iTCM (Dai et al., 2019a), in which 77 commonly used pharmaceutical excipients were, respectively, described by 18 physical properties. The mentioned material properties are as shown in Table 25.2.

Considering the similarity between materials, the material properties matrix (22 × 18) was built by performing the following material matching operations. (1) If a formulation material had the same name and specifications as one pharmaceutical excipient included in the iTCM database, they were assumed to share the same physical properties; (2) in some articles, only the material name was given. The corresponding material properties were regarded as the weighted average of all excipients with the same name but different specifications in the database. (3) If a formulation material was outside the range of recorded materials in the iTCM database, its properties were supplemented with the most similar one in the database. For instance, the API of semi-fine acetaminophen—whose particle sizes D_{10}, D_{50}, and D_{90} were 0.73, 2.75, and 5 μm, respectively (Oka et al., 2015)—was considered to bear the same properties with magnesium stearate that had very similar D_{10}, D_{50} and D_{90} values (i.e., 1.21, 2.82 and 5.32 μm, respectively).

25.2.3 Process parameters processing

The process parameters from different articles were not uniform. In order to make the process parameters comparable, they were processed from two aspects: agitation intensity and granule growth conditions.

25.2.3.1 Calculation of the froude number

The granulator scales recorded in different articles were different, ranging from 0.25 to 600 L. In order to minimize the process variations caused by the scale, the Fr expressing the agitation intensity was calculated as follows (Badawy, Narang, LaMarche, Subramanian, & Varia, 2019; Pandey & Badawy, 2016):

$$Fr = \frac{\omega^2(2R)}{g} \tag{25.1}$$

where R denotes the inner radius of granulator (m). ω is the impeller speed, that is, the revolutions per second (rps) of the impeller, and g is the gravity constant. Considering the fact that the inner radius of the granulator was not given in some articles, it was estimated by using a geometric formula based on an assumption of a cylindrical granulator.

$$R = \sqrt{\frac{V}{\pi H}} \tag{25.2}$$

where V is the granulator scale (L) and H is the height of granulator. If the inner radius of granulator was not given, the equivalent R and H was utilized to calculate R from the granulator scale V. One qualifed article (Pandey et al., 2013) as well as the granulator in our laboratory showed that R was equal to H. In some articles, only the velocity of impeller blade tip was recorded. The impeller speed and the linear velocity of impeller could be transformed into each other with Eq. (25.3):

$$v = 2\pi R\omega \tag{25.3}$$

where v is the velocity of impeller blade tip (m/s).

25.2.3.2 Binder addition-related parameters

In some literature, the binder addition process was described by total volume of water (mL) and total mass of formulation (g). While in other articles, the binder addition process was described by the binder addition rate (mL per minute) or the total addition time (minute). In this chapter, the L/S ratio was calculated in order to make different forms of data comparable. Besides, the theoretical maximum pore saturation (S'_{max}) was calculated to estimate the water saturation in dry powder bed without considering the consolidation of wet granules during the granulation process (Hapgood et al., 2003; Ramachandran et al., 2008).

$$S'_{max} = \frac{w\rho_s(1 - \varepsilon_{min})}{\rho_l \varepsilon_{min}} \tag{25.4}$$

$$\varepsilon_{min} = 1 - \frac{\rho_e}{\rho_p} \tag{25.5}$$

where w is the mass ratio of L/S, ρ_s is the bulk density of the solid particles, and ρ_l is the liquid density. ε_{min} is the minimum porosity the formulation reaches for that particular set of operating conditions, and it is determined by Eq. (25.5). For ungranulated powders, ρ_e is the envelope density of particles and ρ_p is the true density. The true density of the formulation mixtures was estimated by the ideal mixing rule on the basis of every ingredient's true density and mass ratio. In this chapter, the envelope density of particles was unknown, and it was substituted by the tapped density of the formulation mixtures (Iveson & Litster, 1998). In this way, the minimum porosity was estimated by measuring the dry-tapped porosity of the formulation.

25.2.4 Multivariate process modeling

25.2.4.1 Model evaluation indicators

The partial least squares (PLS) was a prevalent data fusion algorithm (Borràs et al., 2016; Gao & Ren, 2010) and was employed to correlate the input and output data matrixes in this chapter. PLS provides a tool by which considerable variations of initial data can be described by a few latent variables (LVs). The PLS regression model was built on the calibration set using SIMCA 13.0 software (Umetrics, Umea, Sweden). Coefficient of determination (R^2), ability of prediction (Q^2), accuracy, and root mean square error of prediction (RMSEP) were used as model quality metrics. Formula of R^2 and Q^2 are as follows:

$$R2 = 1 - SSE/SST \tag{25.6}$$

where SSE is error sum of squares and SST is total sum of squares.

$$Q^2 = \left(1.0 - \prod (PRESS/SS)_a\right) \tag{25.7}$$

where PRESS is predicted residual error sum of square, SS is sum of squared deviations, and subscript a represents a certain variable. Accuracy of PLS model is calculated by the following equation (Han et al., 2018):

$$Accuracy = \frac{Number(|f' - f| \leq Deviation)}{All\ predictions} \tag{25.8}$$

where f' is the prediction value, f is the real value, and "Deviation" is the tolerance limit, and the numerator gives number of acceptable deviations between the predicted value and the real value. "All predictions" are the total number of observations. RMSEP for the prediction set is denoted as follows:

$$\text{RMSEP} = \sqrt{\sum (Y_{\text{obs}} - Y_{\text{pred}})^2 / N} \qquad (25.9)$$

where Y_{obs} and Y_{pred} refers to the predicted residuals for the observations in the prediction set. N is number of samples. *RMSEP* measures the predictive power of the model.

25.2.4.2 Process model derived from literature data

The CQA of the HSWG process (i.e., granule size G_{50}) was at first tried to be predicted by a PLS model established on literature data. The input variables of the intended PLS regression model included 18 formulation material properties and 3 process parameters. For a given formulation, its mixture properties would be repeatedly arranged with related process parameters. Before model building, the outliers were detected to reduce the model prediction error. All 143 pieces of data were at first used to build an initial PLS model correlating the 21 independent variables and the CQA under two LVs. Nine samples were found to be scattered outside the Hotelling T^2 ellipse at 95% confidence. These abnormal samples all used 100% (w/w) HPMC as the formulation material, and the liquid-to-mass ratios were approximate to 2. Besides, the resultant granules were over 6000 μm, which were not common for pharmaceutical applications. The HPMC is a hydrogel-forming polymer. When being reached by the water, the HPMC powders start to agglomerate due to hydrophobic interactions between the substituents on the polymeric chains, forming large aggregates (Simone et al., 2018). At higher HPMC concentrations, such a particular granulation mechanism may bring a big difficulty to provide a precise prediction of the granule size. By removing the outlier formulation No. 1, the rest of 16 formulations were randomly separated into 15 calibration formulations and 1 internal validation formulation (No. 17). The sample size of the calibration set and the internal validation set were 119 and 15, respectively. All variables were centered and scaled before process modeling.

The basic principle of choosing LVs is that when adding an extra LV into the PLS model, no obvious increase of both determinant coefficient (R^2) and prediction ability (Q^2) occur. As a result, four LVs were capable to explain 87.1% of the total variation. The resultant R^2 and Q^2 were 0.743 and 0.718, respectively. Accuracies at different deviations directly showed the prediction performance. The majority (85.0% to be specific) of predictions had predictive errors below 300 μm, and only 36.2% of all predictions had predictive errors within 100 μm. Besides, the RMSEP was 152.6. These results indicated that the process model directly developed from literature data (Model 1) possessed low analysis efficiency. The experiment conditions in different articles varied a lot, which might lead to a model that was not robust enough to provide a precise prediction performance.

25.2.5 Process model derived from integrated data (literature data and laboratory data)

In order to improve the model prediction performance for practical use, a few laboratory data were generated to supplement the literature data.

25.2.5.1 Laboratory wet granulation process method

Granulation operations were performed using the high shear wet granulator (SHK-4, Xi'an Run Tian Pharmaceutical Machinery Co., Ltd., Xi'an, China). A three-blade impeller was located at the bottom of the bowl, and binder liquid was added by gently pouring from the top through a funnel. The total mass of formulation mixture was 150 g in the 1 L granulator, and 300 g in 2 L granulator. Dry mixing under 600 rpm of impeller speed with duration of 3 min was employed prior to wet granulation to ensure uniformity of material mixture. The wet mixing process lasted for 3 min, and the liquid addition and impeller speed were set according to the experimental design. The chopper speed was kept at constant at 1200 rpm, avoiding wet mass caking on the steel wall. The resultant granules were dried at 50°C in an oven overnight. The granule size was determined using the sieve analysis. Six different standard sieves, including 12, 20, 60, 80, 120, and 200 mesh, were employed. After vibrating for 1 min at 30 Hz using a vibration machine (ZNS-300, Beijing Xing Shi Lihe Technology Development Co., Ltd., Beijing, China) particles on each sieve were weighted. The granule size G_{50} was calculated by fitting the curve of mass cumulative size distribution.

25.2.5.2 Simulation data

According to both the experimental design and the formulation properties estimation methods in Section 25.2.2, the mixture properties of each simulation formulation were calculated by the ideal mixing rule, which gave a simulation formulation properties matrix SX (808 × 18). In the formulation matrix mentioned in Section 25.2.2, there were involved three kinds of HPMC, six kinds of lactose, and five kinds of MCC, which were top three frequently used materials. Therefore, two cellulose materials and four lactose materials were selected from the iTCM material library to generate the simulated and experimental formulations. Microcrystalline cellulose (MCC) PH101 (lot No. 1545) was purchased from Asahi Kasei Chemicals Co., Ltd. (Japan). Lactose monohydrate Pharmatose 110M (Batch No. 1009CON), Pharmatose 200M (Batch No. 10095MW), Anhydrous lactose 21AN (Batch No. 1007NX8), and HPMC E15LV (Batch No. D011F6KL01) were purchased from Shanghai Chineway Pharmaceutical Technology Co., Ltd. (Shanghai, China). Lactose granulac 200 (lot No. L1535) was purchased from Molkerei MEGGLE Wasserburg GmbH&Co. KG. (Germany). The deionized water was used as the granulation liquid. The experimental design was realized through a combination of Monte Carlo simulation and random sampling technique. It was assumed that each simulation formulation contained one type of cellulose material and one type of lactose, and eight combinational forms of formulation could be obtained. For every combination, the mass fractions of one material were simulated to be varied from 0% to 100% at 1% increments, and the mass fractions of the other material were varied from 100% to 0% accordingly. As a result, a total of 808 simulation formulations were produced.

Two manipulated process parameters, that is, L/S ratio and impeller speed, were designed to carry out the granulation process. The number of simulated process parameters should match the number of simulated formulations. A total of 808 L/S ratios were simulated randomly in the range from 0.025 to 1.8 g/g, where the upper and lower limits of L/S ratio were figured out from qualified literature data in Section 25.2.1. The process settings must be within the capacity of lab granulator. The maximum and minimum impeller speeds of the lab granulator used were 1200 and 350 s^{-1}, respectively. Hence, 808 impeller speeds as integers were simulated randomly in this range. The Fr was calculated by Eq. (25.1) on the basis of the radius of the granulator and the impeller speed described in Section 25.2.2. A total of 808 values of the Fr were found in the range from 0.6 to 7.75. Due to the limitation of granulator

Table 25.3 Factors and limits in the experimental design.

Operation Boundary	Formulation Variable	Manipulated Process Variable		
	Mass fraction of cellulose material (%)	Liquid to solid ratio (g/g)	Impeller speed (s^{-1})	Granulator scale (L)
Upper limit	100	1.8	1200	2
Lower limit	0	0.025	350	1
Type	Continuous	Continuous	Continuous	Discrete

From Wang et al. (2021).

in our laboratory, the maximum impeller speed of 1200 s^{-1} and the larger granulator radius of 0.087 m resulted the maximum Fr of 7.75, which was lower than the maximum value of Fr (i.e., 16.06) in the collected literature. The corresponding S'_{max} parameter was estimated by Eq. (25.4) on the basis of the simulated parameters. A total of 808 values of the S'_{max} were spread from 2.2 to 410.0. After that a process parameters matrix SP (808 × 3) was built. The operational boundaries of each variable in the experimental design are shown in Table 25.3. The lab granulator had two interchangeable granulation bowls, that is, 1 and 2 L, the radii of which were 0.070 and 0.087 m, respectively. The granulator scale was a discrete variable, and it was randomly allocated to the designed experiments.

The matrix SX and the matrix SP were combined to generate the matrix S (808 × 21). The samples of matrix S could then be projected into the LVs space of PLS Model 1. After the simulations were performed seven granulation experiments for modeling and seven validation experiments were randomly selected from the 808 simulated conditions. Fig. 25.2 shows the LVs space by the first two LVs. The green circles represented calibration samples, and the yellow circles represented the internal validation samples. The gray crosses were 808 simulated samples, and the red triangles were six samples selected for laboratory granulation experiments. It can be seen that laboratory samples are distributed well among the calibration samples. Laboratory samples No. 2 and No. 6 appeared to be nearly overlapped in the PLS scores space. The two samples had the same L/S ratio and consisted of MCC and Lactose with the same mass fractions (i.e., 81% and 19%, respectively, w/w). The different types of Lactose (i.e., Lactose 200 M and Granulac 200) and Fr (i.e., 5.239 and 3.239, respectively) seemed to contribute less variation information to discriminate No. 2 and No. 6.

25.2.5.3 Integrated data model

The laboratory granulation experiments were executed according to Section 2.4. The operational conditions and experimental results are illustrated in Table 25.4. The material mass fractions, the L/S ratio, and the impeller speed showed wide range. The resultant granule sizes had large distributions from 142 to 2017 μm. The sample No.7 failed to produce a granule but got a moist slurry, since the binder liquid was excessive. By adding six valid pieces of experimental data into the calibration set of PLS Model 1, an integrated and augmented calibration set containing 125 samples was established. The PLS Model 2 was built on this integrated dataset. After the optimization of the number of LVs four LVs were capable to explain 86.5% of the total variation. The resultant R^2 and Q^2 were 0.710 and 0.680, respectively. Accuracies at deviations of 300 and 100 μm were 88% and 43%, respectively, which were

Table 25.4 The results of laboratory granulation experiments for model improvement.

No.	Scale (L)	Component 1	Mass Fraction (%)	Component 2	Mass Fraction (%)	Liquid-to-Solid Ratio (L/S)	Froude Number	Impeller Speed (rpm)	S_{max}	G_{50} (μm)
1	1	MCC PH101	64	Lactose 110M	36	0.858	4.212	1030	75.5	1400
2	1	MCC PH101	81	Lactose 200M	19	0.999	5.239	1149	82.7	1394
3	1	MCC PH101	95	Lactose 110M	5	1.333	3.178	895	92.2	2017
4	1	MCC PH101	38	Lactose 200M	62	0.315	0.975	496	41.3	205
5	1	HPMC E15LV	62	Anhydrous lactose	38	0.158	2.352	770	19.1	142
6	2	MCC PH101	81	Granulac 200	19	0.999	3.239	810	82.4	1054
7	1	MCC PH101	13	Lactose 110M	87	1.209	1.557	626	154.2	–

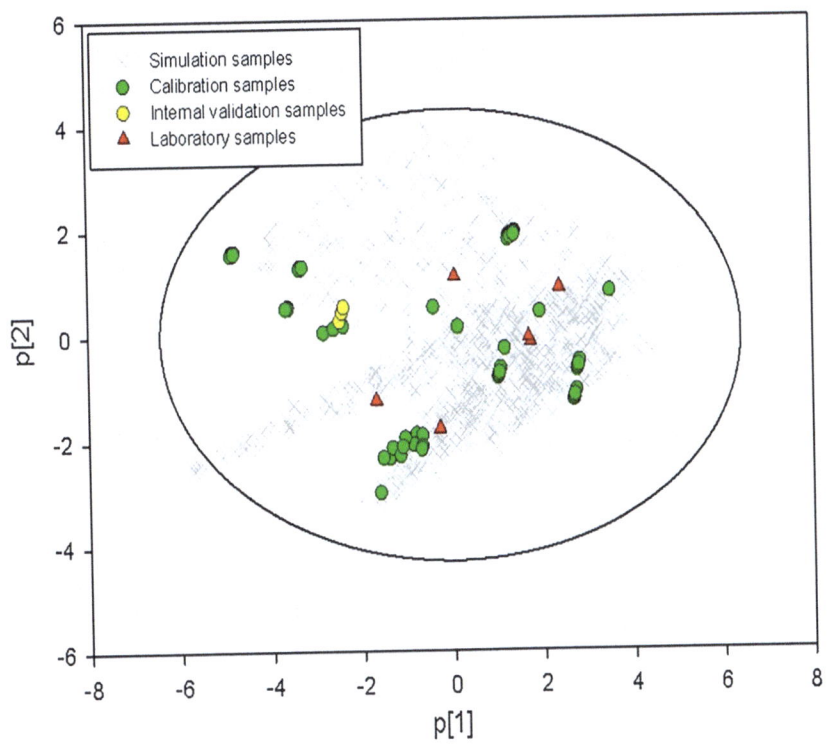

FIGURE 25.2

The LV1 versus LV 2 score scatter plot of PLS Model 1.

higher than that of PLS Model 1. The RMSEP value decreased from 152.6 to 115.7, indicating the mean prediction error of granule size D_{50} was reduced greatly. Although the environment and operating conditions of our experiments were different from that of literature, the HSWG process was considered to follow the same process mechanism. By integration of the literature data and the laboratory data, the underlying features could be extracted through multivariate modeling technique. With the help of a small number of laboratory experiments, the prediction performance of process model could be improved efficiently.

The rationality of modeling results is further analyzed with the help of the variable importance in projection (VIP) plot and the coefficients plot, which are shown in Figs. 25.3 and 25.4, respectively. Variables with VIP values greater than 1 are considered to exert a large impact on the model output. In Fig. 25.3, the first two important variables are L/S ratio and S, with VIP values equaling 1.68 and 1.52, respectively. The two variables also have large positive coefficient values, that is, 0.31 and 0.23, respectively, as shown in Fig. 25.4. Parameter of L/S ratio revealed the amount of liquid binder added into the powder bed. S'_{max} is a measurement of liquid content. Similar to the maximum pore saturation, with the increase of S'_{max}, the granulation experiences nucleation only to rapid growth or even over-wet mass. The rapid growth phase occurred under the conditions of high binder content, leading to larger

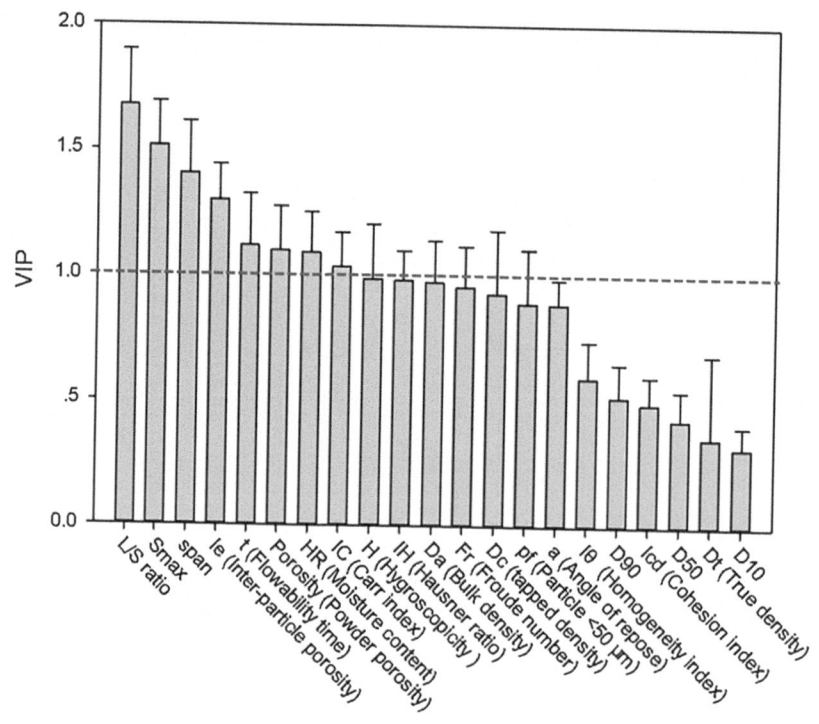

FIGURE 25.3

The variable importance in projection (VIP) plot of PLS Model 2.

granule size. Particularly, when S'_{max} exceeds the upper critical value of steady growth or induction growth process, slurry and mushy mass as the No. 7 experiment in Table 25.4, is likely to be obtained.

Another process parameter, the Fr, was with large negative coefficient value (i.e., −0.28). However, its VIP value was 0.79, which was smaller than 1. The Fr was used to describe the impeller speed mitigating the error influenced by the granulator scale. In this chapter, the Fr was dominated by the impeller speed, and had a negative impact on granule growth. High impeller speed would cause the attrition and breakage process and consequently resulted in the decrease of granule size. However, it should be pointed that the effect of impeller speed on granule size was a two-way regulation, a combination of granule consolidation and breakage dynamic equilibrium (Badawy, Narang, LaMarche, Subramanian, & Varia, 2019; Kristensen et al., 1985). At the early stage of granulation, binder liquid inside the capillary of the wet granule was squeezed out to the surface by shear force, promoting the coalescence and consolidation process. With the impeller speed increasing, shear force sharply rises, leading to intensive breakage of granules due to collisions in the granulator. As a result, the sophisticated mechanism of particle–particle and particle–binder interactions would be difficult to fit.

As shown in Fig. 25.3, VIP values of the majority of material properties are smaller than 1. This indicated that the input material properties had relatively less influence on the granule size, compared with process parameters. *Span* was a vital material property with VIP value of 1.40, and it had a

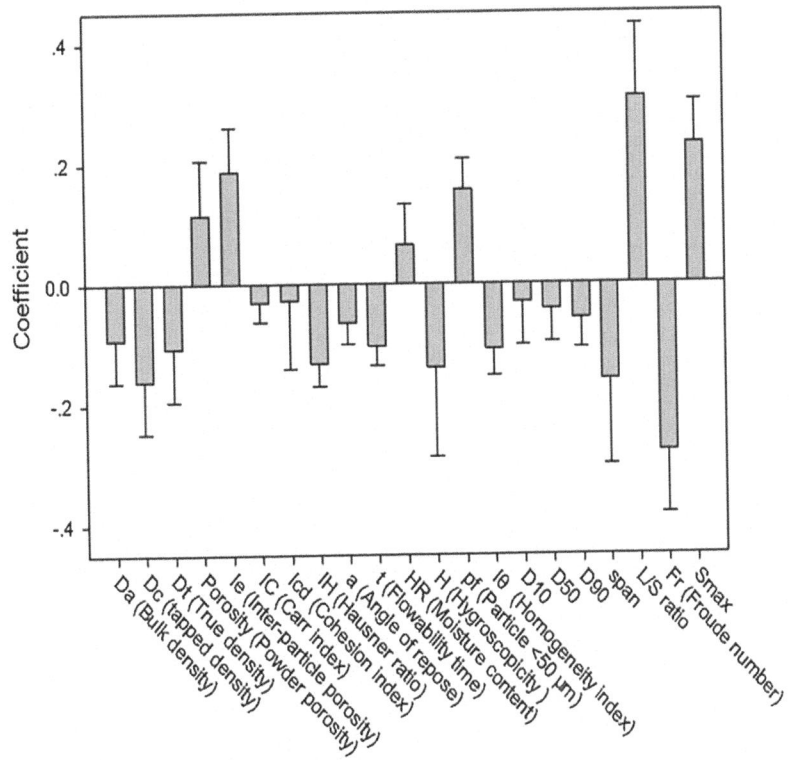

FIGURE 25.4

The coefficients plot of PLS Model 2.

negative impact on granule size since its regression coefficient was −0.16. This tendency was similar to Hounslow et al.'s (2001) work, in which a bimodal particle size distribution would favor coalescence through a layering mechanism of smaller particles onto the surface of larger ones, leading to an uneven growth process.

The porosity attributes of powder bed, such as powder porosity (VIP = 1.10) and interparticle porosity (VIP = 1.16), were also the influential material properties for the granulation process. It has been reported that the penetration time of binder liquid into powder bed would be shorter with porous materials (Hapgood et al., 2002), triggering the nucleation process and promoting the wetting interaction between raw materials and binder.

25.2.5.4 Experimental validation of process model

According to Section 25.2.5.2, the external validation experiments were designed to verify the effectiveness of PLS Model 2. The experimental results are shown in Table 25.5. Similar to the No. 7 experiment in Table 25.2, the No. 14 experiment produced slurry and mushy mass and failed to obtain granules. This could be explained by the extraordinary S'_{max} value (i.e., 107.1) over 100. The RMSEP values of the external validation set calculated from PLS Model 1 and Model 2 were 284.3 and 149.2, respectively.

Table 25.5 The results of external validation experiments.

No.	Scale (L)	Component 1	Mass Fraction 1 (%)	Component 2	Mass Fraction 2 (%)	Liquid-to-Solid Ratio (L/S)	Froude Number	Impeller Speed (rpm)	S_{max}	$Y_{obs}G_{50}$ (μm)	Y_{pre} (μm)
8	2	MCC PH101	64	Granulac 200	36	0.396	1.7	587	39.3	424	458
9	1	MCC PH101	81	Lactose 200M	19	0.862	2.878	852	71.4	1088	774
10	1	MCC PH101	95	Lactose 110M	5	0.882	2.619	812	61.0	978	798
11	1	HPMC E15LV	62	Anhydrous lactose	38	0.296	1.7	655	35.8	207	198
12	1	MCC PH101 (74.3%), HPMC E15LV (25.7%)	70	Anhydrous lactose	30	0.299	5.239	1149	30.6	194	158
13	2	MCC PH101	64	Granulac 200	36	0.658	2.212	670	65.2	641	632
14	1	MCC PH101	38	Lactose 200M	62	0.816	7.001	1200	107.1		

The No. 12 experiment was modified by using 74.3% MCC PH101 and 25.7% HPMC E15LV as Component 1, producing a triple-component formulation to test the robustness of prediction. By using PLS Model 2, the absolute prediction error for No. 12 experiment was 36 μm, which was smaller than RMSEP. These results demonstrated that PLS Model 2 provided improved prediction ability.

Experiments No. 9, No. 10, and No. 13 employed relatively large L/S ratios that were 0.862, 0.882, and 0.658, respectively, and the resulting granule sizes of D_{50} were 1088, 978, and 641 μm, respectively. The average absolute prediction error of the three experiments was 167.6 μm. By contrast, Experiments No. 8, No. 11, and No. 12 employed relatively small L/S ratios that were 0.396, 0.296, and 0.299, respectively, and the resulting granule sizes of G_{50} were 424, 207, and 194 μm, respectively. The average absolute prediction error of the latter three experiments was 26.3 μm. These results confirmed the fact that larger L/S ratios were responsible for larger granule sizes. At a low level of L/S ratio, the relatively small model prediction error would be acquired by PLS Model 2. In addition, the formulations in experiments No. 11 and No. 12 contained 62% and 25.7% of HPMC, respectively. The absolute prediction errors for experiments No. 11 and No. 12 were 9 and 36 μm, respectively, which were smaller than the average prediction error of the improved model. These results indicated that when the concentration of HPMC were not very high, the best prediction performance could be obtained.

25.2.6 Conclusion

In this chapter, the PLS regression was used to build a formulation–process–quality model for the HSWG process. A material database of pharmaceutical excipients was used to estimate physical properties of HSWG formulation, and dimensionless parameters were utilized to reconstruct process variables at different granulator scales. The experimental data of HSWG process from two sources, that is, literature and the authors' laboratory, were fused into a single representation. Results demonstrated that incorporating a small number of laboratory data into the multivariate calibration model could help significantly reduce the prediction error.

25.3 Loss of tabletability after wet granulation based on material library

Nowadays, tablets represent the most widely utilized OSD form in drug manufacturing due to the advantages of self-administration, stability, ease of handling, transportation, and good patient compliance (Sohail Arshad et al., 2021). According to the US Food and Drug Administration Center for Drug Evaluation and Research (CDER) report on novel drug approvals, tablets accounted for 39.6%, 26.4%, and 30.0% of new pharmaceutical products for three consecutive years from 2019 to 2021, respectively, showing that tablets were the first choice of OSD forms in the drug development process (Novel Drugs, 2020; U.S. Food and Drug Administration FDA, 2019; U.S. Food and Drug Administration FDA, 2021). The basic manufacturing routes of tablets mainly include DC, WG, DG, and other technologies. DC is a well-known and simple method in tablet manufacture due to the saving of time, labor, and cost, but it has high requirements and limitations on the physical properties of raw materials, such as flowability, compactability, and die filling of materials (Kása et al., 2009). In practice, it is often necessary to convert fine powders into large agglomerates by granulation processes to improve the compaction and flow properties of the material. According to the research on the manufacturing classification system, 16% of commercial tablet formulations were manufactured using the DC process. Meanwhile, 52% of the commercial tablet formulations were produced by the granulation process, in which 77% were from wet granulation (Leane et al., 2018). This showed that WG was the most popular process option, followed by DC and DG, since WG had a higher tolerance for poor physical properties of materials.

25.3.1 Wet granulation and tableting

The tensile strength (TS) of a tablet is an important quality attribute as the tablet needs to be mechanically strong to withstand further handling, such as film-coating, packaging, transport, and end use by the patient (Sabri et al., 2018). However, a common problem that cannot be ignored in granulation is the loss of tabletability or loss of reworkability, which has a direct effect on tablet TS (Malkowska et al., 1983).

The change of tabletability in wet granulation has not been fully investigated. Various wet granulation techniques, such as HSWG, fluidized bed wet granulation, and twin screw wet granulation (TSWG), as well as the complex interactions between formulation materials and process variables (Thapa et al., 2019), increased the research difficulty. The phenomenon of severe loss of tabletability was also termed as overgranulation in wet granulation, and it often happened on plastic materials like MCC (Osei-Yeboah et al., 2014b). The possible mechanisms were first revealed by Badawy et al. (2006) that the particle porosity was reduced. Later Shi et al. (2010) proved that the MCC granule tabletability was decreased with increasing granule size, and granule size enlargement was the key mechanism for over-granulation in HSWG. In another case, Khorsheed et al., 2019 conducted TSWG with MCC, mannitol, and dibasic calcium phosphate anhydrous as the research objects. The loss of tabletability was observed for MCC and mannitol 100 SD, and it was caused by increased granule strength but was independent of the granule size. By contrast, the brittle materials mannitol C160 and dibasic calcium phosphate anhydrous maintained their tabletability due to the reduced or constant granule size of materials after TSWG. Osei-Yeboah et al. (2014a) showed that the combination of brittle and plastic excipients could solve the problem of tabletability reduction. Except for the investigations on common fillers, such as microcrystalline or lactose, some unusual mechanisms of change of tabletability were found. For instance, the polymorphic transition from delta-mannitol to beta-mannitol occurred during TSWG, which improved the tabletability of delta-mannitol (Vanhoorne et al., 2016a). Another recent example found that the breakage of lengthy crystals of glucose during TSWG significantly increased the tabletability of glucose (Záhonyi et al., 2022). It could be seen that the reported literature on the reduced tabletability of materials in wet granulation was important for designing a robust WG process. However, these experiences were limited to single materials or particular formulations, and a more comprehensive understanding is needed for identifying effective solutions to the loss of tabletability problem in the WG process.

25.3.2 Methodology of the material library

Building of material database for relevant pharmaceutical materials can help understand how material properties influence the process behaviors and the CQAs of finished dosage forms (Hayashi et al., 2021). To establish the material library, 30 materials including 18 pharmaceutical excipients and 12

NPPs were used. All pharmaceutical excipients were purchased commercially, and these excipients played the roles of diluents, binders, and disintegrants in tablet formulations. The selected excipients were suitable for DC and/or granulation processes. According to the deformation characteristics of materials, the excipients could be divided into three types, that is, the plastic, the brittle, and the elastic. MCC PH101 and silicified microcrystalline cellulose (SMCC) were typical plastic excipients (Rojas, Ciro, & Correa, 2014). The brittle excipients included mannitol, calcium phosphate (CaP), dibasic calcium phosphate (DCP), dibasic calcium phosphate anhydrous (DCPA), lactose granulac 200 (Lac G200), and lactose flowlac 100 (Lac F100) (Schönfeld et al., 2022; Paul & Sun, 2017; Yost et al., 2022). The elastic excipients contained pregelatinized starch (PGS), soluble starch, and dextrin (Cho et al., 2021; Picker, 2004; Tanner et al., 2018). Among these excipients, SMCC and lactose cellactose 80 (Lac C80) were co-processed excipients. All NPPs were provided by the Beijing Tcmages Pharmaceutical Co., Ltd. (Beijing, China) and were prepared from 12 medicinal plants that involved commonly used medicinal parts, for example, roots, rhizomes, bark, fruits, flowers, and aerial parts. The manufacturing processes of each NPP included a series of unit operations, such as pretreatment, extraction, filtration, concentration, and spray drying. Compared to excipients, NPPs had the characteristics of multiple compositions, high hygroscopicity, low glass transition temperatures, poor flowability, and compactability (Li et al., 2018).

The powders were characterized to obtain the physical parameters of the powders. The pharmaceutical excipients were sifted using a sieve with an aperture size of 850 μm to remove any lumps. Then, the sieved excipient powders were spread on a tray and were placed in a blast drying oven set at 60°C to dry for 2 days (Tay et al., 2017). The dried powders were then put into a ziplock bag and equilibrated in an environment of relative humidity maintained at 50% and temperature of $25.0 \pm 2.0°C$ for at least 3 days. The NPPs were sensitive to heat and humidity, so they were only sifted by an 850-μm sieve to obtain powders with a homogeneous size before use.

All powdered materials were characterized by 19 physical parameters. Among them, 18 material properties are mentioned as shown in Table 25.2. The solid fraction (SFp) is calculated as $1-\varepsilon$. The testing results of material properties of 30 materials in the material library are achieved in a homemade database named iTCM.

25.3.3 Experimental method for high shear wet granulation
25.3.3.1 Experiment design
As a fact, multiple process factors of HSWG and their interaction could result in complex experiment programs even for simple formulations or single materials. The purpose of the experimental design here was to successfully produce acceptable granules for each material in the material library, but not to fine-tune the process parameters. The arrangement of the experiment design is shown in Table 25.6. The main process parameters considered were the type of wetting agent, the L/S ratio, the impeller speed, and the addition rate of wetting agent. The design principles of the four parameters were as follows.

In order to avoid the introduction of additional substance, deionized water and 95% (v/v) ethanol were chosen as wetting agents because of their volatile properties. NPPs and five starch derivatives (i.e., cold water-soluble starch, cold water-insoluble starch, dextrin, maltodextrin, and PGS) were not easy to wet uniformly, and a sticky wet mass or hard granules could be produced when using water as the wetting agent. For those materials, 95% (v/v) ethanol was used as the wetting agent. Except for the

Table 25.6 The arrangements of experimental design for the high shear wet granulation process.

Run No.	Material No.	Name	Wetting Agent	L/S Ratio (g•g^{-1})	Impeller Speed (rpm)	Addition Rate of Wetting Agent (mL•min^{-1})
1	Z1	Polygoni Multiflori Radix Praeparata	95% Ethanol	0.3	600	50
2	Z2	Angelicae sinensis Radix	95% Ethanol	0.3	600	50
3	Z3	Asari Radix Et Rhizoma	95% Ethanol	0.3	600	50
4	Z4	Menthae Haplocalycis Herba	95% Ethanol	0.3	600	50
5	Z5	Bran-Processed Atractylodis Rhizoma	95% Ethanol	0.3	600	50
6	Z6	Mume Fructus	95% Ethanol	0.3	600	50
7	Z7	Chuanxiong Rhizoma	95% Ethanol	0.3	600	50
8	Z8	Bistortae Rhizoma	95% Ethanol	0.3	600	50
9	Z9	Visci Herba	95% Ethanol	0.3	600	50
10	Z10	Sophorae Flavescentis Radix	95% Ethanol	0.3	600	50
11	Z11	Cinnamomi Cortex	95% Ethanol	0.3	600	50
12	Z12	Buddlejae Flos	95% Ethanol	0.3	600	50
13	E1	Cold water-insoluble starch	95% Ethanol	0.2	300	25
14	E2	Cold water-soluble starch	95% Ethanol	0.2	300	25

(continued on next page)

25.3 Loss of tabletability after wet granulation based on material library

15	E3	Dextrin	95% Ethanol	0.2	300	25
16	E4	Maltodextrin DE20	95% Ethanol	0.2	300	25
17	E5	Pregelatinized starch	95% Ethanol	0.2	300	25
18	E6	Lactose Flowlac® 100	Deionized water	0.1	300	25
19	E7	Lactose Granulac® 200	Deionized water	0.1	300	25
20	E8	Dibasic calcium phosphate anhydrous	Deionized water	0.1	300	25
21	E9	β-cyclodextrin	Deionized water	0.1	300	25
22	E10	Dibasic calcium phosphate	Deionized water	0.2	300	25
23	E11	Mannitol	Deionized water	0.2	300	25
24	E12	Lactose Cellactose® 80	Deionized water	0.3	300	25
25	E13	Calcium phosphate	Deionized water	0.3	300	25
26	E14	Silicified microcrystalline cellulose	Deionized water	0.5	300	25
27	E15	Microcrystalline cellulose PH101	Deionized water	0.5	300	25
28	E16	Polyvinylpolypyrrolidone XL-10	Deionized water	0.6	300	25
29	E17	Low-substituted hydroxypropyl cellulose LH11	Deionized water	0.9	300	25
30	E18	Croscarmellose sodium	Deionized water	1.3	300	25

From Wang et al. (2022).

five starch derivatives, the remaining 13 excipients applied water as the wetting agent to induce their viscosity in forming granules. The L/S ratio was determined for each material based on the results of preliminary experiments (details not shown). The principle of determining the L/S ratio was to avoid the slurry state or formation of large agglomerates.

The impeller speed and the addition rate of wetting agent in this study were referred to the literatures with the same or similar granulator scale (Fayed et al., 2017a,2017b). NPPs have a smaller particle size, stronger interparticle cohesion, and poorer flowability compared to excipients. Therefore, a high impeller speed of 600 rpm was set in the wet granulation of NPPs, which allowed the powders to flow freely and contact sufficiently with the wetting agent. Meanwhile, a low impeller speed of 300 rpm was applied for the wet granulation of pharmaceutical excipients. The chopper speed was kept at 1600 rpm for all experiments. The addition rate of the wetting agent was designed to match the impeller speed. When ethanol was used as the wetting agent, a high-level combination of the impeller speed and the addition rate was used to achieve rapid granulation and to avoid the volatilization of ethanol. In all experiments, the addition time of wetting agent ranged from 1.2 to 15.6 min, depending on the L/S ratio used.

25.3.3.2 Process description

Approximately 300 g of powders were poured into a laboratory scale high shear granulator with a 2 L bowl (SHK-4, Xi'an Runtian Pharmaceutical Machinery Co., Ltd., Xi'an, China). Then, the dry mixing was performed for 1 min at the impeller speed of 300 rpm and without working of the chopper. After dry mixing the wetting agent was added into the granulation bowl using a peristaltic pump (BT00-100 M, Baoding Longer Precision Pump Co., Ltd., Baoding, China) at defined impeller and chopper speeds. The wet massing time lasted for 180 s. At the end of the granulation process, the wet mass was sifted by a 10-mesh standard sieve, and the wet granules were spread out on a tray and transferred into an oven for drying at 55°C. The drying time was controlled until the moisture content of dried granules was close to that of the corresponding powders. Dry granules were separated by using a vibration screen with two standard sieves (ZNS-300, Beijing Kingslh Technology Development Co. Ltd., Beijing, China). After sieving the granules with size fraction of 125 to 250 μm were used for the subsequent tableting process (Bowles et al., 2018; Rezaei et al., 2022; Zhang et al., 2014). The purpose of selecting the same particle size fraction is to facilitate the comparison of the physical and mechanical properties of granules made from different materials.

25.3.3.3 Tableting process

The pretreated powders and the prepared granules were compacted, respectively, into tablets by using a single-punch tablet press machine (C&C600A, Beijing C&C CAMBCAVI Co., Ltd., Beijing, China), which was equipped with a flat-faced punch and die with 10 mm in diameter. The magnesium stearate was used to lubricate the punch surfaces and the die walls before each compaction. After lubrication the powders or granules were manually filled into the die and compacted. Considering the different bulk densities of the materials, the filling mass of the material was set to 300 or 350 mg to ensure the smooth ejection of the tablet. For each material, powdered or granulated, six compaction pressures from 10 to 140 MPa (1 kN = 12.74 MPa) were applied to obtain tablets with different hardness. At least two tablets were obtained under each pressure. The tableting speed was maintained at 25 tablets per minute. The prepared tablets were sealed in a ziplock bag. After being stored for 24 h the weight, diameter, thickness, and diametrical crush force of tablets were measured. The tablet weight was acquired by using an analytical balance (GL124-1SCN, Beijing Sanfu Hezhong Technology Development Co., Ltd., Beijing, China). The diameter and thickness of tablets were measured with a digital calliper (547–401 Digimatic Caliper, Mitutoyo, Japan). The diametrical crush force of tablets was recorded by a tablet hardness tester (YPD-500, Shanghai Huanghai medicine inspection instrument Co., Ltd., Shanghai, China). The TS and the solid fraction (SF) of tablets are calculated by Eqs. (25.10) and (25.12), respectively.

$$TS = \frac{2F}{\pi DH} \tag{25.10}$$

where F (N) is the tablet crush force, D (mm) is the tablet diameter, and H (mm) is the tablet thickness (Fell & Newton, 1970).

$$SF = 1 - \varepsilon \tag{25.11}$$

$$\varepsilon = \frac{\rho_{app}}{\rho true} \tag{25.12}$$

$$\rho_{app} = \frac{m}{\pi \frac{D^2}{4} H} \tag{25.13}$$

where ε is the tablet porosity, ρ_{app} is the apparent tablet porosity, ρ_t (g•cm^{-3}) is the true tablet density, and m (g) is the tablet weight. ρ_t (g•cm^{-3}) is equal to the true density of the material.

The compression models mentioned are shown in Table 25.2. There are nine compression parameters involved in the evaluation: a, b^{-1}, ab, P_y, f, K, k_b, d, and g.

25.3.4 Evaluation of change in tabletability

25.3.4.1 The reworking potential

The reworking potential (RP) index was first proposed by Malkowska et al. (1983). In this method, the tabletability change of the material was calculated by dividing the area under the TS versus compression pressure profile of granules (AUC$_g$) to the area under the TS versus compression pressure profile of powders (AUC$_p$), as shown in Fig. 25.5. The calculation Eq. 25.14) is as follows:

$$\text{Reworking potential} = \frac{\text{AUC}_g}{\text{AUC}_p} \times 100\% \tag{25.14}$$

25.3.4.2 The relative change of tabletability

In order to compare the change of tabletability of different materials based on the concept of material library, this chapter proposed a new index, namely, relative change of tabletability (CoT$_r$). First, the TS versus pressure curves for all powders in the material library and granules prepared from the powders were plotted in order to determine the maximum area under the curve (AUC$_{max}$) and minimum area under the curve (AUC$_{min}$). The maximum change of tabletability (A_{max}) is defined as the difference between AUC$_{max}$ and AUC$_{min}$, as shown in Fig. 25.5. Then, for a particular material, the change of tabletability (C_t) is calculated as the difference between the area under the tabletability curves of the

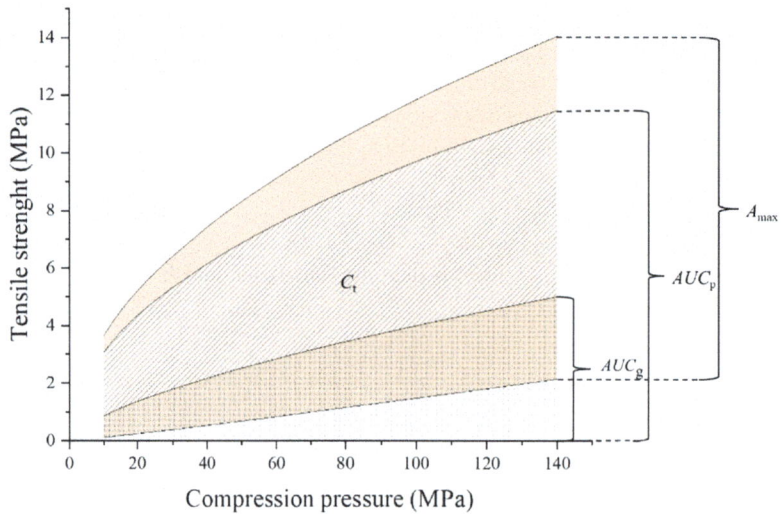

FIGURE 25.5

The schematic diagram for evaluation of the tabletability change.

From Wang et al. (2022).

powder (AUC_p) and the area under the tabletability curves of the granule (AUC_g). Finally, the relative changes of tabletability (CoT_r) of a material could be expressed as dividing C_t by A_{max} in Eq. 25.15):

$CoT_r = [(C_t/A_{max})*(AUC_g - AUC_p)/(AUC_{max} - AUC_{min})] \times 100\%$

$$\frac{C_t}{A_{max}} \frac{AUC_g - AUC_p}{AUC_{max} - AUC_{min}} \times 100\% \tag{25.15}$$

(25.15)

25.3.5 Comparison of compression behavior of powders and granules

25.3.5.1 Compression models

After the tableting and tablet characterization experiments in Section 25.3.3.3 the compression curves interpreting, respectively, the compressibility (i.e., pressure vs. porosity), compactability (i.e., porosity vs. TS), and tabletability (i.e., TS vs. pressure) properties of a material could be plotted. Then, different compression models were used to fit the compression curves under specified pressure ranges in order to acquire the CBCS parameters. When using the Shapiro model, the pressure range was 0 to 50 MPa. For other compression models in the CBCS method, the pressure range was 10 to 140 MPa. Specific equations are given in Table 25.2.

25.3.5.2 Compression model fitting results

Each material, in the form of ungranulated powders or granules, was compressed into tablets with different hardness under the pressure range of 10 to 140 MPa according to the procedures in

Section 25.3.3.3. The curves of porosity–pressure, porosity–TS, and TS–pressure that, respectively, reflected the compressibility, compactability, and tabletability of materials could then be plotted. After that CBCS parameters could be obtained by using different compression models in Table 25.2 to fit the curves. Especially, for CaP (No. E13), whether in the form of powders or granules, the obtained tablets would undergo capping when the applied pressure exceeded 70 MPa. Therefore, this material was not included in the following comparative study of compression behaviors. The goodness of fit of each compression model was evaluated by the determinant coefficient R^2 and the root mean square error. In general, the R^2 values of 339 out of 348 compression models were higher than 0.9, demonstrating the good fitness. The R^2 values of remaining nine compression models were higher than 0.8, revealing the favorite fitness. There are mainly two possible reasons to which the inferior fitting results of the nine compression models can be attributed. The first reason was that the TS versus compression pressure curves of materials, such as β-CD and CWSS, were approximately straight lines parallel to the X-axis when the compression pressure was greater than 60 MPa. Therefore, they could not be fitted well with the Power model. The second reason was that the data points on the TS versus compression pressure curves for materials, for example, CWIS granules, dextrin granules, and PGS granules, were relatively scattered. It was inferred that granules made from starch derivatives were sensitive to the die filling process.

For the sake of comparison, the lowest and highest values of each CBCS parameter are shown in Table 25.7. The parameter a in Kawakita represented the maximal engineering strain of the raw powders or granules. The larger the value of a was, the easier the powder bed of the materials could be compressed. The Kawakita parameter a of ungranulated powder varied from 0.708 (DCP, No. E10) to 1.05 (CWSS, No. E2), and the descriptor a of granules changed between 0.746 (DCPA, No. E'8) and 1.05 (CWIS, No. E'1). This indicated that soluble starch was easier to compress than DCP excipients, which was related to the porous structure of soluble starch and the high density of DCP excipients (Chen et al., 2020; Hentzschel et al., 2011). The parameter ab in the Kawakita model could be used as an indication of the incidence of particle rearrangement during powder compression, and a high ab value corresponded to a high degree of particle rearrangement (Nordström et al., 2009). The values of rearrangement index ab of ungranulated powders ranged from 6.58×10^{-2} (PMRP, No. Z1) to 0.283 (Lac G200, No. E7). The ab values of granules varied from 3.86×10^{-2} (CMC-Na, No. E'18) to 0.288 (β-cyclodextrin, β-CD, No. E'9), suggesting that CMC-Na granules and PMRP powders had nearly no initial particle rearrangement.

The parameter f was used to evaluate the degree of particle fragmentation during the initial compression process. A material with a large f value was prone to particle fragmentation. The Shapiro parameter f of ungranulated powders changed between 8.84×10^{-2} (CWSS, No. E2) and 0.362 (Lac G200, No. E7), and the f values of granules ranged from 3.12×10^{-2} (PGS, No. E'5) to 0.326 (Mannitol, No. E'11). This showed that the two starch excipients had a limited inclination of particle fragmentation. Nevertheless, the lactose and mannitol excipients were easily fragmented during the initial compression process.

The Heckel mean yield pressure, P_y, signified the plastic deformation capacity of powders. The plastic deformation capacity of a material is inversely proportional to the value of its P_y parameter: the lower the P_y value, the greater this capacity. The distribution range of the Heckel parameter P_y for powders was between 41.6 (MHH, No. Z4) and 686 (DCP, No. E10). The P_y value of granules varied from 38.9 (CWIS, No. E'1) to 684 (DCPA, No. E'8). This demonstrated that the DCP excipients had

Table 25.7 The lowest and highest values of each CBCS parameter.

Fitting Relationship	CBCS Parameter	Powders		Granules	
		The Lowest	The Highest	The Lowest	The Highest
Porosity-pressure	a	0.708	1.05	0.746	1.05
	ab	0.0658	0.283	0.0386	0.288
	f	8.84×10^{-2}	0.362	3.12×10^{-2}	0.326
	P_y	41.6	686	38.9	684
	K	6.35	24.5	5.87	24.9
Porosity-TS	k_b	4.54	22.8	5.58	24.9
TS-pressure	d	9.25×10^{-4}	1.00	1.07×10^{-4}	0.517
	g	0.287	1.68	0.398	1.79

From Wang et al. (2022).

poor plastic deformation ability, whereas the starch excipients had good plastic deformation ability. This could be attributed to the high densification degree of DCP and the low densification degree of starch.

The parameter K in the Gurnham equation was related to the compressibility resistance of the powders and granules. The larger the value, the stronger the compression resistance of the material was, and the more difficult it was to compress. The maximum and minimum values of the Gurnham parameter K of the powders were 6.35 (Buddlejae Flos [BF], No. Z12) and 24.5 (DCPA, No. E8), respectively. The largest and the smallest values of the parameter K of granules were 5.87 (CMC-Na, No. E'18) and 24.9 (DCPA, No. E'8).

The parameter k_b in the Ryshkewitch–Duckworth equation represented the bonding capacity of powders and granules. The larger the k_b value was, the worse the bonding capacity acquired of materials was. The values of the Ryshkewitch–Duckworth parameter k_b of powders were distributed between 4.54 (β-CD, No. E9) and 22.8 (DCPA, No. E8). The k_b values of granules varied from 5.58 (PMRP, No. Z'1) to 24.9 (DCPA, No. E'8). These results showed that the bonding capacity of DCPA particles could not be improved after HSWG. This was related to its inherently high degree of densification and poor plastic deformation ability.

Tabletability was the ability of a powdered material to be transformed into a tablet with a certain TS under the applied compaction pressure (Sun, 2016). For ungranulated powders, the descriptors d and g in the power model can represent the tabletability and pressure sensitivity of materials, respectively. The values of descriptor d ranged between 9.25×10^{-4} (MF, No. Z6) and 1.00 (MCC PH101, No. E15). The values of g varied from 0.287 (β-CD, No. E9) to 1.68 (Bistortae Rhizoma (BR), No. Z8 and MF, No. Z6). For granules, the parameters d were distributed between 1.07×10^{-4}–0.517, the range of which was narrower than that of the powders. The lowest and highest values of d could be observed from PGS (No. E'5) and β-CD (No. E'9), respectively. Besides, the values of g varied from 0.398 to 1.79. As with ungranulated powder, the granule of β-CD had the lowest g value of 0.398. Chuanxiong Rhizoma (No. Z'7) had the highest g of 1.79.

Taking all materials in the material library as a whole, the compressibility of materials remained the same after HSWG. In other words, under the same level of mechanical strength, the porosity of the compact was not affected by the granulation process. The compactability of materials would decrease slightly after HSWG, which meant the TS of the powdered tablet was greater than that of the granular tablet at the same tablet porosity. From the d values and the g values, it could be seen that the tabletability of materials would decrease and the pressure sensitivity of materials would increase slightly after HSWG.

25.3.5.3 Principal component analysis

PCA was performed on the data matrix (size 58×9) consisting of 58 samples (29 ungranulated powders and 29 granules). Each sample was featured by nine CBCS parameters. The data matrix was preprocessed by the same method as described in Section 25.3.4.2. The first three PCs cloud explain 89.1% of the total variability, and PC1, PC2, and PC3 accounted for 46.0%, 30.0%, and 13.2%, respectively. The loading plot is shown in Fig. 25.6A. The CBCS parameters associated with PC1 were mainly the compression resistance index K, the rearrangement index ab, the yield pressure b^{-1}, and the fragmentation index f. Among them, K, ab, and f were located on the negative axis of PC1, whereas b^{-1} was situated on the positive axis of PC1. In particular, the parameter ab was close to the parameter f, suggesting that the particle rearrangement capacity and the particle fragmentation capacity were positively correlated within the scope of the material library. The variables contributing

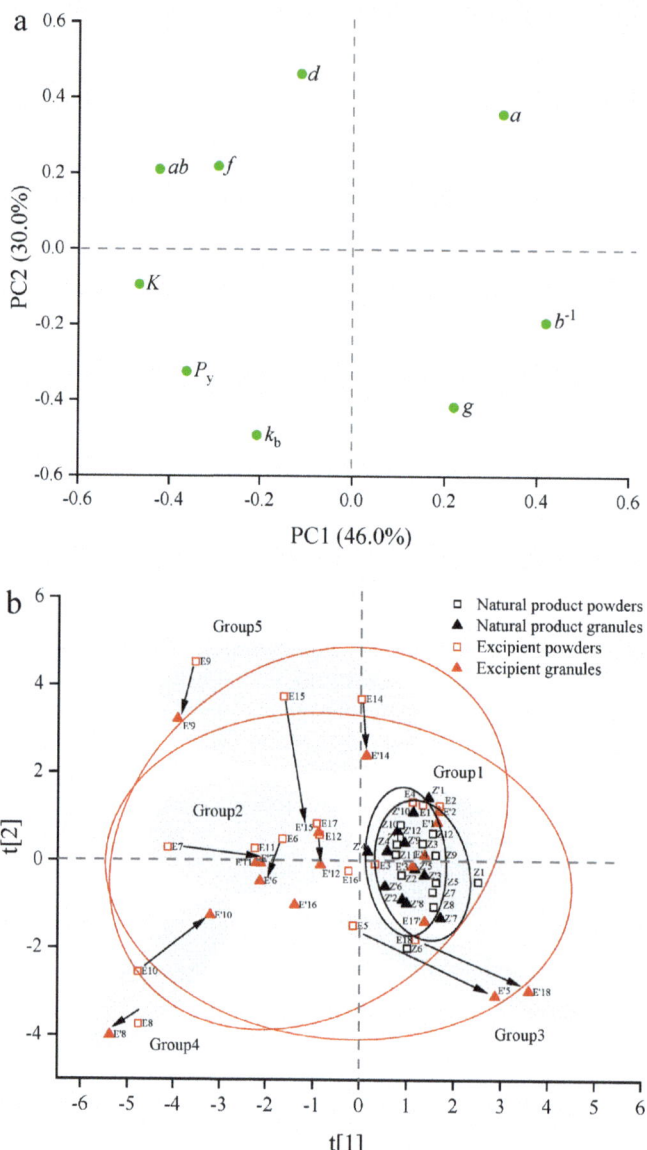

FIGURE 25.6

The PCA analysis of the CBCS parameters data based on the first two principal components: (A) the loading plot; (B) the score plot (the black squares represent NPPs. The black triangles represent NPGs. The red squares represent excipient powders. The red triangles represent excipient granules. Two black circles are 95% confidence ellipses for NPPs and NPGs, respectively. Two red circles are 95% confidence ellipses for excipient powders and excipient granules, respectively. The gray shaded circles represent five groups of materials).

From Wang et al. (2022).

to PC2 were mainly the bonding index k_b, the tabletability index d, and the pressure sensitivity g. The descriptors k_b and g were distributed on the negative side of PC2, and they were on the opposite side of the descriptor d. This showed that materials with good bonding ability and low degree of pressure sensitivity tended to acquire good tabletability. The remaining two parameters, a and P_y, were related to both PC1 and PC2. Parameters a and P_y were located at the upper right and lower left corners of the loading plot, respectively, demonstrating that materials with good plastic deformation capacity were easily compressed.

Fig. 25.6B is the score plot of CBCS parameters; it can be seen there that confidence areas for both excipients and NPPs are not significantly changed before and after HSWG. However, all samples can be roughly divided into five groups according to their locations in the score plot and the changing trend from powders to granules. All NPPs and four starch derivatives (i.e., CWIS, CWSS, dextrin, and maltodextrin) were termed as Group 1 materials, which were mainly distributed on the positive axis of PC1 and were close to the origin. The Group 1 materials were characterized by good plastic deformation ability and weak compression resistance ability, but limited particle rearrangement and particle fragmentation. Moreover, the compression behavior of Group 1 materials did not change before and after granulation. The brittle materials, such as Lac F100 (No. E6), Lac G200 (No. E7), mannitol (No. E11), and Lac C80 (No. E12), were classified as Group 2. The Group 2 materials were situated on the negative axis of PC1, and they had extensive particle rearrangement and fragmentation, strong compression resistance ability, but poor plastic deformation ability. Generally, the Group 1 and Group 2 materials had moderate tabletability, either in the powder or granule forms. Except for Lac G200, the brittle materials mainly suffered from loss of tabletability after HSWG, as shown by the downward directions of arrows in Fig. 25.6B. The material with the largest change in particle rearrangement ability after granulation was Lac G200, whose descriptor ab was reduced from 0.283 to 0.202. The materials of both PGS (No. E5) and CMC-Na (No. E18) belonging to Group 3 were situated at the lower right of the score plot. The changing directions of the two materials were the same, moving downward to the right. This demonstrated that PGS and CMC-Na, especially the granules of the two materials, had poor bonding ability in addition to limited particle rearrangement and particle fragmentation. A synergy of these characteristics might result in the lowest tabletability of the PGS granules ($d = 1.07 \times 10^{-4}$). Group 4 contained two materials, that is, DCPA (No. E8) and DCP (No. E10), and they were located in the lower left of the score plot. This suggested that DCP excipients, especially DCPA, had extensive particle rearrangement and particle fragmentation, but poor plastic deformation ability, weak bonding ability, and strong compressibility resistance. The weak bonding ability of DCPA might be the major reason for its poor tabletability. It was worth noting that the changing trend of the two materials from powders to granules was opposite, resulting in the different plastic deformation ability of granules. The Py of the DCP granules (386) was smaller than that of the ungranulated powders (686), whereas the Py of the DCPA granules (684) was larger than that of the ungranulated powders (615). β-CD (No. E9), SMCC (No. E14), and MCC PH101 (No. E15) were mainly distributed in the upper left of the score plot, and they belonged to Group 5. These materials not only had extensive particle rearrangement and particle fragmentation but also had good bonding ability, all of which might result in good tabletability of these materials. In addition, the changing directions of the three materials from powders to granules were the same, moving from top to bottom. This indicated that the tabletability of the Group 5 materials would reduce significantly after HSWG, but the tabletability of granules was still good. There were two materials, PVPP (No. E16) and L-HPC (No. E17), that did not belong to any of the above five groups. PPVP had the largest loss of compressibility and severe reduction in plastic deformation ability, since its

Table 25.8 The tabletability criteria and classification results for 29 materials.

Category	Criteria	Characteristics	No. of Powders	No. of Granules
1	$d \geq 0.2$, $0 < g < 0.95$	Good tabletability at extreme low-pressure range (20–50 MPa)	7	3
2	$0.002 \leq d < 0.2$ or $d < 0.002$, $g \geq 1.6$	A: Acceptable tabletability at middle pressure (50–100 MPa) and good tabletability at high pressure range (100–140 MPa)	15	13
	$0.002 \leq d < 0.05$ or $d < 0.002$, $g \geq 1.6$	B: Acceptable tabletability at high pressure (100–140 MPa)	6	10
	$0.002 \leq d < 0.005$, $g > 0.95$ or $d < 0.002$, $g \geq 1.6$	C: Unacceptable tabletability over the full pressure range (10–140 MPa)	1	3

From Wang et al. (2022).

descriptor a was reduced from 0.870 to 0.779 and descriptor Py changed from 252 to 444, respectively. As for L-HPC, the HSWG process had the greatest impact on its particle fragmentation, because the f values were decreased from 0.221 to 0.109.

25.3.6 Change of tabletability from powders to granules

25.3.6.1 Qualitative analysis

The TS versus compression pressure curves of 29 powdered materials are shown in Fig. 25.7A. According to the CBCS classification criteria with respect to the tabletability, 29 powders can be divided into two categories, as shown in Table 25.8. Category 1 featured $d \geq 0.2$ and $0 < g < 0.95$. Seven pharmaceutical excipients (i.e., MCC PH101, SMCC, CWSS, CWIS, dextrin, β-CD, and L-HPC) were classified into this category. These cellulose and starch derivatives excipients could produce tablets with good tabletability ($TS \geq 3$ MPa) at the low-pressure range of 20 to 50 MPa. Category 2 was characterized by $0.002 \leq d < 0.2$. Depending on the profile of the TS-pressure curve, it could be further subdivided into Category 2A, 2B, and 2C. In this study, 11 NPPs and four excipients belonged to the Category 2A, and these powders generally had acceptable tabletability ($TS \geq 2$ MPa) at middle pressure range of 50 to 100 MPa and good tabletability at a high pressure range of 100 to 140 MPa. One NPP (i.e., MHH, No. Z4) and five excipients were classified into the Category 2B, which meant they had acceptable TS at high pressure range of 100 to 140 MPa. Only one material, DCPA (No. E8), was classified into the Category 2C that represented unacceptable tabletability ($TS < 2$ MPa) over the investigated pressure range.

It should be pointed out that the tabletability of materials was mainly classified according to the d value, while the influence of the g value on the curve was not considered. In this chapter, it was

25.3 Loss of tabletability after wet granulation based on material library   **899**

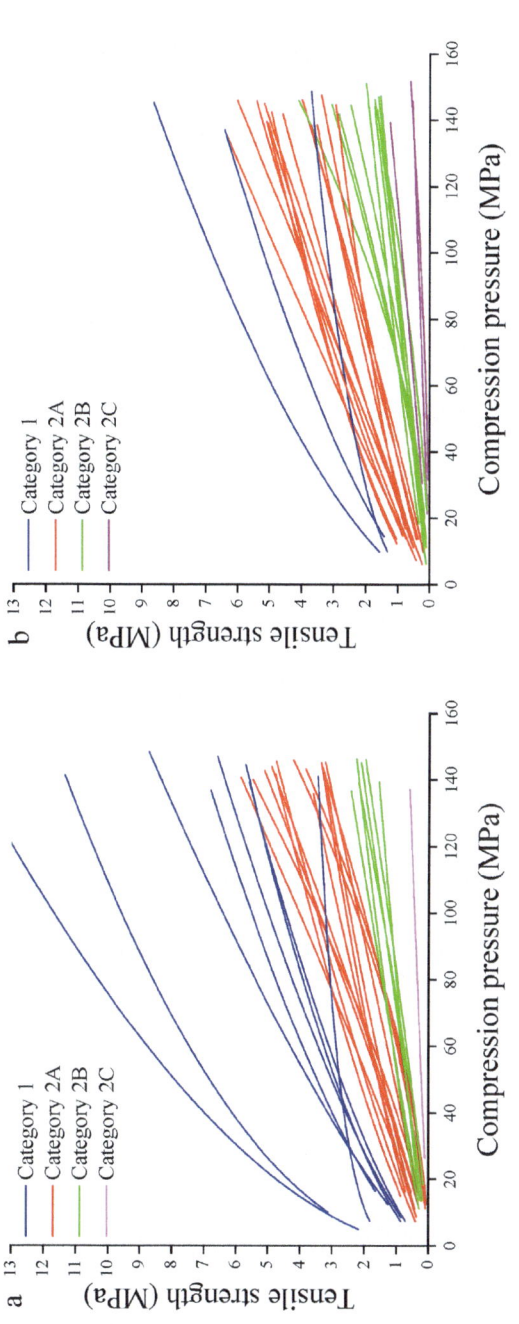

FIGURE 25.7

The tensile strength versus compression pressure curves for 29 materials: (A) the powdered materials; (B) the granular materials.

From Wang et al. (2022).

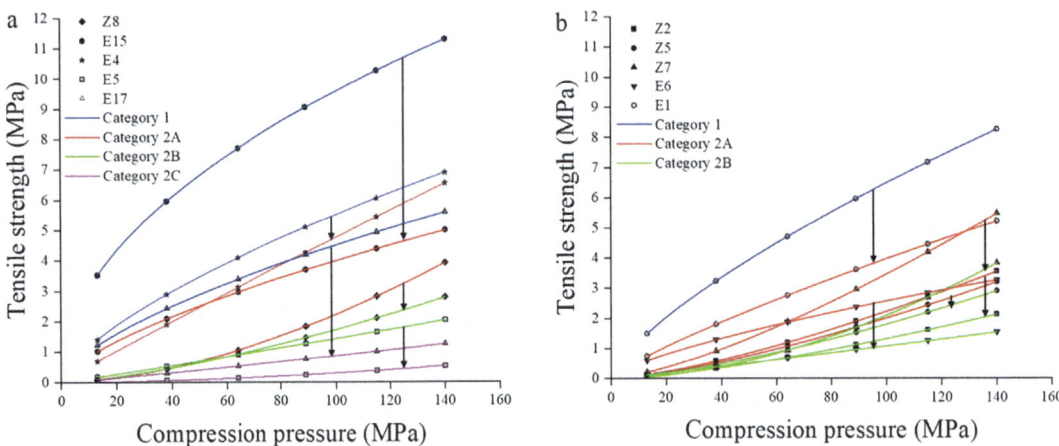

FIGURE 25.8

The change of tabletability for 10 materials. (A) the change of tabletability for five materials (*Bistortae rhizoma* [No. Z8], maltodextrin [No. E4], pregelatinized starch [No. E5], MCC PH101 [No. E15], L-HPC [No. E17]); (B) the change of tabletability for 5 materials (*Angelicae sinensis Radix* [No. Z2], Bran-Processed *Atractylodis rhizoma* [No. Z5], *Chuanxiong rhizoma* [No. Z7], cold water-insoluble starch [No. E1], Lactose Flowlac 100 [No. E6]).

From Wang et al. (2022).

found that the descriptor g had an important influence on the shape of the TS-pressure curve when the d value was lower than 0.002. For example, the d value of Bistortae Rhizoma (No. Z8) was 9.85×10^{-4}, and it might be classified into the Category 3 material. However, the g value of the Z8 material was greater than 1.68, leading to the TS of obtained tablets to be above 2 MPa under the pressure of 50 to 100 MPa; TS of obtained tablets were above 3 MPa under the pressure of 100 to 140 MPa. Therefore, the Bistortae Rhizoma powder belonged to Category 2A. This indicated that the tabletability category should be judged according to the characteristics of the TS versus pressure curve when the g value $\geq$ 1.6 and the d value $<$ 0.002.

The TS versus compression pressure curves and classification of 29 granular materials are shown in Fig. 25.7B. The tabletability classification criteria of granules are consistent with those of powders, and the classification results are shown in Table 25.8. The categories of 19 materials had no change, meaning the tabletability of these materials were not significantly affected by HSWG. The other 10 materials all manifested downgrade trend. The changes of tabletability curves of these 10 materials are plotted in Fig. 25.8. Four excipients that originally belonged to Category 1 were converted into Category 2A (i.e., CWIS, MCC PH101, maltodextrin) or 2C (i.e., L-HPC) after HSWG. The tabletability of five materials (i.e., Lac F100, *A. sinensis* Radix, Bran-Processed Atractylodis Rhizoma, Chuanxiong Rhizoma, Bistortae Rhizoma) had changed from Category 2A to 2B after granulation. The category of PGS (No. E5) was transformed from Category 2B to Category 2C after granulation. Qualitative analysis demonstrated that approximately 34% of materials presented the loss of tabletability after HSWG.

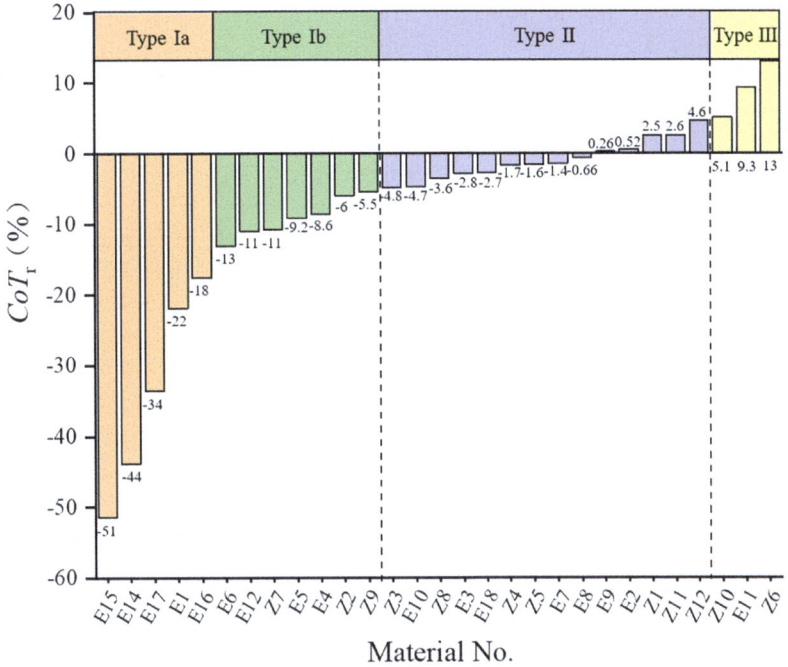

FIGURE 25.9

The tabletability change classification system (TCCS) and the CoT_r values for 29 materials.

25.3.6.2 Quantitative analysis

The reworking potential of materials varied from 13.3% (L-HPC, No. E17) to 183% (MF, No. Z6). Malkowska's method possessed several disadvantages in assessing the change of tabletability. On the one hand, the reworking potential index had the risk of amplifying reduced tabletability of the material belonging to Category 2B and 2C. For example, the reworking potential of DCP (No. E10) was 65.8%, indicating that the material had a serious loss of tabletability. However, the tabletability category of the material before and after granulation had not changed and was maintained as Category 2B. The relative change of tabletability of DCP was −4.74% and belonged to Type a of the following tabletability change classification system (TCCS), which represented the unchanged tabletability. On the other hand, the reworking potential had a poor ability to distinguish materials. For example, the reworking potential of both Lac F100 (No. E6) and MCC PH101 (No. E15) was 40.0%, but the differences between the values of AUC_p and the AUC_g of them were 160 and 626, respectively.

In order to overcome the shortages of the reworking potential index, this study proposed a new index, as described in Section 25.3.4.2. According to Eq. (25.13), the relative change of tabletability of all materials was calculated, and they varied from −51.44% (MCC PH101, No. E15) to 13.07% (MF, No. Z6) after HSWG. The relative change of tabletability of materials can be divided into three types, that is, Type ', Type a, and Type b, as shown in Fig. 25.9. Type ' represented that a material had a loss of tabletability after HSWG. According to the degree of reduced tabletability, Type' could be further classified into two sub-types, that is, Type'a and Type'b. The Type'a was defined as materials with large loss of tabletability, which could be represented by $CoT_r < -15\%$. There were five excipients falling into this type (i.e., MCC PH101, SMCC, CWIS, PVPP and L-HPC). The Type'b was described as materials with moderate loss of tabletability, which was featured by $-15\% \leq CoT_r < -5\%$. There were three NPPs (i.e., *A. sinensis* Radix, *Chuanxiong rhizome*, and *Visci herba*) and four excipients (i.e., Lac F100, Lac C80, maltodextrin, and PGS) belonging to Type'b. Typea stood for no significant change in the CoT_r of materials, which could be characterized by $-5\% \leq CoT_r \leq 5\%$ in this study. There were seven NPPs and seven excipients belonging to the Typea. Typeb was expressed as the increase of tabletability of materials, with the characteristic of $CoT_r > 5\%$. One excipient (i.e., mannitol) and two NPPs (i.e., *Mume fructus* and *Sophorae flavescentis* Radix) were divided into this type.

The established TCCS for materials provided a new means for the initial risk evaluation of materials in the formulation design. In practice, a further evaluation could be performed in conjunction with the roles of materials used in the formulation and the tabletability categories in the CBCS. For instance, the L-HPC belonged to Type 'a and its CoT_r value was -33.60%. When the L-HPC played the role of disintegrant in formulations, its proportion was often small (e.g., 2.5%–5%) (Diós et al., 2015). In such circumstances, the significant loss of tabletability of L-HPC would not be a risk since it would not alter the tabletability of the formulation to a large extent. Similarly, MCC PH101 and SMCC were commonly used fillers in tablet formulation (Solanki et al., 2019; Zhao et al., 2022). The CoT_r values of MCC PH101 and SMCC were -51.44% and -43.80%, respectively, showing that the two excipients were prone to large loss of tabletability after HSWG. However, the tabletability of the two materials after wet granulation may not be a problem, since good or acceptable tablet TS would be achieved. By contrast, the granules of PGS were classified as Category 2C materials after wet granulation. When the PGS was selected as diluent in the tablet formulation, a high risk would be assigned to it.

Mannitol was a frequently used filler in the production of tablets due to its nonhygroscopic character and low drug interaction potential (inertness) (Vanhoorne et al., 2016b). The CoT_r value of mannitol was 9.32% and it belonged to Type b. It had been reported that mannitol had increased or unchanged tabletability properties after wet granulation, which was attributed to the reduction of the primary particle size (Khorsheed et al., 2019). This suggested that mannitol could be used in combination with cellulose excipients having a large loss of tabletability in formulation design, so as to achieve the purpose of balancing the tabletability of the formulation (Kyttä et al., 2020). Besides, mannitol was often used in immediate-release tablets or chewable tablets, due to its water-soluble and sweet taste (Ohrem et al., 2014). The increased tabletability of mannitol might cause problems with the disintegrating of immediate-release tablets or the chewing difficulty of chewable tablets. This showed the increased tabletability of materials also had potential risks during tablet formulation development.

25.4 Conclusion

In this study, the change of tabletability of pharmaceutical materials after HSWG was investigated with the help of the material library approach. Both univariate and multivariate analyses were used to compare the physical properties and the compression behaviors of ungranulated powders and granules. It was proven that HSWG had the advantages of increasing the particle size and improving the flowability of the materials. Five groups of materials with different compression behaviors could be clearly distinguished on the score plot of PCA analysis of the CBCS parameters. The HSWG process did not affect the compressibility of the materials but could decrease the compactability and tabletability of materials. The tabletability categories of 10 materials were downgraded after wet granulation.

The reworking potential and the newly proposed index CoT_r were used to quantify the change of tabletability. The CoT_r had the advantages of mitigating the risk of amplifying the tabletability reduction for materials with weak tabletability. By summarizing the different CoT_r values of materials in the material library, the TCCS was successfully built. It was found that the change of tabletability of the materials could be classified into three types: loss of tabletability, unchanging tabletability, and increase of tabletability, which were defined as Type ', Type a, and Type b. The materials with large and moderate loss of tabletability were further discriminated as Type 'a and' b, respectively. The Type' materials were featured by $CoT_r < -5\%$, and 40% of the materials experinced a loss of tabletability after HSWG. The Type a stood for no significant change in the CoT_r of materials, which could be characterized by $-5\% \leq CoT_r \leq 5\%$, and 50% of the materials belong to Type a. Type b was expressed as the increase of tabletability of materials, with the characteristic of $CoT_r > 5\%$, and 10% of the materials were divided into Type b. The TCCS provided a means for the initial risk evaluation of materials in tablet formulation design.

References

Althnian, A., AlSaeed, D., Al-Baity, H., Samha, A., Dris, A. B., Alzakari, N., & Elwafa, H. K (2021). Impact of dataset size on classification performance: An empirical evaluation in the medical domain. *Applied Sciences, 11*(2), 796.

Alwosheel, A., van Cranenburgh, S., & Chorus, C. G. (2018). Is your dataset big enough? Sample size requirements when using artificial neural networks for discrete choice analysis. *Journal of Choice Modelling, 28,* 167–182. doi:10.1016/j.jocm.2018.07.002.

Badawy, S. I., Narang, A. S., LaMarche, K. R., Subramanian, G. A., & Varia, S. A. (2019). *Handbook of pharmaceutical wet granulation.* Academic Press.

Badawy, S. I. F., Gray, D. B., & Hussain, M. A. (2006). A study on the effect of wet granulation on microcrystalline cellulose particle structure and performance. *Pharmaceutical Research, 23*(3), 634–640. doi:10.1007/s11095-005-9555-z.

Barrasso, D., Eppinger, T., Pereira, F. E., Aglave, R., Debus, K., Bermingham, S. K., & Ramachandran, R. (2015). A multi-scale, mechanistic model of a wet granulation process using a novel bi-directional PBM–DEM coupling algorithm. *Chemical Engineering Science, 123,* 500–513. doi:10.1016/j.ces.2014.11.011.

Bekaert, B., Penne, L., Grymonpré, W., Van Snick, B., Dhondt, J., Boeckx, J., Vogeleer, J., De Beer, T., Vervaet, C., & Vanhoorne, V. (2021). Determination of a quantitative relationship between material properties, process settings and screw feeding behavior via multivariate data-analysis. *International Journal of Pharmaceutics, 602,* 120603. doi:10.1016/j.ijpharm.2021.120603.

Bekaert, B., Van Snick, B., Pandelaere, K., Dhondt, J., Di Pretoro, G., De Beer, T., Vervaet, C., & Vanhoorne, V. (2022). In-depth analysis of the long-term processability of materials during continuous feeding. *International Journal of Pharmaceutics, 614,* 121454. doi:10.1016/j.ijpharm.2022.121454.

Benedetti, A., Khoo, J., Sharma, S., Facco, P., Barolo, M., & Zomer, S. (2019). Data analytics on raw material properties to accelerate pharmaceutical drug development. *International Journal of Pharmaceutics, 563,* 122–134. doi:10.1016/j.ijpharm.2019.04.002.

Bongiorno, V., Gibbon, S., Michailidou, E., & Curioni, M. (2022). Exploring the use of machine learning for interpreting electrochemical impedance spectroscopy data: Evaluation of the training dataset size. *Corrosion Science, 198,* 110119. doi:10.1016/j.corsci.2022.110119.

Borràs, E., Ferré, J., Boqué, R., Mestres, M., & Aceña, L. (2016). Prediction of olive oil sensory descriptors using instrumental data fusion and partial least squares (PLS) regression. *Talanta, 155*, 116–123. doi:10.1016/j.talanta.2016.04.040.

Borràs, E., Ferré, J., Boqué, R., Mestres, M., Aceña, L., & Busto, O. (2015). Data fusion methodologies for food and beverage authentication and quality assessment – A review. *Analytica Chimica Acta, 891*, 1–14. doi:10.1016/j.aca.2015.04.042.

Bostijn, N., Dhondt, J., Ryckaert, A., Szabó, E., Dhondt, W., Van Snick, B., Vanhoorne, V., Vervaet, C., & De Beer, T. (2019). A multivariate approach to predict the volumetric and gravimetric feeding behavior of a low feed rate feeder based on raw material properties. *International Journal of Pharmaceutics, 557*, 342–353. doi:10.1016/j.ijpharm.2018.12.066.

Bowles, B. J., Dziemidowicz, K., Lopez, F. L., Orlu, M., Tuleu, C., Edwards, A. J., & Ernest, T. B. (2018). Co-processed excipients for dispersible tablets—Part 1: Manufacturability. *AAPS PharmSciTech, 19*(6), 2598–2609. doi:10.1208/s12249-018-1090-4.

Braumann, A., Kraft, M., & Mort, P. R. (2010). Parameter estimation in a multidimensional granulation model. *Powder Technology, 197*(3), 196–210. doi:10.1016/j.powtec.2009.09.014.

Cao, J., Shen, H., Zhao, S., Ma, X., Chen, L., Dai, S., Xu, B., & Qiao, Y. (2024). Sample size requirements of a pharmaceutical material library: A case in predicting direct compression tablet tensile strength by latent variable modeling. *Pharmaceutics, 16*(2), 242.

Casian, T., Farkas, A., Ilyés, K., Démuth, B., Borbás, E., Madarász, L., Rapi, Z., Farkas, B., Balogh, A., Domokos, A., Marosi, G., & Tomuță, I. (2019). Data fusion strategies for performance improvement of a process analytical technology platform consisting of four instruments: An electrospinning case study. *International Journal of Pharmaceutics, 567*, 118473. doi:10.1016/j.ijpharm.2019.118473.

CDER Report. Novel Drugs (2020). Summary Available online: https://www.fda.gov/drugs/new-drugs-fda-cders-new-molecular-entities-and-new-therapeutic-biological-products/new-drug-therapy-approvals-2020.

Chen, Y., Dai, G., & Gao, Q. (2020). Preparation and properties of granular cold-water-soluble porous starch. *International Journal of Biological Macromolecules, 144*, 656–662. doi:10.1016/j.ijbiomac.2019.12.060.

Cho, C. H., Kim, J. Y., & Park, E. S. (2021). Utilization of a compaction simulator to formulate mini-tablets containing high dose of acyclovir. *Journal of Drug Delivery Science and Technology, 64*. doi:10.1016/j.jddst.2021.102602.

Dai, S., Xu, B., Shi, G., Liu, J., Zhang, Z., Shi, X., & Qiao, Y. (2019a). SeDeM expert system for directly compressed tablet formulation: A review and new perspectives. *Powder Technology, 342*(3), 517–527. https://doi.org/10.1016/j.powtec.2018.10.027.

Dai, S., Xu, B., Zhang, Z., Yu, J., Wang, F., Shi, X., & Qiao, Y. (2019b). A compression behavior classification system of pharmaceutical powders for accelerating direct compression tablet formulation design. *International Journal of Pharmaceutics, 572*, 118742. doi:10.1016/j.ijpharm.2019.118742.

de Koster, S. A. L., Pitt, K., Litster, J. D., & Smith, R. M. (2019). High-shear granulation: An investigation into the granule consolidation and layering mechanism. *Powder Technology, 355*, 514–525. doi:10.1016/j.powtec.2019.07.076.

Dhondt, J., Bertels, J., Kumar, A., Van Hauwermeiren, D., Ryckaert, A., Van Snick, B., … Vervaet, C. (2022a). A multivariate formulation and process development platform for direct compression. *International Journal of Pharmaceutics, 623*, 121962. doi:10.1016/j.ijpharm.2022.121962.

Dhondt, J., Eeckhout, Y., Bertels, J., Kumar, A., Van Snick, B., Klingeleers, D., … De Beer, T. (2022b). A multivariate methodology for material sparing characterization and blend design in drug product development. *International Journal of Pharmaceutics, 621*, 121801. doi:10.1016/j.ijpharm.2022.121801.

Diós, P., Pernecker, T., Nagy, S., Pál, S., & Dévay, A. (2015). Influence of different types of low substituted hydroxypropyl cellulose on tableting, disintegration, and floating behaviour of floating drug delivery systems. *Saudi Pharmaceutical Journal, 23*(6), 658–666. doi:10.1016/j.jsps.2014.09.001.

Drašković, M., Djuriš, J., Ibrić, S., & Parojčić, J. (2018). Functionality and performance evaluation of directly compressible co-processed excipients based on dynamic compaction analysis and percolation theory. *Powder Technology, 326*, 292–301. doi:10.1016/j.powtec.2017.12.021.

Escotet-Espinoza, M. S., Moghtadernejad, S., Oka, S., Wang, Y., Roman-Ospino, A., Schäfer, E., Cappuyns, P., Assche, I. V., Futran, M., Ierapetritou, M., & Muzzio, F. (2019). Effect of tracer material properties on the residence time distribution (RTD) of continuous powder blending operations. *Powder Technology, 342*, 744–763. doi:10.1016/j.powtec.2018.10.040.

Escotet-Espinoza, M. S., Moghtadernejad, S., Scicolone, J., Wang, Y., Pereira, G., Schäfer, E., Vigh, T., Klingeleers, D., Ierapetritou, M., & Muzzio, F. J. (2018). Using a material property library to find surrogate materials for pharmaceutical process development. *Powder Technology, 339*, 659–676. doi:10.1016/j.powtec.2018.08.042.

Fayed, M. H., Abdel-Rahman, S. I., Alanazi, F. K., Ahmed, M. O., Tawfeek, H. M., & Ali, B. E. (2017). High shear granulation process: Assessing impact of formulation variables on granules and tablets characteristics of high drug loading formulation using design of experiment methodology. *Acta Poloniae Pharmaceutica —Drug Research, 74*(2), 551–564.

Fayed, M. H., Abdel-Rahman, S. I., Alanazi, F. K., Ahmed, M. O., Tawfeek, H. M., & Al-Shdefat, R. I. (2017). New gentle-wing high-shear granulator: Impact of processing variables on granules and tablets characteristics of high-drug loading formulation using design of experiment approach. *Drug Development and Industrial Pharmacy, 43*(10), 1584–1600. doi:10.1080/03639045.2017.1326930.

Fell, J. T., & Newton, J. M. (1970). Determination of tablet strength by the diametral-compression test. *Journal of Pharmaceutical Sciences, 59*(5), 688–691. doi:10.1002/jps.2600590523.

Ferreira, A. P., & Tobyn, M. (2015). Multivariate analysis in the pharmaceutical industry: Enabling process understanding and improvement in the PAT and QbD era. *Pharmaceutical Development and Technology, 20*(5), 513–527. doi:10.3109/10837450.2014.898656.

Gao, L., & Ren, S. (2010). Multivariate calibration of spectrophotometric data using a partial least squares with data fusion. *Spectrochimica Acta Part A: Molecular and Biomolecular Spectroscopy, 76*(3-4), 363–368. doi:10.1016/j.saa.2010.03.024.

Grassmann, P., Sawistowski, H., & Hardbottle, R. (1971). Physical principles of chemical engineering. *AIChE J, 18*(6), 62–82.

Grdešič, P., Sovány, T., & Ilić, I. G. (2018). High-shear granulation of high-molecular weight hypromellose: Effects of scale-up and process parameters on flow and compaction properties. *Drug Development and Industrial Pharmacy, 44*(11), 1770–1782. doi:10.1080/03639045.2018.1496447.

Han, J. K., Shin, B. S., & Choi, D. H. (2019). Comprehensive study of intermediate and critical quality attributes for process control of high-shear wet granulation using multivariate analysis and the quality by design approach. *Pharmaceutics, 11*(6). doi:10.3390/pharmaceutics11060252.

Han, R., Yang, Y., Li, X., & Ouyang, D. (2018). Predicting oral disintegrating tablet formulations by neural network techniques. *Asian Journal of Pharmaceutical Sciences, 13*(4), 336–342. doi:10.1016/j.ajps.2018.01.003.

Hapgood, K. P., Litster, J. D., Biggs, S. R., & Howes, T. (2002). Drop penetration into porous powder beds. *Journal of Colloid and Interface Science, 253*(2), 353–366. doi:10.1006/jcis.2002.8527.

Hapgood, K. P., Litster, J. D., & Smith, R. (2003). Nucleation regime map for liquid bound granules. *AIChE Journal, 49*(2), 350–361. doi:10.1002/aic.690490207.

Hayashi, Y., Nakano, Y., Marumo, Y., Kumada, S., Okada, K., & Onuki, Y. (2021). Application of machine learning to a material library for modeling of relationships between material properties and tablet properties. *International Journal of Pharmaceutics, 609*, 121158. doi:10.1016/j.ijpharm.2021.121158.

Haykin, S., Neural networks and learning machines. Prentice Hall, 3 (2009),

Hentzschel, C. M., Sakmann, A., & Leopold, C. S. (2011). Comparison of traditional and novel tableting excipients: Physical and compaction properties. *Pharmaceutical Development and Technology, 17*(6), 649–653. doi:10.3109/10837450.2011.572897.

Hernández, O. C., Caraballo Rodríguez, I., Bernad Bernad, M. J., & Melgoza Contreras, L. M. (2021). Comparison of the performance of two grades of metformin hydrochloride elaboration by means of the SeDeM system, compressibility, compactability, and process capability indices. *Drug Development and Industrial Pharmacy, 47*(3), 484–497. doi:10.1080/03639045.2021.1892741.

Hildebrandt, C., Gopireddy, S. R., Fritsch, A. K., Profitlich, T., Scherließ, R., & Urbanetz, N. A. (2019). Evaluation and prediction of powder flowability in pharmaceutical tableting. *Pharmaceutical Development and Technology, 24*(1), 35–47. doi:10.1080/10837450.2017.1412462.

Hounslow, M. J., Pearson, J. M. K., & Instone, T. (2001). Tracer studies of high-shear granulation: II. Population balance modeling. *AIChE Journal, 47*(9), 1984–1999. doi:10.1002/aic.690470910.

Ibrahim, A., Kothari, B. H., Fahmy, R., & Hoag, S. W. (2019). Prediction of dissolution of sustained release coated ciprofloxacin beads using near-infrared spectroscopy and process parameters: A data fusion approach. *AAPS PharmSciTech, 20*(6). doi:10.1208/s12249-019-1401-4.

Iveson, S. M., & Litster, J. D. (1998). Growth regime map for liquid-bound granules. *AIChE Journal, 44*(7), 1510–1518. doi:10.1002/aic.690440705.

Iveson, S. M., Litster, J. D., Hapgood, K., & Ennis, B. J. (2001). Nucleation, growth and breakage phenomena in agitated wet granulation processes: A review. *Powder Technology, 117*(1-2), 3–39. doi:10.1016/S0032-5910(01)00313-8.

Kása, P., Bajdik, J., & Zsigmond, Z. (2009). Klára Pintye-Hódi, Study of the compaction behaviour and compressibility of binary mixtures of some pharmaceutical excipients during direct compression. *Chemical Engineering and Processing: Process Intensification, 48*(4), 859–863. doi:10.1016/j.cep.2008.11.002.

Kassem, B. E., Heider, Y., Brinz, T., & Markert, B. (2020). A multivariate statistical approach to analyze the impact of material attributes and process parameters on the quality performance of an auger dosing process. *Journal of Drug Delivery Science and Technology, 60*, 101950. doi:10.1016/j.jddst.2020.101950.

Katz, J. M., Roopwani, R., & Buckner, I. S. (2013). A material-sparing method for assessment of powder deformation characteristics using data collected during a single compression-decompression cycle. *Journal of Pharmaceutical Sciences, 102*(10), 3687–3693. doi:10.1002/jps.23676.

Khorsheed, B., Gabbott, I., Reynolds, G. K., Taylor, S. C., Roberts, R. J., & Salman, A. D. (2019). Twin-screw granulation: Understanding the mechanical properties from powder to tablets. *Powder Technology, 341*, 104–115. doi:10.1016/j.powtec.2018.05.013.

Kristensen, H. G., Holm, P., & Schaefer, T. (1985). Mechanical properties of moist agglomerates in relation to granulation mechanisms Part I. Deformability of moist, densified agglomerates. *Powder Technology, 44*(3), 227–237. doi:10.1016/0032-5910(85)85004-x.

Kulju, T., Paavola, M., Spittka, H., Keiski, R. L., Juuso, E., Leiviskä, K., & Muurinen, E. (2016). Modeling continuous high-shear wet granulation with DEM-PB. *Chemical Engineering Science, 142*, 190–200. doi:10.1016/j.ces.2015.11.032.

Kyttä, K. M., Lakio, S., Wikström, H., Sulemanji, A., Fransson, M., Ketolainen, J., & Tajarobi, P. (2020). Comparison between twin-screw and high-shear granulation—The effect of filler and active pharmaceutical ingredient on the granule and tablet properties. *Powder Technology, 376*, 187–198. doi:10.1016/j.powtec.2020.08.030.

Lamešić, D., Planinšek, O., & Ilić, IG. (2018). Modified equation for particle bonding area and strength with inclusion of powder fragmentation propensity. *European Journal of Pharmaceutical Sciences, 121*, 218–227. doi:10.1016/j.ejps.2018.05.028.

Leane, M., Pitt, K., Reynolds, G. K., Dawson, N., Ziegler, I., Szepes, A., Crean, A. M., Agnol, R. D., Broegmann, B., Charlton, S. T., Davies, C., Gamble, J., Gamlen, M., Hsiao, W. K., Khimyak, Y. Z., Khinast, J., Kleinebudde, P.,

Moreton, C., Oswald, M., … Stone, E. (2018). Manufacturing classification system in the real world: Factors influencing manufacturing process choices for filed commercial oral solid dosage formulations, case studies from industry and considerations for continuous processing. *Pharmaceutical Development and Technology, 23*(10), 964–977. doi:10.1080/10837450.2018.1534863.

Lee, A. R., Kwon, S. Y., Choi, D. H., & Park, E. S. (2017). Quality by Design (QbD) approach to optimize the formulation of a bilayer combination tablet (Telmiduo®) manufactured via high shear wet granulation. *International Journal of Pharmaceutics, 534*(1–2), 144–158. doi:10.1016/j.ijpharm.2017.10.004.

Leuenberger, H. (1982). The compressibility and compactibility of powder systems. *International Journal of Pharmaceutics, 12*(1), 41–55. doi:10.1016/0378-5173(82)90132-6.

Li, Z., Wu, F., Zhao, LiJ, Lin, X., Shen, L., & Feng, Yi (2018). Evaluation of fundamental and functional properties of natural plant product powders for direct compaction based on multivariate statistical analysis. *Advanced Powder Technology, 29*(11), 2881–2894. doi:10.1016/j.apt.2018.08.009.

Li, Z., Yu, Y., Pan, X., & Karim, M. N. (2020). Effect of dataset size on modeling and monitoring of chemical processes. *Chemical Engineering Science, 227*, 115928. doi:10.1016/j.ces.2020.115928.

Malkowska, S., Khan, K. A., Lentle, R., Marchant, J., & Elger, G. (1983). Effect of re-compression on the properties of tablets prepared by moist granulation. *Drug Development and Industrial Pharmacy, 9*(3), 349–361. doi:10.3109/03639048309044679.

Mu, S., Cui, M., & Huang, X. (2020). Multimodal data fusion in learning analytics: A systematic review. *Sensors, 20*(23), 6856. doi:10.3390/s20236856.

Nofrerias, I., Nardi, A., Suñé-Pou, M., Suñé-Negre, J. M., García-Montoya, E., Pérez-Lozano, P., Ticó, J. R., & Miñarro, M. (2019). Comparison between microcrystalline celluloses of different grades made by four manufacturers using the SeDeM diagram expert system as a pharmaceutical characterization tool. *Powder Technology, 342*, 780–788. doi:10.1016/j.powtec.2018.10.048.

Nordström, J., Klevan, I., & Alderborn, G. (2009). A particle rearrangement index based on the Kawakita powder compression equation. *Journal of Pharmaceutical Sciences, 98*(3), 1053–1063. doi:10.1002/jps.21488.

Ohrem, H. L., Schornick, E., Kalivoda, A., & Ognibene, R. (2014). Why is mannitol becoming more and more popular as a pharmaceutical excipient in solid dosage forms? *Pharmaceutical Development and Technology, 19*(3), 257–262. doi:10.3109/10837450.2013.775154.

Oka, S., Kašpar, O., Tokárová, V., Sowrirajan, K., Wu, H., Khan, M., Muzzio, F., Štěpánek, F., & Ramachandran, R. (2015). A quantitative study of the effect of process parameters on key granule characteristics in a high shear wet granulation process involving a two component pharmaceutical blend. *Advanced Powder Technology, 26*(1), 315–322. doi:10.1016/j.apt.2014.10.012.

Ono, T., & Yonemochi, E. (2020). Evaluation of the physical properties of dry surface-modified ibuprofen using a powder rheometer (FT4) and analysis of the influence of pharmaceutical additives on improvement of the powder flowability. *International Journal of Pharmaceutics, 579*, 119165. doi:10.1016/j.ijpharm.2020.119165.

Osei-Yeboah, F., Feng, Y., & Sun, C. C. (2014). Evolution of structure and properties of granules containing microcrystalline cellulose and polyvinylpyrrolidone during high-shear wet granulation. *Journal of Pharmaceutical Sciences, 103*(1), 207–215. doi:10.1002/jps.23776.

Osei-Yeboah, F., Zhang, M., Feng, Y., & Sun, C. C. (2014). A formulation strategy for solving the overgranulation problem in high shear wet granulation. *Journal of Pharmaceutical Sciences, 103*(8), 2434–2440. doi:10.1002/jps.24066.

Pandey, P., & Badawy, S. (2016). A quality by design approach to scale-up of high-shear wet granulation process. *Drug Development and Industrial Pharmacy, 42*(2), 175–189. doi:10.3109/03639045.2015.1100199.

Pandey, P., Tao, J., Chaudhury, A., Ramachandran, R., Gao, J. Z., & Bindra, D. S. (2013). A combined experimental and modeling approach to study the effects of high-shear wet granulation process parameters on granule characteristics. *Pharmaceutical Development and Technology, 18*(1), 210–224. doi:10.3109/10837450.2012.700933.

Paul, S., Chang, S. Y., Dun, J., Sun, W. J., Wang, K., Tajarobi, P., Boissier, C., & Sun, C. C. (2018). Comparative analyses of flow and compaction properties of diverse mannitol and lactose grades. *International Journal of Pharmaceutics, 546*(1–2), 39–49. doi:10.1016/j.ijpharm.2018.04.058.

Paul, S., & Sun, C. C. (2017). Dependence of friability on tablet mechanical properties and a predictive approach for binary mixtures. *Pharmaceutical Research, 34*(12), 2901–2909. doi:10.1007/s11095-017-2273-5.

Perez, P., Sunenegre, J., Minarro, M., Roig, M., Fuster, R., Garciamontoya, E., Hernandez, C., Ruhi, R., & Tico, J. (2006). A new expert systems (SeDeM Diagram) for control batch powder formulation and preformulation drug products. *Journal of Pharmaceutics and Biopharmaceutics, 64*(3), 351–359. doi:10.1016/j.ejpb.2006.06.008.

Picker, K. M. (2004). The 3D model: Explaining densification and deformation mechanisms by using 3D parameter plots. *Drug Development and Industrial Pharmacy, 30*(4), 413–425. doi:10.1081/DDC-120030936.

Pohlman, D. A., & Litster, J. D. (2015). Coalescence model for induction growth behavior in high shear granulation. *Powder Technology, 270*, 435–444. doi:10.1016/j.powtec.2014.07.016.

Ramachandran, R., Poon, J. M.-H., Sanders, CF. W., Glaser, T., Immanuel, C. D., Doyle, F. J., Litster, J. D., Stepanek, F., Wang, F.-Y., & Cameron, I. T. (2008). Experimental studies on distributions of granule size, binder content and porosity in batch drum granulation: Inferences on process modelling requirements and process sensitivities. *Powder Technology, 188*(2), 89–101. doi:10.1016/j.powtec.2008.04.013.

Realpe, A., & Velázquez, C. (2008). Growth kinetics and mechanism of wet granulation in a laboratory-scale high shear mixer: Effect of initial polydispersity of particle size. *Chemical Engineering Science, 63*(6), 1602–1611. doi:10.1016/j.ces.2007.11.018.

Rezaei, L., Meruva, S., & Donovan, M. D. (2022). Effect of manufacturing process on the retention of abuse-deterrent properties of PEO-matrix tablets. *AAPS PharmSciTech, 23*(1). doi:10.1208/s12249-021-02169-6.

Rojas, J., Ciro, Y., & Correa, L. (2014). Functionality of chitin as a direct compression excipient: An acetaminophen comparative study. *Carbohydr. Polym., 103*(4), 134–139.

Rojas, J., Zuluaga, C., & Cadavid, A. (2015). Effect of reprocessing and excipient characteristics on ibuprofen tablet properties. *Journal of Pharmaceutical Research, 14*(7), 1145–1152. doi:10.4314/tjpr.v14i7.4.

Ryckaert, A., Hauwermeiren, J. D, Man, A. F, Djuric, D., Vervaet, C., Nopens, I., & De Beer, T. (2021). TPLS as predictive platform for twin-screw wet granulation process and formulation development. *International Journal of Pharmaceutics, 605*, 120785. doi:10.1016/j.ijpharm.2021.120785.

Sabri, A. H., Hallam, C. N., Baker, N. A., Murphy, D. S., & Gabbott, I. P. (2018). Understanding tablet defects in commercial manufacture and transfer. *Journal of Drug Delivery Science and Technology, 46*, 1–6. doi:10.1016/j.jddst.2018.04.020.

Šantl, M., Ilić, I., Vrečer, F., & Baumgartner, S. (2011). A compressibility and compactibility study of real tableting mixtures: The impact of wet and dry granulation versus a direct tableting mixture. *International Journal of Pharmaceutics, 414*(1-2), 131–139. doi:10.1016/j.ijpharm.2011.05.025.

Schönfeld, B. V., Westedt, U., & Wagner, K. G. (2022). Compression modulus and apparent density of polymeric excipients during compression—impact on tabletability. *Pharmaceutics, 14*(5), 913.

Shi, L., Feng, Y., & Sun, C. C. (2010). Roles of granule size in over-granulation during high shear wet granulation. *Journal of Pharmaceutical Sciences, 99*(8), 3322–3325. doi:10.1002/jps.22118.

Silva, B. M. A., Vicente, S., Cunha, S., Coelho, J. F. J., Silva, C., Reis, M. S., & Simões, S. (2017). Retrospective quality by design (rQbD) applied to the optimization of orodispersible films. *International Journal of Pharmaceutics, 528*(1-2), 655–663. doi:10.1016/j.ijpharm.2017.06.054.

Simone, V. D., Caccavo, D., Lamberti, G., d'Amore, M., & Barba, A. A. (2018). Wet-granulation process: Phenomenological analysis and process parameters optimization. *Powder Technology, 340*, 411–419. doi:10.1016/j.powtec.2018.09.053.

Singh, B. N. (2019). Product development, manufacturing, and packaging of solid dosage forms under QbD and PAT paradigm: DOE case studies for industrial applications. *AAPS PharmSciTech, 20*(8). doi:10.1208/s12249-019-1515-8.

Sohail Arshad, M., Zafar, S., Yousef, B., Alyassin, Y., Ali, R., AlAsiri, A., Chang, M. W., Ahmad, Z., Ali Elkordy, A., Faheem, A., & Pitt, K. (2021). A review of emerging technologies enabling improved solid oral dosage form manufacturing and processing. *Advanced Drug Delivery Reviews, 178*. doi:10.1016/j.addr.2021.113840.

Solanki, N. G., Kathawala, M., & Serajuddin, A. T. M. (2019). Effects of surfactants on itraconazole-hydroxypropyl methylcellulose acetate succinate solid dispersion prepared by Hot Melt Extrusion III: Tableting of extrudates and drug release from tablets. *Journal of Pharmaceutical Sciences, 108*(12), 3859–3869. doi:10.1016/j.xphs.2019.09.014.

Sonnergaard, J. M. (2006). Quantification of the compactibility of pharmaceutical powders. *Journal of Pharmaceutics and Biopharmaceutics, 63*(3), 270–277. doi:10.1016/j.ejpb.2005.10.012.

Sousa, A. S., Serra, J., Estevens, C., Costa, R., & Ribeiro, A. J. (2023). Leveraging a multivariate approach towards enhanced development of direct compression extended release tablets. *International Journal of Pharmaceutics, 646*, 123432. doi:10.1016/j.ijpharm.2023.123432.

Su, J., Zhang, K., Qi, F., Cao, J., Miao, Y., Zhang, Z., Qiao, Y., & Xu, B. (2023). A tabletability change classification system in supporting the tablet formulation design via the roll compaction and dry granulation process. *International Journal of Pharmaceutics, 6*, 100204.

Sun, C. C. (2008). Mechanism of moisture induced variations in true density and compaction properties of microcrystalline cellulose. *International Journal of Pharmaceutics, 346*(1-2), 93–101. doi:10.1016/j.ijpharm.2007.06.017.

Sun, C. C. (2016). A classification system for tableting behaviors of binary powder mixtures. *Journal of Pharmaceutical Sciences, 11*(4), 486–491. doi:10.1016/j.ajps.2015.11.122.

Sun, C. C., & Kleinebudde, P. (2016). Mini review: Mechanisms to the loss of tabletability by dry granulation. *Journal of Pharmaceutics and Biopharmaceutics, 106*, 9–14. doi:10.1016/j.ejpb.2016.04.003.

Tanner, T., Antikainen, O., Ehlers, H., Blanco, D., & Yliruusi, J. (2018). Examining mechanical properties of various pharmaceutical excipients with the gravitation-based high-velocity compaction analysis method. *International Journal of Pharmaceutics, 539*(1–2), 131–138. doi:10.1016/j.ijpharm.2018.01.048.

Tay, J. Y. S., Liew, C. V., & Heng, P. W. S. (2017). Powder flow testing: Judicious choice of test methods. *AAPS PharmSciTech, 18*(5), 1843–1854. doi:10.1208/s12249-016-0655-3.

Thapa, P., Tripathi, J., & Jeong, S. H. (2019). Recent trends and future perspective of pharmaceutical wet granulation for better process understanding and product development. *Powder Technology, 344*, 864–882. doi:10.1016/j.powtec.2018.12.080.

Thoorens, G., Krier, F., Rozet, E., Carlin, B., & Evrard, B. (2015). Understanding the impact of microcrystalline cellulose physicochemical properties on tabletability. *International Journal of Pharmaceutics, 490*(1–2), 47–54. doi:10.1016/j.ijpharm.2015.05.026.

Vanhoorne, V., Bekaert, B., Peeters, E., De Beer, T., Remon, J-P., & Vervaet, C. (2016). Improved tabletability after a polymorphic transition of delta-mannitol during twin screw granulation. *International Journal of Pharmaceutics, 506*(1–2), 13–24. doi:10.1016/j.ijpharm.2016.04.025.

Vanhoorne, V., Van Bockstal, P-J., Van Snick, B., Peeters, E., Monteyne, T., Gomes, P., De Beer, T., Remon, J. P., & Vervaet, C. (2016). Continuous manufacturing of delta mannitol by cospray drying with PVP. *International Journal of Pharmaceutics, 501*(1-2), 139–147. doi:10.1016/j.ijpharm.2016.02.001.

Van Snick, B., Dhondt, J., Pandelaere, K., Bertels, J., Mertens, R., Klingeleers, D., Di Pretoro, G., Remon, J. P., Vervaet, C., De Beer, T., & Vanhoorne, V. (2018). A multivariate raw material property database to facilitate drug product development and enable in-silico design of pharmaceutical dry powder processes. *International Journal of Pharmaceutics, 549*(1–2), 415–435. doi:10.1016/j.ijpharm.2018.08.014.

Wade, J. B., Miesle, J. E., Avilés, S. L., & Sen, M. (2020). Exploring the wet granulation growth regime map – validating the boundary between nucleation and induction. *Chemical Engineering Research and Design, 156*, 469–477. doi:10.1016/j.cherd.2020.02.024.

Wang, T., Ibrahim, A., & Hoag, S. W. (2020). Understanding the impact of magnesium stearate variability on tableting performance using a multivariate modeling approach. *Pharmaceutical Development and Technology, 25*(1), 76–88. doi:10.1080/10837450.2019.1673774.

Wang, Y., Cao, J., Zhao, X., Liang, Z., Qiao, Y., Luo, G., & Xu, B. (2022). Using a material library to understand the change of tabletability by high shear wet granulation. *Pharmaceutics, 14*(12), 2631.

Wang, Y., O'Connor, T., Li, T., Ashraf, M., & Cruz, C. N. (2019). Development and applications of a material library for pharmaceutical continuous manufacturing of solid dosage forms. *International Journal of Pharmaceutics, 569*. doi:10.1016/j.ijpharm.2019.118551.

Wang, Z., Cao, J., Li, W., Wang, Y., Luo, G., Qiao, Y., Zhang, Y., & Xu, B. (2021). Using a material database and data fusion method to accelerate the process model development of high shear wet granulation. *Scientific Reports, 11*(1). doi:10.1038/s41598-021-96097-x.

Yost, E., Mazel, V., Sluga, K. K., Nagapudi, K., & Muliadi, A. R. (2022). Beyond brittle/ductile classification: Applying proper constitutive mechanical metrics to understand the compression characteristics of pharmaceutical materials. *Journal of Pharmaceutical Sciences, 111*, 1984–1991.

Yu, J., Xu, B., Zhang, K., Shi, C., Zhang, Z., & Fu, J. (2019). Using a material library to understand the impacts of raw material properties on ribbon quality in roll compaction. *Pharmaceutics, 11*(12), 662. doi:10.3390/pharmaceutics11120662.

Yu, X., Hounslow, M. J., Reynolds, G. K., Rasmuson, A., Björn, I. N., & Abrahamsson, P. J. (2017). A compartmental CFD-PBM model of high shear wet granulation. *AIChE Journal, 63*(2), 438–458. doi:10.1002/aic.15401.

Záhonyi, P., Szabó, E., Domokos, A., Haraszti, A., Gyürkés, M., Moharos, E., & Nagy, Z. K. (2022). Continuous integrated production of glucose granules with enhanced flowability and tabletability. *International Journal of Pharmaceutics, 626*, 122197. doi:10.1016/j.ijpharm.2022.122197.

Zhang, Y., Binner, J., Rielly, C., & Vaidhyanathan, B. (2014). Comparison of spray freeze dried nanozirconia granules using ultrasonication and twin-fluid atomisation. *Journal of the European Ceramic Society, 34*(4), 1001–1008. doi:10.1016/j.jeurceramsoc.2013.10.033.

Zhang, Y., Cheng, B. C.-Y., Zhou, W., Xu, B., Gao, X., Qiao, Y., & Luo, G. (2019). Improved understanding of the high shear wet granulation process under the paradigm of quality by design using *Salvia miltiorrhiza* granules. *Pharmaceutics, 11*(10), 519. doi:10.3390/pharmaceutics11100519.

Zhang, Y., Xu, B., Wang, X., Dai, S., Shi, X., & Qiao, Y. (2019). Optimal selection of incoming materials from the inventory for achieving the target drug release profile of high drug load sustained-release matrix tablet. *AAPS PharmSciTech, 20*(2). doi:10.1208/s12249-018-1268-9.

Zhao, H., Zhao, L., Lin, X., & Shen, L. (2022). An update on microcrystalline cellulose in direct compression: Functionality, critical material attributes, and co-processed excipients. *Carbohydrate Polymers, 278*, 118968. doi:10.1016/j.carbpol.2021.118968.

U.S. Food and Drug Administration (FDA), 2019. Available online: https://www.fda.gov/drugs/first-generic-drug-approvals/2019-first-generic-drug-approvals.

U.S. Food and Drug Administration (FDA), 2021. Available online: https://www.fda.gov/drugs/new-drugs-fda-cders-new-molecular-entities-and-new-therapeutic-biological-products/novel-drug-approvals-2021.

CHAPTER 26

Emerging paradigms in pharmaceutical wet granulation

Ajit S. Narang[a] and Sherif I.F. Badawy[b]

[a]*Pharmaceutical Sciences, ORIC Pharmaceuticals, Inc., South San Francisco, CA, United States,*
[b]*Drug Product Development, Bristol-Myers Squibb Co, New Brunswick, NJ, United States*

26.1 Introduction

The granulation process transforms the shape, size, surface, and density of powders or powder mixtures to improve their physicochemical properties and handling to enable rapid and robust manufacture on high-speed commercial equipment. A recent review compares various wet granulation technologies such as fluidized bed granulation, extrusion spheronization, spray drying, melt granulation, moisture-activated granulation, thermal adhesive granulation, and foam granulation, with granulation in high- and low-shear mixers (Eremin et al., 2024). The success of a granulation operation depends on optimal achievement of several quality attributes, including enlargement with uniformity of drug and excipient distribution, improved flow, and densification without compromising compactibility, compressibility, tabletability, or drug release. Focusing primarily on pharmaceutical wet granulation, this book has provided an extensive and in-depth understanding of *current* paradigms and practices. The industrial practice of pharmaceutical wet granulation was dominated in the last few decades by high-shear wet granulation (HSWG). Fluid bed granulation is not uncommon in the pharmaceutical industry, but its application has been limited by a number of factors, not the least of which is the ability to uniformly fluidize the cohesive initial blend characteristic of many pharmaceutical formulations. The focus of this book, therefore, has been on the more widely practiced HSWG. Despite its wide utilization, the practice and evolution of HSWG has been loaded with challenges inherent to the fundamental concepts of HSWG. In its fundamental form, HSWG uses a high-shear blender to mix liquid and powder components, creating a time-dependent trajectory of the wet mass attributes until the desired attributes are achieved at the process end point. The concept of "end point" has been key in the practice of wet granulation (Campbell et al., 2011; Landín et al., 1995; Pandey & Badawy, 2016; Sirois & Craig, 2000). Early practice of HSWG emphasized end point determination as the main goal of process development and paid little attention to fundamental process understanding. Wet granulation processes have traditionally been developed using a trial-and-error approach, with the main objective of identifying a "magical" tool to detect granulation end point. A major limitation of the end point concept was the use of methods with only empirical and indirect correlations to the wet mass quality attributes critical to downstream process and product performance. The favorite approach for end point determination in the early days was the "fist test" in which the operator squeezes a sample of the wet mass and arbitrarily decides based on its consistency if the end point is reached. More recently, end point determination methods have attempted to more directly assess critical material attributes such as granule size and density (Huang et al., 2010; Narang et al., 2015, 2017). Although progress has been made in this field, an end point tool which directly assesses these attributes in real time has been elusive and is still the subject of intense research efforts.

26.1.1 End point control

While a reliable end point detection technology could be a valuable part of a successful process control strategy, the scope of modern process development and control in the quality-by-design (QbD) era goes far beyond the end point. Understanding how process dynamics and material properties impact granulation trajectory and end product has been the subject of recent research efforts in HSWG. Physical models of wet granulation have been developed which characterize the rate processes in HSWG, such as nucleation, granule coalescence, and breakage (Iveson et al., 2001). While these models have enhanced the fundamental understanding of wet granulation, they are not directly amenable to the design of industrial processes. These models have been generally developed for a particular rate process using simple systems, and hence, they are not directly applicable to the complex pharmaceutical formulations and wet granulation processes, which involve simultaneous rate processes. Nevertheless, physical models have been valuable to the understanding of the mechanisms involved in the observed effects of process performance and input material properties on process performance and end product attributes (Badawy et al., 2012; Vemavarapu et al., 2009).

26.1.2 Design of experiment approaches

The design of experiment (DoE) approach is now commonly applied to study the effect of HSWG. Early DoE studies were limited to empirical interpretations, and results were not construed in terms of physical mechanisms of granulation (Iskandarani et al., 2001; Ring et al., 2011). Later studies utilized physical models to interpret the effect of process parameters on primary granule properties (such as granule size and density) and on granulation bulk powder properties and final product critical quality attributes (such as flow, compaction properties, and tablet dissolution rate). A similar approach was used for input material properties, where the observed effects of material properties on process performance were linked to the granulation mechanism and physical models of granulation.

26.1.3 Modeling and simulation

Advances have also been made in the computational modeling of HSWG (Chaudhury et al., 2014; Gantt & Gatzke, 2005; Iveson, 2002; Nakamura et al., 2013; Pandey et al., 2013). Predictive numerical simulation of HSWG has evolved over the last two decades with increasing complexity of the models enabled by the leap in available computational power. The various modeling approaches utilized for HSWG include population balance modeling (PBM), discrete element modeling (DEM), PBM coupled with DEM, PBM coupled with computational fluid dynamics (CFD), PBM with a compartmental model and DEM, and PBM coupled with volume of fluid (VoF) models. Early numerical modeling approaches were based on empirical population balance models. The advances in physical models of granulation enabled the evolution of the PBM efforts and the implementation of physically based model parameters. However, the complexity of the HSWG precludes the development of a widely predictive model based on one computational approach. This complexity arises from the number of simultaneous rate processes in the HSWG and the heterogeneity of forces and particle velocities in the granulator. Hybrid modeling approaches such as the combined PBM and DEM modeling, are emerging as promising advances in HSWG process modeling. This handbook presents an in-depth review of the computational modeling of HSWG. Despite the advances in computational models, a fully predictive model which can be used for industrial process control is not yet a reality, and its availability is limited by the high level of complexity that will be required for such a model.

26.1.4 Residual challenges

Taking it all together, substantial progress has been clearly made in the understanding, development, and control of HSWG in the last few decades. Nevertheless, significant challenges remain, which arguably stem from the inherent nature of the process itself. It is becoming more apparent that a paradigm shift may be necessary to address residual challenges, rather than incremental improvements in current paradigms. The current state-of-the-art in HSWG, including advances described above, was covered in-depth in this book. The rest of this chapter provides broad perspectives and overviews of *evolving* paradigms and practices in the field of pharmaceutical wet granulation. These topics include:

- Continuous wet granulation: Among the evolving paradigms, continuous wet granulation is the most likely transformational shift in wet granulation. Unlike HSWG, small quantities of material are continuously loaded, processed, and unloaded. The process operates in a time-invariant dynamic state of control typical of continuous processes in contrast to the time-dependent trajectory characteristic of HSWG. Twin-screw granulation (TSG), which is emerging as the most popular technology for continuous wet granulation, is covered in Chapter 23. TSG overcomes the key challenges of HSWG. The equipment and screws are designed so that the rate processes such as nucleation and coalescence are limited to defined zones in the granulator rather than occurring simultaneously. This makes the process more amenable to computational modeling. This, together with the continuous nature of the process makes advanced process control a reality that is largely lacking in HSWG. Scale-up challenges that hamper HSWG are mostly nonexistent.
- One of the challenges associated with HSWG is the short process time (usually in the range of minutes), which compromises the robustness of end point control and substantially eliminates the possibility of feedback control. Processes at the interface of different technologies such as hot-melt extrusion, coupled with the concepts of wet granulation or dry granulation, coupled with the role of water in wet granulation (e.g., moisture-activated dry granulation), attempt to provide a more robust process end point.
- Processes that concurrently utilize heat and mass flow dynamics, such as steam granulation, melt granulation, and freeze granulation (instead of traditional methodologies that utilize only mass [liquid binder] flow).
- Leaps of granulation technology with improvements and extensions of particular aspects, such as mode of liquid addition (e.g., foam granulation), are evolving to address challenges associated with the uniform distribution of the granulating liquid.

Subsequent sections of this chapter will highlight these evolving paradigms with a critical eye toward both their promise and challenges.

26.2 Moisture-activated granulation

Granulation—whether dry or wet—is a particle agglomeration and size enlargement process that is used to improve the manufacturability of (usually) a mixture of dry powder particles. These manufacturability attributes include flow, density, uniformity of distribution, and compactibility. The granulating liquid for most wet granulation processes is water or an aqueous solution of a hydrophilic polymeric binder. Changes in the concentration of the granulating liquid in the wet granulation process are known to impact the process and outcome of wet granulation. This is a commonly studied variable in QbD development of the wet granulation unit operation. Scientists have explored very low concentrations of water for granulation, calling this process moisture-activated granulation (MAG) or moisture-activated dry granulation (MADG). Water is ubiquitously present on particle surfaces of materials and is associated with plasticization and adhesion. Water associated with molecules could be bound or free water. The bound water, for example, could be the water of crystallization and is not available for reactions. The free water, on the other hand, is mobile, changes with environmental humidity, and is available for reactions in the microenvironment (Narang et al., 2012b). The effect of environmental humidity on the water content of solid materials is assessed using the moisture sorption-desorption studies, typically carried out at a fixed temperature (isothermal). The presence of small amounts of surface water increases particle-particle attractive interactions through the formation of hydrogen bonds and liquid bridge formation (Louati et al., 2017). The MAG generally proceeds by a layering mechanism. A small amount of water is added, usually at a slow addition rate for uniform distribution, to hydrate the binder. The hydrated or "activated" binder becomes sufficiently tacky such that other particles in the formulation are layered on the surface of binder particles. These phenomena require the presence of hydrophilic, water-soluble binders, which create interparticle bonds during the process. Unlike the traditional wet granulation, MAG does not typically involve liquid bridges between particles and does not proceed through the coalescence of the evolving granules. So, the MAG is effective in removing fines that are more readily layered on the binder particles without significant granule growth. Uncontrolled growth is therefore less of a concern when dealing with the MAG process. The moisture-activated granulation process utilizes the same fundamental principles at a liquid level above a critical threshold for particle bonding but below the proportions of water used for the conventional wet granulation processes.

The MAG process could either (1) utilize low quantities of the granulating fluid that would lead to limited particle agglomeration and would still need to be dried in a fluid bed dryer or tray drier (de Jong, 1969) or (2) use an even lower quantity of fluid that the drying step is completely bypassed (Ullah et al., 1987). In another variation of this process, extended storage time of a well-mixed dry binder powder blend (that includes a binder) at high humidity conditions can lead to granule formation even without the mixing operation (Li et al., 2013).

26.2.1 Granulation with different physical states of binder

In exploring the effect of very low quantities of water on the rheological state of the binder, Li et al., 2011a studied the phase behavior, or the physical state, of four different hydrophilic polymeric binders upon exposure to different relative humidity environments at room temperature (Li et al., 2011b). The interaction of binders with water was investigated by studying the water sorption behavior of binders, and the changes in the physical state of binders were studied by dialysis experiments. In addition, glass transition temperatures of the binders were determined in the presence of water. The authors observed that the hydrophilic polymeric binders, such as poly(vinyl pyrrolidone) (PVP) and hydroxypropyl cellulose (HPC), undergo a phase transition from a glassy state to a rubbery/solution state at a certain water content at room temperature. The water content that affects this transition depends on the type of polymer and its molecular weight. For example, PVP K12 undergoes a phase transition from the glassy state to the rubbery/solution state at much lower water content than PVP K29/32 (10% vs. 20%), while the phase transition for HPC occurs with 10% to 15% water (Li et al., 2011b). The authors further studied the effect of physical state of binders on the outcome of the wet granulation process by storing the binary mixtures of calcium carbonate with the binders at humidity conditions that would keep the binders in either the glassy state (60% RH) or the rubbery/solution state (96% RH) (Li et al., 2011a). After exposure to either 60% or 96% RH for 4 weeks, the exposed binary blends were mixed in a Diosna high-shear granulator for 2 min, followed by tray drying in an oven at 50°C for 24 h. The authors observed a significant increase in the particle size and aspect ratio of particles after exposure of the blends to 96% RH, but not 60% RH (Li et al., 2011a).

While these studies are very helpful to improve the fundamental understanding of physical state and attributes of the binder particles on their role in wet granulation, this methodology of prolonged exposure (on the order of weeks) of powder blends to high humidity environments to "activate" the binder is not industrially relevant when compared to traditional granulation approaches that take minutes to hours.

26.2.2 Moisture-activated dry granulation

In a different version of the process, Ullah et al. (1987) prepared granules of powder blends by adding a small quantity of water, ~1% to 4% of the final formula weight of the dry powder blend (which will lead to a higher concentration of the intragranular material), to a high shear mixer during powder mixing. To facilitate uniform distribution of water, the authors further added the least adsorbent materials of the mixture first and the highly adsorbent materials (such as microcrystalline cellulose, [MCC]) last. The authors proved the feasibility and utility of this process by comparing a high drug load formulation of acetaminophen prepared by the MADG and the traditional HSWG process. This formulation utilized PVP as a binder, MCC as a diluent, sodium starch glycollate (SSG) as a disintegrant, and magnesium stearate (MgSt) as a lubricant. A similar profile of granular and product characteristics was observed by the MADG and the HSWG process (Ullah et al., 1987). MCC and silicon dioxide were recommended as preferred excipients for the moisture absorption stage of the MADG process. In another study, Christensen et al. (1994) studied the effect of water content, wet massing time, and dry mixing time using different moisture-absorbing materials on the outcome of MADG—comparing MCC with potato starch or a 1:1 mixture of the two. They observed that the physical properties of the tablets were primarily affected by the water content, the moisture-absorbing material, and the compression force. The authors noted a critical aspect of the amount of water used in this process. The water amount should be low enough to bypass the need for drying. However, too low a water content can compromise flow and aggregation. High water, on the other hand, can increase the strength and tablet disintegration time. The authors observed that an optimum water content was between 1.5% and 2.5%. They further noted that, while it was possible to manufacture MADG formulations of both potato starch and MCC individually as moisture-absorbing materials, the use of a mixture of these two produced tablets with high strength and low disintegration time (Christensen et al., 1994).

Takasaki et al. (2013) further investigated the role of moisture-absorbing materials by studying magnesium aluminometasilicate, pregelatinized starch, crospovidone, and carmellose calcium as alternative moisture absorbents in a lactose monohydrate and PVP-based MADG process. The authors observed that the powder wettability of different moisture adsorbents was key to rapid tablet disintegration. In addition, the overall disintegration time of all tablets produced by the MADG process was significantly lower, compared to the traditional HSWG process. The authors hypothesized that this was primarily because of the better water penetration of MCC in the MADG than in the HSWG process.

The utility of the MADG process for high (>80%) drug load immediate release formulations of cohesive and fluffy drugs was explored by Moravkar et al. (2017) using metformin hydrochloride, acetaminophen, and ferrous ascorbate as model compounds and malto-dextrin, PVP, and HPC as binders. The authors characterized the granules and tablets for a variety of quality attributes and observed that all results were within acceptable ranges. They concluded that the use of MADG allowed for lowering the proportion of excipients used in the formulation to overcome the physicochemical limitations of drug substances to produce high drug-loaded granules for drugs with significantly varying properties and attributes (Moravkar et al., 2017).

While particle adhesion improvements with small amounts of added moisture can be readily demonstrated on most commonly used tableting excipients, the merits of MADG have been shown for a select few drugs. This technology would generally be most beneficial to highly moisture-sensitive compounds, more so than traditional wet granulation. However, the removal of moisture very shortly after water addition allows traditional HSWG technology to be used for some water-sensitive compounds as well (Narang et al., 2012a). Another practical aspect of using MADG is the difficulty in assigning an accurate unit weight composition to the dosage form. Incorporation of water above the equilibrium moisture content of the material blend at operating conditions of product manufacturing indicates that some of that water will be lost during processing. This makes it difficult to identify whether the "added" amount of water should be considered a part of the tablet weight or otherwise.

In an application to a formulation composition that has a high content of absorbent materials, Origoni et al. (2024) studied the use of moisture-activated dry granulation for the development of gastroretentive tablets. The authors highlighted how MADG presents fewer critical process parameters, is energy and time-efficient, and, in their specific application, helped retain the high water absorbent capacity of the excipients in their formulation.

26.3 Technologies with concurrent heat and mass flow dynamics

While traditional wet granulation technologies focus predominantly on mass flow, that is, addition of binder solution to a mixture of solid particles, concurrent use of heat and mass flow to effect granulation incorporates phase transformation of the granulation mass. As much as these methodologies add additional variables and process control requirements, they also bring the promise of process consistency and superior process outcomes. These technologies are exemplified by freeze granulation, melt granulation, thermal adhesion granulation, and steam granulation.

26.3.1 Freeze granulation

Freeze granulation using liquid nitrogen involves spraying a liquid (water or solvent-based) slurry or suspension of particles into liquid nitrogen (which instantaneously freezes the droplets into granules), followed by freezing drying of the frozen droplets. Freeze drying yields granules whose size and shape are controlled by the spraying process parameters, and whose porosity is controlled by the solids concentration in the slurry or suspension. Invented in the ceramics industry (Nyberg et al., 1994), this process has applications to sensitive compounds where size, structure, and homogeneity of primary particles in the suspension should be maintained in the granules and require the production of granules without exposure to process stresses such as shear and heat. Examples of application of this technology could be the production of protein particles and/or particles with consistent shape, size, and porosity for superior aerodynamic performance—such as that required for inhalation drug delivery (Shanmugam, 2015).

Rautenberg and Lamprecht Rautenberg and Lamprecht (2022) highlighted the challenges with the industrial handling of low-density lyophilisates, even as they are technologically useful in improving the processability of thermolabile drugs. In working with mannitol-polyvinylpyrrolidone 25 (PVP 25) mixtures, the authors concluded that these highly porous, low-density microspheres exhibit mechanical stability when their solids content is or exceeds 5%.

In a recent paper, Sartzi et al. (2024) review the principles and the parameters influencing the spray freeze drying process, and discuss the economic and technical barriers to the industrialization of this process.

Freeze granulation using supercritical fluid, also called supercritical freeze granulation, utilizes rapid expansion of supercritical fluid, with liquid carbon dioxide being the preferred supercritical fluid. Sonoda et al. (2009) reported the use of this technology to produce flurbiprofen as a supercritical fluid-soluble drug candidate and lactose as an insoluble carrier. Upon rapid atomization of the supercritical carbon dioxide at room temperature, flurbiprofen fine particles adhering to the surface of lactose particles were obtained. The authors identified this as a valuable technique to reduce the particle size of a highly hydrophobic drug while causing the drug adhesion to hydrophilic particles, thus improving drug dissolution characteristics. Significant improvement in the dissolution rate was observed, compared to the supercritical freeze granulation of the drug alone, which led to agglomeration and loss of dissolution advantage. Supercritical freeze granulation was demonstrated by Sakamoto et al. (2015) to produce porous spherical granules with high flowability and dispersibility, suitable for dry powder inhalation.

These technologies offer promising options for particularly challenging drug molecules (such as proteins), formulation challenges (e.g., segregation), or product requirements (e.g., tight particle size control or aerodynamic properties for inhalation). Nonetheless, the sophistication required (such as handling in pressurized containers) for these technologies is significantly more than traditional wet granulation technologies—making these approaches suitable only for select, rare circumstances where traditional approaches fail to achieve desired outcomes.

26.3.2 Melt granulation

Melt granulation, also called thermoplastic granulation, utilizes binders that melt or soften at relatively low temperatures, or are liquid at room temperature, enabling granulation without the use of water or another granulating liquid. Since the initial studies in the 1990s (Schaefer, 1996), this process has been utilized for both dissolution enhancement of poorly soluble drugs (Passerini et al., 2006) as well as to slow the drug release rate (Voinovich et al., 2000). The binders that can be used in this process are exemplified by hydrophilic binders such as poly(ethylene glycol) (PEG) polymers of different molecular weights, with lower molecular weights existing as liquids at room temperature, and poloxamers; and hydrophobic binders such as waxes, fatty acids, and fatty alcohols. The granulation process is completed with the cooling of the agglomerated powder mass, resulting in the solidification of the binder and formation of robust granules. The melt granulation procedure can be carried out either in situ melt-in (solid binder particles dispersed as dry powder with intragranular ingredients, and then the mixture heats up during processing) or spray-on (in which case liquid or molten binder is sprayed on the intragranular powder mixture, similar to traditional wet granulation).

Melt granulation can be carried out using either a high-shear mixer or a fluid bed granulator. Passerini et al. (2010) compared granulation outcomes of manufacturing a model granulation in these two equipment using the same composition (70%w/w lactose monohydrate as a filler, 20% w/w PEG 6000 as a binder, and 10% w/w ibuprofen or ketoprofen as a drug). Because PEG 6000 is solid at room temperature, this in-situ, melt-in granulation process utilized heating of the fluidization air in a fluid bed granulator and the heated jacket in the high shear granulator to generate adequate temperatures in situ (~57°C–58°C) to melt the binder and effect granulation. The authors evaluated size, morphology, flow, friability, dissolution, and physicochemical properties of the granules manufactured by both processes. The authors observed that while manufacturing with both equipment was feasible, the particle size distribution and morphology of the granules were strongly equipment dependent. Nonetheless, the solid-state characteristics and the dissolution behavior of the drug were equipment independent (Passerini et al., 2010).

The melt granulation process has the advantages of being able to granulate water and heat-sensitive materials, bypassing the heating and drying phases. This process, however, might require elevation of temperature—which may not be uniform throughout the powder bed—and can impart plasticity to the granules, depending on the quantity of the binder used.

26.3.3 Steam granulation

Steam granulation seeks to overcome the issues of granulation fluid distribution by using steam instead of water as the granulating liquid. The high diffusion rate of steam helps readily access the particle surface in the powder bed, while condensation of the steam on the particle surface helps create a layer of binder fluid right at the interface of particle agglomeration and granule nucleation. The uniform distribution of steam on powder particles can help synchronize the initiation of granulation mechanisms (better than traditional wet granulation) and aid the production of granules with not only a more uniform and narrow particle size distribution, but also higher porosity. Steam has a higher diffusion rate into the powder bed than liquid water. Steam also provides a more favorable thermal environment across the granule structure during the drying step. This can help reduce the amount of water necessary for granulation, increase the rate of drying of granules, and produce granules with a larger surface area than those obtained with traditional wet granulation (Vialpando et al., 2013).

Cavallari et al. (2002) compared traditional wet granulation with steam granulation in a Rotolab single pot processor for the granulation of a mixture of piroxicam and β-cyclodextrin. The authors observed higher granule porosity and in vitro rate of drug dissolution with the steam granulated material. The authors hypothesized that the high dissolution rate of wet granulation and steam granulation, compared to physical mixture, could be attributed to the addition of both the mechanical energy (shearing stress produced by the impeller) and/or the thermal energy (steam) to improve the physical interaction between the drug and the cyclodex- trin. They attributed the higher dissolution rate from steam granulation, as compared to wet granulation, to the higher porosity and the larger surface area in the steam granulated material.

This process inherently requires controls for generation, safe handling, and uniform distribution of steam within the granulation mix—as well as for uniform powder bed cooling and temperature regulation. In addition, concomitant application of temperature and humidity in this process provides greater degradation stress to the sensitive APIs.

26.4 Foam granulation

The mode of binder fluid addition during wet granulation is known to impact the process and product outcomes. At small scales of operation, when the absolute quantity of binder fluid to be added is fairly small, the binder fluid is typically added by pouring or pumping using a peristaltic pump. As the process is scaled up, the use of nozzles or spray for the addition of binder fluid is implemented to improve binder distribution and material variability.

Foam granulation technology takes the binder fluid atomization concept to the next level by adding the binder as a foam (Keary & Sheskey, 2004). It utilizes a "foam generator" to introduce the binder liquid. Adding the binder fluid as a foam tremendously increases the surface area per unit volume of the spray solution, increasing the "spread to soak ratio"—the proportion of binder that spreads on the surface of particles versus getting absorbed or soaked into the materials (Shanmugam, 2015). This can result in a narrower granule size distribution and could potentially require less binder and/or granulating fluid than traditional wet granulation while improving the consistency of binder distribution. This process was demonstrated to produce granules and tablets with desired properties at different scales of operation (Sheskey et al., 2007).

26.4.1 Comparison with wet granulation

Cantor et al. (2009) compared the physical and mechanical properties of high drug load formulations prepared by foam granulation to the results produced by traditional wet granulation using 80% drug load formulations of acetaminophen, aspirin, and metformin. The authors measured particle size, surface area, porosity, and Hiestand indices of granulations with granulation liquid delivery using either the spray or foam method. Interestingly, the binder delivery method affected the outcome of granulation differently for different drugs. Thus, while foam granulation showed better granule plasticity for acetaminophen, conventional wet granulation had better plasticity for metformin. The authors concluded that the selection of the granulation process should be based on intrinsic mechanical properties of the drug. Recent studies in foam granulation include studying the effect of binder grade and foam quality in a high drug load formulation. Using hydroxypropyl cellulose as a binder, Koo et al. (2024) found that the granules prepared with foamed binders resulted in higher compact strengths, even though they were not significantly different in compression behavior.

26.4.2 Physics of liquid penetration

Binder distribution mechanism during wet granulation can be very different for foam versus droplet addition of the binder fluid. The phenomenon of nucleation during initial binder contact and distribution in the powder bed can help delineate differences in granulation kinetics and process outcomes between foam and spray granulation. The initial phenomenon of binder contact with a loosely packed powder bed can be a function of several parameters including binder polymer type, mode of addition, powder bed material type, and powder bed porosity. Tan et al. (2009) compared the kinetics of single drop penetration into a loosely packed powder bed, comparing foam and drop methods of addition, HPC and HPMC as the wet granulation binder, and glass ballotini and lactose as loose powder bed. To compare the foam and drop addition methods, the authors defined and studied the specific penetration time (penetration time per unit of binder mass with penetration time defined as the time required for the complete collapse of the foam or the complete disappearance of the drop from the powder surface) and nucleation ratio (the ratio of nuclei granule mass to liquid binder mass). The authors further studied foam quality (by dispensing the foam into a measuring cylinder and allowing it to drain under gravity in time) and liquid fraction (the volume of liquid obtained after the foam has collapsed into liquid in a measuring cylinder). Granule nuclei were collected using a sieve that was larger than the size of the primary particles but smaller than the expected nuclei size. The authors observed that, at a specific penetration time, a much greater mass of binder was absorbed by foam than drop delivery of the binder solution; specific penetration time increased with the binder concentration, as the increasing fluid viscosity slows the velocity of liquid penetration; and that foam granulation was more efficient at nucleation, using less binder liquid to generate the same number of nuclei in the powder bed. Interestingly, no large difference in the nucleation ratios obtained for 3% versus 6% HPC concentration in the foam (Tan et al., 2009).

Foam granulation kinetics and mechanisms are also substrate dependent. For example, Koo et al. (2012) studied foam drainage and collapse times for HPC and HPMC foams using anhydrous lactose or stearic acid as model water-soluble and insoluble substrates, respectively. The authors observed that both foam drainage and collapse times were lower with lactose as a substrate compared to hydrophobic stearic acid. The authors hypothesized that foam-substrate interactions can impact granule formation. The authors speculated that binder distribution in substrates that collapse the foam upon contacting the substrate would predominantly be through mechanical agitation, similar to spray granulation. On the other hand, hydrophobic substrates can maintain the foam long enough to allow mixing of the substrate and the foam, which would make foam properties predominant determinants of the process outcomes.

26.4.3 Liquid marble formation

An interesting advancement of foamed granulation solution addition was shown in the agglomeration of very small particle size hydrophobic powders by solids spreading and nucleating onto a foam bubble, leading to the formation of a powder shell, a hollow granule structure that possesses advantageous processing properties, around a liquid droplet, called a "liquid marble." Hapgood et al. reported the formation of powder shells by placing single fluid droplets onto loosely packed powder beds of hydrophobic powders (Hapgood & Khanmohammadi, 2009). The authors proposed this as a solid spreading nucleation mechanism for the granulation of very small particle size hydrophobic powders. In a subsequent study, Eshtiaghi and Hapgood (2012) studied liquid marble formation with six different mixtures of water and glycerol by releasing droplets onto loosely packed powder beds of fine particle size and hydrophobic materials—colloidal silicon dioxide, glass ballotini, and Teflon. Through this study, the authors revised the conceptual framework of physical flow mechanism involved in glass marble formation and improved the mechanistic understanding of glass marble formation of hydrophobic powders.

The liquid marble and hollow granule formation work, however, is still in very early stages of development, with better understanding and more work required to assess its applicability and scalability.

26.4.4 Advantages and opportunities

Foam granulation offers a useful extension of increasing the surface area of the binder solution to improve solution spreadability and make uniform granules with lower quantities of the granulating fluid and/or polymer. Foam granulation can also be used to achieve a more uniform distribution of small quantities of foam soluble drug. This technology, however, adds another operation to the process—foam generation and delivery—which must be controlled and whose reproducibility and robustness must be established. For example, initial binder distribution during the nucleation stage is a function not only of foam quality but also of the process parameters in a high shear granulator equipment. These parameters interact to potentially lead to distinct wetting and nucleation regimes. In a study of impeller tip speed and foam quality using an HPMC solution foam for the granulation of lactose, Tan and Hapgood (2011) concluded that a low-quality foam tends to induce localized wetting and nucleation under bumping (low impeller tip speed) flow with wetting and nucleation being "foam drainage" controlled, while foam is dispersed by the motion of the agitated powder under roping flow with wetting and nucleation being "mechanical dispersion" controlled.

In addition, operationalizing this technology can be a bit challenging, at least in the initial phases, as multiple critical unit operations must be carried out in a small window of time. In addition, rapid initiation of the particle agglomeration process and shorter time to granulation end point are likely with foam granulation (Keary & Sheskey, 2004), which actually may not be a good thing for process control of high shear wet granulation— already a very short duration process.

26.5 Conclusion and future trends

The understanding and improvement of granulation processes continue to evolve at the interface of material science, data modeling, and process design. This includes continuous improvements and advancements in the currently established wet granulation methodologies (such as low- and high-shear wet granulation, fluid-bed granulation, and twin-screw granulation) toward better process understanding and control using in-line analytical technologies (e.g., process analytical technologies), modeling and simulation, and equipment/process design. The modeling and simulation efforts have started to incorporate the principles of machine learning and automated data processing. In addition, process design modifications that implement concurrent heat and mass flow dynamics and/or hybrid application of principles from different granulation technologies continue to be pursued together. These novel technologies present a new set of opportunities in establishing process robustness and control strategies. The implementation of novel technologies comes with additional regulatory and infrastructure burdens that must be justified by the need for the technology. The traditional technologies are expected to continue to occupy the mainstay of pharmaceutical wet granulation, with the application of advanced and hybrid technologies reserved for special cases where the desired product characteristics cannot be obtained with traditional technologies. A better understanding of the formulation and the process through the application of modeling and data analytics is expected to continue to further improve how the wet granulation processes are developed for the new chemical entities.

References

Badawy, S. I. F., Narang, A. S., Lamarche, K., Subramanian, G., & Varia, S. A. (2012). Mechanistic basis for the effects of process parameters on quality attributes in high shear wet granulation. *International Journal of Pharmaceutics, 439*(1–2), 324–333. https://doi.org/10.1016/j.ijpharm.2012.09.011.

Campbell, G. A., Clancy, D. J., Zhang, J. X., Gupta, M. K., & Choon, K. O. (2011). Closing the gap in series scale up of high shear wet granulation process using impeller power and blade design. *Powder Technology, 205*(1–3), 184–192. https://doi.org/10.1016/j.powtec.2010.09.009.

Cantor, S. L., Kothari, S., & Koo, O. M. Y. (2009). Evaluation of the physical and mechanical properties of high drug load formulations: Wet granulation vs. novel foam granulation. *Powder Technology, 195*(1), 15–24. https://doi.org/10.1016/j.powtec.2009.05.003.

Cavallari, C., Abertini, B., González-Rodríguez, M. L., & Rodriguez, L. (2002). Improved dissolution behaviour of steam-granulated piroxicam. *European Journal of Pharmaceutics and Biopharmaceutics, 54*(1), 65–73. https://doi.org/10.1016/S0939-6411(02)00021-8.

Chaudhury, A., Dana, B., Pandey, P., Wu, H., & Ramachandran, R. (2014). Population balance model development, validation, and prediction of CQAs of a high-shear wet granulation process: Towards QbD in drug product pharmaceutical manufacturing. *Journal of Pharmaceutical Innovation, 9*(1), 53–64. https://doi.org/10.1007/s12247-014-9172-7.

Christensen, L. H., Johansen, H. E., & Schaefer, T. (1994). Moisture-activated dry granulation in a high shear mixer. *Drug Development and Industrial Pharmacy, 20*(14), 2195–2213. 10.3109/03639049409050233.

de Jong, E. J. (1969). The preparation of microgranulates, an improved tabletting technique. *Pharmaceutisch Weekblad, 104*(23), 469–474.

Eremin, V. A., Blynskaya, E. V., Alekseev, K. V., Tishkov, S. V., & Minaev, S. V. (2024). Comparative analysis of innovative wet granulation technologies for pharmaceutical production (A review). *Pharmaceutical Chemistry Journal, 58*(9), 1438–1447. doi:10.1007/s11094-025-03292-5.

Eshtiaghi, N., & Hapgood, K. P. (2012). A quantitative framework for the formation of liquid marbles and hollow granules from hydrophobic powders. *Powder Technology, 223*, 65–76. https://doi.org/10.1016/j.powtec.2011.05.007.

Gantt, J. A., & Gatzke, E. P. (2005). High-shear granulation modeling using a discrete element simulation approach. *Powder Technology, 156*(2–3), 195 212. https://doi.org/10.1016/j.powtec.2005.04.012.

Hapgood, K. P., & Khanmohammadi, B. (2009). Granulation of hydrophobic powders. *Powder Technology, 189*(2), 253–262. https://doi.org/10.1016/j.powtec.2008.04.033.

Huang, J., Kaul, G., Utz, J., Hernandez, P., Wong, V., Bradley, D., Nagi, A., & O'Grady, D. (2010). A PAT approach to improve process understanding of high shear wet granulation through in-line particle measurement using FBRM C35. *Journal of Pharmaceutical Sciences, 99*(7), 3205–3212. https://doi.org/10.1002/jps.22089.

Iskandarani, B., Shiromani, P. K., & Clair, J. H. (2001). Scale-up feasibility in high-shear mixers: Determination through statistical procedures. *Drug Development and Industrial Pharmacy, 27*(7), 651–657.

Iveson, S. M. (2002). Limitations of one-dimensional population balance models of wet granulation processes. *Powder Technology, 124*(3), 219–229. https://doi.org/10.1016/S0032-5910(02)00026-8.

Iveson, S. M., Litster, J. D., Hapgood, K., & Ennis, B. J. (2001). Nucleation, growth and breakage phenomena in agitated wet granulation processes: A review. *Powder Technology, 117*(1–2), 3–39. https://doi.org/10.1016/S0032-5910(01)00313-8.

Keary, C. M., & Sheskey, P. J. (2004). Preliminary report of the discovery of a new pharmaceutical granulation process using foamed aqueous binders. *Drug Development and Industrial Pharmacy, 30*(8), 831–845. https://doi.org/10.1081/DDC-200030504.

Koo, O., Patel, C., & Nikfar, F. (2024). Effect of hydroxy propyl cellulose grade and foam quality on foam granulation of a high drug load formulation. *International Journal of Pharmaceutics, 657*, 124171. https://doi.org/10.1016/j.ijpharm.2024.124171.

Koo, O. M. Y., Ji, J., & Li, J. (2012). Effect of powder substrate on foaml drainage and collapse: Implications to foam granulation. *Journal of Pharmaceutical Sciences, 101*(4), 1385–1390. https://doi.org/10.1002/jps.23053.

Landín, M., Rowe, R. C., & York, P. (1995). Characterization of wet powder masses with a mixer torque rheometer. 3. Nonlinear effects of shaft speed and sample weight. *Journal of Pharmaceutical Sciences, 84*(5), 557–560. https://doi.org/10.1002/jps.2600840508.

Li, J., Tao, L., Buckley, D., Tao, J., Gao, J., & Hubert, M. (2011b). The effect of the physical state of binders on high-shear wet granulation and granule properties: A mechanistic approach towards under- standing high-shear wet granulation process. Part I. Physical characterization of binders. *Journal of Pharmaceutical Sciences, 100*(1), 164–173. https://doi.org/10.1002/jps.22260.

Li, J., Tao, L., Buckley, D., Tao, J., Gao, J., & Hubert, M. (2013). The effect of the physical state of binders on high-shear wet granulation and granule properties: A mechanistic approach to understand the high- shear wet granulation process. Part IV. The impact of rheological state and tip-speeds. *Journal of Pharmaceutical Sciences*. https://doi.org/10.1002/jps.23750.

Li, J., Tao, L., Dali, M., Buckley, D., Gao, J., & Hubert, M. (2011a). The effect of the physical states of binders on high-shear wet granulation and granule properties: A mechanistic approach toward understanding high-shear wet granulation process. Part II. Granulation and granule properties. *Journal of Pharmaceutical Sciences, 100*(1), 294–310. https://doi.org/10.1002/jps.22261.

Louati, H., Oulahna, D., & de Ryck, A. (2017). Effect of the particle size and the liquid content on the shear behaviour of wet granular material. *Powder Technology, 315*, 398–409. https://doi.org/10.1016/j.powtec.2017.04.030.

Moravkar, K. K., Ali, T. M., Pawar, J. N., & Amin, P. D. (2017). Application of moisture activated dry granulation (MADG) process to develop high dose immediate release (IR) formulations. *Advanced Powder Technology, 28*(4), 1270–1280. https://doi.org/10.1016/j.apt.2017.02.015.

Nakamura, H., Fujii, H., & Watano, S. (2013). Scale-up of high shear mixer-granulator based on discrete element analysis. *Powder Technology, 236*, 149–156. https://doi.org/10.1016/j.powtec.2012.03.009.

Narang, A.S., Badawy, S.I.F., Subramanian, G.A., Lamarche, K.R., Bindra, D.S., & Rao, V.M. (2012a). High drug load tablet formulation of brivanib alaninate. USA: Bristol-Myers Squibb Company.

Narang, A. S., Desai, D., & Badawy, S. (2012b). Impact of excipient interactions on solid dosage form stability. *Pharmaceutical Research, 29*(10), 2660–2683. https://doi.org/10.1007/s11095-012-0782-9.

Narang, A. S., Sheverev, V. A., Stepaniuk, V., Badawy, S., Stevens, T., Macias, K., Wolf, A., Pandey, P., Bindra, D., & Varia, S. (2015). Real-time assessment of granule densification in high shear wet granulation and application to scale-up of a placebo and a brivanib alaninate formulation. *Journal of Pharmaceutical Sciences, 104*(3), 1019–1034. https://doi.org/10.1002/jps.24233.

Narang, A. S., Stevens, T., Macias, K., Paruchuri, S., Gao, Z., & Badawy, S. (2017). Application of in-line focused beam reflectance measurement to brivanib alaninate wet granulation process to enable scale-up and attribute-based monitoring and control strategies. *Journal of Pharmaceutical Sciences, 106*(1), 224–233. https://doi.org/10.1016/j.xphs.2016.08.025.

Nyberg, B., Carlstroem, E., & Carlsson, R. (1994). Freeze granulation of liquid phase sintered silicon carbide. *Ceramic Transaction, 42*, 107–114.

Origoni, M. X., Nardi-Ricart, A., Sune-Pou, M., Perez-Lozano, P., Romero-obon, M., Sune-Negre, J. M., Ochoa-Andrade, A. T., & Garcia-Montoya, E. (2024). Exploration of moisture activated dry granulation for the development of gastroretentive tablets aided by SeDeM diagram. *European Journal of Pharmaceutics and Biopharmaceutics, 203*, 114456–114466. https://doi.org/10.1016/j.ejpb.2024.114456.

Pandey, P., & Badawy, S. (2016). A quality by design approach to scale-up of high-shear wet granulation process. *Drug Development and Industrial Pharmacy, 42*(2), 175–189. https://doi.org/10.3109/03639045.2015.1100199.

Pandey, P., Tao, J., Chaudhury, A., Ramachandran, R., Gao, J. Z., & Bindra, D. S. (2013). A combined experimental and modeling approach to study the effects of high-shear wet granulation process parameters on granule characteristics. *Pharmaceutical Development and Technology, 18*(1), 210–224. doi:10.3109/10837450.2012.700933.

Passerini, N., Albertini, B., Perissutti, B., & Rodriguez, L. (2006). Evaluation of melt granulation and ultrasonic spray congealing as techniques to enhance the dissolution of praziquantel. *International Journal of Pharmaceutics, 318*(1–2), 92–102. https://doi.org/10.1016/j.ijpharm.2006.03.028.

Passerini, N., Calogerà, G., Albertini, B., & Rodriguez, L. (2010). Melt granulation of pharmaceutical powders: A comparison of high-shear mixer and fluidised bed processes. *International Journal of Pharmaceutics, 391*(1–2), 177–186. https://doi.org/10.1016/j.ijpharm.2010.03.013.

Rautenberg, A., & Lamprecht, A. (2022). Spray-freeze-dried lyospheres: Solid content and the impact on flowability and mechanical stability. *Powder Technology, 411*, 117905. https://doi.org/10.1016/j.powtec.2022.117905.

Ring, D. T., Oliveira, J. C. O., & Crean, A. (2011). Evaluation of the influence of granulation processing parameters on the granule properties and dissolution characteristics of a modified release drug. *Advanced Powder Technology, 22*(2), 245–252. https://doi.org/10.1016/j.apt.2011.01.006.

Sakamoto, Y., Nakamura, H., & Watano, S. (2015). Development of advanced freeze granulation by supercritical fluid and its application to dry powder inhalation. *Journal of the Society of Powder Technology Japan, 52*(6), 330–336.

Sartzi, M. I., Drettas, D., Stramarkou, M., & Krokida, M. (2024). A comprehensive review of the latest trends in spray freeze drying and comparative insights with conventional technologies. *Pharmaceutics, 16*(12), 1533. doi:10.3390/pharmaceutics16121533.

Schaefer, T. (1996). Melt pelletization in a high shear mixer. X. Agglomeration of binary mixtures. *International Journal of Pharmaceutics, 139*(1–2). https://doi.org/10.1016/0378-5173(96).

Shanmugam, S. (2015). Granulation techniques and technologies: Recent progresses. *BioImpacts, 5*(1), 55–63. https://doi.org/10.15171/bi.2015.04.

Sheskey, P., Keary, C., Clark, D., & Balwinski, K. (2007). Scale-up trials of foam-granulation technology—Higher shear. *Pharmaceutical Technology Europe, 19*(9) 37–38. 41–42, 45–46.

Sirois, P. J., & Craig, G. D. (2000). Scaleup of a high-shear granulation process using a normalized impeller work parameter. *Pharmaceutical Development and Technology, 5*(3), 365–374. https://doi.org/10.1081/PDT-100100552.

Sonoda, R., Hara, Y., Iwasaki, T., & Watano, S. (2009). Improvement of dissolution property of poorly water-soluble drug by supercritical freeze granulation. *Chemical & Pharmaceutical Bulletin (Tokyo), 57*(10), 1040–1044. https://doi.org/10.1248/cpb.57.1040.

Takasaki, H., Yonemochi, E., Messerschmid, R., Ito, M., Wada, K., & Terada, K. (2013). Importance of excipient wettability on tablet characteristics prepared by moisture activated dry granulation (MADG). *International Journal of Pharmaceutics, 456*(1), 58–64. https://doi.org/10.1016/j.ijpharm.2013.08.027.

Tan, M. X. L., & Hapgood, K. P. (2011). Foam granulation: Binder dispersion and nucleation in mixer-granulators. *Chemical Engineering Research and Design, 89*(5), 526–536. https://doi.org/10.1016/j.cherd.2010.07.001.

Tan, M. X. L., Wong, L. S., Lum, K. H., & Hapgood, K. P. (2009). Foam and drop penetration kinetics into loosely packed powder beds. *Chemical Engineering Science, 64*(12), 2826–2836. https://doi.org/10.1016/j.ces.2009.03.008.

Ullah, I., Corrao, R. G., Wiley, G. J., & Lipper, R. A. (1987). Moisture-activated dry granulation: A general process. *Pharmaceutical Technology Europe, 11*(9), 4–54.

Vemavarapu, C., Surapaneni, M., Hussain, M., & Badawy, S. (2009). Role of drug substance material properties in the processibility and performance of a wet granulated product. *International Journal of Pharmaceutics, 374*(1–2), 96–105. https://doi.org/10.1016/j.ijpharm.2009.03.014.

Vialpando, M., Albertini, B., Passerini, N., Bergers, D., Rombaut, P., Martens, J. A., & Mooter, G. V. D. (2013). Agglomeration of mesoporous silica by melt and steam granulation: Part I: A comparison between disordered and ordered mesoporous silica. *Journal of Pharmaceutical Sciences, 102*(11), 3966–3977. doi:10.1002/jps.23700.

Voinovich, D., Moneghini, M., Perissutti, B., Filipovic-Grcic, J., & Grabnar, I. (2000). Preparation high-shear mixer of sustained-release pellets by melt pelletisation. *International Journal of Pharmaceutics, 203*(1–2), 235–244.

Index

Page numbers followed by "*f*" and "*t*" indicate, figures and tables respectively.

A

Abrasion, 822
Acetaminophen (APAP), 229
    model system, 336*t*
    wet granulated, 342*f*
Active pharmaceutical ingredient, 450
Active pharmaceutical ingredients (APIs), 103, 277, 313, 347, 379, 413, 601, 801, 805, 867
    Brivanib formulation
        experimental design for, 103, 103*t*
        regression analysis results for, 107*t*
    final product critical quality attributes
        blend flow, 118
        granule compactibility, 116
        tablet dissolution, 114, 115*f*, 116*f*, 117*f*
    granule and tablet attributes, characterization of, 104
    Pexacerfont formulation
        experimental design for, 103, 105*t*
        regression analysis results for, 108*t*
    process parameters
        impeller speed, 110
        water addition rate, 113
        water amount, 110
        wet massing time, 112
    Razaxabn formulation
        experimental design for, 103, 104*t*
        regression analysis results for, 109*t*
    solid form, 422
Adhesion tension, 332
Aerosolization, 567*f*, 567
Agglomeration, 207, 224, 750
    pharmaceutical powders, 313
    process, 837, 921
    surface tension, 348
    techniques, 3
Amperage, 549
Amylopectin, 290
Amylose, 290
β-anhydrous lactose, 181
Anhydrous lactose, 397*f*
Artificial neural networks, 920,
Aspirin-PVP granulations, 434*t*
Attribute-based scale-up strategy, 612
    acoustic monitoring, 614
    end-point measurements, 612
    FBRM probe, 613
    fill ratio, 615
    granule attributes, 612
    granule densification, 615
    in-line NIR spectroscopy, 614
    in-process granule properties, 612
    Newton power number, 615
    normalization technique, 615
    particle size distribution, 609
    PAT techniques, 614
    process-induced transformations, 613
    pseudo Reynolds number, 615
    wet mass rheological/consistency changes, 613
Avalanching effect, 561
Average particle size (APS), 919

B

Ball milling, 194, 201
Barrel temperature, 817, 852
Base models, 755
Batch melt-granulation, 849
Batch-wise granulation processes, 446
Binders, 285, 286, 418, 843
    acetaminophen tablets, 342*f*
    activation, 371
    adding process, 762
    addition method, 454
    addition-related parameters, 875
    attributes, 457
        addition effects, 359
        binder selection, 351
        critical material attributes, 347
        functional excipients, 347
        granulation outcomes, 348
        granule breakage, 350
        granule formation and growth, 349
        impaction forces, 357
        impact of, 348
        particle velocities, simulation of, 357
        quality attributes of interest, 348
    characteristics, 348, 429*t*
    commercial binder grades, 317*t*
    compatibility considerations, 322
    concentration effect, 484
    considerations for, 322
    copovidone, 319
    distribution mechanism, 919
    efficiency, 321

equilibrium moisture isotherms, 320f
ethyl cellulose, 318
factors affecting, 285, 286, 287
form, 807
granulation mechanisms, 285, 289f
on granulation outcomes, 348
granule strength, 325
gum acacia, 321
hydration, 371
hydrophilic, 843
hydrophobic, 843
hydroxypropyl methylcellulose, 293
hydroxypropylcellulose, 315
hypromellose, 318
level, 847
  threshold, 847
mechanical properties of, 337
methyl cellulose, 315
microcrystalline cellulose, 291
molecular weight effect, 354
particle size effect, 353
performance of, 326
physical-chemical properties, 314
polyvinylpyrrolidone, 293
povidone, 319
properties, 314, 348
regulatory acceptance, 344
selection, 351
sodium carboxymethyl cellulose, 319
solution, 432
solvent, 334
spreading coefficients, 433t
stability, 323
starch, 290
Biological engineering, 756
Blade-particle bed stress, 735, 736f
Blend flow, 118
BMS-540215 (brivanib)
concentration vs. time profile, 647
hydrolysis reaction, 647
level control, 662
rate of formation, 647, 648f
relative humidity, 647, 648f
time profile simulation, 649
Breakage, of wet granules, 210
Brivanib formulation, 500f, 644t
composition, 640t, 662
experimental design for, 639, 641t
manufacturing process, flow diagram, 638f, 654

C

Capillary state granules, 209f, 209

Capillary viscous number, 350
Carbamazepine, 423
Cartesian coordinate system, 824
CBCS parameter
  highest values, 894t
  lowest values, 894t
Chord length distribution (CLD), 353f, 354f, 354, 355f
  comparison of, 505f
  end-point detection algorithm, 501f, 502
  focused beam reflectance measurement
    binder concentration effect, 484
    impeller speed effect, inline, 480
    reproducibility, 491 492 493f–491 492 493f, 490f, 494
    water concentration effect, 489
    water effect, 482
  particle size distributions, 471
  root mean square error (RMSE), 502
  sensitivity of, 484 485 486f–484 485 486f, 487 488 489f–487 488 489f
Circumferential speed, 827
Classical Washburn capillary wetting approach, 329
Coalescence models, 20, 22f
Collisional stress tensor, 731
Collision energy, 740, 741f, 741, 742f, 742
Compactability ratio, 383, 393f, 408
Compaction
  microcrystalline cellulose, 198, 199f, 200t
  properties, 406
Compressibility index, 219, 234
Compression models, 891, 892
Computational fluid dynamics (CFD), 218, 621, 747, 758, 912
  discrete element method, 915
  modeling, 759
  simulation, 756
Computational modeling, 759, 774
Computer aided design (CAD) software, 737
Confirmation batch plan, 246, 248t
Conservation equations, 756
Consolidation, 11
Consolidation rate process, 753
Constant drawing area (CDA) model, 7
Constant porosity, 395
Constant volume Monte Carlo (CVMC), 771
Contact angle, 6, 405
Contact forces, 737
Continuous foam granulation, 803
Continuous melt granulation, 851
Continuous stirred tank reactor (CSTR), 917,
Continuous wet granulation, 830,, 417
Continuum capillary force (CCF) model, 912
Controlled release formulations, 460
Control strategy, 655, 673

at-line release tests, 656, 662, 663f
finished product release tests, 656
in-process attributes control, 656, 659
 film coating, 660, 661f
 fluid bed drying, 656, 659, 660f
 tablet compression, 656, 660
input (raw) material controls, 656, 657, 658f
off-line release tests, 656, 662
process parameters design space
 brivanib alaninate in-process material *vs.* tablet attributes, 656, 658f, 664
 critical wet granulation parameters, 644, 644t, 656
 DOE study, 644
 multivariate process design space, 644t, 657
 wet milling, dry milling, and lubrication parameters, 644
raw materials, 633, 656t
Conventional batch high-shear granulation, 822
Cook's D influence, 236
Cracks, 201
Critical material attributes (CMAs), 228, 269, 344, 558, 669, 672
 granules
  binders, 431
  diluents, 428
  disintegrants, 430
  moisture content, 425
  particle shape and morphology, 425
  particle size, 424
  porosity, 424, 432
  powder density, 428
  solid form, 422
  solubility, 426
  strength, 425
  surface area, 425
  wettability (contact angle), 427
 QbD methodology, 606t, 608f
 wet granulation
  basics, 414
  rate processes, 415
  types, 415
Critical MPs (CMPs), 347
Critical process parameters (CPPs), 344, 420, 558, 669, 672, 693
 correlation between granule porosty and compactability, 607, 608f
 impeller speed, 610f, 617
 potential CPPs, 606t, 607
 scaled-up design space, 609
 spray method, 609
 water addition mode, 609
 of wet granulation, 420
 wet massing time, 619

Critical quality attributes (CQAs), 211, 228, 229, 231, 269, 344, 347, 420, 471, 557, 669, 672, 673, 805, 873
 QbD, 605
Croscarmellose sodium (CCS), 302, 304, 418
Crospovidone (CPVP), 303, 418
Cross-linked poly(acrylic) acid, 191
Crystalline excipients, 423

D

Data-driven models, 913
Data fusion, 873
Data sources, 874
Definitive screening design (DSD), 240
DEM-CFD method
 fluidized bed granulator types, 713
 gas-solid fluidized beds, 712
 Newton's third law, 712
 rotating fluidized bed coater
  coating mass distribution, 712, 714f
  coefficients of variation (CV), temporal change, 712, 715f
  conventional fluidized bed coater, 712, 715f
  schematic illustration, 712, 713f
Design of experiment (DoE), 225, 420, 471, 825, 912
 pharmaceutical QbD
  HS/LS granulation application, 692
  ICH Q8 guidance, 691
  optimization DOE, 692
  product and process development, 691
  screening DOE, 692
DFF sensor response, 522
Dibasic calcium phosphate (DCP), 808
Dicalcium phosphate (DCP), 285
Digital elevation model (DEM), 917
Diluents, 278, 279t, 418, 428
 dicalcium phosphate, 285
 lactose, 278
 mannitol, 283
 microcrystalline cellulose, 280
Dimensionless spray flux, 9
Diosna-boots flow sensor, 512
Diosna high-shear granulator, 357
Direct compression (DC), 670
Discrete element method (DEM), 218, 351, 358f, 618, 621, 747
 calculation algorithm, 707, 708f
 coupling with particle adhesion/coalescence model, 716
 coupling with population balance modeling, 718
 direct numerical simulation, particle transformation, 717f, 719
 granular flows, 725
  Voigt model, 725
 high-shear mixer granulator

agitation torque, 710, 711f
centrifugal tumbling granulator, 709
force acting, 709, 710f
granular flow, 709
impeller geometry, 710
particle-to-particle contact force, 706, 707f
particle-to-wall contact force, 706, 707f
rotational motion, 706
scale-up of high-shear granulation, 726
    advantages, 726
    blade-particle bed stress, 735
    dynamic properties, 731, 731t
    effects of D/d on internal properties in mixers, 733, 734f
    geometric properties, 727, 727t
    kinematic properties, 728, 728t
    relative swept volume per impeller rotation, 733, 735f
simulation, 754, 760, 775
soft sphere model, 706
Disintegrants, 418, 459
    croscarmellose sodium, 302, 304
    crospovidone, 303
    factors affecting, 304
    mechanisms, 300f
    sodium starch glycolate, 301
    starch, 301
    tablet disintegration, 299
Distributive comb mixing element, 141
DMP 754 capsules
    amidine hydrolysis, 167, 168, 169f
    degradants
        concentrations, 173, 174t
        structure, 166f
    encapsulation and compression effects, 167f, 167
    ester hydrolysis, 165, 167, 169f, 173f
    formulation composition, 168, 168t
    lactose blends, 165, 166f
    manufacturing process effects, 171, 172f
    moisture uptake behavior, 171
    pH modifiers, 169, 171f
    physical properties, 173, 174t
    salt form effects, 173
    sodium distribution, 168, 170f
    stability behavior, 165
DMP 755 capsules
    degradant concentrations, 173, 174t
    manufacturing process effects, 174
    physical properties, 173, 174t
    salt form effects, 173
    stability, 173
Drag force
    flow probe, 521
    flow sensor, 513, 548

Droplet-control regime, 753
Droplet state granules, 209f, 209
Drop penetration time, 7
Drug
    binder interactions, 848
    product, 347, 471
    solubility, 849
    stability, 848
Drug substances (DS), 229
    compaction properties, 406
    contact angle, 405
    flow properties, 408
    input material properties, 379
    input variables, 678
    particle size effect
        and granule growth, 401
        and granule porosity, 400
        and wet granule mechanical properties, 394
    physical, chemical, and biological properties, 678, 679t
    solubility, 405
    surface area, 408
    test materials, 381
    water uptake, 408
    wet-granulated product attributes, statistical model, 383
Dry binder addition (DBA), 762
Dry binder powders, 380
Dry granulation (DG), 207, 267, 413, 801, 872
Dry microcrystalline cellulose, 480
Dry mixing models, 371
Dry powder inhaler (DPI), 567
Dual-beam systems, 39
Dye tracer study, 854f
Dynamic similarity, 735

E

Efavirenz, 333t, 337f
End point control, 912
End point determination, 250, 501f, 502
Equilibrium moisture content, 319
Ethanol-water solvent system, 335
Eulerian approach, 754
Eulerian-based continuous model, 769
European Pharmacopeia (pH.Eur.), 315, 348, 418
Excipients, 180, 191, 348
    active substances, role of, 277
    binders, 285, 286
        hydroxylpropyl methylcellulose, 293
        microcrystalline cellulose, 291
        polyvinylpyrrolidone, 293
        starch, 290
    diluents, 278, 279t
        dicalcium phosphate, 285

Index

lactose, 278
mannitol, 283
microcrystalline cellulose, 280
disintegrants
croscarmellose sodium, 302, 304
crospovidone, 303
sodium starch glycolate, 301
starch, 301
functionality, 275, 276
lubricants
magnesium stearate, 298
sodium stearyl fumarate, 299
stearic acid, 297
in silico simulation, 357
variability, 273, 274f
in wet granulation, 417
Experimental design, 887
approaches, 912
arrangements, 888
factors, 879t
limits, 879t
Extruder screw configuration, 851

F

Failure mode and effect analysis (FMEA), 228
Fillers, 451
binder properties, 451
combinations, 808
Fish-bone diagram, 233
Flodex numbers, 383
Flow
function coefficient, 806
properties, 408
ratio, 394f
Fluid bed granulation (FBG), 802
Fluid bed processor, 850
Fluid-bed wet granulation process, 416, 670, 671f
advantages, 216
common input variables, 676, 676t
discrete-element-method, 218
dissolution studies, 220, 220t
fluid-bed granulator, 216, 217f
granules, physical properties of, 220, 220t
lubricated blend, 215, 215t
microcrystalline cellulose, 218
process parameters
amount of granulation fluid (w/w), 684
batch size and volume fill ratio, 687, 687t
equipment type, 686
inlet air condition, 688
pre-heat temperature and pre-heat time, 688
product temperature and exhaust temperature, 689

with report rate and importance ratio, 686f
spray condition, 689
scaleup of, 255
unit operation profile, 675f
Fluidized bed
granulation, 774, 775
granulators, 758
Foam granulation, 919, 921
kinetics, 920
technology, 919
Focused beam reflectance measurement (FBRM), 353f, 354f, 355f, 471, 472
chord length distribution
binder concentration effect, 484
impeller speed effect, inline, 480
reproducibility, 491 492 493f–491 492 493f, 490f, 494
water concentration effect, 489
water effect, 482
C35 probe, 472, 473f, 476f, 477, 496, 502f
dry microcrystalline cellulose, 480
end point indicator algorithm, 501f, 502
granulations
fines, reduction in, 495
particle size, increase in, 483
instrumentation, 475f
probe setup
operation, 473
orientation, 472
position, 472
scaleup, 504
water concentration effect, 489
Food and Drug Administration (FDA), 344, 857
Food Chemicals Codex (FCC), 315, 344
Food powders, 918
Force pulse magnitude, 525
Force pulse magnitude histograms, 525
Formulation properties
estimation, 874
prediction, 874f
Freeze drying, 916
Freeze granulation, 916, 917
Frictional stress models, 770
Froude number, 254, 254t, 615
calculation, 875
Functionality-related characteristics (FRCs), 348, 418
Funicular state granules, 209f, 209

G

Gas-particle interaction, 759
Generally recognized as safe (GRAS) ingredients, 344
Geometric similarity, 727
Geometry

planar, 753
spherical, 753
Gidaspow drag model, 758
Glass bead granule properties, 326t
Glass transition temperature, 320
Goniometer method, 380
Good manufacturing practices (GMP), 344
Granular heat transfer simulation, 754
Granular material, 754
Granular temperature, 729
Granulation, 207, 208f, 267, 372f, 747, 801, 803, 819, 837
    advances in, 270
    attrition and breakage, 210
    binder attributes, impacts, 348
    consolidation and growth, 209
    end-point, 540
    equipment, 922
    focused beam reflectance measurement
        fines, reduction in, 495
        particle size, increase in, 483
    growth kinetics, 497 498 499f–497 498 499f
    mass, 916
    mechanism, 820, 912
    process, 911, 917
    techniques, 748
    wetting and nucleation, 208
Granulators
    and vendors, 225, 226
    volume proportions, 359 360 361 362f–359 360 361 362f
Granule friability test, 42
Granule growth, 379, 383, 391f, 393, 401, 403f, 841
    anhydrous lactose, 403f
    binder viscosity, 324
    drug substance particle size effect, 400
    high-shear wet granulation, 383
    kinetics, 371
Granules
    attrition, 841
    breakage, 136, 350
    consolidation, 13, 373f, 375
        binder hydration, 371
        consolidation rate studies, 372
        fitting parameters, 374t
    critical material attributes
        binders, 431
        diluents, 428
        disintegrants, 430
        growth ratio, 383
        moisture content, 425
        particle shape and morphology, 425
        particle size, 424
        porosity, 424, 432
        powder density, 428
        solid form, 422
        solubility, 426
        strength, 425
        surface area, 425
        wettability (contact angle), 427
    deformation, 14
    densification, 531
    friability tests, 335, 340f
    granulation liquid in nucleation, 43
    growth behavior, 21
    microstructure, 43, 72
    nucleation, 918
    porosity, 372f, 921
    size distribution, 915,, 805
    strength, 72, 841
Granurex, 223, 224f
Growth regime, 750
Growth regime map, 17

H

Hand-squeeze test, 417
Hard-sphere methods, 755
Hausner ratio, 234, 236f
Heckel equation, 913
Hertz-Mindlin contact model, 915,, 755, 756
High-performance computing (HPC), 771
High-shear granulation (HSG), 313, 415, 416f, 473f, 474, 476f, 478f, 479f, 483t, 748, 760f, 766, 821
    bottom-drive, 211, 214, 215t, 216t
    CFD simulation, 768
    coupled simulation, 770
    horizontal shaft, 211, 213f
    mixing effect, 479f, 480f, 481f
    placebo formulation, 483t
    scaleup of
        binder addition rate, 256
        chopper speed, 256
        height of material and batch size, 255
        impeller speed, 258
    top-drive, 211, 214, 215t, 216t
    vertical shaft, 211, 212f
High-shear mixer, 849, 855
High-shear mixer granulator
    agitation torque, 710, 711f
    centrifugal tumbling granulator, 709
    force acting, 709, 710f
    granular flow, 709
    impeller geometry, 710
    mixing/segregation kinetics, 710
    particle kinetic energy, 710, 711f
    PEPT technique, 709

roping flow, 709
High shear wet granulation (HSWG), 93, 125, 268, 367f, 371, 379, 471, 601, 760, 802, 872, 887, 911
    active pharmaceutical ingredients study, 103
    binder, 602
    binder addition, 210
    breakage and attrition rate process, 604
    chopper, 210
    commercial-scale granulators, 620
    consolidation and growth, 604
    continuous wet granulation, 216
    fluid bed drying, 616
    geometric, dynamic, and kinematic similarities, 612
    granule growth, 383
    high-shear granulators
        bottom-drive, 211, 214, 215t, 216t
        horizontal shaft, 211, 213f
        top-drive, 211, 214, 215t, 216t
        vertical shaft, 211, 212f
    ICH Q8, 609
    impellers, 210
    lactose study (Lactose-based formulation study)
    multicomponent pharmaceutical formulation, 604
    pilot-scale granulators, 602
    preblending, 602
    schematic, 602, 603f
    small-scale granulators, 609
    steps, 210
    water-addition, 602
    wet-massing, 602
    wetting and nucleation, 603
Hollow granules, 49, 56
Horizontal shaft granulators, 211, 213f
Hot melt granulation, 413
HS/LS granulations
    common input variables, 675, 675t
    process parameters
        amount of granulation fluid I (w/w) and amount of granulation fluid II (w/w), 684
        batch size and volume fill ratio, 683, 683t, 687
        equipment type, 683
        granulation impeller speed and chopper speed, 685
        jacket/product temperature, 685
        power consumption and torque, 685
        premixing impeller speed and premixing time, 684
        with report frequency, 681f, 681
        risk factor, 682
        solid level of granulation fluid I, 684
        spray rate and spray time, 684
        wet massing time, 685
    unit operation profile, 674f
Hybrid modeling, 912

Hydrophilic polymeric binders, 914
Hydrophobic powders, 920
    blend, 55
    particles, 56
Hydroxyapatite wet granule surface velocity, 19f
Hydroxypropyl acetate succinate (HPMC-AS), 192
Hydroxypropyl cellulose (HPC), 182, 193, 290, 418, 803, 914
Hydroxypropyl methyl cellulose (HPMC), 290, 293, 802
Hygroscopicity, 323

I

Ibuprofen, 329, 334f
Immersional and condensational tests, 329
Immersion mechanism, 819
Impeller
    parameters, 549
    rapid motion, 762
    speed, 110
        effect, 480, 616, 617
    torque, 251, 252t, 511
Inactive Ingredient Database (IID), 344
In-line rheometry, 510
Integrated data model, 879
Interparticle material bridges, 74
Inverse gas chromatography (IGC), 329

J

Japanese Pharmacopeia (JP), 315
Johnson-Kendall-Roberts (JKR)
    model, 755
    theory, 917

K

Kinematic-dynamic scale-up method, 732, 737, 743f, 743t
Kinematic similarity, 733, 739
Kinetic energy, 761
Kinetic theory of granular flow (KTGF), 758
Kneading elements, 811
Kneading time, 234, 236, 239t, 245f

L

Laboratory granulation experiments, 880t
Lactose, 278, 280f, 281f, 451
    granulation, 407f
    monohydrate, 347, 423
Lactose anhydrous
    compactibility, 181, 184, 188t, 189, 192f
    design of experiment, 182, 182t
    fluidized bed drying vs. tray drying, 182
    mercury intrusion porosimetry, 184, 190f
    microcrystalline cellulose, 192, 195t
    to monohydrate, 181

nitrogen adsorption technique, 184
powder Xray diffraction, 184, 187f
regression analysis, 183, 184t
scanning electron micrograph imaging, 189, 191f
sieve analysis, 185
thermogravimetric analysis, 184, 186f
Lactose-based formulation study, 94
compactibility, 97, 99t, 100f
granule size distribution, 96
Plackett-Burman design, 94, 95t, 98t
tapped density, 102
α-lactose monohydrate, 181
6-L Diosna granulator, 473
Linear spring-dashpot (LSD) model, 755
Linear variable displacement transducer (LVDT), 185
Liquid-bound granules, 12f
Liquid marble formation, 920
Liquid-to-solid ratio, 426, 815
Loss-on-drying (LOD), 818
Low-shear wet granulation process, 268, 313, 415, 416f, 674f, 674
planetary mixer, 222
screw type mixer, 222
Lubricants
factors affecting, 296t
magnesium stearate, 298
roles, 297
sodium stearyl fumarate, 299
stearic acid, 297
types, 297
Lump-free dispersion, 315

M

Magnesium stearate, 201, 298, 915
Maillard reaction, 418
Malvern laser diffraction, 474t
Mannitol, 283, 418, 453, 902
Mass flow dynamics, 916
Material attributes, 347
Material libraries, 868
Material sciences tetrahedron, 124
Mathematical models, 912
Mean force pulse magnitude, 544
Mean residence time (MRT), 811
Mechanical mixing, 764
Mechanistic models, 919
Medium-viscosity binders, 59
Melt agglomeration, 849
Melt granulation, 225, 801, 837, 838, 843, 848, 917
applications, 839
critical quality attributes, 842
formulation design, 842

process design, 849
Mercury intrusion porosimetry (MIP), 194, 196f, 201f
Metformin, 335f
Methyl cellulose (MC), 315, 316t
Microcrystalline cellulose, 63, 67, 218, 474, 802, 878, 915
crystallinity, 291
friability of, 283, 284f
functionality, 291
micromeritic changes
compaction, 198, 199f, 200t
moisture uptake, 198f, 198
particle size distribution, 194
porosity and surface area, 194
molecular weight distributions, 283, 284f
particle size distribution, 282f, 282
tabletability, 291f, 291
variability, 295
water absorption for degree of polymerization, 283f, 283
Microscopy techniques, 36
Model-based approach, 754
Model evaluation indicators, 876
Modeling, 754, 912
Modeling techniques
DEM, 621
parameters, 622, 623
PBM, 621
with compartmental model and DEM, 621
coupled with CFD, 621
coupled with DEM, 621
coupled with VoF models, 621
regime-map based modeling, 621
Modern fluid dynamics, 756
Modified capillary number, 14, 350
Mohr's circles analysis, 597
Moist agglomerates, 394, 404
Moisture-assisted dry granulation (MADG), 224, 837, 914, 915
Moisture content, 425
Moisture uptake, of microcrystalline cellulose, 198f, 198
Multiple-impact tester, 42
Multiple linear regression (MLR) analysis, 580
Multiple unit operations, 824
Multiscale model, 773
Multivariate process modeling, 876

N

Naproxen, 329, 334f
National Formulary of the United States Pharmacopeia (USP/NF), 315
Near infrared technology, 916,, 253, 817
New chemical entity (NCE), 323
Newton power number, 254, 254t, 615
Newton's equations of motion, 758

Newton's laws of motion, 757
Nitrogen adsorption technique, 380
Nonionic surfactant, 350
Nonlinear dissolution profiles, 245, 248f
Nucleation
  mechanism, 841
  process, 208, 267, 393, 415
  regime map, 10
Numerical models
  particle assembly systems, 703
  population balance models
    coalescence kernel, 704, 704t
    coalescence of granules, 703, 704f
    DEM coupling with, 716
    hybrid type models, 706
    law of "conservation of numbers,", 703
    multidimensional models, 706
    particle assembly systems, 703
    temporal change, granule size distribution, 703, 705f, 705
  random models, 702f, 702
Numerical resolution methods, 191

O

Optimization DOE, 691
Oral disintegrating tablets (ODT), 873
Oral solid dosage (OSD), 872

P

Paracetamol granulations, 429t
Parametric-based scale-up strategy, 616
  impeller speed, 617
  potential CPPs, 616
  powder bed height/batch size, 616
  process parameters, 616
  spray-related parameters, 616, 620
  water amount, 617
  wet massing time, 616, 619
Partial least squares (PLS), 876
Partially gelatinized starch (PGS), 315
Particle adhesion/coalescence model, DEM coupling with, 716
Particle agglomeration, 495
Particle beam model (PBM), 770, 773
Particle blending model, 771
Particle diffusion coefficient, 730
Particle image velocimetry (PIV), 761
Particle–particle cohesion forces, 359
Particle size distribution (PSD), 353, 356f, 471, 747, 808
  of dry powders, 380
  and granule growth, 401
  and granule porosity, 400
  hydroxypropyl cellulose, 490f
  microcrystalline cellulose, 194
  nested sieve analysis, 500f
  and wet granule mechanical properties, 394
Pendular state granules, 209f, 209
Pexacerfont formulation
  experimental design for, 103, 105t
  regression analysis results for, 108t
Pharmaceutical
  binders, 348
  formulation development, 867
  industry, 747, 748, 802
  powder technology, 314
  Quality by Testing regulatory, 669
  tablet manufacturing process, 917
Pharmaceutical QbD
  application, 669
  DOEs, 691
  elements, 673
  FB wet granulation, 671f
  ICH Q8 (R2) guidance, 669
  PAT, 693
  risk assessment
    fishbone diagram, HS/LS granulation, 692f
    ICH Q9, 691
    modified release (MR) tablet, 691t
  unit operation profile, 672
    FB wet granulation, 675f, 675
    HS/LS wet granulation, 674
    wet granulation development considerations, 673
Physical-based models, 914
Physiologically-based models (PBMs), 920
Pioglitazone hydrochloride, 192
Plackett-Burman design, 94, 95t, 98t
Planetary mixer, 222
Plug flow reactor (PFR), 918
Polymeric excipients, 423
Polymer tablets, 339, 341f
Polymethylmethacrylate (PMMA), 285
Polymorphic conversion, 179
Polyvinylpyrrolidone (PVP), 275, 293, 315
Polyvinylpyrrolidone-vinyl acetate (PVPVA), 192
Population balance equation (PBE), 760
Population balance model, 917,, 367f, 621, 747, 774, 776, 912, 914
  coalescence kernel, 704, 704t
  coalescence of granules, 703, 704f
  with compartment model and DEM, 621
  coupled with CFD, 621
  coupled with DEM, 621
  coupled with VoF models, 621
  DEM coupling with, 716
  hybrid type models, 706

law of "conservation of numbers,", 703
multidimensional models, 706
particle assembly systems, 703
temporal change, granule size distribution, 703, 705f, 705
Porosity
  DPC 963 granulations, 401f
  microcrystalline cellulose, 194
Positron emission particle tracking (PEPT), 618, 769, 818
Post-granulation
  drying, 823
  milling, 823
Powder bed granulation (PBM), 912
Powder bed height/batch size, 616
Powder characterization
  density, 428
  fundamental principles of particles
    cohesion, inter-particulate forces, 569, 571f, 571
    friction, 569f, 569
    gravity, 571, 572f
    liquid bridging, 570f, 570
    mechanical interlocking, 569, 570f
  process environment impact
    aerosolization, 567f, 567
    bulk properties, 566
    cohesive forces, 566
    frictional forces, 566
    intrinsic properties, 566
    mechanical interlocking, 566
    optimum dispersion mechanism, 568
    powder behavior, 567
    powder discharging from hopper, 566
    unit operations, 566
  stylized pharmaceutical tableting process, 560f
Powder feed number (PFN), 825
Powder feed rate, 810
Powder flowability
  adhesion, 562
  cohesion, 562
  compressibility, 562, 563f
  compression, 563
  electrostatic charging, 564, 565f
  flow rate sensitivity, 562
  permeability, 564
  physical changes, 565, 566f
  powder behavior, 560
    external variables, 560
    particle properties, 560, 561f
Powder testing
  biaxial shear cell testing
    advantages and limitations, 577
    data generated, 576
    hopper design methodologies, 576
    measurement procedure, 576
    Mohr's circles analysis, 597
    practicalities, 577
  commercial pressures, 559
  dynamic powder testing
    advantages and limitations, 579
    data generated, 578
    measurement procedure, 578f, 578
    practicalities, 579
    process environment, 578
  hopper and hopper design, 559
  issues evaluation, 558
  method relevancy, 596
  multifaceted powder behavior, 558
  optimizing productivity and quality, 559
  regulatory and economic environment, 559
  relevance concept, 573
  repeatability, 573
  reproducibility, 573
  sensitivity, 573
  traditional powder characterization techniques
    advantages and limitations, 574
    angle of repose, 573
    data generated, 574
    flow through an orifice, 574
    practicalities, 574
    tapped density, 574
    USP/EP guidelines, 573
  uniaxial testing
    advantages and limitations, 576
    data generated, 575
    measurement procedure, 575f, 575
    practicalities, 575
    stages, 575f
    uniform density and homogeneous stress distribution, 575
  US and European pharmacopeia publications, 558
Powder X-ray diffraction pattern, of CPD-1 tablets, 158, 160f
Power consumption, 511, 549
Power consumption measurements, 251, 252t
Practical process relevance
  die filling operation, design space
    aeration data, 586f
    aluminum, 585
    background, 585
    bulk and dynamic properties, 585
    development, 585
    filling performance, 585, 586f
    filling ratio, 585
    large glass beads, 585
    permeability data, 587f
    powders physical properties, 585, 585t

Index

small glass beads, 585
Specific Energy (SE), 585, 587f
tungsten, 585
feeder performance
   average volumetric feed rate, 580, 581t
   calcium citrate, 580
   calcium hydroxide, 580
   cellulose, 580
   DIWE-GLD-87 VR, 580
   Flow Rate Index (FRI), 582
   FT4 Powder Rheometer, 585
   GLD feeder, 584
   maltodextrin, 580
   milk protein, 580
   MLR analysis, 580
   multifaceted powder characterization, 584
   predicted *vs.* actual volumetric feed rate, 582f, 584f
   screw feeders, 579, 580f
   shear thinning behavior, 582
QbD approach
   APAP formulation, 590f, 590
   background, 589
   BFE, 590, 591, 592t
   bulk flow properties, 591, 593f
   correlation, 593
   CQA, 594
   DCP formulation, 590, 591f
   experiments, 589f, 589
   scaling criteria, 594
   tablet hardness *vs.* BFE, 593, 594f
Pregelatinized starch (PGS), 821
Principal component analysis, 855, 895
Principal component analysis (PCA), 253
Process analytical technology, 510
Process analytical technology (PAT), 352, 417, 802, 850
Process capability, 673
Process description, 890
Process fingerprint, 529
Process frequency, 530
Process parameters, 875
Process variables, 852
Product lifecycle management, 673
Pseudodroplet state granules, 209f, 209
Pseudo Reynolds number, 615
PVP-acetaminophen model system, 340t
PVP-acetylsalicylic acid granulations, 337

Q

Qualitative analysis, 898
Quality-by-design, 445
Quality-by-design (QbD), 344, 419, 471, 558
   acetaminophen, 229
brivanib alaninate, 634
calcium carbonate, 228
CMAs, 605, 606t, 608f
CQA, 605
critical material attributes, 228
defined, 225
dissolution rate, resultant tablet, 605
formulation development process, 435f
integrated and overarching application
   design space strategy, 664
   input variables effect, 662
   model-based specification limits, 665
   process robustness by formulation design, 664
methodology, 634, 635f
pharmaceutical development, 419
pharmaceutical drug development, 605
polyethylene glycol, 228
principles, 662
in product development, 229
product quality attributes, 633
quality risk assessments, 633
quality risk management, 419, 605
quality target product profile, 228
Ramipril tablets, development of, 228
risk-based approach, 633
sound science, 605
statistically designed experiments, 633
three-dimensional population balance model, 228
tools
   scale-up, 558
   tableting, 557
TPP, 605
Quality function deployment (QFD), 436
Quality target product profile (QTPP), 228, 229, 230t, 347, 419, 672
Quantitative analysis, 901
Quasistatic diametrical compression, 40

R

Raman spectroscopy, 819
Ramipril tablets, 228
Random coalescence model, 702f, 702
Ranitidine, 285
Razaxabn formulation
   experimental design for, 103, 104t
   regression analysis results for, 109t
Recirculation patterns, 359 360 361 362f–359 360 361 362f, 363 364f–363 364f, 365 366f–365 366f
Reference listed drug (RLD), 229
Relative swept volume, 618
Reproducibility and chord length distribution, 491 492 493f–491 492 493f, 490f, 494

Residence time distribution (RTD), 919,, 817
Residual challenges, 913
Reworking potential, 891
Reynolds number, 254, 254*t*
Risk assessment for drug product, 233, 234*t*, 240*t*
Robust tablets, 342
Roller compaction, 601
Rotating fluidized bed coater (RFBC)
    coating mass distribution, 712, 714*f*
    coefficients of variation (CV), temporal change, 712, 715*f*
    conventional fluidized bed coater, 712, 715*f*
    schematic illustration, 712, 713*f*
Rumpf equation, 405

S

Scanning electron microscope-focused ion beam, 39
Scanning electron microscopy, 38
Screening DOE, 692
Screw design, 812
Screw elements, 811
Screw speed, 811
Seeded granules, 64
Seed granulation, 914,
Shear granulation, 759
Shear stress, 729*f*, 733
Silicified microcrystalline cellulose (SMCC), 886
Simulation, 912
Simulation data, 878
Simulation model, 920
Single impact tester, 42
Single-pot granulation
    Granurex, 223, 224*f*
    heat transfer, 222
    microwave drying, 223
    stripping gas, 222
Sodium carboxymethyl cellulose (NaCMC), 350
Sodium lauryl sulfate (SLS), 335, 809
Sodium starch glycollate (SSG), 301, 418, 915
Sodium stearyl fumarate (SSF), 299
Soft-sphere methods, 755
Solid–liquid surface energy, 327
Solid state transformation
    control of, 191
    formation of high energy metastable form, 180
    hydrate/solvate, 181
    of salt to free form, 180
    to stable polymorphic form, 179
Solid–vapor surface energy, 327
Solubility, 405, 426
Spatial filter velocimetry (SFV), 817
Specific feed load (SFL), 811
Specific mechanical energy (SME), 854

Spray-dried lactose, 280
Spray fluidized bed granulation (SFBG), 921
Spray-related parameters, 620
Stability of formulations
    alkalinizing agent, 161
    CPD-I, 162
    degradation during manufacturing, 155
    DMP 754, 173
    drug-excipient interaction, 156
    experimental design, 157, 157*t*, 158*t*
    pH modifiers, 161, 165*f*, 171*f*, 175
    wet granulation binders, 323
        aldehydes and carboxylic acids, 323
        peroxides, 323
Starch binders, 290, 301, 315, 320
    starch, 301
Statistical model, for wet-granulated product attributes, 383
Steam granulation, 918
Stearic acid, 297
Stokes deformation number, 16, 732
Stress-strain profiles, of wet lactose compacts, 395, 396*f*
Stripping gas, 222
Styl-one evolution compaction simulator, 551
Substrate wetting, 324
Sulfaguanidine transformation, 191
Superficial velocity, 753
Surface energetics, 324
Surface tension, 5, 348, 753
Surfactants, 459, 809
"Swiss-cheese" granules, 67
System parameters, 853

T

Tabletability, 895
    change, 891, 898
    change classification system, 901*f*
    evaluation, 892*f*
Tableting process, 890
Tablets
    hardness, 210, 242, 249*f*
    powder consolidation, 314*f*, 314
Target material properties (TMPs), 347
Target product profile (TPP), 605
Tavares model, 913
Tensile strength (TS), 886
Theophylline monohydrate, 181
Thermogravimetric analysis, 848
Thermoplastic polymer soluplus, 847
Thin powder flow, 521
Three-dimensional population balance model, 228
Timoshenko beam bondmodel (TBBM), 915
Torque, 549

Index

Torque rheometer, 251, 252t
Triazine derivative (CPD-1) tablets
    colloidal silicon dioxide particle effects, 158, 162f
    CPD-II stability, 159, 163f
    formulation-dependent effect, 162, 164f
    pH modifiers, 161, 165f
    powder X-ray diffraction pattern, 158, 160f
    water used for granulation, 158, 161f
    wet granulation process effects, 156, 159f
Turbine mixing elements (TMEs), 820
Turbulence models, 757
Twin-screw extruder, 802, 804, 811, 821, 824
Twin-screw extrusion technology, 802
Twin-screw granulation, 748, 806
    comparison, 822
    couple simulation, 915
    simulation, 917
Twin-screw melt granulation, 837, 851
    advantages, 855
    challenges, 856
Twin-screw mixer (TSM), 916
Twin screw wet granulation (TSWG), 141, 420, 421t, 443, 803, 805, 810, 886
    challenges, 822
    granulation mechanism, 819
    process monitoring, 817
    scale-up, 825

U

Uniaxial compression test, 395
Uniform fluid flow, 519
Unit operations, of wet granulation, 348
Universal testing machine, 339, 341f

V

Vertical high-shear mixer-granulator, 737f, 737
Vertical shaft granulators, 211, 212f
Vibratory transfer systems, 823
Viscosity, 6, 324, 348
Voigt model, 706, 725
Volume of fluid (VoF) models, 621

W

Wall friction angle (WFA), 822
Washburn capillary wetting approach, 329
Water addition rate, 113
Water amount, 617
Water concentration effect, 489
Water holding capacity, 380
Water uptake, 240, 242f, 245f
Wet adhesion, 314
Wet agglomerates, 394
Wet binder addition (WBA), 762
Wet granulation (WG), 155, 268, 414, 748, 801, 872, 885, 886, 919
    advantages, 414t
    API powder properties, 601
    binders
        acetaminophen tablets, 342f
        compatibility considerations, 322
        considerations for, 322
        copovidone, 319
        efficiency, 321
        equilibrium moisture isotherms, 320f
        ethyl cellulose, 318
        gum acacia, 321
        hydroxypropylcellulose, 315
        hypromellose, 318
        methyl cellulose, 315
        physical-chemical properties, 314
        povidone, 319
        properties, 314
        regulatory acceptance, 344
        sodium carboxymethyl cellulose, 319
        solvent, 334
        stability, 323
        starch and modified starches, 320
        supplier reliability, 344
        use levels, 321
        water content, 322
    critical material attributes
        basics, 414
        continuous, 417
        critical process parameters, 420
        end point determination, 417
        excipients, 417
        QbD paradigm, 419
        rate processes, 415
        types, 415
    data fusion, 872
    disadvantages, 414t
    drying and milling, 670
    endpoint detection, 557
    equipment for, 416f
    excipients, 275, 276
    extrusion, 601
    flowability, 557
    fluid bed granulation, 601
    fluidized bed, 269
    granule size, 670
    high shear, 268
    HS/LS granulation, process parameters
        input operating parameters, 680
        process state variables, 680

scale-up consideration, 689
liquid saturation states, 415f
low-shear, 268
melt granulation, 601
methods, 26
modeling, 751
potential process parameters, 421t
process, 3, 4f, 837, 838, 851, 878
process design, QbD quality system
    binder addition method, 679
    classical "trial and error" strategy, 680
    elements, 673
    input variables, drug substance, 678
    intra-granular excipients, 678, 679
    pharmaceutical excipients, 679
    physical, chemical, and biological properties
        drug substance, 678, 679t
    risk assessment and DOEs, 682
    systematic drug-excipient compatibility, 680
    toxicity data of new excipient, 680
product attributes
    material properties, 381
    statistical model, 383
product to unit operation, 672f, 672
purpose, 676
QbD tools
    DOEs, 690
    PAT, 693
    risk assessment, 691, 691t, 692f
quality characterizations, product, 676, 677t
rate process, 749f, 749
rotor granulation, 601
types, 670, 671f
unit operation, 495
Wet granule strength, 15
Wet lactose compacts
    porosity, 399f
    strength of, 398f
    stress-strain profiles, 395, 396f
    yield stress, 399f
Wet massing time, 112, 251, 607, 619
Wet particle agglomeration, 349
Wettability, 324, 427
Wetting process, 208, 267, 415
Wurster process, 268

X

X-ray
    computed tomography, 36
    linear attenuation coefficient, 36

Y

Young's equation, 327, 431
Young's modulus of material, 517